1999 6th Edition

Toys & Prices

Edited by

Sharon Korbeck

Published by

krause publications

700 E. State Street • Iola, WI 54990-0001
Telephone: 715/445-2214

Please call or write for our free catalog of publications.
Our toll-free number to place an order or obtain a free catalog is 800-258-0929 or please use our regular business telephone 715-445-2214 for editorial comment and further information.

Library of Congress Catalog Number: 93-77554
ISBN: 0-87341-654-6

Printed in the United States of America

Contents

Hey! What's That On the Cover?

Farm Set, Nylint, 1960s	$110
Yoda figure, Burger King, 1997	$2
Toy Story Masterpiece Collection, Burger King, 1990s	$3
Assorted Hot Wheels and Matchbox cars	$1-$3 each
Live Action Barbie, Mattel, 1971	$75
Barbie bicycle	$50
Hickory Dickory Dock wind-up tin toy, Mattel	$25
Assorted Fisher-Price Little People	$2
Tweety and Yoda PEZ dispensers	$5
Malibu Barbie Colorforms, Colorforms, 1970s	$15
Yankee Doodles lunch box, King Seeley Thermos, 1975	$40
G.I. Joe Black Adventurer, Hasbro, 1970s	$150
Monkey w/cymbals battery-operated toy	$150
Coca-Cola die-cast bank, Ertl	$30
Ziggy and Twigs Beanie Babies, Ty, 1990s	$15 each
Assorted California Raisins figures, Hardee's	$2-$5 each
KISS action figure, McFarlane Toys, 1997	$15
James Bond Message from M Game, Ideal, 1966	$250
Disney tin top, Chein, 1970s	$20
Popeye gumball machine, 1960s	$20
Darth Vader figure, Kenner	$5
Hey, There, Yogi Bear coloring book, Whitman, 1960s	$25
Sorry board game	$10
Coca-Cola steel truck, Buddy L	$40
Bionic Woman Thermos bottle, Aladdin, 1970s	$25
Ronald McDonald doll	$35
Gumby doll, Knickerbocker, 1970s	$15
Catwoman figure, Batman Returns, Kenner, 1990s	$25
Big Bill Pelican, Fisher-Price, 1960s	$25
Clown battery-operated toy	$200
View-Master viewer	$5
Peanuts dolls (Sally, Linus), Hungerford Plastics, 1950s	$60-$90
Mr. Potato Head, Hasbro, 1990s	$10
Hopalong Cassidy tin target game	$75

Prices listed are for toys in Excellent condition. Toys pictured on cover courtesy of Jon Brecka, Daisy Cain, Joel Edler, Brent Frankenhoff, Greg Krueger, Becca Hansen, Mike Jacquart, Sharon Korbeck, Tom Michael and Steve Rupert.

Acknowledgments

Welcome to *1999 Toys & Prices*!

Every year, as we prepare a new edition of *Toys & Prices*, we face the same challenge. How can we make this book as comprehensive as possible and still keep it manageable?

As we close in on 1,000 pages, we think we've done a pretty good job. While it may be impossible to list every toy ever made and collected, this book hits what we feel are the highlights. I'm sure you'll find not only your favorite toys, but other toys you may have forgotten from your childhood. Remembering them is half the fun — finding them again and buying them complete the experience.

In this our sixth edition, a special thanks goes out to the readers for their interest and our many contributors for their expertise.

Numerous experts contribute their time and knowledge each year to make this book bigger and better. Active collectors may be familiar with their names from ads in *Toy Shop*. And we welcome back some of last year's contributors as well.

Veteran collectors will recognize these people as widely-respected authorities in their various fields, from restaurant premiums and model kits to G.I. Joes and Barbie dolls. They took time from their busy schedules to sift through over 18,000 toys worth — correcting errors, verifying values, adding toy listings and weeding out duplications. It took the labors and long hours of many people to produce this book, and all of them deserve much thanks.

Special thanks goes to the staff of Krause Publications — *Toy Shop* Associate Editor Mike Jacquart; Bonnie Tetzlaff of the book editorial and production staff; Chris Pritchard and Ross Hubbard for cover design/photography; and the production staff for turning copy and photos into a cohesive, coherent package.

Outside contributors/reviewers (I hope I haven't forgotten anyone) include Anthony Balasco, Don Hultzman, Chris Fawcett, Jon Brecka, Carren Kopleton, Dana Johnson, Patricia Long, Jeff Escue, Daniel Yett, Patricia Long, Vincent Santelmo, Dale Womer, Paul Fink, Jeffrey Lowe, John Devlin, John Snyder and Mary Ann and Wolfgang Sell.

The main reason we publish this book is for you — the thousands of toy collectors nationwide and around the world — and we continue to thank you for your support of the book and the hobby.

Here's hoping your year is filled with plenty of toys, good deals and the time to enjoy them both.

I look forward to bringing you the 2000 edition of *Toys & Prices* next year!

Sharon Korbeck
Editor

1999 Toys & Prices

Toy collecting has come a long way in the past decade. We ought to know. We've been there since the beginning.

When *Toy Shop* was first published in 1988, toy collecting was just heating up as a hobby. And while the first issue of Krause Publications' monthly toy magazine had only 48 pages, it has grown into the number one resource in the hobby. Issues now contain almost 200 or more pages every two weeks.

Our experience in the toy collecting hobby has expanded greatly in the past decade — especially in the area of books, particularly price guides.

This book, *Toys & Prices*, is now in its sixth edition. It has added hundreds of pages and dozens of categories since the first edition in 1994. But we've added quite a few titles to our toy publishing stable since then — including Richard O'Brien's popular *Collecting Toys*.

It's hard to believe that next year we'll be working on the 2000 edition of the book! What will become of toy collecting as the millennium approaches?

Toy collecting has changed quite a bit over the years, especially in terms of the type of toys collected. No longer are the oldest "antique" toys generating the most attention. Now new toys, fresh off store shelves, are eagerly anticipated and quickly scarfed up by anxious collectors. Baby boomer toys have hit their heyday, but not yet their peak.

So as we approach 2000, we celebrate the toys that have made so many so happy for so long. This book is for you, the avid toy enthusiast.

But even if you're a casual admirer of the quality and charm of old toys, you may be amazed at what some of those banks, cars, tin toys and other remnants of childhood can sell for today.

It's never too late for you to get started or add to your collection. *Toys & Prices* is designed to help you get the most value for the toys you own now and for the toys you may purchase tomorrow.

From the day we shoot the cover photo to the time the final values are entered into the computer, *Toys & Prices* is a year-long, cooperative process. And like any price guide on the market, it takes the time and talents of dealers, collectors, editors, photographers, designers and marketing specialists to put together a quality product.

When the long task is finally finished, we have a true picture of what we set out to produce — a concise, compact, and current price guide for over 20 toy collecting areas representing over 18,000 toys. No price guide of its nature can be all-inclusive, but we've tried to offer a variety of the main collecting areas.

Look What's New!

Every year, we update the book — adjusting values, proofreading entries and adding new sections.

This year, we've added a section on advertising toys — those popular, lovable, recognizable and highly-collectible characters. Remember Little Sprout and Mr. Peanut? You'll find them and many others in this new section.

We've also updated our vehicle toys section, with two notable additions. Look for listings on the Majorette 200 series and for increased Hot Wheels listings — including recent numbered packs and Treasure Hunts.

Generating price lists for *Toys & Prices* is a year-long process. As we feature articles and price guides in our bi-weekly magazine *Toy Shop*, we keep in mind what toys collectors are most interested in. To do that, we attend toy shows nationwide, study the advertisements in *Toy Shop*, conduct readership surveys and listen to calls and letters from those of you in the collecting trenches.

Keeping Pace in a Hot Hobby

As most of you already know, toy collectors are a fun, albeit sometimes obsessive, bunch. They'll travel miles out of their way to hit a garage sale, destined to uncover a true treasure on the $1 table — and it does happen.

Thousands of people from all walks of life are becoming toy collectors each month, from kids spending their allowances to investment advisors seeking high returns. Toy collectors come in all sizes, ages and nationalities and live all over the globe, but we all share a love of collecting and an appreciation of the artistry, charm, history and just plain fun toys have always offered, and still do.

Toy collecting has been called the hottest collecting field of the 1990s, and with the turn of the century only a few years away, the evidence keeps piling up. Television shows now exist devoted to collecting, museum and library displays spotlight large toy collections, and even some of today's hottest celebrities have admitted their voracious appetites for toys. Toy manufacturers are quickly keeping pace preparing for the secondary market, issuing limited-edition items destined (hopefully) to increase in value.

This hobby has everything going for it — a colorful universe of plentiful and affordable items, the thrill of the hunt, intellectual and emotional gratification, camaraderie and investment potential.

As evidenced in this book, toy collecting is a vast field, discriminating against no one. Novice collectors with little to spend can edge in on the action figure field by purchasing today's "new" collectibles or relatively inexpensive toys such as restaurant premiums and PEZ dispensers. Character toy aficionados are continually bombarded with new characters from Disney, Warner Bros. and the like. More seasoned and affluent collectors can concentrate on some of the older and revered favorites like cast-iron banks or 1930s classic tin.

Where Do Those Prices Come From?

So just where do the prices in this book originate? Our primary sources are print retail ads in *Toy Shop*, dealer price lists, observed prices at toy shows and prices realized from auction houses across the country. Information is compiled by category and entered into databases. These databases are then reviewed for accuracy by the editor of *Toy Shop*.

Each database is then sent out for final review by recognized authorities in the subject fields. These contributing experts are some of the leading collectors and dealers in America, and we are proud to list them on our acknowledgments page.

The final results listed in this book are what we believe to be accurate, current retail prices, listed across a range of grades for each item.

A Cautionary Note

Veteran collectors know that a toy listed for $500 Mint in a price guide may or may not bring such a value — especially since dealers will pay less than book value for items that they will, in turn, mark up. But guides do serve as a starting point for those unfamiliar with the popularity or potential collectibility of particular toys.

Remember, values listed in this book are not offers to buy or sell, but are guidelines as to what you could expect to pay for an item at a show or by mail.

Price guides can be helpful, but they can also be frustrating — both for the editor and the user — if used improperly. Prices for the same item can vary widely from source to source, depending on factors such as geographical differences, personal economics and target market.

Because of this, price guides are meant to be just that — a guide — and not a bible. Feel free to use the values as gauges to aid your purchasing and negotiation, but remember that a dealer's agenda is to get the best price and make a profit too.

The Complete Carry-Along Guide

We've done a lot of revising and reformatting to make this book as easy to use as possible. Sometimes, however, a question may arise as to how to locate a listing in the book.

As a general rule, individual sections override character — if there is a section in the book for the type of toy you're looking for, check there first. For example, if you're looking up a Superman lunch box, look first in the lunch box section. Likewise, a Yogi Bear board game will most likely appear in the postwar games section rather than the character toys section.

The character toys section includes items that do not fit into one of our established sections or are more difficult to classify.

Also, be aware that the words "figure" and "doll" are sometimes used interchangeably, although we have tried to remain consistent.

Overlap is the nature of the beast with toy classification, so if you don't find what you're looking for in one place, check a related section. For example, space-related action figures are most likely to be listed in the action figure section, but you may also find some listed in the space toys section. We're working on ways to streamline such listings to make finding items easier for readers. Your input and suggestions are welcomed.

Are Your Toys in Top Form?

Grading is highly subjective and can often depend on whether you're a buyer or a seller. If you buy a toy in Excellent condition, you might try to call it Near Mint if you are trying to sell, and even slight differences in condition can greatly affect prices. Haggling over condition is often as natural as haggling over price. Perhaps nowhere is that more evident than in toy collecting.

If you are buying this book to estimate the value of your collection before you sell, remember that grading is subjective. Read the definitions of different grades listed in this book, and try to grade your toys as conservatively as you can.

Also note that grading standards vary from class to class. Grading a cast-iron bank or a Victorian board game is a different task than grading model kits or lunch boxes.

Most dealers typically offer between 40 and 60 percent of book value when buying — regardless of condition — as they obviously must be able to turn around and resell toys at a profit in order to stay in business.

While we've proofed and reproofed this book for accuracy, some typos may have slipped in. If you spot a price that seems way off base, don't take it as gospel. It may be a typo, or based on incomplete information. If you do notice what looks like a glaring error, we'd like to know about it so we can be sure to double check it for the next edition.

Likewise, if you own an item not listed here, we would appreciate the chance to add it to our database, especially if you can give us a clean photo of it along with information on its size, maker, year, and any other descriptions that might apply. Photos are welcome, but send only duplicates, as we regret we cannot return them.

The Bug has Bitten

The next time you find yourself at a flea market saying longingly, "I had that doll," or "My brother and I played with that very same truck in our sandbox," you'll know the toy collecting bug is nearby. When you start to negotiate on the price, that bug is moving in . . . fast. When you decide you've got to have that toy at any cost, it's all over — the toy collecting bug has bitten. But it may only sting for a minute . . . then the real fun sets in.

Having fun, after all, should be the ultimate goal of toy collecting. If you aren't having fun, what's the point?

It may take years before you find "the perfect toy" to add to your collection. But that quest and the excitement that fuels it are what keep the hobby going.

A Final Note

For those of you new to toy collecting, WELCOME! Enjoy the ride as you retrieve your past.

To veteran collectors, you already know the emotional and financial benefits this hobby can bring. Share this joy of toys with your children and grandchildren . . . and it just may make you feel like a kid again.

Using The Grading Guides in *Toys and Prices*

Prices listed in *Toys and Prices* are intended only as guidelines of what you might expect to pay for an item in today's market, either at a show or through the mail. The values listed are not offers to buy or sell toys; the publisher does not engage in buying or selling toys, but in compiling information and prices.

Values listed represent our best assessment of current market values at press time and are derived from various printed sources, toy show observations, auction results and the input of expert contributors.

Prices are generally listed for more than one grade of condition. Toy collectors and dealers know grading can be difficult and subjective. It's usually easy to reach agreement on a toy that's Mint In Box, but beyond that, grading becomes a somewhat imprecise science.

In order to provide general guidelines, the editors of *Toys and Prices* have adopted the following grade descriptions. Bear in mind that since various types of toys are looked at in different ways, no single grading system will apply to all of them. Some descriptions may not apply to certain toys, depending on how they are categorized, when they were produced, how they were originally packaged and how they are collected today.

Good, Excellent . . . C10?

Toys & Prices uses a Good, Excellent and Mint grading scale. Many of you, however, may be familiar with the 'C6' or 'C10' numbers used in Richard O'Brien's *Collecting Toys* book.

Don't sweat it; our Good, Excellent and Mint prices roughly figure out to C6, C8 and C10 respectively. Items in less than Good condition may often be considered poor; prices should be lowered accordingly depending on condition.

MIB or MIP (Mint In Box, Mint In Package, C10) — Just like new, in the original package, preferably still sealed. Boxes may have been opened, but any packages inside remain unopened. Blister cards should be intact and unopened. Factory-sealed boxes often command higher prices.

MNP or MNB (Mint No Package, Mint No Box, C10) — This describes a toy typically produced in the 1960s or later in Mint condition, but not in its original package. A toy outside its original package is often referred to as 'loose.'

NM (Near Mint, C9) — A toy that appears like new in overall appearance but exhibits very minor wear and does not have the original box. An exception would be a toy that comes in kit form. A kit toy in Near Mint condition would be expected to have the original box, but the box would display some wear.

EX (Excellent, C8) — A toy that is complete and has been played with. Signs of minor wear may be evident, but the toy is very clean and well cared for.

VG (Very Good, C7) — A toy that obviously has been played with and shows general wear overall. Paint chipping is readily apparent. In metal toys, some minor rust may be evident. In sets, some minor pieces may be missing.

GD (Good, C6) — A toy with evidence of heavy play, dents, chips, and possibly moderate rust. The toy may be missing a major replaceable component, such as a battery compartment door, or may be in need of repair. In sets, several pieces may be missing.

Ultimately, the market is driven by the checkbook, and the bottom line final value of a toy is often the last price at which it was sold. In this sense the toy hobby, like many others, operates in a vacuum, with individual toys setting their own values. Repeated sales of specific items create precedents and establish standards for asking prices across the market. It is by comparison of such multiple transactions that price guides such as this one are created. In the end, the value of a toy is decided by one person, the buyer.

Action Figures

Adventure heroes and superheroes are everywhere, on comic pages, television and movie screens, and, of course, in toy store aisles. The action figure likenesses produced by numerous toy companies are among today's hottest collectibles.

Action figure collecting is one of the fastest growing and potentially largest collectible areas since the baseball card boom of the 1980s. A stroll through the toy section of any store is proof enough. Plus, it is a given that a percentage of today's teen and preteen action figure buyers will become collectors, and their potential numbers are huge. Action figures could bring more collectors into the hobby than G.I. Joe, Hot Wheels and model kits combined.

Hundreds of figures are for sale currently, and they are commonplace in toy stores. In some places, action figures are literally climbing the walls. Why collect them if they can be bought directly from current store shelves? For many collectors, that's exactly how the collecting frenzy begins.

For many action figure collectors, time began in the 1960s. While boys had played with toy soldiers for hundreds of years, these were typically iron or lead figures with no movable parts. The same held true for the hard plastic Marx figures of the 1950s. By definition, however, the term "action figure" was born in the 1960s.

That decade also saw American culture and technologies come of age in ways that changed countless aspects of everyday life, including how toys would be made and sold.

Heroes from the TV Screen

By the late 1950s, television had replaced the dinner table and parlor radio as the family hearth. The sturdy cabinet in the living room captivated with a power only hinted at by radio and which has never been challenged since. It was a working window not only into a wide world of people and places, but also, increasingly, of neat things to buy. Youngsters clustered on the floor, soaking up the names and lore of their new friends and heroes — Wonder Woman, Superman, Batman, G.I. Joe.

From 1961 to 1963, toy makers watched with envy and despair as Mattel's Barbie, aided by TV, took the world of girls' toys by storm. Of course, no one would dream of selling dolls to boys, so this barrier seemed insurmountable. But the wheels of industry would not be easily stopped, and the simple solution to this dilemma ranks as one of the greatest marketing spins of all time. If boys won't play with dolls, why not rename them "action figures?"

Hasbro's first test of G.I. Joe, the male answer to Barbie, debuted at New York's International Toy Fair in early 1964. Toy Fair is where buyers, retailers and manufacturers meet to view upcoming lines — and in the process, make or break a toy's success.

Buyers met the 12-inch G.I. Joe with both hopes and reservations. They wanted to believe that a successful "Barbie for boys" had been created, but as much as Hasbro touted Joe as "America's Movable Fighting Man," the buyers still heard "doll."

Virtually no orders were generated at Toy Fair, so in June, with no fanfare or ad support, Hasbro released the new toy directly into a New York test market. Every test store sold out within a week and the invasion of America was on. By year's end, G.I. Joe had earned Hasbro $17 million, in spite of sales lost to product shortages.

Agent Fox Mulder is one of the figures in McFarlane's popular 1998 X-Files line.

G.I. Joe was the first true fully-articulated action figure for boys, but he wouldn't be alone for long. A.C. Gilbert introduced James Bond figures in 1965, but for the first time in his career, Ian Fleming's super spy failed in his mission. Marx also entered the ring with the Best of the West series, but G.I. Joe had a seemingly limitless arsenal of battle-geared appeal.

The first reasonably successful challenge to G.I. Joe came from Ideal's Captain Action. While Joe's identity was well established, Captain Action was a man of many faces. Ideal designed Captain Action to establish not only his own identity, but also to capitalize on those of many popular superheroes. Joe was just Joe, but Captain Action could become Spider-Man, Batman, the Phantom, Green Hornet and others. Today, Captain Action figures and sets command the second highest prices in the action figure market, second only to classic G.I. Joes.

Ideal's brief foray into the world of superhero action figures paved the way for many to come. While G.I. Joe was forced to temper his image and soften it from the quintessentially military Green Beret Joe of 1967 into the Adventure Team Joe of 1970, superheroes were largely immune to the Vietnam protests that forced Joe's change of mission. By 1969, Ideal tired of Captain Action's complex licensing agreements and discontinued the series, but another company was waiting in the wings. It was Mego.

Mighty Mego

In 1972, Mego released its first superhero series, the six-figure set of Official World's Greatest Super Heroes. These eight-inch tall cloth and plastic figures were joined by 28 others by the time the series ended 10 years later. Mego supplemented this superhero line with licensed film and TV characters from, most notably, *Planet of the Apes*, *Star Trek* and *The Dukes of Hazzard,* as well as historic figures representing the Old West and the World's Greatest Super Knights.

Another milestone in action figure history took place in 1977. Out of nowhere, George Lucas's *Star Wars* had become a worldwide smash, but nobody except Kenner had bothered to secure rights to merchandise toys. When Kenner realized the magnitude of *Star Wars*' potential, it rushed toys through production, but it didn't have time to get action figures on the shelves by Christmas. Instead, Kenner essentially presold the figures as the mail-order Early Bird set.

By Christmas, 1978, the line had grown to 17 figures and the first wave of a deluge of accessories and related toys. The Star Wars figures also established a third standard size for action figures. G.I. Joe and Captain Action were 12-inch figures, Mego figures measured eight inches, and Kenner's *Star Wars* figures were just 3-3/4 inches tall. Their tremendous popularity cemented that size as a new standard that holds to this day.

Next came the six-inch figure, set by Mattel's highly successful and lucrative 1981 Masters of the Universe series. This series was the first to be reverse licensed; in other words, Mattel made the toys first, and then sold the licensing to television and film, not the other way around. Mattel also upped the manufacturing ante by endowing the figures with action features such as punching and grabbing movements, thus enhancing their play value and setting another standard in the process.

Hasbro then scored again with the 1985 introduction of the next level in the evolution of action figures, the transforming figure. The aptly-named

Gene Simmons is part of McFarlane Toys' KISS line.

Transformers did just that, changing from innocuous-looking vehicles into menacing robots with a few deft twists, and then back again. Hasbro's mutating robots also transformed the toy industry, spawning numerous competitors and introducing the element of interchangeability into toy design. Hasbro, however, did not invent the transforming robot. That credit, as far as research shows, goes to a Japanese line called GoDaiKins. But Hasbro perfected the mass merchandising of the concept.

Today's generation of microchip-powered voice simulation and sound effect-laden toys are now the industry standard, but this will undoubtedly be made obsolete by future evolutions of controllability and interaction.

Action figures are big business, and hot series like *Star Trek* and Spawn are now regularly ranked in the top 20 best selling lines by industry trade magazines. An enduring character identity is a key to continued demand and future appreciation. Star Trek has proven itself a worthy long term franchise and is joining the ranks of Star Wars as the blue chip stocks of the action figure market.

The action figure aisles are now attracting more adults, and they are not always buying for their kids. More adults today are buying action figures as collectibles and investments. And those investments will in years hence feed the needs of tomorrow's collectors — the ones who are now sitting on the floor playing with Captain Picard, Batman and Spawn.

The Top 10 Action Figures
(excluding G.I. Joe and Captain Action)
(in Mint in Box condition)

1. Batgirl, Comic Heroines, Ideal, 1967 . . . . $4,500
2. Luke Skywalker w/telescoping saber, Star Wars, Kenner, 1977 3,230
3. Supergirl, Comic Heroines, Ideal, 1967 3,000
4. Mera, Comic Heroines, Ideal, 1967 3,000
5. Wonder Woman, Comic Heroines, Ideal, 1967 3,000
6. Jawa, w/vinyl cape, Star Wars, Kenner, 1977 2,610
7. Scorpio, Major Matt Mason, Mattel, 1967-70 2,250
8. Yak Face, Power of the Force, Kenner, 1985 1350
9. Boba Fett, 3-3/4", Star Wars, Kenner, 1978 700
10. IG-88, 12", Star Wars, Kenner 1980 650

The Top 10 G.I. Joe Figures / Sets
(All made by Hasbro. Prices are for items in Mint in Package condition)

1. Foreign Soldiers of the World, Action Soldiers of the World, 1968 . . . . $5,000
2. G.I. Nurse, Action Girl Series, 1967 4,000
3. Canadian Mountie Set, Action Soldier Series, 1967 4,000
4. Dress Parade Adventure Pack, Action Soldier Series, 1968 3,500
5. Crash Crew Fire Truck Set, Action Pilot Series, 1967 3,500
6. Talking Landing Signal Officer Set, Action Sailor Series, 1968 3,500
7. Talking Shore Patrol Set, Action Sailor Series, 1968 3,500
8. Adventure Pack, Army Bivouac Series, Action Soldier Series, 1968 3,500
9. Military Police Uniform Set, Action Soldier Series, 1967 3,500
10. Shore Patrol, Action Sailor Series, 1967 3,500

Contributors to this section:

Action Figures: Anthony Balasco, Figures, P.O. Box 19482, Johnston, RI 02919

Star Wars Figures: Chris Fawcett, cfawcett@ix.netcom.com

G.I. Joe: Vincent Santelmo, P.O. Box 789, New York, NY 10021; Dale Womer, The Joe Depot, P.O. Box 228, Kulpsville, PA 19443-0228

ACTION FIGURES

A-TEAM (GALOOB, 1984)

NAME	MNP	MIP
3-3/4" FIGURES AND ACCESSORIES		
A-Team Four Figure Set	12	30
Armored Attack Adventure with B.A. Figure	8	20
Bad Guys Figure Set: Viper, Rattle, Cobra, Python	10	25
Combat Attack Gyrocopter	10	25
Command Center Play Set	14	35
Corvette with Face Figure	8	20
Interceptor Jet Bomber with Murdock	10	25
Tactical Van Play Set	6	15
6-1/2" FIGURES AND ACCESSORIES		
Amy Allen	10	30
B.A. Baracus	8	25
Cobra	6	15
Face	6	25
Hannibal	6	25
Murdock	8	25
Off Road Attack Cycle	8	20
Python	6	15
Rattler	6	15
Viper	6	15

ACTION JACKSON (MEGO, 1974)

8" FIGURES		
Action Jackson, Black version	25	60
Action Jackson, blond, brown, or black beard	15	30
Action Jackson, blond, brown, or black hair	15	30
ACCESSORIES		
Parachute Plunge	5	15
Strap-On Helicopter	5	15
Water Scooter	5	15
OUTFITS		
Air Force Pilot	7	15
Army Outfit	7	15
Aussie Marine	7	15
Baseball	7	15
Fisherman	7	15
Football	7	15
Frog Man	7	15
Hockey	7	15
Jungle Safari	7	15
Karate	7	15
Navy Sailor	7	15
Rescue Squad	7	15
Scramble Cyclist	7	15
Secret Agent	7	15
Ski Patrol	7	15
Snowmobile Outfit	7	15
Surf and Scuba Outfit	7	15
Western Cowboy	7	15
PLAY SETS		
Jungle House	40	85
Lost Continent Play Set	40	85
VEHICLES		
Adventure Set	40	85
Campmobile	40	85
Dune Buggy	30	60
Formula Racer	30	60
Mustang	30	60
Rescue Helicopter	40	85

ACTION JACKSON (MEGO, 1974)

NAME	MNP	MIP
Safari Jeep	40	85
Scramble Cycle	20	40
Snowmobile	15	30

ALIENS (KENNER, 1979)

18" FIGURE		
Alien	200	500

ALIENS (KENNER, 1992-94)

1993, SERIES 2		
Alien vs. Predator	15	30
Flying Queen Alien	5	15
Queen Face Hugger	5	15
Snake Alien	5	15
1994		
Arachnid Alien	6	25
Atax	5	20
Clan Leader Predator	5	25
Cracked Tusk Predator	5	15
Invisible Predator (mail-in)	20	40
Kill Krab Alien	5	15
King Alien	10	25
Lasershot Predator (electronic)	15	30
Lava Predator	5	15
Mantis Alien	5	15
Night Cougar Alien	5	15
Night Storm Predator	6	15
Panther Alien	5	15
Rhino Alien	7	15
Spiked Tail Predator	4	15
Stalker Predator	4	15
Swarm Alien (electronic)	15	25
Wild Boar Alien	4	15
5" FIGURES, 1992, SERIES 1		
Apone	5	15
Bull Alien	7	15
Drake	5	15
Gorilla Alien	8	15
Hicks	5	15
Queen Alien	10	25
Ripley	5	15
Scorpion Alien	7	15
ACCESSORIES		
Evac Fighter	7	20
Hovertread	5	20
Power Loader	5	20
Stinger XT-37	5	20

AMERICAN WEST (MEGO, 1973)

8" FIGURES		
Buffalo Bill Cody, boxed	40	75
Buffalo Bill Cody, carded	40	100
Cochise, boxed	40	75
Cochise, carded	40	100
Davy Crockett, boxed	70	110
Davy Crockett, carded	70	140
Shadow (horse), carded	70	140
Sitting Bull, boxed	45	90
Sitting Bull, carded	45	125
Wild Bill Hickok, boxed	40	75
Wild Bill Hickok, carded	40	125

AMERICAN WEST (MEGO, 1973)

NAME	MNP	MIP
Wyatt Earp, boxed	40	75
Wyatt Earp, carded	40	125
PLAY SETS		
Dodge City Play Set, vinyl	100	200

ARCHIES (MARX, 1975)

Archie	15	75
Betty	15	75
Jughead	15	75
Veronica	15	75

ASTRONAUTS (MARX, 1969)

Jane Apollo Astronaut	35	75
Johnny Apollo Astronaut	35	75
Kennedy Space Center Astronaut	35	75

AVENGERS (TOY BIZ, 1998)

FIGURES		
Iron Man	5	10
Loki	5	10
Scarlett Witch	5	10
The Mighty Thor	5	10

BANANA SPLITS (SUTTON, 1970)

Bingo the Bear	45	125
Drooper the Lion	45	125
Fleagle Beagle	45	125
Snorky the Elephant	45	125

BATMAN CRIME SQUAD (KENNER, 1995)

ACCESSORIES		
Attack Jet	7	15
FIGURES		
Air Assault Batman	5	15
Land Strike Batman	5	15
Pirahna Blade Batman	5	15
Sea Claw Batman	5	15
Ski Blast Robin	5	15
Stealthwing Batman	5	15
Torpedo Batman	5	15

BATMAN DARK KNIGHT (KENNER, 1990-91)

FIGURES		
Blast Shield Batman	12	25
Bruce Wayne	7	20
Claw Climber Batman	12	25
Crime Attack Batman	7	15
Iron Winch Batman	7	15
Knockout Joker	25	75
Night Glider Batman	20	35
Power Wing Batman	12	25
Shadow Wing Batman	7	15
Sky Escape Joker	10	30
Thunder Whip Batman	12	25
Wall Scaler Batman	7	15

BATMAN FOREVER (KENNER, 1995)

NAME	MNP	MIP
FIGURES		
Blast Cape Batman	5	15
Fireguard Batman	5	15
Hydro Claw Robin	5	15
Manta Ray Batman	5	15
Night Hunter Batman	5	15
Riddler	10	20
Sonar Sensor Batman	4	10
Street Biker Robin	4	10
Transforming Bruce Wayne	4	10
Transforming Dick Grayson	4	10
Two Face	10	20

BATMAN RETURNS (KENNER, 1992-94)

FIGURES		
Aerostrike Batman	5	15
Air Attack Batman	4	15
Arctic Batman	4	15
Bola Strike Batman	4	15
Bruce Wayne	10	20
Catwoman	10	25
Claw Climber Batman	4	15
Crime Attack Batman	4	15
Deep Dive Batman	5	15
Glider Batman	4	15
High Wire Batman	4	15
Hydrocharge Batman	4	15
Jungle Tracker Batman	4	15
Laser Batman	4	15
Night Climber Batman	4	15
Penguin	15	40
Penguin Commandos	10	25
Polar Blast Batman	4	15
Power Wing Batman	6	15
Robin	10	25
Shadow Wing Batman	4	15
Sky Winch Batman	4	15
Thunder Strike Batman	4	15
Thunder Whip Batman	5	15
VEHICLES		
B.A.T.V. Vehicle	5	15
Bat Cycle	5	25
Bat-Signal Jet	3	15
Batmobile	10	50
Bruce Wayne Custom Coupe	12	50
Camo Attack Batmobile	30	75

BATMAN: THE ANIMATED SERIES (KENNER 1993-95)

ACCESSORIES		
Bat-Signal Jet	3	6
Batmobile	6	20
Hoverbat Vehicle	5	15
Joker Mobile	6	20
Robin Dragster	75	250
Street Jet	15	25
Turbo Batplane	6	20
FIGURES		
Anti-Freeze Batman	4	15
Bane	10	25
Bruce Wayne	10	20
Catwoman	7	25
Clay Face	7	25
Combat Belt Batman	7	40

BATMAN: THE ANIMATED SERIES (KENNER 1993-95)

NAME	MNP	MIP
Dick Grayson	5	15
Ground Assault Batman	5	10
Infrared Batman	5	10
Jet Pack Joker (green face)	10	25
Jet Pack Joker (white face)	10	25
Joker	7	20
Killer Croc	8	30
Knight Star Batman	4	10
Lightning Strike Batman	4	10
Manbat	7	25
Mechwing Batman	4	10
Mr. Freeze	8	20
Ninja Power Pack Batman and Robin	10	25
Ninja Robin	8	15
Parawing Robin	8	15
Penguin	12	85
Phantasm	15	40
Poison Ivy	20	30
Power Vision Batman	8	15
Riddler	10	50
Scarecrow	7	30
Skydive Batman	5	10
Total Armor Batman	4	10
Turbojet Batman	7	15
Two Face	7	20
Ultimate Batman (15")	25	50

BATTLESTAR GALACTICA (MATTEL, 1978-79)

12" FIGURES		
Colonial Warrior	20	55
Cylon Centurian	20	55
3-3/4" FIGURES, 1978, SERIES 1		
Commander Adama	15	30
Cylon Centurian	15	40
Daggit (brown)	15	30
Daggit (tan)	15	30
Imperious Leader	15	30
Ovion	12	25
Starbuck	15	30
3-3/4" FIGURES, 1979, SERIES 2		
Baltar	30	75
Boray	30	75
Cylon Commander	55	110
Lucifer	55	110

BEETLEJUICE (KENNER, 1989-90)

ACCESSORIES		
Creepy Cruiser	5	25
Phantom Flyer	7	15
Snake Mask	7	15
Vanishing Vault	10	20
FIGURES		
Adam Maitland	8	20
Exploding Beetlejuice	5	10
Harry the Haunted Hunter	8	20
Old Buzzard	8	20
Ortho the Obnoxious	8	20
Shipwreck Beetlejuice	5	15
Shish Kabab Beetlejuice	5	15
Showtime Beetlejuice	5	15
Spinhead Beetlejuice	5	15
Street Rat	8	20
Talking Beetlejuice, 12" tall	20	50
Teacher Creature	10	20

BEST OF THE WEST (MARX, 1960s)

NAME	MNP	MIP
FIGURES		
Bill Buck, 1967	100	200
Brave Eagle, 1967	45	90
Buckboard with Horse and Harness	35	75
Chief Cherokee, 1965	45	90
Daniel Boone, 1965	100	200
Davy Crockett	100	200
Fighting Eagle, 1967	45	90
General Custer, 1965	40	80
Geronimo and Pinto	40	80
Geronimo, 1967	45	90
Jamie West, 1967	32	65
Jane West, 1966	40	80
Janice West, 1967	32	65
Jay West, 1967	32	65
Johnny West Covered Wagon, with horse and harness	35	75
Johnny West with Comanche	80	125
Johnny West, 1965	40	80
Josie West, 1967	32	65
Pancho Horse, for 9" figures, 1968	20	40
Princess Wildflower, 1974	50	100
Sam Cobra, 1972	45	90
Sheriff Garrett, 1973	40	80
Thunderbolt Horse	35	75
Zeb Zachary, 1967	40	80

BLACK HOLE (MEGO, 1979-80)

12" FIGURES		
Captain Holland	25	50
Dr. Alan Durant	25	50
Dr. Hans Reinhardt	25	50
Harry Booth	30	60
Kate McCrae	35	80
Pizer	25	50
3-3/4" FIGURES		
Captain Holland, 1979	10	20
Dr. Alan Durant, 1979	10	20
Dr. Hans Reinhardt, 1979	10	25
Harry Booth, 1979	10	25
Humanoid, 1980	70	135
Kate McCrae, 1979	10	25
Maximillian, 1979	17	40
Old B.O.B., 1980	60	120
Pizer, 1979	15	25
S.T.A.R., 1980	60	120
Sentry Robot, 1980	25	60
V.I.N.cent., 1979	30	60

BUCK ROGERS (MEGO, 1979)

12" FIGURES		
Buck Rogers	30	80
Doctor Huer	30	60
Draco	30	60
Draconian Guard	30	60
Killer Kane	30	60
Tiger Man	30	80
Walking Twiki	30	60
3-3/4" FIGURES		
Ardella	6	15
Buck Rogers	20	40
Doctor Huer	6	15
Draco	6	15
Draconian Guard	10	20
Killer Kane	6	15

BUCK ROGERS (MEGO, 1979)

NAME	MNP	MIP
Tiger Man	6	15
Twiki	20	30
Wilma Deering	12	25
3-3/4" PLAY SETS		
Star Fighter Command Center	35	100
3-3/4" VEHICLES		
Draconian Marauder	25	50
Land Rover	20	40
Laserscope Fighter	20	40
Star Fighter	25	50
Starseeker	30	60

CAPTAIN ACTION (IDEAL, 1966-68)

12" FIGURES		
Captain Action, box photo, 1966	200	900
Captain Action, parachute offer on box, 1967	275	1150
Captain Action, photo box, 1967	200	1250
Captain Action, with blue-shirted Lone Ranger on box, 1966	200	900
Captain Action, with red-shirted Lone Ranger on box, 1966	200	900
Dr. Evil, 1967	300	1200
9" FIGURES		
Action Boy, 1967	275	650
Action Boy, with space suit, 1968	350	825
ACCESSORIES		
Action Cave Carrying Case, vinyl, 1967	400	700
Directional Communicator Set, 1966	110	300
Dr. Evil Sanctuary, 1967	600	750
Jet Mortar, 1966	110	300
Parachute Pack, 1966	100	225
Power Pack, 1966	125	250
Quick Change Chamber, Cardboard, Sears Exclusive, 1967	750	900
Silver Streak Amphibian, 1967	500	1000
Silver Streak Garage (with Silver Streak Vehicle, Sears Exclusive)	400	500
Survival Kit, 20 pieces, 1967	125	275
Vinyl Headquarters Carrying Case, Sears Exclusive, 1967	200	500
Weapons Arsenal, 10 pieces, 1966	110	225
ACTION BOY COSTUMES		
Aqualad, 1967	300	900
Robin, 1967	300	1200
Superboy, 1967	300	1000
CAPTAIN ACTION COSTUMES		
Aquaman, 1966	160	600
Aquaman, with flasher ring, 1967	180	750
Batman, 1966	225	1000
Batman, with flasher ring, 1967	250	1100
Buck Rogers, with flasher ring, 1967	450	2700
Captain America, 1966	220	900
Captain America, with flasher ring, 1967	225	900
Flash Gordon, 1966	200	1250
Flash Gordon, with flasher ring, 1967	225	1000
Green Hornet, with flasher ring, 1967	1500	5500
Lone Ranger, blue shirt, with flasher ring, 1967	300	1500
Lone Ranger, red shirt, 1966	170	1350
Phantom, 1966	150	875
Phantom, with flasher ring, 1967	175	900
Sergeant Fury, 1966	200	1000

CAPTAIN ACTION (IDEAL, 1966-68)

NAME	MNP	MIP
Spider-Man, with flasher ring, 1967	550	6000
Steve Canyon, 1966	150	1000
Steve Canyon, with flasher ring, 1967	175	1200
Superman, 1966	200	1200
Superman, with flasher ring, 1967	225	1300
Tonto, with flasher ring, 1967	375	1100

CHiPS (MEGO, 1979)

3-3/4" CARDED FIGURES & ACCESSORIES		
Jimmy Squeaks	5	15
Jon	10	20
Launcher with Motorcycle	25	50
Motorcycle (boxed)	5	30
Ponch	8	20
Sarge	10	25
Wheels Willie	5	15
8" CARDED FIGURES & ACCESSORIES		
Jon	20	50
Motorcycle	30	75
Ponch	15	40
Sarge	25	50

COMIC ACTION HEROES (MEGO, 1975)

3-3/4" FIGURES		
Aquaman	30	60
Batman	20	50
Captain America	20	50
Green Goblin	22	55
Hulk	20	50
Joker	20	50
Penguin	20	50
Robin	20	50
Shazam	20	50
Spider-Man	20	50
Superman	20	45
Wonder Woman	20	40
ACCESSORIES		
Collapsing Tower (w/Invisible Plane & Wonder Woman)	50	125
Exploding Bridge with Batmobile	75	150
Fortress of Solitude with Superman	100	200
Mangler	55	110

COMIC HEROINE POSIN' DOLLS (IDEAL, 1967)

12" BOXED FIGURES		
Batgirl	1000	4500
Mera	600	3000
Supergirl	600	3000
Wonder Woman	600	3000

COMMANDER POWER (MEGO, 1975)

FIGURE WITH VEHICLE		
Commander Power with Lightning Cycle	20	40

DC COMICS SUPER HEROES (TOY BIZ, 1989)

Aquaman	10	30

Exploding Beetlejuice, Kenner, 1989

CHiPS, Ponch, Mego, 1979

World's Greatest Super Heroes,
Iron Man, Mego, 1974

Buck Rogers, Mego, 1979

DC COMICS SUPER HEROES (TOY BIZ, 1989)

NAME	MNP	MIP
Batman	5	15
Bob The Goon	10	20
Flash	7	15
Flash II	10	25
Green Lantern	15	40
Hawkman	15	30
Joker, no Facial Hair	5	15
Joker, w/Facial Hair	10	25
Lex Luthor	5	15
Penguin, long missile	15	25
Penguin, short missile	15	30
Penguin, umbrella-firing	5	15
Riddler	7	15
Superman	20	40
Two Face	15	30
Wonder Woman	8	15

DIE-CAST SUPER HEROES (MEGO, 1979)

6" FIGURES		
Batman	30	75
Hulk	25	65
Spider-Man	30	75
Superman	30	75

DUKES OF HAZZARD (MEGO, 1981-82)

3-3/4" CARDED FIGURES		
Bo Duke	8	15
Boss Hogg	8	20
Cletus	15	30
Cooter	15	30
Coy Duke	15	30
Daisy Duke	12	25
Luke Duke	8	20
Rosco Coltrane	15	30
Uncle Jesse	15	30
Vance Duke	15	30
3-3/4" FIGURES WITH VEHICLES		
Daisy Jeep with Daisy, 1981, boxed	25	50
General Lee Car with Bo and Luke, 1981, boxed	25	50
8" CARDED FIGURES		
Bo Duke	15	30
Boss Hogg	20	40
Coy Duke (card says Bo)	25	50
Daisy Duke	25	50
Luke Duke	15	30
Vance Duke (card says Luke)	25	50

EARTHWORM JIM (PLAYMATES, 1995)

Bob	5	12
Earthworm Jim w/Battle Damage	5	8
Earthworm Jim w/Snott	5	8
Hench Rat w/Evil Cat	5	10
Monstrous Peter Puppy	5	8
Peter Puppy	5	8
Pocket Rocket	5	8
Princess What's Her Name	7	12
Psycrow w/Major Mucus	4	8

FANTASTIC FOUR (TOY BIZ, 1995)

NAME	MNP	MIP
10" BOXED FIGURES		
Dr. Doom	6	12
Human Torch	6	12
Silver Surfer	10	15
5" FIGURES		
Annihilus	3	10
Attuma	3	10
Black Bolt	3	10
Blastaar	3	10
Dr. Doom	3	10
Dragon Man	3	10
Firelord	3	10
Gorgon	3	10
Human Torch	3	10
Invisible Woman	10	20
Mole Man	5	10
Mr. Fantastic	3	10
Namor the Sub-Mariner	3	10
Silver Surfer	6	10
Super Skrull	3	10
Terrax	3	10
Thanos	3	10
Thing	5	10
Thing II	5	10
Triton	3	10
ELECTRONIC 14" FIGURES		
Galactus	10	40
Talking Thing	10	20
VEHICLES		
Fantasticar	7	15
Mr. Fantastic Sky Shuttle	7	15
The Thing's Sky Cycle	7	15

FLASH GORDON (MEGO, 1976)

9" FIGURES		
Dale Arden	35	70
Dr. Zarkow	55	110
Flash Gordon	55	110
Ming	30	60
PLAY SETS		
Flash Gordon Play Set	55	125

FORT APACHE FIGHTERS (MARX, 1960S)

Captain Maddox, 1967	35	70
Fighting Eagle and Comanche	50	100
Fighting Eagle, 1967	35	70
General Custer, 1967	35	70
Geronimo, 1967	35	70

GARGOYLES (KENNER, 1995)

ACCESSORIES		
Gargoyle Castle	20	40
Night Striker	10	20
Rippin' Rider Cycle	7	12
FIGURES		
Battle Goliath	3	15
Broadway	3	15
Bronx	3	15
Brooklyn	4	15
Claw Climber Goliath	3	15

GARGOYLES (KENNER, 1995)

NAME	MNP	MIP
Demona	4	15
Lexington	3	15
Mighty Roar Goliath	5	15
Power Wing Goliath	5	15
Quick Strike Goliath	3	15
Steel Clan Robot	3	15
Stone Armor Goliath	5	15
Strike Hammer Macbeth	3	15
Xanatos	3	15

GHOSTBUSTERS (KENNER, 1986-91)

1986		
Bad to the Bone Ghost	5	15
Banshee Bomber Gooper Ghost with Ecto-Plazm	5	15
Bug-Eye Ghost	5	15
Ecto-1	20	40
Egon Spengler & Gulper Ghost	6	15
Firehouse Headquarters	25	50
Ghost Pooper	5	10
Ghost Zapper	5	15
Gooper Ghost Sludge Bucket	5	15
Gooper Ghost Squisher with Ecto-Plazm	5	15
H2 Ghost	5	15
Peter Venkman & Grabber Ghost	6	15
Proton Pack	20	40
Ray Stantz & Wrapper Ghost	6	15
Slimer Plush Figure, 13"	20	35
Slimer with Pizza	20	40
Stay-Puft Marshmallow Man Plush, 13"	15	30
Winston Zeddmore & Chomper Ghost	7	18
1988		
Brain Blaster Ghost Haunted Human	5	15
Ecto-2 Helicopter	5	15
Fright Feature Egon	5	15
Fright Feature Janine Melnitz	5	15
Fright Feature Peter	5	15
Fright Feature Ray	5	15
Fright Feature Winston	5	15
Gooper Ghost Slimer	12	25
Granny Gross Haunted Human	5	15
Hard Hat Horror Haunted Human	5	15
Highway Haunter	10	20
Mail Fraud Haunted Human	5	15
Mini Ghost Mini-Gooper	5	10
Mini Ghost Mini-Shooter	5	10
Mini Ghost Mini-Trap	5	10
Pull Speed Ahead Ghost	5	15
Terror Trash Haunted Human	5	15
Tombstone Tackle Haunted Human	5	15
X-Cop Haunted Human	5	15
1989		
Dracula	5	15
Ecto-3	5	15
Fearsome Flush	5	10
Frankenstein	3	15
Hunchback	3	15
Mummy	3	15

GHOSTBUSTERS (KENNER, 1986-91)

NAME	MNP	MIP
Screaming Hero Egon	5	15
Screaming Hero Janine Melnitz	5	15
Screaming Hero Peter	5	15
Screaming Hero Ray	5	15
Screaming Hero Winston	5	15
Slimer with Proton Pack, red or blue	15	35
Super Fright Egon with Slimy Spider	5	15
Super Fright Janine with Boo Fish Ghost	5	15
Super Fright Peter Venkman & Snake Head	5	15
Super Fright Ray	5	15
Super Fright Winston Zeddmore & Meanie Wienie	5	15
Wolfman	5	15
Zombie	5	15
1990		
Ecto Bomber with Bomber Ghost	5	15
Ecto-1A with Ambulance Ghost	20	40
Ghost Sweeper	5	15
Gobblin' Goblin Nasty Neck	6	15
Gobblin' Goblin Terrible Teeth	6	15
Gobblin' Goblin Terror Tongue	6	15
Slimed Hero Egon	5	15
Slimed Hero Louis Tully & Four Eyed Ghost	5	15
Slimed Hero Peter Venkman & Tooth Ghost	5	15
Slimed Hero Ray Stantz & Vapor Ghost	5	15
Slimed Hero Winston	5	15
1991		
Ecto-Glow Egon	10	30
Ecto-Glow Louis Tully	10	30
Ecto-Glow Peter	10	30
Ecto-Glow Ray	10	30
Ecto-Glow Winston Zeddmore	10	30

GHOSTBUSTERS, FILMATION (SCHAPER, 1986)

Belfry and Brat-A-Rat	10	15
Bone Troller	10	15
Eddie	10	15
Fangster	10	15
Fib Face	10	15
Futura	10	15
Ghost Popper Ghost Buggy	20	40
Haunter	10	15
Jake	10	15
Jessica	10	15
Mysteria	10	15
Prime Evil	10	15
Scare Scooter Vehicle	10	20
Scared Stiff	6	15
Time Hopper Vehicle	10	15
Tracy	10	15

GI JOE

FIGURE SETS

12" HALL OF FAME FIGURES

NO.	NAME	DESCRIPTION	YEAR	EX	MNP	MIP
6837	Ace	Fighter Pilot	1993	15	20	25
6827	Cobra Commander	Cobra Leader	1992	20	30	35
6839	Destro	Weapons Manufacturer	1993	15	20	25
6019	Duke	Master Sergeant	1991	25	35	45
6826	Duke	Master Sergeant	1992	20	30	35
6111	Grunt	Infantry Squad Leader	1993	5	10	25
6849	Gung-Ho	Dress Marine	1993	15	20	25
6114	Heavy Duty	Heavy Ordinance Specialist	1993	15	20	35
6924	Rapid Fire	Commando	1993	35	50	60
6128	Rock `n Roll	Heavy Weapons Gunner	1993	15	20	25
6828	Snake Eyes	Commando	1992	20	30	35
6829	Stalker	Ranger	1992	20	30	35
6848	Storm Shadow	Ninja	1993	15	20	25
6117	Talking Duke	Talking Battle Commander	1993	25	35	50

3-3/4" SERIES #1, COBRA

NO.	NAME	DESCRIPTION	YEAR	EX	MNP	MIP
6423	Cobra	Infantry Soldier	1982	25	55	110
	Cobra Commander	mail order; Commanding Leader	1982	25	55	110
6424	Cobra Officer	Infantry Officer	1982	25	55	110
6426	Major Bludd	mail order; Mercenary w/card	1982	10	25	50

3-3/4" SERIES #1, GI JOE

NO.	NAME	DESCRIPTION	YEAR	EX	MNP	MIP
6403	Breaker	Communications Officer	1982	20	35	75
6406	Flash	Laser Rifle Trooper	1982	20	35	75
6409	Grunt	Infantry Trooper	1982	20	35	75
6408	Rock `n Roll	Machine Gunner	1982	20	35	75
6407	Scarlett	Counter Intelligence	1982	40	85	175
6402	Short Fuse	Mortar Soldier	1982	20	35	75
6404	Snake Eyes	Commando	1982	40	85	175
6401	Stalker	Ranger	1982	25	35	100
6405	Zap	Bazooka Soldier	1982	20	35	75

3-3/4" SERIES #2, COBRA

NO.	NAME	DESCRIPTION	YEAR	EX	MNP	MIP
6423	Cobra	Reissue	1983	25	50	100
6425	Cobra Commander	Reissue	1983	25	50	100
6424	Cobra Officer	Reissue	1983	25	50	100
6427	Destro	Enemy Weapons Supplier	1983	25	50	75
6426	Major Bludd		1983	15	30	65

3-3/4" SERIES #2, GI JOE

NO.	NAME	DESCRIPTION	YEAR	EX	MNP	MIP
6411	Airborne	Helicopter Assault Trooper	1983	15	35	75
6403	Breaker	Reissue	1983	15	30	65
6415	Doc	Medic	1983	10	20	45
	Duke	mail order; Master Sergeant	1983	10	25	40
6406	Flash	Reissue	1983	15	30	65
6409	Grunt	Reissue	1983	15	30	65
6414	Gung-Ho	Marine	1983	12	30	65
6408	Rock `n Roll	Reissue	1983	15	30	65
6407	Scarlett		1983	30	80	165
6402	Short Fuse	Reissue	1983	15	30	65
6404	Snake Eyes	Reissue	1983	40	80	165
6412	Snow Job	Arctic Trooper	1983	15	25	50
6401	Stalker	Reissue	1983	25	45	90
6413	Torpedo	Navy S.E.A.L.	1983	12	30	65
6410	Tripwire	Mine Detector	1983	15	30	65
6405	Zap	Reissue	1983	15	30	60

3-3/4" SERIES #3, COBRA

NO.	NAME	DESCRIPTION	YEAR	EX	MNP	MIP
6428	Baroness	Intelligence Officer	1983-84	40	85	175
6425	Cobra Commander	mail order; Enemy Leader with Hood	1983-84	10	20	40
6432	Fire Fly	Saboteur	1983-84	40	85	175
6431	Scrap Iron	Anti-Armor Specialist	1983-84	15	30	60
6429	Storm Shadow	Ninja	1983-84	35	75	150

3-3/4" SERIES #3, GI JOE

NO.	NAME	DESCRIPTION	YEAR	EX	MNP	MIP
6421	Blow Torch	Flamethrower	1983-84	10	20	40

FIGURE SETS

3-3/4" SERIES #3, GI JOE

NO.	NAME	DESCRIPTION	YEAR	EX	MNP	MIP
6422	Duke	First Sergeant	1983-84	15	25	50
6416	Mutt	Dog Handler with Dog	1983-84	10	25	50
6420	Recondo	Jungle Trooper	1983-84	15	25	50
6418	Rip-Cord	H.A.L.O. Jumper	1983-84	10	25	50
6419	Road Block	Heavy Machine Gunner	1983-84	15	25	50
6417	Spirit	Tracker with Eagle	1983-84	15	25	50

3-3/4" SERIES #4, COBRA

NO.	NAME	DESCRIPTION	YEAR	EX	MNP	MIP
6433	Buzzer	Mercenary	1984	10	25	50
6450	Crimson Guard	Elite Trooper	1984	15	30	60
6448	Eel	Frogman	1984	15	30	60
6434	Ripper	Mercenary	1984	15	30	60
6449	Snow Serpent	Polar Assault Trooper	1984	15	30	60
6447	Tele-Viper	Communications Trooper	1984	15	30	60
6063	Tomax	Crimson Guard Commander with Xamot	1984	30	65	125
6435	Torch	Mercenary	1984	15	30	60

3-3/4" SERIES #4, GI JOE

NO.	NAME	DESCRIPTION	YEAR	EX	MNP	MIP
6439	Air Tight	Hostile Environment Trooper	1984	20	40	85
6443	Alpine	Mountain Trooper	1984	15	25	50
6445	Barbecue	Fire Fighter	1984	10	25	50
6438	Bazooka	Missile Specialist	1984	10	25	50
6442	Dusty	Desert Trooper	1984	10	25	50
6436	Flint	Warrant Officer	1984	10	25	50
6444	Footloose	Infantry Trooper	1984	10	25	50
6440	Lady Jaye	Covert Operations Officer	1984	25	50	100
6441	Quick Kick	Silent Weapons Martial Artist	1984	20	40	85
6446	Shipwreck	Sailor and Parrot	1984	15	25	50
6437	Snake Eyes	Commando and Wolf	1984	30	60	125
6102	Tripwire	Mine Detector	1984	20	35	75

3-3/4" SERIES #5, COBRA

NO.	NAME	DESCRIPTION	YEAR	EX	MNP	MIP
6456	B.A.T.	Battle Android Trooper	1985	10	15	35
6461	Dr. Mindbender	Master of Mind Control	1985	10	15	35
6460	Monkey Wrench	Mercenary	1985	10	15	35
6473	Viper	Infantry Trooper	1985	10	15	35
6457	Zandar	Zartan's Brother Mercenary	1985	10	15	35
6472	Zarana	Reissue w/earrings	1985	30	65	125
6472	Zarana	Zartan's Sister Mercenary	1985	10	15	35

3-3/4" SERIES #5, GI JOE

NO.	NAME	DESCRIPTION	YEAR	EX	MNP	MIP
6463	Beach Head	Ranger	1985	10	15	30
6471	Dial Tone	Communications Expert	1985	10	15	30
6468	Hawk	Commander	1985	10	15	30
6466	Ice Berg	Snow Trooper	1985	10	15	30
6458	Leather Neck	Marine Gunner	1985	10	15	30
6465	Life Line	Rescue Trooper	1985	10	15	30
6459	Low-Light	Night Spotter	1985	10	15	30
6462	Main Frame	Computer Specialist	1985	10	15	30
6467	Road Block	Heavy Machine Gunner	1985	10	15	30
6469	Sci-Fi	Laser Trooper	1985	10	15	30
	Sgt. Slaughter	Mail order; Drill Instructor	1985	10	15	30
6470	Wet-Suit	Navy S.E.A.L.	1985	15	20	40

3-3/4" SERIES #6, COBRA

NO.	NAME	DESCRIPTION	YEAR	EX	MNP	MIP
6484	Big Boa	Troop Trainer	1986-87	5	10	20
6474	Cobra Commander	Cobra Leader with Battle Armor	1986-87	10	15	30
6154	Cobra-La Team	Three-figure set	1986-87	20	40	75
6487	Crocmaster	Reptile Trainer	1986-87	10	15	30
6479	Crystal Ball	Hypnotist	1986-87	5	10	20
6485	Raptor	Falconer	1986-87	5	10	20
6490	Techno-Viper	Battlefield Technician	1986-87	5	10	20

3-3/4" SERIES #6, GI JOE

NO.	NAME	DESCRIPTION	YEAR	EX	MNP	MIP
6482	Chuckles	Undercover M.P.	1986-87	3	5	10
6475	Crazy Legs	Air Assault Trooper	1986-87	10	15	30
6476	Falcon	Green Beret	1986-87	10	15	30
6488	Fast Draw	Mobile Missile Specialist	1986-87	10	15	25
6486	Gung-Ho	Marine in Dress Blues	1986-87	10	15	25

FIGURE SETS

3-3/4" SERIES #6, GI JOE

NO.	NAME	DESCRIPTION	YEAR	EX	MNP	MIP
6480	Jinx	Ninja Intelligence Officer	1986-87	10	15	25
6478	Law & Order	M.P. with Dog	1986-87	10	15	25
6483	Outback	Survivalist	1986-87	5	10	20
6477	Psych-Out	Deceptive Warfare Trooper	1986-87	5	10	20
	Secret Mission: Brazil	Toys R Us set w/four figures	1986-87	75	150	300
6153	Sgt. Slaughter Set	Three-figure set	1986-87	10	25	50
6491	Sneak Peek	Advanced Recon Trooper	1986-87	10	15	30
6481	Tunnel Rat	Underground Explosive Expert	1986-87	5	10	20

30TH SALUTE SERIES

NO.	NAME	DESCRIPTION	YEAR	EX	MNP	MIP
81271	30th Salute Black Action Soldier		1994	55	100	150
81047	Action Marine		1994	45	60	80
81046	Action Pilot		1994	50	75	125
81048	Action Sailor		1994	60	80	100
81045	Action Soldier		1994	25	50	95
	Green Beret Lt. Joseph Colton	mail order	1994	75	125	175

ACTION GIRL SERIES

NO.	NAME	DESCRIPTION	YEAR	EX	MNP	MIP
8060	G.I. Nurse	Red Cross hat and arm band, white dress, stockings, shoes, crutches, medic bag, stethescope, plasma bottle, bandages and splints.	1967	1750	2000	4000

ACTION MARINE SERIES

NO.	NAME	DESCRIPTION	YEAR	EX	MNP	MIP
7700	Action Marine	Fatigues, green cap, boots, dog tags, insignias and manual	1964	125	145	375
90711	Marine Medic Series	Red Cross helmet, flag and arm bands, crutch, bandages, splints, first aid pouch, stethoscope, plasma bottle, stretcher, medic bag, belt with ammo pouches	1967	325	425	3250
7790	Talking Action Marine		1967	175	200	850
90711	Talking Adventure Pack and Tent Set		1968	275	325	3250
90712	Talking Adventure Pack w/Field Pack Equipment		1968	275	325	3250

ACTION PILOT SERIES

NO.	NAME	DESCRIPTION	YEAR	EX	MNP	MIP
7800	Action Pilot	Orange jumpsuit, blue cap, black boots, dog tags, insignias, manual, catalog and club application	1964	130	165	600
7890	Talking Action Pilot		1967	190	245	1500

ACTION SAILOR SERIES

NO.	NAME	DESCRIPTION	YEAR	EX	MNP	MIP
7600	Action Sailor	White cap, denim shirt and pants, boots, dog tags, navy manual and insignias	1964	125	225	350
7643-83	Navy Scuba Set	Adventure Pack	1968	300	450	3250
7690	Talking Action Sailor		1967	200	330	1250
90621	Talking Landing Signal Officer Set	Talking Adventure Pack	1968	325	350	3500
90612	Talking Shore Patrol Set	Talking Adventure Pack	1968	200	450	3500

ACTION SOLDIER SERIES

NO.	NAME	DESCRIPTION	YEAR	EX	MNP	MIP
7500	Action Soldier	Fatigue cap, shirt, pants, boots, dog tags, army manual and insignias, helmet, belt with pouches, M-1 rifle	1964	100	175	350
7900	Black Action Soldier		1965	450	800	2500
5904	Canadian Mountie Set	Sears' exclusive	1967	850	1500	4000
8030	Desert Patrol Attack Jeep Set	Desert Fighter figure, jeep with steering wheel, spare tire, tan tripod, gun and gun mount and ring, black antenna, tan jacket and shorts, socks, goggles	1967	400	1250	2000
5969	Forward Observer Set	Sears' exclusive	1966	200	375	750
7536	Green Beret	Field radio, bazooka rocket, bazooka, green beret, jacket, pants, M-16 rifle, grenades, camo scarf, belt pistol and holster	1966	275	400	3000

Heavy Artillery Laser, 1983, 3-3/4-inch Series #3

Motorized Battle Tank (MOBAT), 3-3/4-inch Series #3

FIGURE SETS

ACTION SOLDIERS SERIES

NO.	NAME	DESCRIPTION	YEAR	EX	MNP	MIP
7531	Machine Gun Emplacement Set	Sears' exclusive	1965	150	275	1250
7590	Talking Action Soldier		1967	85	135	825
90513	Talking Adventure Pack, Bivouac Equipment		1968	275	325	3000
90517	Talking Adventure Pack, Command Post Equip.		1968	275	375	3000
7557-83	Talking Adventure Pack, Mountain Troop Series		1968	375	650	3500
90532	Talking Adventure Pack, Special Forces Equip.		1968	275	500	3500

ACTION SOLDIERS OF THE WORLD

NO.	NAME	DESCRIPTION	YEAR	EX	MNP	MIP
8105	Australian Jungle Fighter	action figure with jacket, shorts, socks, boots, bush hat, belt, "Victoria Cross" medal, knuckle knife, flamethrower, entrenching tool, bush knife and sheath	1966	250	400	2500
8205	Australian Jungle Fighter	Standard set with action figure uniform, no equipment	1966	150	275	1200
8204	British Commando	Standard set with no equipment	1966	150	275	1750
8104	British Commando	Deluxe set with action figure, helmet, night raid green jacket, pants, boots, canteen and cover, gas mask and cover, belt, Sten sub machine gun, gun clip and "Victoria Cross" medal	1966	300	425	2500
8111-83	Foreign Soldiers of the World	Talking Adventure Pack	1968	750	825	5000
8203	French Resistance Fighter	Standard set with action figure and equipment	1966	125	225	1250
8103	French Resistance Fighter	Deluxe set with figure, beret, short black boots, black sweater, denim pants, "Croix de Guerre" medal, knife, shoulder holster, pistol, radio, sub machine gun and grenades	1966	200	250	2250
8200	German Storm Trooper	Standard set with no equipment	1966	275	325	1300
8100	German Storm Trooper	Deluxe set with figure, helmet, jacket, pants, boots, Luger pistol, holster, cartridge belt, cartridges, "Iron Cross" medal, stick grenades, 9MM Schmeisser, field	1966	275	425	2500
8101	Japanese Imperial Soldier	Deluxe set with figure, Arisaka rifle, belt, cartridges, field pack, Nambu pistol, holster, bayonet, "Order of the Kite" medal, helmet, jacket, pants, short brown	1966	425	675	2700
8201	Japanese Imperial Soldier	Standard set with equipment	1966	300	325	1425
8102	Russian Infantry Man	Deluxe set with action figure, fur cap, tunic, pants, boots, ammo box, ammo rounds, anti-tank grenades, belt, bipod, DP light machine gun, "Order of Lenin" medal, field glasses and case	1966	275	400	2250
8202	Russian Infantry Man	Standard set with no equipment	1966	315	400	1250
5038	Uniforms of Six Nations		1967	750	950	2500

ADVENTURE TEAM

NO.	NAME	DESCRIPTION	YEAR	EX	MNP	MIP
7282	Air Adventurer	with Kung Fu grip	1974	95	125	325
7403	Air Adventurer	includes figure with Kung Fu grip, orange flight suit, boots, insignia, dog tags, rifle, boots, warranty, club insert	1970	120	375	300
7282	Air Adventurer	life-like body figure, uniform and equipment	1976	75	100	200
7283	Black Adventurer	With life-like body and Kung Fu grip	1976	85	125	225
7404	Black Adventurer	includes figure, shirt with insignia, pants, boots, dog tags, shoulder holster with pistol	1970	125	150	375
8026	Bulletman		1976	50	75	150

FIGURE SETS

ADVENTURE TEAM

NO.	NAME	DESCRIPTION	YEAR	EX	MNP	MIP
7278	Eagle Eye Black Commando		1976	85	125	250
7276	Eagle Eye Land Commander		1976	65	80	150
7277	Eagle Eye Man of Action		1976	65	80	165
8050	Intruder Commander		1976	50	75	150
8051	Intruder Warrior		1976	50	75	175
7280	Land Adventurer		1976	35	50	150
7280	Land Adventurer	with life-like body and Kung Fu grip and uniform set	1974	50	65	225
7270	Land Adventurer		1976	20	50	180
7401	Land Adventurer	includes figure, camo shirt and pants, boots, insignia, shoulder holster and pistol, dog tags and team inserts	1970	45	75	200
7500	Man of Action	includes figure, shirt and pants, boots, insignia, dog tags, team inserts	1970	50	75	225
7284	Man of Action	figure with life-like body and Kung Fu grip	1974	45	75	200
7274	Man of Action		1976	25	45	175
8025	Mike Powers/Atomic Man	figure with "atomic" flashing eye, arm that spins hand-held helicopter	1975	20	45	150
7402	Sea Adventurer	includes figure, shirt, dungarees, insignia, boots, shoulder holster and pistol	1970	45	70	245
7271	Sea Adventurer		1976	40	75	250
7281	Sea Adventurer	with life-like body and Kung Fu grip and uniform with equipment	1974	55	75	225
7281	Sea Adventurer		1976	55	85	200
8040	Secret Mountain Outpost		1975	50	85	150
7291	Talking Adventure Team Black Commander	with Kung Fu grip	1974	85	350	750
7406	Talking Adventure Team Black Commander		1973	150	225	600
7400	Talking Adventure Team Commander	includes figure, 2-pocket green shirt, pants, boots, insignia, instructions, dog tag, shoulder holster and pistol	1970	65	125	400
7290	Talking Adventure Team Commander	with Kung Fu grip	1974	75	200	500
7590	Talking Astronaut		1970	90	175	650
7291	Talking Black Commander		1976	125	300	600
7290	Talking Commander		1976	75	115	500
7292	Talking Man of Action		1976	75	120	525
7292	Talking Man of Action	with life-like body and Kung Fu grip	1974	75	200	650
7590	Talking Man of Action	shirt, pants, boots, dog tags, rifle, insignia, instructions	1970	75	125	350

ADVENTURES OF GI JOE

NO.	NAME	DESCRIPTION	YEAR	EX	MNP	MIP
7910	Aquanaut		1969	175	550	3000
7905	Negro Adventurer	Sears' exclusive, includes painted hair figure, blue jeans, pullover sweater, shoulder holster and pistol, plus product letter from Sears	1969	450	750	2750
7980	Sharks Surprise Set w/Frogman	with figure, orange scuba suit, blue sea sled, air tanks, harpoon, face mask, treasure chest, shark, instructions and comic	1969	125	300	750
7615	Talking Astronaut	hard-hand figure with white coveralls with insignias, white boots, dog tags	1969	85	275	1000

BATTLE CORPS

NO.	NAME	DESCRIPTION	YEAR	EX	MNP	MIP
81089	Alley-Viper		1994	3	5	10
81088	Beach-Head		1994	3	5	10
81002	Dialtone		1994	3	5	10
81001	Flint		1994	3	5	10
81004	Ice-Cream Soldier		1994	3	5	10
81007	Life-Line		1994	3	5	10
81012	Major Bludd		1994	3	5	10
81005	Metal-Head		1994	3	5	10
81098	Night Creeper Leader		1994	3	5	10

FIGURE SETS

BATTLE CORPS

NO.	NAME	DESCRIPTION	YEAR	EX	MNP	MIP
81003	Shipwreck		1994	3	5	10
81097	Snow Storm		1994	3	5	10
81008	Stalker		1994	3	5	10
81006	Viper		1994	3	5	10

GI JOE EXTREME

NO.	NAME	DESCRIPTION	YEAR	EX	MNP	MIP
81168	Ballistic		1995	5	6	8
81209	Ballistic		1995	7	8	10
81176	Freight		1995	5	6	8
81179	Inferno		1995	5	6	8
81177	Iron Klaw		1995	5	6	8
81208	Iron Klaw (Deluxe)		1995	7	8	10
81161	Lt. Stone		1995	5	6	8
81207	Lt. Stone (Deluxe)		1995	7	8	10
81295	Lt. Stone vs. Iron Klaw	Super Deluxe	1995	10	15	20
81167	Metalhead		1995	5	6	8
81206	Metalhead (Deluxe)		1995	7	8	10
81162	Sgt. Savage		1995	5	6	8

SHADOW NINJA

NO.	NAME	DESCRIPTION	YEAR	EX	MNP	MIP
81147	Bushido		1994	5	10	25
81146	Night Creeper		1994	5	10	25
81145	Nunchuck		1994	5	10	25
81144	Slice		1994	5	10	25
81141	Snake-Eyes		1994	5	10	25
81142	Storm Shadow		1994	5	10	25
81105	Cobra Astro-Viper		1993	5	10	25
81057	Cobra Blackstar		1994	5	10	25
81056	Cobra Commander		1994	5	10	25
81106	Cobra TARGAT		1993	5	10	25

STAR BRIGADE/GI JOE

NO.	NAME	DESCRIPTION	YEAR	EX	MNP	MIP
81061	Carcass		1994	5	10	25
81102	Countdown		1993	5	10	25
81119	Countdown		1994	5	10	25
81052	Duke		1994	5	10	25
81054	Effects		1994	5	10	25
81058	Lobotomaxx		1994	5	10	25
81103	Ozone		1993	5	10	25
81127	Ozone		1994	5	10	25
81117	Payload		1994	5	10	25
81101	Payload		1993	5	10	25
81059	Predacon		1994	5	10	25
81118	Roadblock		1994	5	10	25
81104	Roadblock		1993	5	10	25
81053	Sci-Fi		1994	5	10	25
81055	Space Shot		1994	5	10	25

SUPER JOE

NO.	NAME	DESCRIPTION	YEAR	EX	MNP	MIP
7510	Gor		1977	40	70	130
7506	Luminos		1977	45	70	130
7503	Super Joe		1977	20	35	70
7504	Super Joe (Black)		1977	35	50	100
7501	Super Joe Commander		1977	25	45	75
7505	The Shield		1977	40	65	125

UNIFORM/EQUIPMENT SETS

3-3/4" SERIES #1, GI JOE

NO.	NAME	DESCRIPTION	YEAR	EX	MNP	MIP
6075	F.L.A.K.	Attack Cannon	1982	20	45	90
6052	H.A.L.	Heavy Artillery Laser with Grand Slam	1982	20	45	90
6071	J.U.M.P.	Jet Pack with Platform	1982	20	45	90
6054	M.M.S.	Mobile Missile System with Hawk	1982	20	45	90

3-3/4" SERIES #2, COBRA

NO.	NAME	DESCRIPTION	YEAR	EX	MNP	MIP
6200	Headquarters Missile-Command Center	with three figures	1983	65	125	250
6083	S.N.A.K.E.	One-Man Battle Armor	1983	10	25	50

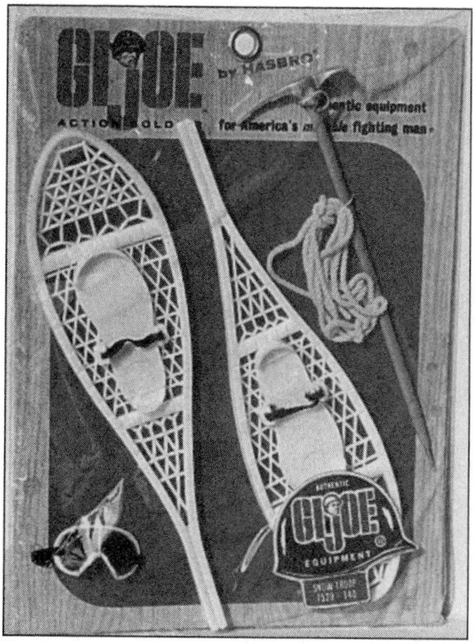

Action Soldier Snow Troop Set, 1966

Green Beret Machine Gun Outpost Set, 1966, Action Soldier Series

Adventure Team Flying Rescue Set, 1971

UNIFORM/EQUIPMENT SETS

3-3/4" SERIES #2, GI JOE

NO.	NAME	DESCRIPTION	YEAR	EX	MNP	MIP
6088	Battle Gear Accessory Pack #1		1983	10	20	40
6020	Headquarters Command Center		1983	40	85	175
6065	Jump	Jet Pack and Platform with Grand Slam	1983	25	50	100
6086-1	Pac/Rats Flamethrower	Remote Control Weapon	1983	10	25	50
6086-2	Pac/Rats Machine Gun	Remote Control Weapon	1983	10	25	50
6086-3	Pac/Rats Missile Launcher	Remote Control Weapon	1983	10	25	50
6074	Whirlwind	Twin Battle Gun	1983	10	25	45

3-3/4" SERIES #3, COBRA

NO.	NAME	DESCRIPTION	YEAR	EX	MNP	MIP
6070	A.S.P.	Assault System Pod	1983	15	30	60
6081-1	C.L.A.W.	Cobra Covert Light Aerial Weapons	1983	15	30	60
6081-2	S.N.A.K.E.	One-Man Armored Suit (white)	1983	10	25	50

3-3/4" SERIES #3, GI JOE

NO.	NAME	DESCRIPTION	YEAR	EX	MNP	MIP
6092	Battle Gear Accessory Pack #2		1983	10	20	40
6125-1	Bivouac	Battle Station	1983	10	25	35
6129-2	Machine Gun Defense Unit		1983	10	20	40
	Manta	mail order; Marine Assault Nautical Air Driven Transport	1983	10	20	35
6129-1	Missile Defense Unit		1983	10	15	25
6129-3	Mortar Defense Unit		1983	10	15	25
6125-3	Mountain Howitzer		1983	10	15	25
	Parachute	mail order; Parachute Pack with Working Parachute	1983	5	10	20
6125-2	Watchtower		1983	10	15	25

3-3/4" SERIES #4, COBRA

NO.	NAME	DESCRIPTION	YEAR	EX	MNP	MIP
6125	Cobra Bunker		1984	5	15	25
6081	Flight Pod	One-Man Bubble Pod	1984	5	15	25
6085	Night Landing	Mini Battlefield Vehicles Assortment	1984	5	15	25
6129	Rifle Range		1984	5	15	25

3-3/4" SERIES #4, GI JOE

NO.	NAME	DESCRIPTION	YEAR	EX	MNP	MIP
6125-2	Air Defense		1984	10	15	25
6129-1	Ammo Dump		1984	10	15	25
6092	Battle Gear Accessory Pack #3		1984	5	10	15
6085-2	Bomb Disposal		1984	10	15	25
6125-1	Check Point		1984	10	15	25
6129-2	Forward Observer		1984	10	15	25
6021	Tactical Battle Platform		1984	10	20	40
6085-1	Weapon Transport	Battlefield	1984	10	15	25

3-3/4" SERIES #5, COBRA

NO.	NAME	DESCRIPTION	YEAR	EX	MNP	MIP
6096	Battle Gear Accessory Pack #4		1985	5	10	15
6130	Surveillance Port Playset		1985	15	30	75
6003	Terror Drome	Armored Headquarters with Fireball Jet and A.V.A.C.	1985	75	150	300

3-3/4" SERIES #5, GI JOE

NO.	NAME	DESCRIPTION	YEAR	EX	MNP	MIP
6130	Outpost Defender Mini Playset		1985	5	10	15

3-3/4" SERIES #6, COBRA

NO.	NAME	DESCRIPTION	YEAR	EX	MNP	MIP
6133-3	Earth Borer		1986	5	10	15
6133-7	Mountain Climber		1986	5	10	15
6133-8	Pom-Pom Gun Pack		1986	5	10	15
6133-5	Rope Crosser		1986	5	10	15

3-3/4" SERIES #6, GI JOE

NO.	NAME	DESCRIPTION	YEAR	EX	MNP	MIP
6133-1	Antiaircraft Gun		1986	5	10	15
6677	Battle Gear Accessory Pack #5		1986	5	10	15

UNIFORM/EQUIPMENT SETS

3-3/4" SERIES #6, GI JOE

NO.	NAME	DESCRIPTION	YEAR	EX	MNP	MIP
6133-2	Helicopter Pack		1986	5	10	15
6006	Mobile Command Center Play Set		1986	25	50	100
6133-4	Rope Walker		1986	5	10	15
6172	S.L.A.M.	Strategic Long-Range Artillery Machine	1986	10	15	30
6098	Vehicle Gear Accessory Pack #1		1986	5	10	15

ACTION MARINE SERIES

NO.	NAME	DESCRIPTION	YEAR	EX	MNP	MIP
7713	Beachhead Assault Field Pack Set	M-1 rifle, bayonet, entrenching shovel and cover, canteen w/cover, belt, mess kit w/cover, field pack, flamethrower, first aid pouch, tent, pegs and poles, tent camo and camo	1964	100	175	325
7711	Beachhead Assault Tent Set	Tent, flamethrower, pistol belt, first-aid pouch, mess kit with utensils and manual	1964	100	200	475
7715	Beachhead Fatigue Pants		1964	15	30	200
7714	Beachhead Fatigue Shirt		1964	20	30	225
7712	Beachhead Field Pack	Cartridge belt, rifle, grenades, field pack, entrenching tool, canteen and manual	1964	40	65	150
7718	Beachhead Flamethrower Set	Reissue	1967	15	30	225
7718	Beachhead Flamethrower Set		1964	15	30	125
7716	Beachhead Mess Kit Set		1964	25	40	275
7717	Beachhead Rifle Set	Reissue	1967	30	50	225
7717	Beachhead Rifle Set	Bayonet, cartridge belt, hand grenades and M-1 rifle.	1964	30	50	150
7703	Communications Field Radio/Telephone Set	Reissue	1967	35	60	275
7703	Communications Field Set		1964	35	50	175
7704	Communications Flag Set	Flags for Army, Navy, Air Corps, Marines and United States.	1964	200	250	475
7702	Communications Poncho		1964	35	50	250
7701	Communications Post and Poncho Set	Field radio and telephone, wire roll, carbine, binoculars, map, case, manual, poncho	1964	125	175	475
7710	Dress Parade Set	Marine jacket, trousers, pistol belt, shoes, hat, M-1 rifle and manual	1964	125	225	450
7710	Dress Parade Set	Reissue	1968	125	225	750
7732	Jungle Fighter Set	Reissue	1968	450	700	2750
7732	Jungle Fighter Set	Bush hat, jacket w/emblems, pants, flamethrower, field telephone, knife and sheath, pistol belt, pistol, holster, canteen w/cover and knuckle knife	1967	450	700	3500
7726	Marine Automatic M-60 Machine Gun Set		1967	35	75	325
7722	Marine Basics Set		1966	55	85	275
7723	Marine Bunk Bed Set		1966	55	80	375
7723	Marine Bunk Bed Set	Reissue	1967	55	80	475
7730	Marine Demolition Set	Reissue	1968	50	100	450
7730	Marine Demolition Set	mine detector and harness, land mine	1966	50	100	350
7721	Marine First Aid Set	First-aid pouch, arm band and helmet	1964	45	85	125
7721	Marine First Aid Set	Reissue	1967	45	85	225
7720	Marine Medic Set	Reissue	1967	25	40	225
7720	Marine Medic Set	with crutch, etc.	1965	25	40	125
7719	Marine Medic Set w/stretcher	First-aid shoulder pouch, stretcher, bandages, arm bands, plasma bottle, stethoscope, Red Cross flag, and manual	1964	175	300	850
7725	Marine Mortar Set		1967	60	80	350
7727	Marine Weapons Rack Set		1967	75	145	625
7708	Paratrooper Camouflage Set	netting and foliage	1964	20	35	65

UNIFORM/EQUIPMENT SETS

ACTION MARINE SERIES

NO.	NAME	DESCRIPTION	YEAR	EX	MNP	MIP
7707	Paratrooper Helmet Set		1964	20	40	85
7709	Paratrooper Parachute Pack		1964	30	80	125
7706	Paratrooper Small Arms Set	Reissue	1967	30	75	225
7731	Tank Commander Set	Reissue	1968	325	500	1525
7731	Tank Commander Set	includes faux leather jacket, helmet and visor, insignia, radio with tripod, machine gun, ammo box	1967	325	500	1750

ACTION PILOT SERIES

NO.	NAME	DESCRIPTION	YEAR	EX	MNP	MIP
7822	Air Academy Cadet Set	deluxe set with figure, dress jacket, shoes, and pants, garrison cap, saber and scabbard, white M-1 rifle, chest sash and belt sash	1967	225	450	1250
7822	Air Academy Cadet Set	Reissue	1968	225	450	1150
7814	Air Force Basics Set	Reissue	1967	30	55	275
7814	Air Force Basics Set		1966	30	55	200
7816	Air Force Mae West Air Vest & Equipment Set		1967	85	125	325
7813	Air Force Police Set	Reissue	1967	70	150	325
7813	Air Force Police Set		1965	70	150	250
7815	Air Force Security Set	Air Security radio and helmet, cartridge belt, pistol and holster	1967	275	350	590
7825	Air/Sea Rescue Set	Reissue	1968	325	550	2500
7825	Air/Sea Rescue Set	includes black air tanks, rescue ring, buoy, depth gauge, face mask, fins, orange scuba outfit	1967	325	550	2500
7824	Astronaut Set	Helmet w/visor, foil space suit, booties, gloves, space camera, propellant gun, tether cord, oxygen chest pack, silver boots, white jumpsuit and cloth cap	1967	100	200	3000
7824	Astronaut Set	Reissue	1968	100	200	1250
7812	Communications Set		1964	55	100	225
7820	Crash Crew Set	fire proof jacket, hood, pants and gloves, silver boots, belt, flashlight, axe, pliers, fire extinguisher, stretcher, strap cutter	1966	125	250	450
7804	Dress Uniform Jacket Set		1964	40	65	250
7805	Dress Uniform Pants		1964	20	35	200
7803	Dress Uniform Set	Air Force jacket, trousers, shirt, tie, cap and manual	1964	225	450	3000
7806	Dress Uniform Shirt & Equipment Set		1964	25	40	200
7823	Fighter Pilot Set	Reissue	1968	400	650	2650
7823	Fighter Pilot Set	working parachute and pack, gold helmet, Mae West vest, green pants, flash light, orange jump suit, black boots	1967	400	650	2500
7812	Scramble Communications Set	Reissue	1967	35	75	250
7812	Scramble Communications Set	Poncho, field telephone and radio, map w/case, binoculars and wire roll	1965	35	75	175
7810	Scramble Crash Helmet	helmet, face mask, hose, tinted visor	1964	65	90	125
7810	Scramble Crash Helmet	Reissue	1967	65	90	225
7808	Scramble Flight Suit	gray flight suit	1964	50	300	225
7808	Scramble Flight Suit		1967	50	75	400
7811	Scramble Parachute Set		1964	20	40	150
7809	Scramble Parachute Set	Reissue	1967	20	40	250
7807	Scramble Set	Deluxe set, gray flight suit, orange air vest, white crash helmet, pistol belt w/.45 pistol, holster, clipboard, flare gun and parachute w/insert	1964	125	225	950
7802	Survival Life Raft Set	Raft with oar and sea anchor	1964	45	90	325
7801	Survival Life Raft Set	Raft with oar, flare gun, knife, air vest, first-aid kit, sea anchor and manual	1964	75	125	550

ACTION SAILOR SERIES

NO.	NAME	DESCRIPTION	YEAR	EX	MNP	MIP
7624	Annapolis Cadet	Reissue	1968	275	375	1350

Military Police Uniform Set,
1967, Action Soldier Series

Destro, 1993,
Hall of Fame Series

UNIFORM/EQUIPMENT SETS

ACTION SAILOR SERIES

NO.	NAME	DESCRIPTION	YEAR	EX	MNP	MIP
7624	Annapolis Cadet	Garrison cap, dress jacket, pants, shoes, sword, scabbard, belt and white M-1 rifle	1967	275	375	1350
7625	Breeches Buoy	Reissue	1968	325	425	1450
7625	Breeches Buoy	yellow jacket and pants, chair and pulley, flare gun, blinker light	1967	325	425	1500
7623	Deep Freeze	Reissue	1968	250	375	1500
7623	Deep Freeze	White boots, fur parka, pants, snow shoes, ice axe, snow sled w/rope and flare gun	1967	250	375	1600
7620	Deep Sea Diver Set	Reissue	1968	325	425	2000
7620	Deep Sea Diver Set	Underwater uniform, helmet, upper and lower plate, sledge hammer, buoy w/rope, gloves, compass, hoses, lead boots and weight belt	1965	325	425	2000
7604	Frogman Scuba Bottoms		1964	20	35	100
7606	Frogman Scuba Tank Set		1964	25	40	100
7603	Frogman Scuba Top Set		1964	25	45	125
7602	Frogman Underwater Demolition Set	Headpiece, face mask, swim fins, rubber suit, scuba tank, depth gauge, knife, dynamite and manual	1964	175	250	1500
7621	Landing Signal Officer	jumpsuit, signal paddles, goggles, cloth head gear, headphones, clipboard (complete), binoculars and flare gun.	1966	225	350	575
7610	Navy Attack Helmet Set	shirt and pants, boots, yellow life vest, blue helmet, flare gun binoculars, signal flags	1964	35	75	150
7611	Navy Attack Life Jacket		1964	20	45	120
7607	Navy Attack Set	life jacket, field glasses, blinker light, signal flags, manual	1964	60	125	425
7609	Navy Attack Work Pants Set		1964	25	40	150
7608	Navy Attack Work Shirt Set		1964	25	40	175
7628	Navy Basics Set		1966	25	55	125
7619	Navy Dress Parade Rifle Set		1965	35	65	125
7619	Navy Dress Parade Set	Billy club, cartridge belt, bayonet and white dress rifle	1964	45	80	175
7626	Navy L.S.O. Equipment Set	helmet, headphones, signal paddles, flare gun	1966	40	80	150
7627	Navy Life Ring Set	U.S.N. life ring, helmet sticker	1966	25	45	150
7618	Navy Machine Gun Set	MG and ammo box	1965	40	80	175
7601	Sea Rescue Set	life raft, oar, anchor, flare gun, first-aid kit, knife, scabbard, manual	1964	95	135	500
7622	Sea Rescue Set	Reissued with life preserver	1966	95	135	500
7612	Shore Patrol	Reissued with radio and helmet and shoes	1967	1000	2000	3500
7612	Shore Patrol	dress shirt, tie and pants, helmet, white belt, .45 and holster, billy club, boots, arm band, sea bag	1964	500	1000	2000
7613	Shore Patrol Dress Jumper Set		1964	75	125	225
7614	Shore Patrol Dress Pant Set		1964	40	75	175
7616	Shore Patrol Helmet and Small Arms Set	white belt, billy stick, white helmet, .45 pistol	1964	40	75	150
7615	Shore Patrol Sea Bag Set		1964	25	50	125

ACTION SOLDIER SERIES

NO.	NAME	DESCRIPTION	YEAR	EX	MNP	MIP
8006.83	Adventure Pack with 12 items	Adventure Pack Footlocker	1968	75	125	600
8005.83	Adventure Pack with 12 items	Adventure Pack Footlocker	1968	75	125	600
8008.83	Adventure Pack with 14 pieces	Adventure Pack Footlocker	1968	75	125	600
8007.83	Adventure Pack with 16 items	Adventure Pack Footlocker	1968	75	125	600

UNIFORM/EQUIPMENT SETS

ACTION SOLDIER SERIES

NO.	NAME	DESCRIPTION	YEAR	EX	MNP	MIP
7549-83	Adventure Pack, Army Bivouac Series		1968	225	450	3500
7813	Air Police Equipment	gray field phone, carbine, white helmet and bayonet	1964	40	95	200
8000	Basic Footlocker	wood tray with cardboard wrapper	1964	35	75	125
7513	Bivouac Deluxe Pup Tent Set	M-1 rifle and bayonet, shovel and cover, canteen and cover, mess kit, cartridge belt, machine gun, tent, pegs, poles, camoflage, sleeping bag, netting, ammo box	1964	115	225	450
7514	Bivouac Machine Gun Set	machine gun set and ammo box	1964	25	40	125
7514	Bivouac Machine Gun Set	Reissue	1967	25	40	225
7515	Bivouac Sleeping Bag	zippered bag	1964	20	30	125
7512	Bivouac Sleeping Bag Set	mess kit, canteen, bayonet, cartridge belt, M-1 rifle, manual	1964	25	30	150
7511	Combat Camouflaged Netting Set	foliage and posts	1964	25	40	85
7572	Combat Construction Set	orange safety helmet, work gloves, jack hammer	1967	325	400	575
7573	Combat Demolition Set		1967	65	100	525
7571	Combat Engineer Set	pick, shovel, detonator, dynamite, tripod and transit with grease gun	1967	125	175	625
7504	Combat Fatigue Pants Set		1964	15	25	110
7503	Combat Fatigue Shirt Set		1964	20	30	125
7505	Combat Field Jacket		1964	45	65	325
7501	Combat Field Jacket Set	Jacket, bayonet, cartridge belt, hand grenades, M-1 rifle and manual	1964	65	100	525
7506	Combat Field Pack & Entrenching Tool		1964	25	45	125
7502	Combat Field Pack Deluxe Set	field jacket, pack, entrenching shovel w/cover, mess kit, first-aid pouch, canteen w/cover	1964	75	125	325
7507	Combat Helmet Set	with netting and foliage leaves	1964	20	35	75
7509	Combat Mess Kit	plate, fork, knife, spoon, canteen, etc.	1964	20	45	85
7510	Combat Rifle and Helmet Set	bayonet, M-1 rifle, belt and grenades	1967	55	100	325
7508	Combat Sandbags Set	three bags per set	1964	10	40	85
7520	Command Post Field Radio and Telephone Set	Field radio, telephone with wire roll and map	1964	35	70	135
7520	Command Post Field Radio and Telephone Set	Reissue	1967	35	70	400
7519	Command Post Poncho	on card	1964	30	45	225
7517	Command Post Poncho Set	Poncho, field radio and telephone, wire roll, pistol, belt and holster, map and case and manual	1964	85	125	400
7518	Command Post Small Arms Set	Holster and .45 pistol, belt, grenades	1964	30	60	100
8009.83	Dress Parade Adventure Pack	Adventure Pack with 37 pieces	1968	750	1250	3500
7533	Green Beret and Small Arms Set		1966	85	110	300
7533	Green Beret and Small Arms Set	Reissue	1967	85	100	425
5978	Green Beret Machine Gun Outpost Set	Sears' exclusive w/two figures and equipment	1966	225	450	1500
7538	Heavy Weapons Set	mortar launcher and shells, M-60 machine gun, grenades, flak jacket, shirt and pants	1967	175	325	1750
7538	Heavy Weapons Set	Reissue	1968	175	325	1500
7523	Military Police Duffle Bag Set		1964	25	40	85
7526	Military Police Helmet and Small Arms Set	Reissue	1967	35	75	250

UNIFORM/EQUIPMENT SETS

ACTION SOLDIER SERIES

NO.	NAME	DESCRIPTION	YEAR	EX	MNP	MIP
7526	Military Police Helmet and Small Arms Set		1964	35	75	125
7524	Military Police Ike Jacket	Jacket with red scarf and arm band	1964	40	60	125
7525	Military Police Ike Pants	Matches Ike jacket	1964	20	30	100
7539	Military Police Uniform Set	includes green or tan uniform, black and gold MP Helmet, billy club, belt, pistol and holster, MP armband and red tunic	1967	450	1650	3500
7539	Military Police Uniform Set	Reissue	1968	450	900	3000
7521	Military Police Uniform Set	includes Ike jacket and pants, scarf, boots, helmet, belt with ammo pouches, .45 pistol and holster, billy club, armband, duffle bag	1964	450	1650	3000
7530	Mountain Troops Set	snow shoes, ice axe, ropes, grenades, camoflage pack, web belt, manual	1964	90	175	350
7516	Sabotage Set	Reissued in photo box	1968	125	250	1700
7516	Sabotage Set	dingy and oar, blinker light, detonator w/strap, TNT, wool stocking cap, gas mask, binoculars, green radio and .45 pistol and holster	1967	125	250	2000
7531	Ski Patrol Deluxe Set	White parka, boots, goggles, mittens, skis, poles and manual	1964	170	350	1250
7527	Ski Patrol Helmet and Small Arms Set		1965	35	75	135
7527	Ski Patrol Helmet and Small Arms Set	Reissue	1967	75	125	250
7529	Snow Troop Set	Reissue	1967	20	45	225
7529	Snow Troop Set	snow shoes, goggles and ice pick	1966	20	45	150
7528	Special Forces Bazooka Set		1966	35	45	225
7528	Special Forces Bazooka Set	Reissue	1967	35	45	325
7532	Special Forces Uniform Set		1966	200	375	1000
7537	West Point Cadet Uniform Set	Dress jacket, pants, shoes, chest and belt sash, parade hat w/plume, saber, scabbard and white M-1 rifle	1967	250	475	1500
7537	West Point Cadet Uniform Set	Reissue	1968	250	375	1200

ACTION SOLDIERS OF THE WORLD

NO.	NAME	DESCRIPTION	YEAR	EX	MNP	MIP
8305	Australian Jungle Fighter Set		1966	25	50	250
8304	British Commando Set	Sten submachine gun, gas mask and carrier, canteen and cover, cartridge belt, rifle, "Victoria Cross" medal, manual	1966	125	200	325
8303	French Resistance Fighter Set	shoulder holster, Lebel pistol, knife, grenades, radio, 7.65 submachine gun, "Croix de Guerra" medal, counter-intelligence manual	1966	25	50	275
8300	German Storm Trooper		1966	125	175	325
8301	Japanese Imperial Soldier Set	field pack, Nambu pistol and holster, Arisaka rifle with bayonet, cartridge belt, "Order of the Kite" medal, counter-intelligence manual	1966	175	275	625
8302	Russian Infantry Man Set	DP light machine gun, bipod, field glasses and case, anti-tank grenades, ammo box, "Order of Lenin" medal, counter-intelligence medal	1966	175	220	325

ADVENTURE TEAM

NO.	NAME	DESCRIPTION	YEAR	EX	MNP	MIP
7490	Adventure Team Headquarters Set	Adventure Team playset	1972	50	125	200
7495	Adventure Team Training Center Set	rifle rack, logs, barrel, barber wire, rope ladder, 3 tires, 2 targets, escape slide, tent and poles, first aid kit, respirator and mask, snake, instructions	1973	75	125	225

Short Fuse, 1982, 3-3/4-inch Series #1

Black Action Soldier, 1965

Astro Locker, 1969, Adventures of G.I. Joe Series

UNIFORM/EQUIPMENT SETS

ADVENTURE TEAM

NO.	NAME	DESCRIPTION	YEAR	EX	MNP	MIP
7345	Aerial Reconnaissance Set	jumpsuit, helmet, aerial recon vehicle with built-in camera	1971	75	125	225
7420	Attack at Vulture Falls	Super Deluxe Set	1975	75	150	275
7414	Black Widow Rendezvous	Super Deluxe Set	1975	125	200	350
7328-5	Buried Bounty	Deluxe Set	1975	10	25	85
7437	Capture of the Pygmy Gorilla Set		1970	100	175	325
8032	Challenge of Savage River	Deluxe Set	1975	100	175	350
7313	Chest Winch Set	Reissue	1974	10	15	75
7313	Chest Winch Set		1972	10	15	40
8033	Command Para Drop	Deluxe Set	1975	200	300	550
7308-3	Copter Rescue Set	blue jumpsuit, red binoculars	1973	15	20	30
7412	Danger of the Depths Set		1970	100	175	325
7338-1	Danger Ray Detection	magnetic ray detector, solar communicator with headphones, two-piece uniform, instructions and comic	1975	45	90	225
7309-2	Dangerous Climb Set		1973	20	35	75
7608-5	Dangerous Mission Set	green shirt, pants, hunting rifle	1973	20	35	75
7370	Demolition Set	armored suit, face shield, bomb, bomb disposal box, extension grips	1971	20	45	125
7371	Demolition Set	with land mines, mine detector and carrying case with metallic suit	1971	75	100	250
7309-5	Desert Explorer Set		1973	20	40	80
7308-6	Desert Survival Set		1973	20	40	80
8031	Dive to Danger	Mike Powers set, orange scuba suit, fins, mask, spear gun, shark, buoy, knife and scabbard, mini sled, air tanks, comic	1975	150	250	450
7328-6	Diver's Distress		1975	35	70	125
7364	Drag Bike Set	three-wheel motorcycle brakes down to backpack size	1971	25	65	125
7422	Eight Ropes of Danger Set		1970	125	225	375
7374	Emergency Rescue Set	shirt, pants, rope ladder and hook, walkie talkie, safety belt, flashlight, oxygen tank, axe, first aid kit	1971	45	75	150
7319-5	Equipment Tester Set		1972	15	20	40
7360	Escape Car Set		1971	30	60	85
7319-1	Escape Slide Set		1972	15	25	40
8028-2	Fangs of the Cobra	Deluxe Set	1975	125	200	375
7423	Fantastic Freefall Set		1970	125	200	375
7431	Fight For Survival Set	with blue parka	1970	300	550	2500
7308-2	Fight for Survival Set	brown shirt and pants, machete	1973	20	30	45
7982	Fight for Survival Set w/Polar Explorer		1969	250	450	850
7351	Fire Fighter Set		1971	20	30	55
7361	Flying Rescue Set		1971	35	60	85
7425	Flying Space Adventure Set		1970	400	600	1000
8000	Footlocker	green plastic with cardboard wrapper	1974	35	70	225
7328-4	Green Danger		1975	30	45	60
7415	Hidden Missile Discovery Set		1970	100	200	450
7308-1	Hidden Treasure Set	shirt, pants, pick axe, shovel	1973	15	25	40
7342	High Voltage Escape Set	net, jumpsuit, hat, wrist meter, wire cutters, wire, warning sign	1971	40	75	150
7343	Hurricane Spotter Set	slicker suit, rain measure, portable radar, map and case, binoculars	1971	55	80	175
7421	Jaws of Death	Super Deluxe Set	1975	325	500	650
7339-2	Jettison to Safety	infrared terrain scanner, mobile rocket pack, two-piece flight suit, instructions and comic	1975	85	200	275
7309-3	Jungle Ordeal Set		1973	15	25	45
7373	Jungle Survival Set		1971	15	25	45
7372	Karate Set		1971	35	70	125
7311	Laser Rescue Set	Reissue	1974	20	35	100
7311	Laser Rescue Set	hand-held laser with backpack generator	1972	20	35	45

UNIFORM/EQUIPMENT SETS

ADVENTURE TEAM

NO.	NAME	DESCRIPTION	YEAR	EX	MNP	MIP
7353	Life-Line Catapult Set		1971	15	25	40
7328-3	Long Range Recon	Deluxe Set	1975	10	20	35
7319-2	Magnetic Flaw Detector Set		1972	10	20	30
7339-3	Mine Shaft Breakout	sonic rock blaster, chest winch, two-piece uniform, netting, instructions, comic	1975	70	125	250
7340	Missile Recovery Set		1971	40	55	85
	Mystery of the Boiling Lagoon	Sears', pontoon boat, diver's suit, diver's helmet, weighted belt and boots, depth gauge, air hose, buoy, nose cone, pincer arm, instructions	1973	150	200	225
7338-2	Night Surveillance	Deluxe Set	1975	35	45	90
7416	Peril of the Raging Inferno	fireproof suit, hood and boots, breathing apparatus, camera, fire extinguisher, detection meter, gaskets	1975	85	150	275
7309-4	Photo Reconnaissance Set		1973	20	30	45
8028-1	Race for Recovery		1975	20	35	125
7341	Radiation Detection Set	jumpsuit with belt, "uranium ore", goggles, container, pincer arm.	1971	30	50	85
7339-1	Raging River Dam Up		1975	60	90	150
7350	Rescue Raft Set		1971	15	20	65
7413	Revenge of the Spy Shark	Super Deluxe Set	1975	50	175	400
7312	Rock Blaster	sonic blaster with tripod, backpack generator, face shield	1972	10	20	35
7315	Rocket Pack Set	Reissue	1974	10	20	50
7315	Rocket Pack Set		1972	10	20	75
7319-3	Sample Analyzer Set		1972	10	20	45
7439.16	Search for the Abominable Snowman Set	Sears, white suit, belt, goggles, gloves, rifle, skis and poles, show shoes, sled, rope, net, supply chest, binoculars, Abominable Snowman, comic book	1973	110	175	350
7375	Secret Agent Set		1971	30	55	175
7328-1	Secret Courier		1975	45	65	135
7309-1	Secret Mission Set		1973	45	65	135
8030	Secret Mission Set	Deluxe Set	1975	65	95	200
7411	Secret Mission to Spy Island Set	comic, inflatable raft with oar, binoculars, signal light, flare gun, TNT and detonator, wire roll, boots, pants, sweater, black cap, camera, radio with earphones, .45 submachine gun	1970	75	125	250
7308-4	Secret Rendezvous Set	parka, pants, flare gun	1973	10	20	35
7319-6	Seismograph Set		1972	10	20	35
7338-3	Shocking Escape	escape slide, chest pack climber, jumpsuit with gloves and belt, high voltage sign, instructions and comic	1975	25	65	125
7362	Signal Flasher Set	large back pack type signal flash unit	1971	15	30	50
7440	Sky Dive to Danger	Super Deluxe Set	1975	90	150	325
7314	Solar Communicator Set	Reissue	1974	10	20	95
7314	Solar Communicator Set		1972	10	20	35
7312	Sonic Rock Blaster Set		1972	10	20	35
7312	Sonic Rock Blaster Set	Reissue	1974	10	20	35
8028-3	Special Assignment	Deluxe Set	1975	30	55	135
7319-4	Thermal Terrain Scanner Set		1972	25	35	50
7480	Three-in-One Super Adventure Set	Danger of the Depths, Secret Mission to Spy Island and Flying Space Adventure Packs	1971	550	975	1250
7480	Three-in-One Super Adventure Set	Cold of the Arctic, Heat of the Desert and Danger of the Jungle	1971	250	400	750
7328-2	Thrust into Danger	Deluxe Set	1975	45	55	175
59289	Trouble at Vulture Pass	Sears' exclusive, Super Deluxe Set	1975	75	125	325
7363	Turbo Copter Set	strap-on one man helicopter	1971	15	35	65
7309-6	Undercover Agent Set	trenchcoat and belt, walkie-talkie	1973	15	30	35
7310	Underwater Demolition Set	hand-held propulsion device, breathing apparatus, dynamite	1972	15	20	40

UNIFORM/EQUIPMENT SETS

ADVENTURE TEAM

NO.	NAME	DESCRIPTION	YEAR	EX	MNP	MIP
7310	Underwater Demolition Set	Reissue	1974	10	20	75
7354	Underwater Explorer Set	self propelled underwater device	1971	15	30	60
7344	Volcano Jumper Set	jumpsuit with hood, belt, nylon rope, chest pack, TNT pack	1971	45	80	250
7436	White Tiger Hunt Set	hunter's jacket and pants, hat, rifle, tent, cage, chain, campfire, white tiger, comic	1970	80	125	275
7353	Windboat Set	back pack, sled with wheels, sail	1971	10	25	55
7309-4	Winter Rescue Set	Replaced Photo Reconnaissance Set	1973	40	75	150

ADVENTURES OF GI JOE

NO.	NAME	DESCRIPTION	YEAR	EX	MNP	MIP
7940	Adventure Locker	Footlocker	1969	80	165	350
7941	Aqua Locker	Footlocker	1969	90	180	375
7942	Astro Locker	Footlocker	1969	90	180	375
7920	Danger of the Depths Underwater Diver Set		1969	140	275	500
7950	Eight Ropes of Danger Set	diving suit, treasure chest, octopus	1969	110	225	525
7951	Fantastic Freefall Set	includes figure with parachute and pack, blinker light, air vest, flash light, crash helmet with visor and oxygen mask, dog tags, orange jump suit, black boots	1969	150	325	675
7982.83	Flight for Survival Set w/o Polar Explorer	Reissue	1969	150	300	500
7952	Hidden Missile Discovery Set		1969	70	135	400
7953	Mouth of Doom Set		1969	125	250	550
7921	Mysterious Explosion Set	basic	1969	60	125	425
7923	Perilous Rescue Set	basic	1969	150	300	500
7922	Secret Mission to Spy Island Set	basic	1969	110	225	450

GI JOE ACTION SERIES, ARMY, NAVY, MARINE AND AIR FORCE

NO.	NAME	DESCRIPTION	YEAR	EX	MNP	MIP
8000	Basic Footlocker		1965	50	75	175
8000.83	Footlocker Adventure Pack	16 pieces	1968	65	135	450
8002.83	Footlocker Adventure Pack	22 pieces	1968	70	145	450
8001.83	Footlocker Adventure Pack	15 pieces	1968	65	135	450
8002.83	Footlocker Adventure Pack	15 pieces	1968	65	135	450

SUPER JOE

NO.	NAME	DESCRIPTION	YEAR	EX	MNP	MIP
7528-1	Aqua Laser		1977	10	20	30
7518-2	Edge of Adventure		1977	10	20	35
7518-3	Emergency Rescue		1977	10	20	30
7528-3	Fusion Bazooka		1977	10	20	30
7538-2	Helipak		1977	10	20	30
7518-1	Invisible Danger		1977	10	20	35
7538-1	Magna Tools		1977	10	20	30
7518-4	Path of Danger		1977	10	20	30
7538-3	Sonic Scanner		1977	10	20	30
7528-2	Treacherous Dive		1977	10	20	30

VEHICLE SETS

3-3/4" SERIES #1, COBRA

NO.	NAME	DESCRIPTION	YEAR	EX	MNP	MIP
	C.A.T.	Motorized Crimson Attack Tank	1982	15	25	55

3-3/4" SERIES #1, GI JOE

NO.	NAME	DESCRIPTION	YEAR	EX	MNP	MIP
6000	M.O.B.A.T.	Motorized Battle Tank with Steeler	1982	30	65	125
6073	R.A.M.	Rapid Fire Motorcycle	1982	15	25	55
6050	V.A.M.P.	Multi-Purpose Attack Vehicle with Clutch	1982	20	35	70

Sky Dive to Danger, 1975 Adventure Team

Official Sea Sled and Frogman Set, 1966, Action Sailor Series

VEHICLE SETS

3-3/4" SERIES #2, COBRA

NO.	NAME	DESCRIPTION	YEAR	EX	MNP	MIP
6097	Cobra Glider	Attack Glider with Viper	1983	35	75	150
6077	F.A.N.G.	Fully Armed Negator Gyro Copter	1983	10	20	45
6051	H.I.S.S.	High Speed Sentry Tank with H.I.S.S.	1983	25	50	100

3-3/4" SERIES #2, GI JOE

NO.	NAME	DESCRIPTION	YEAR	EX	MNP	MIP
6093	A.P.C.	Amphibious Personnel Carrier	1983	15	25	50
4025	Dragon Fly XH-1	Assault Copter with Wild Bill	1983	25	50	100
6097	Falcon	Attack Glider with Grunt	1983	30	75	150
6072	Polar Battle Bear	Sky Mobile	1983	15	25	50
6010	Sky Striker XP-14F	F-14 Jet and Parachute with Ace	1983	35	75	150
6048	Wolverine	Armored Missile Vehicle with Cover Girl	1983	25	50	100

3-3/4" SERIES #3, COBRA

NO.	NAME	DESCRIPTION	YEAR	EX	MNP	MIP
6027	Rattler	Ground Attack Jet with Wild Weasel	1983-84	25	50	100
6055	Stinger	Night Attack Jeep with Cobra Officer	1983-84	20	40	80
6064	Swamp Skier	Chameleon Vehicle with Zartan	1983-84	30	55	125
6058	Water Moccasin	Swamp Boat with Copperhead	1983-84	20	40	80

3-3/4" SERIES #3, GI JOE

NO.	NAME	DESCRIPTION	YEAR	EX	MNP	MIP
7444-3	Attack Cannon (FLAK)		1983	5	10	15
7444-1	Attack Vehicle (VAMP)		1983	15	35	70
7444-4	Battle Tank (MOBAT)		1983	15	25	50
7444-2	Heavy Artillery Laser (HAL)		1983	10	15	25
6005	Killer W.H.A.L.E.	Armored Hovercraft with Cutter	1983	30	65	125
7444-5	Mobile Missile System (MMS)		1983	10	15	30
74450	RAM, HAL & VAMP	three-piece, die-cast set	1983	10	20	40
7444-6	Rapid-Fire Motorcycle (RAM)		1983	15	20	30
6049	S.H.A.R.C.	Submersible High-Speed Attack & Recon Craft with Deep Six	1983	25	50	100
6079	Sky Hawk	V.T.O.L. Jet	1983	10	15	30
6056	Slugger	Self-Propelled Cannon with Thunder	1983	15	25	50
6680	Vamp Jeep with H.A.L.	Attack Vehicle with Heavy Artillery Laser Cannon	1983	20	35	75
6055	Vamp Mark II	Desert Jeep with Clutch	1983	25	40	80

3-3/4" SERIES #4, COBRA

NO.	NAME	DESCRIPTION	YEAR	EX	MNP	MIP
6069	Ferret	All-Terrain Vehicle	1984	10	15	30
6024	Moray	Hydrofoil with Lamprey	1984	15	25	50
6687	Motorized Crimson Attack Tank	MOBAT Tank	1984	30	60	125
6686	Sentry and Missile System	Sears', with H.I.S.S. Tank, Cobra Commander, Officer and Soldier	1984	75	150	250

3-3/4" SERIES #4, GI JOE

NO.	NAME	DESCRIPTION	YEAR	EX	MNP	MIP
6053	A.W.E. Striker	All-Weather Environment Jeep with Crankcase	1984	15	25	50
6078	Armadillo	One-Man Mini-Tank	1984	10	15	35
6023	Bridge Layer	Bridge Laying Trank with Toll Booth	1984	15	25	50
6015	Mauler	Motorized Tank with Heavy Metal	1984	15	25	50
6076	Silver Mirage	Motorcycle with Sidecar	1984	10	15	35
6057	Snowcat	Snow Half-Track Vehicle with Frost-Bite	1984	15	25	50
6001	U.S.S. Flagg	Aircraft Carrier with Admiral Keel Haul	1984	125	250	500

3-3/4" SERIES #5, COBRA

NO.	NAME	DESCRIPTION	YEAR	EX	MNP	MIP
	Air Assault	Air Vehicle	1985	10	15	25
6062	Air Chariot	Vehicle with Serpentor "Cobra Emperor"	1985	15	20	55
	Ground Assault	Land Vehicle	1985	10	20	25
6099-2	Hydro Sled		1985	10	15	25
6099j-1	Jet Pack	One-Man Jet Set	1985	10	15	25
6014	Night Raven S-3P	Surveillance Jet with Drone Pod and Strato Viper	1985	20	35	70
6041	Stun	Split Attack Vehicle with Motor Viper	1985	10	20	35
6068	Swamp Fire	Air/Swamp Transforming Vehicle with Color-Change	1985	10	15	30
6042	Thunder Machine	Compilation Vehicle of Spare Parts with Thrasher	1985	10	20	35

VEHICLE SETS

3-3/4" SERIES #5, GI JOE

NO.	NAME	DESCRIPTION	YEAR	EX	MNP	MIP
6031	Conquest X-30	Super-Sonic Jet with Slip Stream	1985	15	30	65
6066	Devil Fish	High-Speed Attack Boat	1985	10	15	30
6030	H.A.V.O.C.	Heavy Artillery Vehicle Ordinance Carrier with Cross-Country	1985	10	15	35
6067	L.V.C. Recon Sled	Low-Crawl Vehicle Cycle	1985	10	15	30
6022	Tomahawk	Troop Transit Helicopter with Lift Ticket	1985	15	30	65
6061	Triple T	One-Man Tank with Sgt. Slaughter	1985	10	25	50

3-3/4" SERIES #6, COBRA

NO.	NAME	DESCRIPTION	YEAR	EX	MNP	MIP
6087-3	Buzz Boar	Underground Attack Vehicle	1986-87	5	10	15
6070	Dreadnok Air Skiff	Mini-set with Zanzibar	1986-87	10	20	30
6171	Dreadnok Cycle	Compilation Cycle with Gunner Station	1986-87	5	10	20
6029	Maggot	three-in-one tank vehicle with W.O.R.M.S. driver	1986-87	10	15	35
6026	Mamba	Attack Copter with removable pods with Gyro-Viper	1986-87	10	15	30
6170	Pogo	Ballistic Battle Ball	1986-87	5	10	20
6040	Sea Ray	Combination Submarine/Jet with Sea Slug	1986-87	10	20	40
6039	Wolf	Arctic Terrain Vehicle with Ice Viper	1986-87	10	20	45

3-3/4" SERIES #6, GI JOE

NO.	NAME	DESCRIPTION	YEAR	EX	MNP	MIP
6087-2	Coastal Defender	Mini-Vehicle with accessories	1986-87	5	10	15
6004-1	Crossfire-Alfa	Radio Control Vehicle with Rumbler	1986-87	25	50	100
6004-2	Crossfire-Delta	Radio Control Vehicle with Rumbler	1986-87	25	50	100
6002	Defiant Space Shuttle Complex	Space shuttle, space station, crawler	1986-87	75	350	525
6038	Persuader	Laser Tank with Backstop	1986-87	10	15	35
6133-3	Radar Station		1986-87	5	10	15
6087-1	Road Toad	Tow Vehicle with accessories	1986-87	5	10	15

ACTION PILOT SERIES

NO.	NAME	DESCRIPTION	YEAR	EX	MNP	MIP
8040	Crash Crew Fire Truck Set		1967	950	1700	3500
8020	Official Space Capsule Set	space capsule, record, space suit, cloth space boots, space gloves, helmet with visor	1966	175	225	350
5979	Official Space Capsule Set w/flotation	Sears' exclusive with collar, life raft and oars	1966	200	325	700

ACTION SAILOR SERIES

NO.	NAME	DESCRIPTION	YEAR	EX	MNP	MIP
5979	Official Sea Sled and Frogman Set	Sears, with figure and underwater cave, orange scuba suit, fins, mask, tanks, sea sled in orange and black	1966	175	325	650
8050	Official Sea Sled and Frogman Set	without cave	1966	150	300	550

ACTION SOLDIER SERIES

NO.	NAME	DESCRIPTION	YEAR	EX	MNP	MIP
5693	Amphibious Duck	Irwin, 26" long	1967	175	375	700
5397	Armored Car	Irwin, friction powered, 20" long	1967	150	300	500
5395	Helicopter	Irwin, friction powered, 28" long	1967	150	300	500
5396	Jet Fighter Plane	Irwin, friction powered, 30" long	1967	225	475	800
5652	Military Staff Car	Irwin, friction powered, 24" long	1967	200	400	750
5651	Motorcycle and Sidecar	Irwin, 14" long, khaki, with decals	1967	75	150	325
7000	Official Combat Jeep Set	Trailer, steering wheel, spare tire, windshield, cannon, search light, shell, flag, guard rails, tripod, tailgate and hood, without Moto-Rev Sound	1965	200	375	550
7000	Official Jeep Combat Set	With Moto-Rev sound	1965	225	400	650
5694	Personnel Carrier/Mine Sweeper	Irwin, 26" long	1967	300	350	700

ADVENTURE TEAM

NO.	NAME	DESCRIPTION	YEAR	EX	MNP	MIP
	Action Sea Sled	J.C. Penney, 13", Adventure Pack	1973	25	40	85
7005	Adventure Team Vehicle Set		1970	50	75	225
23528	All Terrain Vehicle	14" vehicle	1973	50	75	125
59158	Amphicat	Irwin, scaled to fit two figures	1973	35	55	125

VEHICLE SETS

ADVENTURE TEAM

NO.	NAME	DESCRIPTION	YEAR	EX	MNP	MIP
	Avenger Pursuit Craft	Sears' exclusive	1976	100	175	275
7498	Big Trapper	without action figure	1976	75	105	325
7494	Big Trapper Adventure with Intruder	with action figure	1976	100	150	425
7480	Capture Copter	without action figure	1976	80	175	325
7481	Capture Copter Adventure with Intruder	with action figure	1976	110	200	350
59114	Chopper Cycle	15" vehicle, J.C. Penney's	1973	30	50	100
59751	Combat Action Jeep	18" vehicle, J.C. Penney's	1973	50	65	125
7000	Combat Jeep and Trailer		1976	80	135	550
7439	Devil of the Deep		1974	80	135	325
7460	Fantastic Sea Wolf Submarine		1975	60	100	175
7450	Fate of the Troubleshooter		1974	50	125	225
59189	Giant Air-Sea Helicopter	28" vehicle, J.C. Penney's	1973	50	125	225
7380	Helicopter		1976	50	90	300
7380	Helicopter	14"; yellow; with working winch	1973	50	90	150
7499	Mobile Support Vehicle Set		1972	85	150	325
	Recovery of the Lost Mummy Adventure Set	Sears' exclusive	1971	125	250	425
7493	Sandstorm Survival Adventure		1974	125	200	300
7418	Search for the Stolen Idol Set		1971	120	225	350
7441	Secret of the Mummy's Tomb Set	with Land Adventurer figure, shirt, pants, boots, insignia, pith helmet, pick, shovel, Mummy's tomb, net, gems, vehicle with winch, comic	1970	175	300	600
7442	Sharks Surprise Set w/Sea Adventurer		1970	175	325	550
	Signal All Terrain Vehicle	J.C. Penney's, 12" vehicle	1973	30	65	125
7470	Sky Hawk	5-3/4-foot wingspan	1975	65	100	175
7445	Spacewalk Mystery Set w/Astronaut		1970	225	300	550
79-59301	Trapped in the Coils of Doom	J.C. Penney's exclusive	1974	250	300	550

ADVENTURES OF GI JOE

NO.	NAME	DESCRIPTION	YEAR	EX	MNP	MIP
7980	Sharks Surprise Set w/Frogman		1969	175	325	650
7980.83	Sharks Surprise Set without Frogman		1969	150	300	550
7981	Spacewalk Mystery Set w/Spaceman		1969	150	375	650
7981.83	Spacewalk Mystery Set without Spaceman	Reissue	1969	125	275	550

SUPER JOE

NO.	NAME	DESCRIPTION	YEAR	EX	MNP	MIP
7570	Rocket Command Center		1977	50	100	200
7571	Rocket Command Center	Super Adventure Set including Gor	1977	60	115	225

Action Soldier, 1964

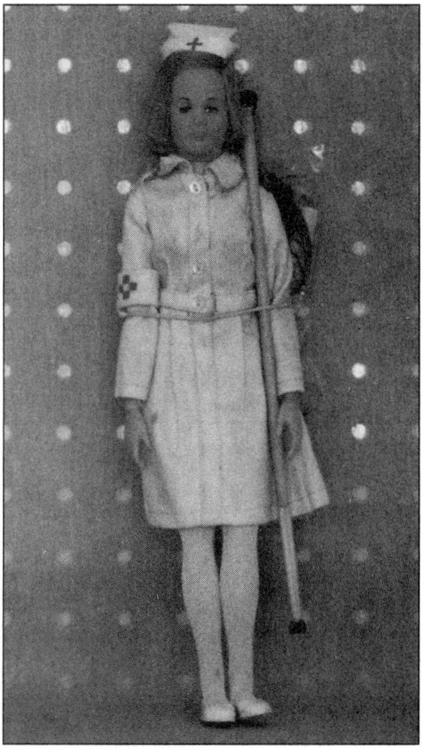

G.I. Nurse, 1967

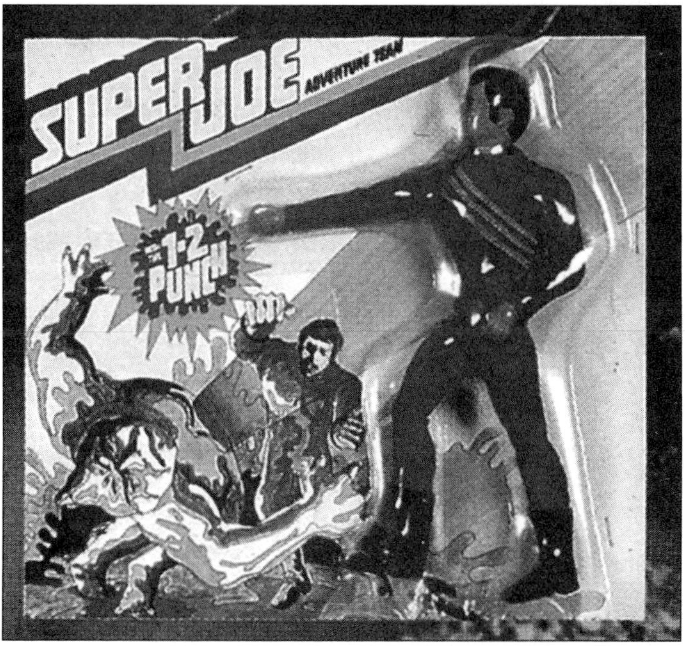

Super Joe, 1977

HAPPY DAYS (MEGO, 1978)

NAME	MNP	MIP
Fonzie, boxed	30	60
Fonzie, carded	30	75
Potsie, carded	30	75
Ralph, carded	30	75
Richie, carded	30	75
PLAY SETS		
Fonzie's Garage Play Set, 1978	60	150
VEHICLES		
Fonzie's Jalopy, 1978	40	80
Fonzie's Motorcycle, 1978	40	80

INDEPENDENCE DAY (TRENDMASTERS, 1996)

FIGURES		
Alien Attacker Pilot	5	10
Alien in Bio Chamber	8	12
Alien Science Officer	5	10
Alien Shock Trooper	5	10
Alien Supreme Commander	15	25
David Levinson	5	10
President Thomas Whitmore	5	10
Steve Hiller	5	10
Ultimate Alien Commander	20	35
Weapons Expert	10	20
Zero Gravity	8	15

INDIANA JONES, ADVENTURES OF (KENNER, 1982-83)

Belloq	10	25
Belloq in Ceremonial Robe, in mailer box	6	15
Belloq in Ceremonial Robe, on card	200	500
Cairo Swordsman	8	20
Convoy Truck	15	35
German Mechanic	15	35
Indiana Jones	50	100
Indiana Jones in German Uniform	20	45
Indiana Jones, 12"	125	250
Map Room Set	20	50
Marion Ravenwood	70	175
Sallah	20	45
Streets of Cairo Set	18	45
Toht	5	15
Well of Souls	30	75

JAMES BOND: MOONRAKER (MEGO, 1979)

12" FIGURES		
Drax	150	200
Holly	150	200
James Bond	125	150
James Bond, deluxe version	350	500
Jaws	400	500

JOHNNY WEST (MARX, 1975)

Jeb Gibson	125	275
Johnny West with Quick Draw	35	70
Sam Cobra with Quick Draw	40	80
Sheriff Garrett	35	70
Thunderbolt, Western Ranch Horse	25	50

JONNY QUEST (GALOOB, 1996)

NAME	MNP	MIP
ACCESSORIES		
Cyber Copter	5	15
Quest Porpoise w/Deep Sea Jonny	5	15
Quest Rover	5	15
QUEST WORLD FIGURES		
Cyber Cycle Jonny Quest	5	10
Cyber Jet Race	5	10
Cyber Suit Hadji	5	10
Cyber Trax Surd	5	10
REAL WORLD FIGURES		
Deep Sea Race Bannon & Hadji	4	8
Jungle Commando Dr. Quest & Ezekiel Rage	4	8
Night Stryker Jonny Quest & Jessie	5	10
Shuttle Pilot Jonny Quest & Race Bannon	4	8
X-Treme Action Jonny Quest & Hadji	4	8

KISS (MCFARLANE, 1997)

6" CARDED FIGURES		
Ace Frehley w/album	5	10
Ace Frehley w/letter stand	7	15
Gene Simmons w/album	5	10
Gene Simmons w/letter base	7	15
Paul Stanley w/album	4	10
Paul Stanley w/letter stand	7	15
Peter Criss w/album	5	10
Peter Criss w/letter stand	7	15

KISS (MEGO, 1978)

12" BOXED FIGURES		
Ace Frehley	100	260
Gene Simmons	110	260
Paul Stanley	100	260
Peter Criss	100	260

LAVERNE AND SHIRLEY (MEGO, 1978)

12" BOXED FIGURES		
Laverne and Shirley	60	125
Lenny and Squiggy	90	175

LEGENDS OF BATMAN (KENNER, 1994)

FIGURES		
Catwoman	8	25
Crusader Batman	4	10
Crusader Robin	4	10
Cyborg Batman	4	10
Dark Rider Batman	10	15
Dark Warrior Batman	4	10
Desert Knight Batman	5	10
Flightpak Batman	5	10
Future Batman	5	10
Joker	10	25
Knightquest Batman	5	10
Knightsend Batman	4	10
Long Bow Batman	4	10
Nightwing Robin	5	10
Power Guardian Batman	5	10
Riddler	10	25
Samurai Batman	4	10
Silver Knight Batman	5	10
Viking Batman	4	10

LEGENDS OF BATMAN (KENNER, 1994)

NAME	MNP	MIP
VEHICLES		
Batcycle	10	20
Batmobile	12	40

LONE RANGER RIDES AGAIN (GABRIEL, 1979)

NAME	MNP	MIP
FIGURES		
Dan Reid	25	50
Little Bear w/Hawk	25	50

LONE RANGER RIDES AGAIN (GABRIEL, 1979)

NAME	MNP	MIP
Lone Ranger	20	40
Red Sleeves	25	50
Tonto	20	40

LONE RANGER, LEGEND OF (GABRIEL, 1982)

NAME	MNP	MIP
FIGURES		
Buffalo Bill Cody	10	25
Butch Cavendish	10	20
General Custer	10	20
Lone Ranger	10	20
Lone Ranger w/Silver	20	50
Scout	10	20
Silver	15	30
Smoke	10	25
Tonto	7	15
Tonto w/Scout	25	50

LOVE BOAT (MEGO, 1981)

NAME	MNP	MIP
4" CARDED FIGURES		
Captain Stubing	10	20
Doc	10	20
Gopher	10	20
Isaac	10	20
Julie	10	25
Vicki	10	25

M*A*S*H (TRISTAR, 1982)

NAME	MNP	MIP
3-3/4" FIGURES AND VEHICLES		
B.J.	5	15
Colonel Potter	5	15
Father Mulcahy	5	15
Hawkeye	5	15
Hawkeye with Ambulance	15	35
Hawkeye with Helicopter	8	20
Hawkeye with Jeep	10	25
Hot Lips	10	20
Klinger	5	15
Klinger in Drag	15	35
M*A*S*H Figures Collectors Set	26	65
Winchester	5	15
8" CARDED FIGURES		
B.J.	15	35
Hawkeye	15	35
Hot Lips	12	30

MAD MONSTER SERIES (MEGO, 1974)

NAME	MNP	MIP
8" FIGURES		
The Dreadful Dracula	80	160
The Horrible Mummy	50	100
The Human Wolfman	75	150
The Monster Frankenstein	45	90
ACCESSORIES		
Mad Monster Castle, vinyl	300	600

MAJOR MATT MASON (MATTEL, 1967-70)

NAME	MNP	MIP
FIGURES		
Callisto, 6"	100	600
Captain Lazer, 12"	125	520
Doug Davis, 6"	100	300
Jeff Long, 6"	100	500
Major Matt Mason, 6"	75	225
Mission Team Four-Pack	175	625
Scorpio, 7"	500	2250
Sergeant Storm, 6"	100	400
VEHICLES AND ACCESSORIES		
Astro-Trak	35	75
Firebolt Space Cannon	35	80
Gamma Ray Guard	30	125
Moon Suit Pak	25	70
Reconojet Pak	25	75
Rocket Launch	25	75
Satellite Launch Pak	25	75
Satellite Locker	30	80
Space Power Suit	30	110
Space Probe Pak	25	75
Space Shelter Pak	25	75
Star Seeker	85	180
Supernaut Power Limbs	30	110
Uni-Tred & Space Bubble	50	125
XRG-1 Reentry Glider	75	200

MARS ATTACKS! (TRENDMASTERS, 1997)

NAME	MNP	MIP
6" FIGURES		
Martian Ambassador	6	12
Martian Leader	6	12
Martian Spy Girl	45	65
Martian Spy Girl, talking	35	85
Martian Trooper	8	15
14" BOXED FIGURES		
Martian Ambassador	10	25
Supreme Commander	10	25

MARVEL FAMOUS COVERS (TOY BIZ, 1997-98)

NAME	MNP	MIP
8" FIGURES		
Aunt May	10	15
Captain America	10	15
Dark Phoenix	10	15
Dr. Doom	10	15
Green Goblin	10	15
Spider-Man	10	15
Storm	12	18
Thor	10	15
Wolverine	10	15

Super Powers, Parademon, Kenner,

DC Comics Super Heroes, Superman, Toy Biz, 1989

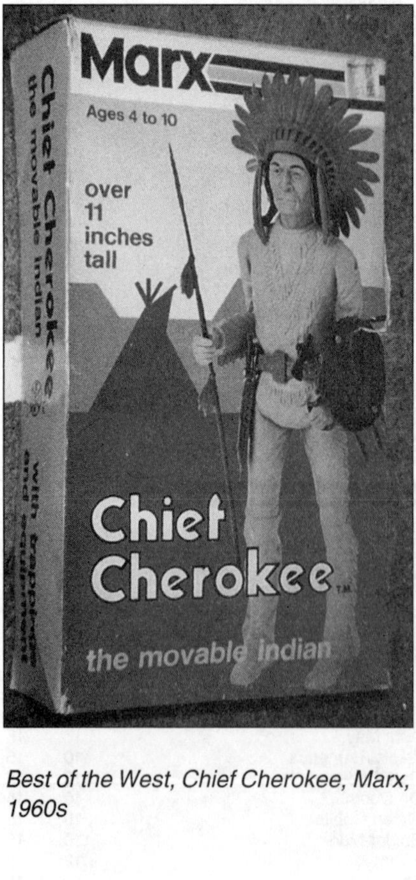

Best of the West, Chief Cherokee, Marx, 1960s

Robin costume for Action Boy, Ideal, 1967

MARVEL SECRET WARS (MATTEL, 1984-85)

NAME	MNP	MIP
4" FIGURES		
Baron Zemo	15	35
Captain America	10	25
Constrictor (foreign release)	30	75
Daredevil	15	35
Doctor Doom	10	20
Doctor Octopus	10	20
Electro (foreign release)	30	75
Falcon	20	40
Hobgoblin	30	60
Ice Man (foreign release)	30	75
Iron Man	20	35
Kang	10	20
Magneto	10	20
Spider-Man, black outfit	25	50
Spider-Man, red and blue outfit	20	40
Three-Figure Set	40	90
Two-Figure Set	25	50
Wolverine, black claws	25	60
Wolverine, silver claws	25	50
ACCESSORIES		
Secret Messages Pack	1	5
Tower of Doom	10	25
VEHICLES		
Doom Copter	10	35
Doom Copter with Doctor Doom	15	55
Doom Cycle	6	20
Doom Cycle with Doctor Doom	10	40
Doom Roller	10	20
Doom Star Glider with Kang	15	30
Freedom Fighter	10	30
Star Dart with Spider-Man (black outfit)	25	50
Turbo Copter	10	40
Turbo Cycle	5	20

MARVEL SUPER HEROES (TOY BIZ, 1990-92)

SERIES 1, 1990		
Captain America	10	25
Daredevil	15	50
Doctor Doom	10	25
Doctor Octopus	10	25
Hulk	5	15
Punisher (cap firing)	5	15
Silver Surfer	10	30
Spider-Man (suction cups)	5	20
SERIES 2, 1991		
Green Goblin (back lever)	15	40
Green Goblin (no lever)	10	25
Iron Man	10	25
Punisher (machine gun sound)	5	15
Spider-Man (web climbing)	15	35
Spider-Man (web shooting)	10	30
Thor (back lever)	15	40
Thor (no lever)	10	25
Venom	10	20
SERIES 3, 1992		
Annihilus	5	15
Deathlok	5	15
Human Torch	5	15

MARVEL SUPER HEROES (TOY BIZ, 1990-92)

NAME	MNP	MIP
Invisible Woman	75	150
Mister Fantastic	5	15
Silver Surfer (chrome)	5	15
Spider-Man (ball joints)	5	15
Spider-Man (web tracer)	5	15
Thing	5	15
Venom (tongue flicking)	15	20
TALKING HEROES		
Cyclops	10	20
Hulk	10	20
Magneto	10	20
Punisher	10	20
Spider-Man	10	20
Venom	10	25
Wolverine	10	20

MASTERS OF THE UNIVERSE (MATTEL, 1981-90)

5-3/4" FIGURES		
Battle Armor He-Man	10	20
Battle Armor Skeletor	5	20
Beast Man	10	20
Blade	10	25
Blast-Attak	5	15
Buzz-Off	5	15
Buzz-Saw Hordak	5	15
Clamp Champ	5	15
Clawful	10	20
Dragstor	5	15
Evil-Lyn	15	30
Extendar	5	15
Faker	15	40
Faker (reissue)	5	15
Fisto	5	15
Grizzlor	5	15
Gwildor	5	15
He-Man, original version	15	30
Hordak	5	15
Horde Trooper	5	15
Jitsu	10	20
King Hiss	5	15
King Randor	10	25
Kobra Khan	5	20
Leech	5	15
Man-At-Arms	10	20
Man-E-Faces	10	20
Mantenna	5	15
Mekaneck	5	15
Mer-Man	5	25
Modulok	5	20
Mosquitor	5	15
Moss Man	5	15
Multi-Bot	5	20
Ninjor	5	20
Orko	5	20
Prince Adam	10	25
Ram Man	15	35
Rattlor	5	15
Rio Blast	5	15
Roboto	5	15
Rokkon	5	15
Rotar	5	15
Saurod	5	20
Scare Glow	15	35
Skeletor, original version	10	25

MASTERS OF THE UNIVERSE (MATTEL, 1981-90)

NAME	MNP	MIP
Snake Face	5	20
Snout Spout	5	15
Sorceress	10	25
Spikor	10	20
SSSqueeze	5	15
Stinkor	5	15
Stonedar	5	15
Stratos, blue wings	10	20
Stratos, red wings	10	20
Sy-Klone	5	15
Teela	10	25
Trap Jaw	10	25
Tri-Klops	10	20
Tung Lashor	5	15
Twistoid	5	15
Two-Bad	5	15
Webstor	5	15
Whiplash	5	15
Zodac	10	20
ACCESSORIES		
Battle Bones Carrying Case	5	10
Battle Cat	10	25
Battle Cat with He-Man (original version)	20	40
Beam-Blaster and Artilleray	15	30
Jet Sled	5	15
Mantisaur	8	15
Megalaser	5	15
Monstroid Creature	15	30
Night Stalker	5	15
Night Stalker with Jitsu	10	25
Panthor (evil cat)	10	25
Panthor with Skeletor (original version)	15	40
Screech	5	15
Screech with Skeletor (original version)	10	25
Stilt Stalkers	5	15
Stridor Armored Horse	5	15
Stridor with Fisto	10	25
Weapons Pak	2	5
Zoar	5	15
Zoar with Teela	15	30
FIFTH ANNIVERSARY FIGURES		
Dragon Blaster Skeletor	10	25
Flying Fists He-Man	10	25
Hurricane Hordak	10	25
Terror Claws Skeletor	10	25
Thunder Punch He-Man	10	25
GRAYSKULL DINOSAUR SERIES		
Bionatops	10	25
Turbodaltyl	10	25
Tyrantisaurus Rex	10	25
METEORBS		
Astro Lion	5	15
Comet Cat	5	15
Cometroid	5	15
Crocobite	5	15
Dinosorb	5	15
Gore-illa	5	15
Orbear	5	15
Rhinorb	5	15
Tuskor	5	15
Ty-Gyr	5	15

MASTERS OF THE UNIVERSE (MATTEL, 1981-90)

NAME	MNP	MIP
PLAY SETS		
Castle GraySkull	25	100
Eternia	100	200
Fright Zone	25	50
Slime Pit	10	20
Snake Mountain	25	50
VEHICLES		
Attak Trak	10	25
Bashasaurus	10	40
Battle Ram	10	40
Blasterhawk	15	40
Dragon Walker	10	25
Fright Fighter	10	40
Land Shark	10	25
Laser Bolt	10	25
Point Dread	10	50
Road Ripper	10	25
Roton	10	30
Spydor	15	40
Wind Raider	10	40

MICRONAUTS (MEGO, 1976-80)

ALIEN INVADERS CARDED		
Antron, 1979	15	30
Centaurus, 1980	35	70
Karrio, 1979	10	20
Kronos, 1980	35	70
Lobros, 1980	35	70
Membros, 1979	15	30
Repto, 1979	13	25
ALIEN INVADERS PLAY SETS		
Rocket Tubes, 1978	23	50
ALIEN INVADERS VEHICLES		
Alphatron	5	10
Aquatron, 1977	10	20
Betatron	5	10
Gammatron	5	10
Hornetroid, 1979	20	40
Hydra, 1976	7	15
Mobile Exploration Lab, 1976	17	35
Solarion, 1978	15	30
Star Searcher, 1978	15	40
Taurion, 1978	11	22
Terraphant, 1979	20	40
BOXED FIGURES		
Andromeda, 1977	10	25
Baron Karza, 1977	15	30
Biotron, 1976	10	25
Force Commander, 1977	10	25
Giant Acroyear, 1977	10	25
Megas, 1981	10	25
Microtron, 1976	5	20
Nemesis Robot, 1978	7	15
Oberon, 1977	10	25
Phobos Robot, 1978	12	25
CARDED FIGURES		
Acroyear II, 1977, red, blue, orange	7	15
Acroyear, 1976, red, blue, orange	10	20
Galactic Defender, 1978, white, yellow	7	15
Galactic Warriors, 1976, red, blue, orange	4	10

Gore, Total Chaos, McFarlane, 1996

Hoof, Total Chaos, McFarlane, 1996

Battlestar Galactica figures, Mattel, 1978-1979

MICRONAUTS (MEGO, 1976-80)

NAME	MNP	MIP
Pharoid with Time Chamber, 1977, blue, red, gray	10	20
Space Glider, 1976, blue, green, orange	5	10
Time Traveler, 1976, clear plastic, yellow, orange	3	10
Time Traveler, 1976, solid plastic, yellow, orange	5	15
MICROPOLIS PLAY SETS		
Galactic Command Center, 1978	20	40
Interplanetary Headquarters, 1978	20	40
Mega City, 1978	20	30
Microrail City, 1978	20	40
PLAY SETS		
Astro Station, 1976	10	20
Stratstation, 1976	15	30
VEHICLES		
Battle Cruiser, 1977	30	60
Crater Cruncher with figure, 1976	5	15
Galactic Cruiser, 1976	7	17
Hydro Copter, 1976	10	25
Neon Orbiter, 1977	6	20
Photon Sled with figure, 1976	5	15
Rhodium Orbiter, 1977	6	20
Thorium Orbiter, 1977	6	20
Ultronic Scooter with figure, 1976	5	15
Warp Racer with figure, 1976	5	15

MIGHTY MORPHIN POWER RANGERS (BANDAI, 1993-95)

3" FIGURES		
Black Ranger	5	15
Blue Ranger	5	15
Pink Ranger	5	15
Red Ranger	5	15
Yellow Ranger	5	15
5" FIGURES W/THUNDER BIKES		
Black Ranger	6	15
Blue Ranger	6	15
Pink Ranger	6	15
Red Ranger	6	15
Yellow Ranger	6	15
5-1/2" CARDED FIGURES		
Black Ranger	3	15
Blue Ranger	3	15
Green Ranger	3	15
Pink Ranger	3	15
Red Ranger	2	15
Yellow Ranger	3	15
8" FIGURES, 1993		
Black Ranger	7	20
Blue Ranger	7	20
Pink Ranger	10	20
Red Ranger	5	20
Yellow Ranger	10	20
8" MOVIE FIGURES, 1995		
Black Ranger	6	15
Blue Ranger	6	15
Pink Ranger	6	15
Red Ranger	6	15
White Ranger	6	15
Yellow Ranger	6	15
ALIENS, 1993		
Baboo	10	20

MIGHTY MORPHIN POWER RANGERS (BANDAI, 1993-95)

NAME	MNP	MIP
Bones	10	20
Finster	10	20
Goldar	10	20
King Sphinx	10	20
Putty Patrol	10	20
Squatt	10	20
ALIENS, 1994		
Evil Eye	7	15
Goo Fish	7	15
Guitardo	7	15
Lord Zed	8	18
Pirantus Head	7	15
Pudgy Pig	7	15
Rhino Blaster	7	15
Socaddillo	7	15
Super Puttys	7	15
KARATE ACTION FIGURES, 1994		
Black Ranger	6	12
Blue Ranger	6	12
Pink Ranger	8	16
Red Ranger	5	10
Yellow Ranger	8	16
POWER RANGERS FOR GIRLS		
Kimberly	10	20
Kimberly/Trini Set	20	40
Trini	10	20
ZORDS, 1993		
Dragon Dagger	20	50
Dragon Zord w/Green Ranger	25	55
MegaZord	15	30
MegaZord Deluxe	20	40
Titanus the Carrier Zord	35	75
ZORDS, 1994		
MegaZord, black/gold, limit. ed.	50	100
Power Cannon	15	35
Power Dome Morphin Set	25	55
Red Dragon Thunder Zord	25	45
Saba (White Sword)	15	30
Thunder Zord Assault Team	25	45
TOR the Shuttle Zord	30	60
Ultra Thunder Zord	35	70
White Tiger Zord w/White Ranger	25	50

MONSTERS (MCFARLANE, 1997-98)

PLAYSETS		
Dracula	7	15
Dr. Frankenstein	7	15
Frankenstein	7	15
Hunchback	7	15
Mummy	7	15
Phantom of the Opera	7	15
Sea Creature	7	15
Werewolf	7	15

NOBLE KNIGHTS (MARX, 1968)

Black Knight	75	190
Bravo Armor Horse	100	130
Gold Knight	60	120
Silver Knight	60	120
Valiant Armor Horse	100	130
Valor Armor Horse	100	130
Victor Armor Horse	100	130

ONE MILLION YEARS, B.C. (MEGO, 1976)

NAME	MNP	MIP
Dimetrodon, 1976, boxed	75	150
Grok, 1976, carded	25	50
Hairy Rino, 1976, boxed	75	150
Mada, 1976, carded	25	50
Orm, 1976, carded	25	50
Trag, 1976, carded	25	50
Tribal Lair Gift Set (Five figures), 1976	70	180
Tribal Lair, 1976	60	120
Tyrannosaur, 1976, boxed	75	150
Zon, 1976, carded	25	50

OUTER SPACE MEN (COLORFORMS, 1968)

CARDED FIGURES		
Alpha 7 / Man from Mars	90	250
Astro-Nautilus / Man from Neptune	90	250
Colossus Rex / Man from Jupiter	100	300
Commander Comet / Man from Venus	90	250
Electron / Man from Pluto	90	250
Orbitron / Man from Uranus	90	250
Xodiac / Man from Saturn	90	250

PLANET OF THE APES (MEGO, 1973-75)

8" FIGURES		
Astronaut Burke, 1975, boxed	50	130
Astronaut Burke, 1975, carded	50	100
Astronaut Verdon, 1975, boxed	50	140
Astronaut Verdon, 1975, carded	50	125
Astronaut, 1973, boxed	50	150
Astronaut, 1975, carded	50	100
Cornelius, 1973, boxed	40	140
Cornelius, 1975, carded	40	100
Dr. Zaius, 1973, boxed	40	150
Dr. Zaius, 1975, carded	40	100
Galen, 1975, boxed	40	140
Galen, 1975, carded	40	100
General Urko, 1975, boxed	50	130
General Urko, 1975, carded	50	100
General Ursus, 1975, boxed	50	120
General Ursus, 1975, carded	50	100
Soldier Ape, 1973, boxed	50	140
Soldier Ape, 1975, carded	50	100
Zira, 1973, boxed	30	150
Zira, 1975, carded	30	100
ACCESSORIES		
Action Stallion, brown motorized, 1975, boxed	50	100
Battering Ram, 1975, boxed	20	40
Dr. Zaius' Throne, 1975, boxed	20	40
Jail, 1975, boxed	20	40
PLAY SETS		
Forbidden Zone Trap, 1975	65	150
Fortress, 1975	60	120
Treehouse, 1975	50	100
Village, 1975	60	130
VEHICLES		
Catapult and Wagon, 1975, boxed	25	50

POCKET SUPER HEROES (MEGO, 1976-79)

NAME	MNP	MIP
3-3/4" FIGURES		
Aquaman, 1976, white card	50	100
Batman, 1976, red card	20	40
Batman, 1976, white card	20	40
Captain America, 1976, white card	50	100
Captain Marvel, 1979, red card	20	40
General Zod, 1979, red card	5	15
Green Goblin, 1976, white card	50	100
Hulk, 1976, white card	15	40
Hulk, 1979, red card	15	30
Joker, 1979, red card	20	40
Jor-El (Superman), 1979, red card	10	20
Lex Luthor (Superman), 1979, red card	10	20
Penguin, 1979, red card	20	40
Robin, 1976, white card	20	40
Robin, 1979, red card	20	40
Spider-Man, 1976, white card	15	40
Spider-Man, 1979, red card	15	30
Superman, 1976, white card	15	30
Superman, 1979, red card	15	30
Wonder Woman, 1979, white card	20	45
ACCESSORIES		
Batcave, 1981	120	300
VEHICLES		
Batmachine, 1979	40	100
Batmobile, 1979, with Batman	80	200
Invisible Jet, 1979	50	125
Spider-Car, 1979, with Spider-Man	30	75
Spider-Machine, 1979	40	100

RAMBO (COLECO, 1985)

FIGURES		
Black Dragon	8	20
Chief	8	20
Colonel Troutman	7	15
Dr. Hyde	10	25
General Warhawk	8	20
Gripper	8	20
K.A.T.	10	20
Mad Dog	8	20
Nomad	10	20
Rambo	10	25
Rambo w/Fire Power	10	25
Sargeant Havoc	8	20
Turbo	8	20
White Dragon	8	20

ROBIN HOOD AND HIS MERRY MEN (MEGO, 1974)

8" FIGURES		
Friar Tuck	25	75
Little John	45	100
Robin Hood	75	150
Will Scarlett	75	150

Major Matt Mason series, Scorpio and Callisto figures, Mattel, 1967-70

CHiPS motorcycle, Mego, 1979

ROBIN HOOD PRINCE OF THIEVES (KENNER, 1991)

NAME	MNP	MIP
ACCESSORIES		
Battle Wagon	15	30
Bola Bomber	5	10
Net Launcher	5	10
Sherwood Forest Play Set	30	60
FIGURES		
Azeem	7	15
Friar Tuck	15	30
Little John	7	15
Robin Hood, Crossbow	5	18
Robin Hood, Crossbow, Costner Head	7	20
Robin Hood, Long Bow	8	17
Robin Hood, Long Bow, Costner Head	10	20
Sheriff of Nottingham	5	15
The Dark Warrior	8	20
Will Scarlett	8	20

ROBOCOP AND THE ULTRA POLICE (KENNER, 1989-90)

ACCESSORIES		
Robo-Glove	25	45
Robo-Helmet	15	40
FIGURES		
Ace Jackson	5	15
Anne Lewis	5	15
Birdman Barnes	8	15
Chainsaw	5	15
Claw Callahan	7	15
Dr. McNamara	5	15
Ed-260	10	25
Headhunter	5	15
Nitro	5	15
RoboCop	9	20
RoboCop Night Fighter	6	20
RoboCop, Gatlin' Gun	15	30
Scorcher	6	15
Sgt. Reed	6	15
Toxic Waster	10	20
Wheels Wilson	6	15
VEHICLES		
Robo-1	10	25
Robo-Command	10	30
Robo-Copter	15	35
Robo-Cycle	5	10
Robo-Hawk	10	35
Robo-Jailer	15	40
Robo-Tank	10	35
Skull-Hog	5	10
Vandal-1	5	20

SHOGUN WARRIORS (MATTEL, 1979)

FIGURES		
Daimos	75	150
Dragun	75	175
Dragun (2nd figure)	75	150
Gaiking	75	150
Godzilla	100	200
Godzilla (2nd figure)	100	250
Mazinga	85	175

SHOGUN WARRIORS (MATTEL, 1979)

NAME	MNP	MIP
Mazinga (2nd figure)	75	150
Raydeen	75	150

SIMPSONS (MATTEL, 1990)

Bart	10	20
Bartman	10	20
Homer	10	20
Lisa	15	30
Maggie	15	30
Marge	10	20
Nelson	10	20
Sofa Set	10	30

SIX MILLION DOLLAR MAN (KENNER, 1975-78)

ACCESSORIES		
Backpack Radio	10	20
Bionic Cycle	10	20
Bionic Mission Vehicle	25	55
Bionic Transport	10	30
Bionic Video Center	25	65
Critical Assignment Arms	15	30
Critical Assignment Legs	15	30
Dual Launch Drag Set	45	80
Flight Suit	15	30
Mission Control Center	25	50
Mission to Mars Space Suit	15	30
OSI Headquarters	30	70
OSI Undercover Blue Denims	15	30
Porta-Communicator	20	50
Tower & Cycle Set	25	50
Venus Space Probe	50	80
FIGURES		
Bionic Bigfoot	75	120
Maskatron	40	100
Oscar Goldman	50	100
Steve Austin	50	100

SPACE: 1999 (MATTEL, 1976)

FIGURES		
Commander Koenig	30	60
Dr. Russell	30	60
Professor Bergman	30	60

SPAWN (MCFARLANE, 1994-97)

ACCESSORIES		
Spawn Alley Play Set	15	50
Spawnmobile	12	40
Violator Monster Rig	12	45
FIGURES		
Alien Spawn	5	15
Angela	5	25
Badrock	8	15
Battleclad Spawn	5	20
Chapel	6	20
Clown	6	20
Clown w/Violator Head	6	20
Commando Spawn	7	20
Cosmic Angela	5	15
Crutch	5	15
Curse, The	5	15

SPAWN (MCFARLANE, 1994-97)

NAME	MNP	MIP
Freak, The	5	15
Future Spawn	10	15
Malebolgia	40	80
Mangler, The	5	15
Medieval Spawn	5	20
Ninja Spawn	7	15
No-Body	5	15
Nuclear Spawn	4	15
Overtkill	7	20
Overtkill II	5	15
Pilot Spawn	10	20
Redeemer	5	15
Sam and Twitch	5	15
Sansker	5	15
Scourge	5	15
Spawn	5	30
Spawn (unmasked, 1st card)	15	30
Spawn II	10	15
Spawn III	7	25
Superpatriot	5	10
Tiffany	5	15
Tremor	8	15
Tremor II	5	10
Vandalizer	5	10
Vertebreaker	10	25
Viking Spawn	5	25
Violator	5	20
Violator II	10	20
Violator w/Chrome Card	10	40
Widow Maker	5	15
Zombie Spawn	5	10

SPAWN, MANGA (MCFARLANE, 1997)

FIGURES

Clown	6	10
Curse	6	10
Goddess	10	15
Manga Ninja Spawn	6	10
Manga Spawn	7	12
Violator	6	12

SPAWN: THE MOVIE (MCFARLANE, 1997)

FIGURES

Al Simmons	4	10
Burnt Spawn	4	10
Clown	4	10
Jason Wynn	4	10
Jessica Priest	4	10
Malebolgia (boxed)	10	30
Spawn (boxed)	10	30
Violator (boxed)	10	30

PLAY SETS

Final Battle	10	20
Graveyard	10	20
Spawn Alley	10	20

SPIDER-MAN: THE ANIMATED SERIES (TOY BIZ, 1994)

10" DELUXE FIGURES

Dr. Octopus	7	15
Hobgoblin	7	15
Kraven	7	15

SPIDER-MAN: THE ANIMATED SERIES (TOY BIZ, 1994)

NAME	MNP	MIP
Lizard	7	15
Spider-Man (hanging)	7	15
Spider-Man w/Suction Cups	7	15
Venom	7	15
Vulture	10	20

15" TALKING FIGURES

Spider-Man	12	25
Venom	12	25

2-1/2" DIE-CAST FIGURES

Spider-Man vs. Carnage	2	4
Spider-Man vs. Dr. Octopus	2	4
Spider-Man vs. Hobgoblin	2	4
Spider-Man vs. Venom	2	4

4-1/2" FIGURES

Alien Spider Slayer	4	8
Carnage	5	15
Dr. Octopus	5	12
Green Goblin	4	15
Hobgoblin	5	12
Kingpin	5	15
Kraven	5	12
Lizard	5	15
Peter Parker	6	15
Rhino	6	25
Scorpion	5	15
Shocker	4	15
Smythe	5	12
Spider-Man (multi-jointed)	3	15
Spider-Man w/Spider Armor	3	15
Spider-Man w/Web Parachute	8	15
Spider-Man w/Web Racer	5	15
Spider-Man w/Web Shooter	5	15
Venom	4	15
Vulture	5	15

ACCESSORIES

Daily Bugle Play Set	10	15

PROJECTORS

Hobgoblin	5	15
Spider-Man	5	15
Venom	5	15

VEHICLES

Hobgoblin Pumpkin Bomber	7	15
Hobgoblin Wing Bomber	10	25

SPIDER-MAN: THE ANIMATED SERIES (TOY BIZ, 1994)

Smythe Battle Chair Attack Vehicle	10	25
Spider-Man Wheelie Cycle	7	15
Spider-Man's Cycle (radio-controlled)	15	30
Tri-Spider Slayer	10	25
Venom Assault Racer	7	15

STAR TREK (MEGO, 1974-80)

12" BOXED FIGURES

Arcturian, 1979	30	60
Captain Kirk, 1979	25	55
Decker, 1979	45	115
Ilia, 1979	25	50
Klingon, 1979	40	85
Mr. Spock, 1979	30	60

G.I. Joe, 3-3/4-Inch Series #1, Scarlett. Photo courtesy of Christie's East

Major Matt Mason, Space Probe Pak, Mattel, 1967-1970

STAR TREK (MEGO, 1974-80)

NAME	MNP	MIP
3-3/4" CARDED FIGURES		
Acturian, 1980	75	150
Betelgeusian, 1980	75	150
Captain Kirk, 1979	10	25
Decker, 1979	10	25
Dr. McCoy, 1979	10	25
Ilia, 1979	10	20
Klingon, 1980	75	150
Megarite, 1980	75	150
Mr. Spock, 1979	10	25
Rigellian, 1980	75	150
Scotty, 1979	10	25
Zatanite, 1980	75	150
8" CARDED FIGURES		
Andorian, 1976	200	400
Captain Kirk, 1974	25	50
Cheron, 1975	75	150
Dr. McCoy, 1974	35	75
Gorn, 1975	80	180
Klingon, 1974	25	50
Lt. Uhura, 1974	50	135
Mr. Spock, 1974	25	50
Mugato, 1976	150	300
Neptunian, 1975	100	225
Romulan, 1976	300	600
Scotty, 1974	35	80
Talos, 1976	165	300
The Keeper, 1975	75	175
PLAY SETS		
Command Bridge (for 3-3/4" figures), 1980	45	105
Enterprise Bridge (for 8" figures), 1976	60	150
Enterprise Bridge with Figures, 1976	95	250
Mission to Gamma VI (for 8" figures), 1976	200	500

STAR TREK COLLECTOR SERIES (PLAYMATES, 1994-95)

9-1/2" BOXED FIGURES		
Borg	10	25
Captain Benjamin Sisko (Command Edition)	10	25
Captain Jean-Luc Picard (Command Edition)	10	25
Captain Jean-Luc Picard (Movie Edition)	10	25
Captain Kirk (Command Edition)	10	25
Captain Kirk (Movie Edition)	10	25
Commander Riker	10	25
Data (Movie Edition)	10	25
Dr. Beverly Crusher	10	25
Geordi La Forge (Movie Edition)	10	25

STAR TREK V (GALOOB, 1989)

BOXED FIGURES		
Captain Kirk	15	75
Dr. McCoy	15	75
Klaa	15	75
Mr. Spock	15	75
Sybok	15	75

STAR TREK: FIRST CONTACT (PLAYMATES, 1996)

NAME	MNP	MIP
5" FIGURES		
Borg	8	15
Data	5	10
Deanna Troi	5	10
Dr. Beverly Crusher	6	12
Geordi LaForge	5	10
Jean-Luc Picard	5	10
Jean-Luc Picard in space suit	6	12
Lily	8	15
William Riker	5	10
Worf	5	10
Zefram Cochrane	5	10
9" FIGURES		
Data	12	18
Jean-Luc Picard	12	18
Jean-Luc Picard in 21st century outfit	15	23
William Riker	12	18
Zefram Cochrane	18	25

STAR TREK: SPACE TALK SERIES (PLAYMATES, 1995)

7" FIGURES		
Borg	5	20
Picard	5	10
Q	5	20
Riker	5	10

STAR TREK: STARFLEET ACADEMY (PLAYMATES, 1996)

5" FIGURES		
Cadet Geordi LaForge	8	15
Cadet Jean-Luc Picard	8	15
Cadet William Riker	8	15
Cadet Worf	8	15

STAR TREK: THE NEXT GENERATION (GALOOB, 1988-89)

3-3/4" FIGURES, SERIES 1		
Data, blue face	70	160
Data, brown face	30	60
Data, flesh face	15	30
Data, spotted face	15	30
Geordi La Forge	5	15
Jean-Luc Picard	5	15
Lt. Worf	5	15
Tasha Yar	10	25
William Riker	5	15
3-3/4" FIGURES, SERIES 2		
Antican	35	100
Ferengi	35	100
Q	35	100
Selay	35	100
ACCESSORIES		
Enterprise	10	35
Ferengi Fighter	15	50
Galileo Shuttle	15	50
Phaser	20	40

STAR TREK: THE NEXT GENERATION (PLAYMATES, 1992-94)

NAME	MNP	MIP
SERIES 1, 1992		
Borg	10	20
Commander Riker	10	18
Data	10	20
Deanna Troi	15	30
Ferengi	10	25
Geordi LaForge	10	25
Gowron the Klingon	12	25
Jean-Luc Picard	10	20
Romulan	15	30
Worf	10	20
SERIES 2, 1993		
Admiral McCoy	5	12
Benzite	7	15
Borg	5	10
Captain Scott (Scotty)	5	10
Commander Riker	6	12
Commander Sela	6	12
Data	6	12
Dathon	7	15
Deanna Troi	6	12
Dr. Beverly Crusher	6	12
Geordi LaForge	7	15
Guinan	7	15
Jean-Luc Picard	6	12
K'Ehleyr	6	12
Locutus	6	12
Lore	7	15
Q	6	12
Spock	6	12
Vorgon	10	20
Wesley Crusher	6	12
Worf	6	12
SERIES 3, 1994		
Barclay	5	15
Beverly Crusher	5	15
Data as Romulan	5	15
Data, dress uniform	5	15
Data, Redemption outfit	75	300
Deanna Troi	3	15

STAR TREK: THE NEXT GENERATION (PLAYMATES, 1992-94)

NAME	MNP	MIP
Dr. Noonian Soong	5	15
Ensign Ro Laren	10	20
Esoqq	20	75
Geordi La Forge	5	15
Gowron	10	40
Guinan	5	15
Hugh Borg	5	15
Lore	3	15
Lwaxana Troi	3	15
Nausicaan	5	15
Picard as Dixon Hill	5	15
Picard as Romulan	3	15
Picard, red uniform	3	15
Q, judge's robes	5	15
Riker, Malcorian	3	15
Riker, red uniform	50	175
Sarek	5	15
Sela	5	15
Spock	5	15
Tasha Yar	5	15
Wesley Crusher	5	15
Worf	8	15

STAR TREK: VOYAGER (PLAYMATES, 1995-96)

5" FIGURES		
B'Elanna Torres	15	25
Chakotay	5	10
Chakotay as a Maquis	8	15
Doctor	8	15
Harry Kim	8	15
Kathryn Janeway	15	25
Kazon	5	10
Kes the Ocampa	12	18
Lt. Carey	5	10
Neelix	5	10
Seska	5	10
Tom Paris	5	10
Torres as a Klingon	5	10
Tuvok	5	10
Vidiian	5	10

Super Powers, Brainiac, Kenner, 1984-1986

Super Powers, Orion, Kenner, 1984-1986

Comic Action Heroes, Robin, Penguin, Mego, 1975

STAR WARS
(Kenner, 1977-1985)

ACCESSORIES

NAME	YEAR	MNP	MIP
DROIDS			
Droids Lightsaber	1985	55	200
EMPIRE STRIKES BACK			
Darth Vader Carrying Case	1982	10	35
Display Arena	1980	25	105
Hoth Wompa	1982	15	55
Laser Pistol	1980	20	85
Lightsaber, red or green	1980	20	55
Lightsaber, yellow	1980	20	70
Mini Figure Case	1980	10	30
Tauntaun, solid belly	1980	20	60
Tauntaun, split belly	1982	19	60
Three-Position Laser Rifle	1980	55	200
EWOKS			
Ewoks Treehouse	1985	18	45
RETURN OF THE JEDI			
Biker Scout Laser Pistol	1984	25	70
C-3PO Carrying Case	1983	17	40
Chewbacca Bandolier Strap	1983	6	15
Darth Vader Carrying Case w/three figures	1983	11	210
Ewok Assault Catapult	1983	11	25
Laser Rifle Carrying Case	1984	15	35
Lightsaber (red or green plastic)		20	50
Rancor Monster	1983	30	70
Tri-Pod Cannon	1983	9	20
Vehicle Maintenance Energizer	1983	8	20
STAR WARS			
24-Figure Carrying Case		15	45
Action Figure Display Stand (Mail-In Premium)	1977	40	300
Han Solo's Laser Pistol		20	115
Inflatable Lightsaber	1977	35	150

ACTION FIGURES

NAME	YEAR	MNP	MIP
12" FIGURES			
Ben (Obi-Wan) Kenobi	1979	135	290
Boba Fett (ESB box)	1979	160	390
Boba Fett (Star Wars box)	1979	160	450
C-3PO	1979	55	150
Chewbacca		75	155
Darth Vader	1978	85	200
Han Solo	1979	210	500
IG-88 (ESB box)	1980	230	650
Jawa	1979	75	195
Luke Skywalker		140	300
Princess Leia Organa	1977	110	240
R2-D2	1979	55	140
Stormtrooper	1979	115	250
DROIDS			
A-Wing Pilot	1985	45	155
Boba Fett	1985	25	475
C-3PO	1985	20	60

ACTION FIGURES

NAME	YEAR	MNP	MIP
Jann Tosh	1985	8	20
Jord Dusat	1985	8	20
Kea Moll	1985	8	20
Kez-Iban	1985	8	20
R2-D2	1985	25	50
Sise Fromm	1985	25	50
Thall Joben	1985	8	20
Tig Fromm	1985	18	50
Uncle Gundy	1985	8	20
EMPIRE STRIKES BACK			
2-1B	1980	9	45
4-LOM		10	140
AT-AT Commander	1980	8	40
AT-AT Driver	1981	8	50
Bespin Security Guard, black	1980	8	45
Bespin Security Guard, white	1980	8	50
Bossk, bounty hunter	1980	10	85
C-3PO, w/removable limbs	1982	8	55
Cloud Car Pilot	1982	14	50
Dengar	1980	9	55
FX-7	1980	8	55
Han Solo in Bespin outfit	1981	14	95
Han Solo in Hoth outfit	1980	12	75
Hoth Rebel Soldier	1980	8	45
IG-88	1980	12	80
Imperial Commander	1981	9	40
Imperial TIE Fighter Pilot	1982	12	75
Lando Calrissian		10	55
Lobot	1981	8	45
Luke Skywalker in Bespin outfit	1980	20	140
Luke Skywalker in Hoth battle gear	1982	14	70
Princess Leia in Bespin gown	1980	19	120
Princess Leia in Hoth outfit	1981	16	80
R2-D2 with sensorscope		10	55
Rebel Commander	1980	8	40
Snowtrooper	1980	14	70
Ugnaught	1981	10	40
Yoda	1981	20	80
Zuckuss	1982	11	70
EWOKS			
Dulok Scout	1985	8	16
Dulok Shaman	1985	8	16
King Gornesh	1985	8	16
Logray	1985	8	18
Urgah	1985	8	17
Wicket	1985	9	20
MAIL-AWAY FIGURES			
Admiral Ackbar		9	18
Anakin Skywalker		30	40
AT-AT Commander		8	15
AT-ST Driver		10	15
C-3PO, removable limbs		8	15
Emperor		11	19
Han Solo in Hoth outfit		12	20

Cantina Adventure Set, 1977, Kenner

Walrus Man, 1978, Kenner

Jawa, 12-inch, 1979, Kenner

ACTION FIGURES

NAME	YEAR	MNP	MIP
Han Solo in trench coat		14	20
Luke Skywalker in Hoth outfit		14	20
Nien Nunb		8	17
Pruneface		9	15
R2-D2 with sensorscope		10	15
POWER OF THE FORCE			
A-Wing Pilot	1985	30	100
Amanaman	1985	85	200
Anakin Skywalker	1985	30	400
AT-AT Driver		8	500
AT-ST Driver		9	65
B-Wing Pilot	1985	8	25
Barada	1985	45	100
Ben (Obi-Wan) Kenobi	1985	15	90
Biker Scout	1985	11	75
C-3PO, w/removable limbs		8	75
Chewbacca		11	75
Darth Vader		15	115
Emperor	1985	11	95
EV-9D9	1985	80	155
Gamorrean Guard		9	230
Han Solo in Carbonite outfit	1985	110	240
Han Solo in Trench Coat		14	360
Imperial Dignitary	1985	40	85
Imperial Gunner	1985	75	130
Jawa	1985	14	80
Lando Calrissian General Pilot	1985	55	120
Luke in X-Wing Fighter Pilot Outfit	1985	13	105
Luke in Battle Poncho		70	110
Luke in Jedi Outfit, green saber		35	190
Luke in Stormtrooper outfit	1985	145	350
Lumat	1985	14	70
Nikto		10	325
Paploo	1985	17	45
R2-D2 with pop-up Lightsaber	1985	75	150
Romba	1985	25	45
Stormtrooper		14	190
Teebo		10	165
Warok	1985	25	50
Wicket	1985	15	140
Yak Face	1985	150	1350
Yoda		20	340
RETURN OF THE JEDI			
8D8	1983	9	30
Admiral Ackbar	1983	9	25
AT-AT Commander		9	25
AT-ST Driver		9	25
B-Wing Pilot		9	25
Bib Fortuna		10	30
Biker Scout	1983	11	30
Chief Chirpa	1983	10	20
Emperor	1983	9	25
Emperor's Royal Guard		11	35
Gamorrean Guard	1983	9	25
General Madine	1983	8	25
Han Solo in trench coat	1984	14	40
Klaatu	1983	8	20
Klaatu in Skiff guard outfit	1983	9	25
Lando Calrissian	1983	12	35

ACTION FIGURES

NAME	YEAR	MNP	MIP
Logray	1983	9	25
Luke Skywalker in Jedi outfit, blue saber	1983	50	160
Luke Skywalker in Jedi outfit, green saber	1983	30	80
Lumat	1983	14	30
Nien Nunb	1983	8	30
Nikto	1984	10	20
Paploo		17	35
Princess Leia in Boushh outfit	1983	20	50
Princess Leia in combat poncho	1984	19	45
Pruneface	1984	9	20
Rancor Keeper	1984	8	20
Rebel Commander	1983	8	25
Rebel Commando	1983	10	25
Ree-Yees	1983	8	20
Squid Head	1983	9	25
Sy Snootles and the Rebo Band, each	1984	25	135
Teebo	1984	10	30
Weequay		9	20
Wicket	1984	15	30
STAR WARS			
Ben Kenobi (original 12)	1977	15	220
Boba Fett	1978	25	700
C-3PO (original 12)	1977	12	150
Chewbacca (original 12)	1977	11	185
Darth Vader (original 12)	1977	15	220
Death Squad Commander (original 12)	1977	12	185
Death Star Droid	1978	12	125
Early Bird Figures (set of four)	1977	220	520
Greedo	1978	11	115
Hammerhead	1978	11	115
Han Solo, large head (original 12)		25	550
Han Solo, small head (original 12)	1977	30	510
Jawa, cloth cape (original 12)	1977	14	165
Jawa, vinyl cape (original 12)	1977	270	2610
Luke Skywalker (original 12)	1977	30	330
Luke Skywalker X-Wing Pilot	1978	13	130
Luke Skywalker, w/telescoping saber (original 12)	1977	150	3230
Power Droid	1978	9	115
Princess Leia Organa (original 12)	1977	35	300
R2-D2 (original 12)	1977	13	160
R5-D4	1978	11	115
Snaggletooth, blue body (Sears Exclusive)	1978	165	250
Snaggletooth, red body	1978	10	115
Stormtrooper (original 12)	1977	14	210
Tusken Raider (original 12)	1977	16	210
Walrus Man	1978	11	115
MICRO SERIES			
Bespin Control Room	1982	12	40
Bespin Freeze Chamber	1982	30	80
Bespin Gantry	1982	12	35
Death Star Compactor	1982	30	70
Death Star Escape	1982	15	45

Cloud City Playset, 1984, Kenner

A-Wing Pilot, 1985, Kenner

Luke Skywalker (Stormtrooper outfit), 1985, Kenner

ACTION FIGURES

NAME	YEAR	MNP	MIP
Death Star World	1982	50	155
Hoth Generator Attack	1982	14	40
Hoth Ion Cannon	1982	19	45
Hoth Turret Defense	1982	15	40
Hoth Wampa Cave	1982	14	40
Hoth World	1982	60	155
Imperial TIE Fighter	1982	20	60
Millennium Falcon	1982	175	390
Snow Speeder	1982	110	250
X-Wing Fighter	1982	20	70

MINI RIGS

NAME	YEAR	MNP	MIP
AST-5	1983	8	20
CAP-2 Captivator	1982	9	25
Desert Sail Skiff	1984	10	25
Endor Forest Ranger	1984	11	20
INT-4 Interceptor	1982	9	25
ISP-6 Imperial Shuttle Pod	1983	8	20
MLC-3 Mobile Laser Cannon	1981	9	25
MTV-7 Multi-Terrain Vehicle	1981	9	30
PDT-8 Personal Deployment Transport	1981	9	30
Radar Laser Cannon	1982	9	20
Security Scout		45	115
Tri-Pod Laser Cannon	1982	9	20
Vehicle Maintenance Energizer	1982	9	20

Play Sets

NAME	YEAR	MNP	MIP
EMPIRE STRIKES BACK			
Cloud City Play Set (Sears Exclusive)	1981	120	390
Dagobah	1982	25	70
Darth Vader's Star Destroyer		50	160
Hoth Ice Planet	1980	45	135
Imperial Attack Base	1980	25	80
Rebel Command Center	1980	65	220
Turret and Probot	1980	45	110
POWER OF THE FORCE			
Jabba's Dungeon	1983	220	340
RETURN OF THE JEDI			
Ewok Village	1983	35	75
Jabba's Dungeon	1983	45	120
STAR WARS			
Cantina Adventure Set (Sears Exclusive)	1977	230	570
Creature Cantina	1977	45	130
Death Star Space Station	1977	90	270
Droid Factory	1977	50	115
Land of the Jawas	1977	45	135

VEHICLES

NAME	YEAR	MNP	MIP
DIE-CAST VEHICLES			
Darth Vader's TIE Fighter	1979	25	65
Land Speeder	1979	25	70
Millennium Falcon	1979	40	155
Slave I	1979	25	85
Snowspeeder	1979	30	90
Star Destroyer	1979	45	175
TIE Bomber	1979	290	830
TIE Fighter	1979	20	60
Twin Pod Cloud Car	1979	25	80
X-Wing	1979	25	75
Y-Wing	1979	45	160
DROIDS			
A-Wing Fighter, Droids Box	1983	240	580
ATL Interceptor	1985	20	60
Imperial Side Gunner	1985	20	55
EMPIRE STRIKES BACK			
AT-AT	1980	105	260
Rebel Transport	1980	40	100
Scout Walker	1982	25	80
Slave I	1980	45	150
Snowspeeder		40	95
Twin-Pod Cloud Car	1980	30	85
EWOKS			
Ewoks Fire Cart	1985	10	25
Ewoks Woodland Wagon	1985	10	25
POWER OF THE FORCE			
Ewok Battle Wagon	1985	45	135
Imperial Sniper Vehicle	1985	50	85
One-Man Sand Skimmer	1985	40	80
Security Scout Vehicle	1985	55	120
Tatooine Skiff		270	670
RETURN OF THE JEDI			
B-Wing Fighter	1984	65	130
Ewok Combat Glider	1984	12	25
Imperial Shuttle	1984	150	310
Speeder Bike	1983	15	35
TIE Interceptor	1984	60	115
Y-Wing Fighter	1983	60	125
STAR WARS			
Darth Vader's TIE Fighter	1977	50	130
Imperial TIE Fighter	1977	40	155
Jawa Sand Crawler (battery-operated)	1977	220	610
Land Speeder (battery-operated)	1977	20	70
Millennium Falcon	1977	85	300
Sonic Land Speeder (J.C. Penney Exclusive)	1977	180	580
X-Wing Fighter	1977	45	165

Teebo, 1984, Kenner

Princess Leia in Boushh outfit, 1983, Kenner

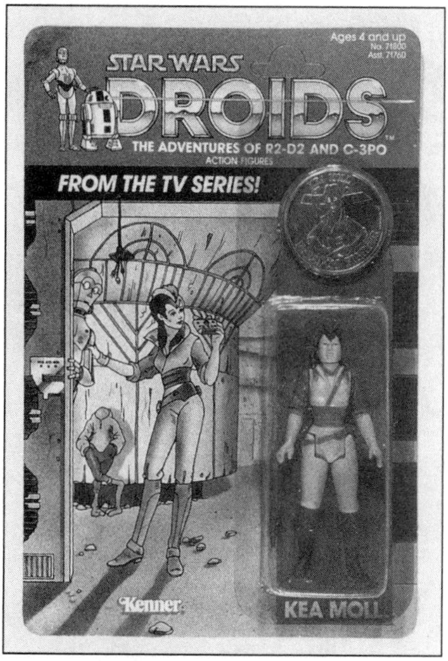

Kea Moll, 1985, Kenner

Imperial Dignitary, 1985, Kenner

Rebel Command Center, 1980, Kenner

Spawn, Cosmic Angela, McFarlane, 1994-1996

Star Trek, Betelgeusian, Mego, 1980

Masters of the Universe, Dragon Walker, Mattel, 1981-1990

STARSKY AND HUTCH (MEGO, 1976)

NAME	MNP	MIP
8" FIGURES & ACCESSORIES		
Captain Dobey	25	50
Car	65	125
Chopper	25	45
Huggy Bear	25	50
Hutch	20	45
Starsky	20	45

SUPER HERO BENDABLES (MEGO, 1972)

5" FIGURES		
Aquaman	50	120
Batgirl	50	120
Batman	35	90
Captain America	35	90
Catwoman	70	175
Joker	60	150
Mr. Mxyzptlk	50	125
Penguin	60	150
Riddler	60	150
Robin	30	75
Shazam	50	125
Supergirl	70	175
Superman	30	75
Tarzan	25	60
Wonder Woman	40	100

SUPER POWERS (KENNER, 1984-86)

5" FIGURES		
Aquaman, 1984	15	35
Batman, 1984	25	55
Braniac, 1984	15	30
Clark Kent, mail-in figure, 1986	100	0
Cyborg, 1986	75	200
Cyclotron, 1986	35	75
Darkseid, 1985	5	15
Desaad, 1985	10	30
Doctor Fate, 1985	25	50
Firestorm, 1985	15	35
Flash, 1984	10	20
Golden Pharoah, 1986	30	65
Green Arrow, 1985	25	55
Green Lantern, 1984	30	60
Hawkman, 1984	25	50
Joker, 1984	15	30
Kalibak, 1985	5	15
Lex Luthor, 1984	5	15
Mantis, 1985	10	30
Martian Manhunter, 1985	10	30
Mr. Freeze, 1986	15	35
Mr. Miracle, 1986	75	200
Orion, 1986	20	40
Parademon, 1985	15	35
Penguin, 1984	20	40
Plastic Man, 1986	40	80
Red Tornado, 1985	25	55
Robin, 1984	25	50
Samurai, 1986	25	50
Shazam (Captain Marvel), 1986	20	40
Steppenwolf, in mail-in bag, 1985	15	0
Steppenwolf, on card, 1985	15	75
Superman, 1984	20	35
Tyr, 1986	25	50
Wonder Woman, 1984	10	20

SUPER POWERS (KENNER, 1984-86)

NAME	MNP	MIP
ACCESSORIES		
Collector's Case, 1984	10	20
PLAY SETS		
Hall of Justice, 1984	30	100
VEHICLES		
Batcopter, 1986	40	75
Batmobile, 1984	40	75
Darkseid Destroyer, 1985	25	50
Delta Probe One, 1985	15	30
Justice Jogger Wind-Up, 1986	10	20
Kalibak Boulder Bomber, 1985	10	25
Lex-Soar 7, 1984	10	20
Supermobile, 1984	15	30

TEEN TITANS (MEGO, 1976)

6-1/2" CARDED FIGURES		
Aqualad	175	350
Kid Flash	175	300
Speedy	300	500
Wondergirl	200	450

TEENAGE MUTANT NINJA TURTLES (PLAYMATES, 1988-92)

1988, SERIES 1		
April O'Neil (no stripe)	60	150
Bebop	3	8
Donatello	8	20
Donatello (w/fan club form)	10	40
Foot Soldier	8	20
Leonardo	8	20
Leonardo (w/fan club form)	10	40
Michaelangelo	8	20
Michaelangelo (w/fan club form)	10	40
Raphael	8	20
Raphael (w/fan club form)	10	40
Rocksteady	8	20
Shredder	8	20
Splinter	8	20
1989, SERIES 2		
Ace Duck (hat off)	5	40
Ace Duck (hat on)	5	15
April O'Neil (blue stripe)	12	30
Baxter Stockman	10	25
Genghis Frog (black belt)	5	15
Genghis Frog (black belt, bagged weapons)	5	30
Genghis Frog (yellow belt)	30	75
Krang	5	15
1989, SERIES 3		
Casey Jones	5	15
General Traag	5	15
Leatherhead	25	50
Metalhead	5	15
Rat King	5	15
Usagi Yojimbo	5	15
1990, SERIES 4		
Mondo Gecko	5	15
Muckman and Joe Eyeball	5	15
Scumbag	5	15
Wingnut & Screwloose	5	15
1990, SERIES 5		
Fugitoid	5	15

World's Greatest Super Heroes, Falcon, Mego, 1974

Battlestar Galactica, Colonial Warrior, Mattel, 1978-1979

Masters of the Universe, Slime Pit, Mattel, 1981-1990

TEENAGE MUTANT NINJA TURTLES (PLAYMATES, 1988-92)

NAME	MNP	MIP
Slash (black belt)	3	25
Slash (purple belt, red "S")	25	75
Triceraton	5	15
1990, SERIES 6		
Mutagen Man	5	15
Napoleon Bonafrog	5	15
Panda Khan	5	15
1991, GIANT TURTLES, 13"		
Donatello	20	40
Leonardo	20	40
Michaelangelo	20	40
Raphael	20	40
1991, SERIES 10		
Grand Slam Raph	5	15
Hose'em Down Don	5	15
Lieutenant Leo	5	15
Make My Day Leo	5	15
Midshipman Mike	5	15
Pro Pilot Don	5	15
Raph the Green Teen Beret	5	15
Slam Dunkin' Don	5	15
Slapshot Leo	5	15
T.D. Tossin' Leonardo	5	15
1991, SERIES 7		
April O'Neil	25	125
April O'Neil ("Press")	10	20
Pizza Face	5	15
Ray Fillet (purple body, red "V")	10	25
Ray Fillet (red body, maroon "V")	10	30
Ray Fillet (yellow body, blue "V")	10	15
1991, SERIES 8		
Don The Undercover Turtle	5	15
Leo the Sewer Samurai	5	15
Mike the Sewer Surfer	5	15
Raph The Space Cadet	5	15
1991, SERIES 9		
Chrome Dome	5	15
Dirt Bag	5	15
Ground Chuck	5	15
Storage Shell Don	5	15
Storage Shell Leo	5	15
Storage Shell Michaelangelo	5	15
Storage Shell Raphael	5	15
1991, WACKY ACTION		
Breakfightin' Raphael	5	15
Creepy Crawlin' Splinter	5	15
Headspinnin' Bebop	5	15
Machine Gunnin' Rocksteady	5	15
Rock & Roll Michaelangelo	5	15
Sewer Swimmin' Don	5	15
Slice 'n Dice Shredder	10	25
Sword Slicin' Leonardo	8	15
Wacky Walkin' Mouser	10	20
1992, GIANT TURTLES, 13"		
Bebop	20	40
Movie Don	20	40
Movie Leo	20	40
Movie Mike	20	40
Movie Raph	20	40
Rocksteady	20	40
1992, SERIES 11		
Rahzer (black nose)	5	15
Rahzer (red nose)	7	25
Skateboard'n Mike	5	15

TEENAGE MUTANT NINJA TURTLES (PLAYMATES, 1988-92)

NAME	MNP	MIP
Super Shredder	5	15
Tokka (brown trim)	9	25
Tokka (gray trim)	5	15
1992, SERIES 12		
Movie Don	5	15
Movie Leo	5	15
Movie Mike	5	15
Movie Raph	5	15
Movie Splinter, no tooth	5	15
Movie Splinter, w/tooth	25	75
VEHICLES/ACCESSORIES		
Flushomatic	4	10
Foot Cruiser	14	35
Foot Ski	4	10
Mega Mutant Killer Bee	3	8
Mega Mutant Needlenose	8	20
Mike's Pizza Chopper Backpack	4	10
Mutant Sewer Cycle with Sidecar	4	10
Ninja Newscycle	5	12
Oozey	4	10
Pizza Powered Sewer Dragster	5	15
Pizza Thrower	14	35
Psycho Cycle	10	25
Raph's Sewer Dragster	6	16
Raph's Sewer Speedboat	5	12
Retrocatapult	4	10
Retromutagen Ooze	2	4
Sewer Seltzer Cannon	4	10
Sludgemobile	7	18
Technodrome, 22"	24	60
Toilet Taxi	5	12
Turtle Blimp, 30" green vinyl	12	30
Turtle Party Wagon	16	40
Turtle Trooper Parachute, 22"	4	10
Turtlecopter	16	40

TOTAL CHAOS (MCFARLANE, 1996)

FIGURES		
Al Simmons	5	15
Dragon Blade	5	15
Gore	5	15
Hoof	5	15
Thorax	5	15
Thresher	5	30

TOTAL JUSTICE (KENNER, 1996)

5" FIGURES		
Aquaman, black armor	5	15
Aquaman, gold armor	7	20
Batman	5	15
Black Lightning	7	20
Darkseid	5	15
Despero	5	15
Flash, The	5	15
Green Arrow	5	20
Green Lantern	5	15
Hawkman	5	15
Huntress	5	20
Parallax	5	20
Robin	5	15
Superman	5	15

Avengers, Iron Man, Toy Biz, 1997

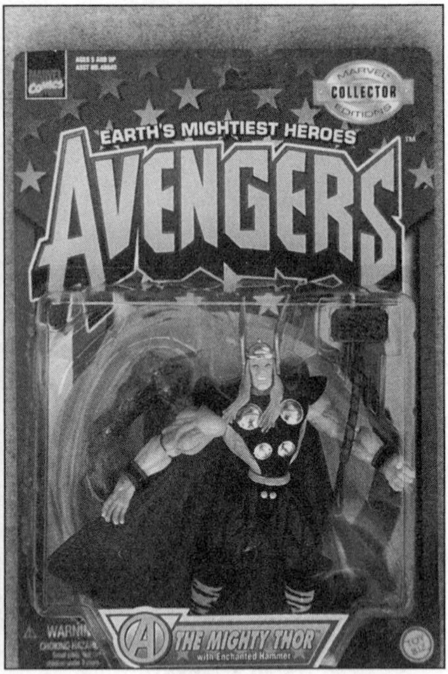

Avengers, The Mighty Thor, Toy Biz, 1997

Best of the West, Fighting Eagle, Marx, 1967

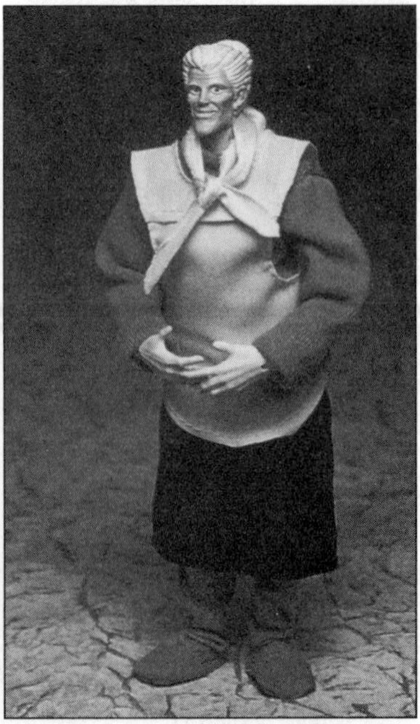

Marvel Famous Covers, Aunt May, Toy Biz, 1997

TOY STORY (THINKWAY, 1996)

NAME	MNP	MIP
5" FIGURES		
Alien	5	15
Boxer Buzz	4	8
Crawling Baby Face	3	9
Fighting Woody	3	9
Flying Buzz (Rocket)	4	8
Hamm	4	8
Karate Buzz	4	8
Kicking Woody	4	8
Quick-Draw Woody	4	8
Rex	4	8
Super Sonic Buzz	3	9
LARGE FIGURES		
Talking Buzz Lightyear	15	50
Talking Woody	15	60

VIKINGS (MARX, 1960s)

Eric the Viking	35	65
Mighty Viking Horse	30	60
Odin the Viking Chieftan	35	65

WALTONS (MEGO, 1975)

8" FIGURES		
Grandma and Grandpa	25	50
John Boy and Ellen	25	50
Mom and Pop	25	50
ACCESSORIES		
Barn	50	100
Country Store	50	100
Truck	40	80
PLAY SETS		
Farm House	50	100
Farm House with Six Figures	0	200

WETWORKS (MCFARLANE, 1995-96)

FIGURES		
Assassin One	5	15
Blood Queen	8	30
Dane	4	15
Delta Commander	5	15
Dozer	4	15
Frankenstein	5	15
Grail	4	15
Mendoza	5	15
Mother One	7	15
Pilgrim	10	30
Vampire	10	15
Werewolf	10	30

WIZARD OF OZ (MEGO, 1974)

4" BOXED FIGURES		
Munchkin Dancer	75	150
Munchkin Flower Girl	75	150
Munchkin General	75	150
Munchkin Lollipop Kid	75	150
Munchkin Mayor	75	150
8" BOXED FIGURES		
Cowardly Lion	25	50
Dorothy with Toto	25	50
Glinda the Good Witch	25	50
Scarecrow	25	50

WIZARD OF OZ (MEGO, 1974)

NAME	MNP	MIP
Tin Woodsman	25	25
Wicked Witch	50	100
Witch's Monkey	75	150
Wizard of Oz	35	100
PLAY SETS		
Emerald City with eight 8" figures	125	350
Emerald City with Wizard of Oz	45	100
Munchkin Land	150	300
Witch's Castle, Sears Exclusive	250	450

WORLD WRESTLING FEDERATION (HASBRO, 1990-94)

FIGURES		
1-2-3 Kid	30	65
Adam Bomb	20	35
Akeem	30	75
Andre the Giant	50	150
Ax	10	30
Bam Bam Bigelow	12	30
Bart Gunn	20	35
Berzerker	6	15
Big Bossman (1990)	8	16
Big Bossman (1992)	7	15
Billy Gunn	20	35
Bret Hart (1992)	10	25
Bret Hart (1993 mail-in)	75	0
Bret Hart (1994)	7	15
British Bulldog	7	15
Brutus the Barber (1990)	12	25
Brutus the Barber (1992)	12	25
Bushwackers (two-pack)	10	20
Butch Miller	5	15
Crush (1993)	12	25
Crush (1994)	6	15
Demolition (two-pack)	20	45
Doink the Clown	7	15
Dusty Rhodes	125	300
Earthquake	15	30
El Matador	6	15
Fatu	5	15
Giant Gonzales	5	15
Greg the Hammer	9	30
Hacksaw Jim Duggan (1991)	5	15
Hacksaw Jim Duggan (1994)	5	15
Honky Tonk Man	25	50
Hulk Hogan (1990)	12	25
Hulk Hogan (1991)	10	20
Hulk Hogan (1992)	10	20
Hulk Hogan (1993, mail-in)	75	100
Hulk Hogan (1993, no shirt)	10	20
I.R.S.	7	15
Jake the Snake Roberts	9	20
Jim Neidhart	6	15
Jimmy Superfly Snuka	12	25
Kamala	10	25
Koko B. Ware	20	60
Legion of Doom (two-pack)	20	40
Lex Luger	15	30
Ludwig Borga	20	35
Luke Williams	5	15
Macho Man (1990)	12	25
Macho Man (1991)	15	35
Macho Man (1992)	15	35
Macho Man (1993)	7	15
Marty Jannetty	5	15
Mountie	6	15

WORLD WRESTLING FEDERATION (HASBRO, 1990-94)

Mr. Perfect (1992)	15	35
Mr. Perfect (1994)	12	25
Nailz	12	25
Nasty Boys (two-pack)	15	80
Owen Hart	20	35
Papa Shango	7	15
Razor Ramon (1993)	15	30
Razor Ramon (1994)	9	18
Repo Man	7	15
Ric Flair	7	15
Rick Martel	5	15
Rick Rude	15	30
Rick Steiner	9	18
Ricky "The Dragon" Steamboat	7	15
Rockers (two-pack)	10	20
Rowdy Roddy Piper	15	30
Samu	5	15
Scott Steiner	8	18
Sgt. Slaughter	15	30
Shawn Michaels (1993)	10	25
Shawn Michaels (1994)	6	15
Sid Justice	6	15
Skinner	6	15
Smash	12	25
Tatanka (1993)	7	15
Tatanka (1994)	7	15
Ted Diabiase (1990)	10	20
Ted Diabiase (1991)	7	15
Ted Diabiase (1994)	7	15
Texas Tornado	10	50
Typhoon	15	30
Ultimate Warrior (1990)	12	25
Ultimate Warrior (1991)	10	20
Ultimate Warrior (1992)	20	40
Undertaker (1992)	9	18
Undertaker (1993, mail-in)	25	50
Undertaker (1994)	15	25
Virgil	7	15
Warlord	6	15
Yokozuna	15	30

WORLD'S GREATEST SUPER HEROES (MEGO, 1972-78)

12-1/2" BOXED FIGURES		
Amazing Spider-Man, 1978	35	75
Batman, 1978	60	125
Captain America, 1978	75	150
Hulk, 1978	30	60
8" FIGURES		
Aquaman, 1972, boxed	50	150
Aquaman, 1972, carded	50	150
Batgirl, 1973, boxed	125	300
Batgirl, 1973, carded	125	250
Batman, fist fighting, 1975, boxed	150	350
Batman, painted mask, 1972, boxed	60	150
Batman, painted mask, 1972, carded	60	100
Batman, removable mask, 1972, boxed	200	350
Batman, removable mask, 1972, Kresge card only	200	450
Bruce Wayne, 1974, boxed, Montgomery Ward exclusive	400	500
Captain America, 1972, boxed	60	200
Captain America, 1972, carded	60	150
Catwoman, 1973, boxed	100	225
Catwoman, 1973, carded	100	225

WORLD'S GREATEST SUPER HEROES (MEGO, 1972-78)

NAME	MNP	MIP
Clark Kent, 1974, boxed, Montgomery Ward exclusive	400	500
Conan, 1975, boxed	120	300
Conan, 1975, carded	120	300
Dick Grayson, 1974, boxed, Montgomery Ward exclusive	400	500
Falcon, 1974, boxed	60	150
Falcon, 1974, carded	60	200
Green Arrow, 1973, boxed	100	250
Green Arrow, 1973, carded	100	400
Green Goblin, 1974, boxed	90	225
Green Goblin, 1974, carded	90	300
Human Torch, Fantastic Four, 1975, boxed	25	90
Human Torch, Fantastic Four, 1975, card	25	50
Incredible Hulk, 1974, boxed	20	100
Incredible Hulk, 1974, carded	20	50
Invisible Girl, Fantastic Four, 1975, boxed	30	150
Invisible Girl, Fantastic Four, 1975, card	30	60
Iron Man, 1974, boxed	75	125
Iron Man, 1974, carded	75	250
Isis, 1976, boxed	75	250
Isis, 1976, carded	75	125
Joker, 1973, boxed	60	150
Joker, 1973, carded	60	150
Joker, fist fighting, 1975, boxed	150	400
Lizard, 1974, boxed	75	200
Lizard, 1974, carded	75	250
Mr. Fantastic, Fantastic Four, 1975, boxed	30	140
Mr. Fantastic, Fantastic Four, 1975, carded	30	60
Mr. Mxyzptlk, open mouth, 1973, boxed	50	75
Mr. Mxyzptlk, open mouth, 1973, carded	50	150
Mr. Mxyzptlk, smirk, 1973, boxed	60	150
Penguin, 1973, boxed	60	150
Penguin, 1973, carded	60	125
Peter Parker, 1974, boxed, Montgomery Ward exclusive	400	500
Riddler, 1973, boxed	100	250
Riddler, 1973, carded	100	400
Riddler, fist fighting, 1975, boxed	150	400
Robin, fist fighting, 1975, boxed	125	350
Robin, painted mask, 1972, boxed	60	150
Robin, painted mask, 1972, carded	60	90
Robin, removable mask, 1972, boxed	250	400
Shazam, 1972, boxed	75	200
Shazam, 1972, carded	75	150
Spider-Man, 1972, boxed	20	100
Spider-Man, 1972, carded	20	40
Supergirl, 1973, boxed	300	450
Supergirl, 1973, carded	300	450
Superman, 1972, boxed	50	125
Superman, 1972, carded	50	100
Tarzan, 1972, boxed	50	150
Tarzan, 1976, Kresge card only	60	225
Thing, Fantastic Four, 1975, boxed	40	150
Thing, Fantastic Four, 1975, carded	40	60
Thor, 1975, boxed	150	300
Thor, 1975, carded	150	300
Wonder Woman, boxed	100	250
Wonder Woman, Kresge card only	100	350

Gene Simmons and Paul Stanley were two of Mego's 1979 KISS figures.

Ideal's Batgirl — one of the Comic Heroines series from 1967 — is one of the most desirable and valuable action figures of all time.

WORLD'S GREATEST SUPER HEROES (MEGO, 1972-78)

NAME	MNP	MIP
Wondergirl	100	240
ACCESSORIES		
Super Hero Carry Case, 1973	40	100
Supervator, 1974	60	120
PLAY SETS		
Aquaman vs. the Great White Shark, 1978	200	500
Batcave Play Set, 1974, vinyl	125	250
Batman's Wayne Foundation Penthouse, 1977, fiberboard	600	1200
Hall of Justice, 1976, vinyl	125	250
SUPERMAN SERIES		
General Zod, 1978	50	100
Jor-El, 1978	50	100
Lex Luthor, 1978	50	100
Superman Play Set, 1978	75	150
Superman, 1978	50	125
VEHICLES		
Batcopter, 1974, boxed	75	150
Batcopter, 1974, carded	55	110
Batcycle, black, 1975, boxed	75	185
Batcycle, black, 1975, carded	60	150
Batcycle, blue, 1974, boxed	75	170
Batcycle, blue, 1974, carded	75	135
Batmobile and Batman	40	100
Batmobile, 1974, boxed	50	125
Batmobile, 1974, carded	50	120
Captain Americar, 1976	100	200
Green Arrowcar, 1976	175	350
Jokermobile, 1976	150	300
Mobile Bat Lab, 1975	125	250
Spidercar, 1976	50	125
WONDER WOMAN SERIES		
Major Steve Trevor, 1978	26	65
Queen Hippolyte, 1978	40	100
Queen Nubia, 1978	40	100
Wonder Woman Play Set, 1978	50	100
Wonder Woman with Diana Prince outfit, 1978	55	80

WORLD'S GREATEST SUPER KNIGHTS (MEGO, 1975)

8" BOXED FIGURES		
Black Knight	80	160
Ivanhoe	60	120
King Arthur	60	120
Sir Galahad	75	150
Sir Lancelot	75	150
ACCESSORIES		
Castle Play Set	80	160
Jousting Horse, battery operated	40	70

WORLD'S GREATEST SUPER PIRATES (MEGO, 1974)

8" BOXED FIGURES		
Blackbeard	70	150
Captain Patch	70	150
Jean LaFitte	80	160
Long John Silver	80	160

X-FILES (MCFARLANE, 1998)

NAME	MNP	MIP
Fireman with Cryolitter	7	25
Mulder with Docile Alien	7	25
Mulder with Human Host/Cryopod Chamber	7	25
Mulder (Arctic Wear) with Human Host/Cryopod Chamber	7	25
Mulder with Victim and Gurney	7	25
Scully with Docile Alien	7	25
Scully with Human Host/Cryopod Chamber	7	25
Scully (Arctic Wear) with Human Host/Cryopod Chamber	7	25
Scully with Victim and Gurney	7	25
Primitive Man with Attack Alien	7	25

X-MEN/X-FORCE (TOY BIZ, 1991-1995)

FIGURES		
Ahab	5	15
Apocalypse #1, 1991	7	15
Apocalypse #1, 1993	4	15
Apocalypse #2	4	15
Archangel	7	15
Archangel #2	4	15
Banshee, 1992	7	15
Banshee, 1993	3	15
Beast	10	20
Bishop	7	15
Black Tom	5	15
Blob	5	15
Bonebreaker	3	15
Bridge	5	15
Bridge	4	15
Brood	4	15
Cable	5	15
Cable #1, 1992	6	15
Cable #1, 1993	4	15
Cable #2	4	15
Cable #3	4	15
Cable #4	3	15
Cable Cyborg	4	15
Caliban	4	15
Cameron Hodge	4	15
Cannonball, Pink	15	35
Cannonball, Purple	10	20
Colossus, 1991	10	20
Colossus, 1993	6	15
Corsair	3	15
Cyclops #1, blue	10	20
Cyclops #1, stripes	5	15
Cyclops #2	5	15
Deadpool	15	35
Deadpool, 1992	15	35
Deadpool, 1995	6	15
Domino	5	15
Forearm	10	20
Forearm	10	20
Forge	10	25
Gambit, 1992	10	20
Gambit, 1993	7	15
Gideon	5	15
Gideon	4	15
Gladiator	3	15
Grizzly	4	15
Havok	4	15
Ice Man #2	4	15

X-MEN/X-FORCE (TOY BIZ, 1991-1995)

NAME	MNP	MIP
Ice Man, 1992	20	45
Ice Man, 1993	10	25
Juggernaut, 1991	10	25
Juggernaut, 1993	4	15
Kane	5	15
Kane #1	7	15
Kane #2	4	15
Killspree	7	15
Krule	3	15
Kylun	3	15
Longshot	7	15
Magneto #1	6	15
Magneto #2	5	15
Maverick	3	15
Morph	10	20
Mr. Sinister	7	15
Nightcrawler	10	25
Nimrod	4	15
Omega Red	6	15
Phoenix	10	25
Professor X	6	15
Pyro	5	15
Quark	3	15
Random	3	15
Raza	3	15
Rictor	3	15
Rogue	10	25
Sabretooth #1	7	15
Sabretooth #2	6	15
Sabretooth #3	4	15
Sauron	3	15
Shatterstar	5	15
Shatterstar #1	5	15
Shatterstar #2	3	15
Silver Samurai	3	15
Slayback	3	15
Spiral	4	15
Storm, 1991	20	40
Storm, 1993	10	20
Strong Guy	3	15
Stryfe	10	25

X-MEN/X-FORCE (TOY BIZ, 1991-1995)

NAME	MNP	MIP
Stryfe	8	15
Sunfire	4	15
Sunspot	3	15
Trevor Fitzroy	3	15
Tusk	5	15
Warpath	10	25
Warpath #1	7	15
Warpath #2	5	15
Warstar	6	15
Wolverine #1, 1991	10	20
Wolverine #1, 1993	7	15
Wolverine #2, 1992	7	15
Wolverine #2, 1993	5	15
Wolverine #3	12	25
Wolverine #4	7	15
Wolverine #5	7	15
Wolverine #6	4	15
Wolverine Fang	4	15
Wolverine, space armor	6	15
Wolverine, street clothes	4	15
X-Cutioner	4	15
X-Treme	4	15

YOUNGBLOOD (MCFARLANE, 1995)

FIGURES

Crypt	5	10
Die Hard	4	8
Dutch	4	8
Sentinel	4	8
Shaft	4	8
Troll	10	20

ZORRO (GABRIEL, 1982)

Amigo	10	20
Captain Ramon	10	20
Picaro	15	35
Sergeant Gonzales	10	20
Tempest	10	25

Advertising Toys

Almost everyone is familiar with advertising characters — how many can you name? Snap, Crackle and Pop; the Pillsbury Doughboy; Speedy Alka Seltzer; the Jolly Green Giant.

What is the purpose of the advertising character? Its primary purpose is to attract the consumer's attention so he or she will buy a product.

Let's face it, many times the store brand item tastes the same or even better than a name brand, so why buy a product that is twice the price? Advertisers hope it's because the character is so familiar and beloved.

Advertising characters are a colorful part of everyday life and can take many forms. The favorite of many collectors are PVC vinyl figures. Such figures are brightly colored, almost unbreakable and make great display pieces. Advertising figures are also found as bendy figures, wind-up toys, radios, telephones and beanbag toys.

Some of the ad characters of today have been around for over 100 years. Aunt Jemima, known for touting maple syrup, is one such example. Aunt Jemima was first used for advertising pancake flour at the World's Fair in the late 1800s. She was a great success; soon after, a paper doll set was used as a promotional item.

Kellogg's has long delighted collectors with their many premiums depicting characters. A 1920s Kellogg's premium was a set of Goldilocks and the three bears dolls. The consumer received a piece of fabric with the image of the doll printed on it. Consumers were supposed to cut, stuff and sew the doll themselves.

Popular cereal pixies Snap, Crackle and Pop were first offered as make-your-own dolls in the 1940s.

Advertising characters have been used to promote everything from tires (remember the Michelin Man?) and shoes (Buster Brown) to fast food (McDonald's Speedee) and dairy products (Elsie the Cow). But cereal characters seem to be among the most popular with collectors.

Recognize these familiar faces?

Quisp is extremely popular, but collectibles are hard to find today. Rare collectibles, like a Quisp bank, can run up to $1,000. Other popular cereal characters include the General Mills monsters (Frankenberry, Boo Berry, Count Chocula and Fruit Brute); Tony the Tiger; and Snap, Crackle and Pop.

Also popular are vintage advertising characters that presented stereotypical (often negative) ethnic characters — think Aunt Jemima and Uncle Mose (pancake mix), Mohawk Carpet Indian and Rastus, the black character used to advertise Cream of Wheat cereal. Because such stereotypes are considered politically incorrect today, such items are extremely collectible.

The Top 10 Advertising Toys
(in Mint condition)

1. Quisp Bank, Quaker Oats, 1960s, 6-1/2" papier mâché $850
2. Esky Store Display, Esquire Magazine, 1940s 375
3. Mr. Peanut Figure, Planters Peanuts, 1930s, 8-1/2" painted wood figure ... 375
4. Speedy Figure, Alka-Seltzer, 1963, 5-1/2" vinyl 350
5. Elsie The Cow Cookie Jar, Borden's, 1950s, 12" 350
6. Good Humor Figure, Good Humor, 1975, 8" vinyl 300
7. Reddy Kilowatt Bobbin' Head, Reddy Communications, 1960s, 6-1/2" 300
8. Vegetable Man Bank, Kraft, 1970s, 8" vegetable man 275
9. Clark Bar Figure, Beatrice Foods, 1960s, 8-1/2" vinyl boy 275
10. Barnum's Animal Crackers Cookie Jar, Nabisco, 1972, ceramic 275

Contributor to this section: Carren Kopleton, Queen's Collection, P.O. Box 235, Clifton, NJ 07011.

ADVERTISING TOYS

NAME	YEAR	DESCRIPTION	GOOD	EX	MINT
7-ELEVEN					
Big Bite Figures	1987	3-1/2" hot dog men holding snacks	7	10	12
AIR INDIA					
Air India Man Figure	1973	4-1/2" statue of man w/flying carpet	20	25	30
ALKA-SELTZER					
Speedy Figure	1963	5-1/2" vinyl boy holding Alka-Seltzer figure	275	300	350
ALPO					
Dan the Dog Cookie Jar	1980s	7" ceramic gray and white sheep dog	35	50	70
AMERICAN HOME FOOD PRODUCTS					
Marky Maypo Figure	1960	9" vinyl figure	35	45	55
AUNT JEMIMA SYRUP					
Aunt Jemima Doll	1950s	13" cloth doll wearing red checkered dress	60	70	90
BASKIN ROBBINS					
Pinky the Spoon Figure	1993	5" bendable figure	4	5	10
BAZOOKA					
Bazooka Joe Doll	1973	19" plush boy pirate	20	25	30
BEATRICE FOODS					
Clark Bar Figure	1960s	8-1/2" vinyl boy holding a Clark bar wearing a striped shirt	200	225	275
BEECH-NUT					
Fruit Stripe Gum Figure	1967	7-1/2" bendable man shaped like a pack of gum riding a motorcycle	150	200	225
BOB'S BIG BOY					
Big Boy Doll	1978	14" cloth boy doll wearing checkerboard outfit	15	25	30
Big Boy Doll	1973	8" plastic jointed doll holding a hamburger	75	125	150
Big Boy Lamp	1960	6-1/2" vinyl figural lamp	40	60	75
Dolly Doll	1978	14" cloth girl doll with Dolly nametag	15	25	30
Nugget Doll	1978	10" cloth dog doll with "Nugget" on collar	15	25	30
BORDEN'S					
Elsie The Cow Cookie Jar	1950s	12"	275	300	350
Elsie The Cow Doll	1950s	12" plush cow that moos when shaken	50	75	85
Elsie The Cow Lamp	1947	electric lamp	150	175	225
BRADFORD HOUSE RESTAURANTS					
Bucky Bradford Figure	1960s	9-1/2" blond pilgrim boy holding dish that reads "It's Yum Yum Time"	25	30	35
BURGER KING					
Burger King Doll	1972	20" plush animated version of king	10	12	15
CAMPBELL					
Campbell Kid Doll	1950s	9-1/2" vinyl, jointed cheerleader wearing a white shirt with "C" in middle	45	65	80
Campbell Kid Doll	1973	16" blonde girl with red hair	25	35	40
Campbell Kid Figure	1970s	7" vinyl Campbell boy with blue overalls	30	40	50
Campbell Kid Figure	1950s	7" boy chef doll with "C"on hat and spoon in his hand	40	50	65

NAME	YEAR	DESCRIPTION	GOOD	EX	MINT
CAMPBELL					
Wizard of O's Figure	1978	7-1/2" vinyl wizard with Spaghetti O's on his hat and bow tie	15	25	30
CHAMPION AUTO STORES					
Champ Man Figure	1991	6" bendable man; head is a flag	7	10	15
CHICKEN DELIGHT INTERNATIONAL					
Chicken Delight Bank	1960s	6" yellow and red chicken holding a tray of biscuits	85	125	150
CHIQUITA					
Chiquita Banana Doll	1974	16" plush dancing banana with fruit on its head	18	20	25
CRACKER JACK					
Cracker Jack Doll	1974	15" plush sailor holding small box of Cracker Jacks snacks	15	20	25
CREST TOOTHPASTE					
Sparkle Telephone	1980s	11" blue and silver snowman-type character	25	35	40
CURAD					
Taped Crusader Figure	1977	7-1/2" male cartoon superhero	65	75	85
CURITY					
Miss Curity Display	1950s	18" plastic store display of a nurse on a base	125	150	185
DEL MONTE					
Clown Bank		colorful, smiling clown	5	10	20
DOMINO'S PIZZA					
Noid Figure	1980s	2-1/2" PVC figure	1	2	3
DOUGLAS OIL COMPANY					
Freddy Fast Figure	1976	7" freckled face boy; hat says "Freddy Fast"	90	110	125
DOW BRANDS					
Scrubbing Bubble Brush	1980s	3-1/2" light blue scrub brush	8	10	12
ESQUIRE MAGAZINE					
Esky Store Display	1940s	11" old man dressed in a tuxedo standing on an Esquire magazine	300	350	375
ESSO					
Esso Tiger Figure	1960s	8-1/2" plastic tiger	20	25	30
EssoTiger Garbage Can	1970s	10" metal garbage can with picture of the Esso tiger	15	20	30
EVEREADY BATTERIES					
Energizer Bunny Plush		22" plush	15	40	65
FACIT ADDING MACHINES					
Facit Man Figure	1964	4" man wearing yellow outfit with black wizard hat	20	25	30
FLORIDA CITRUS DEPARTMENT					
Florida Orange Bird Nodder	1970s	7" plastic nodder	50	60	75
FUNNY FACE DRINK					
Choo Choo Cherry Ramp Walker	1971	3" round, red figure with conductor's hat	70	90	110

ADVERTISING TOYS

NAME	YEAR	DESCRIPTION	GOOD	EX	MINT
FUNNY FACE DRINK					
Funny Face Mugs	1969	3" mugs of Funny Face characters, each	7	10	15
Goofy Grape Pitcher	1973	10" pitcher of smiling, purple character wearing lime green captain's hat	75	95	125
GENERAL MILLS					
Boo Berry Figure	1975	7-1/2" white and light blue ghost with hat and bow tie	75	95	120
Cereal Card Game	1981	card game with different cereal characters	10	15	20
Count Chocula Figure	1975	7-1/2" vinyl vampire	70	85	110
Frankenberry Figure	1975	8" vinyl pink Frankenstein	70	95	120
Fruit Brute Figure	1975	8" vinyl werewolf with striped shirt	75	90	120
Trix Rabbit Figure	1977	9" vinyl white rabbit	30	40	50
GERBER PRODUCTS					
Gerber Boy figure	1985	8" vinyl boy with baseball cap turned sideways that reads "I'm a Gerber Kid"	25	27	30
GOOD HUMOR					
Good Humor Figure	1975	8" vinyl ice cream bar with a bite out of it	260	280	300
GRANDMA'S COOKIES					
Grandma Bank	1988	7-1/2" hard plastic Grandma wearing a blue dress	25	30	35
H.P. HOOD					
Harry Hood Figure	1981	7-1/2" delivery man with "Hood" inscription on chest	55	70	80
HERSHEY'S					
Hershey's Chocolate Figure	1987	4-1/2" bendable candy bar shaped like a man	5	8	10
HUSH PUPPIES					
Hush Puppies Figure	1970s	8" tan and brown basset hound	20	25	35
ICEE					
ICEE Bear Figure	1970s	8" vinyl polar bear drinking an ICEE	25	30	35
INSTY-PRINTS					
Insty-Prints Wizard Figure	1980s	9" vinyl figure of wizard dressed in red outfit with moons and stars	75	100	135
IRON FIREMAN FURNACE					
Iron Fireman Figure	1943	5" metal man shoveling coal	65	75	85
KAHN'S WIENERS					
Beefy Frank Figure	1980	5-1/2" vinyl figural hot dog mustard dispenser	15	18	22
KELLOGG'S					
Dig 'Em Doll	1973	16" smiling frog with baseball hat wearing shirt that reads "Dig Em"	15	20	30
Dig 'Em Slide Puzzle	1979	4" plastic squares that make a scene when put together	8	10	15
Milton The Toaster Figure	1980	5" white, smiling toaster	50	70	90
Snap, Crackle and Pop Figures	1975	7-1/2" vinyl, arms extended, each	25	30	35
Snap, Crackle and Pop Figures	1975	7-1/2" vinyl, arms at side, each	20	25	30
Snap, Crackle and Pop Hand Puppets	1950s	cloth body and vinyl head, each	25	35	40
Tony The Tiger Cookie Jar	1968	7" plastic	70	95	125
Tony The Tiger Figure	1974	7-1/2" vinyl tiger	45	60	75

NAME	YEAR	DESCRIPTION	GOOD	EX	MINT
KELLOGG'S					
Toucan Sam Backpack	1983	12" blue backpack that pictures Toucan Sam sitting on a schoolhouse	15	20	25
Toucan Sam Figure	1984	3" plastic jointed figure with blue body and multi-colored beak	12	15	20
KEN-L-RATION					
Ken-L-Ration Wall Pockets	1960s	3" pair of plastic wall pockets; cat's head and a dog's head	40	50	65
KENTUCKY FRIED CHICKEN					
Colonel Sanders Nodder	1960s	7" plastic nodder of Colonel holding bucket of chicken	70	85	100
KIDDIE CITY TOY STORE					
Kaycee Kangaroo Figure	1980s	vinyl kangaroo with baby in pouch	40	55	65
KRAFT					
Cheesasaurus Rex Figure	1990s	5-1/2" orange dinosaur figures wearing different outfits	3	5	7
Mr. Wiggle Hand Puppet	1966	6" red rubber	150	185	225
Vegetable Man Bank	1970s	8" vegetable man with tomato for a head and a celery body	200	250	275
LABATT'S BREWERY					
Labatt's Beer Man Figure	1972	6" man standing next to wood barrel	20	25	30
MAGIC CHEF					
Magic Chef Figure	1980s	7" chef dressed in tuxedo and chef's hat	10	12	15
MCDONALD'S					
Big Mac Figure	1976	6-1/2" action figure with Big Mac for a head	20	25	30
Mac Tonight Figure	1980s	3-1/2" moon head PVC character wearing a suit	1	3	5
Ronald McDonald Bank	1980s	7-1/2" plastic bank of waving Ronald	15	20	27
MICHELIN TIRES					
Mr. Bib Figure	1980s	12" plastic figure with Michelin sash across chest	30	40	50
NABISCO					
Barnum's Animal Crackers Cookie Jar	1972	ceramic, shaped like a box of animal crackers	175	225	275
Fig Newton Girl	1983	4-1/2" girl with Fig Newton on her head	15	18	20
Nabisco Thing	1996	5" multi-colored bendable figure with Nabisco logo for its head	5	10	15
NATHAN'S					
Nathan's Franksters	1990s	5-1/2" hot dog characters in different outfits	2	3	5
NESTLE'S					
Quik Bunny Doll	1976	24" plush rabbit with letter "Q" on his chest	25	35	40
Quik Bunny Figure	1991	6" bendable brown and tan rabbit with the letter "Q" hanging from his neck	3	7	10
NOVOTEL HOTEL					
Dolphi Figure	1990s	4-1/2" bendable dolphin figures; set of four produced each year	5	8	10
OSCAR MAYER FOODS					
Hot Wheels Wienermobile		die-cast Wienermobile	2	4	6
Little Oscar Puppet		thin plastic, theme song on back	13	3	6

ADVERTISING TOYS

NAME	YEAR	DESCRIPTION	GOOD	EX	MINT
OSCAR MAYER FOODS					
Wienermobile	1950s	11" car; Little Oscar pops up when car is rolled	75	95	135
Wienermobile Bank		plastic	3	7	15
Wienermobile Beanbag	1998	7" plush beanbag	3	5	8
PEPPERIDGE FARM					
Goldfish Plush	1970s	16" plush figure	15	20	25
PILLSBURY					
Biscuit the Cat Puppet	1974	2-1/2" vinyl cat finger puppet	18	25	30
Grandpopper & Grandmommer Figures	1974	5" vinyl pair of figures	80	120	150
PILLSBURY					
Jolly Green Giant Figure	1970s	9-1/2" vinyl green man wearing loincloth of leaves	85	100	125
Little Sprout Figure	1970s	6-1/2" vinyl green figure with leaves on head and body	7	10	15
Little Sprout Inflatable Figure	1976	24" vinyl	25	30	35
Little Sprout Salt and Pepper Shakers	1988	3" ceramic figural shakers	25	30	35
Poppin' Fresh Doll	1972	white velour doll with hat and scarf	20	25	30
Poppin' Fresh Figure	1971	6-1/2" vinyl white baker with blue eyes and blue dot on hat	5	7	10
PIZZA HUT					
Pizza Hut Pete Bank	1969	7-1/2" plastic	20	35	50
PIZZA TIME THEATRE					
Chuck E. Cheese Bank	1980s	6-1/2" plastic	8	12	15
PLANTERS PEANUTS					
Mr. Peanut Doll	1967	21" pillow doll	15	20	25
Mr. Peanut Figure	1992	8-1/2" yellow and black plastic	12	15	20
Mr. Peanut Figure	1930s	8-1/2" painted wood figure with hat and cane	275	300	375
Mr. Peanut Wind-Up	1984	3" yellow peanut man with traditional hat and cane	20	25	30
POST CEREAL					
California Raisins PVC Figure	1987	2" male raisin playing drums	3	5	8
California Raisins Wind-Up	1987	4" female raisin with tambourine	5	7	10
Sugar Bear Doll	1988	12" plush bear with blue "Sugar Bear" shirt	15	20	25
Sugar Bear Doll	1970s	12" plush brown bear with blue "Sugar Bear" shirt	20	25	30
PROCTOR & GAMBLE					
24 Hour Bug	1970s	7" spotted green bug	50	65	80
Hawaiian Punch Doll	1983	15" plush Punchy with red hair and blue and white striped shirt	20	22	25
Hawaiian Punch Radio	1970s	6" figural radio of Punchy	25	35	45
Mr. Clean Figure	1961	8" bald man wearing white clothes with earring; arms are folded	95	120	150
PURINA CHUCK WAGON					
Chuck Wagon	1975	8" vinyl team of horses and checker board covered wagon	25	30	35
QUAKER OATS					
Cap'n Crunch Bank	1969	8" captain wearing blue outfit and sword	45	65	75
Jean LaFoote Bank	1975	8" vinyl pirate wearing green suit and purple hat	60	85	110

NAME	YEAR	DESCRIPTION	GOOD	EX	MINT
QUAKER OATS					
Quake Cereal Doll	1965	9" man with raised arms and letter "Q" across chest	85	95	125
Quisp Bank	1960s	6-1/2" papier mache	500	750	850
Quisp Cereal Doll	1965	10" doll with pink body, green clothes and letter "Q" across stomach	70	85	110
R.J. REYNOLDS					
Joe Camel Can Cooler	1991	4" vinyl	5	10	12
RAID					
Raid Bug	1989	plush green bug	20	25	30
Raid Bug Radio	1980s	Raid bug in standing position with clock on one side and radio on the other	70	200	135
Raid Bug Wind-Up	1983	4" yellow and green angry bug	40	50	70
RALSTON PURINA					
Meow Mix Cat	1976	4-1/2" vinyl yellow cat with black stripes	20	25	30
RECKITT & COLMAN					
Mr. Bubble Figure	1990	8" pink vinyl	20	30	35
REDDY COMMUNICATIONS					
Reddy Kilowatt Bobbin' Head	1960s	6-1/2" Reddy wearing cowboy outfit	225	250	300
Reddy Kilowatt Figure	1961	6" plastic figure with lightbulb for head and lightning bolts for body	150	185	225
RITALIN					
Ritalin Man Statue	1970s	7" smiling plastic statue with hat	55	65	75
SEA HOST					
Clem the Clam Push Puppets	1969	4" fish push puppets; four different fish were issued	15	30	40
SEVEN-UP					
Spot Wind-Up	1990	3" red circular figure with sunglasses	8	12	15
SHOP RITE					
Scrunchy Bear Doll	1970s	16" plush bear with Shop-Rite shirt	20	25	30
SMILE ORANGE DRINK					
Drink Smile Statue	1930s	8" plaster statue of character with orange for head; "Drink Smile" written across base	150	200	250
SMITH KLINE & FRENCH LAB					
Tagamet Figure	1989	2-1/2" bendable pink figure standing on base	15	20	25
SONY					
Sony Boy Figure	1960s	4" vinyl boy wearing yellow "Sony" shirt	150	200	225
SQUIRT BEVERAGE					
Squirt Doll	1961	17" vinyl boy with blond hair and removable clothing; "Squirt" written across shirt	100	150	175
STARKIST					
Charlie the Tuna Doll	1970	15" pull string talking doll	40	50	75
Charlie the Tuna Figure	1973	7-1/2" vinyl Charlie; arms pointed down	50	70	85
Charlie the Tuna Radio	1970	6" radio	50	60	90
Charlie the Tuna Scale	1972	oval shaped bathroom scale with Charlie the Tuna	50	65	80

ADVERTISING TOYS

NAME	YEAR	DESCRIPTION	GOOD	EX	MINT
STERLING DRUG					
Diaparene Baby Doll	1980	5" baby with movable arms and legs; baby wears diaper	35	40	50
TASTYKAKE					
Tastykake Baker Doll	1974	13" pillow doll wearing chef outfit	15	20	25
TONY'S PIZZA					
Mr. Tony Figure	1972	8" vinyl Italian pizza chef	30	35	40
TOYS R US					
Geoffrey Doll	1967	2-1/2" plush Geoffrey wearing red and white striped shirt	20	25	35
TRAVELODGE					
Sleepy Bear Squeeze Toy	1970s	5-1/2" bear wearing pajamas	25	30	35
TROPICANA					
Tropic-Ana Doll	1977	17" pillow doll of Hawaiian girl	15	20	25
U.S. FORESTRY DEPARTMENT					
Woodsy Owl Bank	1970s	8-1/2" ceramic figural bank	100	125	150
U.S. POSTAL SERVICE					
Mr. ZIP Statue	1960s	6-1/2" wood statue with mailbag and pop up hat	150	180	225
WONDER BREAD					
Fresh Guy Figure	1975	4" smiling loaf of bread with polka dots, smiling face and "Wonder" on side	100	150	175

Wienermobile Beanbag, Oscar Mayer, 1998

Banks

Mechanical Banks

Mechanical banks are often considered the royalty of American toys, mainly because they developed at an eventful time in American history, when industrialization was changing every aspect of life, including the change in toys from wood and tin to the newly-discovered cast iron.

The age of mechanical banks began with the end of the Civil War and the dawning of the industrial age and for collecting purposes ended with the beginning of World War II.

The rise of factories during the Civil War laid the groundwork for the coming industrial revolution, and iron was the main metal utilized. Prior to the Civil War, most toys were made of wood, tin or sheet metal. The new process of "casting" iron, with its durability and lower cost, made iron the metal of choice for toy makers. The new process allowed design innovations not previously possible, and toy makers were quick to exploit its potential.

Mikado, 1886, Kyser & Rex. Photo courtesy of Christie's.

Toys of the day reflected the attitudes, activities, personalities and morals of the day. Americans believed in frugality both as a morally-dictated behavior and as a means toward a secure future, and parents of the day strove to instill the virtue in their children. Many of the nursery rhymes and songs of the time espoused the value of thrift, and children were encouraged to save their pennies at every turn. Toy makers saw to it that those parents had clever and colorful allies in the banks themselves.

In 1869, the first patent was issued for a cast-iron mechanical bank, Hall's Excelsior Bank. It was also the first of many banks designed as buildings, a concept which by imitation and variation grew into the largest category in the related field of non-mechanical or still banks.

What's a Mechanical Bank?

Mechanical banks have parts which are set in motion when a coin is deposited. Still banks are receptacles only; depositing coins in them causes no action. In short, if you drop in the coin and something happens, the bank is most likely mechanical. If nothing happens, the bank is classified as still.

The degree of action varies from the simple closing of a mouth or nod of a head to complex multiple figure actions involving music, acrobatics, pratfalls and sports. Along with rarity and subject matter, this complexity of design and action is a primary factor in determining the value of a mechanical bank. From a manufacturing standpoint, intricacy of design and action increased production costs, resulting in higher retail prices and lower overall unit sales. Many years later, that results in rarity.

Some banks never made it into commercial production at all. Bank designers were required to submit working pattern samples with their patent applications, and sometimes these patterns are all that remains of a failed patent bid. Other banks were never intended for commercial production, but were made instead for personal reasons. As one-of-a-kind examples, surviving banks of this type are obviously very rare and command prices in accordance with their rarity.

Think You've Got an Original?

Both mechanical and still cast-iron banks are prime targets for reproduction. Whether intended as forgeries or as honest replicas and marked as such, these items can and do find their way into dealer stock, priced as genuine antique articles. Reproduction labels and imprints can be filed away, metals can be artificially aged, and other telltale areas can be altered in hopes of turning an honest $50 profit into $5,000 or $15,000.

Collectors entering this field are well advised to confine their initial dealings to reputable dealers and auction houses with qualified and impartial expert staffs.

Still Banks

The same companies that made mechanical banks often made still banks as less expensive alternatives. Several banks can be found in both still and mechanical versions. Companies such as Arcade, Ives, Kenton and Stevens are familiar to still and mechanical bank collectors alike.

Again, building-shaped banks are perhaps the single largest type of still banks, with others fashioned as animals, people and busts, and appliances like safes, clocks, mail boxes and globes.

Building-shaped banks span a range from Lincoln's Cabin in pottery to cottages, Victorian houses, mansions and skyscrapers in cast iron. Commemorative banks were particularly popular, such as banks resembling the Washington Monument, National Bank of Los Angeles, Century of Progress building and Eiffel Tower. Other building banks offered variations on general themes such as Home Savings Banks, State Banks and churches.

One notable class of still bank is the registering bank. Often in the shape of a safe or cash register, these banks typically accept only certain coins, such as dimes or nickels. They keep a running tally of deposits and pop open once the bank is filled, typically at $5 or $10. While their delayed reaction mechanism has earned them places in some mechanical collections, they are generally classified as still banks.

A note on restoration: As in many other areas of collecting, restoration of banks is strongly discouraged in the marketplace. Unless undertaken by an experienced professional, the restoration of a bank can result in irreparable damage to its collector value.

While most still banks were made of cast iron, other materials frequently used included glass, pottery and other ceramics, brass, lead, tin, wood, composition and even plastic. This book concentrates on cast-iron banks. Modern ceramic and plastic banks may also be found under their respective character headings in the Character Toys section of this book.

Ertl Die-Cast Banks

Today's market for die-cast banks was pioneered by the Ertl Company of Dyersville, Iowa. Compared to the history of cast-iron banks, die-cast banks were almost literally born yesterday. Ertl's first bank, a replica 1913 Model T Parcel Post Mail Service truck, rolled off the assembly line in 1981. Since then, thousands of banks have followed, in the process creating a rapidly growing, ravenous market. Prices on early and low production models have risen annually, with a few rarities selling today for over $2,000 each.

Ertl's success has spawned both spin-offs and competitors. Currently the farm town of Dyersville is home to three die-cast vehicle companies, all thriving in a booming national market.

Ertl banks represent one major diversion from classic banks of old in that most are different models only from the outside. These banks are collected not by variety of model styles but by variety of corporate sponsors. Ertl offers roughly three dozen body styles in scales of 1:25, 1:30 and 1:43, but these can bear thousands of different corporate imprints such as Amoco, the American Red Cross, Hershey's Chocolate and Texaco.

With several hundred models annually released directly into the collector market, collectors are faced with such variety that they are forced to develop their own collecting specialties. Some collectors confine their efforts to collecting the various banks representing universities. Others collect oil company or beverage company banks like Anheuser-Busch or RC Cola. Harley-Davidson is a particularly popular imprint.

Identification of Ertl banks is potentially confusing. All banks are numbered, but the numbers on the banks are not model numbers. Model numbers are printed only on boxes. The numbers on the banks represent the day and year the bank was built. For example, if a bank is numbered 1255, it was made on the 125th day of 1985. The first three digits are the days and the last is the year. Most collectors and all dealers identify their banks by the number printed on the box, but banks are also easily identifiable by make of vehicle, color scheme and corporate sponsor.

Many Ertl banks are also serial numbered by the corporate sponsor after they leave the Ertl factory. This practice is common and reinforces the limited edition aspect of the banks. Most sponsors will order only one run of banks, again with an eye toward collectibility, but the sizes of individual orders vary by company from as low as 500 to over 100,000.

Trends

Bank collecting, like other hobbies, is ultimately a matter of discretionary income. The prices of many cast-iron mechanical banks have put them in the realm of the wealthy and major investors only. Still banks are also frequently priced beyond the means of casual collectors, but their values are in general a fraction of mechanicals, making them more attractive investment vehicles for collectors of moderate means. Die-cast banks are the last metal banks still affordable to average and beginning collectors, and that fact is key to their explosive success. All three types offer the collector a piece of history and beauty, and each field has its unique attractions and benefits.

Editor's Note: Only a Mint value is given for Ertl bank. Because these banks are of recent vintage, and all were released directly into the collector market, only banks in Mint condition in original packaging are traded regularly.

The Top 10 Mechanical Banks
(in Excellent condition)

1.	Mikado, Kyser & Rex, 1886	$55,000
2.	Jonah and the Whale, Jonah Emerges, Stevens, J.& E., 1880s	55,000
3.	Girl Skipping Rope, Stevens, J.& E., 1890	45,000
4.	Circus Bank, Shepard Hardware, 1888	45,000
5.	Calamity, Stevens, J.& E., 1905	35,000
6.	Rollerskating Bank, Kyser & Rex, 1880s	30,000
7.	Harlequin, Stevens, J.& E., 1907	22,000
8.	Turtle Bank, Kilgore, 1920s	20,000
9.	Picture Gallery Bank, Shepard Hardware, 1885	20,000
10.	Motor Bank, Kyser & Rex, 1889	20,000

The Top 10 Still Banks
(in Excellent condition)

1.	Indiana Paddle Wheeler, unknown manufacturer, 1896	$8,000
2.	Tug Boat, unknown manufacturer	7,500
3.	San Gabriel Mission, unknown manufacturer	7,500
4.	Coin Registering Bank, Kyser & Rex, 1890,	6,500
5.	Hippo Bank, unknown manufacturer	6,000
6.	Dormer Bank, unknown manufacturer	6,000
7.	Chanticleer (Rooster), unknown manufacturer 1911	6,000
8.	Boston State House, Smith & Egge, 1800s	6,000
9.	Electric Railroad, Shimer Toy, 1893	6,000
10.	Camera Bank, Wrightsville Hardware, 1888	5,000

The Top 10 Ertl Die-Cast Banks
(in Mint condition)

1.	Gulf Refining, 1984, #9211	$2,500
2.	Texaco #1 Sampler, 1984, #2128	1,500
3.	Texaco #2 Sampler, 1985, #9238	1,050
4.	Texaco #1, 1986, #2128	950
5.	Texaco #1, 1984, #2128	950
6.	Durona Productions, 1982, #1321	725
7.	Harley-Davidson #1, 1988, #9784	700
8.	Texaco #3 Sampler, 1986, #9396	675
9.	Harley-Davidson #2, 1989, #9135UO	650
10.	Texaco #2, 1985, #9238	650

MECHANICAL BANKS

NAME	COMPANY	YEAR	DESCRIPTION	GOOD	VG	EX
Acrobat	Stevens, J.& E.	1883	deposit coin in opening, press lever which causes the gymnast to kick the clown causing the clown to stand on his head while coin is deposited in bank	2000	4000	9500
Artillery Bank	Stevens, J.& E.	1900s	#24668, 8" long, 6" tall; put coin in barrel, press lever, making soldier drop arm to signal firing, hammer snaps and fires coin through building window	550	900	2200
Artillery Bank	Shepard Hardware	1892	nickel plate version, coin is placed in the cannon, the hammer is pushed back; pressing the thumb piece fires the coin into the fort	500	700	1200
Atlas Bank	Unknown		iron base with white metal figure holding wooden globe with paper litho map; put coin in slot, pulling lever makes coin fall into bank, making globe spin	1000	1500	3500
Bad Accident Mule	Stevens, J.& E.	1890s	10" long, painted; deposit coin under the feet of the driver, press lever, boy jumps into the road, frightening the mule; he rears, the cart and driver are thrown backwards and coin falls in cart	850	1500	3800
Bear and Tree Stump	Judd	1880s	5" tall; put coin on bear's tongue; pressing lever makes tongue lift coin and drop it into bank	500	700	1200
Bear, Slot in Chest	Kenton	1870s	deposit coin in the bear's chest and his mouth opens and closes	750	1000	1500
Bill E. Grin Bank	Judd	1887	4-1/2" tall; dropping coin on top of head makes tongue jut and eyes blink	500	900	2200
Billy Goat Bank	Stevens, J.& E.	1910	deposit coin in slot and pull wire loop forward, goat jumps forward and coin falls in bank	600	1000	1850
Bird on Roof	Stevens, J.& E.	1878	deposit coin in slot on bird's head, pull the wire lever on left side of house, the bird tilts forward, coin rolls into the chimney	650	1000	2000
Bismark Pig	Stevens, J.& E.	1883	lock mechanism, put coin in slot over tail, press the pig's tail, Bismark figure pops up and coin drops	1800	3500	6000
Boy and Bulldog	Judd	1870s	deposit coin between the boy and the dog; pulling lever makes boy lean forward and 'push' coin into bank, the dog moves backwards at the same time and coin falls	600	900	1850
Boy on Trapeze	J. Barton & Smith Co.	1891	9-1/2" tall, painted; place a coin in the slot on the boy's head and he revolves	600	1500	5500
Boy Robbing Bird's Nest	Stevens, J.& E.	1906	8" tall; deposit coin in slot and press the lever on the tree; as the boy falls, the coin disappears in the tree	1000	3000	9500
Boy Scout Camp	Stevens, J.& E.	1912	9-1/2" long; drop coin into tent; pressing lever makes Scout raise flag as coin drops	1500	3500	12000
Boys Stealing Watermelon	Kyser & Rex	1880s	6 1/2" long; put coin in slot on top of dog house; pressing lever makes boy move toward watermelon, dog comes out of house, and coin drops	600	1200	3000
Breadwinners	Stevens, J.& E.	1886	put coin in end of club, cock hammer; pressing button makes 'Labor' hit 'Monopoly', "sending the rascals up" and dropping coin into loaf of bread	3500	7000	15000
Bucking Mule, Miniature	Judd	1870s	put coin in slot; releasing donkey makes him throw man, knocking coin into bank	600	900	1800
Bull and Bear, Single Pendulum	Stevens, J.& E.	1930s	put coin in pendulum on tree stump; pressing lever releases pendulum to swing, dropping coin into either bull or bear	800	1200	1800
Bull Tosses Boy In Well	Unknown		7" long; put coin in boy's hands; pressing lever makes bull spring and boy jumps back, dropping coin in well	750	1200	1700
Bulldog Bank	Stevens, J.& E.	1880s	put coin on dog's nose, pull his tail, the dog opens his mouth and swallows the coin	400	800	2200

Panorama Bank, 1876, Stevens

Stump Speaker Bank, 1886, Shepard Hardware

MECHANICAL BANKS

NAME	COMPANY	YEAR	DESCRIPTION	GOOD	VG	EX
Bulldog Savings Bank	Ives	1878	8-1/2" long; put coin in man's hand, press lever, making dog jump, bite coin, and fall back, dropping coin into bank	1000	2000	3000
Bulldog Standing	Judd	1870s	coin is placed on dog's tongue and tail is lifted; when tail is released the coin is deposited	350	600	1200
Bureau, Five Knob	James A. Serrill	1869	place coin in open drawer; closing drawer makes coin drop	200	350	500
Bureau, Three Knob	James A. Serrill	1869	place coin in open drawer; closing drawer makes coin drop	300	400	550
Butting Buffalo	Kyser & Rex	1888	place coin into tree trunk; pressing lever makes buffalo 'butt' boy up trunk while raccoon flees into tree	1200	2000	4500
Butting Goat	Judd	1870s	deposit coin on holder on tree trunk; lifting the tail causes the goat to spring forward depositing coin in bank. Base length 4-3/4"	600	900	1850
Butting Ram	Ole Storle	1895	put coin on tree limb; pressing lever makes ram butt coin into bank while boy thumbs his nose	2000	6000	12000
Cabin Bank	Stevens, J.& E.	1885	3-1/2" tall, shows darkie in front of cabin; put coin on roof, flip lever to make figure flip and push coin into bank with his feet	300	850	2000
Calamity	Stevens, J.& E.	1905	cock tackles into position, put coin in slot in front of fullback; pressing lever activates tackles and coin drops in the collision	3000	8000	35000
Called Out Bank	Stevens, J.& E.	1900	9" tall; push soldier into bank, drop coin into slot, making soldier pop up and drop coin into bank	4000	10000	20000
Calumet, Large	Calumet Baking Powder	1924	put coin in slot, making box sway back and forth in 'thanks'	150	250	350
Calumet, Small	Calumet Baking Powder	1924	put coin in slot, making box sway back and forth in 'thanks'	175	300	450
Cat and Mouse, Cat Balancing	Stevens, J.& E.	1891	8-1/2" tall; put coin in slot and lock mouse into position; pressing lever makes mouse disappear and kitten appears holding mouse on a ball	1000	1700	3500
Chandlers Bank	National Brass Works	1900s	open the drawer, put in coin, closing drawer drops coin into bank	400	600	1250
Chief Big Moon	Stevens, J.& E.	1899	10" long; put coin in slot in fish tail, push lever, making frog spring up from pond, dropping coin	1200	2200	5000
Chimpanzee Bank	Kyser & Rex	1880	green version; push coin slide toward monkey with log book, making monkey lower head and arm to 'log in' deposit, ringing a bell	1200	2000	3200
Chimpanzee Bank	Kyser & Rex	1880	red variant; push coin slide toward monkey with log book, making monkey lower head and arm to 'log in' deposit, ringing a bell	1400	2200	3200
Cigarette Vending	Unknown (France)	1920s	dispenses real candy cigarettes, illustrated with scenes of children smoking	75	150	250
Circus Bank	Shepard Hardware	1888	place coin on money receptacle, turn the crank and pony goes around the ring and the clown deposits coin	5000	8000	45000
Circus Ticket Collector	Judd	1830	deposit coin on top of barrel and the man's head nods his thanks	550	850	1500
Clever Dick	Saalheimer & Strauss	1920s	German bank; put coin on dog's nose; pressing lever makes him toss coin and swallow it	300	600	850
Clown	Chein	1939	press the lever and the clown sticks out his tongue, place coin on tongue release the lever and he retracts his tongue pulling coin into bank	65	100	150

Motor Bank, 1889, Kyser & Rex

World's Fair Bank, 1893, Stevens

MECHANICAL BANKS

NAME	COMPANY	YEAR	DESCRIPTION	GOOD	VG	EX
Clown Bank, Arched Top	Unknown (England)	1930s	press lever to make tongue jut out, put coin on tongue; lifting lever makes tongue recede, dropping coin	125	275	450
Clown Bust	Chamberlain & Hill	1880s	English bank; put coin in hand; pressing lever makes arm lift coin and clown swallows it	1700	2500	4000
Clown on Globe	Stevens, J.& E.	1890	clown straddles globe on footed base; pressing lever makes globe and clown move and change positions, leaving clown standing on head; 9" tall	900	1500	4500
Clown, Black Face	Unknown (England)	1930s	press lever to jut out tongue, put coin on tongue and release lever to drop coin into bank	150	300	550
Coin Registering Bank	Kyser & Rex	1890s	when the last nickel or dime is deposited totaling $5 the door will pop off and the money can be taken out	1200	2500	5500
Columbian Magic	Introduction	1892	swing open the shelf and place a coin on it, close the shelf and the coin is deposited in bank	150	300	550
Confectionery Bank	Kyser & Rex	1881	8-1/2" tall; depicts lady at candy counter; coin dropped into slot, button pushed, making figure slide to receive candy; bell rings as coin drops	2000	4000	8500
Creedmore Bank	Stevens, J.& E.	1877	9-3/4" long, painted; pull back lever on gun, put coin in slot; pressing man's foot shoots coin into tree	600	900	1500
Cross Legged Minstrel	Unknown (Germany)	1909	put coin in slot; pressing lever makes man tip his hat to you	500	800	1200
Crowing Rooster	Keim & Co. (Germany)	1937	push coin through slot into bank; makes a crowing sound	500	800	1200
Cupola Circular Building	Stevens, J.& E.	1874	push the doorbell lever and the top pops up exposing the cashier, who pivots back and returns to his forward position	4000	8000	12000
Dapper Dan	Marx	1910	deposit coin into the slot and Dapper Dan dances until the spring winds down; to reset wind the key clockwise	300	500	900
Darkie in the Chimney	Unknown		pull the drawer out and the darkie emerges; front door knob when turned counterclockwise allows the trap door in base of bank to be removed	550	1000	1400
Darktown Battery Bank	Stevens, J.& E.		#24670, 10" long, 7-1/4" tall, painted; three lads play ball, red, blue and yellow pitcher's uniform; put coin in pitcher's hand; pressing lever to pitch coin, batter swings and misses, catcher drops	1200	2200	5500
Dentist	Stevens, J.& E.	1880s	9-1/2" long; coin drops into dentist's pocket; press lever at figure's feet, making dentist pull patient's tooth, patient falls backward, dropping coin into gas bag	2000	5500	9500
Dinah	J. Harper & Co.	1911	deposit coin in Dinah's hand and press the lever; she raises her hand, her eyes roll back and her tongue flips in as she swallows the coin	600	900	1450
Dog on Turntable	Judd	1870s	5-1/4" long, 5" tall; turn the handle and dog goes in and deposits penny, coming out of the other door for more	400	800	1200
Dog Tray Bank	Kyser & Rex	1880	place coin on plate; the dog faithfully deposits it in the vault	2000	3500	6000
Eagle and Eaglets	Stevens, J.& E.	1883	#24671, 7-3/4" long, 6" tall, painted; put coin in eagle's beak; pressing lever makes eaglets rise, eagle tilts forward and drops coin into nest	600	950	1650
Electric Safe	Louis	1904	twist the center knob in a clockwise direction until it stops, coin slot becomes operational, rotate dial counterclockwise and the coin falls into the bank	200	450	650

Frog on Round Base, 1872, Stevens

Dog Tray Bank, 1880, Kyser & Rex

Boy and Bulldog, 1870s, Judd

MECHANICAL BANKS

NAME	COMPANY	YEAR	DESCRIPTION	GOOD	VG	EX
Elephant and Three Clowns	Stevens, J.& E.	1883	6" tall, painted; place coin between the rings held by acrobat, move the ball on the feet of the other acrobat and the elephant hits coin with his trunk and coin falls into	600	1000	2250
Elephant Howdah Bank, Pull Tail	Hubley	1934	put coin in elephant's trunk; pulling tail makes animal lift coin over head, dropping coin in howdah	300	500	1250
Elephant Moves Trunk, Large	Williams, A.C.	1905	the trunk of the elephant moves when coin is inserted, trunk automatically closes the slot as soon as coin is deposited, 6-3/4" long	150	250	400
Elephant Three Stars	Unknown	1884	place coin in the elephant's trunk, touch his tail and the coin will be thrown into his head	300	500	850
Elephant, Man Pops Out	Enterprise	1884	push man into howdah or lift trunk to cock mechanism, close howdah lid, put coin in elephant's mouth; pressing lever drops trunk into mouth, dropping coin and man pops up from howdah; 5-3/8" tall	400	800	1250
Feed the Goose	Banker's Thrift	1927	press the tail feathers lever and the goose opens his mouth; toss in a coin and release, the mouth closes, coin is swallowed and his wings rotate	250	400	600
Football	J. Harper & Co.	1895	place a coin on the platform in front of the player's foot, press the lever and he kicks the coin into the goal net	1200	2200	3500
Fortune Teller Savings Bank	Nickel, Baumgarter & Co.	1901	drop a nickel in the slot of the lever give a sharp jerk backwards, the wheel will spin; when it stops pull lever forward and your fortune will appear in window	500	800	1400
Frog on Rock	Kilgore	1920s	press small lever under frog's mouth and he opens to deposit coin	200	400	1100
Frog on Round Base	Stevens, J.& E.	1872	press frog's right foot and put coin in his mouth; release lever and he swallows coin and winks	350	550	1000
Gem Bank	Judd	1878	pull dog back from bank, put coin on tray, lift dog's tail making dog move and drop coin into building	200	550	850
Gem Registering Bank	Stevens, J.& E.	1893	floral embossed rectangular bank with dials at one end; turn thumb piece to top, insert coin, turn thumb piece to bottom, dropping coin into bank	600	1200	3000
Germania Exchange	Stevens, J.& E.	1880s	goat on a keg; put coin on goat's tail; turning faucet makes goat drop coin and lifts up a glass of beer	3500	7000	12000
Giant in Tower	J. Harper & Co.	1892	put coin in slot and giant leans forward	6000	10000	15000
Girl in Victorian Chair	W.S. Reed	1880	girl in blue dress sits with dog in her lap in a highback wicker chair; put coin in chair top and press lever, making coin drop and dog lean forward	2000	3000	4500
Girl Skipping Rope	Stevens, J.& E.	1890	blonde girl in light blue dress jumps rope by means of an ornately housed mechanism	10000	15000	45000
Give Me a Penny	F. W. Smith	1870	bureau with drawer; open drawer and picture rises at back, saying "Give Me A Penny"; putting coin in drawer and closing it makes coin drop and picture fall back into cabinet	1100	2000	3000
Grenadier Bank	J. Harper & Co.	1890s	soldier shoots coin into tree stump	350	650	1100
Guessing Bank, Man's Figure	McLoughlin Brothers	1877	man sits atop a clock with numbers from 1 to 6 repeated around dial; dropping coin makes dial spin and land on a number; bank reads, "Pays Five For One If You Call the Number"	1200	2000	3000
Haley's Elephant	Unknown		8" long, painted gray with red and gold blanket	350	500	650

Dog on Turntable, 1870s, Judd

Mason Bank, 1887, Shepard Hardware

Initiating Bank, First Degree, 1880, Mechanical Novelty Works

MECHANICAL BANKS

NAME	COMPANY	YEAR	DESCRIPTION	GOOD	VG	EX
Hall's Excelsior Bank	Stevens, J.& E.	1869	value varies according to color of bank; 5" tall, yellow version, monkey sits atop building; put coin on tray in monkey's lap, pull string and monkey disappears inside	300	600	850
Hall's Liliput	Stevens, J.& E.	1877	coin laid on plate is carried around by the cashier and placed in the bank; cashier returns to its place and cycle can begin again	350	500	1100
Hall's Liliput, No Tray	Stevens, J.& E.	1877	coin laid on plate is carried around by the cashier and placed in the bank, cashier returns to its place and cycle can begin again	400	550	1100
Harlequin	Stevens, J.& E.	1907	bring figure's hand halfway around to position and place coin in slot and press lever	5000	10000	22000
Hen and Chick	Stevens, J.& E.	1901	place coin in front of hen in slot, raise the lever and as the hen calls, the chicken springs from under her for the coin and disappears	1100	1700	4500
Hindu	Kyser & Rex	1882	deposit coin in mouth and press the lever on the back of the head, his eyes roll down and his tongue swings down, causing the coin to fall into the bank	800	1300	1850
Hold the Fort, Seven Hole	Unknown	1877	pull back the ring until rod is in position, tip the bank, place the coin on target and drop the shot in the cannon; shot follows the coin into the bank and escapes out of the bottom	3000	5000	9500
Home Bank with Dormers	Stevens, J.& E.	1872	pull knob, place penny on its edge in front of the cashier, push the knob to the right, and the coin is deposited in the rear of the bank in vault	800	1200	3500
Home Bank with Tickets, Morrison's	Wm. Morrison	1900s	place 50 tickets in bank, deposit coin in slot and pull desk in front of cashier as far as it will go and coin is deposited in the vault; don't release until you receive your ticket	300	400	750
Home Bank, No Dormers	Wm. Morrison	1872	6" tall, 5" wide, no dormers, bank teller in cage; put coin in side of building and pull tray in front of teller to get a receipt	650	1000	3000
Hoop-La Bank	J. Harper & Co.	1895	place coin in dog's mouth and press the lever, the dog jumps thru the hoop and deposits the coin in the barrel	800	1200	5500
Horse Race, Straight Base	Stevens, J.& E.	1870	pull cord to start spring, place the horses' heads opposite the star, deposit the coin in the opening and the race will begin	3500	5500	8000
Humpty Dumpty Bank	Shepard Hardware	1882	8-1/2" tall; put coin in Humpty's hand, press lever, making him drop coin into bank	800	2200	4000
I Always Did 'Spise a Mule--Jockey	Stevens, J.& E.	1879	#24672, 10-1/2" long, 8" tall; put coin in jockey's mouth; pressing lever makes mule kick, throwing jockey which drops coin into bank	800	1200	2800
I Always Did 'Spise a Mule-Bench	Stevens, J.& E.	1897	put the mule and boy into position; when the knob is touched, the base causes the mule to kick the boy over, throwing the coin from the bench into the receptacle below	800	1200	2800
Independence Hall	Enterprise	1875	semi-mechanical bronze finish bank; drop coin in tower and pull lever to make bell ring	350	550	1050
Indian Shooting Bear	Stevens, J.& E.	1883	10-3/8" long; put coin on rifle barrel; pressing level makes Indian shoot coin into bear	900	2500	4500
Initiating Bank, First Degree	Mechanical Novelty Works	1880	10-1/2" long, Eddy's patent on base; place coin on boy's tray, push lever, the goat butts boy forward and the frog moves upward as the coin slides from the tray into the frog's mouth	5000	7000	9000
Initiating Bank, Second Degree	Mechanical Novelty Works	1880	goat is pressed down to lock mechanism; put coin on man's tray; pressing lever makes goat butt man, dropping coin into frog's mouth	3000	5000	8000

Hen and Chick, 1901, Stevens

Elephant Three Stars, 1884

Mule Entering Barn, 1880, Kyser & Rex

MECHANICAL BANKS

NAME	COMPANY	YEAR	DESCRIPTION	GOOD	VG	EX
Japanese Ball Tosser Safe	Weeden	1888	tin windup; put coin in slot, wind mechanism, making figure move arms, appearing to juggle balls	1000	2200	5000
Joe Socko Novelty Bank	Straits Mfg.	1930s	Joe Palooka is turned clockwise as coin is deposited into slot making him turn quickly, swinging his right arm and knocking down his opponent. Base length 3-1/2"	250	400	650
Jolly Joe Clown, with Verse	Saalheimer & Strauss	1930s	push lever down, making tongue jut and eyes close; put coin on tongue; lifting lever makes tongue recede into mouth, dropping coin, and eyes open	350	600	850
Jolly Nigger in High Hat	Starkie (England)	1920	put coin in hand; pressing lever makes figure swallow coin while eyes roll and ears wiggle	250	400	650
Jolly Nigger in High Hat	J. Harper & Co.	1880s	put coin in hand; pressing lever makes arm raise, dropping coin into mouth, eyes roll and tongue moves back into mouth as coin drops	250	400	650
Jolly Nigger, String Tie	Unknown (England)	1890s	put coin in his hand; pressing lever makes him lift coin and swallow it, tongue pulling back and eyes rolling	175	300	500
Jonah and the Whale	Shepard Hardware	1890s	put coin on Jonah's back; pressing lever makes Jonah turn toward whale's mouth, dropping coin into mouth.	1200	2500	4500
Jonah and the Whale, Jonah Emerges	Stevens, J.& E.	1880s	deposit coin in side of whale; pull Jonah into position by pulling tail backwards; press the lever, the coin is deposited and Jonah will appear	12000	20000	55000
Juke Box Bank	Haji	1950s	marked "100 Select-O-Matic" on both sides; insert coin, wind up, turntable turns, music plays	95	200	275
Keene Savings Bank	Kingsbury Mfg.	1902	drop coin in bank, press lever to show amount	350	600	850
Kick Inn	Melvisto Novelty	1921	place coin on the ledge and push the lever on edge of base, the donkey kicks the ledge with his hind feet; ledge moves upward and tosses the coin into the side of the inn	200	500	900
Kiltie	J. Harper & Co.	1931	deposit coin in Scotchman's hand and press the lever; he raises his arm, lowers his eyes and deposits coin in his shirt pocket	750	1200	3000
Leap Frog Bank	Shepard Hardware	1891	7-1/2" long; put standing boy behind stooping one, put coin in slot; pressing lever makes standing boy leap over other one, who pushes lever on tree, dropping coin into bank	750	1500	4500
Lighthouse	Unknown	1891	two slots, one is a still bank, other on top of tower takes nickels; the button on top of tower will permit the bank to open only after 100 nickels are deposited	800	1200	3000
Lion and Two Monkeys	Kyser & Rex	1883	9" long; put coin in monkey's hand; pressing lever makes monkey lower hand and drop coin into lion's mouth	650	1200	2200
Lion Hunter	Stevens, J.& E.	1911	hunter figure shoots coin into lion's mouth	2000	5000	10000
Little Jocko	Ferdinand Strauss	1912	drop penny into cup, turn the crank, the monkeys dance and music plays	500	1200	2200
Little Joe (Darkie Bank)	J. Harper & Co.	1910	put coin in Joe's hand; pressing lever makes him lift and swallow coin	150	350	575
Little Miss Muffet	Unknown	1930s	drop coin into bank and spider appears in window	150	400	675
Little Moe	Chamberlin & Hill	1931	place coin in Moe's hand, press the lever on his left shoulder, he raises his right arm, his tongue flips in and he swallows the coin as he lowers his arm, tipping his hat	250	400	600
Lucky Wheel Money Box	Jacob & Co.	1929	deposit coin in top of bank; the wheel spins and stops at one of twelve fortune messages	300	500	700

Cat and Mouse: Cat Balancing, 1891, Stevens

Hold the Fort, Seven Hole, 1877

MECHANICAL BANKS

NAME	COMPANY	YEAR	DESCRIPTION	GOOD	VG	EX
Magic Bank	Stevens, J.& E.	1873	6" tall; open door to find cashier, put coin on his tray; pressing lever makes cashier disappear and drop coin in building	800	1200	2200
Magician Bank	Stevens, J.& E.	1901	8" tall; put coin on table; pressing lever makes magician lower hat over coin, dropping coin into bank while magician nods head	2500	3500	6500
Mama Katzenjammer	Kenton	1908	deposit coin in Mama's back in slot; her eyes roll up and return to their original position	3000	5000	7500
Mammy and Child	Kyser & Rex	1884	put coin on apron; pressing lever makes Mammy lower spoon to baby, Mammy's head lowers and baby's leg lifts as coin drops	3000	4500	6500
Mason Bank	Shepard Hardware	1887	7-1/4" long; drop coin onto hod, press lever, making hod move and drop coin into brick wall	2000	4000	7500
Memorial Money Bank	Enterprise	1876	slide the lever forwards to expose the coin slot, deposit coin, release the lever and it snaps back to ring the Liberty Bell	600	1200	1850
Merry-Go-Round	Kyser & Rex	1888	put coin in slot, turn the handle and chimes will ring, the figures will revolve and the attendant turns, raises stick, and coin drops	7500	11000	15000
Mikado	Kyser & Rex	1886	put coin under right hat and turn crank, making coin mysteriously move to left hat, which is lifted to display coin	15000	25000	55000
Milking Cow	Stevens, J.& E.	1885	deposit coin in cow's back, the lever under the cow's throat is pressed, the cow will kick up its hind leg, upset the boy and the milk pail and deposit coin	3000	5000	7000
Minstrel Bank	Saalheimer & Strauss	1928	7-1/2" tall; press lever making tongue stick out, put coin on tongue, lift lever to make coin retract on tongue into bank	350	600	900
Monkey and Coconut	Stevens, J.& E.	1886	8-1/2" tall; put coin in monkey's hand; pressing lever makes monkey drop coin into coconut	350	900	2500
Monkey and Parrot	Saalheimer & Strauss	1925	put coin into monkey's hands; pressing lever makes monkey toss coin into parrot's mouth	350	450	650
Monkey Bank	Hubley	1920s	9" long, painted dark green base, put coin in monkey's mouth; pressing lever makes monkey spring forward, dropping coin into organ	600	1000	1600
Monkey Tips Hat	Chein	1940s	drop a coin into bank, making monkey tip his hat	100	200	400
Monkey With Tray	German	1900s	unclothed monkey squats on square base with zoo scenes; put coin on tray; pressing tail makes monkey raise tray, head tilt back and swallow coin	300	450	650
Monkey With Tray, Uniformed	German	1900s	uniformed monkey squats on square base with animal scenes; put coin on tray; pressing tail makes monkey raise tray, head tilt back and swallow coin	350	500	750
Mosque Bank	Judd	1880s	9" tall, electroplated; put coin on tray on gorilla's head, turning lever makes gorilla turn, dropping coin into bank	600	1200	2000
Motor Bank	Kyser & Rex	1889	wind the rod with key, drop a coin in slot and the trolley car is set in motion	10000	15000	20000
Mule Entering Barn	Kyser & Rex	1880	8-1/2" long, gray barn version; put coin between mule's hind legs; pushing lever makes mule kick coin into barn and dog appears.	600	1200	3500
Musical Savings Bank, Regina	Regina Music Box	1894	bank is styled like a mantle clock; wind up mechanism, drop in coin, music plays	2500	4000	6000

I Always Did 'Spise a Mule-Bench, 1897, Stevens

Leap Frog Bank, 1891, Shepard Hardware

MECHANICAL BANKS

NAME	COMPANY	YEAR	DESCRIPTION	GOOD	VG	EX
National Bank	Stevens, J.& E.	1873	place coin on door ledge and push doorbell; the door revolves, slinging the coin into the bank, the man behind the window of the door quickly moves to the right to get out of the way	2000	3500	6000
National, Your Savings	Unknown	1900s	replica cash register, with one slot and key each for pennies, nickels, dimes and quarters; dropping coin and pressing key rings a bell	275	550	850
New Creedmore Bank	Stevens, J.& E.	1891	place coin on barrel of rifle, press right foot and coin is shot into the bull's eye of the target; as coin enters it strikes gong bell	550	950	1850
Novelty Bank	Stevens, J.& E.	1873	open door and put coin on tray, release door which closes by a spring and teller turns and drops coin into vault	550	1000	1850
Octagonal Fort Bank	Unknown	1890	also called Fort Sumter, cock mechanism, put coin in barrel end; pressing lever makes coin fire into tower; 10-3/4" long	1200	2200	3500
Organ Bank	Kyser & Rex	1881	6" tall, brown organ, monkey in blue pants and coat, yellow hat; turn handle and a chime of bells will ring while monkey deposits coins placed in tambourine, tipping his hat in thanks	350	500	850
Organ Bank with Boy and Girl	Kyser & Rex	1882	7-1/2" tall; put coin on tray; turning crank makes monkey lower tray, dropping coin into bank while boy and girl turn	500	850	1650
Organ Bank with Cat and Dog	Kyser & Rex	1882	#24663, 8-1/2" tall; put coin on tray; turning crank makes monkey lower tray dropping coin into bank while cat and dog rotate	500	850	1650
Organ Bank, Miniature	Kyser & Rex	1890s	put coin in slot; turning crank makes bells ring, monkey turn, and coin drop	500	850	1800
Organ Grinder and Dancing Bear	Stevens, J.& E.	1890s	6-7/8" long, 5-1/2" tall, 4-3/4" wide, painted windup; put coin in slot; pushing button makes grinder deposit coin, play organ, and bear dance	2300	3500	7500
Owl, Slot in Book	Kilgore	1926	5-3/4"; deposit a coin in the slot and Blinkey's eyes will roll down and back	350	650	950
Owl, Slot in Head	Kilgore	1926	5-5/8", deposit coin in slot, eyes roll forward	350	650	950
Owl, Turns Head	Stevens, J.& E.	1880	7-1/2" tall, brown bird with yellow highlights, glass eyes; insert coin in branch; pressing lever makes owl turn head as coin drops	250	500	900
Paddy and the Pig	Stevens, J.& E.	1882	8-1/2" tall; put coin in pig's nose; pressing lever makes pig kick coin into Paddy's mouth	900	2500	4000
Panorama Bank	Stevens, J.& E.	1876	place coin in slot on roof and next picture appears in the window	2500	4500	6500
Patronize the Blind Man	Stevens, J.& E.	1878	place coin in the blind man's hand, the dog takes the coin and deposits it in the bank and returns to his position	2000	3000	4500
Peg-Leg Beggar	Judd	1875s	insert coin in slot on hat and the beggar nods his thanks	750	1200	1700
Pelican Bank, Baseball Player	Stevens, J.& E.	1878	8" tall; close bird's beak, put coin into top of head which makes beak open revealing a baseball player's head	2000	3500	5000
Picture Gallery Bank	Shepard Hardware	1885	place coin in hand of center figure and he deposits coin; all the letters of the alphabet and numbers 1 to 26 are shown in rotation; also 26 animals or objects with short word for each letter	4500	7500	20000
Pig in a High Chair	Stevens, J.& E.	1897	5-1/4" tall; put coin on tray; pressing lever makes tray bring coin to pig's mouth	350	550	1400
Pistol Bank	Richard Elliot	1909	5-1/2" long; pull trigger halfway back, a hook appears from end of the barrel, place a dime on hook and pull trigger; coin is snatched and pistol fires when deposited	450	750	1650

Bad Accident Mule, 1890s, Stevens

Professor Pug Frog's Great Bicycle Feat, 1886, Stevens

MECHANICAL BANKS

NAME	COMPANY	YEAR	DESCRIPTION	GOOD	VG	EX
Popeye Knockout Bank with Box	Straits Mfg.	1929	turn Popeye to the right, lift his opponent to his feet and when the coin is deposited Popeye will deliver knockout punch	250	650	950
Presto Bank	Kyser & Rex	1894	pull drawer open and deposit coin, release drawer and coin is deposited in bank	250	525	950
Professor Pug Frog's Great Bicycle Feat	Stevens, J.& E.	1886	place coin on rear wheel; turning crank makes frog spin, dropping coin into bank	1500	3500	8500
Punch and Judy	Banks & Sons	1929	pressing lever makes figures rise, put coin in slot; releasing lever makes figures fall back into bank as coin drops	2700	5000	7000
Punch and Judy, Large Letters	Shepard Hardware	1884	7-1/2" tall; put coin on Judy's tray, press lever and Judy deposits coin while Punch tries to hit her with stick	600	2500	4500
Punch and Judy, Small Letters	Shepard Hardware	1884	7-1/2" tall; put coin on Judy's tray, press lever and Judy deposits coin while Punch tries to hit her with stick	600	1500	2500
Queen Victoria Bust	J. Harper & Co.	1887	drop coin into slot in crown, making eyes roll; also made in brass	3000	5000	7500
Rabbit in Cabbage	Kilgore	1925	4-1/2" long, white with a green base; press coin into the slot and ears will rise and then flop back down	200	650	1100
Rabbit Standing, Small	Lockwood Mfg.		put coin in rabbit's paws; pressing tail moves ears and drops coin	450	850	1250
Reclining Chinaman	Stevens, J.& E.	1882	8-1/4" long; put coin in slot on log, press lever to make figure raise hand, showing hand of cards and saluting the depositor; as coin falls, rat runs out of the end of the log	1500	2500	6500
Red Riding Hood	W.S. Reed	1880s	put coin in slot in pillow; moving lever makes Grandma's mask shift, revealing the wolf; Red turns her head in 'fear' and coin drops	8500	14000	18000
Registering Dime Savings Bank	Ives	1890	deposit dime in chute and pull lever to right, then left and the dime will be deposited and door locked, amount registered on dial; when $10 is deposited door will unlock and pop out	800	1600	2500
Rollerskating Bank	Kyser & Rex	1880s	place coin in roof slot, press lever, the skaters glide to the rear of the rink as the coin is deposited in bank, the man turns to give a little girl a wreath	12000	18000	30000
Rooster Bank	Kyser & Rex	1900s	6-1/4", put coin on rooster's tail; pressing lever causes the rooster to move his head in a crowing position, money is deposited	600	900	1500
Safety Locomotive	Edward J. Colby	1887	weight of money dropped on cab loosens the smokestack, then it can be lifted out and the money poured from the opening	650	1400	2200
Saluting Sailor	Unknown (Germany)		pressing lever makes sailor lift left arm, exposing coin slot while saluting with right arm; dropping coin and releasing lever completes cycle	300	600	900
Santa at the Chimney	Shepard Hardware	1889	6" tall, painted; put coin in his hand; pressing lever makes him drop coin into chimney	800	1800	4000
Santa at the Desk	S & E		deposit coin in large phone and dial Santa's number, the phone rings and he will pick it up and nod his head, as he writes "Merry Xmas" on a lighted piece of paper	250	450	675
Schley Bottling Up Cervera	Unknown	1899	bottle neck shows two portraits, Schley and Cervera; coins can only be dropped when Cervera is visible; shake bank to pull up Cervera picture, dropping coin will make Schley's picture return	4500	8000	18000
Scotchman	Saalheimer & Strauss	1930s	German bank; lifting lever makes eyes blink and tongue jut; put coin on tongue and lower lever to make tongue recede and drop coin into bank	300	500	700

Uncle Remus, 1891, Kyser & Rex

Kick Inn, 1921, Melvisto Novelty

MECHANICAL BANKS

NAME	COMPANY	YEAR	DESCRIPTION	GOOD	VG	EX
Smyth X-Ray Bank	Henry C. Hart	1898	set coin in the path of the scope slot and look into the eyepiece; you will see through the coin and out the other end of the bank; press lever and the coin is deposited in the bank	900	1800	3200
Speaking Dog	Shepard Hardware	1885	deposit coin on girl's plate; when thumb piece is pressed the girl's arm moves, depositing coin through trap door on bench and dog wags his tail and moves his mouth.	850	1500	2800
Sportsman, Fowler	Stevens, J.& E.	1892	place a coin in slot, set the trap, place the bird on the trap and push the lever; bird rises in the air and the sportsman fires, downing bird; bank can use paper caps	7500	12000	18000
Springing Cat	Charles Bailey	1882	lock cat in place, put coin in slot; pulling lever makes cat move toward coin as a mouse appears, knocks coin into bank, and escapes back into bank, leaving cat with open jaws	3500	6000	8000
Squirrel and Tree Stump	Mechanical Novelty Works	1881	pressing lever makes squirrel move and drop coin into stump	850	1250	2250
Starkie's Aeroplane	Starkie (England)	1919	move plane up pole and lock in place, put coin in slot on plane; pressing lever makes plane coast down pole, dropping coin in bank base	1200	1800	3500
Stollwerck Hand Shadows	Stollwerck (Germany)	1880s	candy dispenser; deposit coin in slot on top, pull drawer handle to dispense candy; decorated with art of a woman's arm making shadow animal	100	250	400
Stollwerck Postman	Stollwerck (Germany)	1900s	candy vending machine decorated with images on a postal theme; drop coin in slot, pull handle to receive candy	200	350	500
Stollwerck Red Riding Hood	Stollwerck (Germany)	1900s	candy vending machine decorated with images of Red Riding Hood; drop coin in slot, pull handle to receive candy	175	300	450
Stump Speaker Bank	Shepard Hardware	1886	9-3/4" tall, painted; place coin in hand, press the small knob on top of the box, which lowers the arm and opens the satchel to deposit the coin; release lever and mouth moves up and down	850	2200	4500
Sweet Thrift	Beverly Novelty	1928	candy dispenser; drop coin in slot and open drawer to receive candy; 5-3/4" tall; red, green or yellow	100	250	450
Tabby Bank	Unknown	1887	cat sits atop large egg, waiting for chick to hatch; drop coin in slot in cat's back and the chick moves its head	250	650	1050
Tammany Bank	Stevens, J.& E.	1873	6" tall; fat politician sits in chair, yellow vest, brown jacket, blue pants; put coin in his hand and he drops it into his pocket	400	800	3000
Tank and Cannon	Starkie (England)	1919	cannon fires coin into tank bank	650	900	1500
Teddy and the Bear	Stevens, J.& E.	1907	10" long, painted; cock gun and put coin in it, push bear into tree and close cover; pressing lever makes Teddy lower head in aim, gun fires coin into tree and bear pops	1000	2200	4500
Thrifty Animal Bank	Buddy L	1940	a registering bank with two coin slots, one takes dimes which become visible as acorns on the tree and the second slot is a still bank; the top of bank pops off when full	250	500	850
Thrifty Tom's Jigger Bank	Ferdinand Strauss	1910	drop coin into the slot and Thrifty Tom does a dance that lasts until the spring winds down	200	450	850
Toad on Stump	Stevens, J.& E.	1886	press lever to open toad's mouth, put coin on mouth and release lever, dropping coin into bank	550	875	1650
Tommy Bank	J. Harper & Co.	1914	cock the rifle, lay a coin in front of the launcher and press the lever on top of soldier's left side; the coin is shot into the tree as his head rises	2500	3500	6500

MECHANICAL BANKS

NAME	COMPANY	YEAR	DESCRIPTION	GOOD	VG	EX
Treasure Chest Musical Bank	Faith Mfg.	1930	bronze or silver finish domed chest; wind mechanism, dropping coin in slot makes music play	300	500	700
Trick Dog Bank (Six-Part Base)	Hubley	1888	deposit coin in dog's mouth; pressing lever makes dog jump through clown's hoop and coin is deposited in his barrel	500	800	1800
Trick Dog Bank, Solid Base	Hubley	1920s	deposit coin in dog's mouth; pressing lever makes dog jump through clown's hoop and coin is deposited in his barrel	200	450	1000
Trick Pony	Shepard Hardware	1885	7" long, 8" tall; put coin in horse's mouth, pulling lever makes pony drop coin in trough	900	1250	3500
Trick Savings Bank	Unknown		5-1/2", deposit coin, coin disappears with drawer closed	55	125	200
Turtle Bank	Kilgore	1920s	press a coin in the slot and Pokey's neck extends and then returns	7500	12000	20000
Uncle Remus	Kyser & Rex	1891	5-3/4" long, painted; deposit coin on roof and press the chicken's head, the policeman moves toward Uncle Remus who slams the door to prevent getting caught	1700	2300	3500
Uncle Sam with Carpet Bag	Shepard Hardware	1886	11-1/2" tall; put coin in Sam's hand; pressing lever lowers coin into his carpet bag	1000	2800	8000
Uncle Tom, No Star, Lapels	Kyser & Rex	1882	put coin on tongue; pressing lever makes Tom swallow coin and move eyes	250	650	1250
Uncle Tom, With Lapels and Star	Kyser & Rex	1882	put coin on tongue; pressing lever makes Tom swallow coin and move eyes	250	650	1250
United States and Spain	Stevens, J.& E.	1898	U.S. cannon faces Spanish ship; cock cannon and insert paper cap; pressing lever fires cap and shot which strikes ship's mast while coin drops	2000	4500	7500
United States Safe Bank	Stevens, J.& E.	1880	drop coin in slot, making top flip up showing a small bank book in which to write entry of deposit	800	2200	4000
Victorian Money Box	Unknown (England)		dropping coin into box makes girl in doorway curtsy	800	1300	1800
Volunteer Bank	J. Harper & Co.	1885	length 10"; cock rifle and put coin in slot; pressing lever makes man fire rifle, shooting coin into tree stump	750	1250	1850
Watch Dog Safe	Stevens, J.& E.	1890s	drop coin in top of bank, lift lever, coin falls into the bank as the dog opens his mouth and barks; release and mouth closes	350	675	1250
Weeden's Plantation Darkie Bank	Weeden	1888	5-1/2" tall, windup bank; putting coin in slot at side of house makes banjo player play and other figure dance	750	1250	1850
William Tell	Stevens, J.& E.	1896	10-1/2" long; lock lever on gun, lower head to aim; lowering boy's arm reveals apple, put coin on gun, press shooter's foot to fire coin into castle, knocking down apple, ringing a gong	600	1250	2500
Wimbledon Bank	J. Harper & Co.	1885	cock rifle in reclining redcoat's hands, put coin on barrel; pressing lever shoots coin into tree as soldier's head rises	3500	5500	8000
Wireless Bank	Hugo Mfg.	1926	battery operated; putting coin on roof of bank and clapping hands makes cover swing over, dropping coin into bank	350	650	850
World's Fair Bank, with Lettering	Stevens, J.& E.	1893	deposit coin on Columbus' feet; pressing lever makes the Indian Chief popup from log, offering peace pipe as Columbus salutes him	850	1250	1850
Zoo Bank	Kyser & Rex	1890s	building bank; put coin in slot; pressing monkey's face makes coin drop and shutters open on lower windows, and faces of lion and tiger appear through windows	850	1650	2250

STILL BANKS

NAME	COMPANY	YEAR	DESCRIPTION	GOOD	EX
$100,000 Money Bag	Unknown		3-5/8" tall, silver gray finish	300	650
1 Pounder Shell Bank	Grey Iron Casting	1918	8" artillery shell, "1 Pounder Bank"	25	95
1876 Bank, Large	Judd, H.L.	1895	3-3/8" tall, building bank with bronze/copper finish	75	250
1926 Sesquicentennial Bell	Grey Iron Casting	1926	3-3/4" x 3-7/8" diam.	75	200
A.A.O.S.M.S. Shriner's Fez	Allen Mfg.	1920s	2-3/8" red fez with tassle and gold lettering	250	650
Administration Building	Magic Introduction	1893	5", unpainted	250	650
Air Mail Bank on Base	Dent	1920	6-3/8" tall, red	375	1000
Alamo	Alamo Iron Works	1930s	1-7/8" tall, 3-3/8" wide, unpainted bronze finish	200	450
Alphabet Bank	Unknown		3-1/2", octagonal	1200	3000
Amherst Buffalo	Unknown	1930s	5-1/4" tall, 8" long	150	350
Amish Boy	Wright, John	1970	5" tall, painted	10	65
Amish Boy in White Shirt	Wright, John	1971	5" tall, blue coveralls, black hat	10	65
Amish Girl	Wright, John	1970	5" tall, painted	10	65
Andy Gump	Arcade	1928	4-3/8" tall, Andy sits reading a paper, painted	500	950
Apollo 8	Wright, John	1968	4-1/4", red, white and blue	20	75
Apollo, Plain	Wright, John	1968	4-1/4", unpainted	20	65
Apple	Kyser & Rex	1882	5-1/4" tall, painted apple on twig with leaves	600	1450
Arabian Safe	Kyser & Rex	1882	4-9/16" x 4-1/4"	100	300
Armoured Car	Williams, A.C.	1900s	3-3/4" tall, 6-3/4" long, red car on gold wheels	650	2500
Art Deco Elephant	Unknown		4-3/8" tall, red	100	225
Aunt Jemima	Williams, A.C.	1900s	5-7/8", also called Mammy with Spoon	125	325
Auto	Williams, A.C.	1910?	5-3/4" long, black, red wheels, four passengers	500	1200
Baby in Cradle	Unknown	1890s	3-1/4" tall, rocking cradle	500	1400
Bank of Columbia	Arcade	1800s	4-7/8", unpainted, "Bank of Columbia"	150	375
Bank of England Safe	Kyser & Rex	1882	identical to Egyptian Safe except front is embossed Bank of England	350	650
Barrel	Judd, H.L.	1873	2-3/4" tall	100	225
Baseball on Three Bats	Hubley	1914	5-1/4"	350	1400
Baseball Player	Williams, A.C.	1909	5-3/4 inches, gold	100	325
Baseball Player	Williams, A.C.	1910s	5-3/4" tall, several colors	200	475
Basket Puzzle Bank	Nicol	1894	2-3/4" tall, 3-1/2" wide, unpainted	300	650
Basket Registering Bank, Woven	Braun, Chas. A.	1902	2-7/8" x 3-3/4"	50	125
Basset Hound	Unknown		3-1/8", bronze finish	650	1500
Battleship Maine	Grey Iron Casting	1800s	5-1/4" tall, 6-5/8" long, "Maine"	650	3000
Battleship Maine	Stevens, J. & E.	1901	6" tall, 10-1/4" long, white	500	3500
Battleship Oregon	Stevens, J. & E.	1890s	4-7/8" long, silver finish	200	450
Be Wise Owl	Williams, A.C.	1900s	4-7/8" x 2-1/2"	150	375
Bean Pot	Unknown		3", red cooking pot, nickel registering	150	450
Bear Seated on Log	Unknown		7"	400	950
Bear Stealing Pig	Ober	1913	5-1/2" tall, painted	400	1000
Bear with Honey Pot	Hubley		6-1/2" tall, painted	75	175
Bear, Begging	Williams, A.C.	1900s	5-3/8", bronze finish	75	150
Beehive Bank	Kyser & Rex	1882	2-3/8"	250	500
Beehive Registering Savings Bank	Unknown	1891	5-3/8" x 6-1/2"	200	425
Beehive with Brass Top	Gobeille, W.M.		5-1/2" tall on base, unpainted	350	750

BANKS

STILL BANKS

NAME	COMPANY	YEAR	DESCRIPTION	GOOD	EX
Bethel College Administration Building	Service Foundry	1935	2-7/8" x 5-1/4"	175	350
Bicentennial Bell	Unknown	1976	4" x 4"	25	45
Billiken	Williams, A.C.	1909	4-1/4" tall, on square base, bronze finish, red cap	55	125
Billiken on Throne	Williams, A.C.	1909	6-1/2" tall	65	175
Billy Bounce	Hubley	1900s	4-11/16" tall, silver painted body	375	900
Billy Possum ("Possum & Taters")	Harper, J.M.	1909	3" x 4-3/4", on base "Billy Possum"	1200	4500
Bird Bank Building	Unknown		5-7/8" unpainted cupola building with bird on top, "Bank New York"	650	2800
Bird Cage Bank	Arcade	1900s	3-7/8" tall, similar to Crystal Bank #926, but glass is replaced by open mesh	50	125
Bismark Bank, (Pig)	Unknown	1883	3-3/8", "Bismark Bank"	100	300
Bismark Pig with Rider	Unknown	1880s	7-1/4" tall, 6-1/2" long, bronze finish	1000	3500
Boss Tweed	Unknown	1870s	3-7/8" tall	1500	3500
Boston State House	Smith & Egge	1800s	6-3/4" tall, painted	3000	6000
Boxer Bulldog	Hubley	1900s	4-1/2", seated, bronze finish	125	225
Boy Scout	Williams, A.C.	1910s	5-7/8" tall, brown finish	50	125
Boy with Large Football	Hubley	1914	5-1/8" tall, brown	2000	3200
Buckeye (SBCCA)	Filler, Lou	1973	3-1/2", painted "Ohio The Buckeye State", "SBCC 1973"	25	150
Buffalo Bank	Williams, A.C.	1900s	3-1/8" x 4-3/8", gold	50	175
Buffalo Nickel	Knerr, George	1970s	3-7/8"	35	100
Building with Belfry	Kenton		8" tall, in browns	550	3500
Bull on Base	Unknown		4" tall, unpainted	200	450
Bull with Long Horns	Unknown		3-11/16" tall, painted	50	125
Bulldog, Large	Wright, John	1960s	6", painted	25	65
Bulldog, Seated	Hubley	1928	3-7/8"	200	400
Bulldog, Standing	Arcade	1900s	2-1/4", painted	250	450
Bungalow Bank	Grey Iron Casting	1900s	3-3/4" x 3", white cottage with green roof	225	425
Bust of Man	Unknown		5"	100	350
Buster Brown & Tige	Williams, A.C.	1900s	5-1/2"	100	275
Cadet	Hubley	1905	5-3/4" tall, blue uniform with gold trim	300	750
Camel, Kneeling	Kyser & Rex	1889	2-1/2" tall, 4-3/4" long	350	750
Camel, Large	Williams, A.C.	1900s	7-1/4" x 6-1/4"	200	425
Camel, Small	Hubley	1920s	4-3/4" x 3-7/8"	100	225
Camera	Wrightsville	1888		1000	5000
Camera Bank	Wrightsville Hdw.	1800s	4-5/16" tall, bronze finish bellows camera on tripod	2500	5000
Campbell Kids	Williams, A.C.	1900s	3-5/16" x 4-1/8"	150	350
Cannon	Hubley	1914	3" tall, 6-7/8" long, black cannon on red wheels	2500	5000
Capitalist, The (Everett True)	Ober	1913	5" tall, painted	1200	1800
Capitol Bank	Riverside Foundry	1981	5-1/8"	25	50
Captain Kidd	Unknown	1900s	5-5/8" tall, Kidd stands by tree trunk with shovel, base reads "Captain Kidd"	275	450
Carpenter Safe	Harper, J.M.	1907	4-3/8"	2500	5000
Cash Register Savings Bank	Hubley	1906	4-3/4", unpainted, "Cash Register Savings Bank"	500	750
Cash Register Savings Bank	Unknown	1880s	5-5/8" tall, round face on three claw foot feet, "Cash Register Savings Bank"	350	850

STILL BANKS

NAME	COMPANY	YEAR	DESCRIPTION	GOOD	EX
Cash Register with Mesh	Arcade	1900s	3-3/4" tall, red finish with gold-bronze mesh	50	125
Castle Bank, Small	Kyser & Rex	1882	3" x 2-13/16"	200	450
Cat on Tub	Williams, A.C.	1920s	4-1/8" tall, bronze finish	100	200
Cat with Ball	Williams, A.C.	1900s	2-1/2" tall	200	375
Cat with Bow	Hubley	1930s	4-1/8"	275	575
Cat with Bow, Seated	Grey Iron Casting	1922	4-3/8" tall, brown finish	225	475
Cat with Bow, Seated	Wright, John		4-3/8" x 2 7/8", painted, white body, red bow	25	50
Cat with Long Tail	Grey Iron Casting	1910s	4-3/8" tall, 6-3/4" long	375	875
Cat with Soft Hair, Seated	Arcade	1900s	4-1/4" x 2 7/8"	85	225
Century of Progress Building	Arcade	1933	4-1/2" x 7", white, "A Century Of Progress" building from Chicago World's Fair	800	1800
Champion Heater	Unknown		4-1/8" green and black, "Champion"	125	400
Chanticleer (Rooster)	Unknown	1911	4-5/8", bronze finish, painted face and comb	850	6000
Chicken Feed Bag	Knerr, George	1973	4-5/8", "Chicken Feed"	35	300
Chipmunk with Nut	Unknown		4-1/16", black	300	950
Church Towers	Unknown		6-3/4"	850	1900
Church Window Safe	Shimer Toy	1890s	3 1/16"	50	125
City Bank with Chimney	Unknown	1870s	6-3/4" tall, painted	650	2200
City Bank with Crown	Unknown	1870s	5-1/2" tall, painted with red "crown" on roof	600	1800
City Bank with Teller	Judd, H.L.		5-1/2", bronze finish	400	700
Clown	Williams, A.C.	1908	6-1/4", gold and red with tall curved hat	125	250
Clown Bust	Knerr, George	1973	4-7/8", painted	100	350
Coca-Cola Bank	Unknown		3-3/8" tall, red and green with logo	600	1500
Coin Registering Bank	Kyser & Rex	1890	6-3/4" with red doors and dome	2500	6500
Colonial House with Porch, Large	Williams, A.C.	1900s	4" tall, white	100	275
Colonial House with Porch, Small	Williams, A.C.	1910s	3" tall, brown finish with red, green or gold roof	75	200
Columbia	Kenton		4-1/2" tall, silver finish building bank	600	900
Columbia Bank	Kenton	1890s	5-3/4" tall, unpainted silver finish	300	700
Columbia Bank	Kenton	1890s	8-3/4" tall, bronze finish	600	1000
Columbia Magic Savings Bank	Magic Introduction	1892	5", unpainted, "Columbia Magic Savings Bank"	300	700
Columbia Tower	Grey Iron Casting	1897	6-7/8", unpainted three-story tower	450	950
Covered Bridge	Wright, John	1960s	2-1/2" tall, 6-1/8" long, white with red roof	35	75
Covered Wagon	Wilton Products		6-5/8" long, unpainted	10	25
Cow	Williams, A.C.	1920	3-3/8" x 5-1/4", brown or red finish	75	350
Crosley Radio, Large	Kenton	1930s	5-1/8" tall, green with gold highlights	650	1500
Crosley Radio, Small	Kenton	1930s	4-5/16" tall, green	150	600
Cross	Unknown		9-1/4" tall, dark finish, "God Is Love" on base	750	2200
Crown Bank on Legs, Small	Unknown		4-5/8", painted	600	850
Cupola Bank	Vermong Novelty Works	1869	5-1/2" tall, painted building with center roof cupola	300	650
Cupola Bank	Stevens, J. & E.	1872	4-1/4" x 3-3/8", red and gray	100	365
Cupola Bank	Stevens, J. & E.	1870s	3-1/4" tall, black	75	225
Cutie Dog	Hubley	1914	3-7/8", painted	65	175

Dutch Girl, Grey Iron Casting

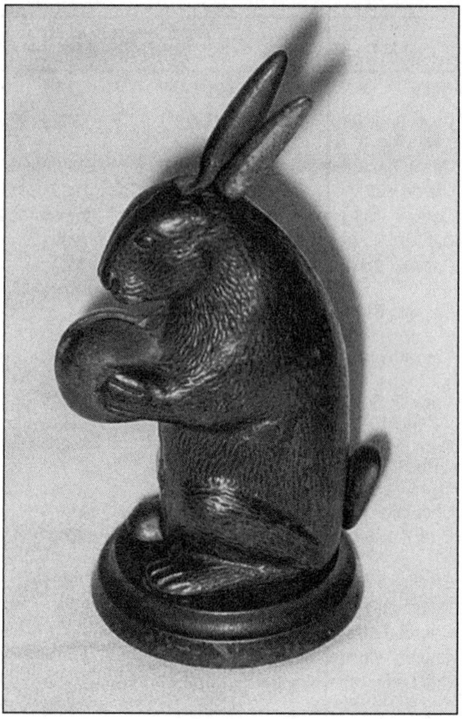

Rabbit Standing, Large, 1908, A.C. Williams

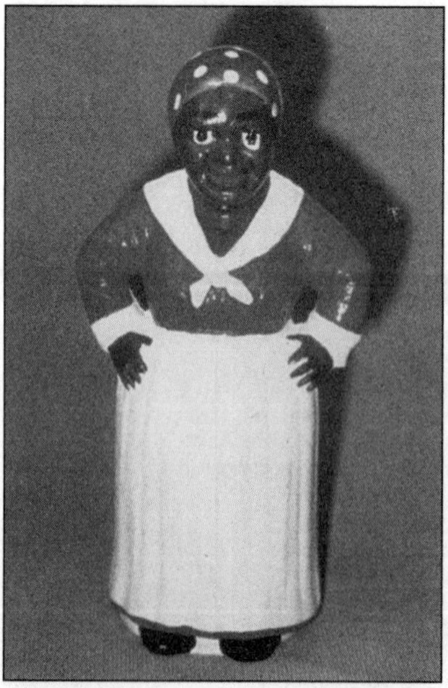

Mammy with Hands on Hips, 1900s, Hubley

Multiplying Bank, 1883, Stevens

STILL BANKS

NAME	COMPANY	YEAR	DESCRIPTION	GOOD	EX
Daisy	Shimer Toy	1899	2-1/8" tall, red safe bank	50	175
Darkey Sharecropper	Williams, A.C.	1900s	5-1/2" tall, toes visible on one foot	75	375
Decker's Iowana, (Pig)	Unknown		2-5/16", unpainted	75	200
Derby	Unknown		1-5/8" tall, 3-1/8" long, "Pass Around the Hat"	100	325
Dime Registering Coin Barrel	Kyser & Rex	1889	4" x 2-1/2", unpainted	125	225
Dime Savings	Shimer Toy	1899	2-1/2" safe, "Dime Savings"	200	425
Dog on Tub	Williams, A.C.	1920s	4-1/16" x 2" diam., bronze finish	125	200
Dog Smoking Cigar	Hubley		4-1/4", painted, white body, red bow tie	450	850
Dolphin Boat Bank	Grey Iron Casting	1900s	4-1/2" tall, sailor boy in boat holds anchor	500	850
Domed Bank	Williams, A.C.	1899	3" tall	20	95
Domed Mosque Bank	Grey Iron Casting	1900s	4-1/4" tall, gold/bronze finish	85	175
Domed Mosque Bank	Grey Iron Casting	1900s	3-1/8" tall, bronze finish	65	145
Donkey	Unknown		3-1/4" tall, black with red yoke	100	200
Donkey "I Made St. Louis Famous"	Arcade	1903	4-11/16" tall, gray finish	800	1500
Donkey on Base	Unknown		6-9/16" tall	250	650
Donkey with Blanket	Kenton	1930s	3-7/8" tall, painted, gray with red blanket	450	950
Donkey, Large	Williams, A.C.	1920s	6-13/16" tall, painted	150	275
Donkey, Small	Arcade	1910s	4-1/2" tall, blue, gold or gray finish	85	200
Dormer Bank	Unknown		4-3/4" tall, painted building bank with red roof	3500	6000
Double Door	Williams, A.C.	1900s	5-7/16" building with two doors, painted white with gold highlights	200	375
Doughboy	Grey Iron Casting	1919	7" tall, painted World War I soldier	350	850
Dry Sink	Wright, John	1970	3" x 2-3/4", dark finish	25	45
Duck	Hubley	1930s	4-3/4", white painted body	150	275
Duck Bank	Williams, A.C.	1900s	4-7/8", unpainted	150	275
Duck on Tub "Save for a Rainy Day"	Hubley	1930s	5-3/8"	95	450
Duck, Round	Kenton	1930s	4" tall, painted, yellow body, red beak and top of head	225	450
Dutch Boy	Grey Iron Casting		6-3/4" tall	600	850
Dutch Boy	Unknown		8-1/4" tall, doorstop conversion	150	275
Dutch Boy on Barrel	Hubley	1930s	5-5/8"	75	275
Dutch Girl	Grey Iron Casting		6-1/2" tall, bronze finish	600	850
Dutch Girl Holding Flowers	Hubley	1930s	5-1/2" tall, painted, iron trap in base	100	275
Eagle Bank Building	Unknown		9-3/4" tall, painted building with gold eagle on roof	450	950
Eagle with Ball, Building	Unknown		10-3/4" tall, building with eagle and ball on roof	850	4500
Edison Bust	Blevins, Charlotte	1972	5-5/16"	35	65
Eggman (Wm. Howard Taft)	Arcade	1910	4-1/8" tall	850	2500
Egyptian Tomb	Kyser & Rex	1882	6-1/4" square safe on base, decorated with Sphinx and obelisk on front, sides show, pyramid, walled ruins and urn with flowers, gold	450	750

STILL BANKS

NAME	COMPANY	YEAR	DESCRIPTION	GOOD	EX
Electric Railroad	Shimer Toy	1893	8-1/4" long	2500	6000
Elephant on Bench on Tub	Williams, A.C.	1920s	3-7/8"	125	225
Elephant on Tub	Williams, A.C.	1920s	5-3/8", in bronze finish	100	185
Elephant on Tub, Decorated	Williams, A.C.	1920s	5-3/8", painted version of #483	125	200
Elephant on Wheels	Williams, A.C.	1920s	4" tall, unpainted	150	300
Elephant Trumpeting	Wright, John	1971	7-1/4" tall, black finish	15	35
Elephant with Bent Knee	Kenton	1904	3-1/2", tan finish	200	375
Elephant with Chariot, Large	Hubley	1900s	4-3/4" tall, also made without chariot	2000	3000
Elephant with Chariot, Small	Hubley	1906	7" long, gray elephant, red chariot, yellow wheels	1400	2200
Elephant with Howdah, Large	Williams, A.C.	1900s	6-3/4", gold	85	150
Elephant with Howdah, Large	Williams, A.C.	1900s	4-7/8" x 6-3/8"	65	125
Elephant with Howdah, Short Trunk	Hubley	1910	3-3/4" tall, painted gray with red belt	125	275
Elephant with Howdah, Small	Williams, A.C.	1900s	3-1/2" x 5"	65	125
Elephant with Raised Slot	Unknown		4-1/2" tall, gray body, gold blanket	150	350
Elephant with Swivel Trunk	Unknown		2-1/2", black finish with gold swivel trunk	125	250
Elephant with Tin Chariot	Wing	1900s	8" long, red chariot	1000	1600
Elephant with Tucked Trunk	Arcade	1900s	2-3/4" x 4-5/8", red or green	65	125
Elephant with Turned Trunk, Seated	Unknown		4-1/4", unpainted	450	950
Elephant, "GOP 1936"	Hubley	1936	3-1/2" tall, "GOP 1936"	750	1200
Elephant, Circus	Hubley	1930s	3-7/8", painted, with lavender pants and red dotted white shirt	150	350
Elf	Unknown		10" tall, painted, converted doorstop	150	450
English Setter	Wright, John	1970	8-1/2" tall, black	125	275
Fidelity Safe, Large	Kyser & Rex	1880	3-5/8" tall, green with gold trim, "Fidelity Safe"	150	300
Fidelity Trust Vault, Lord Fauntleroy	Barton Smith Co.	1890	6-1/2" x 5-7/8"	300	650
Fido	Hubley	1914	5", painted, white body, black eyes and ears, red collar	60	145
Fido on Pillow	Hubley	1920s	7-3/8" long, painted	100	200
Finial Bank	Kyser & Rex	1887	5-3/4" tall, 4-3/8" wide, building bank with single finial on roof	275	850
Flags Bank (SBCCA)	Littlestown Harware	1976	3-1/4" tall, 6" square white pyramid with color US flags	75	125
Flat Iron Building Bank	Kenton	1900s	5-1/2" tall, silver	135	350
Floral Safe (National Safe)	Stevens, J. & E.	1898	4-5/8" x 4-1/8"	125	350
Football Player	Williams, A.C.	1910s	5-7/8" tall, bronze finish	250	450
Foreman	Grey Iron Casting	1951	4-1/2", painted	175	350
Fort	Unknown	1910s	4-1/8", unpainted bronze finish	125	275
Fort Mt. Hope	Unknown		2-7/8" tall	125	425
Four Tower	Ohio Foundry	1949	5-3/8", painted white building with red roof	35	85
Four Tower	Stevens, J. & E.		5-3/4", unpainted with gold highlights	125	375

STILL BANKS

NAME	COMPANY	YEAR	DESCRIPTION	GOOD	EX
Foxy Grandpa	Hubley	1920s	5-1/2" tall, painted	150	375
Frog	Iron Art	1973	4-1/8", deep green finish	75	125
Frowning Face	Unknown		5-5/8" tall, hanging bank, chin drops below surface level	850	1750
G.E. Radio Bank	Arcade	1930s	3-3/4" tall, brown cabinet radio on four legs	125	325
G.E. Refrigerator, Small	Hubley	1930s	3-3/4", blue	75	225
Gas Pump	Unknown		5-3/4" tall, red	275	650
Gem Stove	Abendroth Bros.		4-3/4", brown finish	75	175
General Butler	Stevens, J.& E.	1880s	cast iron	1800	3500
General Butler	Stevens, J. & E.	1880s		1000	3500
General Butler	Stevens, J. & E.	1884	6-1/2" tall, painted head on frog body	1500	3500
General Pershing Bust	Grey Iron Casting	1918	7-3/4" tall, bronze finish	75	150
General Sheridan on Base	Arcade	1910s	6" tall, General seated on rearing horse	250	650
George Washington Bust on Safe	Harper, J.M.	1903	5-7/8" tall	1000	2500
Gettysburg Bank	Wilton Products	1960	4-3/4" x 7-1/4" gray monument with reclining soldier	75	200
Give Me A Penny	Hubley	1900s	5-1/2" tall Black figure in hat, painted	200	450
Globe Bank With Eagle	Enterprise Mfg.	1875	5-3/4", red with eagle on globe	125	350
Globe on Arc	Grey Iron Casting	1900s	5-1/4" tall, red	100	300
Globe on Claw Feet	Kenton		6"	175	375
Globe on Hand	Unknown	1893	4", bronze finish	375	1275
Globe on Wire Arc	Arcade	1900s	4-5/8" tall, painted spinning globe, red continents	125	450
Globe Safe with Hinged Door	Kenton	1900s	5"	100	250
Globe Savings Fund Bank	Kyser & Rex	1889	7-1/8", painted "Globe Savings Fund 1888"	1800	3000
Gold Eagle	Wright, John	1970	5-3/4"	5	20
Good Luck Horseshoe	Arcade	1908	4-1/4" tall, Buster Brown & Tige with horse inside horseshoe	150	400
Goose Bank	Arcade	1920s	3-3/4", unpainted	85	175
Graf Zeppelin	Williams, A.C.	1920s	6-5/8" long, silver gray finish	85	375
Graf Zeppelin on Wheels	Williams, A.C.	1934	7-3/4" long silver pulltoy bank	150	475
Grandpa's Hat	Unknown		2-1/4" tall, 3-7/8" wide, top hat	225	450
Grenade with Pin	Bartlett Mayward		4-1/4"	85	175
Gunboat	Kenton		8-1/2" long, blue hull, white top, twin masts	650	1800
Hall Clock	Hubley	1900s	5-1/4" tall, brown finish, paper face	275	475
Hall Clock	Arcade	1923	5-5/8" tall, dark finish with gold highlights	300	700
Hall Clock with Cast Face	Hubley	1920s	5-3/26" tall	275	425
Hanging Mailbox	Williams, A.C.	1920s	5-1/8" tall, green, wall mount mailbox replica, gold lettering	65	175
Hanging Mailbox on Platform	Unknown	1800s	7-1/4" tall, red box hangs on post in platform base	650	1500
Hard Hat	Knerr, George	1970s	1-15/16" tall, white with red lettering	100	250
Harleysville Bank	Unicast Foundry	1959	2-5/8" tall, 5-1/4" long, white with gray roof	75	225
Hen on Nest	Unknown	1900s	3", bronze finish with red highlights	100	1750
High Rise Building	Kenton		7" tall	200	550
High Rise, Tiered	Kenton		5-3/4"	125	350

World's Fair Administration Building, 1893

Captain Kidd, 1900s

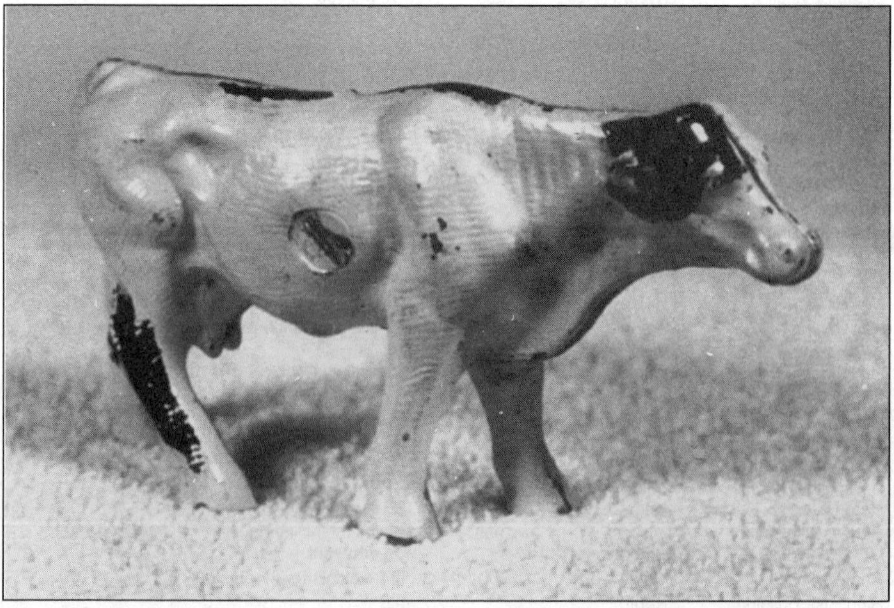

Cow, 1920, A.C. Williams

STILL BANKS

NAME	COMPANY	YEAR	DESCRIPTION	GOOD	EX
Hippo	Unknown		2" tall, 5-3/16" long, bronze with red highlights	3500	6000
Holstein Cow	Arcade	1910s	2-1/2" tall, 4-5/8" long, black finish	125	350
Home Bank	Judd, H.L.	1890s	4" x 3-1/2", dark finish	175	500
Home Bank with Crown	Stevens, J. & E.	1872	5-1/4", painted, "Home Bank"	475	1400
Home Savings Bank	Unknown		10-1/2" painted, "Property of Peoples Savings Bank, Grand Rapids, Mich."	175	650
Home Savings Bank	Shimer Toy	1899	5-7/8", painted	150	525
Home Savings Bank	Unknown		9-5/8" tall, painted	175	650
Home Savings Bank with Dog Finial	Stevens, J. & E.	1891	5-3/4" tall	125	450
Home Savings Bank with Finial	Stevens, J. & E.	1891	3-1/2" tall, mustard finish	125	375
Honey Bear	Unknown		2-1/2", silver finish unpainted bear sits eating honey	675	1200
Hoover/Curtis Elephant "GOP"	Hubley	1928	3-3/8", ivory finish	675	1600
Horse on Tub, Decorated	Williams, A.C.	1920s	5-5/6"	135	300
Horse on Wheels	Williams, A.C.	1920	4-1/4", deep red finish	150	450
Horse, "Beauty"	Arcade	1900s	4-1/8" x 4-3/4", black with raised "Beauty" on side	85	175
Horse, Prancing	Arcade	1910s	4-1/4" tall, black with gray hooves	55	150
Horse, Prancing with Belly Band	Unknown		4-1/2", light bronze finish	175	375
Horse, Prancing, Large	Williams, A.C.	1910s	7-3/16" tall, bronze finish	75	165
Horse, Rearing on Oval Base	Williams, A.C.	1920s	5-1/8" x 4-7/8"	95	250
Horse, Rearing on Pebbled Base	Unknown		7-1/4" x 6-1/2", gold finish	85	165
Horseshoe with Mesh	Williams, A.C.		Horse head inside horseshoe that forms end of mesh coin cage, bronze finish	65	145
Hot Point Electric Stove	Arcade	1925	6", white, on legs	350	975
House with Basement	Ohio Foundry Co.	1893	4-5/8" square, painted	850	1600
House with Bay Window	Unknown	1874	5-5/8" tall, painted	900	2200
House with Chimney lot	Unknown		2-7/8" x 2-13/16", painted	275	850
House with Knight	Unknown		7-1/4" unpainted "Savings Bank" with knight figure on roof peak	375	950
Hub	Magic Introduction	1892	5" x 5-1/4" x 1-5/8"	300	850
Humphrey-Muskie Donkey	Unknown	1968	4-1/2" tall, pale silver finish, "Humphrey Muskie 68"	10	35
Humpty Dumpty	Unknown	1930s	5-1/2" tall, painted, white egg, red brick wall	375	850
Humpty Dumpty, Seated	Russell, Edward K.	1974	5-3/8" tall, painted	75	125
Husky	Grey Iron Casting	1910s	5"	200	550
I Made Chicago Famous, Large Pig	Harper, J.M.	1902	2-5/8" x 5-5/16"	250	550
I Made Chicago Famous, Small Pig	Harper, J.M.	1902	2-1/8" x 4-1/8"	200	400
Ice Box	Arcade		4-1/4" tall, white, "Save For Ice"	175	650
Independence Hall	Unknown	1875	8-1/8" tall, 15-1/2" long, mustard building on base with bell tower	1800	3500
Independence Hall	Enterprise Mfg.	1875	10" tall, deep red/brown finish	450	1150

BANKS

STILL BANKS

NAME	COMPANY	YEAR	DESCRIPTION	GOOD	EX
Independence Hall Tower	Enterprise Mfg.	1876	9-1/2"	225	525
Indian Chief Bust	Unknown	1978	4-7/8", unpainted	35	85
Indian Family	Harper, J.M.	1905	3-5/8" x 5-1/8", unpainted	850	2200
Indian Head Penny	Knerr, George	1972	3-1/4" diam.	35	75
Indian Seated on Log	Ouve, A.	1970s	3-5/8" tall, unpainted	85	150
Indian with Tomahawk	Hubley	1900s	5-7/8"	175	550
Indiana Paddle Wheeler	Unknown	1896	7-1/8" long, black with red trim	4000	8000
International Eagle on Globe	Unknown		8" x 8", unpainted	1200	2500
Ironmaster's House	Kyser & Rex	1884	4-1/2", unpainted	600	1250
Japanese Safe	Kyser & Rex	1882	5-3/8" tall	100	300
Japanese Safe	Kyser & Rex	1883	5-1/2" tall, painted	125	375
Jarmulowsky Building	Stevens, J. & E.		7-3/4" tall, bronze finish building bank	1200	2000
Jewel Safe	Stevens, J. & E.	1907	5-3/8", unpainted	125	350
John Brown Fort	Unknown		3" tall, red with white cupola	85	135
Junior Cash Register, Small	Stevens, J. & E.	1920s	5-1/4" x 4-5/8" elaborate cast with slot at top	175	375
Kelvinator Bank	Arcade	1930s	#832, 4-1/2" tall, white with grey trim replica refrigerator	150	375
Key	Somerville, W.J.	1905	5-1/2" long, silver finish skeleton key	250	650
Key, St. Louis World's Fair	Unknown	1904	5-3/4" long, dark finish	275	700
King Midas	Hubley	1930s	4-1/2" tall, painted	1250	2500
Kitty Bank	Hubley	1930s	4-3/4" tall, painted, white body with blue bow	65	150
Klondyke	Unknown		3-1/4" cube	650	1400
Kodak Bank	Stevens, J. & E.	1905	4-1/4" tall, 5" wide, "Kodak Bank"	200	450
L'il Tot	Watkins, Bob	1982	5-7/8"	125	175
Labrador Retriever	Unknown		4-1/2" black finish with gold collar	125	375
Lamb	Wright, John	1970	3-1/4" tall, painted white with black highlights	35	75
Lamb, Small	Unknown		3-3/16", painted white	200	375
Laughing Pig	Hubley		2-1/2", painted	125	275
Liberty Bell	Harper, J.M.	1905	3-3/4"	275	550
Liberty Bell with Yoke	Arcade	1920s	3-1/2"	25	65
Liberty Bell, Miniature	Penncraft		3-1/2" x 1-3/4"	20	35
Lighthouse	Lane Art	1950s	9-1/2" tall, "Light of the World"	125	250
Lighthouse	Unknown	1891	10-1/4" tall, red tower rises from unpainted base	1200	3000
Limousine	Arcade	1921	same as #1478, but with steel wheels	1200	2800
Limousine	Arcade	1920s	8-1/16" long, black with white rubber tires	750	2500
Limousine Yellow Cab	Arcade	1921	repaint of #1478	1400	2800
Lincoln High Hat	Unknown	1880s	2-3/8" tall, black finish, "Pass Around the Hat"	125	225
Lion on Tub, Decorated	Williams, A.C.	1920s	5-1/2" tall	125	225
Lion on Tub, Plain	Williams, A.C.	1920s	7-1/2" tall, bronze finish	100	200
Lion on Tub, Small	Williams, A.C.	1920s	4-1/8" tall, brown or green finish	85	175
Lion on Wheels	Williams, A.C.	1920s	4-1/2" x 5-1/2", gold	145	225
Lion, Ears Up	Williams, A.C.	1930s	3-5/8" x 4-1/2"	75	125
Lion, Small	Williams, A.C.	1934	2-1/2" x 3-5/8"	85	150
Lion, Tail Between Legs	Unknown		3" x 5-1/4"	85	145
Lion, Tail Left	Hubley	1910s	3-3/4" tall, bronze finish	100	175
Lion, Tail Right	Williams, A.C.	1900s	5-1/4" tall, bronze finish	55	150
Lion, Tail Right	Arcade	1900s	4" tall	55	100
Lion, Tail Right	Williams, A.C.	1920s	3-1/2" x 4-15/16"	55	100

STILL BANKS

NAME	COMPANY	YEAR	DESCRIPTION	GOOD	EX
Little Red Riding Hood Safe	Harper, J.M.	1907	5-1/16" tall, painted	2000	4000
Log Cabin	Kyser & Rex	1882	2-1/2" x 3-1/4", painted	175	425
Lost Dog	Judd H.L.	1890s	5-3/8", unpainted	275	850
Lucky Cabin	Wright, John	1970	4-1/8" tall, painted with horseshoe over door	35	65
Mailbox on Legs, Large	Hubley	1920s	5-1/2" tall, green street corner box replica	85	225
Mailbox on Legs, Small	Hubley	1928	3-3/4" tall, green replica street corner mailbox	35	100
Main Street Trolley with People	Williams, A.C.	1920s	3" x 6-3/4" bronze finish	175	475
Main Street Trolley without People	Williams, A.C.	1920s	6-3/4" long	175	400
Majestic Radio Bank	Arcade	1930s	4-1/2" tall, mahogany finish replica of a floor standing radio on fou r legs, coin slot in back, with key	125	200
Majestic Refrigerator Bank	Arcade	1930s	4-1/2" tall, in red, green or blue with gold trim, replica of single door fridge on four legs, coin slot in back, with key lock	375	600
Mammy	Unknown	1970s	8-1/4" tall, doorstop conversion, red dress, white apron	10	25
Mammy with Hands on Hips	Hubley	1900s	5-1/4" tall, red dress, white apron	85	400
Man in Barrel	Stevens, J. & E.	1890s	3-3/4" tall, painted	175	375
Man on Cotton Bale	US Hardware	1898	4-7/8" tall, painted darkie sits on hay bale, red scarf, yellow pants	1500	2500
Marietta Silo	Unknown		5-1/2" gray finish	275	650
Marshall Stove	Unknown		3-7/8", red	125	225
Mary & Little Lamb	Unknown	1901	4-3/8" tall, painted white with red trim	350	1000
Mascot	Hubley	1914	5-3/4" tall, boy stands on baseball	850	1850
McKinley/Teddy Elephant	Unknown	1900	2-1/2" tall, bronze finish	350	650
Mean Standing Bear	Hubley		5-1/2"	100	225
Mellow Furnace	Liberty Toy		3-9/16" x 3-1/8", brown finish	125	225
Mermaid Boat	Grey Iron Casting	1900s	4-1/2" tall, companion piece to Dolphin, girl in boat holds fish	350	850
Merry-Go-Round	Grey Iron Casting	1920s	4-5/8" tall, unpainted	175	550
Metropolitan Bank	Stevens, J. & E.	1872	5-7/8", "Metropolitan Bank"	125	275
Mickey Mouse	Wright, John	1970s	5" x 3-3/4" bookend bank, painted	85	150
Mickey Mouse, Hands on Hips	Unknown		9" tall, painted	125	450
Middy with Clapper	Unknown	1887	5-1/4" brown finish	150	350
Minuteman	Hubley	1905	6" tall, painted	200	525
Model T Ford	Arcade	1920s	4" tall, black	650	1250
Moody & Sankey	Smith & Egge	1870	5" painted, two oval portraits on front	800	1850
Mosque, Large, Three-Story	Williams, A.C.	1920s	3-1/2" tall	45	125
Mosque, Small, Two-Story	Unknown		2-7/8" tall	35	115
Mother Hubbard Safe	Harper, J.M.	1907	4-1/2" tall	1500	5000
Mulligan Policeman (Keystone Cop)	Williams, A.C.	1900s	5-3/4", painted	175	400
Multiplying Bank	Stevens, J. & E.	1883	6-1/2" painted building	700	2200
Mutt & Jeff	Williams, A.C.	1900s	4-1/4" x 3-1/2", gold	75	275
National Safe	Stevens, J. & E.	1800s	3-3/8" tall, unpainted	65	125
Nest Egg	Smith & Egge	1873	3-3/8" tall on base, bronze finish egg on side, "Horace"	450	850

Santa Claus with Tree, 1910s, Hubley

Woolworth Building, small, 1915, Kenton

Bear, Begging, 1900s, A.C. Williams

STILL BANKS

NAME	COMPANY	YEAR	DESCRIPTION	GOOD	EX
Nesting Doves Safe	Harper, J.M.	1907	5-1/4", bronze finish	1500	3500
New Heatrola Bank	Kenton	1920s	4-1/2" tall, green finish with red trim	85	275
Newfoundland Dog	Arcade	1930s	3-5/8" x 5-3/8", blue or green finish	100	225
Newfoundland Dog with Pack	Unknown		4-11/16" tall	85	175
Nixe	Unknown		4-1/2" tall, silver boy in boat, "Nixe"	350	1450
Nixon Bust	Blevins, Charlotte	1972	5-5/16"	45	85
Nixon/Agnew Elephant	Unknown	1968	2-5/8"	15	35
North Pole Bank	Grey Iron Casting	1920s	4-1/4", unpainted, "Save Your Money And Freeze It"	375	775
Oak Stove	Shimer Toy	1899	2-3/8" tall, unpainted	125	475
Old Abe with Shield, Eagle	Unknown	1880	3-7/8", unpainted	450	1300
Old South Church	Unknown		10" tall, bronze finish	2000	5000
One Car Garage	Williams, A.C.	1920s	2-1/2", painted	125	250
One Story House	Grey Iron Casting	1900s	3" tall	65	175
Oregon Gunboat	Kenton		11" long, blue hull, gray guns, black and red stacks, "Oregon"	850	1800
Organ Grinder	Hubley		6-3/16" x 2-1/8", painted	125	350
Oriental Boy on Pillow	Hubley	1920s	5-1/2" tall, painted	85	200
Oriental Camel	Unknown		3-3/4" tall, on rockers	300	875
Ornate Hall Clock	Hubley	1900s	5-7/8" tall, tan finish, paper face	200	425
Osborn Pig	Unknown		2" x 4", "You can bank on the Osborn..."	100	350
Oscar the Goat	Unknown		7-3/4" tall, black with silver hooves and horns	75	175
Owl	Vindex Toys	1930	4-1/4", painted	75	325
Owl on Stump	Unknown		3-5/8", red	65	125
Ox	Kenton		4-3/8", painted	85	150
Palace	Ives	1885	7-1/2" tall, 8" wide	850	3000
Park Bank Building	Unknown		4-3/8" painted	450	1450
Parlor Stove	Unknown		6-7/8", gray and black	275	425
Parrot on Stump	Unknown		6-1/4", painted	125	450
Pavillion	Kyser & Rex	1880	3-1/8" x 3"	225	500
Pay Phone Bank	Stevens, J. & E.	1926	7-3/16", unpainted	450	1800
Pearl Street Bank	Unknown		4-1/4", unpainted, silver finish	350	850
Peg Legged Pirate	Unknown		5-1/4", unpainted	25	85
Pelican	Hubley	1930s	4-3/4", painted white	350	1000
Penny Register Pail	Kyser & Rex	1889	2-3/4", unpainted	125	250
Penthouse Building	Williams, A.C.		5-7/8" tall, silver finish	350	850
Peters Weatherbird	Arcade		4-1/4" tall	750	2500
Phoenix Dime Register Trunk	Piaget	1890	3-3/4" x 5" steamer trunk	125	250
Pig, A Christmas Roast	Unknown		3-1/4" x 7-1/8"	85	250
Pig, Seated	Williams, A.C.	1900s	3" x 4-9/16"	35	125
Plymouth Rock 1620	Unknown		3-7/8" long, "1620"	650	1850
Polar Bear, Begging	Arcade	1900s	5-1/4", white	275	450
Policeman Bank	Arcade	1930s	5-5/8" tall, blue with aluminum finish on gloves and star, gold buttons, black shoes, flesh face and hands	250	1000
Policeman Safe	Harper, J.M.	1907	5-1/4"	1250	4500
Polish Rooster	Unknown		5-1/2"	850	2500
Polish Rooster	Unknown		5-1/2" tall, painted	850	2200
Pooh Bank	Unknown		5" x 4-7/8"	5	15
Possum	Arcade	1910s	2-3/8" tall, 4-3/8" long, silver finish	125	575

STILL BANKS

NAME	COMPANY	YEAR	DESCRIPTION	GOOD	EX
Postal Savings Mailbox	Nicol	1920s	6-3/4"	85	275
Pot Bellied Stove	Knerr, George	1968	5-3/4" tall, flat black finish	25	65
Potato	Martin, Mary A.	1897	5-1/4" long, "Bank"	850	1650
Presto Bank	Williams, A.C.	1900s	3-5/8" tall, silver finish with gold dome	85	175
Presto Bank	Unknown		3-1/4" tall, silver finish, "Bank"	65	150
Presto Bank	Unknown		4-1/4" tall building, silver with gold dome	85	175
Presto Trick Bank	Kyser & Rex	1892	4-1/2" tall, red doors and roof	250	850
Professor Pug Frog Bank	Williams, A.C.	1900s	3-1/4"	275	550
Pugdog, Seated	Kyser & Rex	1889	3-1/2", painted	250	475
Puppo	Hubley	1920s	4-7/8" tall, painted bee on body	125	250
Puppo on Pillow	Hubley	1920s	5-5/8" x 6", painted brown, cream, black, pink	150	275
Put Money in Thy Purse	Unknown	1886	2-3/4" tall change purse, black	625	950
Puzzle Try Me	Unknown	1868	2-11/16" tall, safe, "Puzzle Try Me"	475	975
Quadrafoil House	Several Makers	1900s	3-1/8" tall	125	225
Queen Stove	Wright, John	1975	3-3/4" to cook top, "Queen" on oven door	25	65
Quilted Lion	Unknown		3-3/4" tall, 4-3/4" long, bronze finish	185	450
Rabbit Lying Down	Unknown		2-1/8" x 5-1/8", unpainted	175	575
Rabbit Standing, Large	Williams, A.C.	1908	6-1/4" tall, brown metal finish	125	325
Rabbit with Carrot	Knerr, George	1972	3-3/8", painted white, orange and green carrot	85	200
Rabbit, Begging	Williams, A.C.	1900s	5-1/8"	85	275
Rabbit, Large, Seated	Hubley	1900s	4-5/8" tall, painted white with pink highlights	125	375
Rabbit, Small, Seated	Arcade	1910s	3-5/8" tall	125	325
Radio Bank	Hubley	1928	3-5/16" tall, metallic blue	100	375
Radio Bank with Three Dials	Kenton	1920s	3" tall, 4-5/8" long, red	100	350
Radio with Combination Door	Kenton	1930s	4-1/2" red, metal sides and back	125	375
Reclining Cow	Unknown		2-1/8" tall, 4" long, black	100	400
Recording Bank	Unknown		6-5/8" x 4-1/4"	200	575
Red Ball Safe	Unknown		3", red ball on base	175	425
Red Goose Shoes on Base	Arcade	1920s	5-1/2", on pedestal w/base	300	750
Red Goose Shoes on Pedestal	Unknown		4-7/16" red goose on bronze base	175	350
Red Goose Shoes, Squatty	Arcade	1920s	4" tall, red body, yellow feet	275	500
Reindeer on Base	Wright, John	1973	10" x 8"	75	125
Reindeer, Large	Williams, A.C.	1900s	9-1/2" tall, bronze finish	125	250
Reindeer, Small	Williams, A.C.		6-1/4" tall, bronze finish	75	150
Reliable Parlor Stove	Schneider & Trenkramp		6-1/4"	425	850
Republic Pig	Wilton Products	1970s	7" tall, painted pig in business suit	35	85
Rhesus Monkey	Unknown		8-1/2" converted doorstop, painted	35	125
Rhino	Arcade	1910s	2-5/8" tall, 5" long, gold	225	650
Rochester Clock	Unknown		5" tall with working clock	225	750
Rocking Chair	Manning, C.J.	1898	6-3/4" tall, brown finish	1500	2750
Rocking Horse	Knerr, George	1975	5-5/8", white with red saddle, "SBCC"	350	550
Roller Safe	Kyser & Rex	1882	3-11/16" x 2-7/8"	125	245
Roof Bank	Stevens, J. & E.	1887	5-1/4" x 3-3/4"	125	350
Roof Bank	Grey Iron Casting	1900s	5-1/4"	125	300

STILL BANKS

NAME	COMPANY	YEAR	DESCRIPTION	GOOD	EX
Rooster	Hubley/Williams	1910s	4-3/4", brown finish with red comb and wattle	125	300
Rooster	Arcade	1910s	4-5/8", black with red comb	125	350
Rooster, Large	Unknown	1913	6-3/4", unpainted except for red comb and wattle	550	1250
Rumplestiltskin	Unknown	1910s	6" x 2-1/4"	200	500
Saddle Horse	Grey Iron Casting	1928	4-3/8" tall	375	650
Safe Deposit	Shimer Toy	1899	3-5/8", "Safe Deposit"	85	150
Safety Locomotive	Unknown	1887	3-1/4" tall, gray	1250	2200
Sailor, Medium	Hubley	1910s	5-1/4" tall	225	475
San Gabriel Mission	Unknown		4-5/8" x 3-3/4", painted, musical building	2000	7500
Santa Claus	Hubley	1900s	5-3/4", painted with arms folded in front	450	950
Santa Claus with Tree	Hubley	1910s	5-3/4" tall with arms folded in front, tree at back, painted	450	950
Santa with Wire Tree	Ives	1890s	7-1/4" tall, with removable ornate tree	875	1500
Scottie, Seated	Hubley	1930s	4-7/8" x 6", black finish, red collar	125	300
Scrollwork Safe	Unknown	1900s	2-3/4" tall	85	225
Seal on Rock	Arcade	1900s	3-1/2", black	175	500
Security Safe	Unknown	1894	4-1/2" tall, red door	125	275
Security Safe Deposit	Unknown	1881	3-7/8" tall	95	150
Shell Out	Stevens, J. & E.	1882	4-3/4" long, conch shell on base, off white	225	700
Show Horse	Lane Chair	1973	5-7/8" tall	75	150
Six Sided Building, Two Story	Unknown		3-3/8" tall	100	275
Six-Sided Building	Unknown		2-3/8" tall, unpainted	225	650
Skyscraper Bank	Williams, A.C.	1900s	5-1/2" tall, silver building, four gold posts	85	150
Skyscraper Bank	Williams, A.C.	1900s	4-3/8" tall, silver building, four gold posts	85	125
Skyscraper with Six Posts	Williams, A.C.	1900s	6-1/2" tall, silver building, gold posts	125	450
Songbird on Stump	Williams, A.C.	1900s	4-3/4", bronze finish	300	800
Space Heater with Bird	Chamberlain & Hill	1890s	English, 6-1/2" tall	175	375
Space Heater with Flowers	Unknown	1890s	English, 6-1/2" tall, Far East motif, red finish	175	375
Spaniel, Large	Wright, John	1960s	10-1/2" long, painted	65	125
Spitz	Grey Iron Casting	1928	4-1/4", bronze finish	225	575
Squirrel with Nut	Unknown		4-1/8"	425	1250
St. Bernard with Pack, Large	Williams, A.C.	1900s	5-1/2" x 7-3/4"	125	225
St. Bernard with Pack, Small	Williams, A.C.	1900s	3-3/4" x 5-1/2"	85	175
Star Safe	Kyser & Rex	1882	2-5/8" tall	150	450
State Bank	Kenton	1890s	3" tall, unpainted building bank	95	200
State Bank	Kenton	1900	8" x 7"	550	1200
State Bank	Kyser & Rex	1890s	5-1/2" tall, bronze building bank	125	325
State Bank	Arcade	1910s	4-1/8" tall, bronze finish	85	175
Statue of Liberty	Williams, A.C.		6-3/8" tall	85	125
Statue of Liberty	Kenton	1900s	6-1/16" tall	85	125
Statue of Liberty	Kenton	1900s	6-3/8" tall, silver finish with gold highlights	100	175
Statue of Liberty, Large	Kenton	1900s	9-1/2" tall, silver gray finish, gold highlights	350	1200

STILL BANKS

NAME	COMPANY	YEAR	DESCRIPTION	GOOD	EX
Steamboat	Williams, A.C.	190s	7-5/8" long, brown finish	125	375
Steamboat with Small Wheels	Kenton		7-7/16" long, silver finish	175	425
Stop Sign	Dent	1920	5-5/8" tall, green with red and gold highlights	325	850
Stork Safe	Harper, J.M.	1907	5-1/2"	850	1750
Street Car	Grey Iron Casting	1891	4-1/2" long, painted	250	650
Sun Dial	Arcade	1900s	4-5/16" tall	650	2000
Sunbonnet Sue	Unknown	1970	7-1/2", painted	65	165
Tabernacle Savings	Keyless Lock Co.		2-1/4" x 5", unpainted	850	1250
Taft-Sherman Bust	Harper, J.M.	1908	4" tall, one side Smiling Jim, other side Peaceful Bill	1000	1850
Tank Bank 1918, Large	Williams, A.C.	1920s	3" tall x 3-11/16" long, gold finish	100	200
Tank Bank 1918, Small	Williams, A.C.	1920s	2-3/8"long, gold finish	65	150
Tank Bank 1919	Unknown		3" x 5-1/2", silver finish, "1919"	125	350
Tank Savings Bank	Ferrosteel	1919	9-1/2" long, "Tank Savings Bank"	175	525
Teddy Bear	Arcade	1900	2-1/2" x 3-7/8"	125	350
Teddy Roosevelt Bust	Williams, A.C.	1919	5" tall	175	450
Templetone Radio	Arcade	1930s	4-1/2", red	275	575
Thoroughbred	Hubley	1946	5-1/4", bronze finish	75	150
Three Wise Monkeys	Williams, A.C.	1900s	3-1/4" tall, 3-1/2" wide	225	550
Time Is Money Clock Bank	Williams, A.C.	1910s	3-1/2" tall, alarm clock shaped, gold finish, "Time Is Money"	125	200
Time Safe	Roche, E.M. Co.		7" tall, 3-3/4" wide, unpainted	375	750
Tower	Kenton	1915	4-1/8", unpainted	175	375
Tower Bank	Harper, J.M.	1900s	9-1/4" tall, unpainted, brown finish	175	375
Tower Bank	Kyser & Rex	1890	6-7/8" building with tower rising from roof, "Tower Bank 1890"	1200	2200
Town Hall Bank	Kyser & Rex	1882	4-5/8", red, "Town Hall Bank"	375	950
Toy Soldier	Worley, Laverne A.	1982	7-1/2" tall, painted, "SBCCA"	15	65
Treasure Chest	Wright, John	1970	2-3/4" x 4", smaller version is #928	60	35
Triangular Building	Hubley	1914	6" tall, "Bank"	325	675
Trick Buffalo	Unknown		5-1/2" tall, black	750	1500
Trolley Car	Kenton	1900s	5-1/4" long, painted silver	225	650
Trunk on Dolly	Piaget	1890	2-5/8" x 3-9/16"	175	350
Trust Bank	Stevens, J. & E.	1800s	7-1/4"	1800	3000
Tug Boat	Unknown		5-1/2" long, red, pulltoy	4500	7500
Turkey, Large	Williams, A.C.	1900s	4-1/4" x 4", painted wattle	250	550
Turkey, Small	Williams, A.C.	1900s	3-3/8" tall, red head and wattle	150	275
Turtle Bank	Unknown		1" tall, 3-7/16" long	2000	3500
Two Car Garage	Williams, A.C.	1920s	2-1/2", painted	125	350
Two Goats Butting	Harper, J.M.		4-1/2", two goats on tree stump, "Two Kids" on base	950	2000
Two Story House	Williams, A.C.	1930s	3-1/16" tall, brown finish with red roof	75	150
Two-Faced Black Boy, Large	Williams, A.C.	1900s	4-1/8" tall	125	350
Two-Faced Black Boy, Small	Williams, A.C.	1900s	3-1/8" x 2-3/4"	85	300
Two-Faced Devil	Williams, A.C.	1004	4-1/4" tall, deep red	550	1250
Two-Faced Indian	Williams, A.C.	1900s	4-5/16" tall, bronze finish with painted highlights	1500	2750
U.S. Bank, Eagle Finial	Unknown	1890s	9-1/4" tall, green with gold trim	850	1500
U.S. Mail	Kenton	1900s	4-3/4" tall, silver gray with red lettering	100	375
U.S. Mail Bank with Combination Lock	Fish, O.B.	1903	6-7/8" tall, silver gray with red lettering	225	775
U.S. Mail with Eagle	Kenton	1930s	4-1/8" x 3-1/2"	85	175

STILL BANKS

NAME	COMPANY	YEAR	DESCRIPTION	GOOD	EX
U.S. Mail with Eagle	Hubley	1906	4" x 4"	175	325
U.S. Mail, Small	Kenton	1900s	3-5/8" x 2-3/4", silver or green mail box with red lettering	75	150
U.S. Navy Akron Zeppelin	Williams, A.C.	1930	6-5/8" long, silver finish, "US Navy Akron"	175	500
U.S. Treasury Bank	Grey Iron Casting	1920s	3-1/4", painted	250	475
Ulysses S. Grant Bust	Unknown	1976	5-1/2" tall	125	250
Ulysses S. Grant Bust on Safe	Harper, J.M.	1903	5-5/8" tall	1750	3000
Uncle Sam Hat	Knerr, George		2" x 3", red, white and blue	125	350
United Banking and Trust, Building Bank	Williams, A.C.		3" tall, bronze finish	225	450
Victorian House	Stevens, J. & E.	1892	4-1/2", unpainted deep gray finish	175	375
Victorian House	Unknown		3-1/4" tall, gray metallic finish	150	275
Villa	Kyser & Rex	1894	5-9/16" unpainted except for red finial	375	850
Villa Bank	Kyser & Rex	1882	3-7/8" x 3-3/8", "1882"	375	700
Vindex Bulldog	Vindex Toys	1931	5-1/4" tall, painted, "Vindex Toys"	125	275
Washington Bell with Yoke	Grey Iron Casting	1932	2-3/4", red	125	325
Washington Monument	Williams, A.C.	1900s	6" tall	150	325
Washington, George, Bust	Grey Iron Casting	1920s	8" tall, bronze finish	850	1450
Watch Dog Safe	Unknown		5-1/8", with brass handle, dog stands guard on front	1850	4000
Water Spaniel with Pack (I Hear A Call)	Harper, J.M.	1900	5-3/8" x 7-7/8"	225	450
Weaver Hen	Unknown	1970s	6", white with red comb and wattle, "Weaver"	20	50
Westside Presbyterian Church	Unknown	1916	3-3/4" x 3-5/8", silver finish	350	950
Whale of a Bank	Knerr, George	1975	2-3/4" x 5-3/16", "A Whale of a Bank"	85	200
Whippet on Base	Unknown		3-1/2" tall, gold finish	75	125
White City Barrel #1 on Cart	Nicol	1894	5" long, unpainted, "White City Puzzle Savings Bank, A Barrel of Money"	275	475
White City Barrel, Large	Nicol	1893	5-1/8" tall, silver finish barrel	175	275
White City Pail	Nicol	1893	2-5/8" tall, silver finish pail with handle	125	225
White City Puzzle Safe #10	Nicol	1893	4-5/8", unpainted	125	225
White City Puzzle Safe #12	Nicol	1893	4-7/8", unpainted	150	325
White Horse on Base	Knerr, George	1973	9-1/2" tall	125	225
Wirehaired Terrier	Hubley	1920s	4-5/8", painted	125	275
Wisconsin Beggar Boy	Unknown		6-7/8" tall, "Help the Crippled Children of Wisconsin"	525	900
Wisconsin War Eagle	Unknown	1880	2-7/8"	675	1500
Wise Pig, The	Hubley	1930s	6-5/8" tall, painted off white pig holding plaque	85	225
Woolworth Building	Kenton	1915	5-3/4" x 1-1/4"	85	150
Woolworth Building	Kenton	1915	7-7/8" tall, bronze finish	100	225
Work Horse on Base	Unknown		9" tall, painted white	75	125
Work Horse with Flynet	Arcade	1910s	4" tall	300	800
World's Fair Administration Building	Unknown	1893	6" x 6", painted	1400	2250
Yellow Cab	Arcade	1921	7-7/8" long, orange and black, rubber tires	1500	2400
York Stove	Abendroth Bros.		4" tall, unpainted, "York Stove"	225	525
Young America	Kyser & Rex	1882	4-3/8" x 3-1/8" safe	125	275

ERTL BANKS

NAME	NO.	YEAR	DESCRIPTION	MINT
4-H Clubs of America	*9701*	1988	1917 Model T Van, white body, trim	30
4-H Clubs of America	*9848*	1988	1905 Ford Delivery Van, white body, green trim	35
4-H Clubs of America	*9379*	1987	1913 Model T Van, white body, blue trim	35
A.J. Seibert Co.	*1323*	1987	1913 Model T Van, white body, red trim	125
AC Rochester #1 - United Auto Workers	*9746*	1989	1950 Chevy Panel, red body, black trim	110
ACE Hardware	*9019*	1989	1918 Runabout, red body, black trim	75
ACE Hardware #1	*9038*	1989	1918 Runabout, red/white body, black trim	35
ACE Hardware #2	*7697EO*	1990	1926 Mack w/crates, red body, black/brown trim	35
ACE Hardware #3-Marked "3rd Edition"	*9459*	1989	1932 Ford Panel, red body, white trim	30
Achenbach's Pastry Shop	*9643*	1989	Step Van, white body, white trim	30
Agway #1	*9444*	1986	1913 Model T Van, white body, red trim	300
Agway #2	*9195*	1987	1918 Runabout, black body, black trim	40
Agway #3	*9743*	1988	1905 Ford Delivery Van, black body, black trim	30
Agway #4 - Ltd. Ed	*9687*	1989	1932 Ford Panel, black body, silver trim	35
Agway #5 Ltd. Ed W/Spare	*7514*	1990	1917 Model T Van, blue body, blue trim	35
Alberta	*9218*	1985	1913 Model T Van 1 of 10 Canadian Provinces, white body, brown trim	30
Alex Cooper Auctioneers	*9201*	1984	1913 Model T Van, white body, red trim	60
Alka-Seltzer #1	*9155*	1987	1918 Runabout, blue body, beige/blue trim	140
Alka-Seltzer #2	*9791*	1988	1917 Model T Van, white body, blue trim	45
Allerton, Illinois Centennial	*9460*	1986	1913 Model T Van, white body, red trim	35
Allied Can Lines #1	*1369*	1983	1913 Model T Van, orange body, black trim	80
Allied Van Lines #2	*2136*	1984	Horse & Carriage, orange body, black trim	65
Allied Van Lines #3	*2119*	1985	1917 Model T Van, orange body, black trim	65
Allied Van Lines #4	*9776*	1988	1937 Ford Tractor/Trailer, orange body, black trim	65
Allied Van Lines #5	*7517UO*	1990	1947 IH Tractor/Trailor, orange body, trim	40
Allis-Chalmers "A-C"	*1201*	1984	1926 Mack Truck, tan body, black trim	40
Allis-Chalmers "A-C"	*2226EO*	1989	1918 Runabout, orange body, black trim	30
Alzheimer's Association #2-Ltd. Ed	*9594UO*	1990	1905 Ford Delivery Van, white body, purple trim	30
Alzheimer's Association-#1 Ltd. Ed	*9680*	1989	1913 Model T Van, white body, purple trim	95
American Red Cross #1	*9294*	1987	1913 Model T Van, black body, white trim	75
American Red Cross #2	*9294*	1988	1913 Model T Van, black body, white trim	35
American Red Cross #3 W/Spare Tire	*9294*	1989	1913 Model T Van, white body, red trim	50
American Red Cross #4-Gold Spokes L.E.	*9294*	1989	1913 Model T Van, white body, red trim	45
American Red Cross #5 Ltd. Ed.	*9685*	1989	1905 Ford Delivery Van, black body, trim	45
American Red Cross #6 Ltd. Ed.	*2984UO*	1990	1950 Chevy Panel, red body, black trim	43
American Red Cross #7-Ltd. Ed.	*7616*	1990	1926 Mack w/crates, red body, brown trim	35
Amoco	*1333*	1988	1905 Ford Delivery Van, white body, red trim	90
Amoco	*9150*	1987	1913 Model T Van, white body, red trim	155
Amoco #1	*9373*	1986	1926 Mack Tanker, silver body, red trim	250
Amoco #2	*9173*	1987	1926 Mack Tanker, white body, red trim	225
Amoco #3	*9447*	1987	1926 Mack Tanker, silver body, black trim	350
Amoco (Certicare)	*7668UA*	1990	1932 Ford Panel, white body, black trim	45
Amoco (Certicare)	*9151*	1987	1913 Model T Van, white body, red trim	50
Amoco - Atlas Auto Products	*9496*	1988	1905 Ford Delivery Van, white body, red trim	45
Amoco 100th Anniversary - Ltd. Ed.	*9745*	1989	1917 Model T Van, white body, blue trim	35
Amoco 100th Anniversary-Ltd. Ed.	*9660*	1989	1918 Runabout, white body, black/blue trim	55
Amoco Stanolind #1 Polarine Oil & Grease	*9673*	1989	1917 Model T Van, orange body, orange trim	500
Amoco Stanolind #2 Polarine Ltd. Ed.	*9060*	1989	1926 Mack Tanker, orange body, black trim	165
Amoco Stanolind #3 - Polarine Lubricants	*9383*	1989	1926 Mack Tanker, orange body, black trim	150
Amoco Stanolind #4 Ltd. Ed.	*7657UA*	1990	1932 Ford Panel, dark green body, black trim	155
Amoco-Red Crown Gas-Stand. Oil Ltd. Ed.	*9563UA*	1990	Horse Team & Tanker, black body, red trim	25
Amsouth	*9454*	1986	1913 Model T Van, white body, blue trim	65
Andrews Toy Shop Ltd. Ed.	*1322UA*	1990	1913 Model T Van, white body, blue trim	25

ERTL BANKS

NAME	NO.	YEAR	DESCRIPTION	MINT
Anheuser-Busch #2	*9047*	1989	1926 Mack w/crates, red body, black trim	60
Anheuser-Busch #3	*7574EO*	1990	1931 Hawkeye Truck, red body, black trim	55
Anheuser-Busch (Chrome) 1st Issue	*9766*	1988	1918 Barrel Runabout, red body, black trim	135
Anheuser-Busch (Chrome) Reissue	*9766*	1990	1918 Barrel Runabout, red body, black trim	75
Anthracite Battery	*9264*	1987	1905 Ford Delivery Van, white body, red trim	45
Arkansas 150th Anniversary	*9367*	1986	1913 Model T Van, white body, red trim	40
Arkansas Razorbacks	*9353*	1985	1913 Model T Van, white body, red trim	45
Arm & Hammer	*9486*	1987	1913 Model T Van, yellow body, red trim	105
Arm & Hammer	*9938UO*	1989	1932 Ford Panel, yellow body, red trim	70
Arm & Hammer	*9828*	1988	1905 Ford Delivery Van, yellow body, red trim	30
Arm & Hammer (Decal)	*9828*	1989	1905 Ford Delivery Van, yellow body, red trim	75
Arm & Hammer W/Spare	*7553UO*	1990	1932 Ford Panel, yellow body, red trim	40
Armour Food	*9891*	1988	1913 Model T Van, white body, red trim	400
Arrow Distributing #1-Ltd. Ed.	*9270*	1987	1932 Ford Panel, white body, blue trim	85
Arrow Distributing #2	*9725*	1988	1950 Chevy Panel, silver body, trim	125
Arrow Distributing #3	*9328*	1989	1918 Runabout, blue body, white trim	30
Arrow Distributing #4	*7542UO*	1990	1905 Ford Delivery Van, white body, red trim	30
Artworks (Donneckers)	*7550UO*	1990	1905 Ford Delivery Van, white body, red trim	35
Associated Grocers of Colorado	*9212*	1984	1913 Model T Van, white body, trim	65
Atlanta Falcons	*1248*	1984	1913 Model T Van, silver body, red trim	50
Atlas Van Lines #1	*9514*	1987	1926 Mack Truck, white body, blue trim	100
Atlas Van Lines #2-Ltd. Ed.	*9771*	1988	1932 Ford Panel, white body, red trim	35
Atlas Van Lines #3 W/Spare-Ltd. Ed.	*9577*	1989	1913 Model T Van, white body, blue trim	35
Atlas Van Lines #4-Ltd. Ed.	*7612UA*	1990	1937 Ford Tractor/Trailer, white body, blue trim	35
Baker Oil Tools	*9210UP*	1990	1926 Mack Truck, yellow body, blue trim	35
Baltimore Gas and Electric	*9153*	1987	1932 Ford Panel, black body, light brown trim	180
Baltimore Gas and Electric #2	*9870*	1988	1918 Runabout, black body, trim	30
Baltimore Gas and Electric #3	*9752*	1989	1950 Chevy Panel, gold body, black trim	45
Baltimore Gas and Electric #4	*2102UO*	1990	Step Van, beige body, blue trim	40
Barq's Root Beer #1	*9826*	1988	1913 Model T Van, metalized silver body, trim	60
Barq's Root Beer #2	*9072*	1989	1932 Ford Panel w/spare tire, metalized silver body, trim	55
Barq's Root Beer #3	*9054UO*	1990	1918 Barrel Runabout, silver body, black trim	50
Barrett Jackson Car Auction	*9361UP*	1990	1950 Chevy Panel, black body, white trim	40
Barrick's Farm Sales	*9271*	1987	1913 Model T Van, white body, red trim	25
Basehor, Kansas	*9007*	1989	1905 Ford Delivery Van, white body, red trim	25
Beckman High School #1	*9311*	1989	1905 Ford Delivery Van, green body, gold trim	30
Beckman High School #2	*1656UO*	1990	1913 Model T Van, green body, yellow trim	25
Bell System	*9803*	1988	1932 Ford Panel, black body, trim	35
Bell System	*9801*	1988	Horse & Carriage, dark brown body, light brown trim	30
Bell Telephone #1	*9203*	1984	1950 Chevy Panel, olive body, trim	60
Bell Telephone #1	*9298*	1987	1918 Runabout w/o ladder, olive body, black trim	40
Bell Telephone #1	*2141*	1984	Horse & Carriage, dark blue body, black trim	80
Bell Telephone #2	*9800*	1988	1918 Runabout w/ladder, dark green body, black trim	35
Bell Telephone #2	*9203*	1985	1950 Chevy Panel, olive body, trim	40
Bell Telephone 70th Anniversary	*9695*	1981	1913 Model T Van, gray body, black trim	100
Bell Telephone AT&T	*7610IU*	1990	1905 Ford Delivery Van, black body, trim	25
Bell Telephone Canada	*7609UO*	1990	1905 Ford Delivery Van, black body, trim	30
Bell Telephone of America	*9802*	1988	1937 Ford Tractor/Trailer, white body, blue trim	40
Bell Telephone of Canada	*1327*	1982	1913 Model T Van, black body, trim	65
Bell Telephone Pioneers of America	*2117*	1985	1913 Model T Van, black body, trim	25
Bell Telephone System	*9646*	1981	1913 Model T Van, black body, trim	65
Bell Telephone Yellow Pages	*2142*	1984	1926 Mack Truck, yellow body, black trim	55
Ben Franklin	*9688*	1989	1905 Ford Delivery Van, gray body, red trim	35
Ben Franklin	*1319*	1989	1918 Runabout, gray body, red trim	25
Big "A" - Wagner Brake	*1366UA*	1990	1918 Runabout, black body, red trim	115
Big "A" Auto Parts	*9482*	1987	1905 Ford Delivery Van, white body, red trim	55

ERTL BANKS

NAME	NO.	YEAR	DESCRIPTION	MINT
Big "A" Auto Parts	*9094*	1989	1926 Mack Truck, black body, red trim	25
Big "A" Auto Parts-Ltd. Ed.	*9772*	1988	1917 Model T Van, black body, trim	35
Big Bear Family Center	*9981*	1988	1918 Runabout, white/black body, brown trim	25
Big Bear Family Center	*9006*	1989	1905 Ford Delivery Van, white body, red trim	30
Biglerville Hose Co.	*9760*	1988	1913 Model T Van, white body, red trim	35
Binkley-Hurst Bros. 50th Anniversary	*9626*	1989	1913 Model T Van, white body, red trim	30
BJR Auto Radiator Service	*9059*	1989	1932 Ford Panel, black body, silver trim	80
BJR Auto Radiator Service	*9500*	1986	1913 Model T Van, silver body, blue trim	75
BJR Auto Radiator Service	*7614UO*	1990	1950 Chevy Panel, red body, black trim	100
Bookmobile (Coos Bay)	*9257*	1985	1913 Model T Van, white body, red trim	40
Boone Co. Fair	*9716*	1988	1905 Ford Delivery Van, white body, blue trim	30
Borg Warner #1 - Ltd. Ed.	*9346*	1985	1913 Model T Van, white body, red trim	75
Borg Warner #2 - Ltd. Ed.	*9390*	1986	1913 Model T Van, white body, red trim	55
Bost Bakery	*9029*	1988	1917 Model T Van, white body, trim	35
Bost Bakery #1 (Gold Spokes)	*9235*	1985	1913 Model T Van, white body, red trim	210
Bost Bakery #1 (Red Spokes) Ltd. Ed.	*9235*	1985	1913 Model T Van, white body, red trim	35
Bost Bakery #2	*9437*	1986	1913 Model T Van, white body, red trim	40
Bost Bakery Ltd. Ed.	*9170*	1987	1926 Mack Tanker, white body, red trim	45
Brendle's (Gold Spokes)	*9823*	1988	1917 Model T Van, white body, blue trim	35
Brendle's (Red Spokes)	*9823*	1988	1917 Model T Van, white body, blue trim	125
Breyer's Ice Cream	*9028*	1988	1905 Ford Delivery Van, black body, trim	75
Breyer's Ice Cream	*9617*	1989	1905 Ford Delivery Van, cream body, black trim	45
Briggs & Stratton	*9509*	1989	1937 Ford Tractor/Trailer, white body, trim	50
Briggs & Stratton	*9986*	1988	1918 Runabout, white body, black trim	145
British Columbia (1 of 10 Canadian Prov)	*9221*	1985	1913 Model T Van, white body, pink trim	30
Broadlands Centennial	*9880*	1988	1905 Ford Delivery Van, white body, blue trim	25
Brownberry Bakeries	*9441*	1986	1913 Model T Van, white body, trim	60
Buckeye, Arizona	*9287*	1987	1905 Ford Delivery Van, white body, blue trim	55
Budweiser	*1315*	1983	1913 Model T Van, white body, red trim	195
Bush's Pork & Beans	*1357*	1990	1905 Ford Delivery Van, white body, red trim	30
Bussmann Fuses 75th Anniversary	*9333*	1989	1918 Runabout, white body, black trim	65
C.R.'s Friendly Market	*9699UO*	1989	1917 Model T Van, white body, orange trim	30
Campbell's Pork & Beans	*9184*	1987	1918 Runabout, white body, red trim	75
Campbell's Pork & Beans	*9394*	1986	1905 Ford Delivery Van, red body, black trim	75
Canada	*9226*	1985	1913 Model T Van, white body, red trim	25
Canada Dry Ginger Ale	*2133*	1985	1913 Model T Van, green body, trim	135
Canada Dry Ginger Ale	*7680UO*	1990	1918 Barrel Runabout, green/white body, black trim	40
Cape Canaveral Range Operations Truck	*1996*		blue pickup truck, limited to 500	30
Cardinal Foods	*2139*	1984	1913 Model T Van, white body, red trim	95
Carl Buddig Meats	*2106*	1984	1913 Model T Van, red body, trim	80
Carl's Chicken Barbeque	*9089*	1989	Step Van, white body, trim	25
Carlisle H.S. Thundering Herd	*2140UP*	1990	1937 Ford Tractor/Trailer, white body, trim	35
Carlisle H.S. Thundering Herd	*9682*	1989	1913 Model T Van, white body, green trim	25
Carlisle H.S. Thundering Herd	*9937UA*	1989	1918 Runabout, green body	25
Carlisle Productions - Fall Carlisle	*7570UO*	1990	1950 Chevy Panel, red body	70
Carnation	*9178*	1987	1913 Model T Van, white body, red trim	45
Carnation	*9179*	1987	1926 Mack Tanker, white body, red trim	75
Castrol #1	*9463*	1986	1913 Model T Van, white body, green trim	75
Castrol #1	*9464*	1986	1926 Mack Tanker, white body, green trim	140
Castrol #2	*9464UP*	1987	1926 Mack Tanker, white body, green trim	75
Castrol #2	*9463*	1987	1913 Model T Van, white body, green trim	55
Celotex	*9475*	1987	1913 Model T Van, white body, red trim	95
Celotex	*9317*	1987	1926 Mack Truck, red body, black trim	550
Central Hawkeye Gas Engine	*9196*	1987	1905 Ford Delivery Van, white body, red trim	30
Champion Sparkplug	*9067OU*	1990	1918 Runabout, white body, black trim	60
Charter Oak Centennial	*9031*	1989	1918 Runabout, white body, black trim	20
Chemical Bank	*1662UP*	1990	1905 Ford Delivery Van, white body, blue trim	30
Chevrolet #1 - Heartbeat of America	*9873*	1989	1950 Chevy Panel, white body, black trim	60

Anheuser-Busch 1918 Barrel Runabout

Atlas Van Lines 1913 Ford Model T Van

ERTL BANKS

NAME	NO.	YEAR	DESCRIPTION	MINT
Chevrolet #1 - Today's Truck	*9561UO*	1989	1950 Chevy Panel, black body, trim	35
Chevrolet #2 - Today's Truck	*9561UP*	1990	1950 Chevy Panel, white body, black trim	25
Chevrolet #2 - Today's Truck	*9873UP*	1990	1950 Chevy Panel, white body, black trim	25
Chevrolet Barrel 1/43 Dime Bank	*9931*	1990	1930 Chevy Stake Truck, blue body, graphic trim	10
Chevrolet Heartbeat of America	*9048*	1989	1950 Chevy Tractor Trailer, white on white	35
Chicago Cubs	*7545*	1990	1926 Mack Truck, white body, blue trim	30
Chicago Tribune	*9017*	1989	Horse & Carriage, black on black	55
Chicago Tribune	*2150*	1989	1917 Model T Van, spare tire, black on black	45
Chicago Tribune	*9102*	1988	1917 Model T Van, black on black	70
Chicago Tribune	*9386*	1987	Step Van, white on white	110
Chipco	*9882*	1988	1905 Ford Delivery Van, black body, green trim	35
Chiquita Bananas	*9662*	1989	1913 Model T Van, yellow body, blue trim	70
Christmas-Happy Holidays 1989	*9584*	1989	1913 Model T Van, white body, red trim	35
Christmas-Happy Holidays 1990	*7575DO*	1990	1905 Ford Delivery Van, red body, green trim	20
Chrome King-American Bumper	*9825*	1988	1913 Model T Van, silver on silver	30
Cincinnati Bengals	*1249*	1984	1913 Model T Van, white body, orange trim	55
Cintas	*7666UP*	1990	Step Van, white	80
Citgo #1 Lubricants	*9307*	1988	1926 Mack Tanker, white body, black trim	425
Citgo #2 Lubricants	*9456EA*	1990	1918 Barrel Runabout, black body, white trim	145
Citgo #3	*7537*	1989	1913 Model T Van, spare tire, white body, red trim	75
Classic Motorbooks 25th Anniversary	*7567UO*	1990	1950 CHevy Panel, blue body, silver trim	60
Clemson University	*9523*	1989	1918 Runabout, white body, orange, black trim	55
Clemson University Limited Edition	*9775*	1988	1913 Model T Van, white body, orange trim	55
CO-OP, The Farm Store	*9245*	1985	1913 Model T Van, tan body, green trim	45
Coast to Coast	*9049*	1989	1926 Mack Truck with crates, white body, black, brown trim	25
Coast to Coast	*9188*	1987	1913 Model T Van, white body, black trim	45
Coast to Coast	*9742*	1988	1905 Ford Delivery Van, white body, black trim	25
Coast to Coast Hardware Store	*2105EO*	1990	1918 Runabout, white body, black trim	25
Cohen & Sons, William	*9339*	1989	1926 Mack Truck, red body, tan trim	65
Comet Cleanser	*7507UO*	1990	1905 Ford Delivery Van, metallic green body, gold trim	30
Conoco #1	*9750*	1989	1926 Mack Truck, silver body, green trim	250
Conoco #2	*7523UA*	1990	Horse Team, Tanker, black body, white trim	45
Coos Bay, House of Books	*9256*	1985	1013 Model T Van, white body, red trim	30
Country General	*1345UO*	1990	1918 Runabout, white body, red trim	25
Country Store-Reiman First Edition	*7564UO*	1990	1926 Mack Truck, yellow body, black trim	30
Crescent Electric Supply	*9008*	1989	1913 Model T Van, white body, blue trim	80
Cumberland Valley Tractor Pullers 1988	*1324*	1988	1926 Mack Tanker, silver body, red trim	35
Cumberland Valley Tractor Pullers 1989	*9657*	1989	1932 Ford Panel, silver body, black trim	40
Cumberland Valley Tractor Pullers 1990	*9761UO*	1990	1937 Ford Tractor Trailer, silver body, blue trim	35
Currie's	*7529UO*	1990	1905 Ford Delivery Van, white body, red trim	30
Cycle-AM Motocross	*9204*	1984	1913 Model T Van, white body, red trim	95
Daily Press	*9056*	1989	1905 Ford Delivery Van, white body, red trim	60
Daily Press, Newport News, Virginia	*9521*	1987	1913 Model T Van, white body, red trim	160
Dairy Farm	*9525*	1987	1913 Model T Van, black on black	495
Dairy Queen	*9034*	1989	1932 Ford Panel, white body, blue trim	95
Dairy Queen	*9033*	1989	1918 Runabout, red body, black trim	85
Dairy Queen	*9284*	1988	1937 Ford Tractor Trailer, white on white	95
Dairy Queen	*9144*	1987	1913 Model T Van, white body, red trim	125
Dairy Queen Limited Edition	*9285*	1988	1917 Model T Van, white on white	85
Dallas Cowboys	*1247*	1984	1913 Model T Van, silver body, blue trim	115
Decorah, Iowa	*9762*	1988	1932 Ford Panel, white body, blue trim	30
Decorah, Iowa	*9424*	1986	1905 Ford Delivery Van, white body, red trim	30
Decorah, Iowa	*9143*	1987	1918 Runabout, blue body, beige trim	30
Decorah, Iowa, Chamber of Commerce	*9255*	1985	1913 Model T Van, white body, red trim	45

ERTL BANKS

NAME	NO.	YEAR	DESCRIPTION	MINT
Decorah, Iowa, Chamber of Commerce	*9677*	1989	1926 Mack Truck, white body, red trim	30
Delaval	*9681*	1989	Step Van, white on white	110
Delaware Valley Old Time Power & Equip.	*9492*	1987	1918 Runabout, white body, black trim	30
Delaware Valley Old Time Power & Equip.	*9522*	1986	1905 Ford Delivery Van, white body, red trim	30
Democratic Party, Election '88	*9806*	1988	1905 Ford Delivery Van, white body, blue trim	105
Detroit News	*1667*	1983	1913 Model T Van, red body, blue trim	75
Deutz-Allis	*2217*	1989	1905 Ford Delivery Van, white body, black trim	20
Deutz-Allis	*2209*	1987	1913 Model T Van, white body, blue trim	25
Diamond Crystal Salt	*9438*	1986	1926 Mack Truck, red body, black trim	260
Diamond Crystal Salt	*9414*	1987	1913 Model T Van, white body, red trim	75
Dixie Brewing	*9728*	1988	1937 Ford Tractor Trailer, both doors labelled, white on white	200
Dixie Brewing	*9073*	1989	1918 Barrel Runabout, green body, black, white trim	50
Dixie Brewing Limited Edition	*9728*	1988	1937 Ford Tractor Trailer, white on white	75
Dobyns-Bennett High School, Kingsport,Tennessee	*7516UO*	1990	1950 Chevy Panel, maroon body, white trim	90
Dolly Madison	*9206UP*	1990	Step Van, white	65
Domtar Gypsum	*9824*	1989	1913 Model T Van, white body, blue trim	25
Double "J" Limited Edition	*9215*	1985	1913 Model T Van, white body, red trim	25
Dr. Pepper	*7572*	1990	1926 Mack Truck, red body, white trim	60
Dr. Pepper	*7573UO*	1990	1918 Runabout, white body, red trim	60
Dr. Pepper Special Edition	*9739*	1988	1905 Ford Delivery Van, white body, red trim	50
Drake Hotel	*2113*	1984	1913 Model T Van, white body, red trim	130
Drake, The (Hilton Hotels)	*7672UO*	1990	1932 Ford Panel, white body, blue trim	90
Dubuque G&CC Invitational Golf #1	*1657*	1990	1917 Model T Van, white body, blue trim	165
Dubuque, Iowa	*9503*	1986	1905 Ford Delivery Van, white body, red trim	30
Durona Productions	*9313*	1986	1932 Ford Panel, white body, blue trim	345
Durona Productions	*1321*	1982	1913 Model T Van, cream on cream	725
Dyersville Historical Society	*7571UO*	1990	1918 Barrel Runabout, red, black body, white trim	25
Dyersville Historical Society	*9037*	1989	1932 Ford Panel, white body, red trim	85
Dyersville Historical Society	*9529*	1986	1905 Ford Delivery Van, white body, red trim	35
Dyersville Historical Society	*9490*	1987	1913 Model T Van, white body, red trim	35
Dyersville Historical Society	*9883*	1988	1918 Runabout, red body, black trim	35
East Buchanan, Iowa	*9360*	1985	1913 Model T Van, white body, red trim	30
East Tennessee University	*7627*	1990	1913 Model T Van, white body, blue trim	40
Eastview Pharmacy	*1317UP*	1990	1950 Chevy Panel, blue body, silver trim	95
Eastview Pharmacy, Limited Edition	*9671*	1989	1913 Model T Van, white body, blue trim	125
Eastwood Company #1	*9325*	1990	1950 Chevy Panel, blue on blue	600
Eastwood Company #1	*9325*	1989	1950 Chevy Panel, blue on blue	600
Eastwood Company #2	*9562UO*	1990	1932 Ford Panel with spare tire, tan body, maroon trim	375
Eastwood Company #3	*2985UO*	1990	1931 Hawkeye Truck, green body, black trim	100
Eastwood Company #4	*7664UO*	1990	1937 Ford Tractor Trailer, red body, green trim	225
Eastwood Company #5	*2141UP*	1990	1930 Diamond T Tanker, blue on blue	100
Edy's Ice Cream	*9644*	1989	1905 Ford Delivery Van, cream body, black trim	30
Elma, Iowa	*9399*	1986	1913 Model T Van, white body, blue trim	225
Elmira Maple Festival Ltd. Ed.	*9759UA*	1990	1905 Ford Delivery Van,white body,red trim	30
Elmira Syrup Festival	*9656*	1989	1913 Model T Van, white body, blue trim	25
Entenmann's	*9780*	1988	Step Van, white body, trim	100
Entenmann's	*9455*	1986	1913 Model T Van, white body, blue trim	80
Entenmann's	*1317*	1987	Step Van, white body, trim	90
Ephrata Fair 1989	*9141*	1989	1913 Model T Van, black body, blue trim	30
Ephrata Fair 1990	*7541UO*	1990	1950 Chevy Panel, red body, black trim	30
Ertl Collector's Club	*9064*	1989	1950 Chevy Panel, gold body, white trim	95
Ertl Collector's Club	*1668*	1983	1913 Model T Van, white body, red trim	150
Ertl Collector's Club Ltd. Ed.	*1660PA*	1990	Step Van, black body	45

ERTL BANKS

NAME	NO.	YEAR	DESCRIPTION	MINT
Ertl N.Y. Premium Incentive Show Ltd. Ed.	*9737*	1988	1905 Ford Delivery Van, silver body, black trim	50
Ertl Safety Award	*7554UA*	1990	1913 Model T Van, white body, blue trim	100
Evers Toy Store	*9566UO*	1990	Horse Team & Tanker, white body, black trim	25
F-D-R Associates Ltd. Ed.	*9378*	1987	1905 Ford Delivery Van, silver body, black trim	45
Fanny Farmer	*2104*	1983	1913 Model T Van, white body, brown trim	35
Farm Bureau Co-Op	*7622*	1990	1913 Model T Van, white body, red trim	25
Farm Toy Capital of The World #3 L.E.	*9189*	1987	1926 Mack Tanker, silver body, red trim	70
Farm Toy Capital of The World #4 L.E.	*9779*	1988	1932 Ford Panel, black body, silver trim	50
Farm Toy Capital of The World #5 L.E.	*9107*	1989	1905 Ford Delivery Van, green body, black trim	30
Farm Toy Capital of The World #6 L.E.	*1664UP*	1990	1931 Hawkeye Truck, red body, black trim	35
Farm Toy Capitol of The World #1 L.E.	*9233*	1986	1913 Model T Van, white body, green trim	95
Farm Toy Capitol of The World #2 L.E.	*9510*	1986	1918 Runabout, blue body, blue/beige trim	85
Federal Express	*9334*	1989	Step Van, white body, trim	45
Felix Grundy Days Ltd. Ed.	*6125*	1989	1913 Model T Van, white body, red trim	25
Field of Dreams-Universal Studios Ltd. Ed.	*7617UA*	1990	1905 Ford Delivery Van, white body, blue trim	50
Fina	*9043*	1989	1905 Ford Delivery Van, white body, blue trim	35
Fina	*9186*	1987	1926 Mack Tanker, white body, blue trim	90
Fina - Ltd. Ed.	*9456*	1989	1917 Model T Van, white body, trim	150
Fina - Ltd.. Ed.	*9407*	1987	1913 Model T Van, white body, blue trim	65
Firehouse Films (Durona)	*9369*	1990	1950 Chevy Panel, red body, black trim	165
First National Bank (Oklahoma) Ltd. Ed.	*2988UO*	1990	1918 Runabout, white body, gold trim	50
First Tennessee Bank #1 Ltd. Ed.	*1318*	1988	1917 Model T Van, white body, blue trim	165
First Tennessee Bank #2 Ltd. Ed.	*9331*	1989	1918 Runabout, white body, black/blue trim	30
Flav-O-Rich Ltd. Ed.	*9044*	1989	1913 Model T Van, white body, red trim	30
Flint Piston Service - U.A.W. #2	*7551UO*	1990	Step Van, white body	65
Food City - Ltd. Ed.	*9857*	1988	1905 Ford Delivery Van, silver body, black trim	30
Food Lion - Ltd. Ed.	*9279*	1987	1913 Model T Van, gold body, blue trim	30
Ford	*0865*	1986	1905 Ford Delivery Van, white body, blue trim	25
Ford	*0837EO*	1987	1918 Runabout, white body, blue trim	25
Ford #1	*1334*	1981	1913 Model T Van, white body, blue trim	35
Ford #2 (Nat'l Truck Dlrs) Ltd. Ed.	*1322*	1983	1913 Model T Van, white body, blue trim	110
Ford Motorsports #1	*9871*	1988	1905 Ford Delivery Van, white body, blue trim	55
Ford Motorsports #2	*2151*	1989	1918 Runabout, white body, blue trim	30
Ford Motorsports #3	*1658*	1990	1913 Model T Van, white body, blue trim	35
Ford New Holland #5	*0374*	1990	1917 Model T Van, white body, blue trim	20
Franco-American	*9302*	1986	1926 Mack Truck, red body, green trim	60
Freihofer Baking Co.	*9710*	1988	Step Van, red body, trim	50
Frito-Lay	*9633*	1989	1950 Chevy Panel, orange body, trim	40
Frito-Lay	*9634*	1989	Step Van, white body, tan trim	45
Frito-Lay	*9632*	1989	1913 Model T Van, white body, red trim	30
Fuller Brush Co.	*9085*	1989	1905 Ford Delivery Van, white body, red trim	30
Future Farmers of America	*9531*	1987	1913 Model T Van, white body, blue trim	35
Gateway Toy Show #1 -9th Anniversary	*9598UO*	1989	1950 Chevy Tractor/Trailer, white body, trim	55
Genstar (Gypsum Products Co.)	*9358*	1985	1913 Model T Van, white body, blue trim	25
Georgia Tech	*9251*	1985	1932 Ford Roadster, metallic gold body, white trim	125
Gilbertville, Iowa	*9368*	1986	1932 Ford Panel, white body, red trim	65
Gilbertville, Iowa 3rd Annual	*9246*	1985	1913 Model T Van, white body, red trim	50
Glaxo	*1353UO*	1990	1932 Ford Panel, white body, blue trim	30
Glendale Medical Center	*9266*	1987	1913 Model T Van, white body, blue trim	60
Global Van Lines	*1655*	1983	1913 Model T Van, light blue body, black trim	45
Global Van Lines	*1655UO*	1990	1913 Model T Van, light blue body, black trim	25
Golden Flake	*9118*	1987	Step Van, white body, trim	70

Flav-O-Rich 1913 Model T Van

Unique Gardens 1926 Mack Tanker, 1987

New Holland 1905 Ford Delivery Van, 1987

ERTL BANKS

NAME	NO.	YEAR	DESCRIPTION	MINT
Good (J.F. Good Co.)	*9524*	1986	1913 Model T Van, white body, trim	125
Good (J.F. Good Co.)	*9332*	1989	1926 Mack Truck, white body, red trim	20
Good (J.F. Good Co.)	*9603*	1988	1918 Runabout, white body, brown trim	40
Goshen, H. & W. Dairy	*9146*	1987	Horse & Carriage, white body, orange trim	70
Grauer's Paint	*2139UO*	1990	1932 Ford Panel, white body, red trimn	40
Gulf - That Good Gulf Gasoline (Reissue)	*7652UO*	1990	1926 Mack Tanker, orange body, blue trim	45
Gulf Ohio Gas Marketing	*9443*	1984	1932 Ford Panel, white body, red trim	410
Gulf Ohio Pipeline	*9443*	1984	1932 Ford Panel, white body, red trim	465
Gulf Refining	*9211*	1984	1950 Chevy Panel, orange body, black trim	2500
H.E. Butts	*1365*	1983	1913 Model T Van, white body, red trim	25
Hamm's Beer	*7619UO*	1990	1926 Mack Truck, white body, red trim	70
Hamm's Beer	*2145*	1984	1913 Model T Van, white body, blue trim	85
Hardware Hank	*7635EO*	1990	1917 Model T Van, red body, black trim	20
Harley-Davidson #1	*9784*	1988	1918 Runabout, olive body, black trim	700
Harley-Davidson #2	*9135UO*	1989	1926 Mack w/crates, red body, brown trim	650
Harley-Davidson #3	*7525UA*	1990	1932 Ford Panel, black body, trim	350
Hartford Provisions	*2108*	1983	1913 Model T Van, white body, red trim	265
Hawkeye Tech	*9533*	1986	1913 Model T Van, white body, red trim	30
Heatcraft - Lennox Ltd. Ed.	*7562UA*	1990	1926 Mack Truck, white body, red trim	30
Heating Alternatives Ltd.	*1312*	1987	1913 Model T Van, white body, red trim	30
Heilig Meyers	*9749*	1989	1926 Mack Truck, green body, trim	25
Heilig Meyers 75th Ann. (1913-1988)	*9700*	1988	1913 Model T Van, green body, trim	30
Heineken Beer #1	*9570UO*	1989	1918 Barrel Runabout, green/white body, black trim	175
Heinz "57"	*1345*	1981	1913 Model T Van, white body, trim	75
Hemmings Motor News #1 (Irish Green)	*9669*	1989	1932 Ford Panel, light green body, black trim	225
Hemmings Motor News #2 (British Green)	*9669*	1989	1932 Ford Panel, green body, black trim	45
Hemmings Motor News #2 (British Green)	*9669*	1990	1932 Ford Panel, green body, black trim	25
Henderson Centennial	*9370*	1986	1913 Model T Van, white body, blue trim	30
Henny Penny	*9890*	1988	1905 Ford Delivery Van, white body, red trim	40
Henny Penny	*9946UO*	1989	1932 Ford Panel, white body, red trim	40
Henny Penny	*9945UO*	1989	1918 Runabout, white body, black trim	25
Henny Penny	*9889*	1988	1913 Model T Van, white body, red trim	30
Hershey Auto Club	*9799*	1988	1917 Model T Van, white body, trim	90
Hershey Auto Club	*7640UO*	1990	1926 Mack Truck, white body, black trim	35
Hershey Auto Club	*9084*	1989	1918 Barrel Runabout, tan body, brown trim	75
Hershey Auto Club (Regional)	*7639UO*	1990	1905 Ford Delivery Van, white body, maroon trim	40
Hershey's Chocolate Milk	*1349UO*	1990	1926 Mack Tanker, brown body, trim	45
Hershey's Cocoa	*9665*	1989	1905 Ford Delivery Van, white body, brown trim	70
Hershey's Golden Almond	*2129*	1990	1913 Model T Van 1000+, goldplate body, metalllic trim	100
Hershey's Kisses	*2126UO*	1990	1950 Chevy Panel 1000+, chrome body, metalized trim	135
Hershey's Milk Chocolate	*1350UO*	1990	1913 Model T Van, brown body, trim	25
Hershey's Milk Chocolate With Almonds	*1351UO*	1990	1913 Model T Van, brown body, trim	25
Hills Department Stores	*9768*	1988	1913 Model T Van, white body, red trim	25
Hinckley & Schmitt	*9427*	1986	1913 Model T Van, white body, blue trim	25
Hoffman Laroche	*9974*	1988	1905 Ford Delivery Van, white body, blue trim	145
Hoffman Laroche	*9601*	1988	1913 Model T Van, white body, red trim	155
Holiday Wholesale	*9470*	1987	Step Van, white body, blue trim	30
Holly Cliff Farms #1	*9477*	1987	1926 Mack Tanker, silver body, red trim	50
Holly Cliff Farms #2	*9972*	1989	1926 Mack Truck, white body, black trim	40
Holt Mfg. #1 (Caterpillar)	*7709DO*	1989	1905 Ford Delivery Van, black on black	20
Home Federal Savings	*2149*	1984	1926 Mack Truck, white body, black trim	35
Home Hardware #1	*1356*	1982	1913 Model T Van, yellow body, black trim	185
Home Hardware #2	*2109*	1984	1926 Mack Truck, yellow body, black trim	125
Home Hardware #3	*9250*	1985	1932 Ford Panel, yellow body, black trim	60

NAME	NO.	YEAR	DESCRIPTION	MINT
Home Hardware #4	*9401*	1986	1905 Ford Delivery Van, yellow body, black trim	65
Home Hardware #5	*9145*	1987	1918 Ford Runabout, yellow body, black trim	45
Home Hardware #6	*9819*	1988	1950 Chevy Panel, yellow body, black trim	35
Home Hardware #7	*9012*	1989	1926 Mack Truck with crates, yellow body, black trim	35
Home Hardware #8	*9011*	1989	1917 Model T Van, yellow body, black, tan trim	30
Home Savings & Loan	*9846*	1988	1918 Ford Runabout, white body, black trim	25
Home Savings & Loan	*9292*	1989	1926 Mack Truck, white body, red trim	40
Home Savings & Loan	*9309*	1989	1921 Ford Panel, white body, blue trim	40
Home Savings & Loan	*9845*	1988	1913 Model T Van, white body, red trim	30
Home Savings & Loan	*9844*	1988	1905 Ford Delivery Van, white body, blue trim	20
Homestead Collectibles	*9651*	1989	1905 Ford Delivery Van, white body, red trim	35
Hostess Cakes #1	*1661*	1982	1913 Model T Van, white body, black trim	50
Hostess Cakes #2	*9422*	1986	1913 Model T Van, white body, blue trim	30
Howard Brand Discounts	*1366*	1983	1913 Model T Van, white body, red trim	30
Hudson Bay Company	*9781*	1988	1913 Model T Van, black on black	45
Husker Harvest Days	*1346UO*	1990	1918 Ford Runabout, white body, red trim	25
I.B.C.	*9610*	1988	Step Van, white on white	65
Idaho Centennial 1890-1990	*9139*	1989	1913 Model T Van, white body, red trim	30
Ideal Trucking	*2963UO*	1990	1937 Ford Tractor Trailer, white	35
IDED	*9851*	1988	1918 Ford Runabout, white body, black trim	85
IDED (Ertl Logo)	*9849*	1988	1918 Ford Runabout, white body, black trim	115
IGA #1	*1651*	1983	1913 model T Van, white body, red trim	55
IGA #2	*2138*	1984	1926 Mack Truck, white body, red trim	45
IGA #3	*2126*	1984	1913 Model T Van, white body, red trim	45
IGA #4 (60th Annv.)	*9350*	1985	1932 Ford Panel, white body, red trim	50
IGA #5	*9120*	1987	1905 Ford Delivery Van, white body, red trim	25
IGA #6	*9023*	1988	1918 Ford Runabout, red body, black trim	25
IGA #7	*9015*	1989	1950 Chevy Panel, red body, white trim	25
IGA #8	*7696EO*	1990	1931 Hawkeye Truck, white body, red trim	25
IGA Credit Union	*9794*	1988	1905 Ford Delivery Van, white body, blue trim	25
Imperial Palace #1	*2107*	1983	1913 Model T Van, gray body, black trim	40
Imperial Palace #2	*9943UO*	1989	1926 Mack Truck, white body, blue trim	40
Independence, Iowa (4th of July, 1985)	*9253*	1985	1913 Model T Van, white body, red trim	30
Independence, Iowa (4th of July, 1986)	*9248*	1986	1913 Model T Van, white body, blue trim	30
Independence, Iowa (4th of July, 1987)	*9194*	1987	1918 Ford Runabout, white body, black trim	25
Independence, Iowa, Chamber of Commerce	*9236*	1984	1913 Model T Van, white body, red trim	35
Independence, Iowa, Christmas	*9303*	1986	1905 Ford Delivery Van, white body, red trim	30
Independence, Iowa, Christmas	*9499*	1987	1905 Ford Delivery Van, white body, blue trim	35
Independence, Iowa, Christmas	*9888*	1988	1918 Ford Runabout, red on red	30
Independence, Iowa, Christmas	*9359*	1985	1913 Model T Van, white body, red trim	650
Independence, Iowa, Christmas	*9359*	1985	1913 Model T Van, white body, red trim	35
Independence, Iowa, Lion's Club	*9507*	1986	1913 Model T Van, gray body, blue trim	70
Independence, Iowa, Lion's Club	*9288*	1987	1917 Model T Van, white body, blue trim	35
Indianapolis 500	*9813*	1988	1918 Ford Runabout, white body, black trim	45
Iowa Fireman's Assn. 105th	*1346*	1983	1913 Model T Van, white body, red trim	195
Iowa Fireman's Assn. 106th	*2137*	1984	1913 Model T Van, white body, red trim	40
Iowa Fireman's Assn. 107th	*9237*	1985	1913 Model T Van, white body, red trim	60
Iowa Fireman's Assn. 109th	*9165*	1987	1005 Ford Delivery Van, white body, red trim	55
Iowa Hawkeyes #1	*1311*	1983	1913 Model T Van, yellow body, black trim	55
Iowa Hawkeyes #10 (10th Anniversary Edition)	*1665UO*	1990	1931 Hawyeye with crates, yellow body, black trim	30
Iowa Hawkeyes #2	*1351*	1982	1913 Model T Van, yellow body, black trim	30
Iowa Hawkeyes #3	*1355*	1983	1913 Model T Van, black body, yellow trim	55
Iowa Hawkeyes #4	*1663*	1983	1926 Mack Truck, yellow body, black trim	60
Iowa Hawkeyes #6	*2135*	1984	1926 Mack Truck, yellow body, black trim	35
Iowa Hawkeyes #7	*9180*	1987	1918 Ford Runabout, yellow body, black trim	35

ERTL BANKS

NAME	NO.	YEAR	DESCRIPTION	MINT
Iowa Hawkeyes #8	*9810*	1988	1905 Ford Delivery Van, black body, yellow trim	35
Iowa Hawkeyes #9	*9748*	1989	1932 Ford Panel, yellow body, black trim	40
Iowa Jaycee Express	*9701*	1990	1913 Model T Van, gray body, black trim	90
Iowa Jaycee Express Limited Edition	*9457UO*	1990	1918 Ford Runbaout, beige body, blue trim	25
Iowa State Cyclones #1	*9259*	1985	1913 Model T Van, yellow body, red trim	30
Iowa State Cyclones #2	*9834*	1988	1905 Ford Delivery Van, yellow body, red trim	35
Iowa State Cyclones #3	*9127*	1989	1950 Chevy Panel. red body, yellow trim	45
Iowa State Cyclones #4	*1312UP*	1990	1918 Ford Runabout, red body, yellow trim	30
J.C. Penney	*2976UO*	1990	1918 Ford Barrel Runabout, red body, tan trim	25
J.C. Penney	*2975UO*	1990	1932 Ford Panel with spare tire, beige body, red trim	35
J.C. Penney	*9640*	1989	1050 Chevy Panel, yellow on yellow	30
J.C. Penney	*9641*	1989	1926 Mack Truck, white body, blue trim	30
J.C. Penney	*1328*	1988	1918 Ford Runabout, white body, blue trim	35
J.C. Penney	*2977UO*	1990	1926 Mack Truck with crates, yellow body, black trim	20
J.C. Penney	*9232*	1985	Horse, Carriage, tan body, black trim	65
J.C. Penney	*9234*	1985	1932 Ford Panel, beige body, black trim	75
J.C. Penney #1	*1354*	1983	1913 Model T Van, white body, red trim	155
J.C. Penney #2	*1354*	1983	1913 Model T Van, gray body, black trim	95
J.C. Penney #3	*1354*	1985	1913 Model T Van, yellow body, black trim	40
J.C. Penney (Golden Rule)	*9639*	1989	1917 Model T Van, green body, black trim	25
J.C. Penney (Golden Rule)	*1326*	1988	1905 Ford Delivery Van, orange body, black trim	35
J.I. Case	*0668*	1988	1913 Model T Van, red on red	20
J.I. Case	*0216*	1984	1926 Mack Truck, tan body, black trim	35
J.I. Case	*0286*	1990	1913 Model T Van	25
J.I. Case	*0401*	1989	1905 Ford Delivery Van, red body, black trim	25
J.I. Case	*0699*	1987	1905 Ford Delivery Van, tan body, gray trim	20
J.L. Kraft	*2147*	1985	1913 Model T Van, yellow body, black trim	90
J.T. General Store	*1321*	1982	1913 Model T Van, beige on beige	35
J.T. General Store (90th Anniversary)	*9674*	1989	1905 Ford Delivery Van, beige body, brown trim	25
Jack Daniels	*9077*	1989	1918 Ford Runabout with barrels, black on black	50
Jack Daniels	*9852*	1988	1905 Ford Delivery Van, black on black	75
Jackson Brewery	*9050*	1989	1905 Ford Delivery Van, white body, red trim	35
Janesville, Iowa	*9344*	1985	1913 Model T Van, white body, red trim	25
Jesup, Iowa, Chamber of Commerce	*9258*	1985	1913 Model T Van, white body, red trim	30
Jim Beam #116 (Northern Ohio)	*7661UO*	1990	1932 Ford Panel, white body, blue trim	65
Jim Beam (District #1)	*1313UO*	1990	1905 Ford Delivery Van, white body, blue trim	45
Jim Beam (District #10)	*9989*	1988	1926 Mack Truck, white body, black trim	35
Jim Beam (District #5) #1	*9412*	1986	1913 Model T Van, white body, red trim	175
Jim Beam (District #5) #2	*9387*	1987	1932 Ford Panel, white body, blue trim	95
Jim Beam (District #5) #3	*9729*	1988	1918 Ford Runabout, white body, black trim	85
Jim Beam (District #5) #4	*9676*	1989	1905 Ford Delivery Van, yellow body, red trim	85
Jim Beam (District #5) #5	*2964UO*	1990	1918 Ford Runabout with barrels, red body, beige trim	75
Jim Beam (District #6)	*7442*	1990	1932 Ford Panel	30
Jim Beam (District #8)	*9683*	1989	1913 Model T Van, white body, red trim	35
Jim Beam (District #9)	*9647*	1989	1917 Model T Van, white body, blue trim	35
Jim Beam (District #9) Susquehanna	*1316*	1990	1950 Chevy Panel, blue body, silver trim	75
Jim Beam, Sugar River Beamers	*2125UO*	1990	1905 Ford Delivery Van, white body, red trim	30
Jim's Auto Sales	*9818*	1988	1913 Model T Van, white body, red trim	25
John Deere #1	*0531*	1984	1926 Mack Truck, green body, yellow trim	115
John Deere #2	*5534*	1986	1926 Mack Truck, green body, yellow trim	70
John Deere #3	*5564EO*	1989	1926 Mack Truck, yellow body, green trim	40
John Deere I	*5621*	1989	1950 Chevy Panel, green body, yellow trim	35
John Deere II	*5621*	1989	1950 Chevy Panel, Wide Tire Version, green body, yellow trim	25

Flint Piston Service Step Van

Riverview Nursery 1937 Ford Tractor Trailer

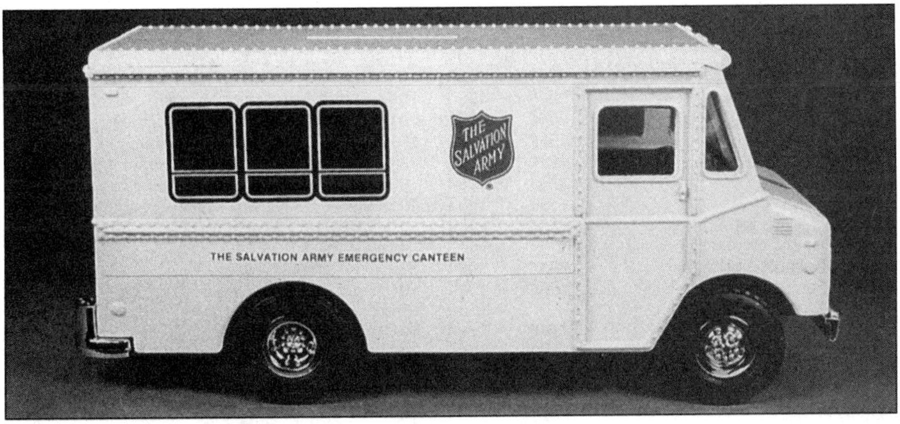

Salvation Army Step Van, 1987

ERTL BANKS

BANKS

NAME	NO.	YEAR	DESCRIPTION	MINT
Johnson Wax	*9459*	1987	1913 Model T Van, white body, red trim	165
Kansas State Fair	*9272*	1987	1913 Model T Van, white body, blue trim	215
Kansas State Fair	*9272*	1987	1913 Model T Van, white body, red trim	75
Kauffman's Orchard Apple Farm	*7560UO*	1990	1918 Ford Runabout, white body, black trim	30
Kerr McGee Oil	*9130*	1989	1913 Model T Van, gray body, black trim	45
Kerr McGee Oil	*9773*	1988	1926 Mack Tanker, gray body, black trim.T	50
Kerr McGee Oil	*7662UA*	1990	1932 Ford Panel with spare tire, gray body, black trim	35
Key Federal Bank	*9702*	1988	1918 Ford Runabout, white body, black trim	35
Key Federal Bank	*9703*	1988	1932 Ford Panel, white body, blue trim	35
Key-Aid	*9485*	1988	1905 Ford Delivery Van, white body, red trim	30
Key-Aid	*9175*	1987	1913 Model T Van, white body, red trim	70
Key-Aid	*9944*	1989	1932 Ford Panel, black body, silver trim	30
Key-Aid Limited Edition	*1332UA*	1990	1931 Hawkeye with crates, green body, black trim	30
Kidde	*9351*	1985	1913 Model T Van, white body, red trim	55
King Edward Cigars	*9854*	1988	1913 Model T Van, red body, black trim	40
Kingsport, Tennessee #1	*9174*	1987	1918 Ford Runabout, white body, black trim	110
Kingsport, Tennessee (Citivan)	*7682UA*	1990	1932 Ford Panel, black body, silver trim	40
Kodak #1	*9985*	1987	1905 Ford Delivery Van, gold spokes, yellow body, red trim	295
Kodak #2	*9985*	1987	1905 Ford Delivery Van, red spokes, yellow body, red trim	90
Kraft Dairy Group	*9675*	1989	1917 Model T Van, white on white	35
Kroger Foods	*9511*	1986	1913 Model T Van, white body, blue trim	65
Kuiken Brothers	*9362*	1985	1913 Model T Van, white body, red trim	30
Lake Odessa, Michigan, Centennial	*9519*	1986	1913 Model T Van, white body, blue trim	30
Leidy's	*9578UO*	1989	1937 Ford Tractor Trailer, white	30
Leinenkugel's	*7569UO*	1990	1918 Ford Runabout with barrels, red, black body, white trim	80
Lennox #1	*9461*	1986	1913 Model T Van, white body, red trim	135
Lennox #2	*9192*	1987	1932 Ford Panel, white body, red trim	210
Lennox #3	*9793*	1988	1918 Ford Runabout, red body, white trim	45
Lennox #4 Limited Edition	*9323*	1989	1905 Ford Delivery Van, white body, red trim	25
Lennox #5 Limited Edition	*7561UA*	1990	1926 Mack Truck, white body, red trim	30
LePage Glue	*2120*	1984	1913 Model T Van, yellow body, red trim	35
Light Commercial Vehicle Assn.	*2123UO*	1990	1913 Model T Van, white body, green trim	35
Link-Belt	*2107UP*	1990	1918 Ford Runabout, red body, black trim	40
Lion Coffee #1	*9306*	1988	1913 Model T Van, red on red	65
Lion Coffee #2 (125th Annv.)	*9434*	1989	1905 Ford Delivery Van, white body, green trim	35
Lipton Tea #1	*7505*	1989	1913 Model T Van, white body, red trim	75
Little Debbie #1	*9377*	1987	Step Van, white on white	90
Little Debbie #2	*9377UO*	1990	Step Van, white on white	70
Lolli Pups	*2146*	1984	1913 Model T Van, yellow body, brown trim	40
Longview, Illinois	*9704*	1988	1917 Model T Van, white body, blue trim	20
Loras College, Ltd. Ed.	*9679*	1989	1913 Model T Van, white body, purple trim	35
Los Angeles Times	*7667UA*	1990	1917 Model T Van, blue on blue	45
Louisiana State University	*9516*	1986	1913 Model T Van, white body, purple trim	50
MAC Tools	*9608*	1988	1950 Chevy Panel, gray on gray	300
Mace Brothers	*9391*	1986	1913 Model T Van, white body, red trim	55
Madison Electric, Ltd. Ed.	*9589UA*	1990	1913 Model T Van, blue body, white trim	105
Manitoba, Canada	*9220*	1985	1913 Model T Van, white body, green trim	30
Marathon Oil	*9783*	1988	1926 Mack Truck, silver body, red trim	350
Marshall Fields	*1650*	1983	1913 Model T Van, white body, green trim	125
Marshfield, Wisconsin, Chamber of Commerce	*9658*	1989	1926 Mack Truck, white body, green trim	25
Martin's Potato Chips	*9856*	1988	Step Van, white on white	45
Massey Ferguson	*1089*	1989	1918 Ford Runabout, red body, black, yellow trim	25
Massey Ferguson	*1348*	1990	1913 Model T Van	15
Massey Ferguson	*1122*	1984	1926 Mack Truck, tan body, black trim	35
Massey-Harris	*1092*	1987	1913 Model T Van, gold body, red trim	25
Matco	*9659*	1989	Step Van	65
Maurice's	*9476*	1987	1932 Ford Panel, gray body, maroon trim	35
Mayer's Well Drilling	*9267*	1987	1913 Model T Van, white body, blue trim	35

ERTL BANKS

NAME	NO.	YEAR	DESCRIPTION	MINT
McGlynn Bakery	*9495*	1987	Step van, white on white	85
McGlynn Bakery #1	*2112*	1984	1913 Model T Van, red spokes, white body, black trim	45
McGlynn Bakery #2	*2112*	1986	1913 Model T Van, black spokes, white body, black trim	35
McLain Boiler - 1 of 3 Piece Set	*2122R*	1990	1905 Ford Delivery Van, white body, red trim	30
McLain Boiler - 1 of 3 Piece Set	*2122RO*	1990	1917 Model T Van, white body, blue trim	30
Meijer Foods	*9352*	1985	1913 Model T Van, red body, white trim	35
Mellon Bank	*9318*	1987	1913 Model T Van, white body, green trim	25
Mellon Bank	*1358UO*	1990	1950 Ford Delivery Van, white body, green trim	35
Merit Oil	*9980*	1988	1926 Mack Tanker, white body, red trim	115
Merita Bread (American Bakeries)	*1316*	1987	1917 Model T Van, white on white	35
Merita Bread (American Bakeries)	*9316*	1987	1913 Model T Van, white body, red trim	35
Meyer's Funeral Home	*9568UA*	1990	1932 Ford Panel, black body, silver trim	40
Michigan Milk Producers Assn.	*1325*	1983	1913 Model T Van, white body, red trim	95
Michigan Sesquicentennial	*9172*	1987	1913 Model T Van, white body, blue trim	35
Michigan State University, Rose Bowl	*9026*	1988	1917 Model T Van, white on white	35
Mike's Trainland #1	*9850*	1988	1937 Ford Tractor Trailer, white on white	70
Mike's Trainland #2	*9411*	1989	1913 Model T Van, white body, red trim	30
Mike's Trainland #3	*9221UO*	1990	1905 Ford Delivery Van, blue on blue	30
Minnesota Industrial Tools	*9308UO*	1990	Step Van, silver	35
Minnesota Vikings	*1246*	1984	1913 Model T Van, white body, purple trim	55
Missouri Tourism, "Wake Up To Missouri"	*2143UO*	1990	1905 Ford Delivery Van, blue	25
Monroe Shocks, "America Rides Monroe"	*7511UR*	1990	1913 Model T Van, yellow body, blue trim	35
Monroe Shocks, "Monroe"	*7511UO*	1990	1913 Model T Van, yellow body, blue trim	35
Monroe Shocks, "The World Rides Monroe"	*7511UP*	1990	1913 Model T Van, yellow body, blue trim	35
Montana Centennial	*9747*	1989	1913 Model T Van, white body, red trim	25
Montgomery Ward #1	*9542*	1981	1913 Model T Van, green body, black trim	200
Montgomery Ward #2	*9052*	1982	1917 Model T Van, brown on brown	100
Montgomery Ward #3	*1363*	1983	1926 Mack Truck, yellow body, black trim	60
Montgomery Ward #4	*1367*	1983	Horse, Carriage, blue body, black trim	60
Montgomery Ward #5	*2110*	1984	1932 Ford Panel, yellow body, green trim	65
Montgomery Ward #6	*9230*	1985	1905 Ford Delivery Van, red body, black trim	60
Monticello, Iowa	*9364*	1986	1913 Model T Van, white body, red trim	40
Moorman Mfg. Co.	*9585UO*	1990	1905 Ford Delivery Van, white body, red trim	30
Morton Salt	*9787*	1988	1905 Ford Delivery Van, blue on blue	55
Mountain Dew	*7670UO*	1990	1950 Chevy Panel, green body, white trim	40
Mrs. Baird's	*9474*	1987	Horse, Carriage, white body, blue trim	120
Mutual Savings & Loan	*9187*	1987	1905 Ford Delivery Van, white body, blue trim	25
N.E.W. Hobby	*9479*	1987	Step Van, white on white	25
Nabisco Almost Home Cookies	*1653*	1984	1913 Model T Van, tan body, red trim	65
Nabisco Premium Saltine Crackers	*9699*	1981	1913 Model T Van, yellow body, maroon trim	300
NASA	*9467*	1989	Step Van, white on white	40
Nash Finch	*9347*	1985	1913 Model T Van, white body, red trim	155
National Street Rod Association	*7504*	1990	1937 Ford Tractor Trailer, white on white	80
National Van Lines #1	*9505*	1986	1913 Model T Van, white body, blue trim	65
National Van Lines #2	*9119*	1987	1932 Ford Panel, white body, blue trim	95
National Van Lines #3	*1342*	1988	1905 Ford Delivery Van, white body, blue trim	60
Neilson's Ice Cream	*2131*	1984	1913 Model T Van, white on white	30
New Brunswick, Canada	*9223*	1985	1913 Model T Van, white body, purple trim	25
New Holland	*9415*	1986	1913 Model T Van, yellow body, red trim	150
New Holland	*0379*	1987	1905 Ford Delivery Van, yellow body, red trim	15
New Holland Tractors	*9397*	1986	1913 Model T Van, yellow body, red trim	30
Newfoundland, Canada	*9224*	1985	1913 Model T Van, white body, purple trim	25
Nittany Machinery Association	*9642*	1989	1913 Model T Van, white body, red trim	30
Norand Data Systems	*9207UO*	1990	1926 Mack Truck with crates, white body, black trim	25

NAME	NO.	YEAR	DESCRIPTION	MINT
North American Van Lines	*9030*	1988	1937 Ford Tractor Trailer, red, tan body, red trim	65
North Dakota Centennial	*9045*	1989	Horse, Carriage, tan body, black trim	30
North Dakota Centennial	*9690*	1989	1913 Model T Van, white body, brown trim	25
Northern Electric	*9517*	1986	1905 Ford Delivery Van, black on black	40
Nova Scotia, Canada	*9222*	1985	1913 Model T Van, white body, red trim	25
Oakwood Mobile Homes	*9822*	1988	1917 model T Van, white body, blue trim	30
Oelwein, Iowa, Chamber of Commerce	*9361*	1985	1913 Model T Van, white body, red trim	25
Old Country	*9478*	1987	Step Van, white on white	95
Old El Paso	*7636UO*	1990	1905 Ford Delivery Van, yellow body, red trim	45
Olivet Union, Nazarene University	*2115UP*	1990	1905 Ford Delivery Van, yellow body, purple trim	30
Ontario, Canada	*9217*	1985	1913 Model T Van, white body, green trim	25
Orkin #1	*9389*	1987	1913 Model T Van, white body, red trim	35
Orkin #2	*9842*	1988	1905 Ford Delivery Van, white body, red trim	30
Oroweat	*9286*	1987	Step Van, white on white	60
Otasco #1	*1359*	1982	1913 Model T Van, yellow body, black trim	105
Otasco #2 (65th Annv.)	*1368*	1983	1913 Model T Van, yellow body, black trim	65
Otasco #3	*2134*	1984	1926 Mack Truck, yellow body, black trim	100
Otasco #4	*9342*	1985	1950 Chevy Panel, yellow on yellow	110
Otasco #5	*9371*	1986	1932 Ford Panel, yellow body, black trim	110
Otasco #6	*9168*	1987	1918 Ford Runabout, yellow body, black trim	75
Otasco #7 (70th Anniversary)	*9777*	1988	1918 Ford Runabout, white body, blue, black trim	70
Our Own Hardware	*9767*	1988	1913 Model T Van, white body, red trim	25
Overnite Trucking	*9142*	1987	1913 Model T Van, silver body, blue trim	250
Overnite Trucking	*9068*	1989	1937 Ford Tractor Trailer, blue cab, silver trailer	80
P.J. Valves	*7519Uo*	1990	1905 Ford Delivery Van, white body, blue trim	30
Parcel Post Mail Service	*9467*	1981	1913 model T Van, black on black	140
Peavey	*1357*	1982	1913 Model T Van, white body, red trim	25
Penn State	*9512*	1989	1918 Ford Runabout, white body, blue trim	75
Penn State	*9263*	1987	1917 Model T Van, white body, blue trim	160
People's National Bank	*9581UO*	1990	1918 Ford Runabout, blue body, white trim	45
People's National Bank	*9877*	1988	1913 Model T Van, white body, green trim	145
Pepsi-Cola	*9637*	1989	1932 Ford Panel, white body, blue trim	75
Pepsi-Cola	*6936*	1989	1918 Ford Runabout, blue body, black, white trim	55
Pepsi-Cola	*9635*	1989	1950 Chevy Panel, narrow tires, blue body, white trim	75
Pepsi-Cola	*9736*	1988	1905 Ford Delivery Van, white body, blue trim	80
Pepsi-Cola	*1314*	1987	1917 Model T Van, white body, blue trim	125
Pet Milk	*1652*	1983	1913 Model T Van, orange spokes, white body, black trim	85
Pet Milk	*1652*	1984	1913 Model T Van, red spokes, white body, black trim	65
Petty Enterprises #1	*9573UO*	1989	1913 Model T Van, blue body, red trim	200
Petty Enterprises #2	*9574UO*	1989	1913 Model T Van, blue on blue	410
Philadelphia Cream Cheese	*9835*	1988	1913 Model T Van, silver on silver	45
Phillips 66 #1 (Old Logo)	*9407UP*	1990	1931 Hawkeye Tanker, orange body, black trim	160
Pittsburgh Steelers	*1298*	1984	1913 Model T Van, white body, black trim	55
Pizza Today Magazine	*9933UO*	1990	1932 Ford Panel, yellow body, black trim	35
Pocono Antique Bazaar	*2917*	1990	1913 Model T Van, white body, brown trim	25
Poynors Home & Auto	*9273*	1987	1932 Ford Panel, white body, blue trim	30
Preston Trucking Co.	*9579UO*	1990	1937 Ford Tractor Trailer, orange	45
Prince Edward Island, Canada	*9219*	1985	1913 Model T Van, white body, orange trim	25
Protivin, Iowa, Holy Trinity	*9784*	1988	1918 Ford Runabout, white body, black trim	40
Publix	*9698*	1989	1950 Chevy Tractor Trailer with Reefer, white body, green trim	35
Publix	*9431*	1986	1950 Chevy Panel, white body, green trim	95
Publix	*9436*	1986	1905 Ford Delivery Van, white body, green trim	35

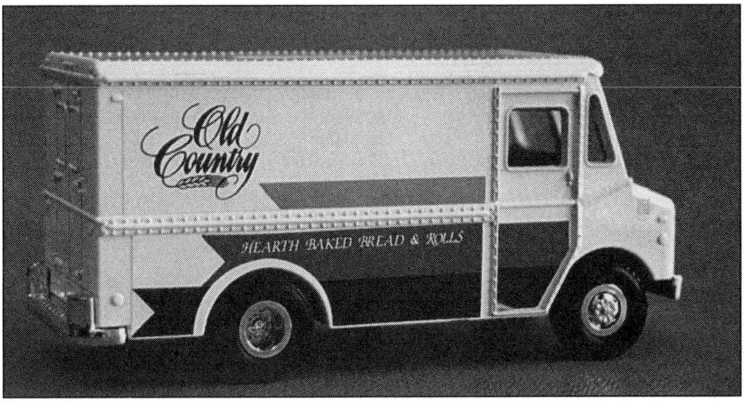

Old Country Step Van, 1987

Sasco Aloe Vera 1917 Model T, 1985

Pocono Antique Bazaar 1913 Model T Van

ERTL BANKS

NAME	NO.	YEAR	DESCRIPTION	MINT
Publix	*9693*	1989	1926 Mack Truck with crates, green body, brown trim	30
Publix	*9718*	1988	1937 Ford Tractor Trailer, white body, green trim	40
Publix #1	*9248*	1985	1932 Ford Panel, white body, green trim	45
Publix #1	*1337*	1981	1913 Model T Van, white body, green trim	45
Publix #1	*2115*	1984	1926 Mack Truck, black tires, red spokes, white body, green trim	125
Publix #1	*9149*	1987	1918 Ford Runabout, white body, green trim	55
Publix #2	*9719*	1988	1926 Mack Tanker, white cab, body, green trim	35
Publix #2	*2115*	1985	1926 Mack Truck, white tires, green spokes, white body	100
Publix #2	*1337*	1984	1913 Model T Van, red wheels, white body, green trim	35
Publix #2 (Pleasure)	*9723*	1988	1918 Ford Runabout, white body, green trim	25
Publix #2 Food & Pharmacy	*9147*	1987	1932 Ford Panel, white body, green trim	35
Publix #3 (The Deli)	*9694*	1989	1918 Ford Runabout, white body, brown trim	25
Publix #3 Limited Edition	*1337*	1985	1913 Model T Van, white body, green trim	95
Publix Danish Bakery	*9152*	1987	1905 Ford Delivery Van, white body, brown trim	30
Publix Danish Bakery	*9249*	1985	1913 Model T Van, white body, brown trim	35
Publix Danish Bakery	*9722*	1988	1950 Chevy Panel, brown body, white trim	35
Publix Danish Bakery	*9697*	1989	1917 Model T Van, white body, brown trim	25
Publix Danish Bakery	*9435*	1986	1932 Ford Panel, white body, brown trim	30
Publix Danish Bakery #2	*7690DO*	1990	1932 Ford Panel, with spare tire, white body, orange trim	35
Publix Dari-Fresh	*9695*	1989	1937 Ford Tanker Trailer, white body, green trim	35
Publix Dari-Fresh	*9001*	1989	1905 Ford Delivery Van, white body, green trim	25
Publix Dari-Fresh	*9430*	1986	1926 Mack Tanker, white body, green cab, trim	45
Publix Dari-Fresh	*7686DO*	1990	1931 Hawkeye Tanker, white body, green trim	25
Publix Deli	*7689DO*	1990	1905 Ford Delivery Van, orange body, brown trim	20
Publix Floral	*7687DO*	1990	1918 Ford Runabout, white body, green trim	20
Publix Food & Pharmacy	*9692*	1989	1950 Chevy Panel, green body, white trim	35
Publix Food & Pharmacy	*9721*	1988	1905 Ford Delivery Van, white body, black trim	25
Publix Food & Pharmacy	*7688DO*	1990	1913 Model T Van, green body, white trim	20
Publix Produce	*9696*	1989	1932 Ford Panel, green body, white trim	45
Publix Produce	*9720*	1988	1926 Mack Truck, white body, green trim	30
Publix Produce	*7685DO*	1990	1931 Hawkeye Truck, white body, green trim	20
Publix Produce Limited Edition	*9148*	1987	1917 Model T Van, white body, green trim	35
Quakertown National Bank #1	*9515*	1986	1913 Model T Van, white body, blue trim	75
Quakertown National Bank #2	*9291*	1987	1932 Ford Panel, white body, blue trim	55
Quakertown National Bank #3	*9979*	1988	1905 Ford Delivery Van, white body, blue trim	70
Quakertown National Bank #4	*9417*	1989	1918 Ford Runabout, blue body, white trim	45
Quakertown National Bank #5	*1370UP*	1990	1917 Model T Van, white body, blue trim	35
Quality Farm & Fleet #1	*9491*	1987	1913 Model T Van, white body, red trim	30
Quality Farm & Fleet #2	*9609*	1988	1905 Ford Delivery Van	30
Quality Farm & Fleet #3	*9609*	1989	1918 Ford Runabout, white body, red trim	25
Quebec, Canada	*9216*	1985	1913 Model T Van, white body, blue trim	25
R.C. Cola	*9827*	1988	1917 Model T Van, white body, blue trim	40
Ragrai XVII	*9108*	1989	1905 Ford Delivery Van, white body, blue trim	25
RCA #1	*9314*	1987	1913 Model T Van, white body, black trim	60
RCA #2	*9315*	1987	1905 Ford Delivery Van, white body, red trim	55
RCA #3	*9275*	1987	1926 Mack Truck, white body, red trim	40
RCA #4	*1344*	1988	1918 Ford Runabout, black body, white trim	35
RCA #5	*1343*	1988	1917 Model T Van, white body, red trim	35
RCA #6	*9621*	1989	1932 Ford Panel, white body, black trim	35
Red Crown Gasoline #1 (1st Run)	*7654UO*	1990	1931 Hawkeye Tanker, red on red	75
Red Crown Gasoline #1 (2nd Run)	*7654UO*	1990	1931 Hawkeye Tanker, red on red	60

ERTL BANKS

NAME	NO.	YEAR	DESCRIPTION	MINT
Red Rose Tea	*2130*	1984	1913 Model T Van, red on red	40
Renninger's Antique Market, Adamstown, PA	*7556UO*	1990	1932 Ford Panel, white body, red trim	30
Renninger's Antique Market, Adamstown, PA	*9712*	1988	1905 Ford Delivery Van, white body, red trim	35
Renninger's Antique Market, Adamstown, PA	*9894*	1989	1918 Ford Runabout with barrels, white body, red trim	35
Renninger's Antique Market, Kutztown, PA	*9714*	1988	1905 Ford Delivery Van, black body , red trim	35
Renninger's Antique Market, Kutztown, PA	*9895*	1989	1918 Ford Runabout with barrels, black body, red trim	35
Renninger's Antique Market, Kutztown, PA	*7555*	1990	1932 Ford Panel, black body, red trim	30
Renninger's Antique Market, Mt. Dora, FL	*9713*	1988	1905 Ford Delivery Van, brown on brown	35
Renninger's Antique Market, Mt. Dora, FL	*9896*	1989	1918 Ford Runabout with barrels, tan body, brown trim	35
Renninger's Antique Market, Mt. Dora, FL	*7557UO*	1990	1932 Ford Panel, beige body, brown trim	35
Republican Central Comm., Douglas County	*9369*	1986	1913 Model T Van, white body, red trim	75
Republican Party, Election 1988	*9805*	1988	1905 Ford Delivery Van, white body, red trim	150
Reynolds Aluminum	*9731*	1988	1917 Model T Van, white on white	55
Reynolds Wrap	*9162*	1987	1913 Model T Van, silver body, blue trim	285
Richlandtown	*2969UO*	1990	1905 Ford Delivery Van, white body, red trim	30
Richlandtown	*9290*	1987	1926 Mack Tanker, white body, red trim	65
Ringling Brothers Circus	*9726*	1988	1937 Ford Tractor Trailer, white body, yellow trim	145
Ringling Brothers Circus	*9027*	1988	1913 Model T Van, red on red	175
Riverview Nursery	*9121*	1987	1926 Mack Tanker, white body, red trim	65
Riverview Nursery (Purina)	*9804*	1988	1937 Ford Tractor Trailer, white on white	45
Rogersville, Tennessee	*9082*	1989	1932 Ford Panel with spare tire, black body, silver trim	80
Rogersville, Tennessee, Heritage Days	*2108UP*	1990	1918 Ford Runabout, red body, black trim	35
Round Top Arms	*9418*	1989	1918 Ford Runabout, white body, black trim	25
Rubschlager Bakery	*9446*	1986	1913 Model T Van, white body, red trim	40
S&F Toys #1	*9205*	1984	1913 Model T Van, gray body, maroon trim	20
S&F Toys #2	*9205*	1989	1913 Model T Van, gray body, maroon trim	20
Sacred Heart Church	*9340*	1985	1913 Model T Van, white body, red trim	35
Safety Kleen	*9289*	1987	Step Van, yellow on yellow	55
Safety Kleen (Dry Cleaner)	*9506*	1986	1913 Model T Van, yellow body, black trim	45
Safety Kleen (Parts Cleaner Service)	*9506*	1986	1913 Model T Van, yellow body, black trim	45
Saia Trucking	*9462*	1986	1913 Model T Van, white body, red trim	415
Salvation Army	*9109*	1987	1913 Model T Van, white body, red trim	100
Salvation Army	*2101*	1987	Step Van, white on white	110
San Francisco 49ers	*1297*	1984	1913 Model T Van, gold body, red trim	105
Sara Lee	*9941*	1989	1950 Chevy Panel, maroon body, black trim	70
Sasco Aloe Vera	*9354*	1985	1917 Model T Van, gray body, maroon trim	25
Saskatchewan, Canada	*9225*	1985	1913 Model T Van, white body, orange trim	20
Schneider Meats	*9445*	1986	1950 Chevy Panel, orange body, black trim	150
Schneider Meats	*9228*	1985	1926 Mack Truck, orange body, blue trim	525
Schneider Meats	*9229*	1985	1932 Ford Panel, orange body, blue trim	115
Schneider Meats	*1332*	1984	1913 Model T Van, orange body, blue trim	50
Schwan's Ice Cream	*9210*	1984	1950 Chevy Panel, brown on brown	165
Scott Tissue #1	*9652*	1989	1917 Model T Van, gold spokes, white body, blue trim	85
Scott Tissue #2	*9652UP*	1990	1917 Model T Van, silver spokes, white body, blue trim	50
Sealed Power Piston Rings	*2121UO*	1990	1905 Ford Delivery Van, white body, red trim	25
Sealed Power Speed-Pro	*1663UP*	1990	1905 Ford Delivery Van, black body, red trim	25
Sears	*2129*	1984	1913 Model T Van, black body, white trim	30
Servistar	*7510UA*	1990	1918 Ford Runabout, gray body, black, red trim	20
Servistar	*9036*	1989	1913 Model T Van, gray body, blue trim	25
Servistar	*9817*	1988	1905 Ford Delivery Van, gray body, red trim	45

ERTL BANKS

NAME	NO.	YEAR	DESCRIPTION	MINT
Seven-Up	*1662*	1988	1913 Model T Van, white body, green trim	180
Shelby Life Insurance	*1347*	1988	1905 Ford Delivery Van, white body, blue trim	85
Shopko	*9039*	1989	1913 Model T Van, white body, blue trim	30
Shoprite	*9711*	1988	1905 Ford Delivery Van, yellow body, red trim	15
Shoprite	*9666*	1989	1926 Mack Truck with crates, yellow body, red trim	25
Shoprite	*9163*	1987	1918 Ford Runabout, yellow body, red trim	30
Shoprite	*9426*	1986	1913 Model T Van, yellow body, red trim	35
Sidney Fire	*1335*	1988	1913 Model T Van, white body, red trim	20
Silver Springs Flea Market	*9619*	1988	1918 Ford Runabout, white body, black trim	30
Silver Springs Speedway	*2972UO*	1990	1932 Ford Panel, white body, red trim	55
Sinclair	*2120UO*	1990	1926 Mack Tanker, white body, green trim	55
Sinclair	*2119UP*	1990	1926 Mack Tanker, white body, green trim	80
Smoke Craft	*9493*	1987	1918 Ford Runabout, white body, black trim	30
Smoke Craft	*9494*	1987	1905 Ford Delivery Van, white body, red trim	30
Smoke Craft #1	*9164*	1987	1913 Model T Van, white body, blue trim	55
Smoke Craft Meats	*9654*	1989	1932 Ford Panel, white body, blue trim	35
Smoke Craft Meats	*9653*	1989	Step Van, white on white	40
Smokey The Bear #1	*9124*	1989	1913 Model T Van, white body, green trim	40
Smokey The Bear #2	*9123*	1989	1918 Ford Runabout, green, white body, black trim	45
Sohio Gas	*9269*	1987	1926 Mack Tanker, white body, red trim	300
South Dakota Centennial	*9879*	1988	Horse, Carriage, blue body, black trim	35
Southern States Oil	*9797*	1988	1926 Mack Truck, gray body, red trim	50
Southern States Oil	*7628UO*	1990	1937 Ford Tractor Trailer, red body, black trim	45
Southern States Oil	*9322*	1989	1918 Ford Runabout, red body, black trim	60
Southern States Oil	*9199*	1987	1926 Mack Tanker, silver body, red trim	275
Sparklettes Water	*9741*	1988	1926 Mack Truck, green on green	30
Spartan Food Stores	*9247*	1985	1913 Model T Van, green on green	30
St. Columbkille	*9304*	1986	1913 Model T Van, white body, blue trim	45
Steamtown, USA	*9167*	1987	1926 Mack Tanker, white body, red trim	100
Steelcase	*9041*	1989	1926 Mack Truck with crates, black body, brown trim	40
Steelcase	*9604*	1988	1905 Ford Delivery Van, blue body, silver trim	30
Steelcase	*7502UO*	1990	1937 Ford Tractor Trailer, white body, black trim	95
Steelcase	*7503UO*	1990	1932 Ford Panel, blue body, silver trim	95
Steelcase #1	*1657*	1982	1913 Model T Van, red spokes, blue body, silver trim	95
Steelcase #2	*1657*	1989	1913 Model T Van, chrome spokes, blue body, silver trim	45
Steelcase #2	*9265*	1987	1918 Ford Runabout, blue body, black trim	45
Steelcase #3	*1657*	1989	1913 Model T Van, red spokes with chrome trim, blue body, silver	100
Stevens Brothers Cartage (Bekins)	*9841*	1988	1926 Mack Truck, white body, black trim	35
Still Transfer Company	*9224*	1990	1937 Ford Tractor Trailer, white body, red trim	25
Strawberry Point, Iowa	*9353*	1990	1913 Model T Van, white body, red trim	45
Stroh's Beer	*7679UO*	1990	1918 Ford Runabout with barrels, red, black body, red trim	55
Sunbeam Bread	*9631*	1989	1913 Model T Van, white body, red trim	45
Sunbeam Bread	*1330*	1990	1932 Ford Panel with spare tire, blue body, yellow trim	35
Sunbeam Bread	*9518*	1986	1913 Model T Van, white body, red trim	100
Sunbeam Bread Step Van #1	*9638*	1989	Step Van, yellow body, blue trim	60
Sunholidays Travel	*9618*	1988	1918 Ford Runabout, white body, black trim	35
Sunmaid Raisins	*9575*	1989	1905 Ford Delivery Van, red on red	30
Sunmaid Raisins	*9576*	1990	Step Van, red on red	40
Super Valu #1	*9663EO*	1990	1932 Ford Panel with spare tire, white body, red trim	25
Support Your Local Fire Department	*7613UO*	1990	1950 Chevy Panel, red on red	30
Support Your Local Police	*7506UO*	1990	1950 Chevy Panel, black on black	25
Support Your Local Sheriff	*7540UO*	1990	1950 Chevy Panel, black on black	30

NAME	NO.	YEAR	DESCRIPTION	MINT
Sussex County Farm & Horse Show (Coors)	*7632UO*	1990	1905 Ford Delivery Van, white body, red trim	60
Swiss Valley #1	*2140*	1984	1913 Model T Van, white body, red trim	85
Swiss Valley #3	*9847*	1988	1095 Ford Delivery Van, white body, orange trim	35
Tabasco (McIlhenny)	*9078*	1989	1918 Ford Runabout with barrels, orange body, black/white trim	30
Tabasco (McIlhenny)	*9878*	1988	1905 Ford Delivery Van, white body, orange trim	25
Tennessee Homecoming 1986	*9420*	1986	1913 Model T Van, white body, red trim	160
Terminix International #1	*9465*	1986	1913 Model T Van, white body, orange trim	45
Terminix International #2	*9840*	1988	1917 Model T Van, white on white	25
Terminix International #3	*9086*	1989	1905 Ford Delivery Van, white body, orange trim	30
Texaco #1	*2128*	1986	1913 Model T Van, Silk Screened, white body, red trim	950
Texaco #1	*2128*	1984	1913 Model T Van, Silk Screened, white body, red trim	950
Texaco #1 Sampler	*2128*	1984	1913 Model T Van, Applied Label, white body, red trim	1500
Texaco #2	*9238*	1985	1926 Mack Tanker, white body, red trim	650
Texaco #2 Sampler	*9238*	1985	1926 Mack Tanker, white body, red trim	1050
Texaco #3	*9396*	1986	1932 Ford Panel, white body, red trim	450
Texaco #3 Sampler	*9396*	1986	1932 Ford Panel, white body, red trim	675
Texaco #4	*9321*	1987	1905 Ford Delivery Van, black body, red trim	165
Texaco #4 Sampler	*9376*	1987	1905 Ford Delivery Van, white body, red trim	575
Texaco #5	*9740*	1988	1918 Ford Runabout, black body, red trim	100
Texaco #5 Sampler	*9740*	1988	1918 Ford Runabout, gold spokes, black body, red trim	250
Texaco #6	*9040VO*	1989	1926 Mack Truck, Marker 1925 Mack, red body, black trim	85
Texaco #7	*9330VO*	1990	1930 Diamond T Tanker, embossed, red body, black trim	75
Texaco #7 Sampler	*9330*	1990	1930 Diamond T Tanker, No #7 mark on base, red body, black trim	185
Thompson Trucking	*9613*	1989	1937 Ford Tractor Trailer, white body, yellow trim	30
Thunderhills #1	*9268*	1987	1913 Model T Van, white body, blue trim	350
Thunderhills #2	*9792*	1988	1905 Ford Delivery Van, white body, red trim	285
Thunderhills #3	*9326*	1989	1926 Mack Truck, white body, black trim	265
Thunderhills #4	*7566UO*	1990	1918 Ford Runabout, white body, black trim	225
Tide Soap	*7509*	1990	1913 Model T Van, orange	55
Tioga County, NY, Bicentennial	*1321*	1990	1913 Model T Van, white body, blue trim	30
Tisco	*9983UO*	1990	1917 Model T Van, white body, black trim	50
Tisco	*9949UO*	1989	1918 Ford Runabout, white body, black trim	30
Tisco	*9948UO*	1989	1926 Mack Truck, white body, green trim	100
Tisco	*9649UO*	1990	Step Van, white body, red trim	35
Tisco	*9983UA*	1990	1905 Ford Delivery Van, white body, red trim	115
Titleist Golf Balls	*9489*	1987	1913 Model T Van, white body, red trim	65
Tom's Snack Foods	*1337UO*	1990	Step Van, Old Logo, tan body, red trim	30
Tom's Snack Foods	*1338UP*	1990	Step Van, New Logo, tan body	30
Toy Farmer #1	*1664*	1983	1913 Model T Van, white body, blue trim	40
Toy Farmer #2 (10th Anniversary)	*9483*	1987	1913 Model T Van, red body, blue trim	25
Toy Shop #1	*9442*	1989	1926 Mack Truck, white body, red trim	40
Toy Shop #2	*2118*	1990	1913 Model T Van, white body, red trim	25
Toy Tractor Times #1	*9480*	1988	1937 Ford Tractor Trailer, white on white	30
Toy Tractor Times #2	*9480*	1988	1937 Ford Tractor/Trailer, white body, trim	30
Toymaster	*2105*	1984	1913 Model T Van, yellow body, trim	35
Tractor Supply Company #1	*1349*	1982	1913 Model T Van, white body, red trim	40
Tractor Supply Company #1	*9355*	1986	1905 Ford Delivery Van, red body, white trim	25
Tractor Supply Company #10	*9133*	1989	1926 Mack w/crates, red body, black trim	35
Tractor Supply Company #2	*1349*	1983	1913 Model T Van, red body, white trim	40
Tractor Supply Company #3	*2121*	1984	1926 Mack Truck, red body, white trim	55
Tractor Supply Company #4	*9208*	1984	1932 Ford Panel, white body, red trim	60
Tractor Supply Company #5	*9207*	1985	1950 Chevy Panel, white body, trim	65
Tractor Supply Company #6	*9355*	1986	1917 Model T Van, red body, white trim	30

ERTL BANKS

NAME	NO.	YEAR	DESCRIPTION	MINT
Tractor Supply Company #6	*9357*	1986	1905 Ford Delivery Van, white body, red trim	25
Tractor Supply Company #7	*9356*	1986	1917 Model T Van, white body, red trim	25
Tractor Supply Company #8	*9530*	1986	1918 Runabout, white body, red trim	30
Tractor Supply Company #9	*2100*	1987	1926 Mack Truck, white body, red trim	40
Traer, Iowa Lions Club	*9416*	1986	1913 Model T Van, white body, purple trim	45
Trappey "Bull Brand" Hot Sauce	*1311*	1990	1905 Ford Delivery Van, cream body, red trim	35
Tremont Area Ambulance Association	*9754*	1989	1913 Model T Van, white body, blue trim	30
Tremont Area Ambulance Association	*7665UO*	1990	1932 Ford Panel, white body, blue trim	40
Tropicana Orange Juice	*7637UO*	1990	1905 Ford Delivery Van, white body, orange trim	75
Trucklite	*9898*	1988	1937 Ford Tractor/Trailer, white body, trim	30
True Value #1	*1348*	1982	1913 Model T Van, white body, blue trim	225
True Value #2	*1362*	1983	1926 Mack Truck, red body, black trim	105
True Value #3	*1296*	1984	1950 Chevy Panel (Marked 1948 Chevy Panel), light brown body	80
True Value #4	*9232*	1985	1932 Ford Panel, white body, blue trim	65
True Value #5	*9366*	1986	1918 Runabout, white body, blue trim	40
True Value #6	*9301*	1987	1905 Ford Delivery Van, white body, red trim	35
True Value #7	*9105*	1988	1926 Mack w/crates, brown body, red trim	25
True Value #8	*9623*	1989	1918 Barrel Runabout, red body, black trim	25
True Value #9	*7625EO*	1990	Horse Team & Tanker, red body, black trim	25
Trustworthy Hardware #1 - Ltd. Ed.	*9260*	1985	1917 Model T Van, white body, brown trim	195
Trustworthy Hardware #2	*9395*	1986	1905 Ford Delivery Van, white body, brown trim	55
Trustworthy Hardware #3 - Ltd. Ed.	*9375*	1987	1926 Mack Truck, white body, brown trim	40
Trustworthy Hardware #4 - Ltd. Ed.	*9774*	1988	1918 Runabout, brown body, black trim	30
Trustworthy Hardware #5 W/Spare	*9744*	1989	1913 Model T Van, white body, brown trim	25
Trustworthy Hardware #6 Ltd. Ed. (Sample)	*9100YA*	1990	1932 Ford Panel, brown body, silver trim	75
Trustworthy Hardware #6 W/Spare Ltd. Ed.	*9100UA*	1990	1932 Ford Panel, white body, brown trim	35
Turner Hydraulics	*2106UP*	1990	1931 Hawkeye Truck, white body, black trim	30
U.S. Mail	*2136*	1990	1932 Ford Panel, white body, red/white/black trim	10
U.S. Mail	*1659*	1988	1905 Ford Delivery Van, white body, blue trim	95
U.S. Mail	*9727*	1988	1937 Ford Tractor/Trailer, white body, trim	40
U.S. Mail	*9730*	1988	1917 Model T Van, white body, trim	40
U.S. Mail	*9532*	1989	1913 Model T Van, white body, trim	25
U.S. Mail	*9209UO*	1990	Step Van, white body, red/white/black trim	110
U.S. Mail	*9051*	1989	1932 Ford Panel, white body, dark blue trim	90
U.S. Mail #1 (Limited Edition)	*9532*	1987	1913 Model T Van, white body, trim	35
U.S. Mail #2 Ltd. Ed. W/Spare	*9843*	1988	1918 Runabout, white body, trim	45
U.S. Mail #3 Ltd. Ed. W/Spare	*9052*	1989	1932 Ford Panel, white body, blue trim	25
U.S. Mail #4 Ltd. Ed.	*7641UA*	1990	1905 Ford Delivery Van, blue body, red trim	30
U.S. Mail (Express Mail)	*9169*	1987	1926 Mack Truck, white body, black trim	115
U.S. Mail (Express Mail)	*9893*	1988	1937 Ford Tractor/Trailer, white body, trim	85
U.S. Mail (With Reversed Eagle)	*9532*	1986	1913 Model T Van, white body, trim	285
U.S. Mail - W/O Spare	*9296*	1987	1918 Runabout, white body, black trim	85
U.S.A. Baseball Team	*9795*	1988	1905 Ford Delivery Van, white body, navy blue trim	60
Unique Gardens	*9202*	1984	1913 Model T Van, white body, red trim	45
Unique Gardens	*9122*	1987	1926 Mack Tanker, white body, red trim	65
United Hardware	*9299*	1987	1950 Chevy Panel, red body, white trim	50
United Van Lines #1	*2100*	1984	1913 Model T Van, white body, black trim	55
United Van Lines #2	*9227*	1985	Horse & Carriage, white body, black trim	40
United Van Lines #3	*9393*	1986	1905 Ford Delivery Van, white body, black trim	25
United Van Lines #4 - Ltd. Ed.	*9715*	1988	1917 Model T Van, white body, black trim	30
United Van Lines #5 - Ltd. Ed.	*9096*	1989	1918 Runabout, white body, black trim	25
University of Florida Gators	*9821*	1988	1905 Ford Delivery Van, white body, orange trim	55
University of Indiana	*9497*	1987	1913 Model T Van, white body, red trim	30

NAME	NO.	YEAR	DESCRIPTION	MINT
University of Kansas - '88 Nat'l Champs	*9816*	1988	1905 Ford Delivery Van, white body, blue trim	40
University of Kansas Jayhawks	*9605*	1988	1913 Model T Van, white body, red trim	40
University of Michigan Wolverines	*9513*	1989	1913 Model T Van, yellow body, blue trim	40
University of Nebraska (Go Big Red)	*1330*	1982	1913 Model T Van, white body, red trim	70
University of Northern Iowa	*9300*	1986	1913 Model T Van, white body, trim	400
University of Wisconsin	*9655*	1989	1913 Model T Van, white body, red trim	30
V & S Variety Stores #1	*9622*	1989	1905 Ford Delivery Van, white body, orange trim	35
V & S Variety Stores #2	*7625EO*	1990	1918 Runabout, black body, orange trim	25
Valley Forge	*9616*	1989	1926 Mack Truck, white body, red trim	100
W.R. Meadows Inc.	*2968*	1990	1916 Model T Van, green body, white trim	90
W.W. Irwin Gasoline Maintenance Company	*2142UP*	1990	1905 Ford Delivery Van, light blue body	30
Washington State Centennial	*9137*	1989	1913 Model T Van, white body, red trim	25
Washington Suburban Sanitary Commission	*7522UO*	1990	1926 Mack Tanker, black body, red trim	30
Watkin's Inc.	*9786*	1988	1905 Ford Delivery Van, black body, red trim	55
Weber's Supermarket	*9276*	1987	1913 Model T Van, white body, red trim	30
Weil-McLain Boiler - 1 of 3 Piece Set	*2122RO*	1990	1926 Mack Truck, red body, black trim	30
Western Auto	*1328*	1981	1913 Model T Van, red body, black trim	110
Wheatbelt Stores - Five Point	*9481*	1987	1913 Model T Van, tan body, brown trim	25
Wheelers	*1358*	1982	1913 Model T Van, white body, red trim	30
White House (National Fruit Products)	*9004*	1989	1913 Model T Van, yellow body, green trim	235
White House (National Fruit Products)	*9005*	1989	1913 Model T Van, yellow body, green trim	210
Wilson Foods	*9897*	1988	1918 Ruanbout, red body, black trim	35
Winn-Dixie #1 (Red Spokes)	*1364*	1983	1913 Model T Van, white body, green trim	55
Winn-Dixie #1 (White Spokes)	*1364*	1984	1913 Model T Van, white body, green trim	45
Winn-Dixie #10 - Ltd. Ed.	*9014*	1989	1926 Mack w/ crates, white body, green/brown trim	30
Winn-Dixie #11 - Ltd. Ed.	*9013*	1989	1950 Chevy Panel, white body, green trim	35
Winn-Dixie #12 W/Spare Ltd. Ed.	*7694*	1990	1913 Model T Van, white body, green trim	25
Winn-Dixie #13 Ltd. Ed.	*7693*	1990	1931 Hawkeye Truck, white body, green trim	25
Winn-Dixie #2	*2125*	1984	1926 Mack Truck, green body, white trim	60
Winn-Dixie #3	*9341*	1985	1932 Ford Panel, white body, green trim	55
Winn-Dixie #4	*9423*	1986	1932 Ford Panel, green body, white trim	45
Winn-Dixie #5	*9392*	1986	1905 Ford Delivery Van, white body, green trim	55
Winn-Dixie #6	*9116*	1987	1918 Runabout, green body, white trim	30
Winn-Dixie #7	*9117*	1987	1905 Ford Delivery Van, green body, white trim	30
Winn-Dixie #8 - Ltd. Ed.	*9706*	1988	1937 Ford Tractor/Trailer, white body, green trim	30
Winn-Dixie #9 (W/Spare Tire)	*9707*	1988	1932 Ford Panel, white body, green trim	35
Wonder Bread #1	*1660*	1982	1913 Model T Van, white body, black trim	75
Wonder Bread #2	*9421*	1986	1913 Model T Van, white body, black trim	40
Wonder Bread #3	*9161*	1987	1913 Model T Van, white body, black trim	50
Wood Heat	*9498*	1987	1905 Ford Delivery Van, white body, blue trim	35
Wood Heat	*9978*	1989	1937 Ford Tractor/Trailer, white body, trim	55
Worldwide Products	*9114*	1987	Step Van, blue/white body	75
Worthington Firehouse	*7539UO*	1990	1905 Ford Delivery Truck, white body, red trim	40
Yelton Trucking	*9402*	1986	1913 Model T Van, white body, red trim	30
Yelton Trucking - Limited Edition	*9171*	1987	1926 Mack Tanker, white body, red trim	40
Yoder Popcorn	*9627*	1989	Step Van, white body, trim	55
York Peppermint Patties #1 W/ Spare	*9069*	1990	1932 Ford Panel, silver body, trim	105
York Peppermint Patties #3	*9071UO*	1990	1937 Ford Tractor/Trailer, blue body, silver trailer	105
Yuengling Beer	*9770*	1988	1937 Ford Tractor/Trailer	155
Yuengling Beer	*9429*	1986	1913 Model T Van, white body, red trim	160
Yuengling Beer	*9176*	1987	1905 Ford Delivery Van, white body, red trim	135

Barbie Dolls, Fashions and Accessories

It's hard to believe that in 1999, America's best known doll will turn 40.

First made in 1959, Mattel's Barbie doll is still one of the most popular toys of all time. Mattel was innovative in the early 1960s, extensively using television to advertise its merchandise, particularly the Barbie doll. A tremendous marketing campaign was launched which displayed Barbie dressed for dating, school and a variety of careers. The seeds were planted in children's minds, and imaginations shifted into overdrive. There was a quick response to these television commercials, and the toy became quite a phenomenon.

Modern American culture has been chronicled through the doll's beautifully detailed wardrobe. Styles range from the early 1960s Jacqueline Kennedy suits with pillbox hats to wild, mod 1970s fashions like polyester disco dresses. Designs by Bob Mackie continue to be favorites with their upscale, glittering glamour.

Today, the Barbie doll, her friends and accessories, whether new or old, are highly collectible. One reason for the popularity of collecting and selling Barbie dolls is that they were favorite playthings for most baby boomers. These adults are now attempting to replace their discarded toys and recapture a part of their childhood. Many collectors are also interested in the investment potential of Barbie paraphernalia. Mattel estimates collectors now number in excess of 100,000.

Hot Prospects

The 1988 Holiday Barbie is a perfect example of the effect that production run has on price appreciation. First in a series, the doll has rapidly increased in value both because of its beauty and limited issue. Due to the success of the 1988 doll, the 1989-1992 Holiday Barbie dolls were released in large quantities. While a very popular series, subsequent dolls have not matched the meteoric rise of the 1988 Holiday Barbie which retailed for $29.95 and now fetches $700 or more on the secondary market.

The Teen Talk Barbie is another doll worthy of mention. Mattel issued an estimated 350,000 of these talking dolls in 1992 for $34.99. Approximately one percent have a voice chip that says, among other things, "Math is Tough." This phrase was met with disdain by many, and use of the chip was halted. In fact, Mattel offered a refund for any doll voicing the offending phrase.

Returning that Barbie, however, would have been a costly mistake, for its current value on the secondary market is approximately $400. New dolls that become valuable very quickly are one reason why this collectible is so popular.

Not all Barbie dolls increase in value, however. Dolls made in the 1970s and later have been slower to increase in value than vintage Barbie dolls from the 1960s. Limited-edition dolls released in the 1980s and 1990s have shown mixed results depending upon the size of the production run.

The 1996 Pink Splendor doll is an example of a high-end collector's edition doll that received a lukewarm reception by collectors. Retailing for $900, the doll was quickly offered for slightly less on the secondary mar-

1966 American Girl

Looking Out for #1

The first Ponytail Barbies all look somewhat similar to the non-collector, but there are subtle differences. The first Barbie (1959), known to collectors as Ponytail Barbie #1, featured:

- holes in the bottom of the feet with copper tubes
- zebra-stripe one-piece swimsuit
- blonde or brunette hair with soft curly bangs
- red fingernails, toenails and lips
- gold hoop earrings
- white irises and severely pointed black eyebrows
- heavy, black facial paint
- pale, almost white, ivory skin tone
- body markings: Barbie T.M./ Pats. Pend. / ©MCMLVIII/ by/Mattel/Inc.

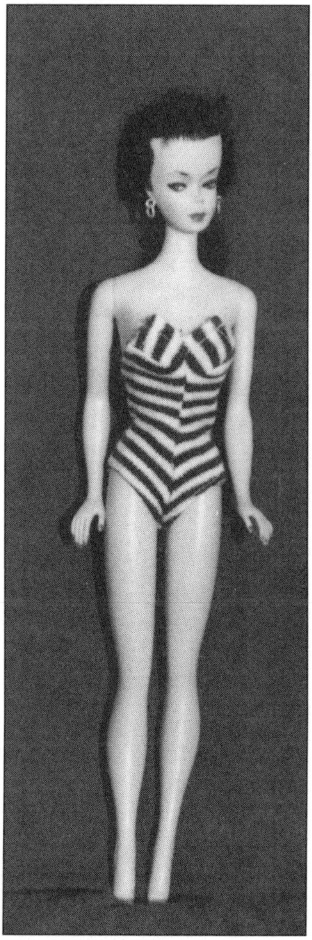

Ponytail Barbie, #1

ket. Collectors believed the high cost was unjustified. And while collectors still love Mattel's high-end collector's edition Barbie dolls (especially those with attire created by Bob Mackie and other fashion designers), Mattel may have pushed Pink Splendor's price out of many collectors' hands.

How Do You Price a Barbie Doll?

Values listed in this chapter are for dolls Mint In Box (MIB, Mint condition, never removed from original box with all accessories) and Mint No Box (MNB, Mint condition dolls and outfits without original packaging).

The earliest Barbie dolls were packaged in two-piece boxes and held in place by a cardboard liner. The doll could be taken out of the liner with no difficulty. Thus, the term Never Removed From Box (NRFB) frequently is accurate only for later issue dolls which were sealed in the box.

For consistency, the MIB term is used throughout this portion of the price guide, and it indicates a doll in the condition that it left the Mattel factory. Expect to receive between 50 and 70 percent of book value when selling to a dealer. Fifty percent is the norm with the higher percentage reserved for items of great value and scarcity.

Pricing Barbie dolls that are not Mint is quite subjective. A doll that has been played with may have a myriad of flaws such as green ears, neck splits, missing fingers, fading skin tone or hair cut or missing.

Remember that MIB and MNB condition dolls and outfits are rare, and prices reflect that scarcity. Dolls and fashions that are "played with" are plentiful and command a very small percentage of the MIB or Mint price.

Trends

Today's Barbie collectors concentrate on two main areas — vintage 1960s dolls and collector dolls from the 1980s and 1990s. Some of these dolls have experienced a rapid increase in value. For example, the 1990 Gold

Bob Mackie retailed for around $200 and is now selling for $1,000 on the secondary market. Also, vintage Barbie dolls and clothing in Mint condition continue to increase in value at a respectable rate each year. FAO Schwarz exclusives often enjoy increased value on the secondary market.

Promising recent issues include the Little Bo Peep (Children's Collector Series), Barbie Loves Elvis gift set and X-Files Barbie and Ken.

Don't overlook accessories such as paper dolls, vinyl cases or vehicles. Many of these items are still rather affordable and most likely will increase in value.

Regular issue (or "play") Barbie dolls from 1970s to present have shown flat returns on investment. This is projected to continue for many years because of the poor quality of the 1970s dolls and the enormous production runs of the 1980s.

Don't Be Fooled by Dates!

Think you have a vintage 1960s doll? Don't be fooled!

The dates and markings found on the buttocks of Barbie dolls don't necessarily tell you how old the doll is. Often the date only refers to the date the doll's body style was copyrighted.

That means even dolls from the 1990s may have a 1960s date on them.

The Top 10 Barbie Dolls
(in Mint in Box condition)

1. Ponytail Barbie #1, brunette, 1959 . . . . . $9,000
2. Ponytail Barbie #1, blonde, 1959 8,000
3. Ponytail Barbie #2, brunette, 1959 7,000
4. Ponytail Barbie #2, blonde, 1959 6,000
5. American Girl Side-Part Barbie, 1965 4,000
6. Color Magic Barbie, midnight black hair, 1966 3,200
7. American Girl Barbie, brunette, titian, 1966 2,700
8. American Girl Barbie with Color Magic Face, 1966 2,700
9. Color Magic Barbie, blonde, plastic box, 1966 2,400
10. Pink Jubilee Barbie, only 1,200 made, 1989 2,200

The Top 10 Barbie Fashions
(in Never-Removed-From-Box condition)

1. Roman Holiday, #968 . . . . . $4,800
2. Pan American Stewardess, #1678 4,500
3. Easter Parade, #971 4,000
4. Gay Parisienne, #964 3,800
5. Beautiful Bride, #1698 2,100
6. Shimmering Magic, #1664 1,800
7. Here Comes The Groom, #1426 1,600
8. Campus Sweetheart, #1616 1,500
9. Commuter Set, #916 1,500
10. Gold n' Glamour, #1647 1,500

Contributor to this section: Patricia Long, 5608 S. Kingshighway, St. Louis, MO 63109.

BARBIE & FRIENDS

NO.	NAME	YEAR	MNB	MIB
9423	All American Barbie	1991	4	30
9425	All American Christie	1991	4	25
9424	All American Ken	1991	4	25
9427	All American Kira	1991	4	35
9426	All American Teresa	1991	4	30
3553	All Star Ken	1981	7	25
9099	All Stars Barbie	1989	5	25
9352	All Stars Christie	1989	5	20
9361	All Stars Ken	1989	5	20
9360	All Stars Midge	1989	5	30
9353	All Stars Teresa	1989	5	30
1010	Allan, bendable leg	1966	150	550
1000	Allan, straight leg	1964	55	130
4930	American Beauty Mardi Gras Barbie	1987	40	120
3137	American Beauty Queen	1991	5	45
3245	American Beauty Queen, black doll	1991	5	35
1070	American Girl Barbie "Color Magic Face"	1966	1200	2700
1070	American Girl Barbie, all blondes	1965	800	1800
1070	American Girl Barbie, brunette, titian	1966	1200	2700
1070	American Girl Side-Part Barbie, brunette, blonde, titian	1965	2225	4000
5640	Angel Face Barbie	1982	8	40
4828	Animal Lovin' Barbie, black doll	1989	5	75
1350	Animal Lovin' Barbie, white version	1989	5	40
1395	Animal Lovin' Ginger Giraffe	1989	7	20
1351	Animal Lovin' Ken	1989	5	20
1352	Animal Lovin' Nikki	1989	7	20
1393	Animal Lovin' Zizi Zebra	1989	7	20
1207	Astronaut Barbie, black doll	1985	30	40
2449	Astronaut Barbie, white version	1985	25	75
15204	Autumn Glory, Enchanted Seasons Collection	1996	40	90
9434	Babysitter Courtney	1991	4	10
9433	Babysitter Skipper	1991	4	10
1599	Babysitter Skipper, black doll	1991	4	10
9000	Baggy Casey, blonde (Baggie)	1975	25	50
9613	Ballerina Barbie on Tour, gold, 1st version	1976	45	125
9093	Ballerina Barbie, 1st version	1976	20	65
9528	Ballerina Cara	1976	25	65
9805	Barbie & Her Fashion Fireworks	1976	20	60
3751	Barbie & the Beat	1990	5	30
2752	Barbie & the Beat Christie	1990	5	15
2754	Barbie & the Beat Midge	1990	6	20
1144	Barbie with Growin' Pretty Hair	1971	100	300
9601	Bathtime Fun Barbie	1991	3	20
9603	Bathtime Fun Barbie, black doll	1991	3	15
3237	Beach Blast Barbie	1989	3	20
3253	Beach Blast Christie	1989	4	15
3238	Beach Blast Ken	1989	4	15
3244	Beach Blast Miko	1989	4	15
3242	Beach Blast Skipper	1989	4	15
3251	Beach Blast Steven	1989	4	15
3249	Beach Blast Teresa	1989	5	15
9907	Beautiful Bride Barbie	1978	75	165
1290	Beauty Secrets Barbie, 1st issue	1980	20	65
1295	Beauty Secrets Christie	1980	20	55
1018	Beauty, Barbie's Dog	1979	12	30
9404	Benetton Barbie	1991	6	45
9407	Benetton Christie	1991	6	35
9409	Benetton Marina	1991	6	35

BARBIE & FRIENDS

NO.	NAME	YEAR	MNB	MIB
1293	Black Barbie, 1st issue black	1980	15	80
1142	Brad, bendable leg	1970	50	250
850	Bubblecut Barbie	1962	300	500
850	Bubblecut Barbie Brownette	1961	900	1800
850	Bubblecut Barbie Sidepart, all hair colors	1961	350	800
850	Bubblecut Barbie, all blondes	1961	150	385
850	Bubblecut Barbie, black haired	1961	200	425
850	Bubblecut Barbie, brunette	1961	200	425
850	Bubblecut Barbie, titian	1961	200	425
850	Bubblecut Barbie, white ginger	1962	450	1000
3577	Buffy and Mrs. Beasley	1968	125	275
3311	Busy Barbie	1972	150	295
3313	Busy Francie	1972	225	450
13675	Busy Gal Nostalgic Barbie	1994	25	55
3314	Busy Ken	1971	75	150
3312	Busy Steffie	1971	175	350
4547	Calgary Olympic Skating Barbie	1987	25	60
4439	California Dream Barbie	1988	5	20
4443	California Dream Christie	1988	6	15
4441	California Dream Ken	1988	8	15
4442	California Dream Midge	1988	3	15
4440	California Dream Skipper	1988	13	20
4403	California Dream Teresa	1988	15	20
7377	Carla, European exclusive	1976	65	140
1180	Casey, brunette, blonde, Twist and Turn	1967	85	300
1180	Casey, titian, Twist and Turn	1967	150	400
3570	Chris, titian, blonde, brunette	1967	35	250
1150	Color Magic Barbie, blonde, plastic box	1966	800	2400
1150	Color Magic Barbie,midnight hair	1966	1200	3200
15469	Cool Shavin' Ken	1996	10	15
3022	Cool Times Barbie	1989	5	30
3217	Cool Times Christie	1989	5	25
3219	Cool Times Ken	1989	5	25
3216	Cool Times Midge	1989	7	25
3218	Cool Times Teresa	1989	9	25
7079	Cool Tops Courtney	1989	7	20
9351	Cool Tops Kevin	1989	5	20
4989	Cool Tops Skipper	1989	7	15
5441	Cool Tops Skipper, black doll	1989	5	15
7123	Costume Ball Barbie	1991	6	25
7134	Costume Ball Barbie, black doll	1991	6	15
7154	Costume Ball Ken	1991	6	30
7160	Costume Ball Ken, black doll	1991	6	20
4859	Crystal Barbie, black doll	1984	10	25
4598	Crystal Barbie, white version	1984	10	35
9036	Crystal Ken, black doll	1983	15	25
4898	Crystal Ken, white version	1983	8	30
3509	Dance Club Barbie	1989	5	45
3513	Dance Club Devon	1989	5	40
3512	Dance Club Kayla	1989	5	85
3511	Dance Club Ken	1989	5	40
4836	Dance Magic Barbie	1990	7	25
7080	Dance Magic Barbie, black doll	1990	7	25
7081	Dance Magic Ken	1990	6	20
7082	Dance Magic Ken, black doll	1990	6	20
7945	Day-to-Night Barbie, black doll	1985	10	35
7944	Day-to-Night Barbie, Hispanic version	1985	17	40
7929	Day-to-Night Barbie, white version	1985	10	40

BARBIE & FRIENDS

NO.	NAME	YEAR	MNB	MIB
9018	Day-to-Night Ken, black doll	1984	8	20
9019	Day-to-Night Ken, white version	1984	8	25
9217	Deluxe Quick Curl Barbie	1976	20	75
9219	Deluxe Quick Curl Cara	1976	20	50
9218	Deluxe Quick Curl P.J.	1976	20	50
9428	Deluxe Quick Curl Skipper	1976	20	50
	Dentist Barbie	1997	15	30
3850	Doctor Barbie	1988	8	45
4118	Doctor Ken	1988	5	40
1116	Dramatic New Living Barbie, all hair colors	1970	65	275
1117	Dramatic New Living Skipper	1970	50	100
1623	Dream Bride	1992	10	40
9180	Dream Date Barbie	1982	7	30
4077	Dream Date Ken	1982	5	30
5869	Dream Date P.J.	1982	8	40
4817	Dream Date Skipper	1990	5	15
2242	Dream Glow Barbie, black doll	1986	12	25
1647	Dream Glow Barbie, Hispanic version	1986	25	70
2248	Dream Glow Barbie, white version	1986	12	45
2421	Dream Glow Ken, black doll	1986	13	20
2250	Dream Glow Ken, white version	1986	13	15
9180	Dream Time Barbie, pink	1985	10	25
9180	Dream Time Barbie, two versions	1985	10	25
2290	Earring Magic Ken	1993	15	45
7093	Fabulous Fur Barbie	1983	20	65
5313	Fashion Jeans Barbie	1981	15	65
5316	Fashion Jeans Ken	1982	12	25
2210	Fashion Photo Barbie, two versions	1978	20	75
2324	Fashion Photo Christie	1978	20	75
2323	Fashion Photo P.J.	1978	35	85
7193	Fashion Play Barbie	1983	10	30
9429	Fashion Play Barbie	1990	2	30
4835	Fashion Play Barbie	1987	10	25
9629	Fashion Play Barbie	1991	2	20
5953	Fashion Play Barbie, black doll	1991	2	15
5954	Fashion Play Barbie, Hispanic version	1990	2	15
870	Fashion Queen Barbie	1963	145	500
1189	Feelin' Fun Barbie, two versions, white, 1st issue	1988	5	20
3421	Feelin' Groovy Barbie	1985	100	175
9916	Flight Time Barbie, black doll	1990	5	20
9584	Flight Time Barbie, white version	1990	5	30
9600	Flight Time Ken	1990	5	20
1143	Fluff	1971	75	200
1122	Francie Hair Happenin's	1970	150	400
1170	Francie Twist and Turn, bendable leg, all hair colors, long hair	1967	150	400
	Francie Twist and Turn, flip hair	1969	200	400
1129	Francie with Growin' Pretty Hair	1971	75	225
1170	Francie, bendable leg, black doll	1967	900	1800
1130	Francie, bendable leg, white version, blonde, brunette	1966	150	350
1140	Francie, straight leg, brunette, blonde	1966	200	450
7270	Free Moving Barbie	1974	50	100
7283	Free Moving Cara	1974	40	80
7280	Free Moving Ken	1974	30	70
7281	Free Moving P.J.	1974	30	80
7668	Fun to Dress Barbie, black doll	1987	3	15
1373	Fun to Dress Barbie, black doll	1988	3	15
4939	Fun to Dress Barbie, black doll	1989	2	15

BARBIE & FRIENDS

NO.	NAME	YEAR	MNB	MIB
7373	Fun to Dress Barbie, Hispanic version	1989	3	15
4808	Fun to Dress Barbie, white version	1989	3	15
4558	Fun to Dress Barbie, white version	1987	3	20
4372	Fun to Dress Barbie, white version	1988	3	15
1739	Funtime Barbie, black doll	1987	5	25
1738	Funtime Barbie, white version	1987	5	25
7194	Funtime Ken	1987	7	20
1953	Garden Party Barbie	1989	8	25
1922	Gift Giving Barbie	1986	5	30
1205	Gift Giving Barbie	1989	5	30
7262	Gold Medal Olympic Barbie Skater	1975	20	100
7264	Gold Medal Olympic Barbie Skier	1975	20	100
7263	Gold Medal Olympic P.J. Gymnast	1975	20	85
7261	Gold Medal Olympic Skier Ken	1975	20	85
7274	Gold Medal Olympic Skipper	1975	20	85
3533	Golden Dreams Barbie Glamorous Night	1981	25	65
1974	Golden Dreams Barbie, two versions	1981	8	60
3249	Golden Dreams Christie	1981	10	65
15121	Got Milk? Barbie	1996	15	30
15122	Got Milk? Barbie, black	1996	15	30
7834	Great Shapes Barbie, black doll	1984	5	25
7025	Great Shapes Barbie, white version	1984	5	35
7025	Great Shapes Barbie, with Walkman	1984	12	40
7310	Great Shapes Ken	1984	5	25
7417	Great Shapes Skipper	1984	5	25
4253	Groom Todd	1982	15	45
9222	Growing Up Ginger	1977	30	75
7259	Growing Up Skipper	1977	10	55
9561	Happy Birthday Barbie	1991	8	30
1922	Happy Birthday Barbie	1984	8	35
1922	Happy Birthday Barbie	1981	8	45
9561	Happy Birthday Barbie, black doll	1991	8	30
15816	Happy Holidays	1996	30	45
12155	Happy Holidays, gold gown	1994	30	125
12156	Happy Holidays, gold gown, black doll	1994	30	80
10824	Happy Holidays, gold/red	1993	30	140
10911	Happy Holidays, gold/red gown, black doll	1993	30	90
1871	Happy Holidays, green gown	1991	40	200
2696	Happy Holidays, green gown, black doll	1991	40	100
14123	Happy Holidays, green/silver gown	1995	20	65
14124	Happy Holidays, green/silver gown, black doll	1995	20	45
4098	Happy Holidays, pink gown	1990	25	200
4543	Happy Holidays, pink gown, black doll	1990	20	80
1703	Happy Holidays, red gown	1988	100	800
17832	Happy Holidays, red/white/metallic	1997	5	10
17833	Happy Holidays, red/white/metallic, black doll	1997	5	10
1429	Happy Holidays, silver gown	1992	30	165
2396	Happy Holidays, silver gown, black doll	1992	30	100
3253	Happy Holidays, white gown	1989	50	300
7470	Hawaiian Barbie	1975	25	65
7470	Hawaiian Barbie	1977	30	80
5040	Hawaiian Fun Barbie	1991	3	20
5044	Hawaiian Fun Christie	1991	3	20
9294	Hawaiian Fun Jazzie	1991	3	20
5041	Hawaiian Fun Ken	1991	3	15
5043	Hawaiian Fun Kira	1991	3	15
5042	Hawaiian Fun Skipper	1991	3	15
5045	Hawaiian Fun Steven	1991	3	15

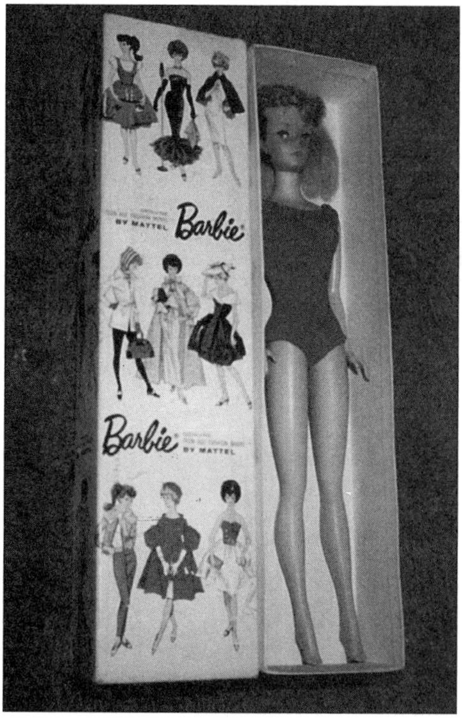

Ponytail Barbie #5, blond

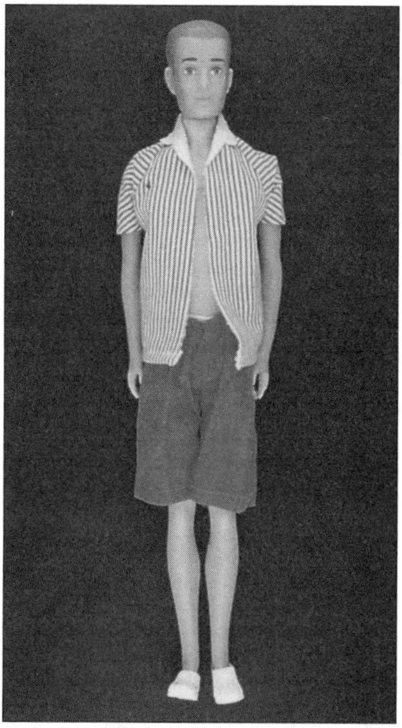

Early Ken doll, 1960s

American Girl Barbie, titian

Skipper, bendable leg, 1965

NO.	NAME	YEAR	MNB	MIB
7495	Hawaiian Ken	1984	7	30
2960	Hawaiian Ken	1979	13	50
3698	High School Chelsie	1989	5	20
3600	High School Dude, Jazzie's boyfriend	1989	5	20
3635	High School Jazzie	1989	5	20
3636	High School Stacie	1989	5	20
1292	Hispanic Barbie	1980	15	70
2249	Home Pretty Barbie	1990	8	30
2390	Homecoming Queen Skipper, black doll	1988	8	20
1952	Homecoming Queen Skipper, white version	1988	12	25
1757	Horse Lovin' Barbie	1983	10	40
3600	Horse Lovin' Ken	1983	8	25
5029	Horse Lovin' Skipper	1983	8	25
7927	Hot Stuff Skipper	1984	5	18
7365	Ice Capades Barbie, 50th Anniversary	1990	5	35
7348	Ice Capades Barbie, black doll, 50th Anniversary	1990	5	25
7375	Ice Capades Ken	1990	5	25
15473	Inline Skating Barbie	1996	5	25
15474	Inline Skating Ken	1996	5	25
15475	Inline Skating Midge	1996	5	25
	International Barbie Arctic	1997	20	30
3626	International Barbie Australian	1993	10	35
9094	International Barbie Brazillian	1990	15	75
4928	International Barbie Canadian	1988	15	95
11180	International Barbie Chinese	1994	10	35
7330	International Barbie Czechoslovakian	1991	30	110
11104	International Barbie Dutch	1994	10	35
4973	International Barbie English	1992	12	80
3898	International Barbie Eskimo	1982	50	150
9844	International Barbie Eskimo, reissue	1991	8	75
	International Barbie French	1997	10	25
3188	International Barbie German	1987	45	125
12598	International Barbie German, reissue	1995	10	25
2997	International Barbie Greek	1986	30	80
3189	International Barbie Iceland	1987	30	125
3897	International Barbie India	1982	50	175
14451	International Barbie India, reissue	1995	12	25
7517	International Barbie Irish	1984	45	150
12998	International Barbie Irish, reissue	1994	10	20
1601	International Barbie Italian	1980	65	225
2256	International Barbie Italian, reissue, two box versions	1993	10	40
4647	International Barbie Jamaican, silver earrings	1992	12	85
9481	International Barbie Japanese	1985	55	125
14163	International Barbie Japanese, reissue	1986-87	12	150
11181	International Barbie Kenyan	1994	10	35
4929	International Barbie Korean	1988	15	100
7329	International Barbie Malaysian	1991	10	75
1917	International Barbie Mexican	1989	15	75
	International Barbie Mexican, reissue	1995	12	25
1753	International Barbie Native American I, two box versions	1993	12	35
11609	International Barbie Native American II	1995	10	30
12699	International Barbie Native American III	1995	10	30
7376	International Barbie Nigerian	1990	15	80
14450	International Barbie Norwegian, pink flowers	1996	12	25
3262	International Barbie Oriental	1981	55	200
1600	International Barbie Parisienne	1980	65	200
9843	International Barbie Parisienne, reissue	1991	8	75
2995	International Barbie Peru	1986	30	100

BARBIE & FRIENDS

NO.	NAME	YEAR	MNB	MIB
12700	International Barbie Polynesian	1995	10	30
	International Barbie Puerto Rican	1997	15	30
1602	International Barbie Royal	1980	65	225
	International Barbie Russian	1997	20	30
1916	International Barbie Russian	1989	20	100
3263	International Barbie Scottish	1981	50	200
9845	International Barbie Scottish, reissue	1991	8	75
4031	International Barbie Spanish	1983	40	175
4963	International Barbie Spanish, reissue	1992	12	40
4032	International Barbie Swedish	1983	35	85
7451	International Barbie Swiss	1984	35	150
4061	Island Fun Barbie	1988	3	20
4092	Island Fun Christie	1988	3	20
4060	Island Fun Ken	1988	3	15
4064	Island Fun Skipper	1988	3	15
4093	Island Fun Steven	1988	3	15
4117	Island Fun Teresa	1988	3	15
3633	Jazzie Workout	1989	5	12
1756	Jewel Secrets Barbie, black doll, two box versions	1987	6	55
1737	Jewel Secrets Barbie, white doll, two box versions	1987	6	35
3232	Jewel Secrets Ken, black doll	1987	6	25
1719	Jewel Secrets Ken, rooted hair	1987	6	25
3133	Jewel Secrets Skipper	1987	6	25
3179	Jewel Secrets Whitney	1987	8	25
750	Ken, bendable leg, brunette, blonde	1965	155	350
1124	Ken, bendable leg, talking Ken	1970	60	125
750	Ken, flocked hair, brunette, blonde	1961	100	350
750	Ken, painted hair, brunette, blonde	1962	50	150
9325	Kevin	1991	5	10
2597	Kissing Barbie, two versions	1979	8	60
2955	Kissing Christie	1979	10	60
9725	Lights & Lace Barbie	1991	4	30
9728	Lights & Lace Christie	1991	4	30
9727	Lights & Lace Teresa	1991	4	25
1155	Live Action Barbie	1971	60	150
1152	Live Action Barbie Onstage	1971	75	250
1175	Live Action Christie	1971	60	250
1159	Live Action Ken	1971	55	150
1172	Live Action Ken on Stage	1971	40	150
1156	Live Action P.J.	1971	65	250
1153	Live Action P.J. on Stage	1971	75	175
1116	Living Barbie	1970	65	250
7072	Lovin' You Barbie	1983	20	100
3989	Magic Curl Barbie, black doll	1982	8	25
3856	Magic Curl Barbie, white version	1982	10	35
3137	Magic Moves Barbie, black doll	1985	15	35
2126	Magic Moves Barbie, white doll	1985	15	35
1067	Malibu Barbie	1971	15	60
1067	Malibu Barbie	1975	15	40
7745	Malibu Christie	1975	10	40
7745	Malibu Christie	1977	15	40
1068	Malibu Francie	1971	15	45
1088	Malibu Ken	1976	8	25
1087	Malibu P.J.	1975	5	45
1069	Malibu Skipper	1977	8	45
1080	Midge, bendable leg, blonde, titian	1965	325	600
1080	Midge, bendable leg, brunette	1965	325	600
860	Midge, straight leg, blonde	1964	75	200

BARBIE & FRIENDS

NO.	NAME	YEAR	MNB	MIB
860	Midge, straight leg, brunette, titian	1964	100	225
860	Midge, with teeth, all hair colors	1965	165	325
1060	Miss Barbie (Sleep-eye)	1964	700	1400
4224	Mod Hair Ken	1973	45	65
9988	Music Lovin' Barbie	1985	15	45
2388	Music Lovin' Ken	1985	15	45
2854	Music Lovin' Skipper	1985	20	75
9942	My First Barbie	1991	3	25
1875	My First Barbie, aqua and yellow dress	1981	10	30
9944	My First Barbie, black doll	1990	4	20
9943	My First Barbie, black doll	1991	3	20
9944	My First Barbie, Hispanic version	1991	3	20
9943	My First Barbie, Hispanic version	1990	5	25
1875	My First Barbie, pink checkered dress	1983	5	35
1801	My First Barbie, pink tutu, black doll	1987	5	20
1788	My First Barbie, pink tutu, white version	1987	5	20
9858	My First Barbie, white dress, black doll	1984	7	25
1875	My First Barbie, white dress, white version	1984	5	30
1281	My First Barbie, white tutu, black doll	1988	6	15
1282	My First Barbie, white tutu, Hispanic version	1988	6	20
1280	My First Barbie, white tutu, white version	1988	5	20
9942	My First Barbie, white version	1990	4	20
9940	My First Ken	1990	4	15
9940	My First Ken	1991	3	15
1389	My First Ken, 1st issue	1989	4	15
9342	New Look Ken	1976	23	65
7807	Newport Barbie, two versions	1974	25	140
1170	No Bangs Francie	1971	900	1800
1127	Nurse Julia, 1-piece outfit	1969	100	200
1127	Nurse Julia, 2-piece outfit, talking	1969	100	200
4405	Nurse Whitney	1987	20	45
15430	Ocean Friends Barbie, Ocean Magic	1996	5	17
15430	Ocean Friends Ken, Ocean Magic	1996	5	17
15431	Ocean Friends Kira, Ocean Magic	1996	5	17
15123	Olympic Gymnast	1996	10	25
4885	Party Treats Barbie	1989	8	25
9516	Peaches n' Cream Barbie, black doll	1984	8	40
7926	Peaches n' Cream Barbie, white version	1984	8	45
4869	Pepsi Spirit Barbie	1989	10	40
4867	Pepsi Spirit Skipper	1989	10	35
4555	Perfume Giving Ken, black doll	1989	6	25
4554	Perfume Giving Ken, white version	1989	6	25
4552	Perfume Pretty Barbie, black doll	1989	8	25
4551	Perfume Pretty Barbie, white version	1989	8	25
4557	Perfume Pretty Whitney	1987	8	35
3551	Pink n' Pretty Barbie	1982	15	40
3554	Pink n' Pretty Christie	1982	8	35
5336	Playtime Barbie	1984	15	20
850	Ponytail Barbie #1, blonde	1959	4000	8000
850	Ponytail Barbie #1, brunette	1959	4500	9000
850	Ponytail Barbie #2, blonde	1959	3500	6000
850	Ponytail Barbie #2, brunette	1959	4000	7000
850	Ponytail Barbie #3, blonde	1960	600	1200
850	Ponytail Barbie #3, brunette	1960	800	1500
850	Ponytail Barbie #4, blonde	1960	300	600
850	Ponytail Barbie #4, brunette	1960	350	800
850	Ponytail Barbie #5, blonde	1961	175	400
850	Ponytail Barbie #5, brunette	1961	250	500

BARBIE & FRIENDS

NO.	NAME	YEAR	MNB	MIB
850	Ponytail Barbie #5, titian	1961	250	500
850	Ponytail Barbie #6, blonde	1962	200	300
850	Ponytail Barbie #6, brunette	1962	200	350
850	Ponytail Barbie #6, titian	1962	250	400
850	Ponytail Swirl Style Barbie, blonde	1964	350	600
850	Ponytail Swirl Style Barbie, brunette	1964	350	600
850	Ponytail Swirl Style Barbie, platinum	1964	500	1000
850	Ponytail Swirl Style Barbie, titian	1964	350	600
1117	Pose n' Play Skipper (baggie)	1973	20	55
2598	Pretty Changes Barbie	1978	8	45
7194	Pretty Party Barbie	1983	12	30
4220	Quick Curl Barbie	1973	20	80
7291	Quick Curl Cara	1975	20	60
4222	Quick Curl Francie	1973	20	55
4221	Quick Curl Kelley	1973	20	75
8697	Quick Curl Miss America, blonde	1974	35	75
8697	Quick Curl Miss America, brunette	1973	45	175
4223	Quick Curl Skipper	1973	20	50
1090	Ricky	1965	55	160
1140	Rocker Barbie, 1st issue	1986	7	40
3055	Rocker Barbie, 2nd issue	1987	7	25
1196	Rocker Dana, 1st issue	1986	7	40
3158	Rocker Dana, 2nd issue	1987	7	20
1141	Rocker Dee-Dee, 1st issue	1986	7	30
3160	Rocker Dee-Dee, 2nd issue	1987	7	20
2428	Rocker Derek, 1st issue	1986	7	30
3173	Rocker Derek, 2nd issue	1987	7	20
2427	Rocker Diva, 1st issue	1986	7	30
3159	Rocker Diva, 2nd issue	1987	7	20
3131	Rocker Ken, 1st issue	1986	7	30
1880	Rollerskating Barbie	1980	8	60
1881	Rollerskating Ken	1980	8	40
4973	Safari Barbie	1983	8	30
1019	Scott	1979	15	60
9109	Sea Lovin' Barbie	1984	8	35
9110	Sea Lovin' Ken	1984	8	30
4931	Sensations Barbie	1987	5	12
4977	Sensations Becky	1987	5	12
4976	Sensations Belinda	1987	5	12
4967	Sensations Bopsy	1987	5	12
4960	Sensations Bopsy Bibops	1987	15	95
7799	Show and Ride Barbie	1988	8	35
7511	Ski Fun Barbie	1991	6	15
7512	Ski Fun Ken	1991	6	15
7513	Ski Fun Midge	1991	6	25
1030	Skipper, bendable leg, brunette, blonde, titian	1965	100	325
950	Skipper, straight leg, brunette, blonde, titian	1964	50	225
950	Skipper, straight leg, reissues, brunette, blonde, titian	1971	125	400
1120	Skooter, bendable leg, brunette, blonde, titian	1965	100	400
1040	Skooter, straight leg, brunette, blonde, titian	1965	55	175
15408	Snowboard Barbie	1996	10	20
13534	Solo in the Spotlight, blonde	1995	12	30
13820	Solo in the Spotlight, brunette	1995	12	30
15419	Sparkle Barbie	1996	10	20
1294	Sport n' Shave Ken	1980	8	40
1190	Standard Barbie, blonde, brunette	1970	300	500
1190	Standard Barbie, blonde, brunette	1967	300	500
1190	Standard Barbie, titian	1967	500	1000

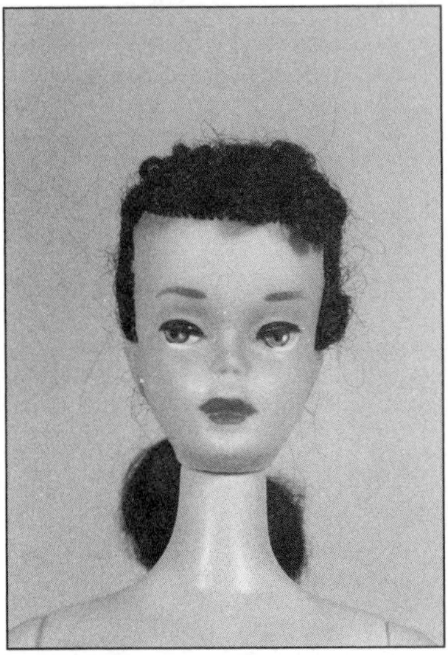

Ponytail Barbie #4, brunette

Ponytail Barbie #3, blonde

Twist and Turn Barbie, 1967, dressed in 1969 Fab City ensemble

NO.	NAME	YEAR	MNB	MIB
1190	Standard Barbie, titian	1970	500	1000
1281	Starr Kelley	1979	8	35
1283	Starr Shaun	1979	8	40
1280	Starr Starr	1979	8	35
1282	Starr Tracy	1979	8	35
3360	Stars & Stripes Air Force Barbie	1990	15	50
3966	Stars & Stripes Army Barbie	1989	10	40
9693	Stars & Stripes Navy Barbie	1991	10	40
9694	Stars & Stripes Navy Barbie, black doll	1991	10	35
1283	Style Magic Barbie	1989	5	20
1288	Style Magic Christie	1989	5	20
1915	Style Magic Skipper	1989	10	20
1290	Style Magic Whitney	1989	5	20
7027	Summit Barbie	1990	8	30
7029	Summit Barbie, Asian version	1990	10	30
7028	Summit Barbie, black doll	1990	8	30
7030	Summit Barbie, Hispanic version	1990	10	35
7745	Sun Gold Malibu Barbie, black doll	1983	5	15
4970	Sun Gold Malibu Barbie, Hispanic version	1985	3	20
1067	Sun Gold Malibu Barbie, white version	1983	5	15
3849	Sun Gold Malibu Ken, black doll	1983	3	15
	Sun Gold Malibu Ken, Hispanic version	1985	3	20
1088	Sun Gold Malibu Ken, white version	1983	3	15
1187	Sun Gold Malibu P.J.	1983	5	15
1069	Sun Gold Malibu Skipper	1983	5	15
1067	Sun Lovin' Malibu Barbie	1978	5	20
1088	Sun Lovin' Malibu Ken	1978	5	20
1187	Sun Lovin' Malibu P.J.	1978	5	20
1069	Sun Lovin' Malibu Skipper	1978	5	20
7806	Sun Valley Barbie	1974	20	130
7809	Sun Valley Ken	1974	20	100
4970	Sunsational Hispanic Barbie	1984	20	35
1067	Sunsational Malibu Barbie	1982	6	25
4970	Sunsational Malibu Barbie, Hispanic version	1982	8	25
7745	Sunsational Malibu Christie	1982	6	20
3849	Sunsational Malibu Ken, black doll	1982	15	35
1187	Sunsational Malibu P.J.	1982	6	30
1069	Sunsational Malibu Skipper	1982	5	35
7745	Sunset Malibu Christie	1973	20	65
1068	Sunset Malibu Francie	1971	25	65
1088	Sunset Malibu Ken	1972	15	50
1187	Sunset Malibu P.J.	1971	10	50
1069	Sunset Malibu Skipper	1971	20	50
3296	Super Hair Barbie, black doll	1987	8	20
3101	Super Hair Barbie, white version	1987	8	25
5839	Super Sport Ken	1982	8	20
2756	Super Teen Skipper	1978	7	20
9828	Supersize Barbie	1977	75	200
9839	Supersize Christie	1977	75	200
2844	Supersize Super Hair Barbie	1979	85	175
4983	Superstar Ballerina Barbie, 1st version	1976	20	60
9720	Superstar Barbie	1977	15	95
1605	Superstar Barbie 30th Anniversary, black doll	1989	6	35
1604	Superstar Barbie 30th Anniversary, white version	1989	8	25
9950	Superstar Christie	1977	20	75
2211	Superstar Ken	1978	17	50
1550	Superstar Ken 30th Anniversary, black doll	1989	5	30
1535	Superstar Ken 30th Anniversary, white version	1989	7	50

NO.	NAME	YEAR	MNB	MIB
1067	Superstar Malibu Barbie	1977	10	35
	Superstar Promotional	1976	50	110
7796	Sweet 16 Barbie	1974	25	75
7635	Sweet Roses Barbie	1989	7	25
7455	Sweet Roses P.J.	1983	15	25
2064	Tahiti, Barbie's Pet Parrot	1985	3	15
1115	Talking Barbie, all hair colors (nape curls)	1968	200	400
1114	Talking Brad	1970	65	175
1195	Talking Busy Barbie	1972	150	300
1196	Talking Busy Ken	1972	80	160
1186	Talking Busy Steffie	1972	175	350
1126	Talking Christie	1968	65	175
1128	Talking Julia	1968	65	175
1111	Talking Ken	1969	40	200
1113	Talking P.J.	1969	65	175
1125	Talking Stacey, blonde, titian	1968	200	400
1107	Talking Truly Scrumptious	1969	300	600
13915	Teacher Barbie, black doll, painted on panties	1996	15	20
13914	Teacher Barbie, painted on panties	1996	15	25
3634	Teen Dance Jazzie	1989	7	35
5893	Teen Fun Skipper Cheerleader	1987	5	15
5899	Teen Fun Skipper Party Teen	1987	5	15
5889	Teen Fun Skipper Workout	1987	5	15
3634	Teen Jazzie (Teen Dance)	1989	4	35
3631	Teen Looks Jazzie Cheerleader	1989	4	20
3633	Teen Looks Jazzie Workout	1989	4	20
5507	Teen Scene Jazzie, two box versions	1991	5	35
4855	Teen Sweetheart Skipper	1988	5	25
5745	Teen Talk Barbie "Math is Tough"	1992	50	275
5745	Teen Talk Barbie, two box versions	1992	15	45
1950	Teen Time Courtney	1988	5	10
1951	Teen Time Skipper	1988	5	10
1760	Tennis Barbie	1986	5	25
1761	Tennis Ken	1986	5	25
3590	Todd	1967	50	200
4103	Tracey Bride	1983	8	45
1022	Tropical Barbie, black doll	1986	3	15
1017	Tropical Barbie, white version	1986	3	20
1023	Tropical Ken, black doll	1986	3	15
4060	Tropical Ken, white version	1986	3	20
2056	Tropical Miko	1986	3	20
4064	Tropical Skipper	1986	3	20
1108	Truly Scrumptious, non-talking	1969	250	500
3580	Tutti, all hair colors, floral dress with yellow ribbon	1967	45	130
3550	Tutti, all hair colors, pink, white gingham suit, hat	1966	50	180
8128	Tutti, Germany	1978	40	110
1185	Twiggy	1967	150	350
5723	Twirley Curls Barbie, black doll	1983	8	30
5724	Twirley Curls Barbie, Hispanic version	1982	10	40
5579	Twirley Curls Barbie, white version	1982	8	30
1160	Twist and Turn Barbie, blondes and brunettes	1967	150	350
1160	Twist and Turn Barbie, titian	1967	200	450
1119	Twist and Turn Christie	1969	100	275
1160	Twist and Turn Flip Barbie, all hair colors	1969	150	350
1118	Twist and Turn P.J., blonde	1970	75	200
1105	Twist and Turn Skipper, all hair colors	1969	50	300
1165	Twist and Turn Stacey, blonde, titian	1968	200	400
4774	UNICEF Barbie, Asian version	1989	17	40

BARBIE & FRIENDS

NO.	NAME	YEAR	MNB	MIB
4770	UNICEF Barbie, black doll	1989	7	30
4782	UNICEF Barbie, Hispanic version	1989	7	40
1920	UNICEF Barbie, white version	1989	7	30
	University Barbie	1997	12	18
1675	Vacation Sensations Barbie, blue	1986	10	32
1675	Vacation Sensations Barbie, pink	1988	12	45
1182	Walk Lively Barbie	1972	75	200
1184	Walk Lively Ken	1972	40	175
3200	Walk Lively Miss America Barbie, brunette	1972	60	150
1183	Walk Lively Steffie	1972	75	275
7011	Wedding Fantasy Barbie, black doll	1989	7	25
2125	Wedding Fantasy Barbie, white version	1989	7	40
9607	Wedding Party Alan	1991	7	20
9608	Wedding Party Barbie	1991	7	20
9852	Wedding Party Kelly & Todd	1991	15	35
9609	Wedding Party Ken	1991	7	20
9606	Wedding Party Midge	1991	7	15
3469	Western Barbie, three hair styles	1981	8	40
2930	Western Fun Barbie, black doll	1989	5	15
9932	Western Fun Barbie, white version	1989	5	15
9934	Western Fun Ken	1989	5	15
9933	Western Fun Nia	1989	5	15
3600	Western Ken	1981	7	40
5029	Western Skipper	1982	8	30
4103	Wet n' Wild Barbie	1990	8	35
4121	Wet n' Wild Christie	1989	3	15
4104	Wet n' Wild Ken	1989	3	15
4120	Wet n' Wild Kira	1989	3	15
4138	Wet n' Wild Skipper	1989	3	15
4137	Wet n' Wild Steven	1989	3	15
4136	Wet n' Wild Teresa	1989	3	15
1009	Wig Wardrobe Midge	1965	200	500
7808	Yellowstone Kelley	1974	100	275

COLLECTORS' EDITIONS, STORE EXCLUSIVES, GIFT SETS

17341	After the Walk, Coca-Cola	1997	40	75
3712	All American Barbie & Starstepper	1991	12	35
17313	American Stories American Indian	1997	10	20
14612	American Stories Civil War Nurse Barbie	1996	10	25
12578	American Stories Colonial Barbie	1995	10	25
14715	American Stories Indian and Papoose	1996	10	25
17312	American Stories Patriot Barbie	1997	10	20
12577	American Stories Pilgrim Barbie	1995	10	25
12680	American Stories Pioneer Barbie	1995	10	25
14756	American Stories Shopkeeper Barbie	1996	10	25
5854	Ames Country Looks Barbie	1993	12	30
2452	Ames Denim 'N Lace Barbie	1992	10	30
5756	Ames Hot Looks Barbie	1992	10	30
	Ames Lady Bug Barbie	1997	5	15
2909	Ames Party in Pink	1991	10	35
18667	Angels of Music Harpist Angel	1998	45	90
17603	Anne Klein Barbie	1997	35	70
3406	Applause Barbie Holiday	1991	20	35
5313	Applause Style Barbie	1990	10	45
17690	Avon Mrs. P.F.E. Albee	1997	35	75
15202	Avon Spring Blossom, black	1996	10	20
15201	Avon Spring Blossom, white	1996	10	25

BARBIE

COLLECTORS' EDITIONS, STORE EXCLUSIVES, GIFT SETS

NO.	NAME	YEAR	MNB	MIB
15587	Avon Winter Velvet, black	1996	10	25
15571	Avon Winter Velvet, white	1996	17	35
11182	B Mine Barbie	1993	7	25
3208	Back To School	1993	15	35
9613	Ballerina on Tour Gift Set	1976	25	125
17763	Ballroom Beauties Moonlight Waltz	1997	40	70
4431	Barbie and Friends: Ken, Barbie, P.J.	1983	25	75
	Barbie and Ken Camping Out	1983	25	65
892	Barbie and Ken Tennis Gift Set	1962	450	1000
3303	Barbie Beautiful Blues Gift Set	1967	1600	3000
17450	Barbie Loves Elvis Gift Set	1997	35	75
16079	Barbie Millicent Roberts Gift Set	1996	30	65
19433	Barbie Millicent Roberts Green Thumb Fashion	1998	15	30
16076	Barbie Millicent Roberts Homecoming Fan, Goin' to the Game	1996	30	65
17567	Barbie Millicent Roberts Perfectly Suited	1987	30	70
16077	Barbie Millicent Roberts Picnic in the Park	1996	30	40
19791	Barbie Millicent Roberts Pinstripe Power Gift Set	1998	30	65
19772	Barbie Millicent Roberts Snow Chic, So Chic Fashion	1998	15	30
1013	Barbie's Round the Clock Gift Set, Bubblecut	1964	1200	3000
1011	Barbie's Sparkling Pink Gift Set, Bubblecut	1964	1000	2400
1017	Barbie's Wedding Party Gift Set	1964	1000	2400
9601	Bathtime	1991	5	15
1702	Beauty Secrets Barbie Pretty Reflections Gift Set	1979	40	100
17040	Bill Blass Barbie	1997	90	150
17641	Billions of Dreams Barbie	1997	100	300
	Billy Boy's	1987	125	250
19365	Birds of Beauty Collection The Peacock	1998	50	100
11589	Birthday Fun at McDonald's Gift Set	1994	15	35
9385	BJ's Children's Palace, black	1990	15	45
4385	BJ's Children's Palace, white	1990	20	45
15987	BJ's Rose Bride Barbie	1996	15	45
14452	Bloomingdale's Donna Karan, blonde	1995	45	110
14595	Bloomingdale's Donna Karan, blonde	1994	35	100
14452	Bloomingdale's Donna Karan, brunette	19965	35	135
15950	Bloomingdale's Ralph Lauren Barbie	1997	40	80
12152	Bloomingdale's Savvy Shopper	1994	50	150
5405	Bob Mackie Designer Gold	1990	350	800
4247	Bob Mackie Empress Bride Barbie	1992	500	1200
SA415	Bob Mackie Fantasy Goddess of Asia	1998	120	240
14056	Bob Mackie Goddess of the Sun	1995	75	180
15522	Bob Mackie Jewel Essence Amethyst Aura	1997	40	80
15519	Bob Mackie Jewel Essence Diamond Dazzle	1997	40	80
15521	Bob Mackie Jewel Essence Emerald Embers	1997	40	80
15520	Bob Mackie Jewel Essence Ruby Radiance	1997	40	80
15523	Bob Mackie Jewel Essence Sapphire Splendor	1997	40	80
17934	Bob Mackie Madame du Barbie	1997	100	250
10803	Bob Mackie Masquerade Ball Barbie	1993	250	400
14105	Bob Mackie Moon Goddess	1996	75	160
4248	Bob Mackie Neptune Fantasy Barbie	1992	500	1200
2703	Bob Mackie Platinum	1991	300	700
12046	Bob Mackie Queen of Hearts Barbie	1994	125	235
2704	Bob Mackie Starlight Splendor	1991	300	700
13966	Caroling Fun Barbie	1995	8	20
3304	Casey Goes Casual Gift Set	1967	800	1800
4385	Child's World Disney Barbie	1990	15	35
9835	Child's World Disney Barbie, black doll	1990	15	35
16900	Children's Collector Series Cinderella	1997	25	40

Twist and Turn Christie, 1969

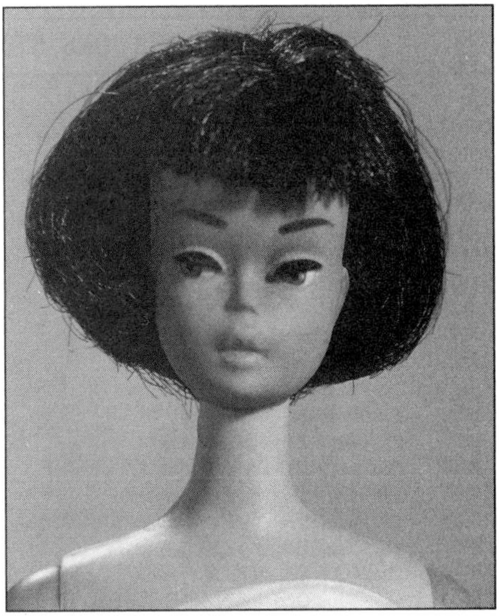

American Girl Barbie, 1965

Malibu Barbie, 1971

COLLECTORS' EDITIONS, STORE EXCLUSIVES, GIFT SETS

NO.	NAME	YEAR	MNB	MIB
14960	Children's Collector Series Little Bo Peep	1995	25	120
13016	Children's Collector Series Rapunzel	1994	25	50
	Children's Collector Series Sleeping Beauty	1998	20	40
16013	Christian Dior 50th Anniversary Barbie	1997	60	175
13168	Christian Dior Barbie	1994	60	150
14615	Chuck E. Cheese	1996	15	30
19367	City Seasons Autumn in Paris	1998	25	50
	City Seasons Spring in Tokyo	1998	25	50
	City Seasons Summer in Rome	1998	25	50
19429	City Seasons Winter in New York	1998	25	50
	Classic Ballet Sugar Plum Fairy	1997	15	35
18509	Classic Ballet Swan Lake	1998	15	35
1521	Classique Benefit Ball Barbie	1992	50	165
10149	Classique City Style Barbie	1993	30	110
11622	Classique Evening Extravaganza	1994	25	65
11638	Classique Evening Extravaganza, black doll	1994	25	65
19361	Classique Evening Sophisticate	1998	30	60
12999	Classique Midnight Gala	1995	25	70
10148	Classique Opening Night Barbie	1993	30	95
17136	Classique Romantic Interlude Barbie	1997	35	55
17137	Classique Romantic Interlude Barbie, black	1997	35	55
15461	Classique Starlight Dance	1996	25	60
15819	Classique Starlight Dance, black doll	1996	25	60
11623	Classique Uptown Chic Barbie	1994	25	75
2366	Club-Doll-Jewel Jubilee	1991	10	25
4893	Cool City Blues: Barbie, Ken, Skipper	1989	20	45
15528	Couture Portrait in Taffeta	1996	65	135
17572	Couture Serenade in Satin	1997	65	135
4917	Dance Club Barbie Gift Set	1989	25	60
5409	Dance Magic Gift Set Barbie & Ken	1990	15	35
9058	Dance Sensation Barbie Gift Set	1984	15	40
2996	Deluxe Tropical Barbie	1986	10	30
11276	Easter Fun Barbie	1994	10	20
12793	Easter Party	1995	20	45
15846	Empress Sissy Barbie	1996	60	150
14992	Enchanted Evening Barbie, blonde	1996	25	35
15407	Enchanted Evening Barbie, brunette	1996	25	35
15948	Escada Barbie	1996	60	150
	Evergreen Princess Barbie	1994	75	150
3196	Fantastica Barbie	1992	20	40
	FAO Schwarz Barbie as George Washington (American Beauty Series)	1996	30	75
13257	FAO Schwarz Circus Star Barbie	1994	35	140
7734	FAO Schwarz Golden Greetings Barbie	1989	65	265
14061	FAO Schwarz Jeweled Splendor (125th Anniversary)	1995	100	225
1539	FAO Schwarz Madison Ave. Barbie	1992	100	200
2921	FAO Schwarz Night Sensation	1991	75	175
2017	FAO Schwarz Rockettes Barbie	1993	75	200
11652	FAO Schwarz Silver Screen Barbie	1994	65	200
14684	FAO Schwarz Statue of Liberty Barbie (American Beauty Series)	1995	35	75
	FAO Schwarz Summer in San Francisco	1998	30	75
5946	FAO Schwarz Winter Fantasy	1990	100	245
17382	Fashion Luncheon	1997	25	50
863	Fashion Queen Barbie & Her Friends	1964	1000	2400
864	Fashion Queen Barbie & Ken Trousseau Gift Set	1964	1200	2800
19632	Fashion Savvy Uptown Chic #2	1998	25	50
10339	Festiva Barbie	1993	15	35
1194	Francie Rise n' Shine Gift Set	1971	600	1200

COLLECTORS' EDITIONS, STORE EXCLUSIVES, GIFT SETS

NO.	NAME	YEAR	MNB	MIB
1042	Francie Swingin' Separates Gift Set	1966	700	1500
17601	Francie Wild Bunch	1997	20	55
14808	Francie, 30th Anniversary	1996	35	65
3826	Fun to Dress Barbie Gift Set	1993	8	15
18547	Gap Barbie and Kelly Gift Set	1997	15	35
	Gap Barbie and Kelly Gift Set, black	1997	15	35
16450	Gap Barbie, black	1997	20	45
16449	Gap Barbie, white	1997	30	60
	General Mills Winter Dazzle, black	1997	5	10
18456	General Mills Winter Dazzle, white	1997	5	10
12009	Gold Jubilee Barbie	1994	500	1200
	Graduation, Class of 1996	1996	5	15
	Graduation, Class of 1997	1997	5	10
17782	Grand Ole Opry Country Rose	1997	40	80
16708	Great Eras Chinese Empress	1997	35	50
11397	Great Eras Egyptian Queen	1994	20	140
12792	Great Eras Elizabethan Queen	1994	25	65
4063	Great Eras Flapper	1993	25	200
16707	Great Eras French Lady	1997	35	70
3702	Great Eras Gibson Girl	1993	25	140
15005	Great Eras Grecian Goddess	1995	35	45
12791	Great Eras Medieval Lady	1994	25	65
11478	Great Eras Southern Belle	1994	20	145
14900	Great Eras Victorian Lady	1995	35	50
19631	Great Fashions of the 20th Century, Dance 'til Dawn	1998	30	60
18630	Great Fashions of the 20th Century, Promenade in the Park	1998	30	60
16295	Great Villains as Cruella DeVil, Power in Pinstripes	1996	35	75
	Great Villains as Cruella DeVil, Ruthless in Red	1997	35	75
14108	Hallmark Holiday Memories Barbie	1994	20	45
17094	Hallmark Holiday Traditions	1997	20	45
17094	Hallmark Holiday Traditions	1997	25	50
14880	Hallmark Valentine Barbie (Sweet Valentine Barbie)	1996	20	45
12579	Hallmark Victorian Skater Barbie (Victorian Elegance Barbie)	1994	75	150
15621	Hallmark Yuletide Romance Barbie	1996	20	45
9519	Happy Birthday Barbie Gift Set	1984	35	85
0175	Harvey Nichols Special Edition (250 made)	1995	500	1500
7470	Hawaiian Barbie	1982	25	100
1879	Hills Blue Elegance Barbie	1992	12	40
3274	Hills Evening Sparkle	1990	10	40
3549	Hills Moonlight Rose	1991	7	40
4843	Hills Party Lace Barbie	1989	15	45
12412	Hills Polly Pocket Barbie	1994	12	50
13940	Hills Sea Pearl Mermaid Barbie	1995	8	35
12192	Holiday Dreams Barbie	1994	10	20
10280	Holiday Hostess Barbie	1993	25	50
15581	Holiday Season	1996	5	15
10928	Hollywood Hair Deluxe Gift Set	1993	15	35
12701	Hollywood Legends Dorothy (Wizard of Oz)	1995	20	180
15498	Hollywood Legends Eliza Doolittle (My Fair Lady), green	1996	30	75
15500	Hollywood Legends Eliza Doolittle (My Fair Lady), lace ball gown	1996	30	75
15501	Hollywood Legends Eliza Doolittle (My Fair Lady), pink	1996	30	75
15497	Hollywood Legends Eliza Doolittle (My Fair Lady), white ascot	1996	30	75
14901	Hollywood Legends Glinda (Wizard of Oz)	1996	35	75
16573	Hollywood Legends Ken as Cowardly Lion (Wizard of Oz)	1996	35	75
15499	Hollywood Legends Ken as Henry Higgins (My Fair Lady)	1996	25	75
	Hollywood Legends Ken as Rhett (Gone with the Wind)	1993	25	65
	Hollywood Legends Ken as Scarecrow (Wizard of Oz)	1996	35	65

COLLECTORS' EDITIONS, STORE EXCLUSIVES, GIFT SETS

NO.	NAME	YEAR	MNB	MIB
14902	Hollywood Legends Ken as Tin Man (Wizard of Oz)	1996	35	65
13676	Hollywood Legends Maria (Sound of Music)	1994	25	55
17451	Hollywood Legends Marilyn Monroe, pink	1997	35	55
17452	Hollywood Legends Marilyn Monroe, red	1997	35	55
17155	Hollywood Legends Marilyn Monroe, white	1997	35	55
12045	Hollywood Legends Scarlett (Gone with the Wind), green curtain	1993	25	55
12997	Hollywood Legends Scarlett (Gone with the Wind), green/white dress	1993	25	65
13254	Hollywood Legends Scarlett (Gone with the Wind), New Orleans dress	1993	25	65
12815	Hollywood Legends Scarlett (Gone with the Wind), red dress	1993	25	65
1865	Home Shopping Club Evening Flame	1992	70	150
18667	Illusion	1998	50	100
17031	In the Limelight	1997	40	80
12043	International Gift Set	1994	30	60
	Japanese Living Eli	1970	700	1400
1596	JC Penney Barbie Pink Premier Gift Set	600	1500	
2702	JCPenney Enchanted Evening (Evening Elegance Series)	1991	40	110
7057	JCPenney Evening Elegance (Evening Elegance Series)	1990	40	80
1278	JCPenney Evening Sensation (Evening Elegance Series)	1992	12	80
10684	JCPenney Golden Winter (Evening Elegance Series)	1993	12	65
12191	JCPenney Night Dazzle (Evening Elegance Series)	1994	15	50
14010	JCPenney Royal Enchantment Barbie	1995	25	55
2366	Jewel Jubilee	1991	25	65
	Julia Simply Wow Gift Set	1976	400	1000
	K-B Toys Glamour Barbie	1997	10	20
	K-B Toys Glamour Barbie, black	1997	10	20
4870	K-Mart Peach Pretty Barbie	1989	10	30
3117	K-Mart Pretty in Purple	1992	12	25
3121	K-Mart Pretty in Purple, black doll	1992	12	25
	Keepsake Treasures Barbie and the Tale of Peter Rabbit	1998	15	35
2977	Kissing Barbie Gift Set	1978	25	65
10309	Kool-Aid Wacky Warehouse Barbie I	1992	25	65
11763	Kool-Aid Wacky Warehouse Barbie II	1994	25	65
	Kraft Treasures Barbie	1992	30	75
	L.E. Festival Holiday Barbie (540 made)	1994	700	1200
10123	Little Debbie Barbie	1993	25	45
1585	Living Barbie Action Accents Gift Set	1970	500	1000
1117	Living Skipper	1970	40	250
7583	Loving You Barbie Gift Set	1983	45	100
1703	Malibu Barbie "The Beach Party", with case	1979	17	35
1248	Malibu Ken Surf's Up Gift Set	1971	75	200
	Mattel Festival 35th Anniversary (3,500 made)	1994	250	500
11591	Mattel Festival 35th Anniversary Gift Set (975 made)	1994	400	800
	Mattel Festival Banquet, blonde	1994	175	325
	Mattel Festival Banquet, brunette	1994	125	250
	Mattel Festival Banquet, redhead	1994	225	475
11160	Mattel Festival Doctor, brunette (1,500 made)	1994	75	175
11921	Mattel Festival Gymnast (1,500 made)	1994	75	175
12155	Mattel Festival Happy Holiday	1994	270	540
	Mattel Festival Haute Couture, rainbow (500 made)	1994	200	400
	Mattel Festival Haute Couture, red velvet (480 made)	1994	200	400
12191	Mattel Festival Night Dazzle, brunette (420 made)	1994	150	300
12905	Mattel Festival Snow Princess, brunette (285 made)	1994	600	1200
10051	Meijers Shopping Fun	1993	10	20
863	Meijers Something Extra	1992	10	20
4983	Mervyns Ballerina Barbie	1983	30	75
7093	Mervyns Fabulous Fur	1986	25	65

COLLECTORS' EDITIONS, STORE EXCLUSIVES, GIFT SETS

NO.	NAME	YEAR	MNB	MIB
1012	Midge's Ensemble Gift Set	1964	1200	3200
17780	Midnight Princess	1997	30	60
857	Mix n' Match Gift Set	1962	800	1800
3210	Montgomery Wards (mail order box)	1972	30	700
3210	Montgomery Wards Barbie (pink box)	1972	350	800
2483	My First Barbie Gift Set	1991	8	20
5386	My First Barbie Gift Set, pink tutu	1987	18	40
1979	My First Barbie Gift Set, pink tutu	1986	15	35
1875	My First Barbie, pink tutu, Zayre's Hispanic	1987	8	45
20442	NASCAR Barbie, Pro Sports Series 50th Anniversary	1998	20	40
	National Convention, Barbie and the Bandstand	1996	225	475
	National Convention, Barbie Around the World Festival	1985	125	250
	National Convention, Barbie Convention 1980	1980	125	250
	National Convention, Barbie Forever Young	1989	125	250
	National Convention, Barbie Loves a Fairytale	1991	150	300
	National Convention, Barbie Loves New York	1984	125	250
	National Convention, Barbie Ole	1995	225	475
	National Convention, Barbie Wedding Dreams	1992	50	100
	National Convention, Barbie's Pow Wow	1983	125	250
	National Convention, Beach Blanket Barbie	1997	225	475
	National Convention, Christmas With Barbie	1987	125	250
	National Convention, Come Rain or Shine	1988	125	250
	National Convention, Deep in the Heart of Texas	1990	125	250
	National Convention, Michigan Entertains Barbie	1982	125	250
	National Convention, The Magic of Barbie	1994	175	375
	National Convention, You've Come a Long Way	1993	225	475
	National Convention, Barbie's Reunion	1986	125	250
16289	Nicole Miller	1996	30	65
11591	Nostalgic 35th Anniversary Gift Set	1994	75	150
11590	Nostalgic 35th Anniversary, Blonde	1994	25	50
11782	Nostalgic 35th Anniversary, Brunette	1994	50	80
	Nostalgic Wedding Day Party	1997	25	40
5472	Nutcracker Barbie	1992	85	250
16442	Ocean Friends Gift Set	1996	20	45
16443	Olympic Barbie Gift Set	1996	15	30
1014	On Parade Gift Set, Barbie, Ken, Midge	1964	800	1700
3803	Osco Picnic Pretty	1993	20	40
14024	Oshagatsu Barbie	1995	40	80
1588	P.J.'s Swinging Silver Gift Set	1970	700	1500
10926	Paint 'N Dazzle Deluxe Gift Set	1993	17	35
2001	Party Premier	1992	15	30
9025	Party Sensation	1990	30	70
7009	Peach Blossom Barbie	1992	15	30
13598	Peppermint Princess Barbie	1995	35	85
5239	Pink & Pretty Barbie Gift Set	1982	35	90
3756	Pink Jubilee Barbie, only 1200 made	1989	800	2200
16091	Pink Splendor	1996	200	600
15280	Poodle Parade	1996	25	55
2598	Pretty Changes Barbie Gift Set	1978	35	85
2901	Pretty Hearts Barbie	1992	7	15
1135	Pretty Pairs Angie N' Tangie	1970	125	250
1133	Pretty Pairs Lori N' Rori	1970	125	250
1134	Pretty Pairs Nan N' Fran	1970	125	250
25192	Radio Shack Earring Magic	1991	20	65
3161	Red Romance Barbie	1993	7	15
1858	Royal Romance	1992	20	45
	Russell Stover Candie Barbie	1997	5	15
14617	Russell Stover Easter (with Easter basket)	1996	10	25

Quick Curl Barbie, 1973

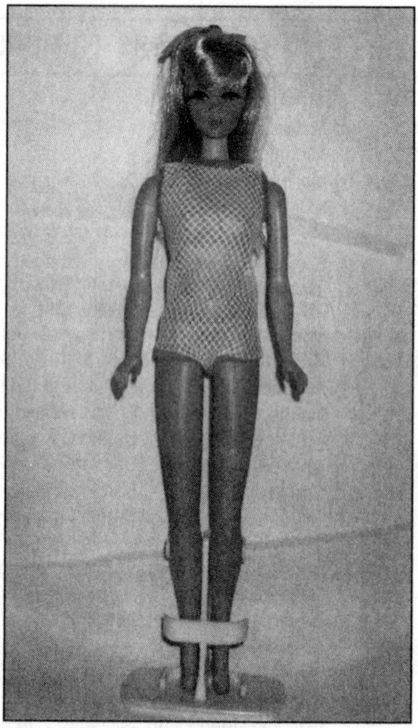

Twist and Turn Barbie, 1967

Twiggy, 1967

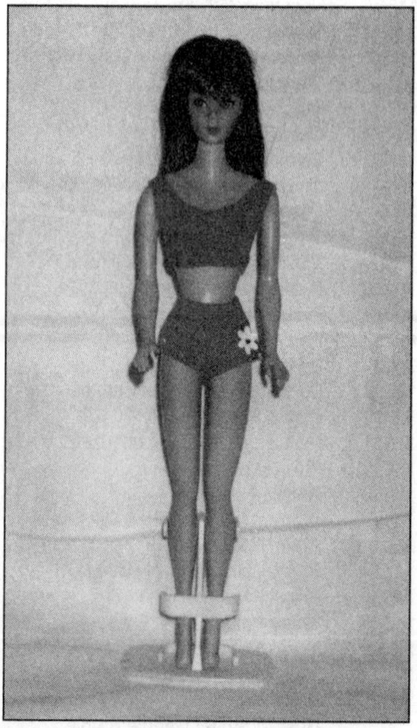

Standard Barbie, brunette, 1967

COLLECTORS' EDITIONS, STORE EXCLUSIVES, GIFT SETS

NO.	NAME	YEAR	MNB	MIB
14956	Russell Stover Easter (with software package)	1996	10	25
	Russell Stover Little Bo Peep	1997	10	20
	Sam's 1950s Fun	1996	15	30
17693	Sam's Sixties Fun Barbie, redhead	1997	10	20
17252	Sam's Sixties Fun, blonde	1997	10	20
	Sam's Sweet Moments Barbie	1997	25	50
13741	Schooltime Fun	1995	15	30
2998	Sears 100th Celebration Barbie	1986	20	85
1866	Sears Barbie Twinkle Town Set	1969	800	1600
3817	Sears Blossom Beautiful Barbie	1992	100	250
2306	Sears Dream Princess	1992	25	50
10292	Sears Enchanted Princess	1994	35	75
3596	Sears Evening Enchantment Barbie	1989	10	35
5588	Sears Lavender Surprise	1990	8	30
9049	Sears Lavender Surprise, black doll	1990	8	25
7669	Sears Lilac and Lovely Barbie	1988	10	50
1193	Sears Perfectly Plaid Gift Set	1971	800	1500
13011	Sears Ribbons and Roses Barbie	1995	10	50
12410	Sears Silver Sweetheart Barbie	1994	17	70
2586	Sears Southern Belle	1991	10	50
4550	Sears Star Dream Barbie	1987	10	70
9042	Sears Winter Sports	1975	65	115
12384	Season's Greetings Barbie	1994	50	85
10929	Secret Hearts Gift Set	1993	17	35
17572	Serenade in Satin	1997	65	135
1364	Service Merchandise Blue Rhapsody	1991	125	225
12005	Service Merchandise City Sophisticate	1994	20	75
13612	Service Merchandise Ruby Romance Barbie	1995	25	65
1886	Service Merchandise Satin Nights, two earring versions	1992	20	80
10994	Service Merchandise Sparkling Splendor	1993	17	80
10143	Sharin Sisters Gift Set	1993	12	25
5716	Sharin Sisters Gift Set	1992	12	25
3142	Shopko/Venture Blossom Beauty	1991	10	60
1876	Shopko/Venture Party Perfect	1992	12	45
15951	Silver Royale	1996	25	65
1021	Skipper Party Time Gift Set	1964	100	650
1172	Skipper Swing 'a' Rounder Gym Gift Set	1972	100	400
1036	Skooter Cut n' Button Gift Set (Sears)	1967	150	650
2262	Snap 'N Play Gift Set (Snap 'N Play Deluxe)	1992	10	30
	Snow Princess Barbie	1994	75	150
15762	Soda Fountain Sweetheart, Coca-Cola	1996	25	150
4116	Spiegel Regal Reflections Barbie	1992	75	325
10969	Spiegel Royal Invitation Barbie	1993	35	110
14009	Spiegel Shopping Chic Barbie	1995	35	100
3347	Spiegel Sterling Wishes	1991	45	150
12077	Spiegel Theatre Elegance	1994	150	200
17441	Spiegel Winners Circle	1997	35	65
	Splash 'N Color Barbie Gift Set	1997	10	20
3477	Spring Bouquet Barbie	1993	10	20
7008	Spring Parade Barbie	1992	15	25
2257	Spring Parade Barbie, black doll	1992	15	25
15006	Star Trek Barbie and Ken	1996	35	50
14070	Starlight Waltz Barbie	1995	35	85
19366	Sunflower Barbie, Artist Series, Van Gogh	1998	40	80
1648	Swan Lake Barbie	1991	65	200
3208	Sweet Spring Barbie	1992	10	20
17860	Tangerine Twist	1997	30	60
4583	Target Baseball Date Barbie	1993	10	25

COLLECTORS' EDITIONS, STORE EXCLUSIVES, GIFT SETS

NO.	NAME	YEAR	MNB	MIB
2954	Target Cute'n Cool	1991	8	40
3203	Target Dazzlin' Date Barbie	1992	10	35
7476	Target Gold and Lace Barbie	1989	10	40
2587	Target Golden Evening	1991	6	45
10202	Target Golf Date Barbie	1993	10	20
5955	Target Party Pretty Barbie	1990	6	35
5413	Target Pretty in Plaid Barbie	1992	15	35
14110	Target Steppin' Out Barbie	1995	8	15
7801	Tennis Star Barbie & Ken	1988	18	40
1199	Tiff Pose N' Play	1972	175	375
19633	Timeless Sentiments Angel of Joy	1998	25	50
19364	Together Forever Collection Romeo & Juliet	1998	50	100
	Toys R Us 35th Anniversary Midge	1998	25	50
12149	Toys R Us Astronaut Barbie (Career Collection)	1994	15	40
12150	Toys R Us Astronaut Barbie, black doll	1994	15	35
10217	Toys R Us Back to School Barbie	1993	10	20
3722	Toys R Us Barbie for President	1992	17	65
3940	Toys R Us Barbie for President, black doll	1992	17	35
7970	Toys R Us Bath Time Skipper	1992	12	25
9342	Toys R Us Beauty Pagent Skipper	1991	10	25
1490	Toys R Us Cool 'N Sassy	1992	10	20
4110	Toys R Us Cool 'N Sassy, black doll	1992	10	20
	Toys R Us Crystal Splendor	1996	12	25
10712	Toys R Us Dream Wedding Gift Set	1993	20	65
10713	Toys R Us Dream Wedding Gift Set, black doll	1993	22	65
12322	Toys R Us Emerald Elegance (Society Style Collection)	1995	15	65
12323	Toys R Us Emerald Elegance, ethnic doll	1994	15	30
17443	Toys R Us Emerald Enchantment Barbie	1997	25	50
2721	Toys R Us Fun School	1991	6	40
17692	Toys R Us Harley-Davidson Barbie	1997	30	60
	Toys R Us Harley-Davidson Barbie, 2nd version	1998	30	60
10507	Toys R Us Love to Read Barbie	1993	12	35
4581	Toys R Us Malt Shop Barbie	1993	10	20
10608	Toys R Us Moonlight Magic Barbie	1993	15	70
10609	Toys R Us Moonlight Magic Barbie, black doll	1993	15	65
13233	Toys R Us POG Barbie	1995	7	15
10688	Toys R Us Police Officer Barbie	1993	15	70
10689	Toys R Us Police Officer Barbie, black doll	1993	15	65
13555	Toys R Us Purple Passion	1995	10	30
13554	Toys R Us Purple Passion, ethnic doll	1995	10	25
1276	Toys R Us Radiant in Red Barbie	1992	12	80
4113	Toys R Us Radiant in Red Barbie, black doll	1992	12	65
13256	Toys R Us Sapphire Dream Barbie (Society Style Collection)	1995	50	75
	Toys R Us Sapphire Sophisticate	1997	12	25
2721	Toys R Us School Fun Barbie	1992	10	20
4411	Toys R Us School Fun Barbie, black doll	1992	10	20
10682	Toys R Us School Spirit Barbie	1993	10	45
10683	Toys R Us School Spirit Barbie, black doll	1993	10	30
7799	Toys R Us Show n' Ride Barbie	1989	10	37
10491	Toys R Us Spots 'N Dots Barbie	1993	12	25
10885	Toys R Us Spots 'N Dots Teresa	1993	12	40
2917	Toys R Us Sweet Romance	1991	8	30
1433	Toys R Us Totally Hair Courtney	1992	10	20
1430	Toys R Us Totally Hair Skipper	1992	10	20
7735	Toys R Us Totally Hair Whitney	1992	20	40
5949	Toys R Us Winter Fun Barbie	1990	10	40
2783	Trail Blazin'	1991	10	25
2996	Tropical Barbie Deluxe Gift Set	1985	20	45

COLLECTORS' EDITIONS, STORE EXCLUSIVES, GIFT SETS

NO.	NAME	YEAR	MNB	MIB
3556	Tutti and Todd Sundae Treat Set	1966	150	350
3554	Tutti Me n' My Dog	1966	150	350
3553	Tutti Nighty Night Sleep Tight	1965	100	300
4097	Twirley Curls Barbie Gift Set	1982	30	85
18941	Twist and Turn Smasheroo	1998	25	50
12675	Valentine Barbie	1994	7	20
774505	Valerie	1978	125	225
19788	Vera Wang Bridal Barbie	1998	80	158
1859	Very Violet Barbie	1992	10	20
4589	Wal-Mart 25th Year Pink Jubilee Barbie	1987	20	50
2282	Wal-Mart Anniversary Star Barbie (30th Anniversary)	1992	15	35
3678	Wal-Mart Ballroom Beauty	1991	8	40
13014	Wal-Mart Country Bride	1995	8	15
13015	Wal-Mart Country Bride, black doll	1995	8	15
13016	Wal-Mart Country Bride, Hispanic	1995	8	15
	Wal-Mart Country Star Barbie, white, Hispanic, or black	1994	7	20
7335	Wal-Mart Dream Fantasy	1990	8	45
1374	Wal-Mart Frills and Fantasy Barbie	1988	7	50
3963	Wal-Mart Lavender Look Barbie	1989	7	20
	Wal-Mart Skating Dreams	1997	5	15
15510	Wal-Mart Skating Star Barbie	1996	10	20
10592	Wal-Mart Superstar Barbie	1993	15	35
11645	Wal-Mart Tooth Fairy Barbie	1994	7	15
	Wal-Mart Tooth Fairy, 3rd edition	1997	5	15
1247	Walking Jamie Strollin' in Style Gift Set	1972	300	600
17783	Water Lily (Artist Series, Monet)	1997	40	75
17119	Wedding Day Barbie, blonde	1997	15	35
17120	Wedding Day Barbie, redhead	1997	15	35
10924	Wedding Fantasy Gift Set	1993	50	125
9852	Wedding Party Gift Set, Six Dolls	1991	45	125
15186	Wessco Carnival Cruise Barbie	1997	25	50
13912	Wessco International Traveler #1	1995	35	75
16158	Wessco International Traveler #2	1996	25	50
5408	Western Fun Gift Set Barbie & Ken	1989	12	30
5408	Western Fun Gift Set Barbie & Ken	1990	12	30
7637	Winn Dixie Party Pink Barbie	1989	7	30
5410	Winn Dixie Pink Sensation	1990	6	25
3284	Winn Dixie Southern Beauty	1991	10	20
	Winter Fantasy Assortment	1997	10	20
	Winter Fantasy, black	1997	10	20
15334	Winter Fantasy, blonde	1996	20	45
15530	Winter Fantasy, brunette	1996	30	65
10655	Winter Princess Barbie	1993	100	400
10658	Winter Royale	1993	20	40
13513	Winter's Eve Barbie	1995	15	35
2583	Woolworth's Special Expressions, black doll, blue dress	1991	4	10
5505	Woolworth's Special Expressions, black doll, pink dress	1990	4	10
3200	Woolworth's Special Expressions, black doll, salmon dress	1992	5	12
7326	Woolworth's Special Expressions, black doll, white dress	1989	5	12
2582	Woolworth's Special Expressions, blue dress	1991	4	15
10048	Woolworth's Special Expressions, blue dress, pastel print	1993	5	15
3197	Woolworth's Special Expressions, peach dress	1992	5	20
5504	Woolworth's Special Expressions, pink dress	1990	3	20
4842	Woolworth's Special Expressions, white version	1989	5	20
	Workin' Out Barbie Gift Set	1997	10	25

PORCELAIN BARBIES

1110	30th Anniversary Ken	1991	75	150

PORCELAIN BARBIES

NO.	NAME	YEAR	MNB	MIB
7957	30th Anniversary Midge	1993	65	140
11396	30th Anniversary Skipper	1994	50	110
5475	Benefit Performance Barbie	1988	200	475
1708	Blue Rhapsody Barbie	1986	300	600
16962	Blushing Orchid Bride Barbie	1997	75	175
1553	Crystal Rhapsody, blonde	1992	100	200
10201	Crystal Rhapsody, brunette	1993	100	200
3415	Enchanted Evening Barbie	1987	175	400
9973	Gay Parisienne Blonde (Disney)	1991	225	400
9973	Gay Parisienne Brunette	1991	150	225
9973	Gay Parisienne Redhead (Disney)	1991	225	650
10246	Gold Sensation	1993	175	350
18326	Holiday Ball	1997	75	175
15760	Holiday Caroler	1996	75	175
14311	Holiday Jewel	1996	.55	175
14479	Mattel's 50th Anniversary Barbie	1995	300	600
5351	Plantation Belle	1992	100	200
7526	Plantation Belle Barbie (red and blonde Disney)	1992	100	500
14541	Romantic Rose Bride	1995	75	175
10950	Royal Splendor	1993	100	200
1249	Silken Flame Barbie (brunette)	1993	100	200
11099	Silken Flame, blonde	1993	250	500
11875	Silver Starlight	1994	175	350
7613	Solo in the Spotlight	1990	100	225
5313	Sophisticated Lady	1990	125	225
10050	Special Expressions, pastel print	1993	5	15
10049	Special Expressions, pastel print	1993	6	18
3198	Special Expressions, salmon dress	1992	5	15
12953	Star Lily Bride Barbie (Wedding Flower Series)	1994	125	175
2641	Wedding Day Barbie	1989	300	600

Barbie in Switzerland #0822

Little Theatre, 1964, Mattel

BARBIE VINTAGE FASHIONS 1959-66

NO.	OUTFIT	MNP	MIP
1631	Aboard Ship	150	450
934	After Five	55	175
984	American Airlines Stewardess	70	200
917	Apple Print Sheath	40	150
0874	Arabian Nights	125	350
989	Ballerina	50	200
953	Barbie Baby-Sits, 1st version	80	250
1605	Barbie in Hawaii	75	180
0823	Barbie in Holland	85	200
0821	Barbie in Japan	225	500
0820	Barbie in Mexico	85	225
0822	Barbie in Switzerland	75	225
1634	Barbie Learns to Cook	110	500
1608	Barbie Skin Diver	40	100
962	Barbie-Q Outfit	70	180
1651	Beau Time	65	350
1698	Beautiful Bride	650	2100
1667	Benefit Performance	400	1300
1609	Black Magic	100	375
947	Bride's Dream	90	300
1628	Brunch Time	100	350
981	Busy Gal	140	450
956	Busy Morning	100	250
1616	Campus Sweetheart	350	1500
0889	Candy Striper Volunteer	125	350
954	Career Girl	70	350
1687	Caribbean Cruise	80	200
0876	Cheerleader	80	200
0872	Cinderella	140	475
1672	Club Meeting	140	400
1670	Coffee's On	80	175
916	Commuter Set	450	1500
1627	Country Club Dance	125	425
1603	Country Fair	40	140
1604	Crisp'n Cool	80	150
918	Cruise Stripes	55	125
1626	Dancing Doll	110	250
1666	Debutante Ball	350	1100
946	Dinner At Eight	55	165
1633	Disc Date	150	375
1613	Dog n' Duds	110	325
1669	Dreamland	70	150
0875	Drum Majorette	80	175
971	Easter Parade	1500	4000
983	Enchanted Evening	125	500
1695	Evening Enchantment	300	500
1660	Evening Gala	130	350
961	Evening Splendor	110	275
1676	Fabulous Fashion	200	500
943	Fancy Free	30	80
1635	Fashion Editor	350	700
1656	Fashion Luncheon	350	1300
1691	Fashion Shiner	90	220
1696	Floating Gardens	220	500

BARBIE VINTAGE FASHIONS 1959-66

NO.	OUTFIT	MNP	MIP
921	Floral Petticoat	40	125
1697	Formal Occasion	190	500
1638	Fraternity Dance	195	500
979	Friday Night Date	75	200
1624	Fun At The Fair	140	300
1619	Fun n' Games	140	300
931	Garden Party	50	175
1606	Garden Tea Party	70	225
1658	Garden Wedding	150	250
964	Gay Parisienne	1500	3800
1647	Gold n' Glamour	650	1500
992	Golden Elegance	125	325
1610	Golden Evening	105	300
911	Golden Girl	70	175
1645	Golden Glory	165	425
945	Graduation	40	65
0873	Guinevere	140	275
1665	Here Comes The Bride	450	1000
1639	Holiday Dance	165	550
942	Icebreaker	55	100
1653	International Fair	125	400
1632	Invitation To Tea	190	500
0819	It's Cold Outside, brown	50	110
0819	It's Cold Outside, red	65	165
1620	Junior Designer	140	300
1614	Junior Prom	190	600
1621	Knit Hit	80	165
1602	Knit Separates	55	140
957	Knitting Pretty, blue	80	190
957	Knitting Pretty, pink	140	350
978	Let's Dance	55	200
0880	Little Red Riding Hood & The Wolf	220	500
1661	London Tour	150	350
1600	Lunch Date	40	150
1649	Lunch On The Terrace	80	300
1673	Lunchtime	90	275
1646	Magnificence	350	650
944	Masquerade	60	200
1640	Matinee Fashion	350	525
1617	Midnight Blue	250	800
1641	Miss Astronaut	330	700
1625	Modern Art	190	650
940	Mood For Music	80	200
933	Movie Date	125	100
1633	Music Center Matinee	250	600
965	Nightly Negligee	55	125
1644	On The Avenue	330	500
985	Open Road	140	400
987	Orange Blossom	55	165
1650	Outdoor Art Show	190	500
1637	Outdoor Life	110	275
1601	Pajama Party	20	80
1678	Pan American Stewardess	1000	4500

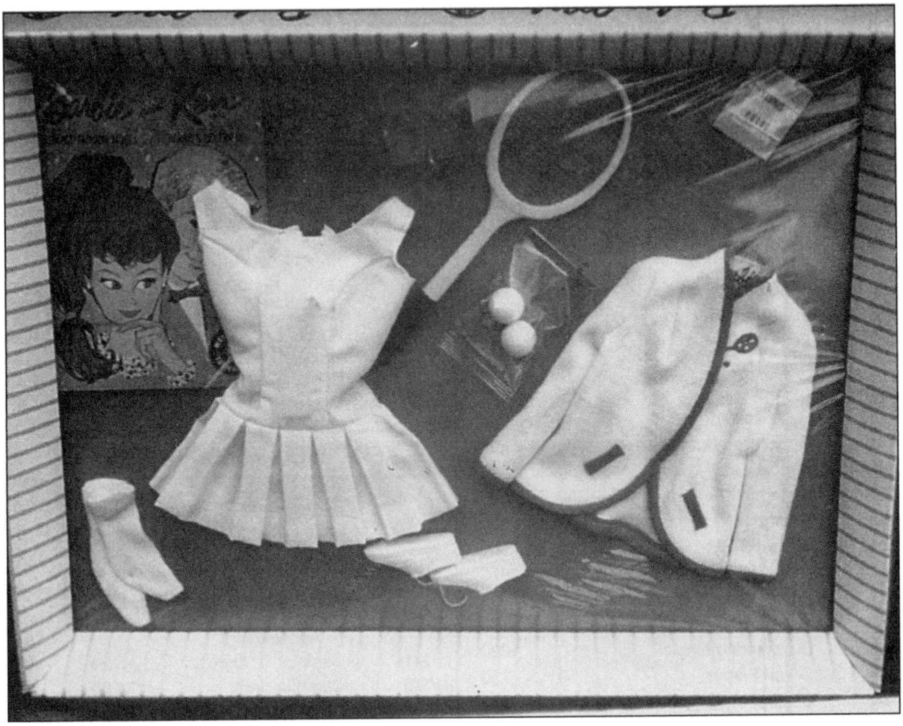

Tennis Anyone, #941

Barbie's Austin Healy Convertible, 1962, Irwin

BARBIE

BARBIE VINTAGE FASHIONS 1959-66

NO.	OUTFIT	MNP	MIP
958	Party Date	80	300
1692	Patio Party	140	325
915	Peachy Fleecy	70	150
1648	Photo Fashion	140	425
967	Picnic Set	125	275
1694	Pink Moonbeams	110	300
966	Plantation Belle	200	650
1643	Poodle Parade	250	800
1652	Pretty As A Picture	110	375
1686	Print Aplenty	140	275
949	Rain Coat	50	80
1654	Reception Line	190	525
939	Red Flare	40	150
991	Registered Nurse	85	250
963	Resort Set	50	150
1668	Riding In The Park	350	525
968	Roman Holiday	1500	4800
1611	Satin n' Rose	90	350
1615	Saturday Matinee	400	875
951	Senior Prom	65	225
986	Sheath Sensation	45	150
1664	Shimmering Magic	550	1800
977	Silken Flame	50	150
988	Singing In The Shower	50	140
1629	Skater's Waltz	140	365
948	Ski Queen	80	200
1636	Sleeping Pretty	100	325
1674	Sleepytime Gal	105	220
1642	Slumber Party	105	350
982	Solo In The Spotlight	125	325
993	Sophisticated Lady	165	400
937	Sorority Meeting	55	225
1671	Sporting Casuals	70	250
0949	Stormy Weather	50	140
1622	Student Teacher	165	400
1690	Studio Tour	110	250
969	Suburban Shopper	85	250
1675	Sunday Visit	275	500
1683	Sunflower	140	245
976	Sweater Girl	65	175
973	Sweet Dreams, pink	165	400
973	Sweet Dreams, yellow	55	150
955	Swingin' Easy	55	250
941	Tennis Anyone	55	120
959	Theatre Date	70	225
1612	Theatre Date	50	225
1688	Travel Togethers	110	240
1655	Under Fashions	200	600
919	Undergarments	40	80
1685	Underprints	55	225
1623	Vacation Time	80	200
972	Wedding Day Set	140	400
1607	White Magic	100	300
975	Winter Holiday	55	250

FRANCIE FASHIONS 1966

NO.	OUTFIT	MNP	MIP
1259	Checkmates	80	150
1258	Clam Diggers	80	175
1256	Concert In The Park	55	180
1257	Dance Party	55	310
1260	First Formal	80	265
1252	First Things First	55	110
1254	Fresh As A Daisy	55	125
1250	Gad-About	55	185
1251	It's A Date	55	150
1255	Polka Dots N' Raindrops	55	125
1261	Shoppin' Spree	80	150
1253	Tuckered Out	55	125

KEN VINTAGE FASHIONS 1961-66

0779	American Airlines Captain, two versions	110	350
797	Army and Air Force	75	150
1425	Best Man	800	1200
1424	Business Appointment	600	1300
1410	Campus Corduroys	20	65
770	Campus Hero	30	100
0782	Casuals, striped shirt	55	90
782	Casuals, yellow shirt	30	70
1416	College Student	125	500
1400	Country Clubbin'	55	110
793	Dr. Ken	60	150
785	Dreamboat	30	100
0775	Drum Major	75	200
1407	Fountain Boy	95	180
1408	Fraternity Meeting	30	100
791	Fun On Ice	50	110
1403	Going Bowling	30	90
1409	Going Huntin'	50	120
795	Graduation	20	50
1426	Here Comes The Groom	800	1600
1412	Hiking Holiday	100	210
1414	Holiday	125	210
780	In Training	20	60
1420	Jazz Concert	80	325
1423	Ken A Go Go	300	700
0774	Ken Arabian Nights	75	175
1404	Ken In Hawaii	70	125
0777	Ken In Holland	75	150
0778	Ken In Mexico	75	200
0776	Ken In Switzerland	75	200
1406	Ken Skin Diver	30	60
0773	King Arthur	150	280
794	Masquerade (Ken)	60	125
1427	Mountain Hike	140	275
1415	Mr. Astronaut	300	700
1413	Off To Bed	110	200
792	Play Ball	50	130
788	Rally Day	35	85

Barbie Goes Travelin' Case, 1965, SPP

Barbie & Francie Case, 1967, SPP

KEN VINTAGE FASHIONS 1961-66

NO.	OUTFIT	MNP	MIP
1405	Roller Skate Date, with hat	40	150
1405	Roller Skate Date, with slacks	40	150
1417	Rovin' Reporter	125	325
796	Sailor	60	125
786	Saturday Date	30	125
1421	Seein' The Sights	100	400
798	Ski Champion	60	125
0781	Sleeper Set, blue	55	100
781	Sleeper Set, brown	30	65
1401	Special Date	75	180
783	Sport Shorts	20	60
1422	Summer Job	150	575
784	Terry Togs	30	75
0772	The Prince	150	400
789	The Yachtsman, no hat	55	110
0789	The Yachtsman, with hat	100	500
790	Time For Tennis	30	100
1418	Time To Turn In	80	175
799	Touchdown	40	110
787	Tuxedo	65	150
1419	TV's Good Tonight	80	200
1411	Victory Dance	40	150

RICKY FASHIONS 1965-66

1506	Let's Explore	55	75
1501	Lights Out	40	65
1504	Little Leaguer	40	80
1502	Saturday Show	40	65
1505	Skateboard Set	55	80
1503	Sunday Suit	40	75

SKIPPER VINTAGE FASHIONS 1964-66

1905	Ballet Class	30	90
1923	Can You Play?	40	130
1926	Chill Chasers	40	75

SKIPPER VINTAGE FASHIONS 1964-66

NO.	OUTFIT	MNP	MIP
1912	Cookie Time	55	125
1933	Country Picnic	165	400
1911	Day At The Fair	90	150
1929	Dog Show	90	275
1909	Dreamtime	30	100
1906	Dress Coat	30	90
1904	Flower Girl	30	125
1920	Fun Time	65	100
1919	Happy Birthday	95	425
1934	Junior Bridesmaid	165	425
1917	Land & Sea	50	110
1935	Learning To Ride	140	225
1932	Let's Play House	90	275
1930	Loungin' Lovelies	35	125
1903	Masquerade (Skipper)	40	115
1913	Me N' My Doll	90	200
1915	Outdoor Casuals	55	90
1914	Platter Party	40	100
1916	Rain Or Shine	30	80
1928	Rainy Day Checkers	55	275
1901	Red Sensation	30	80
1907	School Days	40	225
1921	School Girl	80	275
1918	Ship Ahoy	60	150
1902	Silk N' Fancy	30	100
1908	Skating Fun	40	110
1936	Sledding Fun	75	210
1910	Sunny Pastels	40	100
1924	Tea Party	75	210
1922	Town Togs	70	120
1900	Under-Pretties	30	55
1925	What's New At The Zoo?	55	85

TUTTI FASHIONS 1966

3601	Puddle Jumpers	30	55
3603	Sand Castles	30	80
3602	Ship Shape	30	80
3604	Skippin' Rope	30	80

Barbie Record, 1965, Columbia

Barbie Fashion Stage

BARBIE & FRIENDS ACCESSORIES

NAME	COMPANY	YEAR	EX	MINT
Colorforms Sets				
Barbie Lace & Dress Dancing Doll	Colorforms	1975	4	8
Barbie Sport Fashion	Colorforms	1975	4	8
Barbie's 3-D Fashion Theatre	Colorforms	1970	9	18
Dress-Up Kit New Living Barbie	Colorforms	1970	10	35
Malibu Barbie	Colorforms	1972	7	15
Miscellaneous				
Barbie & Ken Wipe Away Cloths		1964	85	250
Barbie and Ken Hangers	SPP	1960s	40	45
Barbie Beauty Kit	Roclar	1961	145	250
Barbie Bubble Bath	Roclar	1961	60	90
Barbie Carry-All Wallet	SPP	1963	145	255
Barbie Disco Record Player		1976	85	160
Barbie Dresser Accessories		1962	175	300
Barbie Ge-Tar	Mattel	1963	200	700
Barbie Hair Fair	Mattel	1966	75	145
Barbie Nurse Kit	Pressman	1962	100	325
Barbie Powder Mitt		1961	75	140
Barbie Pretty Up Time		1960s	50	100
Barbie Record	Columbia Records	1965	70	120
Barbie Store Display	Mattel	1980	10	30
Barbie Wig Wardrobe	Mattel	1960s	75	150
Barbie's Dog Snowball	Arco/Mattel	1990s	20	35
Barbie, Ken, Midge Pencil Case	SPP	1964	100	260
Jack and Jill magazine advertisements		1960s	15	35
Jigsaw Puzzle	Whitman	1963-65	30	60
Jumbo Trading Cards	Dynamic	1962	200	345
Record Player	Emenee	1961	700	950
Record Tote	Ponytail	1961	75	180
Vanity Fair Transistor Radio	Vanity Fair	1963	900	1200
Wine Set		1986	40	75
Paper Dolls				
Barbie and Ken Cut-Outs	Whitman	1962	40	60
Barbie and Skipper Stand-Up Dolls	Whitman	1964	45	90
Barbie Costume Dolls	Whitman	1964	65	120
Barbie Doll Cut-Outs	Whitman	1962	60	120
Barbie Has a New Look Paper Dolls	Whitman	1967	40	70
Barbie, Ken and Midge Paper Dolls	Whitman	1963	50	90
Francie and Casey Paper Dolls	Whitman	1967	40	70
Malibu Barbie Paper Dolls	Whitman	1972	25	30
Meet Francie Paper Dolls	Whitman	1966	50	90
Midge Cut-Outs	Whitman	1963	75	90
Skipper Paper Dolls	Whitman	1965	65	120
Skooter Paper Dolls	Whitman	1965	65	120
Structures				
Barbie Fashion Stage	Mattel	1971	60	125
Barbie Goes to College	Mattel	1964	200	650
Barbie's 2nd Dream House	Mattel	1964	100	250
Barbie's Dream House	Mattel	1962	30	300
Barbie's Dream Kitchen	Mattel	1965	300	500
Fashion Shop	Mattel	1962	100	600
Francie House		1965	90	140
Jamie's Penthouse (Sears)		1971	220	475
Little Theatre	Mattel	1964	100	550
Quick Curl Boutique		1973	40	80
Skipper Jeweled Bed		1965	125	250
Skipper's Dream Room	Mattel	1965	200	600
Skipper's Schoolroom	Mattel	1965	200	650
Timepieces				
Barbie Personal Photo Clock	Bradley/Elgin	1964	520	675
Barbie Starbright Boudoir Clock	Bradley/Elgin	1964	640	775

Barbie and Ken Wipe Away Cloths, 1963

Skipper Jeweled Bed

BARBIE & FRIENDS ACCESSORIES

NAME	COMPANY	YEAR	EX	MINT
Brokn' Heart Pendant	Bradley/Elgin	1965	425	575
Brokn' Heart Wristwatch	Bradley/Elgin	1964	325	425
Curly Bangs Pendant	Bradley/Elgin	1963-64	175	350
Curly Bangs Wristwatch	Bradley/Elgin	1963-64	300	450
Midge Wristwatch	Bradley/Elgin	1964	375	525
Skipper Wristwatch	Bradley/Elgin	1964	380	515
Swirl Ponytail Wristwatch	Bradley/Elgin	1964	250	325

Vehicles

NAME	COMPANY	YEAR	EX	MINT
Allan's Mercedes Roadster	Irwin	1964	75	575
Barbie's Airplane	Irwin	1964	400	3500
Barbie's Austin Healy Convertible	Irwin	1962	100	450
Barbie's Austin Healy, lavender	Irwin	1964	800	1000
Barbie's Speedboat	Irwin	1964	100	1800
Barbie's Sport Plane	Irwin	1964	2000	3800
Ken's Hot Rod	Irwin	1963	100	950
Skipper's Speedboat	Irwin	1965	1450	1850

Vinyl Cases

NAME	COMPANY	YEAR	EX	MINT
Barbie & Francie Case	SPP	1967	35	50
Barbie & Stacey Sleep & Keep Case		1969	75	100
Barbie Double Case	SPP	1963	12	25
Barbie Double Case	Ponytail	1961	15	30
Barbie Goes Travelin' Case	SPP	1965	100	325
Barbie Single Case	SPP	1963	7	15
Barbie Single Case	Ponytail	1961	10	20
Barbie Train Case	SPP	1961	45	70
Fashion Queen Case	SPP	1963	75	150
Hatbox-Style Cases	Mattel	1961	20	50
Ken Cases (U.S. versions)	Ponytail	1961	12	25
Midge Cases (U.S. versions)		1963	20	45
Miss Barbie Case	SPP	1964	75	150
Skipper Cases (U.S. versions)		1964	12	25

Barbie & Stacey Sleep 'N Keep Case, Sears Exclusive, top; Jamie's Penthouse, 1971

Drum Majorette, #0875

Battery-Operated Toys

Armies of colorful, playful tin toys are frequently found at toy shows and flea markets. Among the most popular are battery-operated toys.

The heyday of battery-operated toys began in the 1940s, with Japanese companies leading the charge. Popular wind-up and friction toys were soon replaced with longer-lasting battery-operated versions.

Thousands of designs were made with toys featuring a variety of creative characters — from bubble-blowing monkeys to robots to cigar-smoking clowns. Disney creations and other cartoon characters like Popeye were also popular. The ingenuity of the Japanese created some truly unique novelty toys, valued today for their motion, design and humorous appeal.

Toys were often complex and could replicate several motions, such as walking, lifting or drumming. Tin lithography was usually quite interesting and detailed.

It is often difficult to identify the manufacturer of a battery-operated toy. Many companies used only initials to mark toys; some didn't mark them at all. Major manufacturers of battery-operated toys from the 1940s-1960s include Marx, Linemar (Marx's Japanese subsidiary), Alps, Marusan, Yonezawa, Bandai, Asahi Toy and Modern Toy (MT).

Manufacture of tin toys dropped off in the 1960s in favor of cheaper methods and materials.

Determining Value

As in other collecting areas, toys with character ties generally command more money. Space-related battery-operated toys are also popular.

It's difficult to find most battery-operated toys in Mint condition in original boxes. Battery-operated toys are highly susceptible to rust, corrosion and yellowing of fabric/plush.

Having the original box greatly increases the toy's desirability. Instructions were often printed on the box, and the name of the toy on the box often didn't match the toy exactly.

The complexity of the toy also determines value. A toy with three or more actions will generally be worth more than a toy that performs only one action. Even though they may be nice display pieces, toys that don't work command much less.

This book lists toys in Good, Excellent and Mint in Box conditions.

To be graded in Mint condition, a toy should be operational, clean, free of rust or corrosion, and have the original box. Note that those in Good condition will exhibit wear, but will still be operational. Lesser condition toys and those that are not operational are not graded here. Some collectors acquire them for parts.

Dancing Merry Chimp, 1960s, Kuramochi

The Top 10 Battery-Operated Toys (in Mint condition)

1.	Mickey the Magician, Linemar, 1960s	$2,500
2.	Bubble Blowing Popeye, Linemar, 1950s	2,500
3.	Gypsy Fortune Teller, Ichida, 1950s	2,200
4.	Big Loo, Marx, 1960s	2,200
5.	Smoking Spaceman, Linemar, 1950s	1,600
6.	Kooky-Spooky Whistling Tree, Marx, 1950s	1,600
7.	Drumming Mickey Mouse, Linemar, 1950s	1,600
8.	Nutty Nibs, Linemar, 1950s	1,400
9.	Tarzan, Marusan, 1960s	1,000
10.	Super Susie, Linemar, 1950s	1,000

Contributor to this section: Don Hultzman, 5026 Sleepy Hollow Rd., Medina, OH 44256.

Cragstan Crapshooter, 1950s

BATTERY-OPERATED

Battery-Operated Toys

NAME	COMPANY	YEAR	DESCRIPTION	GOOD	EX	MIB
ABC Fairy Train	MT	1950s	14-1/2" long	80	95	160
Accordion Player Bunny	Alps	1950s	12" tall	150	300	425
Air Cargo Prop-Jet Airplane	Marx	1960s	Seaboard World Airlines, 12" long	150	300	425
Air Control Tower	Bandai	1960s	11" tall, 37" span	210	315	450
Air Defense Pom-Pom Gun	Linemar	1950s	14" long	115	175	250
Aircraft Carrier	Marx	1950s	20" long	275	450	625
Alley, the Exciting New Roaring Stalking Alligator	Marx	1960s	17-1/2" long	145	225	310
American Airlines DC-7	Linemar	1960s	17-1/2" long, 19" wingspan	175	300	425
American Airlines Electra	Linemar	1950s	18" long, 19-1/2" wingspan	175	300	410
American Airlines Flagship Caroline	Linemar	1950s	18" long, 19-1/2" wingspan	175	275	385
Anti-Aircraft Unit No. 1	Linemar	1950s	12-1/2" long	140	225	310
Army Radio Jeep -- J1490	Linemar	1950s	7-1/4" long	75	125	165
Arthur A-Go-Go	Alps	1960s	10" tall	135	200	295
Atomic Rocket X-1800	MT	1960s	9" long	145	225	325
B-58 Hustler Jet	Marx	1950s	21" long, 12" wingspan	425	650	925
Ball Playing Dog	Linemar	1950s	9"	125	170	250
Barber Bear	Linemar	1950s	9-1/2" tall	175	300	400
Barking Boxer Dog	Marx	1950s	7" long	45	65	85
Barney Bear Drummer	Alps	1950s	11" tall	110	185	250
Barnyard Rooster	Marx	1950s	10" tall	85	135	185
Bartender	TN	1960s	11-1/2" tall	50	75	100
Bear the Cashier	MT	1950s	7-1/2" tall	180	280	375
Bengali -- The Exciting New Growling, Prowling Tiger	Linemar	1961	18-1/2" tall	75	125	175
Big John	Alps	1960s	12" tall	55	75	125
Big John the Indian Chief	TN	1960s	12-1/2" tall	75	130	175
Big Loo	Marx	1960s	38" tall	950	1500	2200
Big Max Robot	Remco	1958	7" tall	80	125	175
Big Parade	Marx	1963	15" long	300	450	600
Bimbo the Clown	Alps	1950s	9-1/4" tall	190	325	400
Blushing Gunfighter	Y Co.	1960s	11" tall	120	200	235
Blushing Willie	Y Co.	1960s	10" tall	50	75	100
Bobby the Drumming Bear	Alps	1950s	10" tall	175	275	380
Bongo Player	Alps	1960s	10" tall	75	110	160
Bongo the Drumming Monkey	Alps	1960s	9-1/2" tall	75	115	165
Brave Eagle	TN	1950s	11" tall	85	140	190
Brewster the Rooster	Marx	1950s	9-1/2" tall	110	175	235
Bubble Blowing Bear	MT	1950s	9-1/2" tall	130	200	275
Bubble Blowing Monkey	Alps	1950s	10" tall	100	120	200
Bubble Blowing Popeye	Linemar	1950s	11-3/4" tall	750	1200	2500
Bubbling Bull	Linemar	1950s	8" tall	75	110	155
Bunny the Magician	Alps	1950s	14-1/2" tall	200	275	400
Burger Chef	Y Co.	1950s	9" tall	150	225	300
Busy Housekeeper	Alps	1950s	8-1/2" tall	150	230	325
Busy Secretary	Linemar	1950s	7-1/2" tall	125	200	300
Busy Shoe Shining Bear	Alps	1950s	10" tall	110	175	240
Cabin Cruiser with Outboard Motor	Linemar	1950s	12" long	100	135	200
Calypso Joe	Linemar	1950s	11" tall	190	310	400
Camera Shooting Bear	Linemar	1950s	11" tall	350	475	700
Cappy the Baggage Porter Dog	Alps	1960s	12" tall	75	130	175
Caterpillar	Alps	1950s	16" long	80	125	175
Central Choo Choo	MT	1960s	15" long	30	40	60
Charlie the Drumming Clown	Alps	1950s	9-1/2" tall	135	200	275

Battery-Operated Toys

NAME	COMPANY	YEAR	DESCRIPTION	GOOD	EX	MIB
Charlie Weaver	TN	1962	12"	60	90	120
Charm the Cobra	Alps	1960s	6" tall	90	130	175
Chee Chee Chihuahua	Mego	1960s	8" tall	25	40	60
Chef Cook	Y Co.	1960s	11-1/2" tall	100	200	300
Chippy the Chipmunk	Alps	1950s	12" long	75	120	155
Circus Fire Engine	MT	1960s	11" long	110	175	235
Clancy the Great	Ideal	1960s	19-1/2" tall	85	135	200
Climbing Donald Duck on Friction Fire Engine	Linemar	1950s	12" long	300	525	700
Clown Circus Car	MT	1960s	8-1/2" long	100	175	235
Clown on Unicycle	MT	1960s	10-1/2" tall	180	280	375
Colonel Hap Hazard Robot	Marx	1968	11-1/4" tall	300	440	600
Coney Island Penny Machine	Remco	1950s	13" tall	90	130	300
Coney Island Rocket Ride	Alps	1950s	13-1/2" tall	300	450	600
Cragstan Beep Beep Greyhound Bus	Cragstan	1950s	20" long	100	150	220
Cragstan Crapshooter	Y Co.	1950s	9-1/2" tall	90	130	180
Cragstan Crapshooting Monkey	Alps	1950s	9" tall	60	100	140
Cragstan Mother Goose	Y Co.	1960s	8-1/4" tall	85	125	165
Cragstan Playboy	Cragstan	1960s	13" tall	85	130	185
Cragstan Roulette, A Gambling Man	Y Co.	1960s	9" tall	130	200	275
Crawling Baby	Linemar	1940s	11" long	45	75	100
Daisy, the Jolly Drumming Duck	Alps	1950s	9" tall	120	180	240
Dancing Merry Chimp	Kuramochi	1960s	100	150	200	
Dandy, the Happy Drumming Pup	Alps	1950s	8-1/2" tall	85	145	200
Dennis the Menace	Rosko	1950s	9" tall, with xylophone	120	180	240
Disney Acrobats	Linemar	1950s	9" tall; Mickey, Donald, and Pluto	300	625	800
Disney Fire Engine	Linemar	1950s	11" long	425	650	870
Disneyland Fire Engine	Linemar	1950s	18" long	295	440	600
Donald Duck	Linemar	1960s	8" tall	175	275	380
Doxie the Dog	Linemar	1950s	9" long	20	25	45
Drinking Captain	S & E	1960s	100	150	200	
Drummer Bear	Alps	1950s	10" tall	130	200	275
Drumming Mickey Mouse	Linemar	1950s	10" tall	800	1000	1600
Drumming Polar Bear	Alps	1960s	12"	75	120	165
Ducky Duckling	Alps	1960s	8"	35	55	85
El Toro, Cragstan Bullfighter	TN	1950s	9-1/2" long	90	145	200
Feeding Bird Watcher	Linemar	1950s	9" tall	300	350	600
Fido the Xylophone Player	Alps	1950s	8-3/4" tall	115	170	240
Flintstone Yacht	Remco	1961	17" long	90	145	200
Frankenstein Monster	TN	1960s	14" tall	130	200	275
Frankie the Rollerskating Monkey	Alps	1950s	12"	90	150	200
Fred Flintstone Bedrock Band	Alps	1962	9-1/2" tall	400	500	900
Fred Flintstone Flivver	Marx	1960s	7" long	400	600	900
Friendly Jocko, My Favorite Pet	Alps	1950s	8" tall	110	175	245
Gino, Neapolitan Balloon Blower	Tomiyama	1960s	10" tall	80	130	180
Girl with Baby Carriage	TN	1960s	8" tall	100	150	200
Godzilla Monster	Marusan	1970s	11-1/2" tall	125	160	250
Good Time Charlie	MT	1960s	12" tall	90	130	175
Grandpa Panda Bear	MT	1950s	9" tall	100	175	245

Telephone Bear, 1950s, Linemar

Picnic Bear, 1950s, Alps

McGregor, 1960s, Rosko

Battery-Operated Toys

NAME	COMPANY	YEAR	DESCRIPTION	GOOD	EX	MIB
Great Garloo	Marx	1960s	23" tall green monster	300	450	600
Green Caterpillar	Daiya	1950s	19-1/2" long	150	250	350
Gypsy Fortune Teller	Ichida	1950s	12" tall, 20 cards	1100	1700	2200
Happy & Sad Face Cymbal Clown	Y Co.	1960s	10" tall	120	180	200
Happy Fiddler Clown	Alps	1950s	9-1/2" tall	200	300	425
Happy Naughty Chimp	Daishin	1960s	9-1/2" tall	50	80	130
Happy Santa One-Man Band	Alps	1950s	9" tall	100	185	245
Hippo Chef	Y Co.	1960s	10" tall	100	160	600
Hobo Clown with Accordion	Alps	1950s	10-1/2" tall	185	280	400
Hoop Zing Girl	Linemar	1950s	11-1/2" tall	115	185	245
Hoopy the Fishing Duck	Alps	1950s	10" tall	275	375	525
Hooty the Happy Owl	Alps	1960s	9" tall	65	100	145
Hungry Cat	Linemar	1960s	9" tall	250	450	700
Hungry Hound Dog	Y Co.	1950s	9-1/2" tall	150	275	300
Ice Cream Baby Bear	MT	1950s	200	300	400	
Indian Joe	Alps	1960s	12" tall	55	90	120
Jo-Jo the Flipping Monkey	TN	1970s	10" tall	35	55	80
Jocko the Drinking Monkey	Linemar	1950s	11" tall	70	110	155
Jolly Bambino	Alps	1950s	9" tall	200	300	400
Jolly Bear Peanut Vendor	TN	1950s	8" tall	250	285	500
Jolly Daddy	Marusan	1950s	8-3/4" tall	165	250	350
Jolly Drummer Chimpy	Alps	1950s	9" tall	75	100	140
Jolly Pianist	Marusan	1950s	8" tall	90	145	200
Jolly Santa on Snow	Alps	1950s	12-1/2" tall	100	200	300
Josie the Walking Cow	Daiya	1950s	14" long	100	150	200
Jumbo the Bubble Blowing Elephant	Y Co.	1950s	7-1/4" tall	55	85	115
Jungle Trio	Linemar	1950s	8" tall	400	600	800
King Zor	Ideal	1962	blue plastic dinosaur, 30" long	100	250	550
Kissing Couple	Ichida	1950s	10-3/4" long	145	200	300
Knitting Grandma	T-N	1950s	8-1/2" tall	150	225	325
Kooky-Spooky Whistling Tree	Marx	1950s	14-1/4" tall	450	760	1600
Lambo Elephant	Alps	1950s	16" long with trailer	260	385	525
Linemar Music Hall	Linemar	1950s	8" tall	100	120	200
Lion	Linemar	1950s	9" long	70	100	145
Loop the Loop Clown	T-N	1960s	10" tall	60	120	175
Mac the Turtle	Y Co.	1960s	8" tall	85	135	185
Magic Man Clown	Marusan	1950s	11" tall	250	385	525
Magic Snowman	MT	1950s	11-1/4" tall	125	190	265
Main Street	Linemar	1950s	19-1/2" long	480	950	970
Major Tooty	Alps	1960s	14" tall	85	130	175
Mambo the Jolly Drumming Elephant	Alps	1950s	9-1/2" tall	90	150	200
Marching Bear	Alps	1960s	10" tall	150	225	300
Marshal Wild Bill	Y Co.	1950s	10-1/2" tall	180	280	375
Marvelous Mike	Saunders	1950s	17" long	100	175	245
Maxwell Coffee-Loving Bear	TN	1960s	10" tall	100	150	200
McGregor	Rosko	1960s	12" tall	90	140	185
Mickey the Magician	Linemar	1960s	10" tall	785	1200	2500
Mighty Kong	Marx	1950s	11" tall	250	325	500
Mischievous Monkey	MT	1950s	18" tall	200	300	400
Miss Friday the Typist	TN	1950s	8" tall	115	180	245
Mr. MacPooch	SAN	1950s	8" tall	130	195	280
Mr. Traffic Policeman	A-I	1950s	14" tall	175	270	365
Mumbo Jumbo	Alps	1960s	9-3/4" tall	85	120	160

Battery-Operated Toys

NAME	COMPANY	YEAR	DESCRIPTION	GOOD	EX	MIB
Musical Bear	Linemar	1950s	10" tall	200	300	400
Musical Jackal	Linemar	1950s	10" tall	150	225	600
Musical Marching Bear	Alps	1950s	11" tall	115	170	600
Nutty Mad Indian	Marx	1960s	12" tall	80	120	175
Nutty Nibs	Linemar	1950s	11-1/2" tall	700	850	1400
Ol' Sleepy Head RIP	Y Co.	1950s	9" tall	150	250	320
Panda Bear	MT	1970s	10" long	30	45	65
Pat the Roaring Elephant	Y Co.	1950s	9" long	100	160	225
Peppermint Twist Doll	Haji	1950s	12" tall	150	225	300
Peppy Puppy	Y Co.	1950s	8" long	45	70	100
Pet Turtle	Alps	1960s	7" long	65	100	145
Pete the Space Man	Bandai	1960s	5" tall	60	90	130
Peter the Drumming Rabbit	Alps	1950s	13" tall	145	200	275
Picnic Bear	Alps	1950s	10" tall	75	120	165
Picnic Bunny	Alps	1950s	10" tall	80	130	180
Picnic Monkey	Alps	1950s	10" tall	100	150	200
Picnic Poodle	STS	1950s	7" long	40	60	80
Pierrot Monkey Cycle	M-T	1950s	8" tall	325	460	650
Pinkee the Farmer	MT	1950s	9-1/2" long	100	155	200
Pipie the Whale	Alps	1950s	12" long	90	100	180
Pistol Pete	Marusan	1950s	10-1/4" tall	250	270	500
Playful Puppy	MT	1950s	5" tall	100	150	200
Polar Bear	Alps	1970s	8" long	40	60	85
Popcorn Eating Bear	MT	1950s	9" tall	100	150	200
Rambling Ladybug	M-T	1960s	8" long	60	90	125
Reading Bear	Alps	1950s	9" tall	85	150	210
Rembrandt the Monkey Artist	Alps	1950s	8" tall	170	240	365
Ricki the Begging Poodle	Rock Valley	1950s	9" long	35	45	65
Roarin' Jungle Lion	Marx	1950s	16" long	140	200	285
Rock 'N Roll Monkey	Rosko	1950s	13" tall	140	200	280
Rocking Chair Bear	MT	1950s	10" tall	100	140	175
Roller Skater	Alps	1950s	12" tall	90	135	180
Root Beer Counter	K Co.	1960s	8" tall	100	160	230
Sam the Shaving Man	Plaything Toy	1960s	11-1/2" tall	125	180	245
Sammy Wong the Tea Totaler	TN	1950s	10" tall	140	200	285
Santa Claus on Handcar	MT	1960s	10" tall	100	140	200
Santa Claus on Reindeer Sleigh	M-T	1950s	17" long	400	600	825
Santa Claus on Scooter	MT	1960s	10" tall	100	140	180
Santa Copter	MT	1960s	8-1/2" high	75	100	140
Saxophone Playing Monkey	Alps	1950s	9-1/2" tall	200	300	400
Serpent Charmer	Linemar	1950s	7" tall	175	300	425
Shaggy the Friendly Pup	Alps	1960s	8" long	35	45	65
Shoe Maker Bear	TN	1960s	8-1/2" tall	100	165	225
Shoe Shine Bear	T-N	1950s	9" tall	130	195	275
Shoe Shine Joe	Alps	1950s	11" tall	125	195	265
Shoe Shine Monkey	TN	1950s	9" tall	135	195	260
Shooting Gorilla	MT	1950s	12" tall	155	235	325
Shutterbug Photographer	T-N	1950s	420	620	820	
Skating Circus Clown	TPS	1950s	6" tall	450	595	800
Skiing Santa	MT	1960s	12" tall	160	230	300
Skipping Monkey	TN	1960s	9-1/2" tall	45	60	80
Slalom Game	T-N	1960s	15-1/4" long	100	160	225
Sleeping Baby Bear	Linemar	1950s	9" long	195	285	400
Sleeping Pup	Alps	1960s	9" long	45	75	100
Slurpy Pup	TN	1960s	6-1/2" long	45	60	100

Bongo the Drumming Monkey, 1960s, Alps

Saxophone Playing Monkey, 1950s, Alps

Battery-Operated Toys

NAME	COMPANY	YEAR	DESCRIPTION	GOOD	EX	MIB
Smokey Bear	SAN	1950s	9" tall	250	300	500
Smoking Bunny	SAN	1950s	10-1/2" tall	100	140	185
Smoking Elephant	Marusan	1950s	8-3/4" tall	115	165	225
Smoking Spaceman	Linemar	1950s	12" tall	750	1200	1600
Sneezing Bear	Linemar	1950s	9" tall	180	270	365
Snoopy Sniffer	MT	1960s	8" long	40	60	80
Spanking Bear	Linemar	1950s	9" tall	160	200	320
Strutting Sam	Haji	1950s	10-1/2" tall	200	300	430
Sunday Driver	MT	1950s	10" long	55	90	125
Super Susie	Linemar	1950s	9" tall	500	750	1000
Suzette the Eating Monkey	Linemar	1950s	8-3/4" tall	300	450	625
Switchboard Operator	Linemar	1950s	7-1/2" tall	250	380	800
Tarzan	Marusan	1960s	13" tall	480	765	1000
Teddy Bear Swing	T-N	1950s	17" tall	300	380	600
Teddy the Boxing Bear	Y Co.	1950s	9" tall	115	190	240
Teddy the Rhythmical Drummer	Alps	1960s	11" tall	85	120	165
Teddy-Go-Kart	Alps	1960s	10-1/2" long	75	125	160
Telephone Bear	Linemar	1950s	7-1/2" tall	180	280	380
Television Spaceman	Alps	1960s	14-1/2" tall	400	600	825
Tinkling Trolley	M-T	1950s	10-1/2" long	110	170	230
Tom and Jerry Choo Choo	M-T	1960s	10-1/4" long	125	185	255
Tom and Jerry Handcar	M-T	1960s	7-3/4" long	135	200	270
Tom-Tom Indian	Y Co.	1960s	10-1/2" tall	80	120	160
Topo Gigio Playing the Xylophone	TN	1960s		265	440	525
Traveler Bear	Linemar	1950s	8" tall	85	125	400
Trumpet Playing Bunny	Alps	1950s	10" tall	145	200	280
Trumpet Playing Monkey	Alps	1950s	9" tall	150	225	300
Tubby the Turtle	Y Co.	1950s	7" long	50	75	100
Tumbles the Bear	Yanoman	1960s	8-1/2" tall	75	125	160
Twirly Whirly	Alps	1950s	13-1/2" tall	290	450	600
Walking Bear with Xylophone	Linemar	1950s	10" tall	175	270	365
Walking Elephant	Linemar	1950s	8-1/2" tall	80	120	155
Walking Esso Tiger	Marx	1950s	11-1/2" tall	200	300	400
Walking Itchy Dog	Alps	1950s	9" long	45	75	100
Western Locomotive	MT	1950s	10-1/2" long	45	65	80
Windy the Elephant	TN	1950s	9-3/4" tall	120	180	245
Yeti the Abominable Snowman	Marx	1960s	12" tall	250	335	500
Yo-Yo Clown	Alps	1960s	9" tall	150	190	300
Yo-Yo Monkey	Alps	1960s	9" tall	100	185	245
Yo-Yo Monkey	YM	1960s	12" tall	85	135	180
Yummy Yum Kitty	Alps	1950s	9-1/2" tall	170	270	325
Zero Fighter Plane	Bandai	1950s	15" wingspan	150	230	300

Beanie Babies

Beanbag toys have been around for years, so have plush animals. But never have plush beanbag animals caused such a commotion as the Beanie Babies craze of 1997.

The inexpensive ($5 retail) toys by Ty, Inc. have been around since 1994, but have recently sparked a collecting craze of manic proportions. Adults as well as children are clamoring for the tiny plush pets with creative names like Tabasco the bull, Web the spider and Chocolate the moose.

There are over 100 Beanie Babies animals (and some non-animals like a ghost and a snowman) in existence, but variations in tags, colors and styles doubles the number collectors can look for. Each Beanie Baby is labeled with its name, a poem and its birth date on a heart-shaped tag. Old tags are paper "To/From" style tags or a tag sewn into the animal.

Retired Beanie Babies (ones that the company no longer produces) are especially popular, commanding incredible prices on the secondary market — some up to $1,000. Adding to the toys' appeal is that they are only available through smaller retailers such as gift shops.

What's New?

The biggest Beanie Babies splash of 1998 was Princess, a purple bear sporting a rose on its chest. Princess, with an appropriate poem, was dedicated to the memory of Diana, Princess of Wales. Immediately upon issue, the bears were selling for $200-$500; many were auctioned, and the proceeds were donated to charities.

Beanie Babies in the most demand are those that have been retired, those with variations, and those that contain mistakes (such as misspelled names). Babies with "old" tags tend to be more popular than those with new ones.

Beanie Babies bears have seen incredible increases in popularity, especially those with embroidered logos on their chests, like Peace (with a peace sign), Princess (with a rose) and Erin (with a shamrock).

Also in demand are Teenie Beanie Babies, miniature Beanie Babies offered as McDonald's restaurant premiums in 1997 and 1998.

The long-term investment potential of Beanie Babies is still questionable, but the past several years have proven strong and sure for collectors. Prices have increased for most retired Beanies, and even new issues debuted last year for more than the usual $5 retail price.

Kiwi, Garcia and Righty (left to right) are three Beanie Babies captivating collectors.

The Top 10 Beanie Babies
(All listed are retired. Prices are for items in Mint condition)

1.	Peanut the Elephant, dark blue	$4,200
2.	Bongo the Monkey, with "Nana" tag	4,000
3.	Punchers the Lobster	4,000
4.	Derby the Horse, old tag, fine yarn	3,800
5.	Cubbie [Brownie] the Bear, old "Brownie" tag	3,700
6.	Teddy the Bear, old face	2,500
7.	Quackers the Duck, without wings	2,000
8.	Humphrey the Camel	1,800
9.	Slither the Snake	1,750
10.	Teddy the Cranberry Bear, old face	1,600

Tuffy, Bones and Weenie are three Beanie Babies dogs. Bones and Weenie are retired.

BEANIE BABIES

NAME	DESCRIPTION	MINT
Ants the Anteater		7
Batty the Bat		8
Bernie the St. Bernard		9
Blackie the Bear	old tag	32
Blackie the Bear	new tag	8
Bongo the Monkey	new tag, brown or tan tail	9
Britannia the Bear	European release	165
Bruno the Terrier		9
Chip the Calico Cat		8
Chocolate the Moose	old tag	35
Chocolate the Moose	new tag	9
Claude the Crab	tie-dyed	18
Congo the Gorilla		9
Crunch the Shark		9
Cubbie the Bear	Chicago Cubs promo	250
Cubbie the Bear	new tag	20
Curly the Bear		20
Daisy the Cow	new tag	8
Daisy the Cow	old tag	40
Derby the Horse	old tag, coarse yarn	30
Derby the Horse	new tag, coarse yarn	10
Derby the Horse	dot on forehead	12
Doby the Doberman		8
Dotty the Dalmatian		8
Early the Robin		7
Erin the Bear	green w/shamrock on chest	150
Fetch the Golden Retriever		7
Fleece the Lamb		9
Fortune the Panda		7
Freckles the Leopard		9
Gigi the Poodle		7
Glory the Bear		75
Gobbles the Turkey		9
Hissy the Snake		9
Iggy the Iguana		15
Jabber the Parrot		7
Jake the Mallard Duck		7
Kuku the Cockatoo		7
Lucky the Ladybug	w/21 glued on dots; discontinued	500
Maple the Bear	Olympic version	400
Maple the Bear	w/Pride tag	375
Maple the Bear	Canadian exclusive	265
Mel the Koala		10
Mystic the Unicorn	iridescent horn	10
Nanook the Husky		10
Nuts the Squirrel		8
Peace the Tie-Dyed Bear		50
Pinky the Flamingo	old tag	50
Pinky the Flamingo	new tag	8
Pouch the Kangaroo		8
Pounce the Cat		8
Prance the Cat	striped	9
Princess the Bear	Princess Diana commemorative, purple bear w/rose	75
Puffer the Puffin		8
Pugsly the Pug Dog		10
Rainbow the Chameleon		13
Righty the Elephant	no flag	350
Ringo the Raccoon	old tag	50
Ringo the Raccoon	new tag	8
Roary the Lion		8
Rocket the Blue Jay		7
Scoop the Pelican		8
Seaweed the Otter	old tag	30
Seaweed the Otter	new tag	12
Sly the Fox	new tag, white belly	12
Smoochy the Frog		9
Snip the Cat	Siamese	8
Snort the Bull	red	8
Snort the Bull	with "Tabasco" tag	75
Spike the Rhinoceros		5
Spinner the Spider		8
Spunky the Cocker Spaniel		15
Stinger the Scorpion		7
Stinky the Skunk	new tag	8
Stinky the Skunk	old tag	30
Stretch the Ostrich		15
Strut the Rooster		10
Tracker the Bassett Hound		7
Tuffy the Terrier		8
Valentino the Bear	new tag	22
Valentino the Bear	old tag	50
Whisper the Deer		7
Wise the Owl		7
Wrinkles the Dog		8

Retired

Ally the Alligator	old tag	100
Ally the Alligator	new tag	55
Baldy the Eagle		16
Bessie the Cow	old tag	50
Bessie the Cow	new tag	60
Blizzard the Snow Tiger		15
Bones the Dog	new tag	16
Bones the Dog	old tag	55
Bongo the Monkey	brown tail	75
Bongo the Monkey	tan tail	50
Bongo the Monkey	with "Nana" tag	4000

Beanie Babies

Retired

NAME	DESCRIPTION	MINT
Bronty the Brontosaurus		820
Bubbles the Fish	old tag	175
Bubbles the Fish	new tag	140
Bucky the Beaver	old tag	60
Bucky the Beaver	new tag	35
Bumble the Bee	old tag	550
Bumble the Bee	new tag	500
Caw the Crow		550
Chilly the Polar Bear		1600
Chops the Lamb	old tag	200
Chops the Lamb	new tag	160
Coral the Fish	old tag, retired	200
Coral the Fish	new tag	170
Cubbie [Brownie] the Bear	old "Brownie" tag	3700
Cubbie the Bear	new tag	20
Cubbie the Bear	old tag	150
Derby the Horse	old tag, fine yarn	3800
Digger the Crab	red, old tag	275
Digger the Crab	red, new tag	100
Digger the Crab	orange	650
Doodle the Rooster		45
Ears the Bunny	old tag	40
Ears the Bunny	new tag	20
Echo the Dolphin		20
Flash the Dolphin	old tag	200
Flash the Dolphin	new tag	100
Flip the Cat	old tag	70
Flip the Cat	new tag	30
Floppity the Bunny	lavender	20
Flutter the Butterfly		550
Garcia the Bear	tie-dyed, new tag	175
Garcia the Bear	tie-dyed, old tag	200
Goldie the Goldfish	old tag	200
Goldie the Goldfish	new tag	45
Gracie the Swan		16
Grunt the Razorback	new tag	150
Grunt the Razorback	old tag	200
Happy the Hippo	grey	700
Happy the Hippo	lavender, old tag	75
Happy the Hippo	lavender, new tag	20
Hippity the Bunny	mint green	20
Hoot the Owl	new tag	45
Hoot the Owl	old tag	70
Hoppity the Bunny	pink	20
Humphrey the Camel		1800
Inch the Worm	new tag, felt antennas	100
Inch the Worm	old tag, felt antennas	170
Inch the Worm	new tag, yarn antennas	16
Inky the Octopus	beige, w/mouth	600

Retired

NAME	DESCRIPTION	MINT
Inky the Octopus	new tag, pink	15
Inky the Octopus	no mouth	600
Inky the Octopus	old tag, pink	50
Jolly the Walrus		20
Kiwi the Toucan	new tag	170
Kiwi the Toucan	old tag	220
Lefty the Donkey		305
Legs the Frog	new tag	20
Legs the Frog	old tag	200
Libearty the Bear	without flag	375
Libearty the Bear	with flag	375
Lizzy the Lizard	blue w/black spots, new tag	22
Lizzy the Lizard	blue w/black spots, old tag	225
Lizzy the Lizard	tie-dye	750
Lucky the Ladybug	w/11 spots on fabric	25
Lucky the Ladybug	old tag, seven dots	300
Lucky the Ladybug	new tag, seven dots	100
Magic the Dragon	new tag, light pink stitching	38
Magic the Dragon	old tag, light pink stitching	70
Magic the Dragon	new tag, regular pink stitching	160
Manny the Manatee	old tag	200
Manny the Manatee	new tag	150
Mystic the Unicorn	old tag, fine mane	375
Mystic the Unicorn	new tag, fine mane	175
Nip the Cat	gold w/pink ears	760
Nip the Cat	white face and belly	400
Nip the Cat	new tag, white paws	15
Nip the Cat	old tag, white paws	300
Patti the Platypus	raspberry	820
Patti the Platypus	deep fuchsia	900
Patti the Platypus	new tag, fuchsia	20
Patti the Platypus	magenta	650
Patti the Platypus	old tag, fuchsia	50
Peanut the Elephant	new tag, light blue	16
Peanut the Elephant	dark blue	4200
Peanut the Elephant	old tag, light blue	300
Peking the Panda		1600
Pinchers the Lobster	old tag, red	40
Pinchers the Lobster	new tag, red	16
Pinchers the Lobster	old tag, orange	110
Punchers the Lobster		4000
Quackers the Duck	new tag, with wings	20
Quackers the Duck	without wings	2000
Radar the Bat	new tag	150

Bongo the Monkey

Hoot the Owl

Ally the Alligator

Retired

NAME	DESCRIPTION	MINT
Radar the Bat	old tag	200
Rex the Tyrannosaurus		700
Righty the Elephant	with flag	325
Rover the Dog		25
Scottie the Terrier		18
Seamore the Seal	new tag	150
Seamore the Seal	old tag	250
Slither the Snake		1750
Sly the Fox	new tag, brown belly	125
Snowball the Snowman		25
Sparky the Dalmatian		110
Speedy the Turtle	new tag	22
Speedy the Turtle	old tag, retired	200
Splash the Whale	new tag	100
Splash the Whale	old tag	200
Spook the Ghost	old tag, not "Spooky"	225
Spooky the Ghost	old tag	50
Spooky the Ghost	new tag	25
Spot the Dog	without spot	1600
Spot the Dog	new tag, with spot	60
Spot the Dog	old tag, with spot	70
Squealer the Pig	old tag	50
Squealer the Pig	new tag	20
Steg the Stegosaurus		750
Sting the Ray	old tag	200
Stripes the Tiger	tan	20
Stripes the Tiger	orange, stripes on tail	275
Tabasco the Bull	new tag	190
Tabasco the Bull	old tag	225
Tank the Armadillo	nine lines w/shell	65
Tank the Armadillo	nine lines, no shell	230
Tank the Armadillo	seven lines, old or new tags	150
Teddy the 1997 Holiday Bear		40
Teddy the Bear	brown, new tag, new face	100
Teddy the Bear	old face	2500
Teddy the Bear	brown. old tag, new face	200
Teddy the Cranberry Bear	old face	1600
Teddy the Cranberry Bear	new face	1550
Teddy the Jade Bear	new face	1550
Teddy the Jade Bear	old face	1500
Teddy the Magenta Bear	old face	1500

Retired

NAME	DESCRIPTION	MINT
Teddy the Magenta Bear	new face	1550
Teddy the Teal Bear	old face	1500
Teddy the Teal Bear	new face	1550
Teddy the Violet Bear	new face	1550
Teddy the Violet Bear	old face	1500
Trap the Mouse		1300
Tusk the Walrus	old tag	150
Tusk the Walrus	new tag	110
Twigs the Giraffe	new tag	15
Twigs the Giraffe	old tag	40
Velvet the Panther	new tag	25
Velvet the Panther	old tag	65
Waddle the Penguin	new tag	20
Waddle the Penguin	old tag	40
Waves the Whale		20
Web the Spider		1000
Weenie the Dog	old tag	40
Weenie the Dog	new tag	20
Ziggy the Zebra	old tag	50
Ziggy the Zebra	new tag	20
Zip the Cat	all black	1500
Zip the Cat	black w/white paws	440
Zip the Cat	old tag, white paws	40

McDonald's Teenie Beanies, 1997-98

Bones the Dog	1998 set	10
Bongo the Monkey	1998 set	10
Chocolate the Moose	First set	10
Chops the Lamb	First set	10
Doby the Doberman	1998 set	10
Goldie the Goldfish	First set	10
Happy the Hippo	1998 set	10
Inch the Worm	1998 set	10
Lizz the Lizard	First set	10
Mel the Koala	1998 set	10
Patti the Platypus	First set	15
Peanut the Elephant	1998 set	10
Pinchers the Lobster	1998 set	10
Pinky the Flamingo	First set	15
Quacks the Duck	First set	10
Scoop the Pelican	1998 set	10
Seamore the Seal	First set	12
Snort the Bull	First set	12
Speedy the Turtle	First set	10
Twigs the Giraffe	1998 set	10
Waddles the Penguin	1998 set	10
Zip the Cat	1998 set	10

Character Toys

Everybody loves a character — right? In the field of toy collecting, characters are those familiar faces created by comic books, television cartoons, movies and more.

Character toy collecting remains the broadest field in the hobby. If time and space would permit, we could easily fill this book with character toys and still only skim its colorful surface.

From Atom Ant to Mickey Mouse, from Popeye to Zorro and everywhere in between, the choices are mind boggling. Collectors are restricted only by their budgets, as the variety even within single categories like Mickey Mouse or Popeye is wide enough to comprise entire collections. Tin wind-ups, bisques, dolls, books, puppets, play sets, puzzles, models, coloring books, games — it never ends.

Beginning collectors are soon faced with either narrowing their collecting into specialties or being overwhelmed by choices, and the options are so tempting that many collectors never specialize at all. Flash Gordon and Mickey Mouse not only look just fine together, they belong together. Dick Tracy, Li'l Abner and Donald Duck look great side by side on a shelf, and whatever configuration a developing toy collection takes, it will tell a fascinating story of its time and its relationship to other toys around it. For this reason, those diverse and eclectic collections can be the most satisfying of all.

For those who are already engaged in or are considering specialization, a few words of introduction to some of the major classes in this section are in order.

Disney Mania!

Walt Disney has often been called the single greatest contributor to American culture. Disneyana is one of the largest and most vigorous areas of toy collecting, and it shows no signs of slowing.

Stan Laurel and Oliver Hardy Banks, 1974, Play Pal

No comic creations have ever been as honored (or as frequently pirated) as Disney characters. From the rat-snouted mouse of 1928 to the virtually human version of the 1990s, the various stages of Mickey Mouse have adorned more toys and trinkets than any other image in modern history. And he is hardly alone.

The wide world of Walt has grown full, is richly populated, and the theory of cosmic expansion holds true — each new Disney film brings new planets of characters into being. And across our planet, millions of budding little collectors line up to be first for their movie tickets, popcorn, and every piece of merchandise their allowances will buy.

The implication is clear. The market demand for the classic toys of Mickey, Donald, Pinocchio, and other early Disney characters is secure and will likely stay that way. We can only guess what those first 1930s Disney toys will be worth when they finally turn 100 years old, or what *A Bug's Life* and *Mulan* toys will be worth in 2098.

Just How Popular?

Not all toys endlessly escalate in value. Enduring popularity is a key element in predicting future demand for pop culture collectibles. How popular was it when it was new? How long did it stay at the top? These and other factors all impact on future collectibility.

As collectibles mature and their active collecting public increases, their prices can be expected to rise. But once the collector base begins to erode, once it gets old or loses interest or gets interrupted by a war or replaced by the next big thing, then values can begin to slide. When collectors go away, sometimes an entire hobby goes with them. The best insurance against this is the continued popularity of some aspect of their identity, particularly if it is character based.

Lily Munster Doll, Remco, 1964

Another prime example of this longevity is Popeye, who made his first appearance in 1929 in a comic strip called Thimble Theatre, created by E.C. Segar. The spinach-eating sailor rapidly became one of the most popular characters of his day, bringing fame to his cartoon companions as well.

That fame manifested itself in some of the most beautiful and well-designed toys of the 1930s. Classic Marx toys like Popeye the Champ and Popeye and Olive Oyl Jiggers have continually topped price lists not only because they are superb examples of the toy makers' art, but also because their characters continue to charm to this day.

Durability translates into collectibility, and the power to draw new collectors from each maturing generation. The universal and continuing appeal of Mickey and Minnie Mouse, Superman, Snoopy, Popeye, Batman and others are assurances that future generations of collectors will seek out their toys and place them proudly on their shelves, someday perhaps right next to their 50-year-old Buzz Lightyear action figures.

Trends

Character toys see little chance of waning popularity, with new characters cropping up with every movie, TV show and book. Especially hot in 1998 were toys related to blockbuster movies, like *Godzilla* and *The X-Files*, and the ever-popular rock group KISS. A renewed interest in 1960s and 1970s pop culture generated lots of interest in vintage games, lunch boxes, dolls and other toys from that era — particularly those that remembered TV shows like *The Brady Bunch*, *The Dukes of Hazzard* and *The Partridge Family*.

Among newer toys, Batman and Superman toys failed to generate as much interest as vintage toys bearing those superheroes. And while new Superman toys don't excited collectors much today, Superman premiums from the 1940s continue to lead lists of the top-selling character collectibles.

Demand for a particular character may ebb and flow according to times and market whims, but overall this is one of the strongest and most reliable areas of toy collecting. Some Good prices were lowered, accounting for the increasing desire for toys in Excellent or better condition.

The Top 10 Character Collectibles
(in Mint condition)

1. Action Comics #1, DC Comics, 1938, first appearance of Superman . . . $100,000
2. Superman Member Ring, 1940 100,000
3. Superman Gum Ring, Gum, Inc., 1940 40,000
4. Superman Candy Ring, Leader Novelty Candy, 1940 22,000
5. Superman Trading Cards, Gum Inc., 1940, set of 72 10,000
6. Superman Patch, 1939, Supermen of America, Action Comics 10,000
7. Superman-Tim Club Ring, 1940s 10,000
8. Captain Marvel Sirocco Figurine, Fawcett, 1940s 8,500
9. Donald Duck Bicycle, Shelby, 1949 6,000
10. Superman Patch, 1940s 5,500

Contributor to this section: John Snyder, Jr., Diamond International Galleries, 1966 Greenspring Dr., Suite 401, Timonium, MD 21093.

Nightmare Before Christmas, Lock, Shock and Barrel dolls, Applause, 1990s

CHARACTER TOYS

101 DALMATIANS

TOY	COMPANY	YEAR	DESCRIPTION	GOOD	EX	MIB
101 Dalmatians Snow Dome	Marx	1961	3" x 5" x 3-1/2" tall	35	65	100
101 Dalmatians Wind-Up	Linemar	1959		60	115	375
Dalmatian Pups Figures	Enesco	1960s	4-1/2" tall, set of three	12	23	125
Lucky Figure	Enesco	1960s	4" tall	35	65	100
Lucky Squeeze Toy	Dell		7" tall, squeakers in the bottom	5	10	35

ADDAMS FAMILY

TOY	COMPANY	YEAR	DESCRIPTION	GOOD	EX	MIB
Gomez Hand Puppet	Ideal	1965		50	100	210
Lurch Figure	Remco	1964		75	150	320
Morticia Figure	Remco	1964		100	200	620
Morticia Halloween Costume	Ben Cooper	1964	with hair	50	125	250
Morticia Halloween Costume	Ben Cooper	1964	painted hair	30	90	180
Morticia Hand Puppet	Ideal	1965		55	100	230
Thing Bank		1964	plastic, battery-operated	35	75	200
Uncle Fester Figure	Remco	1964		100	250	600

ALICE IN WONDERLAND

TOY	COMPANY	YEAR	DESCRIPTION	GOOD	EX	MIB
Adventures in Costumeland Game	Walt Disney World	1980s	created for Disney costume division members, game board and pieces in a small vinyl garment bag	75	125	200
Alice and White Rabbit Mug	TDL/Daiichi Seimei	1991	promo piece	20	35	50
Alice Bank	Leeds	1950s	figural	50	80	125
Alice Cookie Jar	Regal		13-1/2" tall	950	1600	2500
Alice Cookie Jar	Leeds	1950s	printed in relief	70	115	175
Alice Costume	Ben Cooper	1950s	costume made until 1970s	20	35	65
Alice Disneykin	Marx	1950s	hard plastic	8	16	30
Alice Disneykin	Marx	1950s	unpainted, soft plastic	10	16	25
Alice Doll	Duchess	1951	#739 7-1/2" tall	30	65	120
Alice Doll	Duchess	1951	12-1/2" tall	40	80	160
Alice Doll	Gund	1950s	flat vinyl, stuffed	15	30	60
Alice Doll	Horsman	1970s	Alice has a castle on her apron	15	30	60
Alice Doll	Pedigree	1970s	Great Britain	15	20	60
Alice Doll	Horsman	1970s	#1071, Walt Disney Classics	15	30	60
Alice Figure	Shaw	1951		100	200	350
Alice Figure	Bully	1984	Germany, PVC, wearing a blue or red dress, each	4	8	15
Alice Figure	Hagen-Renaker	1956		100	250	400
Alice Figure	Marx	1950s	painted with "Holland" stamped on the bottom	10	20	40
Alice Figure	Aldon Industries	1950s	plastic, cut-out standup	25	50	100
Alice Figure	Sears	1980s	Magic Kingdom Collection, bone china	8	15	30
Alice in Wonderland and Cinderella Book	Collins	1950s	Great Britain version	15	40	50
Alice in Wonderland Big Coloring Book	Whitman	1951	#301 Big Golden	15	35	75
Alice in Wonderland Book	Whitman	1951	#426, Big Golden Book	10	30	65
Alice in Wonderland Book	Whitman	1950s	#10426, Big Golden Book, gold foil backing	10	25	50

ALICE IN WONDERLAND

TOY	COMPANY	YEAR	DESCRIPTION	GOOD	EX	MIB
Alice in Wonderland Book	Whitman	1950s	#10426, without gold backing	5	15	30
Alice in Wonderland Book	Whitman	1950s	#2074 Cozy Corner Book, green endpapers	7	15	40
Alice in Wonderland Book	Dell Publishing	1951	#331	20	40	85
Alice in Wonderland Book	Whitman	1951	Sandpiper Book with dust jacket	10	20	60
Alice in Wonderland Card Game	Thos. De LaRue	1980s		2	5	10
Alice in Wonderland Classic Series with Disney Book	Whitman	1950s	#2140 Lewis Carroll text with Disney dust jacket	10	20	40
Alice in Wonderland Coloring Book	Whitman	1974	#1049	5	10	30
Alice in Wonderland Paint Book	Whitman	1951	#2167	10	25	75
Alice in Wonderland Punch-Out Book	Whitman	1951	#2164	25	60	150
Alice in Wonderland Sticker Fun	Whitman	1950s	#2193 stencil and coloring book	15	25	60
Alice Marionette	Peter Puppet	1950s	comes in two different boxes, one "Alice in Wonderland" the other Peter Puppet Disney	30	65	200
Alice Meets the White Rabbit Book	Whitman	1951	#D-19, Little Golden Book	10	20	50
Alice Mug	Disney	1970s		10	20	35
Alice Paper Dolls	Whitman	1976	#1948	8	20	30
Alice Paper Dolls	Whitman	1972	#4712	10	20	30
Alice Snow Dome	Marx	1961	3" tall featuring Alice & the White Rabbit in front of tree	12	25	60
Alice Stationery and Notepad	Pak-Well	1970s	#77065 with a fan card cover	8	15	25
Alice Wristwatch	U.S. Time	1950s	Alice peeking through pink flowers and a plastic statue	75	165	350
Alice Wristwatch	Alba	1990s	Japan, gold tone face	10	25	40
Alice Wristwatch	U.S. Time	1950s	came with ceramic statue	75	165	350
Balloons	Oak Rubber	1951	four different designs	10	16	25
Balloons	Eagle Rubber	1951		10	16	25
Blocks	Chad Valley	1950s	five cylindrical tin blocks with color graphics	50	120	200
Bread Labels	NBC	1950s	12 different styles, each	8	16	30
Bread Seal Poster	NBC	1950s		10	25	50
Bread Stickers	Continental Baking/ Wonder	1974	five styles, each	3	5	10
Bridge Card Game	Whitman	1950s	two decks of cards	30	60	120
Canasta Card Game	Whitman	1950s	two Canasta decks of White Rabbit cards	25	60	100
Candy Tin	Edward Sharp & Sons	1950s	Great Britain English Toffee Tin	50	75	150
Cards	Royal Desserts	1951	16 different cards on the back of dessert packages, each	8	15	30
Caterpillar Figure	Hagen-Renaker	1956		100	250	400
Cereal Box with Record	General Mills	1956	Wheaties	60	150	250
Cheshire Cat Costume Pattern	McCall's	1951		20	35	60
Cheshire Cat Doll	Disneyland	1970s	plush	8	20	35
Child's Vanity	Neevel	1950s	illustrated with film scenes	40	75	150
Clock Radio	General Electric	1970s	features characters on face	15	35	60
Disneykin Play Set	Marx	1950s		150	325	600
Disneyland Figures Set	United China	1960s	large set, each piece	10	25	40

ALICE IN WONDERLAND

TOY	COMPANY	YEAR	DESCRIPTION	GOOD	EX	MIB
Disneyland Figures Set	United China	1970s	small figures, each piece	8	15	25
Dormouse Doll	Lars/Italy	1950s	stuffed	150	325	500
Dormouse Figure	Shaw	1951		120	200	350
Fan Card	Walt Disney	1951	original release with 1973 invitation to studio screening	30	60	75
Fan Card	Walt Disney	1951	premium sent to fans who wrote letters to studio	10	25	50
Film Viewer	Tru-Vue	1950s	viewer and filmstrip	25	50	75
Glass	Pepsi Cola	1970s	featuring Alice, part of Wonderful World of Disney set	8	16	25
Glasses	Libbey	1951	eight styles released for film opening, each	35	60	100
Handbag	Salient	1950s	child's pink vinyl shoulder bag, Alice with rocking fly	20	35	50
Handbag	ACME Briefcase	1950s	child's red leather shoulder bag	20	35	50
Hatbox	Neevel	1950s	Caterpillar or Tea Party graphics	50	80	125
Jingle Ball	Vanguard	1951		10	20	35
Looking Glass Cookie Jar	Fred Roberts		raised characters on the body of the jar with a mirror lid	125	260	500
Mad Hatter Costume	Ben Cooper	1950s	costume made until 1970s	20	35	60
Mad Hatter Costume Pattern	McCall's	1951		15	30	50
Mad Hatter Disneykin	Marx	1950s		20	40	75
Mad Hatter Disneykin	Marx	1950s	unpainted, soft plastic	10	16	25
Mad Hatter Doll	Gund	1950s	plush	100	225	350
Mad Hatter Doll	Gund	1950s	flat vinyl, stuffed	15	35	60
Mad Hatter Figure	Sydney Pottery	1950s	large size, sold only in Australia	195	325	500
Mad Hatter Figure	Marx	1950s	painted with "Holland" stamped on the bottom	10	16	30
Mad Hatter Figure	Shaw	1951		60	150	300
Mad Hatter Figure	Schmid	1980s	playing xylophone	15	35	60
Mad Hatter Figure	Hagen-Renaker	1956		100	225	350
Mad Hatter Hand Puppet		1960s	hand puppet with cloth body	15	25	40
Mad Hatter Marionette	Peter Puppet	1950s		40	80	150
Mad Hatter Nodder	Marx	1950s		25	75	150
Mad Hatter Snap Eeze	Marx	1950s		10	20	30
Mad Hatter Snow Dome	New England Collectors Society	1980s	crystal, Mad Hatter, St. Patrick's Day	15	25	40
Mad Hatter Teapot	Regal	1950s		450	975	1500
Mad Hatter's Tea Party Book	Whitman	1951	#D-23 Little Golden Book	15	20	25
Mad Hatter/March Hare Figure	TDL	1980s	Mad Hatter, March Hare with saxophone	20	35	50
Mad Tea Party/Cheshire Cat Mug	Applause	1988	Cheshire Cat handle with Tea Party on the mug	10	16	25
Magic Picture Kit Set	Jiffy Pop	1974	set of four	10	16	30
Make-Up Kit	Hasbro	1951		20	50	75
March Hare Costume	Ben Cooper	1950s	costume made until 1970s	15	35	60
March Hare Costume Pattern	McCall's	1951	pattern	10	25	50
March Hare Disneykin	Marx	1950s	unpainted and soft plastic	10	16	25
March Hare Doll	Gund	1950s	flat vinyl, stuffed	20	40	60
March Hare Doll	Gund	1950s	plush	150	325	500
March Hare Figure	Hagen-Renaker	1956		100	225	400
March Hare Figure	Shaw	1951		150	300	450
March Hare Figure	Marx	1950s	painted with "Holland" stamped on the bottom	10	16	30

ALICE IN WONDERLAND

TOY	COMPANY	YEAR	DESCRIPTION	GOOD	EX	MIB
March Hare Figure	Sydney Pottery	1950s	large size, sold only in Australia	150	325	500
March Hare Marionette	Peter Puppet	1950s		30	65	120
March Hare Snap Eeze	Marx	1950s		15	20	30
March Hare Twistoy	Marx	1950s		10	16	30
Molding Set	Model Craft	1951		40	65	120
Molding Set	Great Britain	1952	similar to Model Craft set	40	90	150
Music Box	TDL	1980s	ceramic teacup	40	65	125
Music Box	TDL	1980s	plastic, tea cup rotates	40	65	125
Music Box	Disneyland	1980s	Wooden box features Alice and White Rabbit "I'm Late"	25	40	75
Picture Frame	Dexter-Mahnke	1970s	cloth picture frame, featuring Mad Tea Party	15	30	40
Pitcher	Regal	1950s	King of Hearts	150	325	500
Plaque	Disneyland	1970s	wooden, with Alice and live flowers	10	20	35
Poster	Kraft	1980	Disneyland 25th Anniversary Family Reunion	10	25	50
Poster	Disneyland	1958	Alice attraction	150	500	800
Poster	Walt Disney World	1980s	costume division poster featuring Cheshire Cat	10	25	50
Puppet Theatre	Peter Puppet	1950s		80	130	225
Puzzle	Jaymar	1951	Alice and Rabbit	15	50	75
Puzzle	Jaymar	1951	Alice under a tree	15	50	75
Puzzle	Jaymar	1951	Croquet cast scene	15	50	75
Puzzle	Stafford/ England	1979	wooden, Great Britain Tea Party	8	20	35
Puzzle	TDL	1980s	#18, Mad Tea Party and Cast	5	10	20
Puzzle	Jaymar	1951	Tea Party scene	15	50	75
Queen of Hearts Card Game	Edu-Cards	1975		15	25	40
Queen of Hearts Disneykin	Marx	1950s		25	50	75
Queen of Hearts Disneykin	Marx	1950s	unpainted, soft plastic	15	20	30
Queen of Hearts Doll	Gund	1950s	flat vinyl, stuffed	20	50	75
Queen of Hearts Figure	Marx	1950s	painted with "Holland" stamped on the bottom	8	16	25
Queen of Hearts Figure	Sears	1980s	Magic Kingdom Collection, bone china	10	20	30
Ramp Walker	Marx	1950s	Mad Hatter and White Rabbit	15	35	60
Record Player	RCA Victor	1951	45 rpm	75	125	150
Record/Little Nipper Giant Storybook	RCA Victor	1951	LY-437, 33, 45, or 78 rpm, each	50	100	150
Rubber Stamp Set	All Night Media	1989		5	10	20
Rubber Stamp Set	Multiprint	1970s	Italy #177	15	30	50
Salt and Pepper Shakers	Regal	1950s	blue or white featuring Alice	150	300	500
Salt and Pepper Shakers	Regal	1950s	featuring TweedleDee and TweedleDum	150	300	500
School Bag	ACME Briefcase	1950s	fabric and leather	25	65	100
Sewing Cards	Whitman	1951		25	50	100
Sewing Kit	Hasbro	1951	7" sewing machine and 5" doll, all plastic	35	75	125
Soap Set		1951		30	60	125
Tea Cake Box	TDL	1980s	with a Mad Tea Party lid	5	10	25
Tea Cup and Saucer	TDL	1986	ceramic, Mad Tea Party	60	120	200
Tea Cup Wristwatch	U.S. Time	1950s	picture of Mad Hatter with an overlay	150	300	500
The Unbirthday Party Book	Whitman	1974	#22, Walt Disney Showcase	4	15	20

ALICE IN WONDERLAND

TOY	COMPANY	YEAR	DESCRIPTION	GOOD	EX	MIB
Thimble	New England Collectors Society	1980s	Mad Hatter and Alice	5	13	20
Ticket	Disneyland	1970s	employee screening ticket featuring Cheshire Cat	5	10	20
TV Scene, White Rabbit and March Hare	Marx	1950s		30	65	100
TweedleDee and TweedleDum Dolls	Gund	1950s	flat vinyl, stuffed, each	20	40	60
TweedleDee and TweedleDum Dolls	TDL	1980s	plush, each	15	30	50
TweedleDee Figure	Shaw	1951		60	130	200
TweedleDee Figure	Sydney Pottery	1950s	large size, sold only in Australia	150	325	500
TweedleDum Figure	Shaw	1951		60	130	200
TweedleDum Figure	Sydney Pottery	1950s	large size, sold only in Australia	150	325	500
Vase	Enesco	1960s	featuring Alice's head	50	100	150
View-Master Set	GAF	1970s	three reels, Disney	8	13	20
Wall Decor	Dolly Toy	1951	#260 contains Alice, Mad Hatter, March Hare and a lamp	50	100	175
Wall Plaque	Leisuramics	1974	bisque, oval shape, featuring Mad Hatter	15	35	50
Wall Plaque	Leisuramics	1974	bisque, oval shape, featuring TweedleDee and TweedleDum	15	35	50
Wallet	Salient	1950s	vinyl, featuring Mad Tea Party	15	35	50
Wallet	Salient	1950s	vinyl, featuring White Rabbit	15	35	50
Walrus Doll	Lars/Italy	1950s	stuffed	150	325	500
Walrus Figure	Shaw	1951		100	225	350
Walrus Figure	Sydney Pottery	1950s	large size, sold only in Australia	150	325	500
White Rabbit Creamer	Regal	1950s		100	250	400
White Rabbit Disneykin	Marx	1950s		25	50	75
White Rabbit Disneykin	Marx	1950s	unpainted and soft plastic	10	20	30
White Rabbit Doll	Sears	1970s	plush with waistcoat and umbrella	10	20	30
White Rabbit Doll	Buena Vista/ Disney	1974	plush	50	100	150
White Rabbit Doll	TDL	1980s	plush	10	25	50
White Rabbit Doll	Gund	1950s		100	225	350
White Rabbit Doll	Gund	1950s	flat vinyl, stuffed	15	35	50
White Rabbit Doll	Disneyland	1970s	small with black spectacles	8	15	25
White Rabbit Doll	Disneyland	1970s	large, plush with yellow spectacles	10	20	30
White Rabbit Ears	TDL	1980s		5	15	25
White Rabbit Figure	Sears	1980s	Magic Kingdom Collection, bone china	8	16	30
White Rabbit Figure	Shaw	1951		60	130	200
White Rabbit Figure	Marx	1950s	painted with "Holland" stamped on the bottom	8	16	25
White Rabbit Figure	Italy		5-1/2" tall, ceramic	15	35	50
White Rabbit Rolykin	Marx	1950s		15	35	50
White Rabbit Sugar Bowl	Regal	1950s		150	325	500

AMOS & ANDY

TOY	COMPANY	YEAR	DESCRIPTION	GOOD	EX	MIB
Amos & Andy Card Party	A.M. Davis	1930	6" x 8", score pads & tallies	20	75	150
Amos & Andy Fresh Air Taxi	Marx	1930s	5" x 8" long, tin wind-up	500	1000	1750
Amos Wind-Up	Marx	1930	12" tall, tin	500	1000	1750
Puzzle	Pepsodent	1932	8-1/2" x 10", pictured Amos, Andy and other characters	25	100	175

ANDY GUMP

TOY	COMPANY	YEAR	DESCRIPTION	GOOD	EX	MIB
Andy Gump Automobile	Arcade		7" x 6", cast iron with a large figure	350	650	1200
Brush & Mirror			4" diameter, red on ivory colored surface of brush	20	35	75
Chester Gump Playstone Funnies Mold Set		1940s		50	100	150
Chester Gump/Herby Nodders		1930s	ceramic 2-1/4" string nodders, each	50	150	250

ARCHIES

TOY	COMPANY	YEAR	DESCRIPTION	GOOD	EX	MIB
Archie Halloween Costume	Ben Cooper	1969		15	50	75
Archies Paper Dolls	Whitman	1969		15	35	50
Jalopy	Marx	1975	12", plastic	25	50	75
Puzzle	Jaymar	1960s	"Swinging Malt Shop"	10	30	50

ATOM ANT & FRIENDS

TOY	COMPANY	YEAR	DESCRIPTION	GOOD	EX	MIB
Atom Ant Kite	Roalex	1960s		20	75	125
Atom Ant Punch-Out Set	Whitman	1966		30	100	175
Atom Ant Push Puppet	Kohner	1960s		20	50	75
Atom Ant Puzzle	Whitman	1966		15	40	75
Atom Ant Soaky	Purex	1966		15	50	75
Morocco Mole Bubble Club Soaky	Purex	1960s	7" hard plastic	20	50	100
Squiddly Diddly Bubble Club Soaky	Purex	1960s	10-1/2" hard plastic	20	50	100
Winsome Witch Bubble Club Soaky	Purex	1960s	10-1/2" hard plastic	20	50	100

BABES IN TOYLAND

TOY	COMPANY	YEAR	DESCRIPTION	GOOD	EX	MIB
Babes in Toyland Go Mobile Friction Car	Linemar	1961	4" x 5"x 6"	60	115	375
Babes in Toyland Hand Puppets	Gund		Silly Dilly Clown, Soldier, or Gorgonzo, each	35	65	100
Babes in Toyland Twist 'N Bend Toy	Marx	1963	4" tall flexible toy with Private Valiant holding a baton	10	25	45
Babes in Toyland Wind-Up Toy	Linemar	1950s	tin	75	135	375
Cadet Doll	Gund		15-1/2" tall, fabric	15	30	75
Puzzle	Jaymar	1961		15	25	40

BAMBI

TOY	COMPANY	YEAR	DESCRIPTION	GOOD	EX	MIB
Bambi Book	Grosset & Dunlap	1942	black & white illustrations	15	25	45
Bambi Prints	New York Graphic Society	1947	11" x 14" framed	25	50	100
Bambi Soaky	Colgate-Palmolive			10	25	45
Flower Bank		1940s	5" x 5"x 7" tall, plaster	45	90	150
Lamp			Bambi and Thumper	20	50	125
Throw Rug		1960s	21" x 39", Bambi and Thumper	25	50	75
Thumper Ashtray	Goebel	1950s	4" tall	25	50	125
Thumper Bank	Leeds	1950s	ceramic, figural	40	80	150
Thumper Book	Grosset & Dunlap	1942	color & black/white illustrations	10	25	45
Thumper Doll			16" tall, plush	12	25	35

BAMBI

TOY	COMPANY	YEAR	DESCRIPTION	GOOD	EX	MIB
Thumper Pull Toy	Fisher-Price	1942	#533, 7-1/2" x 12", wood & metal, Thumper's tail rings the bell	25	60	125
Thumper Soaky	Colgate-Palmolive	1960s		15	50	75

BANANA SPLITS

TOY	COMPANY	YEAR	DESCRIPTION	GOOD	EX	MIB
Banana Band	Larami	1973	horn, sax, mouth harp	20	50	75
Banana Buggy Model Kit	Aurora	1968		50	150	275
Banana Splits Bingo Costume	Ben Cooper	1968		50	125	250
Banana Splits Doll	Sutton	1960s	12" tall Drooper	30	100	175
Banana Splits Kut-Up Kit	Larami	1973		15	40	60
Banana Splits Mug		1969	plastic yellow dog mug	10	40	60
Banana Splits Record	Kellogg's	1969		15	50	75
Paint-By-Number Set	Hasbro	1969		40	75	125
Puzzle	Whitman	1969	frame tray	15	50	65
Talking Telephone	Hasbro	1969		60	150	275

BARNEY GOOGLE

TOY	COMPANY	YEAR	DESCRIPTION	GOOD	EX	MIB
Barney Google & Spark Plug	Nifty	1920s	7-1/2" tall, wind-up	350	750	1500
Barney Google & Spark Plug Figurines			3" x 3", bisque, on white bisque pedestal	40	125	175
Barney Google Doll	Schoenhut	1922	8-1/2" tall, wood & wood composition	175	400	850
Spark Plug Doll	Schoenhut	1922	9" long x 6-1/2" tall, jointed wood construction with fabric	175	400	850
Spark Plug in Bathtub		1930	5" long die-cast, white	90	200	300
Spark Plug Pull Toy			10" x 8" tall, wood	80	145	225
Spark Plug Squeaker Toy		1923	5" long, rubber with squeaker in mouth	20	60	125
Spark Plug Toy		1920s	5" tall, wood, on wheels	45	125	200

BATMAN

TOY	COMPANY	YEAR	DESCRIPTION	GOOD	EX	MIB
Bat Bomb	Mattel	1966		30	75	150
Bat Cycle	Toy Biz	1989		5	10	15
Bat Machine	Mego	1979		25	50	100
Bat Ring		1966	yellow plastic, originally for a gumball machine	25	50	75
Bat-Troll Doll	Wish-Nik	1966	vinyl, dressed in a blue felt Batman outfit with cowl and cape	75	150	300
Batboat	Duncan	1987		10	20	35
Batboat Pullstring Toy	Eidai		made in Japan	50	100	175
Batcave Play Set	Mego	1974	vinyl	125	250	450
Batcoin Lot	Space Magic Limited	1966	Four 1-1/2" diameter metal coins, each depicting a scene featuring Batman and Robin battling villains	20	40	75
Batman "Punch-O" Drink Mix		1966	small paper packet	20	30	50
Batman & Robin Society Membership Button	Button World	1966	full color litho metal button featuring Batman and Robin and the words "Charter Member-Batman and Robin Society"	10	20	30

CHARACTER

BATMAN

TOY	COMPANY	YEAR	DESCRIPTION	GOOD	EX	MIB
Batman & Superman Record Album	Wonderland	1969-71	45 rpm record Batman theme song from the 1966 television show and "The Superman Song"	15	50	75
Batman 3-D Comic Book	DC Comics	1966	9" x 11" comic with 3-D pages and glasses	35	100	250
Batman and Robin Hand Puppets	Ideal	1966	12" soft vinyl head, plastic body, each	35	125	175
Batman and Robin Valentine		1966		5	15	25
Batman Annual		1965-66	8" x 10" hardback annual contains reprinted stories from 1950s Batman and Detective Comics	15	30	50
Batman Arcade Game	Bluebox	1989	electronic	50	100	150
Batman Bank		1989	figural bank given away with Batman Cereal	4	7	10
Batman Bank		1966	7" tall glazed china figural bank depicts Batman with hands on hip	45	90	125
Batman Batarang Toss	Pressman	1966		125	250	400
Batman Bendy Figure	Diener	1960s	on card	30	60	80
Batman Candy Box	Phoenix Candy	1966	2-1/2" x 3-1/2" x 1", several color scenes	25	75	150
Batman Cartoon Kit	Colorforms	1966		20	50	80
Batman Cast and Paint Set		1960s	plaster casting mold and paint set	35	75	150
Batman Cereal Box	Ralston	1989		15	50	75
Batman Comic Book and Record Set	Golden Records	1966	33-1/3 rpm record, full size Batman comic book, official Batman membership card with secret Batman code on back	30	60	120
Batman Crazy Foam		1974		20	40	80
Batman Crusader Sundae Fudgesicle	Popsicle	1966	7" white and brown paper wrapper	5	10	20
Batman Dinner Set		1966	ceramic; three pieces	35	70	140
Batman Drinking Glass	Pepsi	1976	7" tall glass tumbler, all Batman characters, each	10	20	25
Batman Figure	Ideal	1966	3" yellow plastic, detachable gray plastic cape	10	25	45
Batman Figure	Takara/ Japan	1989		40	80	150
Batman Figure	Presents	1989	15-1/2" tall vinyl and cloth figure on base, 1970s logo, metal stand	20	40	75
Batman Figure	Applause	1988	15" tall with stand	10	20	40
Batman Figure	Bully	1989	7" bendy	5	10	20
Batman Figure	Billiken	1989	8" on card	5	10	20
Batman Figure and Parachute	CDC	1966	11" x 9" card, metallic blue figure of Batman and working parachute	20	40	80
Batman Figure with Flyaway Action	Mego	1976	12-1/2" tall	50	100	200
Batman Flying Copter	Remco	1966	12" plastic with guide-wire control	35	80	150
Batman Flying Figure on String	Ben Cooper	1973	6" rubber figure of Batman with rubber cape	10	20	35
Batman Fork	Imperial	1966	6" stainless steel with embossed figure of Batman, with "Batman" engraved towards the bottom	8	25	50
Batman Halloween Costume	Ben Cooper	1965	plastic Halloween mask and purple and yellow cape, several versions, some feature logo on chest	15	30	60

BATMAN

TOY	COMPANY	YEAR	DESCRIPTION	GOOD	EX	MIB
Batman Helmet and Cape Set	Ideal	1966	blue hard plastic cowl shaped helmet and soft blue vinyl cape with drawstring	100	225	450
Batman Inflated Gliding Figure	Ideal	1966	16" soft plastic inflatable Batman with free flowing cape and hard plastic cable rail	25	50	100
Batman Kite	Hiflyer	1982		5	10	20
Batman Lamp	Vanity Fair		Made in Taiwan	40	70	140
Batman Lobby Display	Warner Bros.	1989	Michael Keaton cardboard stand up	75	125	175
Batman Lucky Charm Display Card		1966	4" x 4" paper display card used in bubble gum machines, card shows Bat logo and red "Be protected--Get your Batman lucky charm now"	10	20	35
Batman Meets Blockbuster Coloring Book	Whitman	1966	40 pages	15	30	60
Batman on a String Figure	Fun Things	1966	4" rubber, flexible arms, legs and removable cape	15	30	60
Batman Paint by Number Set	Hasbro	1965	five pre-numbered sketches, ten oil paint vials and brush	40	85	175
Batman Pencil Box	Empire Pencil	1966	gun-shaped pencil box with set of Batman pencils	20	50	100
Batman Pillow		1966	10" x 12" with 1940s logo	20	40	75
Batman Pinball Game	Marx	1960s	tin litho with plastic casing	30	75	150
Batman Play Set	Ideal	1966	11 pieces including characters and vehicles	300	600	1200
Batman Postcards	Dexter Press	1966	set of eight postcards with Carmine Infantino artwork	30	60	120
Batman Postcards	Dexter Press	1966	three full color postcards, each taken from a comic panel from Batman comics, each	5	10	20
Batman Push Puppet	Kohner	1966	3" plastic with push button on bottom	20	45	85
Batman Radio Belt and Buckle		1966		25	75	150
Batman Record	SPC	1966	45 rpm, sleeve shaped like Batman's head; also available in Robin, Joker, Penguin, Riddler and Batmobile versions	15	35	60
Batman Returns Display	Warner Bros.	1992		10	50	75
Batman Returns Lobby Display	Warner Bros.	1992	Michael Keaton life-size cardboard stand up	30	75	150
Batman Returns Watch	Consort	1989	gray or yellow Bat logo	10	20	35
Batman Road Race Set		1960s	slot car racing set	100	250	350
Batman Slot Car	Magicar (England)	1966	5" long Batmobile being driven by Batman and Robin in illustrated display window box	100	200	400
Batman Soundtrack Record	20th Century Fox	1966	mono and stereo versions, each	50	140	250
Batman String Puppet	Madison	1977		35	75	150
Batman Super Powers Stain & Paint Set		1984		5	15	25
Batman Superfriends Lite Brite Refill Pack		1980		5	15	25
Batman Superhero Stamp Set		1970s		10	20	40
Batman Switch & Go Play Set	Mattel	1966	9" plastic Batmobile, 40 feet of track, figures, etc.	125	250	500
Batman Target Game	Hasbro	1966	tin litho target with plastic revolver and rubber-tipped darts	40	80	160

BATMAN

TOY	COMPANY	YEAR	DESCRIPTION	GOOD	EX	MIB
Batman Trace-a-Graph	Emenee	1966		30	75	150
Batman Utility Belt	Ideal	1960s		300	1000	1500
Batman vs. the Joker Book	Signet	1966	160-page paperback	5	10	15
Batman Wastepaper Basket		1966	10" tall, color tin litho	30	60	125
Batman Wind-Up	Billiken	1989	Tin litho	25	50	100
Batman Wristwatch	Quintel	1991	digital	6	12	18
Batman Yo-Yo	SpectraStar	1989		3	6	12
Batman/Robin Flicker-Flasher Ring	Vari-Vue	1966	silver plastic base	10	20	25
Batmobile	Aoshinu (Japan)	1980s	motorized	25	45	90
Batmobile	AHI	1972	11" long tin litho battery-operated mystery action car with blinking light and jet engine noise	65	130	260
Batmobile	Bandai	1980s	pullback vehicle with machine guns	20	50	100
Batmobile	Matsushiro		radio-controlled	50	100	200
Batmobile	Apollo (Japan)		radio-controlled	50	100	200
Batmobile	Azrak-Hamway	1974	battery operated	35	75	150
Batmobile	Toy Biz	1989	remote control	10	25	50
Batmobile	Rich Man's Toys	1989	remote control	85	160	350
Batmobile	Simms	1960s	plastic car on card	15	30	60
Batmobile	Duncan	1977	12" x 8" on card	20	45	90
Batmobile AM Radio	Bandai	1970s		35	80	150
Batmobile Display Sign	Burry's	1969	34" x 48" die-cut 3-D plastic story display, raised images of Batman, Robin, and Batmobile, bright orange with yellow lettering	400	750	1500
Batmobile Motorized Kit	Aoshinu (Japan)	1980s	smaller snap kit	15	30	60
Batphone	Marx	1966		60	120	225
Batscope Dart Launcher	Tarco	1966		20	40	80
Batwing	Toy Biz	1980s		5	15	35
Beach Towel		1966	34" x 58" white, Batman hitting a crook	40	95	180
Bread Wrapper	New Century Bread	1966	plastic	20	45	60
Button Display Card		1966	full color display card used in bubble gum machines which offered Batman buttons	10	35	50
Cake Decoration		1960s	2" hard plastic figure of Robin or Batman, each	8	15	20
Cake Decorations	Space Magic Limited	1966	4" plastic one dimensional figures of Batman, Robin and old 1940s Batman logo	15	30	60
Candy Cigarettes		1960s	made in England	15	30	60
Catwoman Iron-On Patch		1966	Catwoman with the words "Batkids Fan Club"	10	50	75
Catwoman Returns	Horizon	1990s	vinyl model kit	10	20	40
Catwoman Watch	Consort	1991	Batman Returns	7	15	30
Catwoman Watch	Quintel	1991	digital	5	10	18
Catwoman's Revenge Record	Power Records	1975	33-1/3 rpm story record	5	10	18
Cave Tun-L	New York Toy	1966	26" x 26" x 2" tunnel	500	1100	2100
Charm Bracelet		1966	on card	25	50	100
Child's Belt		1960s	elastic with bronze logo buckle	20	40	60
Child's Dinner Plate	Boontonware	1966	7" plastic with image of Batman and Robin	15	30	40

BATMAN

TOY	COMPANY	YEAR	DESCRIPTION	GOOD	EX	MIB
Child's Mittens		1973	children's blue plastic vinyl, raised illustration of Batman and logo	10	20	40
Child's Pajamas	Wormser	1966	light blue, two piece pajamas, full color Batman logo on chest	200	450	900
Chocolate Milk Carton	Reiter & Hart	1966	one-quart carton in yellow, red and brown, features front and back panels of Batman in action poses	150	300	660
Christmas Ornament	Presents	1989		5	10	18
Coffee Mug	Anchor-Hocking	1966	milk glass, action pose of Batman on one side and the Bat logo on the opposite side	15	30	50
Coins	Transogram	1966	plastic, set	20	40	75
Costume Patterns	McCalls	1960s	patterns for making Batman, Robin, and Superman costumes, paper envelope, each	15	30	60
Costume Store Poster	Ben Cooper	1966	12" x 24", yellow	25	100	150
Dot-To-Dot & Coloring Book	Vasquez Brothers	1967	Batman with Robin the Boy Wonder, printed in the Phillipines, 20 pages	20	40	75
Drinking Glass		1989	5", made in France	10	20	35
Escape Gun	Lincoln	1966	red plastic spring-loaded pistol with Batman decal, two separate firing barrels	30	60	120
Flicker Pictures Display Card		1966	bubble gum machine display card	10	25	45
Frame Tray Puzzle	Whitman	1966	11" x 14", Batman and Robin thwarting the Joker	15	30	60
From Alfred to Zowie! Book	Golden Press	1966		10	20	35
Give-A-Show Projector Cards	Kenner	1960s	four slide cards in box	10	25	50
Glow-in-the Dark Poster	Ciro Art	1966	18" x 14" poster of Batman and Robin swinging across Gotham City	10	50	75
Gotham City Stunt Set	Tonka	1989		15	30	60
Inflatable TV Chair		1982		10	20	35
Jelly Jar	W.H. Marvin	1966	5"-6" glass jar with color label, "Bat" Pure Apple Jelly	250	500	700
Joker Bank	Mego	1974	plastic	25	50	100
Joker Cereal Bowl	Sun Valley	1966	5" hard plastic	10	20	40
Joker Figure	Presents		15" vinyl figure	10	20	35
Joker Figure	Ideal	1966	3" blue plastic	10	20	40
Joker Figure	Applause	1988	vinyl with stand	10	20	35
Joker Record	SPC	1966	45 rpm, sleeve shaped like Joker's head	20	40	75
Joker Van	Ertl	1989	die-cast vehicle on card	4	8	12
Joker Wind-up	Billiken	1989		40	80	150
Joker Wristwatch	Fossil	1980s		25	50	75
Joker Wristwatch	Quintell	1989	digital	10	15	25
Joker Yo-Yo	SpectraStar	1989		5	10	18
Lapel Pin	Mamsell	1966	2" bat-shaped metal, black with yellow eyes	15	30	50
Life Magazine		1966	March 11, 1966 issue, Adam West as Batman on cover	15	50	75
Magic Magnetic Gotham City Play Set	Remco	1966	cardboard city, character figures	200	450	850
Mug		1966	5" clear plastic; color wrap around sheet	30	60	100
Official Bat-Signal Stickers	Alan-Whitney	1966		10	20	35
Paint-By-Number Book	Whitman	1966		20	40	75

BATMAN

TOY	COMPANY	YEAR	DESCRIPTION	GOOD	EX	MIB
Party Hat	Amscan/ Canadian	1972	7" child's cardboard hat depicts Batman and Robin	8	15	25
Penguin Returns Model Kit	Horizon			10	20	40
Pennant		1966	11 x 29" white felt, illustration of the Dynamic Duo swinging on ropes with the Bat-signal in the background	15	75	100
Projector Gun	Toy Biz	1989		8	15	25
Puppet Theater Stage	Ideal	1966	marketed by Sears, 19" x 11" x 20" cardboard stage with hand puppets	75	200	350
Ray Gun		1960s	7" long blue and black futuristic space gun with bat sights and bats on handgrip	80	250	375
Riddler/Batman Punching Riddler Flicker-Flasher Ring	Vari-Vue	1966	silver plastic base	10	25	40
Robin Character Sponge	Epic	1966	5"	20	50	100
Robin Figure	Palitoy	1970s	8" figure on card	15	40	80
Robin Figure	Presents		cloth and vinyl, on base	10	20	35
Robin Figure	Ideal	1966	3" plastic, detachable yellow plastic cape	10	20	40
Robin Figure	Applause	1988	vinyl with stand	10	20	35
Robin Iron-on Patch		1966	2-1/2" diameter patch, Batkids Fan Club	8	35	50
Robin on a String Figure	Ben Cooper		4" tall, rubber	10	20	35
Robin Ornament	Presents	1989		5	13	20
Robin Placemat		1966	13" x 18" vinyl	10	25	40
Robin Push Puppet	Kohner	1966	3", plastic, push button on bottom	25	60	90
Robin Shuttle	Mego	1979	sized for British-made Mego figures, in box	20	40	80
Robin/Dick Grayson Flicker-Flasher Ring	Vari-Vue	1966	silver plastic base	10	20	40
Rubber Stamp Set	Kellogg's	1966	2" x 5" hard black plastic case, set of six plastic stamps plus ink pad: Batman, Robin, Batmobile, Joker, Riddler and Penguin	50	100	200
Shooting Arcade	AHI	1970s	graphics of Joker, Catwoman and Penguin	25	75	130
Sip-A-Drink Cup		1966	British, 6" tall, white plastic	50	100	200
Slam Bang Ice Cream Carton	Cabarrus Creamery	1966	features Batman and Robin on side panels	15	40	55
Sparkle Paint Set	Kenner	1966	paint and six pre-numbered sketches of Batman	30	75	140
Sticker Fun with Batman Book	Watkins-Strathmore	1966	8" x 11" softbound with stickers	15	40	80
Super Accelerator Batmobile	AHI	1970s	on card	15	40	80
Talking Alarm Clock	Janex	1975	plastic clock with Bat logo on face	40	80	150
Three Villains of Doom Book	Signet	1966	160 pages	10	25	50
Turbine-Sound Batmobile	Toy Biz	1989		10	15	20
TV Guide	TV Guide	1966	March 26-April 1 issue, photo cover of Adam West as Batman	45	150	200
Video Game Watch	Tiger	1989	with alarm	8	15	30

BEANY & CECIL

TOY	COMPANY	YEAR	DESCRIPTION	GOOD	EX	MIB
Beany & Cecil & Their Pals Record Player	Vanity Fair	1961		75	150	300

BEANY & CECIL

TOY	COMPANY	YEAR	DESCRIPTION	GOOD	EX	MIB
Beany & Cecil Carrying Case		1960s	9" diameter with strap, vinyl-covered cardboard	40	85	90
Beany & Cecil Gun	Mattel	1961	with propeller disks	25	100	150
Beany & Cecil Puzzle	Playskool	1961	wooden frame tray	25	55	85
Beany & Cecil Skill Ball		1960s	colorful tin with wood frame	30	125	110
Beany & Cecil Travel Case		1960s	square, 4-1/2" x 3-1/2" x 3" red vinyl, carrying strap, illustrated with characters	30	70	90
Beany & Cecil Travel Case		1960s	8" tall, round, red vinyl with zipper and strap	25	55	85
Beany and His Magic Set Book		1953	Tell-a-Tale Book	8	19	35
Beany Doll	Mattel	1963	15", non-talking	20	75	100
Beany Figure	Caltoy	1984	8" tall	5	15	30
Beany Talking Doll	Mattel	1950s	17" tall, stuffed cloth, vinyl head with pull string	60	150	300
Bob Clampetts' Beany Coloring Book	Whitman	1960s		15	50	80
Captain Huffenpuff Puzzle		1961	large	20	60	100
Cecil and His Disguise Kit	Mattel	1962	17" tall plush Cecil with disguise wigs, mustaches, etc.	30	75	150
Cecil in the Music Box	Mattel	1961	jack-in-the-box	65	200	350
Cecil Soaky			8-1/2" tall, plastic	20	75	100
Leakin' Lena Boat	Irwin	1962	plastic and wood	50	100	200
Leakin' Lena Pound 'N Pull Toy	Pressman	1960s	wood	50	100	200

BETTY BOOP

TOY	COMPANY	YEAR	DESCRIPTION	GOOD	EX	MIB
Betty Boop Delivery Truck	Schylling	1990	tin litho	15	30	45
Betty Boop Doll	M-Toy	1986	12" vinyl jointed	10	20	40
Betty Boop Doll		1930s	wood with composition head	100	350	600
Betty Boop Doll Clothing	M-Toy	1986	outfits for 12" dolls high fashion boutique, each	5	10	20
Betty Boop Figure	NJ Croce	1988	9" bendy	4	8	12
Betty Boop Figure		1980s	3" PVC figure, eight different poses and outfits, each	2	3	5

BLONDIE & DAGWOOD

TOY	COMPANY	YEAR	DESCRIPTION	GOOD	EX	MIB
Blondie Figure		1940s	2-1/2" tall, lead	10	20	30
Blondie Paint Book	Whitman	1947		20	75	100
Blondie Paint Set	American Crayon	1946		15	75	100
Blondie Paper Dolls	Whitman	1944		50	150	200
Blondie Paper Dolls	Whitman	1955		20	75	100
Blondie's Peg Board Set	King Features	1934	9" x 15-1/2", multi-colored pegs, hammer, cut-outs of Dagwood, Blondie, etc.	35	100	200
Blondie's Presto Slate	Presto	1944	10" x 13" illustration of Blondie & Dagwood and other characters	15	50	75
Dagwood & Kids Figures	King Features	1944	Dagwood, Alexander, and Cookie, each	25	60	120
Dagwood Marionette		1945	14"	50	100	200
Dagwood's Solo Flight Airplane	Marx	1935	12" wingspan, plane 9" in length	250	500	1000
Lucky Safety Card		1953	2" x 4" cards, Dagwood offers safety tips	10	55	50
Puzzle		1930s	Featured Funnies	20	75	100

BOZO

TOY	COMPANY	YEAR	DESCRIPTION	GOOD	EX	MIB
Bozo the Clown Beach Towel		1960s	16" x 24"	8	15	25
Bozo the Clown Doll		1970s		10	20	40
Bozo the Clown Figure		1970s	vinyl, 5" tall	5	10	15
Bozo the Clown Push Button Marionette	Knickerbocker	1962		20	40	80
Bozo the Clown Puzzle	Whitman	1965	#4516	10	20	35
Bozo the Clown Slide Puzzle		1960s		10	25	50
Bozo the Clown Soaky	Palmolive	1960s		10	40	60
Bozo Trick Trapeze		1960s	red base	15	35	60

BRADY BUNCH

TOY	COMPANY	YEAR	DESCRIPTION	GOOD	EX	MIB
Brady Bunch Halloween Costume	Collegeville	1970s	smock reads "One of The Brady Bunch"	40	90	150
Brady Bunch Kite Fun Book	Pacific Gas & Electric	1976		15	30	50
Brady Bunch Paper Dolls Cut-Out Book	Whitman	1973		25	55	90
Brady Bunch Puzzle			frame tray	25	45	80
Brady Bunch Trading Cards	Topps	1971	55 cards	250	550	800
Kitty Karry-All Doll	Remco	1969		75	150	300

BUGS BUNNY

TOY	COMPANY	YEAR	DESCRIPTION	GOOD	EX	MIB
Bugs Bunny Bank	Dakin	1971	on a basket of carrots	15	30	50
Bugs Bunny Bank		1940s	5-3/4" x 5-1/2", pot metal, figure on base	40	80	175
Bugs Bunny Bendy	Applause	1980s	4" tall	5	10	15
Bugs Bunny Charm Bracelet		1950s	brass charms of Bugs Bunny, Tweety, Sniffles, Fudd, etc.	20	40	80
Bugs Bunny Chatter Chum	Mattel	1982		10	20	40
Bugs Bunny Clock	Litech	1972	12" x 14"	30	75	150
Bugs Bunny Colorforms Set	Colorforms	1958		15	40	60
Bugs Bunny Costume	Collegeville	1960s	mask and costume	9	16	30
Bugs Bunny Figure	Dakin	1976	yellow globes in "Cartoon Theater" box	10	20	40
Bugs Bunny Figure	Warner Bros.	1975	5-1/2" tall, ceramic, holding carrot	15	35	60
Bugs Bunny Figure	Dakin	1971	10" tall	10	25	50
Bugs Bunny Figure	Warner Bros.	1975	2-3/4" tall, ceramic	10	20	35
Bugs Bunny in Uncle Sam Outfit	Dakin	1976	distributed through Great America Theme Park, Illinois	20	40	75
Bugs Bunny Mini Snow Dome	Applause	1980s		5	10	15
Bugs Bunny Musical Ge-Tar	Mattel	1977		10	25	40
Bugs Bunny Night Light	Applause	1980s		5	10	15
Bugs Bunny Soaky			soft rubber	9	16	30
Bugs Bunny Talking Alarm Clock	Janex	1974	battery-operated	35	75	150
Bugs Bunny Talking Doll	Mattel	1971		25	50	100
Bugs Bunny Wristwatch	Lafayette	1978		20	40	80

CALIFORNIA RAISINS

TOY	COMPANY	YEAR	DESCRIPTION	GOOD	EX	MIB
California Raisins Chalkboard	Rose Art	1988		5	10	15

Dick Tracy Target Game, 1940s, Marx

Snoopy Matic Instant Load Camera, 1970s, Helm Toy

Illya Kuryakin Special Secret Lighter Gun, 1966, Ideal

CALIFORNIA RAISINS

TOY	COMPANY	YEAR	DESCRIPTION	GOOD	EX	MIB
California Raisins Clay Factory	Rose Art	1988		10	30	40
California Raisins Colorforms Play Set	Colorforms	1987		5	10	25
California Raisins Crayon By Number	Rose Art	1988		10	30	40
California Raisins Wind-Up Walkers	Rasta	1987		4	10	15

CAPTAIN AMERICA

TOY	COMPANY	YEAR	DESCRIPTION	GOOD	EX	MIB
Captain America Figure	Lakeside	1970s	bendy	30	60	125
Captain America Rocket Racer	Buddy L	1984	Secret Wars remote controlled battery operated car	30	100	150

CAPTAIN MARVEL

TOY	COMPANY	YEAR	DESCRIPTION	GOOD	EX	MIB
Adventures of Captain Marvel Ink Blotter/Ruler	Republic/ Fawcett	1940s	6" blotter with ruler advertises the 12-part serial, theatre premium	200	600	850
Boy Who Never Heard of Captain Marvel Mini Comic	Bond Bread	1940s		35	75	150
Captain Marvel and Billy's Big Game Mini Comic		1940s		90	180	375
Captain Marvel Beanbags		1940s	Captain Marvel, Mary Marvel or Hoppy, each	50	100	150
Captain Marvel Beanie		1940s	cap shows image of Captain Marvel flying toward word "Shazam," blue	70	200	350
Captain Marvel Beanie		1940s	cap shows image of Captain Marvel flying toward word "Shazam," pink	200	600	850
Captain Marvel Booklet	Fawcett	1940s		25	45	300
Captain Marvel Brunch Bag		1940s	red rectangular vinyl with strap handle	50	150	200
Captain Marvel Button		1940s	celluloid, pinback	25	65	100
Captain Marvel Buzz Bomb	Fawcett	1950s	paper airplane in envelope	20	50	75
Captain Marvel Club Button		1941	tin litho, showing Captain Marvel in bust 3/4 view, with "Shazam" in lightning bolts at bottom	25	65	100
Captain Marvel Club Felt Shoulder Patches	Fawcett	1940s	Captain Marvel diving towards Earth, yellow	50	175	350
Captain Marvel Club Felt Shoulder Patches	Fawcett	1940s	Captain Marvel diving towards Earth, blue	25	75	150
Captain Marvel Club Membership Card	Fawcett	1940s		15	50	75
Captain Marvel Code Finder		1943		125	500	650
Captain Marvel Comic Hero Punch-Outs	Lowe	1942	cardboard figures	75	150	300
Captain Marvel Felt Pennant	Fawcett	1940s	yellow, shows Captain Marvel flying	50	125	225
Captain Marvel Felt Pennant	Fawcett	1940s	blue, shows Captain Marvel flying	35	80	150
Captain Marvel Film Viewer Gun		1940s	gun-shaped movie viewer with film strips from Paramount series	100	200	375
Captain Marvel Flannel Patch	Fawcett	1940s		25	65	125
Captain Marvel Glow Pictures	Fawcett	1940s	set of four	225	600	850
Captain Marvel Iron-Ons	Fawcett	1950s	sheet	15	30	55

CAPTAIN MARVEL

TOY	COMPANY	YEAR	DESCRIPTION	GOOD	EX	MIB
Captain Marvel Jr. Ski Jump	Reed & Associates	1947	paper, in envelope	10	20	50
Captain Marvel Jr. Statuette	Fawcett	1940s	hand-painted plastic	500	1750	3600
Captain Marvel Jr. Wristwatch		1940s	blue band, round dial with blue costumed Jr.	300	500	850
Captain Marvel Key Chain	Fawcett	1940s		35	100	150
Captain Marvel Lightning Wind-Up Race Car	Fawcett	1947	4" long tin wind-up in green, yellow, orange or blue, each	45	150	225
Captain Marvel Magic Dime Register Bank	Fawcett	1948	available in three colors	55	105	250
Captain Marvel Magic Flute		1940s	on die-cut card, shows Captain Marvel on side	40	100	150
Captain Marvel Magic Lightning Box	Fawcett	1940s		50	100	165
Captain Marvel Magic Membership Card	Fawcett	1940s		15	50	75
Captain Marvel Magic Picture	Reed & Associates	1940s	paper, shows Billy Batson "transforming" into Captain Marvel	15	50	75
Captain Marvel Magic Whistle	Fawcett	1948	seed company premium, picture of Captain Marvel on both sides, on card	40	100	150
Captain Marvel Meets the Weatherman Mini Comic	Bond Bread	1940s	Bond Bread premium	45	90	150
Captain Marvel Neck Tie	Fawcett	1940s		30	75	150
Captain Marvel Overseas Cap		1940s		75	200	300
Captain Marvel Paint Set		1940s	paint set with five chalk figurines	150	400	600
Captain Marvel Paper Horn	Fawcett	1940s		15	30	50
Captain Marvel Patch	Fawcett	1940s		25	75	125
Captain Marvel Pinback	Fawcett	1940s		25	50	75
Captain Marvel Pinback Pattern	Fawcett	1940s	pattern for original pinback	15	25	75
Captain Marvel Portrait	Whiz Comics/ Fawcett	1940s		45	100	175
Captain Marvel Portrait	Republic	1940s	different version than Whiz Comics portrait	45	100	175
Captain Marvel Power Siren	Fawcett	1940s		45	90	150
Captain Marvel Puzzle	Reed & Associates	1940s	in envelope	35	75	125
Captain Marvel Puzzle	Fawcett	1941		25	45	75
Captain Marvel Secret Code Sheet	Fawcett	1940s		10	20	40
Captain Marvel Sirocco Figurine	Fawcett	1940s		2000	6000	8500
Captain Marvel Skull Cap		1940s		65	200	350
Captain Marvel Soap	Fawcett	1947	three illustrated bars in box	70	175	350
Captain Marvel Stationary	Fawcett	1940s	paper and envelopes in box	100	200	325
Captain Marvel Statuette	Fawcett	1940s	hand-painted plastic, shows Captain Marvel standing with arms crossed, on base with name engraved	1000	3000	5200
Captain Marvel Suspenders	Fawcett	1940s		40	75	125
Captain Marvel Sweater		1940s	white or off-white, red Captain Marvel logo	55	150	200
Captain Marvel Tattoo Transfers	Fawcett	1940s		25	75	100
Captain Marvel Tie Bar		1940s	on card	30	60	100
Captain Marvel Wristwatch		1948	in box, shows Captain Marvel holding an airplane	195	600	1000

CAPTAIN MARVEL

TOY	COMPANY	YEAR	DESCRIPTION	GOOD	EX	MIB
Captain Marvel, Jr. Booklet	Fawcett	1940s		25	45	75
Fawcett's Comic Stars Christmas Tree Ornaments	Fawcett	1940s	metal star-shaped ornaments with art of Captain Marvel and Hoppy	25	40	65
Giveaway Comics #1		1950		100	500	900
Giveaway Comics #2		1950		75	450	800
Giveaway Comics #3		1950		75	450	800
Mary Marvel Figurine			5"	500	1600	2600
Mary Marvel Illustrated Soap	Fawcett	1947	three soap bars in box	90	225	350
Mary Marvel Patch	Fawcett	1940s		80	225	350
Mary Marvel Pinback	Fawcett	1940s		15	30	45
Mary Marvel Stationery	Fawcett	1940s	boxed	60	150	250
Mary Marvel Statuette	Fawcett	1940s	hand-painted plastic	700	2000	3500
Mary Marvel Wristwatch		1940s	in box	175	450	900
Membership Secret Code Card	Fawcett	1940s		17	50	75
Rocket Raider	Fawcett	1940s	paper airplane in envelope	10	20	35

CAPTAIN MIDNIGHT

TOY	COMPANY	YEAR	DESCRIPTION	GOOD	EX	MIB
Air Heroes Stamp Album		1930s	twelve stamps	25	75	150
Captain Midnight Cup			plastic, 4" tall, "Ovaltine-The Heart of a Hearty Breakfast"	25	65	125
Captain Midnight Map	Skelly Oil	1940	11" x 17"	125	500	750
Captain Midnight Medal		1930s	gold, wings and words "Flight Commander"	80	175	325
Captain Midnight Membership Manual		1930s	Secret Squadron official code and manual guide	50	120	200
Captain Midnight Secret Society Decoder		1949		35	90	160

CARTOON/COMIC CHARACTERS

TOY	COMPANY	YEAR	DESCRIPTION	GOOD	EX	MIB
Alfred E. Neuman Figurine		1960s	base says "What Me Worry?"	40	85	165
Andy Panda Bank	Walter Lantz	1977	7" tall, hard plastic	10	25	40
Beetle Bailey Comic Strip Stamper Set	Ja-Ru	1981	7 stampers, book, crayon	5	10	20
Beetle Bailey Gun Set	Ja-Ru	1981	cord gun and target	5	20	30
Bloom County Opus Doll		1986	10" tall, plush, penguin Opus wearing a Santa Claus cap	5	20	30
Breezley Soaky	Purex	1967	9" tall, plastic	25	60	110
Cadbury the Butler Figure	DFC	1981	3-1/2" figure from Richie Rich	7	13	20
Chilly Willy Doll	Walter Lantz	1982	plush	5	15	20
Daffy Dog Poster			10" x 13", The Morning After	9	16	25
Dan Dunn Pinback Button		1930s	1-1/4"	25	75	150
Doggie Daddy Metal Trivet		1960s	says "You have to work like a dog to live like one"	12	30	45
Dudley Do-Right Doll	Wham-O	1972	bendy	10	30	60
Dudley Do-Right Jigsaw Puzzle	Whitman	1975	Dudley and Snidley	10	30	50
Easy Show Movie Projector Films	Kenner	1965	numerous cartoon characters, each film	8	15	30
Favorite Funnies Printing Set		1930s	#4004, Orphan Annie, Herby, and Dick Tracy, six stamps, pad, paper and instructions	40	125	200
Geoffrey Jack-in-the-Box	Toys R Us	1970s	jack-in-the-box	15	35	65
Hagar the Horrible Doll		1983	12" tall	10	20	35
Hair Bear Bunch Mug		1978	Square Bear figural mug	5	13	18
Hair Bear Bunch Wristwatch		1972	medium gold tone case, base metal back, articulated hands, red leather snap down band	30	70	125

CHARACTER

CARTOON/COMIC CHARACTERS

TOY	COMPANY	YEAR	DESCRIPTION	GOOD	EX	MIB
Harold Teen Playstone Funnies Mold Set		1940s		25	75	150
Henry on Trapeze Toy	G. Borgfeldt		6" x 9", celluloid, wind-up, jointed Henry suspended from trapeze	155	425	850
Herman and Katnip Punch Out Kite	Saalfield	1960s	folds into a kite	10	20	35
Josie & the Pussycats Paper Doll Book	Whitman	1971		20	40	75
Katzenjammer Kids Jigsaw Puzzle		1930s	9-1/2" x 14", Featured Funnies	25	75	150
King Leonardo Doll	Holiday Fair	1960s	cloth plush dressed in royal robe	30	70	125
Little Audrey Dress Designer Kit	Saalfield	1962	die cut doll and accessories in illustrated box	15	30	60
Little Audrey Shoulder Bag Leathercraft Kit	Jewel	1961		20	75	100
Little Lulu Bank			8" tall hard plastic with black fire hydrant	20	45	80
Little Lulu Dish		1940s	5-1/2" hand painted ceramic, pictures of Lulu, Tubby and her friends	50	125	200
Little Lulu Paint Book	Whitman	1944		20	75	150
Little Lulu Puzzles	Whitman	1973	four frame tray puzzles	20	50	90
Mush Mouse Pull Toy	Ideal	1960s	pull toy with vinyl figure	40	85	180
Nancy Music Box	United Feature	1968	ceramic	30	70	125
Peter Potamus Soaky			11" tall	15	35	60
Rosie's Beau Puzzle		1930s	9-1/2" x 14", Featured Funnies	15	50	75
Scrappy Bank			3" x 3-1/2" metal, embossed illustration of Scrappy and his dog	35	100	200
Smilin' Jack Puzzle		1930s	9-1/2" x 14", Featured Funnies	20	75	125
Supercar Molding Color Kit	Sculptorcraft	1960s	set of rubber plaster casting models of vehicle and show characters, including Mike Mercury, Beaker and Popkiss, Jimmy and Mitch, Masterspy	35	125	200
Touche Turtle Soaky		1960s	standing	25	60	110
Touche Turtle Soaky		1960s	lying down	15	30	55
Winnie Winkle Playstone Funnies Mold Set		1940s		25	50	100
Wonder Woman Figure	Presents		14" tall cloth and vinyl figure on base	9	25	50
Yipee Pull Toy	Ideal	1960s	with vinyl figures of Yipee, Yapee & Yahoee	40	80	150

CASPER THE FRIENDLY GHOST

TOY	COMPANY	YEAR	DESCRIPTION	GOOD	EX	MIB
Casper Costume	Collegeville		"Ghostland" costume, #216	25	40	75
Casper Doll	Sutton & Sons	1972	rubber squeeze doll with logo	10	20	40
Casper Doll		1960s	15" cloth	20	40	80
Casper Doll			7-3/4" tall squeeze doll holds black spotted puppy	25	50	85
Casper Figure Lamp	Archlamp	1950	17" tall	25	75	125
Casper Halloween Costume	Collegeville	1960s	mask and costume	12	25	45
Casper Hand Puppet		1960s	8" tall, cloth and plastic head	10	30	60
Casper Jigsaw Puzzle	Ja-Ru	1988		2	4	6
Casper Light Shade		1960s		10	25	50
Casper Night Light	Duncan	1975	6-1/2" tall	15	35	60
Casper Soaky	Colgate-Palmolive			15	50	75

CASPER THE FRIENDLY GHOST

TOY	COMPANY	YEAR	DESCRIPTION	GOOD	EX	MIB
Casper Spinning Top		1960s	blue top with figure of Casper inside	15	30	45
Casper the Friendly Ghost Talking Doll	Mattel	1961	15" tall, terry cloth, plastic head with a pull string voice box	30	65	110
Casper Wind-Up Toy	Linemar	1950s	tin	100	275	500
Wendy the Good Witch Soaky				15	50	75

CHARLIE CHAPLIN

TOY	COMPANY	YEAR	DESCRIPTION	GOOD	EX	MIB
Charlie Chaplin Doll			11-1/2" tall, wind-up	200	500	900
Charlie Chaplin Figure			8-1/2" tall, tin with cast iron feet, wind-up	300	600	1250
Charlie Chaplin Figure			2-1/2" tall, lead	15	75	150
Charlie Chaplin Pencil Case			8" long	20	45	85
Charlie Chaplin Toy			4" tall, spring mechanism tips his hat when string is pulled	30	100	175
Charlie Chaplin Wristwatch	Bradley	1985	oldies series, quartz, large black plastic case and band, sweep seconds, shows Chaplin as Little Tramp	15	35	75
Charlie Chaplin Wristwatch	Bubbles/ Cadeaux	1972	Swiss, large chrome case, black and white dial, articulated sweep cane second hand, black leather band	40	80	175

CHARLIE'S ANGELS

TOY	COMPANY	YEAR	DESCRIPTION	GOOD	EX	MIB
Charlie's Angels Paper Dolls	Toy Factory	1977	Farrah, Kate or Jaclyn sets, each	10	25	40
Charlie's Angels Pendant	Fleetwood Toys	1977	4" plastic figure of Farrah hangs from pendant	10	25	40
Cheryl Ladd Doll	Mattel	1978	12" tall	15	30	50
Farrah Fawcett Doll	Mattel	1977	12"	20	60	100
Kate Jackson Doll	Mattel	1978	12"	15	35	60
Kelly Doll	Hasbro	1977	8"	10	25	50
Kris Doll	Hasbro	1977	8"	10	25	45
River Race Outfits	Palitoy	1977		10	30	45
Sabrina Doll	Hasbro	1977	8"	10	50	75
Sabrina, Kelly, and Kris Gift Set	Hasbro	1977		15	60	80
Slalom Caper Outfits	Palitoy			10	20	40
Underwater Intrigue Outfits	Palitoy			10	20	40

CHIPMUNKS

TOY	COMPANY	YEAR	DESCRIPTION	GOOD	EX	MIB
Alvin Doll	Knickerbocker	1963	14" tall plush with vinyl head	10	30	50
Chipmunks Soaky		1960s	10" tall, Alvin, Simon, or Theodore, each	8	30	50
Chipmunks Toothbrush		1984	battery-operated	8	20	35
Chipmunks Wallet		1959	vinyl	8	20	30

CINDERELLA

TOY	COMPANY	YEAR	DESCRIPTION	GOOD	EX	MIB
Cinderella & Prince Wind-Up Toy	Irwin		5" tall, plastic	30	100	225
Cinderella Alarm Clock	Westclox		2-1/2" x 4-1/2" x 4" tall	30	70	150

CINDERELLA

TOY	COMPANY	YEAR	DESCRIPTION	GOOD	EX	MIB
Cinderella Bank		1950s	ceramic, Cinderella holding magic wand	15	50	95
Cinderella Charm Bracelet		1950	golden brass link with five charms, Cinderella, Fairy Godmother, slipper, pumpkin coach and Prince	20	45	80
Cinderella Doll	Horsman		8" tall in illustrated box	15	60	100
Cinderella Doll			11" tall, blue stain ballgown with white bridal gown, glass slippers, holding Little Little Golden Book	15	35	75
Cinderella Figurine			5" tall, ceramic	10	35	75
Cinderella Figurine			5" tall, plastic	5	15	25
Cinderella Molding Set	Model Craft	1950s	set of character molds in illustrated box	40	75	130
Cinderella Musical Jewelry Box			mahogany music box plays "So This Is Love"	15	35	75
Cinderella Paper Dolls	Whitman	1965		20	45	75
Cinderella Puzzle	Jaymar	1960s		10	20	35
Cinderella Soaky		1960s	11" tall, blue	15	25	40
Cinderella Wind-Up Toy	Irwin	1950	5" tall, Cinderella & Prince dancing	45	95	180
Cinderella Wristwatch	Timex	1958	Cinderella and castle, pink leather band	35	75	140
Cinderella Wristwatch	US Time	1950		40	125	250
Fairy Godmother Pitcher			7" tall figural pitcher	20	45	80
Gus Doll	Gund	1950s	13" tall, gray doll with dark red shirt & green felt hat	50	100	185
Gus/Jaq Serving Set	Westman	1960s	creamer, pitcher and sugar bowl	35	70	130
Prince Charming Hand Puppet	Gund	1959	10" tall	20	45	85

CRUSADER RABBIT

TOY	COMPANY	YEAR	DESCRIPTION	GOOD	EX	MIB
Crusader Rabbit Book	Wonder Book	1958		10	30	50
Crusader Rabbit in Bubble Trouble Book	Whitman	1960		8	20	40
Crusader Rabbit Paint Set		1960s	13" x 19"	15	50	100
Crusader Rabbit Soaky	Purex	1960s		25	75	125
Crusader Rabbit Trace & Color Book	Whitman	1959		15	50	75

DANGER MOUSE

TOY	COMPANY	YEAR	DESCRIPTION	GOOD	EX	MIB
Danger Mouse Doll	Russ	1988	15" tall	15	40	60
Danger Mouse ID Set	Gordy	1985		5	10	15
Danger Mouse Pendant Necklace	Gordy	1986		5	15	25

DENNIS THE MENACE

TOY	COMPANY	YEAR	DESCRIPTION	GOOD	EX	MIB
Dennis the Menace & Ruff Book Ends		1974	ceramic	30	65	125
Dennis the Menace and Ruff Book	Whitman	1959	Little Golden Book	8	16	30
Dennis the Menace Colorforms Set	Colorforms	1961		15	45	80
Dennis the Menace Giant Mischief Kit	Hasbro	1950s		35	85	175
Dennis the Menace Paint Set	Pressman	1954	paints, crayons, brush and trays	20	45	90
Dennis the Menace Tiddley Winks	Whitman	1961		15	32	60

DENNIS THE MENACE

TOY	COMPANY	YEAR	DESCRIPTION	GOOD	EX	MIB
Dennis the Menace TV Show Puzzle	Whitman	1960		15	30	55

DEPUTY DAWG

TOY	COMPANY	YEAR	DESCRIPTION	GOOD	EX	MIB
Deputy Dawg Doll	Ideal	1960s	14" tall, cloth with plush arms & vinyl head	20	65	100
Deputy Dawg Figure	Dakin	1977	6 " tall, plastic body with vinyl head	25	40	75
Deputy Dawg Soaky		1966	9-1/2" tall, plastic	10	40	60

DICK TRACY

TOY	COMPANY	YEAR	DESCRIPTION	GOOD	EX	MIB
45 Special Water Handgun	Tops Plastics	1950s	plastic	30	65	100
Ace Detective Book	Whitman	1943		10	40	60
Adventures of Dick Tracy and Dick Tracy Jr. Book	Whitman	1933	320 pages, hardcover Big Little Book	70	300	500
Adventures of Dick Tracy the Detective Book	Whitman	1933	first of Big Little Book series, hardcover	150	500	750
Air Detective Cap	Quaker	1938		45	150	250
Air Detective Member Badge		1938	brass, wing shape	30	100	150
Auto Magic Picture Gun		1950s	6-1/2" x 9" metal picture gun & filmstrip	30	75	150
Automatic Police Station	Marx	1950s	tin litho police station and car	125	400	750
Automatic Target Range Gun	Marx	1967	BB gun mounted in an enclosed plastic shooting gallery	50	100	210
B.O. Plenty Figure	Marx	1950s	Famous Comic Figures series, waxy cream, pink, 60mm tall	15	25	45
B.O. Plenty Wind-Up	Marx	1940s	8-1/2" tall holding baby Sparkle, litho tin, walks, hat tips up and down when key is wound	110	225	450
Baby Sparkle Plenty Coloring Book	Saalfield	1948	#1015, cover has Baby Sparkle sitting in a chair	20	50	90
Baby Sparkle Plenty Paper Dolls	Saalfield	1948	#1510, on cover, Baby Sparkle is standing by a clothes line	20	50	90
Baking Set	Pillsbury	1937	cookie cutter, six press-out sheets with pictures of Dick Tracy and his pals	50	100	200
Big Boy Figure	Playmates	1990		4	10	20
Black Light Magic Kit	Stroward	1952	ultra-violet bulb, cloth, invisible pen, brushes & fluorescent dyes	50	100	200
Bonny Braids Coloring Book	Saalfield		#1174, Dick Tracy's New Daughter	20	40	75
Bonny Braids Doll	Ideal	1951	14" tall with toothbrush	85	175	350
Bonny Braids Doll	Ideal	1952	8" tall, crawls when wound	75	150	300
Bonny Braids Doll		1950s	6" tall, plastic, walking wobble doll	20	55	100
Bonny Braids Paper Dolls	Saalfield	1951	#1559, Dick Tracy's new daughter and Tess	20	75	125
Bonny Braids Pin	Charmore	1951	1-1/4" figure plastic pin on full color card	20	60	100
Bonny Braids Store Contest Card		1951	5-1/2" x 5-1/2"	20	60	100
Bonny Braids Stroll Toy	Charmore	1951	tin litho, Bonny doll in carriage	35	75	150
Booklet	Big Thrill Chewing Gum	1934	five different premium books, eight pages, each	25	60	120
Breathless Mahoney Figure	Applause	1990	14" tall	4	7	10

DICK TRACY

TOY	COMPANY	YEAR	DESCRIPTION	GOOD	EX	MIB
Camera Dart Gun	Larami	1971	8mm camera-shaped toy with dart-shooting viewer	20	50	90
Candy Box	Novel Package	1940s	box with cartoons and story on back; comic strips on bottom	30	175	250
Christmas Tree Light Bulb		1930s	early painted figure of Dick Tracy	30	70	125
Coloring Set	Hasbro	1967	six pre-sketched, numbered pictures to color, with pencils	30	65	110
Convertible Squad Car	Marx	1948	20", friction power with flashing lights	100	300	600
Copmobile	Ideal	1963	24" long, white and blue plastic, battery-operated with a microphone with amplified speaker on top	50	100	195
Crimestopper Club Kit	Chicago Tribune	1961	premium kit containing badge, whistle, decoder, magnifying glass, fingerprinting kit, ID card, crimestopper textbook	20	40	75
Crimestopper Play Set	Hubley	1970s	Dick Tracy cap gun, holster, handcuffs, wallet, flashlight, badge and magnifying glass	30	65	125
Crimestoppers Set	Larami	1973	handcuffs, nightstick & badge	10	25	45
Decoder Card	Post Cereal		cereal premium, red or green	15	40	75
Detective Badge	Quaker	1938		40	85	160
Detective Button		1930s	celluloid pinback with portrait, newspaper premium	25	50	90
Detective Club Belt		1937	leather with secret pouch	75	200	400
Detective Club Crime Stoppers Badge	Guild	1940s		25	50	90
Detective Club Pin		1942	yellow, tab back	25	50	90
Detective Dick Tracy and the Spider Gang Book	Whitman	1937	240 pages, Big Little Book	20	45	90
Detective Kit		1944	Dick Tracy Junior Detective Manual, Secret Decoder, ruler, Certificate of Membership and badge	150	400	600
Dick Tracy & Little Orphan Annie Button	Genung Promo			250	750	1000
Dick Tracy and His G-Men Book	Whitman	1941	432 pages, Big Little Book, hardcover with flip pictures	20	45	90
Dick Tracy and the Bicycle Gang Book	Whitman	1948	288 pages, Big Little Book, hardcover	20	40	80
Dick Tracy and the Boris Arson Gang Book	Whitman	1935	432 pages, Big Little Book, hardcover	25	55	100
Dick Tracy and the Hotel Murders Book	Whitman	1937	432 pages, hardcover Big Little Book	20	50	95
Dick Tracy and the Invisible Man Book	Whitman	1939	Quaker premium, 132 pages, softcover Big Little Book	40	150	200
Dick Tracy and the Mad Killer Book	Whitman	1947	288 pages, hardcover Big Little Book	20	50	95
Dick Tracy and the Mystery of the Purple Cross Book	Whitman	1938	320 pages, Big Big Book, hardcover	80	160	300
Dick Tracy and the Phantom Ship Book	Whitman	1940	432 pages, hardcover Big Little Book	20	45	90
Dick Tracy and the Racketeer Gang Book	Whitman	1936	432 pages, hardcover Big Little Book	25	55	100
Dick Tracy and the Stolen Bonds Book	Whitman	1934	320 pages, hardcover, Big Little Book	25	55	100
Dick Tracy and the Tiger Lilly Gang Book	Whitman	1949	288 pages, hardcover Big Little Book	20	45	90
Dick Tracy and the Wreath Kidnapping Case Book	Whitman	1945	432 pages, hardcover Big Little Book	20	45	90
Dick Tracy and Yogee Yamma Book	Whitman	1946	352 pages, hardcover Big Little Book	20	45	90

DICK TRACY

TOY	COMPANY	YEAR	DESCRIPTION	GOOD	EX	MIB
Dick Tracy Bingo, Lock Them Up in Jail and Harmonize with Tracy Game		1940s	object of each is to roll BBs into different holes on the face of a glass framed game card for points	40	85	150
Dick Tracy Braces	Deluxe	1940s	Chicago Tribune premium, suspenders on colorful card	20	75	150
Dick Tracy Braces for Smart Boys and Girls	Deluxe	1950s	Police badge, metal handcuffs, whistle, suspenders with a Dick Tracy badge as a holder and magnifying glass	30	75	150
Dick Tracy Candid Camera	Seymour Sales	1950s	with 50mm lens, plastic carrying case & 127 film	40	80	150
Dick Tracy Car	Marx	1950s	6-1/2" long, light blue with machine gun pointing out of the front window	75	200	350
Dick Tracy Cartoon Kit	Colorforms	1962		20	45	85
Dick Tracy Comic Book	Popped Wheat Cereal	1947	premium	6	10	20
Dick Tracy Crime Lab	Ja-Ru	1980s	click pistol, fingerprint pad, badge and magnifying glass, available in orange and bright yellow	8	15	30
Dick Tracy Crime Stopper Badge		1960s	star shape giveaway badge from WGN "9 Official Dick Tracy Crimestopper" TV Badge	30	65	125
Dick Tracy Crime Stopper Game	Ideal	1963	workstation contains crime indicator dial, decoder knobs, criminal buttons, clue cards and holders and clue windows	40	90	180
Dick Tracy Crime Stoppers Laboratory	Porter Chemical	1955	60 power microscope, fingerprint pack, glass slides and magnifying glass and textbook	85	175	350
Dick Tracy Detective Club Belt Badge			premium	30	75	150
Dick Tracy Detective Club Wrist Radios	Gaylord	1945		80	165	300
Dick Tracy Detective Set	Pressman	1930s	color graphics of Junior and Dick Tracy, ink roller, glass plate, and Dick Tracy fingerprint record paper	80	160	300
Dick Tracy Doll		1930s	13" tall, composition, grey trench coat with moveable head and mouth that operates with back pull string, gray or yellow coat	130	275	550
Dick Tracy Encounters Facey Book	Whitman	1967	260 pages, hardcover Big Little Book, cover price 39 cents	8	15	25
Dick Tracy Figure	Lakeside		bendy	15	30	50
Dick Tracy Figure	Professional Art	1940s	7" unpainted, detailed white chalk figure	75	150	280
Dick Tracy Figures	Marx	1950s	Famous Comic Figures series, several characters, each	40	75	150
Dick Tracy From Colorado to Nova Scotia Book	Whitman	1933	320 pages, hardcover Big Little Book	25	50	90
Dick Tracy Hand Puppet	Ideal	1961	10-1/2", fabric & vinyl, with record	35	80	150
Dick Tracy in 3-D Comic Book	Blackthorne	1986	Ocean Death Trap	2	3	5
Dick Tracy in Action Model Kit	Aurora	1968	plastic	80	175	350
Dick Tracy in Chains of Crime Book	Whitman	1936	432 pages, hardcover, Big Little Book	25	55	100

DICK TRACY

TOY	COMPANY	YEAR	DESCRIPTION	GOOD	EX	MIB
Dick Tracy Jr. Bombsight	Miller Bros. Hat	1940s	cardboard	35	65	120
Dick Tracy Jr. Click Pistol #78	Marx	1930s	aluminum	30	75	150
Dick Tracy Jr. Detective Agency Tie Clasp		1930s	silver or brass, each	25	50	90
Dick Tracy Junior Detective Kit Book	Golden Press	1962	punchout book of Tracy tools, including badges, revolver, wrist radio	25	50	100
Dick Tracy Lamp		1950s	painted ceramic bust of Tracy in black coat, yellow hat and red tie	700	1500	3000
Dick Tracy Little Golden Book	Golden Press	1962	features characters from the TV show	15	20	40
Dick Tracy Mask	Philadelphia Inquirer	1933	paper	70	200	350
Dick Tracy Meets a New Gang Book	Whitman	1939	Quaker premium, 132 pages, softcover Big Little Book	55	175	250
Dick Tracy Monogram Ring	Quaker	1938	ring shows initials only, no Tracy name or picture	200	600	1200
Dick Tracy Nodder		1960s	6-1/2" tall, ceramic nodding head bust	200	600	1100
Dick Tracy on the High Seas Book	Whitman	1939	432 pages, hardcover Big Little Book	25	55	100
Dick Tracy on the Trail of Larceny Lu Book	Whitman	1935	432 pages, hardcover Big Little Book	30	60	110
Dick Tracy on Voodoo Island Book	Whitman	1944	352 pages, hardcover Big Little Book	20	40	75
Dick Tracy Original Radio Broadcast Album	Coca-Cola	1972	presents the cast from "The Case of the Firebug Murders" radio show	20	45	85
Dick Tracy Out West Book	Whitman	1933	300 pages, hardcover, Big Little Book	30	60	100
Dick Tracy Paint Book	Saalfield	1930s	96 pages	60	130	250
Dick Tracy Picture	Pillsbury	1940s	part of set of eight, each 7" x 10" in mat, shows Tracy and Junior	60	120	230
Dick Tracy Pinball Game	Marx	1967	14 x 24", shows characters from TV show pilot	40	90	175
Dick Tracy Play Set	Ideal	1973	contains 18 cardboard figures that measure 3-1/2" to 5" tall, with carrying case	50	120	225
Dick Tracy Play Set	Placo	1982	plastic dart gun, targets of different villians and a set of handcuffs	10	30	50
Dick Tracy Pop-Pop Game	Ja-Ru	1980s	Diet Smith and Flattop are targets	10	15	25
Dick Tracy Puzzle		1952	11" x 14" frame tray	20	50	95
Dick Tracy Returns Book	Whitman	1939	432 pages, hardcover Big Little Book, Republic movie serial tie-in	20	45	85
Dick Tracy Ring	Miller Bros. Hat	1940s	enameled portrait	50	140	250
Dick Tracy Service Patrol Ring		1966	premium	15	35	60
Dick Tracy Soaky	Colgate-Palmolive	1965	10" tall	20	50	90
Dick Tracy Solves the Penfield Mystery Book	Whitman	1934	320 pages, hardcover, Big Little Book	25	55	100
Dick Tracy Sparkle Paints	Kenner	1963	paints, brushes and six pictures to paint	25	50	100
Dick Tracy Special FBI Operative Book	Whitman	1943	432 pages, hardcover, Big Little Book	20	45	75
Dick Tracy Special Ray Gun	Larami	1964	Remington .41 derringer with a metal Dick Tracy New York Police Detective Badge	20	55	90

DICK TRACY

TOY	COMPANY	YEAR	DESCRIPTION	GOOD	EX	MIB
Dick Tracy Super Detective Book	Whitman	1941		20	40	75
Dick Tracy Target	Marx	1941	10" square tin litho with "Recovery" and "Rescuing" points on front and bullseye target on back	40	110	200
Dick Tracy Target Game	Marx	1940s	17" circular cardboard target, with dart gun and box	65	175	350
Dick Tracy Target Set	Larami	1969	red, green or blue; shoots rubber bands	15	30	55
Dick Tracy the Man with No Face Book	Whitman	1938	432 pages, hardcover, Big Little Book	20	45	90
Dick Tracy the Super Detective Book	Whitman	1939	432 pages, hardcover Big Little Book	20	45	90
Dick Tracy Two-Way Wristwatch	Playmates	1990	watch with no radio function	4	7	15
Dick Tracy vs. Crooks in Disguise Book	Whitman	1939	352 pages, hardcover, Big Little Book with flip pictures	25	50	100
Dick Tracy Wing Bracelet	Quaker	1938	cereal premium, girl's bracelet with airplane and wording "Dick Tracy Air Detective"	50	150	200
Dick Tracy Wristwatch	Bradley	1959		50	100	200
Dick Tracy Wristwatch	Omni	1981	digital; police car box	15	40	80
Dick Tracy Wristwatch	New Haven	1937	oblong, round, or square face, in box	75	300	500
Dick Tracy Wristwatch with Animated Gun	New Haven	1951		50	175	350
Dick Tracy's Ghost Ship Book	Whitman	1939	Quaker premium, 132 pages, softcover Big Little Book	30	125	200
Dick Tracy's Two-in-One Mystery Puzzle	Jaymar	1958	one puzzle shows the crime and the other the solution	30	60	110
Dinnerware Set	Zak Designs	1980s	plate, cup, bowl	10	15	30
Dinnerware Set	Homer Laughlin	1950s	bowl, dinner plate, and mug	70	150	290
Famous Funnies Deluxe Printing Set		1930s	14 stamps, paper and stamp pad, in illustrated box	40	90	180
Favorite Funnies Printing Set	Stampercraft	1935	features Tracy and other cartoon characters	30	65	125
Film Strip Viewer	Acme	1948	viewer and two films in colorful illustrated box	40	90	175
Film Strip Viewer	Acme	1964	viewer with two boxes of film, on card, jumbo movie style	20	45	85
Film Viewer	Larami	1973	mini color televiewer with two paper filmstrips	10	20	40
Fingerprint Set	Pressman	1933	microscope, fingerprint pad, magnifying glass and badge	95	200	400
Flashlight	Quaker	1939	3" pen light, black	30	60	120
Flashlight	Bantam Lite	1961	metal wrist light	20	50	100
Flashlight	Quaker	1939	red, green, and black; bullet shaped with shield tag, pocket size	35	75	150
Flattop Story Double Record Set	Mercury Records	1947	record, book, comics	50	110	200
Get Away Car	Playmates	1990		10	20	40
Gravel Gertie Figure	Marx	1950s	Famous Comic Figures series	10	15	30
Handcuffs	John Henry	1946	metal toy handcuffs on display header card	25	55	110
Hat	Miller Bros. Hat	1940s	wool fedora, blue/gray	40	100	175
Hemlock Holmes Hand Puppet	Ideal	1961	includes record	40	90	175
Hingees "Dick Tracy and his Friends to Life" Punch-outs	Reed & Associates	1944	6-1/2" tall figures, Tess Trueheart, Chief Brandon, Junior, Pat Patton and Tracy	25	50	100
Joe Jitsu Hand Puppet	Ideal	1961	10-1/2", fabric & vinyl, includes record	40	90	175

DICK TRACY

TOY	COMPANY	YEAR	DESCRIPTION	GOOD	EX	MIB
Junior Detective Kit	Sweets Company	1944	certificate, secret code dial, wall chart, file cards & tape measure	55	115	225
Junior Dick Tracy Crime Detection Folio		1942	radio premium, contained detective's notebook, decoder with three mystery sheets, and puzzle	60	130	250
Little Honey Moon Doll	Ideal	1965	16" space baby, bubble helmet and outfit with white pigtails, doll sitting on half a moon with stars in the background	70	160	300
Luger Water Gun	Larami	1971		15	35	65
Mobile Commander	Larami	1973	toy telephone with plastic connecting tube, plastic gun and badge	20	40	75
Motorola Presents Dick Tracy Comic Book	Motorola	1953	premium comic book with paper mask and vest	40	80	150
Offical Holster Outfit	Classy Products	1940s	leather holster with painted Tracy profile	50	200	300
Pep B.O. Plenty Pin	Kellogg's	1945	tin litho button	15	25	45
Pep Chief Brandon Pin	Kellogg's	1945	tin litho button	10	15	25
Pep Dick Tracy Pin	Kellogg's	1945	tin litho button	20	35	50
Pep Flattop Pin	Kellogg's	1945	tin litho button	20	35	50
Pep Flintheart Pin	Kellogg's	1945	tin litho button	10	20	30
Pep Gravel Gertie Pin	Kellogg's	1945	tin litho button	15	30	45
Pep Junior Tracy Pin	Kellogg's	1945	tin litho button	10	20	30
Pep Pat Patten Pin	Kellogg's	1945	tin litho button	10	20	30
Pep Tess Trueheart Pin	Kellogg's	1945	tin litho button	12	20	30
Playstone Funnies Kasting Kit	Allied	1930s	molds for casting figures of Tracy and other characters	40	80	150
Police Car	Linemar	1949	tin, battery-operated remote control	175	350	600
Police Car	Marx		9" long, green litho tin, friction drive when wond up siren on the side of station turns rapidly	75	150	250
Police Whistle No. 64	Marx		tin	20	40	80
Pop Gun	Tip Top Bread	1944	7-1/2" x 4-1/2" paper pop gun, premium	35	100	175
Power Jet Squad Gun	Mattel	1962	29" long cap and water rifle	45	100	200
Puzzle, "Dick Tracy's New Daughter"	Saalfield	1951	tray puzzle of Bonny Braids	20	60	100
Puzzle, "The Bank Holdup"	Jaymar	1960s	triple-thick interlocking pieces featuring the TV cartoon	15	60	100
Puzzles, Dick Tracy Big Little Book Picture Puzzles	Whitman	1938	8" x 10" x 2" contains two puzzles of BLB scenes	50	150	250
Rapid Fire Tommy Gun	Parker Johns	1940s	20" long tommy gun with Tracy on stock	70	175	300
Repeater Cap Gun	Larami	1972		15	35	65
Riot Car	Marx	1950s	friction, sparkling, 7-1/2" long, 1946	80	200	375
Secret Code Book	Quaker	1938	premium	20	60	125
Secret Code Writer and Pencil		1939		35	80	200
Secret Compartment Ring	Quaker	1938	removable cover picturing Tracy and good luck symbols	80	175	375
Secret Detective Methods and Magic Tricks Book	Quaker	1939	cereal premium	20	75	150
Secret Detector Kit	Quaker	1938	Secret Formula Q-11 & negatives	40	150	300
Secret Service Patrol Badge	Quaker	1938	brass Captain badge	55	120	250

DICK TRACY

TOY	COMPANY	YEAR	DESCRIPTION	GOOD	EX	MIB
Secret Service Patrol Badge	Quaker	1938	brass girl's division badge	20	50	90
Secret Service Patrol Badge	Quaker	1938	Inspector General	175	600	1000
Secret Service Patrol Badge	Quaker	1938	Lieutenant	40	150	250
Secret Service Patrol Badge	Quaker	1938	2nd year chevron	10	25	45
Secret Service Patrol Badge	Quaker	1938	Sergeant	30	75	150
Secret Service Patrol Bracelet	Quaker	1938	chain bracelet with small head of Dick Tracy & Junior & four leaf clover	70	150	250
Secret Service Patrol Leader Pin	Quaker	1938	litho bar pin, Patrol Leader	250	600	1000
Secret Service Patrol Member Button	Quaker	1938	1-1/4" blue & silver, pinback	20	45	80
Secret Service Phones	Quaker	1938	cardboard phones, walkie talkie type	45	150	250
Shoulder Holster Set	J. Hapern	1950s	leather holster with Dick Tracy's profile	45	90	175
Siren Pistol	Marx	1930s	pressed steel, 8-1/2" long	70	175	350
Space Coupe	Aurora	1968	assembly required, all plastic	80	200	400
Sparkle Plenty Bank	Jayess	1940s	12" tall, base features a medallion of Dick Tracy as Godfather	120	250	500
Sparkle Plenty Christmas Tree Lights	Mutual Equipment	1940s	seven-light set	30	65	125
Sparkle Plenty Christmas Tree Lights	Mutual Equipment	1940s	15-light set	40	90	175
Sparkle Plenty Doll	Ideal	1947	12" tall	100	220	425
Sparkle Plenty Figure	Marx	1950s	Famous Comic Figures series	10	15	30
Sparkle Plenty Islander Ukette	Styron	1950	musical instrument, junior size, with instruction book	60	125	250
Sparkle Plenty Washing Machine	Kalon Radio	1940s	12" tall tin litho, pictured outside on tub is Gravel Gertie doing the wash as B.O. Plenty holds baby Sparkle	70	150	300
Squad Car	Marx	1950s	20" long, green litho tin, convertible with friction drive with a battery operated flashing light to the driver's side	120	250	475
Squad Car No. 1	Marx	1950s	11" long, with characters painted on windows, Dick Tracy badge on door, equipped with red light and machine gun	90	195	375
Squad Car No. 1 Pedal Car	Murray	1950s	deep green with yellow markings and white plastic light on hood	550	1100	2200
Steve the Tramp Figure	Playmates	1990		5	20	35
Sub-Machine Gun	Tops Plastics	1950s	12" long, red, green or blue, water gun "holds over 500 shots on one filling," Dick Tracy decal on magazine	65	130	250
Sunday Funnies Board Game	Ideal	1972		20	50	95
Super 8 Color Film	Republic	1965	b/w cartton "Trick or Treat"	10	25	50
Talking Phone	Marx	1967	green with ivory handle, battery-operated with 10 different phrases	30	65	120
The Blank Figure	Playmates	1990	figure with gun and hat with featureless face attached	70	115	175
The Capture of Boris Arson Book	Pleasure Books	1935	pop-up book	80	185	375

Mickey Mouse Fire Truck, Sun Rubber

Popeye in the Music Box, 1957, Mattel

DICK TRACY

TOY	COMPANY	YEAR	DESCRIPTION	GOOD	EX	MIB
Transistor Radio Receivers	American Doll & Toy	1961	shoulder holster and secret ear plug with two transistor radio receivers	35	75	150
Two-Way Electronic Wrist Radios	Remco	1950s	2-1/2" x 9 1/2" x 13-1/2", plastic battery-operated wrist radios	55	110	200
Two-Way Wrist Radios	Ertl	1990	battery-operated	10	15	25
Two-Way Wrist Radios	American Doll & Toy	1960s	plastic with power pack, battery-operated	40	90	175
Wall Clock		1990s	16" x 20" battery power quartz, face shows Disney movie Tracy talking into wrist radio	12	20	30
Wallpaper Section		1950s	shows comic strip scenes of Tracy and seven other characters	15	40	75
Wrist Band AM Radio	Creative Creations	1976	with earphone and two mercury batteries; box shows Tracy and Flattop	30	65	125
Wrist Radio	Da-Myco Products	1947	crystal set with receiver on a leather band, 30" wires and connectors for aerial and ground, no batteries, no tubes and no electric	130	275	525
Wrist TV	Ja-Ru	1980s	paper roll of cartoon strips are threaded through the TV viewer	8	15	30

DISNEY

TOY	COMPANY	YEAR	DESCRIPTION	GOOD	EX	MIB
Aristocats Thomas O'Malley Figure	Enesco	1967	8" tall ceramic	30	65	125
Black Hole Puzzle	Whitman	1979	jigsaw puzzle, V.I.N.C.E.N.T. or Cygnus	5	10	15
Carousel	Linemar		7" tall with 3" figures, wind-up	150	300	500
Casey Jr. Disneyland Express Train	Marx	1950s	12" long, tin, wind-up	65	130	250
Character Molding & Coloring Set		1950s	red rubber molds of Bambi, Thumper, Dumbo, Goofy, Flower and Jose Carioca to make plaster figures	35	75	150
Color Television Bank	Linemar	1950s	4" x 4-1/2", tin, Mickey or Donald on side panels, litho screen rotates	200	350	500
Disney "Sea Scouts" Puzzle	Williams Ellis	1930s		16	50	75
Disney Figure Golf Balls			set of twelve	12	23	35
Disney Filmstrips	Craftman's Guild	1940s	13 color filmstrips	95	180	275
Disney Rattle	Noma	1930s	4" tall with Mickey & Minnie, Donald & Pluto carrying a Christmas tree	60	115	275
Disney Shooting Gallery	Welso Toys	1950s	8" x 12" x 1-1/2" tin target with molded figures of Mickey, Donald, Goofy and Pluto	60	115	225
Disney Tin Tray	Ohio Art		8" x 10", pictures Mickey & Minnie Mouse, Goofy, Horace, Pluto, Donald Duck & Clarabelle	23	45	65
Disney Treasure Chest Set	Craftman's Guild	1940s	red plastic film viewer and filmstrips in blue box designed like a chest	65	125	190
Disney World Globe	Rand McNally	1950s	6-1/2" metal base, 8" diameter with Disney characters	25	50	175
Disneyland Ashtray		1950s	5" diameter, china, with Tinker Bell & castle	11	20	125
Disneyland Auto Magic Picture Gun & Theater		1950s	battery-operated metal gun with oval filmstrip	45	85	130

DISNEY

TOY	COMPANY	YEAR	DESCRIPTION	GOOD	EX	MIB
Disneyland Bagatelle	Wolverine	1970s	large game with Disneyland graphics	15	25	100
Disneyland Electric Light	Econlite	1950s	picture of Disney characters leaving a bus on a drum base	45	80	225
Disneyland F.D. Fire Truck	Linemar		18" long, battery-operated, moveable, Donald Duck fireman climbs the ladder	70	130	250
Disneyland Felt Banner	Disney	1960s	"The Magic Kingdom," 24-1/2" red/white/blue coat of arms	25	50	75
Disneyland Give-A-Show Projector Color Slides		1960s	112 color slides	60	115	200
Disneyland Haunted House Bank	Japanese	1960s		35	65	225
Disneyland Metal Craft Tapping Set	Pressman	1950s		20	35	90
Disneyland Miniature License Plates	Marx	1966	2" x 4" plates with Mickey, Minnie and Pluto, or Snow White, Donald and Goofy, each	12	23	35
Disneyland Pen		1960s	6" long with a picture of a floating riverboat	15	25	40
Disneyland Puzzle	Whitman	1956	frame tray	15	50	75
Duck Tales Travel Tote		1980s	travel agency premium	4	7	10
Early Settlers Log Set	Halsam	1960s	log building set based on Disneyland's Tom Sawyer's Island	25	45	125
Fantasia Bowl	Vernon Kilns	1940	12" diameter & 2-1/2" tall, pink bowl with a winged nymph from Fantasia	115	210	325
Fantasia Cup & Saucer Set	Vernon Kilns	1940	6-1/4" diameter saucer & 2" tall cup	60	115	175
Fantasia Figure	Vernon Kilns		half-woman, half-zebra centaur	80	145	300
Fantasia Musical Jewelry Box	Schmid Bros.	1990	box features Mickey and plays "The Sorcerer's Apprentice"	30	55	85
Fantasia Unicorn Figure	Vernon Kilns	1940s	ceramic black-winged unicorn	45	80	175
Figural Light Switch Plates	Monogram		hand-painted switch plates: Goofy, Donald, Mickey, on card, each	5	10	20
Happy Birthday/Pepsi Placemats	Pepsi	1978	set of four mats	5	20	30
Horace Horsecollar Hand Puppet	Gund	1950s		25	50	125
Johnny Tremain Figure & Horse	Marx	1957	plastic, 9-1/2" tall horse and 5-1/2" tall Johnny	60	115	175
Jose Carioca Figure	Marx	1960s	5-1/2" tall plastic, wire arms and legs	45	80	150
Jose Carioca Figure	Marx		2" tall, plastic	30	55	50
Jose Carioca Wind-Up Toy	France	1940s	3-1/2" x 5" x 7-1/2" tall	90	210	325
Lap Trays	Hasko	1960s	set of four: Donald, Goofy & Pluto, Peter Pan, and the Seven Dwarfs	35	65	150
Nautilus Expanding Periscope	Pressman	1954	inspired by 20,000 Leagues Under the Sea, 19" long	35	100	175
Nautilus Wind-Up Submarine	Sutcliffe/ England	1950s		80	145	250
Official Santa Fe & Disneyland R.R. Scale Model Train	Tyco	1966	HO scale, electric	155	295	450
Pecos Bill Wind-Up Toy	Marx	1950s	10" tall, riding his horse Widowmaker and holding a metal lasso	35	100	225
Robin Hood of the Range Better Little Book	Whitman	1942		7	20	50

DISNEY

TOY	COMPANY	YEAR	DESCRIPTION	GOOD	EX	MIB
Sand Pail & Shovel	Ohio Art	1930s	features pie-eyed Mickey selling cold drinks to Pluto, Minnie & Clarabelle	30	95	225
Second National Duck Bank	Chein		3-1/2" tall x 7" long	85	145	225
Silly Symphony Lights	Noma		set of eight	55	100	350
Toby Tyler Circus Playbook	Whitman	1959	punch-out character activity book	25	50	75
Walt Disney Movie Viewer & Cartridge	Action Films	1972	#9312 with the cartridge "Lonesome Ghosts"	9	16	25
Walt Disney's Character Scramble	Plane Facts	1940s	10 cardboard figures, 6" tall	20	40	125
Walt Disney's Clock Cleaners Book	Whitman	1938	linen-like illustrated book	40	100	150
Walt Disney's Game/ Parade/Academy Award Winners	American Toy Works		15 games for all ages	60	115	175
Walt Disney's Realistic Noah's Ark	W.H. Greene	1940s	6" x 7" x 18" Ark, 101 2" animals, and 4" human figures on cardboard	125	345	600
Walt Disney's Silly Symphony Bells	Noma		Christmas tree bells	40	100	250
Walt Disney's Snap-Eeze Set	Marx	1963	12-1/2" x 15" x 1" box with 12 flat plastic figures	60	115	275
Walt Disney's Television Car	Marx		8" long, friction toy lights up a picture on the roof when motor turns	125	300	500

DONALD DUCK

TOY	COMPANY	YEAR	DESCRIPTION	GOOD	EX	MIB
Carpet Sweeper	Ohio Art	1930s	3" with wooden handle, Donald Duck and Minnie Mouse	40	100	225
Crayon Box	Transogram	1946	tin, pictures Donald and Mickey	25	50	75
Daisy Duck Wristwatch	US Time	1948		100	195	475
Donald Duck & Pluto Car	Sun Rubber		6-1/2" long, hard rubber	25	60	170
Donald Duck Alarm Clock	Glen Clock/ Scotland	1950s	5-1/2" x 5-1/2" x 2", Donald pictured with blue bird on his hand	95	225	450
Donald Duck Alarm Clock	Bayard	1960s	2" x 4-1/2" x 5"	80	225	450
Donald Duck Bank	Crown Toy	1938	6" tall, composition, movable head	55	100	425
Donald Duck Bank		1940s	4-1/2" x 4-1/2" x 6-1/2" tall, ceramic, Donald holding a rope with a large brown fish by his side	35	65	165
Donald Duck Bank		1940s	5-1/2" x 6" x 7-1/2" tall, china, Donald seated holding a coin in one hand	70	130	275
Donald Duck Bicycle	Shelby	1949		2000	4000	6000
Donald Duck Camera	Herbert-George	1950s	3" x 4" x 3"	25	75	175
Donald Duck Car	Sun Rubber	1950s	2-1/2" x 3-1/2" x 6-1/2", rubber	35	65	200
Donald Duck Doll	Knickerbocker	1938	17" tall, red jacket, black plush hat, Donald as drum major	90	400	950
Donald Duck Doll	Mattel	1976	talking doll	25	75	100
Donald Duck Driving Pluto Toy			9" long, wind-up, celluloid	275	500	1200
Donald Duck Dump Truck	Linemar	1950s		80	145	375
Donald Duck Figure	Japanese	1930s	celluloid, long-billed Donald	55	100	450
Donald Duck Figure	Seiberling		6" tall, hollow rubber with squeaker in the base	35	65	125
Donald Duck Figure	Seiberling		6" tall, solid rubber with movable head	40	70	125
Donald Duck Figure	Seiberling		5" tall, rubber	45	80	125

DONALD DUCK

TOY	COMPANY	YEAR	DESCRIPTION	GOOD	EX	MIB
Donald Duck Figure	Dell	1950s	7" tall, rubber	35	65	125
Donald Duck Fun-E-Flex Figure	Fun-E-Flex	1930s	wooden Donald on red sled with rope	40	80	275
Donald Duck Funnee Movie Set	Transogram	1940	box features Donald, Mickey, and the nephews	45	150	250
Donald Duck Funnee Movie Set	Irwin	1949	hand-crank viewer and four films in box	80	145	225
Donald Duck Lamp	Dolly Toy	1970s	Donald on a tug boat	25	60	150
Donald Duck Lamp		1940s	9" tall, china, Donald holding an axe standing next to a tree trunk	55	100	225
Donald Duck Light Switch Cover	Dolly Toy	1976	plastic, Donald on a boat	4	7	10
Donald Duck Marionette	Peter Puppet	1950s	6-1/2" tall	25	60	150
Donald Duck Music Box	Anri	1971	Donald with guitar, plays "My Way"	35	65	125
Donald Duck Nodder		1960s	5-1/2" tall on green base	20	75	150
Donald Duck Paint Box	Transogram	1938		20	75	200
Donald Duck Projector	Stephens	1950s	projector with four films	50	120	200
Donald Duck Puppet	Pelham Puppets	1960s	10" tall, hollow composition	25	50	175
Donald Duck Push Puppet	Kohner	1950s	sailor Donald	25	65	150
Donald Duck Push Toy	Gong Bell	1950s		30	55	85
Donald Duck Puzzle	Jaymar	1940s	frame tray	10	25	40
Donald Duck Ramp Walker	Marx	1950s	1-1/4" x 3-1/2" x 3" tall, Donald is pulling red wagon with his nephews	100	125	250
Donald Duck Sand Pail	Ohio Art	1939	4-1/2" tall, tin, Donald at beach playing tug-of-war with his two nephews	30	100	175
Donald Duck Sand Pail	Ohio Art	1950s	3-1/2" tall, tin, Donald in life preserver fighting off seagulls	30	55	85
Donald Duck Scooter	Marx	1960s	tin wind-up, 4" x 4" x 2"	80	140	225
Donald Duck Skating Rink Toy	Mettoy	1950s	4" diameter, Donald is skating while other Disney characters circle the rink	35	65	100
Donald Duck Sled	S.L. Allen	1935	36" long wooden slat and metal runner sled with character decals, Donald and nephews	90	260	500
Donald Duck Snow Shovel	Ohio Art		wood, tin, litho	50	115	175
Donald Duck Soaky		1950s	7" tall	10	40	80
Donald Duck Soap	Disney		figural castile soap	35	65	85
Donald Duck Squeeze Toy	Dell	1960s	8" tall, rubber	10	25	45
Donald Duck Tea Set	Ohio Art		7-1/2" tray, 2-1/4" diameter cups, saucers	40	80	165
Donald Duck Telephone Bank	N.N. Hill Brass	1938	5" tall with cardboard figure	70	130	275
Donald Duck the Drummer	Marx	1940s	5" x" 7 x 10" tall, Donald beats on a metal drum	150	300	400
Donald Duck Toothbrush Holder		1930s	bisque, figural	35	60	150
Donald Duck Toy	Schuco		5-1/2" tall, wind-up with a bellows, quacking sound	300	550	850
Donald Duck Toy Raft	Ideal	1950s	2" tall, blue plastic raft with yellow sail with Donald looking through a telescope	35	65	100
Donald Duck Trapeze Toy	Linemar		5" tall, celluloid, wind-up	55	105	450
Donald Duck Umbrella Handle		1930s	3-1/4" tall	35	60	125
Donald Duck Watering Can	Ohio Art	1938	6" tall, tin litho	20	35	135
Donald Duck Wind-Up	Durham Plastic	1972	6-1/2" tall, hard plastic	14	25	50

DONALD DUCK

TOY	COMPANY	YEAR	DESCRIPTION	GOOD	EX	MIB
Donald Duck Wristwatch	US Time	1940s		125	300	450
Donald Duck WWI Pencil Box	Dixon		5" x 8-1/2" x 1-1/4" deep, Donald flying a plane, holding a tomahawk	35	60	125
Donald Tricycle Toy	Linemar	1950s	tin	235	440	675
Huey, Dewey and Louie Dolls	Gund	1950s	set of three 8" tall dolls	60	120	325
Louie Doll	Gund	1940s	8" tall, white/light green plush	35	65	100
Ludwig Von Drake Figure	Marx	1961	3" tall, from the "Snap-Eeze" collection	9	16	35
Ludwig Von Drake in Go Cart	Linemar	1960s	tin and plastic	100	245	450
Ludwig Von Drake Mug		1961	3-1/2" white china	15	20	45
Ludwig Von Drake Squeeze Toy	Dell	1961	8" tall, rubber	10	20	45
Ludwig Von Drake Tiddly Winks	Whitman	1961		10	16	45
Ludwig Von Drake Wonderful World of Color Pencil Box	Hasbro	1961	box shows Ludwig and the nephews	25	50	75
Uncle Scrooge Wallet		1970s	3" x 4", Uncle Scrooge tossing coins	10	16	25

DR. DOOLITTLE

TOY	COMPANY	YEAR	DESCRIPTION	GOOD	EX	MIB
Dr. Doolittle Figure	Mattel	1967	7" tall	15	35	75
Dr. Doolittle Figure	Mattel	1967	5" tall with parrot	15	30	50
Dr. Doolittle Giraffe-in-the-Box	jack-in-the-box			35	75	150

DR. SEUSS

TOY	COMPANY	YEAR	DESCRIPTION	GOOD	EX	MIB
Cat in the Hat Doll	Coleco	1983	plush	10	30	60
Cat in the Hat Ge-Tar	Mattel	1970		45	75	150
Cat in the Hat Jack-in-the-Box	Mattel	1969		45	100	190
Cat in the Hat Riding Toy	Coleco	1983		25	50	125
Cat in the Hat Rocking Toy	Coleco	1983		25	50	125
Cat in the Hat Talking Puppet	Mattel	1970	vinyl head	35	75	225
Grinch Doll	Coleco	1983		30	55	150
Horton the Elephant Doll	Coleco	1983		20	40	75
Lorax Doll	Coleco	1983		20	40	75
Mattel-O-Phone	Mattel	1970		35	50	150
Star-Bellied Sneetch Doll	Coleco	1983		30	50	100
Talking Cat in the Hat Doll	Mattel	1970		45	100	250
Talking Hedwig Doll	Mattel	1970		50	100	200
Talking Horton Doll	Mattel	1970		50	100	300
Thidwick the Moose Doll	Coleco	1983		10	25	60
Yertle the Turtle Doll	Coleco	1983	12" plush	10	30	60

DUMBO

TOY	COMPANY	YEAR	DESCRIPTION	GOOD	EX	MIB
Dumbo Christmas Ornament			2" porcelain bisque, 50th anniversary	5	10	15
Dumbo Doll			12" plush	10	16	25
Dumbo Figure	Dakin			10	20	40
Dumbo Milk Pitcher		1940s	6" tall	25	50	125
Dumbo Roll Over Wind-Up Toy	Marx	1941	4" tall, tin with tumbling action	150	250	500
Dumbo Squeeze Toy	Dell	1950s	5" tall	10	20	45

CHARACTER

ELMER FUDD

TOY	COMPANY	YEAR	DESCRIPTION	GOOD	EX	MIB
Elmer Fudd Figure	Dakin	1971	in a red hunting outfit	30	65	125
Elmer Fudd Figure		1950s	metal 5", on green base with his name embossed, Elmer in hunting outfit	50	115	225
Elmer Fudd Figure	Dakin	1977	Fun Farm	15	30	55
Elmer Fudd Figure	Dakin	1968	8" tall	15	45	75
Elmer Fudd Mini Snow Dome	Applause	1980s		7	15	30
Elmer Fudd Mug	Applause	1980s	figural	5	10	20
Elmer Fudd Pull Toy	Brice Toys	1940s	9", wooden, Elmer in fire chief's car	60	130	250

FAMILY AFFAIR

TOY	COMPANY	YEAR	DESCRIPTION	GOOD	EX	MIB
Buffy Halloween Costume	Ben Cooper	1970		20	40	80
Buffy Make-Up and Hairstyling Set	Amsco	1971		20	40	80
Buffy with Mrs. Beasley Dolls	Mattel	1967	6" Buffy with smaller Mrs. Beasley	25	60	110
Family Affair Cartoon Kit	Colorforms	1970		10	25	45
Family Affair Puzzle	Whitman	1970		10	25	45
Mrs. Beasley Paper Dolls	Whitman	1970s	several variations	10	25	45
Mrs. Beasley Rag Doll	Mattel	1973	14"	10	25	40
Talking Mrs. Beasley Doll	Mattel	1967		45	90	200

FELIX THE CAT

TOY	COMPANY	YEAR	DESCRIPTION	GOOD	EX	MIB
Felix Cartoon Lamp Shade			6" tall	25	60	130
Felix Doll		1920s	13" tall, jointed arms	110	225	450
Felix Doll		1920s	8" tall, wood, fully jointed	75	250	500
Felix Figure	Schoenhut	1920s	4" tall, wood, leather ears, stands on a white wood base	40	200	400
Felix Flashlight		1960s	contains whistle	20	175	300
Felix Manual Dexterity Game	German	1920s	2", round, metal	60	125	230
Felix on a Scooter	Nifty	1924	tin wind-up	175	350	700
Felix Pencil Case		1950s		25	60	110
Felix Punching Bag		1960s	11" tall inflatable bobber	25	60	110
Felix Sip-a-Drink Cup			5" tall	15	25	50
Felix Soaky		1960s	10" tall, red plastic	20	60	100
Felix Soaky		1960s	10" tall, blue plastic	20	60	100
Felix Soaky		1960s	10" tall, black plastic	15	50	90
Felix Squeak Toy		1930s	6" tall, soft rubber	20	60	100
Felix the Cat Pull Toy	Nifty	1920s	5-1/2" tall, 8" long, tin, cat is chasing two red mice on the front of the cart, litho pictures of Felix on side	150	325	650
Felix Wristwatch		1960s		60	125	250

FERDINAND THE BULL

TOY	COMPANY	YEAR	DESCRIPTION	GOOD	EX	MIB
Ferdinand Doll	Knickerbocker	1938	5" x 9" x 8-1/2" tall, joint composition with cloth tail & flower stapled in his mouth	95	180	275
Ferdinand Figure			9" tall, plastic	16	30	125
Ferdinand Figure	Delco	1938	4-1/2" tall, ceramic, bull seated with a purple garland around his neck	35	65	125
Ferdinand Figure	Seiberling	1930s	3" x 5-1/2" x 4" tall, rubber	25	50	125
Ferdinand Figure			3-1/2" bisque	11	20	85

FERDINAND THE BULL

TOY	COMPANY	YEAR	DESCRIPTION	GOOD	EX	MIB
Ferdinand Figure	Disney	1940s	composition	70	130	275
Ferdinand Hand Puppet	Crown Toy	1938	9-1/2" tall	45	80	175
Ferdinand Savings Bank	Crown Toy		5" tall, wood composition with silk flower with metal trap door	20	40	175
Ferdinand the Bull Book	Whitman	1938	linen picture book	18	35	125
Ferdinand the Bull/The Matador Wind-Up Toy	Marx	1938	5-1/2" high by 8" long, wind-up action between the matador & Ferdinand "bull fight"	225	425	750
Ferdinand Toy	Knickerbocker	1940	wood composition with jointed head & legs with flower in his mouth	40	75	250
Ferdinand Wind-Up Toy	Marx	1938	6" long, tin, wind-up, when wound the wire tail spins & causes him to jump around	150	350	450

FLINTSTONES

TOY	COMPANY	YEAR	DESCRIPTION	GOOD	EX	MIB
Baby Puss Figure	Knickerbocker	1961	10" tall, vinyl	30	70	125
Bamm Bamm Bank		1960s	11" tall, hard plastic figure sitting on turtle	15	35	60
Bamm Bamm Bubble Pipe	Transogram	1963	figural pipe on illustrated card	12	25	50
Bamm Bamm Doll	Ideal	1962	15" tall	45	100	200
Bamm Bamm Figure	Dakin	1970	7" tall	15	35	70
Bamm Bamm Finger Puppet	Knickerbocker	1972		5	15	20
Bamm Bamm Soaky	Purex	1960s		15	30	45
Barney Bank		1973	solid plastic, Barney holding a bowling ball	12	30	50
Barney Doll		1962	6" tall, soft vinyl doll, movable arms & head	20	40	75
Barney Figure	Flintoys	1986		4	7	10
Barney Figure	Dakin	1970	7-1/4" tall	15	35	65
Barney Figure	Knickerbocker	1961	10" tall, vinyl	35	50	150
Barney Finger Puppet	Knickerbocker	1972		5	13	20
Barney Night Light	Electricord	1979	figural	5	10	20
Barney Policeman Figure	Flintoys	1986		4	8	12
Barney Riding Dino Toy	Marx	1960s	8" long, metal & vinyl, wind-up	110	250	500
Barney Soaky	Roclar	1970s		5	7	20
Barney Wind-Up Toy	Marx	1960s	3-1/2" tall figure, tin	80	175	350
Barney's Car	Flintoys	1986		7	13	20
Betty Figure	Knickerbocker	1961	10" tall, vinyl	40	95	180
Betty Figure	Flintoys	1986		4	7	10
Dino Bank			china, Dino carrying a golf bag	40	90	175
Dino Bank		1973	hard vinyl, blue with Pebbles on his back	15	35	60
Dino Bath Puppet Sponge		1973	bath mitt	10	18	35
Dino Doll			movable head and arms	10	20	35
Dino Figure	Flintoys	1986		4	7	10
Dino Figure	Dakin	1970	7-3/4" tall	20	50	90
Dino Wind-Up Toy	Marx	1960s	3-1/2" tall, tin	80	175	350
Fang Figure	Dakin	1970	7" tall	25	50	90
Flintmobile	Flintoys	1986		9	16	25
Flintmobile with Fred Figure	Flintoys	1986		16	30	45
Flintstones Ashtray		1960	ceramic with Wilma	20	55	95
Flintstones Bank		1971	19" tall with Barney & Bamm Bamm	20	45	75
Flintstones Car	Remco	1964	battery operated car with Barney, Fred, Wilma and Betty	80	175	350
Flintstones Figure Set	Spoontiques	1981	eight figures	30	45	85

FLINTSTONES

TOY	COMPANY	YEAR	DESCRIPTION	GOOD	EX	MIB
Flintstones Figures	Empire	1976	three-inch solid figures of Fred, Barney, Wilma and Betty	10	40	75
Flintstones Figures	Imperial	1976	eight acrylic figures: Fred, Barney, Wilma, Betty, Pebbles, Bamm Bamm, Dino and Baby Puss	15	35	65
Flintstones House	Flintoys	1986		12	25	35
Flintstones Lamp			9-1/2" tall, plastic Fred with lampshade picturing characters	40	85	175
Flintstones Paint Box	Transogram	1961		15	35	55
Flintstones Party Place Set	Reed	1969	tablecloth, napkins, plates, cups	8	15	35
Flintstones Roto Draw	British	1969		25	65	95
Flintstones Tru-Vue Film Card	Tru-Vue	1962	#T-37, with strips of Fred	25	65	95
Fred Bubble Blowing Pipe			soft vinyl with curved stem	5	10	15
Fred Doll		1960	13" soft vinyl doll with movable head	40	90	190
Fred Doll	Perfection Plastic	1972	11" tall	12	30	50
Fred Figure	Knickerbocker	1960	15" tall	35	80	150
Fred Figure	Dakin	1970	8-1/4" tall	20	45	80
Fred Figure	Flintoys	1986		4	8	12
Fred Figure	Knickerbocker	1961	10" tall, vinyl	30	70	130
Fred Finger Puppet	Knickerbocker	1972		5	13	20
Fred Flintstone's Bedrock Bank	Alps	1962	9", tin & vinyl, battery operated	150	300	600
Fred Flintstone's Lithograph Wind-Up	Marx	1960s	3-1/2" tall figure, metal	80	165	350
Fred Gumball Machine		1960s	plastic, shaped like Fred's head	15	30	50
Fred Loves Wilma Bank			ceramic	45	100	175
Fred Night Light		1970	figural	5	10	20
Fred Policeman Figure	Flintoys	1986		4	8	12
Fred Push Puppet	Kohner	1960s		10	25	40
Fred Riding Dino	Marx	1962	18" long battery operated with Fred in Howdah	150	300	600
Fred Riding Dino	Marx	1962	8" long, tin & vinyl, wind-up	150	300	600
Great Big Punch-Out Book	Whitman	1961		15	35	70
Motorbike	Flintoys	1986		5	10	15
Pebbles Bank			9" tall vinyl with Pebbles sitting in chair	9	20	35
Pebbles Doll	Ideal	1963	15" tall	50	110	200
Pebbles Doll	Mighty Star	1982	vinyl head, arms and legs, cloth stuffed body 12" tall	15	25	45
Pebbles Figure	Dakin	1970	8" tall with blonde hair and purple velvet shirt	20	40	75
Pebbles Finger Puppet	Knickerbocker	1972		5	13	20
Pebbles Flintstone Cradle	Ideal	1963	for a 15" doll	35	70	125
Pebbles Soaky	Purex	1960s		15	30	45
Police Car	Flintoys	1986		7	13	20
Wilma Figure	Knickerbocker	1961	10" tall, vinyl	45	90	180
Wilma Figure	Flintoys	1986		4	7	10
Wilma Friction Car	Marx	1962	metal	80	160	350

FLIPPER

TOY	COMPANY	YEAR	DESCRIPTION	GOOD	EX	MIB
Flipper Halloween Costume	Collegeville	1964		20	40	80
Flipper Magic Slate	Lowe	1960s		8	15	30
Flipper Model Kit	Revell	1965		15	30	60

FLIPPER

TOY	COMPANY	YEAR	DESCRIPTION	GOOD	EX	MIB
Flipper Puzzle	Whitman	1960s	several variations, each	8	18	25
Flipper Ukelele	Mattel	1968		20	20	30

FLYING NUN

Flying Nun Chalkboard	Screen Gems	1967		20	40	75
Flying Nun Doll	Hasbro	1960s	4"	20	100	175
Flying Nun Doll	Hasbro	1967	11"	30	70	150
Flying Nun Halloween Costume	Ben Cooper	1967		20	60	100
Flying Nun Paint-By-Number Set	Hasbro	1960s	two scenes and 10 paint vials	15	25	50
Flying Nun Paper Doll Set	Saalfield	1969	five dolls and costumes	15	30	50

FOGHORN LEGHORN

Foghorn Leghorn Figure	Dakin	1970	6-1/4" tall	20	40	80
Foghorn Leghorn Figure	Applause	1980s	PVC	3	50	75
Foghorn Leghorn Hand Puppet		1960s	9", fabric with vinyl head	10	60	100

FONTAINE FOX

Powerful Katrinka Toy		1923	5-1/2" tall, wind-up, pushing Jimmy in a wheelbarrow	225	475	950
Toonerville Trolley			4" tall, red pot metal	150	500	1100
Toonerville Trolley		1922	7-1/2" tall, wind-up	250	500	1000
Toonerville Trolley			3" tall	80	190	375
Toonerville Trolley	Nifty		miniature, 2" tall	90	180	375

GARFIELD

Garfield 3-D Light Switch Plate	Prestigeline	1978		175	600	1200
Garfield Chair Bank	Enesco	1981		10	20	35
Garfield Easter Figure	Enesco	1978		5	10	20
Garfield Figure	Enesco	1978	Garfield as graduate	5	10	15
Garfield Figure Bank	Enesco	1981	4-3/4"	10	20	35
Garfield Music Box	Enesco	1981	Garfield dancing	15	30	55

GASOLINE ALLEY

Skeezix Comic Figure		1930s	6" chalk statue	10	20	40
Skeezix Stationery		1926	6" x 8-1/2"	8	15	30
Uncle Walt & Skeezix Figure Set			bisque, Uncle Walt, Skeezix, Herby & Smitty	50	100	200
Uncle Walt & Skeezix Pencil Holder	F.A.S.		5" tall, bisque	40	175	300

GILLIGAN'S ISLAND

Gilligan's Floating Island Play Set	Playskool	1977		75	190	350
Gilligan's Island Notepad	Whitman	1965	Gilligan and Skipper on cover	12	30	55
Gilligan's Island Trading Cards	Topps	1965	set of 55 cards	400	800	1550
New Adventures of Gilligan Dip Dots Painting Set	Kenner	1975	book with paints and brush	220	570	1200

GIRL FROM U.N.C.L.E.

TOY	COMPANY	YEAR	DESCRIPTION	GOOD	EX	MIB
1967 British Annual Book	World Distributors	1967	hardcover, 95 pages, photo cover	10	20	35
1968 British Annual Book	World Distributors	1968	hardcover, 95 pages, photo cover	10	20	35
1969 British Annual Book	World Distributors	1969	hardcover, 95 pages, photo cover	10	20	35
Costume	Halco	1967	transparent or painted mask, dress-style costume has show logo and silhouette image of Girl spy holding smoking gun, in illustrated window box	50	110	225
Garter Holster	Lone Star	1966	metal pistol fires small plastic bullets from metal shells, checker design vinyl holster and bullet pouch, on card	60	125	250
Girl From U.N.C.L.E. Doll	Marx	1967	11" tall with 30 accessories in illustrated box	250	500	1000
Music from the Television Series	M.G.M. Records	1966	photo cover shows Stephanie against a wall	9	18	35
Secret Agent Wristwatch	Bradley	1966	watch has pink face with April Dancer image, in case	80	175	350

CHARACTER

GOOFY

TOY	COMPANY	YEAR	DESCRIPTION	GOOD	EX	MIB
Backwards Goofy Wristwatch	Helbros	1972		265	490	750
Backwards Goofy Wristwatch	Pedre		silver case 2nd edition	35	65	150
Goofy Car	Spain		vinyl head Goofy drives tin litho car	25	30	45
Goofy Doll		1970s	5" x 6" x 13" tall, fabric & vinyl, laughing doll	25	50	75
Goofy Figure	Arco		bendy	5	10	20
Goofy Figure	Marx		Snap-Eeze figure	12	35	50
Goofy Lil' Headbobber	Marx			16	30	75
Goofy Night Light	Horsman	1973	green, figural	16	30	45
Goofy Rolykin	Marx			25	50	75
Goofy Safety Scissors	Monogram	1973	on card	4	7	15
Goofy Toothbrush	Pepsodent	1970s		4	7	15
Goofy Twist'N Bend Figure	Marx	1963	4" tall	11	20	30
Goofy with Bump 'N Go Action Lawn Mower	Illfelder	1980s	3-1/2" x 10" x 11", plastic figure pushing lawn mower with silver handle	35	65	100

GREEN HORNET

TOY	COMPANY	YEAR	DESCRIPTION	GOOD	EX	MIB
Assistant Badge	Don Howard Associates	1966		45	95	200
Bike Badge	Burry Cookies	1966	premium; w/Vari-Vue flasher	85	175	350
Black Beauty Balloon Toy	Oak Rubber	1966		70	145	275
Black Beauty Slot Car	Aurora	1966	clear box w/insert	80	175	350
Black Beauty Slot Car	BZ Industries	1966	large scale	150	375	750
Captain Action Flasher Ring	Vari-Vue	1966	blue	5	15	25
Captain Action Flasher Ring	Vari-Vue	1966	chrome	12	20	35
Charm	Cracker Jack	1966	hornet-shaped	10	30	50
Charms	Folz Vending	1966	hornet-shaped	10	15	25
Comic Strip Stickers	Folz Vending	1966	7" long, from vending machines, each	25	50	100
Electric Drawing Set	Lakeside	1966		90	190	275
Flasher Button	Vari-Vue	1966	no pinback, 7"	25	45	80

GREEN HORNET

TOY	COMPANY	YEAR	DESCRIPTION	GOOD	EX	MIB
Flasher Button	Vari-Vue	1966	no pinback, 3"	10	20	35
Flasher Button	Vari-Vue	1966	pinback, 3"	15	30	45
Flasher Rings	Vari-Vue	1960s	chrome base, each	8	15	20
Flasher Rings	Vari-Vue	1960s	blue plastic base, each	5	8	15
Flashlight Whistle	Bantamlight	1966		35	75	150
Frame Tray Puzzles	Whitman	1966	box of four	40	80	150
Green Hornet Bendy Figure	Lakeside	1966		35	75	150
Green Hornet Bubble Gum Ring	Frito Lay		rubber ring, in cello pack	20	45	90
Green Hornet Candy/Toy Box	Phoenix Candy	1966	several variations	35	50	100
Green Hornet Charm Bracelet		1966	gold finish chain with five charms: Hornet, Van, Kato, Pistol, Black Beauty, on 3" x 7-1/2" illustrated card	45	90	180
Green Hornet Colorforms Set	Colorforms	1966		60	125	250
Green Hornet Dashboard	Remco	1966		300	1000	2000
Green Hornet Mini Walkie Talkies	Remco	1966		75	150	300
Green Hornet Print Putty	Colorforms	1966		20	50	95
Green Hornet Seal Ring	General Mills	1940	cereal premium	350	750	1500
Green Hornet Soundtrack Record	20th Century Fox	1966		250	75	150
Green Hornet Troll Figure	Damm	1966	3" tall	45	75	150
Green Hornet Troll Figure	Uneeda Wishnik	1966	7" tall	55	100	250
Green Hornet TV Guide		1966	Cover features Van Williams and Bruce Lee	45	90	180
Green Hornet Utensils	Imperial Knife	1966	fork and spoon	25	60	100
Green Hornet Walkie Talkies	Remco	1966		175	75	150
Green Hornet Wallet		1966	green vinyl, Hornet or Kato	20	45	90
Green Hornet Wrist Radios	Remco	1966	battery-operated	125	250	500
Halloween Costume	Ben Cooper	1966	several variations	80	175	350
Hand Puppet	Ideal	1966	with hat	70	150	275
Inflatable Raft	Ideal	1966		160	350	700
Instant Squeeze Candy	Dre's Inc.	1966	toothpaste-type container w/hornet-shaped plug	40	90	175
Kite	Roalex	1966		30	75	150
Magic Eyes Movie Viewer Slides	Sawyers	1966		75	150	300
Magic Rub-On Set	Whitman	1966		70	150	275
Magic Slate	Watkins-Strathmore	1966	three variations	35	75	150
Mini Movie Viewer	Chemtoy	1971	with filmstrips	20	45	80
Mini Movie Viewer	Acme/ Chemtoy	1966	with filmstrips	45	90	180
Numbered Pencil and Paint Set	Hasbro	1966		70	150	275
Paint By Number Set	Hasbro	1966		60	125	250
Pencil Case	Hasbro	1966		30	65	120
Pencils	Empire Pencil	1966	five on card	30	80	150
Pennant	RMS	1966	blue or orange	20	45	90
Playing Cards	Ed-U-Cards	1966		10	75	150
Punch-Out Book	Whitman	1966		75	175	350
Secret Agent Badge	Don Howard Associates	1966		15	30	45
Stardust Craft Kit	Hasbro	1966		35	75	150

GREEN HORNET

TOY	COMPANY	YEAR	DESCRIPTION	GOOD	EX	MIB
The Case of the Disappearing Doctor Book	Whitman	1966		15	25	40
The Green Hornet Cracks Down Book	Whitman	1942	Better Little Books	20	45	90
The Green Hornet Returns Book	Whitman	1941	Better Little Books	20	45	90
The Green Hornet Strikes Book	Whitman	1940	Better Little Books	20	45	90
Thingmaker Mold and Accessories	Mattel	1966		75	150	300
Trading Cards	Donruss	1966	set of 44	85	100	325
Trading Cards Display Box	Donruss	1966		100	250	500
Trading Cards Wrapper	Donruss	1966		15	30	60
Trading Stickers	Topps	1966	set of 44	85	160	325
Trading Stickers Display Box	Topps	1966		100	250	500
Trading Stickers Wrapper	Topps	1966		10	25	50
Wrist Signal Light	Bantamlight	1966		40	80	175

GUMBY

TOY	COMPANY	YEAR	DESCRIPTION	GOOD	EX	MIB
Adventures of Gumby Electric Drawing Set	Lakeside	1966		20	40	70
Gumby Adventure Costume	Lakeside	1960s	fireman, cowboy, knight and astronaut, each	23	45	80
Gumby Colorforms Set	Colorforms	1988		8	10	25
Gumby Figure	Applause	1980s	12" tall, poseable	9	16	25
Gumby Figure	Applause	1980s	5-1/2 " bendy, three styles	4	8	15
Gumby Hand Puppet	Lakeside	1965	9" tall with vinyl head	10	20	50
Gumby Modeling Dough	Chemtoy	1960s		20	45	75
Gumby's Jeep	Lakeside	1960s	yellow tin litho, Gumby & Pokey's names are printed on seat	80	175	265
Pokey Figure	Lakeside	1960s		15	30	65
Pokey Modeling Dough	Chemtoy	1960s		20	45	75

HAPPY HOOLIGAN

TOY	COMPANY	YEAR	DESCRIPTION	GOOD	EX	MIB
Happy Hooligan Nesting Toys	Anri		4" tall, wooden set of four nesting pieces	50	95	190
Happy Hooligan Toy	Chein	1932	6" tall, wind-up, walking figure	130	275	550

HECKLE & JECKLE

TOY	COMPANY	YEAR	DESCRIPTION	GOOD	EX	MIB
Heckle & Jeckle Figures			7" soft foam figures	12	23	35
Heckle & Jeckle Story Book	Wonder Book	1957		9	16	25
Heckle and Jeckle Skooz-It Game	Ideal	1963	cylindrical container	25	30	50
Little Roquefort Figure		1959	8-1/2" tall, wood	15	35	65

HONEY WEST

TOY	COMPANY	YEAR	DESCRIPTION	GOOD	EX	MIB
Formal Outfit	Gilbert	1965		25	50	100
Honey West Accessory Set	Gilbert	1965	cap-firing pistol, binoculars, shoes and glasses	20	40	75
Honey West Accessory Set	Gilbert	1965	telephone purse, lipstick, handcuffs and telescope lens necklace	20	40	75

HONEY WEST

TOY	COMPANY	YEAR	DESCRIPTION	GOOD	EX	MIB
Honey West Doll	Gilbert	1965	12" tall with black leotards, belt, shoes, binoculars and gun	75	150	300
Karate Outfit	Gilbert	1965		20	45	90
Pet Set with Ocelot	Gilbert	1965		30	60	90
Secret Agent Outfit	Gilbert	1965		20	45	85

HOWDY DOODY

TOY	COMPANY	YEAR	DESCRIPTION	GOOD	EX	MIB
Clarabell Bank	Strauss	1976	flocked plastic, 9"	15	35	65
Clarabell Jumping Toy	Linemar	1950s	7" tall tin litho, squeeze lever to make figure hop forward and squeak	200	400	800
Clarabell Marionette doll	Peter Puppet	1950s		100	200	400
Flub-a-Dub Figure	TeeVee Toys	1950s	4" x 4" painted plastic, movable mouth	35	75	135
Flub-a-Dub Flip A Ring Game	Flip-A-Ring	1950s	9", ring toss game	20	40	75
Flub-a-Dub Marionette	Peter Puppet	1950s		90	200	425
Flub-a-Dub Puppet	Gund	1950s		35	70	120
Howdy Doody & Clarabell Book	Simon & Schuster	1952	Little Golden Book	10	20	35
Howdy Doody Acrobat	Arnold	1950s	tin, plastic, Howdy swings on high bar	100	25	50
Howdy Doody Air-O-Doodle Circus Train	Plasticraft/ Kagran	1950s	red/yellow plastic train, boat and plane toy on card with cut out character passengers	40	80	150
Howdy Doody and Clarabell Coloring Book	Whitman	1955		15	100	175
Howdy Doody and Clarabell Puppet Mitten Kit	Connecticut Leather	1950s		25	50	100
Howdy Doody and his Magic Hat Book	Whitman	1953	Little Golden Book	10	20	30
Howdy Doody and Mr. Bluster Book	Whitman	1954	Little Golden Book	10	20	30
Howdy Doody and the Musical Forest Record	RCA	1950s	45 rpm	18	35	75
Howdy Doody and the Princess Book	Whitman	1952	Little Golden Book	10	20	35
Howdy Doody and You Record	RCA	1950s	45 rpm	18	35	50
Howdy Doody Bank	Vandor		ceramic figural head	20	50	80
Howdy Doody Bank	Strauss	1976	flocked plastic, 9"	15	35	65
Howdy Doody Bank		1950s	7" tall, ceramic, Howdy riding a pig	60	150	275
Howdy Doody Bubble Pipe	Lido	1950s	4" long, Howdy or Clarabell	25	150	275
Howdy Doody Button	New York Sunday News	1949	reads "New Color Comic -- Sunday News"	25	50	90
Howdy Doody Coin	Kellogg's	1950s	plastic, silver, raised bust on Howdy on front	15	30	60
Howdy Doody Color TV Set	American Plastic		plastic, with films	125	250	475
Howdy Doody Coloring Books	Whitman	1955	boxed set of six	35	100	200
Howdy Doody Comic Book	Dell	1949	Issue No. 1	100	400	1000
Howdy Doody Cookbook	Welch's	1952		25	75	150
Howdy Doody Cookie-Go-Round	Luce/Krispy Kan	1950s	lithographed cookie tin	70	150	275
Howdy Doody Costume	Collegeville	1950s		45	80	180

HOWDY DOODY

TOY	COMPANY	YEAR	DESCRIPTION	GOOD	EX	MIB
Howdy Doody Crayon Set	Milton Bradley	1950	16 crayons with pictures	40	80	150
Howdy Doody Doll		1950s	7" tall vinyl squeeze toy, Howdy in blue pants and red shirt	35	75	150
Howdy Doody Doll	Ideal	1950s	eyes and mouth move	35	75	150
Howdy Doody Doll	Applause	1988	11" cloth doll	8	15	30
Howdy Doody Doll	Goldberger	1970s	30", vinyl ventriloquist doll	35	80	150
Howdy Doody Doll	Goldberger	1976	12" vinyl ventriloquist doll	20	40	75
Howdy Doody Dominoes		1950s		30	30	50
Howdy Doody Figure	Stahlwood		5" x 7" rubber squeeze figure on airplane	170	350	700
Howdy Doody Fingertronic Puppet Theater	Sutton	1970s		20	40	75
Howdy Doody Flasher Rings	Nabisco	1950s	set of eight plastic character rings	85	170	325
Howdy Doody Flicker Ring	Nabisco	1950s		9	18	35
Howdy Doody in Funland Book	Whitman	1953	Little Golden Book	10	20	35
Howdy Doody Kiddie Pool	Ideal	1950s	40" diameter, yellow/blue vinyl	80	150	275
Howdy Doody Marionette	Peter Puppet	1950s		80	175	350
Howdy Doody Mug	Ovaltine	1950s	red plastic w/Howdy decal (Be Keen, Drink Chocolate Flavored Ovaltine)	35	45	70
Howdy Doody Music Box	Vandor		Howdy playing piano	35	50	80
Howdy Doody Night Light	Leco	1950s	figural, Howdy's face	35	75	140
Howdy Doody Outdoor Sports Box		1950s	tin litho box w/colorful graphics	30	60	110
Howdy Doody Paint Set	Marx	1950s	plaster figures, paint	30	100	175
Howdy Doody Paint Set	Milton Bradley	1950s		35	75	150
Howdy Doody Pencil Case		1950s	vinyl; smiling Howdy on front	18	90	150
Howdy Doody Phono Doodle	Sharatone Products			100	250	500
Howdy Doody Pumpmobile	Nylint		tin vehicle	100	250	500
Howdy Doody Puppet Show Set		1950s	includes plastic figures of Howdy, Clarabell, Mr. Bluster, Flub, Dilly Dally	80	170	325
Howdy Doody Puzzle	Whitman	1950s	frame tray, Howdy Goes Fishing	20	45	65
Howdy Doody Puzzle Set	Milton Bradley	1950s	set of three	40	50	75
Howdy Doody Ranch House Tool Box	Liberty Steel	1950s	14" x 6" x 3" illustrated steel box with handle	35	100	175
Howdy Doody Salt and Pepper Shakers	Peter Puppet	1950s	shape of Howdy's head; removable blue vinyl neckerchief	75	160	300
Howdy Doody Sand Forms	Ideal/Kagran	1952	on card	40	80	150
Howdy Doody Songs Record	Take Two	1974	record, cut-outs, coloring book	15	30	55
Howdy Doody Sticker Fun Book	Whitman	1952		15	30	65
Howdy Doody Swim Ring	Ideal	1950s	inflatable, 20" diameter	20	40	75
Howdy Doody Talking Alarm Clock	Janex	1974		25	90	180
Howdy Doody Television	Lido	1950s	filmstrips w/TV box	20	100	200
Howdy Doody Ukulele	Emenee	1950s	plastic, white or yellow, 17"	35	60	100
Howdy Doody Umbrella	Hollander	1950s	Howdy head for handle	30	80	150
Howdy Doody Wall Walker	Tigrett			20	45	90
Howdy Doody Wristwatch	Ingraham	1954	deep blue band with blue and white dial showing character faces	130	200	400
Howdy Doody Wristwatch	Ever Tick/ Kagran	1950s	glow-in-the-dark	75	150	300

Dick Tracy on Voodoo Island Book, 1944, Whitman

Charlie's Angels Paper Dolls, Toy Factory

Above: Mickey Mouse watch and stand, 1950s, Timex. Below: Mickey Mouse alarm clock, 1975, Bradley

HOWDY DOODY

TOY	COMPANY	YEAR	DESCRIPTION	GOOD	EX	MIB
Howdy Doody Xylo-Doodle		1950s	yellow plastic piano/ xylophone w/colorful graphics	60	350	650
Howdy Doody's Animal Friends Book	Whitman	1956	Little Golden Book	10	20	35
Howdy Doody's Circus Book	Whitman	1950	Little Golden Book	10	20	35
Howdy Doody's Electric Carnival Game	Harett-Gilmar	1950s		50	110	200
Howdy Doody's Laughing Circus Record Set	RCA	1950s	two 78 rpm records	30	65	120
Howdy Doody's Lucky Trip Book	Whitman	1953	Little Golden Book	10	20	35
Howdy Doody's One-Man Band	Trophy Products/ Kagran		musical instruments	90	200	375
Mr. Bluster Bank	Strauss	1976	flocked plastic, 9"	15	30	60
Princess Summerfall Winterspring Doll	Beehler Arts	1950s	8", hard plastic, braided black hair	75	175	325
Princess Summerfall Winterspring Sewing Cards	Milton Bradley	1950s	four cards, thread, plastic needle	25	60	125
Puppets	Gund	1950s	Howdy, Bluster, Clarabell, Dilly or Princess	20	50	80
Sparkle Gun	Ja-Ru	1987	plastic gun	5	10	20
Spinning Top	Lorenz Bolz	1970s	tin top w/characters	25	55	100

HUCKLEBERRY HOUND

TOY	COMPANY	YEAR	DESCRIPTION	GOOD	EX	MIB
Hokey Wolf Figure	Marx	1961	TV-Tinykin	10	25	50
Hokey Wolf Figure	Dakin	1970		25	70	130
Huckleberry Hound Bank	Knickerbocker	1960	10" tall, hard plastic figural	10	25	35
Huckleberry Hound Bank	Dakin	1980	5" tall figural bank of Huck sitting	10	25	45
Huckleberry Hound Doll	Knickerbocker	1959	18" plush, vinyl hands & face	15	35	65
Huckleberry Hound Figure		1960s	6" tall, glazed china	15	30	60
Huckleberry Hound Figure	Marx	1961	TV-Tinykins	12	25	45
Huckleberry Hound Figure	Dakin		8" tall	15	35	70
Huckleberry Hound Figure Set	Marx	1961	TV-Tinykins	30	70	100
Huckleberry Hound Go-Cart	Linemar	1960s	6-1/2" tall, friction	75	150	300
Huckleberry Hound Wind-Up Toy	Linemar	1962	4" tall, tin	70	150	300
Huckleberry Hound Wristwatch	Bradley	1965	chrome case, wind-up mechanism, gray leather band, face shows Huck in full view	30	65	135
Mr. Jinks Doll	Knickerbocker	1959	13" tall, plush, vinyl face	20	45	150
Mr. Jinks Soaky	Purex	1960s	10" tall, Pixie & Dixie, hard plastic	10	25	40
Pixie & Dixie Dolls	Knickerbocker	1960	12" tall, each	20	40	50
Pixie & Dixie Magic Slate		1959		10	20	40

HUMPTY DUMPTY

TOY	COMPANY	YEAR	DESCRIPTION	GOOD	EX	MIB
Humpty Dumpty Bubble Bath	Avon	1960s	figural plastic container	5	15	50
Humpty Dumpty Crib Toy	Mattel	1980s	plastic, sits on crib rail	3	6	12
Humpty Dumpty Figural Soap	Avon	1990s		2	5	10
Humpty Dumpty Figure	Marx	1950s	Fairy Tale series, plastic	3	7	10
Humpty Dumpty Game	Orchard Toys	1981	British matching game	5	12	25

HUMPTY DUMPTY

TOY	COMPANY	YEAR	DESCRIPTION	GOOD	EX	MIB
Humpty Dumpty Magazine		1950s, 1960s		1	3	5
Humpty Dumpty Musical Toy	Alladin Plastics	1960s	plastic, record inside toy	10	25	40
Humpty Dumpty Potato Chip Tin		1990s	large blue and gold tin	7	15	35
Humpty Dumpty Pull Toy	Fisher-Price	1970s	plastic, several color variations	1	2	5
Humpty Dumpty Puzzle			wood frame tray puzzles, several varieties	3	5	10

I DREAM OF JEANNIE

TOY	COMPANY	YEAR	DESCRIPTION	GOOD	EX	MIB
I Dream of Jeannie Costume	Ben Cooper	1970s		7	10	20
I Dream of Jeannie Doll	Remco	1977	6"	25	45	80
I Dream of Jeannie Doll	Ideal	1965	18"	35	80	150
I Dream of Jeannie Play Set	Remco	1977	with 6" doll	20	100	175

INDIANA JONES

TOY	COMPANY	YEAR	DESCRIPTION	GOOD	EX	MIB
Indiana Jones 3-D View-Master Gift Set	View-Master			15	30	60
Indiana Jones Backpack	Pepsi			20	45	85
Indiana Jones The Legend Mug				5	10	15
Last Crusade Button	Pepsi		retailer button	5	10	15
Temple of Doom Calendar				5	10	15
Temple of Doom Storybook	hardbound			7	15	25

JAMES BOND

TOY	COMPANY	YEAR	DESCRIPTION	GOOD	EX	MIB
007 Bullet Hole Stickers		1987	simulated bullet holes, magnetic license holder	6	15	25
007 Dart Gun	Imperial	1984	photo of Roger Moore	9	25	45
007 Exploding Cigarette Lighter	Coibel			6	15	25
007 Exploding Coin	Coibel		on blister pack	6	15	25
007 Exploding Pen	Coibel		on blister pack	6	15	25
007 Exploding Spoon	Coibel		on blister pack	6	15	25
007 Flicker Rings		1967	set of 11	30	65	110
007 Radio Trap	Multiple Toys	1966	radio with secret business cards	40	85	150
007 Submachine Gun	Imperial	1984	photo of Roger Moore	10	25	45
007 Toy Pistol	Edgemark			5	15	25
007 Video Disc Promo Kit	RCA	1980s	disc, poster, pamphlet	15	30	50
Bond Golf Tees		1987	British, six tees and pencil in leather pouch	8	15	25
Bond Pocket Diary		1988	British leather bound	5	15	25
Electric Drawing Set	Lakeside	1965	plastic tracing board, pencils, sharpener, adventure sheets	40	90	180
Goldfinger II Role Playing Game				6	10	20
Goldfinger Puzzle	Milton Bradley	1965	Bond's Bullets Blaze	20	40	75
Goldfinger Puzzle	Milton Bradley	1965	Bond and Golden Girl	25	50	100
James Bond .380 Pistol	Imperial	1984	with photo of Roger Moore	10	25	40
James Bond Alarm Clock			painting of Roger Moore in center	18	35	75

CHARACTER

JAMES BOND

TOY	COMPANY	YEAR	DESCRIPTION	GOOD	EX	MIB
James Bond Aston Martin	Gilbert	1965	12", battery operated	125	250	500
James Bond Aston Martin Slot Car	Gilbert	1965		50	100	200
James Bond B.A.R.K. Attache Case	Multiple Toys	1965	007 luger, missles, hideaway pistol and rocket launching device, case and gun bears 007 logo	75	150	300
James Bond Computer Game "The Stealth Affair"		1990	based on Licence to Kill	12	25	50
James Bond Disguise Kit	Gilbert	1965		35	75	150
James Bond Disguise Kit #2	Gilbert	1965		35	75	150
James Bond Doll	Gilbert	1965	12"	80	165	325
James Bond Doll Accessory Set	Gilbert	1965	trench coat, mask, shoes, binoculars, glasses, grenade, pants and hat	45	90	350
James Bond Doll Action Apparel Set	Gilbert	1965	scuba tanks, spear gun, knife	20	45	225
James Bond Hand Puppet	Gilbert	1965		35	75	100
James Bond Harpoon Gun (Thunderball)	Lone Star	1960s	box illustrated with undersea fight scene graphics	35	75	150
James Bond Hideaway Pistol	Coibel	1985		12	25	45
James Bond Jr. Action Ninja Figure	Hasbro			5	10	55
James Bond Jr. Bath Towel				10	20	20
James Bond Jr. Buddy Mitchell Figure	Hasbro			5	10	15
James Bond Jr. Capt. Walker D. Plank figure	Hasbro			3	6	10
James Bond Jr. CD Player/Weapons Kit	Hasbro			10	20	25
James Bond Jr. Corvette				10	20	30
James Bond Jr. Crime Fighter Set			includes handcuffs, watch and walkie-talkie	5	10	15
James Bond Jr. Dr. Derange figure	Hasbro			3	6	10
James Bond Jr. Dr. No. figure	Hasbro			3	6	10
James Bond Jr. Figure	Hasbro		with scuba gear	3	6	10
James Bond Jr. Figure	Hasbro		with pistol	3	6	10
James Bond Jr. Gordo Leiter figure	Hasbro			3	6	10
James Bond Jr. I.Q. figure	Hasbro		with weapon device	3	6	10
James Bond Jr. Jaws Figure	Hasbro			5	10	15
James Bond Jr. Karate Punch Target Game			gun with boxing glove and targets of villain	8	15	15
James Bond Jr. Ninja Play Set			throw stars, nunchukas and badge	8	15	20
James Bond Jr. Ninja Wrist Weapon Set				5	10	15
James Bond Jr. Odd Job figure	Hasbro			3	6	10
James Bond Jr. Scum Shark Mobile				8	12	20
James Bond Jr. Sticker Book				3	6	10
James Bond Jr. Sub Cycle				5	12	20
James Bond Jr. Vehicles	Ertl		S.C.U.M. Helicopter, Bond's car and van; set of three	15	30	60
James Bond Parachute Set	Imperial	1984	orange and green parachutist figures	10	20	30
James Bond Pocket Watch		1981	Roger Moore on face	20	40	75

JAMES BOND

TOY	COMPANY	YEAR	DESCRIPTION	GOOD	EX	MIB
James Bond Ring		1964	glass oval w/photo of Sean Connery	10	25	35
James Bond Roadrace Set	Gilbert	1965	scenery tracks, Aston Martin and other cars	150	300	575
James Bond Secret Attache Case	MPC	1965		125	260	600
James Bond Secret Service Game	Spears	1965		30	75	525
James Bond Sting Pistol	Coibel		8-shot cap pistol and booklet from On Her Majesty's Secret Service	15	30	55
James Bond Tarot Game		1973		10	25	50
James Bond Video Game	Coleco	1983		15	25	45
James Bond Wall Clock		1981	large clock with painting of Roger Moore in action	25	45	80
Jaws Figure	Mego	1979		130	270	525
Licence to Kill Car Set	Matchbox		jeep, seaplane, oil tanker, and copter	20	40	75
Living Daylights German Pistol	Wicke	1980s	25-shot cap gun	10	25	45
Living Daylights Record Mobile		1980s		5	15	25
May Day Pistol		1985		10	20	35
Moonraker Drax Copter	Corgi			15	30	55
Moonraker Halloween Costume		1979	mask of Roger Moore and space suit costume	20	50	60
Moonraker Jr. Set	Corgi		Aston Martin and Lotus	15	35	125
Moonraker Jr. Shuttle	Corgi			10	20	35
Moonraker Shuttle & Drax Copter Jr.'s	Corgi		set of two	15	30	55
Moonraker Space Shuttle	Corgi			25	50	100
Moonraker Spanish Card Game			33 cards with different color stills	12	25	50
Multi-Action Aston Martin		1960s	friction operated car for Agent 711	65	125	240
Octopussy Gift Set	Corgi		plane, truck and horse trailer	30	60	120
Odd Job Doll	Gilbert	1965	12" doll in karate outfit, with derby	125	250	500
Odd Job Puppet	Gilbert		plastic hand puppet	50	300	550
Scuba Outfit #2	Gilbert	1965		15	30	60
Scuba Outfit #3	Gilbert	1965		15	30	70
Scuba Outfit #4	Gilbert	1965		15	30	60
Scuba Outfit Deluxe	Gilbert	1965		30	65	70
Ski Outfit	Gilbert	1965		40	90	180
Spy Who Loved Me Helicopter Jr.	Corgi			10	20	30
Spy Who Loved Me Jr. Set	Corgi		Lotus and Copter	15	30	60
Spy Who Loved Me Puzzle	Milton Bradley		several scenes, each	15	35	65
Stromberg Helicopter	Corgi			25	50	75
The James Bond Box		1965	rare game played with dice	35	70	130
Thunderball Balloon Target Game		1965		40	80	150
Thunderball Puzzle	Milton Bradley	1965	several pictures, each	20	40	75
Thunderball Set	Gilbert	1965		40	90	175
Tuxedo Outfit	Gilbert	1965		45	90	180
View to a Kill Michelin Button			Bond on large size pinback with Midas man	5	10	15

JETSONS

Elroy Toy	Transogram	1963		35	75	20

JETSONS

TOY	COMPANY	YEAR	DESCRIPTION	GOOD	EX	MIB
Jetson Figures	Applause	1990	10" tall; Judy, George, Elroy, Rosie, each	5	100	175
Jetsons Birthday Surprise Book	Whitman	1963	Tell-A-Tale Book	10	20	35
Jetsons Colorforms Kit	Colorforms	1963		35	70	140
Puzzle	Whitman	1962	70 pieces	20	80	150

JUNGLE BOOK

TOY	COMPANY	YEAR	DESCRIPTION	GOOD	EX	MIB
Baloo Doll			12" tall, plush	9	50	90
Jungle Book Carrying Case	Ideal	1966	5" x 14" x 8"	25	50	90
Jungle Book Dinner Set			vinyl placemat, bowl, plate and cup	7	15	25
Jungle Book Fun-L Tun-L	New York Toy	1966	108" x 24"	35	65	120
Jungle Book Magic Slate	Watkins-Strathmore	1967		9	20	30
Jungle Book Sand Pail and Shovel	Chein	1966	tin litho, illustrated with Jungle Book characters	20	45	75
Jungle Book Tea Set	Chein	1966	tin litho, plates, saucers, tea cups and serving tray	20	45	85
Jungle Book Utensils			fork and spoon with melamine handles	4	7	15
Mowgli Figure	Holland Hill	1967	8", vinyl	15	35	65
Mowgli's Hut Mobile & Figures	Multiple Toymakers	1968	2" x 3" x 3" mobile with Baloo & King Louis figures	30	65	110
Mowgli/Baloo Wristwatch			digital, clear plastic band	5	10	15
Shere Kahn Figure	Enesco	1965	5" tall, ceramic	15	15	20

LADY & THE TRAMP

TOY	COMPANY	YEAR	DESCRIPTION	GOOD	EX	MIB
Lady and Tramp Figures	Marx	1955	Lady is 1-1/2" tall & white, Tramp is 2" tall & tan	20	45	80
Lady Doll	Woolikin	1955	5" x 8" x 8-1/2", light tank with burnt orange accents on face, ears, stomach & tail, plastic eyes, nose & a white silk ribbon around neck	40	50	90
Modeling Clay	Pressman	1955		20	100	175
Perri Doll	Steiff	1950s	6" tall, plush	35	65	125
Puzzle	Whitman	1954	11" x 15", frame tray	15	30	50
Toy Bus	Modern Toys/ Japan	1966	3-1/2" x 4" x 14" long	100	40	60
Tramp Doll	Schuco	1955	8" tall, brown with a white underside & face, hard plastic eyes & nose	50	100	200

LAUREL & HARDY

TOY	COMPANY	YEAR	DESCRIPTION	GOOD	EX	MIB
Laurel & Hardy Die Cut Puppets	Larry Harmon	1970s	moveable, each	20	125	225
Laurel & Hardy Die-Cut Puppets	Dell	1982	soft vinyl, each	10	50	100
Laurel and Hardy TV Set		1976	with paper filmstrips	15	30	60
Oliver Hardy Bank	Play Pal	1974	13-1/2" tall, plastic	15	30	45
Oliver Hardy Doll	Dakin		5" tall wind-up dancing/ shaking vinyl doll	35	40	50
Oliver Hardy Figure	Dakin	1974	7-1/2" tall	25	90	150
Stan Laurel Bank	Play Pal	1974	13-1/2" tall, plastic	10	30	45
Stan Laurel Bank		1972	15" tall, vinyl figural bank	15	35	60
Stan Laurel Figure	Dakin	1974	8" tall	25	50	100

CHARACTER

LITTLE MERMAID

TOY	COMPANY	YEAR	DESCRIPTION	GOOD	EX	MIB
Ariel Doll			9" tall, in gown	9	75	125
Ariel Doll	PVC			9	16	25
Ariel Jewelry Box			5-3/4" x 4-1/2", musical	9	16	25
Ariel Toothbrush			battery operated with holder	9	16	25
Eric Doll			9-1/2" tall in full dress uniform	9	16	25
Flounder & Ariel Faucet Cover			plastic	5	10	15
Flounder Doll			15" plush fish	9	16	25
Flounder Pillow			14" x 24" shaped like Flounder	7	13	20
Little Mermaid Figures	Applause		several PVC characters, each	2	4	5
Little Mermaid Purse			6" diameter, vinyl, canteen style purse	5	10	15
Little Mermaid Snow Globe			4" water globe	9	16	25
Scuttle Doll			15" seagull, plush	11	20	30
Sebastian Doll			16" crab, plush	9	16	25
Under the Sea Jewelry Box			4" mahogany, musical	18	35	50

LITTLE ORPHAN ANNIE

TOY	COMPANY	YEAR	DESCRIPTION	GOOD	EX	MIB
Annie Doll	Knickerbocker	1982	10" tall, with two dresses and removable heart locket	5	10	15
Annie Doll	Knickerbocker	1982	10" doll without locket	2	5	7
Annie Figures	Knickerbocker	1982	six figures, 2" tall, each	2	3	5
Beetleware Mug	Ovaltine	1935		30	45	85
Beetleware Mug	Ovaltine	1933		20	60	100
Daddy Warbucks Doll	Knickerbocker	1982		9	50	85
ID Bracelet	Ovaltine	1934		25	45	85
Little Orphan Annie & Chizzler Book		1930s	Big Little Book	20	60	100
Little Orphan Annie & Sandy Ashtray		1930s	3" tall, ceramic	40	80	150
Little Orphan Annie & Sandy Dolls	Famous Artists Syndicate	1930	9-3/4" tall, each	40	80	150
Little Orphan Annie & Sandy Toothbrush Holder		1930s	bisque	30	65	120
Little Orphan Annie & the Gooneyville Mystery Book	Whitman	1947		15	30	150
Little Orphan Annie & the Haunted House Comic Book	Cupples & Leon	1928		35	110	225
Little Orphan Annie Bucking the World Comic Book	Cupples & Leon	1929	hardcover	35	125	250
Little Orphan Annie Clothespins	Gold Medal	1938	clothesline & pulley	15	35	75
Little Orphan Annie Colorforms Set	Colorforms	1970s		8	18	35
Little Orphan Annie Costume		1930s	mask & slip-over paper dress	25	50	100
Little Orphan Annie Cut-Outs	Miller Toys	1960s	Sandy, Grunts the Pig, Pee Wee the Elephant	20	100	200
Little Orphan Annie Doll	Well Toy	1973	7" tall	15	20	45
Little Orphan Annie Figure		1940s	1-1/2" tall, lead	12	25	35
Little Orphan Annie in the Circus Comic Book	Cupples & Leon	1927	9" x 7", 86 pages	35	110	225
Little Orphan Annie Light Up the Candles Game	Ovaltine		3-1/2" x 5", premium	15	125	250
Little Orphan Annie Mug	Ovaltine	1930	Uncle Wiggily	20	45	80
Little Orphan Annie Music Box	N.Y. News	1970	figural	25	60	100

LITTLE ORPHAN ANNIE

TOY	COMPANY	YEAR	DESCRIPTION	GOOD	EX	MIB
Little Orphan Annie Punch-Outs	King, Larson, McMahon	1944	3-D punch-outs of Annie, Sandy, Punjab & Daddy Warbucks	25	50	100
Little Orphan Annie Rummy Cards	Whitman	1937	5" x 6", colored silhouettes of Annie on the back	15	75	150
Little Orphan Annie Shipwrecked Comic Book	Cupples & Leon	1931	9" x 7", 86 pages	30	90	200
Little Orphan Annie Stove		1930s	electric version, 8" x 9", gold metal, litho plates, functional oven doors & back burner	60	135	260
Little Orphan Annie Stove		1930s	non-electric model, gold-brass lithographed labels of Annie & Sandy, oven doors functional	50	100	225
Little Orphan Annie Wind-Up Toy	Marx	1930s	5" tall, tin	175	375	725
Miss Hannigan Doll	Knickerbocker	1982		5	10	15
Molly Doll	Knickerbocker	1982		5	10	15
Punjab Doll	Knickerbocker	1982		5	15	20
Puzzle	Novelty Dist.		Famous Comics	18	35	50
Radio Annie's Secret Society Booklet		1936	6" x 9"	30	60	120
Radio Annie's Secret Society Manual		1938	6" x 9", 12 pages	45	75	150
Round Decoder Pin	Ovaltine	1936		20	100	175
Sandy Figure		1940s	3/4" tall, lead	15	50	90
Sandy Wind-Up Toy	Marx	1930s	4" tall, tin	75	25	45
Secret Compartment Decoder Pin	Ovaltine	1936		20	150	300
Shake-Up Mug	Ovaltine	1931		25	60	100
Sunburst Decoder Pin	Ovaltine	1937		25	60	90

LOONEY TUNES

TOY	COMPANY	YEAR	DESCRIPTION	GOOD	EX	MIB
Bugs Bunny Hand Puppet	Zany	1940s	rubber head	30	70	125
Daffy Duck Bank	Applause	1980s	figural	8	15	25
Daffy Duck Figure	Applause	1980s	4" bendy	8	15	25
Daffy Duck Figure	Dakin	1968	8-1/2" tall	15	25	45
Daffy Duck Figure	Dakin	1970s		15	30	55
Elmer Fudd Hand Puppet	Zany	1940s	rubber head	30	65	125
Foghorn Leghorn Hand Puppet	Zany	1940s	rubber head	30	55	100
Pepe Le Pew Figure	Dakin	1971	8" tall	25	50	100
Pepe Le Pew Goofy Gram	Dakin	1971		25	50	100
Speedy Gonzales Figure	Dakin	1970	7-1/2" tall, vinyl	15	25	45
Speedy Gonzales Figure	Dakin		5", vinyl	10	20	30
Sylvester & Tweety Bank		1972	vinyl	14	25	40
Sylvester & Tweety Figures	Warner Bros.	1975	6" tall	9	16	25
Sylvester Figure	Dakin	1976	Cartoon Theater	9	16	25
Sylvester Figure	Dakin	1971	Sylvester on a fish crate	15	25	40
Sylvester Figure	Dakin	1969		11	20	30
Sylvester Figure	Oak Rubber	1950	6", rubber	16	30	90
Sylvester Hand Puppet	Zany	1940s	rubber head	30	65	125
Sylvester Soaky	Colgate			9	16	25
Tasmanian Devil Bank	Applause	1980s		5	10	15
Tasmanian Devil Doll	Mighty Star	1971	13" tall, plush	9	16	25
Tasmanian Devil Figure	Superior	1989	7", plastic, on base	3	5	8
Tweety Doll	Dakin	1969	6", moveable head & feet	9	16	25
Tweety Figure	Dakin	1976	Cartoon Theater	9	16	25
Tweety Figure	Dakin	1971	on bird cage	14	25	40

CHARACTER

LOONEY TUNES

TOY	COMPANY	YEAR	DESCRIPTION	GOOD	EX	MIB
Tweety Figure	Dakin	1971	Goofy Gram, holding red heart	15	25	40
Tweety Hand Puppet	Zany	1940s	rubber head	30	65	125
Tweety Soaky	Colgate	1960s	8-1/2", plastic	9	16	25

MAGILLA GORILLA

TOY	COMPANY	YEAR	DESCRIPTION	GOOD	EX	MIB
Droop-A-Long Coyote Soaky	Purex	1960s	12", plastic	20	45	75
Droop-A-Long Hand Puppet	Ideal		vinyl head	20	45	85
Magilla Gorilla Book	Golden	1964	Big Golden Book	10	20	35
Magilla Gorilla Cannon	Ideal	1964		30	70	135
Magilla Gorilla Cereal Bowl	MB Inc.			7	35	50
Magilla Gorilla Doll		1960s	11" tall, cloth body, hard arms & legs, hard plastic head	25	50	150
Magilla Gorilla Doll	Ideal	1966	18-1/2" plush with vinyl head	50	100	250
Magilla Gorilla Plate		1960s	8" diameter, plastic	7	16	25
Magilla Gorilla Pull Toy	Ideal	1960s	with vinyl figure	35	75	150
Magilla Gorilla Puppet	Ideal			35	75	150
Magilla Gorilla Push Puppet	Kohner	1960s	brown plastic figure in pink shorts and shoes, yellow base	20	45	85
Punkin' Puss Soaky	Purex	1960s	11-1/2" tall, plastic	15	35	60
Ricochet Rabbit Hand Puppet	Ideal	1960s	11", vinyl head	30	40	75
Ricochet Rabbit Soaky	Purex	1960s	10-1/2" tall, plastic	30	65	125

MAN FROM U.N.C.L.E.

TOY	COMPANY	YEAR	DESCRIPTION	GOOD	EX	MIB
1966 British Annual	World Distributors	1966	hardcover, 95 pages, photo cover	15	75	70
1967 British Annual	World Distributors	1967	hardcover, 95 pages, photo cover	12	25	50
1968 British Annual	World Distributors	1968	hardcover, 95 pages, photo cover	12	25	50
1969 British Annual	World Distributors	1969	hardcover, 95 pages, photo cover	12	25	50
Action Figure Apparel Set	Gilbert	1965	bullet proof vest, three targets, three shells, binoculars, and bazooka	45	90	180
Action Figure Armament Set	Gilbert	1965	for 12" figures: jacket, cap firing pistol with barrel extension, bipod stand, telescopic sight, grenade belt, binoculars, accessory pouch and beret	40	80	160
Action Figure Arsenal Set #1	Gilbert	1965	tommy gun, bazooka, three shells, cap firing pistol & attachments, in shallow window box	40	80	160
Action Figure Arsenal Set #2	Gilbert	1965	cap firing THRUSH rifle with telescopic sight, grenade belt and four grenades, on wrapped header card	40	80	160
Action Figure Jumpsuit Set	Gilbert	1965	for 12" figures: jumpsuit with boots, helmet with chin strap, 28" parachute & pack, cap firing tommy gun with scope, instructions	50	100	200
Action Figure Pistol Conversion Kit	Gilbert	1965	binoculars and pistol with attachments, for 12" figures, on wrapped header card	20	40	80

MAN FROM U.N.C.L.E.

TOY	COMPANY	YEAR	DESCRIPTION	GOOD	EX	MIB
Action Figure Scuba Set	Gilbert	1965	for 12" Gilbert dolls: swim trunks, air tanks, tank bracket, tubes, scuba jacket and knife	60	125	250
Affair of the Gentle Saboteur Book	Whitman	1966	hardcover	8	15	20
Affair of the Gunrunners' Gold Book	Whitman	1967	hardcover	8	15	20
Alexander Waverly Figure	Marx	1966	blue plastic, 5-3/4" tall, stamped with character's name and U.N.C.L.E. logo on the bottom of base	6	12	25
Arcade Cards		1960s	postcards with b/w photo fronts, Napoleon or Illya	4	9	18
Attache Case	Lone Star	1966	small cardboard briefcase, contains die-cast Mauser & parts to assemble U.N.C.L.E. Special	125	225	450
Attache Case, British	Lone Star	1966	cardboard covered in vinyl, 9mm automatic luger, shoulder stock, sight, silencer, belt, holster, secret wrist holster & pistol that fires cap & cork, grenade, wallet with passport, play money	250	500	900
Attache Case, British	Lone Star	1965	15" x 8" x 2" vinyl case with a pistol, holster, walkie talkie, cigarette box gun, U.N.C.L.E. badge, international passport, invisible cartridge pen and handcuffs	225	450	850
Bagatelle Game	Hong Kong	1966	8" x 14" pinball game	75	150	300
Bicycle License Plates	Marx	1967	four different, metal: Man from U.N.C.L.E., The Girl from U.N.C.L.E., Napoleon Solo, Illya Kuryakin, each	15	30	45
Calcutta Affair Book	Whitman	1967	254 pages, Big Little Book	4	8	15
Candy Cigarette Box	Cadet Sweets	1966	candy and trading card, illustrated box	20	45	90
Candy Cigarette Counter Display Box	Cadet Sweets	1966	holds 72 candy cigarette boxes, illustrated	250	500	125
Coin of El Diablo Affair Book	Wonder Books	1965	softcover, 48 pages	8	15	25
Counter Spy Water Gun	Hong Kong	1960s	luger water gun with unlicensed Napoleon Solo illustration header card	5	15	25
Counterspy Outfit	Marx	1966	contains trench coat with secret pockets, pistol, shoulder holster, launcher barrel, silencer, scope sight, two pair of glasses, beards, eye patch, badge case, etc., in box	110	225	450
Counterspy Outfit Store Display	Marx	1966	35" x 36" wide cardboard display with one piece of each item in Counterspy Outfit	300	600	1200
Crime Buster Gift Set	Corgi	1966	set includes Man from U.N.C.L.E. car, James Bond Aston Martin and Batmobile with Batboat on trailer, in window box	250	500	1000
Die-Cast Car	Playart	1968	2-3/4" long, die-cast metal, metallic purple	80	175	350
Die-Cast Metal Gun	Lone Star	1965	die-cast automatic cap pistol with plastic grips, plus cut-out badge, on card	60	125	250

MAN FROM U.N.C.L.E.

TOY	COMPANY	YEAR	DESCRIPTION	GOOD	EX	MIB
Diving Dames Affair Book	Souvenir Press/ England	1967	#10 in series	4	8	275
Doomsday Affair Book	Souvenir Press	1965	#2 in series	4	8	15
Fingerprint Kit		1966	ink pad, roller, code book, magnifier, fingerprint records and pressure plate, in illustrated window box	125	250	500
Flicker Ring		1966	blue plastic with "changing portrait" of Napoleon or Illya, each	5	12	18
Flicker Ring		1965	silver plastic ring with b/w photos, each	8	15	20
Foto-Fantastiks Coloring Set	Eberhard Faber	1965	six colored pencils, paint brush, and six 8" x 10" photos, came in four different versions, each	35	80	160
Generic Spies Figures	Marx	1966	six different solid plastic, unpainted figures 5-3/4" tall, each	6	12	18
Handkerchief	England	1966	U.N.C.L.E. logo, Illya & Napoleon	30	63	125
Headquarters Transmitter	Cragstan	1965	molded gold colored plastic transmitter, amplifier and under cover case, silver ID card, 20-foot wire, in box	75	150	300
Illya Kuryakin Action Puppet	Gilbert	1965	13" tall, soft vinyl hand puppet of Illya holding a communicator, on 10" x 16" card	75	150	300
Illya Kuryakin Costume	Halco	1967	painted mask, rayon costume in three colors showing Illya holding a gun, in illustrated window box	40	85	175
Illya Kuryakin Doll	Gilbert	1965	12" tall, plastic, black sweater, pants and shoes, spring loaded arm for firing cap pistol, folding badge, ID card and instruction sheet, in photo box	75	150	300
Illya Kuryakin Figure	Marx	1966	blue or gray plastic figure, 5-3/4" tall, stamped with character's name and U.N.C.L.E. logo on the bottom of base	10	25	45
Illya, That Man From U.N.C.L.E. Book	Pocket Books	1966	6" x 9" paperback, 100 pages of David McCallum	8	18	35
Invisible Writing Cartridge Pen	Platinum/ England	1965	pen, two vials of ink and two invisible ink vials	110	225	450
Magic Slates	Watkins-Strathmore	1965	9" x 14" slate with two punch-out figures of either Napoleon or Illya, each	35	75	150
Man from the U.N.C.L.E. Record	Capitol Records	1965	45 rpm with The Man from U.N.C.L.E. theme song and "The Vagabond"	25	50	100
Man from U.N.C.L.E. and other TV Themes Record	Metro Records	1965	photo cover, has three songs from U.N.C.L.E. plus theme songs from Dr. Kildare, Mr. Novak, Bonanza and other shows	8	15	30
Man from U.N.C.L.E. Button	Button World	1965	3-1/2" diam. round button with portrait of Napoleon or Illya, each	8	15	25
Man from U.N.C.L.E. Card Game	Japan	1966	small artwork cards in illustrated box	35	75	150
Man from U.N.C.L.E. Code Board		1966	chalkboard with line art illustrations	70	150	300

MAN FROM U.N.C.L.E.

TOY	COMPANY	YEAR	DESCRIPTION	GOOD	EX	MIB
Man from U.N.C.L.E. Finger Puppets	Dean	1966	vinyl; THRUSH agent, Solo, Kuryakin, Waverly and two female agents; window box	135	275	550
Man from U.N.C.L.E. Playing Cards	Ed-U-Cards	1965	standard 54-card deck with action photo illustrations, on card	15	30	60
Man from U.N.C.L.E. Playing Cards Display Box	Ed-U-Cards	1965	holds 12 packs	125	250	500
Man from U.N.C.L.E. Puzzles	Jaymar	1965	frame tray; three versions; each	20	40	70
Man from U.N.C.L.E. Record	Crescendo Records	1965	by the Challengers, cover shows blonde female spy with gun	5	10	20
Man from U.N.C.L.E. Record	Union/Japan	1966	45 rpm with photo sleeve	35	75	150
Man from U.N.C.L.E. Sheet Music	Hastings Music Corp.	1964	six pages, theme song and a brief description of the TV show	10	20	40
Man from U.N.C.L.E. Trading Cards	Topps	1965	set of 55 b/w photo cards	40	85	150
Man from U.N.C.L.E. Trading Cards	Cadet Sweets	1966	set of 50 cards, color photos, set	20	40	80
Man from U.N.C.L.E. Trading Cards	ABC/ England	1966	25 cards	20	40	80
Mystery Jigsaw Series Puzzles	Milton Bradley	1965	14" x 24" puzzle, 250 pieces plus story booklet, The Loyal Groom, The Vital Observation, The Impossible Escape, The Micro-Film Affair, each	20	45	85
Napoleon Solo Costume	Halco	1965	transparent plastic "mystery mask," costume has line art shirt, tie, shoulder holster and U.N.C.L.E. logo, in illustrated box	45	90	175
Napoleon Solo Credentials and Passport Set	Ideal	1965	silver ID card, badge, identification wallet, slide window passport, on header card	30	60	120
Napoleon Solo Credentials and Secret Message Sender	Ideal	1965	message sender, badge, and silver ID, on card	35	75	150
Napoleon Solo Doll	Gilbert	1965	11" tall, plastic, white shirt, black pants and shoes, spring loaded arm for firing cap pistol, folding badge, ID card and instruction sheet	70	145	290
Napoleon Solo Figure	Marx	1966	blue or gray plastic figure, 5-3/4" tall stamped with character's name and U.N.C.L.E. logo on the bottom of base	10	25	45
Pinball Affair Game	Marx	1966	12" x 24" tin litho pinball game	65	140	275
Pistol Cane Gun	Marx	1966	25" long, cap firing, bullet shooting aluminum cane with eight bullets and one metal shell, on illustrated card	125	250	500
Power Cube Affair Book	Souvenir	1968	#15 in series, British	4	8	15
Puzzle	Milton Bradley	1966	10" x 19", 100 pieces, Illya's Battle Below	15	35	65
Puzzle	Milton Bradley	1966	10" x 19", 100 pieces, Illya Crushes THRUSH	15	35	65
Puzzles	England	1966	four 11" x 17" puzzles, each with 340 pieces: The Getaway, Solo in Trouble, The Frogman Affair, Secret Plans, each	40	80	160

MAN FROM U.N.C.L.E.

TOY	COMPANY	YEAR	DESCRIPTION	GOOD	EX	MIB
Secret Agent Wristwatch	Bradley	1966	gray watch face shows Solo holding a communicator, came with either plain "leather" or "mod" watch band, in case	110	225	450
Secret Code Wheel Pinball	Marx	1966	10" x 22" x 6" tin litho pinball game	75	150	300
Secret Message Pen	American Character	1966	6-1/2" long double tipped pen for writing invisible messages, on header card	90	180	300
Secret Print Putty	Colorforms	1965	putty in a gun shaped container, print paper, display cards of Kuryakin & Solo and a book of spy & weapons illustrations, on card	20	45	90
Secret Service Gun	Ideal	1965	pistol, holster, badge and silver ID card, in window box	150	300	600
Secret Service Pop Gun		1960s	bagged Luger pop gun on header card with unlicensed illustration of Illya and Napoleon on header	5	10	20
Secret Weapon Set	Ideal	1965	clip loading cap firing pistol, holster, ID wallet, silver ID card, U.N.C.L.E. badge, two demolition grenades and holster, in window	180	375	750
Shirt		1965	has secret pocket, glow-in-the-dark badge & ID, photo package	175	350	700
Shoot Out! Game	Milton Bradley	1965	skill and action game for two players, plastic marble game in illustrated box	75	150	300
Shooting Arcade Game	Marx	1966	tin litho arcade with mechanical wind-up THRUSH agent targets for pellet shooting pistol, scope and stock attachments	200	400	800
Shooting Arcade Game	Marx	1966	smaller version with THRUSH spinner targets	125	250	500
Spy Magic Tricks	Gilbert	1965	mystery gun, Illya playing cards, tricks	125	250	500
Television Picture Story Book	P.B.S. Limited	1968	hardcover, 62 pages, Gold Key reprints	10	18	35
THRUSH Agent Figures	Marx	1966	three different blue plastic figures, 5-3/4" tall stamped with titles and U.N.C.L.E. logo on the bottom of each base, each	8	18	25
THRUSH Ray-Gun Affair Game	Ideal	1966	four U.N.C.L.E. agent pieces, Area Decoder cards, 3-D THRUSH hideouts, THRUSH vehicles, crayons, dice and a rotating "ray gun," in illustrated box	50	100	200
THRUSH-Buster Display Box	Corgi	1966	large display box with graphics, holds 12 cars	160	325	650
U.N.C.L.E. Badges Store Display	Lone Star	1965	illustrated card holds 12 triangular black plastic badges with gold lettering, with badges	45	90	180

MARY POPPINS

TOY	COMPANY	YEAR	DESCRIPTION	GOOD	EX	MIB
Mary Poppins Doll	Gund	1964	11-1/2" tall, bendable	30	85	175
Mary Poppins Figure				20	40	75
Mary Poppins Manicure Set	Tre-Jur	1964		20	40	75

MARY POPPINS

TOY	COMPANY	YEAR	DESCRIPTION	GOOD	EX	MIB
Mary Poppins Paper Dolls	Whitman	1973	with magic tote bag	15	35	65
Mary Poppins Pencil Case		1964	vinyl with zipper top	10	18	35
Mary Poppins Puzzle	Jaymar	1964	frame tray	6	14	24
Mary Poppins Tea Set	Chein	1964	tin, creamer, plates, place settings, cups, serving tray	30	60	100

MICKEY & MINNIE MOUSE

TOY	COMPANY	YEAR	DESCRIPTION	GOOD	EX	MIB
50 Years with Mickey Wristwatch	Bradley	1983	small round chrome case, inscription and serial number on back	45	80	150
Crayons	Transogram	1946		20	40	60
Mickey & Donald Alarm Clock	Jerger/ Germany	1960s	2-1/2" x 5" x 7"; metal case, dark brass finish, 3-D plastic figures of Mickey & Donald on either side	55	100	150
Mickey & Donald Jack-in-the-Box	Lakeside	1966	Donald pops out	30	50	145
Mickey & Minnie & Donald Throw Rug			26" x 41", Mickey & Minnie in an airplane with Donald parachuting	60	115	175
Mickey & Minnie Carpet			27" x 41", Peg Leg Pete is lassoed by Mickey; all characters in western outfits	115	210	325
Mickey & Minnie Dolls	Gund	1940s	13" tall, each	100	210	550
Mickey & Minnie Flashlight	Usalite Co.	1930s	6" long, Mickey leading Minnie through the darkness guided by flashlight & Pluto	35	65	100
Mickey & Minnie Sand Pail	Ohio Art		5" x 5", tin, Mickey, Minnie, Pluto & Donald in boat looking across water at castle, with swivel handle	40	80	200
Mickey & Minnie Sled	S.L. Allen	1935	32" long, wooden slat sled, metal runners, "Mickey Mouse" decal on steering bar	100	260	425
Mickey & Minnie Snow Dome	Monogram	1970s	3" x 4" x 3" tall, Mickey & Minnie with a pot of gold at the end of the rainbow	12	23	35
Mickey & Minnie Tea Set	Ohio Art	1930s	5" x 8", pitcher pictures Mickey at piano & cups picture Mickey, Pluto & Minnie	45	80	200
Mickey & Minnie Toothbrush Holder			4-1/2" tall, bisque, toothbrush holes are located behind their heads	55	100	175
Mickey & Minnie Toothbrush Holder			2-1/2" x 4" x 3-1/2", Mickey & Minnie on sofa with Pluto at their feet	70	130	250
Mickey & Minnie Trash Can	Chein	1970s	13" tall tin litho, Mickey and Minnie fixing a flat tire on one side, other side shows Mickey feeding Minnie soup	25	50	125
Mickey & Minnie Tray	Ohio Art	1930s	5-1/2" x 7-1/4", tin, Mickey & Minnie in rowboat	20	40	130
Mickey & Pluto Ashtray			3" x 4" x 3" tall, ceramic, Mickey & Pluto playing banjos while sitting on the edge of the ashtray	115	210	325
Mickey & Pluto Wristwatch	Bradley	1980	LCD quartz, black vinyl band	19	35	55
Mickey & Three Pigs Spinning Top	Lackawanna		9" diameter, tin	25	65	150
Mickey Mouse Activity Book	Whitman	1936	40 pages	20	35	55
Mickey Mouse Alarm Clock	Bradley	1975	travel alarm, large red cube case, shut off button on top, separate alarm wind, in sleeve box	30	55	85

MICKEY & MINNIE MOUSE

TOY	COMPANY	YEAR	DESCRIPTION	GOOD	EX	MIB
Mickey Mouse Alarm Clock	House Martin	1988	5" x 7" x 2"	11	20	30
Mickey Mouse Alarm Clock	Bayard	1960s	2" x 4-1/2" x 4-1/2" tall, 1930s style Mickey with movable head that ticks off the seconds	95	180	275
Mickey Mouse Ashtray			3-1/2" tall, wood composition figure of Mickey	35	65	125
Mickey Mouse Baby Gift Set		1930s	silver-plated cup, fork, spoon, cup and napkin holder	90	165	250
Mickey Mouse Band Drum		1936	7" x 14" diameter, cloth mesh & paper drum heads	50	90	250
Mickey Mouse Band Leader Bank	Knickerbocker		7-1/2" tall, plastic	9	16	25
Mickey Mouse Band Sand Pail	Ohio Art	1938	6" tall, tin, pictures of Mickey Mouse, Minnie, Horace, Pluto, & Clarabelle the cow parading down street	20	60	200
Mickey Mouse Band Spinning Top			9" diameter	25	50	125
Mickey Mouse Bank		1930s	2" x 3" x 2-1/4", tin, shaped like a treasure chest with Mickey & Minnie on the "Isle of the Thrift"	150	295	450
Mickey Mouse Bank		1930s	2-1/2" x 6", shaped like a mailbox with Mickey holding an envelope	55	100	275
Mickey Mouse Bank	Fricke & Nacke	1978	3" x 5" x 7" tall, embossed image of Mickey in front, side panels have Minnie, Goofy, Donald and Pluto	12	23	150
Mickey Mouse Bank	Crown Toy	1938	6" tall, composition, key locked trap door on base, with figure standing next to treasure chest & head is movable	55	100	350
Mickey Mouse Bank		1950s	5" x 5-1/2" x 6", china, shaped like Mickey's head with slot between his ears	25	50	125
Mickey Mouse Bank	German	1934	2-1/2" x 3", tin, bright yellow bank shaped like a beehive, Mickey approaching door holding a honey jar in one arm & key to open the door in the other	175	340	525
Mickey Mouse Bank	Transogram	1970s	5" x 7-1/2" x 19" tall, plastic with Mickey standing on a white chest	12	23	35
Mickey Mouse Bank	Wolverine	1960s	1-1/2" x 5-1/2" x 1" tall, plastic	50	75	
Mickey Mouse Bean Bag Game	Marks Bros.	1930s		90	125	200
Mickey Mouse Beanie		1950s	blue/yellow felt hat with Mickey on front	16	60	150
Mickey Mouse Boxed Lantern Slides	Ensign	1930s	5" x 6" x 2", cartoons: Traffic Troubles, Gorilla Mystery, Cactus Kid, Castaway, Delivery Boy, Fishin' Around, Firefighters, Moose Hunt & Mickey Steps Out	150	275	425
Mickey Mouse Bubble Buster Gun	Kilgore	1936	8" long, cork gun	55	100	150
Mickey Mouse Bump-N-Go Spaceship	Matsudaya	1980s	battery operated tin litho with clear dome, has six flashing lights, rotating antenna	35	65	100
Mickey Mouse Camera	Helm Toy	1970s	2" x 5" x 4-1/2", Mickey in engineer's uniform riding on top of the train	11	20	30
Mickey Mouse Camera	Ettelson	1960s	3" x 3" x 5"	18	35	50

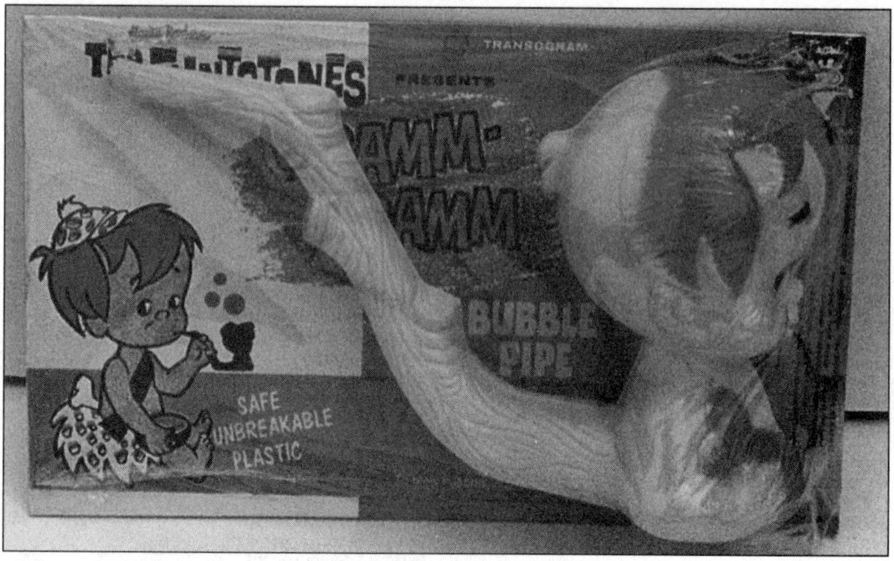

Bamm-Bamm Bubble Pipe, 1963, Transogram

Dick Tracy Squad Car No.1, 1950s, Marx

MICKEY & MINNIE MOUSE

TOY	COMPANY	YEAR	DESCRIPTION	GOOD	EX	MIB
Mickey Mouse Camera	Child Guidance	1970s	4" x 7" x 7"	9	16	25
Mickey Mouse Car	Polistil	1970s	2" x 4" x 1-1/2" tall plastic car with rubber figure, Mickey in driver's seat	25	50	75
Mickey Mouse Cardboard House	O.B. Andrews	1930s	14" x 12" x 13" tall	115	210	325
Mickey Mouse Chatty Chums	Mattel	1979		6	12	25
Mickey Mouse Christmas Lights	Noma	1930s	eight lamps with holiday decals of characters	115	210	325
Mickey Mouse Clock	Elgin	1970s	10" x 15" x 3", electric wall clock	11	20	30
Mickey Mouse Club Bank	Play Pal Plastics	1970s	4-1/2" x 6" x 11-1/2" tall, vinyl	16	30	45
Mickey Mouse Club Coffee Tin		1950s	illustrated lid promotes the club, tin included MM badge	30	55	85
Mickey Mouse Club Fun Box	Whitman	1957	box includes stamp book, club scrapbook, six coloring books and four small gameboards	40	70	110
Mickey Mouse Club Magic Divider	Jacmar	1950s	arithmetic game	25	50	75
Mickey Mouse Club Magic Kit	Mars Candy	1950s	two 8" x 20" punch-out sheets	25	50	100
Mickey Mouse Club Magic Subtractor	Jacmar	1950s	arithmetic game	11	20	30
Mickey Mouse Club Marionette		1950s	3" x 6-1/2" x 13-1/2" tall, composition figure of a girl with black felt hat & mouse ears	50	150	250
Mickey Mouse Club Mouseketeer Doll	Horsman	1960s	8" tall, denim outfit	25	45	90
Mickey Mouse Club Mousketeer Ears	Kohner		7" x 12-1/2"	11	20	30
Mickey Mouse Club Mousketeer Handbag	Connecticut Leather		leather/vinyl crafting set	45	60	75
Mickey Mouse Club Newsreel with Sound	Mattel	1950s	4" x 4-1/2" x 9" tall, orange box, plastic projector with two short filmstrips, record, cardboard screen, cartoons "Touchdown Mickey" and "No Sail"	75	165	275
Mickey Mouse Club Plastic Plate	Arrowhead	1960s	9" diameter, clubhouse with Goofy, Pluto & Donald wearing mouse ears and sweaters with club emblems	16	30	45
Mickey Mouse Club Rhythm Makers Set	Emenee			65	100	300
Mickey Mouse Club Toothbrush	Pepsodent	1970s		4	7	10
Mickey Mouse Colorforms Set	Colorforms	1976	8" x 12-1/2" x 1", Spirit of '76	10	16	60
Mickey Mouse Cup	Cavalier	1950s	3" tall, silver-plated cup with a 2-1/2" opening	25	50	75
Mickey Mouse Dart Gun Target	Marks Brothers	1930s	10" target, dart gun & suction darts	30	55	225
Mickey Mouse Dinner Set	Empresa Electro	1930s	china, 2" creamer, 3-4-1/2" plates and 3-5" plates, 7" long dish, two oval platters	130	245	375
Mickey Mouse Doll		1930s	4" x 7" x 10-1/2" tall, movable arms & legs, swivel head, black velveteen body & red felt pants	300	575	875
Mickey Mouse Doll	Schuco	1950s	10" tall	100	210	350
Mickey Mouse Doll	Gund	1960s	12", Mickey as fireman	25	50	75

CHARACTER

MICKEY & MINNIE MOUSE

TOY	COMPANY	YEAR	DESCRIPTION	GOOD	EX	MIB
Mickey Mouse Doll	Knickerbocker	1935	11" tall, stuffed with removable shoes & jointed head, red shorts	115	210	325
Mickey Mouse Doll	Hasbro	1970s	4" x 5-1/2" x 7-1/2" tall, talks	18	35	50
Mickey Mouse Doll	Knickerbocker		5" x 7" x 12" tall, fabric, Mickey in checkered shorts & green jacket with white felt flower stapled to it	175	340	525
Mickey Mouse Doll	Horsman	1972	3" x 10" x 12" tall, talking doll, says five different phrases	18	35	50
Mickey Mouse Dominoes	Halsam		Mickey and Pluto on dominoes	75	95	150
Mickey Mouse Drum	Ohio Art	1930s	6" diameter	30	55	175
Mickey Mouse Electric Casting Set	Home Foundary	1930s	9-1/2" x 16" x 2"	60	115	175
Mickey Mouse Electric Table Radio	General Electric	1960s	4-1/2" x 10-1/2" x 6" tall	25	50	125
Mickey Mouse Figure			4" tall, bisque, dressed in a green nightshirt	35	65	100
Mickey Mouse Figure	Seiberling	1930s	6-1/2" tall, rubber	65	120	185
Mickey Mouse Figure	Goebel	1930s	3-1/2" tall, in a hunting outfit reading a book	35	65	100
Mickey Mouse Figure			composition, part of the Lionel Circus Train Set	55	100	150
Mickey Mouse Figure	Seiberling	1930s	3-1/2" tall, black hard rubber	25	50	75
Mickey Mouse Figure		1930s	3" tall, bisque, holding a parade flag & sword	35	60	95
Mickey Mouse Figure		1930s	3" tall, bisque, plays saxophone	35	60	95
Mickey Mouse Figure	Seiberling	1930s	3-1/2" tall, latex	55	100	150
Mickey Mouse Figure	Marx	1970	6" tall, vinyl	9	16	35
Mickey Mouse Fire Engine	Sun Rubber		7" long, rubber, push toy	40	80	135
Mickey Mouse Fire Truck with Figure	Sun Rubber		2-1/2" x 6-1/2" x 4", Mickey driving & mold-in image of Donald standing on the back holding onto his helmet	45	80	125
Mickey Mouse Fun-E-Flex Figure	Fun-E-Flex	1930s	7", bendy	245	455	700
Mickey Mouse Gumball Bank	Hasbro	1968		15	30	45
Mickey Mouse Has a Busy Day Book	Whitman	1937	16 pages	25	35	55
Mickey Mouse in Giantland Book	David McKay	1934	45 pages, hardcover	45	80	125
Mickey Mouse Jack-In-the-Box		1970s	5-1/2" square tin litho box shows Mickey, Pluto, Donald and Goofy, Mickey pops out	23	45	65
Mickey Mouse Lamp	Soreng-Manegold	1935	10-1/2" tall	70	130	225
Mickey Mouse Lionel Circus Train	Lionel	1935	five cars w/Mickey, 30" train, 84 inches of track, circus tent, Sunoco station, truck, tickets, Mickey composition statue	550	1050	1750
Mickey Mouse Lionel Circus Train Handcar	Lionel		metal, 9" long with 6" tall composition/rubber figures of Mickey & Minnie	235	440	675
Mickey Mouse Magic Slate	Watkins-Strathmore	1950s	8-1/2" x 14" tall	15	30	45
Mickey Mouse Map of the United States	Dixon	1930s	9-1/4" x 14"	35	65	100
Mickey Mouse Marbles	Monarch		marbles and Mickey bag, on card	4	7	15

MICKEY & MINNIE MOUSE

TOY	COMPANY	YEAR	DESCRIPTION	GOOD	EX	MIB
Mickey Mouse Mechanical Pencil		1930s	head of Mickey on one end and decal of Mickey walking on other side of pencil	35	65	100
Mickey Mouse Mechanical Robot	Gabriel			65	120	135
Mickey Mouse Mousegetar		1960s	10" x 30" x 2-1/2" black plastic	35	75	100
Mickey Mouse Movie Projector	Keystone	1934	5-1/2" x 11-1/2" x 11" tall for 8mm movies	140	260	400
Mickey Mouse Movie-Fun Shows	Mastercraft	1940s	7-1/2" square by 4" deep, animated action movies	95	180	275
Mickey Mouse Music Box	Anri	1971	5" x 3-1/2", plays "If I Were Rich Man"	45	80	125
Mickey Mouse Music Box	Schmid	1970s	3-1/2" x 5-1/2", plays "Mickey Mouse Club March," ceramic, Mickey in western clothes standing next to a cactus	30	55	85
Mickey Mouse Music Box			4-1/2" x 5" x 7", plays "It's a Small World," Mickey in conductor's uniform standing on cake	18	35	50
Mickey Mouse Music Box	Japan	1970s	4" x 6" tall, china, plays "Side by Side," Mickey is brushing a kitten in a washtub	12	23	55
Mickey Mouse Music Box	Schmid		3" x 6" x 7-1/2" tall, bisque, Spirit of '76, plays "Yankee Doodle;" Mickey, Goofy and Donald are dressed as Revolutionary War Minutemen	60	115	175
Mickey Mouse Musical Money Box		1970s	3" x 6", tin box with Mickey, Pluto, Donald and Goofy	30	55	85
Mickey Mouse Night Light	Disney	1938	4" tall, tin	90	165	250
Mickey Mouse Old Timers Fire Engine	Matsudaya	1980s	red, tin and plastic fire truck with Mickey at the wheel	35	65	100
Mickey Mouse Pencil Box	Dixon	1937	5-1/2" x 9" x 3/4", Mickey, Goofy & Pluto riding a rocket	35	65	120
Mickey Mouse Pencil Box	Dixon	1937	5" x 8-1/2" x 1-1/4", Mickey is a circus ringmaster & Donald riding a seal	35	65	120
Mickey Mouse Pencil Box	Dixon	1930s	Mickey ready to hitch Horace to a carriage in which Minnie is sitting	35	65	120
Mickey Mouse Pencil Box	Dixon	1930s	5-1/2" x 10-1/2" x 1-1/4", Mickey in a gymnasium	35	65	150
Mickey Mouse Pencil Holder	Dixon	1930s	4-1/2" tall	60	115	135
Mickey Mouse Pencil Sharpener			3" tall, celluloid, sharpener located on base	55	100	225
Mickey Mouse Pencil Sharpener	Hasbro	1960s	shape of Mickey's head, pencil goes into mouth	12	23	35
Mickey Mouse Picture Gun	Stephens	1950s	6-1/2" x 9-1/2" x 3", metal, lights to show filmstrips	55	100	250
Mickey Mouse Pitcher	Germany	1930s	2" x 3" diameter, china, white with green shading around the base, Mickey on each side	60	115	175
Mickey Mouse Play Tiles	Halsam	1964	336 tiles	15	25	40
Mickey Mouse Pocket Watch	Bradley	1976	3-1/2" x 4-1/2" x 3/4", Mickey in his Bicentennial outfit	55	500	950
Mickey Mouse Pocket Watch	Ingersoll	1930s	2" diameter	195	350	850
Mickey Mouse Pocket Watch	Bradley	1970s		35	65	100
Mickey Mouse Pocket Watch	Lorus	1988	#2202, quartz, small gold bezel, gold chain and clip fob, articulated hands	20	40	60

MICKEY & MINNIE MOUSE

TOY	COMPANY	YEAR	DESCRIPTION	GOOD	EX	MIB
Mickey Mouse Presents a Silly Symphony Book	Whitman	1934	Big Little Book	20	45	65
Mickey Mouse Print Shop Set	Fulton Specialty	1930s	6-1/2" x 6-1/2", ink pad, stamper, metal tweezers, wooden tray	70	130	200
Mickey Mouse Pull Toy	N.N. Hill Brass	1935	14" tall, wood & metal	130	245	375
Mickey Mouse Pull Toy	Toy Kraft		7" x 22" x 8" tall, horse cart drawn by wooden horses	300	550	850
Mickey Mouse Puppet Forms	Colorforms	1960s		10	20	50
Mickey Mouse Radio	Emerson	1934	wood composition cabinet with designs of Mickey Mouse playing various musical instruments	335	700	3000
Mickey Mouse Radio	Philgee	1970s		16	30	45
Mickey Mouse Record Player	General Electric	1970s	playing arm is the design of Mickey's arm	55	100	150
Mickey Mouse Riding Toy	Mengel	1930s	6" x 17" x 16" tall	385	715	1100
Mickey Mouse Rodeo Rider	Matsudaya	1980s	plastic wind-up, cowboy Mickey rides a bucking bronco	20	40	60
Mickey Mouse Rolykin	Marx		1-1/2" tall, ball bearing action	10	16	40
Mickey Mouse Rub 'N Play Magic Transfer Set	Colorforms	1978		15	25	50
Mickey Mouse Safety Blocks	Halsam	1930s	nine blocks	80	145	225
Mickey Mouse Sand Pail	Ohio Art	1938	6" tall, tin, Mickey, Donald & Goofy playing golf	25	80	225
Mickey Mouse Sand Pail	Ohio Art	1938	3" tall, tin	25	70	125
Mickey Mouse Sand Shovel	Ohio Art		10" long, tin	15	50	100
Mickey Mouse Saxophone		1930s	16" tall	115	210	325
Mickey Mouse Scissors	Disney		3" long, child's scissors with Mickey figure	20	35	55
Mickey Mouse Serving Tray		1960s	11" diameter, tin	11	20	30
Mickey Mouse Sewing Cards	Colorforms	1978	7-1/2" x 12" cut-out card designs of Mickey, Minnie, Pluto, Clarabelle, Donald Duck and Horace	10	20	70
Mickey Mouse Sled	Flexible Flyer	1930s	18" x 30" x 6" tall, wood	100	210	750
Mickey Mouse Squeeze Toy	Dell	1960s	8" tall	15	30	45
Mickey Mouse Squeeze Toy	Dell	1950s	rubber, Mickey as hitchhiking hobo	35	60	95
Mickey Mouse Squeeze Toy	Sun Rubber	1950s	10" tall, rubber	20	40	60
Mickey Mouse Stamp Pad		1930s	3" long	25	50	75
Mickey Mouse Steamboat	Matsudaya	1988	wind-up plastic steamboat with Mickey as Steamboat Willie, runs on floor as smokestacks go up and down, box says "60 Years With You"	30	55	125
Mickey Mouse Tea Set		1930s	3" saucer, 2" pitcher, 2-1/2" sugar bowl each piece shows Mickey & Minnie in a rowboat	25	50	225
Mickey Mouse Tea Set	Wolverine		plastic	40	110	150
Mickey Mouse Telephone Bank	N.N. Hill Brass	1938	5" tall with cardboard figure of Mickey	70	130	275
Mickey Mouse Throw Rug	Alex. Smith Carpet	1935	26" x 42"	125	225	350
Mickey Mouse Tractor	Sun Rubber		5" long, rubber	25	50	90
Mickey Mouse Transistor Radio	Gabriel	1950s	6-1/2" x 7" x 1-1/2"	35	65	125

MICKEY & MINNIE MOUSE

TOY	COMPANY	YEAR	DESCRIPTION	GOOD	EX	MIB
Mickey Mouse Tricycle Toy	Steiff	1932	8-1/2" x 7", wood & metal frame, action movement	325	780	3000
Mickey Mouse Twirling Tail Toy	Marx	1950s	3" x 5-1/2" x 5-1/2" tall, with a built-in key, metal tail spins around as the toy vibrates	115	210	325
Mickey Mouse Utensils	Wm. Rogers & Son	1947	6" fork and 5-1/2" spoon	45	80	125
Mickey Mouse Viewer	Craftsmen's Guild	1940s	film viewer with 12 films	65	120	185
Mickey Mouse Wall Clock	Elgin	1978	9" diameter dial, 15" long, shaped like oversize watchband, "50 Happy Years" logo on dial	25	50	75
Mickey Mouse Wash Machine	Ohio Art		8" tin litho with Mickey & Minnie Mouse pictured doing their wash	40	80	250
Mickey Mouse Water Globes		1970s	3" x 4-1/2" x 5" tall, three dimensional plastic figures of Mickey seated with a plastic water globe between his legs	25	50	75
Mickey Mouse Watering Can	Ohio Art	1938	6" tin litho	35	120	225
Mickey Mouse Wind-Up Musical Toy	Illco	1970s	6" tall, plays "Lullaby & Goodnight," 3-D figure of Mickey in red pants & yellow shirt	20	35	50
Mickey Mouse Wind-Up Toy	Gabriel	1978	plastic transparent figure of Mickey with visible metal gears	9	15	35
Mickey Mouse Wind-Up Trike	Korean	1960s	tin litho trike with plastic Mickey with flag and balloon on handle, bell on back	80	145	225
Mickey Mouse Wristwatch			two-gun Mickey, saddle tan western style band	23	45	65
Mickey Mouse Wristwatch	Timex	1958	electric	115	210	325
Mickey Mouse Wristwatch	Bradley	1970s	2-1/2" x 6" x 2-1/2", plastic case, white dial with Mickey playing tennis, the second hand has a tennis ball on the end of it	35	65	100
Mickey Mouse Wristwatch	Ingersoll	1939	rectangular with standard second hand between Mickey's legs	200	375	600
Mickey Mouse Wristwatch	Timex	1960s	large round case, stainless back, articulated hands, red vinyl band	45	80	75
Mickey Mouse Wristwatch	Bradley	1984	medium white case, articulated hands, black face, sweep seconds, white vinyl band, in plastic window box	20	35	55
Mickey Mouse Wristwatch	Bradley	1983	medium black octagonal case, articulated hands, no numbers on face, in plastic window box	20	250	375
Mickey Mouse Wristwatch	Bradley	1970s	white plastic case, watch on pendant, bubble crystal, articulated hands, gold chain	25	50	75
Mickey Mouse Wristwatch	Bradley	1978	commemorative edition	60	115	175
Mickey Mouse Yarn Sewing Kit	Marks Bros.	1930s	seven cards, yarn, needle	35	150	250
Mickey's Air Mail Plane	Sun Rubber	1940s	3-1/2" x 6" long, 5" wingspan, rubber	35	65	150
Minnie Mouse Alarm Clock	Bradley	1970s	pink metal electric two-bell clock with articulated hands	23	45	65
Minnie Mouse Car	Matchbox	1979		5	10	20

MICKEY & MINNIE MOUSE

TOY	COMPANY	YEAR	DESCRIPTION	GOOD	EX	MIB
Minnie Mouse Choo-Choo Train Pull Toy	Linemar	1940s	3" x 8-1/2" x 7" tall, green metal base & green wooden wheels	70	125	250
Minnie Mouse Clock	Phinney-Walker	1970s	8" diameter by 1-1/2" deep, plastic, "Behind Every Great Man, There is a Woman!"	35	65	100
Minnie Mouse Doll	Knickerbocker	1935	14" tall, stuffed, cloth, polka-dot skirt & lace pantaloons	105	195	475
Minnie Mouse Doll	Petz	1940s	10" tall	115	210	325
Minnie Mouse Figure	Ingersoll	1958	5-1/2" tall, plastic	30	55	85
Minnie Mouse Fun-E-Flex Figure	Fun-E-Flex	1930s	5", bendy	125	225	350
Minnie Mouse Hand Puppet		1940s	11" tall, white on red polka-dot, fabric hard cover and a pair of black & white felt hands	50	95	160
Minnie Mouse Music Box	Schmid	1970s	3-1/2" diameter, plays "Love Story"	11	20	50
Minnie Mouse Rocker	Marx	1950s	tin wind-up, rocker moves back & forth with gravity motion of her head & ears	250	450	695
Minnie Mouse Wristwatch	Timex	1958	small round chrome case, stainless back, articulated hands, yellow vinyl band	55	100	150
Minnie Mouse Wristwatch	Bradley	1978	gold case, sweep seconds, red vinyl band, articulated hands	16	30	45
Minnie with Bump-n-Go Action Shopping Cart	Illfelder	1980s	4" x 9-1/2" x 11-1/2" tall, plastic, battery operated Minnie pushing cart	25	50	75
Mouseketeer Cut-Outs	Whitman	1957	figures and accessories	23	45	65
Mouseketeer Fan Club Typewriter		1950s	lithographed tin	45	85	150
Puzzle	Marks Brothers	1930s	10" x 12", Mickey polishing the boiler on his "Mickey Mouse R.R." train engine and Minnie waving from the cab	40	70	110
Puzzle	Whitman	1957	Adventureland, 11" x 15" frame tray; Mickey, Minnie, Donald & his nephew in boat surrounded by jungle beasts	16	30	45
Puzzle	Jaymar	1960s	Pluto's Wash and Scrub Service	12	23	50
Spinning Top	Chein	1950s	tin litho, features Mickey in cowboy outfit and other characters	30	100	225
Spinning Top	Fritz Bueschel	1930s	7" diameter, 7" tall, Mickey, Minnie, a nephew, Donald & Horace playing a musical instrument	50	100	250
The Adventures of Mickey Mouse Book	David McKay	1931	full color illustrations, softcover	45	80	125

MIGHTY MOUSE

TOY	COMPANY	YEAR	DESCRIPTION	GOOD	EX	MIB
Charm Bracelet		1950s	brass charms of Gandy Goose, Terry Bear, Mighty Mouse and other Terrytoon characters	20	40	75
Mighty Mouse Ball Game	Ja-Ru	1981		4	7	15
Mighty Mouse Cinema Viewer	Fleetwood	1979	with four strips	7	15	25
Mighty Mouse Doll	Ideal	1950s	14" tall, stuffed, cloth	30	60	120
Mighty Mouse Dynamite Dasher	Takara	1981		5	100	200
Mighty Mouse Figure	Dakin	1977	hard & soft vinyl figure	20	40	70
Mighty Mouse Figure	Dakin	1978	Fun Farm	20	45	75

MIGHTY MOUSE

TOY	COMPANY	YEAR	DESCRIPTION	GOOD	EX	MIB
Mighty Mouse Figure		1950s	9-1/2" rubber, squeaks	20	40	70
Mighty Mouse Flashlight	Dyno	1979	3-1/2" figural light	7	60	90
Mighty Mouse Hit the Claw Target Set	Parks			50	100	190
Mighty Mouse Make-a-Face Sheet	Towne	1958	with dials to change face parts	15	30	50
Mighty Mouse Mighty Money	Fleetwood	1979		2	50	75
Mighty Mouse Money Press	Ja-Ru	1981	stampers, pads and money	5	10	20
Mighty Mouse Movie Viewer	Chemtoy	1980		5	10	20
Mighty Mouse Picture Play Lite	Janex	1983		6	11	20
Mighty Mouse Sneakers	Randy Co.	1960s	children's, graphics on box, picture on sneakers	20	40	100
Mighty Mouse Wallet	Larami	1978		5	10	15
Mighty Mouse Wristwatch	Bradley	1979	chrome case	20	15	25
Puzzle	Fleetwood	1979	Mighty Mouse/Heckle and Jeckle	6	75	100
Puzzle			tray puzzle; Mighty Mouse and his TV Pals	6	12	25

MISCELLANEOUS CHARACTERS

TOY	COMPANY	YEAR	DESCRIPTION	GOOD	EX	MIB
A-Team Rocket Ball Target Set		1983	gumballs with gun and target	4	9	17
A-Team Shrinky Dinks Set		1980s		4	8	16
Amy Carter Paper Dolls	Toy Factory	1970s	14" cardboard doll w/ accessories	15	25	45
Ben Casey Pencils	Hassenfeld Bros.	1962	10 red/white pencils on card	15	30	60
Bewitched Samantha Doll	Ideal	1967	12-1/2" tall	175	350	700
Bewitched Tabitha Paper Doll Set	Magic Wand	1966	11" cardboard doll, clothes	25	50	100
Bonzo Scooter Toy			7" scooter with 6" Bonzo, wind-up	125	250	475
Bruce Lee Figure	Largo	1986	8" with weapon	15	35	65
Captain Kangaroo Presto Slate	Fairchild	1960s	slate on illustrated card, several versions	15	25	35
Captain Kangaroo Puzzle	Whitman	1960	#4446, frame tray	10	30	50
Captain Video Rocket Launcher	Lido	1952		60	175	300
Daktari Puzzle	Whitman	1967	100 pieces	10	20	35
Diamond Jim Figure		1930s	5-1/2" tall	30	65	125
Dr. Kildare Photo Scrapbook		1962		10	20	40
Dragnet Badge 714	Knickerbocker	1955	2-1/2" bronze finish badge in yellow box with illustration of Jack Webb, box bottom has ID card	10	40	75
Evel Knievel Stunt Stadium	Ideal	1974	large vinyl case with accessories	35	75	150
George Bush Figure			7" tall	8	15	30
Hardy Boys Dolls	Kenner	1979	12" tall Joe Hardy (Shaun Cassidy) or Frank Hardy (Parker Stevenson)	15	40	75
Holly Hobbie Wristwatch	Bradley	1982	small gold tone case, base metal back, yellow plastic band	5	10	35
Jimmy Carter Radio			peanut-shaped transistor radio	20	42	85
Jimmy Carter Wind-Up Walking Peanut			5" tall	10	20	40

CHARACTER

MISCELLANEOUS CHARACTERS

TOY	COMPANY	YEAR	DESCRIPTION	GOOD	EX	MIB
Joan Palooka Stringless Marionette	Nat'l Mask & Puppet	1952	12-1/2" tall "daughter of Joe Palooka" doll comes with pink blanket and birth certificate	50	100	200
Kennedy Kards		1960s	red, white, and blue playing cards w/cartoon artwork	10	15	30
Knight Rider Impossibles Stunt Set	LJN	1982		45	95	175
Knight Rider Wrist Communicator	Larami	1982		9	19	35
Komic Kamera Film Viewer Set		1950s	5" long, Dick Tracy, Little Orphan Annie, Terry & the Pirates & The Lone Ranger	35	75	145
Kukla, Fran and Ollie Puppet Show	Milton Bradley	1962	cardboard stage, puppets, props	45	150	250
Little King Lucky Safety Card	New York Journal	1953	2" x 4" cards, Little King warns of safety	8	15	30
Lyndon Johnson Figure	Remco	1960s		17	37	75
Mork and Mindy Colorforms	Colorforms	1979		10	15	25
Mr. & Mrs. Potato Head Set	Hasbro	1960s	cars, boats, shopping trailer, etc.	22	45	90
Mr. Ed Talking Horse Puppet	Mattel	1962		32	75	110
Mr. Potato Head Frankie Frank	Hasbro	1966	companion to Mr. Potato Head, with accessories	20	40	80
Mr. Potato Head Frenchy Fry	Hasbro	1966	companion to Mr. Potato Head, with accessories	20	40	80
Mr. Potato Head Ice Pops	Hasbro	1950s	plastic molds for freezing treats, in box	16	30	60
Patton Figure	Excel Toy		poseable doll with clothing and accessories	20	40	80
Prince Charles of Wales Figure	Goldberger	1982	13" tall, dressed in palace guard uniform	18	35	70
Red Ranger Ride 'Em Cowboy	Wyandotte	1930s	tin wind-up rocker	110	225	550
Rin-Tin-Tin Magic Erasable Pictures	Transogram	1955		45	125	225
Ringling Bros. & Barnum & Bailey Circus Play Set		1970s	vinyl, with animals, trapeze personnel, clowns and assorted circus equipment	20	40	80
Sandra Dee Paper Dolls	Saalfield	1959	two cardboard dolls w/outfits	20	45	75
Sgt. Preston of the Yukon Punch-Out Cards	Quaker	1950s	"Big Game Trophy" cardboard cut-outs, set of nine	25	75	150
Sir Reginald Play-N-Save Bank		1960s	7" tall plastic lion & hunter, on a 15" green plastic base, the hunter fires the coin into the lion's mouth	45	100	150
Soupy Sales Card Game	Jaymar	1960s	Slap Jack, Old Maid, Funny Rummy, or Hearts/Crazy 8s	15	35	50
Soupy Sales Doll	Remco	1960s	5" doll	40	125	200
Soupy Sales Doll	Knickerbocker	1966	12" plush, vinyl head	25	60	100
Space Ghost Puzzle	Whitman	1967		20	50	75
Starsky & Hutch Puzzle	HG Toys	1970s		15	30	45
Starsky & Hutch Shoot-Out Target Set	Berwick	1970s		23	45	90
Sylvester Stallone Rambo Figure		1986	18" tall, poseable figure	7	13	25
Tuesday Weld Paper Dolls	Saalfield	1960	two cardboard dolls w/outfits	25	45	75
Uncle Don's "Puzzy & Sizzy" Membership Card		1950s		10	20	35
Willie Whopper Pencil Case		1930s	green with illustrations of Willie, Pirate and his gal	40	75	120

MOON MULLINS

TOY	COMPANY	YEAR	DESCRIPTION	GOOD	EX	MIB
Moon Mullins & Kayo Railroad Handcar Toy	Marx	1930s	6" long, wind-up, both figures bendable arms & legs	225	450	900
Moon Mullins & Kayo Toothbrush Holder			4" tall, bisque	30	65	130
Moon Mullins Figure Set			bisque, Uncle Willie, Kayo, Moon Mullins & Emmy, 2-1/4" to 3-1/2"	80	250	450
Moon Mullins Playstone Funnies Mold Set		1940s		35	75	150
Puzzle		1930s	9-1/2" x 14", Featured Funnies	15	50	75

MR. MAGOO

TOY	COMPANY	YEAR	DESCRIPTION	GOOD	EX	MIB
Mr. Magoo Car	Hubley	1961	7-1/2" x 9" long, metal, battery-operated	100	225	450
Mr. Magoo Doll	Ideal	1962	5" tall, vinyl head with cloth body	40	90	175
Mr. Magoo Doll	Ideal	1970	12" tall	20	45	85
Mr. Magoo Drinking Glass		1962	5-1/2" tall	10	25	45
Mr. Magoo Figure	Dakin		7" tall	25	75	150
Mr. Magoo Hand Puppet		1960s	vinyl head, cloth body	25	50	100
Mr. Magoo Puzzle	Warren	1978	frame tray	7	15	25
Mr. Magoo Soaky	Palmolive	1960s	10" tall, vinyl & plastic	15	50	75

MUNSTERS, THE

TOY	COMPANY	YEAR	DESCRIPTION	GOOD	EX	MIB
Grandpa Doll	Remco	1964		150	300	600
Herman Munster Doll	Remco	1964		150	300	700
Lily Baby Doll	Ideal	1965	unlicensed "monster baby"	40	80	160
Lily Doll	Remco	1964		150	300	600
Puzzle	Whitman	1960s	frame tray	30	50	100
Puzzle	Whitman	1965	100 pieces, boxed	35	75	150
The Last Resort Book	Whitman	1964		13	30	40

MUTT & JEFF

TOY	COMPANY	YEAR	DESCRIPTION	GOOD	EX	MIB
Mutt & Jeff Bank			5" tall, cast iron, two piece construction held together by screw in the back	40	90	175
Mutt & Jeff Dolls		1920s	8" x 6-1/2" tall, composition hands & heads with heavy cast iron feet, movable arms & legs, fabric clothing; pair	125	250	500
Mutt & Jeff Figures	A. Steinhardt & Bros	1911	ceramic, with a coin inserted into base; pair	75	150	300

NIGHTMARE BEFORE CHRISTMAS

TOY	COMPANY	YEAR	DESCRIPTION	GOOD	EX	MIB
Bandanna	Fashion Victim		two styles: Jack or Lock, Shock & Barrel	4	5	6
Baseball Caps	Fashion Victim		three styles: Fishbone with metal keychain; Lock, Shock & Barrel; or Jack "Bone Daddy"	8	10	12
Beach Towel	Fashion Victim			22	29	30
Bionic Airwalker Balloon	Anagram		featuring Jack Skellington; over 7" tall	15	20	30
Bookmarks	OSP Publishing		set of eight styles including wallet cards	10	15	20
Boxer Shorts	Stanley DeSantis		Lock, Shock & Barrel or Jack styles; cotton	6	8	10
Brass Keychain	Disney		Jack Skellington's Tombstone	3	4	5

CHARACTER

NIGHTMARE BEFORE CHRISTMAS

TOY	COMPANY	YEAR	DESCRIPTION	GOOD	EX	MIB
Buttons			set of 12 featuring logo, characters, etc.	15	20	30
Cardboard Store Display	Applause		over 6" wide	50	60	125
Comforter	Wamsutta		twin size featuring Lock, Shock & Barrel	45	55	75
Cookie Jar	Treasure Craft		ceramic; Jack on Tombstone	40	60	130
Drawstring Bag	Jack Skellington on front			10	15	20
Gift Bags	Cleo		five styles available	6	8	10
Glow Oogie Boogie Figure	Applause		PVC figure	7	12	30
Greeting Cards			set of six featuring various characters	15	20	30
Handheld Video Game	Tiger			25	30	40
Jack Figure	Hasbro		bendy on card	5	10	20
Jack in Coffin Figure	Applause		12"	30	50	225
Jack Soaky			glow head	7	10	35
Kaleidoscope	C. Bennett Scopes			16	18	20
Lock, Shock & Barrel Dolls	Applause		set of three; 6" cloth and vinyl dolls	25	35	60
Lock, Shock & Barrel Figures	Applause		set of three, 3" PVC figures	10	15	35
Magic Action Figures	Applause		set of three; 4" figures has its own action when rolled	20	30	40
Mayor Figure	Hasbro		bendy on card	10	25	40
Mayor Wind-Up Music Box	Schmid		head spins while music plays "What's This?"	40	70	100
Mug	Selandia		10 oz. acrylic mug with floating snowflakes and glitter	5	7	18
Mug	Selandia		16 oz. acrylic mug with floating snowflakes and glitter	5	7	18
Mylar Balloons	Anagram		five variations, each	10	15	25
Notepad	Beach		75-sheet pad featuring Lock, Shock & Barrel	3	4	5
Oogie Boogie Doll	Applause		16"; makes farting noise when squeezed	15	25	50
Partyware	Beach		65 pieces	14	16	20
Pencil			Whirly Sally or Whirly Jack	4	10	12
Pin	Oopsa Daisy		several character styles, pewter	10	18	20
Pop-Up Book				15	30	45
Postcard Book			30 full-color postcards	5	10	15
Purse			multi-compartment with mirror	5	10	20
PVC Figure on Drinking Straws			Jack or Sally	3	7	8
Rhinestone Pin			white stones featuring Jack Skellington	30	55	65
Rhinestone Pin	Oopsa Daisy		red stones featuring Jack Skellington	40	65	75
Sally Doll	Hasbro		removable limbs	40	75	300
Sally Figure	Hasbro		bendy on card	15	25	50
Sally in Coffin Figure	Applause		12"	50	100	350
Santa Claus Doll	Applause		10" plush	10	15	35
Santa Jack Figure	Hasbro		bendy on card	6	8	10
Santa Puppet Doll	Hasbro			20	30	60
Silk Necktie			many styles/patterns	15	20	30
Sky Floater Kite	Spectra Star			5	9	12
Stickers	Gibson		featuring movie characters	2	4	6
Sunglasses	Jet Vision Limited			12	15	18
Talking Jack Doll	Hasbro			45	70	275
Temporary Tattoos	US Kids			2	4	6

NIGHTMARE BEFORE CHRISTMAS

TOY	COMPANY	YEAR	DESCRIPTION	GOOD	EX	MIB
Tumbler	Selandia		7 oz., acrylic	10	12	14
Vest	Fashion Victim		Lock, Shock & Barrel or Jack styles	12	15	20
Video Release Poster	Touchstone		24" x 36"	15	18	30
Wooden Ornaments	Kurt S. Adler		set of 10; 4" to 6" tall; individually packaged	55	60	70
Wristwatch	Timex		six styles available, each	20	30	50
Wristwatch	Burger King		four styles with plastic bands; digital time	10	15	30
Yo-Yo	Spectra Star			5	7	10

PEANUTS

TOY	COMPANY	YEAR	DESCRIPTION	GOOD	EX	MIB
Batter-Up Snoopy Colorforms	Colorforms	1979		15	35	40
Big Quart-O-Snoopy Bubbles	Chemtoy	1970s		9	13	15
Camp Kamp Play Set	Child Guidance	1970s	rubber camp building with characters	45	65	75
Can You Catch It, Charlie Brown? Game	Ideal	1976		75	110	125
Charlie Brown Costume	Collegeville		with mask	12	20	30
Charlie Brown Deluxe View-Master Gift Pak	GAF/View-Master	1970s	cylindrical container holds seven reels and viewer	35	50	60
Charlie Brown Doll	Determined	1970s	plastic; wearing baseball gear	45	65	75
Charlie Brown Doll	Ideal	1976	removable clothing	25	40	45
Charlie Brown Doll	Hungerford Plastics	1958	8-1/2" plastic	45	77	90
Charlie Brown Nodder	Japanese	1960s	5-1/2" tall, bobbing head	20	75	150
Charlie Brown Pocket Doll	Boucher	1968	7"	20	30	35
Charlie Brown Punching Bag	Determined	1970s		25	40	45
Charlie Brown Push Puppet	Ideal	1977		20	50	55
Charlie Brown Wristwatch	Determined	1970s	Charlie Brown in baseball gear, yellow face, black band	110	160	200
Charlie Brown's All-Star Dugout Play Set	Child Guidance	1970s		30	40	50
Charlie Brown's Backyard Play Set	Child Guidance	1970s		20	30	35
Chirping Woodstock	Aviva	1977	plastic with electronic sound	18	26	30
Dolls	Simon Simple	1960s	7-1/2", Charlie Brown, Lucy, or Linus	20	40	50
Dress Me Belle Doll	Knickerbocker	1983	Belle wearing pink dress with blue dots	25	35	40
Dress Me Snoopy Doll	Knickerbocker	1983	Snoopy wearing blue jeans and red/yellow shirt	25	35	40
Electronic Snoopy Playmate	Romper Room/ Hasbro	1980		84	120	140
Express Station Set	Aviva	1977	Snoopy riding locomotive	36	50	60
Formula-1 Racing Car	Aviva	1978	11-1/2" plastic with Woodstock or Snoopy	75	110	125
Joe Cool Punching Bag	Ideal	1976		20	30	40
Joe Cool Push Puppet	Ideal	1977		20	35	50
Kaleidorama	Determined	1979		12	17	20
Linus Doll	Ideal	1976	removable clothing	50	100	150
Linus Doll	Hungerford Plastics	1958	8-1/2" plastic	50	75	100
Linus Music Box	Anri		wood, Linus in pumpkin patch on cover, plays "Who Can I Turn To?"	85	120	140
Linus Pocket Doll	Boucher	1968	7"	15	26	35

PEANUTS

TOY	COMPANY	YEAR	DESCRIPTION	GOOD	EX	MIB
Lucy and Charlie Brown Music Box	Anri	1971	4", each character beside large mushroom, plays "Rose Garden"	100	150	175
Lucy Candlestick Holder	Hallmark		7-1/4" figural composition	15	21	25
Lucy Doll	Ideal	1976	removable clothing	25	50	60
Lucy Doll	Hungerford Plastics	1958	8-1/2" plastic	45	75	90
Lucy Music Box	Anri	1971	5", Lucy with mushrooms on ground, plays "Love Story"	135	190	225
Lucy Music Box	Anri	1969	6-1/2", Lucy behind psychiatrist booth, plays "Try to Remember"	135	190	225
Lucy Nurse Push Puppet	Ideal	1977		20	35	45
Lucy Pocket Doll	Boucher	1968	7" open mouth	15	25	30
Lucy Pocket Doll	Boucher	1968	7" smiling	15	30	35
Lucy Tea Party Game	Milton Bradley	1972		25	35	40
Lucy Wristwatch	Timex	1970s	small chrome case, articulated arms, sweep seconds, white vinyl band	20	50	75
Lucy's Watch Wardrobe	Determined	1970s	white face, comes with blue, white, and pink bands	80	140	165
Lucy's Winter Carnival Colorforms	Colorforms	1973		20	40	45
Official Peanuts Baseball	Wilson	1969	illustrated with characters	50	95	110
Parade Drum	Chein	1969	large tin drum features characters in director's chairs	90	160	185
Peanuts Banks	United Features	1970	set of five	45	80	125
Peanuts Coloring Books	Saalfield	1960s	boxed set of five, cover features baseball scene	30	70	80
Peanuts Deluxe Play Set	Determined	1975	includes Lucy's psychiatrist booth and three action figures	90	200	235
Peanuts Drum	Chein	1974	tin, features characters with instruments	70	100	175
Peanuts Kindergarten Rhythm Set	Chein	1972	four percussion instruments	90	160	200
Peanuts Magic Catch Puppets	Synergistics	1978	four characters with Velcro balls	10	20	30
Peanuts Music Box	Schmid	1985	ceramic, characters piled on car, "Clown Capers", plays "Be a Clown"	70	105	120
Peanuts Music Box	Schmid	1984	8", ceramic, characters revolve around Christmas tree, plays "Joy to the World"	70	170	200
Peanuts Music Box	Schmid	1972	8" wooden ferris wheel box/ bank, plays "Spinning Wheel"	165	235	275
Peanuts Pelham Puppets	Pelham/ Tiderider	1979	7"-8" Charlie Brown, Snoopy or Woodstock	45	75	100
Peanuts Projects	Determined	1963	activity book	30	45	50
Peanuts Puppets Display	Pelham/ Tiderider	1979	display theater	235	335	390
Peanuts Show Time Finger Puppets	Ideal	1977	several rubber character puppets	24	34	40
Peanuts Skediddler Clubhouse Set	Mattel	1970	three rubber skediddlers, Snoopy, Lucy, Charlie Brown	126	180	210
Peanuts Stackables	Determined	1979	four hard rubber figures	18	26	30
Peanuts Tea Set		1961	one tray, two plates, two cups, four small plates	25	50	75
Peanuts Trace and Color	Saalfield	1960s	five book set	36	50	60

PEANUTS

TOY	COMPANY	YEAR	DESCRIPTION	GOOD	EX	MIB
Peanuts: A Book to Color	Saalfield		cover features Snoopy and Charlie Brown on skateboard	30	45	50
Peppermint Patty Doll			14" tall, cloth	7	13	20
Peppermint Patty Doll	Ideal	1976	removable clothing	27	40	45
Piano	Ely	1960s	wood with characters on top	162	230	270
Picture Maker	Mattel	1971	plastic character stencils	45	65	75
Pigpen Doll	Hungerford Plastics	1958	8-1/2" plastic	65	89	105
Playmate Snoopy Doll	Determined	1971	6" plush	15	21	25
Push 'N' Fly Snoopy	Romper Room/ Hasbro	1980	pull toy featuring Snoopy the Flying Ace	12	17	25
Push and Play with the Peanuts Gang	Child Guidance	1970s	plastic with rubber characters	30	50	55
Puzzle	Playskool	1979	six pieces, Snoopy leaning on bat	5	7	8
Puzzle	Determined	1971	eight-panel cartoon strip, 1,000 pieces	15	25	30
Puzzle	Determined	1971	four "Love Is" scenes, 1,000 pieces	25	40	45
Puzzle	Milton Bradley	1973	Schroeder, Charlie Brown, Snoopy, and Lucy on baseball mound	10	20	25
Rowing Snoopy	Mattel	1981		20	40	45
Sally Doll	Hungerford Plastics	1958	6-1/2" plastic	60	90	110
Schroeder and Piano Doll	Hungerford Plastics	1958	7" plastic	100	235	275
Schroeder Music Box	Anri	1971	5", Schroeder at the piano, plays "Beethoven's Emperor's Waltz"	90	125	150
Schroeder's Piano	Child Guidance	1970s		50	95	110
Schroeder's Piano	Aviva			25	40	45
See 'N' Say Snoopy Says	Mattel	1969		40	80	95
Snoopy Action Toys	Aviva	1977	wind-up Snoopy as drummer or boxer	25	45	50
Snoopy and Charlie Brown Copter	Aviva/Hasbro	1979	plastic	15	25	30
Snoopy and his Bugatti Race Car Model Kit	Monogram/ Mattel	1971		50	105	120
Snoopy and his Flyin' Doghouse	Mattel	1974		60	95	110
Snoopy and his Motorcycle Model Kit	Monogram/ Mattel	1971		60	85	100
Snoopy and his Sopwith Camel Model Kit	Monogram/ Mattel	1971		60	105	120
Snoopy and Woodstock in Wagon	Aviva		green die-cast with white wheels	10	14	16
Snoopy and Woodstock on Skateboard	Aviva			10	14	16
Snoopy and Woodstock Radio	Determined	1970s	plastic, two-dimensional doghouse	20	35	40
Snoopy as Astronaut Doll	Ideal	1977	14" plush with helmet and space suit	90	150	175
Snoopy as Astronaut Doll	Determined	1969	9", rubber head, plastic body	30	50	55
Snoopy as Astronaut Doll	Knickerbocker		5" vinyl	30	50	60
Snoopy as Astronaut Music Box	Schmid	1970s		20	35	55
Snoopy as Beagle Scout in Bus	Hasbro	1983	die-cast	4	5	6
Snoopy as Flying Ace Costume	Collegeville		with mask	15	20	25
Snoopy as Flying Ace in Wagon	Aviva		yellow die-cast with red wheels	10	15	18

PEANUTS

TOY	COMPANY	YEAR	DESCRIPTION	GOOD	EX	MIB
Snoopy as Joe Cool in Wagon	Aviva		purple die-cast with orange wheels	10	15	18
Snoopy as Magician Doll	Ideal	1977	14" plush with cape, hat, and mustache	90	150	175
Snoopy as Rock Star Doll	Ideal	1977	14" plush with wig, shoes and microphone	105	150	175
Snoopy Autograph Doll	Determined	1971	10-1/2"	20	25	30
Snoopy Bank	United Feature	1968	7" figural bank	14	25	40
Snoopy Bank Radio	Concept 2000	1978	plastic, Snoopy dancing	48	70	80
Snoopy Biplane	Aviva	1977	die-cast, Snoopy as Flying Ace	15	26	30
Snoopy Color 'N' Recolor	Avalon	1980		30	50	55
Snoopy Copter Pull Toy	Romper Room	1980	sound and action toy	6	9	10
Snoopy Costume	Collegeville		with mask	12	17	20
Snoopy Deep Diver Submarine	Knickerbocker	1980s	plastic	35	55	65
Snoopy Doghouse Radio	Determined	1970s	plastic	35	55	65
Snoopy Doll	Determined	1970s	plastic jointed	25	40	45
Snoopy Doll	Hungerford Plastics	1958	7" plastic	45	75	95
Snoopy Doll	Determined	1971	15" plush, felt eyes, eyebrows, and nose; red tag around neck	20	35	40
Snoopy Doll	Ideal		7" rag doll	7	10	20
Snoopy Drive-In Movie Theater	Kenner	1975		90	145	170
Snoopy Emergency Set	Hasbro		Snoopy in three vehicles	33	50	55
Snoopy Express	Aviva	1982	mechanical wind-up train, plastic	20	30	35
Snoopy Express	Aviva	1977	mechanical wind-up train, wood, includes track, tunnel, and signs	30	55	65
Snoopy Family Car	Aviva	1978		40	65	75
Snoopy Family Car	Aviva	1977	die-cast convertible, 2-1/4"	35	50	60
Snoopy Gravity Raceway	Aviva	1977		40	60	80
Snoopy Gyro Cycle	Aviva/Hasbro	1982	plastic friction toy	20	40	50
Snoopy Handfuls	Hasbro		twin pack; characters in die-cast racers	15	30	50
Snoopy Hero Time Watch	Determined	1970s	Snoopy in dancing pose, red band	55	100	145
Snoopy Hi-Fi Radio	Determined	1977	three-dimensional Snoopy wearing headphones, plastic	35	50	70
Snoopy High Wire Act	Monogram/ Mattel	1973		25	45	50
Snoopy in the Music Box	Mattel	1969	metal jack-in-the-box	15	30	55
Snoopy in Tow Truck	Hasbro	1983	die-cast	5	7	10
Snoopy is Joe Cool Model Kit	Monogram/ Mattel	1971	Snoopy rides surfboard	40	75	90
Snoopy Jack-in-the-Box	Romper Room/ Hasbro	1980	plastic doghouse jack-in-the-box	7	10	20
Snoopy Magician Push Puppet	Ideal	1977		15	30	40
Snoopy Marionette	Pelham/ Tiderider	1979	27", Pelham Puppets	200	445	525
Snoopy Movie Viewer	Kenner	1975		15	25	30
Snoopy Music Box	Aviva	1974	6", Snoopy with hobo pack w/ Woodstock, plays "Born Free"	30	45	50

Mad Hatter marionette, 1950s, Peter Puppet

Man from U.N.C.L.E. Illlya Kuryakin gun set, 1966, Ideal

PEANUTS

TOY	COMPANY	YEAR	DESCRIPTION	GOOD	EX	MIB
Snoopy Music Box	Aviva	1982	ceramic heart-shaped base with Snoopy and Woodstock hugging on top, plays "Love Makes the World Go Round"	30	55	65
Snoopy Music Box	Aviva	1979	8", Snoopy on doghouse shaped box, roof is removable lid, plays "Candy Man"	35	75	85
Snoopy Music Box	Quantasia	1984	ceramic Snoopy and musical note on base, plays "Fur Elise"	35	40	55
Snoopy Music Box	Quantasia	1985	plastic, Snoopy in boat inside waterglobe, plays "Blue Hawaii"	30	40	50
Snoopy Music Box	Schmid	1986	6" ceramic, Snoopy as Lion Tamer, plays "Pussycat, Pussycat"	45	65	90
Snoopy Music Box	Schmid	1986	7-1/2" ceramic, Snoopy next to Christmas tree, plays "O, Tannenbaum"	90	140	165
Snoopy Music Box	Schmid	1984	7" ceramic, Snoopy and Woodstock on seesaw, plays "Playmates"	70	105	120
Snoopy Musical Ge-tar	Mattel	1969	crank handle	25	40	55
Snoopy Musical Guitar	Aviva	1980	plastic, crank handle	15	30	45
Snoopy Nodder	Japanese	1960s	5-1/2" tall, bobbing head	20	35	65
Snoopy Paint-by-Number Set	Craft House	1980s	12" x 16"	10	20	30
Snoopy Paper Dolls	Determined	1976	with 10 outfits	20	40	45
Snoopy Phonograph	Vanity Fair	1979	features picture of dancing Snoopy	70	105	120
Snoopy Playhouse	Determined	1977	plastic doghouse with furniture, Snoopy, and Woodstock	40	65	80
Snoopy Playland	Aviva	1978	Snoopy in bus and six other characters	45	65	75
Snoopy Pocket Doll	Boucher	1968	7" with Flying Ace outfit	20	30	40
Snoopy Racing Car Stickshifter	Aviva	1978		90	140	165
Snoopy Radio	Determined	1975	plastic two-dimensional	30	60	70
Snoopy Radio	Determined	1970s	figural radio, Snoopy on green grass, plastic	20	30	35
Snoopy Radio	Determined	1977	plastic square radio with Snoopy pointing to dial	30	65	75
Snoopy Radio Controlled Fire Engine	Aviva	1980	with Woodstock transmitter	48	70	80
Snoopy Radio-Controlled Doghouse	Aviva	1980		30	55	65
Snoopy Sheriff Push Puppet	Ideal	1977		15	25	35
Snoopy Sign Mobile	Avalon	1970s		45	75	85
Snoopy Skediddler and His Sopwith Camel	Mattel	1969	with carrying case	100	205	250
Snoopy Slot Car Racing Set	Aviva	1977		50	85	100
Snoopy Slugger	Playskool	1979	ball, bat and cap	15	20	30
Snoopy Snack Attack Game	Gabriel	1980		20	35	40
Snoopy Snippers Scissors	Mattel	1975	plastic	25	40	50
Snoopy Soaper	Kenner	1975	gold soap dispenser with Snoopy on top	15	40	45
Snoopy Tea Set	Chein	1970	metal, features tray, plate, cups, and saucers	45	145	190
Snoopy the Critic	Aviva	1977	Snoopy and Woodstock on doghouse with microphone	90	190	225
Snoopy the Flying Ace Push Puppet	Ideal	1977		20	30	40

PEANUTS

TOY	COMPANY	YEAR	DESCRIPTION	GOOD	EX	MIB
Snoopy Toothbrush	Kenner	1972	Snoopy on doghouse holder	15	30	35
Snoopy Wristwatch	Timex	1970s	gold bezel, tennis ball circles Snoopy on clear disk, articulated hands holding racket, denim background and band	25	50	75
Snoopy Wristwatch	Determined	1969	Snoopy in dancing pose, silver or gold case, various colors	60	90	115
Snoopy Wristwatch	Lafayette Watch	1970s	Snoopy dancing, Woodstock is the second hand, silver case with red face and black band	60	111	130
Snoopy's 'Lectric Comb and Brush	Kenner	1975		30	45	50
Snoopy's Beagle Bugle	Child Guidance	1970s	plastic	50	75	90
Snoopy's Bubble Blowing Bubble Tub	Chemtoy	1970s		16	40	45
Snoopy's Dog House	Romper Room/ Hasbro	1978	Snoopy walks on roof	20	40	45
Snoopy's Dream Machine	DCS	1980	small version, no blinking lights, laminated cardboard	50	77	90
Snoopy's Dream Machine	DCS	1979	with blinking lights	90	150	175
Snoopy's Fantastic Automatic Bubble Pipe	Chemtoy	1970s		5	10	12
Snoopy's Good Grief Glider	Child Guidance	1970s	spring load launcher	30	65	75
Snoopy's Pencil Sharpener	Kenner	1974		15	40	45
Snoopy's Pound-A-Ball Game	Gabriel/Child Guidance	1980		30	65	75
Snoopy's Shape Register	Gabriel/Child Guidance	1980	plastic cash register	25	45	50
Snoopy's Soft House	Knickerbocker	1980	soft cloth house	20	35	40
Snoopy's Spaceship AM Radio	Concept 2000	1978	plastic	50	95	110
Snoopy's Stunt Spectacular	Child Guidance	1978	Snoopy on motorcycle	30	50	60
Snoopy's Swim and Sail Club	Child Guidance	1970s	characters and water vehicles	45	77	90
Snoopy's Take-a-Part Doghouse	Gabriel/Child Guidance	1980		15	20	25
Snoopy, Woodstock, and Charlie Brown Radio	Concept 2000	1970s	two-dimensional plastic	20	30	40
Snoopy-Matic Instant Load Camera	Helm Toy	1970s	uses 110 film	90	150	175
Speak Up, Charlie Brown Talking Storybook	Mattel	1971	cardboard with vinyl pages	75	115	135
Spinning Top	Ohio Art		5", Snoopy and the Gang	5	15	30
Spinning Top	Chein	1960s	faces of Snoopy, Charlie Brown, Lucy, and Linus	30	50	70
Stack-Up Snoopy	Romper Room/ Hasbro	1980		10	13	15
Super Cartoon Maker	Mattel	1970	molds to make character figures	60	115	135
Swimming Snoopy	Concept 2000	1970s		20	40	45
Table Top Snoopy Game	Nintendo	1980s		60	120	140
Tabletop Hockey	Munro Games	1972		60	110	125
Talking Peanuts Bus	Chein	1967	metal, chracters seen in windows	100	320	525
Tell Time Clock	Concept 2000	1980s	three styles: Snoopy, Woodstock or Charlie Brown	18	26	30

PEANUTS

TOY	COMPANY	YEAR	DESCRIPTION	GOOD	EX	MIB
Tell Us a Riddle, Snoopy Game	Colorforms	1974		20	50	60
Tub Time Snoopy Doll	Knickerbocker	1980s	rubber	15	30	35
Vaporizer/Humidifier	Milton Bradley		plastic, 13" x 16", Snoopy on doghouse	40	70	90
Woodstock Climbing String Action	Aviva/Hasbro	1977	plastic	15	25	30
Woodstock Costume	Collegeville		with mask	12	15	20
Woodstock in Ice Cream Truck	Aviva		friction vehicle	6	9	10
Yankee Doodle Snoopy	Colorforms	1975		20	35	40

PEE WEE HERMAN

Ball Dart Set				5	10	15
Billy Baloney Doll	Matchbox	1988	18" tall	10	15	40
Chairry Figure	Matchbox	1988	5" tall	3	5	10
Chairry Figure	Matchbox		15" tall	10	15	40
Conky Wacky Wind-Up	Matchbox	1988		3	5	10
Cowboy Curtis Figure	Matchbox			5	10	30
Globey with Randy	Matchbox	1988		5	10	30
King of Cartoons Figure	Matchbox	1988	5" tall	5	10	30
Magic Screen Figure	Matchbox	1988	5" tall poseable	5	10	30
Magic Screen Wacky Wind-Up	Matchbox	1988	6" tall	3	5	10
Miss Yvonne Doll	Matchbox	1988	poseable 5" tall	5	10	40
Pee Wee Herman Deluxe Colorforms	Colorforms	1980s		7	10	35
Pee Wee Herman Doll	Matchbox		15" tall, non talking	7	15	40
Pee Wee Herman Doll	Matchbox	1988	poseable 5" tall	4	8	20
Pee Wee Herman Play Set	Matchbox	1989	20" x 28" x 8" for use with 5" figures, Pee Wee's bike, folds into large carrying case	10	25	50
Pee Wee Herman Slumber Bag	Matchbox	1988		10	20	35
Pee Wee Herman Ventriloquist Doll	Matchbox	1980s		25	60	100
Pee Wee with Scooter and Helmet	Matchbox	1988		4	7	10
Pee Wee Yo-Yo				3	5	10
Pterri Doll	Matchbox		13" tall	15	25	45
Pterri Wacky Wind-Ups	Matchbox	1988		3	5	10
Reba Figure	Matchbox	1988	poseable	5	10	30
Ricardo Figure	Matchbox	1988		5	10	30
Vance the Talking Pig Figure	Matchbox	1987		15	30	80
View-Master Gift Set	View-Master	1980s		5	10	15

PETER PAN

Captain Hook Figure			8" tall, plastic	9	15	25
Captain Hook Hand Puppet	Gund	1950s	9" tall	30	60	100
Peter Pan Baby Figure	Sun Rubber	1950s		30	65	125
Peter Pan Charm Bracelet		1974		15	25	45
Peter Pan Doll	Duchess Doll	1953	11-1/2" tall, brown trim fabric shoes, green mesh stockings with flocked outfit, hat with a large red feather, shiny silver white metal dagger in belt, eyes, arms & head move	135	275	425

PETER PAN

TOY	COMPANY	YEAR	DESCRIPTION	GOOD	EX	MIB
Peter Pan Doll	Ideal	1953	18" tall	95	190	275
Peter Pan Hand Puppet	Oak Rubber	1953	rubber	25	50	95
Peter Pan Map of Neverland		1953	18" x 24", collectors issue for users of Peter Pan Beauty Bar	30	60	125
Peter Pan Nodder		1950s	6" tall	65	130	250
Peter Pan Paper Dolls	Whitman	1952	11 die-cut cardboard figures	25	75	150
Peter Pan Push Puppet	Kohner	1950s	6" tall, green & flesh colored beads, plastic head, light green plastic hat	20	45	80
Peter Pan Sewing Cards	Whitman	1952		15	25	45
Puzzle	Jaymar	1950s	frame tray puzzle shows Peter, Wendy, John and Michael flying over Neverland	15	25	45
Tinker Bell Doll	Duchess Doll	1953	8" tall, flocked green outfit with a pair of large white fabric wings with gold trim, eyes open & close, jointed arms & head moves	85	170	325
Tinker Bell Figure	Sutton	1960s	7" tall, plastic & rubber figure	20	45	75
Tinker Bell Pincushion		1960s	with 1-1/2" tall Tinker Bell figure, in clear plastic display can	20	35	65
Wendy Doll	Duchess Doll	1953	8" tall, full purple length skirt with purple bow in back of dress, eyes open & close, jointed arms & head moves	85	170	325

PHANTOM

TOY	COMPANY	YEAR	DESCRIPTION	GOOD	EX	MIB
Oil Paint by Numbers Set	Hasbro	1967		200	300	500
Phantom 2040 Carded Toys	Ja-Ru	1995	sword, crossbow, gun or whistle light	7	10	15
Phantom 2040 Rubber Ball	Unice	1995		10	15	25
Phantom Archery Set	Larami	1976	bow, arrows, animals	75	100	150
Phantom Binoculars	Larami	1976		100	150	200
Phantom Candy Jar	KFS	1996	plastic w/figural lid	10	20	30
Phantom Club Rubber Stamper Skull Ring		1950s		800	1100	1500
Phantom Costume	Character Costumes	1989		35	50	60
Phantom Costume	Ben Cooper	1950s		150	200	300
Phantom Costume	Collegeville	1970s		65	85	125
Phantom Dagger Set	Larami	1976	knife and sheath	75	100	150
Phantom Desert Survival Kit	Larami	1976	canteen and Thermos bottle	75	100	150
Phantom Figure		1980s	6" dark blue PVC figure	20	30	40
Phantom Figure		1990s	6" purple PVC figure	10	20	30
Phantom Figure	Street Players	1996	4" action figure w/skull throne or horse	10	15	20
Phantom Figures	Spain	1990	3-1/2", two pieces, purple PVC	10	15	20
Phantom Giant Games Book	World Distributors	1968		75	100	150
Phantom Jungle Play Set	Larami	1976	figure, palm trees, animals	100	150	200
Phantom Pathfinder Set	Larami	1976	compass, canteen, binoculars, case	75	100	150
Phantom Puffy Magnet	Hanna-Barbera	1975		35	45	60
Phantom Safari Set	Larami	1976	figure, truck, trailer, animals	125	175	250
Phantom Squirt Camera	Larami	1976		75	100	150
Phantom Water Pistol	Nasta	1974	with holster	65	75	90
Phantom Water Pistol	Nasta	1974		75	100	125

PHANTOM

TOY	COMPANY	YEAR	DESCRIPTION	GOOD	EX	MIB
Playing Cards	John Sands	1990		15	20	25
Playing Cards	Bulls Dist.	1990	Swedish	15	20	25
Rub-On Transfer Set	Hasbro	1967		75	100	150
Syrocco Figure	Pillsbury Mills	1944	brown	1000	1250	1700
Syrocco Figure	Pillsbury Mills	1944	purple	500	750	1000

PINK PANTHER

TOY	COMPANY	YEAR	DESCRIPTION	GOOD	EX	MIB
Pink Panther & The Fancy Party Book	Golden			5	10	15
Pink Panther & The Haunted House Book	Golden			5	10	15
Pink Panther at Castle Kreep Book	Whitman			4	7	10
Pink Panther at The Circus Sticker Book	Golden	1963		5	15	30
Pink Panther Coloring Book	Whitman	1976	cover shows Pink Panther roasting hot dogs	5	15	25
Pink Panther Figure	Dakin	1971	8" tall, with legs closed	15	30	60
Pink Panther Figure	Dakin	1971	8" tall, with legs open	15	30	60
Pink Panther Memo Board			write on/wipe off memo board	5	10	15
Pink Panther Motorcycle			2-1/2" plastic	4	7	10
Pink Panther Music Box	Royal Orleans	1982	Christmas limited edition	25	60	100
Pink Panther Music Box	Royal Orleans	1984	Christmas limited edition	25	60	100
Pink Panther Music Box	Royal Orleans	1983	Christmas limited edition	25	60	100
Pink Panther One-Man Band	Illco	1980	10" tall, battery-operated, plush body with vinyl head	25	50	75
Pink Panther Pool Game	Ja-Ru	1980s		4	7	15
Pink Panther Putty	Ja-Ru	1970s		4	7	10
Pink Panther Wind-Up			3" tall, plastic, walking wind-up with trench coat & glasses	7	13	20
Puzzle	Whitman	1960s	100 pieces, several pictures available, each	8	20	30

PINOCCHIO & JIMINY CRICKET

TOY	COMPANY	YEAR	DESCRIPTION	GOOD	EX	MIB
Figaro Figure	Multi-Wood Products	1940	3" tall, wood composition	30	75	150
Figaro Figure	Knickerbocker	1940s	composition, movable limbs and head	90	150	240
Figaro Friction Toy	Linemar	1960s	1-1/2" x 3" x 1-1/2", tin	50	120	250
Figaro Roll Over Wind-Up Toy	Marx	1940	5" long with ears & tail, tin	125	225	400
Gepetto Figure	Multi-Wood Products	1940	5-1/2" tall, wood composition	50	95	145
Gideon Figure	Multi-Wood Products		5" tall	35	65	100
Honest John Figure	Multi-Wood Products		2-1/2" x 3" base with a 7" tall figure	45	80	150
Jiminy Cricket Doll	Ideal	1940	wooden jointed	125	250	450
Jiminy Cricket Figure	Ideal	1940s	wood jointed, hand painted	150	200	350
Jiminy Cricket Figure	Marx		3-1/2" x 4-3/4" tall, Snap-Eeze, white plastic base with movable arms & legs	25	50	75
Jiminy Cricket Hand Puppet	Gund	1950s	11" tall	15	40	100
Jiminy Cricket Marionette	Pelham Puppets	1950s	3" x 6" x 10" tall, dark green head with, large eyes, gray felt hat	70	130	275
Jiminy Cricket Ramp Walker	Marx	1960s	1" x 3" x 3" tall, pushing a bass fiddle	75	125	250
Jiminy Cricket Soaky			7" tall bottle	10	40	60

PINOCCHIO & JIMINY CRICKET

TOY	COMPANY	YEAR	DESCRIPTION	GOOD	EX	MIB
Jiminy Cricket Toothbrush Set	Dupont	1950s	plastic wall hanging Jiminy holds a toothbrush	25	50	75
Jiminy Cricket Wristwatch	US Time	1948	Birthday Series	50	150	275
Lampwick Figure	Multi-Wood Products	1940	5-1/2" tall, wood composition	30	75	125
Pin the Nose on Pinocchio Game	Parker Brothers	1939	15-1/2" x 20"	30	75	125
Pinocchio & Jiminy Cricket Dolls	Knickerbocker	1962	6" tall, vinyl, titled "Knixies", each	20	45	90
Pinocchio & Jiminy Push Puppet	Marx	1960s	2-1/2" x 5" x 4" tall, double puppet	30	45	100
Pinocchio Bank	Play Pal Plastics	1970s	7" x 7" x 10" tall, vinyl, 3-D molded head of Pinocchio	10	20	35
Pinocchio Bank	Play Pal Plastics	1960s	11-1/2" tall, plastic	12	23	35
Pinocchio Bank	Crown Toy	1939	5" tall, wood composition with metal trap door on back	60	120	250
Pinocchio Book		1940	Big Little Book	23	45	65
Pinocchio Book	Grosset & Dunlap	1939	9-1/2" x 13", laminated cover	25	50	75
Pinocchio Book	Whitman	1939	96 pages	16	30	45
Pinocchio Book Set	Whitman	1940	8-1/2" x 11-1/2", set of six books, 24 pages each	75	175	300
Pinocchio Color Box	Transogram		also known as paint box	15	30	60
Pinocchio Crayon Box	Transogram	1940s	4-1/2" x 5-1/2" x 1/2" deep, tin	15	30	60
Pinocchio Cut-Out Book	Whitman	1940		30	70	125
Pinocchio Doll	Ideal	1940	8" tall, wood composition head, others are jointed wood	50	105	250
Pinocchio Doll	Knickerbocker	1940	3-1/2" x 4" x 9 1/2" tall, jointed composition doll with movable arms & head	175	400	750
Pinocchio Doll	Ideal	1939	12" tall with wire mesh arms & legs	80	180	400
Pinocchio Doll	Ideal	1940	19-1/2" tall, composition	225	425	750
Pinocchio Doll	Ideal	1940	10" tall, wood composition head, jointed arms & legs attached to body	60	130	300
Pinocchio Figure	Crown Toy		9-1/2" tall, jointed arms	40	90	180
Pinocchio Figure	Multi-Wood Products	1940	5" tall, wood composition	40	80	200
Pinocchio Hand Puppet	Knickerbocker	1962		20	45	75
Pinocchio Hand Puppet	Crown Toy		9" tall, composition	20	50	45
Pinocchio Hand Puppet	Gund	1950s	10" tall, with squeaker	25	50	125
Pinocchio Music Box			plays "Puppet on a String"	15	30	60
Pinocchio Paint Book	Disney	1939	11" x 15", heavy paper cover	25	35	85
Pinocchio Paperweight & Thermometer	Plastic Novelties	1940		25	50	75
Pinocchio Plastic Cup	Safetyware	1939	2-3/4" tall, plastic	25	50	100
Pinocchio Push Puppet	Kohner	1960s	5" tall	15	25	65
Pinocchio Snow Dome	Disney	1970s	3" x 4-1/2" x 5" tall, Pinocchio holds plastic dome between hands & feet	15	40	75
Pinocchio Soaky				10	25	50
Pinocchio Tea Set	Ohio Art	1939	tin tray, plates, saucers, serving platter, cups, bowls & smaller plates	50	90	225
Pinocchio the Acrobat Wind-Up Toy	Marx	1939	2-1/2" x 11" x 17" tall, turning on a trapeze-like frame	275	550	900
Pinocchio Walker	Marx	1939	9" tall, tin, animated eyes, rocking action	200	400	750
Pinocchio Wind-Up Toy	Linemar		6" tall, tin wind-up, arms & legs move	75	175	350
Puzzle	Jaymar	1960s	5" x 7", "Pinocchio's Expedition"	11	20	30

PINOCCHIO & JIMINY CRICKET

TOY	COMPANY	YEAR	DESCRIPTION	GOOD	EX	MIB
Walt Disney Tells the Story of Pinocchio Book	Whitman	1939	4-1/4" x 6-1/2" paperback, 144 pages	25	50	75
Walt Disney's Pinocchio Book	Random House	1939	8-1/2" x 11-1/2", hardcover	15	40	60

PLUTO

TOY	COMPANY	YEAR	DESCRIPTION	GOOD	EX	MIB
Pluto Alarm Clock	Allied	1955	4" x 5-1/2" x 10" tall, eyes and hands shaped like dog bones, glow in the dark	60	125	250
Pluto Bank	Animal Toys Plus	1970s	9" tall vinyl, Pluto standing in front of a doghouse	15	30	45
Pluto Bank	Disney	1940s	4" x 4-1/2" x 6-1/2", ceramic	35	65	110
Pluto Figure	Seiberling	1930s	3-1/2" long, rubber	30	65	120
Pluto Figure	Seiberling	1930s	7" tall, rubber	30	65	120
Pluto Fun-E-Flex Figure	Fun-E-Flex	1930s	wood	25	50	100
Pluto Hand Puppet	Gund	1950s	9" tall	15	35	75
Pluto Lantern Toy	Linemar	1950s		125	275	450
Pluto Pop-A-Part Toy	Multiple Toymakers	1965	9" long, plastic	15	30	45
Pluto Pop-Up Critter Figure	Fisher-Price	1936	wooden, Pluto standing on base 10-1/2" long	80	145	175
Pluto Purse	Gund	1940s	9" x 14" x 2"	25	50	85
Pluto Push Toy	Fisher-Price	1936	8" long, wood	70	130	250
Pluto Rolykin	Marx		1" x 1" x 1-1/2" tall, ball bearing action	15	30	60
Pluto Sports Car	Empire		2" long	9	16	25
Pluto the Acrobat Trapeze Toy	Linemar		10" tall, metal, celluloid, wind-up	60	200	250
Pluto the Drum Major	Marx/ Linemar	1950s	tin, mechanical	175	325	700
Pluto Toy	Linemar		4" long, friction motor makes tongue wag	30	55	200
Pluto Toy	Linemar		9" friction toy, Pluto pulling red wagon	60	125	250
Pluto Tricycle Toy	Linemar	1950s	tin	150	325	650
Pluto Watch Me Roll Over	Marx	1939	8" long, tin, Pluto turns over as his tail passes beneath him	150	275	425

POPEYE

TOY	COMPANY	YEAR	DESCRIPTION	GOOD	EX	MIB
60th Anniversary Bank	Presents	1988	metal, P5988	5	15	20
60th Anniversary Candle Box	Presents	1989	metal, heart shaped, #P5979	5	10	15
60th Anniversary Collection Book	Hawk Books	1990		20	35	50
Adventures of Popeye Book	Saalfield	1934		25	55	100
Adventures of Popeye Game	Transogram	1957		35	70	140
Apprentice Printer	MSS	1970s		6	15	30
Ball & Jacks Set	MSS			6	10	20
Ball & Paddle	BC			6	10	15
Balloon Pump		1957	inflato-pump	25	50	100
Barber Shop	Larami	1970s		5	15	20
Baseball	Ja-Ru	1983		5	10	20
Beach Boat	H.G. Industries	1980	red or yellow	5	10	15
Bell	Vandor	1980	Popeye on top, ceramic	8	15	25
Belt Buckle	Lee	1980	Popeye with spinach	8	15	20
Belt Buckle	U.S. Spinach Growers		Strength thru Spinach	8	15	20
Belt Buckle	Pyramid Belt	1973	Popeye with sailor hat	10	18	35
Biffbat-Fly Back Paddle		1935		15	30	55

POPEYE

TOY	COMPANY	YEAR	DESCRIPTION	GOOD	EX	MIB
Big Surprise Book	Wonder Books	1976		4	8	15
Billion Bubbles	Larami	1984		4	8	15
Blackboard	Bar Zim	1962		15	30	60
Blinky Cup	Beacon Plastics			8	15	25
Bluto Button	Lisa Frank	1979	2", Bluto getting socked	2	3	5
Bluto Dippy Dumper	Marx	1935	wind-up	200	400	800
Bluto Figure	Cristallerie Antonio		Italian crystal	10	20	30
Bluto Wind-Up Toy	Marx	1938	celluloid, with Brutus, horse, and cart	325	500	750
Bookends	Vandor	1980	ceramic, Popeye and Brutus	20	45	75
Bop Bag	Miner Industries	1981		6	12	25
Bowl	National Home Products	1979	plastic	5	10	18
Bowl	Deka	1971	oval, plastic	8	18	25
Bowl	Vandor	1980	ceramic, 1 of 3	8	15	25
Boxing Game	Harmony	1981		5	10	15
Boxing Gloves	Everlast	1960s		25	50	100
Brutus Button	Mini Media	1983	1", "I'm Mean"	2	3	5
Brutus Button	Strand	1985	3", "Ya Little Runt"	2	3	5
Brutus Button	Strand	1985	3", "Gonna Eat You for Breakfast"	2	3	5
Brutus Button	Factors	1980	3", movie	2	3	5
Brutus Dog Toy	Petex	1986		4	10	15
Brutus Doll	Presents	1985	large	9	18	35
Brutus Doll	Presents	1985	small	5	10	20
Brutus Figure	Presents	1990	PVC	2	3	5
Brutus Figure	Comic-Spain	1984	Brutus with club	4	7	10
Brutus Figure	Bully	1981	pink shirt	5	13	25
Brutus Figure	Japan Olympics	1962	wood, Brutus in barrel	75	150	300
Brutus Figure	KFS-Hearst	1991	wood	4	7	10
Brutus Figure Painting Kit	Avalon	1980		6	10	20
Brutus Hand Puppet	Gund	1960s		15	25	65
Brutus Hi-Pop Ball	Ja-Ru	1981		6	10	18
Brutus Hookies	Tiger	1977		5	10	15
Brutus in Jeep		1950s	tiny plastic car	15	30	45
Brutus in Steamroller	Lesney/ Matchbox	1980	Matchbox	8	13	25
Brutus Jump-Up	Imperial	1970s		8	15	30
Brutus Mini Bank	KFS	1979		10	20	30
Brutus Music Box	Presents	1989	#P5984	8	15	25
Brutus Music Box	KFS	1980	Brutus dancing	10	15	30
Brutus Painting Kit	Avalon	1980		8	15	20
Brutus Soaky	Colgate-Palmolive	1960s		15	35	60
Brutus Sports Car		1950s	tiny plastic car	10	25	40
Brutus Wind-Up Toy	Durham	1980		5	10	20
Bubble 'N Clean	Woolfoam	1960s		20	40	75
Bubble Blower	Transogram	1958		15	30	55
Bubble Blower Boat	Larami	1984		6	10	15
Bubble Blowing Popeye	Linemar	1950s		275	475	1000
Bubble Blowing Train	Hong Kong	1970s	pink	8	18	30
Bubble Pipe	Ja-Ru	1985		5	10	15
Bubble Pipe	KFS	1960s	yellow with red end	10	25	40
Bubble Set	Transogram	1936	two wooden pipes, tray, soap in 5" x 7-1/2" box	20	40	70
Bubble Shooter	Ja-Ru	1980s	orange or yellow body	5	10	18

CHARACTER

POPEYE

TOY	COMPANY	YEAR	DESCRIPTION	GOOD	EX	MIB
Bubbleblaster	Carlin Playthings	1980		8	15	20
Bubbles Blaster	Larami	1984		5	10	15
Bubbles with Dip Pow Bubbles	MSS	1986		5	10	15
Button	Offset Gravure	1930s	1", New York Evening Journal	15	30	55
Button	Mini Media	1983	1", "No Wimps"	2	3	5
Cabinet			mirrored	25	50	100
Candy	Alberts	1980	bonbons	6	10	15
Candy Box	Phoenix Candy	1960	Popeye & his pals	15	25	40
Candy Cigarettes	Primrose Confectionery - England	1959		10	15	25
Candy Rings	Alberts	1989		6	10	15
Candy Sticks	Hearst	1989	red box	6	10	15
Candy Sticks	World Candies	1990	48 count	6	10	15
Cap Gun	Ja-Ru	1981		6	10	15
Captain George Presents Popeye Book	Memory Lane	1970		5	10	25
Cereal Bowl	National Home Products	1979		5	10	15
Cereal Box	Cocoa-Puffs	1987	with gum	3	7	10
Chain Bubbles Maker	Larami	1984	red Popeye	6	10	15
Chalk	American Crayon	1936	white, 18 pieces	8	15	30
Change Purse	Sanrio	1990		3	7	10
Character Figures	Popeye's Chicken	1991	blue plastic, several characters available	1	3	5
Character Figures	Spoontiques	1981	pewter, three 1" figures: Olive w/hands clasped, Popeye with muscles, Jeep standing, each	10	20	30
Character Figures	Spoontiques	1980	two 1" figures: Jeep lifting tail, Swee'Pea with feet showing, each	10	20	30
Character Figures	Spoontiques	1980	2" figures: Popeye with barbell, Popeye with parrot, Olive walking, Popeye flexing muscles, Popeye with spinach, each	10	20	30
Charm Bracelet	Peter Brams	1990	silver or gold	10	18	25
Checker Board	Ideal	1959		15	30	50
Chimes Doll	J. Swedlin	1950s	gray plush body, chimes	15	40	60
Chinese Jump Rope	MSS			4	10	15
Chocolate Mold		1940s	metal, Popeye	35	80	150
Chocolate Mold	Turmic Plastics	1991	plastic, Popeye	3	6	10
Christmas Lamp Shades	General Electric	1930s	set of ten	60	125	250
Christmas Light Covers	General Electric Textolite	1929	"Cheers"	30	60	125
Christmas Ornament	Presents	1987	Alice the Goon, Swee'Pea, Wimpy, Popeye, Olive Oyl or Brutus, each	4	8	15
Christmas Ornament	Bully	1981	Bluto	8	15	25
Christmas Ornament	Bully	1981	Dufus	8	15	25
Christmas Ornament	Presents	1989	Season's Greetings	4	8	15
Christmas Tree Lamp Set	General Electric	1935		70	140	250
Circus Man Film	Brumberger	1950s	8mm	10	20	30

POPEYE

TOY	COMPANY	YEAR	DESCRIPTION	GOOD	EX	MIB
Clothes Brush	KFS	1929	wooden, black or brown	20	40	75
Color Markers	Sanrio	1990	six	2	5	8
Color-Me Stickers	Diamond Toymakers	1983		4	8	15
Color-Vue Pencil-by-Numbers	Hasbro	1979		8	18	28
Colorforms Birthday Party Set	Colorforms	1961		20	45	90
Colorforms Movie Version	Colorforms	1980		7	13	20
Comb & Brush	KFS	1979		7	15	25
Construction Trucks	Larami	1981		4	11	16
Cook's Catch-All	KFS	1980	ceramic, Wimpy	10	20	30
Cookie Jar	McCoy	1965	ceramic white suited Popeye	50	100	200
Crayons	Dixon	1958	12 giant crayons	10	18	35
Crayons	American Crayon	1950s		10	17	30
Crayons	American Crayon	1933	12 giant crayons	10	18	35
Cup	New Zealand	1940s	Popeye on skis, ceramic	18	40	70
Daily Dime Bank	KFS	1956		15	30	75
Daily Quarter Bank	Kalon	1950s	4-1/2" tall, metal	60	120	200
Danger Ahoy! Book	Whitman	1969	Big Little Book	5	10	18
Deep Sea Danger Book	Whitman	1980	Big Little Book	2	4	8
Dice		1990	with Popeye head	1	3	5
Dime Register Bank	KFS	1929	square, window shows total deposits	35	75	175
Dish	New Zealand	1940s	ceramic, Popeye and Olive	20	45	75
Dish Set	Boontonware	1964	three piece plastic	20	40	75
Dockside Presto Magix	APC	1980		3	6	11
Double Action Water Gun Set	MSS			4	8	15
Drawing Board	KFS	1978	slate with rope attached	8	16	25
Drawing Desk	Carlin Playthings	1980		10	20	30
Duck Shoot	Ja-Ru	1980s		5	10	15
Dufus Figure	Bully	1981	with hand on stomach	8	15	25
Egg Cup	Japan	1940s	Popeye sitting at table with spinach	30	65	100
Egg Cup & Mug	Magna	1989	Great Britain	15	40	75
Erase-O-Board & Magic Screen Set	Hassenfeld Bros.	1957		20	45	90
Film Card	Tru-Vue	1959	T-28	8	18	30
Film Projector	Cinexin-Spain		8mm with 13 movies	65	130	200
Finger Puppet Family	Denmark Plastics	1960s		20	40	80
Flashlight	Larami	1983	blue, yellow, or red	5	10	18
Flashlight	Bantam-Lite	1960s	three color, wrist light	8	15	25
Fleas A Crowd Record	Peter Pan	1962	78 rpm	6	10	20
Foto-Fun Printing Kit	Fun Bilt	1958		20	40	80
Freezicles	Imperial	1980		5	12	18
Fun Booklets	Spot-O-Gold	1980	set of 10	10	25	50
Funny Color Foam	Creative Aerosol	1983		5	12	18
Funny Face Maker	Jaymar	1962		15	30	60
Funny Films Viewer	Acme	1940s		20	40	80
Funny Fire Fighters	Marx	1930s	celluloid figures, Popeye on ladder & Bluto drives fire truck--both figures wear boxing gloves	250	950	1900
Ghost Ship to Treasure Island Book	Whitman	1967	Big Little Book	5	12	15
Giant 24 Big Picture Coloring Book	Merrigold Press	1981		3	7	13

CHARACTER

POPEYE

TOY	COMPANY	YEAR	DESCRIPTION	GOOD	EX	MIB
Giant Paint Book	Whitman	1937	blue or red	25	45	85
Give-A-Show Projector	Kenner		projector with slides	35	70	130
Great Big Popeye Paint and Crayon Book	McLoughlin Bros.	1937		25	45	85
Gumball Dispenser	Superior Toys	1983	pocket pack	5	10	15
Gumball Machine	Hasbro	1968	6", shape of Popeye's head	10	18	35
Gumball Machine	Superior Toys	1983	Popeye eating spinach	7	15	30
Gumball Machine	Superior Toys	1983	Popeye gives Olive flowers	7	15	30
Hag of the Seven Seas Pop-Up Book	Blue Ribbon Books	1935		45	90	175
Halloween Bucket	Renz	1979	shaped like Popeye's head, red, yellow or blue	10	15	25
Harmonica	Larami	1973		10	15	25
Hat & Pipe	Empire Plastics	1950s		18	35	60
Holster Set	Halco	1960s		25	50	100
Horseshoe Magnets	Larami	1984		5	10	15
House that Popeye Built Book	Wonder Books	1960		5	12	18
Hunting Knife	Larami	1973		10	20	35
ID Set	Gordy	1982		5	12	16
In a Sock for Susan's Sake Book	Whitman	1940	Big Little Book	20	40	75
In Quest of Poopdeck Pappy Book	Whitman	1937	Big Little Book	20	40	75
Indian Fighter Film	Atlas Films	8mm		5	10	15
Jack-in-the-Box	Mattel	1961		40	80	150
Jack-in-the-Box	Nasta	1979		15	35	60
Jack-in-the-Box	Nasta	1983		10	20	35
Jackknife	Imperial	1940s	green Popeye on pearl handle	30	70	130
Jeep Doll	Presents	1985	two sizes	8	13	20
Jeep Figure	KFS-Hearst	1991	wood	4	7	10
Jeep Lucky Spinner	KFS	1936		30	65	110
Jeep Wall Plaque			ceramic	4	7	15
Jiffy Pop Fun 'N Games Booklet	Spot-O-Gold	1980		3	6	10
Jumbo Card Game	House of Games	1978		8	15	20
Jumbo Trading Card Game	Dynamic Toy	1960s		8	18	35
Kaleidoscope	Larami	1979		6	13	18
Kazoo & Harmonica	Larami	1979		6	13	18
Kazoo Pipe	Northwestern Products	1934		20	35	65
Kazoo Pipe	Peerless Playthings	1960s	yellow	10	20	35
King of the Jungle Film	Atlas Films	1960s	8mm	5	10	20
Kite	Sky-Way	1980	inflatable	6	12	17
Kite	Sky-Way	1980	regular	4	8	13
Knapsack	Fabil	1979		10	15	25
Knockout Bank	Straits	1935		190	400	800
Kooky Straw	Imperial	1980		4	8	15
Lamp		1940s	boat with Popeye light bulb	200	400	800
Lantern	Linemar	1950s	7-1/2" tall, battery operated, light in belly	125	225	400
Life Raft	KFS	1979	large, blue	9	18	25
Little Pops the Ghost Book	Random House	1981		4	8	16
Little Pops the Magic Flute Book	Random House	1981		4	8	16

TOY	COMPANY	YEAR	DESCRIPTION	GOOD	EX	MIB
Little Pops the Spinach Burgers Book	Random House	1981		4	8	16
Little Pops the Treasure Hunt Book	Random House	1981		4	8	16
Mae West & Popeye Book			eight pages	5	10	15
Magic Eyes Film Card	Tru-Vue	1962	set of three	12	25	45
Magic Glow Putty	FC Famous Toys			6	10	15
Magic Play Around Game	Amsco	1960s		20	45	85
Magic Slate	Lowe	1959		10	20	40
Magic Slate Paper Saver	Whitman	1981		4	8	15
Make-A-Picture Premium	Quaker	1934		10	20	40
Marble		1940s	1" blue/white with black/white or red Popeye	8	15	20
Marble Set	Akro Agate	1935	#116	100	200	450
Marble Set	Imperial	1980		5	10	15
Marble Shooter		1940s	milk glass container	15	30	45
Metal Tapping Set	Carlton Dank	1950s		25	50	100
Metal Target Set	Ja-Ru	1983		8	15	20
Metal Whistle	Ja-Ru	1981		5	10	15
Micro-Movie	Fascinations	1990	Popeye-Ali Baba	4	8	12
Mini Hurricane Lamp	Presents	1989	P5981-1993, 60th year	4	8	12
Mini Lunch Box	Sanrio	1990	plastic	4	8	12
Mini Memo Board	Freelance	1980		5	10	12
Miniature Train Set	Larami	1980		6	11	18
Mirror	Freelance	1978	Popeye lifting weights	8	13	20
Mirror	Freelance	1979	Olive with mirror	7	13	18
Mirror Rattle	Cribmates	1979		8	13	20
Mix or Match Storybook	Random House	1981		4	9	18
Model Kit	Carto	1970s	Popeye & Olive Oyl	25	50	100
Modeling Clay	American Crayon	1936		20	45	90
Motor Friend	Nasta	1976		10	20	30
Mug	Schmid	1950s	ceramic	10	20	30
Muscle Builder Bluto	Carlin Playthings	1980		6	12	18
Muscle Builder Popeye	Carlin Playthings	1980		6	12	18
Music Box	Vandor	1980	Wimpy on top of hamburger, ceramic	18	35	65
Music Box	Vandor	1980	revolving Olive with Popeye dancing, ceramic	18	35	65
Music Box	Vandor	1980	revolving Popeye spanks Swee'Pea, ceramic	18	35	65
Music Lovers Film	Atlas Films	1960s	8mm	8	13	18
Musical Mug	KFS	1982	ceramic	10	20	30
Musical Rattle	Cribmates	1979		8	13	20
My Popeye Coloring Kit	American Crayon	1957		20	40	80
Official Popeye Pipe		1958	5" stem with 2" bowl, battery operated, "It lites, it toots"	20	60	100
Old Time Wild West Train	Larami	1984		5	10	15
Olive Oyl & Swee'Pea Hot Water Bottle	Duarry	1970		30	75	150
Olive Oyl & Swee'Pea Snow Globe	Presents	1989	several styles	4	8	15
Olive Oyl & Swee'Pea Telephone Shoulder Rest	Comvu	1982		5	10	20
Olive Oyl & Swee'Pea Thermometer	KFS	1981		8	13	18
Olive Oyl & Swee'Pea Wash Up Book	Tuffy Books	1980		3	6	12

POPEYE

TOY	COMPANY	YEAR	DESCRIPTION	GOOD	EX	MIB
Olive Oyl Bank		1940s	cast iron	60	120	200
Olive Oyl Bike Bobbers	KFS	1960s		15	25	45
Olive Oyl Button	Factors	1980	3"	1	2	4
Olive Oyl Button	Pep	1946		8	20	35
Olive Oyl Button	Mini Media	1983	1", "More than just a pretty face"	1	2	4
Olive Oyl Costume	Ben Cooper	1976		8	20	35
Olive Oyl Costume	Collegeville	1950s		15	35	70
Olive Oyl Cup	Coke	1977	Coke Kollect-A-Set	4	8	15
Olive Oyl Doll	Dakin	1960s	8" tall	18	35	70
Olive Oyl Doll	Dakin	1970s	hard plastic w/removable clothes	15	35	70
Olive Oyl Doll	Rempel	1950s	small	15	40	75
Olive Oyl Doll	Presents	1990	small molded plastic, musical # P5948	8	13	20
Olive Oyl Doll	Dakin	1970s	Cartoon Theatre, in box	15	40	75
Olive Oyl Doll	Presents	1991	small molded plastic, # P5966	4	7	10
Olive Oyl Doll	Presents	1985	small	8	15	20
Olive Oyl Doll	Presents	1985	Christmas, large	14	25	35
Olive Oyl Doll	Presents	1991	Christmas, small	8	13	20
Olive Oyl Doll			9" vinyl sqeeze doll, Olive with Swee'Pea	12	25	40
Olive Oyl Doll	Uneeda	1979	removable clothing	10	25	38
Olive Oyl Doll	Toy Toons	1990		4	7	10
Olive Oyl Figure	Amscan	1980	large bendy	6	12	18
Olive Oyl Figure	KFS	1940	8" tall	30	75	150
Olive Oyl Figure	Ben Cooper	1974	rubber	10	20	35
Olive Oyl Figure	Comics Spain	1986	6" bendy	4	8	12
Olive Oyl Figure	Jesco	1988	small bendy	2	4	8
Olive Oyl Figure	KFS-Hearst	1991	wood	4	7	10
Olive Oyl Figure	Mexico	1990	ceramic	5	10	18
Olive Oyl Figure	Chester	1990	10", Olive with rolling pin	5	10	18
Olive Oyl Figure		1940s	5" wooden jointed	30	90	175
Olive Oyl Figure	Cristallerie Antonio		Italian crystal	10	15	25
Olive Oyl Figure	Presents	1990	PVC	1	2	4
Olive Oyl Figure	Comics Spain	1984	PVC, Olive with flower	3	7	12
Olive Oyl Figure	Multiple Toymakers	1950s	2" tall	7	13	20
Olive Oyl Figure	Bully	1981	holding flower	8	15	25
Olive Oyl Figure	Bully	1981	with hands clasp	8	15	25
Olive Oyl Figure	KFS	1980	arms clamped together, hanging figure	5	10	15
Olive Oyl Figure		1940s	lead	10	20	30
Olive Oyl Figure	Presents	1990	3" tall, plastic	1	2	4
Olive Oyl Figure	Jesco	1988	large bendy	3	6	10
Olive Oyl Figure Painting Kit	Avalon	1980		6	12	18
Olive Oyl Foam Toy	Cribmates	1979		8	15	25
Olive Oyl Hairbrush	Cribmates	1979	musical	8	15	25
Olive Oyl Hand Puppet	Gund	1960s	comic strip body	20	45	75
Olive Oyl Hi-Pop Ball	Ja-Ru	1981		5	10	15
Olive Oyl Hookies	Tiger	1977		4	10	15
Olive Oyl in a Sports Car	Lesney/ Matchbox	1980		8	15	25
Olive Oyl in Airplane	Corgi	1970s		8	15	25
Olive Oyl Jump-Up	Imperial			5	10	15
Olive Oyl Marionette	Gund	1950s	11-1/2" tall	25	70	130
Olive Oyl Mini Bank	KFS	1979		10	20	30
Olive Oyl Mug	Vandor	1980	ceramic	8	15	23

CHARACTER

POPEYE

TOY	COMPANY	YEAR	DESCRIPTION	GOOD	EX	MIB
Olive Oyl Mug	Schmid	1950s	musical ceramic	15	40	70
Olive Oyl Music Box	Presents	1989	#P5983	8	15	25
Olive Oyl Music Box	KFS	1980	Olive dancing	8	15	25
Olive Oyl on Troubled Waters Record	Peter Pan	1976	45 rpm	4	8	15
Olive Oyl Painting Kit	Avalon	1980		4	8	15
Olive Oyl Push Puppet	Kohner	1960s	4" tall, plastic	15	30	50
Olive Oyl Sports Car		1950s	tiny plastic car	10	15	25
Olive Oyl Squeak Toy	Cribmates	1979	on a stick	5	13	18
Olive Oyl Squeeze Toy	Rempel	1950s	vinyl	15	35	70
Olive Oyl Swim Ring	Wet Set-Zee Toys	1979		5	10	15
Olive Oyl Tiles	Italy	1970s	3" x 5" with stand	12	25	40
Olive Oyl Toboggan	KFS	1979		6	15	20
Olive Oyl Wall Plaque			ceramic	3	7	10
Original Radio Broadcasts Record	Golden Age	1977	33 rpm	5	10	15
Paint 'N Puff Set	Art Award	1979	two versions	6	12	18
Paint with Water Book	Whitman	1981		3	6	10
Painting & Crayon Book	England	1960		13	23	45
Paper Party Blowouts	Gala/James River	1988		2	4	7
Paperweight		1937	"Popeye Eats Del Monte Spinach"	35	70	125
Pencil Case		1930s		15	35	70
Pencil Case	Sanrio	1990		4	8	13
Pencil Case	Eagle	1936	beige, #9027	20	45	75
Pencil Case	Hassenfeld Bros.	1950s	red	20	45	75
Pencil Sharpener	KFS	1929	orange celluloid	15	30	60
Pick-Up Sticks	Lido	1957		12	30	50
Picture Disc Record	Peter Pan	1982	33 rpm	6	10	15
Picture Disc Record	Record of America	1948	78 rpm	15	30	60
Pig for a Friend Mug	Vandor	1980	ceramic	6	10	15
Pin	JCPenney	1935	Back to School Days with Popeye	12	20	35
Pirate Island Presto Magix	American Pub.	1980		4	8	15
Plane & Parachute	Fleetwood	1980		6	12	18
Play Money		1930s	color bucks-framed	10	25	50
Play Money	The Toy House	1970		5	10	20
Playing Cards	Presents	1988	metal box, #P5998, two decks, 60th year	4	8	15
Pocket Pin Ball	Nintendo/Ja-Ru	1983	cups	5	10	15
Pocket Pin Ball	Nintendo/Ja-Ru	1983	holes	5	10	15
Pollution Solution Record	Peter Pan	1970s	45 rpm	4	8	13
Pool Table	Larami	1984		4	8	13
Poopdeck Pappy Doll	Presents	1985		15	25	35
Pop Maker & Son	Ja-Ru	1987		2	4	8
Pop Pistol	Larami	1984		4	8	13
Popeye & Betty Boop Film	Exclusive Films	1935	8mm film	11	25	45
Popeye & Brutus Punch Me Bop Bag	Dartmore	1960s		15	30	50
Popeye & Cast Cigar Box				10	15	25
Popeye & Friends Record	Merry Records	1981	33 rpm	3	6	10
Popeye & Olive Oyl Button	Lisa Frank	1980	1", cowboy Popeye and Indian Olive	2	3	5

Popeye Soaky, 1980s, KFS

Green Hornet hand puppet, 1966, Ideal

3-D Adventures of Superman comic book, 1950s, DC Comics

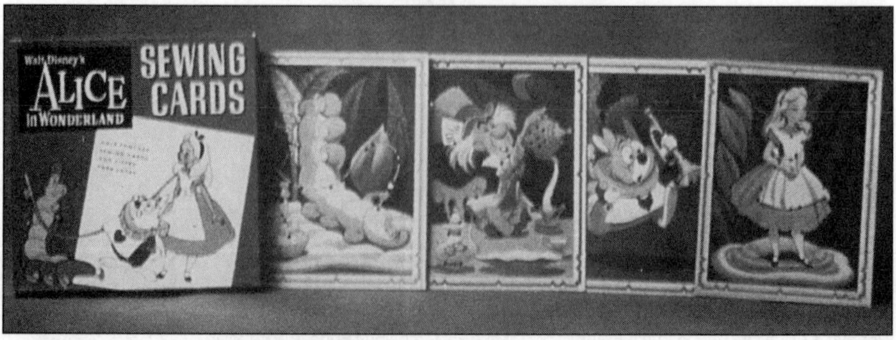

Alice sewing cards, 1951, Whitman

POPEYE

TOY	COMPANY	YEAR	DESCRIPTION	GOOD	EX	MIB
Popeye & Olive Oyl Button		1970s	1-1/2", "I Love You"	2	3	5
Popeye & Olive Oyl Music Box	Schmid		8-1/4" figural box	40	80	150
Popeye & Olive Oyl Sand Set	Peer Products	1950s	bucket, shovel	15	40	70
Popeye & Olive Oyl Suspenders	KFS	1979	blue	6	12	18
Popeye & Olive Oyl Toy Watch	Unknown	1970s	flicker	5	10	15
Popeye & Oscar Flicker Ring			blue	8	15	20
Popeye & Shark Swim Ring	Laurel Star-Japan	1960s		10	25	35
Popeye & Swee'Pea Coloring Book	Whitman	1970	1056-31	6	13	20
Popeye & Swee'Pea Flicker Badge	Varivue	1960s		8	15	25
Popeye & Swee'Pea Flicker Ring			blue	8	15	25
Popeye & Swee'Pea Snow Globe	Presents	1989	several varieties	3	7	12
Popeye & the Deep Sea Mystery Big Little Book	Whitman	1939		20	35	65
Popeye & Wimpy Flicker Badge	Varivue	1960s		8	15	25
Popeye & Wimpy Flicker Ring			blue	8	15	25
Popeye & Wimpy Walk-A-Way Toy	Marx	1964		25	50	95
Popeye Activity Coloring Book	Grosset & Dunlap	1978		5	10	15
Popeye Activity Pad	Merrigold Press	1982		3	7	13
Popeye Air Mattress	Zee Toys	1979		8	20	25
Popeye Alarm Clock	Smiths	1967	British	75	150	300
Popeye All Picture Comic Book	Whitman	1942	Big Little Book	18	65	50
Popeye and his Jungle Pet Book	Whitman	1937		20	45	85
Popeye and the Jeep Book	Whitman	1937	Big Little Book	20	45	85
Popeye and the Pet Book	Peter Haddock	1987	book three of four	2	4	8
Popeye and the Time Machine Book	Quaker	1990	mini comic	1	2	4
Popeye Apron	Chester	1990		3	6	12
Popeye Arcade	Fleetwood	1980		6	12	25
Popeye Arcade Game	Parker Bros.	1980	card game	5	10	15
Popeye at the Wheel	Woolnough	1950s	musical	125	250	500
Popeye Ball			rubber kick ball	7	13	20
Popeye Ball Toss Game	KFS	1950s		30	50	75
Popeye Bank	Play Pal	1972	shape of Popeye's head, plastic	15	30	45
Popeye Bank	Sanrio	1990	with padlock	4	8	15
Popeye Bank		1940s	9" cast iron	60	125	200
Popeye Bank		1940s	Popeye with life preserver	30	60	100
Popeye Bank	Play Pal	1970s	plastic, Popeye sitting on rope	10	30	40
Popeye Bank	Renz	1979	beige bust	8	20	35
Popeye Bank	Presents	1991	vinyl, Popeye with removable pipe	6	12	18
Popeye Bank	Leonard	1980	silver, Popeye sitting	15	25	45
Popeye Bank	Vandor	1980	ceramic, Popeye sitting on rope	10	20	30
Popeye Bank			ceramic, Popeye in light blue cap	5	10	15
Popeye Bank	Mexico	1990	ceramic bust	10	25	40

POPEYE

TOY	COMPANY	YEAR	DESCRIPTION	GOOD	EX	MIB
Popeye Bank	Vandor	1980	ceramic	125	250	500
Popeye Bathtub Toy	Stahlwood	1960s	floating boat	15	30	55
Popeye Beach Set	Peer Products/ KFS	1950s	plastic rowboat, accessories	15	35	50
Popeye Bend-I-Face	Lakeside	1967		15	35	60
Popeye Bingo	Nasta	1980		6	10	15
Popeye Bingo	Bar Zim	1929		15	45	80
Popeye Blow Me Down Airport	Marx	1935		250	500	1000
Popeye Book	Random House	1980	hardcover, based on movie	2	4	8
Popeye Break-A-Plate Game	Combex	1963		30	65	125
Popeye Bubble Liquid	M. Shimmel Sons	1970s	shaped like Popeye with necktie similar to a sailor's knot	8	15	25
Popeye Button	KFS	1950s	1", Famous Studios	12	20	30
Popeye Button	Pep	1946		15	35	50
Popeye Button	KFS	1989	3", marine conservation	2	3	5
Popeye Button	Factors	1980	3", movie	2	3	5
Popeye Button	KMOX TV	1960s	1", S.S. Popeye	4	7	10
Popeye Button	Lowe	1959	sew-on card	12	20	30
Popeye Button	Strand	1985	3", several styles	2	3	5
Popeye Button	S. Cruz	1990	2", Santa Cruz boardwalk	2	3	5
Popeye Calls on Olive Oyl Book	Whitman	1937	8-1/2" x 9-1/2"	20	45	85
Popeye Car	Vandor	1980	Popeye & Olive in blue or pink car	20	40	75
Popeye Carnival	Toymaster	1965		50	100	195
Popeye Charm			solid gold	100	150	200
Popeye Charm		1930s	celluloid	10	20	30
Popeye Charm			silver with dangly parts	10	15	20
Popeye Climbs a Mountain Book	Wonder Books	1983		2	4	7
Popeye Color & Recolor Book	Jack Built	1957	color, wipe & color again	15	25	45
Popeye Coloring Set	Hasbro	1960s	numbered, with pencils	12	25	40
Popeye Costume	Collegeville	1980s		6	10	16
Popeye Costume	Ben Cooper	1984		4	8	12
Popeye Costume	Collegeville	1950s		15	35	50
Popeye Cup	Deca Plastics	1979	plastic	4	7	15
Popeye Cup	Popeye Picnic	1989	plastic	1	2	4
Popeye Cup & Saucer	Japan	1930s		40	70	130
Popeye Dog Toy	Petex	1986		3	6	10
Popeye Doll	Presents	1985	small	8	13	20
Popeye Doll	Presents	1985	small doll with pipe molded into hand	8	13	20
Popeye Doll	Sears/ Cameo	1957	13" in box	150	310	600
Popeye Doll	Presents	1990	small molded plastic, musical #P5949	8	13	20
Popeye Doll	Rempel	1950s	small	20	40	75
Popeye Doll	Dakin	1970s	Cartoon Theatre, in box	25	50	95
Popeye Doll	Dakin	1974	squeaks	15	30	55
Popeye Doll	Chad Valley	1950s	7" tall, squeaks	20	40	75
Popeye Doll	Etone	1983	8" plush	6	10	15
Popeye Doll	Lakeside	1968	12" tall, sponge rubber	10	25	45
Popeye Doll	Uneeda	1979		15	35	70
Popeye Doll	Woolikin	1950s	white plush	20	40	75
Popeye Doll	Stack	1936	12" tall, wood jointed, with pipe	80	175	350

POPEYE

TOY	COMPANY	YEAR	DESCRIPTION	GOOD	EX	MIB
Popeye Doll		1960s	9" vinyl squeeze doll, Popeye with Swee'Pea	12	30	50
Popeye Doll	Toy Toons	1990		4	7	10
Popeye Doll	Gund	1958	20" tall	30	70	130
Popeye Doll			23" china	75	150	300
Popeye Doll	Chicago Herald American	1950s		20	60	100
Popeye Doll	Uneeda	1979	16" tall	15	35	60
Popeye Doll	Cameo	1935		95	195	375
Popeye Doll	Quaker	1960s	12" cloth	10	20	40
Popeye Doll	Dakin	1970s	hard plastic w/removable clothes	15	40	85
Popeye Figure		1970s	ceramic, removeable head Popeye	20	45	75
Popeye Figure	Sirocco-KFS	1944	5", wood	35	75	100
Popeye Figure	Presents	1990	PVC	1	2	4
Popeye Figure	England	1950s	7", bendy, yellow pants	25	50	100
Popeye Figure		1940s	celluloid with wooden feet	20	40	75
Popeye Figure		1950s	plastic, Popeye on four wheels with telescope	20	40	75
Popeye Figure	Japan	1950s	celluloid	15	30	55
Popeye Figure	Cristallerie Antionio Imperatore		Italian crystal	10	16	25
Popeye Figure	Comics Spain	1984	PVC, Popeye with spinach	4	7	12
Popeye Figure	Bully	1981	several variations	9	18	30
Popeye Figure		1940s	lead	15	25	35
Popeye Figure	Presents	1990	3" tall, plastic	2	3	5
Popeye Figure	Japan Olympics	1962	wood, Popeye at bat	125	250	500
Popeye Figure	Bronco	1978	bendy	6	10	15
Popeye Figure	Chester	1990	10", ceramic, Popeye with spinach	6	10	15
Popeye Figure	KFS-Hearst	1991	wood	6	10	15
Popeye Figure		1930s	5" tall, wood jointed, held together with string	40	90	175
Popeye Figure	Dakin		8" tall with spinach can	15	30	60
Popeye Figure	Imperial	1979		6	10	15
Popeye Figure	Mexico	1990	ceramic	7	15	20
Popeye Figure	Duncan	1970	8" tall	15	30	55
Popeye Figure	Ben Cooper	1974	rubber	8	15	25
Popeye Figure	Combex	1960s	rubber, Popey with a can of spinach	15	20	40
Popeye Figure		1930s	12" tall, chalk, with pipe and hat, ashtray base	40	85	150
Popeye Figure	Lakeside	1968	miniflex	9	18	35
Popeye Figure	Lakeside	1969	superflex	10	20	40
Popeye Figure	Comics Spain	1986	6" bendy, white pants	3	6	12
Popeye Figure	Jesco	1988	small bendy	2	4	8
Popeye Figure	Amscan	1980	small bendy	4	7	10
Popeye Figure	Amscan	1980	large bendy	6	12	16
Popeye Figure	Jesco	1988	bendy	6	10	15
Popeye Figure Painting Kit	Avalon	1980		4	8	13
Popeye Finger Rings	Post Toasties	1949		18	35	70
Popeye Fishing Game	Fleetwood	1980		5	10	15
Popeye Fishing Game	Transogram	1962	magnetic	15	30	55
Popeye Flicker Badge	Varivue	1960s	Popeye eating spinach	8	15	20
Popeye Flickers	Sonwell	1960s		8	15	20
Popeye French Record	Polygram	1981	45 rpm	6	10	15

CHARACTER

POPEYE

TOY	COMPANY	YEAR	DESCRIPTION	GOOD	EX	MIB
Popeye Galley Steward Figure	KFS	1980	ceramic	9	18	30
Popeye Games	Ed-U-Card	1960s	set of four games	10	25	40
Popeye Getar	Mattel	1960s	14" long, shaped like Popeye's face, plays "I'm Popeye the Sailor Man"	20	45	90
Popeye Glass	Coke	1977	Coke Kollect-A-Set	4	7	10
Popeye Glow Putty	Larami	1984		3	6	10
Popeye Goes Swimming Colorforms	Colorforms	1963		15	35	60
Popeye Goes to School Television	Zaboly	1950s		15	30	55
Popeye Hammer Game	Holgate	1960s		60	125	250
Popeye Hand Puppet	Gund	1950s	plush	15	30	50
Popeye Hand Puppet	Gund	1960s	Popeye's head on cloth body	20	45	80
Popeye Hi-Pop Ball	Ja-Ru	1981		4	8	13
Popeye Hot Water Bottle	Duarry	1970		40	90	175
Popeye How to Draw Cartoons Book	Joe Musial/D. McKay	1939		20	45	90
Popeye in a Spinach Truck	Lesney/ Matchbox	1980		8	13	25
Popeye in Boat	Corgi	1970s		8	20	35
Popeye in Puddleburg Book	Saalfield	1934	Big Little Book	20	40	75
Popeye in the Movies Record	Peter Pan		33 rpm with book	6	10	15
Popeye Jump-Up	Imperial	1970s		8	15	30
Popeye Lamp	Alan Jay	1959	Popeye with legs folded holding spinach	35	75	135
Popeye Lamp		1940s	telescope with Popeye at base	40	90	175
Popeye Lamp	Idealite	1935	Popeye base & boat scene	250	500	900
Popeye Lantern	Linemar	1960s	7-1/2" tall, metal, battery operated	70	150	290
Popeye Launches His New Song Hits Record	Peter Pan	1958	45 rpm	8	20	30
Popeye Learn & Play Activity Book	Allen Canning	1985		2	4	6
Popeye Magic Play Around	Amsco	1950s	characters with magnetic bases that slide across play set	30	65	110
Popeye Marionette	Gund	1950s	11-1/2" tall	45	95	200
Popeye Marionette	Create-Japan		wood	85	175	325
Popeye Mechanical Pencil	Eagle	1929	10-1/2" long illustrated pencil with box	25	50	100
Popeye Meets his Rival Book	Whitman	1937	8-1/2" x 11-1/2"	20	45	85
Popeye Menu Pinball Game	Durable Toy & Novelty	1935		45	90	175
Popeye Mini Bank	KFS	1979		4	8	16
Popeye Mini Tennis Game	Nordic	1970s		10	15	25
Popeye Mini Winder	Durham	1980		4	8	15
Popeye Model Kit	Tokyo Plamo	1964	#808	35	75	150
Popeye Nail-On Game	Colorforms	1963		20	40	75
Popeye Night Light	Arrow Plastic			8	13	20
Popeye on Parade/Strike Me Pink Record	Cricket	1950s	45 rpm	10	20	35
Popeye on Rocket Coloring Book	Whitman - France	1980		8	13	20
Popeye on Safari Book	Quaker	1990	mini comic	2	3	5
Popeye on Tricycle	Linemar		4-1/2", tin wind-up with celluloid arms & legs, bell rings behind Popeye	100	200	450
Popeye One Man Band	Larami	1980s		6	10	15
Popeye Paddle Ball	Larami	1984	with color photo of Popeye	6	10	15

POPEYE

TOY	COMPANY	YEAR	DESCRIPTION	GOOD	EX	MIB
Popeye Paint and Crayon Set	Milton Bradley	1934		25	50	100
Popeye Paint Book	McLoughlin Bros.	1932	blue	25	40	60
Popeye Paint By Numbers	Hasbro	1960s		12	25	50
Popeye Paint Coloring Book	Whitman	1951		15	30	50
Popeye Paint Set	American Crayon	1933	6", tin	25	45	65
Popeye Paint-By-Numbers	Hasbro	1981		8	13	20
Popeye Painting Kit	Avalon	1980		6	10	15
Popeye Party Game	Whitman	1937	posters, paper pipes, game box	20	40	75
Popeye Pencil-By-Numbers	Hasbro	1979		10	25	35
Popeye Peppy Puppet	Kohner	1970		10	20	30
Popeye Picture	KFS-Sears		silver foil	10	15	25
Popeye Pin			Popeye at steering wheel, stick pin	4	7	10
Popeye Pinball Game	Ja-Ru	1983		6	10	15
Popeye Pipe	Edmonton Pipe	1970	figural head	15	25	40
Popeye Pipe	Harmony	1980		6	10	15
Popeye Pipe	Micro-Lite-KFS	1958		10	25	50
Popeye Pipe	KFS	1970s	plastic kazoo, red & blue	6	20	35
Popeye Pipe	MSS	1970s	plastic, white	8	20	35
Popeye Pipe		1940s	red wooden	15	25	35
Popeye Pipe Toss Game	Rosebud Art	1935	small version with wooden pipe	20	45	90
Popeye Pistol	Delcast		Super mini cap with 24 caps No. 807-BB	6	20	30
Popeye Pistol	Marx	1935		50	150	195
Popeye Play Set	Cribmates	1979	Popeye, Olive Oyl, and Swee'Pea squeak toys, mirror, rattle and pillow	15	30	45
Popeye Playing Card Game	Whitman	1934	5" x 7", green box	15	30	50
Popeye Playing Card Game	Whitman	1938	blue box	13	25	45
Popeye Playing Card Game	Parker Bros.	1983		6	10	15
Popeye Pop-Up Book	Random House	1981		6	10	15
Popeye Popcorn	Purity Mills	1949	in can	20	35	50
Popeye Presto Paints	Kenner	1961		15	50	65
Popeye Pull Toy	Metal Masters	1950s	10-1/2" x 11-1/2", xylophone, wood with paper litho labels, metal wheels	100	225	425
Popeye Punch-Out Play Book	Whitman	1961		12	20	30
Popeye Puppet	Kohner		pull string, Popeye jumps	10	15	25
Popeye Puppet Show Book	Pleasure Books	1936		25	40	60
Popeye Push Puppet	Kohner	1960	4" tall	20	40	80
Popeye Record	Peter Pan	1977	33 rpm, four stories, #1114	6	10	15
Popeye Ring Toss Game	Fleetwood	1980		6	10	15
Popeye Ring Toss Game	Transogram	1957		30	50	75
Popeye Sailboat	KFS	1976		8	13	20
Popeye Sees the Sea Book	Whitman	1936	Big Little Book	20	35	65
Popeye Service Station	Larami	1979		6	15	25
Popeye Shipwreck Game	Einson-Freeman	1933		40	65	100

POPEYE

TOY	COMPANY	YEAR	DESCRIPTION	GOOD	EX	MIB
Popeye Snow Globe	KFS	1960s	Popeye holds globe between legs	25	40	75
Popeye Soaky	Colgate-Palmolive	1960s		12	30	55
Popeye Soaky	KFS	1987	British	12	20	35
Popeye Song Folio Book	Famous Music	1936		20	50	85
Popeye Sparkler	Chein	1959		90	125	175
Popeye Speed Boat	Harmony	1981		10	16	25
Popeye Speedboard Pull Toy		1960s		175	350	675
Popeye Spinach Eater Pull Toy	Fisher-Price	1939	#488, standing Popeye drums on spinach can drum	105	225	425
Popeye Spinach Target Game	Gardner	1960s		40	65	100
Popeye Sports Car	Linemar	1950s		225	375	600
Popeye Squeeze Toy	Rempel	1950s	8" tall, vinyl	15	35	65
Popeye Squeeze Toy	Cribmates	1979	Popeye on a stick	8	13	20
Popeye Stay in Shape Book	Tuffy Books	1980		6	10	15
Popeye Supergyro	Larami	1980s		6	10	15
Popeye Surprise Present Book	Peter Haddock	1987		4	7	10
Popeye the Juggler Bead Game	Bar-Zim	1929	3-1/2" x 5", covered with glass	20	35	50
Popeye the Ladies Man Record			33 rpm	8	13	20
Popeye the Movie Book	Avon Printing	1980		6	10	15
Popeye the Movie Soundtrack Record	Paramount	1980		6	10	15
Popeye the Pilot	Chein	1940s	tin wind-up airplane, 8" long, 8" wingspan	310	525	800
Popeye the Sailor Man & His Friends Record	Golden	1960s	33 rpm	6	10	15
Popeye the Sailor Man Book	Whitman	1942	Big Little Book	20	35	65
Popeye the Sailor Man Record	Diplomat Records	1960	33 rpm	6	10	15
Popeye the Sailor Man Record	Peter Pan	1976	33 rpm	6	10	15
Popeye the Sailor Man Record	Golden	1959	45 rpm	6	10	15
Popeye the Weatherman Colorforms	Colorforms	1959		20	50	90
Popeye Thimble		1990		4	7	10
Popeye Tiles	Italy	1970s	3" x 5" with stand	15	35	65
Popeye Toboggan	KFS	1979		6	10	15
Popeye Toothbrush Holder	Vandor		5" tall, figural	10	15	25
Popeye Toothbrush Set	Nasta	1980s	holds two toothbrushes	6	15	20
Popeye Train Pull Toy	Larami			135	225	350
Popeye Train Set	Larami	1973		6	15	25
Popeye Transit Company Moving Van	Linemar	1950	tin	300	650	1200
Popeye Tricky Trapeze	Kohner	1970		15	35	65
Popeye Tricky Walker	Jaymar	1960s	plastic	10	25	50
Popeye Tricycling	Linemar	1950s		250	500	900
Popeye Tug & Dingy Pull Toy	Fisher-Price	1950s	wood	135	225	350
Popeye Tugboat	Ideal	1961	inflatable	15	35	55
Popeye TV Cartoon Kit	Colorforms	1966		20	40	75
Popeye TV Magic Putty	MSS	1970s		6	10	15
Popeye Utensils	Arrow Plastic	1970s	spoon and fork	8	13	20
Popeye Video Game	Nintendo	1983		12	20	30

POPEYE

TOY	COMPANY	YEAR	DESCRIPTION	GOOD	EX	MIB
Popeye vs. Bluto the Bad Book	Quaker	1990	mini comic	2	3	5
Popeye Wall Hanging	Amscan	1979	42", jointed	8	13	20
Popeye Water Colors	American Crayon	1933		15	35	60
Popeye Water Sprinkler	KFS	1960s	with rubbber head	15	25	40
Popeye with his Friends Book	Whitman	1937		20	45	80
Popeye Wristwatch	Bradley	1979		60	100	150
Popeye Wristwatch	Armitron	1991		20	35	50
Popeye Wristwatch	Bradley	1964	#308, green case	115	195	300
Popeye Wristwatch	KFS-Japan	1990	Popeye in ship's wheel	40	65	100
Popeye Wristwatch	Unique	1987	Popeye's head pops open, digital	12	20	30
Popeye Wristwatch	New Haven	1938		125	300	600
Popeye Writing Tablet	KFS	1929	Popeye with spinach hypo	20	35	50
Popeye's Adventure Book	Purnell	1958	England	12	20	30
Popeye's Ark Book	Saalfield	1936	Big Little Book	20	35	65
Popeye's Favorite Sea Shanties Record	RCA Camden	1960		10	20	30
Popeye's Favorite Sea Songs Record	Peter Pan	1959	45 rpm	10	20	30
Popeye's Favorite Stories Record	RCA Camden	1960	33 rpm	10	20	30
Popeye's Gang Pinball Game	MSS	1970s		13	25	45
Popeye's Lucky Jeep Game	Northwestern Products	1936		40	65	120
Popeye's Official Wallet	KFS	1959		10	25	45
Popeye's Peg Board Game	Bar Zim	1934		45	90	180
Popeye's Sliding Boards & Ladders	Warren Built-Rite	1958		12	25	50
Popeye's Songs About...... Record	Golden	1961	33 rpm	8	18	25
Popeye's Spinach Hunt Game	Whitman	1976		10	20	35
Popeye's Submarine	Larami	1973		10	25	40
Popeye's Three Game Set	Built-Rite	1956		15	30	60
Popeye's Tiddly Winks	Parker Bros.	1948		20	35	60
Popeye's Treasure Map Game	Whitman	1977		12	20	35
Popeye's Where's Me Pipe Game				30	45	70
Popeye/Olive Oyl/ Swee'Pea Lamp Shade		1950s		40	65	100
Popeye/Olive Oyl/Wimpy Decals	IGS Stores	1935		8	13	20
Popeye/Olive Oyl/Wimpy Skill Games	Lido	1965		10	16	25
Popsicle Harmonica	Czech	1929		60	125	225
Punch Ball	National Latex	1970s		10	20	35
Punch'Em Talking Rattle Toy	Sanitoy	1950s		15	35	70
Punching Bag	Dartmore	1960s		15	30	50
Punching Bag Film	Brumberger	1950s	8mm	8	15	25
Puppetforms	Colorforms	1950s		25	40	60
Puzzle	Illco	1987	11 pieces, Popeye 3-D	4	7	10
Puzzle	Saalfield	1932	Popeye in Four	35	55	85
Puzzle	Tower Press	1962	wood, Popeye	10	16	25
Puzzle	Roalex	1960s	tile	20	35	50
Puzzle	Opera Mundi	1977	tile, "Popeye's Riddle"	10	16	25
Puzzle	Ja-Ru	1987	Popeye & Son TV show	6	10	15

POPEYE

TOY	COMPANY	YEAR	DESCRIPTION	GOOD	EX	MIB
Puzzle	Ja-Ru	1989	"Birthday Cake and Ice Cream"	6	10	15
Puzzle	Ja-Ru	1981	comic	6	10	15
Puzzle	Illco	1987	11 pieces, Olive 3-D	6	10	15
Puzzle	Jaymar	1991	63 pieces, jumbo, Popeye's boat	6	10	15
Puzzle	Ja-Ru	1989	boating and dancing	6	10	15
Puzzle	Jaymar	1945	22" x 13-1/2", Popeye	15	35	60
Puzzle	Jaymar	1991	63 pieces, Popeye blowing candles	6	10	15
Puzzle	Jaymar	1991	63 pieces, jumbo, Popeye and Olive surfing	4	7	10
Puzzle	American Pub. Corp.	1976	5-1/2" round can	10	16	25
Puzzle	Jaymar	1991	Christmas scene, inlaid, 12 pieces	4	7	10
Puzzle	Jaymar	1991	12 pieces, Popeye holding turkey	4	7	10
Puzzle	Jaymar	1991	12 pieces, Popeye gang swimming	4	7	10
Puzzle	England	1959	120 pieces, "What a Catch"	25	40	60
Puzzle	Waddington's House of Games	1978	27" x 18" floor puzzle	8	13	20
Puzzle	Jaymar	1991	63 pieces, jumbo, Popeye rescues Olive	6	10	15
Puzzle	Larami	1973	magnetic	6	10	15
Puzzle Game	Waddington's House of Games	1978		10	16	25
Puzzle Party Book	Cinnamon House	1979		6	10	15
Quest for the Rainbird Book	Whitman	1943	Big Little Book	20	35	65
Race Set	Ja-Ru	1989		6	10	15
Race to Pearl Peak Book	Golden	1982		6	10	15
Rain Boots	KFS	1950s	spinach power	13	30	60
Record Player	Emerson	1960s	Dynamite Music Machine	30	65	130
Ring the Bell with Hammer Game	Harett-Gilmar	1960s		25	40	60
Ring Toss Stand-Up Game	Transogram	1958		15	30	55
Road Building Set	Larami	1979		6	10	15
Roller Skating Popeye	Linemar	1950s		250	525	100
Roly Poly & Cork Gun Game	Knickbocker	1958		50	80	125
Roly Poly Popeye		1940s	with beaded arms & celluloid	75	130	200
Rub 'N Win Party Game	Spot-O-Gold	1980		5	10	15
S.S. Funboat Coloring Book	Merrigold Press	1981		4	7	10
Sailboats	Larami	1981		6	10	15
Sailor & the Spinach Stalk Coloring Book	Whitman	1982	1150-1	4	7	10
Sailor's Knife	Ja-Ru	1981		6	10	15
Scott Fun 'N Games Booklet	Spot-O-Gold	1980	set of five	10	15	25
Screen-A Show Projector	Denys Fisher	1973		35	80	140
Sea Hag Doll	Presents	1985		15	25	35
Sea Hag Hand Puppet	Presents	1987		8	13	20
Secret Message Pen	Gordy	1981		6	10	15
Shaving Kit	Larami	1979		6	10	15
Six Popeye Songs Record	Wonderland Records	1950s	45 rpm	8	15	25
Skeet Shoot Game	Irwin	1950		40	80	150
Sketchbook	Japan	1960		12	25	40

POPEYE

TOY	COMPANY	YEAR	DESCRIPTION	GOOD	EX	MIB
Skoozit Pick-A-Puzzle Game	Ideal	1966	8-1/4" round can	15	35	70
Sleeping Bag	KFS	1979		15	35	65
Sling Darts	KFS			20	35	50
Soap Dispenser	Woolfoam	1970s		15	25	35
Soap on a Rope	KFS		white, shaped like Popeye's head	15	25	40
Soap Set	Kerk Guild	1930s	Olive Oyl, Swee'Pea, & Popeye soap figures	35	75	145
Soapy Popeye Boat	Kerk Guild	1950s		20	40	75
Song & Story Skin Diver Record	KFS	1964		8	15	20
Songs of Health Record	Golden	1960s	45 rpm	6	10	15
Songs of Safety Record	Golden	1960s	45 rpm	6	10	15
Spinach Can Bank	KFS	1975	blue can with raised characters	15	25	40
Squirt Face	Ja-Ru	1981		6	10	15
Stationery	Presents	1989	metal box, heart-shaped note paper, #P5976	6	10	15
Stitch-A-Story	KFS			12	20	30
Storage Box	Sanrio	1990	smoke colored	6	10	15
Sun-Eze Pictures	Tillman Toy	1962		10	20	35
Sunday Funnies Soda Can	Flavor Valley	1970s		8	13	20
Sunglasses	Larami	1980s	red, yellow, or blue	6	10	15
Super Race with Launcher	Fleetwood	1980		6	10	15
Surf Rider	Wet Set-Zee Toys	1979		10	20	35
Suspenders		1970s	red, white & blue with plastic emblems	8	18	30
Swee'Pea Bank	Vandor	1980	6-1/2" figural	15	25	45
Swee'Pea Bean Bag	Dakin	1974		8	18	30
Swee'Pea Button	Lisa Frank	1980	1", Swee'Pea with Jeep	2	3	5
Swee'Pea Cup	Coke	1977	Coke Kollect-A-Cup	4	7	10
Swee'Pea Doll	Presents	1985	small	8	13	20
Swee'Pea Doll	Presents	1985	large	10	16	25
Swee'Pea Doll	Presents	1991	Christmas, small	8	13	20
Swee'Pea Doll	Presents	1991	small molded plastic, # P5968	8	13	20
Swee'Pea Doll	Uneeda	1979		8	13	20
Swee'Pea Egg Cup	Vandor	1980	ceramic	10	16	25
Swee'Pea Figure			ceramic, one of five	4	7	10
Swee'Pea Figure	Cristallerie Antionio Imperatore		Italian crystal	10	16	25
Swee'Pea Figure	Presents	1984	PVC	2	3	5
Swee'Pea Figure	Comics Spain	1984	PVC, Swee'Pea with cake	4	7	10
Swee'Pea Figure	Presents	1990	3" tall, plastic	4	7	10
Swee'Pea Hand Puppet	Gund	1960s	bonnet on head, cloth body decorated with baby lambs	13	25	50
Swee'Pea Hand Puppet	Presents	1987		8	13	20
Swee'Pea Hi-Pop Ball	Ja-Ru	1981		6	10	15
Swee'Pea Mini Bank	KFS	1979		8	18	35
Swee'Pea Music Box	KFS	1980	Swee'Pea dancing	8	13	20
Swee'Pea Music Box	Presents	1989	#P5986	8	13	20
Swee'Pea Night Light	Presents	1978	bone china	10	20	40
Swee'Pea Snow Globe	Presents	1989	several styles	4	7	10
Swee'Pea Squeak Toy		1970s	frowning or smiling	12	20	30
Swee'Pea Wall Plaque		1950s	ceramic	12	30	45
Swee'Pea's Lemonade Stand Television Film	Zaboly	1950s		10	20	35
Swirler Flying Barrel Toy	Imperial	1980		6	12	20
Talking View-Master Set	GAF	1962	old type	12	25	35

POPEYE

TOY	COMPANY	YEAR	DESCRIPTION	GOOD	EX	MIB
Tambourine	Larami	1980s		6	10	15
Tambourine	Santa Cruz	1990		6	10	15
Target Ball	Ja-Ru	1981		6	10	15
Telephone	Comvu	1982		15	25	35
Telescope	Larami	1973		6	15	25
The Outer Space Zoo Book	Golden	1980		4	7	10
Thimble Theatre Book	Whitman	1935	Big Big Book	30	65	100
Thimble Theatre Cut-Outs	Aldon	1950s		25	45	65
Toy Watch	Sekonda-Japan	1987	comic strip band	30	50	75
Trace & Color	Fleetwood	1980		6	10	15
Training Cup	Deka	1971		8	13	20
Transistor Radio	Philgee	1960s		20	35	50
Trash Can	KFS			10	15	25
Trinket Box	Vandor	1980	Popeye laying on top, ceramic	10	16	25
Trinket Box	Vandor	1980	Popeye's head in preserver, ceramic	10	16	25
Trumpet Bubble Blower	Larami	1984	pink or yellow pan	6	10	15
Tube-A-Loonies	Larami	1973	five small tubes on card	6	10	15
Turn-A-Scope	Larami	1979		6	10	15
TV Set with Three Film Scrolls		1957		12	20	30
TV Tray	KFS	1979		8	13	20
Umbrella	KFS	1979	blue & white	12	20	30
View-Master Set	Sawyers	1959	three reels	12	20	35
View-Master Set	GAF	1962	three pack	8	13	20
View-Master Set	GAF	1959	"The Fish Story"	6	10	15
View-Master Set	GAF	1959	"The Hunting Bird"	6	10	15
Wagon Works Film	Atlas Films	1960s	8mm	6	10	15
Wallet	Larami	1978		8	13	20
Wallet	Sanrio	1990		6	10	15
Wallet	Presents	1991	P-5432, tri-fold	6	10	15
Water Ball Game	Nintendo	1983	one basket	6	10	15
Whale of a Tale Record	Peter Pan	1981	45 rpm	6	10	15
What! No Spinach? Book	Golden	1981		6	10	15
Whistle Candy	Alberts	1989		6	10	15
Whistling Flashlights	Bantam-Lite	1960s	six with display	135	225	350
Whistling Wing Ding	Mego	1950s		20	35	50
Wimpy Button	Strand	1985	3", "Must go Home and Water the Ducks"	2	3	5
Wimpy Button	Lisa Frank	1979	2"	2	3	5
Wimpy Button	Pep	1946		10	16	25
Wimpy Cup	Coke	1977	Coke Kollect-A-Set	4	7	10
Wimpy Dog Toy	Petex	1986		8	13	20
Wimpy Doll	Presents	1985	holding a hamburger	14	23	35
Wimpy Doll	KFS	1950s	rubber	20	35	50
Wimpy Figure	Bully	1981	yellow hat	6	12	25
Wimpy Figure		1940s	lead	12	20	35
Wimpy Figure	Sirocco-KFS	1944	5", wood	30	50	75
Wimpy Figure	Comics Spain	1984	Wimpy with hamburger	4	7	10
Wimpy Figure	Presents	1990	PVC	2	3	5
Wimpy Figure	Cristallerie Antonio Imperatore		Italian crystal	10	15	25
Wimpy Figure	Buitoni	1950s	premium	25	55	110
Wimpy Hand Puppet	Gund	1950s	fabric hand cover, vinyl squeaker head & voice	15	35	60
Wimpy Hand Puppet	Gund	1960s	cloth body	13	25	50
Wimpy Hand Puppet	Presents	1987		8	13	20

POPEYE

TOY	COMPANY	YEAR	DESCRIPTION	GOOD	EX	MIB
Wimpy in Back to his First Love Book			eight pages	6	10	15
Wimpy Magnet				6	10	15
Wimpy Music Box	Presents	1989	#P5985	8	13	20
Wimpy Rings	Post Toasties	1949		8	25	50
Wimpy Squeeze Toy	Rempel	1950s	vinyl	20	35	50
Wimpy Squeeze Toy	Cribmates	1979		12	20	30
Wimpy the Hamburger Eater Book	Whitman	1938	Big Little Book	20	35	65
Wimpy Thermometer	KFS	1981		8	13	20
Wimpy Tricks Popeye & Roughhouse Book	Whitman	1937		35	55	85
Wimpy Tricks Popeye Book	Whitman	1937		40	65	100
Wimpy Tugboat	Ideal	1961	inflatable	20	35	50
Wimpy What's Good to Eat? Book	Tuffy Books	1980		6	10	15
Wood Slate	Ja-Ru	1983		6	10	15
Write on/Wipe Off Board	Freelance	1979		6	10	15

PORKY PIG

TOY	COMPANY	YEAR	DESCRIPTION	GOOD	EX	MIB
Porky & Petunia Figures	Warner Bros.	1975	4-1/2" tall	9	16	25
Porky Pig Bank		1930s	bisque, orange, blue, and yellow	55	100	150
Porky Pig Doll	Mattel	1960s	17", cloth, vinyl head	15	30	45
Porky Pig Doll	Gund	1950	14" tall	45	80	125
Porky Pig Doll	Dakin	1968	7-3/4" tall in black velvet jacket	11	20	30
Porky Pig Soaky				11	30	50
Porky Pig Umbrella		1940s	hard plastic 3" figure on end, Porky & Bugs printed in red cloth	40	75	115

QUICK DRAW MCGRAW

TOY	COMPANY	YEAR	DESCRIPTION	GOOD	EX	MIB
Auggie Doggie Doll	Knickerbocker	1959	10" tall, plush with vinyl face	16	30	45
Auggie Doggie Soaky	Purex	1960s	10" tall, plastic	15	40	65
Baba Looey Bank	Knickerbocker	1960s	9" tall, vinyl, plastic head	16	30	35
Baba Looey Doll	Knickerbocker	1959	20" tall, plush with vinyl donkey ears & sombrero	35	65	100
Blabber Doll	Knickerbocker	1959	15" tall, plush with vinyl face	25	50	75
Blabber Soaky	Purex	1960s	10-1/2" tall, plastic	16	35	50
Quick Draw McGraw Bank		1960	9-1/2", plastic, orange, white, & blue	20	35	55
Quick Draw McGraw Book	Whitman	1960		15	25	40
Quick Draw McGraw Doll	Knickerbocker	1959	16", plush with vinyl face, cowboy hat	45	80	125
Quick Draw McGraw Moving Target Game	Knickerbocker	1960s		95	125	200
Quick Draw Mold & Model Cast Set		1960		25	50	75
Scooper Doll	Knickerbocker	1959	20", plush with vinyl face	30	55	85

RAGGEDY ANN & ANDY

TOY	COMPANY	YEAR	DESCRIPTION	GOOD	EX	MIB
Camel with the Wrinkled Knees Book		1924		75	100	150
Raggedy Andy Bank	Play Pal		11"	15	20	25
Raggedy Andy Figure		1970s	rubber/wire, 4" tall	5	10	15
Raggedy Andy Puppet	Dakin	1975	cloth	15	25	30
Raggedy Ann Bank	Play Pal	1974	11"	15	20	25

RAGGEDY ANN & ANDY

TOY	COMPANY	YEAR	DESCRIPTION	GOOD	EX	MIB
Raggedy Ann Coloring Book		1968		5	10	15
Raggedy Ann Puppet	Dakin	1975	cloth	15	25	30
Raggedy Ann/Andy Paper Doll Book	Whitman	1974	#1962	10	20	25
Raggedy Ann/Andy Record Player			square, cardboard	45	55	65
Raggedy Ann/Andy Record Player		1950s	plastic, heart shaped	80	90	100
Raggedy Ann/Andy Wastebasket		1972	tin	8	15	20

ROAD RUNNER & WILE E. COYOTE

TOY	COMPANY	YEAR	DESCRIPTION	GOOD	EX	MIB
Road Runner & Coyote Lamp		1977	12-1/2", figures standing on base	20	35	125
Road Runner Bank			standing on base	5	10	15
Road Runner Costume	Collegeville	1960s		20	30	55
Road Runner Doll	Mighty Star	1971	13" tall	9	16	25
Road Runner Figure	Dakin		plastic, Cartoon Theater	9	16	65
Road Runner Figure	Dakin	1971	Goofy Gram	14	25	40
Road Runner Figure	Dakin	1968	8-3/4" tall	12	23	35
Road Runner Hand Puppet	Japanese	1970s	10", vinyl head	4	7	10
Wile E. Coyote & Road Runner Figures	Royal Crown	1979	7" tall, each	11	20	30
Wile E. Coyote Doll	Mighty Star	1971	18" tall, plush	9	16	25
Wile E. Coyote Doll	Dakin	1970	on explosive box	15	30	45
Wile E. Coyote Figure	Dakin	1976	Cartoon Theater	9	16	25
Wile E. Coyote Figure	Dakin	1971	Goofy Gram, fused bomb in right hand	14	25	40
Wile E. Coyote Figure	Dakin	1968	10" tall	12	23	35
Wile E. Coyote Hand Puppet	Japanese	1970s	10" vinyl head	4	7	10
Wile E. Coyote Night Light	Applause	1980s		12	23	35

ROCKY

TOY	COMPANY	YEAR	DESCRIPTION	GOOD	EX	MIB
Apollo Creed Doll	Phoenix Toys	1983	8" tall	4	5	10
Clubber Lang Doll	Phoenix Toys	1983	8" tall	5	10	15
Rocky Balboa Doll	Phoenix Toys	1983	8" tall	5	10	15

ROCKY & BULLWINKLE

TOY	COMPANY	YEAR	DESCRIPTION	GOOD	EX	MIB
Bullwinkle & Rocky Clock Bank	Larami	1969	4-1/2" tall, plastic	25	50	80
Bullwinkle & Rocky Wastebasket		1961	11" tall, metal with Jay Ward cast pictured	35	65	135
Bullwinkle Bank		1960s	6" tall, glazed china	60	115	360
Bullwinkle Bank	Play Pal	1972	11-1/2", plastic	35	50	90
Bullwinkle Cartoon Kit	Colorforms	1962		35	50	150
Bullwinkle Dinner Set	Boonton Molding	1960s	plate and cup pictures Bullwinkle and the Cheerios Kid	25	45	75
Bullwinkle Figure	Dakin	1976	Cartoon Theater, 7-1/2" tall, plastic	25	50	90
Bullwinkle Flexy Figure	Larami	1970		10	20	50
Bullwinkle for President Bumper Sticker		1972		10	16	30
Bullwinkle Jewelry Hanger		1960s	5" tall, suction cup on back	12	23	40
Bullwinkle Magic Slate		1963		15	30	50
Bullwinkle Make Your Own Badge Set	Larami	1960s		20	35	50

ROCKY & BULLWINKLE

TOY	COMPANY	YEAR	DESCRIPTION	GOOD	EX	MIB
Bullwinkle Paintless Paint Book	Whitman	1960		15	30	50
Bullwinkle Spell & Count Board		1969		9	16	35
Bullwinkle Stamp Set	Larami	1970		11	20	35
Bullwinkle Stickers		1984	Bullwinkle, Sherman and Peabody, Snidely Whiplash	5	10	20
Bullwinkle Talking Doll	Mattel	1970		25	40	120
Bullwinkle Travel Adventure Board Game	Transogram	1960s		30	55	90
Bullwinkle Travel Game	Larami	1971	magnetic	15	30	50
Bullwinkle's Circus Time Toy		1969	Bullwinkle on a elephant	20	35	60
Bullwinkle's Circus Time Toy		1969	Rocky on a circus horse	20	35	60
Bullwinkle's Double Boomerangs	Larami	1969	set of two on illustrated card	12	23	40
Dudley Do-Right Figure	Dakin	1976	Cartoon Theater	15	30	50
Dudley Do-Right Figure	Wham-O	1972	5" tall, flexible	10	20	50
Dudley Do-Right Puzzle	Whitman	1975		10	15	30
Mr. Peabody Bank		1960s	6" tall, glazed china	80	145	300
Mr. Peabody Figure	Wham-O	1972	4" tall, flexible	10	20	40
Natasha Figure	Wham-O	1972		10	20	40
Rocky & Bullwinkle Bank		1960	5" tall, glazed china	80	145	290
Rocky & Bullwinkle Coloring Book	Watkins-Strathmore	1962		15	30	50
Rocky & Bullwinkle Movie Viewer		1960s	#225, red and white plastic viewer with three movies	23	45	70
Rocky & Bullwinkle Presto Sparkle Painting Set	Kenner	1962	six cartoon pictures & two comic strip panels	25	55	100
Rocky & Bullwinkle Puzzle	Whitman	1972	boxed	15	30	60
Rocky & Bullwinkle Toothpaste Holder		1960s	glazed china	70	175	365
Rocky & His Friends Book	Whitman	1960s	Little Golden Book	9	16	30
Rocky Bank		1950s	5" tall, slot in large tail, glazed china	60	115	275
Rocky Figure	Dakin	1976	Cartoon Theater, 6-1/2" tall, plastic	20	50	100
Rocky Flexy Figure	Larami	1970		10	20	45
Rocky Soaky			10-1/2" tall, plastic	12	23	40
Rocky the Flying Squirrel Coloring Book	Whitman	1960		15	30	50
Sherman Figure	Wham-O	1972	4" tall, flexible	10	25	50
Snidely Whiplash Figure	Wham-O	1972	5" tall, flexible	10	20	40

ROGER RABBIT

TOY	COMPANY	YEAR	DESCRIPTION	GOOD	EX	MIB
Animates		1988	Doom, Roger, Eddie & Smart Guy, each	4	7	10
Baby Herman & Roger Rabbit Mug	Applause			4	7	10
Baby Herman Figure	LJN	1988	ceramic	7	13	20
Baby Herman Figure	LJN	1987	6" figure on card	9	16	25
Benny the Cab	LJN			20	40	60
Benny the Cab Doll	Applause	1988	6" long	5	10	15
Boss Weasel Figure	LJN	1988	4" bendable	3	5	8
Deluxe Color Activity Book	Golden		#5523	4	7	10
Dip Flip Game	LJN			7	13	20
Eddie Valiant Animate	LJN	1988	6" tall, bendable	3	5	8
Eddie Valiant Figure	LJN	1988	4", flexible	3	5	8
Jessica License Plate				5	10	15
Jessica Zipper Pull				5	10	15
Judge Doom Animate	LJN	1988	6" tall poseable	4	7	10

ROGER RABBIT

TOY	COMPANY	YEAR	DESCRIPTION	GOOD	EX	MIB
Judge Doom Figure	LJN	1988	4" bendable	3	5	8
Paint with Water Book	Golden		#1702	5	10	15
Paint-A-Cel Set			Benny the Cab and Roger pictures	5	10	15
Roger Rabbit Animate	LJN	1988	6" poseable	4	7	10
Roger Rabbit Blow-Up Buddy			36" tall	5	10	15
Roger Rabbit Bullet Hole Wristwatch	Shiraka	1987	white case and leather band, in plastic display box	20	35	55
Roger Rabbit Doll	Applause		17" tall	5	10	15
Roger Rabbit Doll	Applause		8-1/2" tall	4	7	10
Roger Rabbit Figure	LJN	1988	4" bendable	3	5	8
Roger Rabbit Silhouette Wristwatch	Shiraka	1987	large gold case, black band	18	35	50
Roger Wacky Head Puppets	Applause		hand puppets	4	7	10
Roger Wind-Up	Matsudaya	1988		18	35	50
Smart Guy Figure	LJN	1988	flexible	3	5	8
Talking Roger in Benny the Cab			17" tall	11	20	30
Trace & Color Book	Golden		#2355	4	7	10
View-Master Gift Set	View-Master			7	13	20

ROOKIES, THE

TOY	COMPANY	YEAR	DESCRIPTION	GOOD	EX	MIB
Rookie Chris Doll	LJN	1973	8" tall	10	20	35
Rookie Mike Doll	LJN	1973	8" tall	10	20	35
Rookie Terry Doll	LJN	1973	8" tall	10	20	35
Rookie Willy Doll	LJN	1973	8" tall	10	20	35

ROOTIE KAZOOTIE

TOY	COMPANY	YEAR	DESCRIPTION	GOOD	EX	MIB
Rootie Kazootie Club Button		1950s	1" tin litho	15	40	50
Rootie Kazootie Drum		1950s	8" diameter drum, Rootie on the drum head	20	45	65
Rootie Kazootie Fishing Tackle Box	RK Inc.	1950s	tin litho, bright graphics	15	25	40

RUFF & REDDY

TOY	COMPANY	YEAR	DESCRIPTION	GOOD	EX	MIB
Ruff & Reddy Draw Cartoon Set Color	Wonder Art			45	80	125
Ruff & Reddy Go To A Party Tell-A-Tale Book	Whitman	1958		15	30	45
Ruff & Reddy Magic Rub-Off Picture Set	Transogram	1958		45	80	125

SCOOBY DOO

TOY	COMPANY	YEAR	DESCRIPTION	GOOD	EX	MIB
Scooby Doo and the Pirate Treasure Book	Golden	1974	Little Golden Book	5	10	15
Scooby Doo Hand Puppet	Ideal	1970s	vinyl head	20	35	55
Scooby Doo Paint with Water Book		1984		5	10	15
Scooby Doo Squeak Toy	Sanitoy	1970s	6" tall	20	35	40

SECRET SQUIRREL

TOY	COMPANY	YEAR	DESCRIPTION	GOOD	EX	MIB
Secret Squirrel Bubble Club Soaky	Purex	1960s		15	25	55

Mickey Mouse steamboat, 1988, Masudaya

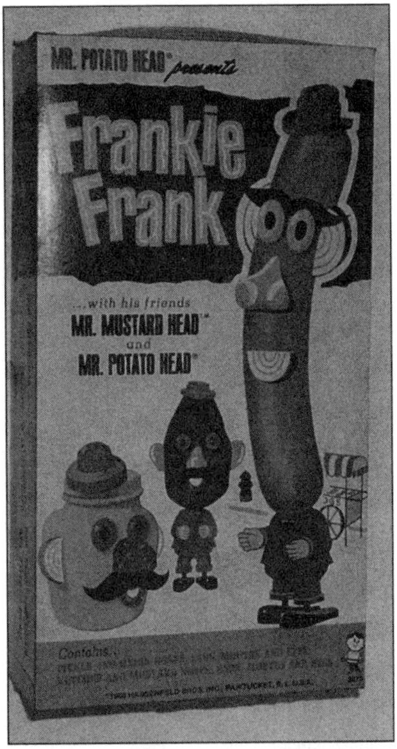

Mr. Potato Head, Frankie Frank, 1966, Hasbro

Superman Scrap Book, 1940, Saalfield

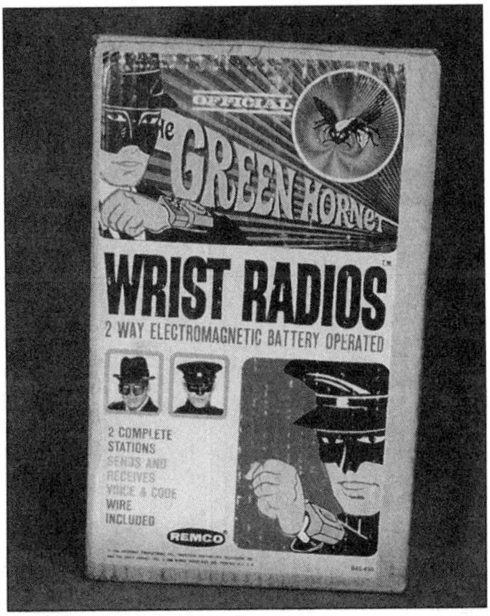

Green Hornet Wrist Radios, 1966, Remco

SECRET SQUIRREL

TOY	COMPANY	YEAR	DESCRIPTION	GOOD	EX	MIB
Secret Squirrel Push Puppet	Kohner	1960s	plastic figure in white coat, blue hat holding binoculars	15	35	55
Secret Squirrel Puzzle		1967	frame tray	14	25	40
Secret Squirrel Ray Gun				14	25	40

SHIRLEY TEMPLE

TOY	COMPANY	YEAR	DESCRIPTION	GOOD	EX	MIB
Baby Take A Bow Dress		1934	red cotton, red polka dots on collar and reverse on the bottom with silk ribbons, fits 16" doll	85	95	100
Ballerina Outfit		1957-58	blue/green nylon and tulle, flower hair piece; fits 12" vinyl doll	80	85	100
Blue Bird Doll		1939	18", felt skirt and vest, organdy blouse and apron with blue bird appliques	900	950	1000
Blue Bird Doll		1939	20" composition	800	900	1000
Bridge Cards	U.S. Playing Card	1934		35	40	45
Bright Eyes Doll		1936	13" composition, plaid dress, leather shoes	600	650	700
Bright Eyes Dress		1934	fits 16" dolls, several colors variations	85	90	100
Bright Eyes Outfit			white corduroy coat and hat with original pin, fits 16" doll	100	150	200
Captain January Doll		1936	16", green pique with silk ribbons	650	700	725
Captain January Doll	Ideal		22", dark blue sailor suit with white trim	800	850	900
Captain January Doll	Ideal	1957	12", vinyl head, dark blonde hair, plastic body, arms and legs	175	200	250
Captain January Doll	Ideal	1936	20", blue floral print, cotton school dress with ruffled collar	700	750	800
Captain January Doll		1936	13", white sailor suit, red bow tie, white hat, original pin	600	650	700
Captain January Doll		1936	18", cotton sailor suit with anchor applique and red silk tie	650	700	800
Captain January Outfit		1936	red or green pleated pique, silk ribbons; fits a 16" doll	85	95	100
Coat and Hat	Ideal	1958	red corduroy coat and hat, fits 12" doll	60	70	75
Coat and Hat			velveteen coat and hat with red buttons; fits 16" doll	80	90	100
Composition Book	Western	1935		25	30	35
Curly Top Doll		1935	11", dotted swiss dress, blue silk ribbons	600	650	700
Curly Top Doll		1935	16" composition; different versions available	650	700	750
Curly Top Doll		1935	18" composition, mohair hair, hazel eyes	600	650	750
Curly Top Doll		1935	18", pleated dotted swiss with blue silk ribbons, red or lavender	600	650	750
Curly Top Doll	Ideal	1935	11", pink organdy dress	800	900	1000
Curly Top Dress		1935	striped cotton dress, fits 20" doll	85	95	100
Dimples Outfit		1936	heavy felt jacket with red trim; fits 16" doll	85	95	100
Doll Buggy	F.A. Whitney Carriage	1935	wicker, hubcaps are labeled Shirley Temple	300	350	400
Doll Trunk			18" tall with "Our Little Girl" decal on the side, photo of Shirley	100	150	200

SHIRLEY TEMPLE

TOY	COMPANY	YEAR	DESCRIPTION	GOOD	EX	MIB
Doll Wardrobe Trunk			20" heavy cardboard/wood steamer trunk for 18" doll, leather strap, metal latch with lock & key, two drawers, and four cardboard hangers	100	150	200
Dress	Ideal	1958	nylon with loop details; fits 12" doll	60	70	75
Dress			light blue organdy; fits 16" doll	85	95	100
Dress, Jacket, and Purse		1959	nylon	65	70	75
Hair Ribbon and Band	Ribbon Mills	1934	Several colors/styles	15	20	35
Hanger			cardboard; picture of Shirley Temple, came with all outfits	15	20	30
Heidi Doll		1937	18", striped cotton skirt, black velveteen top with red braid, apron, mohair pigtails with red ribbons	950	1000	1200
Heidi Doll		1960-61	17"	250	300	350
Heidi Doll	Ideal	1957	12", vinyl head, dark blonde hair and plastic body, arms and legs	175	200	250
Jumper and Blouse	Ideal	1959	blue velveteen jumper with floral applique, cotton blouse	60	70	75
Jumpsuit			red/white checkered	35	40	50
Jumpsuit			red with white flowers	80	85	100
Little Colonel Doll	Alexander		13", pink hat with ruffle, pink dress and bloomers	450	500	600
Little Colonel Doll	Ideal	1934	13", smiling, pink hat with feather and dress and bloomers	650	750	850
Little Colonel Doll			13", variations in pantaloons, bonnets and shoes	750	800	850
Little Colonel Doll			13", taffeta outfit, variations in the collar, ruffles and pantaloons	750	800	850
Little Colonel Doll			20", blue organdy, bonnet with pink feather	900	950	1000
Little Colonel Doll		1934	13", light blue dress with bonnet, pantaloons and shoes with buckles	750	800	850
Little Colonel Doll		1934	18", pink organdy dress, bonnet, white pantaloons	750	800	850
Littlest Rebel Doll		1935	13", yellow/brown dress, white ruffled collar, yellow shoes	750	800	850
Littlest Rebel Doll		1935	16", polka dot outfit, pantaloons, gray felt hat	600	650	675
Littlest Rebel Doll		1935	18", cotton and organdy outfit	650	700	750
Littlest Rebel Doll		1935	22", red/white cotton dress with organdy collar	750	800	850
Littlest Rebel Dress		1935	yellow/brown, fits an 18" doll	125	150	200
Littlest Rebel Outfit			checkered dress, rick rack ribbon on sleeves, lace collar and apron	85	95	100
My Life and Times by Shirley Temple Book	Saalfield	1936	Big Little Book #116	65	70	75
Nightcoat and Cap	Ideal	1958-59	flannel; fits 12" doll	60	70	75
Our Little Girl Outfit		1935	blue or white pique, white dog appliques, fits 16" doll	85	95	100
Our Little Girl Outfit		1935	red or blue dress with music appliques, matching hat, fits 16" doll	85	95	100
Paper Doll Book	Whitman	1976	#1986	8	10	20
Paper Dolls	Saalfield	1934	four 8" dolls and 30 outfits; first licensed set, #2112	50	100	250
Paper Dolls	Saalfield	1958	#4435	35	40	75

SHIRLEY TEMPLE

TOY	COMPANY	YEAR	DESCRIPTION	GOOD	EX	MIB
Paper Dolls	Saalfield	1959	18" folding doll with easel, costumes, and accessories, #5110	7	20	45
Pen and Pencil Set	David Kahn	1930s		85	100	150
Pin	Reliable		"The World's Darling/Genuine Shirley Temple, A Reliable Doll"	60	70	80
Polka Dot Dancing Dress		1935	blue organdy with matching hat and sunsuit	85	90	100
Poor Little Rich Girl Doll		1936	13", blue silk pajamas	300	650	800
Poor Little Rich Girl Doll		1936	18", pique sunsuit with matching tam	250	600	800
Poor Little Rich Girl Doll		1936	13", sailor dress with white trim and matching tam	250	350	650
Poor Little Rich Girl Outfit		1935	pique and organdy dress, fits 16" doll	85	95	100
Poor Little Rich Girl Outfit		1936	pleated red plaid with white collar, fits 16" doll	85	95	100
Poor Little Rich Girl Outfit		1935	blue sailor dress, fits 16" doll	85	95	100
Promo Photo			8" x 10" photo of Shirley with facsimile autograph, came with all composition dolls and outfits	25	30	35
Purse		1950s	red, white, and black with Shirley Temple lettering	25	30	35
Rain Cape and Umbrella			plaid red or blue rain cape with hood and matching umbrella; fits 18" doll	80	100	125
Rebecca of Sunnybrook Farm Doll		1957	12" vinyl, blue bib overalls, blue polka dot blouse, straw hat	100	200	300
Rebecca of Sunnybrook Farm Doll		1957	12" vinyl, red felt jumper and plastic purse	150	200	300
Rebecca of Sunnybrook Farm Doll		1957	12" vinyl, blue bib overalls with plaid blouse, black low shoes	150	200	200
Rebecca of Sunnybrook Farm Doll	Ideal	1957	12", vinyl head, dark blonde hair, plastic body, arms and legs	175	200	300
Scrap Book	Saalfield	1935	#1714	60	65	75
Shirley Standing Doll	Saalfield	1935	cardboard doll on platform and different outfits, #1719	30	65	150
Shirley Temple A Great Big Book to Color	Saalfield	1936	#1717	35	40	45
Shirley Temple at Play Book	Saalfield	1935	#1712	40	45	55
Shirley Temple Baby	Ideal		18", painted hair, chubby toddler body with dimpled cheeks, flirty eyes, dressed in pink organdy with silk ribbons	500	750	1000
Shirley Temple Baby			20", composition head, arms and legs, cloth body	650	800	1100
Shirley Temple Baby			16", mohair wig, flirty eyes	500	700	1000
Shirley Temple Doll	Ideal	1958-59	17" vinyl, brown twinkle eyes, pink/blue dress	200	300	400
Shirley Temple Doll			27" composition with pink taffeta, bonnet and pantaloons	800	1100	1500
Shirley Temple Doll			22" composition with facial molding	550	750	850
Shirley Temple Doll			20" composition with facial molding	450	700	800
Shirley Temple Doll			18" composition with facial molding	500	650	750
Shirley Temple Doll			16", black velveteen coat and hat	175	300	750

SHIRLEY TEMPLE

TOY	COMPANY	YEAR	DESCRIPTION	GOOD	EX	MIB
Shirley Temple Doll			16", light blue organdy with pink hemstitching and silk ribbons	250	500	700
Shirley Temple Doll			16" composition	350	600	700
Shirley Temple Doll			13", pleated pique with white applique	250	500	650
Shirley Temple Doll	Reliable (Ideal/ Canada)		13", yellow organdy dress, silk ribbon	450	750	900
Shirley Temple Doll	Ideal/Hong Kong	1972	16", vinyl, red polka dot dress, Stand Up and Cheer	150	180	200
Shirley Temple Doll	Ideal/Hong Kong	1972	15" vinyl manufactured for Montgomery Ward's	185	200	225
Shirley Temple Doll	Ideal		22" jointed, composition body, blonde mohair, hazel glass eyes, open mouth, red/white polka dot dress	200	400	850
Shirley Temple Doll	Ideal	1960s	17", yellow nylon dress and Twinkle Eyes wrist tag	200	300	400
Shirley Temple Doll	Made in Japan		8" composition with pink silk undies	175	200	275
Shirley Temple Doll	Ideal	1958	15" vinyl, yellow nylon dress trimmed with lace and ribbon around skirt	150	300	400
Shirley Temple Doll	Ideal	1934	18", pleated pale green, pink, or blue dress, embroidered collar and pink silk ribbon	500	600	800
Shirley Temple Doll	Ideal	1934	18", light pink organdy	450	600	800
Shirley Temple Doll	Blossom		18" cloth with Little Colonel dress	200	300	400
Shirley Temple Doll		1960s	17", yellow party dress, white purse with Shirley Temple lettering	200	300	400
Shirley Temple Doll		1960	15" vinyl, blue jumper with red/ white gingham blouse and pocket facing	100	200	300
Shirley Temple Doll		1958	15" vinyl, red nylon dress with floral detailing at collar, hair ribbon, purse	100	200	300
Shirley Temple Doll		1957	12", vinyl, molded hands and feet, synthetic rooted wig, pink slip trimmed with lace	100	200	300
Shirley Temple Doll		1957	12" vinyl, molded hands and feet, synthetic rooted wig, two piece slip/undies	100	200	300
Shirley Temple Doll		1957	12" vinyl, pink/blue nylon dress with daisy appliques, hat, purse	100	200	350
Shirley Temple Doll		1935	18" facial molding doll with mohair wig (parted in the center), light complexion, outfit from "Curly Top"	250	450	900
Shirley Temple Doll	Ideal		19" vinyl, twinkle eyes, dressed in pink nylon, black purse	175	300	400
Shirley Temple Dolls and Dresses	Saalfield	1960	Two dolls with different outfits, #1789	10	15	30
Shirley Temple Five Books About Me	Saalfield	1936	Just a Little Girl, Twinkletoes, On the Movie Lot, In Starring Roles, and Little Playmate, #1730	100	110	125
Shirley Temple in The Littlest Rebel	Saalfield	1935	Big Little Book, #1595	15	20	30
Shirley Temple My Book to Color	Saalfield	1937	#1768	30	35	45

SHIRLEY TEMPLE

TOY	COMPANY	YEAR	DESCRIPTION	GOOD	EX	MIB
Shirley Temple Pastime Box	Saalfield	1937	Four activity books: Favorite Puzzles, Favorite Games, Favorite Sewing Cards, and Favorite Coloring Book, #1732	50	100	125
Shirley Temple Scrap Book	Saalfield	1936	#1722	55	65	75
Shirley Temple Scrap Book	Saalfield	1937	#1763	65	70	75
Shirley Temple Story Book	Saalfield	1935	#1726	20	25	35
Shirley Temple Treasury Book	Random House	1959		20	25	35
Shirley Temple with Lionel Barrymore in The Little Colonel	Saalfield	1935	Big Little Book, #1095	15	20	25
Shirley Temple's Blue Bird Coloring Book	Saalfield	1939		30	35	45
Shirley Temple's Busy Book	Saalfield	1958	Activity book, #5326	40	45	55
Shirley Temple's Favorite Poems	Saalfield	1936	#1720	15	25	30
Stand Up and Cheer Doll		1934	11", dotted red, blue, or green organdy dress with silk ribbon	400	600	750
Stand Up and Cheer Doll		1934	13" composition, dotted organdy green dress	100	150	200
Stand Up and Cheer Doll			11" composition, short rayon dress with blue polka dots	650	850	900
Stand Up and Cheer Doll Trunk		1934		100	150	200
Stowaway Doll		1936	20" , two-piece linen with brass buttons	550	700	900
Stowaway Doll		1936	25", pink taffeta, mohair hair	600	800	1000
Stowaway Outfit		1936	red or blue pique, fits 16" doll	85	95	100
Texas Ranger Doll			17", plaid shirt, leather vest with trim	500	600	850
Texas Ranger Doll			20", plaid shirt, leather vest with trim, chaps, holster, and metal gun	600	800	1100
Texas Ranger/Cowgirl Doll			11", plaid cotton shirt, leather vest, chaps, holster, metal gun, felt 10 gallon hat with "Ride 'Em Cowboy" printed band	350	600	950
Texas Ranger/Cowgirl Gun			Came with Texas Ranger outfit	30	40	65
The Story of Shirley Temple Book	Saalfield	1934	Big Little Book, #1089	15	20	30
This Is My Crayon Book	Saalfield	1936	#1711	45	50	55
Wee Willie Winkie Doll		1937	18", long sleeve cotton jacket, two pockets and six brass buttons, plaid wool skirt, tan belt with brass buckle	600	900	1000
Wee Willie Winkie Doll	Ideal	1957	12", vinyl head, dark blonde hair, plastic body, arms and legs	175	200	250
Wee Willie Winkie Outfit		1937	pique outfit and tam, slip/undies; fits 16" doll	100	125	150
Wool Coat			fits 18" dolls, from Little Miss Marker	50	75	100

SLEEPING BEAUTY

TOY	COMPANY	YEAR	DESCRIPTION	GOOD	EX	MIB
Fairy Godmother Hand Puppets		1958	set of three: 10-1/2" tall, Flora, Merryweather, & Fauna, each	50	70	200

SLEEPING BEAUTY

TOY	COMPANY	YEAR	DESCRIPTION	GOOD	EX	MIB
King Huber/King Stefan Hand Puppets	Gund	1956	10" tall, molded rubber heads with fabric hand cover	25	50	125
Puzzle	Whitman	1958	11-1/2" x 14-1/2", Sleeping Beauty with forest animals	15	30	45
Puzzle	Whitman	1958	11-1/2" x 14-1/2", Sleeping Beauty with Prince Phillip and Three Good Fairies circling	15	30	45
Puzzle	Whitman	1958	11-1/2" x 14-1/2", Three Good Fairies circling around a baby in a crib	15	25	40
Sleeping Beauty Alarm Clock	Phinney-Walker	1950s	2-1/2" x 4" x 4-1/2" tall, Sleeping Beauty surrounded by three birds and petting a rabbit	50	75	125
Sleeping Beauty Doll Crib Mattress		1960s	9" x 17", Sleeping Beauty and the fairies	12	23	35
Sleeping Beauty Jack-In-The-Box	Enesco	1980s	Princess Aurora, wooden box, plays "Once Upon A Dream"	35	60	135
Sleeping Beauty Magic Paint Set	Whitman			25	45	65
Sleeping Beauty Squeeze Toy	Dell	1959	4" x 4" x 5" tall, rubber, Sleeping Beauty with rabbit	25	45	65
Sleeping Beauty Sticker Fun Book	Whitman	1959		12	25	35

SMOKEY BEAR

TOY	COMPANY	YEAR	DESCRIPTION	GOOD	EX	MIB
Smokey Bank			6" tall, china	12	23	85
Smokey Bear and the Campers Book	Golden	1961		7	15	20
Smokey Bear Coloring Book	Whitman	1958		7	15	25
Smokey Bear Record	Peter Pan		45 rpm	2	5	8
Smokey Bobbing Head Figure		1960s	6-1/4" tall	15	45	175
Smokey Doll	Ideal	1950s	15" plush, vinyl face	40	65	125
Smokey Figure	Dakin	1971	figure on a tree stump	25	50	125
Smokey Soaky		1960s	9" tall, plastic	5	10	25
Smokey Wristwatch	Hawthorne	1960s		35	90	175

SNOW WHITE & THE SEVEN DWARFS

TOY	COMPANY	YEAR	DESCRIPTION	GOOD	EX	MIB
Baby Rattle	Krueger	1938	Snow White at piano, Dwarfs playing instruments	130	245	375
Bashful Doll	Ideal	1930s		55	100	200
Dime Register Bank	Disney	1938	holds up to five dollars	50	125	250
Doc Doll	Ideal	1930s		55	100	200
Doc Lamp	LaMode Studios	1938	8" tall, plaster	45	80	175
Dopey Bank	Crown Toy	1938	7-1/2" tall, wood composition	55	100	175
Dopey Dime Register Bank	Disney	1938	holds up to $5	45	100	225
Dopey Doll	Knickerbocker	1938	11" tall composition	90	165	375
Dopey Doll	Krueger		14" tall	90	165	250
Dopey Doll	Chad Valley	1938	cloth body	50	95	195
Dopey Doll	Ideal	1930s		55	100	225
Dopey Figure		1960s	ceramic figure and barrel	15	30	75
Dopey Lamp		1940s	9" tall, ceramic base with Dopey	35	65	175
Dopey Rolykin	Marx		2"	30	50	100
Dopey Soaky		1960s	10" tall	11	30	40
Dopey Ventriloquist Doll	Ideal	1938	18" tall	115	210	500
Dopey Walker	Marx	1938	9" tall, tin, rocking walker	275	525	800

SNOW WHITE & THE SEVEN DWARFS

TOY	COMPANY	YEAR	DESCRIPTION	GOOD	EX	MIB
Grumpy Doll	Knickerbocker	1938	11" tall, composition	90	165	350
Happy Doll		1930s	5-1/2" tall, composition, holding a silver pick with a black handle	35	65	175
Happy Toy	YS Toys (Taiwan)		battery operated, Happy fries eggs	40	75	225
Ironing Board	Wolverine		tin board and cover	15	25	40
Pencil Box	Venus Pencil		3" x 8" x 1"	60	115	175
Puzzle	Jaymar	1960s	11" x 14"	18	35	50
Puzzles	Whitman	1938	set of two	50	100	175
Radio	Emerson	1938	8" x 8" with characters on cabinet	325	800	2000
Refrigerator	Wolverine	1970s	15", tin, single door, white and yellow depicting Snow White	15	25	85
Safety Blocks	Halsam	1938	7-1/2" x 14-1/2"	35	65	100
Sand Pail	Ohio Art	1938	8", tin, Snow White plays hide-n-seek with the dwarfs	25	60	200
Seven Dwarfs Figures	Seiberling	1938	5-1/2", rubber	115	210	450
Seven Dwarfs Target Game	Chad Valley	1930s	6-1/2" x 11-1/2" target, spring locked gun	115	210	325
Sled	S. L. Allen	1938	40" long wood slat and metal runner sled with character decals	140	260	400
Sneezy Doll	Krueger		14" tall	115	210	325
Snow White & the Seven Dwarfs Book	Whitman	1938	Big Little Book	15	40	60
Snow White and the Seven Dwarfs Dolls	Deluxe	1940s	22" Snow White and 7" dwarfs	250	475	1200
Snow White Doll	Horsman		8", in illustrated box	12	23	75
Snow White Doll	Knickerbocker	1939	3" x 7" x 3-1/2", composition with movable arms and legs	60	115	300
Snow White Doll	Ideal	1938	3-1/2" x 6-1/2" x 16" tall, fabric face and arms, red/white dress with dwarf and forest animal design	245	450	750
Snow White Doll	Knickerbocker	1940	12" tall, composition	80	145	400
Snow White Lamp	LaMode Studios	1938	8-1/2" tall	35	65	150
Snow White Marionette	Tony Sarg/ Alexander	1930s	12-1/2" tall	60	115	225
Snow White Mirror		1940s	9-1/2", plastic handle	23	45	65
Snow White Model Making Set	Sculptorcraft	1930s		60	115	175
Snow White Paper Dolls	Whitman	1938	10" x 15" x 1-1/2"	80	145	225
Snow White Sewing Set	Ontex	1940s		25	50	75
Snow White Sink	Wolverine	1960s	tin	15	25	45
Snow White Soaky				12	35	50
Snow White Table Quoits	Chad Valley	1930s	9-1/2" x 21" x 1-1/4" deep	115	210	325
Tea Set	Marx	1960s	teapot, five saucers, large plates and tea cups	35	65	145
Tea Set	Ohio Art	1937	plates, cups, tray, saucers	45	120	250
Tea Set	Wadeheath	1930s	white china, teapot, cups, saucers, and creamer	115	210	325

SPORTS

Dorothy Hamill Doll	Ideal	1975	11-1/4" tall	25	50	75
Evel Knievel Doll	Ideal		6" tall	15	30	50
Julius Erving (Dr. J) Doll		1974		20	40	60
Muhammed Ali Wristwatch	Bradley	1980	chrome case, sweep seconds, brown leather band, face shows Ali in trunks and gloves, with signature beneath	25	50	75

SPORTS

TOY	COMPANY	YEAR	DESCRIPTION	GOOD	EX	MIB
O.J. Simpson Doll		1974		30	75	250
Wayne Gretzky Doll	Mattel		12" tall	25	50	75

STEVE CANYON

TOY	COMPANY	YEAR	DESCRIPTION	GOOD	EX	MIB
Steve Canyon Costume	Halco	1959		25	50	150
Steve Canyon's Interceptor Station Punch Out	Golden	1950s		30	75	150
Steve Canyon's Membership Card & Badge			1/2" x 4" Milton Caniff membership card for the Airagers, Morse code on back, 3" tin litho color badge with gold feathers with Steve's face centered	70	130	200

SUPERHEROES

TOY	COMPANY	YEAR	DESCRIPTION	GOOD	EX	MIB
Amazing Spider-Car	Mego	1976	10" long, red plastic	20	35	50
Aquaman Halloween Costume	Ben Cooper	1967		20	55	80
Aquaman Jigsaw Puzzle	Whitman	1968	100 pieces; Aquaman and Mera	15	35	50
Aquaman Scourge of the Sea Book	Whitman	1968	Big Little Book	10	20	35
Captain America Bendy Figure	Lakeside	1966	6", rubber	45	80	125
Captain America Halloween Costume	Ben Cooper	1967		50	100	200
Captain America Rocket Racer	Buddy L	1984	Secret Wars, remote control battery-operated car	15	25	75
Captain America Scooter	Marx	1967	4", yellow plastic friction toy with figure	75	200	300
Comic Book Tattoos	Topps	1967	Aquaman, Wonder Woman, Superman, or Batman	15	25	35
Flash Glass	Pepsi	1978		15	35	50
Green Lantern Halloween Costume	Ben Cooper	1967		75	150	250
Hawkman Button	Button World	1966	3", "Hawkman Superhero Club"	15	25	30
Justice League of America Display Card	Fleer	1970	cardboard display from inside gumball machine	25	40	60
Marvel Superheroes Card Game	Milton Bradley	1978		15	20	35
Marvel Superheroes Colorforms Set	Colorforms	1983		7	15	25
Marvel Superheroes Easy Show Projector	Kenner	1967	projector and three cartridges	50	75	175
Marvel Superheroes Puzzle	Milton Bradley	1967	100 pieces	35	65	150
Marvel Superheroes Sparkle Paint Set	Kenner	1967		50	75	150
Marvel World Adventure Play Set	Amsco	1975	with stand-up scenes and figures	50	75	150
Mr. Bubble Superfriends Box		1984	bubble bath box features Superman, Wonder Woman, Batman, and Robin	15	25	50
Spider-Man Button	Button World	1966	3", "Superhero Club"	15	35	50
Spider-Man Coloring Book	Marvel Books	1983	oversized, "The Arms of Doctor Octopus"	7	15	25
Spider-Man Crazy Foam	American Aerosol	1974		10	30	45
Spider-Man Friction Vehicle	Marx	1968	tin litho w/Spider-Man figure at wheel	75	150	350
Spider-Man Hand Puppet	Imperial	1976	9", vinyl head, plastic body	10	25	30
Wonder Woman Glass	Pepsi	1978	6"	10	25	35

SUPERHEROES

TOY	COMPANY	YEAR	DESCRIPTION	GOOD	EX	MIB
Wonder Woman Record	Peter Pan Records	1977	33-1/3" rpm record w/comic book	10	25	30

SUPERMAN

TOY	COMPANY	YEAR	DESCRIPTION	GOOD	EX	MIB
3-D Adventures of Superman Comic Book	DC Comics	1950s	with 3-D goggles	90	250	650
Action Comics #1	DC Comics	1938	first appearance of Superman	30000	80000	100000
Adventures of Superman Book	Random House	1942	6-1/2" x 9" hardcover, 220 pages, author George Lowther, full color dust jacket	500	750	1200
Adventures of Superman Book	Random House	1942	6-1/2" x 9" hardcover by George Lowther, no dust jacket	150	200	250
Adventures of Superman Book	Random House	1942	4" x 5-1/2" armed services edition	100	200	400
Bicycle Siren	Empire	1970s		10	20	30
Book and Record Set	Musette Records	1947	The Magic Ring	75	125	200
Book and Record Set	Musette Records	1947	The Flying Train	75	100	150
Book, With Superman at the Gilbert Hall of Science	Gilbert	1948	32-page promo catalog for Gilbert's Erector Sets and other toys, illustrated with Superman	50	100	200
Bubble Gum Badge	Fo-Lee Gum Corp.	1948	shield-shaped brass finish badge shows a variant of the chain breaking pose inside a sun burst pattern ringed with stars	1000	3500	5000
Children's Dish Set	Boontonware	1966	7" plate, 5-1/2" bowl, 3-1/2" cup, all white plastic with Superman image	25	50	75
Cinematic Picture Pistol	Daisy	1940	non-electric, film is viewed through view in back of gun, metal gun with one pre-loaded 28 scene Superman film	300	750	1500
Crayon-by-Numbers Set	Transogram	1954	16 crayons & 44 action scenes	65	90	125
Daily Planet Jetcopter	Corgi	1979		20	40	60
Dangle Dandies Mobile	Kellogg's	1955	set of eight cut outs on boxes of Rice Krispies & Corn Flakes	75	100	150
Energized Superman Figure	Remco	1979	12" tall battery operated hard body figure, in box	35	65	100
Fan Card	National Comics	1950s	5" x 7" promo b/w post card with signature "Best Wishes, George Reeves"	65	80	100
Fan Card		1942	7" x 10", shows full color Superman in hands on hips pose, reads, "Best wishes from your friend Superman"	250	500	700
Film Viewer		1947	1-1/2" x 4" x 6-1/2" wide boxed set of hand-held viewer and six films	150	200	350
Film Viewer		1965	plastic hand viewer with two boxes of film, on illustrated card	25	40	65
Film Viewer	Acme	1947	1-1/2" x 6-1/2" wide boxed set of hand-held viewer & two films	70	130	200
Flying Noise Balloon	Van Dam	1966	oversized balloon makes noise in flight, Superman illustration on balloon and card	12	23	35
Flying Superman	Kellogg's	1950s	plastic premium, 5" x 6-1/2", toy only	35	100	150

CHARACTER

SUPERMAN

TOY	COMPANY	YEAR	DESCRIPTION	GOOD	EX	MIB
Flying Superman	Kellogg's	1950s	premium, 5" x 6 1/2", rubber band propelled, with instruction sheet & mailer	50	150	250
Flying Superman	Transogram	1955	12-1/2" tall molded plastic figure propelled by "super flight launcher", a rubber band attached to a pistol grip holder, on illustrated	55	100	150
Hair Brush	Avon	1976	Superman handle, illustrated box	15	25	40
Hair Brush	Monarch	1940	wood, full length decal of Superman, brush came in handle and no-handle styles with box	50	150	300
Jumbo Movie Viewer	Acme	1950s	blue/yellow plastic viewer with 35 mm "theatre size" film, on illustrated card, with one filmstrip	90	165	250
Junior Defense League of America Membership Certificate	Superman, Inc.	1940s	Superman Bread premium, red/blue print and Superman bust and logo on paper, "signed" by Clark Kent	250	500	750
Krypto-Raygun Film Viewer	Daisy	1940	battery-operated metal projector gun, seven Superman filmstrips, illustrated box	500	1000	2500
Kryptonite Rock		1970s	glow-in-the-dark rocks sold as kryptonite chunks, in illustrated box	10	15	25
Mini Comic Book	Kellogg's	1955	cereal premium, #1-B, Supershow of Metropolis	90	175	300
Mini Comic Book	Kellogg's	1955	cereal premium, #1, The Superman Time Capsule	125	230	350
Mini Comic Book	Kellogg's	1955	cereal premium, #1-A, Duel in Space	90	175	300
Movie Viewer	Acme Plastics	1948	red plastic viewer and three individually boxed Superman films, in small full color die-cut box	75	350	450
Movie Viewer	Acme Plastics	1947	black plastic viewer, white knob and two individually boxed Superman films, in red/blue die-cut box	100	400	500
Movie Viewer	Acme	1950s	black/red plastic viewer with two individually boxed Superman films	70	130	200
Movie Viewer	Acme Plastics	1940	tortoise shell plastic, and three individually boxed Superman films, in large full color die-cut box	150	500	750
Official Magic Kit	Bar-Zim	1956	magic balls, disappearing cards, multiplying corks, vanishing trick, shell game, balancing belt & directions, in illustrated box	250	500	750
Official Superman Costume	Ben Cooper	1954	blue/red suit with red/yellow monogram & belt, in box	140	260	400
Official Superman Krypto-Raygun Film Viewer	Daisy	1940	includes raygun, battery, bulb, lens and one film strip, in illustrated box	500	1000	1500
Official Superman Playsuit	Ben Cooper	1970	cloth suit in illustrated box	20	40	60
Official Superman Playsuit with Beanie	Funtime Playwear	1954	rayon outfit, red cap with screened Superman image, navy & red suit with 5" gold monogram & belt, flannel beanie with screened Superman images, plastic propellor	300	800	1000

SUPERMAN

TOY	COMPANY	YEAR	DESCRIPTION	GOOD	EX	MIB
Official Superman Two-Piece Kiddie Swim Set		1950s	set of rubber swim fins & goggles with Superman's image or "S" symbol, in box	50	150	225
Original Radio Broadcasts Record	Nostalgia Lane	1977	old Superman radio teleplays, in illustrated sleeve showing chain breaking pose in color and b/w strip panels	10	25	50
Paint-by-Numbers Watercolor Set	Transogram	1954	16 watercolors & 44 action scenes	75	200	250
Patch		1939	5-1/2" diameter fabric premium patch, shows 3/4 profile of Superman breaking chains off chest, Supermen of America -- Action Comics	1500	7000	10000
Patch		1970s	triangular orange cloth patch with red border shows Superman flying over desert scene	5	10	15
Patch		1973	diamond-shaped cloth patch shows Superman standing against vertical red/white stripes, wide yellow border has stars and reads "Superman Junior Olympics"	5	10	15
Patch	Superman Bread	1942	cardboard shield	200	500	1000
Patch		1970s	rectangular white cloth patch with green border shows full color Supergirl flying	5	10	15
Patch		1940s	3-1/2" round patch shows chain breaking pose	750	2000	5500
Patch		1970s	rectangular white patch with green border shows full color Superboy running toward viewer	5	10	15
Patch		1970s	cloth diamond-shaped patch of "S" logo in gold/red, several sizes, each	5	10	15
Pen	Jaffe	1947	red/blue pen on illustrated card	100	150	250
Pencil Box	Mattel	1966		10	15	25
Pencil Holder	Superman, Inc.	1940s	hollow holder in shape of large pencil, illustrated on shaft with red and blue on white images and Superman-Tim Club logos	400	800	1000
Pennant		1973	35th anniversary item, felt pennant has Amazing World of Superman logo and reads "Metropolis, Illinois, Home of Superman", came in two sizes, each	12	25	35
Pennant		1940s	yellow pennant with Superman image and raised logo	300	800	1000
Pillow		1960s	12" square felt pillow with color art of flying Superman	35	65	100
Pin	Kellogg's	1940s	7/8" round pin with black/red/blue bust of Superman, most common pin in Pep Cereal series of late 1940s	12	25	35
Play-Doh Set	Play-Doh	1985	three cans of clay, play mat, Batmobile and molds for making eight characters	4	7	10
Puzzle	Saalfield	1940	300 pieces, 12 x 16", Superman the Man of Tomorrow	150	350	500

SUPERMAN

TOY	COMPANY	YEAR	DESCRIPTION	GOOD	EX	MIB
Puzzle		1940	set of two Superman puzzles	300	600	800
Puzzle	Whitman	1966	150 pieces, 14" X 18", Superman	12	25	35
Puzzle	APC	1973		10	15	20
Puzzle	Saalfield	1940	300 pieces, 12 x 16", Superman with Muscles Like Steel	150	350	500
Puzzle	Saalfield	1940	500 pieces, 16 x 20", Superman Stands Alone	200	500	750
Puzzle	Saalfield	1940	300 pieces, 12 x 16", Superman Shows his Super Strength	150	350	500
Puzzle	Saalfield	1940	300 pieces, 12 x 16", Superman Saves the Streamliner	150	350	500
Puzzle	Saalfield	1940	500 pieces, 16 x 20", Superman Saves a Life	200	500	750
Puzzle	Whitman	1966	two frame tray puzzles: one shows Superman fighting space robot, other shows him flying past manned rocket ship in outer space, each	25	50	75
Puzzles	Saalfield	1940	set of six small puzzles with 42 pieces each	130	245	375
Record & Club Membership Kit		1966	33-1/3 rpm record of the original comic, Superman Club card, shoulder patch & 1" tin litho club button, in 12" square illustrated box	100	150	200
Super Babe Doll	Imperial Crown Toy	1947	15" tall, rubber skin, movable arms and legs, sleep eyes, composition head	500	1000	2000
Super Candy & Toy	Phoenix Candy	1967	boxed candy with a small toy inside each box	25	50	75
Super Heroes String Puppets	Madison	1978	string controlled cloth and vinyl marionette	25	40	65
Superman 3-D Cut-Out Picture	Kellogg's	1950s	4-1/2 x 6-1/2" premium framed cut-out of Superman from the back of cereal box, reads "Best Wishes From Your Friend	60	80	100
Superman Action Game	American Toy Works	1940s	wood and cardboard wartime game, Superman holds tottering bridge and kids toss bean bags at tanks on bridge	750	1500	2000
Superman and Supergirl Push Puppets	Kohner	1968	set of two: 5-1/4" on bases, in window box	100	150	250
Superman Back-a-Wack	Dell	1966	blue plastic paddle with gold imprinted "S" logo and name, elastic string and red ball, on illustrated card	15	30	45
Superman Balloon		1966	small balloon with centered image	5	10	15
Superman Bank		1949	9-1/2" painted ceramic shows youthful looking Superman standing on a cloud, possibly an unlicensed import or carnival prize	100	300	500
Superman Bank		1974	bust of Superman	20	30	50
Superman Beanie		1940s	hat with two-color Superman embossed images on brim	500	700	1000
Superman Belt	Pioneer	1940s	brown leather with Superman images and rectangular buckle, in box	100	200	300
Superman Belt	Kellogg's	1950s	28" long red plastic, aluminum "S" symbol buckle in red/ yellow	100	250	300

SUPERMAN

TOY	COMPANY	YEAR	DESCRIPTION	GOOD	EX	MIB
Superman Belt	Pioneer	1940s	clear plastic with color images and round brass buckle in box	100	200	350
Superman Belt Buckle		1940s	square metal buckle shows red/blue chain breaking pose	60	115	175
Superman Book and Record Set	Peter Pan	1970s	two stories with record	5	10	15
Superman Button	WABC Radio	1966	radio premium button for "It's a Bird, It's a Plane..." production, shows faceless Superman with "WABC 77" across chest	100	150	200
.Superman Candy	Leader Novelty Candy	1940	boxed candy with punch-out trading cards on box back and coupons redeemable for Superman items, red box, front shows chain-breaking pose	500	750	1000
Superman Candy Ring	Leader Novelty Candy	1940	gold finish hidden compartment ring, face is embossed with "M", lightning bolt and eyeball symbol, compartment shows Superman decal image	1500	7500	22000
Superman Christmas Card		1940s	4" x 5", Superman Brings You Christmas Greetings, shows him flying with small tree in hands	50	150	200
Superman Cigarette Lighter	Dunhill	1940s	battery-operated table-top lighter has chrome finish figure standing on black base	500	1000	1500
Superman City Game	Remco	1966	board game with magnetic figures and buildings	150	300	500
Superman Club Button		1966	3-1/2" celluloid button, shows 3/4 profile thigh-up view of Superman in hands on hips pose, reads, "Official Member Superman	10	20	30
Superman Coloring Book	Saalfield	1941	11" x 15", cover shows Superman flying over farm house with Lois Lane	100	350	500
Superman Coloring Book	Saalfield	1940	cover shows Superman carrying bomb down to submarine on water's surface	100	350	500
Superman Coloring Book	Saalfield	1955	cover shows Superman standing with hands on hips on left and three inset action scenes on right	30	100	150
Superman Coloring Book	Saalfield	1958	11" x 15", cover shows Superman standing circus lion cage with arm in lion's mouth	30	100	150
Superman Coloring Book	Saalfield	1940	11" x 15", cover shows Superman breaking through brick wall	100	350	500
Superman Coloring Book	Whitman	1966	8" x 11", shows Superman standing hands on hips in front of comic panels	10	20	30
Superman Costume		1950s	red pants & tie-on cape, blue shirt with red, blue & yellow "S" emblem on chest, yellow belt	100	260	400
Superman Crazy Foam	American Aerosol	1970s	spray bath soap in full color illustrated can	15	25	40

CHARACTER

SUPERMAN

TOY	COMPANY	YEAR	DESCRIPTION	GOOD	EX	MIB
Superman Crusader Ring		1940s	brass or silver finish ring shows forward facing bust of Superman	50	160	300
Superman Cup	Burger King	1984	one of four in set with figural handles, others are Batman, Wonder Woman and Darkseid	5	7	10
Superman Cut-Outs	Saalfield	1940	perforated figures, heavy stock paper, blue cover	500	1500	2500
Superman Cut-Outs	Saalfield	1940	red cover, lighter paper than blue book, non-perforated cut-outs	500	1500	2500
Superman Dime Register Bank		1940s	1/2" x 2-1/2" x 2 1/2" yellow tin, front shows full color Superman breaking chains off chest, held $5 in dimes	100	250	375
Superman Doll	Knickerbocker		20" tall plush in box	12	25	35
Superman Doll	Toy Works	1977	25-1/2" tall, cloth, with cape	12	25	35
Superman Doll	Ideal	1940	13-1/2" tall, wood jointed body with composition head, movable head, arms and legs, ball knob hands	500	1000	2000
Superman Electronic Question & Answer Quiz Machine	Lisbeth Whiting Co.	1966	battery operated quiz game in full color illustrated box	45	80	125
Superman Figure	Chemtoy		rubber, three different poses, on card, each	10	15	25
Superman Figure	Presents		15" vinyl/cloth figure on base	10	15	25
Superman Figure	Fun Things		6" rubber figure on card	9	16	25
Superman Figure	Japan	1979	plastic body with soft vinyl head, movable arms and head, in illustrated window box	20	40	60
Superman Figure	Palitoy		8" figure on card	35	65	100
Superman Figurine	Craft Master	1984	solid figurine and paint set, on illustrated card	10	20	30
Superman Figurine	Ideal	1966	3" tall hard plastic painted or unpainted figure on base, removable cape, part of Justice League series	35	65	85
Superman Golden Muscle Building Set	Peter Puppets Playthings	1954	handles, springs, hand grippers, jump rope, wall hooks, measuring tape, progress chart, membership certificate & button, illustrated box	700	1000	1500
Superman Gum Ring	Gum, Inc.	1940	gold finish hidden compartment ring, face is embossed with "S", lightning bolt and Superman bust, compartment shows Superman image	4000	15000	40000
Superman Hand Puppet	Ideal	1966	11", cloth body, vinyl head	25	40	65
Superman Hood Ornament	Lee	1940s	chrome finish, shows Superman in stylized running pose with box	500	2000	4500
Superman II Trading Cards	Topps	1981	set of 88 cards, complete set	5	10	15
Superman II Trading Cards	Costa Rican	1981	set of 88 cards, complete set	18	35	50
Superman II View-Master Set			three reels, based on film	4	8	12
Superman III Trading Cards	Topps	1983	set of 99 cards, complete set	5	10	15
Superman III View-Master Set			three reels, based on film	4	7	10

Captain Kangaroo Presto Slate, 1960s, Fairchild

Popeye crayons and chalk, 1930s, American Crayon

SUPERMAN

TOY	COMPANY	YEAR	DESCRIPTION	GOOD	EX	MIB
Superman Junior Defense League Pin		1940s	die cut pin in shape of flying Superman holding banner aloft, gold finish pin with red/ white/blue detailing	35	125	250
Superman Junior Horseshoe Set	Super Swim	1950s	four rubber horseshoes, two rubber bases and two wood pegs and Official Sports Club card & rules for sportsmanship	50	100	175
Superman Junior Swim Goggles	Super Swim	1950s	plastic lenses, rubber goggles with red strap, "S" logo and membership card for Superman Safety Swim Club	35	80	150
Superman Kite	Pressman	1966		15	30	45
Superman Kite	Hiflyer	1982		3	5	8
Superman Krypto-Raygun Filmstrips	Daisy	1940s	extra boxed films for Krypto-Raygun, each	15	30	45
Superman Krypton Rocket	Park Plastics	1956	2" x 9" x 9-1/2" water powered rocket with Krypton generating pump, reserve fuel tank & Krypton Rocket, in illustrated box, same as Kellogg's rocket but in mass market packaging with added fuel tank	150	300	400
Superman Krypton Rocket	Kellogg's	1954	2" x 9" x 9-1/2" water powered rocket with "Krypton generating pump," in mailer box	150	300	400
Superman Mechanical Super Heroes Toy	Marx	1968	5-1/2" tall by 3-1/2" wide tin wind-up figure, part of four piece set, others include Batman, Spider-Man and Captain America	35	65	100
Superman Member Ring		1940		7500	30000	100000
Superman Moccasins	Penobscot Shoe	1940s	leather moccasins with Superman chain breaking pose on toe upper	150	600	800
Superman Mug		1950s	left handed mug, shows Superman on front and name across cape in back, handle has arrow and star	65	115	175
Superman Mug		1966	white glass, red & blue logo with Superman image, reverse picture is Superman breaking chain	10	20	30
Superman Muscle Building Club Button	Peter Puppets Playthings	1954	part of Golden Muscle Building set, full color bust in sunburst circle in white button, reads "Superman Muscle Building Club"	50	250	500
Superman Necktie		1940s		35	100	150
Superman Necktie Set		1940s	boxed set of two ties, small tie shows Superman standing with arms crosses, larger tie shows Superman landing	250	750	1000
Superman of Metropolis Award Certificate		1973	premium given out during Metropolis, Illinois' 1973 35th anniversary of Superman celebration	9	16	25
Superman Official Eight-Piece Junior Quoit Set		1940s	game in illustrated box includes wood and rubber game pieces, instruction booklet & membership card for "Superman Official Sports Club"	75	100	150

SUPERMAN

TOY	COMPANY	YEAR	DESCRIPTION	GOOD	EX	MIB
Superman Paint Set	American Toy Works	1940s	box shows Superman flying up toward upper right corner of box, with pallet and brushes at lower right, "Paint Set" inside pallet	250	750	1000
Superman Paint Set	American Toy Works	1940s	larger version, shows Superman in front of pallet background	250	750	1000
Superman Paint Set	American Toy Works	1940	three brushes, small palette, water cup, 14 different paints and four b/w cartoon panels, in 11" x 14-1/2" color illustrated box	250	750	1000
Superman Paint-by-Number Book	Whitman	1966	11" x 13-1/2", 40 pictures plus coloring guide on back cover	15	25	40
Superman Phone Booth Radio	Vanity Fair	1978	battery operated AM radio of green British-style booth has color bas-relief Superman exiting	35	65	100
Superman Pinball Game	Bally	1978	full-sized arcade game	200	500	750
Superman Planter		1970s	3" diameter painted ceramic	5	10	15
Superman Play Set	Ideal	1973	self-contained vinyl covered full color snap-close case opens to three backdrops, Fortress of Solitude, Daily Planet and villain's hideout, for staging action scenes with supplied color punch-outs	25	50	85
Superman Playsuit	Fishback	1940	red and blue imprinted blue smock shirt and pants with tie-on red cape	150	500	900
Superman Pogo Stick		1977	48" with a vinyl bust on top	25	50	100
Superman Pop-Up Book	Random House	1979	hardcover, full color	10	16	25
Superman Press-Out Book	Whitman	1966	punch out, assemble and hang scenes and characters	20	35	50
Superman Push Puppet	Kohner	1966	5-1/4" on base, in window box	25	40	65
Superman Radio		1973	transistor radio made in punch-out shape of Superman from waist up	35	65	100
Superman Rain Cape	Firestone	1950s	26" plastic, "S" logo on back	12	25	35
Superman Record Player		1978	latching box briefcase type record player illustrated on all sides in full color, also features b/w origin strip on back	55	100	150
Superman Ring	Nestle's	1978	premium ring, gold finish with white circle center and yellow/red diamond "S" logo in middle	15	50	60
Superman Roller Skates	Larami	1975	plastic with color bust of Superman shaped around front of each skate, in illustrated window box	20	40	60
Superman Rub-Ons	Hasbro	1966	magic picture transfers in box illustrated with picture of Superman flying	20	40	60
Superman School Bag	Acme	1950s	red/blue fold-over clasp vinyl bag, screened full color Superman figure, black plastic handle and shoulder strap	45	80	125
Superman Scrap Book	Saalfield	1940	cover shows Superman flying over mountains toward stylized S.O.S. transmission	150	400	600

SUPERMAN

TOY	COMPANY	YEAR	DESCRIPTION	GOOD	EX	MIB
Superman Senior Rubber Horseshoe Set		1950s	in box	55	100	150
Superman Senior Swim Goggles	Super Swim	1950s	plastic lenses, rubber goggles with red strap, "S" logo on bridge, membership card for Superman Safety Swim Club	40	80	150
Superman Sky Hero	Marx	1977	rubber band glider with color Superman image, on card	25	40	65
Superman Soaky	Avon	1978	9-1/2" bubble bath bottle, Superman stands atop building	30	40	50
Superman Soaky	Colgate Palmolive	1965	10" soap bottle, shows him standing with hands at sides	20	40	60
Superman Song Record	A.A. Records	1950s	6" two-song, 45 rpm record in sleeve, other song is "Tarzan Song"	15	30	45
Superman Space Satellite Launcher Set	Kellogg's	1950s	premium set of generic plastic gun with firing "satellite wheel" and illustrated instruction sheet, in mailer box	150	450	750
Superman Stamp Set		1965	set of six wood-backed character stamps	9	16	25
Superman Statue	Syracuse Ornament	1942	5-1/2" tall composition statue of Superman in hands on hips pose, finished in brown patina with red/black highlights	1000	2750	4500
Superman Statue		1940s	15" tall, crude painted plaster carnival prize	75	200	350
Superman Super Watch	Toy House	1967	plastic toy watch with moveable hands, watch "case" is large "S" chest symbol, on illustrated card	15	30	45
Superman Supertime Wristwatch	National Comics	1950s	gray band, stamped red "S" logo, silver western style buckle, full color hands on hips pose inside chrome finish case, second hand, even-hour numbers around face, in full color box	250	500	750
Superman Suspenders	Pioneer	1948	elastic, illustrated box	350	400	600
Superman Tank	Linemar	1958	large battery operated tin, 3-D Superman with a cloth cape, in illustrated box	350	1500	2500
Superman Telephone	ATE	1979	plastic phone with large figure of Superman in hands on hips pose standing over key pad, receiver hangs up into back of his cape, illustrated box	150	750	1000
Superman the Movie Trading Cards	OPC	1979	set of 132 cards	12	25	35
Superman the Movie Trading Cards	Topps	1979	set of 88 cards, second issue	12	25	35
Superman the Movie Trading Cards	French	1979	set of 180 cards	18	35	50
Superman the Movie Trading Cards	Topps	1978	set of 77 cards, first issue	7	13	20
Superman the Movie View-Master Set		1979	three reels, based on film	5	10	15
Superman Tilt Track	Kohner	1966	marble skill game, in illustrated window box	65	95	150
Superman To The Rescue Coloring Book	Whitman	1964	cover shows Superman rescuing woman	12	25	35
Superman Toothbrush	Janex	1970s	figural, battery operated	20	35	50
Superman Towels	G.H. Wood	1970s	children's sponge towels, illustrated	12	25	35

SUPERMAN

TOY	COMPANY	YEAR	DESCRIPTION	GOOD	EX	MIB
Superman Trading Cards	Topps	1966	set of 66 cards, George Reeves TV series scenes	90	165	250
Superman Trading Cards	Gum Inc.	1940	2-1/2" x 3-1/4" cards, each	75	125	250
Superman Trading Cards	Topps	1965	24 packages with b/w picture cards with scenes from the television series, per card	4	7	10
Superman Trading Cards	Gum Inc.	1940	2-1/2" x 3-1/4" cards, set of 72	2500	8000	10000
Superman Trading Cards	Topps	1966	set of 66 cards, shows George Reeves TV series scenes, each card	2	4	6
Superman Trading Cards	Topps	1965	24 packages with b/w picture cards with scenes from the television series, set	105	195	300
Superman Trading Cards Display Box	Topps	1966	2" x 4" x 8" display box of 24 unopened packs, box shows George Reeves bust	125	225	350
Superman Trading Cards Wrapper	Gum Inc.	1940	4-1/2" x 6" waxed paper	300	500	750
Superman Turnover Tank	Linemar	1950s	tin wind-up, flat tin Superman wears yellow suit and red cape, olive drab tank is marked M-25, rare in box	500	1000	2000
Superman Turnover Tank	Marx	1940	2-1/2 x 3" x 4" long tin wind-up	500	1000	2000
Superman Utility Belt	Remco	1979	illustrated window box, decoder glasses, kryptonite detector, nonworking watch, handcuffs, ring, decoder map, press card and secret message	25	50	75
Superman View-Master Set	GAF		three reels, based on cartoon series	5	10	15
Superman Wall Clock	New Haven	1978	plastic and cardboard battery operated framed wall clock showing Superman fighting alien shaceship	25	40	65
Superman Wallet	Croyden	1950s	brown, color embossed flying Superman and logo	50	150	200
Superman Wallet		1960s	brown leather	25	40	65
Superman Water Gun	Multiple Toymakers	1967	6" plastic	30	100	150
Superman Workbook	DC Comics	1940s	English grammar workbook	100	450	700
Superman Wristwatch		1977	gold bezel, stainless back, blue leather band, face shows Superman flying upward from below	35	65	100
Superman Wristwatch	Bradley	1959	dial shows Superman flying over city, second hand, chrome finish case	600	750	1200
Superman Wristwatch	Una-Donna	1986	plastic case and band, several color and face illustrations, each	7	13	20
Superman Wristwatch	New Haven Clock	1940s	squared-oval faced watch, leather band, dial shows Superman standing, hands on hips, in illustrated box	525	975	1500
Superman Wristwatch	New Haven Clock	1939	flattened oval face, shows color image of standing Superman from knees up, leather band	500	1200	1700
Superman's Christmas Adventure Comic Book	Macy's	1940	1940 Macy's holiday premium	700	1500	3500
Superman's Christmas Adventure Record	Decca	1940s	set of three 78 rpm records in illustrated sleeves	150	300	500
Superman's Christmas Play Book		1944	department store premium	200	500	750

SUPERMAN

TOY	COMPANY	YEAR	DESCRIPTION	GOOD	EX	MIB
Superman-Tim Club Button	Superman, Inc.	1940s	two different, both say Superman-Tim Club and have red/blue lettering and images on white background	75	100	150
Superman-Tim Club Membership Card	Superman, Inc.	1940s	blue/red or red/black card	750	125	200
Superman-Tim Club Press Card		1940s	blue/red card for identifying self as an Official Reporter for club	100	150	200
Superman-Tim Club Redbacks	Superman, Inc.	1940s	red on white imprinted coupons styled to look like money, denominations of $1, $5 and $10 "redbacks", each	10	20	30
Superman-Tim Club Ring		1940s	bronze finish metal ring with embossed image of Superman in flight, with initials S and T near his feet	1000	5000	10000
Superman-Tim Magazine		1940s	5" x 7" monthly store premium, each	25	50	75
Supermen of America Membership Certificate		1948	8-1/2" x 11", signed by "Clark Kent"	75	150	200
Toy Wristwatch	Germany	1950s	non-working toy watch, blue plastic band, rectangular case with full color full standing pose on white dial	35	65	100
Trick Picture Sun Camera	Made in Japan	1950s	when left in the sun for two minutes, the camera "develops" a picture of Superman fighting a space monster	75	115	175
Utensil Set	Imperial Knife	1966	stainless steel spoon and fork set with Superman on the handles, on 4-1/2 x 10" illustrated card	50	75	100
Wall Banner		1966	16" x 25" with hanging rod at top, shows large central picture of Superman in front of city skyline and two lower panels of him smashing rocks and flying through space	40	70	110

TARZAN

Kala Ape Figure	Dakin	1984	3" tall	4	7	10
Tarzan Flasher Ring	Vari-Vue	1960s		7	15	25
Tarzan Party Set	Amscan	1977		7	15	20
Young Tarzan Figure	Dakin	1984	4", bendable	4	8	12

THREE LITTLE PIGS

Big Bad Wolf Pocket Watch	Ingersoll	1934		175	500	750
Puzzle	Jaymar	1940s	7" x 10" x 2"	25	50	75
Three Little Pigs Bracelet		1930s	1/2" x 2-1/4", wolf blowing down a house with pig running away	55	150	200
Three Little Pigs Sand Pail	Ohio Art	1930s	4-1/2", tin	45	80	125
Three Little Pigs Soaky Set	Drew Chemical	1960s	8" tall each: Three Little Pigs and the Big Bad Wolf	60	115	175
Three Little Pigs Wind-Up Toy	Schuco	1930s	4-1/2" pigs playing fiddle, fife & drum	245	450	700
Who's Afraid of the Big Bad Wolf Game	Parker Brothers	1930s		60	250	400

TOM & JERRY

TOY	COMPANY	YEAR	DESCRIPTION	GOOD	EX	MIB
Jerry Figure	Marx	1973	4" tall	12	23	35
Puzzles	Whitman		four frame tray puzzles	15	25	40
Tom & Jerry Bank	Gorham	1980	6" tall	18	35	50
Tom & Jerry Figure Set		1975	walking, Tom, Jerry, & Droopy	15	30	45
Tom & Jerry Go Kart	Marx	1973	plastic, friction drive	25	50	75
Tom & Jerry on Scooter	Marx	1971	plastic friction drive	12	23	35
Tom & Jerry Wristwatch	Bradley	1985	quartz, oldies series, small white plastic case and band, sweep seconds, face shows Tom squirting Jerry with hose	10	16	25
Tom Figure	Marx	1973	6" tall	12	23	35

TOP CAT

TOY	COMPANY	YEAR	DESCRIPTION	GOOD	EX	MIB
Top Cat Figure	Marx	1961	TV-Tinykins, plastic	15	30	45
Top Cat Soaky		1960s	10" tall, vinyl	15	30	50
Viewmarx Micro-Viewer	Marx	1963	plastic	15	30	45

UNDERDOG

TOY	COMPANY	YEAR	DESCRIPTION	GOOD	EX	MIB
Kite Fun Book	Pacific Gas and Electric	1970s		5	25	75
Puzzle	Whitman	1975	100 pieces	9	16	25
Underdog Costume	Ben Cooper	1969		50	75	115
Underdog Dot Funnies Kit	Whitman	1974		9	16	25
Underdog Figure	Dakin	1976	plastic, Cartoon Theater	25	75	125

WELCOME BACK, KOTTER

TOY	COMPANY	YEAR	DESCRIPTION	GOOD	EX	MIB
Halloween Costume	Collegeville	1976	several styles	10	15	20
Sweathogs Dolls	Mattel	1976	Epstein, Washington, Barbarino, Kotter, Horshack; each	5	10	25
Sweathogs Grease Machine Cars	Ahi	1977	3" long, plastic cars; various styles, each	10	25	40
Welcome Back, Kotter Classroom	Mattel	1976	play set	20	40	60
Welcome Back, Kotter Colorforms Set	Colorforms	1976		5	10	20

WINKY DINK & YOU

TOY	COMPANY	YEAR	DESCRIPTION	GOOD	EX	MIB
When Winky Winks at You Record	Decca	1956		40	50	75
Winky Dink Book	Golden	1956	Little Golden Book	9	16	25
Winky Dink Comic Book	Dell	1950s	#663	15	20	30
Winky Dink Magic Crayons		1960s		16	30	45
Winky Dink Official TV Game Kit		1950s		25	45	70
Winky Dink Secret Message Game	Lowell	1950s		70	125	195
Winky Dink Winko Magic Kit		1950s		15	30	45

WINNIE THE POOH

TOY	COMPANY	YEAR	DESCRIPTION	GOOD	EX	MIB
Jack-In-The-Box	Carnival Toys	1960s		15	30	45

WINNIE THE POOH

TOY	COMPANY	YEAR	DESCRIPTION	GOOD	EX	MIB
Kanga and Roo Squeak Toy	Holland Hill	1966	vinyl	12	23	35
Lamp	Dolly Toy	1964	7" tall	25	50	75
Magic Slate	Western Publishing	1965	8-1/2" x 13-1/2"	15	30	45
Puzzle	Whitman	1964	frame tray	7	13	20
Radio	Thilgee	1970s	5" x 6" x 1-1/2" tall	25	50	75
Winnie the Pooh & Christopher Robin Dolls	Horsman	1964	Winnie the Pooh 3-1/2" tall and Christopher 11" tall, set	60	115	200
Winnie the Pooh Button		1960s	3-1/2" celluloid	5	10	15
Winnie the Pooh Doll		1960s	12" tall	12	23	35
Winnie the Pooh Snow Globe			5-1/2", musical	15	30	45

WIZARD OF OZ

TOY	COMPANY	YEAR	DESCRIPTION	GOOD	EX	MIB
Carpet Sweeper	Bissell	1939	child-sized	100	165	250
Cast 'N Paint Set		1975	makes six 6" figures	15	25	35
Chalk Board	Roth American	1975	wood frame, steel stand with chalk, chalk holder and eraser	16	27	40
Christmas Ornaments	Presents	1989	cloth and vinyl: Dorothy, Scarecrow, Tin Man, Cowardly Lion, Glinda & Wicked Witch, each	8	10	15
Christmas Ornaments	Bradford Novelty	1977	4-1/2" tall, Dorothy, Scarecrow, Tin Man, Cowardly Lion, each	5	8	10
Cookie Jar	Clay Art	1990	white with relief figures of characters	20	35	50
Cowardly Lion Bank		1960s	ceramic, red nose	25	40	65
Cowardly Lion Costume	Collegeville	1989	deluxe	17	30	45
Cowardly Lion Costume	Ben Cooper	1975	costume & mask	12	20	30
Cowardly Lion Costume	Collegeville	1989	plastic mask and vinyl bodysuit	6	10	15
Cowardly Lion Doll	M-D Tissue	1971	cloth, light brown body with white snout	6	10	15
Cowardly Lion Doll	Presents	1988	vinyl	17	30	45
Cowardly Lion Doll	Largo Toys	1989	rag doll	8	13	20
Cowardly Lion Doll	Ideal	1984	9", Character Dolls series	17	30	45
Cowardly Lion Mask	Newark Mask Company	1939	linen, hand painted	50	80	125
Cowardly Lion Mask	Don Post Studios	1983	rubber	35	55	85
Cowardly Lion Music Box	Schmid	1983		17	30	45
Cowardly Lion Wind-Up	Durham Industries	1975	on illustrated card	10	16	25
Crayon Box	Cheinco	1975	rectangular, metal	5	8	12
Cut & Make Masks	Dover	1982	eight cut-out color masks	4	7	10
Dandy Lion Doll	Artistic	1962	14" tall, cloth & vinyl	30	50	75
Decoupage Kit		1975	two wooden plaques, scenes based on film	12	20	30
Doodle Dolls	Whiting	1979	three dolls: cardboard parts, yarn, Styrofoam balls, fabric	6	10	15
Dorothy and Friends Visit Oz Book	Curtis Candy	1967	candy premium	5	8	12
Dorothy and Toto Doll	Ideal	1984	9", Character Dolls series	17	30	45
Dorothy Bank		1960s	ceramic, blue dress with brown wicker basket	25	40	65
Dorothy Costume	Collegeville	1989	plastic mask & vinyl bodysuit	6	10	15
Dorothy Costume	Collegeville	1989	deluxe, includes red metallic glitter chips for shoes	17	30	45
Dorothy Costume	Ben Cooper	1975	costume & mask	12	20	30
Dorothy Doll	M-D Tissue	1971	cloth, stuffed, yellow hair, orange jumper	6	10	15

WIZARD OF OZ

TOY	COMPANY	YEAR	DESCRIPTION	GOOD	EX	MIB
Dorothy Doll	Madame Alexander	1991	8", blue checked jumper with white blouse, basket with Toto and red shoes, Storyland Dolls series	20	35	50
Dorothy Doll	Largo Toys	1989	Judy Garland	8	13	20
Dorothy Doll	Presents	1988	vinyl, blue checkered jumper, white blouse, red slippers, yellow brick road base	17	30	45
Dorothy Doll	Sears	1939	15-1/2", Judy Garland, red or blue checked jumper with black pin curls	375	620	950
Dorothy Doll	Ideal	1939	18", Judy Garland, blue checked jumper, open and close brown eyes	425	715	1100
Dorothy Doll	Ideal	1939	15-1/2", Judy Garland, blue checked jumper, open and close brown eyes	325	550	850
Dorothy Doll	Ideal	1939	13", Judy Garland, blue checked jumper, open and close brown eyes	250	425	650
Dorothy Doll	Effanbee	1984	14-1/2" tall, vinyl, Judy Garland, blue dress and hair ribbons, ruby slippers, Legend Series	35	60	95
Dorothy Meets the Wizard Book	Curtis Candy	1967	candy premium	6	10	15
Dorothy Music Box	Schmid	1983	plays "Over the Rainbow"	20	35	50
Dorothy Squeak Toy	Burnstein	1939	7", hollow rubber	70	115	175
Erasers	Applause	1989	set of six: figural, Scarecrow, Cowardly Lion, Dorothy, Wicked Witch, Tin Man, Glinda, set	10	15	25
Fun Shades	Multi Toys	1989	children's sunglasses with character images	3	5	7
Game of The Wizard of Oz	Whitman	1939		90	145	225
Give-A-Show Projector Slides	Kenner	1968	35 color slides, five different shows	15	25	40
Glinda Doll	Presents	1989	vinyl, pink dress with pink crown and wand, yellow brick road base	17	30	45
Glinda Squeak Toy	Burnstein	1939	7", hollow rubber	75	130	200
Glinda's Magic Wand	Multi Toys	1989	battery operated wand with red glitter star on end, lights up, on illustrated card	4	7	10
Jack Pumpkinhead and the Sawhorse of Oz Book	Rand McNally	1939	hardcover, also contains "Tik Tok and the Gnome King of Oz"	30	50	75
Jack Pumpkinhead Doll	Oz Doll & Toy	1924	13"	390	650	1000
Jack Pumpkinhead Figure	Heart & Heart	1985	Return to Oz, 3"-4" tall, plastic jointed	30	50	75
Little Dorothy & Toto of Oz Book	Rand McNally	1939	hardcover, also contains "The Cowardly Lion and the Hungry Tiger"	30	50	75
Little Golden Book Series		1951	The Road to Oz, The Emerald City of Oz, & The Tin Woodman of Oz, each	6	10	15
Lollipop Guild Boy Doll	Presents	1989	vinyl, plaid shirt, green shorts and striped socks, on a yellow brick road base	17	30	45
Lullabye League Girl Doll	Presents	1989	vinyl, pink ballerina dress and slippers with hat on a yellow brick road base	17	30	45
Magic Picture Kit	Jiffy Pop Popcorn	1968		4	7	10
Magic Slate	Lowe	1961	art based on animated TV show	10	15	25
Magic Slate	Whitman	1976		4	7	10
Magic Slate	Western	1985	Return to Oz	4	7	10

WIZARD OF OZ

TOY	COMPANY	YEAR	DESCRIPTION	GOOD	EX	MIB
Magic Slate	Western	1989		3	5	7
Magic Story Cloth	Raco	1978	38" x 44" plastic sheet, eight crayons and sponge	6	10	15
Magnets	Vanderbilt Products	1989	several characters available, each	2	3	5
Magnets	Grynnen Barrett	1987	six character magnets in box	6	10	15
Mask Book	Watermill Press	1990	four paper masks	3	5	7
Mayor of Munchkinland Doll	Presents	1989	vinyl, black suit and shoes on a yellow brick road base	17	30	45
Munchkins Figures	Presents	1988	PVC, 1-3/4" to 2-3/4": Mayor, Lollipop Guild Boy, Sleepyhead Girl, Lady, Soldier and Ballerina, each	3	5	8
Off to See the Wizard Colorforms	Colorforms	1967		20	30	55
Off to See the Wizard Dancing Toys	Marx	1967	mechanical, dancing Tin Man, the Cowardly Lion and the Scarecrow, Montgomery Ward's exclusive, each	50	80	150
Off to See the Wizard Flasher Rings	Vari-Vue	1967	gumball machine prizes, silver painted resin, gold painted, dark or light blue plastic, each	6	10	20
Off to See the Wizard Hand Puppet	Mattel	1968	talking, four vinyl heads on finger tips, Toto and Cowardly Lion on thumb pad, ten phrases	25	40	70
Oz Time Wristwatch	Macy's	1989	50th anniversary premium, round face with Emerald City, black plastic band	25	35	75
Oz-Kins Figures	Aurora	1967	plastic Burry Biscuit premium: set of 10	35	60	95
Pails	Swift & Company	1950s	Oz Peanut Butter, red & yellow and red, yellow & white tin	20	35	50
Paint by Number 'N Frame Set	Hasbro	1969	16" x 18", two plastic frames, 18 watercolors, brush and eight pictures to paint	15	25	35
Paint by Number Set	Hasbro	1973	six oil paint vials and brush	6	10	15
Paint by Number Set	Craft House	1979	two 10" x 14" panels, 15 colors, brush and instructions	10	15	25
Paint by Number Set	Craft Master	1968	six paints, brush, picture of Tin Man, Cowardly Lion, or the Scarecrow	15	25	40
Paint by Number Set	Art Award	1989	three different versions	3	5	8
Paint with Crayons Set	Art Award	1989	four pictures based on MGM film characters, in illustrated box	3	5	8
Paper Dolls	The Toy Factory	1975	Dorothy, Tin Man, Scarecrow, Cowardly Lion & Toto, clothes and accessories	8	13	20
Patchwork Girl Doll	Oz Doll & Toy	1924	13"	70	115	175
Playing Cards	Presents	1988	tin holds two decks	5	8	12
Puppet Theatre	Proctor & Gamble	1965	cardboard theater designed for P & G puppets	30	50	75
Puzzle	Whitman	1976	frame tray	3	5	7
Puzzle	Western	1989	frame tray, 100 pieces, Glinda and Dorothy in Munchkinland	4	7	10
Puzzle	Milton Bradley	1990	1000 piece jigsaw featuring the 1989 Norman James Company poster	6	10	15
Puzzle	Crisco Oil	1985	Return to Oz, mail away premium, 200 piece puzzle	6	10	15

WIZARD OF OZ

TOY	COMPANY	YEAR	DESCRIPTION	GOOD	EX	MIB
Puzzle	Reilly & Lee	1932	#1, two softcover editions of the Scarecrow & the Tin Man, Ozma & the Little Wizard, plus two puzzles, in box	175	295	450
Puzzle	Haret-Gilmar	1960s	10" x 14" puzzle in canister	6	10	15
Puzzle	American Puzzle Company	1976	200 piece puzzle in canister	4	7	10
Puzzle	Effanbee	1984	in canister	8	13	20
Puzzles	Reilly & Lee	1932	set #2, two softcover editions of Tik-Tok & Jack Pumpkinhead & the Sawhorse, plus 25 piece puzzles	175	295	450
Puzzles	Whitman	1967	set of three: Peter Pan, Alice in Wonderland and The Wizard of Oz, in box	10	16	25
Puzzles	Jaymar	1960s	set of four, 100 pieces each, each	10	16	25
Puzzles	Doug Smith	1977	17" x 22" each, frame tray	10	16	25
Puzzles	Golden Press	1985	frame tray, Return to Oz characters	2	3	4
Puzzles	Jaymar	1960s	frame tray	10	16	25
Return to Oz Game	Golden Press	1985		8	13	20
Return to Oz Hand Puppets	Welch's Jelly	1985	Scarecrow, Gump, or Tik-Tok, Return to Oz promotion	10	16	25
Return To Oz Little Golden Books	Western	1985	Dorothy Returns to Oz, Escape from the Witch's Castle, Dorothy in the Ornament Room, Dorothy Saves the Emerald City, each	2	3	4
Rubber Stamps		1989	18 character stamps	10	16	25
Rubber Stamps		1989	set of 11 characters in plastic case	8	13	20
Rubber Stamps	Multi Toys	1989	12 figural stampers	3	5	8
Rusty the Tin Man Doll	Artistic Toy Company	1962	14" tall, cloth & vinyl	30	50	75
Scarecrow & the Tin Man Book	G. W. Dillingham	1904		115	195	300
Scarecrow & the Tin Man Book	Perks Publishing	1946	black & yellow pictures	15	25	35
Scarecrow Bank		1960s	ceramic	25	40	65
Scarecrow Costume	Ben Cooper	1968	battery-operated light-up mask	15	25	40
Scarecrow Costume	Ben Cooper	1967		15	25	35
Scarecrow Costume	Collegeville	1989	plastic mask and vinyl bodysuit	6	10	15
Scarecrow Costume	Collegeville	1989	deluxe, includes straw	17	30	45
Scarecrow Doll	Oz Doll & Toy	1924	13"	390	650	1000
Scarecrow Doll	M-D Tissue	1971	cloth, stuffed, frowning, blue pants and red and white plaid jacket	6	10	15
Scarecrow Doll	Presents	1988	vinyl, brown pants, green shirt, and black hat & shoes, on a yellow brick road base	17	30	45
Scarecrow Doll	Largo Toys	1989	rag doll	8	13	20
Scarecrow Doll	Ideal	1984	9", Character Dolls series	17	30	45
Scarecrow Figure	Dalen Products	1984	6' inflatable	10	15	25
Scarecrow Figure	Artisans Studio	1939	4", wood composition	60	100	150
Scarecrow Figure		1968	15", ceramic, painted or unpainted	15	25	35
Scarecrow Figure	Heart & Heart	1985	Return to Oz, 3"-4" tall, plastic jointed	25	40	65

WIZARD OF OZ

TOY	COMPANY	YEAR	DESCRIPTION	GOOD	EX	MIB
Scarecrow Mask	Newark Mask	1939	linen, hand painted	50	80	125
Scarecrow Mask	Don Post Studios	1983	rubber	35	55	85
Scarecrow Music Box	Schmid	1983		17	30	45
Scarecrow Night Light	Hamilton Gifts	1989	7", unpainted bone china	10	15	25
Scarecrow Talkin' Patter Pillow Doll	Mattel	1968	cloth, pull-string, says 10 phrases, dark blue pants and sleeves, white gloves, black boots	30	50	75
Scarecrow Wind-Up	Durham Industries	1975	on illustrated card	10	16	25
Scarecrow-in-the-Box	Mattel	1967	jack-in-the-box	15	25	40
Showboat Play Set	Remco	1962	pink plastic showboat with oversized central stage area, four different plays, scenery, players & scripts	35	55	85
Snack 'N Sip Pals	Multi Toys	1989	12 red & white striped straws with detachable character figures	3	5	7
Socrates the Scarecrow Doll	Artistic	1962	14" tall, cloth & vinyl	30	50	75
Stand-Up Rub-Ons	Hasbro	1968	three full color transfer sheets, character outline sheets of 10 characters	15	25	40
Stationery	Whitman	1939	10 sheets & envelopes, with character illustations	70	115	175
Stitch a Story Set	Hasbro	1973	two framed pictures, thread and embroidery needle	8	13	20
Strawman Doll	Ideal	1939	17", Ray Bolger, tan or pink pants, black or navy jacket	295	495	750
Tales of the Wizard of Oz Coloring Book	Whitman	1962	art from animated TV show	10	15	25
Tales of the Wizard of Oz Comic Book	Dell	1962	#1306	8	13	20
Tea Set	Ohio Art	1970s	30-piece set, red and yellow plastic	25	40	65
The Tin Woodsman and Dorothy Book	Curtis Candy	1967	candy premium	5	8	12
The Wonderful Cut-Outs of Oz Book	Crown	1985	35 figures to cut out	6	10	15
Tik Tok Figure	Heart & Heart	1985	Return to Oz, 3"-4" tall, plastic jointed	30	50	75
Tin	Multi Toys	1989	8" x 10" x 2", illustrated with Emerald City and characters	6	10	15
Tin Man Bank		1960s	ceramic, silver	25	40	65
Tin Man Costume	Ben Cooper	1968		15	25	35
Tin Man Costume	Collegeville	1989	deluxe	17	30	45
Tin Man Costume	Ben Cooper	1975	costume and mask	12	20	30
Tin Man Costume	Halco	1961	costume & mask	17	30	45
Tin Man Costume	Collegeville	1989	plastic mask and vinyl bodysuit	6	10	15
Tin Man Doll	Oz Doll & Toy	1924	13"	390	650	1000
Tin Man Doll	M-D Tissue	1971	cloth, stuffed, gray body, blue eyes, red heart	6	10	15
Tin Man Doll	Presents	1988	vinyl, silver body with a heart clock on chaint, on a yellow brick road base	17	30	45
Tin Man Doll	Ideal	1984	9", Character Dolls series	15	25	40
Tin Man Doll	Largo Toys	1989	rag doll	8	13	20
Tin Man Figure	Artisans Studio	1939	4", wood composition	60	100	150
Tin Man Figure		1968	15", ceramic, painted or unpainted	15	25	35
Tin Man Figure		1985	Return to Oz, 3"-4" tall, plastic jointed	25	40	65
Tin Man Mask	Newark Mask	1939	linen, hand painted	50	80	125

WIZARD OF OZ

TOY	COMPANY	YEAR	DESCRIPTION	GOOD	EX	MIB
Tin Man Mask	Don Post Studios	1983	rubber	35	55	85
Tin Man Music Box	Schmid	1983		17	30	45
Tin Man Robot	Remco	1969	21-1/2" tall, battery operated, lifts legs and swings arms as he walks	70	115	175
Tin Man Wind-Up	Durham Industries	1975		10	16	25
Toto Doll	Presents	1988	5-1/2" plush	8	13	20
Toy Watch		1940s	tin, Scarecrow and the Tin Man on either side of non-working dial	15	25	35
Trash Can	Chein	1975	oval metal	17	30	45
Vinyl Stick-On Play Set	Multi Toys	1989	10 vinyl stickers with Emerald City background, on header card	3	5	8
Wall Decorations	Shepard Press	1967	20 punch-out decorations	17	30	45
Water Guns	Durham	1976	heads of Scarecrow, Tin Man, or Cowardly Lion, water squirts out of nose, each	10	16	25
Wicked Witch Doll	Presents	1989	vinyl, black dress & hat, green face & hands holding broom on a yellow brick road base	20	35	50
Wicked Witch Mask	Ben Cooper	1975		6	10	15
Wicked Witch Squeak Toy	Burnstein	1939	7", hollow rubber	90	145	225
Wizard of Oz Book	Western	1975	#310-32, Little Golden Book	2	3	4
Wizard of Oz Christmas Book	Gimbel's	1968	New York department store premium	10	15	25
Wizard of Oz Color-By-Number Book	Karas Publishing	1962	#A-116, Twinkle Books series	8	13	20
Wizard of Oz Coloring Book	Swift	1955	advertised Oz Peanut Butter	10	15	25
Wizard of Oz Coloring Book	Playmore	1970s	#A400-10	3	5	7
Wizard of Oz Coloring Book	Saalfield	1957		10	15	25
Wizard of Oz Comic Book	Dell	1956	Dell Junior Treasury, #5	10	15	25
Wizard of Oz Comic Book	Dell	1957	Classic Illustrated Jr., #535	10	15	25
Wizard of Oz Costume	Halco	1961	costume & mask	17	30	45
Wizard of Oz Dart Game	Dart Board Equipment	1939	board illustrated with yellow brick road and circular targets of Oz characters, with three darts	195	325	500
Wizard of Oz Dolls	Effanbee	1985	11-1/2", Dorothy, Scarecrow, Tin Man, Cowardly Lion, each	12	20	30
Wizard of Oz Figures	Multiple Toymakers	1967	6" tall, bendy, several characters, on card	15	25	35
Wizard of Oz Figures	Presents	1988	3-3/4": Dorothy, Scarecrow, Tin Man, Cowardly Lion, Wicked Witch, Glinda, each	2	3	5
Wizard of Oz Figures	Multi Toys	1989	4" poseable figures, set of six	12	20	30
Wizard of Oz Figures	Just Toys	1989	several characters, bendy, each	3	5	8
Wizard of Oz Figures	Presents	1989	six figures on musical bases, each	10	12	15
Wizard of Oz Hand Puppets	Proctor & Gamble	1965	plastic	8	13	20
Wizard of Oz Hand Puppets	Presents	1989	several characters available, each	10	15	20
Wizard of Oz Hand Puppets	Multi Toys	1989	set of six on blister cards, each	6	10	15
Wizard of Oz Live! Dolls	Applause		cloth, three different sizes, each	8	13	20
Wizard of Oz Lunch Box	Aladdin	1989	plastic, bottle shows ruby slippers on yellow brick road	6	10	15

WIZARD OF OZ

TOY	COMPANY	YEAR	DESCRIPTION	GOOD	EX	MIB
Wizard of Oz Paint Book	Whitman	1939		25	40	65
Wizard of Oz Paper Dolls	Whitman	1976		4	7	10
Wizard of Oz Pocket Watch	Westclock	1980s	silver finish case, four characters on dial	17	30	45
Wizard of Oz Squeak Toy	Burnstein	1939	7" tall, hollow rubber, several characters	70	115	175
Wizard of Oz Sticker Fun Book	Whitman	1976		3	5	7
Wizard of Oz Wind-Ups	Multi Toys	1989	50th anniversary editions, several characters, each	3	5	8
Wizard of Oz Wristwatch	EKO	1989	quartz, illustrated face showing Emerald City, black plastic band	10	16	25
Wizard of Oz Wristwatch	EKO	1989	child's LCD, red face in round yellow case, plastic band shows yellow brick road and Emerald City	6	10	15
Wizard Squeak Toy	Burnstein	1939	7", hollow rubber	75	130	200

WOODY WOODPECKER

TOY	COMPANY	YEAR	DESCRIPTION	GOOD	EX	MIB
Alarm Clock	Columbia Time	1959	Woody's Cafe	60	115	175
Lamp		1971	20" tall, plastic	11	20	30
Paper Dolls	Saalfield	1968	Woody Woodpecker and Andy Panda	15	30	45
Playing Cards		1950s	two decks in a carrying case	30	55	85
Woody Woodpecker Hand Puppet	Mattel	1963	pull-string voice box	30	60	90
Woody Woodpecker Nodder		1950s	plastic	50	100	150
Woody Woodpecker's Fun-o-Rama Punch-Out Book		1972		9	16	25

YOGI BEAR

TOY	COMPANY	YEAR	DESCRIPTION	GOOD	EX	MIB
Boo Boo Doll	Knickerbocker	1960s	9-1/2" tall, plush	35	60	50
Bubble Pipe	Transogram	1963	Yogi Bear figural pipe	15	30	45
Character Figures		1960	12" tall, Yogi, Boo Boo, and Ranger Smith, each	15	30	45
Cindy Bear Doll	Knickerbocker	1959	16", plush with vinyl face	35	65	60
Coat Rack	Wolverine	1979	48", red wood, Yogi and Boo Boo cut out in front, growth chart on back	35	65	95
Hot Water Bottle		1966		25	50	75
Magic Slate		1963		11	20	30
Safety Scissors	Monogram	1973	on card	4	7	10
Snagglepuss Figure	Dakin	1970		30	55	85
Snagglepuss Soaky	Purex	1960s	9" tall, vinyl/plastic	15	25	55
Snagglepuss Sticker Fun Book	Whitman	1963		11	20	30
Yogi Bear and Cindy Push Puppet Set	Kohner	1960s	boxed set of two	45	75	200
Yogi Bear and Pixie & Dixie Game Car	Whitman		7-1/2" pile on game in car	12	23	35
Yogi Bear Bank	Dakin	1980	7", figural	5	10	15
Yogi Bear Bank	Knickerbocker	1960s	22", figural	15	30	45
Yogi Bear Cartoonist Stamp Set	Lido	1961		25	50	75
Yogi Bear Doll	Knickerbocker	1959	10" tall	35	65	100
Yogi Bear Doll		1962	6", soft vinyl with movable arms & head	20	40	60
Yogi Bear Doll	Knickerbocker	1960s	19" plush	25	50	75

YOGI BEAR

TOY	COMPANY	YEAR	DESCRIPTION	GOOD	EX	MIB
Yogi Bear Doll	Knickerbocker	1959	16" plush with vinyl face	35	65	100
Yogi Bear Figure	Knickerbocker	1960s	9" tall, plastic	23	45	65
Yogi Bear Figure	Dakin	1970	7-3/4" tall	15	30	45
Yogi Bear Figure	Marx	1961	TV-Tinykins	15	30	45
Yogi Bear Friction Toy		1960s	Yogi in yellow tie and green hat, illustrated red box	35	65	95
Yogi Bear Ge-Tar	Mattel	1960s		50	75	150
Yogi Bear Hand Puppet	Knickerbocker			15	30	45
Yogi Bear Paint 'em Pals	Craft Master	1978	paint-by-number set	10	15	25
Yogi Bear Push Puppet	Kohner	1960s		12	23	35
Yogi Bear Stuff & Lace Doll	Knickerbocker	1959	items to make a 13" x 5" doll	20	40	60
Yogi Bear Wristwatch		1963		35	65	100
Yogi Score-A-Matic Ball Toss Game	Transogram	1960		40	75	115
Yogi Squeeze Doll	Sanitoy	1979	12", vinyl	9	16	25
Yogi vs. Magilla for President Coloring Book	Whitman	1964		15	40	65
Yogi Wristwatch	Bradley	1967	medium base metal case, shows Yogi with hobo sack on stick, black vinyl band	35	100	175

YOSEMITE SAM

TOY	COMPANY	YEAR	DESCRIPTION	GOOD	EX	MIB
Mini Snow Dome	Applause	1980s		9	16	25
Musical Snow Dome	Applause	1980s		15	30	45
Yosemite Sam Figure	Dakin	1978	Fun Farm	11	20	30
Yosemite Sam Figure	Dakin	1968	7" tall	12	23	35
Yosemite Sam Figure	Dakin	1971	on treasure chest	14	25	40
Yosemite Sam Nodder		1960s	6-1/4", bobbing head & spring mounted head	35	100	175

Coloring Books

Paper goods have always been popular among collectors, whether it be tickets, postcards, Valentines, display pieces or other ephemera. Coloring books can be added to that category, although collectors of specific characters are already familiar with adding these items to their collections.

Coloring books have been around for more than a century. As early as the 1880s, the McLoughlin Brothers, a 19th century giant in publishing children's books, produced The Little Folks Painting Book illustrated by Kate Greenaway. The firm continued to publish books of this type through the 1920s when it became a part of the Milton Bradley Company.

Some of the first advertising premium coloring and painting books appeared in the 1890s including Hood's Sarsaparilla Painting Book in 1894.

During the early 1900s, the Stokes Company issued some of the first true coloring books of the century with Buster's Paint Book and the Buster Brown Stocking Paint Box, a cardboard folder containing packets of pictures to color. Both booklets, starring Buster Brown, were the work of Yellow Kid comic strip artist Richard Outcault.

Among the major coloring book producers of the 20th century were Saalfield, Whitman, and Merrill Publishing. Saalfield was founded in Akron, Ohio, in the early 1900s and achieved fame with the popular Billy Whiskers series and later with the immortal Shirley Temple coloring books. Whitman Publishing was founded in Racine, Wis., in the early 1900s, and today as Western Publishing is one of the country's largest producers of children's books. Merrill Publishing came along in the 1930s, but for the next 20 years they provided a major challenge to Whitman in the field of coloring books.

Walt Disney's immortal characters were naturals for the coloring book medium, and that went for Disney's TV heroes as well, like Fess Parker as Daniel Boone.

In the early 1940s, Western film heroes burst into the scene with their own books. Gene Autry and Roy Rogers joined fictional heroes like Red Ryder and Hopalong Cassidy. Cowboy heroes continued as coloring book stars into the 1950s, as did a host of comic strip and comic book characters like Captain Marvel, Superman, Blondie, Bugs Bunny and Little Lulu. Adventures in space got an early start with coloring books too. Buck Rogers may have beat Flash Gordon to publication as a newspaper

Nancy and Sluggo, 1970, Whitman

COLORING BOOKS

comic strip, but Flash Gordon appeared first in a 1934 painting book, followed a year later by Whitman's Buck Rogers Paint Book.

The 1960s could be called the golden age for coloring books. Television provided a wealth of worthy subjects, from sitcom to cartoon characters.

Trends

Aside from Disney and other comic classics of the 1930s, the coloring books of the 1960s tend to be among the most collectible because of the heavy following of character collectors.

Condition is very important when grading coloring books. A Mint condition coloring book should not be colored in, and the cover and pages should be in prime condition (no creases, tears, or marks). Character-related books — especially Disney, Western and superheroes — tend to be the most highly sought. Generic titles generate little collector interest.

The Top 10 Coloring Books
(in Mint condition)

1.	Courageous Cat and Minute Mouse, Artcraft	$175
2.	John Wayne Coloring Book, Saalfield, 1951	150
3.	Hood's Sarsaparilla Painting Book, Hood, 1894	150
4.	Gone with the Wind Paint Book, Merrill, 1940	150
5.	Funny Company, Whitman, 1966	130
6.	Touche Turtle, Whitman	125
7.	Shirley Temple Crosses the Country, Saalfield, 1939	125
8.	Draw & Paint Tom Mix, Whitman, 1935	125
9.	Charlie Chaplin Up in the Air, M.A. Donahue & Co., 1914	125
10.	Walt Disney's Mickey Mouse Paint Book, Whitman, 1937	110

Coloring Books

NAME/COMPANY/YEAR	NO.	GOOD	EX	MINT
101 Dalmatians (Watkins-Strathmore, 1960)	1864	5	15	25
ABC-TV Discovery (Saalfield)	9659	10	20	30
Addams Family, The (Artcraft)	4331	40	65	100
Adventures of Batman (Whitman, 1966)		15	30	45
Alvin and the Chipmunks (Watkins-Strathmore, 1966)	1878-G	5	7	10
Alvin and the Chipmunks (Whitman)	1157	10	30	35
Amazing Chan & the Chan Clan, The (Whitman)	1063	10	25	35
Andy Griffith Show (Saalfield)	5361	15	25	50
Annie Oakley (Whitman, 1957)	1756	20	40	65
Apollo, Man on the Moon (Artcraft, 1969)		10	20	25
Archie's Pal Jughead (Whitman, 1972)		10	15	25
Archies, The (Whitman)	1135	10	15	20
Astro Boy (Saalfield)	4551	25	80	100
Astronut (Treasure, 1963)		20	30	35
Atom Ant (Whitman, 1966)	1113	15	30	70
Attack! Fighting Men in Action (Whitman, 1964)		15	20	30
Baba Looey (Watkins-Strathmore)	1853	10	20	40
Banana Splits (Whitman)	1062	30	50	70
Barney Google/Snuffy Smith (Lowe, 1950s)	2831	10	20	35
Bat Masterson (Saalfield, 1959)	4634	20	40	50
Beany & Cecil (Whitman)	203425	15	30	55
Beany & Cecil (Whitman)	1648	10	30	55
Beany's Coloring Book (Whitman)	112015	10	30	55
Beatles (Saalfield)	5240	30	60	75
Beatles Official Coloring Book (Peerless, 1964)	5240	30	45	75
Beaver's Big Book (Saalfield)	5327	20	45	65
Beetle Bailey (Lowe, 1961)	2860	20	30	35
Ben Casey (Saalfield)	9532	10	15	35
Ben-Hur (Lowe)	2851	20	30	40
Bette Davis (Merrill, 1942)	4817	25	45	75
Betty Grable (Merrill, 1951)	1501	25	40	65
Beverly Hillbillies (Watkins-Strathmore, 1964)	1883	15	30	40
Beverly Hillbillies (Whitman, 1963)	1137	15	30	35
Bewitched Fun and Activity Book (Treasure)	8908	25	50	65
Bing Crosby (Saalfield)	4840	15	25	45
Blackbeard's Ghost (Whitman, 1968)		15	20	25
Blondie (Saalfield, 1968)	9961	15	25	40
Bob Hope (Saalfield, 1954)	1257	20	40	55
Bonanza (Saalfield, 1960s)	1617	12	30	35
Bonny Braids Coloring Book (Saalfield, 1951)	2366	15	25	55
Boo Boo Bear (Watkins-Strathmore)	1882	15	20	30
Bozo the Clown (Whitman)	1179	10	20	25
Brady Bunch (Whitman)	1004	20	30	40
Brady Bunch (Whitman)	1061	20	35	45
Brady Bunch (Whitman, 1974)	1657	15	30	40
Brenda Starr (Saalfield, 1964)	9675	25	40	50
Buccaneer, The (Saalfield)	4633	10	15	25
Buffy and Jody Coloring Book (Whitman)	1640	5	10	15
Bugs Bunny (Whitman)	1147	10	30	40
Bugs Bunny's Big Busy Color & Fun (Whitman)	2017	10	30	40
Bullwinkle the Moose (Whitman, 1960)		25	45	50
Buster Brown Coloring Book (1950s)		10	15	35
Camelot (Whitman)	1157	15	20	25
Captain America (Whitman, 1966)	1181	15	30	35
Captain Kangaroo (Lowe, 1977)	4967	10	20	30
Captain Kangaroo Trace and Color (Whitman, 1960)	1413	10	15	20
Car 54, Where are You? (Whitman, 1962)	1157	20	40	55
Carmen Miranda (Saalfield)	2370	30	50	60
Casey Jones (Saalfield)	4618	10	25	35
Charge - G.I. Joe (Watkins, 1965)		25	40	45
Charlie Chan (Saalfield, 1941)	2355	30	65	90
Charlie Chaplin Up in the Air (M.A. Donahue & Co., 1914)	317	30	85	125
Chatter (Winter)	4106	10	25	35

Coloring Books

NAME/COMPANY/YEAR	NO.	GOOD	EX	MINT
Chitty Chitty Bang Bang (Whitman, 1968)	1654	5	10	20
Choo Choo (Watkins-Strathmore)	1857	10	20	30
Circus Boy (Whitman)	1198	10	20	30
Cisco Kid (Saalfield, 1950)	2078	15	20	35
Clutch Cargo (Artcraft)	9547	50	60	75
Clutch Cargo (Whitman, 1965)		50	90	95
Combat (Saalfield, 1963)		10	30	35
Convoy (Artcraft)	9637	5	10	20
Courageous Cat and Minute Mouse (Artcraft)	9540	60	140	175
Cuffy and Captain Gallant (Lowe)	2521	15	25	35
Curiosity Shop (Artcraft)	5353	10	20	25
Daktari (Whitman)	1649	15	25	35
Danger! with the Green Hornet (Watkins-Strathmore, 1966)		20	35	55
Dastardly and Muttley (Whitman)	1023	20	40	45
Dennis the Menace (Whitman)	1135	12	30	40
Dick Tracy (Saalfield, 1946)	2536	20	60	75
Dick Van Dyke (Artcraft)	9557	30	75	100
Dick Van Dyke (Saalfield/Artcraft, 1963)	9557	25	40	45
Dino the Dinosaur (Whitman)	1117	10	15	20
Disneyland (Whitman, 1961)	2975	5	10	15
Diver Dan (Saalfield)	4512	30	60	75
Doctor Dolittle and His Animals (Watkins-Strathmore, 1967)	1871-4	10	15	20
Donald Duck (Whitman, 1959)		15	20	25
Donna Reed (Artcraft)	2218	15	35	40
Donny and Marie (Whitman)	1641	5	12	25
Doris Day "A Warner Brothers Star" (Whitman, 1955)	1751	20	35	50
Doris Day (Whitman)	1143	10	25	30
Doris Day (Whitman, 1952)	1138	20	35	50
Double Deckers (Artcraft)	3936	10	20	25
Double Deckers (Saalfield)	3836	10	20	25
Dr. Kildare Play Book (Lowe)	3092	10	15	35
Draw & Paint Tom Mix (Whitman, 1935)		30	80	125
Edgar Bergen's Charlie McCarthy Paint Book (Whitman, 1938)	690	25	40	85
Electro Man (Lowe)	4971	7	25	35
Elizabeth Taylor (Whitman, 1950)	1119	30	60	80
Elizabeth Taylor (Whitman, 1954)	1144	30	50	75
Elmer Fudd & Friends (Watkins-Strathmore, 1962)	1849	10	20	30
Elmer Fudd (Western)	1878	12	25	35
Elvis (TN Mfg. & Dist. Co., 1983)		10	20	45
Emmett Kelly as Willy the Clown (Saalfield)	4533	10	20	30
Esther Williams (Merrill, 1950)	1591	20	40	70
Eve Arden Coloring Book (Saalfield, 1953)	2310	10	20	45
F-Troop (Saalfield)	9560	15	30	45
Family Affair (Whitman)	1640	7	15	25
Family Affair (Whitman, 1968)		7	20	25
Fantastic Osmonds: A Coloring Book, The (Osbro, 1973)	4622	15	35	50
Felix the Cat (Saalfield, 1959)	4655	15	35	60
Fireball XL5 (Golden, 1963)		35	50	90
Flintstones Gismos and Gadgets (Whitman, 1961)	1117	10	20	30
Flintstones Hoppy the Hopparoo, The (Whitman)	1117	15	25	35
Flintstones Pebbles & Bamm-Bamm, The (Whitman)	1004	10	20	25
Flintstones with Pebbles & Bamm-Bamm, The (Whitman)	1117	10	15	25
Flintstones, The (Whitman)	1138	10	20	30
Flintstones, The (Whitman, 1964)		10	20	25
Flipper Paint with Water (Whitman)	1340	5	15	25
Flying Nun (Artcraft)	4672	12	25	35
Foreign Legionnaire (Abbott)	2613	10	20	30
Frankenstein, Jr. (Whitman)	1115	20	40	50
Freddie (Whitman)	B400	10	15	20
Funky Phantom (Whitman)	1003	12	30	40
Funny Company (Whitman, 1966)		45	125	130
Fury (Whitman, 1958)	1199	10	20	30
G.I. Joe (Whitman)	1412	15	30	50

Coloring Books

NAME/COMPANY/YEAR	NO.	GOOD	EX	MINT
G.I. Joe Action Coloring Book (Whitman)	1156	10	25	40
Gabby Hayes (Abbott, 1954)		10	25	35
Garrison's Gorillas (Whitman, 1968)	1149	10	20	35
Gene Autry (Whitman, 1949)	1157	25	45	70
Gene Autry (Whitman, 1951)	1153	25	45	60
Gene Autry Cowboy Adventures to Color (Merrill, 1941)	4803	25	60	90
Gentle Ben (Whitman, 1968)	1642	10	25	30
George of the Jungle (Whitman, 1967)		25	50	100
Get Smart (Saalfield)	4519	20	40	50
Get Smart (Saalfield)	9562	20	40	50
Gilligan's Island (Whitman, 1965)	1135	15	30	50
Godzilla (Resource Pub.)	630	20	40	45
Gone with the Wind Paint Book (Merrill, 1940)	3403	30	50	150
Grace Kelly (Whitman, 1956)	1752	30	60	75
Green Acres (Whitman)	1188	25	45	60
Green Hornet (Watkins, 1966)		15	20	50
Green Hornet (Whitman, 1966)	1190	20	35	50
Greer Garson (Merrill, 1944)	3480	25	50	70
Grizzly Adams (NcNally)	06528	5	10	20
Grizzly Adams and Ben (McNally)	06534	5	10	30
Gumby and Pokey (Whitman, 1966)	1141	10	20	40
Gunsmoke (Whitman, 1958)	1184	20	30	45
H.R. Pufnstuf (Whitman, 1970)	1093	10	20	50
Harlem Globetrotters (Whitman, 1971)	1085	7	10	25
Hee Haw (Saalfield, 4538)	1970	10	20	30
Hercules (Lowe)	2838	20	40	75
Hi Mr. Jinks (Watkins-Strathmore)	1832	10	15	20
Hi! I'm Mrs. Beasley Color and Read (Whitman/Western, 1972)	1364	10	20	30
Hong Kong Phooey (Saalfield)	#H1852	10	20	30
Hood's Sarsaparilla Painting Book (Hood, 1894)		30	50	150
Hopalong Cassidy (Abbott)	1311	30	50	75
Hopalong Cassidy Starring William Boyd (Lowe, 1950)	1200	30	55	75
Hoppity Hopper (Whitman, 1965)		20	45	50
Hot Wheels (Whitman, 1969)		20	30	50
Howdy Doody (Whitman, 1950)	2093	15	25	60
Howdy Doody (Whitman, 1952)	217625	20	35	60
Howdy Doody Fun Book (Whitman, 1951)		20	30	40
Huckleberry Hound (Watkins-Strathmore)	1883	10	20	50
Huckleberry Hound (Whitman, 1959)	1130	10	20	40
Huckleberry Hound (Whitman, 1959)		15	35	45
I Love Lucy Coloring Book (Whitman, 1954)		25	50	75
I Love Lucy, Lucille Ball, Desi Arnaz, Little Ricky (Western/Dell, 1955)		25	45	70
Incredible Hulk at the Circus (Whitman, 1977)	1040	5	10	15
It's About Time (Whitman)	1134	15	30	50
Jack Webb's Safety Squad (Lowe)	1525	20	35	45
Jackie Gleason's TV Show (Abbott, 1956)	2614	35	45	60
Jambo (Whitman)	1173	5	10	20
Jetsons, The (Whitman)	1114	15	20	40
Jetsons, The (Whitman, 1962)	1135	20	45	60
Jetsons, The (Whitman, 1963)		15	30	35
JFK (Kanrom, 1962)		10	15	35
Jimmy Durante (Funtime)	F5035	30	60	75
Jimmy Durante Cut-Out Coloring Book (Pocket Books, 1952)	F5035	20	40	55
John Wayne (Saalfield, 1951)	1238	40	100	150
Johnny Lightning (Whitman)	1057	10	20	35
Jonny Quest (Whitman)	1091	20	50	75
Jonny Quest (Whitman)	1111	20	50	75
Josie and the Pussy Cats (Saalfield)	H1881	15	30	50
Journey to the Center of the Earth (Whitman)	1137	7	15	30
June Allyson (Whitman, 1952)	1862	20	40	55
Jungle Book (Saalfield)	9524	10	20	25
Kato's Revenge (Watkins-Strathmore)	1824	20	30	40

Yogi Bear, Whitman

Steve Canyon, 1952, Saalfield

Hopalong Cassidy starring William Boyd, 1950, Lowe

Adventures of Batman, 1966, Whitman

Coloring Books

NAME/COMPANY/YEAR	NO.	GOOD	EX	MINT
Kimba the White Lion (Saalfield)	4575	35	60	90
King Kong (Whitman)	1097	30	60	75
King Leonardo (Whitman)	1138	25	40	60
Korg (Saalfield)	H1853	5	10	30
Lad A Dog (Artcraft)	4526	12	20	40
Lady and the Tramp (Whitman, 1954)	1139	7	15	25
Lancelot Link (Whitman)	1146	10	20	25
Land of the Giants, The (Whitman, 1969)	1138	15	30	55
Land of the Lost (Whitman, 1975)	1045	15	25	40
Lariat Sam (Whitman, 1962)	1133	30	60	75
Lassie (Whitman)	1114	15	25	35
Lassie (Whitman, 1958)	1039	15	25	30
Lassie (Whitman, 1969)	1656	10	20	30
Lassie (Whitman, 1969)	1642	15	25	30
Laugh-In (Saalfield)	4633	15	30	50
Laurel & Hardy (Saalfield, 1972)	3883	10	15	25
Leave It To Beaver (Saalfield/Artcraft, 1958)	5662	20	50	60
Leave It To Beaver: A Book to Color (Saalfield, 1963)	5662	15	30	60
Lennon Sisters, The (Watkins-Strathmore)	1833	10	20	45
Lennon Sisters, The (Whitman, 1958)	158	10	20	45
Li'l Abner & Daisy Mae (Saalfield, 1942)	2391	25	60	100
Li'l Abner (Saalfield, 1941)	121	30	50	75
Liddle Kiddles (Whitman)	1411	20	30	55
Liddle Kiddles (Whitman)	1648	20	40	60
Lieutenant, The (Saalfield)	9577	10	20	30
Little Annie Rooney Paint Book (Whitman, 1930s)		25	40	75
Little Lulu & Tubby (Whitman, 1959)		20	30	50
Little Lulu (Watkins-Strathmore, 1950s)	1852	15	25	50
Little Orphan Annie (Saalfield, 1974)	4689	15	30	45
Little Rascals (Saalfield)	4546	10	20	35
Lone Ranger (Whitman, 1951)	1117-15	25	40	50
Loopy de Loop (Whitman)	2946	15	25	40
Loretta Young (Saalfield, 1956)	1108	25	40	50
Lucille Ball/Desi Arnaz (Whitman, 1953)	2079	25	40	60
Lucy Show Cut-Out Coloring Book, The (Golden, 1963)		25	40	75
Lucy Show, The (Funtime)	GF227	15	30	50
Magilla Gorilla (Whitman)	1113	20	30	50
Man Called Flintstone, The (Whitman)	1090	7	10	25
Man from U.N.C.L.E. (Watkins-Strathmore)	1855	15	25	50
Man from U.N.C.L.E. (Western)	1855	15	30	45
Man from U.N.C.L.E. (Whitman, 1965)	1095	15	30	40
Margaret O'Brien Paint Book (Whitman, 1947)	1155	25	40	75
Milton the Monster (Whitman)	1139	10	20	40
Mister Ed the Talking Horse (Whitman, 1963)	1135	15	30	45
Mister Magoo Cut-Out Coloring Book (Golden, 1961)	GF186	12	20	30
Moby Dick (Whitman)	1170	5	10	20
Monkees Big Beat Fun Book (Young World, 1986)		20	35	50
Monster Squad (McNally)	06428	7	10	25
Munsters, The (Whitman)	1149	20	40	75
Munsters, The (Whitman)	1648	30	65	75
Mushmouse and Punkin' Puss (Whitman)	1342	20	30	45
My Favorite Martian (Whitman)	1148	15	20	35
My Mother the Car (Saalfield)	4518	20	40	50
My Three Sons (Whitman, 1967)	1113	10	20	30
N.W. Passage (Lowe)	2852	10	20	30
Nancy and Sluggo (Whitman, 1970s)		15	25	40
Nancy Drew #1 (Treasure)	16003	5	10	15
Nancy Drew #2 (Treasure)	16004	5	10	15
Nanny and the Professor (Artcraft)	9620	7	10	25
Nanny and the Professor (Saalfield, 1971)	3829	7	10	25
National Velvet (Whitman)	2975	10	20	45
National Velvet (Whitman)	1186	10	20	40
New York World's Fair (Spertus, 1960s)		5	10	20

Coloring Books

NAME/COMPANY/YEAR	NO.	GOOD	EX	MINT
New Zoo Revue, The (Saalfield)	C1854	5	10	20
Our Gang Coloring Book (Saalfield, 1933)	966	10	20	35
Ozzie & Harriet, David & Ricky (Saalfield)	125910	25	40	50
Partridge Family (Artcraft, 1970)	3939	20	35	50
Partridge Family, The (Artcraft)	3937	20	30	40
Partridge Family, The (Saalfield)	4537	20	35	45
Patty Duke (Whitman)	1141	15	25	40
Patty Duke (Whitman, 1964)	1122	15	25	35
Peanuts (Saalfield)	5615	15	20	25
Pebbles Flintstone (Whitman)	1117	5	10	20
Peter Potamus (Whitman, 1964)	1139	10	20	35
Petticoat Junction (Whitman)	1111	20	40	60
Pinky Lee's Health and Safety (Funtime)	DF104	30	40	55
Pinocchio (Whitman, 1961)	1184	10	15	25
Planet of the Apes Cut and Color Book (Saalfield)	C2434	10	15	35
Popeye (Lowe, 1958)	2834	20	35	45
Prince Valiant (Saalfield, 1954)		15	20	30
Quick Draw McGraw (Whitman, 1959)		20	35	45
Raggedy Ann & Andy (Saalfield, 1944)	2498	20	40	75
Ramar (Saalfield)	4881	10	20	30
Ramar of the Jungle (Saalfield)	2208	15	20	30
Range Rider, The (Abbott, 1956)		15	20	35
Rat Patrol (Saalfield)	2273	20	35	50
Red Ryder (Whitman, 1952)	1155	15	30	50
Ricochet Rabbit (Whitman)	1142	20	40	50
Rin Tin Tin (Whitman, 1959)		10	25	30
Ripcord (Artcraft)	9529	20	30	40
Rita Hayworth (Merrill, 1942)	3483	30	60	75
Road Runner, The (Whitman)	1133	10	15	25
Rocky & Bullwinkle (Whitman, 1961)		12	20	45
Rocky the Flying Squirrel (Whitman, 1960)	1134	10	20	35
Roger Ramjet (Whitman)	1115	20	50	75
Ronny Howard of the Andy Griffith Show (Artcraft)	5644	10	35	50
Roy Rogers & Dale Evans (Whitman, 1951)	2171	20	40	55
Roy Rogers and Dale Evans' Big Book to Color (Whitman)	1184	15	30	50
Roy Rogers Paint Book (Whitman, 1944)	668	25	45	65
Roy Rogers' Pal Pat Brady (Whitman, 1955)	1255	10	25	30
Ruff and Reddy (Watkins-Strathmore)	1856	10	20	30
Sabrina (Whitman)	1071	10	20	40
Samson and Goliath (Whitman)	1113	10	15	25
Seahunt (Artcraft)	4558	15	35	45
Sergeant Preston (Whitman, 1953)	1329	10	15	25
Sgt. Bilko (Treasure, 1959)	330	15	35	45
Shadow (Saalfield, 1974)	4636	10	15	25
Shari Lewis (Artcraft)	5654	10	20	30
Shari Lewis (Saalfield)	5335	10	20	30
Shari Lewis and Her Puppets (Saalfield)	4527	15	25	35
Shazan (Whitman)	1084	12	20	30
Shirley Temple (Saalfield, 1935)	1738	20	50	60
Shirley Temple Crosses the Country (Saalfield, 1939)	1779	30	110	125
Shirley Temple Drawing & Coloring Book (Saalfield, 1936)	1724	30	60	100
Sigmund and the Sea Monsters (Saalfield, 1974)		10	30	40
Six Million Dollar Man (Saalfield, 1974)	C1832	10	15	20
Six Million Dollar Man Activity Book (Rand McNally, 1977)	C2471	10	15	20
Skippy (Whitman, 1970)	1116	5	10	20
Smitty Color Book (McLoughlin Bros., 1931)		20	40	75
Snagglepuss & Yakky Doodle (Whitman, 1962)	1209	15	25	40
Sonja Henie Coloring Book (Merrill, 1939)	3476	20	50	75
Soupy Sales (Treasure)	8907	10	15	25
Space Angel (Artcraft)	9528	30	50	100
Space Ghost (Whitman)	1082	10	30	40
Space:1999 (Saalfield)	C1881	10	20	30
Speed Buggy (Saalfield)	H1860	15	30	35

Coloring Books

NAME/COMPANY/YEAR	NO.	GOOD	EX	MINT
Star Trek (Saalfield, 1975)	C1856	5	10	20
Star Trek (Saalfield, 1975)	C1862	10	15	25
Steve Canyon (Saalfield, 1952)	123410	20	35	60
Stingray (Whitman, 1965)	1133	15	35	40
Super Circus (Whitman)	1251	15	25	35
Super Six (Whitman)	1181	15	30	60
Superboy (Whitman, 1967)		10	25	30
Superman (Whitman, 1964)		20	35	40
Tales of the Texas Rangers (Saalfield, 1958)	2071	10	15	25
Tales of the Vikings (Saalfield)	4566	10	15	25
Tales of Wells Fargo (Whitman, 1957)		15	25	35
Tarzan (Whitman)	1157	10	30	40
Tarzan (Whitman, 1968)	1157	10	30	40
Tennessee Tuxedo (Whitman)	1011	15	40	60
That Girl (Saalfield)	4510	10	20	30
That Girl (Saalfield, 1967)	4513	10	20	30
Three Stooges Funny Coloring Book (Norman Maurer, 1960)	2855	20	40	55
Three Stooges, The (Lowe)	2822	25	50	80
Three Stooges, The (Whitman)	1135	10	20	35
Time Tunnel (Saalfield, 1967)		25	50	80
Tom and Jerry (Whitman, 1959)		10	20	30
Tom Corbett (Saalfield, 1950)		15	30	55
Tonto (Whitman, 1957)	2953	15	25	40
Top Cat (Whitman, 1961)	1185	15	30	40
Touche Turtle (Whitman)	1185	20	50	125
Tweety Coloring Book (Whitman, 1955)	2953	10	15	25
Twelve O'Clock High (Artcraft, 1960s)	9636	10	20	35
Uncle Martin the Martian (Funtime)	GF233	15	30	40
Underdog (Whitman)	1666	15	25	40
Underdog (Whitman)	1010	20	30	35
Voyage to the Bottom of the Sea (Whitman, 1965)	1851-B	15	30	50
Wacky Races (Whitman)	1067	15	30	40
Wacky World of the Great Grape Ape, The (McNally)	06440	10	15	20
Wally Gator (Whitman)	2975	30	50	70
Walt Disney Presents Winnie the Pooh (Whitman)	1058	5	10	15
Walt Disney's Alice in Wonderland Paint Book (Whitman, 1951)	2167	25	50	60
Walt Disney's Annette (Whitman, 1964)	1145	10	15	40
Walt Disney's Mickey Mouse Club (Whitman)	1059	7	10	15
Walt Disney's Mickey Mouse Paint Book (Whitman, 1937)	1069	30	60	110
Walt Disney's Peter Pan (Whitman, 1952)	2186	15	30	55
Walt Disney's Snow White and the Seven Dwarfs Paint Book (Whitman, 1938)		30	50	110
Walt Disney's Spin & Marty (Whitman, 1956)	1194	5	10	20
Walt Disney's Swamp Fox (Whitman, 1961)	1134	10	20	25
Walt Disney's Zorro (Whitman)	1158	15	30	50
Welcome Back Kotter (Whitman, 1977)	1081	5	10	15
Where's Huddles? (Whitman)	1089	5	10	25
Whirlybirds (Whitman)	1151	15	35	60
Wild Bill Hickok and Jingles (Saalfield, 1953)	120910	15	30	45
Wild Kingdom (Whitman)	1004	10	15	20
Woody Woodpecker (Whitman, 1950s)	1416	15	20	35
Wyatt Earp Coloring Book (Whitman, 1958)		20	35	55
Yakky Doodle and Chopper (Whitman)	1800	15	25	35
Yogi Bear (Whitman)	1111	15	20	35
Yosemite Sam (Watkins, 1963)		20	25	30

A.C. Gilbert Erector Sets

What adult doesn't recall the joy of using toys to vicariously experience adult tasks — whether it be driving a toy truck or playing house? Construction toys were designed to make kids òfeel bigó and exercise their imagination and creativity.

Those boxes of parts of many sizes, shapes, colors, and purpose taught basic mechanics, logic, fine motor coordination and a host of other skills — all while providing hours of fun.

Most boys can recall the colors, heft and smell of their first construction set, even if they may not remember when they received it or who gave it to them. Those unfortunates who never got one no doubt made do with sets of their own design — including wood scraps, string, old wheels and nails.

Few joys of childhood rival that of laboriously constructing some mechanical wonder, watching it work, and then gleefully tearing it all apart.

While many other construction sets have been made through the years, none has received the most collectors' attention and praise like the original A.C. Gilbert Erector sets.

A Set by Any Other Name isn't the Same

Alfred Carlton Gilbert, born in 1884, was both a product of and producer for his time, the industrial coming of age in America. He inspired legions of bright young tinkerers to grow up to become architects of the future — engineers, scientists and craftsmen of all types.

As a child, Gilbert loved magic tricks, a fascination that would later pay off almost like magic. Also an aspiring sports star, he decided on a career in health education, and worked his way through Yale medical school by performing magic shows. At Yale he

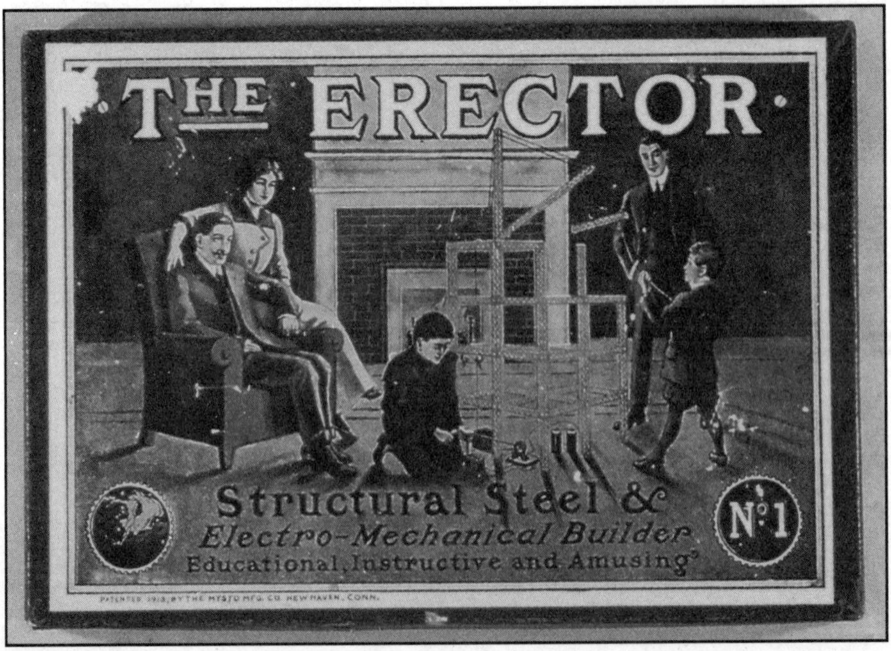

Set #1, 1913

Set #10042, 1959-'61

also won a berth on the 1908 U.S. Olympic team and won the gold medal in the pole vault.

While at Yale, Gilbert began giving magic lessons and became frustrated by the lack of available magic props for his students. A fellow amateur magician named John Petrie was also a mechanic, and he and Gilbert began making small magic kits and selling them.

After the Olympics, Gilbert returned to medical school and the magic business, which soon grew into a mail order magic company, supplying both amateurs and professionals. Gilbert and Petrie formed the Mysto Manufacturing Company.

By 1909, the newly degreed Dr. Gilbert left his medical career to pursue his magic business — and business was good. In his promotional travels across America, Gilbert began selling to toy store buyers and soon realized the need for quality American-made toys. Deep in thought, he went home, and the result was the birth of the Erector set.

The origin of the first Erector set is the stuff of debate and myth, and even Gilbert told several versions of it during his life. What is fact is that Petrie was not impressed with Gilbert's new toy idea, and this eventually led Gilbert to buy out Petrie's interest in the Mysto Manufacturing Company.

Gilbert showed his new Mysto-Erector toy at the 1913 Toy Fair and retail orders poured in, threatening to swamp his new company. Gilbert rolled out a major national promotion for his new line, including an ingenious model building contest that provided him hundreds of new and free model configurations he promoted in future instruction manuals.

The 1913 holiday sales season was a resounding success, and Gilbert turned his sights on refining and improving the product line, upgrading the girders and providing factory-assembled motors for the 1914 kits. He also continued the model building contests with huge success, offering real cars as prizes, and those new model ideas were incorporated into instruction manuals.

A new plant in 1915 allowed Gilbert to expand the line into better motors for his Erector sets, tin toys and electric fans to fill the off-season work void.

In 1916, Gilbert renamed his firm The A.C. Gilbert Company, but kept the Mysto name for his magic sets. He also created the Gilbert Engineering Institute for Boys, a national club which became a highly successful promotional tool.

In 1917, World War I threatened to sideline the entire toy industry. Gilbert was called by the fledgling Toy Manufacturers Association, which he helped create, to lobby Congress for permission for the toy industry to continue making toys. Armed with his own toys, Gilbert triumphed and was hailed nationwide as the "Man Who Saved Christmas."

Gilbert introduced his now-famous chemistry sets in 1917 and proceeded to dramatically expand the Gilbert line in the years to come, introducing tool chests, a line of power tools and books.

It was a year of great change in 1920, with a near total overhaul of the Gilbert lines and the introduction of a series of more advanced scientific, mechanical and radio sets. That year also saw the introduction of the No. 10 Deluxe Erector sets, now by far the most highly desirable of all Gilbert construction sets.

A true overhaul came in 1924, when Gilbert introduced the new Erector line. He completely redesigned the Erector system with thinner girders and debuted curved girders and numerous other advancements. The 1924 line also saw streamlining across all other divisions, with many toys dropped from production.

In 1927, Gilbert expanded his control over the construction toy market by buying Trumodel, a line of high quality toys that were redesigned and rolled into his deluxe Erector sets.

Gilbert's promotional activities took another turn in 1929 when he entered into a number of exclusive contracts with Sears, Macy's, J.C. Penney and other national retailers. Each company was allowed to market low-end Gilbert products, but not under the Erector name. They were sold under the names Trumodel, Steel-Tech and Little Jim among others, with each retailer having a different line. These sets have since become collectible in their own right.

Also in 1929, Gilbert essentially monopolized the American construction toy market by acquiring his chief rival, Meccano. For 1930, the Erector and Meccano lines were partially merged in the New American Meccano line, now commonly called Erector-Meccano.

Set #10062, 1958

The Great Depression finally hit the Gilbert company, but not before the Erector series hit its all-time pinnacle with the Erector Hudson Locomotive model, which was included in the top-of-the-line sets. In 1931, Gilbert's most expansive No. 10 set was produced. In addition to containing parts to build all models from lower numbered sets (which included the White Truck model, the Steam Shovel, and the Zeppelin), it also included parts for both the Hudson Locomotive and its tender. Gilbert would never offer an Erector set this grand again.

The Depression forced major downsizing at Gilbert, but the company would survive. The Erector line was acquired by Gabriel Industries in 1967. As previously noted, the Erector line has been revived with resounding success.

Trends

While still a relatively small subset of the toy collecting field, construction set collecting, particularly of Erector and Meccano products, is enjoying moderate and steady growth. Original Gilbert sets command the highest premiums, particularly the earliest ones bearing the company's previous Mysto Manufacturing name.

Not all collecting fields have their day in the sun, but this one, rich as it is in history and variety, presents a strong case for higher visibility.

Note: All sets listed are made by The A.C. Gilbert Company. Collectors tend to identify sets by number and year, rather than name; in this book, sets are listed by set number, then by year.

Erector sets are rarely found in Mint condition, especially with all the pieces and the box intact. This section lists two grades for sets collectors may likely find — Good and Excellent. Sets that have been refurbished and reassembled as a whole may be found and can command a price 40 to 75 percent higher than a similar, unrefurbished set.

Values given are averages for the time period noted. As a general rule, older sets are worth more than newer ones; larger sets are worth more than smaller.

The Top 10 A.C. Gilbert Erector Sets
(in Excellent condition)

1. Set #10, Complete Erector in all its Glory, 1929-31, oak chest $7,875
2. Set #10, Complete Deluxe Set, 1928, nine-drawer oak chest 7,875
3. Set #10, Erector Deluxe in all its Glory, 1927, eight-drawer oak chest 7,875
4. Set #10, 1920-26, wood box 7,875
5. Set #9, Mechanical Wonders Set
   with 110-Volt Motor, 1929-32, wood box 4,610
6. Set #8-1/2, 1931-32, wood box 3,150
7. Set #7-1/2, Motorized Erector, 1914, wood box 1,575
8. Set #12-1/2, 1956-57, metal box 1,365
9. Set #12-1/2, 1948-50, metal box 1,365
10. Set #10093, 1959, metal box 1,265

Contributor to this section: Daniel Yett, 1900 Penbrooke Tr., Dayton, OH 45459-3432.

A. C. GILBERT ERECTOR SETS

Set #0

YEAR	BOX TYPE	GOOD	EX
1913	cardboard	80	130
1914	cardboard	60	100
1915	cardboard	120	290

Set #1

YEAR	BOX TYPE	GOOD	EX
1913	cardboard	105	210
1914-34	cardboard	80	130
1945	cardboard	80	130

Set #1-1/2

YEAR	BOX TYPE	GOOD	EX
1935-42	cardboard	40	55
1950-56	cardboard	40	55

Set #2

YEAR	BOX TYPE	GOOD	EX
1913	cardboard	110	220
1914-21	cardboard	75	115

Set #2-1/2

YEAR	BOX TYPE	GOOD	EX
1936-42	cardboard	45	65
1946-56	cardboard	45	65

YEAR	BOX TYPE	GOOD	EX
1913	cardboard	195	315
1914-34	cardboard	95	155
1945	cardboard	95	155

Set #3-1/2

YEAR	BOX TYPE	GOOD	EX
1935	cardboard	125	165
1936-42	cardboard	65	85

Set #4

YEAR	BOX TYPE	GOOD	EX
1913	cardboard	220	420
1914	cardboard	130	210
1915	wood	130	210
1916-34	cardboard	130	210
1945	metal	105	185

Set #4-1/2

YEAR	BOX TYPE	GOOD	EX
1935	metal	140	235
1936-42	cardboard	50	85
1946-57	metal	140	235

Set #5

YEAR	BOX TYPE	GOOD	EX
1913	cardboard	220	525
1914	cardboard	180	290
1915	wood	145	265
1916	cardboard	145	265
1923	wood	550	895

Set #5

YEAR	BOX TYPE	GOOD	EX
1930	cardboard	120	290

Set #5-1/2

YEAR	BOX TYPE	GOOD	EX
1936-38	metal	95	140

Set #6

YEAR	BOX TYPE	GOOD	EX
1913		630	840
1914	wood	185	290
1915-16	wood	155	260
1920	wood	155	260
1921	wood	155	260
1931	cardboard	100	260
1945	metal	190	260

Set #6-1/2

YEAR	BOX TYPE	GOOD	EX
1935-42	metal, not made in 1936 or 1937	50	95
1946-57	metal	50	95

Set #7

YEAR	BOX TYPE	GOOD	EX
1913	wood	315	470
1914-32	wood	160	260
1933-34	metal	160	260
1945	metal	210	315

Set #7-1/2

YEAR	BOX TYPE	GOOD	EX
1914	wood	655	1575
1928-32	wood	260	420
1934-37	metal	90	155
1938-57	metal	75	130

Set #8

YEAR	BOX TYPE	GOOD	EX
1913	wood	575	945
1914-31	wood	575	945
1933-34	metal	575	945
1945	metal	260	350

Set #8-1/2

YEAR	BOX TYPE	GOOD	EX
1931-32	wood	1900	3150
1933	metal	190	240
1935-57	metal	105	180

Set #9

YEAR	BOX TYPE	GOOD	EX
1929-32	wood	2650	4610
1945	metal	210	315

Set #9-1/2

YEAR	BOX TYPE	GOOD	EX
1935	metal	290	390

A. C. GILBERT ERECTOR SETS

Set #9-1/2

YEAR	BOX TYPE	GOOD	EX
1936-49	metal	210	315

Set #10

YEAR	BOX TYPE	GOOD	EX
1920-26	wood	4610	7875
1927	eight drawer oak chest	4610	7875
1928	nine-drawer oak chest	4610	7875
1929-31	oak	4610	7875

Set #10-1/2

YEAR	BOX TYPE	GOOD	EX
1936-42	metal	315	470
1949-57	metal	235	395

Set #12-1/2

YEAR	BOX TYPE	GOOD	EX
1948-50	metal	735	1365
1956-57	metal	735	1365

Erector Junior

YEAR	BOX TYPE	GOOD	EX
1943-47	cardboard, #5	100	225
1943-47	cardboard, #3	70	185
1943-47	cardboard, #1	40	115

Junior Erector

YEAR	BOX TYPE	GOOD	EX
1949-55	cardboard, #4	45	100
1949-55	cardboard, #2	30	75
1949-55	cardboard, #10	85	225

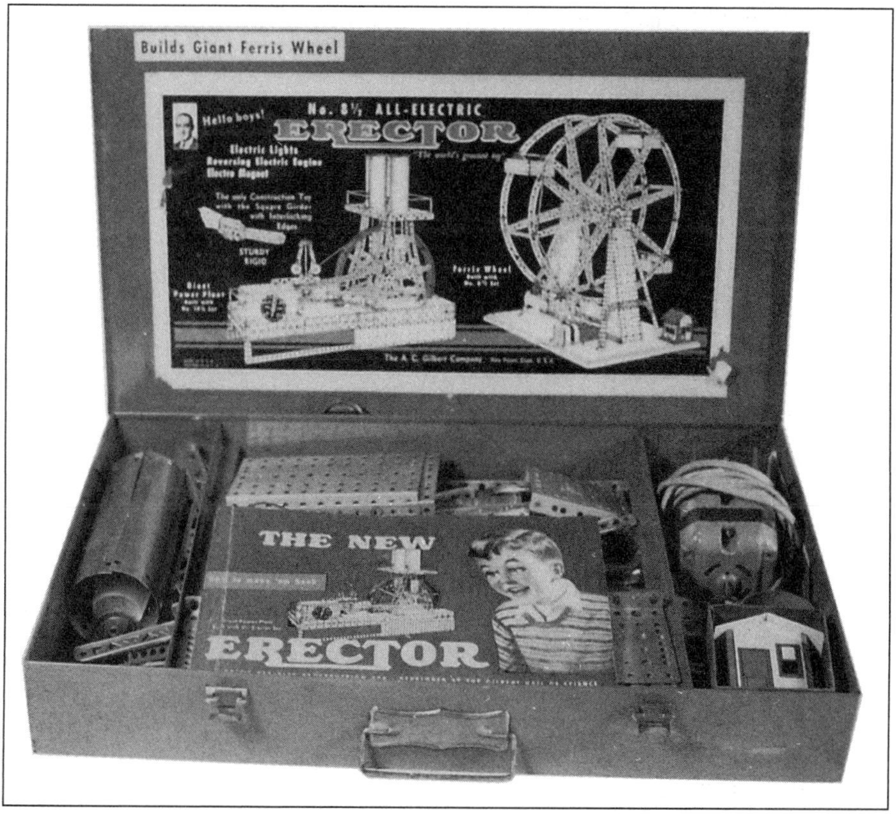

Set No. 8-1/2, late 1930s

Post 1957 Sets

NO.	YEAR	BOX TYPE	GOOD	EX
10026	1958	cardboard	45	85
10082	1958	metal	315	525
10072	1958	metal	185	290
10062	1958	metal	77	145
10052	1958	metal	55	95
10041	1958	metal	45	85
10031	1958	cardboard	65	105
10092	1958	metal	785	1260
10011	1958-61	cardboard tube	45	70
10021	1958-61	cardboard tube	55	95
10093	1959	metal	735	1265
10073	1959	metal	185	290
10083	1959	metal	315	525
10063	1959-61	metal	80	150
10032	1959-61	cardboard tube	70	115
10042	1959-61	metal	45	85
10037	1960	cardboard tube	135	210
18040	1960-61	tube	105	150
18010	1960-61	tube	55	80
10074	1960-61	metal	85	150
18020	1960-61	tube	65	95
18030	1960-61	tube	80	115
10084	1960-61	metal	160	295
10094	1960-62	metal	420	685
10201	1962	metal	60	95
10211	1962	metal	85	160
10221	1962	metal	100	185
10231	1962	metal	210	315
10161	1962-63	carton	55	85
10171	1962-63	carton	65	105
10181	1962-63	metal	50	80
10128	1963-64	metal	120	160
10129	1963-64	metal	210	315
10127	1963-64	metal	95	135
10253	1964	cardboard	35	45
10254	1964	cardboard	40	55
10252	1964	cardboard	30	40
10251	1964	cardboard	22	35
10354	1965-66	cardboard	45	55
10353	1965-66	cardboard	35	45
10352	1965-66	cardboard	30	40
10351	1965-66	cardboard	22	35
10601	1965-66	cardboard	115	210
10606	1966	cardboard	180	265

Fisher-Price Toys

The familiar sturdy wood toys with paper lithography made by Fisher-Price are colorful remnants of how toys used to be made. Collectors especially value their quality and nostalgia.

While the rest of the world suffered through the Depression, Fisher-Price set up shop in 1930 in East Aurora, N.Y. Founded by Herman Fisher, Irving Price and Helen Schelle, the company hoped to bring toys of matchless charm and inherent quality to children.

Sixteen toys were introduced in the 1931, the first production year, including Granny Doodle and Doctor Doodle, a pair of charming ducks.

Most collectors are familiar with the look of vintage Fisher-Price. Early toys featured a unique, colorful, crisp paper lithography on wood. Finding such charm today is as rare as some of the early toys.

It didn't take long for Fisher-Price to realize the value of licensing. In 1935, the company issued the Walt Disney Mickey Mouse Band, featuring Mickey and Pluto. That piece alone can command around $2,000 in Mint in Box condition. Other Disney characters and Popeye also became popular Fisher-Price toys.

Popeye the Sailor, 1936

Donald Duck Drummer, 1949

Musical toys, especially items featuring bells or xylophones, became perennial favorites.

By the late 1950s, plastic was beginning to appear as accents on toys, but by the late 1960s, most of the toys' bodies were made entirely of plastic.

The common Fisher-Price Play Family Little People appeared as early as 1959 in a yellow wooden Safety School Bus. Made of wood, they could be removed from their vehicles. Their body shapes and compositions have changed throughout the years. Today the Little People are about three times the size of the originals and are made of plastic.

Today, Fisher-Price is a subsidiary of toy giant Mattel. The company continues to make Little People play sets in a variety of themes, but has expanded its line to include outdoor playground and riding toys, apparel and books.

Trends

Collecting vintage Fisher-Price, especially pre-1950s pieces, can be costly, since it is difficult to find older pieces in premium condition. Many 1930s-1940s wooden toys can command $400-$2,000 or more in Mint in Box condition.

But collectors shouldn't pass on more modern, plastic Fisher-Price toys. Many toys from the 1960s and later are gaining favor with younger collectors, especially Little People play sets and musical radios, clocks, and televisions. The musical toys are still fairly easy to find and can range from $5-$30 in general.

Some of the more common and easy-to-find toys include the Chatter Telephone, Pull-a-Tune Xylophone, Corn Popper and numerous bees and Snoopy dogs. Because of their inherent play value and lengthy history, many were and still are used as toys and are often found in less than Mint condition.

Editor's Note: Many Fisher-Price toys had several variations, which often means several different values. Also, the date found on the toy may only be a copyright date, which may be earlier than the actual date of manufacture.

The Top 20 Fisher-Price Toys
(in Mint in Box condition)

1. Skipper Sam, 1934, #155 .... $2,800
2. Popeye the Sailor, 1936, #703 2,400
3. Popeye, 1935, #700 2,200
4. Woodsy-Wee Circus, 1931, #201 2,100
5. Mickey Mouse Band, 1935, #530 2,100
6. Popeye Cowboy, 1937, #705 2,100
7. Doughboy Donald, 1942, #744 2,000
8. Drummer Bear, 1931, #102 2,000
9. Doctor Doodle, 1931, #100 2,000
10. Granny Doodle, 1931, #101 2,000
11. Granny Doodle & Family, 1933, #101 2,000
12. Barky Puppy, 1931, #103 2,000
13. Chubby Chief, 1932, #110 2,000
14. Dumbo Circus Racer, 1941, #738 2,000
15. Doc & Dopey Dwarfs, 1938, #770 2,000
16. Dog Cart Donald, 1936, #149 2,000
17. Raggedy Ann & Andy, 1941, #711 1,975
18. Walt Disney's Easter Parade, 1936, #475 1,900
19. Pushy Doddle, 1933, #507 1,900
20. Scotty Dog, 1933, #710 1,900

Squeaky the Clown, 1958

Queen Buzzy Bee, 1962

FISHER-PRICE TOYS

NAME	NO.	YEAR	GOOD	EX	MINT
Allie Gator	653	1960	40	80	160
Amusement Park	932	1963	50	100	200
Baby Chick Tandem Cart	50	1953	45	65	125
Barky	462	1958	45	75	125
Barky Buddy	150	1934	400	900	1800
Barky Puppy	103	1931	550	1000	2000
Big Bill Pelican	794	1961	45	75	125
Big Performing Circus	250	1932	450	900	1800
Blackie Drummer	785	1939	400	650	1000
Bonny Bunny Wagon	318	1959	30	45	80
Boom Boom Popeye	491	1937	550	800	1600
Bossy Bell	656	1959	35	45	70
Bouncing Bunny Wheelbarrow	727	1939	450	750	1300
Bouncy Racer	8	1960	40	65	90
Bucky Burro	166	1955	95	200	550
Buddy Bronc	430	1938	200	400	800
Buddy Bullfrog	728	1959	40	65	125
Bunny Basket Cart	301	1957	30	45	80
Bunny Basket Cart	303	1960	30	45	95
Bunny Bell Cart	520	1941	125	250	550
Bunny Bell Cart	604	1954	45	75	150
Bunny Cart	5	1948	75	100	145
Bunny Drummer	512	1942	150	225	550
Bunny Drummer	505	1946	150	225	550
Bunny Egg Cart	28	1950	50	70	125
Bunny Egg Cart	404	1949	50	70	125
Bunny Engine	703	1954	45	75	135
Bunny Racer	474	1942	125	225	550
Busy Bunny Cart	719	1936	150	300	750
Butch the Pup	333	1951	45	75	125
Buzzy Bee	325	1950	40	60	80
Cash Register	972	1960	35	75	125
Chatter Monk	798	1957	45	65	145
Chatter Telephone	747	1962	30	45	65
Choo-Choo Local	517	1936	350	700	1500
Chubby Chief	110	1932	450	950	2000
Chuggy Pop-Up	616	1955	40	65	135
Circus Wagon	156	1942	125	300	650
Concrete Mixer Truck	926	1959	150	300	625
Corn Popper	785	1957	35	50	115
Corn Popper	788	1963	35	50	85
Cotton Tail Cart	525	1940	200	400	800
Cowboy Chime	700	1951	150	300	600
Dandy Dobbin	765	1941	150	325	675
Dashing Dobbin	742	1938	250	400	850
Ding-Dong Duckey	724	1949	175	225	575
Dinkey Engine	642	1959	35	50	95
Dizzy Dino	407	1931	250	400	800
Dizzy Donkey	433	1939	45	80	160
Doc & Dopey Dwarfs	770	1938	550	1000	2000
Doctor Doodle	477	1940	150	300	600
Doctor Doodle	100	1931	600	1000	2000
Dog Cart Donald	149	1936	600	1000	2000

Fisher-Price Toys

NAME	NO.	YEAR	GOOD	EX	MINT
Doggy Racer	7	1942	150	300	600
Donald Cart	469	1940	450	900	1800
Donald Duck & Nephews	479	1941	350	700	1400
Donald Duck Cart	544	1942	200	350	750
Donald Duck Cart	605	1954	125	250	500
Donald Duck Cart	500	1937	350	600	1500
Donald Duck Choo-Choo	450	1942	225	450	900
Donald Duck Choo-Choo	450	1940	350	600	1200
Donald Duck Delivery	715	1936	350	750	1500
Donald Duck Drum Major	432	1948	150	300	600
Donald Duck Drum Major	550	1940	150	300	600
Donald Duck Drummer	454	1949	175	225	550
Donald Duck Pop-Up	425	1938	325	650	1200
Donald Duck Xylophone	185	1938	325	750	1000
Donald Duck Xylophone	177	1946	150	350	700
Dopey Dwarf	770	1939	350	650	1700
Doughboy Donald	744	1942	550	1000	2000
Drummer Bear	102	1931	550	1000	2000
Ducky Cart	51	1950	45	85	145
Ducky Cart	6	1948	45	85	145
Ducky Daddles	14	1941	35	65	135
Ducky Daddles	148	1942	175	300	600
Dumbo Circus Racer	738	1941	450	1000	2000
Easter Bunny	490	1936	165	300	675
Elsie's Dairy Truck	745	1948	300	700	1400
Farm Truck	845	1954	145	300	600
Farmer in Dell Music Box	763	1962	30	40	65
Farmer in the Dell TV Radio	166	1963	25	35	75
Fido Zilo	707	1955	35	65	135
Fire Truck	630	1959	35	50	95
Fuzzy Fido	444	1941	125	200	550
Gabby Duck	767	1952	55	80	160
Gabby Goofies	777	1963	35	40	65
Gabby Goofies	775	1956	25	45	75
Gabby Goose	120	1936	35	65	95
Galloping Horse & Wagon	737	1948	150	350	750
Go'N Back Jumbo	360	1931	450	900	1800
Gold Star Stagecoach	175	1954	150	425	975
Golden Gulch Express	191	1961	65	95	165
Granny Doodle	101	1931	600	1000	2000
Granny Doodle & Family	101	1933	650	1500	2000
Happy Helicopter	498	1953	150	225	425
Happy Hippo	151	1962	35	65	145
Hot Diggety	800	1934	450	900	1800
Howdy Bunny	757	1939	300	650	1100
Huckleberry Hound Xylo	711	1961	175	225	475
Huffy Puffy Train	999	1958	55	80	175
Humpty Dump Truck	145	1963	40	65	125
Humpty Dumpty	757	1957	165	225	550
Husky Dump Truck	145	1961	30	45	95
Jack-n-Jill TV Radio	148	1959	25	45	85
Jingle Giraffe	472	1956	55	175	375

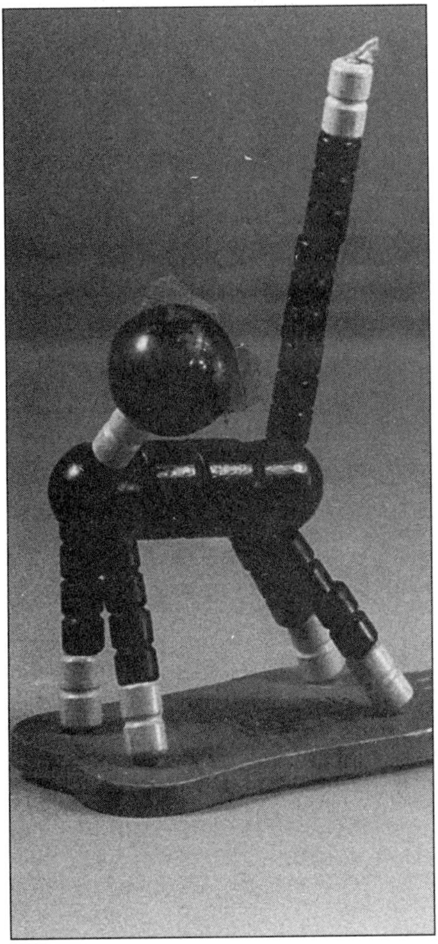

Tailspin Tabby Pop-Up, 1947

Lop-Ear Looie, 1934

Fisher-Price Toys

NAME	NO.	YEAR	GOOD	EX	MINT
Jolly Jumper	793	1963	25	50	85
Jolly Jumper	450	1954	35	75	165
Juggling Jumbo	735	1958	150	225	425
Jumbo Jitterbug	422	1940	65	150	325
Jumbo Rolo	755	1951	65	150	295
Junior Circus	902	1963	75	200	395
Katy Kackler	140	1954	45	80	160
Kicking Donkey	175	1937	175	285	725
Kitty Bell	499	1950	65	150	275
Lady Bug	658	1961	30	45	75
Leo the Drummer	480	1952	175	225	450
Little Snoopy		1970s	3	7	10
Looky Chug-Chug	161	1949	75	175	325
Looky Fire Truck	7	1950	65	100	250
Lop-Ear Looie	415	1934	150	300	600
Lucky Monk	109	1932	350	850	1800
Merry Mousewife	662	1962	35	50	80
Merry Mutt	473	1949	45	75	145
Mickey Mouse Band	530	1935	750	1500	2100
Mickey Mouse Choo-Choo	432	1938	450	850	1800
Mickey Mouse Drummer	476	1941	150	300	625
Mickey Mouse Puddle Jumper	310	1953	150	225	395
Mickey Mouse Safety Patrol	733	1956	165	300	625
Mickey Mouse Xylophone	798	1942	275	500	925
Mickey Mouse Xylophone	798	1939	300	600	975
Mickey Mouse Zilo	714	1963	165	325	575
Mini Bus		1970s	1	3	5
Molly Moo-Moo	190	1956	65	200	400
Moo-oo Cow	155	1958	40	80	160
Mother Goose Cart	784	1955	35	70	145
Music Box Barn	764	1960	25	45	90
Music Box Sweeper	131	1961	40	65	165
Musical Elephant	145	1948	165	325	700
Musical Push Chime	722	1950	35	50	95
Musical Sweeper	225	1953	40	65	170
Musical Sweeper	100	1950	65	135	275
Musical Tick Tock Clock	997	1962	25	45	90
Nosey Pup	445	1956	45	75	155
Patch Pony	616	1963	25	40	85
Perky the Pot	686	1958	40	65	135
Peter Bunny Cart	472	1939	150	325	725
Peter Bunny Engine	721	1949	75	200	425
Peter Bunny Engine	715	1941	70	225	475
Peter Pig	479	1959	35	50	110
Pinky Pig	695	1958	35	65	145
Pinky Pig	695	1956	40	70	160
Pinocchio	494	1939	105	400	975
Pinocchio	720	1939	200	400	940
Playful Puppy	625	1961	35	60	120
Playland Express	192	1962	40	80	165
Pluto Pop-Up	440	1936	35	75	160
Pony Chime	137	1962	30	65	135

Fisher-Price Toys

NAME	NO.	YEAR	GOOD	EX	MINT
Pony Express	733	1941	65	200	445
Poodle Zilo	739	1962	30	65	145
Pop 'N Ring	809	1959	40	65	125
Popeye	700	1935	700	1500	2200
Popeye Cowboy	705	1937	650	1000	2100
Popeye Spinach Eater	488	1939	225	650	1300
Popeye the Sailor	703	1936	600	1200	2400
Prancing Horses	766	1937	175	400	900
Pudgy Pig	478	1962	35	50	110
Puffy Engine	444	1951	35	75	145
Pull-A-Tune Xylophone	870	1957	30	45	95
Pushy Doddle	507	1933	350	800	1900
Pushy Elephant	525	1934	225	600	1700
Pushy Piggy	500	1932	350	800	1800
Quacko Duck	300	1939	135	250	600
Quacky Family	799	1946	30	50	85
Queen Buzzy Bee	444	1962	25	50	80
Rabbit Cart	52	1950	35	70	140
Racing Bunny Cart	723	1938	150	350	700
Racing Ponies	760	1936	200	500	850
Racing Pony	705	1933	325	1000	1475
Racing Rowboat	730	1952	105	225	475
Raggedy Ann & Andy	711	1941	400	900	1975
Rattle Ball	682	1959	20	30	50
Riding Horse	254	1940	400	900	1475
Rock-A-Bye Bunny Cart	788	1940	175	425	825
Rock-A-Stack	627	1960	15	25	40
Roller Chime	123	1953	35	70	140
Rolling Bunny Basket	310	1961	35	65	135
Rooster Cart	469	1938	95	225	575
Safety School Bus	990	1962	50	100	175
Safety School Bus	983	1959	150	300	600
Scotty Dog	710	1933	350	850	1900
Shaggy Zilo	738	1960	40	80	160
Skipper Sam	155	1934	650	1300	2800
Sleepy Sue (Turtle)	495	1962	25	45	90
Smokie Engine	642	1960	30	40	85
Snoopy Sniffer	180	1958	50	100	195
Snoopy Sniffer	180	1938	65	200	475
Snorky Fire Engine	168	1960	50	95	195
Sonny Duck Cart	410	1941	75	200	475
Space Blazer	750	1953	175	400	850
Squeaky the Clown	777	1958	125	225	550
Stake Truck	649	1960	40	80	160
Streamline Express	215	1935	350	700	1700
Strutter Donald Duck	510	1941	125	300	700
Struttin' Donald Duck	900	1939	205	450	875
Sunny Fish	420	1955	75	200	425
Suzie Seal	460	1961	25	40	85
Tabby Ding Dong	730	1939	200	400	800
Tailspin Tabby	400	1931	125	350	700
Tailspin Tabby Pop-Up	600	1947	100	200	400

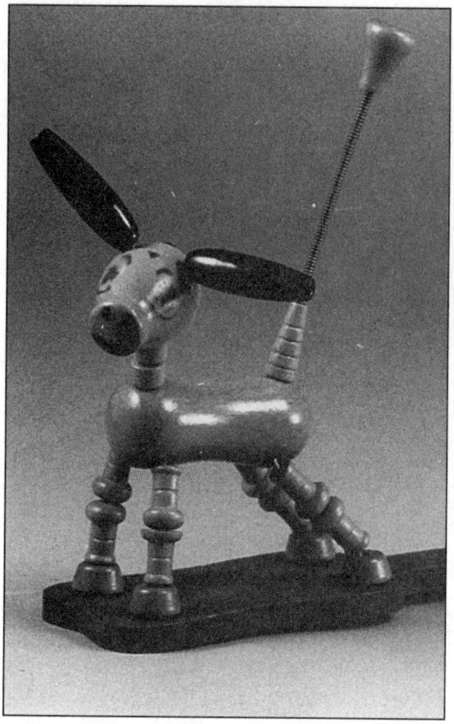

Dizzy Donkey, 1939

Leo the Drummer, 1952

Fuzzy Fido, 1941

Fisher-Price Toys

NAME	NO.	YEAR	GOOD	EX	MINT
Talk-Back Telephone	747	1961	35	70	140
Talking Donald Duck	765	1955	35	70	140
Talky Parrot	698	1963	50	100	195
Tawny Tiger	654	1962	30	65	165
Teddy Bear Parade	195	1938	300	700	1700
Teddy Bear Zilo	777	1950	35	50	165
Teddy Choo-Choo	465	1937	200	400	825
Teddy Drummer	775	1936	300	600	1200
Teddy Station Wagon	480	1942	150	300	650
Teddy Tooter	712	1957	100	225	525
Teddy Tooter	150	1940	200	400	895
Teddy Trucker	711	1949	100	225	575
Teddy Xylophone	752	1948	175	225	475
Ten Little Indians TV Radio	159	1961	25	50	75
This Little Pig	910	1963	25	40	45
Thumper Bunny	533	1942	225	400	800
Timber Toter	810	1957	50	100	175
Timmy Turtle	150	1953	40	80	160
Tiny Teddy	634	1955	40	75	145
Tiny Tim	496	1957	35	70	135
Tip-Toe Turtle	773	1962	30	45	95
Toot Toot Engine	641	1962	30	45	90
Tow Truck	615	1960	45	70	155
Toy Wagon	131	1951	125	250	500
Trotting Donald Duck	741	1937	300	650	1450
Tuggy Turtle	139	1959	45	75	165
Uncle Timmy Turtle	125	1956	50	100	195
Waggy Woofy	437	1942	100	250	500
Walt Disney's Carnival	207	1936	325	600	1200
Walt Disney's Donald Duck	208	1936	250	400	900
Walt Disney's Easter Parade	475	1936	450	1000	1900
Walt Disney's Mickey Mouse	209	1936	250	400	800
Walt Disney's Pluto-the-Pup	210	1936	225	450	800
Whistling Engine	617	1957	40	80	160
Wiggily Woofer	640	1957	40	95	195
Winky Blinky Fire Truck	200	1954	45	100	195
Woodsy-Wee Circus	201	1931	500	900	2100
Woofy Wagger	447	1947	45	75	185
Woofy Wowser	700	1940	100	325	825
Ziggy Zilo	737	1958	50	100	175

Games

Just a few years ago, except for a handful of visionaries, no one considered board and card games worthy of collecting, which in hindsight is itself a strong indicator of their future collectibility. Games were playthings only and were treated casually compared to thousands of other now highly-collectible items. Their entire value was in the playing, in how much fun and social interaction they afforded. They were casually played, their parts cheerfully smudged, crumpled and misplaced. Once their fun quotient was exhausted, they were just as casually tossed in closets, attics or in the trash.

While the idea of games as collectibles is a recent one in the world of antiques, the field is making up for that oversight with a vengeance. This is particularly true in the three most visible groupings — Victorian games, sports games and character/TV-related games.

Victorian games are loosely defined as those produced between the Civil War and World War I. During this period, manufacturing technologies advanced from the hand-tinted games of the pre-1860s to chromolithography, which ushered in the age of mass production.

The premier marque of Victorian games belongs to McLoughlin Bros., which progressed from hand tinting to lithography in grand style, producing many of the most beautifully illustrated — and most valuable — games ever made. In many cases, simply the McLoughlin name on the box will make a game considerably more valuable than the identical item produced by a competitor.

McLoughlin was purchased by Milton Bradley in 1920, which reissued many games under the Milton Bradley name. As would be expected, these reissues have only a fraction of the value of McLoughlin originals.

Another noteworthy game maker was Parker Brothers, which since 1883 has produced many games rivaling the beauty and collectibility of McLoughlin, Bliss, Clark & Sowdon and J.H. Singer are other names to watch for, as they also held to high standards of creativity and artistry in the execution of their games.

However, if you do come across a Victorian game in your travels, most likely it will be by Milton Bradley, by far the most prolific manufacturer of the age. While its games were generally neither as beautiful nor as valuable as those of its competitors, they are still collected for both their artwork and subject matter.

Sports games, of which there are many, have enjoyed immense gains in value along with the boom in other sports memorabilia which essentially began in the late 1970s. This area also illustrates the issue of relative value across collecting fields.

In general, sports collectors assign a higher value to sports games than game collectors do, and asking prices for the same game can vary widely from a toy show to a sports show, or even from one table to another at the same show. But the point is repeatedly proven that collectors within a specialized field will often pay more for a particular item than generalized collectors in a related field.

Still, no hobby operates in a vacuum. Increasing demand by sports collectors will also drive up prices for game collectors, just as demand by Christmas collectors for Santa games will drive up prices for both Christmas and game collectors. Game collectors might always assign a lower value to the same game than either of these more specialized groups, but whatever its origin, demand drives value. As collectors grow more sophisticated and knowledgeable about various sources of items in their special areas, these cross-field price variances begin to fade, but they will probably never disappear altogether, particularly when one field can experience a boom time when another may not.

Sports games are collected either by individual sport, like baseball or horse racing, or by personality, such as Babe Ruth's Baseball Game (1926) or Mickey Mantle's Big League Baseball (1955). Personality-based games typically have higher values than their generic counterparts, and older generic games are normally more valuable than newer ones.

Postwar Games Also Popular

Many collectors consider World War II as the breaking point between antique and modern games. However, this turning point can be defined perhaps just as logically as either before or after the advent of TV. The fact that World War II and the rise of television took place in the same decade is a largely convenient coincidence. Which event wrought the most lasting changes on American life can make for a rousing evening's debate, but TV's impact on modern marketing and consumer habits is hard to overestimate.

Motor Cycle Game, 1905, Milton Bradley

Character games in particular lend themselves to the theory of television classification. Pre-TV games can be seen as evolving from Victorian nursery rhyme games like McLoughlin's Little Goldenlocks and The Three Bears (1890), numerous variations on Mother Goose and Milton Bradley's Little Jack Horner (1910) to name a few.

These were supplemented by popular comic characters like Winnie Winkle, The Katzenjammer Kids, Chester Gump, Dick Tracy and also by literary entries such as Parker Brother's classic Wonderful Game of Oz (1921) and Kerk Guild's A. A. Milne's Winnie The Pooh Game (1931).

The 1930s also saw the tapping of a limitless well of Disney character games, beginning with numerous Mickey Mouse games and following with Ferdinand The Bull, Snow White, Alice In Wonderland and many more. Disney's merchandising grew more sophisticated with each new film, setting precedents that have become today's standard practice.

Television games of the 1950s saw a general decline in the quality of materials and execution and an increasing reliance on character affiliation. Games like Lowell's Gunsmoke (1950), Milton Bradley's Annie Oakley (1950) and Parker Brothers' Bing Crosby — Call Me Lucky (1954) set a new, albeit lower, standard for game production and marketing.

TV-based games of the 1950s and '60s are currently the most eagerly sought-after of all modern games, and their values reflect that demand.

Trends

Today's TV, film and cartoon character games — those from the 1960s and newer — are strong bets for future appreciation. Comic character games from the 1940s and earlier, however, are losing their nostalgic value and have to stand on artistic merit. Values of these games have been declining.

Strong bets for future investment include sports games, particularly if they are celebrity-based, and games which address current historical or cultural affairs. Also worth seeking out are the limited run products of smaller players in the game field, again, especially those with character affiliation.

Word games, generic strategy games and non-character related games have historically demonstrated little or no appreciation, and their heritage as poor investments will likely be inherited by their modern counterparts. Exceptions are games in genres like science fiction, all of which have better than normal potential as long as sci-fi remains in vogue. Another exception is in games which utilize unique, complex or intricate playing pieces and apparatus. A good example of this can be seen in Transogram's Ka-Bala (1965). This game would be moderately collectible simply by virtue of its fortune telling theme, but its board-dominating, glow-in-the-dark eyeball centerpiece is thrillingly ghoulish.

The burgeoning field of game collecting offers a wealth of benefits including investment potential, history and artistry. But the games can also be played. While most Victorian and modern character games are sought primarily for box and board art, many have been designed to fulfill the mission of fun.

Editor's Note: Since experts agree that so few Mint examples of prewar games exist, the prewar section in this edition lists only two grades, Good and Excellent, as Excellent is the highest grade generally applied to prewar games in auction catalogs and dealer price lists. This is particularly true of Victorian games.

The postwar game section lists three grades of value — Good, Excellent and Mint — as enough examples of Mint condition modern games exist to have established sales histories on which to base market values.

Games in both sections are listed alphabetically by name. Letters listed after the game denote the type of game: B - Board Games, C - Card Games, S - Skill/Action Games.

Lippy the Lion Game, 1963, Transogram

The Top 10 Prewar Games
(in Excellent condition)

1. Bulls and Bears, McLoughlin Bros., 1896 .... $13,000
2. New Parlor Game of Baseball, Sumner, 1896 10,000
3. Champion Baseball Game, Schultz, 1889 6,800
4. Little Fireman Game, McLoughlin Bros., 1897 6,000
5. Zimmer Baseball Game, McLoughlin Bros., 1885 6,000
6. Teddy's Ride from Oyster Bay to Albany, Jesse Crandall, 1899 5,500
7. Golf, Schoenhut, 1900 5,000
8. Egerton R. Williams Popular Indoor Baseball Game, Hatch, 1886 5,000
9. Great Mails Baseball Game, Walter Mails Baseball Game, 1919 4,100
10. Darrow Monopoly, Charles Darrow, 1934 4,000

The Top 10 Postwar Games
(in Mint condition)

1. Elvis Presley Game, Teen Age Games, 1957 .... $1,000
2. Red Barber's Big League Baseball Game, G & R Anthony, 1950s 900
3. Win A Card Trading Card Game, Milton Bradley, 1965 900
4. Munsters Drag Race Game, Hasbro, 1965 800
5. Munsters Masquerade Game, Hasbro, 1965 800
6. Creature From The Black Lagoon, Hasbro, 1963 750
7. Strike Three, Tone Products, 1948 725
8. Jonny Quest Game, Transogram, 1964 700
9. Munsters Picnic Game, Hasbro, 1965 700
10. Gilligan's Island, Game Gems, 1965 600

Contributors to this section: Postwar Games, Jeff Lowe's extravaGAMEza, 5005 Tamara Lane, West Des Moines, IA 50265, 515-267-8768; **Prewar Games,** Paul Fink, Paul Fink's Fun and Games, P.O. Box 488, 59 South Kent Road, Kent, CT 06757.

PREWAR GAMES

NAME	TYPE	YEAR	COMPANY	GOOD	EX
21st Century Football	B	1930s	Kerger	55	90
400 Game, The	B	1890s	J.H. Singer	100	150
400, Aristocrat of Games, The	S	1933	Morris Systems	15	25
A&P Relay Boat Race Coast-to-Coast	B	1930s	A&P	30	40
ABC	C	1900s	Parker Brothers	25	40
ABC Baseball Game	B	1910s		430	715
ABC, Game of	B	1914		60	100
Abcdarian, The	B	1899	Chaffee & Selchow	40	100
Across the Channel	B	1926	Wolverine	50	85
Across the Continent	B	1892	Parker Brothers	100	175
Across the Continent	B	1922	Parker Brothers	150	300
Across the Sea Game	B	1930	Gabriel	60	100
Across the Yalu	B	1905	Milton Bradley	75	175
Add-Too	B	1940	All-Fair	10	15
Admiral Byrd's South Pole Game Little America	B	1930s	Parker Brothers	125	300
Admirals, The Naval War Game	B	1939	Merchandisers	75	120
ADT Delivery Boy	B	1890	Milton Bradley	120	200
ADT Messenger Boy (Small Version)	B	1915	Milton Bradley	40	70
Advance And Retreat, Game of	B	1900s	Milton Bradley	95	175
Aero Ball	S	1940s	Game Makers	30	75
Aero-Chute	B	1940	American Toy Works	75	150
Aeroplane Race	B	1922	Wolverine	60	95
After Dinner	B	1937	Frederick H. Beach (Beachcraft)	10	20
Air Base Checkers	B	1942	Einson-Freeman	20	30
Air Mail, The	B	1930	Archer Toy	75	125
Air Mail, The Game of	B	1927	Milton Bradley	95	100
Air Ship Game, The	B	1912	McLoughlin Bros.	150	300
Air Ship Game, The	B	1904	McLoughlin Bros.	300	700
Airplane Speedway Game	B	1941	Lowe	20	30
Airship Game, The	C	1916	Parker Brothers	30	50
Akins Real Baseball	B	1915	Akins	450	750
Aldjemma	B	1944	Corey Games	30	45
Alee-Oop	B	1937	Royal Toy	30	50
Alexander's Baseball	B	1940s		245	400
Alice in Wonderland	B	1930s	Parker Brothers	60	150
Alice In Wonderland, Game of	B	1923	Stoll & Edwards	50	85
All American Basketball	B	1941	Corey Games	55	65
All American Football	B	1935		35	55
All-American Big Boy Baseball Game	B	1920s	Rosensteel-Pulich	200	500
All-American Football	B	1925	Parker Brothers	100	165
All-Star Baseball Game	B	1935	Whitman	100	165
Allegrando	C	1884	Theodore Presser	40	60
Allie-Patriot Game	C	1917	McDowell And Mellor	30	50
Alpha Baseball Game	B	1930s	Redlich	100	150
Alpha Football Game	B	1940s	Replica	70	115
Amateur Golf	B	1928	Parker Brothers	145	245
Ambuscade, Constellations And Bounce	B	1877	McLoughlin Bros.	150	300
America's Football	B	1939	Trojan Games	55	90
America's Yacht Race	B	1904	McLoughlin Bros.	450	750
American Boy Game	B	1920s	Milton Bradley	75	125
American Derby	B	1931	Henschel	55	90
American Football Game	B	1930	Ace Leather Goods	70	115
American History, The Game of	C	1890s	Parker Brothers	50	90

GAMES

PREWAR GAMES

NAME	TYPE	YEAR	COMPANY	GOOD	EX
American League Fan Craze Card Game	C	1904	Fan Craze	1950	3250
American National Game Baseball	C	1909	American National Game Co.	75	200
American Revolution, The New Game of The	B	1844	Lorenzo Burge	960	1600
American Sports	B	1880s		110	180
Amusing Game of Conundrums	C	1853	John McLoughlin	750	1300
Amusing Game of Innocence Abroad, The	B	1888	Parker Brothers	135	300
Amusing Game of the Corner Grocery	C	1890s		100	150
Anagrams	C	1885	Peter G. Thompson	50	90
Ancient Game of the Mandarins, The	B	1923	Parker Brothers	45	75
Andy Gump, His Game	B	1924	Milton Bradley	60	100
Anex-A-Gram	B	1938	Embossing	20	40
Animal & Bird Lotto	B	1926	All-Fair	15	20
Animal Bingo	S		Baldwin Manufacturing	40	125
Apple Pie	C	1895	Parker Brothers	40	60
Arena	B	1896	Bliss	120	200
Armstead's Play Ball	C	1910s	Austin	200	500
Astronomy	C	1905	Cincinnati Game	40	60
Athletic Sports	B	1900	Parker Brothers	145	245
Attack, Game of	B	1889	Bliss	500	700
Auction Letters	C	1900	Parker Brothers	40	60
Authors	B	1890s	J.H. Singer	40	80
Authors	B	1861	Whipple & Smith	100	300
Authors Illustrated	C	1893	Clark & Sowdon	40	60
Authors, Game of Standard	C	1890s	McLoughlin Bros.	25	50
Authors, The Game of	C	1890s	Parker Brothers	15	35
Auto Game, The	B	1906	Milton Bradley	100	200
Auto Race Electro Game	B	1929	Knapp Electric & Novelty	125	210
Auto Race Game	B	1925	Milton Bradley	200	325
Auto Race Jr.	B	1925	All-Fair	150	200
Auto Race, Army, Navy, Game Hunt (Four game set)	B	1920s	Wilder	110	180
Auto Race, Game Of	B	1920s	Orotech	105	175
Auto-Play Baseball Game	B	1911	Auto-Play	425	700
Automobile Race, Game of the	B	1904	McLoughlin Bros.	500	1000
Avilude	C	1873	West & Lee	65	100
Aydelott's Parlor Baseball	B	1910		195	325
Babe Ruth National Game of Baseball	B	1929	Keiser-Fry	550	910
Babe Ruth's Baseball Game	B	1926	Milton Bradley	200	600
Babe Ruth's Official Baseball Game	B	1940s	Toytown	430	715
Baby Barn Yard	B	1940s	B.L. Fry Products	15	25
Bag of Fun	S	1932	Rosebud Art	15	20
Bagatelle, Game of	B	1898	McLoughlin Bros.	200	875
Bagdad, The Game of The East	B	1940	Clover Games	25	40
Balance The Budget	C	1938	Elten Game	45	75
Balloonio	S	1937	Frederick H. Beach	20	40
Bally Hoo	C	1931	Gabriel	30	65
Bambino	S	1934	Bambino Products	75	125
Bambino (Baseball, Chicago World's Fair)	B	1933	Johnson Store Equipment	295	490
Bambino Baseball Game	B	1940	Mansfield-Zesiger	145	250
Bamboozle, or The Enchanted Isle	B	1876	Milton Bradley	175	250
Bang Bird	S	1924	Doremus Schoen	20	30

PREWAR GAMES

NAME	TYPE	YEAR	COMPANY	GOOD	EX
Bang, Game of	B	1903	McLoughlin Bros.	100	200
Banner Lye Checkerboard	B	1930s	Geo E. Schweig & Son	15	20
Barage	B	1941	Corey Games	50	100
Barber Pole	S	1908	Parker Brothers	50	75
Barn Yard Tiddledy Winks	S	1910s	Parker Brothers	50	85
Barney Google and Spark Plug Game	B	1923	Milton Bradley	100	200
Baron Munchausen Game, The	B	1933	Parker Brothers	35	50
Base Hit	B	1944	Games	55	90
Baseball	B	1942	Lowe	15	25
Baseball & Checkers	B	1925	Milton Bradley	75	150
Baseball Dominoes	B	1910	Evans	250	400
Baseball Game	B	1930	All-Fair	125	250
Baseball Game & G-Man Target Game	B	1940	Marks Brothers	100	165
Baseball Game, New	B	1885	Clark & Martin	165	275
Baseball Wizard Game	B	1916	Morehouse	265	450
Baseball, Game of	B	1886	J.H. Singer	325	550
Baseballitis Card Game	C	1909	Baseballitis Card	125	205
Bases Full	S	1930		45	70
Basilinda	B	1890	Horsman	105	175
Basket Ball	S	1929	Russell	150	350
Basketball	B	1942	Lowe	15	25
Basketball Card Game	C	1940s	Warren/Built-Rite	20	35
Basketball Game, Official	B	1940	Toy Creations	55	90
Batter Up, Game of	C	1908	Fenner Game	150	200
Battle Checkers	B	1925	Pen Man	15	50
Battle Game, The	B	1890s	Parker Brothers	120	200
Battle of Ballots	B	1931	All-Fair	55	125
Battle of Manila	B	1899	Parker Brothers	400	650
Battles, or Fun For Boys, Game of	S	1889	McLoughlin Bros.	600	1000
Bean-Em	S	1931	All-Fair	250	600
Bear Hunt, Game of	B	1923	Milton Bradley	45	70
Beauty And The Beast, Game of	B	1905	Milton Bradley	45	75
Bee Gee Baseball Dart Target	B	1935s	Bee Gee	70	115
Bell Boy Game, The	B	1898	Chaffee & Selchow	425	700
Belmont Park	B	1930	Marks Brothers	75	125
Bengalee	B	1940s	Advance Games	20	35
Benny Goodman Swings	B	1930s	Toy Creations	65	125
Benson Football Game, The	B	1930s	Benson	85	140
Betty Boop Coed Bridge	C	1930s		50	75
Bible ABCs and Promises	C	1940s	Judson Press	15	25
Bible Authors	C	1895	Evangelical Pub.	20	35
Bible Boys	B	1901	Zondervan	10	15
Bible Characters	B	1890s	Decker & Decker	10	25
Bible Cities	C	1920s	Nellie T. Magee	15	25
Bible Lotto	B	1933	Goodenough and Woglom	10	15
Bible Quotto	B	1932	Goodenough and Woglom	6	10
Bible Rhymes	B	1933	Goodenough and Woglom	10	15
Bicycle Cards	C	1898	Parker Brothers	150	200
Bicycle Game	B	1896	Donaldson Brothers	205	350
Bicycle Game, The New	B	1894	Parker Brothers	400	700
Bicycle Race	B	1910	Milton Bradley	85	140
Bicycle Race Game, The	B	1898	Chaffee & Selchow	430	715

PREWAR GAMES

NAME	TYPE	YEAR	COMPANY	GOOD	EX
Bicycle Race, A Game for the Wheelmen	B	1891	McLoughlin Bros.	700	1000
Bicycling, The Merry Game of	B	1900	Parker Brothers	100	165
Big Apple	B	1938	Rosebud Art	30	50
Big Bad Wolf Game	B	1930s	Parker Brothers	75	125
Big Business	B	1936	Parker Brothers	75	90
Big Business	B	1937	Transogram	20	50
Big League Baseball Card Game	C	1940s	State College Game Lab	35	60
Big League Basketball	B	1920s	Baumgarten	145	245
Big Six: Christy Mathewson Indoor Baseball Game	B	1922	Piroxloid	150	500
Big Ten Football Game	B	1936	Wheaties	55	90
Bike Race Game, The	B	1930s	Master Toy	35	100
Bild-A-Word	B	1929	Educational Card & Game	20	35
Billy Bump's Visit To Boston	C	1888	Parker Brothers	25	55
Billy Whiskers	B	1923	Saalfield	45	100
Billy Whiskers	B	1924	Russell	45	75
Bilt-Rite Miniature Bowling Alley	B	1930s	Atwood Momanus	70	115
Bing Miller Base Ball Game	B	1932	Ryan	750	2000
Bingo	B	1925	Rosebud Art	15	25
Bingo	S	1929	All-Fair	15	25
Bingo or Beano	B	1940s	Parker Brothers	10	15
Bird Center Etiquette	C	1904	Home Game	55	90
Bird Lotto	B	1940s	Gabriel	20	35
Birds, Game of	C	1899	Cincinnati Game	55	90
Black Beauty	B	1921	Stoll & Edwards	40	65
Black Cat Fortune Telling Game, The	C	1897	Parker Brothers	95	150
Black Falcon of The Flying G-Men, The	B	1939	Ruckelshaus	175	400
Black Sambo, Game of	B	1939	Gabriel	90	200
Blackout	B	1939	Milton Bradley	60	150
Block	C	1905	Parker Brothers	15	20
Blockade	B	1941	Corey Games	50	95
Blondie Goes To Leisureland	B	1935	Westinghouse	20	35
Blondie Playing Card Game	C	1941	Whitman	55	90
Blow Football Game	B	1912		30	50
Blox-O	B	1923	Lubbers & Bell	15	25
Bluff	B	1944	Games of Fame	15	25
Bo Bang & Hong Kong	B	1890	Parker Brothers	275	450
Bo McMillan's Indoor Football	B	1939	Indiana Game	60	125
Bo Peep Game	B	1895	McLoughlin Bros.	195	325
Bo Peep, The Game of	B	1890	J.H. Singer	200	300
Boake Carter's Star Reporter	B	1937	Parker Brothers	105	225
Bobb, Game of	S	1898	McLoughlin Bros.	200	400
Bomb The Navy	B	1940s	Pressman	20	50
Bombardment, Game of	B	1898	McLoughlin Bros.	100	350
Bomber Ball	S	1940s	Game Makers	75	125
Bombs Away	B	1944	Toy Creations	75	175
Bookie	B	1931	Bookie Games	55	90
Boston Baseball Game	B	1906	Boston Game	495	825
Boston Globe Bicycle Game of Circulation	B	1895	Boston Globe	55	90
Boston-New York Motor Tour	B	1920s	American Toy	90	150
Botany	C	1900s	G.H. Dunston	50	100
Bottle Imps, Game of	S	1907	Milton Bradley	300	500
Bottle-Quoits	B	1897	Parker Brothers	50	85

PREWAR GAMES

NAME	TYPE	YEAR	COMPANY	GOOD	EX
Bottoms Up	B	1934	Embossing	20	45
Bourse, or Stock Exchange	C	1903	Flinch Card	25	40
Bow-O-Winks	S	1932	All-Fair	60	120
Bowl 'em	B	1930s	Parker Brothers	20	35
Bowling Alley	S	1921	N.D. Cass	20	35
Bowling Board Game	B	1896	Parker Brothers	350	575
Box Hockey	B	1941	Milton Bradley	35	60
Boxing Game, The	B	1928	Stoll & Edwards	85	140
Boy Hunter, The	S	1925	Parker Brothers	60	100
Boy Scouts	B	1910s	McLoughlin Bros.	125	200
Boy Scouts in Camp	B		McLoughlin Bros.	200	400
Boy Scouts Progress Game	B	1924	Parker Brothers	150	200
Boy Scouts, The Game of	B	1926	Parker Brothers	200	350
Boy Scouts, The Game of	C	1912	Parker Brothers	75	125
Boys Own Football Game	B	1900s	McLoughlin Bros.	325	750
Bradley's Circus Game	B	1882	Milton Bradley	60	150
Bradley's Telegraph Game	B	1900s	Milton Bradley	85	145
Bradley's Toy Town Post Office	B	1910s	Milton Bradley	90	150
Bringing Up Father Game	B	1920	Embee Distributing	75	175
Broadway	B	1917	Parker Brothers	150	250
Brownie Auto Race	B	1920s	Jeanette Toy & Novelty	115	195
Brownie Character Ten Pins Game	S	1890s		125	200
Brownie Horseshoe Game	B	1900s	M.H. Miller	30	50
Brownie Kick-In Top	S	1910s	M.H. Miller	35	60
Brownie Ring Toss	B	1920s	M.H. Miller	30	50
Buck Rogers and His Cosmic Rocket Wars Game	B	1934		200	350
Buck Rogers In The 25th Century	C	1936	All-Fair	160	350
Buck Rogers Siege of Gigantica Game	B	1934		400	600
Bucking Bronco	B	1930s	Transogram	30	50
Buffalo Bill, The Game of	B	1898	Parker Brothers	200	400
Buffalo Hunt	B	1898	Parker Brothers	175	250
Bugle Horn or Robin Hood	C	1850s	McLoughlin Bros.	600	1000
Bugle Horn or Robin Hood, Game of	B	1895	McLoughlin Bros.	350	650
Bugville Games	B	1915	Animate Toy	75	125
Bula	S	1943	Games Of Fame	30	45
Bull In The China Shop	S	1937	Milton Bradley	20	40
Bulls and Bears	B	1936	Parker Brothers	150	250
Bulls and Bears	B	1896	McLoughlin Bros.	10000	13000
Bunco	C	1904	Home Game	35	55
Bunker Golf	B	1932		115	195
Bunny Rabbit, Or Cottontail & Peter, The Game of	B	1928	Parker Brothers	85	145
Buried Treasure, The Game of	B	1930s	Russell	35	60
Buster Brown at Coney Island	B	1890s	J. Ottmann Lith.	225	350
Buster Brown at the Circus	C	1900s	Selchow & Righter	100	250
Buster Brown Hurdle Race	B	1890s	J. Ottmann Lith.	330	550
Buster Brown, Pin The Tail On The Tiger Game	S	1900s		60	100
Busto	S	1931	All-Fair	100	200
Buying and Selling Game	B	1903	Milton Bradley	100	200
Buzzing Around	S	1924	Parker Brothers	40	65
Cabby	B	1940	Selchow & Righter	40	75
Cabin Boy	B	1910	Milton Bradley	60	100
Cadet Game, The	B	1905	Milton Bradley	100	150
Cake Walk Game, The	B	1900s	Parker Brothers	750	1500

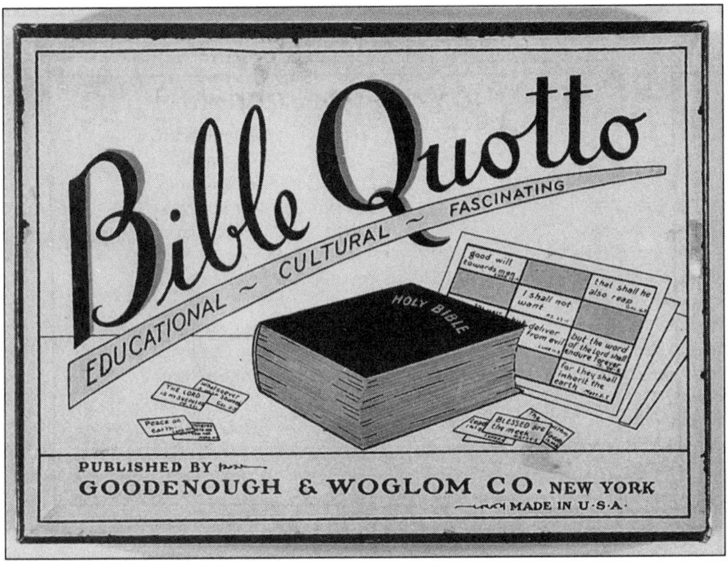

Bible Quotto, 1932, Goodenough and Woglom

Pop the Hat, 1930s, Milton Bradley

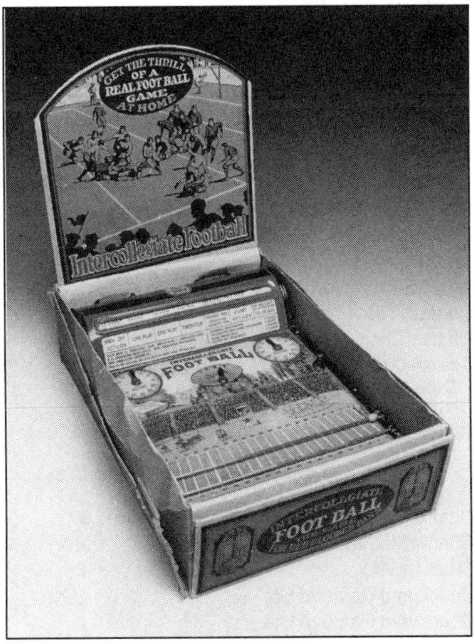

Intercollegiate Football, 1923, Hustler Toy

PREWAR GAMES

NAME	TYPE	YEAR	COMPANY	GOOD	EX
Cake Walk, The	B	1900s	Anglo American	600	1300
Calling All Cars	B	1938	Parker Brothers	45	75
Camelot	B	1950s	Parker Brothers	20	35
Camouflage, The Game of	C	1918	Parker Brothers	35	75
Canoe Race	B	1910	Milton Bradley	35	60
Capital Cities Air Derby, The	B	1929	All-Fair	200	350
Captain and the (Katzenjammer) Kids	B	1940s	Milton Bradley	50	100
Captain Hop Across Junior	B	1928	All-Fair	100	200
Captain Jinks	C	1900s	Parker Brothers	30	55
Captain Kidd And His Treasure	B	1896	Parker Brothers	195	350
Captain Kidd Junior	B	1926	Parker Brothers	50	75
Captive Princess	B	1880	McLoughlin Bros.	135	225
Captive Princess	B	1899	McLoughlin Bros.	65	150
Captive Princess, Tournament And Pathfinders, Games of	B	1888	McLoughlin Bros.	100	200
Capture The Fort	B	1914	Valley Novelty Works	45	75
Car Race & Game Hunt	B	1920s	Wilder	100	165
Cargo For Victory	B	1943	All-Fair	65	95
Cargoes	B	1934	Selchow & Righter	30	75
Carl Hubbell Mechanical Baseball	B		Gotham	200	500
Carnival, The Show Business Game	B	1937	Milton Bradley	50	100
Casey on the Mound	B	1940s	Kamm Games	100	200
Cat	B	1915	Carl F. Doerr	15	25
Cat And Witch	S	1940s	Whitman	25	65
Cat, Game of	B	1900	Chaffee & Selchow	270	450
Catching Mice, Game of	B	1888	McLoughlin Bros.	200	325
Cats And Dogs	B	1929	Parker Brothers	50	150
Cavalcade	B	1930s	Selchow & Righter	40	60
Cavalcade Derby Game	S	1930s	Wyandotte	50	85
Century Ride	B	1900	Milton Bradley	60	150
Century Run Bicycle Game, The	B	1897	Parker Brothers	210	350
Champion Baseball Game, The	B	1889	Schultz	4100	6800
Champion Game of Baseball, The	B	1910s	Proctor Amusement	60	100
Champion Road Race	B	1934	Champion Spark Plugs	100	165
Championship Baseball Parlor Game	B	1914	Grebnelle Novelty	150	250
Championship Fight Game	B	1940s	Frankie Goodman	20	50
Champs, The Land of Brawno	B	1940	Selchow & Righter	30	50
Characteristics	B	1845	Ives	180	300
Characters, A Game of	C	1889	Decker & Decker	25	40
Charge, The	B	1898	E.O. Clark	180	300
Charlie Chan Game	C	1939	Whitman	30	75
Charlie Chan, The Great Charlie Chan Detective Game	B	1937	Milton Bradley	150	250
Charlie McCarthy Game of Topper	B	1938	Whitman	20	45
Charlie McCarthy Put and Take Bingo Game	B	1938	Whitman	30	45
Charlie McCarthy Question and Answer Game	C	1938	Whitman	40	65
Charlie McCarthy Rummy Game	C	1938	Whitman	20	35
Charlie McCarthy's Flying Hats	B	1938	Whitman	25	40
Chasing Villa	B	1920	Smith, Kline & French	65	180
Checkered Game of Life	B	1911	Milton Bradley	150	250
Checkered Game of Life	B	1860	Milton Bradley	200	300
Checkered Game of Life	B	1866	Milton Bradley	130	200
Checkers & Avion	B	1925	American Toy Works	30	50
Chee Chow	B	1939	Gabriel	15	25

PREWAR GAMES

NAME	TYPE	YEAR	COMPANY	GOOD	EX
Cheerios Bird Hunt	B	1930s	General Mills	15	25
Cheerios Hook The Fish	B	1930s	General Mills	15	25
Chessindia	B	1895	Clark & Sowdon	55	95
Chester Gump Game	B	1938	Milton Bradley	65	125
Chester Gump Hops over the Pole	B	1930s	Milton Bradley	65	95
Chestnut Burrs	C	1896	Fireside Game	35	65
Chevy Chase	B	1890	Hamilton-Myers	75	125
Chicago Game Series Baseball	B	1890s	Doan	1175	1950
Chin-Chow and Sum Flu	B	1925	Novitas Sales	6	10
China	B	1905	Wilkens Thompson	50	85
Chinaman Party	S	1896	Selchow & Righter	75	130
Ching Gong	B	1937	Gabriel	20	40
Chiromagica, or The Hand of Fate	B	1901	McLoughlin Bros.	225	350
Chivalrie Lawn Game	B	1875		60	100
Chivalry	B	1925	Parker Brothers	75	125
Chivalry, The Game of	S	1888	Parker Brothers	100	175
Chocolate Splash	B	1916	Willis G. Young	51	85
Christmas Goose	B	1890	McLoughlin Bros.	500	1000
Christmas Jewel, Game of the	B	1899	McLoughlin Bros.	250	400
Christmas Mail	B	1890s	J. Ottmann Lith.	390	650
Chutes And Ladders	B	1943	Milton Bradley	15	35
Cinderella	C	1895	Parker Brothers	55	90
Cinderella	C	1921	Milton Bradley	15	25
Cinderella	B	1923	Stoll & Edwards	40	95
Cinderella	C	1905	Milton Bradley	50	80
Cinderella or Hunt the Slipper	C	1887	McLoughlin Bros.	75	105
Circus Game	B	1914		75	125
Citadel	B	1940	Parker Brothers	40	60
Cities	B	1932	All-Fair	25	40
City Life, or The Boys of New York, The Game of	C	1889	McLoughlin Bros.	65	250
City of Gold	B	1926	Zulu Toy	100	200
Classic Derby	B	1930s	Doremus Schoen	30	50
Click	S	1930s	Akro Agate	50	85
Clipper Race	B	1930	Gabriel	25	75
Clown Tenpins Game	B	1912		60	100
Clown Winks	S	1930s	Gabriel	15	25
Coast To Coast	B	1940s	Master Toy	15	25
Cock Robin	C	1895	Parker Brothers	40	75
Cock Robin and His Tragical Death, Game of	C	1885	McLoughlin Bros.	65	105
Cock-A-Doodle-Doo Game	B	1914		60	100
Cocked Hat, Game of	B	1892	J.H. Singer	150	250
College Baseball Game	B	1890s	Parker Brothers	350	700
College Boat Race, Game of	B	1896	McLoughlin Bros.	350	650
Colors, Game of	B	1888	McLoughlin Bros.	150	225
Columbia's Presidents and our Country, Game of	C	1886	McLoughlin Bros.	250	400
Columbus	B	1892	Milton Bradley	775	1300
Combination B	B	1922	Wilder	40	65
Combination Tiddledy Winks	S	1910	Milton Bradley	40	65
Comic Conversation Cards	C	1890	J. Ottmann Lith.	100	200
Comic Leaves of Fortune-The Sibyl's Prophecy	C	1850s	Charles Magnus	500	900
Comical Animals Ten Pins	B	1910	Parker Brothers	185	310
Comical Game of "Who", The	C	1910s	Parker Brothers	35	50

PREWAR GAMES

NAME	TYPE	YEAR	COMPANY	GOOD	EX
Comical Game of Whip, The	C	1920s	Russell	20	45
Comical History of America	C	1924	Parker Brothers	40	75
Comical Snap, Game of	C	1903	McLoughlin Bros.	35	55
Commanders of our Forces, The	C	1863	E.C. Eastman	150	225
Commerce	C	1900s	J. Ottmann Lith.	40	100
Competition, or Department Store	C	1904	Flinch Card	25	35
Cones & Corns	S	1924	Parker Brothers	39	65
Conette	S	1890	Milton Bradley	45	75
Coney Island Playland Park	B	1940	Vitaplay Toy	54	90
Conflict	B	1942	Parker Brothers	150	200
Conquest of Nations, or Old Games With New Faces, The	C	1853	Willis P. Hazard	100	175
Construction Game	B	1925	Wilder	150	200
Contack	S	1939	Parker Brothers	5	15
Coon Hunt Game, The	B	1903	Parker Brothers	450	1200
Corn & Beans	B	1875	E.G. Selchow	50	85
Corner The Market	B	1938	Whitman	25	40
Cortella	B	1915	Atkins	22	35
Costumes and Fashions, Game of	C	1881	Milton Bradley	80	225
Cottontail and Peter, The Game of	B	1922	Parker Brothers	75	150
Country Club Golf	B	1920s	Hustler Toy	75	125
Country Store, The	B	1890s	J.H. Singer	75	125
County Fair, The	C	1891	Parker Brothers	45	75
Cousin Peter's Trip to New York, Game of	C	1898	McLoughlin Bros.	45	75
Covered Wagon	B	1927	Zulu Toy	55	85
Cowboy Game, The	B	1898	Chaffee & Selchow	210	425
Cows In Corn	S	1889	Stirn & Lyon	150	250
Crash, The New Airplane Game	B	1928	Nucraft Toys	30	70
Crazy Traveller	S	1920s	Parker Brothers	40	50
Crazy Traveller	B	1908	Parker Brothers	40	60
Crickets In The Grass	S	1920s	Madmar Quality	35	50
Crime & Mystery	B	1940s	Frederick H. Beach (Beachcraft)	15	25
Criss Cross Words	B	1938	Alfred Butts	90	150
Crooked Man Game	B	1914		45	75
Cross Country	B	1941	Lowe	20	30
Cross Country Marathon	B	1920s	Milton Bradley	50	125
Cross Country Marathon Game	B	1930s	Rosebud Art	100	200
Cross Country Racer	B	1940	Automatic Toy	45	75
Cross Country Racer (w/wind-up cars)	B	1940s		75	130
Crossing the Ocean	B	1893	Parker Brothers	87	175
Crow Cards	C	1910	Milton Bradley	9	15
Crow Hunt	B	1904	Parker Brothers	40	60
Crow Hunt	S	1930	Parker Brothers	50	85
Crows In The Corn	S	1930	Parker Brothers	45	75
Crusade	B	1930s	Gabriel	27	45
Crusaders, Game of the	B	1888	McLoughlin Bros.	150	300
Cuckoo, A Society Game	B	1891	J.H. Singer	40	100
Curly Locks Game	B	1910	United Game	60	100
Cycling, Game of	B	1910	Parker Brothers	100	165
Daisy Clown Ring Game	B	1927	Schacht Rubber	9	15
Daisy Horseshoe Game	B	1927	Schacht Rubber	9	15
Danny McFayden's Stove League Baseball Game	B	1920s	National Game	295	490
Darrow Monopoly	B	1934	Charles Darrow	1500	4000

PREWAR GAMES

NAME	TYPE	YEAR	COMPANY	GOOD	EX
Day at the Circus, Game of	B	1898	McLoughlin Bros.	300	400
Deck Derby	B	1920s	Wolverine	36	60
Deck Ring Toss Game	S	1910		30	50
Decoy	B	1940	Selchow & Righter	45	75
Defenders of the Flag	C	1922	Stoll & Edwards	35	65
Defenders of The Flag Game	B	1920s		24	40
Democracy	B	1940	Toy Creations	50	85
Department Store, Game of Playing	B	1898	McLoughlin Bros.	520	1300
Derby Day	C	1900s	Parker Brothers	30	50
Derby Day	B	1930	Parker Brothers	45	75
Derby Steeple Chase	B	1888	McLoughlin Bros.	150	250
Detective, The Game of	B	1889	Bliss	1200	2000
Dewey at Manila	C	1899	Chaffee & Selchow	100	150
Dewey's Victory	B	1900s	Parker Brothers	120	200
Diamond Game of Baseball, The	B	1894	McLoughlin Bros.	1275	2150
Diamond Heart	B	1902	McLoughlin Bros.	150	225
Diceball	B	1938	Ray-Fair	90	145
Dicex Baseball Game, The	B	1925	Chester S. Howland	195	325
Dick Tracy Detective Game	B	1937	Whitman	40	100
Dick Tracy Detective Game	B	1933	Einson-Freeman	45	150
Dick Tracy Playing Card Game	C	1934	Whitman	50	75
Dick Tracy Super Detective Mystery Card Game	C	1937	Whitman	65	90
Dig	S	1940	Parker Brothers	2	5
Dim Those Lights	S	1932	All-Fair	200	400
Din	C	1905	Horsman	25	45
Discretion	B	1942	Volume Sprayer	30	45
Disk	S	1900s	Madmar Quality	35	55
District Messenger Boy, Game of	B	1886	McLoughlin Bros.	250	400
District Messenger Boy, Game of	B	1904	McLoughlin Bros.	80	200
Diving Fish	S	1920s	C.E. Bradley	20	30
Dixie Land, Game of	C	1897	Fireside	60	90
Doctor Busby Card Game	C	1910		35	55
Doctor Quack, Game of	C	1922	Russell	25	40
Doctors and the Quack	C	1890s	Parker Brothers	45	75
Dodging Donkey, The	S	1920s	Parker Brothers	45	75
Dog Race	B	1937	Transogram	20	40
Dog Show	B	1890s	J.H. Singer	85	140
Dog Sweepstakes	B	1935	Stoll & Einson	45	75
Donald Duck Game	C	1930s	Whitman	15	30
Donald Duck Playing Game	C	1941	Whitman	25	40
Donald Duck's Own Party Game	B	1938	Parker Brothers	80	200
Donkey Party	S	1887	McLoughlin Bros.	75	150
Double Eagle Anagrams	C	1890	McLoughlin Bros.	35	55
Double Flag Game, The	C	1904	McLoughlin Bros.	75	125
Double Game Board (Baseball)	B	1925	Parker Brothers	50	75
Double Header Baseball	B	1935	Redlich	145	250
Down and Out	S	1928	Milton Bradley	50	75
Down the Pike with Mrs. Wiggs at the St. Louis Exposition	C	1904	Milton Bradley	40	75
Dr. Busby	C	1937	Milton Bradley	20	50
Dr. Busby	C	1900s	J. Ottmann Lith.	40	60
Dr. Busby	C	1890s	J.H. Singer	60	75
Dr. Fusby, Game of	C	1890s	McLoughlin Bros.	65	125
Dreamland Wonder Resort Game	B	1914	Parker Brothers	275	700

PREWAR GAMES

NAME	TYPE	YEAR	COMPANY	GOOD	EX
Drive 'n Putt	B	1940s	Carrom Industries	50	90
Drummer Boy Game, The	B	1890s	Parker Brothers	100	250
Dubble Up	B	1940s	Gabriel	15	25
Dudes, Game of the	B	1890	Bliss	225	375
Durgin's New Baseball Game	B	1885	Durgin & Palmer	425	700
E-E-YAH Base Ball Game	B	1900s	National Games	600	1800
Eagle Bombsight	B	1940s	Toy Creations	75	150
East is East and West is West	B	1920s	Parker Brothers	80	200
Easy Money	B	1936	Milton Bradley	35	55
Ed Wynn The Fire Chief	B	1937	Selchow & Righter	45	75
Eddie Cantor's Tell It To The Judge	B	1930s	Parker Brothers	30	75
Egerton R. Williams Popular Indoor Baseball Game	C	1886	Hatch	2500	5000
Election	B	1896	Fireside	20	35
Electric Baseball	B	1935	Einson-Freeman	35	60
Electric Football	B	1930s	Electric Football	55	90
Electric Magnetic Baseball	B	1900		175	295
Electric Questioner	B	1920	Knapp Electric & Novelty	20	35
Electric Speed Classic	B	1930	Pressman	390	650
Electro Gameset	B	1930	Knapp Electric & Novelty	30	45
Elementaire Musical Game	B	1896	Theodore Presser	20	35
Elite Conversation Cards	C	1887	McLoughlin Bros.	35	75
Ella Cinders	B	1944	Milton Bradley	40	80
Elmer Layden's Scientific Football Game	B	1936	Cadaco	40	80
Elsie the Cow Game, The	B	1941	Selchow & Righter	50	100
Enchanted Forest Game	B	1914		120	200
Endurance Run	B	1930	Milton Bradley	60	150
Errand Boy, The	B	1891	McLoughlin Bros.	200	350
Ethan Allen's All-Star Baseball Game	B	1942	Cadaco-Ellis	60	150
Evening Parties, Game of	B	1910s	Parker Brothers	180	300
Excursion to Coney Island	C	1880s	Milton Bradley	55	75
Excuse Me!	C	1923	Parker Brothers	15	35
Faba Baga or Parlor Quiots	S	1883	Morton E. Converse	40	65
Fairies' Cauldron Tiddledy Winks Game, The	S	1925	Parker Brothers	35	50
Fairyland Game	B	1880s	Milton Bradley	60	95
Famous Authors	C	1910	Parker Brothers	40	65
Famous Authors	C	1943	Parker Brothers	10	15
Fan Craze Card Game	C	1904	Fan Craze	175	295
Fan-i-Tis	B	1913	C.W. Marsh	110	180
Fan-Tel	B	1937	Schoenhut	20	30
Farmer Jones' Pigs	B	1890	McLoughlin Bros.	165	200
Fascination	S	1890	Selchow & Righter	35	50
Fashionable English Sorry Game, The	B	1934	Parker Brothers	20	50
Fast Mail Game	B	1910	Milton Bradley	105	225
Fast Mail Railroad Game	B	1930s	Milton Bradley	50	85
Favorite Art, Game of	C	1897	Parker Brothers	40	65
Favorite Steeple Chase	B	1895	J.H. Singer	100	200
Ferdinand Card Game	C	1938	Whitman	25	65
Ferdinand The Bull Chinese Checkers Game	B	1930s		60	100
Fibber McGee	B	1936	Milton Bradley	25	40
Fibber McGee and The Wistful Vista Mystery	B	1940	Milton Bradley	30	45

Alice in Wonderland, 1930s, Parker Brothers

Peter Coddle Tells of his Trip to Chicago, 1890s, Parker Brothers

GAMES

PREWAR GAMES

NAME	TYPE	YEAR	COMPANY	GOOD	EX
Fiddlestix	S	1937	Plaza	10	15
Fig Mill	B	1916	Willis G. Young	25	80
Finance	B	1937	Parker Brothers	25	40
Finance And Fortune	B	1936	Parker Brothers	35	50
Fire Alarm Game	B	1899	Parker Brothers	1300	2300
Fire Department	B	1930s	Milton Bradley	80	125
Fire Fighters Game	B	1909	Milton Bradley	120	200
Fish Pond	S	1890	E.O. Clark	50	125
Fish Pond	S	1920s	Wilder	25	60
Fish Pond Game, Magnetic	S	1891	McLoughlin Bros.	175	300
Fish Pond, Game of	S	1910s	Wescott Brothers	30	50
Fish Pond, New and Improved	S	1890s	McLoughlin Bros.	75	125
Fish Pond, The Game of	S	1890	McLoughlin Bros.	150	250
Fishing Game	S	1899	Martin	30	65
Five Hundred, Game of	C	1900s	Home Game	20	35
Five Little Pigs	S	1890s	J.H. Singer	50	100
Five Wise Birds, The	S	1923	Parker Brothers	25	50
Flag Travelette	B	1895	Archarena	45	100
Flags	C	1899	Cincinnati Game	55	90
Flap Jacks	S	1931	All-Fair	80	225
Flapper Fortunes	B	1929	Embossing	30	45
Flash	S	1940s	Pressman	20	50
Flight To Paris	B	1927	Milton Bradley	150	250
Flinch	C	1902	Flinch Card	10	15
Fling-A-Ring	B	1930s	Wolverine	20	35
Flip It	B	1925	American Toy Works	40	60
Flip It	B	1940	Deluxe Game	20	30
Flip It, Auto Race & Transcontinental Tour	B	1920s	Deluxe Game	35	90
Flitters	S	1899	Martin	45	75
Flivver	B	1927	Milton Bradley	150	250
Floor Croquet Game	S	1912		60	100
Flowers, Game of	B	1899	Cincinnati Game	45	75
Flying Aces	B	1940s	Selchow & Righter	50	100
Flying the Beam	B	1941	Parker Brothers	75	125
Flying the United States Airmail	B	1929	Parker Brothers	200	400
Fobaga (football)	B	1942	American Football	50	80
Follow the Stars	B	1922	Watts	225	375
Foolish Questions	C	1920s	Wallie Dorr	45	85
Foot Race, The	B	1900s	Parker Brothers	60	100
Football	B	1930s	Wilder	50	80
Football Game	B	1898	Parker Brothers	295	495
Football Knapp Electro Game Set	B	1929	Knapp Electric & Novelty	125	205
Football, The Game of	B	1895	George A. Childs	75	125
Football-As-You-Like-It	B	1940	Wayne W. Light	85	145
Fore Country Club Game of Golf	B	1929	Wilder	175	295
Fortune	B	1938	Parker Brothers	40	70
Fortune Teller, The	B	1905	Milton Bradley	20	50
Fortune Telling	C	1920s	All-Fair	40	75
Fortune Telling & Baseball Game	B	1889		85	140
Fortune Telling Game	C	1930s	Stoll & Edwards	35	55
Fortune Telling Game	B	1934	Whitman	50	100
Fortune Telling Game, The	C	1890s	Parker Brothers	40	65
Fortunes, Game of	C	1902	Cincinnati Game	55	95

PREWAR GAMES

NAME	TYPE	YEAR	COMPANY	GOOD	EX
Forty-Niners Gold Mining Game	B	1930s	National Games	25	40
Foto World	B	1935	Cadaco	90	150
Foto-Electric Football	B	1930s	Cadaco	35	50
Foto-Finish Horse Race	B	1940s	Pressman	30	45
Four and Twenty Blackbirds	S	1890s		650	1100
Four Dare Devils, The	S	1933	Marx, Hess & Lee	40	65
Fox and Geese	B	1903	McLoughlin Bros.	60	150
Fox and Geese, The New	C	1888	McLoughlin Bros.	45	75
Fox and Hounds	B	1900	Parker Brothers	85	140
Fox Hunt	B	1905	Milton Bradley	40	65
Fox Hunt	B	1930s	Lowe	20	35
Foxy Grandpa at the World's Fair	C	1904	J. Ottmann Lith.	150	200
Foxy Grandpa Hat Party	B	1906	Selchow & Righter	55	90
Fractions	C	1902	Cincinnati Game	25	45
Frank Buck's Bring 'em Back Alive Game	C	1937	All-Fair	50	100
Frisko	B	1937	Embossing	20	30
Frog He Would a Wooing Go, The	B	1898	McLoughlin Bros.	800	1200
Frog School Game	B	1914		45	75
Frog Who Would a Wooing Go, The	B	1920s	United Game	60	125
Fun at the Circus	B	1897	McLoughlin Bros.	360	600
Fun at the Zoo, A Game	B	1902	Parker Brothers	120	200
Fun Kit	B	1939	Frederick H. Beach (Beachcraft)	15	20
Fut-Ball	B	1940s	Fut-Bal	35	60
G-Men	C	1936	Milton Bradley	35	55
G-Men Clue Games	B	1935	Whitman	60	150
Game of Baseball	B	1886	McLoughlin Bros.	750	1600
Game of Friendly Fun	B	1939	Milton Bradley	50	145
Games You Like To Play	B	1920s	Parker Brothers	75	100
Gamevelope	C	1944	Morris Systems	20	35
Gang Busters Game	B	1939	Whitman	50	100
Gang Busters Game	B	1938	Lynco	150	250
Gavitt's Stock Exchange	C	1903	W.W. Gavitt	35	65
Gee-Wiz Horse Race	S	1928	Wolverine	50	85
General Headquarters	B	1940s	All-Fair	50	100
Genuine Steamer Quoits	S	1924	Milton Bradley	15	25
Geographical Cards	C	1883	Peter G. Thompson	35	65
Geographical Lotto Game	B	1921		20	30
Geography Game	B	1910s	A. Flanagan	15	25
Geography up to Date	C	1890s	Parker Brothers	35	55
George Washington's Dream	C	1900s	Parker Brothers	45	75
Ges It Game	B	1936	Knapp Electric & Novelty	20	45
Get The Balls Baseball Game	B	1930		20	30
Glydor	B	1931	All-Fair	50	125
Go Bang	B	1898	Milton Bradley	100	300
Go to the Head of the Class	B	1938	Milton Bradley	15	35
Goat, Game of	C	1916	Milton Bradley	15	35
Going To The Fire Game	B	1914		90	150
Gold Hunters, The	B	1900s	Parker Brothers	105	175
Gold Rush, The	C	1930s	Cracker Jack	20	50
Golden Egg	C	1850s	McLoughlin Bros.	450	750
Golden Egg, The	C	1845	R.H. Pease	165	275
Goldenlocks & The Three Bears	B	1890	McLoughlin Bros.	320	800
Golf	B	1900	Schoenhut	2700	5000

PREWAR GAMES

NAME	TYPE	YEAR	COMPANY	GOOD	EX
Golf Tokalon Series, The Game of	B	1890s	E.O. Clark	350	575
Golf, A Game of	B	1930	Milton Bradley	145	245
Golf, Game of	B	1896	McLoughlin Bros.	425	715
Golf, The Game of	B	1905	Clark & Sowdon	325	550
Golf, The Game of	B	1898	J.H. Singer	100	300
Golliwogg	C	1907	Milton Bradley	200	300
Gonfalon Scientific Baseball	B	1930	Pioneer Game	110	180
Good Old Aunt, The	B	1892	McLoughlin Bros.	150	250
Good Old Game of Corner Grocery, The	C	1900s	Parker Brothers	40	60
Good Old Game of Dr. Busby	C	1920s	United Game	25	50
Good Old Game of Dr. Busby	C	1900s	Parker Brothers	55	85
Good Old Game of Innocence Abroad, The	B	1888	Parker Brothers	180	300
Good Things To Eat Lotto	B	1940s	Gabriel	15	25
Goose Goslin Scientific Baseball	B	1935	Wheeler Toy	250	400
Goose, The Jolly Game of	B	1851	J.P. Beach	750	1250
Goosey Gander, Or Who Finds the Golden Egg, Game of	B	1890	J.H. Singer	675	1150
Goosy Goosy Gander	B	1896	McLoughlin Bros.	300	500
Graham McNamee World Series Scoreboard Baseball Game	B	1930	Radio Sports	250	400
Grand National Sweepstakes	B	1937	Whitman	20	50
Grande Auto Race	B	1920s	Atkins	75	150
Grandma's Game of Useful Knowledge	C	1910s	Milton Bradley	20	50
Grandmama's Improved Arithmetical Game	C	1887	McLoughlin Bros.	30	50
Grandmama's Improved Geographical Game	C	1887	McLoughlin Bros.	55	90
Grandmama's Sunday Game: Bible Questions, Old Testament	C	1887	McLoughlin Bros.	35	65
Graphic Baseball	B	1930s	Northwestern Products	165	275
Great American Baseball Game, The	B	1906	William Dapping	145	250
Great American Flag Game, The	B	1940	Parker Brothers	35	60
Great American Game	B	1910	Neddy Pocket Game	145	250
Great American Game of Baseball, The	B	1907	Pittsburgh Brewing	145	250
Great American Game, Baseball, The	B	1923	Hustler Toy	150	225
Great American Game, The	B	1925	Frantz	110	180
Great American War Game	B	1899	J.H. Hunter	600	1000
Great Battlefields	C	1886	Parker Brothers	150	250
Great Composer, The	C	1901	Theodore Presser	35	55
Great Family Amusement Game, The	B	1889	Einson-Freeman	20	35
Great Horse Race Game, The	B	1925	Selchow & Righter	70	115
Great Mails Baseball Game	C	1919	Walter Mails Baseball Game	2475	4100
Gregg Football Game	B	1924	Albert A. Gregg	175	285
Greyhound Racing Game	B	1938	Rex Manufacturing	15	25
Guess Again, The Game of	C	1890s	McLoughlin Bros.	75	120
Gumps at the Seashore, The	B	1930s	Milton Bradley	65	135
Gym Horseshoes	B	1930	Wolverine	30	45
Gypsy Fortune Telling Game	C	1909	McLoughlin Bros.	100	175
Gypsy Fortune Telling Game, The	B	1895	Milton Bradley	75	125
H.M.S. Pinafore	C	1880	McLoughlin Bros.	300	500
Halma	B	1885	Milton Bradley	30	45
Halma	B	1885	Horsman	35	60
Hand of Fate	B	1901	McLoughlin Bros.	1200	2000
Happitime Bagatelle	S	1933	Northwestern Products	30	45

PREWAR GAMES

NAME	TYPE	YEAR	COMPANY	GOOD	EX
Happy Family, The	B	1910	Milton Bradley	15	25
Happy Hooligan Bowling Type Game	B	1925		60	100
Happy Landing	S	1938	Transogram	30	45
Hardwood Ten Pins Wooden Game	B	1889		60	100
Hare & Hound	B	1895	Parker Brothers	245	400
Hare and Hounds	B	1890	McLoughlin Bros.	150	250
Harlequin, The Game of The	B	1895	McLoughlin Bros.	150	300
Harold Teen Game	B	1930s	Milton Bradley	75	150
Have-U It?	C	1924	Selchow & Righter	15	25
Heads and Tails	C	1900s	Parker Brothers	30	50
Heedless Tommy	B	1893	McLoughlin Bros.	240	400
Hel-Lo Telephone Game	B	1898	J.H. Singer	100	200
Helps to History	B	1885	A. Flanagan	20	35
Hen that Laid the Golden Egg, The	B	1900	Parker Brothers	105	175
Hendrik Van Loon's Wide World Game	B	1935	Parker Brothers	35	65
Hening's In-Door Game of Professional Baseball	B	1889	Inventor's	525	875
Hens and Chickens, Game of	C	1875	McLoughlin Bros.	150	225
Heroes of America	B	1920	Educational Card & Game	20	35
Hey What?	C	1907	Parker Brothers	20	35
Hi-Way Henry	B	1928	All-Fair	600	1500
Hialeah Horse Racing Game	B	1940s	Milton Bradley	40	65
Hickety Pickety	B	1924	Parker Brothers	20	50
Hidden Titles	C	1908	Parker Brothers	20	35
Hide and Seek, Game of	B	1895	McLoughlin Bros.	300	1000
Hippodrome Circus Game	B	1895	Milton Bradley	200	300
Hippodrome, The	B	1900s	E.O. Clark	150	200
Historical Cards	C	1884	Peter G. Thompson	55	75
History up to Date	C	1900s	Parker Brothers	55	65
Hit and Run Baseball Game	B	1930s	Wilder	150	290
Hit That Line	B	1930s	LaRue Sales	100	165
Hockey, Official	B	1940	Toy Creations	50	75
Hokum	C	1927	Parker Brothers	25	35
Hold The Fort	B	1895	Parker Brothers	125	225
Hold Your Horses	B	1930s	Klauber Novelty	10	20
Hollywood Movie Bingo	C	1937	Whitman	50	75
Home Baseball Game	B	1897	McLoughlin Bros.	900	1700
Home Defenders	B	1941	Saalfield	15	25
Home Diamond	C	1913	Phillips	150	225
Home Diamond, The Great Baseball Game	B	1925	Phillips	175	295
Home Games	B	1900s	Martin	105	175
Home History Game	C	1910s	Milton Bradley	55	75
Home Run King	B	1930s	Selrite	275	450
Home Run with Bases Loaded	B	1935	T.V. Morrison	205	350
Honey Bee Game	B	1913	Milton Bradley	50	85
Hood's Spelling School	B	1897	C.I. Hood	20	35
Hood's War Game	C	1899	C.I. Hood	40	65
Hoop-O-Loop	B	1930	Wolverine	20	30
Hoot	C	1926	Saalfield	55	85
Hop-Over Puzzle	S	1930s	Pressman	20	35
Hornet	B	1941	Lowe	30	45
Horse Race	B	1943	Lowe	10	20
Horse Racing	B	1935	Milton Bradley	35	60
Horses	B	1927	Modern Makers	45	75

NAME	TYPE	YEAR	COMPANY	GOOD	EX
Hounds & Hares	B	1894	J.W. Keller	35	60
House that Jack Built	C	1900s	Parker Brothers	55	75
House that Jack Built, The	C	1887	McLoughlin Bros.	55	85
Household Words, Game of	C	1916	Household Words Game	50	85
How Good Are You?	B	1937	Whitman	10	15
How Silas Popped the Question	C	1915	Parker Brothers	35	50
Howard H. Jones Collegiate Football	B	1932	Municipal Service	40	100
Huddle All-American Football Game	B	1931		100	165
Hungry Willie	S	1930s	Transogram	40	70
Hunting Hare, Game of	B	1891	McLoughlin Bros.	205	350
Hunting in the Jungle	S	1920s	A. Gropper	30	45
Hunting the Rabbit	B	1895	Clark & Sowdon	70	115
Hunting, The New Game of	B	1904	McLoughlin Bros.	360	600
Hurdle Race	B	1905	Milton Bradley	75	150
Hymn Quartets	B	1933	Goodenough and Woglom	10	15
I Doubt It	C	1910	Parker Brothers	35	55
Ice Hockey	B	1942	Milton Bradley	50	75
Illustrated Mythology	C	1896	Cincinnati Game	55	75
Improved Geographical Game, The	B	1890s	Parker Brothers	60	100
Improved Historical Cards	C	1900	McLoughlin Bros.	35	55
In and Out the Window	B	1940s	Gabriel	20	35
In-Door Baseball	B	1926	E. Bommer Foundation	100	180
India	B	1940	Parker Brothers	15	20
India Bombay	B	1910s	Cutler & Saleeby	25	40
India, An Oriental Game	B	1890s	McLoughlin Bros.	100	125
India, Game of	B	1910s	Milton Bradley	15	40
Indianapolis 500 Mile Race Game	B	1938	Shaw	350	575
Indians and Cowboys	B	1940s	Gabriel	40	65
Indoor Football	B	1919	Underwood	145	250
Indoor Golf Dice	S	1920s	W.P. Bushell	20	35
Indoor Horse Racing	B	1924	Man-O-War	70	115
Industries, Game of	C	1897	A.W. Mumford	25	55
Inside Baseball Game	B	1911	Popular Games	300	500
Intercollegiate Football	B	1923	Hustler Toy	125	245
International Automobile Race	B	1903	Parker Brothers	875	1450
International Spy, Game of	B	1943	All-Fair	75	125
Ivanhoe	C	1886	Parker Brothers	55	75
Jack and Jill	B	1890s	Parker Brothers	55	95
Jack and Jill	B	1909	Milton Bradley	60	100
Jack and the Bean Stalk	B	1895	Parker Brothers	110	175
Jack and the Bean Stalk, The Game of	B	1898	McLoughlin Bros.	550	900
Jack Spratt Game	B	1914		45	75
Jack Straws, The Game of	S	1901	Parker Brothers	25	35
Jack-Be-Nimble	B	1940s	Embossing	20	35
Jackie Robinson Baseball Game	B	1940s	Gotham	425	725
Jackpot	B	1943	B.L. Fry Products	15	25
Jamboree	S	1937	Selchow & Righter	75	150
Japan, The Game of	B	1903	J. Ottmann Lith.	180	300
Japanese Ball Game	S	1930s	Girard	35	65
Japanese Games of Cash and Akambo	B	1881	McLoughlin Bros.	150	250
Japanese Oracle, Game of	C	1875	McLoughlin Bros.	200	325
Japanola	S	1928	Parker Brothers	35	60
Jaunty Butler	S	1932	All-Fair	75	150

GAMES

Glydor, 1931, All-Fair

Captive Princess, 1899, McLoughlin Brothers

PREWAR GAMES

NAME	TYPE	YEAR	COMPANY	GOOD	EX
Jav-Lin	S	1931	All-Fair	105	225
Jeep Board, The	B	1944	Lowe	15	35
Jeffries Championship Playing Cards	B	1904		35	55
Jig Chase	B	1930s	Game Makers	35	50
Jig Race	B	1930s	Game Makers	35	50
Jockey	B	1920s	Carrom Industries	35	60
Joe "Ducky" Medwick's Big League Baseball Game	C	1930s	Johnson-Breier	125	200
John Gilpin, Rainbow Backgammon and Bewildered Travelers	B	1875	McLoughlin Bros.	175	300
Johnny Get Your Gun	B	1928	Parker Brothers	60	100
Johnny's Historical Game	C	1890s	Parker Brothers	55	95
Jolly Clown Spinette	S	1932	Milton Bradley	30	45
Jolly Pirates	B	1938	Russell	20	35
Jolly Robbers	S	1929	Wilder	50	75
Journey to Bethlehem, The	B	1923	Parker Brothers	95	160
Jumping Frog, Game of	C	1890	J.H. Singer	75	125
Jumping Jupiter	S	1940s	Gabriel	30	50
Jumpy Tinker	B	1920s	Toy Creations	20	30
Jungle Hunt	B	1940s	Rosebud Art	30	50
Jungle Hunt	S	1940	Gotham Pressed Steel	30	45
Jungle Jump-Up Game	S	1940s	Judson Press	30	45
Junior Baseball Game	B	1915	Benjamin Seller	100	165
Junior Basketball Game	B	1930s	Rosebud Art	55	90
Junior Bicycle Game, The	B	1897	Parker Brothers	250	400
Junior Combination Board	B	1905	McLoughlin Bros.	30	50
Junior Football	B	1944	Deluxe Game	30	45
Junior Motor Race	B	1925	Wolverine	40	70
Just Like Me, Game of	C	1899	McLoughlin Bros.	55	95
Kan-Oo-Win-It	B	1893	McLoughlin Bros.	345	575
Kate Smith's Own Game America	B	1940s	Toy Creations	40	65
Katzenjammer Kids Hockey	S	1940s	Jaymar	40	65
Katzy Party	S	1900s	Selchow & Righter	70	120
Keeping Up with the Jones'	B	1921	Parker Brothers	50	85
Keeping Up with the Jones', The Game of	B	1921	Phillips	75	100
Kellogg's Baseball Game	B	1936	Kellogg's	25	35
Kellogg's Boxing Game	B	1936	Kellogg's	35	55
Kellogg's Football Game	B	1936	Kellogg's	25	40
Kellogg's Golf Game	B	1936	Kellogg's	25	40
Kentucky Derby Racing Game	B	1938	Whitman	20	40
Kilkenny Cats, The Amusing Game of	B	1890	Parker Brothers	100	200
Kindergarten Lotto	S	1904	Strauss	50	80
King's Quoits, New Game of	B	1893	McLoughlin Bros.	300	450
Kings	B	1931	Akro Agate	55	95
Kings, The Game of	C	1845	Josiah Adams	150	300
Kitty Kat Cup Ball	B	1930s	Rosebud Art	75	150
Klondike Game	B	1890s	Parker Brothers	345	575
Knockout	B	1937	Scarne Games	55	90
Knockout Andy	S	1926	Parker Brothers	35	90
Knute Rockne Football Game, Official	B	1930	Radio Sports	250	400
Ko-Ko the Clown	B	1940	All-Fair	20	30
Komical Konversation Kards	C	1893	Parker Brothers	50	75
Kriegspiel Junior	B	1915	Parker Brothers	50	80
Kuti-Kuts	S	1922	Regensteiner	20	30

PREWAR GAMES

NAME	TYPE	YEAR	COMPANY	GOOD	EX
La Haza	B	1923	Supply Sales	10	20
Lame Duck, The	B	1928	Parker Brothers	60	100
Land and Sea War Games	B	1941	Lowe	40	65
Lasso the Jumping Ring	B	1912		60	100
Lawson's Baseball Card Game	C	1910		100	175
Lawson's Patent Game of Baseball	C	1884	Lawson's Card	500	1000
Le Choc	B	1919	Milton Bradley	50	85
League Parlor Base Ball	B	1889	Bliss	600	1000
Leap Frog Game	B	1900	McLoughlin Bros.	165	275
Leap Frog, Game of	B	1910	McLoughlin Bros.	45	75
Leaping Lena	S	1920s	Parker Brothers	100	200
Lee at Havana	B	1899	Chaffee & Selchow	75	150
Leslie's Baseball Game	B	1909	Perfection Novelty	145	250
Let's go to College	B	1944	Einson-Freeman	30	45
Let's Play Games, Golf	B	1939	American Toy Works	50	80
Let's Play Polo	B	1940	American Toy Works	45	75
Letter Carrier, The	B	1890	McLoughlin Bros.	100	200
Letters	B	1878	Horsman	30	45
Letters Improved for the Logomachist	C	1878	Noyes & Snow	50	100
Letters or Anagrams	B	1890s	Parker Brothers	30	50
Lew Fonseca Baseball Game, The	B	1920s	Carrom Industries	525	875
Library of Games	B	1938	American Toy Works	15	25
Library of Games	C	1939	Russell	55	75
Lid's Off, The	S	1937	American Toy Works	35	75
Life in the Wild West	B	1894	Bliss	300	500
Life of the Party	B	1940s	Rosebud Art	35	50
Life's Mishaps & Bobbing 'Round the Circle	B	1891	McLoughlin Bros.	400	500
Light Horse H. Cooper Golf Game	B	1943	Trojan Games	175	295
Limited Mail & Express Game, The	B	1894	Parker Brothers	200	375
Lindy Flying Game	C	1927	Parker Brothers	35	65
Lindy Flying Game, The New	C	1927	Nucraft Toys	45	75
Lindy Hop-Off	B	1927	Parker Brothers	200	400
Lion & the Eagle, or the Days of '76	C	1883	E.H. Snow	75	125
Literature Game	B	1897	L.J. Colby	15	25
Little Black Sambo, Game of	B	1934	Einson-Freeman	75	150
Little Bo-Beep Game	B	1914		60	100
Little Boy Blue	B	1910s	Milton Bradley	50	85
Little Colonel	B	1936	Selchow & Righter	75	100
Little Cowboy Game, The	B	1895	Parker Brothers	200	450
Little Fireman Game	B	1897	McLoughlin Bros.	4000	6000
Little Jack Horner Golf Course	B	1920s		145	250
Little Jack Horner, A Game	B	1910s	Milton Bradley	45	75
Little Nemo Game	B	1914		100	200
Little Orphan Annie Bead Game	S	1930s		20	35
Little Orphan Annie Game	B	1927	Milton Bradley	125	250
Little Orphan Annie Rummy Cards	C	1937	Whitman	45	75
Little Orphan Annie Shooting Game	S	1930s	Milton Bradley	30	50
Little Red Riding Hood	B	1900	McLoughlin Bros.	200	500
Little Shoppers	B	1915	Gibson Game	150	250
Little Soldier, The	B	1900s	United Game	95	160
London Bridge	B	1899	J.H. Singer	100	150
London Game, The	B	1898	Parker Brothers	165	275
Lone Ranger Game, The	B	1938	Parker Brothers	40	80

PREWAR GAMES

NAME	TYPE	YEAR	COMPANY	GOOD	EX
Lone Ranger Hi-Yo Silver!! Target Game	S	1939	Marx	70	115
Looping the Loop	B	1940s	Advance Games	25	40
Los Angeles Rams Football Game	B	1930s	Zondine	175	295
Lost Diamond, Game of	B	1896	McLoughlin	200	300
Lost Heir, Game of the	C	1910	Milton Bradley	30	50
Lost Heir, The Game of	C	1893	McLoughlin Bros.	65	85
Lost in the Woods	B	1895	McLoughlin Bros.	660	900
Lotto	B	1932	Milton Bradley	6	10
Lou Gehrig's Official Playball	B	1930s	Christy Walsh	525	875
Lowell Thomas' World Cruise	B	1937	Parker Brothers	100	300
Luck, The Game of	B	1892	Parker Brothers	60	100
Lucky 7th Baseball Game	B	1937	All-American	100	250
Mac Baseball Game	B	1930s	Mc Dowell	145	250
Macy's Pirate Treasure Hunt	B	1942	Einson-Freeman	20	35
Madrap, The New Game of	B	1914		45	75
Magic Race	B	1942	Habob	55	90
Magnetic Jack Straws	B	1891	Horsman	27	45
Magnetic Treasure Hunt	B	1930s	American Toy Works	15	25
Mail, Express Or Accommodation, Game of	B	1920s	Milton Bradley	135	225
Mail, Express or Accommodation, Game of	B	1895	McLoughlin Bros.	450	1000
Major League Ball	B	1921	National Game Makers	325	550
Major League Base Ball Game	B	1912	Philadelphia Game	650	1000
Major League Baseball Game	C	1900s	Parker Brothers	50	150
Make A Million	C	1934	Rook Card	15	30
Mammoth Conette	S	1898	Milton Bradley	90	150
Man Hunt	B	1937	Parker Brothers	150	250
Man in the Moon	B	1901	McLoughlin Bros.	2100	3500
Mansion of Happiness	B	1864	Ives	180	300
Mansion of Happiness	B	1843	Ives	300	400
Mansion of Happiness, The	B	1895	McLoughlin Bros.	510	1000
Mar-Juck	S	1923	Regensteiner	20	30
Marathon Game, The	B	1930s	Rosebud Art	75	150
Marble Muggins	S	1920s	American Toy	100	200
Marriage, The Game of	B	1899	J.H. Singer	200	300
Match 'em	B	1926	All-Fair	12	20
Mathers Parlor Baseball Game	B	1909	McClurg	30	50
Mayflower, The	C	1897	Fireside Game	55	85
Meet the Missus	B	1937	Fitzpatrick Brothers	45	75
Mental Whoopee	B	1936	Simon & Schuster	10	15
Merry Hunt, The	B		Singer	250	400
Merry Steeple Chase	B	1890s	J. Ottmann Lith.	45	70
Merry-Go-Round	B	1898	Chaffee & Selchow	1500	2000
Messenger Boy Game	B	1910	J.H. Singer	100	200
Messenger, The	B	1890	McLoughlin Bros.	75	125
Meteor Game	S	1916	A.C. Gilbert	25	45
Mexican Pete - I Got It	B	1940s	Parker Brothers	20	40
Mickey Mouse Baseball	B	1936	Post Cereal	55	90
Mickey Mouse Big Box of Games & Things To Color	B	1930s		45	75
Mickey Mouse Bridge Game	C	1935	Whitman	35	55
Mickey Mouse Circus Game	B	1930s	Marks Brothers	180	300
Mickey Mouse Coming Home Game	B	1930s	Marks Brothers	80	200
Mickey Mouse Miniature Pinball Game	S	1930s	Marks Brothers	20	35

PREWAR GAMES

NAME	TYPE	YEAR	COMPANY	GOOD	EX
Mickey Mouse Old Maid Game	C	1930s	Whitman	35	65
Mickey Mouse Roll'em Game	B	1930s	Marks Brothers	90	150
Mickey Mouse Shooting Game	S	1930s	Marks Brothers	120	200
Mickey Mouse Skittle Ball Game	S	1930s	Marks Brothers	60	100
Mickey Mouse Soldier Target Set	S	1930s	Marks Brothers	60	100
Midget Auto Race	B	1930s	Cracker Jack	10	15
Midget Speedway, Game of	B	1942	Whitman	55	90
Miles at Porto Rico	B	1899	Chaffee & Selchow	75	125
Miniature Golf	B	1930s	Miniature Golf	35	60
Miss Muffet Game	B	1914		60	100
Mistress Mary, Quite Contrary	B	1905	Parker Brothers	60	105
Modern Game Assortment	B	1930s	Pressman	25	40
Moneta: "Money Makes Money", Game of	B	1889	F.A. Wright	90	150
Monkey Shines	B	1940	All-Fair	20	30
Monopolist, Mariner's Compass And Ten Up	B	1878	McLoughlin Bros.	400	750
Monopoly	B	1935	Parker Brothers	35	75
Monopoly Jr. Edition	B	1936	Parker Brothers	15	30
Moon Mullins Automobile Race	B	1927	Milton Bradley	60	150
Mother Goose Bowling Game	B	1884	Charles M. Crandall	510	850
Mother Goose, Game of	B	1921	Stoll & Edwards	30	50
Mother Hubbard	C	1875	McLoughlin Bros.	75	125
Mother Hubbard Game	B	1914		60	100
Motor Boat Race, An Exciting	B	1930	American Toy Works	110	180
Motor Cycle Game	B	1905	Milton Bradley	300	600
Motor Race	B	1922	Wolverine	90	150
Movie Inn	B	1917	Willis G. Young	45	75
Movie Millions	B	1938	Transogram	100	200
Movie-Land Keeno	C	1929	Wilder	150	250
Movie-Land Lotto	B	1920s	Milton Bradley	45	75
Moving Picture Game, The	B	1920s	Milton Bradley	70	120
Mr. Ree	B	1937	Selchow & Righter	60	100
Mumbly Peg	S	1920s	All-Fair	25	50
Musical Lotto	C	1936	Tudor Metal Products	35	65
Mutuels	B	1938	Mutuels	85	150
My Word, Horse Race	B	1938	American Toy Works	60	100
Mythology	C	1900	Cincinnati Game	55	75
Mythology, Game of	B	1884	Peter G. Thompson	50	90
Napoleon LaJoie Baseball Game	C	1913	Parker Brothers	150	500
Napoleon, Game of	B	1895	Parker Brothers	775	1300
National American Baseball Game	C	1910	Parker Brothers	130	200
National Baseball Game, The	C	1913	National Baseball Playing Card	700	1200
National Derby Horse Race	B	1938	Whitman	20	35
National Game of Baseball, The	B	1900s		500	875
National Game of the American Eagle, The	B	1844	Ives	2000	4000
National Game, The	B	1889	National Game	875	1450
National League Ball Game	B	1890	Yankee Novelty	350	575
Nations or Quaker Whist, Game of	C	1898	McLoughlin Bros.	75	110
Nations, Game of	C	1908	Milton Bradley	25	45
Naughty Molly	C	1905	McLoughlin Bros.	75	100
Naval Maneuvers	B	1920	McLoughlin Bros.	600	900
Navigator	B	1938	Whitman	45	75
Navigator Boat Race	B	1890s	McLoughlin Bros.	115	195

NAME	TYPE	YEAR	COMPANY	GOOD	EX
Nebbs on the Air, A Radio Game	B	1930s	Milton Bradley	50	125
Nebbs, Game of the	B	1930s	Milton Bradley	40	75
Neck and Neck	B	1929	Embossing	25	40
Neck and Neck	B	1930	Wolverine	50	80
Nellie Bly	B	1898	J.H. Singer	150	300
New Baseball Game	B	1885	Clark & Martin	165	275
New Parlor Game of Baseball	B	1896	Sumner	5000	10000
New York Recorder Newspaper Supplement Baseball Game	B	1896		430	715
Newsboy, Game of the	B	1890	Bliss	1200	2000
NFL Strategy	B	1935	Tudor	45	70
Nine Men Morris	B	1930s	Milton Bradley	25	45
Ninteenth Hole Golf Game	B	1930s	Einson-Freeman	85	140
Nip & Tuck Hockey	B	1928	Parker Brothers	115	195
No-Joke	B	1941	Volume Sprayer	12	20
Nok-Out Baseball Game	B	1930	Dizzy & Daffy Dean	350	575
North Pole Game, The	B	1907	Milton Bradley	50	125
Nosey, The Game of	C	1905	McLoughlin Bros.	225	400
Object Lotto	B	1940s	Gabriel	15	25
Obstacle Race	B	1930s	Wilder	70	115
Ocean to Ocean Flight Game	B	1927	Wilder	55	125
Office Boy, The	B	1889	Parker Brothers	150	300
Official Radio Baseball Game	B	1930s	Toy Creations	50	100
Official Radio Basketball Game	B	1939	Toy Creations	45	75
Official Radio Football Game	B	1940	Toy Creations	45	75
Oh, Blondie!	C	1940s		30	45
Old Curiosity Shop	C	1869	Novelty Game	200	300
Old Hunter & His Game	B	1870		175	295
Old Maid	B	1898	Chaffee & Selchow	25	60
Old Maid	C	1890s	J.H. Singer	25	50
Old Maid & Old Bachelor, The Merry Game of	B	1898	McLoughlin Bros.	180	400
Old Maid Card Game	C	1889		20	50
Old Maid Fun Full Thrift Game	C	1940s	Russell	30	50
Old Maid or Matrimony, Game of	B	1890	McLoughlin Bros.	150	250
Old Maid, Game of	C	1870	McLoughlin Bros.	75	125
Old Maid, with Characters from Famous Nursery Rhymes	C	1920s	All-Fair	30	50
Old Mother Goose	B	1898	Chaffee & Selchow	105	175
Old Mother Hubbard, Game of	B	1890s	Milton Bradley	60	100
Old Mrs. Goose, Game of	B	1910	Milton Bradley	50	85
Old Time Shooting Gallery	S	1940	Warren-Built-Rite	15	25
Oldtimers	B	1940	Frederick H. Beach (Beachcraft)	15	20
Oliver Twist, The Good Old Game of	C	1888	Parker Brothers	150	225
Ollo	B	1944	Games Of Fame	15	25
Olympic Runners	B	1930	Wolverine	125	175
On the Mid-Way	B	1925	Milton Bradley	45	75
One Two Button Your Shoe	B	1940s	Master Toy	15	25
Open Championship Golf Game	B	1930s	Beacon Hudson	45	75
Opportunity Hour	B	1940	American Toy Works	20	35
Ot-O-Win Football	B	1920s	Ot-O-Win Toys & Games	55	90
Ouija	B	1920	William Fuld	35	70
Our Bird Friends	C	1901	Sarah H. Dudley	20	30
Our Defenders	B	1944	Master Toy	40	65
Our Gang Tipple Topple Game	S	1930	All-Fair	160	400

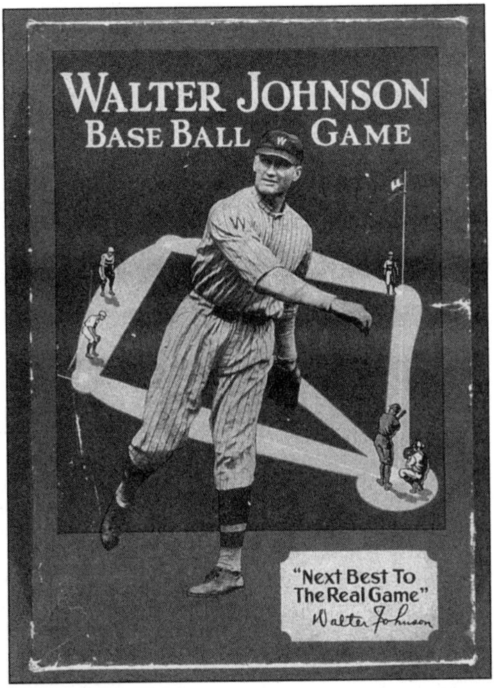

Walter Johnson Baseball Game, 1930s

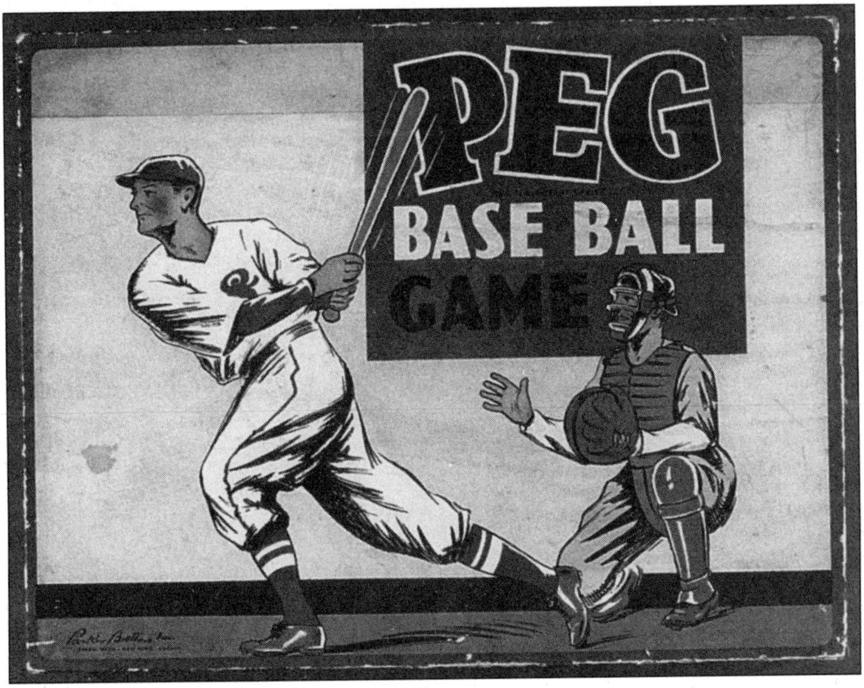

Peg Baseball, 1920s, Parker Brothers

GAMES

PREWAR GAMES

NAME	TYPE	YEAR	COMPANY	GOOD	EX
Our National Ball Game	B	1887	McGill & DeLany	425	650
Our National Life	C	1903	Cincinnati Game	35	65
Our No. 7 Baseball Game Puzzle	B	1910	Satisfactory	110	180
Our Union	B	1896	Fireside Game	25	40
Outboard Motor Race, The	B	1930s	Milton Bradley	35	75
Overland Limited, The	B	1920s	Milton Bradley	60	115
Owl and the Pussy Cat, The	B	1900s	E.O. Clark	210	350
Pan-Cake Tiddly Winks	B	1920s	Russell	55	125
Pana Kanal, The Great Panama Canal Game	B	1913	Chaffee & Selchow	65	110
Panama Canal Game	B	1910	Parker Brothers	200	400
Par Golf Card Game	B	1920	National Golf Services	115	195
Par, The New Golf Game	B	1926	Russell	115	195
Parcheesi	B	1880s	H.B. Chaffee	50	75
Parker Brothers Post Office Game	B	1910s	Parker Brothers	105	175
Parlor Base Ball, Game of	B	1892	McLoughlin	1750	2900
Parlor Baseball Game	B	1908	Mathers	200	325
Parlor Croquet	B	1940	Pressman	20	35
Parlor Football Game	B	1890s	McLoughlin Bros.	525	875
Parlor Golf	B	1897	Chaffee & Selchow	55	90
Pat Moran's Own Baseball Game	B	1919	Smith, Kline & French	325	400
Patch Word	C	1938	All-Fair	20	35
Patent Parlor Bowling Alley	B	1899	Thomas Kochka	70	115
Paws & Claws	C	1895	Clark & Sowdon	60	90
Pe-Ling	B	1923	Cookson & Sullivan	25	45
Pedestrianism	B	1879		350	575
Peeza	S	1935	Toy Creations	25	45
Peg at my Heart	B	1914	Willis G. Young	20	45
Peg Baseball	B	1915	Parker Brothers	145	250
Peg Baseball	B	1924	Parker Brothers	50	100
Peg'ity	B	1925	Parker Brothers	10	25
Peggy	B	1923	Parker Brothers	35	55
Pegpin, Game of	B	1929	Stoll & Edwards	25	45
Pennant Puzzle	B	1909	L.W. Hardy	250	400
Pennant Winner	B	1930s	Wolverine	175	295
Penny Post	B	1892	Parker Brothers	150	250
Pepper	C	1906	Parker Brothers	20	35
Peter Coddle and his Trip to New York	C	1890s	J.H. Singer	35	65
Peter Coddle tells of his Trip to Chicago	C	1890	Parker Brothers	30	50
Peter Coddle's Trip to New York	C	1925	Milton Bradley	20	35
Peter Coddle's Trip to New York, The Game of	C	1888	Parker Brothers	45	75
Peter Coddle's Trip to the World's Fair	C	1939	Parker Brothers	50	75
Peter Coddle, Improved Game of	C	1900	McLoughlin Bros.	35	55
Peter Coddles	C	1890s	J. Ottmann Lith.	25	55
Peter Pan	B	1927	Selchow & Righter	100	200
Peter Peter Pumpkin Eater	B	1914	Parker Brothers	60	125
Peter Rabbit Game	B	1910	Milton Bradley	55	95
Peter Rabbit Game	B	1940s	Gabriel	45	75
Philadelphia Inquirer Baseball Game, The	B	1896		145	250
Philo Vance	B	1937	Parker Brothers	70	175
Phoebe Snow, Game of	B	1899	McLoughlin Bros.	150	250
Piggies, The New Game	B	1894	Selchow & Righter	330	550
Pigskin	B	1940	Parker Brothers	20	50

PREWAR GAMES

NAME	TYPE	YEAR	COMPANY	GOOD	EX
Pigskin, Tom Hamilton's Football Game	B	1934	Parker Brothers	25	60
Pike's Peak or Bust	S	1890s	Parker Brothers	75	150
Pilgrim's Progress, Going To Sunday School, Tower of Babel	B	1875	McLoughlin Bros.	120	250
Pinafore	B	1879	Fuller Upham	45	75
Pinch Hitter	B	1930s		110	180
Pines, The	B	1896	Fireside Game	15	25
Ping Pong	S	1902	Parker Brothers	100	175
Pinocchio Pitfalls Marble Game	B	1940		30	50
Pinocchio Playing Card Game	C	1939	Whitman	55	95
Pinocchio Ring The Nose Game	B	1940		20	30
Pinocchio Target Game	S	1938	American Toy Works	90	150
Pinocchio the Merry Puppet Game	S	1939	Milton Bradley	55	95
Pioneers of the Santa Fe Trail	B	1935	Einson-Freeman	25	40
Pirate & Traveller	B	1936	Milton Bradley	25	50
Pirate Ship	B	1940	Lowe	15	25
Pitch Em, The Game of Indoor Horse Shoes	S	1929	Wolverine	25	35
Pla-Golf Board Game	B	1938	Pla-Golf	775	1300
Play Ball	B	1920	National Game	145	250
Play Football	B	1934	Whitman	55	90
Play Hockey Fun with Popeye & Wimpy	B	1935	Barnum	205	350
Pocket Baseball	B	1940	Toy Creations	15	25
Pocket Edition Major League Baseball Game	B	1943	Anderson	85	140
Pocket Football	B	1940	Toy Creations	25	40
Polar Ball Baseball	B	1940	Bowline Game	90	145
Pool, Game of	B	1898		700	1175
Pop the Hat	S	1930s	Milton Bradley	35	50
Posting, A Merry Game of	B	1890s	J.H. Singer	180	300
Pro Baseball	B	1940		70	115
Professional Game of Base Ball	B	1896	Parker Bros.	500	100
Psychic Baseball	C	1927	Psychic Baseball	60	150
Psychic Baseball Game	B	1935	Parker Brothers	175	295
Quarterback	B	1914	Littlefield	100	165
Rabbit Hunt, Game of	B	1870	McLoughlin Bros.	175	295
Race for the Cup	B	1910s	Milton Bradley	125	200
Race, The Game of the	B	1860s		425	715
Races, The Game of the	B	1844	William Crosby	850	1400
Racing Stable, Game of	B	1936	D & H Games	115	195
Radio Game	B	1926	Milton Bradley	40	100
Raggedy Ann's Magic Pebble Game	B	1941	Milton Bradley	50	100
Rainy Day Golf	B	1920	Selchow & Righter	75	150
Rambles	B	1881	American	195	275
Ranger Commandos	S	1942	Parker Brothers	35	60
Razz-O-Dazz-O Six Man Football	B	1938	Gruhn & Melton	60	100
Real Baseball Card Game	C	1900	National Baseball	180	275
Realistic Baseball	B	1925	Realistic Game & Toy	205	350
Realistic Golf	B	1898	Parker Brothers	875	1450
Red Riding Hood and the Wolf, The New Game	C	1887	McLoughlin Bros.	75	120
Red Riding Hood, Game of	B	1898	Chaffee & Selchow	200	350
Red Ryder "Whirli-Crow" Target Game	S	1940s	Daisy	150	250
Red Ryder Target Game	B	1939	Whitman	75	150
Rex	C	1920s	J. Ottmann Lith.	35	65

PREWAR GAMES

NAME	TYPE	YEAR	COMPANY	GOOD	EX
Rex and the Kilkenny Cats Game	B	1892	Parker Brothers	45	75
Ride 'em Cowboy	S	1939	Gotham Pressed Steel	25	45
Ring My Nose	S	1926	Milton Bradley	60	100
Ring Scaling	S	1900	Martin	25	45
Ring-A-Peg	B	1885	Horsman	25	45
Rip Van Winkle	B	1890s	Clark & Sowdon	175	225
Rival Policemen	B	1896	McLoughlin Bros.	2000	4000
Road Race, Air Race (Two-game set)	B	1928	Wilder	145	250
Robinson Crusoe for Little Folks, Game of	C	1900s	E.O. Clark	35	65
Robinson Crusoe, Game of	B	1909	Milton Bradley	30	75
Roll-O Football	B	1923	Supply Sales	35	60
Roll-O Golf	B	1923	Supply Sales	35	60
Roll-O Junior Baseball Game	B	1922	Roll-O	325	550
Roll-O-Motor Speedway	B	1922	Supply Sales	65	110
Roly Poly Game	B	1910		30	50
Roodles	C	1912	Flinch Card	30	50
Rook	C	1906	Rook Card	10	15
Roosevelt at San Juan	C	1899	Chaffee & Selchow	100	200
Rose Bowl Championship Football Game	B	1940s	Lowe	40	80
Rough Riders, The Game of	B	1898	Clark & Sowdon	400	600
Roulette Baseball Game	B	1929	W. Barthonomae	115	195
Round the World Game	B	1914	Milton Bradley	150	200
Round the World with Nellie Bly	B	1890	McLoughlin Bros.	210	350
Royal Game of Kings and Queens	B	1892	McLoughlin Bros.	375	650
Rube Bressler's Baseball Game	B	1936	Bressler	130	215
Rube Walker & Harry Davis Baseball Game	B	1905		875	1450
Rummy Football	B	1944	Milton Bradley	35	60
Runaway Sheep	B	1892	Bliss	165	275
Sabotage	C	1943	Games Of Fame	25	45
Saratoga Horse Racing Game	B	1920	Milton Bradley	40	100
Saratoga Steeple Chase	B	1900	J.H. Singer	120	400
Scout, The	B	1900s	E.O. Clark	105	175
Scouting, Game of	B	1930s	Milton Bradley	200	300
Scrambles	B	1941	Frederick H. Beach (Beachcraft)	15	20
Shadow Game, The	B	1940s	Toy Creations	500	900
Shopping, Game of	B	1891	Bliss	1500	2000
Shuffle-Board, The New Game of	S	1920	Gabriel	50	85
Shufflebug, Game of	B	1921		20	30
Siege of Havana, The	B	1898	Parker Brothers	180	300
Simba	S	1932	All-Fair	75	150
Sippa Fish	B	1936	Frederick H. Beach (Beachcraft)	20	30
Skating Race Game, The	B	1900	Chaffee & Selchow	500	750
Skeezix and the Air Mail	B	1930s	Milton Bradley	65	125
Skeezix Visits Nina	B	1930s	Milton Bradley	65	125
Ski-Hi New York to Paris	B	1927	Cutler & Saleeby	100	150
Skippy, A Card Game	C	1936	All-Fair	45	75
Skippy, Game of	B	1932	Milton Bradley	75	150
Skirmish at Harper's Ferry	B	1891	McLoughlin Bros.	330	550
Skit Scat	C	1905	McLoughlin Bros.	50	85
Skor-It Bagatelle	B	1930s	Northwestern Products	145	250
Sky Hawks	B	1931	All-Fair	120	200

PREWAR GAMES

NAME	TYPE	YEAR	COMPANY	GOOD	EX
Skyscraper	B	1937	Parker Brothers	250	400
Slide Kelly! Baseball Game	B	1936	B.E. Ruth	70	115
Slugger Baseball Game	B	1930	Marks Brothers	110	180
Smitty Game	B	1930s	Milton Bradley	200	350
Smitty Speed Boat Race Game	B	1930s	Milton Bradley	50	125
Smitty Target Game	S	1930s	Milton Bradley	40	75
Snake Game	B	1890s	McLoughlin Bros.	150	200
Snap	C	1883	Horsman	20	50
Snap Dragon	B	1903	H.B. Chaffee	135	225
Snap, Game of	C	1892	McLoughlin Bros.	40	65
Snap, Game of	C	1910s	Milton Bradley	20	35
Snap, The Game of	C	1905s	Parker Brothers	25	40
Snap-Jacks	S	1940s	Gabriel	15	25
Sniff	B	1940s	The Embossing Co.	20	30
Snow White and the Seven Dwarfs	S	1938	American Toy Works	150	250
Snow White and the Seven Dwarfs	B	1938	Parker Brothers	125	200
Snow White and the Seven Dwarfs	B	1938	Milton Bradley	125	150
Snug Harbor	B	1930s	Milton Bradley	50	85
Socko the Monk, The Game of	B	1935	Einson-Freeman	15	25
Soldier Boy Game	B	1914	United Game	60	150
Soldier's Cavalry	B		McLouglin Bros.	150	300
Soldiers on Guard	B			300	500
Speculation	B	1885	Parker Brothers	40	65
Speed Boat	B	1920s	Parker Brothers	85	140
Speed Boat Race	B	1926	Wolverine	70	115
Speed King, Game Of	B	1922	Russell	115	195
Speedem Junior Auto Race Game	B	1929	All-Fair	70	175
Speedway Motor Race	B	1920s	Smith, Kline & French	145	250
Spider's Web, Game of	B	1898	McLoughlin Bros.	75	150
Spin 'em Target Game	S	1930s	All Metal Product	25	40
Spin It	S	1910s	Milton Bradley	20	50
Squails	B	1877	Milton Bradley	35	60
Squails	B	1870s	Adams	100	150
Stage	C	1904	C.M. Clark	100	150
Stak, International Game of	S	1937	Marks Brothers	30	60
Stanley in Africa	B	1891	Bliss	900	1800
Star Baseball Game	C	1941	W.P. Ulrich	70	115
Star Basketball	B	1926	Star Paper Products	125	205
Star Ride	B	1934	Einson-Freeman	15	25
Stars on Stripes Football Game	B	1941	Stars & Stripes Games	55	90
Stax	S	1930s	Marks Brothers	12	20
Steeple Chase	B	1890	J.H. Singer	100	200
Steeple Chase & Checkers	B	1910	Milton Bradley	55	90
Steeple Chase, Game of	B	1900s	E.O. Clark	60	100
Steeple Chase, Game of	B	1910s	Milton Bradley	40	65
Steeple Chase, Improved Game of	B	1890s	McLoughlin Bros.	195	325
Steps to Health Coke Game	B	1938	CDN	40	70
Sto-Auto Race	B	1920s	Stough	65	110
Sto-Quoit	B	1920s	Stough	10	15
Stock Exchange	B	1936	Parker Brothers	50	100
Stock Exchange, The Game of	B	1940s	Stox	40	65
Stop & Go	B	1936	Einson-Freeman	25	50
Stop and Go	B	1928	All-Fair	100	150
Stop and Shop	B	1930	All-Fair	100	150
Stop, Look, and Listen, Game of	B	1926	Milton Bradley	40	100

PREWAR GAMES

NAME	TYPE	YEAR	COMPANY	GOOD	EX
Strat: The Great War Game	B	1915	Strat Game	25	45
Strategy, Game of	B	1891	McLoughlin Bros.	240	400
Strategy, Game of Armies	B	1938	Corey Games	40	75
Stratosphere	B	1936	Whitman	100	200
Stratosphere	B	1930s	Parker Brothers	50	125
Street Car Game, The	B	1890s	Parker Brothers	200	400
Strike Out	B	1920s	All-Fair	175	295
Strike-Like	B	1940s	Saxon Toy	55	90
Stubborn Pig, Game of the	B		Milton Bradley	100	200
Stunt Box	B	1941	Frederick H. Beach (Beachcraft)	12	20
Submarine Drag	B	1917	Willis G. Young	75	125
Substitute Golf	B	1906	John Wanamaker	300	600
Suffolk Downs	B	1930s	Corey Game	85	140
Superman Action Game	S	1940	American Toy Works	60	100
Superman, Adventures of	B	1942	Milton Bradley	300	400
Susceptibles, The	B	1891	McLoughlin Bros.	325	550
Sweep	B	1929	Selchow & Righter	25	40
Sweeps	B	1930s	E.E. Fairchild	25	60
Sweepstakes	B	1930s	Haras	55	90
Swing A Peg	B	1890s	Milton Bradley	30	50
T.G.O. Klondyke	B	1899	J.H. Singer	150	250
Table Croquet	S	1890s	Milton Bradley	30	75
Table Golf	B	1909	McClurg	15	25
Tackle	B	1933	Tackle Game	70	115
Tactics	S	1940	Northwestern Products	25	45
Tait's Table Golf	B	1914	John Tait	350	575
Tak-Tiks, Basketball	B	1939	Midwest Products	15	25
Take It And Double	B	1943	Frederick H. Beach (Beachcraft)	20	35
Take It or Leave It	B	1942	Zondine Game	25	45
Take-Off	C	1930s	Russell	25	45
Teddy's Bear Hunt	B	1907	Bowers & Hard	375	650
Teddy's Ride from Oyster Bay to Albany	B	1899	Jesse Crandall	3000	5500
Tee Off	B	1935	Donogof	115	195
Telegrams	B	1941	Whitman	40	70
Telegraph Boy, Game of the	B	1888	McLoughlin Bros.	250	350
Telegraph Messenger Boy, Game of the	C	1886	McLoughlin Bros.	55	95
Telepathy	B	1939	Cadaco-Ellis	65	110
Tell Bell, The	B	1928	Knapp Electric & Novelty	75	100
Ten Pins	B	1920	Mason & Parker	35	60
Tennis & Baseball	B	1930		70	115
Terry and the Pirates	B	1930s	Whitman	50	100
Tete-A-Tete	B	1892	Clark & Sowdon	40	100
They're Off, Race Horse Game	B	1930s	Parker Brothers	15	30
Thorobred	B	1940s	Lowe	55	90
Thorton W. Burgess Animal Game	B	1925	Saalfield	70	115
Three Bears	B	1910s	Milton Bradley	20	50
Three Bears, The	C	1922	Stoll & Edwards	30	50
Three Blind Mice, Game of	B	1930s	Milton Bradley	25	45
Three Little Kittens	B	1910s	Milton Bradley	60	100
Three Little Pigs Game	B	1933	Einson-Freeman	55	95
Three Little Pigs, The Game of the	B	1933	Kenilworth Press	100	165
Three Men in a Tub	B	1935	Milton Bradley	40	65

Animal Bingo, Baldwin Manufacturing

The Air Ship Game, early 1900s, McLoughlin Brothers

PREWAR GAMES

NAME	TYPE	YEAR	COMPANY	GOOD	EX
Three Men on a Horse	B	1936	Milton Bradley	20	50
Three Merry Men	C	1865	Amsdan	75	100
Three Point Landing	B	1942	Advance Games	35	55
Thrilling Indoor Football Game	B	1933	Cronston	70	115
Through The Clouds	B	1931	Milton Bradley	75	150
Through the Locks to the Golden Gate	B	1905	Milton Bradley	125	225
Ticker	B	1929	Glow Products	50	100
Tiddledy Wink Tennis	S	1890	E.I. Horsman	45	75
Tiddledy Winks, Game of	S	1910s	Parker Brothers	35	55
Tiddley Golf Game	B	1928	Milton Bradley	80	135
Tiddley Winks Game	B	1920s	Wilder	15	35
Tiger Hunt, Game of	B	1899	Chaffee & Selchow	270	450
Tiger Tom, Game of	B	1920s	Milton Bradley	30	75
Ting-A-Ling, The Game of	B	1920	Stoll & Edwards	25	45
Tinker Toss	S	1920s	Toy Creations	25	40
Tinkerpins	B	1916	Toy Creations	55	90
Tip the Bellboy	S	1929	All-Fair	120	200
Tip Top Fish Pond	S	1930s	Milton Bradley	20	45
Tip-Top Boxing	B	1922	LaVelle	200	350
Tipit	B	1929	Wolverine	15	20
Tit for Tat Indoor Hockey	B	1920s	Lemper Novelty	45	75
Tit-Tat-Toe	B	1929	The Embossing	15	25
Tit-Tat-Toe, Three in a Row	B	1896	Austin & Craw	45	75
To the Aid of your Party	B	1942	Leister Game	15	25
Tobagganing at Christmas, Game of	B	1899	McLoughlin Bros.	1000	1500
Toboggan Slide	B	1890s	Hamilton-Myers	225	385
Toboggan Slide	B	1890s	J.H. Singer	200	325
Toll Gate, Game of	B	1890s	McLoughlin Bros.	280	700
Tom Barker Card Game	C	1913		1400	2300
Tom Hamilton's Pigskin	B	1935	Parker Brothers	50	100
Tom Sawyer and Huck Finn, Adventures of	B	1925	Stoll & Edwards	70	120
Tom Sawyer on the Mississippi	B	1935	Einson-Freeman	40	100
Tom Sawyer, The Game of	B	1937	Milton Bradley	45	75
Toonerville Trolley Game	B	1927	Milton Bradley	175	300
Toonin Radio Game	B	1925	All-Fair	150	300
Toot	C	1905	Parker Brothers	35	65
Top-Ography	B	1941	Cadaco	35	45
Topsy Turvey, Game of	B	1899	McLoughlin Bros.	150	250
Tortoise and the Hare	B	1922	Russell	45	85
Toss-O	S	1924	Lubbers & Bell	15	35
Totem	C	1873	West & Lee	45	75
Toto, The New Game	B	1925	Baseball Toto Sales	55	90
Touchdown	B	1937	Cadaco	65	110
Touchdown	S	1930s	Milton Bradley	150	250
Touchdown Football Game	B	1920s	Wilder	100	165
Touchdown or Parlor Football, Game of	B	1897	Union Mutual Life	85	140
Touchdown, The New Game	B	1920	Hartford	70	115
Touring	C	1926	Parker Brothers	20	40
Touring	C	1906	Wallie Dorr	35	50
Tourist, A Railroad Game	B	1900s	Milton Bradley	100	225
Tournament	B	1858	Mayhew & Baker	180	300
Town Hall	B	1939	Milton Bradley	20	30
Toy Town Bank	B	1910	Milton Bradley	90	150

PREWAR GAMES

NAME	TYPE	YEAR	COMPANY	GOOD	EX
Toy Town Conductors Game	B	1910	Milton Bradley	105	175
Toy Town Target with Repeating Pistol	S	1911	Milton Bradley	55	95
Toy Town Telegraph Office	B	1910s	Parker Brothers	75	150
Trackle-Lite	B	1940s	Saxon Toy	55	90
Traffic Hazards	B	1939	Trojan Games	30	75
Trailer Trails	B	1937	Offset Gravure	35	60
Train for Boston	B	1900	Parker Brothers	500	1500
Traits, The Game of	C	1933	Goodenough and Woglom	20	35
Transatlantic Flight, Game of the	B	1925	Milton Bradley	125	300
Transport Pilot	B	1938	Cadaco	25	40
Trap-A-Tank	B	1920s	Wolverine	40	70
Traps & Bunkers	B	1926?	Milton Bradley	115	195
Traps and Bunkers, A Game of Golf	S	1930s	Milton Bradley	25	40
Travel, The Game of	B	1894	Parker Brothers	225	375
Treasure Hunt	B	1940	All-Fair	20	50
Treasure Island	B	1923	Stoll & Edwards	75	125
Treasure Island	B	1934	Stoll & Einson	60	90
Treasure Island, Game of	B	1923	Gem	40	65
Triangular Dominoes	B	1885	Frank H. Richards	35	60
Trilby	B	1894	E.I. Horsman	270	450
Trip Around the World, A	B	1920s	Parker Brothers	25	45
Trip Round the World, Game of	B	1897	McLoughlin Bros.	300	500
Trip through our National Parks: Game of Yellowstone, A	C	1910s	Cincinnati Game	35	65
Trip to Washington	B	1884	Milton Bradley	100	175
Triple Play	B	1930s	National Games	12	20
Trips of Japhet Jenkens & Sam Slick	C	1871	Milton Bradley	20	35
Trolley	C	1904	Snyder Brothers	45	75
Trolley Came Off, The	C	1900s	Parker Brothers	45	75
Trolley Ride, The Game of the	B	1890s	Hamilton-Myers	210	350
Trunk Box Lotto Game	B	1890s	McLoughlin Bros.	15	35
Tumblin Five Acrobats	B	1925	Doremus Schoen	12	20
Turn Over	B	1908	Milton Bradley	75	100
Turnover	B	1898	Chaffee & Selchow	50	85
Tutoom, Journey to the Treasures of Pharoah	B	1923	All-Fair	150	225
Twentieth Century Limited	B	1900s	Parker Brothers	90	150
Twenty Five, Game of	C	1925	Milton Bradley	10	15
Ty Cobb's Own Game of Baseball	B	1920s	National Novelty	350	650
U-Bat-It	B	1920s	Schultz Star	70	115
U.S. Postman Game	B	1914		60	100
Uncle Jim's Question Bee	B	1938	Kress	20	30
Uncle Sam at War with Spain, Great Game of	B	1898	Rhode Island Game	325	550
Uncle Sam's Baseball Game	B	1890	J.C. Bell	525	875
Uncle Sam's Mail	B	1893	McLoughlin Bros.	210	400
Uncle Wiggily's New Airplane Game	B	1920s	Milton Bradley	50	175
United States Air Mail Game, The	S	1930s	Parker Brothers	50	85
United States History, The Game of	C	1903	Parker Brothers	25	35
Van Loon Story of Mankind Game, The	B	1931	Kerk Guild	50	85
Vanderbilt Cup Race	B	1906	Bowers & Hard	600	1200
Varsity Football Game	B	1942	Cadaco-Ellis	45	75
Varsity Race	B	1899	Parker Brothers	425	725
Vassar Boat Race, The	B	1899	Chaffee & Selchow	420	700

PREWAR GAMES

NAME	TYPE	YEAR	COMPANY	GOOD	EX
Venetian Fortune Teller, Game of	C	1898	Parker Brothers	125	165
Verborum	C	1883	Peter G. Thompson	35	65
Vest Pocket Checker Set	B	1929	Embossing	15	25
Vest Pocket Quoits	B	1944	Colorful Creations	25	45
Victo	B	1943	Spare Time	15	25
Victory	B	1920s	Klak New Haven	105	175
Vignette Author	B	1874	E.G. Selchow	35	75
Visit of Santa Claus, Game of the	B	1899	McLoughlin Bros.	300	700
Visit to the Farm	B	1893	Bliss	300	500
Vox-Pop	B	1938	Milton Bradley	25	45
Voyage Around the World, Game of	B	1930s	Milton Bradley	105	175
Wa-Hoo Pick-Em Up Sticks	S	1936	Doremus Schoen	15	25
Wachter's Parlor Base Ball (bagatelle)	B	1925	Wachter	145	250
Walk the Plank	B	1925	Milton Bradley	50	95
Walking the Tightrope	B	1920	Milton Bradley	50	100
Walking the Tightrope	B	1897	McLoughlin Bros.	125	250
Walt and Skeezix Gasoline Alley Game	C	1927	Milton Bradley	100	175
Walt Disney's Game Parade	B	1930s		30	50
Walt Disney's Ski Jump Target Game	S	1930s	American Toy Works	175	295
Walt Disney's Uncle Remus Game	S	1930s	Parker Brothers	60	125
Walter Johnson Baseball Game	B	1930s		195	325
Waner's Baseball Game	B	1930s	Waner's Baseball Game	350	575
Wang, Game of	B	1892	Clark & Sowdon	25	65
War and Diplomacy	C	1899	Chaffee & Selchow	85	125
War of Nations	B	1915	Milton Bradley	40	65
War of Words	C	1910	McLoughlin Bros.	35	60
Ward Cuff's Football Game	B	1938	Continental Sales	125	200
Washington's Birthday Party	S	1911	Russell	55	95
Watch on De Rind	B	1931	All-Fair	200	400
Waterloo	B	1895	Parker Brothers	325	550
Watermelon Frolic	B	1900	Horsman	135	225
Watermelon Patch	B	1940s	Craig Hopkins	25	45
Watermelon Patch Game	S	1896	McLoughlin Bros.	900	2000
Way to the White House, The	B	1927	All-Fair	75	150
We, The Magnetic Flying Game	B	1928	Parker Brothers	100	200
West Point	B	1902		90	150
What Would You Do?	C	1933	Geo E. Schweig & Son	15	25
What's My Name?	B	1920s	Jaymar	15	25
When My Ship Comes In	C	1888	Parker Brothers	55	75
Where do you Live?	C	1890s	J.H. Singer	55	75
Where's Johnny?	C	1885	McLoughlin Bros.	65	110
Which is It? Speak Quick or Pay	C	1889	McLoughlin Bros.	65	110
Whip, The Comical Game of	C	1930	Russell	30	50
Whippet Race	B	1940s	Pressman	20	35
Whirlpool Game	B	1890s	McLoughlin Bros.	65	150
White Wings	B	1930s	Glevum Games	45	70
Who is the Thief?	C	1937	Whitman	35	55
Wide Awake, Game of	B	1899	McLoughlin Bros.	150	250
Wide World and a Journey Round It	B	1896	Parker Brothers	165	275
Wild West Cardboard Game	B	1914		90	150
Wild West, Game of the	B	1889	Bliss	750	1500
Wilder's Football Game	B	1930s	Wilder	55	90
Win, Place & Show	B	1940s	3M	25	40
Winko Baseball	B	1940	Milton Bradley	45	70

PREWAR GAMES

NAME	TYPE	YEAR	COMPANY	GOOD	EX
Winnie Winkle Glider Race Game	B	1930s	Milton Bradley	65	150
Winnie-The-Pooh Game	B	1933	Parker Brothers	100	200
Winnie-The-Pooh Game, A. A. Milne's	B	1931	Kerk Guild	50	85
Witzi-Wits	B	1926	All-Fair	100	200
Wizard, The	B	1921	Fulton Specialty	15	25
Wogglebug Game of Conundrums, The	C	1905	Parker Brothers	200	500
Wonder Tiddley Winks	S	1899	Martin	20	35
Wonderful Game of Oz (pewter pieces)	B	1921	Parker Brothers	600	1300
Wonderful Game of Oz (wooden pieces)	B	1921	Parker Brothers	200	400
Wordy	B	1938	Pressman	25	45
World Flyers, Game of the	B	1926	All-Fair	150	300
World Series Baseball Game	B	1940s	Radio Sports	205	350
World Series Parlor Baseball	B	1916	Clifton E. Hooper	150	250
World's Championship Baseball	B	1910	Champion Amusement	175	295
World's Championship Golf Game	B	1930s	Beacon Hudson	145	250
World's Columbian Exposition, Game of the	B	1893	Bliss	450	750
World's Educator Game	B	1889	Reed	45	75
World's Fair Game	B	1939		90	150
World's Fair Game, The	B	1892	Parker Brothers	800	1400
Worth While	C	1907	Doan	25	40
WPA, Work, Progress, Action	B	1935	All-Fair	100	200
Wyhoo!	C	1906	Milton Bradley	40	60
Wyntre Golf	B	1920s	All-Fair	175	400
X-Plor-US	B	1922	All-Fair	100	150
Ya-Lo Football Card Game	B	1930s		85	140
Yacht Race	B	1930s	Pressman	150	250
Yacht Race	B	1890s	Clark & Sowdon	200	350
Yachting	B	1890	J.H. Singer	70	115
Yale Harvard Football Game	B	1922	LaVelle	200	300
Yale Harvard Game	B	1890	McLoughlin Bros.	1050	1775
Yale-Princeton Football Game	B	1895	McLoughlin Bros.	575	1100
Yankee Doodle!	B	1940	Cadaco-Ellis	30	50
Yankee Doodle, A Game of American History	B	1895	Parker Brothers	285	475
Yankee Pedlar, Or What Do You Buy	C	1850s	John McLoughlin	725	1200
Yankee Trader	B	1941	Corey Games	35	80
Yellowstone, Game of	C	1895	Fireside Game	75	100
You're Out! Baseball Game	B	1941	Corey Games	35	65
Young Athlete, The	B	1898	Chaffee & Selchow	425	700
Young Folks Historical Game	C	1890s	McLoughlin Bros.	45	85
Young Peddlers, Game of the	C	1859	Mayhew & Baker	125	175
Young People's Geographical Game	C	1900s	Parker Brothers	35	55
Yuneek Game	B	1889	McLoughlin Bros.	450	750
Zimmer Baseball Game	B	1885	McLoughlin Bros.	3000	6000
Zip-Top	B	1940	Deluxe Game	35	55
Zippy Zepps	B	1930s	All-Fair	350	600
Zoo Hoo	B	1924	Lubbers & Bell	125	200
Zoom	C	1941	Whitman	35	55
Zoom, Original Game of	B	1940s	All-Fair	45	85
Zulu Blowing Game	B	1927	Zulu Toy	50	100

Dr. Kildare, 1962, Ideal

Melvin the Moon Man, 1960s, Remco

POSTWAR GAMES

NAME	TYPE	YEAR	COMPANY	GOOD	EX	MINT
$25,000 Pyramid	B	1980s	Cardinal Industries	3	10	15
$64,000 Question Quiz Game	B	1955	Lowell	20	45	75
1-2-3 Game Hot Spot!	B	1961	Parker Brothers	6	15	25
1863, Civil War Game	B	1961	Parker Brothers	20	40	65
2 For The Money	B	1955	Lowell	20	30	55
20,000 Leagues Under the Sea	B	1950s	Gardner	25	55	100
25 Ghosts	B	1969	Lakeside	15	25	40
300 Mile Race	B	1955	Warren	15	35	60
4000 A.D. Interstellar Conflict Game	B	1972	House of Games	8	20	35
77 Sunset Strip	B	1960	Lowell	35	50	75
77 Sunset Strip	B	1960	Warner	40	65	100
A-Team	B	1984	Parker Brothers	5	10	15
Abbott & Costello Who's On First?	B	1978	Selchow & Righter	5	10	15
ABC Monday Night Football	B	1972	Aurora	8	30	40
ABC Monday Night Football Roger Staubach Edition	B	1973	Aurora	8	20	35
ABC Sports Winter Olympics	B	1987	Mindscape	10	15	25
Acquire (plastic tiles)	B	1968	3M	12	30	45
Acquire (wood tiles)	B	1963	3M	20	40	75
Across the Board Horse Racing Game	B	1975	MPH	8	20	35
Across the Continent (cars)	B	1960	Parker Brothers	20	40	65
Across the Continent (trains)	B	1952	Parker Brothers	30	50	80
Action Baseball	B	1965	Pressman	35	45	50
Addams Family	B	1965	Ideal	65	100	175
Addams Family	C	1965	Milton Bradley	25	55	75
Addams Family	B	1973	Milton Bradley	20	40	65
Admirals	B	1960s	Parker Brothers	25	55	75
Advance To Boardwalk	B	1985	Parker Brothers	5	15	20
Adventure in Science, An	B	1950	Jacmar	20	30	50
Agent Zero-M Spy Detector	B	1964	Mattel	30	50	80
Aggravation	B	1970	Lakeside	5	10	15
Air Assault on Crete	B	1977	Avalon Hill	10	15	25
Air Empire	B	1961	Avalon Hill	40	100	160
Air Race Around the World	B	1950s	Lido	12	30	50
Airways	S	1950s	Lindstrom Tool & Toy	30	50	75
Alfred Hitchcock "Why?"	B	1965	Milton Bradley	10	20	30
Alfred Hitchcock Presents Mystery Game	B	1958	Milton Bradley	20	35	55
Alien	B	1979	Kenner	15	45	70
All American Football	B	1969	Cadaco	5	10	20
All In The Family	B	1972	Milton Bradley	10	20	30
All My Children	B	1985	TSR	5	10	20
All Pro Baseball	B	1969	Ideal	20	35	55
All Pro Basketball	B	1969	Ideal	20	35	50
All Pro Football	B	1967	Ideal	8	20	30
All Star Baseball	B	1960s	Cadaco-Ellis	20	50	75
All Star Baseball	B	1989	Cadaco	5	12	20
All Star Baseball	B	1970s	Cadaco-Ellis	10	25	45
All Star Baseball Fame	B	1962	Cadaco-Ellis	15	25	40
All Star Basketball	B	1950s	Gardner	55	90	135

POSTWAR GAMES

NAME	TYPE	YEAR	COMPANY	GOOD	EX	MINT
All Star Electric Baseball & Football	B	1955	Harett-Gilmar	35	60	90
All Star Football	B	1950	Gardner	55	90	135
All The King's Men	B	1979	Parker Brothers	6	10	15
All Time Greats Baseball Game	B	1971	Midwest Research	15	35	50
American Derby, The	B	1951	Cadaco-Ellis	20	45	70
Angry Donald Duck Game	S	1970s	Mexico	40	65	100
Animal Crackers	B	1970s	Milton Bradley	4	7	11
Annette's Secret Passage	B	1958	Parker Brothers	25	55	80
Annie Oakley (larger)	B	1955	Milton Bradley	20	50	75
Annie Oakley (smaller game)	B	1950s	Milton Bradley	12	30	40
Annie, The Movie Game	B	1981	Parker Brothers	4	6	10
Anti-Monopoly	B	1973	Anti-Monopoly	8	20	30
APBA Baseball Master Game	B	1975	APBA	20	50	75
APBA League Football	B	1980s	APBA	12	20	30
APBA Pro League Football	B	1964	APBA	25	60	85
APBA Saddle Racing Game	B	1970s	APBA	15	25	40
Apple's Way	B	1974	Milton Bradley	15	40	60
Archie Bunker's Card Game	C	1972	Milton Bradley	6	15	25
Archies, The	B	1969	Whitman	25	45	65
Arnold Palmer's Inside Golf	B	1961	D.B. Remson	50	85	130
Around The World In 80 Days	B	1957	Transogram	25	40	65
Art Lewis Football Game	B	1955	Morgantown Game	70	115	175
Art Linkletter's House Party	B	1968	Whitman	12	30	50
Art Linkletter's People are Funny Party Game	C	1954	Whitman	10	25	35
As The World Turns	B	1966	Parker Brothers	12	30	50
ASG Baseball	B	1989	3W (World Wide Wargames)	12	20	30
ASG Major League Baseball	B	1973	Gerney Games	25	60	100
Assembly Line	B	1953	Selchow & Righter	25	65	85
Atom Ant Game	B	1966	Transogram	45	70	110
Aurora Pursuit! Game	B	1973	Aurora	8	25	40
Autograph Baseball Game	B	1948	Philadelphia Inquirer	110	180	275
B-17 Queen of The Skies	B	1983	Avalon Hill	6	10	15
B.T.O. (Big Time Operator)	B	1956	Bettye-B	20	45	70
Bali	C	1954	I-S Unlimited	8	20	30
Ballplayer's Baseball Game	B	1955	Jon Weber	30	50	75
Bamboozle	B	1962	Milton Bradley	20	45	65
Banana Tree	B	1977	Marx	10	15	25
Barbapapa Takes A Trip	B	1977	Selchow & Righter	3	5	8
Barbie's Little Sister Skipper Game	B	1964	Mattel	15	40	60
Barbie, Queen of The Prom	B	1960	Mattel	30	60	85
Baretta	B	1976	Milton Bradley	10	25	40
Barnabas Collins Game	B	1969	Milton Bradley	20	50	80
Barney Miller	B	1977	Parker Brothers	10	15	25
Barnstormer	B	1970s	Marx	20	35	55
Bart Starr Quarterback Game	B	1960s	175	295	450	
Baseball	S	1960s	Tudor	25	40	60
Baseball Card All Star Game	C	1987	Captoys	5	12	15
Baseball Card Game	C	1950s	Ed-U-Cards	15	40	60
Baseball Card Game, Official	C	1965	Milton Bradley	20	55	80
Baseball Challenge	B	1980	Tri-Valley Games	15	25	35

POSTWAR GAMES

NAME	TYPE	YEAR	COMPANY	GOOD	EX	MINT
Baseball Game, Official	B	1969	Milton Bradley	100	165	250
Baseball Game, The	B	1988	Horatio	12	20	30
Baseball Strategy	B	1973	Avalon Hill	6	15	25
Baseball, A Sports Illustrated Game	B	1975	Time	25	40	65
Baseball, Football & Checkers	B	1957	Parker Brothers	35	60	90
Bash!	S	1967	Ideal	10	15	25
Basketball Strategy	B	1974	Avalon Hill	10	15	25
Bat Masterson	B	1958	Lowell	45	75	120
Batman	B	1978	Hasbro	15	30	50
Batman and Robin Game	B	1965	Hasbro	45	75	120
Batman Batarang Toss	S	1966	Pressman	150	250	400
Batman Card Game	C	1966	Ideal	20	65	85
Batman Game	B	1966	Milton Bradley	45	75	100
Batman Pin Ball	S	1966	Marx	55	95	150
Bats in the Belfry	S	1964	Mattel	20	55	80
Batter Up	B	1946	M. Hopper	30	50	75
Batter Up Card Game	C	1949	Ed-U-Cards	10	25	45
Batter-Rou Baseball Game (Dizzy Dean)	B	1950s	Memphis Plastic	100	165	250
Battle Cry	B	1962	Milton Bradley	25	60	85
Battle Line	B	1964	Ideal	25	50	75
Battle of The Planets	B	1970s	Milton Bradley	15	25	35
Battleship	B	1965	Milton Bradley	10	15	25
Battlestar Galactica	B	1978	Parker Brothers	10	15	25
Beany & Cecil Match It	B	1960s	Mattel	30	50	80
Beat Inflation	B	1975	Avalon Hill	8	20	30
Beat The Buzz	B	1958	Kenner	10	15	25
Beat The Clock	B	1960s	Milton Bradley	6	10	16
Beat The Clock	B	1954	Lowell	35	60	95
Beatles Flip Your Wig Game	B	1964	Milton Bradley	60	125	175
Beetle Bailey, The Old Army Game	B	1963	Milton Bradley	40	70	90
Behind The 8 Ball Game	B	1969	Selchow & Righter	8	20	35
Ben Casey MD Game	B	1961	Transogram	15	30	45
Bermuda Triangle	B	1976	Milton Bradley	10	15	25
Betsy Ross and the Flag	B	1950s	Transogram	18	45	70
Beverly Hillbillies Game	C	1963	Milton Bradley	10	25	40
Beverly Hillbillies Game	B	1963	Standard Toykraft	40	65	100
Bewitched	B	1965	T. Cohn	50	80	120
Bewitched Stymie Game	C	1960s	Milton Bradley	15	40	60
Bible Baseball	B	1950s	Standard	50	75	150
Bible Quiz Lotto	C	1949	Jack Levitz	5	12	20
Big 5 Poosh-M Up	S	1950s	Knickerbocker	25	40	60
Big Foot	B	1977	Milton Bradley	5	12	20
Big Game Hunt, The	S	1947	Carrom Industries	15	25	40
Big League Baseball	B	1959	Saalfield	55	90	135
Big League Baseball Game	B	1966	3M	15	25	40
Big League Manager Football	B	1965	BLM	20	35	50
Big Payoff	B	1984	Payoff Enterprises	6	10	15
Big Six Sports Game	B	1950s	Gardner	125	295	450
Big Sneeze Game, The	S	1968	Ideal	10	25	40
Big Time Colorado Football	B	1983	B.J. Tall	6	10	15
Big Town	B	1954	Lowell	50	80	125
Billionaire	B	1973	Parker Brothers	5	15	20

POSTWAR GAMES

NAME	TYPE	YEAR	COMPANY	GOOD	EX	MINT
Bing Crosby's Game, Call Me Lucky	B	1954	Parker Brothers	35	60	85
Bingo-Matic	B	1954	Transogram	5	10	20
Bionic Crisis	B	1975	Parker Brothers	6	15	25
Bionic Woman	B	1976	Parker Brothers	6	15	25
Bird Brain	B	1966	Milton Bradley	10	25	40
Bird Watcher	B	1958	Parker Brothers	18	50	75
Birdie Golf	B	1964	Barris	30	75	100
Black Ball Express	B	1957	Schaper	10	25	40
Black Beauty	B	1957	Transogram	25	50	65
Black Box	B	1978	Parker Brothers	5	10	15
Blade Runner	B	1982		25	60	85
Blast Off	B	1953	Selchow & Righter	40	100	150
Blast, The Game of	B	1973	Ideal	8	20	35
Blitzkrieg	B	1965	Avalon Hill	5	15	25
Blockhead	B	1954	Russell	10	15	25
Blondie	B	1970s	Parker Brothers	6	10	15
Blondie and Dagwood's Race for the Office	B	1950	Jaymar	30	45	70
Blue Line Hockey	B	1968	3M	10	25	40
Bluff	B	1964	Saalfield	10	25	40
BMX Cross Challenge Action Game	B	1988	Cross Challenge	6	10	15
Bob Feller's Big League Baseball	B	1949	Saalfield	75	150	200
Bobbsey Twins	B	1957	Milton Bradley	8	20	30
Bobby Shantz Baseball Game	B	1955	Realistic Games	80	150	225
Body Language	B	1975	Milton Bradley	5	12	20
Boggle	B	1976	Parker Brothers	3	8	12
Bonanza Michigan Rummy Game	B	1964	Parker Brothers	30	45	70
Booby Trap	S	1965	Parker Brothers	10	15	25
Boom or Bust	B	1951	Parker Brothers	50	100	135
Booth's Pro Conference Football	B	1977	Sher-Co	10	15	25
Boots and Saddles	B	1960	Chad Valley	35	65	100
Bop The Beetle	S	1963	Ideal	20	35	55
Boris Karloff's Monster Game	B	1965	Gems	75	125	200
Boston Marathon Game, Official	B	1978	Perl Products	15	25	35
Boston Red Sox Game	C	1964	Ed-U-Cards	15	35	55
Bottoms Up	B	1970s		3	5	8
Bowl & Score	B	1974	Lowe	6	10	15
Bowl And Score	B	1962	Lowe	6	10	15
Bowl Bound!	B	1973	Sports Illustrated	15	25	40
Bowl-A-Matic	S	1963	Eldon	45	75	120
Brain Waves	B	1977	Milton Bradley	6	15	25
Branded	B	1966	Milton Bradley	30	60	90
Brass Monkey Game, The	B	1973	U.S. Game Systems	10	25	40
Break Par Golf Game	B	1950s	Warren/Built-Rite	8	20	30
Break The Bank	B	1955	Bettye-B	20	50	75
Breaker 1-9	B	1976	Milton Bradley	5	10	15
Brett Ball	B	1981	9th Inning	15	30	45
Bride Bingo	B	1957	Leister Game	5	12	20
Broadside	B	1962	Milton Bradley	20	45	70
Bruce Jenner Decathlon Game	B	1979	Parker Brothers	4	7	11

Merry Milkman, 1955, Hasbro

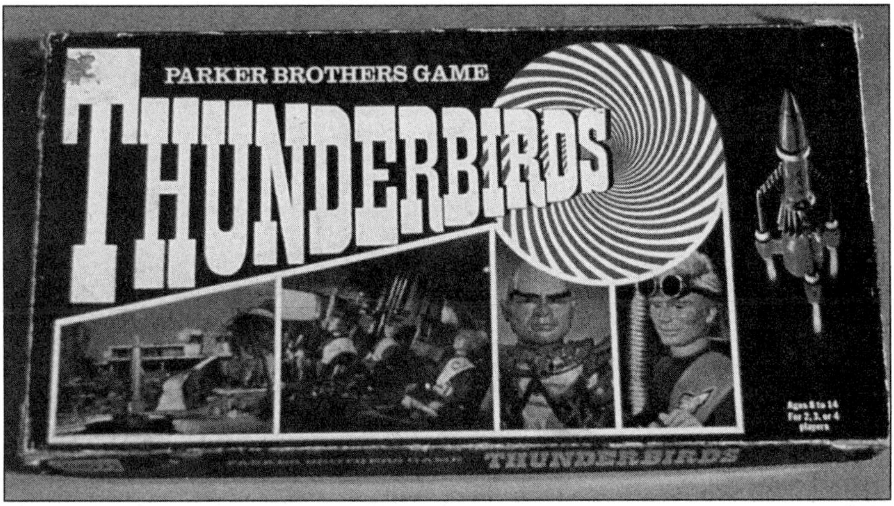

Thunderbirds, 1965, Parker Brothers

GAMES

POSTWAR GAMES

NAME	TYPE	YEAR	COMPANY	GOOD	EX	MINT
Buck Fever	B	1984	L & D Robton	12	20	30
Buckaroo	B	1947	Milton Bradley	20	35	55
Bucket Ball	B	1972	Marx	10	15	25
Bug-A-Boo	B	1968	Whitman	10	15	20
Bugaloos	B	1971	Milton Bradley	15	35	60
Bugs Bunny Game (Bagatelle)	S	1975	Ideal	15	25	40
Bugs Bunny Under The Cawit Game	B	1972	Whitman	15	25	40
Building Boom	B	1950s	Kohner	10	20	35
Built-Rite Swish Basketball Game	B	1950s	Warren/Built-Rite	10	20	35
Bullwinkle Card Game	C	1962	Ed-U-Cards	10	30	50
Bullwinkle Hide & Seek Game	B	1961	Milton Bradley	25	40	75
Bullwinkle's Super Market Game	B	1970s	Whitman	15	25	60
Buster Brown Game and Play Box	B	1950s	Buster Brown Shoes	40	75	125
C&O/B&O	B	1969	Avalon Hill	20	50	95
Cabbage Patch Kids	B	1984	Parker Brothers	5	10	15
California Raisins Board Game	B	1987	Decipher	4	16	25
Call It Golf	B	1966	Strauss	15	25	40
Call It!	S	1978	Ideal	5	12	20
Call My Bluff	B	1965	Milton Bradley	15	20	30
Calling All Cars	B	1950s	Parker Brothers	45	100	150
Calling Superman	S	1955	Transogram	25	85	125
Calvin & The Colonel High Spirits	B	1962	Milton Bradley	12	30	50
Camelot	B	1955	Parker Brothers	20	30	50
Camouflage	B	1961	Milton Bradley	5	20	35
Camp Granada Game, Allan Sherman's	B	1968	Milton Bradley	18	45	65
Camp Runamuck	B	1965	Ideal	15	30	65
Campaign	B	1971	Waddington	10	30	45
Campaign	B	1966	Campaign Game	8	25	40
Campaign: The American "Go" Game	B	1961	Saalfield	15	45	65
Can You Catch It Charlie Brown?	B	1976	Ideal	10	15	20
Can't Stop	B	1980	Parker Brothers	5	12	20
Candid Camera Game	B	1963	Lowell	15	40	65
Candid Camera Target Shot	S	1950s	Lindstrom Tool & Toy	35	60	90
Candyland	B	1960s	Milton Bradley	10	25	45
Candyland	B	1949	Milton Bradley	25	50	90
Cannonball Run, The	B	1981	Cadaco	4	16	25
Caper	B	1970	Parker Brothers	20	30	50
Capital Punishment	B	1981	Hammerhead	25	45	75
Captain America	B	1977	Milton Bradley	10	15	25
Captain America	B	1966	Milton Bradley	40	65	105
Captain Caveman and the Teen Angels	B	1981	Milton Bradley	8	20	30
Captain Gallant Desert Fort Game	B	1956	Transogram	20	40	65
Captain Kangaroo	B	1956	Milton Bradley	40	85	125
Captain Video Game	B	1952	Milton Bradley	75	125	200
Car Travel Game	B	1958	Milton Bradley	5	16	25
Cardino	B	1970	Milton Bradley	10	15	25
Careers	B	1965	Parker Brothers	6	15	25

POSTWAR GAMES

NAME	TYPE	YEAR	COMPANY	GOOD	EX	MINT
Careers	B	1957	Parker Brothers	8	20	30
Careful: The Toppling Tower	S	1967	Ideal	10	15	25
Cargoes	B	1958	Selchow & Righter	8	25	40
Carl Hubbell Mechanical Baseball	B	1950	Gotham	100	200	300
Carl Yastrzemski's Action Baseball	B	1968	Pressman	90	145	195
Carrier Strike	B	1977	Milton Bradley	15	25	45
Cars 'n Trucks Build-A-Game	B	1961	Ideal	15	45	80
Case of The Elusive Assassin, The	B	1967	Ideal	20	50	75
Casey Jones	B	1959	Saalfield	30	45	70
Casper The Friendly Ghost Game	B	1959	Milton Bradley	5	10	20
Casper The Friendly Ghost Game	B	1974	Schaper	8	18	30
Cat & Mouse	B	1964	Parker Brothers	7	15	20
Catchword	B	1954	Whitman	3	10	15
Catfish Bend Storybook Game	B	1978	Selchow & Righter	15	20	35
Cattlemen, The	B	1977	Selchow & Righter	10	15	25
Cavalcade	B	1953	Selchow & Righter	15	45	65
Caveat Emptor	B	1971	Plan B	5	16	25
Centipede	B	1983	Milton Bradley	6	10	15
Century of Great Fights	B	1969	Research Games	40	75	110
Challenge Golf at Pebble Beach	B	1972	3M	15	25	35
Challenge the Yankees	B	1960s	Hasbro	150	325	450
Challenge Yahtzee	B	1974	Milton Bradley	7	15	20
Championship Baseball	B	1966	Championship Games	10	20	30
Championship Basketball	B	1966	Championship Games	10	20	30
Championship Golf	B	1966	Championship Games	10	20	30
Changeover: The Metric Game	B	1976	John Ladell	8	20	30
Changing Society	B	1981	Phil Carter	5	16	25
Chaos	B	1965	Amsco Toys	5	16	25
Charge It!	C	1972	Whitman	5	15	25
Charlie Brown's All Star Baseball Game	B	1965	Parker Brothers	35	60	90
Charlie's Angels	B	1977	Milton Bradley	6	10	20
Charlie's Angels (Farrah Fawcett box)	B	1977	Milton Bradley	10	20	45
Chase, The	B	1966	Cadaco	12	30	45
Chaseback	B	1962	Milton Bradley	5	16	25
Checkpoint: Danger!	B	1978	Ideal	5	16	25
Cherry Ames' Nursing Game	B	1959	Parker Brothers	25	65	90
Chess	B	1977	Milton Bradley	3	5	8
Chevyland Sweepstakes	B	1968	Milton Bradley	25	50	100
Chex Ches Football	B	1971	Chex Ches Games	15	25	40
Cheyenne	B	1958	Milton Bradley	25	60	95
Chicago Sports Trivia Game	B	1984	Sports Trivia	6	10	15
Chicken In Every Pot, A	B	1980s	Animal Town Game	20	30	50
Chicken Lotto	S	1965	Ideal	8	25	40
Children's Hour, The	B	1946	Parker Brothers	10	30	45
CHiPS	B	1981	Ideal	7	10	20

POSTWAR GAMES

NAME	TYPE	YEAR	COMPANY	GOOD	EX	MINT
CHiPS Game	B	1977	Milton Bradley	4	7	10
Chit Chat Game	B	1963	Milton Bradley	6	10	15
Chopper Strike	B	1976	Milton Bradley	10	25	40
Chug-A-Lug	B	1969	Dynamic	5	16	25
Chute-5	B	1973	Lowe	3	5	8
Chutes & Ladders	B	1956	Milton Bradley	10	15	25
Chutes Away!	S	1978	Gabriel	15	45	65
Chutzpah	B	1967	Middle Earth	25	50	75
Chutzpah	B	1967	Cadaco	10	15	25
Cimarron Strip	B	1967	Ideal	50	85	135
Circle Racer Board Game	B	1988	Sport Games USA	6	10	15
Circus Game	B	1947		15	25	40
Cities Game, The	B	1970	Psychology Today	8	25	40
Civil War	B	1961	Avalon Hill	15	45	65
Civilization	B	1982	Avalon Hill	7	10	20
Clash of the Titans	B	1981	Whitman	10	25	45
Class Struggle	B	1978	Bernard Ollman	5	16	25
Clean Sweep	B	1960s	Schaper	20	50	75
Clean Water	B	1972	Urban Systems	8	25	40
Clickety-Clak	S	1950s	Milton Bradley	10	30	45
Cloak & Dagger	B	1984	Ideal	8	25	40
Close Encounters of the Third Kind	B	1977	Parker Brothers	7	15	20
Clue	B	1949	Parker Brothers	25	40	65
Clue	B	1972	Parker Brothers	4	7	11
Clunk-A-Glunk	S	1968	Whitman	8	25	40
Code Name: Sector	B	1977	Parker Brothers	10	30	45
Collector, The	B	1977	Avalon Hill	5	16	25
College Basketball	B	1954	Cadaco-Ellis	20	30	55
Columbo	B	1973	Milton Bradley	6	10	15
Combat	B	1963	Ideal	25	45	70
Combat	C	1964	Milton Bradley	10	25	40
Comin' Round The Mountain	B	1954	Einson-Freeman	30	50	80
Computer Baseball	B	1966	Epoch Playtime	25	40	65
Computer Basketball	B	1969	Electric Data	45	75	115
Computerized Pro Football	B	1971	Data Prog.	15	25	40
Concentration (25th Anniversary Ed.)	B	1982	Milton Bradley	6	10	16
Concentration (3rd Ed.)	B	1960	Milton Bradley	12	20	35
Conestoga	B	1964	Washburne Research	25	65	95
Coney Island Penny Pitch	S	1950s	Novel Toy	33	55	88
Coney Island, The Game of	B	1956	Selchow & Righter	35	75	125
Conflict	B	1960	Parker Brothers	15	45	65
Confucius Say	B	1960s	Pressman	10	30	45
Conquer	B	1979	Whitman	5	16	25
Conquest of the Empire	B	1984	Milton Bradley	45	100	150
Consetta and Her Wheel of Fate	B	1946	Selchow & Righter	30	75	110
Conspiracy	B	1982	Milton Bradley	4	7	11
Contigo	B	1974	3M	7	20	35
Cootie	B	1949	Schaper	10	25	35
Count Coup	B	1979	Marcian Chronicles	10	30	45
Count Down Space Game	B	1960	Transogram	15	45	65

POSTWAR GAMES

NAME	TYPE	YEAR	COMPANY	GOOD	EX	MINT
Countdown	B	1967	Lowe	25	45	75
Counter Point	B	1976	Hallmark	10	15	25
Cowboy Roundup	B	1952	Parker Brothers	15	25	40
Cowboys & Indians	C	1949	Ed-U-Cards	8	20	35
Cracker Jack Game	B	1976	Milton Bradley	5	16	25
Crazy Clock Game	S	1964	Ideal	35	75	100
Creature Features	B	1975	Athol	20	50	75
Creature From The Black Lagoon	B	1963	Hasbro	150	400	750
Cribb Golf	B	1980s		10	25	40
Crosby Derby, The	B	1947	Fishlove	50	90	135
Cross Up	B	1974	Milton Bradley	5	16	25
Crosswords	B	1954	National Games	12	20	32
Crusader Rabbit Game	B	1960s		50	100	175
Cub Scouting, The Game of	B	1987	Cadaco	5	16	25
Curious George Game	B	1977	Parker Brothers	4	6	10
Curse of the Cobras Game	B	1982	Ideal	12	20	32
Cut Up Shopping Spree Game	B	1968	Milton Bradley	6	10	16
Dallas (TV Role Playing)	B	1980	SPI	4	7	11
Dallas Game	C	1980	Mego	3	10	15
Danger Pass	B	1964	Game Partners	15	45	65
Daniel Boone Trail Blazer	B	1964		25	60	85
Daniel Boone Wilderness Trail	C	1964	Transogram	10	40	65
Dark Crystal Game, The	B	1982	Milton Bradley	7	20	35
Dark Shadows Game	B	1968	Whitman	18	40	60
Dark Tower	B	1981	Milton Bradley	50	120	175
Dastardly & Muttley	B	1969	Milton Bradley	25	45	65
Dating Game, The	B	1967	Hasbro	15	25	40
Davy Crockett Adventure Game	B	1956	Gardner	45	75	120
Davy Crockett Frontierland Game	B	1955	Parker Brothers	20	50	75
Davy Crockett Radar Action Game	B	1955	Ewing Mfg. & Sales	51	85	136
Davy Crockett Rescue Race Game	B	1950s	Gabriel	20	50	75
Dawn of The Dead	B	1978	SPI	15	40	65
Daytona 500 Race Game	B	1989	Milton Bradley	10	25	35
Dead Pan	B	1956	Selchow & Righter	8	25	40
Dealer's Choice	B	1972	Parker Brothers	20	40	60
Dear Abby	B	1972	Ideal	7	20	35
Decathalon	B	1972	Sports Illustrated	15	25	35
Decoy	B	1956	Selchow & Righter	12	40	60
Deduction	B	1976	Ideal	2	7	12
Deluxe Wheel of Fortune	B	1986	Pressman	5	8	13
Dennis The Menace Baseball Game	B	1960		20	50	70
Denny McLain Magnetik Baseball Game	B	1968	Gotham	115	195	295
Deputy Dawg Hoss Toss	S	1973		15	25	40
Deputy Dawg TV Lotto	B	1961		21	35	56
Deputy Game, The	B	1960	Milton Bradley	30	50	80
Derby Day	B	1959	Parker Brothers	21	35	56
Derby Downs	B	1973	Great Games	10	30	45

Bugaloos, 1971, Milton Bradley

Intrigue, 1954, Milton Bradley

POSTWAR GAMES

NAME	TYPE	YEAR	COMPANY	GOOD	EX	MINT
Detectives Game, The	B	1961	Transogram	30	50	80
Dick Tracy Crime Stopper	B	1963	Ideal	57	95	152
Dick Tracy Playing Card Game	C	1937	Whitman	20	45	75
Dick Tracy Playing Card Game	C	1939	Esquire Novelty	20	45	75
Dick Tracy Playing Card Game	C	1934	Whitman	25	60	100
Dick Tracy The Master Detective Game	B	1961	Selchow & Righter	30	50	80
Dick Van Dyke Board Game	B	1964	Standard Toykraft	45	100	175
Diet	B	1972	Dynamic	5	16	25
Diner's Club Credit Card Game, The	B	1961	Ideal	15	45	65
Dinosaur Island	B	1980	Parker Brothers	5	16	25
Diplomacy	B	1961	Games Research	21	35	56
Diplomacy	B	1976	Avalon Hill	15	25	40
Direct Hit	B	1950s	Northwestern Products	40	70	110
Dirty Water-The Water Pollution Game	B	1970	Urban Systems	8	20	30
Disney Dodgem Bagatelle	S	1960s	Marx	30	65	90
Disney Mouseketeer	B	1964	Parker Brothers	40	65	100
Disneyland Game	B	1965	Transogram	15	35	50
Dispatcher	B	1958	Avalon Hill	20	50	75
Dobbin Derby	B	1950	Cadaco-Ellis	8	25	40
Doctor Who	B	1980s	Denys Fisher	35	75	100
Dogfight	B	1962	Milton Bradley	25	60	85
Dollar A Second	B	1955	Lowell	20	50	75
Dollars & Sense	B	1946	Sidney Rogers	100	150	200
Domain	B	1983	Parker Brothers	2	3	5
Don Carter's Strike Bowling Game	B	1964	Saalfield	55	90	135
Don't Break the Ice	S	1960s	Schaper	5	16	25
Don't Spill the Beans	S	1967	Schaper	5	16	25
Donald Duck Big Game Box	B	1979	Whitman	10	15	20
Donald Duck Pins & Bowling Game	B	1955s	Pressman	35	60	90
Donald Duck Tiddley Winks Game	B	1950s		6	10	15
Donald Duck's Party Game	B	1950s	Parker Brothers	10	35	50
Dondi Potato Race Game	B	1950s	Hasbro	15	35	50
Donkey Party Game	B	1950	Saalfield	15	25	40
Donny & Marie Osmond TV Show Game	B	1977	Mattel	8	25	40
Double Cross	B	1974	Lakeside	5	16	25
Double Trouble	B	1987	Milton Bradley	3	5	8
Doubletrack	B	1981	Milton Bradley	2	7	12
Dr. Kildare	B	1962	Ideal	25	40	65
Dracula Mystery Game	B	1960s	Hasbro	80	160	225
Dracula's "I Vant To Bite Your Finger" Game	B	1981	Hasbro	10	20	30
Dragnet	B	1955	Parker Brothers	70	115	185
Dragnet	B	1955	Transogram	35	60	95
Dragon's Lair	B	1983	Milton Bradley	7	10	20
Dragonmaster	C	1981	Lowe	6	15	25

POSTWAR GAMES

NAME	TYPE	YEAR	COMPANY	GOOD	EX	MINT
Driver Ed	B	1973	Cadaco	6	10	20
Duell	B	1976	Lakeside	5	16	25
Dukes of Hazzard	B	1981	Ideal	6	15	25
Dunce	B	1955	Schaper	7	20	35
Dune	B	1984	Parker Brothers	6	10	15
Dune, Frank Herbert's	B	1979	Avalon Hill	15	35	50
Dungeon Dice	B	1977	Parker Brothers	5	8	15
Dungeons & Dragons	B	1980	Mattel	6	10	15
Duplicate Ad-Lib	B	1976	Lowe	5	10	15
Duran Duran Game	B	1985	Milton Bradley	15	25	40
Dynamite Shack Game	S	1968	Milton Bradley	10	25	45
Dynomutt	B	1977	Milton Bradley	10	15	25
E.T. The Extra-Terrestrial	B	1982	Parker Brothers	6	10	15
Earl Gillespie Baseball Game	B	1961	Wei-Gill	25	40	65
Earth Satellite Game	B	1956	Gabriel	35	75	125
Ecology	B	1970	Urban Systems	7	20	35
Egg and I, The	B	1947	Capex	30	50	80
El Dorado	B	1977	Invicta	7	20	35
Electra Woman and Dyna Girl	B	1977	Ideal	10	25	45
Electric Sports Car Race	B	1959	Tudor	35	60	90
Electronic Detective Game	B	1970s	Ideal	20	30	45
Electronic Lightfight	B	1981	Milton Bradley	10	30	45
Electronic Radar Search	B	1967	Ideal	10	15	25
Eliot Ness and the Untouchables	B	1961	Transogram	30	70	100
Ellsworth Elephant Game	B	1960	Selchow & Righter	30	45	70
Elvis Presley Game	B	1957	Teen Age Games	300	650	1000
Emenee Chocolate Factory	B	1966		6	10	15
Emergency	B	1974	Milton Bradley	10	15	25
Emily Post Popularity Game	B	1970	Selchow & Righter	15	20	35
Emperor of China	B	1972	Dynamic	6	18	30
Empire Auto Races	B	1950s	Empire Plastics	20	30	50
Empire Strikes Back, Hoth Ice World	B	1977	Kenner	10	20	30
Enemy Agent	B	1976	Milton Bradley	15	20	30
Energy Quest	B	1977	Weldon	5	16	25
Engineer	B	1957	Selchow & Righter	8	25	40
Entertainment Trivia Game	B	1984	Lakeside	5	10	15
Entre's Fun & Games In Accounting	B	1988	Entrepreneurial Games	4	7	12
Ergo	B	1977	Invicta	5	16	25
Escape From New York	B	1980	TSR	10	15	25
Escape from the Casbah	B	1975	Selchow & Righter	7	20	35
Escape From the Death Star	B	1977	Kenner	10	20	30
Espionage	B	1973	MPH	6	18	30
Everybody's Talking!	B	1967	Watkins-Strathmore	8	25	40
Executive Decision	B	1971	3M	5	16	25
Expanse	B	1949	Milton Bradley	15	45	65
Extra Innings	B	1975	J. Kavanaugh	15	25	35
Eye Guess	B	1960s	Milton Bradley	15	20	35
F-Troop	B	1965	Ideal	40	100	155
F.B.I.	B	1958	Transogram	35	55	90
F/11 Armchair Quarterback	B	1964	James R. Hock	15	25	40
Fact Finder Fun	B	1963	Milton Bradley	10	15	25

POSTWAR GAMES

NAME	TYPE	YEAR	COMPANY	GOOD	EX	MINT
Facts In Five	S	1967	3M	5	15	25
Fall Guy, The	B	1981	Milton Bradley	10	15	25
Falls	B	1950s	National	85	140	215
Family Affair	B	1967	Whitman	25	40	65
Family Feud	B	1977	Milton Bradley	4	12	20
Family Ties Game, The	B	1986	Apple Street	10	20	25
Famous 500 Mile Race	B	1988		8	13	20
Fang Bang	B	1966	Milton Bradley	10	15	25
Fangface	B	1979	Parker Brothers	5	8	13
Fantastic Voyage Game	B	1968	Milton Bradley	15	25	40
Fantasy Island Game	B	1978	Ideal	7	20	35
Fascination	S	1962	Remco	15	35	50
Fast 111s	B	1981	Parker Brothers	5	16	25
Fast Golf	C	1977	Whitman	8	13	20
Fastest Gun, The	B	1974	Milton Bradley	10	35	50
Fat Albert	B	1973	Milton Bradley	15	25	40
Fearless Fireman	B	1957	Hasbro	35	100	150
Feed the Elephant!	S	1952	Cadaco-Ellis	10	35	50
Feeley Meeley Game	S	1967	Milton Bradley	7	15	20
Felix the Cat Dandy Candy Game	B	1957		10	15	25
Felix the Cat Game	B	1968	Milton Bradley	8	20	30
Felix the Cat Game	B	1960	Milton Bradley	10	25	45
Feudal	B	1967	3M	15	25	40
Fighter Bomber	B	1977	Cadaco	15	25	40
Finance	B	1962	Parker Brothers	10	20	35
Fire Chief	B	1957	Selchow & Righter	8	25	40
Fire Fighters!	B	1957	Russell	15	25	40
Fire House Mouse Game	B	1967	Transogram	10	35	50
Fireball XL-5	B	1963	Milton Bradley	40	100	145
Fireball XL-5 Magnetic Dart Game	S	1963	Magic Wand	75	125	200
First Class Farmer	B	1965	F & W Publishing	7	20	35
First Down	B	1970	TGP Games	50	80	125
Fish Pond	B	1950s	National Games	10	30	45
Fishbait	B	1965	Ideal	40	60	75
Flagship Airfreight: The Airplane Cargo Game	B	1946	Milton Bradley	40	70	115
Flash Gordon	B	1970s	House Of Games	15	20	35
Flash: The Press Photographer Game	B	1956	Selchow & Righter	18	50	75
Flea Circus Magnetic Action Game	S	1968	Mattel	15	25	40
Flight Captain	B	1972	Lowe	8	25	40
Flintstones	B	1980	Milton Bradley	15	20	35
Flintstones	B	1971	Milton Bradley	15	25	40
Flintstones Animal Rummy	C	1960	Ed-U-Cards	7	25	45
Flintstones Brake Ball	S	1962	Whitman	45	75	120
Flintstones Cut-Ups Game	C	1963	Whitman	15	45	75
Flintstones Dino The Dinosaur Game	B	1961	Transogram	45	75	120
Flintstones Hoppy The Hopperoo Game	B	1964	Transogram	45	75	120
Flintstones Mechanical Shooting Gallery	S	1962	Marx	75	125	200
Flintstones Mitt-Full Game	B	1962	Whitman	40	65	100

POSTWAR GAMES

NAME	TYPE	YEAR	COMPANY	GOOD	EX	MINT
Flintstones Stone Age Game	B	1961	Transogram	20	45	65
Flip 'N Skip	B	1971	Little Kennys	5	10	15
Flip Flop Go	B	1962	Mattel	6	10	15
Flipper Flips	B	1960s	Mattel	30	50	80
Flying Nun Game, The	B	1968	Milton Bradley	20	30	50
Flying Nun Marble Maze Game, The	S	1967	Hasbro	20	30	50
Fonz Game, The	B	1976	Milton Bradley	15	25	40
Fooba-Roo Football Game	B	1955	Memphis Plastic	25	40	65
Football Fever	B	1985	Hansen	20	35	50
Football Strategy	B	1972	Avalon Hill	3	10	15
Football Strategy	B	1962	Avalon Hill	8	25	40
Football, Baseball, & Checkers	B	1948	Parker Brothers	15	45	65
Fore	B	1954	Artcraft Paper	20	35	50
Formula One Car Race Game	B	1968	Parker Brothers	15	40	60
Fortress America	B	1986	Milton Bradley	25	40	65
Fortune 500	B	1979	Pressman	8	25	40
Foto-Electric Baseball	B	1951	Cadaco-Ellis	15	35	65
Four Lane Road Racing	B	1963	Transogram	15	45	65
Fox & Hounds, Game of	B	1948	Parker Brothers	10	30	45
Frank Cavanaugh's American Football	B	1955	F. Cavanaugh	25	60	90
Frankenstein Game	B	1962	Hasbro	80	160	225
Frisky Flippers Slide Bar Game	B	1950s	Warren/Built-Rite	5	10	15
Frontier Fort Rescue Game	B	1956	Gabriel	10	35	50
Frontier-6	B	1980	Rimbold	8	25	40
Fu Manchu's Hidden Hoard	B	1967	Ideal	20	50	75
Fugitive	B	1966	Ideal	75	150	200
Funky Phantom Game	B	1971	Milton Bradley	10	15	25
Funny Bones Game	C	1968	Parker Brothers	5	7	11
G.I. Joe	B	1982	International Games	15	25	40
G.I. Joe Adventure	B	1982	Hasbro	20	30	50
G.I. Joe Bagatelle Gun Action Game	S	1970s	Hasbro	7	12	15
G.I. Joe Card Game	B	1965	Whitman	10	15	25
G.I. Joe Marine Paratrooper	B	1965	Milton Bradley	30	45	70
Gambler's Golf	B	1975	Gammon Games	10	16	25
Games People Play Game, The	B	1967	Alpsco	7	20	35
Gammonball	B	1980	Fun-Time Products	10	16	25
Gang Way For Fun	B	1964	Transogram	25	40	65
Gardner's Championship Golf	B	1950s	Gardner	10	35	50
Garfield	B	1981	Parker Brothers	4	6	10
Garrison's Gorillas	B	1967	Ideal	45	75	120
Gay Puree	B	1962		25	40	65
Gene Autry's Dude Ranch Game	B	1950s	Warren/Built-Rite	30	70	100
General Hospital	B	1980s	Cardinal	15	20	30
General Hospital	B	1974	Parker Brothers	10	15	25
Generals, The	B	1980	Ideal	7	20	35
Gentle Ben Anima l Hunt Game	B	1967	Mattel	30	95	100
Geo-Graphy	B	1954	Cadaco-Ellis	5	15	25
George of the Jungle Game	B	1968	Parker Brothers	40	90	125

Monkees Game, 1968, Transogram

Bewitched Stymie Game, 1960s, Milton Bradley

Monster Old Maid, 1964, Milton Bradley

NAME	TYPE	YEAR	COMPANY	GOOD	EX	MINT
Get Beep Beep: The Road Runner Game	B	1975	Whitman	15	25	35
Get Smart Game	B	1966	Ideal	40	95	130
Get That License	B	1955	Selchow & Righter	10	35	50
Get the Message	B	1964	Milton Bradley	5	16	25
Get the Picture	B	1987	Worlds Of Wonder	6	10	15
Gettysburg	B	1960	Avalon Hill	5	16	25
Ghosts	B	1985	Milton Bradley	8	20	35
Giant Wheel Thrills 'n Spills Horse Race	B	1958	Remco	30	50	75
Gidget	C	1966	Milton Bradley	15	35	50
Gil Hodges' Pennant Fever	B	1970	Research Games	65	100	150
Gilligan's Island	B	1965	Game Gems	175	350	600
Gilligan, The New Adventures of	B	1974	Milton Bradley	15	35	50
Gingerbread Man	B	1964	Selchow & Righter	10	35	50
Globe-Trotters	B	1950	Selchow & Righter	8	25	40
Globetrotter Basketball, Official	B	1950s	Meljak	60	100	150
Gnip Gnop	S	1971	Parker Brothers	5	16	25
Go For Broke	B	1965	Selchow & Righter	10	15	25
Go for the Green	B	1973	Sports Illustrated	10	25	40
Go Go Go	C	1950s	Arco Playing Card	6	20	35
Goal Line Stand	B	1980	Game Shop	12	20	30
Godfather	B	1971	Family Games	6	10	15
Godzilla	B	1960s	Ideal	75	250	400
Godzilla	B	1978	Mattel	30	50	80
Going to Jerusalem	B	1955	Parker Brothers	8	25	40
Going, Going, Gone!	B	1975	Milton Bradley	9	15	25
Gold!	B	1981	Avalon Hill	5	16	25
Golden Trivia Game	B	1984	Western	6	10	15
Goldilocks	B	1955	Cadaco-Ellis	8	25	40
Gomer Pyle Game	B	1960s	Transogram	20	45	65
Gong Hee Fot Choy	C	1948	Zondine Game	20	35	55
Gong Show Game	B	1975	Milton Bradley	15	25	40
Gong Show Game	B	1977	American Publishing	15	25	40
Good Guys 'N Bad Guys	B	1973	Cadaco	5	16	25
Good Ol' Charlie Brown Game	B	1971	Milton Bradley	15	20	30
Goofy's Mad Maze	B	1970s	Whitman	6	10	15
Goonies	B	1980s	Milton Bradley	10	15	20
Gooses Wild	B	1966	CO-5	2	7	12
Gotham Professional Basketball	B	1950s	Gotham	35	50	70
Gotham's Ice Hockey	S	1960s	Gotham	25	60	85
Grab A Loop	S	1968	Milton Bradley	7	12	20
Grabitz	C	1979	International Games	2	5	7
Grand Master of Martial Arts	B	1986	Hoyle	6	10	15
Gray Ghost, The	B	1958	Transogram	30	70	100
Great Escape, The	B	1967	Ideal	8	25	40
Great Grape Ape Game, The	B	1975	Milton Bradley	15	25	40
Green Acres Game, The	B	1960s	Standard Toykraft	25	65	90
Green Ghost Game	B	1965	Transogram	75	150	200
Green Hornet Quick Switch Game	B	1966	Milton Bradley	180	300	475
Gremlins	B	1984	International Games	10	15	25
Greyhound Pursuit	B	1985	N/N Games	8	13	20
Grizzly Adams	B	1978	Waddington's House of Games	15	25	40

POSTWAR GAMES

NAME	TYPE	YEAR	COMPANY	GOOD	EX	MINT
Groucho's TV Quiz Game	B	1954	Pressman	30	65	100
Groucho's You Bet Your Life	B	1955	Lowell	50	100	140
Group Therapy	B	1969	Group Therapy Assn.	3	12	20
Guinness Book of World Records Game, The	B	1979	Parker Brothers	5	9	15
Gulf Strike	B	1983	Victory Games	5	16	25
Gunsmoke Game	B	1950s	Lowell	40	65	100
Gusher	B	1946	Carrom Industries	35	75	100
Half-Time Football	B	1979	Lakeside	5	9	15
Handicap Harness Racing	B	1978	Hall of Fame Games	15	25	35
Hands Down	S	1965	Ideal	10	15	25
Hang On Harvey	B	1969	Ideal	15	20	30
Hangman	B	1976	Milton Bradley	5	8	15
Hank Aaron Baseball Game	B	1970s	Ideal	50	80	125
Hank Aaron Bases Loaded	B	1976	Twentieth Century Enterprises	45	70	100
Hank Bauer's "Be a Manager"	B	1960s	Barco Games	75	125	175
Happiness	B	1972	Milton Bradley	10	25	45
Happy Days	B	1976	Parker Brothers	8	20	30
Happy Little Train Game, The	B	1957	Milton Bradley	3	12	20
Hardy Boys Mystery Game, Secret of Thunder Mountain	B	1978	Parker Brothers	7	15	25
Hardy Boys Mystery Game, The	B	1968	Milton Bradley	15	25	40
Hardy Boys Treasure	B	1960	Parker Brothers	35	30	85
Harlem Globetrotters Game	B	1971	Milton Bradley	12	30	50
Harlem Globetrotters Official Edition Basketball	B	1970s	Cadaco-Ellis	45	75	115
Harpoon	B	1955	Gabriel	10	35	50
Harry Lorayne Memory Game, The	B	1976	Reiss	7	20	35
Harry's Glam Slam	C	1962	Harry Obst	35	60	90
Hashimoto San	B	1963	Transogram	30	45	70
Haul the Freight	B	1962	Bar-Zim	15	50	75
Haunted House Game	B	1963	Ideal	65	100	175
Haunted Mansion	B	1970s	Lakeside	30	50	80
Have Gun Will Travel Game	B	1959	Parker Brothers	30	75	100
Hawaii Five-O	B	1960s	Remco	30	90	120
Hawaiian Eye	B	1960	Transogram	95	160	250
Hawaiian Punch Game	B	1978	Mattel	5	16	25
Hector Heathcote	B	1963	Transogram	50	85	135
Hex: The Zig-Zag Game	B	1950	Parker Brothers	10	35	50
Hi Pop	S	1946	Advance Games	20	35	55
Hi-Ho! Cherry-O	B	1960	Whitman	6	10	15
Hide 'N' Thief	B	1965	Whitman	7	20	35
Hide-N-Seek	B	1967	Ideal	10	15	25
High-Bid	B	1965	3M	7	20	35
Hip Flip	B	1968	Parker Brothers	6	10	15
Hippety Hop	B	1947	Corey Game	25	40	65
Hippopotamus	B	1961	Remco	8	25	40
Hit The Beach	B	1965	Milton Bradley	35	75	100
Hobbit Game, The	B	1978	Milton Bradley	15	35	60
Hoc-Key	S	1958	Cadaco-Ellis	10	35	50
Hock Shop	B	1975	Whitman	3	12	20
Hocus Pocus	B	1960s	Transogram	30	45	70

POSTWAR GAMES

NAME	TYPE	YEAR	COMPANY	GOOD	EX	MINT
Hog Tied	B	1981	Selchow & Righter	3	12	20
Hogan's Heroes Game	B	1966	Transogram	35	85	155
Holiday	B	1958	Replogle Globes	40	65	100
Hollywood Awards Game	B	1976	Milton Bradley	8	25	40
Hollywood Go	B	1954	Parker Brothers	30	45	75
Hollywood Squares	B	1980	Milton Bradley	4	6	10
Hollywood Squares	B	1974	Ideal	6	10	15
Home Court Basketball	B	1954		145	250	375
Home Game	B	1950s	Pressman	30	50	80
Home Stretch Harness Racing	B	1967	Lowe	35	60	90
Home Team Baseball Game	B	1957	Selchow & Righter	35	60	90
Honey West	B	1965	Ideal	50	85	135
Honeymooners Game, The	B	1986	TSR	7	12	20
Hoodoo	B	1950	Tryne	7	12	20
Hookey Go Fishin'	B	1974	Cadaco	10	16	25
Hopalong Cassidy Bean Bag Toss Game	S	1950s		20	40	60
Hopalong Cassidy Chinese Checkers Game	B	1950s		20	50	75
Hopalong Cassidy Game	B	1950s	Milton Bradley	40	100	145
Hoppity Hooper Pin Ball Game	S	1965	Lido	40	75	115
Horse Play	B	1962	Schaper	10	35	50
Horseshoe Derby Game	B	1950s	Built-Rite	8	20	30
Hot Property!	B	1980s	Take One Games	8	25	40
Hot Rod	B	1953	Harett-Gilmar	30	50	75
Hot Spot	B	1961	Parker Brothers	10	20	30
Hot Wheels Game	B	1982	Whitman	8	13	20
Hot Wheels Wipe-Out Game	B	1968	Mattel	30	50	75
Hotels	B	1987	Milton Bradley	8	25	40
Houndcats Game	B	1970s	Milton Bradley	8	15	25
House Party	B	1968	Whitman	10	25	35
Houston Astros Baseball Challenge Game	B	1980	Croque	15	25	35
How To Succeed In Business Without Really Trying	B	1963	Milton Bradley	10	15	25
Howard Hughes Game, The	B	1972	Family Games	10	35	50
Howdy Doody Adventure Game	B	1950s	Milton Bradley	45	75	100
Howdy Doody Card Game	C	1954	Russell	20	50	75
Howdy Doody Dominoes Game	S	1950s	Ed-U-Cards	60	100	200
Howdy Doody Quiz Show	B	1950s	Multiple Products	40	75	
Howdy Doody's Electric Carnival Game	B		Harrett-Gilmar	15	20	35
Howdy Doody's Own Game	B	1949	Parker Brothers	75	125	200
Howdy Doody's Three Ring Circus	B	1950	Harett-Gilmar	45	75	120
Howdy Doody's TV Game	B	1950s	Milton Bradley	35	50	100
Huckleberry Hound	B	1981	Milton Bradley	10	20	35
Huckleberry Hound Bumps	B	1960	Transogram	20	50	75
Huckleberry Hound Spin-O-Game	B	1959		45	75	120
Huckleberry Hound Tiddly Winks	B	1959	Milton Bradley	15	30	60

POSTWAR GAMES

NAME	TYPE	YEAR	COMPANY	GOOD	EX	MINT
Huckleberry Hound Western Game	B	1959	Milton Bradley	25	40	65
Huckleberry Hound's Huckle Chuck Target Game	S	1961	Transogram	25	50	100
Huggin' The Rail	S	1948	Selchow & Righter	45	65	100
Hullabaloo	B	1965	Remco	25	65	90
Humor Rumor	B	1969	Whitman	10	20	30
Humpty Dumpty Game	B	1950s	Lowell	6	10	15
Hunch	B	1956	Happy Hour	7	20	35
Hungry Ant, The	B	1978	Milton Bradley	5	16	25
Hungry Henry	S	1969	Ideal	7	20	35
Hunt For Red October	B	1988	TSR	5	15	25
Hurry Up	B	1971	Parker Brothers	5	16	25
I Dream of Jeannie Game	B	1965	Milton Bradley	30	75	105
I Spy	B	1965	Ideal	50	95	150
I Survived New York!	C	1981	City Enterprises	4	7	12
I Wanna Be President	B	1983	J.R. Mackey	5	16	25
I'm George Gobel, And Here's The Game	B	1955	Schaper	40	60	80
I-Qubes	S	1948	Capex	10	15	25
Identipops	B	1969	Playvalue	75	175	250
Image	B	1972	3M	3	12	20
Incredible Hulk	B	1978	Milton Bradley	6	10	15
Indiana Jones: Raiders of The Lost Ark	B	1981	Kenner	20	35	55
Indianapolis 500 75th Running Race Game	B	1991	International Games	8	13	20
Input	B	1984	Milton Bradley	4	7	15
Inside Moves	B	1985	Parker Brothers	3	12	20
Inspector Gadget	B	1983	Milton Bradley	15	25	40
Instant Replay	B	1987	Parker Brothers	8	13	20
Intercept	B	1978	Lakeside	7	20	35
International Grand Prix	B	1975	Cadaco	30	50	75
Interpretation of Dreams	B	1969	Hasbro	10	15	25
Interstate Highway	B	1963	Selchow & Righter	20	50	75
Intrigue	B	1954	Milton Bradley	20	40	60
Inventors, The	B	1974	Parker Brothers	5	12	20
Ipcress File	B	1966	Milton Bradley	20	50	75
Ironside	B	1976	Ideal	55	95	150
Is the Pope Catholic?!	B	1986	Crowley Connections	10	25	45
Isolation	B	1978	Lakeside	2	7	12
Itinerary	B	1980	Xanadu Leisure	3	12	20
Jace Pearson's Tales of The Texas Rangers	B	1955	E.E. Fairchild	50	75	125
Jack & Jill Target Game	S	1948	Cadaco-Ellis	7	20	35
Jack and The Beanstalk	B	1946	National Games	30	45	75
Jack and The Beanstalk Adventure Game	B	1957	Transogram	25	50	75
Jack Barry's Twenty One	B	1956	Lowe	20	30	50
Jackie Gleason's and AW-A-A-A-Y We Go!	B	1956	Transogram	75	125	200
Jackie Gleason's Story Stage Game	B	1955	Utopia Enterprises	100	200	300
Jackpot	B	1975	Milton Bradley	7	11	20
Jacmar Big League Electric Baseball	B	1950s	Jacmar	100	175	250

Flipper Flips, 1960s, Mattel

Mighty Mouse Rescue Game, 1960s, Harett-Gilmar

Hopalong Cassidy Game, 1950s, Milton Bradley

Swayze, 1954, Milton Bradley

POSTWAR GAMES

NAME	TYPE	YEAR	COMPANY	GOOD	EX	MINT
James Bond 007 Goldfinger Game	B	1966	Milton Bradley	35	75	100
James Bond 007 Thunderball Game	B	1965	Milton Bradley	35	75	100
James Bond Live and Let Die Tarot Game	C	1973	US Games Systems	20	55	85
James Bond Message From M Game	B	1966	Ideal	100	250	350
James Bond Secret Agent 007 Game	B	1964	Milton Bradley	15	35	50
James Bond You Only Live Twice	B	1984	Victory Games	5	8	13
James Clavell's Noble House	B	1987	FASA	7	20	35
James Clavell's Shogun	B	1983	FASA	7	20	35
James Clavell's Tai-Pan	B	1987	FASA	7	20	35
James Clavell's Whirlwind	B	1986	FASA	7	20	35
Jan Murray's Charge Account	B	1961	Lowell	25	40	65
Jan Murray's Treasure Hunt	B	1950s	Gardner	10	25	35
Jaws, The Game of	S	1975	Ideal	5	16	25
JDK Baseball	B	1982	JDK Baseball	20	45	65
Jeanne Dixon's Game of Destiny	B	1968	Milton Bradley	7	12	20
Jeopardy	B	1964	Milton Bradley	10	15	25
Jerry Kramer's Instant Replay	B	1970	EMD Enterprises	15	25	40
Jet World	B	1975	Milton Bradley	8	25	40
Jetsons Fun Pad Game	B	1963	Milton Bradley	40	100	145
Jetsons Game	B	1985	Milton Bradley	5	10	15
Jetsons Out of this World Game	B	1963	Transogram	85	140	225
Jimmy the Greek Oddsmaker Football	B	1974	Aurora	15	25	35
Jockette	B	1950s	Jockette	25	40	60
Joe Palooka Boxing Game	B	1950s	Lowell	35	65	100
John Drake Secret Agent	B	1966	Milton Bradley	30	45	70
Johnny Apollo Moon Landing Bagatelle	S	1969	Marx	20	35	55
Johnny Ringo	B	1959	Transogram	75	125	200
Johnny Unitas Football Game	B	1970	Pro Mentor	35	65	100
Joker's Wild	B	1973	Milton Bradley	5	10	15
Jonathan Livingston Seagull	B	1973	Mattel	8	20	30
Jonny Quest Game	B	1964	Transogram	200	500	700
Jose Canseco's Perfect Baseball Game	B	1991	Perfect Game	8	13	20
Jubilee	B	1950s	Cadaco	10	25	35
Jumbo Jet	B	1963	Jumbo	10	25	35
Jumpin'	B	1964	3M	7	20	35
Jumping DJ	B	1962	Mattel	15	50	75
Junior Bingo-Matic	B	1968	Transogram	6	10	15
Junior Executive	B	1963	Whitman	7	15	30
Junior Quarterback Football	B	1950s	Warren/Built-Rite	7	20	35
Justice	B	1954	Lowell	30	45	75
Justice League of America	B	1967	Hasbro	70	150	250
Ka Bala	B	1965	Transogram	40	75	120
KaBoom!	S	1965	Ideal	10	15	25
Kar-Zoom	B	1964	Whitman	15	20	35
Karate, The Game of	B	1964	Selchow & Righter	7	20	35
Kardball	C	1946	Ajak	20	30	40

POSTWAR GAMES

NAME	TYPE	YEAR	COMPANY	GOOD	EX	MINT
Karter Peanut Shell Game	B	1978	Morey & Neely	7	20	35
Kennedys, The	B	1962	Transogram	30	65	100
Kentucky Derby	B	1960	Whitman	15	25	40
Kentucky Jones	B	1964	T. Cohn	25	40	65
Ker-Plunk	S	1967	Ideal	5	10	15
Keyword	B	1954	Parker Brothers	5	10	15
Kick Back	S	1965	Schaper	5	16	25
Kick-Off Soccer	B	1978	Camden Products	7	20	35
Kimbo	S	1950s	Parker Brothers	10	15	25
King Arthur	S	1950s	Northwestern Products	25	40	65
King Kong Game	B	1966	Milton Bradley	10	25	40
King Kong Game	B	1976	Ideal	8	20	30
King Kong Game	B	1963	Ideal	75	200	325
King Leonardo and His Subjects Game	B	1960	Milton Bradley	30	50	80
King of the Sea	B	1975	Ideal	8	25	40
King Oil	B	1974	Milton Bradley	7	20	35
King Pin Deluxe Bowling Alley	S	1947	Baldwin Mfg.	10	20	30
King Tut's Game	B	1978	Cadaco	5	16	25
King Zor, The Dinosaur Game	B	1964	Ideal	25	65	100
Kismet	B	1971	Lakeside	45	100	150
KISS On Tour Game	B	1978	Aucoin	20	30	50
Klondike	B	1975	Gamma Two	20	55	75
Knight Rider	B	1983	Parker Brothers	7	12	20
Knockout, Electronic Boxing Game	S	1950s	Northwestern Products	90	150	240
Know Your States	C	1955	Garrard Press	10	25	45
Kojak	B	1975	Milton Bradley	7	12	20
Kommisar	B	1960s	Selchow & Righter	20	30	40
Kooky Carnival	B	1969	Milton Bradley	10	35	50
Korg 70,000 BC	B	1974	Milton Bradley	10	15	25
Kreskin's ESP	B	1966	Milton Bradley	10	20	30
Krokay	S	1955	Transogram	4	10	15
Krull	B	1983	Parker Brothers	6	10	15
KSP Baseball	B	1983	Koch Sports Products	15	25	35
Kukla & Ollie	B	1962	Parker Brothers	20	45	65
Lancer	B	1968	Remco	70	120	190
Land of The Giants	B	1968	Ideal	60	100	160
Land of The Lost	B	1975	Milton Bradley	20	50	75
Land of The Lost Pinball	S	1975	Larami	10	20	30
Landslide	B	1971	Parker Brothers	5	16	25
Laramie	B	1960	Lowell	45	85	175
Las Vegas Baseball	B	1987	Samar Enterprises	8	13	20
Laser Attack Game	B	1978	Milton Bradley	7	20	35
Lassie Game	B	1965	Game Gems	10	35	50
Last Straw	B	1966	Schaper	5	10	15
Laugh-In's Squeeze Your Bippy Game	B	1968	Hasbro	60	100	160
Laurel & Hardy Game	B	1962	Transogram	12	30	40
Laverne & Shirley Game	B	1977	Parker Brothers	9	15	25
Leapin' Letters	S	1969	Parker Brothers	5	16	25

POSTWAR GAMES

NAME	TYPE	YEAR	COMPANY	GOOD	EX	MINT
Leave It To Beaver Ambush Game	B	1959	Hasbro	20	45	70
Leave It To Beaver Money Maker	B	1959	Hasbro	20	45	70
Leave It To Beaver Rocket To The Moon	B	1959	Hasbro	20	45	70
Lee Vs Meade: Battle of Gettysburg	B	1974	Gamut of Games	12	35	50
Legend of Jesse James Game, The	B	1965	Milton Bradley	30	70	100
Lemans	B	1961	Avalon Hill	25	40	65
Let's Bowl a Game	B	1960	DMR	20	35	50
Let's Go to the Races	B	1987	Parker Brothers	7	20	35
Let's Make A Deal	B	1970s	Ideal	10	15	25
Let's Play Basketball	C	1965	D.M.R.	12	20	35
Let's Play Golf "The Hawaiian Open"	B	1968	Burlu	25	40	60
Let's Play Safe Traffic Game	B	1960s	X-Acto	25	50	90
Let's Play Tag	B	1958	Milton Bradley	5	16	25
Leverage	B	1982	Milton Bradley	4	6	10
LF Baseball	B	1980	Len Feder	12	20	30
Li'l Abner's Spoof Game	C	1950	Milton Bradley	65	95	135
Lie Detector Game	B	1961	Mattel	20	50	75
Lieutenant, The	B	1963	Transogram	30	70	100
Life, The Game of	B	1960	Milton Bradley	10	15	25
Limit Up	B	1980	Willem	6	18	30
Line Drive	B	1953	Lord & Freber	35	75	100
Linebacker Football	B	1990	Linebacker	12	20	30
Linkup	B	1972	American Greetings	5	16	25
Linus the Lionhearted Uproarious Game	B	1965	Transogram	50	85	135
Lion and the White Witch, The	B	1983	David Cook	5	16	25
Lippy the Lion Game	B	1963	Transogram	25	45	70
Little Black Sambo	B	1952	Cadaco-Ellis	60	120	200
Little Boy Blue	B	1955	Cadaco-Ellis	6	18	30
Little Creepies Monster Game	B	1974	Toy Factory	6	10	15
Little House On The Prairie	B	1978	Parker Brothers	8	20	30
Little League Baseball Game	B	1950s	Standard Toykraft	25	45	70
Little Orphan Annie	B	1981	Parker Brothers	10	20	30
Little Red Schoolhouse	B	1952	Parker Brothers	20	35	55
Lobby	B	1949	Milton Bradley	8	25	40
Lone Ranger and Tonto Spin Game, The	S	1967	Pressman	15	25	40
Long Shot	B	1962	Parker Brothers	45	75	125
Longball	B	1975	Ashburn Industries	30	50	75
Look All-Star Baseball Game	B	1960	Progressive Research	35	60	90
Looney Tunes Game	B	1968	Milton Bradley	20	40	65
Lord of the Rings, The	B	1979	Milton Bradley	25	60	85
Los Angeles Dodgers Baseball Game	B	1964	Ed-U-Cards	30	50	80
Lost Gold	B	1975	Parker Brothers	7	20	35
Lost In Space Game	B	1965	Milton Bradley	45	75	120
Lost Treasure	B	1982	Parker Brothers	7	20	35
Lottery Game	B	1972	Selchow & Righter	6	18	30
Love Boat World Cruise	B	1980	Ungame	5	15	20

POSTWAR GAMES

NAME	TYPE	YEAR	COMPANY	GOOD	EX	MINT
Loving Game, The	B	1987	R.J.E. Enterprises	4	6	10
Lucan, The Wolf Boy	B	1977	Milton Bradley	5	10	15
Lucky Break	B	1975	Gabriel	10	20	30
Lucky Strike	B	1972	International Toy	5	10	15
Lucky Town	B	1946	Milton Bradley	15	50	75
Lucy Show Game, The	B	1962	Transogram	60	100	160
Lucy's Tea Party Game	B	1971	Milton Bradley	20	35	55
Ludwig Von Drake Ball Toss Game	B	1960		6	10	15
Luftwaffe	B	1971	Avalon Hill	3	8	15
M Squad	B	1958	Bell Toys	35	75	125
M*A*S*H* Game	B	1981	Milton Bradley	15	35	50
MacDonald's Farm	B	1948	Selchow & Righter	20	45	65
MAD Magazine Game, The	B	1979	Parker Brothers	4	12	20
MAD, What Me Worry?	B	1987	Milton Bradley	6	10	15
Madame Planchette Horoscope Game	B	1967	Selchow & Righter	6	18	30
Magilla Gorilla	B	1964	Ideal	40	65	100
Magnetic Fish Pond	S	1948	Milton Bradley	15	25	65
Magnetic Flying Saucers	B	1950s	Pressman	21	35	55
Magnificent Race	B	1975	Parker Brothers	10	25	40
Mail Run	B	1960	Quality Games	15	50	75
Main Street Baseball	B	1989	Main St. Toy	20	35	55
Major League Baseball	B	1965	Cadaco	10	20	30
Major League Baseball Magnetic Dart Game	B	1958	Pressman	55	90	135
Man from U.N.C.L.E. Illya Kuryakin Card Game	C	1966	Milton Bradley	10	25	40
Man from U.N.C.L.E. Napoleon Solo Game	B	1965	Ideal	25	45	70
Man from U.N.C.L.E. Pinball Game	S	1966		80	135	215
Man from U.N.C.L.E. Target Game	S	1966	Marx	130	275	550
Man from U.N.C.L.E. THRUSH Ray Gun Affair Game	B	1966	Ideal	50	85	135
Manage Your Own Team	B	1950s	Warren	10	30	40
Management	B	1960	Avalon Hill	12	40	60
Mandinka	B	1978	Lowe	3	12	20
Manhunt	B	1972	Milton Bradley	5	16	25
Maniac	B	1979	Ideal	7	15	20
Marathon Game	S	1978	Sports Games	10	20	35
Marblehead	S	1969	Ideal	5	16	25
Margie, The Game of Whoopie	B	1961	Milton Bradley	12	30	50
Mark "Three"	S	1972	Ideal	6	18	30
Marlin Perkins' Zoo Parade	B	1965	Cadaco-Ellis	15	35	60
Martin Luther King Jr.	B	1980	Cadaco	6	10	15
Marx-O-Matic All Star Basketball	S	1950s	Marx	150	250	400
Mary Hartman, Mary Hartman	B	1976	Reiss Games	10	25	35
Mary Poppins Carousel Game	B	1964	Parker Brothers	12	30	50
Masquerade Party	B	1955	Bettye-B	45	75	100
Mastermind	B	1970s	Invicta	5	10	15
Masterpiece, The Art Auction Game	B	1971	Parker Brothers	10	20	30
Match	C	1953	Garrard Press	10	15	25

GAMES

Time Tunnel, 1966, Ideal

Winky Dink Official TV Game Kit, 1950s

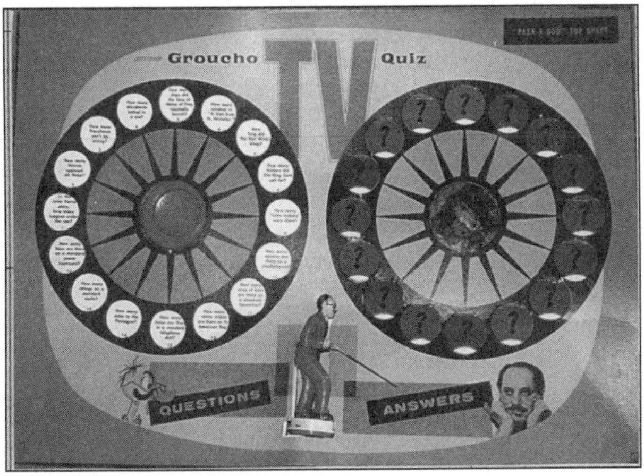

Groucho's TV Quiz Game, 1954, Pressman

POSTWAR GAMES

NAME	TYPE	YEAR	COMPANY	GOOD	EX	MINT
Match Game (3rd Ed.), The	B	1963	Milton Bradley	15	25	40
Matchbox Traffic Game	B	1960s		25	45	70
McDonald's Game, The	B	1975	Milton Bradley	10	20	30
McHale's Navy Game	B	1962	Transogram	30	50	80
McMurtle Turtle	B	1965	Cadaco-Ellis	8	25	40
Mechanical Shooting Gallery	S	1950s	Wyandotte	70	135	195
Meet The Presidents	B	1953	Selchow & Righter	15	25	40
Melvin The Moon Man	B	1960s	Remco	50	85	135
Men Into Space	B	1960	Milton Bradley	20	60	90
Mentor	S	1960s	Hasbro	20	70	100
Merger	B	1965	Universal Games	6	18	30
Merry Milkman	B	1955	Hasbro	35	85	125
Merv Griffin's Word For Word	B	1963	Mattel	5	16	25
Miami Vice: The Game	B	1984	Pepperlane	10	25	35
Mickey Mantle's Action Baseball	B	1960	Pressman	50	125	175
Mickey Mantle's Big League Baseball	B	1958	Gardner	125	250	325
Mickey Mouse	B	1976	Parker Brothers	6	10	15
Mickey Mouse	B	1950	Jacmar	35	55	90
Mickey Mouse Basketball	B	1950s	Gardner	55	90	135
Mickey Mouse Canasta Jr.	C	1950	Russell	20	55	85
Mickey Mouse Haunted House Bagatelle	S	1950s		210	350	550
Mickey Mouse Jr. Royal Rummy	C	1970s	Whitman	5	15	30
Mickey Mouse Library of Games	C	1946	Russell	30	65	125
Mickey Mouse Lotto Game	B	1950s	Jaymar	10	15	25
Mickey Mouse Pop Up Game	B	1970s	Whitman	7	15	20
Mickey Mouse Slugaroo	B	1950s		20	30	50
Mid Life Crisis	B	1982	Gameworks	5	15	20
Mighty Comics Super Heroes Game	B	1966	Transogram	20	70	100
Mighty Hercules Game	B	1963	Hasbro	125	300	500
Mighty Heroes on the Scene Game	B	1960s	Transogram	35	75	120
Mighty Mouse	B	1978	Milton Bradley	15	20	30
Mighty Mouse Rescue Game	B	1960s	Harett-Gilmar	35	60	125
Mighty Mouse Target Game	S	1960s	Parks	30	60	90
Mille Bornes	C	1962	Parker Brothers	2	7	12
Milton The Monster	B	1966	Milton Bradley	25	45	70
Mind Maze Game	S	1970	Parker Brothers	5	16	25
Mind Over Matter	B	1968	Transogram	10	15	25
Miss America Pageant Game	B	1974	Parker Brothers	5	16	25
Miss Popularity Game	B	1961	Transogram	20	35	55
Missing Links	B	1964	Milton Bradley	6	18	35
Mission: Impossible	B	1967	Ideal	40	100	135
Mission: Impossible	B	1975	Berwick	10	15	25
Mister Ed Game	B	1962	Parker Brothers	20	50	75
Mister Football	B	1951	Alkay	30	75	100
Mob Strategy	B	1969	NBC-Hasbro	6	18	30
Monday Morning Quarterback	B	1963	Zbinden	20	35	50
Money! Money! Money!	B	1957	Whitman	6	18	30

NAME	TYPE	YEAR	COMPANY	GOOD	EX	MINT
Monkees Game	B	1968	Transogram	30	75	105
Monkeys and Coconuts	B	1965	Schaper	10	15	20
Monster Game	B	1977	Ideal	45	75	120
Monster Game, The	B	1965	Milton Bradley	15	25	40
Monster Lab	S	1964	Ideal	175	275	450
Monster Mansion	B	1981	Milton Bradley	7	12	20
Monster Old Maid	C	1964	Milton Bradley	10	40	55
Monster Squad	B	1977	Milton Bradley	10	25	40
Monsters of the Deep	B	1976	Whitman	5	16	25
Moon Shot	B	1960s	Cadaco	15	35	60
Moon Tag, Game of	B	1957	Parker Brothers	75	125	200
Mork and Mindy	B	1978	Milton Bradley	10	15	25
Mostly Ghostly	B	1975	Cadaco	15	20	30
Mouse Trap	S	1963	Ideal	25	45	70
Movie Moguls	B	1970	RGI	6	10	15
Movie Studio Mogul	B	1981	International Mktg.	8	25	40
Mr. Bug Goes To Town	B	1955	Milton Bradley	15	35	60
Mr. Doodle's Dog	B	1940s	Selchow & Righter	20	45	70
Mr. Machine Game	B	1961	Ideal	40	65	100
Mr. Mad Game	S	1970	Ideal	12	40	60
Mr. Magoo Maddening Misadventures Game, The	B	1970	Transogram	45	75	120
Mr. Magoo Visits The Zoo	B	1961	Lowell	25	45	70
Mr. President	B	1967	3M	8	25	40
Mr. Ree	B	1957	Selchow & Righter	10	30	45
Mr. T Game	C	1983	Milton Bradley	5	15	25
Mt. Everest	B	1955	Gabriel	15	35	55
Mug Shots	B	1975	Cadaco	7	12	20
Munsters Card Game	C	1966	Milton Bradley	20	45	65
Munsters Drag Race Game	B	1965	Hasbro	250	600	800
Munsters Masquerade Game	B	1965	Hasbro	250	600	800
Munsters Picnic Game	B	1965	Hasbro	200	500	700
Muppet Show	B	1977	Parker Brothers	10	15	25
Murder on the Orient Express	B	1967	Ideal	20	50	75
Murder She Wrote	B	1985	Warren	4	6	10
Mushmouse & Punkin Puss	B	1964	Ideal	45	75	120
MVP Baseball, The Sports Card Game	B	1989	Ideal	8	13	20
My Fair Lady	B	1960s	Standard Toykraft	15	35	50
My Favorite Martian	B	1963	Transogram	50	90	125
My First (Walt Disney Character) Game	B	1963	Gabriel	20	35	60
Mystery Checkers	B	1950s	Creative Designs	15	20	30
Mystery Date	B	1965	Milton Bradley	60	125	175
Mystery Mansion	B	1984	Milton Bradley	10	25	35
Mystic Skull The Game of Voodoo	B	1965	Ideal	30	50	80
Mystic Wheel of Knowledge	B	1950s	Novel Toy	15	25	40
Name That Tune	B	1959	Milton Bradley	20	30	50
Nancy Drew Mystery Game	B	1957	Parker Brothers	35	75	110
NASCAR Daytona 500	B	1990	Milton Bradley	10	25	35
National Football League Quarterback, Official	B	1965	Standard Toykraft	20	45	60
National Inquirer	B	1991	Tyco	10	15	20
National Lampoon's Sellout	B	1970s	Cardinal	6	15	25

POSTWAR GAMES

NAME	TYPE	YEAR	COMPANY	GOOD	EX	MINT
National Pro Football Hall of Fame Game	B	1965	Cadaco	15	25	40
National Pro Hockey	B	1985	Sports Action	15	25	40
National Velvet Game	B	1950s	Transogram	20	40	65
NBA Basketball Game, Official	B	1970s	Gerney Games	25	55	80
NBC Game of the Week	B	1969	Hasbro	15	35	50
NBC Peacock	B	1966	Selchow & Righter	20	50	75
NBC Pro Playoff	B	1969	Hasbro	10	25	45
NBC TV News	B	1960	Dadan	15	35	50
Nebula	B	1976	Nebula	4	6	10
Neck & Neck	B	1981	Yaquinto	8	13	20
Negamco Basketball	B	1975	Nemadji Game	10	16	25
New Avengers Shooting Game	B	1976	Denys Fisher	165	275	440
New Frontier	B	1962	Colorful Products	30	50	80
New York World's Fair	B	1964	Milton Bradley	20	50	75
New York World's Fair Children's Game	C	1964	Ed-U-Cards	15	35	55
Newlywed Game (1st Ed.)	B	1967	Hasbro	10	14	20
NFL Armchair Quarterback	B	1986	Trade Wind	8	13	20
NFL Football Game, Official	S	1968	Ideal	20	30	50
NFL Franchise	B	1982	Rohrwood	10	16	25
NFL Game Plan	B	1980	Tudor	6	10	15
NFL Quarterback	B	1977	Tudor	14	23	35
NFL Strategy	B	1976	Tudor	5	16	25
NHL All-Pro Hockey	B	1969	Ideal	10	16	25
NHL Strategy	B	1976	Tudor	10	35	50
Nibbles 'N Bites	S	1964	Schaper	5	16	25
Nieuchess	B	1961	Avalon Hill	12	40	60
Nightmare On Elm Street	B	1989	Cardinal	15	30	45
Nile	B	1967	Lowe	6	18	30
Nixon Ring Toss	S	1970s		20	30	50
No Respect, The Rodney Dangerfield Game	B	1985	Milton Bradley	15	25	40
No Time for Sergeants Game	B	1964	Ideal	12	40	60
Noah's Ark	B	1953	Cadaco-Ellis	8	25	40
Nok-Hockey	B	1947	Carrom	20	35	55
Noma Party Quiz	B	1947	Noma Electric	20	35	50
Northwest Passage	B	1969	Impact	12	20	30
Nuclear War	C	1965	Douglas Malewicki	25	40	75
Number Please TV Quiz	B	1961	Parker Brothers	15	25	40
Numble	B	1968	Selchow & Righter	12	20	30
Numeralogic	B	1973	American Greetings	6	18	30
Nurses, The	B	1963	Ideal	10	35	50
Nutty Mads Bagatelle	S	1963	Marx	25	45	85
Nutty Mads Target Game	S	1960s	Marx	35	70	125
NY Mets Baseball Card Game, Official	C	1961	Ed-U-Cards	20	45	65
O.J. Simpson See-Action Football	B	1974	Kenner	50	125	175
Obsession	B	1978	Mego	5	10	15
Octopus	B	1954	Norton Games	12	40	60
Off To See The Wizard	B	1968		10	15	25
Oh Magoo Game	B	1960s	Warren	15	30	45

POSTWAR GAMES

NAME	TYPE	YEAR	COMPANY	GOOD	EX	MINT
Oh What a Mountain	B	1980	Milton Bradley	5	16	25
Oh, Nuts! Game	B	1968	Ideal	18	45	65
Oh-Wah-Ree	B	1966	3M	4	7	12
Oil Power	B	1980s	Antfamco	10	35	50
Old Shell Game, The	B	1974	Selchow & Righter	7	20	35
Oldies But Goodies	B	1987	Orig. Sound Record	6	18	30
On Guard	B	1967	Parker Brothers	6	10	15
On Target	S	1973	Milton Bradley	12	40	60
Operation	B	1965	Milton Bradley	10	15	25
Option	B	1983	Parker Brothers	2	7	12
Orbit	B	1959	Parker Brothers	10	30	50
Organized Crime	B	1974	Koplow Games	6	18	30
Orient Express	B	1985	Just Games	5	16	25
Original Home Jai-Alai Game, The	B	1984	Design Origin	15	25	35
Oscar Robertson's Pro Basketball Strategy	B	1964	Research Games	70	115	175
Our Gang Bingo	B	1958		50	85	135
Outdoor Survival	B	1972	Avalon Hill	6	10	15
Outer Limits	B	1964	Milton Bradley	110	180	275
Outlaw Trail	B	1972	Dynamic	6	18	30
Outwit	B	1978	Parker Brothers	5	10	15
Overboard	B	1978	Lakeside	3	12	18
Overland Trail Board Game	B	1960	Transogram	50	100	135
Ozark Ike's Complete 3 Game Set	B	1956	Builtrite	55	90	135
P.T. Boat 109 Game	B	1963	Ideal	30	50	90
Pac-Man	B	1980	Milton Bradley	5	15	20
Pan American World Jet Flight Game	B	1960	Hasbro	10	35	50
Panic Button	B	1978	Mego	7	15	20
Panzer Blitz	B	1970	Avalon Hill	6	12	20
Panzer Leader	B	1974	Avalon Hill	8	15	25
Par '73	B	1961	Big Top Games	15	25	40
Par Golf	B	1950s	National Games	20	50	75
Par-A-Shoot Game	S	1947	Baldwin	15	25	40
Parcheesi (Gold Seal Ed.)	B	1964	Selchow & Righter	7	12	20
Pari Horse Race Card Game	B	1959	Pari Sales	20	35	50
Paris Metro	B	1981	Infinity Games	10	16	25
Park & Shop	B	1952	Traffic Game	25	60	85
Park and Shop Game	B	1960	Milton Bradley	10	35	50
Parker Brothers Baseball Game	B	1955	Parker Brothers	45	70	110
Partridge Family	B	1974	Milton Bradley	10	25	40
Pass It On	B	1978	Selchow & Righter	3	12	20
Password	B	1963	Milton Bradley	6	15	25
Pathfinder	B	1977	Milton Bradley	5	15	20
Patty Duke Game	B	1963	Milton Bradley	25	40	75
Paul Brown's Football Game	B	1947	Trikilis	110	180	275
Paydirt	B	1979	Avalon Hill	5	12	18
Paydirt!	B	1973	Sports Illustrated	10	16	25
Payoff Machine Game	B	1978	Ideal	5	16	25
Peanuts: The Game of Charlie Brown And His Pals	B	1959	Selchow & Righter	20	30	50
Pebbles Flintstone Game	B	1962	Transogram	20	35	55
Pee Wee Reese Marble Game	B	1956	Pee Wee Enterprises	175	295	450

Wanted Dead or Alive, 1959, Lowell

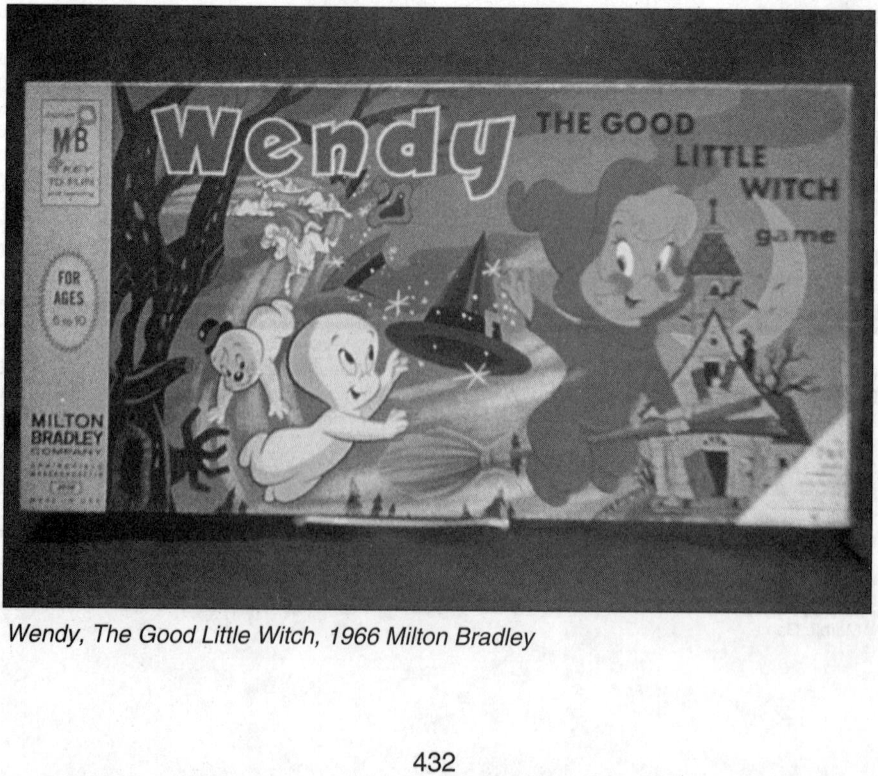

Wendy, The Good Little Witch, 1966 Milton Bradley

POSTWAR GAMES

NAME	TYPE	YEAR	COMPANY	GOOD	EX	MINT
Pennant Chasers Baseball Game	B	1946	Craig Hopkins	25	45	70
Pennant Drive	B	1980	Accu-Stat Game	8	13	20
People Trivia Game	B	1984	Parker Brothers	7	11	20
Perquackey	B	1970	Lakeside	3	5	8
Perry Mason Case of The Missing Suspect Game	B	1959	Transogram	20	30	50
Personalysis	B	1957	Lowell	15	25	40
Peter Gunn Dectective Game	B	1960	Lowell	25	60	90
Peter Pan	B	1953	Transogram	15	35	60
Peter Potamus Game	B	1964	Ideal	30	50	80
Peter Principle Game	B	1981	Avalon Hill	5	16	25
Peter Principle Game	B	1973	Skor-Mor	6	18	30
Petropolis	B	1976	Pressman	10	35	50
Petticoat Junction	B	1963	Standard Toykraft	35	55	90
Phalanx	B	1964	Whitman	10	30	45
Phantom Game, The	B	1965	Transogram	95	160	255
Phantom's Complete Three Game Set, The	B	1955	Built-Rite	30	75	125
Phil Silvers' You'll Never Get Rich Game	B	1955	Gardner	30	75	105
Philip Marlowe	B	1960	Transogram	20	50	75
Phlounder	B	1962	3M	6	18	30
Pigskin Vegas	B	1980	Jokari/US	6	10	15
Pinbo Sport-o-Rama	B	1950s		35	60	90
Pinhead	B	1959	Remco	10	35	50
Pink Panther Game	B	1977	Warren	10	25	40
Pink Panther Game	B	1981	Cadaco	4	7	12
Pinky Lee and the Runaway Frankfurters	B	1950s	Lisbeth Whiting	30	75	105
Pinocchio	B	1977	Parker Brothers	5	8	15
Pinocchio Board Game, Disney's	B	1960	Parker Brothers	10	15	25
Pinocchio, The New Adventures of	B	1961	Lowell	25	45	70
Pirate and Traveller	B	1953	Milton Bradley	10	30	50
Pirate Raid	B	1956	Cadaco-Ellis	8	25	40
Pirate's Cove	B	1956	Gabriel	15	35	65
Pitchin' Pal	S	1952	Cadaco-Ellis	7	20	35
Pizza Pie Game	B	1974	Milton Bradley	5	16	25
Planet of the Apes	B	1974	Milton Bradley	18	45	65
Play Ball! A Baseball Game of Skill	B	1940s	Rosebud Art	50	75	150
Play Basketball with Bob Cousy	B	1950s	National Games	115	195	295
Play Your Hunch	B	1960	Transogram	8	25	40
Playoff Football	B	1970s	Crestline	20	35	50
Plaza	B	1947	Parker Brothers	7	20	35
Ploy	B	1970	3M	5	16	25
Plus One	B	1980	Milton Bradley	6	18	30
Pocket Size Bowling Card Game	B	1950s	Warren/Built-Rite	15	25	40
Pocket Whoozit	B	1985	Trivia	4	7	10
Point of Law	B	1972	3M	4	14	20
Pole Position	B	1983	Parker Brothers	8	13	20
Police Patrol	B	1955	Hasbro	40	100	150
Politics, Game of	B	1952	Parker Brothers	15	45	65
Pony Express, Game of	B	1947	Polygon	12	40	60

POSTWAR GAMES

NAME	TYPE	YEAR	COMPANY	GOOD	EX	MINT
Pony Polo	S	1960s	Remco	10	20	35
Pooch	B	1956	Hasbro	10	35	50
Poosh-em-up Slugger Bagatelle	B	1946	Northwestern Products	30	75	105
Pop Yer Top!	B	1968	Milton Bradley	8	25	40
Pop-Up Store Game	B	1950s	Milton Bradley	12	40	60
Popeye, Adventures of	B	1957	Transogram	50	75	125
Population	B	1970	Urban Systems	6	18	30
Postman	B	1957	Selchow & Righter	8	25	40
Pothole Game, The	B	1979	Cadaco	10	25	40
Pow, The Frontier Game	B	1955	Selchow & Righter	12	40	60
Power 4 Car Racing Game	B	1960s	Manning	15	50	75
Power Play Hockey	B	1970	Romac	12	35	50
Prediction Rod	B	1970	Parker Brothers	5	16	25
Presidential Campaign	B	1979	John Hansen	6	18	30
Prince Caspian	B	1983	David Cook	5	16	25
Prince Valiant (Harold Foster's)	B	1950s	Transogram	12	40	60
Prize Property	B	1974	Milton Bradley	8	25	40
Pro Baseball Card Game	C	1980s	Just Games	6	10	15
Pro Bowl Live Action Football	S	1960s	Marx	35	65	95
Pro Draft	B	1974	Parker Brothers	15	25	40
Pro Football	B	1980s	Strat-O-Matic	6	10	15
Pro Foto-Football	B	1977	Cadaco	6	15	25
Pro Franchise Football	B	1987	Rohrwood	10	16	25
Pro Golf	B	1982	Avalon Hill	7	11	17
Pro Quarterback	B	1964	Tod Lansing	20	40	60
Probe	B	1964	Parker Brothers	2	7	12
Products and Resources, Game of	B	1962	Selchow & Righter	6	18	30
Profit Farming	B	1979	Foster Enterprises	6	18	30
Prospecting	B	1953	Selchow & Righter	25	65	90
Public Assistance	B	1980	Hammerhead	12	40	60
Pug-i-Lo	B	1960	Pug-i-Lo Games	55	90	135
Pursue the Pennant	B	1984	Pursue the Pennant	25	40	60
Pursuit!	B	1973	Aurora	8	25	40
Push Over	B	1981	Parker Brothers	3	12	18
Put and Take	B	1956	Schaper	5	16	25
Puzzling Pyramid	B	1960	Schaper	6	18	30
Quad-Ominos	B	1978	Pressman	2	6	10
Quarterback Football Game	B	1969	Transogram	25	40	65
Qubic	B	1965	Parker Brothers	3	10	15
Quick Shoot	S	1970	Ideal	6	18	30
Quinto	B	1964	3M	6	18	30
Quiz Panel	B	1954	Cadaco-Ellis	6	18	30
Race-A-Plane	B	1947	Phon-O-Game	12	40	60
Race-O-Rama	B	1960	Warren/Built-Rite	10	25	45
Raceway	B	1950s	B & B Toy	30	50	75
Radaronics	B	1946	ARC	20	65	90
Raggedy Ann	B	1956	Milton Bradley	6	18	30
Rainy Day Golf	B	1980	Bryad	6	18	30
Raise the Titanic	B	1987	Hoyle	12	30	40
Rat Patrol Game	B	1966	Transogram	40	70	100
Rat Patrol Spin Game	S	1967	Pressman	30	65	85
Rawhide	B	1959	Lowell	75	175	300

POSTWAR GAMES

NAME	TYPE	YEAR	COMPANY	GOOD	EX	MINT
Raymar of The Jungle	B	1952	Dexter Wayne	40	100	135
Razzle	B	1981	Parker Brothers	2	6	10
Razzle Dazzle Football Game	B	1954	Texantics Unlimited	50	80	125
React-Or	B	1979		15	25	40
Real Action Baseball Game	B	1966	Real-Action Games	20	35	50
Real Baseball Card Game	B	1990	National Baseball	110	180	275
Real Ghostbusters, The	B	1986	Milton Bradley	6	18	30
Real Life Basketball	B	1974	Gamecraft	10	16	25
Realistic Football	B	1976	Match Play	15	25	35
Rebel, The	B	1961	Ideal	50	85	135
Rebound	B	1971	Ideal	5	16	25
Record Game, The	B	1984	The Record Game	8	25	40
Red Barber's Big League Baseball Game	B	1950s	G & R Anthony	350	575	900
Red Herring	B	1945	Cadaco-Ellis	12	40	60
Red Rover Game, The	B	1963	Cadaco-Ellis	5	16	25
Reddy Clown 3-Ring Circus Game	B	1952	Parker Brothers	15	25	55
Reese's Pieces Game	B	1983	Ideal	5	8	15
Reflex	B	1966	Lakeside	5	16	25
Regatta	B	1968	3M	8	25	40
Regatta	B	1946		40	70	110
Replay Series Baseball	B	1983	Bond Sports	6	10	15
Restless Gun	B	1950s	Milton Bradley	15	25	60
Return To Oz Game	B	1985	Western	10	15	25
Reward	B	1958	Happy Hour	10	35	50
Rich Farmer, Poor Farmer	B	1978	McJay Game	5	16	25
Rich Uncle The Stock Market Game	B	1955	Parker Brothers	18	40	60
Rich Uncle, Game of	B	1946	Parker Brothers	18	40	60
Richie Rich	B	1982	Milton Bradley	3	5	10
Rickenbacker Ace Game	B	1946	Milton Bradley	60	150	200
Ricochet Rabbit Game	B	1965	Ideal	45	75	120
Rifleman Game	B	1959	Milton Bradley	20	50	75
Rin Tin Tin Game	B	1950s	Transogram	20	50	65
Ringmaster	B	1947	Cadaco-Ellis	12	40	60
Rio, The Game of	B	1956	Parker Brothers	10	35	50
Ripley's Believe It Or Not	B	1979	Whitman	6	10	15
Risk	B	1959	Parker Brothers	25	50	75
Riverboat Game	B	1950s	Parker Brothers/ Disney	20	35	55
Road Runner Game	B	1968	Milton Bradley	20	40	60
Road Runner Pop Up Game	B	1982	Whitman	20	30	50
Robert Schuller's Possibility Thinkers Game	B	1977	Selchow & Righter	3	5	10
Robin Hood	B	1955	Harett-Gilmar	30	50	80
Robin Hood Game	B	1970s	Parker Brothers	7	15	30
Robin Hood, Adventures of	B	1956	Bettye-B	45	75	120
Robin Roberts Sports Club Baseball Game	B	1960	Dexter Wayne	100	165	250
Robocop VCR Game	B	1988	Spinnaker	7	20	35
Robot Sam the Answer Man	B	1950	Jacmar	25	45	70
Rock 'N' Roll Replay	B	1984	Baron-Scott	7	20	35
Rock the Boat Game	B	1978	Milton Bradley	6	18	30
Rock Trivia	B	1984	Pressman	5	10	15

POSTWAR GAMES

NAME	TYPE	YEAR	COMPANY	GOOD	EX	MINT
Rocket Patrol Magnetic Target Game	S	1950s	American Toy Products	45	75	120
Rocket Race	B	1958	Stone Craft	30	75	125
Rocket Race To Saturn	B	1950s	Lido	15	20	35
Rocket Sock-It Rifle and Target Game	S	1960s	Kenner	20	35	75
Rodeo, The Wild West Game	B	1957	Whitman	12	40	60
Roger Maris' Action Baseball	B	1962	Pressman	50	125	175
Rol-A-Lite	B	1947	Durable Toy & Novelty	45	75	120
Rol-It	B	1954	Parker Brothers	4	12	20
Roll And Score Poker	B	1977	Lowe	3	7	12
Roll-A-Par	B	1964	Lowe	15	25	40
Roman X	B	1964	Selchow & Righter	7	20	35
Route 66 Game	B	1960	Transogram	75	125	200
Roy Rogers Game	B	1950s		20	35	55
Rribit, Battle of the Frogs	B	1982	Genesis Enterprises	6	18	30
Ruffhouse	B	1980	Parker Brothers	3	10	15
Rules of the Road	B	1977	Cadaco	5	16	25
Run to Win	B	1980	Cabela	6	18	30
Russian Campaign, The	B	1976	Avalon Hill	3	5	10
S.O.S.	B	1947	Durable Toy & Novelty	40	70	110
S.W.A.T. Game	B	1970s	Milton Bradley	10	15	25
Sabotage	B	1985	Lakeside	2	6	10
Saddle Racing Game	B	1974	APBA	12	30	50
Safari	B	1950	Selchow & Righter	12	40	60
Safecrack	B	1982	Selchow & RIghter	4	15	20
Sail Away	B	1962	Howard Mullen	15	45	75
Salvo	B	1961	Ideal	15	25	40
Samsonite Basketball	B	1969	Samsonite	15	35	60
Samsonite Football	B	1969	Samsonite	12	30	50
Sandlot Slugger	B	1960s	Milton Bradley	35	60	90
Saratoga: 1777	S	1974	Gamut of Games	10	25	50
Save the President	B	1984	Jack Jaffe	12	30	45
Say When!	B	1961	Parker Brothers	6	18	30
Scan	C	1970	Parker Brothers	3	10	15
Scarne's Challenge	C	1947	John Scarne Games	10	25	35
Scavenger Hunt	B	1983	Milton Bradley	3	5	10
Scooby Doo and Scrappy Doo	B	1983	Milton Bradley	3	8	16
Scoop	B	1956	Parker Brothers	40	60	100
Score Four	B	1968	Funtastic	3	10	15
Scotland Yard	B	1985	Milton Bradley	3	10	15
Scott's Baseball Card Game	C	1989	Scott's Baseball Cards	12	20	30
Scrabble	B	1953	Selchow & Righter	5	12	20
Screaming Eagles	B	1987	Milton Bradley	4	15	20
Screwball The Mad Mad Mad Game	B	1960	Transogram	30	75	105
Scribbage	B	1963	Lowe	3	10	15
Scrimmage	B	1973	SPI	5	15	25
Scruples	B	1986	Milton Bradley	5	10	15

GAMES

Fugitive, 1966, Ideal

National Velvet, 1950s, Transogram

POSTWAR GAMES

NAME	TYPE	YEAR	COMPANY	GOOD	EX	MINT
Sea World Treasure Key	B	1983	International Games	4	15	20
Sealab 2020 Game	B	1973	Milton Bradley	6	10	15
Seance	B	1972	Milton Bradley	18	60	80
Secrecy	B	1965	Universal Games	6	18	30
Secret Agent Man	B	1966	Milton Bradley	25	40	65
Secret of NIMH	B	1982	Whitman	6	10	15
Secret Weapon	B	1984	Selchow & Righter	5	16	25
Seduction	B	1966	Createk	20	30	45
See New York 'Round the Town	B	1964	Transogram	8	25	40
Sergeant Preston Game	B	1950s	Milton Bradley	15	35	50
Set Point	B	1971	XV Productions	8	20	30
Seven Keys	B	1961	Ideal	18	40	60
Seven Seas	B	1960	Cadaco-Ellis	30	50	80
Seven Up	B	1960s	Transogram	7	12	20
Sha-ee, the Game of Destiny	B	1963	Ideal	20	65	90
Shadowlord!	B	1983	Parker Brothers	3	10	15
Sharpshooter	S	1962	Cadaco-Ellis	12	40	60
Shazam, Captian Marvel's Own Game	B	1950s	Reed & Associates	30	60	95
Shenanigans	B	1964	Milton Bradley	15	40	60
Sheriff of Dodge City	B	1966	Parker Brothers	8	25	40
Sherlock Holmes	B	1950s	National Games	12	40	60
Sherlock Holmes	B	1980	Whitman	5	16	25
Sherlock Holmes Game, The	B	1974	Cadaco	6	18	30
Shifty Checkers	B	1973	Aurora	8	25	40
Shifty Gear Game	B	1962	Schaper	5	16	25
Shindig	B	1965	Remco	25	65	90
Shmo	B	1960s	Remco	10	35	50
Shogun	B	1986	Milton Bradley	18	45	75
Shopping	B	1973	John Ladell	5	16	25
Shotgun Slade	B	1960	Milton Bradley	20	45	65
Show-Biz	B	1950s	Lowell	15	50	75
SI: The Sporting Word Game	B	1961	Time	10	16	25
Siege Game	B	1966	Milton Bradley	10	35	50
Silly Carnival	B	1969	Whitman	7	12	20
Silly Safari	B	1966	Topper	40	65	105
Simpsons Mystery of Life, The	B	1990	Cardinal	5	7	10
Sinbad	B	1978	Cadaco	20	30	50
Sinking of The Titanic, The	B	1976	Ideal	35	75	100
Sir Lancelot, Adventures of	B	1960s	Lisbeth Whiting	40	70	110
Situation 4	B	1968	Parker Brothers	5	16	25
Situation 7	B	1969	Parker Brothers	8	25	40
Six Million Dollar Man	B	1975	Parker Brothers	5	10	15
Skatebirds Game	B	1978	Milton Bradley	7	20	35
Skatterbug, Game of	B	1951	Parker Brothers	30	50	80
Skedaddle	B	1965	Cadaco-Ellis	8	25	40
Skeeter	C	1950s	Arco Playing Card	6	20	35
Ski Gammon	B	1962	American Publishing	10	16	25
Skill-Drive	B	1950s	Sidney Tarrson	10	35	50
Skins Golf Game, Official	B	1985	O'Connor Hall	12	20	30

NAME	TYPE	YEAR	COMPANY	GOOD	EX	MINT
Skip Bowl	S	1955	Transogram	6	18	30
Skip-A-Cross	B	1953	Cadaco	5	12	20
Skipper Race Sailing Game	B	1949	Cadaco-Ellis	25	60	90
Skirmish	B	1975	Milton Bradley	20	35	55
Skirrid	B	1979	Kenner	4	15	20
Skudo	B	1949	Parker Brothers	8	25	40
Skully	B	1961	Ideal	3	5	10
Skunk	B	1950s	Schaper	7	12	20
Sky Lanes	B	1958	Parker Brothers	20	65	90
Sky's The Limit, The	B	1955	Kohner	15	25	40
Sla-lom Ski Race Game	B	1957	Cadaco-Ellis	20	50	75
Slap Stick	S	1967	Milton Bradley	7	20	35
Slap Trap	B	1967	Ideal	8	25	40
Slapshot	B	1982	Avalon Hill	6	10	15
Slip Disc	B	1980	Milton Bradley	5	16	25
Smack-A-Roo	S	1964	Mattel	6	18	30
Smess, The Ninny's Chess	B	1970	Parker Brothers	10	35	50
Smog	B	1970	Urban Systems	8	20	30
Smokey: The Forest Fire Prevention Bear	B	1961	Ideal	40	65	105
Smurf Game	B	1984	Milton Bradley	4	7	12
Snafu	B	1969	Gamescience	7	20	35
Snagglepuss Fun at the Picnic Game	B	1961	Transogram	40	60	100
Snake Eyes	B	1957	Selchow & Righter	15	50	75
Snake's Alive, Game of	B	1967	Ideal	6	18	30
Snakes & Ladders	B	1974	Summmerville/ Canada	4	7	10
Snakes In The Grass	B	1960s	Kohner	10	15	25
Snappet Catch Game with Harmon Killebrew	B	1960	Killebrew	55	90	135
Snob, A Fantasy Shopping Spree	B	1983	Helene Fox	7	20	35
Snoopy & The Red Baron	B	1970	Milton Bradley	15	35	50
Snoopy Come Home Game	B	1973	Milton Bradley	5	16	25
Snoopy Game	B	1960	Selchow & Righter	25	45	70
Snoopy's Doghouse Game	B	1977	Milton Bradley	5	16	25
Snow White and the Seven Dwarfs	B	1970s	Cadaco	6	10	15
Snuffy Smith Game	B	1970s	Milton Bradley	10	30	45
Sod Buster	B	1980	Santee	10	16	25
Solarquest	B	1986	Western	6	10	15
Solid Gold Music Trivia	B	1984	Ideal	6	10	15
Solitaire (Lucille Ball)	B	1973	Milton Bradley	5	15	20
Sons of Hercules Game, The	B	1966	Milton Bradley	30	70	100
Soupy Sales Sez Go-Go-Go Game	B	1960s	Milton Bradley	55	95	150
Southern Fast Freight Game	B	1970	American Publishing	8	25	40
Space Angel Game	B	1966	Transogram	25	65	85
Space Game	B	1953	Parker Brothers	25	65	85
Space Pilot	B	1951	Cadaco-Ellis	35	75	100
Space Shuttle 101	B	1978	Media-Ungame	8	20	30
Space Shuttle, The	B	1981	Ungame	8	20	30
Space: 1999 Game	B	1975	Milton Bradley	10	15	25
Special Agent	B	1966	Parker Brothers	10	25	40
Special Detective/Speedway	B	1959	Saalfield	15	40	60

POSTWAR GAMES

NAME	TYPE	YEAR	COMPANY	GOOD	EX	MINT
Speed Circuit	S	1971	3M	15	30	45
Speedorama	B	1950s	Jacmar	30	50	80
Speedway, Big Bopper Game	B	1961	Ideal	35	60	90
Spider and the Fly	B	1981	Marx	12	20	30
Spider's Web Game, The	B	1969	Multiple Plastics	7	12	20
Spider-Man Game, The Amazing	B	1967	Milton Bradley	20	45	85
Spider-Man with The Fantastic Four	B	1977	Milton Bradley	10	15	25
Spin Cycle Baseball	B	1965	Pressman	25	40	65
Spin The Bottle	B	1968	Hasbro	6	10	15
Spin Welder	B	1960s	Mattel	7	12	20
Spiro T. Agnew American History Challenge Game	B	1971	Gabriel	20	35	55
Sporting News Baseball	B	1986	Mundo Games	8	13	20
Sports Arena No. 1	B	1954	Rennoc Games & Toys	35	60	90
Sports Illustrated Baseball	B	1972	Sports Illustrated	25	40	65
Sports Illustrated College Football	B	1971	Sports Illustrated	8	25	40
Sports Illustrated Decathlon	B	1972	Time	8	25	40
Sports Illustrated Handicap Golf	B	1971	Sports Illustrated	15	25	35
Sports Illustrated Pro Football	B	1970	Time	15	25	40
Sports Illustrated, All Time All Star Baseball	B	1973	Sports Illustrated	35	75	100
Sports Trivia Game	B	1984	Hoyle	6	10	15
Sports Yesteryear	B	1977	Skor-Mor	15	25	35
Spot Cash	B	1959	Milton Bradley	7	15	20
Spy vs. Spy	B	1986	Milton Bradley	10	15	25
Square Mile	B	1962	Milton Bradley	12	40	60
Square-It	B	1961	Hasbro	5	16	25
Squares	B	1950s	Schaper	4	15	20
Squatter: The Australian Wool Game	B	1960s	John Sands/Australia	15	25	45
St. Louis Cardinals Baseball Card Game	B	1964	Ed-U-Cards	20	40	65
Stadium Checkers	B	1954	Schaper	7	12	20
Stagecoach West Game	B	1961	Transogram	35	75	125
Stampede	B	1956	Gabriel	8	25	40
Star Reporter	B	1950s	Parker Brothers	30	70	110
Star Team Battling Spaceships	B	1977	Ideal	10	15	25
Star Trek Adventure Game	B	1985	West End Games	8	20	30
Star Trek Game	B	1960s	Ideal	40	100	135
Star Trek: The Next Generation	B	1993	Classic Games	15	25	50
Star Wars Adventures of R2D2 Game	B	1977	Kenner	12	20	30
Star Wars Battle at Sarlacc's Pit	B	1983	Parker Brothers	10	15	25
Star Wars Escape from Death Star	B	1977	Kenner	10	15	25
Star Wars Monopoly	B	1997	Parker Brothers	15	20	35
Star Wars ROTJ Ewoks Save The Trees	B	1984	Parker Brothers	10	15	25
Star Wars Wicket the Ewok	B	1983	Parker Brothers	7	12	20
Star Wars X-Wing Aces Target Game	B	1978		20	30	50
Starship Troopers	B	1976	Avalon Hill	5	16	25
Starsky & Hutch	B	1977	Milton Bradley	5	16	25
State Capitals, Game of	B	1952	Parker Brothers	12	20	30
States, Game of the	B	1975	Milton Bradley	4	6	10
Statis Pro Football	B	1970s	Statis-Pro	25	40	65

POSTWAR GAMES

NAME	TYPE	YEAR	COMPANY	GOOD	EX	MINT
Stay Alive	B	1971	Milton Bradley	4	6	10
Steps of Toyland	B	1954	Parker Brothers	20	35	55
Steve Allen's Qubila	B	1955	Lord & Freber	8	25	40
Steve Canyon	B	1959	Lowell	20	50	85
Steve Scott Space Scout Game	B	1952	Transogram	30	75	105
Stick the IRS!	B	1981	Courtland Playthings	5	16	25
Sting, The	B	1976	Ideal	10	25	40
Stock Car Race	B	1950s	Gardner	20	50	75
Stock Car Racing Game	B	1956	Whitman	18	45	70
Stock Car Racing Game (w/Petty/Yarborough)	B	1981	Ribbit Toy	17	30	45
Stock Car Speedway, Game of	B	1965	Johnstone	55	90	135
Stock Market Game	B	1970	Avalon Hill	7	12	20
Stock Market Game	B	1955	Gabriel	20	30	50
Stock Market Game	B	1963	Whitman	6	18	30
Stock Market Specialist	B	1983	John Hansen	6	18	30
Stoney Burk	B	1963	Transogram	30	70	100
Stop Thief	B	1979	Parker Brothers	5	16	25
Straight Arrow	B	1950	Selchow & Righter	25	45	70
Straightaway	B	1961	Selchow & Righter	45	70	110
Strat-O-Matic Baseball	B	1961	Strat-O-Matic	100	165	250
Strat-O-Matic College Football	B	1976	Strat-O-Matic	12	30	50
Strat-O-Matic Hockey	B	1978	Strat-O-Matic	12	30	50
Strat-O-Matic Sports "Know-How"	B	1984	Strat-O-Matic	6	10	15
Strata 5	B	1984	Milton Bradley	5	16	25
Strategic Command	B	1950s	Transogram	25	45	70
Stratego (wood pieces)	B	1961	Milton Bradley	20	45	70
Strategy Manager Baseball	B	1967	McGuffin-Ramsey	2	6	12
Strategy Poker Fine Edition	C	1967	Milton Bradley	5	12	18
Strato Tac-tics	B	1972	Strato-Various	8	25	40
Stretch Call	B	1986	Sevedeo A. Vigil	12	20	30
Strike Three	B	1948	Tone Products	275	475	725
Stuff Yer Face	B	1982	Milton Bradley	6	20	35
Stump the Stars	B	1962	Ideal	8	25	40
Sub Attack Game	B	1965	Milton Bradley	7	20	35
Sub Search	B	1973	Milton Bradley	8	25	40
Sub Search	B	1977	Milton Bradley	6	18	30
Sudden Death!	B	1978	Gabriel	5	16	25
Suffolk Downs Racing Game	B	1947	Corey Game	60	100	150
Sugar Bowl	B	1950s	Transogram	20	45	65
Summit	B	1961	Milton Bradley	20	45	65
Sunken Treasure	B	1976	Milton Bradley	7	12	20
Sunken Treasure	B	1948	Parker Brothers	15	25	45
Super Coach TV Football	B	1974	Coleco	25	40	65
Super Market	B	1953	Selchow & Righter	10	35	50
Super Powers	B	1984	Parker Brothers	15	25	40
Super Spy	B	1971	Milton Bradley	15	25	40
Superboy Game	B	1960s	Hasbro	45	75	135
Supercar Road Race	B	1962	Standard Toykraft	45	100	140
Supercar to the Rescue Game	B	1962	Milton Bradley	40	90	125
Superheroes Card Game	C	1978	Milton Bradley	15	30	55
Superman & Superboy	B	1967	Milton Bradley	40	65	105

POSTWAR GAMES

NAME	TYPE	YEAR	COMPANY	GOOD	EX	MINT
Superman Game	B	1966	Merry Manufacturing	35	55	90
Superman Game	B	1965	Hasbro	45	75	120
Superman Game	C	1966	Whitman	35	75	110
Superman II	B	1981	Milton Bradley	10	20	35
Superman III	B	1982	Parker Brothers	4	12	20
Superman Spin Game	S	1967	Pressman	40	65	105
Superman, Adventures of	B	1940s	Milton Bradley	100	250	375
Superstar Baseball	B	1966	Sports Illustrated	30	50	75
Superstar Pro Wrestling Game	B	1984	Super Star Game	8	13	20
Superstar TV Sports	B	1980	ARC	6	10	15
Superstition	B	1977	Milton Bradley	10	15	25
Sure Shot Hockey	B	1970	Ideal	15	25	40
Surfside 6	B	1961	Lowell	30	75	125
Surprise Package, Your	B	1961	Ideal	25	60	85
Survive!	B	1982	Parker Brothers	3	10	15
Suspense	S	1950s	Northwestern Products	15	20	30
Swahili Game	B	1968	Milton Bradley	15	20	30
Swap, the Wheeler-Dealer Game	B	1965	Ideal	7	20	35
Swat Baseball	B	1948	Milton Bradley	20	35	50
Swayze	B	1954	Milton Bradley	20	35	55
Swish	B	1948	Jim Hawkers Games	45	75	115
Swoop	B	1969	Whitman	7	12	20
Sword In The Stone Game	B	1960s	Parker Brothers	15	35	50
Swords and Shields	B	1970	Milton Bradley	10	35	50
Syllable	C	1948	Garrard Press	10	15	25
T.V. Bingo	B	1970	Selchow & Richter	3	5	10
Tabit	B	1954	John Norton	25	35	75
Tactics II	B	1984	Avalon Hill	6	10	15
Taffy's Party Game	B	1960s	Transogram	10	15	25
Tales of Wells Fargo	B	1959	Milton Bradley	40	65	105
Talking Baseball	B	1971	Mattel	25	60	85
Talking Football	B	1971	Mattel	20	50	75
Talking Monday Night Football	B	1977	Mattel	10	30	50
Tally Ho!	B	1950s	Whitman	25	40	65
Tangle	B	1964	Selchow & Righter	6	18	30
Tank Battle	B	1975	Milton Bradley	15	35	50
Tank Command	B	1975	Ideal	7	20	35
Tantalizer	B	1958	Northern Signal	15	30	55
Tarzan	B	1984	Milton Bradley	5	10	15
Tarzan To The Rescue	B	1976	Milton Bradley	10	15	25
Taxi!	B	1960	Selchow & Righter	15	35	50
Tee Off by Sam Snead	B	1973	Glenn Industries	35	75	110
Teed Off!	B	1966	Cadaco	6	15	25
Teeko	B	1948	John Scarne Games	20	25	30
Telephone Game, The	B	1982	Cadaco	7	20	35
Television	B	1953	National Novelty	35	75	100
Tell It To The Judge	B	1959	Parker Brothers	12	35	60
Temple of Fu Manchu Game, The	B	1967	Pressman	20	30	50
Ten-Four, Good Buddy	B	1976	Parker Brothers	5	7	12
Tennessee Tuxedo	B	1963	Transogram	75	125	200
Tennis	B	1975	Parker Brothers	10	16	25
Tension	B	1970	Kohner	7	12	20
Terrytoons Hide N' Seek Game	B	1960	Transogram	35	95	100

GAMES

Jan Murray's Charge Account, 1961, Lowell

Veda, The Magic Answer Man, 1974, Milton Bradley

Battle of the Planets, 1970s, Milton Bradley

GAMES

POSTWAR GAMES

NAME	TYPE	YEAR	COMPANY	GOOD	EX	MINT
Test Driver Game, The	B	1956	Milton Bradley	25	75	125
Texas Millionaire	B	1955	Texantics	25	65	95
Texas Rangers, Game of	B	1950s	All-Fair	12	40	60
That's Truckin'	B	1976	Showker	7	20	35
They're at the Post	B	1976	MAAS Marketing	25	40	65
Thing Ding Robot Game	B	1961	Schaper	115	195	310
Think Twice	B	1974	Dynamic	4	15	20
Think-Thunk	B	1973	Milton Bradley	6	18	30
Thinking Man's Football	B	1969	3M	10	20	30
Thinking Man's Golf	B	1966	3M	10	20	30
Third Reich	B	1974	Avalon Hill	7	12	20
Thirteen	B	1955	Cadaco-Ellis	5	16	25
This Is Your Life	B	1954	Lowell	15	35	50
Three Little Pigs	B	1959	Selchow & Righter	7	20	35
Three Musketeers	B	1958	Milton Bradley	35	55	90
Three Stooges Fun House Game	B	1950s	Lowell	100	275	400
Thunder Road	B	1986	Milton Bradley	10	16	25
Thunderbirds Game	B	1965	Waddington/England	40	70	115
Tic-Tac Dough	B	1957	Transogram	20	30	45
Tickle Bee	S	1956	Schaper	15	20	35
Tiddle Flip Baseball	B	1949	Modern Craft	20	35	50
Tiddle-Tac-Toe	B	1955	Schaper	3	10	15
Tight Squeeze	S	1967	Mattel	7	20	35
Tilt Score	B	1964	Schaper	4	15	20
Time Bomb	S	1965	Milton Bradley	35	50	75
Time Machine	B	1961	American Toy	75	150	250
Time Tunnel Game, The	B	1966	Ideal	90	150	240
Time Tunnel Spin Game, The	S	1967	Pressman	60	100	160
Tiny Tim Game of Beautiful Things, The	B	1970	Parker Brothers	25	65	90
Tip-It	S	1965	Ideal	5	16	25
Tipp Kick	S	1970s	Top Set	15	25	40
Tom & Jerry	B	1977	Milton Bradley	10	15	25
Tom & Jerry Adventure In Blunderland	B	1965	Transogram	25	45	70
Tom Seaver Game Action Baseball	B	1969	Pressman	50	125	175
Tomorrowland Rocket To Moon	B	1956	Parker Brothers	20	50	75
Toot! Toot!	B	1964	Selchow & Righter	8	25	40
Tootsie Roll Train Game	B	1969	Hasbro	20	30	50
Top Cat Game	B	1962	Transogram	50	120	175
Top Cop	B	1961	Cadaco-Ellis	35	65	105
Top Pro Basketball Quiz Game	B	1970	Ed-U-Cards	6	15	25
Top Pro Football Quiz Game	B	1970	Ed-U-Cards	6	15	25
Top Scholar	B	1957	Cadaco-Ellis	5	16	25
Top Ten College Basketball	B	1980	Top Ten Game	10	16	25
Top-ography	B	1951	Cadaco-Ellis	4	15	20
Topple	B	1979	Kenner	4	15	20
Topple Chairs	S	1962	Eberhard Faber	6	18	30
Tornado Bowl	B	1971	Ideal	5	16	25
Total Depth	B	1984	Orc Productions	12	40	60
Touch	C	1970	Parker Brothers	5	16	25
Touche Turtle Game	B	1964	Ideal	45	100	175
Tournament Labyrinth	S	1980s	Pressman	10	15	25

POSTWAR GAMES

NAME	TYPE	YEAR	COMPANY	GOOD	EX	MINT
Town & Country Traffic Game	B	1950s	Ranger Steel	75	125	200
Track Meet	B	1972	Sports Illustrated	15	25	35
Trade Winds: The Caribbean Sea Pirate Treasure Hunt	B	1960	Parker Brothers	18	45	65
Traffic Game	B	1968	Matchbox	35	55	90
Traffic Jam	B	1954	Harett-Gilmar	40	60	80
Trail Blazers Game	B	1964	Milton Bradley	10	35	50
Trail Drive	C	1950s	Arco Playing Card	15	35	65
Trails to Tremble By	B	1971	Whitman	6	18	30
Trap Door	B	1982	Milton Bradley	3	10	15
Trap-em!	B	1957	Selchow & Righter	8	25	40
Trapped	B	1956	Bettye-B	18	45	75
Traps, The Game of	B	1950s	Traps	75	125	200
Travel America	B	1950	Jacmar	15	25	40
Travel-Lite	B	1946	Saxon Toy	45	75	120
Treasure Island	B	1954	Harett-Gilmar	30	50	80
Tri-Ominoes, Deluxe	B	1978	Pressman	3	5	10
Tribulation, The Game of	B	1981	Whitman	3	10	15
Triple Play	B	1978	Milton Bradley	5	10	15
Triple Yahtzee	B	1972	Lowe	4	6	10
Tripoley Junior	B	1962	Cadaco-Ellis	5	16	25
Trivial Pursuit	B	1981	Selchow & Righter	3	10	15
Troke (Castle Checkers)	B	1961	Selchow & Righter	6	18	30
Tru-Action Electric Baseball Game	B	1955	Tudor	25	50	80
Tru-Action Electric Basketball	S	1965	Tudor	25	50	75
Tru-Action Electric Harness Race Game	S	1950s	Tudor	15	40	65
Tru-Action Electric Sports Car Race	B	1959	Tudor	20	50	75
Trump, the Game	B	1989	Milton Bradley	5	16	25
Trust Me	B	1981	Parker Brothers	5	8	13
Truth or Consequences	B	1962	Lowell	10	35	50
Truth or Consequences	B	1955	Gabriel	12	40	60
Try-It Maze Puzzle Game	S	1965	Milton Bradley	7	12	20
TSG I: Pro Football	B	1971	TSG	35	55	85
Tumble Bug	B	1950s	Schaper	6	18	30
Turbo	B	1981	Milton Bradley	4	15	20
TV Guide Game	B	1984	Trivia	8	13	20
Twelve O'Clock High	B	1965	Ideal	25	60	85
Twelve O'Clock High	C	1966	Milton Bradley	20	40	65
Twiggy, Game of	B	1967	Milton Bradley	40	60	85
Twilight Zone Game	B	1960s	Ideal	75	100	200
Twinkles Trip to the Star Factory	B	1960	Milton Bradley	45	75	120
Twister	B	1966	Milton Bradley	10	15	25
Twixt	B	1962	3M	5	16	25
Two For The Money	B	1950s	Lowell	10	25	40
U.N. Game of Flags	B	1961	Parker Brothers	12	20	30
U.S. Air Force, Game of	B	1950s	Transogram	25	45	70
Ubi	B	1986	Selchow & Righter	6	18	30
Ultimate Golf	B	1985	Ultimate Golf	20	35	55
Uncle Wiggly	B	1979	Parker Brothers	4	7	12
Undercover: The Game of Secret Agents	B	1960	Cadaco-Ellis	20	35	55

POSTWAR GAMES

NAME	TYPE	YEAR	COMPANY	GOOD	EX	MINT
Underdog	B	1964	Milton Bradley	20	60	100
Underdog Save Sweet Polly	B	1972	Whitman	25	45	70
Undersea World of Jacques Cousteau	B	1968	Parker Brothers	30	45	60
Ungame	B	1975	Ungame	4	6	10
United Nations, A Game about the	B	1961	Payton Products	7	20	35
Universe	B	1966	Parker Brothers	6	18	30
Untouchables, The	S	1950s	Marx	95	160	250
Ur, Royal Game of Sumer	B	1977	Selchow & Righter	3	10	15
Uranium Rush	B	1955	Gardner	30	70	100
USAC Auto Racing	B	1980	Avalon Hill	25	65	85
Vagabondo	B	1979	Invicta	5	16	25
Vallco Pro Drag Racing Game	B	1975	Zyla	15	25	35
Valvigi Downs	B	1985	Valvigi	6	18	30
Vaquero	B	1952	Wales Game Systems	15	50	75
Varsity	B	1955	Cadaco-Ellis	12	30	50
VCR Basketball Game	B	1987	Interactive VCR Games	6	10	15
VCR Quarterback Game	B	1986	Interactive VCR Games	8	13	20
Veda, The Magic Answer Man	B	1960s	Pressman	20	30	45
Vegas	B	1974	Milton Bradley	4	15	20
Verbatim	B	1985	Lakeside	3	10	15
Verdict	B	1959	Avalon Hill	15	50	75
Verdict II	B	1961	Avalon Hill	12	40	60
Verne Gagne World Champion Wrestling	B	1950	Gardner	70	115	175
Vice Versa	B	1976	Hallmark Games	5	16	25
Video Village	B	1960	Milton Bradley	10	30	40
Vietnam	B	1984	Victory Games	10	15	25
Vince Lombardi's Game	B	1970	Research Games	35	70	100
Virginian, The	B	1962	Transogram	35	80	125
Visit To Walt Disney World Game	B	1970	Milton Bradley	15	20	35
Voice of The Mummy	B	1960s	Milton Bradley	25	60	90
Voodoo Doll Game	B	1967	Schaper	30	45	65
Voyage of the Dawn Treader	B	1983	David Cook	5	16	25
Voyage to Cipangu	B	1979	Heise-Cipangu	7	20	35
Wackiest Ship In The Army	B	1964	Ideal	40	65	105
Wacky Races Game	B	1970s	Milton Bradley	15	25	40
Wagon Train	B	1960	Milton Bradley	25	40	65
Wahoo	B	1947	Zondine	15	20	30
Wally Gator Game	B	1963	Transogram	50	85	130
Walt Disney's 101 Dalmatians	B	1960	Whitman	20	35	55
Walt Disney's 20,000 Leagues Under The Sea	B	1954	Jacmar	45	75	120
Walt Disney's Jungle Book	B	1967	Parker Brothers	15	25	45
Walt Disney's Official Frontierland	B	1950s	Parker Brothers	25	45	70
Walt Disney's Sleeping Beauty Game	B	1958	Whitman	30	50	80
Walt Disney's Swamp Fox Game	B	1960	Parker Brothers	30	50	80

POSTWAR GAMES

NAME	TYPE	YEAR	COMPANY	GOOD	EX	MINT
Waltons	B	1974	Milton Bradley	5	15	20
Wanted Dead or Alive	B	1959	Lowell	50	75	125
War At Sea	B	1976	Avalon Hill	10	20	30
War of the Networks	B	1979	Hasbro	6	18	30
Watergate Scandal, The	B	1973	American Symbolic	15	20	30
Waterloo	B	1962	Avalon Hill	10	20	45
Waterworks	C	1972	Parker Brothers	5	12	18
Weird-Ohs Game, The	B	1964	Ideal	85	145	230
Welcome Back, Kotter	C	1976	Milton Bradley	5	16	25
Welcome Back, Kotter	B	1977	Ideal	10	25	40
Welfare	B	1978	Jedco	12	40	60
Wendy, The Good Little Witch	B	1966	Milton Bradley	85	145	230
West Point Story, The	B	1950s	Transogram	10	35	50
What Shall I Be?	B	1966	Selchow & Righter	10	15	25
What Shall I Wear?	B	1969	Selchow & Righter	5	16	25
What's My Line Game	B	1950s	Lowell	25	45	65
Whatzit?	B	1987	Milton Bradley	5	16	25
Wheel of Fortune	B	1985	Pressman	5	8	13
Where's The Beef?	B	1984	Milton Bradley	6	10	15
Which Witch?	B	1970	Milton Bradley	25	60	85
Whirl Out Game	B	1971	Milton Bradley	6	18	30
Whirl-A-Ball	B	1978	Pressman	10	15	25
Whirly Bird Play Catch	B	1960s	Innovation Industries	20	35	50
White Shadow Basketball Game, The	B	1980	Cadaco	10	25	40
Who Can Beat Nixon?	B	1971	Dynamic	10	35	50
Who Framed Roger Rabbit?	B	1987	Milton Bradley	20	35	55
Who What Or Where?	B	1970	Milton Bradley	5	8	13
Who, Game of	B	1951	Parker Brothers	30	50	80
Whodunit	B	1972	Selchow & Righter	10	15	25
Whodunit?	B	1959	Cadaco-Ellis	8	25	40
Whosit?	B	1976	Parker Brothers	6	10	15
Wide World	B	1962	Parker Brothers	15	25	40
Wide World of Sports Golf	B	1975	Milton Bradley	12	30	45
Wide World Travel	B	1957	Parker Brothers	25	45	50
Wil-Croft Baseball	B	1971	Wil-Croft	10	16	25
Wild Bill Hickock	B	1955	Built-Rite	45	75	125
Wild Kingdom Game	B	1977	Teaching Concepts	20	35	50
Wildlife	B	1971	Lowe	15	35	50
Willie Mays "Say Hey"	B	1954	Toy Development	200	350	525
Willie Mays "Say Hey" Baseball	B	1958	Centennial Games	190	295	450
Willie Mays Push Button Baseball	B	1965	Eldon	175	295	450
Willow	B	1988	Parker Brothers	6	10	15
Win A Card Trading Card Game	B	1965	Milton Bradley	350	575	900
Win, Place & Show	B	1966	3M	6	20	30
Wine Cellar	B	1971	Dynamic	5	16	25
Wing-Ding	S	1951	Cadaco-Ellis	8	25	40
Winko Baseball	B	1940s	Milton Bradley	45	70	95
Winky Dink Official TV Game Kit	B	1950s		20	30	50
Winnie The Pooh Game	B	1959	Parker Brothers	30	50	80
Winnie The Pooh Game	B	1979	Parker Brothers	5	8	13
Winning Ticket, The	B	1977	Ideal	6	20	30
Wiry Dan's Electric Baseball Game	B	1953	Harett-Gilmar	25	40	65

POSTWAR GAMES

NAME	TYPE	YEAR	COMPANY	GOOD	EX	MINT
Wiry Dan's Electric Football Game	B	1953	Harett-Gilmar	15	40	65
Wise Old Owl	C	1950s	Novel Toy	20	35	55
Witch Pitch Game	B	1970	Parker Brothers	15	25	40
Wizard of Oz Game	B	1962	Lowe	20	30	50
Wizard of Oz Game	B	1974	Cadaco	10	15	25
Wolfman Mystery Game	B	1963	Hasbro	80	160	225
Woman & Man	B	1971	Psychology Today	4	15	20
Wonder Woman Game	B	1967	Hasbro	20	35	65
Wonderbug Game	B	1977	Ideal	8	25	35
Woody Woodpecker Game	B	1959	Milton Bradley	60	95	145
Woody Woodpecker's Crazy Mixed Up Color Factory	B	1972	Whitman	12	20	30
Woody Woodpecker's Moon Dash Game	B	1976	Whitman	12	20	30
Woody Woodpecker, Travel With	B	1950s	Cadaco-Ellis	35	65	100
World Bowling Tour	B	1979	World Bowling Tour	7	20	35
World Champion Wrestling Official Slam O' Rama	B	1990	International Games	5	8	12
World of Micronauts	B	1978	Milton Bradley	10	15	25
World of Wall Street	B	1969	NBC-Hasbro	4	15	20
World's Fair Game, The Official New York	B	1964	Milton Bradley	25	45	85
World's Greatest Baseball Game	B	1977	J. Woodlock	30	50	80
Wow! Pillow Fight For Girls Game	S	1964	Milton Bradley	10	20	30
Wrestling Superstars	B	1985	Milton Bradley	8	13	20
WWF Wrestling Game	B	1991	Colorforms	5	8	12
Wyatt Earp Game	B	1958	Transogram	35	60	85
Xaviera's Game	B	1974	Dynamic	7	20	35
Yacht Race	B	1961	Parker Brothers	40	85	125
Yahtzee	B	1956	Lowe	6	10	15
Yertle, The Game of	B	1960	Revell	55	95	150
Yogi Bear Break A Plate Game	B	1960s	Transogram	50	80	130
Yogi Bear Cartoon Game	B	1950s		3	5	10
Yogi Bear Game	B	1971	Milton Bradley	20	35	65
Yogi Bear Go Fly A Kite Game	B	1961	Transogram	25	60	85
Yogi Berra Pitch Kit	S	1963	Ross Products	35	50	100
Your America	B	1970	Cadaco	2	7	12
Yours For a Song	B	1962	Lowell	20	35	55
Zaxxon	B	1982	Milton Bradley	6	10	15
Zig Zag Zoom	B	1970	Ideal	10	20	30
Ziggy Game, A Day With	B	1977	Milton Bradley	15	20	30
Zingo	B	1950s	Empire Plastics	15	20	30
Zip Code Game	B	1964	Lakeside	35	55	90
Zomax	B	1988	Zomax	8	25	40
Zoography	B	1972	Amway	5	16	25
Zorro Game, Walt Disney's	B	1966	Parker Brothers	30	65	95
Zorro Target Game W/Dart Gun	B	1950s	Knickerbocker	20	30	50
Zowie Horseshoe Game	S	1947	James L. Decker	20	35	55

Get Up and Go!

(Prices listed are for toys Mint in Box)

Chef Cook, 1960s, Y Co.
$300

Bubble Blowing Monkey, 1950s, Alps
$200

Drinking Captain, 1960s, S&E
$200

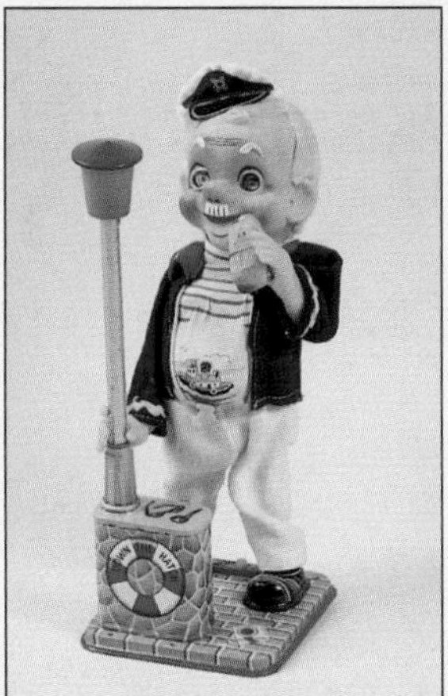

Charlie Weaver, 1962, TN
$120

Big Loo, 1960s, Marx
$2,200

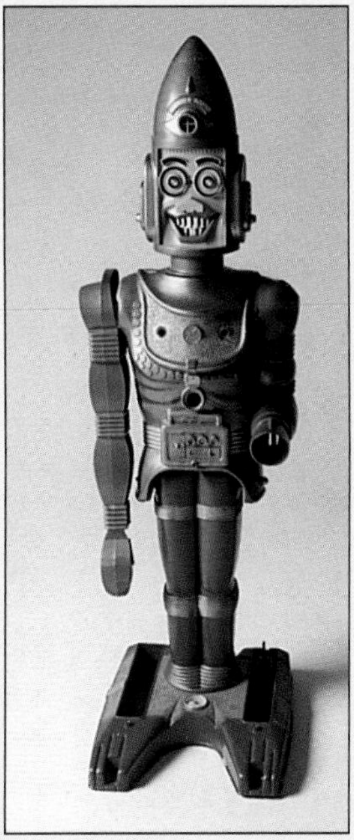

Still Queen of Dolls

(All toys made by Mattel. All photos courtesy of Mattel. Prices listed are for dolls Mint in Box)

Sunflower Barbie, ▲
1998
$80

Happy Holidays, ▲
1997
$10

Pink Splendor, ▲
1997
$600

NASCAR Barbie, ▲
1998
$40

Bob Mackie Fantasy Goddess of Asia, 1998
$240

Angel of Joy, 1998
$50

Grand Ole Opry Country Rose, 1997
$80

Dance 'til Dawn, 1998
$60

Fashion Savvy Uptown Chic, 1998
$50

Boxes of Fun

(Prices listed are for games in Mint Condition)

Partridge Family, 1974, Milton Bradley

$40

Mystery Date, 1965, Milton Bradley

$175

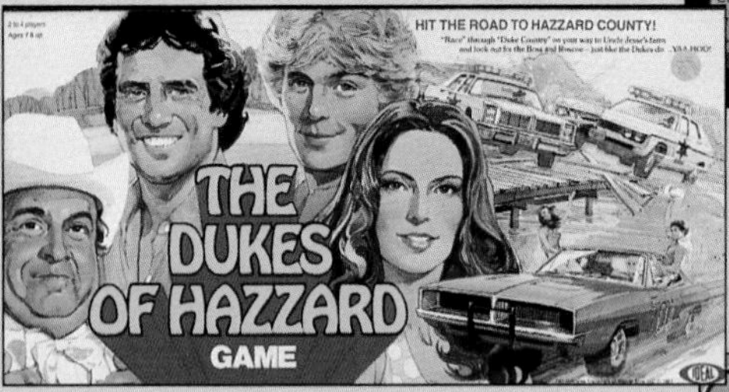

Dukes of Hazzard, 1981, Ideal

$25

Barney Miller, 1977, Parker Brothers

$25

Board games are collected for their historic value and colorful graphics. Pictured are three desirable prewar games, each valued at $100 or more in top condition.

Fast Food Fun

(Prices listed are for toys in Mint Condition)

Hunchback of Notre Dame Puppets, 1996, Burger King

$4 each

BK Kids Club Glo-Force, 1996, Burger King

$2 each

Funny Fry Friends, 1990, McDonald's

$4 each

Berenstain Bears, 1987, McDonald's

$4 each

Land Before Time Puppets, 1991, Pizza Hut

$8 each

Today's Hottest Collectible

(All toys made by Ty Inc. Prices listed are for toys in Mint Condition with original hang tags)

Pounce the Cat,
$8

Twigs the Giraffe,
$15

Speedy the Turtle,
$10

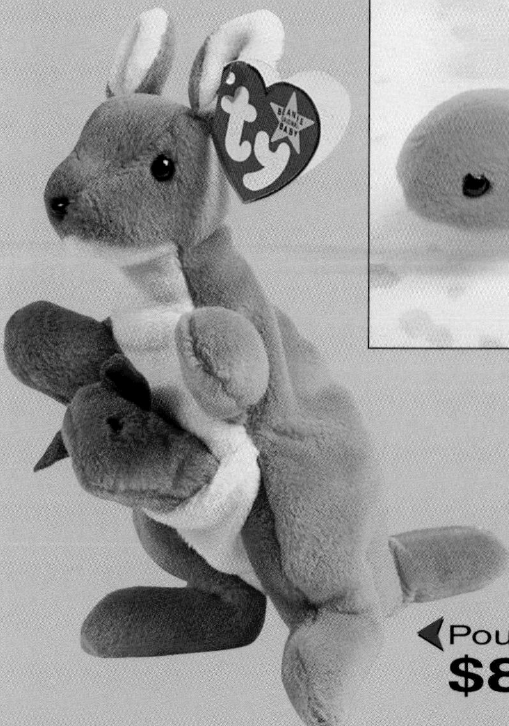

Pouch the Kangaroo,
$8

Lizzy the Lizard,
$22

Happy the Hippo,
$20

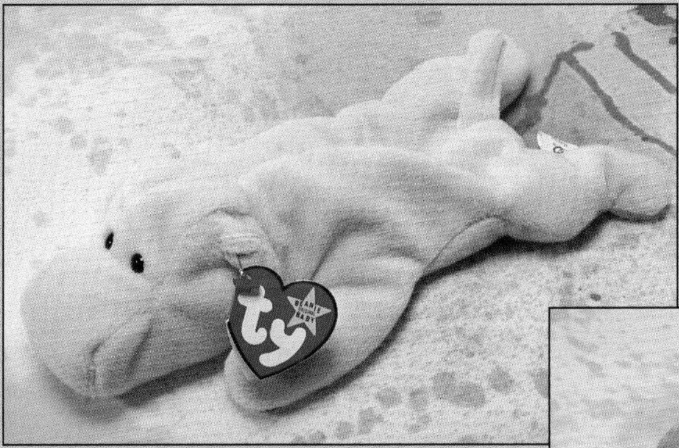

Goldie the Goldfish,
$45

Bessie the Cow,
$60

Iggy the Iguana,
$15

Toys with a Slogan

(Prices listed are for toys in Mint Condition)

Raid Bug, 1989, Raid
$30

Quik Bunny Doll, 1976, Nestle's Foods
$40

Wienermobile Mail-Away Bank, Oscar Mayer
$15

Energizer Bunny, Eveready
$65

Clown Mail-Away Bank, Del Monte
$20

Toy Guns

Remember the toy guns of your childhood? Baby boomers who grew up watching TV Westerns and space operas have been largely responsible for the increased interest in toy gun collecting. This, plus the increasing interest in related Western and cowboy hero collectibles, has brought prices and demand into new realms.

The first cap gun patent dates back to 1860. The preferred material of toy construction until World War II was cast iron. Most early cap guns, exploders and figural guns were made of cast iron, though variations of plastic, tin and stamped metal also appear.

The late 1930s was a golden age of cast-iron guns. Many guns of this period have realistic revolving cylinders, fine nickel finishes, colorful plastic grips and complicated mechanisms.

Metal toy gun production ceased around 1940. The companies that survived after 1946 experimented with a variety of materials and methods of production, the most successful of which was die casting. This involved injecting an alloy mixture into a mold, which resulted in fine detail, lower costs and a much lighter gun.

Die-cast guns dominated the 1950s and '60s marketplace. Television brought little buckaroos daily installments of the thrilling adventures of cowboy heroes such as Roy Rogers, Gene Autry and Hopalong Cassidy. Manufacturers such as the George Schmidt Company, Classy Products, Marx, Hubley, Nichols, Kilgore and Wyandotte all scrambled to arm these cowpokes and compete for a piece of the market. This, by some standards, was the golden age of die-cast guns.

Mattel, a relative newcomer to the business, introduced its first Western-style cap gun, the Fanner 50, in 1957. This shiny, oversized gun would, by its incredible success, drive many manufacturers out of the toy gun market and some out of business entirely.

The bottom line of toy making has always been cost. The less money spent on production, the greater the profit. Mattel produced a line of low priced plastic and die-cast cap guns, rifles and machine guns that arrived on the market at a time when TV Western heroes such as Maverick, The Lawman, Sugarfoot, Bronco and Matt Dillon were replacing the older serial cowboys.

Cap guns would never be the same. All major producers added plastic to their lines, and the race was on to lower production costs and remain competitive. As far as most collectors are concerned, things went downhill from there. Each year toy guns seemed to decrease in quality and detail.

While cap guns sold in the American market were largely produced by American manufacturers in both the prewar and postwar periods, other toy guns came from overseas producers. Tin lithography became an art form in postwar Japan. The same Japanese companies that produced the robots and tin cars so widely sought by today's collectors also produced space ray guns, cowboy, G-men and military-style tin litho guns for the American market.

While prewar Japanese toy guns are rare, postwar sparkling ray guns, pop guns, water and clicker guns abound. These guns were generally cheaper than their American competition and provided a profitable low end product for American retailers.

As the postwar Japanese economy grew and workers demanded more money, toy gun manufacturing moved from Japan to Hong Kong to Taiwan to Korea and eventually to mainland China, currently one of the major producers of toy guns and caps.

The Last Gunfighter

By the late 1960s, the majority of TV Westerns had been canceled. Cowboy heroes were forgotten, and toy guns in general were seen in a more negative light. Fads of spy guns and space guns revived the market for a while, but the golden age was over. The eventual end of this tale can be found today in toy stores. Now found in the toy gun aisles are numerous multi-colored Super-Soakers and noise-making space guns, but Roy Rogers holster sets or even Fanner 50s are nearly extinct.

Hawkeye Cap Pistol, Kilgore, 1950s

Walking through any toy or collectibles show or picking up a toy publication today will bring you in contact with collectible toy guns. The prices, conditions and varieties are confusing, to say the least. To make sense of toy gun collecting, you must first understand that there are all sorts of toy guns. There are cap guns, water guns, dart guns, cork guns, clicker guns, BB guns, air guns, pellet guns, guns that shoot peas, and some that can even fire four or five types of ammunition.

Where to Begin Collecting?

What do you want to collect, and why do you want to collect it? Most toy gun collectors are buying back a bit of their childhood. They played with these guns in countless childhood fantasies, and holding them again unlocks memories. Some collectors are searching only for the toy guns they had (or wish they had) as children. These folks may actually fire their guns and may not be as interested in condition and packaging as collectors on the other side of the spectrum who collect for the investment. Investment collectors seek only the highest grade Mint in Box guns that will appreciate in value. They never play with their guns, and the idea of firing caps in a gun gives them shivers. Most collectors fall between these two extremes.

Some collectors amass guns by a specific manufacturer, like Hubley, Mattel or Nichols, while others search for the shootin' irons that bear the names of their cowboy heroes. Some specialize in military-style toy weapons, bullet-firing machine guns, rocket-firing bazookas and cap-firing hand grenades. Recently, space guns have increased in popularity. Collectors needn't limit collections to guns only. Some people collect toy gun advertising, catalogs, store displays or other accessories.

Another point to consider is condition. It may be a Holy Grail quest to collect only Mint In Box pieces. Many times guns came off the assembly line in less than perfect condition, due to bad batches of metal, lack of quality control or poor design. Toy guns are, after all, just that — toys. They were meant to be played with, used and then discarded. To expect them to exist in pristine condition for decades is often wishful thinking.

The high acid content of the paper pulp used to make toy gun boxes has damaged the finish of many guns. Mattel's Fanner guns are especially prone to this problem. Some guns have inherent problems that surface with age, such as flaking or lifting of their finishes. This occurs because the gun was not properly prepared before being plated or the particular batch of plating wasn't properly mixed. The big Nichols Stallion .45s suffered from this problem. Sometimes a better quality of nickel just wasn't available and the production line had to use what was on hand. Any grading system used should be flexible and take this into consideration.

Watch for Variations

Variations enhance quality and expand the depth of a collection, but they can also drive collectors crazy. Forced by economic considerations, most companies would continue production of a popular gun for years with only minor changes to the molds or packaging. Each of these changes, no matter how minor, is technically a variation.

A good example of production variations can be found in Hubley's Texan cap guns. The Texan began life in the golden age of cast iron, the mid-1930s. It was a large, well-designed, sturdy pistol that fired roll caps and had a revolving cylinder and a Western look. The initial cast-iron version was offered in a cap-firing version and a dummy version. (Most manufacturers offered dummy versions in states and cities where caps were prohibited as fireworks.) The dummy version had a different hammer than the standard version, so it couldn't fire caps.

For a while, Hubley boosted sales appeal by obtaining a license to use the famous Colt firearms rearing horse logo instead of the traditional Hubley star. This variation is a must for any complete Texan collection. After World War II, the Texan reappeared in a nearly identical die-cast version that was offered in nickel, gold, gray or blue finishes with both standard or dummy hammers.

To further complicate matters, Hubley redesigned the gun and issued it as a Texan, offering it in nickel, gold and metal finishes. All these guns had the familiar Hubley longhorn steer head plastic grips. These vary by being either all white or white with a black steer.

Hubley also produced a smaller version of the Texan, called the Texan Jr., in many variations and box styles. As far as variations go, an enthusiast could amass quite a collection of variations on just one gun. This is why deciding what you want to collect is so important.

Beware of Reproductions

Reproductions and misrepresentations are common in any collecting field and a hazard for both the novice and experienced collector. Reproductions of highly-sought toy guns have yet to appear on the scene, mainly because the production and tooling costs for the potential market demand would be too high. But this has not stopped many small-time entrepreneurs from producing reproduction hammers, grips and laser-copy boxes.

Toy restoration and repair is not a concern as imperative as reproduction. As certain pieces become more difficult to find, restoration is a logical approach to fill the collecting demand. But restored or repaired pieces should always be presented as such and should be priced accordingly lower than an unrestored version.

Most laser-copy boxes are easy to spot. The inside cardboard is usually whiter than an original and many times flaws, tears and dirt from the original are visible.

Reproduction hammers are usually easy to spot, especially when placed next to an original. The original usually tends to be cleaner and more detailed. Clues such as a Mint hammer and pitted pan or anvil (striking surface) are also dead giveaways.

Misrepresentations are sometimes the most difficult situations to unmask. Many dealers misrepresent pieces simply from lack of knowledge and experience, and many beginning collectors fall victim for the same reasons. Simple mistakes include listing incorrect dates or manufacturer. Beware the dealer who doesn't know the difference between cast iron and die cast.

Experience will show you that an "H" in a diamond or oval on the side of a gun is usually the trademark of either Leslie-Henry (the Diamond H brand) or the Halco (J.

Halpern) Company, not Hubley. Hubley guns are usually marked "Hubley" somewhere on the gun. Knowing trademarks and distinguishing styles is learned through experience.

What Gun...What Holster?

One area of major confusion seems to be centered around the question of "What gun went in what holster?" Holster makers were often not the same companies that made the guns. The cowboy cap gun craze of the 1940s through the 1950s supported the growth of many small production leather holster manufacturers. Companies such as Halco, Classy Products and Pilgrim Leather Goods produced holsters for major manufacturers and at the same time produced empty holsters. These empty sets were sold to jobbers who would buy quantities of guns — the cheaper the better — from different manufacturers to fill them. They would then sell the gun/holster sets to individual retailers.

Large chain stores such as Toys R Us and Kmart didn't exist at this time. Every local department store, hardware store, hobby shop, and sporting goods store was a potential customer. Years later, it's not uncommon to find Leslie Henry Gene Autry Pistols in a Lasso 'Em Bill Holster set by Keyston Bros. What is this set? It's not really a Gene Autry set, no matter how nice it looks, and it shouldn't be represented as such. Most character holster sets were produced by major companies that paid hefty licensing fees to use cowboy names on both guns and holsters. Though there were a number of exceptions, most of these sets were sold with matching guns and holsters. The Gunsmoke set would have Marshal Matt Dillon guns in it, the Have Gun, Will Travel set would have Paladin guns, and so forth. Beware of the Mint In Box character set with a photo of the cowboy star on the box cover, his name on the holsters paired with common, no-name guns.

Trends

Toy guns are among the most passionately sought collectibles in the world. There are collectors who would trade or give anything for that gun they had as a child. Unfortunately, there are not enough of some guns to go around. Others exist in great surplus and will probably not increase in value until well into the next century. This shortage of desirable pieces and a glut of less desirable one has created occasional lapses of sanity on the part of both buyers and sellers.

Auction and show prices for classic Western holster sets such as Hopalong Cassidy, Lone Ranger and Roy Rogers have remained strong and have spurred many older collectors (who obtained their sets at a fraction of these prices) to consider liquidating parts of their collections.

As these guns begin to go on the open market again, most will obtain healthy profits for their owners while improving the quality of new owners' collections. This fuels the market.

On the other hand, many of the same toy guns can be found year after year at toy shows. These shunned pieces have become permanent fixtures in dealer displays for various reasons. Broken hammers, rust, cracked grips, worn or incorrect holster and unjustifiably high prices are the primary reasons these toys are not bought.

The last year has seen increased interest in space guns. Classics such as the Hubley Atomic Disintegrator, Daisy Buck Rogers XZ-31 Rocket Pistol and BCM Space Outlaw have shared the spotlight with more common plastic clicker and water guns from the 1950s and 1960s. Lost in Space guns have been listed with astronomical auction prices, but these prices may reflect the extremely small supply, rather than actual desirability.

With the increased interest in classic plastic of the 1950s dime store variety, the best money's on plastic water, clicker, flashlight, and dart guns by companies such as Palmer, Renwal, Irwin, Park, Remco, Marx and others.

Mattel guns in Good condition are getting harder to find. The most desirable continue to be the Fanner .45 and the Fanner 50. All Mattel guns suffer from similar aging problems. Their gleaming finishes rarely survive well outside the box. There will never be enough Mattel buckle or tommy-burst machine guns to go around.

The Top 10 Toy Guns / Sets
(in Mint in Box condition)

1.	Man From U.N.C.L.E. THRUSH Rifle, Ideal, 1966	$2,500
2.	Lost In Space Roto-Jet Gun Set, Mattel, 1966	2,000
3.	Man From U.N.C.L.E. Attache Case, Ideal, 1965	1,500
4.	Cap Gun Store Display, Nichols, 1950s	950
5.	Showdown Set with Three Shootin' Shell Guns, Mattel, 1958	950
6.	Man From U.N.C.L.E. Attache Case, Lone Star, 1966	850
7.	Man From U.N.C.L.E. Napoleon Solo Gun Set, Ideal, 1965	800
8.	Lost In Space Helmet and Gun Set, Remco, 1967	800
9.	Roy Rogers Double Gun & Holster Set, Classy, 1950s	800
10.	Man From U.N.C.L.E. Attache Case, Lone Star, 1966	750

Buck Rogers XZ-31 Rocket Pistol, Daisy, 1934

GUNS

Detective/Spy

NAME	COMPANY/DESCRIPTION	YEAR	GOOD	EX	MIP
	Benton Harbor Novelty				
Dick Cap Pistol	4-3/4", automatic, side loading, black finish	1950s	20	30	45
	Edison				
Sharkmatic Cap Gun	6" automatic, style pistol, fires "Supermatic System" strip caps	1980s	10	15	20
	Hubley				
Dick Cap Pistol	die cast, 4-1/4" automatic style, side loading with nickel finish	1950s	25	40	65
Dick Cap Pistol	4-1/8" cast iron automatic style, side loading, nickel finish, Dick oval in red paint	1930	75	125	195
	Ideal				
Man From U.N.C.L.E. Attache Case	15" x 10" x 2-1/2", comes with cap firing pistol and clip, ID card, wallet, cap grenade, badge, passport and secret message sender	1965	525	975	1500
Man From U.N.C.L.E. Illya K. Special Lighter Gun	cigarette lighter gun shoots caps, has radio compartment concealed behind fake cigs, in window box	1966	125	225	350
Man From U.N.C.L.E. Illya Kuryakin Gun Set	includes clip loading, cap firing plastic pistol, badge, wallet, ID card, in window box	1966	210	390	600
Man From U.N.C.L.E. Napoleon Solo Gun Set	clip loading, cap firing plastic pistol with rifle attachments, badge, ID card, in window box	1965	280	520	800
Man From U.N.C.L.E. Pistol and Holster	7" long pistol and plastic holster, both with orange ID sticker	1965	45	85	125
Man From U.N.C.L.E. Stash Away Guns	three cap firing guns, holsters, two straps, ID card and badge, in window box	1966	210	390	600
Man From U.N.C.L.E. THRUSH Rifle	36" long	1966	875	1600	2500
	Knickerbocker				
Dragnet Badge 714 Triple Fire Comb. Game	tin litho stand-up target has four plastic spinners, includes two 6" black plastic guns, one fires darts, the other cork gun, includes four plastic darts & three corks	1955	90	150	225
Dragnet Snub Nose Cap Pistol	plastic/die cast works, 6-3/4", black plastic with gold "Dragnet" and "714" badge on grip, on card	1960s	30	55	75
Dragnet Water Gun	6" black plastic, .38 Special-style with gold "714" shield on grip	1960s	15	20	35
	Lone Star				
Girl From U.N.C.L.E. Garter Holster	"gang buster" metal gun fires plastic bullets from metal shells, checker design vinyl holster and bullet pouch, on card	1966	80	145	225
Man From U.N.C.L.E. Attache Case	vinyl covered cardboard, 9mm automatic Luger, shoulder stock, sight, silencer, belt, holster, secret wrist holster and pistol, grenade, wallet, passport, money	1966	300	550	850
Man From U.N.C.L.E. Attache Case	vinyl case with pistol, holster, walkie talkie, cigarette box gun, badge, passport, invisible cartridge pen, handcuffs	1966	265	475	750

Detective/Spy

NAME	COMPANY/DESCRIPTION	YEAR	GOOD	EX	MIP
	Lone Star				
Man From U.N.C.L.E. Attache Case	small cardboard briefcase, contains die cast Mauser and parts to assemble U.N.C.L.E. Special	1966	140	260	400
	Marx				
G-Man Gun Wind-Up Machine Gun	23" tin-litho, red, black, orange & gray litho, round drum magazine, wind-up mechanism makes sparks from muzzle, uses cigarette flint, wooden stock	1948	145	200	275
Man From U.N.C.L.E. Pistol Cane	25" long, cap firing, bullet shooting aluminum cane with eight bullets and one metal shell, on card	1966	175	325	500
	Mattel				
Agent Zero M Snap-Shot Camera Pistol	7-1/2" extended, camera turns into pistol at press of a button	1964	25	50	75
Agent Zero W Potshot	3" die cast potshot derringer, gold finish, has brown vinyl armband holster holds two Shootin' Shell cartridges, gold buckle with Agent Zero W logo	1965	40	85	125
Official Detective Shootin' Shell Snub-Nose	.38 die cast chrome with brown plastic grips, black vinyl shoulder holster, wallet, badge, ID card, Pistol Range Target & bullets	1959	95	175	250
Official Detective Shootin' Shell Snub-Nose	.38 die cast 7" chrome finish, gold cylinder, brown plastic grips, Private Detective badge & Shootin' Shell bullets	1960	60	100	165
Official Dick Tracy Shootin' Shell Snub-Nose .38	die cast chrome .38 with brown plastic grips, chrome finish with Shootin' Shell bullets & Stick-m caps	1961	75	125	185
	Nichols				
Detective Shell Firing Pistol	5-1/2" snub-nose pistol chambers & fires six 3-pc. cap cartridges, cut-out badge, bullet cartridges, extra red plastic bullet heads	1950s	100	175	275
	Unknown				
Man From U.N.C.L.E. Secret Service Pop Gun	bagged luger pop gun on header card with unlicensed art of Napoleon and Illya	1960s	4	7	10

Military/Automatics

NAME	COMPANY/DESCRIPTION	YEAR	GOOD	EX	MIP
	Buddy L				
Spitfire Cap Firing Machine Gun	biped stand attached to muzzle, black plastic stock & grip with cap or clicker firing	1950	80	130	175
	Coibel				
Official James Bond 007 Thunderball Pistol	4-1/2" Walther PPK style, single shot fires plastic caps, Secret Agent ID	1985	15	25	45
	Daisy				
Model 12 SoftAir Gun	machine gun style, loads "SoftAir" pellets in plastic cartridge, 10 rounds, spring fired, can be cocked by barrel grip or bolt	1990	35	50	75

Military/Automatics

NAME	COMPANY/DESCRIPTION	YEAR	GOOD	EX	MIP
Daisy					
SA Automatic Burp Gun	10" with stock, black plastic, burp-gun style, removeable clip, loads & fires white plastic bullets	1970s	20	35	50
Edison					
Matic 45 Cap Gun	24", plastic gun with stock, fires "Supermatic System" strip caps	1980s	10	15	20
Esquire Novelty					
7580 UZI Automatic	10-1/2", battery operated, black plastic uses 250-shot roll caps, shoulder strap	1986	20	45	60
Hubley					
Army .45 Cap Pistol	6-1/2" automatic, dark gray finish, white plastic grips, pop-up cap	1950s	45	90	165
Automatic Cap Pistol No. 290	die cast, 6-1/2", nickel finish, brown checkered grips, magazine pops up when slider is pulled back		85	150	225
Larami					
9mm Z-Matic Uzi Cap Gun	8" replica, removeable cap storage magazine, black finish, small orange plug in barrel	1984	5	10	15
Maco					
Molotov Cocktail Tank Buster Cap Bomb	6" plastic & die cast, insert caps in head & throw	1964	10	15	25
MP Holster Set	plastic pistol has removeable magazine, loads & ejects bullets, white leather belt & holster	1950s	65	120	175
Paratrooper Carbine	24" carbine, removeable magazine, fires plastic bullets, bayonet & target	1950s	75	115	165
USA Machine Gun	12" tripod-mounted gun fires plastic bullets, red & yellow plastic	1950s	65	130	200
Main Machine					
Mustang Toy Machine Gun	25" long chrome and hard plastic paper firing gun	1950s	75	125	175
Marx					
Anti Aircraft Gun	mechanical, sparks, tin litho, 16-1/2" long	1941	35	100	195
Army Automatic Pistol	2-1/2" automatic, (ACP style), black with white grips, small leather holster with flaps (Marx Miniature)	1950s	10	15	25
Army Pistol with Revolving Cylinder	tin litho		35	60	95
Army Sparking Pop Gun		1940-50	45	75	125
Desert Patrol Machine Pistol	plastic, 11" long		30	45	75
Green Beret Tommy Gun	sparkles, trigger action, on card	1960s	35	55	95
Mini-M.A.G. Combat Gun	cap pistol, miniature scale, die cast		20	30	40
Siren Sparkling Airplane Pistol	heavy-gauge enamel steel, 9-1/2" wingspan, 7" long		50	75	100

Military/Automatics

NAME	COMPANY/DESCRIPTION	YEAR	GOOD	EX	MIP
	Mattel				
Sparkling Siren Machine Gun	26" long	1949	45	75	110
Special Mission Tommy Gun			20	35	45
Tommy Gun	sparks and makes noise	1939	65	95	150
Burp Gun	17" plastic with die cast works, perforated roll caps fired by cranking the handle	1957	25	50	75
Mattel-O-Matic Air Cooled Machine Gun	16" machine gun fires perforated roll caps by cranking handle, plastic with die cast, tripod-mounted, plastic/die cast works, red and black plastic, box	1955	45	85	125
Wind-Up Burp Gun	plastic/pressed steel, 24", "Grease Gun", fires perforated roll caps, fold-over wire stock, cap storage in magazine, boxed	1955	50	90	150
	National				
Automatic Cap Pistol	4-1/4" automatic style, grip swivels to load, black finish	1925	45	75	135
Automatic Cap Pistol	6-1/2" silver finish with simulated walnut grip	1950s	35	65	90
	Nichols				
Army 45 Automatic	4-1/4" all metal, side loading automatic, olive	1959	15	20	35
	Parris				
M-1 Kadet Training Rifle	32" wood/metal M-1 carbine, clicker action, metal barrel, trigger guard, bolt	1960s	20	35	50
	Redondo				
Revolver Mauser Cap Pistol	6-1/4", Mauser style automatic pop-up magazine, silver finish, brown plastic grips	1960s	4	7	10
	Stevens				
Spitfire Automatic Cap Pistol	4-5/8" cast iron, side loading, silver finish, "flying airplanes" white plastic grips	1940	60	110	165
	Topper/Deluxe				
Johnny Seven One Man Army-OMA	36" multi-purpose seven guns in one, removeable pistol fires caps, rifle fires white plastic bullets, bolt spring fired machine gun "tommy gun" sound, rear launcher fires grenades, forward diff. shell	1964	125	225	350
	Unknown				
MM Automatic Carbine	24" recoil red slide in muzzle & flashing light, brown & black plastic	1960s	15	30	45

Miscellaneous

NAME	COMPANY/DESCRIPTION	YEAR	GOOD	EX	MIP
	Ambrit Industries				
Spud Gun	case aluminum, pneumatic all-metal gun shoots pellets	1950s	15	35	50

Miscellaneous

NAME	COMPANY/DESCRIPTION	YEAR	GOOD	EX	MIP
	Atomic Industries				
Dynamic Automatic Repeating Bubble Gun	8" black plastic pistol projects bubbles		25	40	65
	Buddy L				
Paper Cracker Rifle	26" steel & machined aluminum mechanism, barrel, trigger & operating lever, uses 1000 shot paper roll, brown plastic stock	1940s	95	165	225
	Daisy				
Buzz Barton Special, No. 195	blue metal finish, wood stock with ring sight	1930s	85	125	200
Jack Armstrong Shooting Propeller Plane Gun	5-1/2" gun, shoots flying disc, pressed tin	1933	35	60	95
Model No. 25 Pump Action BB Gun	plastic stock	1960s	35	65	125
Targeteer No. 18 Target Air Pistol	10" gun metal finish, push barrel to cock, Daisy BB tin with special BBs, spinner target	1949	50	85	125
Water Pistol No. 17	5-1/4" tin	1940	25	40	65
Water Pistol No. 8	5-1/4" tin	1930s	25	50	80
	EMU Rififi				
Automatic Sparkling Pistol	6-1/2", plastic/metal, uses cigarette lighter flints to make sparks, available in green-red, yellow-green, red or white colors	1960s	15	30	45
	Esquire Novelty				
Hideaway Derringer	3-1/2" single shot, loads solid metal bullet, grip is removeable to store two more bullets, gold finish, white plastic grips	1950s	65	90	110
	Hong Kong				
Potato Gun	Spud Gun, plastic, pneumatic action fires potato pellet from muzzle	1991	3	7	10
	Hubley				
Midget Cap Pistol	5-1/2" long, die cast, all metal flintlock with silver finish	1950s	20	35	50
Pirate Cap Pistol	9-1/2" side-by-side flintlock style with die cast frame & cast double hammers & trigger, chrome finish, white plastic grips feature Pirate in red oval	1950	50	100	175
Tiger Cap Pistol	6 7/8", single action, mammoth caps, metal finish	1935	35	65	100
Trooper Cap Pistol	6-1/2" all metal, pop up cap magazine, nickel finish, black grips	1950	30	50	85
Winner Cap Pistol	4-3/8" automatic style, pop-up magazine release in front of trigger guard, nickel finish	1940	45	85	135
	Kilgore				
Clip 50 Cap Pistol	4-1/4", unusual automatic style, black Bakelite plastic frame, removeable cap magazine clip	1940	75	110	175
Mountie Automatic Cap Pistol	6", double action, automatic style with pop-up magazine, unusual nickel finish, black plastic grips		20	30	45

Flintlock Pistol, 1954, Hubley 1954,

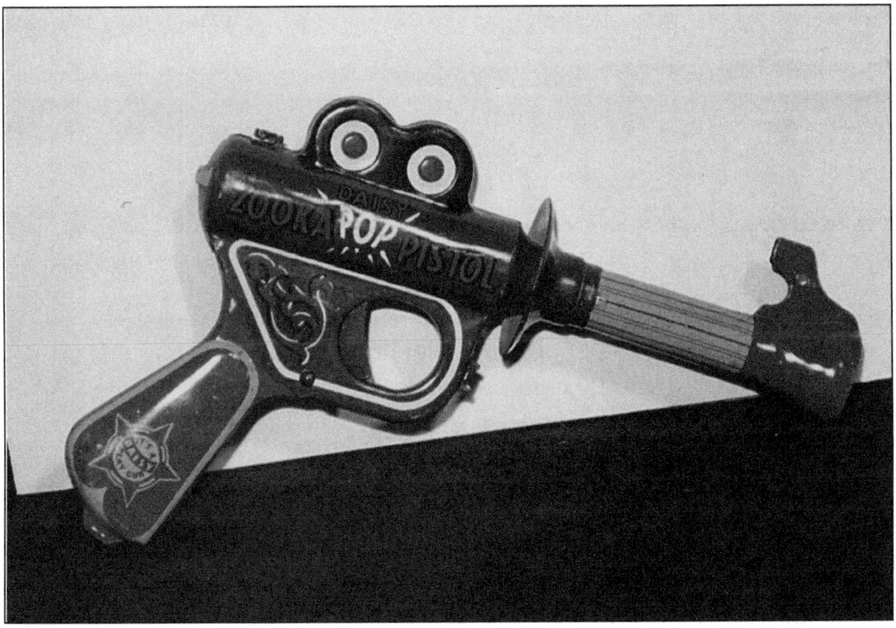

Zooka Pop Pistol, 1954, Daisy

NAME	COMPANY/DESCRIPTION	YEAR	GOOD	EX	MIP
	Kilgore				
Presto Cap Pistol	5-1/8", pop up cap magazine nickel finish brown plastic grips	1940	65	120	185
Rex Automatic Cap Pistol	3-7/8", blue metal finish cast iron, small size automatic style, side loading, white pearlized grips	1939	45	80	125
	Langson				
Nu-Matic Paper Popper Gun	pressed steel, 7" squeeze grip trigger, mechanism pops roll of paper (reel at top of gun) to make loud noise, black finish	1940s	30	50	75
	Marx				
Automatic Repeater Paper Pop Pistol	7-3/4" long		35	65	95
Blastaway Cap Gun	50 shooter repeater		25	35	50
Burp Gun	20" battery operated, green & black plastic	1960s	20	35	50
Click Pistol	pressed steel, 7-3/4" long		30	45	60
Click Pistol	tin litho		20	35	50
Famous Firearms Deluxe Edition Collectors Album	set of four rifles, five pistols & four holsters miniature series, includes: Mare's Laig, Thompson machine gun, Sharps rifle, Winchester saddle rifle, Derringer, .38 snub-nose, Civil War pistol, six shooter/Flint	1959	50	95	150
Marx Miniatures Famous Gun Sets	Four-gun set features tommy gun, Civil War revolver, "Mare's Laig" & western saddle rifle	1958	25	45	65
Marxman Target Pistol	plastic, 5-1/2" long		35	50	70
Popeye Pirate Click Pistol	tin litho, 10" long	1930s	100	150	200
Repeating Cap Pistol	aluminum		20	30	50
Sparkling Pop Gun			25	40	65
Streamline Siren Sparkling Pistol	tin litho		25	45	65
	Meldon				
P-38 Clicker Pistol	7-1/2" black finish, automatic style	1950s	35	50	75
	Midwest				
Long Tom Dart Gun	11" pressed steel	1950s	35	50	75
	Palmer Plastics				
Airplane Clicker Pistol	4-1/2" yellow & black plane, red pilot & guns	1950s	25	45	65
	Park Plastics				
Atomee Water Pistol	4-1/4" black plastic	1960s	10	20	35
	Rosvi				
Revolver Aquila Pop Pistol	10" green finish, pop gun breaks to cock, fires cork from barrel, sparks from mechanism under barrel	1960s	15	20	35
	Stevens				
25 Jr. Cap Pistol	4-1/8" automatic, side loading, silver finish	1930	25	50	75

Miscellaneous

NAME	COMPANY/DESCRIPTION	YEAR	GOOD	EX	MIP
	Stevens				
6 Shot Cap Pistol	6-3/4", six separate triggers revolve to deliver caps to hammer, metal finish		85	135	175
Model 25-50	4-1/2" nickel finish	1930	35	65	95
Pluck Cap Pistol	3-1/2" cast iron, single shot single action	1930	15	30	45
	Unknown				
Double Holster Set	black leather, large size, steer head conches, lots of red jewels, fringe, holsters only, no guns	1950s	65	100	150
	Welco				
Spud Gun (Tira Papas)	6" all-metal gun	1960s	10	20	35

Police

NAME	COMPANY/DESCRIPTION	YEAR	GOOD	EX	MIP
	Acme Novelty				
G-Boy Pistol	7" automatic (ACP) style entire left rear side of gun swings down to load	1950s	15	35	50
	Hubley				
Mountie Automatic Cap Pistol	die cast, 7-1/4" automatic style, pop up lever-release magazine, blue finish	1960s	20	35	55
	Kilgore				
Machine Gun Cap Pistol	5-1/8", long cast iron crank-fired gun	1938	100	175	250
	Marx				
.38 Cap Pistol	with caps, miniature scale, die cast		35	75	125
Detective Snub-Nose Special	die cast, 5-3/4" top release break, unusual revolving cylinder, fires Kilgore style disc caps, chrome finish with black plastic grips	1950s	65	100	150
Dick Tracy Click Pistol			35	50	75
Dick Tracy Jr. Click Pistol	aluminum	1930s	100	150	200
Dick Tracy Siren Pistol	pressed steel, 8-1/2" long	1935	40	75	125
Dick Tracy Sparkling Pop Pistol	tin litho		35	55	95
G-Man Automatic Silent Arm Pistol	tin		35	55	70
G-Man Automatic Sparkling Pistol	pressed steel, 4" long	1930s	45	65	110
G-Man Machine Gun	tin miniature, wind-up	1940s	25	35	65
G-Man Sparkling Sub-Machine Gun			50	75	100
G-Man Tommy Gun	sparkles when wound	1936	55	80	110
Gang Buster Crusade Against Crime Sub-Machine Gun	litho, metal with wooden stock, 23" long	1938	95	150	225
Marx Miniatures Detective Set	miniature cap firing brown and gray tommy gun, chrome pistol and holster on card with wood grain frame border	1950s	25	45	65

Police

NAME	COMPANY/DESCRIPTION	YEAR	GOOD	EX	MIP
	Marx				
Sheriff Signal Pistol	plastic, 5-1/2" long	1950	20	30	50
Siren Sparkling Pistol	tin litho		20	35	45
	Mattel				
Official Dick Tracy Tommy Burst Machine Gun	25" Thompson style machine gun fire perforated roll caps, single shot or in full burst when bolt is pulled back, brown plastic stock & black plastic body, lift up rear sight, Dick Tracy decal on stock	1960s	100	175	275
	Pilgrim Leather				
Peter Gunn Private Eye Revolver & Holster Set	36" die cast Remington with six two-pc. bullets, badge & wallet, Peter Gunn business cards, black leather shoulder holster	1959	200	325	450
	T. Cohn				
Sparkling "Sure-Shot" Machine Gun	long body tin multicolored red/yellow/blue tin noise making gun, great box graphics show boy shooting sparks as pigtailed blonde girl looks	1950s	50	75	125

Space

NAME	COMPANY/DESCRIPTION	YEAR	GOOD	EX	MIP
	Ahi				
Lost In Space Saucer Gun	disc shooting gun	1977	30	50	75
	Arco				
Ro-Gun "It's A Robot"	Shogun-type robot transforms into a rifle, in window box	1984	8	13	20
	Arliss				
Space Dart Gun	4" solid color plastic gun, shoots standard rubber tipped darts	1950s	12	20	30
X100 Mystery Dart Gun	3-3/4" long, yellow or gray plastic gun on cardboard display card, with two yellow and blue talcum impregnated darts which create a smoke effect when striking any target	1956	25	40	60
	Asahitoy/Japan				
Space Navigator Gun	3-1/2" long, tin, looks like sawed off military .45, colorfully trimmed blue body with smiling spaceman, blasting winged rocketship and "Space Navigator" logo on grips, planets and star on body	1953	35	60	95
	Aviva				
Star Trek:TMP Water Pistol	gray plastic, early pistol-grip phaser design	1979	10	20	35
	Azrak-Hamway				
Star Trek Water Pistol	white plastic, shaped like U.S.S. Enterprise	1976	15	30	45

Space

NAME	COMPANY/DESCRIPTION	YEAR	GOOD	EX	MIP

B.C.M.

NAME	COMPANY/DESCRIPTION	YEAR	GOOD	EX	MIP
Space Outlaw Ray Gun	10" long, chrome plated, die cast metal, recoiling barrel action, "Cosmic", "Sonic" or "Gamma" power levels, large red clear plastic teardrop shaped window	1965	115	250	450

Beaver Toys

NAME	COMPANY/DESCRIPTION	YEAR	GOOD	EX	MIP
Bee-Vo Bell Gun	#204, 6-1/2" long, red plastic, fires trapped marble at bell in muzzle, in box	1950s	25	40	60

Budson

NAME	COMPANY/DESCRIPTION	YEAR	GOOD	EX	MIP
Flash Gordon Air Ray Gun	10" long, unusual air blaster, handle on top cocks mechanism, shoots blast of air, red and silver pressed steel	1950s	175	295	450

Chein

NAME	COMPANY/DESCRIPTION	YEAR	GOOD	EX	MIP
Atomic Flash Gun	7-1/2" long, tin, sparkling action seen through tinted elongated oval plastic muzzle, with yellow and red on turquoise body with red lettered "Atomic Flash" over trigger	1955	40	65	100

Chemtoy

NAME	COMPANY/DESCRIPTION	YEAR	GOOD	EX	MIP
Pop Gun	4-1/2" long red hard plastic gun with space designs on handle	1967	25	35	50

Daisy

NAME	COMPANY/DESCRIPTION	YEAR	GOOD	EX	MIP
Buck Rogers Holster for U-238 Atomic Pistol	leather holster only	1946	40	65	100
Buck Rogers Holster for XZ-35 Pop Gun	embossed leather, attached to belt by two short riveted straps	1934	40	65	100
Buck Rogers in the 25th Century Pistol Set	holster is red, yellow & blue leather gun is 9-1/2" pressed steel pop gun	1930s	200	300	425
Buck Rogers in the 25th Century XZ-35 Rocket Pistol	holster is red, yellow & blue leather, gun is 9-1/2" pressed steel Rocket Pistol with single cooling fin at barrel base	1934	155	260	400
Buck Rogers U-235 Atomic Pistol	9-1/2" long, pressed steel, makes pop noise and flash in window when trigger is pulled	1946	105	180	275
Buck Rogers XZ-31 Rocket Pistol	10-1/2" long, heavy blued metal, grip pumps the action, gun pops when trigger is pulled	1934	135	225	350
Buck Rogers XZ-35 Space Gun	7" long, heavy blued metal ray gun, the grip pumps the action and the gun pops when trigger is pulled, single cooling fin at barrel base, also called "Wilma Gun"	1934	105	180	275
Buck Rogers XZ-38 Disintegrator Pistol	10-1/2" long, polished copper or blued finish, four flutes on barrel, spark is produced in the window on top of the gun when the trigger is pulled	1936	115	195	300
Buck Rogers XZ-44 Liquid Helium Water Gun	7-1/2" long, red and yellow lightning bolt design stamped metal body with a leather bladder to hold water. A later version was available in copper finish	1936	145	245	375
Daisy Rocket Dart Pistol	7" long, red, blue and yellow sheet metal gun with blue body, blue grips with yellow trim, blue and yellow barrel stripes, same body as Zooka Pop Pistol but with connecting rod from gun to barrel	1954	60	100	150

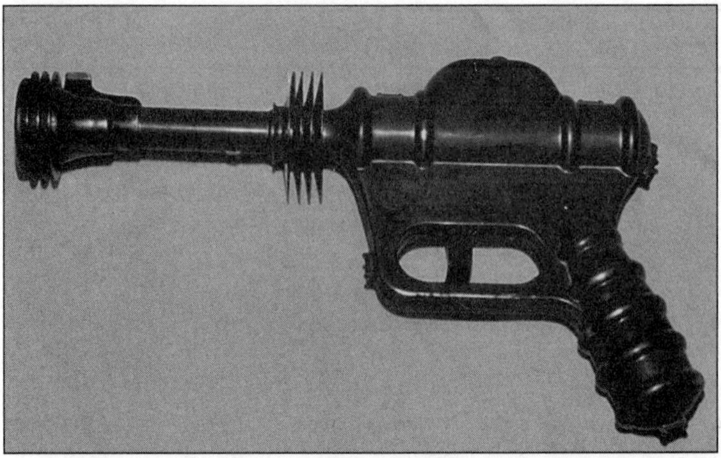

Buck Rogers XZ-38 Disintegrator Pistol, Daisy, 1930s

Mattel-O-Matic Air Cooled Machine Gun, 1955, Mattel

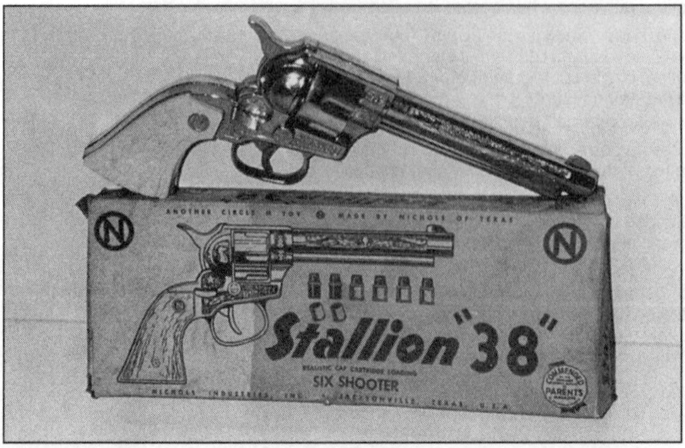

Stallion .38 Cap Pistol, Nichols, 1950s

NAME	COMPANY/DESCRIPTION	YEAR	GOOD	EX	MIP
	Daisy				
Daisy Zooka Pop Pistol	7" long, colorful red, blue and yellow sheet metal gun with blue body, red grips with yellow trim and litho star reading "It's a Daisy Play Gun," yellow barrel with red stripes, and wide red muzzle, handle cock	1954	80	130	200
Star Trek:III Phaser	white & blue plastic gun with light & sound effects	1984	30	50	75
	Daiya/Japan				
Baby Space Gun	6" friction siren & spark action	1950s	35	60	95
Space Gun	6" long, tin, sparkling action, metallic teal finish with red grooves and muzzle, green spaceship on body above "Space Gun," small Daiya logo inside red/yellow burst on grip with "577001" at bottom of grip	1957	35	60	95
Super Sonic Space Gun	7-1/2" long, tin litho, metallic gray body with red gunsight fin, friction siren and sparkling action, large oval center art with outstanding lunar scene of rockets, mountains and Earth in sky, red helmeted spaceman on grip	1957	40	65	100
	Endoh/Japan				
Super Sonic Gun	9" long, tin, sparkling action with three red plastic spark windows and clear red plastic barrel, blue body with red lightning bolt beneath yellow "Super Sonic" on rounded gun body, small ENDOH logo printed above grips	1957	40	65	100
	Futuristic Products				
Strato Gun	9" long, gray finish die cast, cap firing, internal hammer, top of gun lifts to load	1950s	70	115	175
Strato Gun	9" long, chrome finish die cast, red cooling fins, cap firing, internal hammer, top of gun lifts to load	1950s	100	165	300
	Gilbert				
Moon McDare Space Gun Set		1966	25	40	65
	H.G. Toys				
Alien Blaster Target Game	set features large free standing cardboard Alien target and plastic dart shooting rifle, gun has large block letters "Alien" on side, based on the movie	1979	55	95	145
	H.Y. Mfg./Hong Kong				
Razer Ray Gun	plastic bronze finish body with five large cooling fins near red plastic barrel, friction sparkling action, chrome finish muzzle tip, "Razer Ray Gun" embossed on rear of barrel	1972	10	15	25
	Haji				
Atomic Gun	9" long, red, gray and yellow tin litho gun with plastic muzzle, friction sparkling action, large hollow letter "ATOMIC GUN" on body	1969	15	30	45
Over and Under Ray Gun	8-1/2" long, red, yellow, white and black tin litho gun with two over and under reciprocating plastic muzzles, friction sparkling action	1960s	30	50	75

Space

NAME	COMPANY/DESCRIPTION	YEAR	GOOD	EX	MIP
	Hasbro				
Jet Plane Missile Gun	jet shaped handgun shoots darts, targets supplied on box back	1968	35	60	95
	Hero Toy/Japan				
Space Gun	7" long, tin litho, friction sparkling action, yellow body with blue and red trim, small Hero Toy logo by trigger	1960	35	55	85
	Hiller				
Atom Ray Gun	5-1/2" long, sleek red body gun of aluminum and brass with bulbous water reservoir on top of gun, reads "Atom Ray Gun" between two lightning bolts on reservoir	1949	135	225	350
	Hong Kong				
Ratchet Water Pistol Ray Gun	6-1/2" unusual pull back mechanism loads pistol, ratchet forces water out when trigger is pulled	1960s	25	40	65
Satellite & Rocket Pistol	5" long, green plastic gun fires either yellow plastic darts or saucers, on card	1960s	12	20	30
Visible Sparkling Ray Gun	8-1/2" long, plastic, mechanism visible, bagged with header card		15	25	40
	Hubley				
Atomic Disintegrator Ray Gun	8" in long, die cast metal with red handles, ornately embellished with dials and other equipment outcroppings, shoots caps	1954	150	300	650
	Ideal				
Ideal Flash Gun	9" long, plastic 3-color flashlight gun with red or blue body and bulbous contrasting-color blue or red rimmed flash unit, trigger switch and tail battery compartment cover, with color switch at top of flash unit	1957	80	130	200
Ratchet Sound Space Gun	7" long, red plastic with silver trim, flywheel ratchet on top of gun	1950s	30	50	75
Star Team Ionization Nebulizer	9" water gun fires water mist, red, white, blue & black plastic, Star Team decal	1969	30	50	75
	Irwin				
Clicker Ray Gun	9" long, red plastic with deep blue cooling fins on barrel base	1960	30	45	70
Space Ship Flashlight Gun	7-1/4", blue plastic ray gun has cockpit with orange spaceman, nose unscrews for AAA batteries, pulling trigger lights nose & moves guns & spaceman	1950s	60	100	150
	Jak-Pak				
Rocket Gun	7" hard yellow/green plastic with spring loaded plunger that shoots corks up to 50 feet	1958	8	13	20
	Japan				
888 Space Gun	3" long, tin, shoots caps, painted blue body and grip with stars, planets and spaceship, red barrel with "888" above grip	1955	30	50	75
Atomic Gun	5" long, gold, blue, white and red tin litho, friction sparkling action, "Atomic Gun" on body sides	1960s	20	35	50

Space

NAME	COMPANY/DESCRIPTION	YEAR	GOOD	EX	MIP
	Japan				
Jet Gun	6" long, tin, sparkling action, red body with three small red tinted spark windows near muzzle, grip shows silver-suited astronaut in modern helmet and wording "JET GUN" at top of grip near trigger	1957	35	60	90
Ray Gun	6-1/2" long, tin, sparkling action with two red tinted plastic tapered rectangle windows at muzzle, "Ray Gun" in red at top of body with rocket exhaust encircling green/blue planet against deep blue star studded background	1957	30	50	80
S-58 Space Gun	12" long, tin litho, deep metallic blue body with friction sparkling action, "S-58" on muzzle, with ringed planet graphic on front sight	1957	35	55	85
Space Atomic Gun	5-1/2" long, tin, sparkling action seen through red tinted plastic window, two-tone blue body with red/white atomic symbol on grip, "Space Atomic Gun" letters around oval spaceship-and-stars logo above trigger	1955	30	50	75
Space Gun	9" long, friction sparkling action with three red tinted plastic spark windows and clear red plastic barrel, body in metallic blue with large red "SPACE GUN" letters on yellow	1957	35	55	85
Super Space Gun	6" long, tin litho, friction sparkling action, blue on blue body with white/yellow/red highlights, large red on white "SUPER SPACE" lettering on side	1960	25	40	65
	Kenner				
Star Wars Han Solo Laser Pistol	flat dark gray finish Luger-like design with realistic scope sight and barrel extensions, black and white "Star Wars" decal on side	1978	30	50	75
Star Wars:ESB Han Solo Laser Pistol	same as original Han Solo pistol but with "ESB" decal on side	1980	20	35	50
Star Wars:ESB Laser Pistol		1980	20	35	55
Star Wars:ESB Three-Position Laser Rifle		1980	55	90	135
Star Wars:ROTJ Biker Scout Laser Pistol		1983	20	35	50
	Knickerbocker				
4-Barrel Waist Space Dart Gun Belt	11" wide gun system on belt, designed to be worn on waist or chest and aimed with periscope sight, red plastic belt	1950s	30	50	75
Space Jet Water Pistol	4" long, black plastic with white "Space Jet" lettering and spaceship line art on sides, fill plug in gunsight	1957	15	30	45
	KO/Japan				
Space Jet Gun	9" long, tin, sparkling action with black body, orange "Space Jet" on body with orange and red atomic symbol on grip, clear green plastic finned barrel base, clear blue plastic finned muzzle	1957	35	60	90
	Larami				
Dick Tracy Special Ray Gun	derringer with metal Dick Tracy New York Police Detective Badge	1964	30	50	75

Space

NAME	COMPANY/DESCRIPTION	YEAR	GOOD	EX	MIP
	Lido				
Space X-Ray Gun	#46598, 8-1/2" long, plastic, friction sparkling action, same body as Razer Ray Gun but with more futuristic handgrip and noisemaker at rear, sold in bag with header card	1970s	15	25	35
	LJN				
Dune Fremen Tarpel Gun	8" long, battery operated with internal light, light beam and chirping sound, plastic	1984	25	40	65
Dune Sardaukar Laser Gun	7" black plastic with flashing lights, battery operated	1984	20	35	50
	M & L Toy				
Space Rocket Gun	9" gray plastic, modern police-style pistol grip and shell chamber body with oversized barrel and muzzle sights, spring loaded, shoots rocket projectiles, in box with two "rockets"	1950s	55	95	145
	Marx				
Atomic Ray Gun	30" long,"Captain Space Solar Scout," blue plastic with oversized telescope sight flashlight and "electric buzzer" sound	1957	75	130	200
Cherilea Space Gun	miniature scale, die cast		25	40	60
Flash Gordon Radio Repeater Clicker Pistol	10" long, silver and red long-barrel gun with "Radio Repeater" in red beneath barrel and image and Flash Gordon name in side of grips	1937	175	295	450
Flash Gordon Signal Pistol	7" long, green bulbous teardrop sheet metal body with red flared sight and muzzle, prominent Flash Gordon decal on body, siren sounds when trigger is pulled	1930s	195	325	500
Flash Gordon Water Pistol	7-1/2" long, plastic with whistle in handle	1940s	80	130	200
Rex Mars Atomic Pistol Flashlite	plastic, battery operated	1950s	40	65	100
Rex Mars Planet Patrol 45 Caliber Machine Gun	22" long, tin and plastic, wind-up	1950s	60	100	150
Rocket Signal Pistol	same bulbous teardrop metal body as Flash Gordon Signal Pistol and Siren sparkling Airplane Pistol but without siren hole or wings; same rear fin, red with litho of three horizontally stacked finned orange/yellow bombs	1930s	135	225	350
Space Patrol Atomic Flashlight Pistol	gold/bronze finish pistol with seven large cooling fins on barrel and three smaller ones at back of gun, large clear plastic diffuser on muzzle, white "Official Space Patrol" on	1950s	135	225	350
Sparking Atom Buster Pistol	aluminum		30	50	75
Sparking Space Gun Rifle			50	80	125
Tom Corbett Official Space Cadet Gun	poorly designed composite rifle with modern military plastic stock and front grip at ends of long tin litho gun body with litho bombs and "Ray Adjuster" scale	1950s	135	225	350
Tom Corbett Space Cadet Atomic Flashlight Pistol	identical to Space Patrol Atomic Flashlight Pistol except for body colors and "Tom Corbett Space Cadet" printed upside down on handgrip	1950s	135	225	350

Space

NAME	COMPANY/DESCRIPTION	YEAR	GOOD	EX	MIP
	Marx				
Tom Corbett Space Cadet Gun	10-1/2" long, sheet metal clicker based on Flash Gordon Radio repeater molds, red body, blue barrel reads "Space Cadet," handgrips show bust of Tom in front of planet with rocket ship symbol above	1952	125	210	325
	Matchbox				
Robotech Water Pistol		1985	6	10	15
	Mattel				
Battlestar Galactica Lasermatic Pistol		1978	15	30	45
Battlestar Galactica Lasermatic Rifle		1978	25	40	65
Lost In Space Roto-Jet Gun Set	TV tie-in, modular gun can be reconfigured into different variations, shoots discs	1966	775	1300	2000
Space:1999 Astro Popper Gun	on card	1976	6	10	15
	Mercury Toys				
Planet Clicker Bubble Gun	8" long, plastic, red body with yellow accents, dip the barrel in bubble solution and pull trigger to make bubbles and produce click sound, in illustrated box	1953	40	65	100
	Metamol/Spain				
Jack Dan Space Gun	7-1/2" long, in black, red or blue painted die cast metal cap gun with "Jack Dan" over trigger	1959	105	180	275
	Mil Jo				
Space Scout Spud Gun	7" black & white plastic	1960s	15	30	45
	Nasta				
Flash Gordon Space Water Gun	water ray gun on illustrated card	1976	6	10	15
Flash Gordon Three Color Ray Gun	battery operated	1976	8	13	20
Space Water Pistol		1976	6	10	15
Sparkling Ray Gun		1976	6	10	15
	Nomura/Japan				
Cosmic Gun	12" long, plastic, battery operated with a small electric motor that runs reciprocating light in clear red plastic barrel, dark blue body, red and orange lettered "COSMIC GUN" decal	1970	35	55	85
Space Control Space Gun	3" long, tin sparkling gun with green body, red sights, decorated all over with stars and planets, red and yellow "Space Control" letters over trigger and spacemen firing gun and rocket flying overhead on grips	1954	30	50	75
	Norton-Honer				
Buck Rogers Sonic Ray Flashlight Gun	7-1/4" black, green & yellow plastic with code signal screw	1955	70	115	175

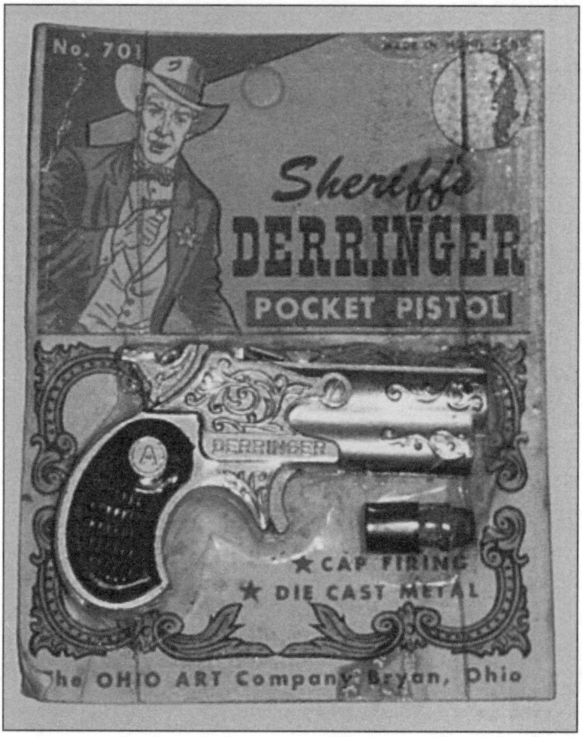

Sheriff's Derringer Pocket Pistol, 1960s, Ohio Art

Anti-Aircraft Gun, 1941, Marx

Space

NAME	COMPANY/DESCRIPTION	YEAR	GOOD	EX	MIP
	Nu-Age Products				
Smoke Ring Gun	large, sleek gray finished breakfront pistol with red barrel and muzzle ring, used rocket shaped matches to produce smoke, trigger fired smoke rings, small engraved "Smoke Ring Gun" logo on gunsight fin	1950s	175	295	450
	Ohio Art				
Astro Ray Laser Lite Beam Dart Gun	10" red & white plastic flashlight lights target with four darts	1960s	70	115	175
	Palmer Plastics				
Ray Gun Water Pistol	5-1/2" many color variations: green, orange, translucent blue, royal blue, black, yellow & red	1950s	10	15	20
Ray Gun Water Pistols	5-1/2" many color variations green, orange, translucent blue, royal blue, black, yellow & red	1950s	8	13	20
Space Water Gun	5-1/2" long, clear red plastic body with embossed Ringed planet and star, four cooling fins at barrel base, hollow telescope sight, yellow plastic trigger, white plastic stopper attached by loop to red knob at gun back	1957	15	30	45
	Park Plastics				
Atomee Water Pistol	4-1/4" black plastic	1960s	15	25	35
Space Water Gun	6" long, red transparent plastic, stopper at rear of gun, finned trigger guard, zeppelin-shaped reservoir with single embossed lightning bolt running its length, tiny "Park Plastics" imprinted along lateral reservoir fin	1960	15	25	35
	Power House Candy				
Captain Video Rite-O-Lite Flashlight Gun	3" long, red plastic gun with bulb, space map, paper, directions and order form, in mailing envelope	1950s	20	35	55
	Quisp Cereal/Quaker				
Quisp Powered Sugar Space Gun	7" long, red, mail away premium		150	250	400
	Ranger Steel Products				
Cosmic Ray Gun	9" long, tin body with plastic barrel, boldly painted in blue, yellow and red lightning bolts	1954	50	80	125
Cosmic Ray Gun #249	8" long, plastic, blue body, yellow barrel, red tip, in box showing two space kids in bubble helmets and backpacks shooting at spaceships	1953	40	65	100
Space Pistol	large yellow and orange flint gun		40	65	100
	Rayline				
Star Trek Tracer Gun	6-1/2" plastic firing tracer gun	1966	40	65	100
Star Trek Tracer Scope	rifle with discs	1968	50	80	125
	Remco				
Jupiter 4 Color Signal Gun	9" long black, red and yellow plastic gun that lights up in four colors, red telescoping sight	1950s	35	55	85
Lost In Space Helmet and Gun Set	child size helmet with blue flashing light and logo decals, blue and red molded gun	1967	310	525	800

NAME	COMPANY/DESCRIPTION	YEAR	GOOD	EX	MIP
	Remco				
Space:1999 Utility Belt Set	with disc shooting stun gun, watch and compass	1976	12	20	30
Star Trek Phaser	black plastic shaped like pistol, electronic sound, flashlight projects target	1975	30	50	75
Star Trek Phaser	Astro Buzz-Ray Gun with three color flash beam	1967	50	80	125
	Royal Plastics				
Flash-O-Matic, The Safe Gun	7" long red and yellow plastic battery operated light beam gun	1950s	60	100	150
	S. Horikawa/Japan				
Floating Satellite Target Game	6-1/2" x 9", battery operated, includes a pistol and three rubber tipped darts, a blower supports the styrofoam ball on a column of air and the players shoot darts to knock it down	1958	175	295	450
	San/Japan				
Space Gun	3-1/4" long, tin, sparkling action, aqua blue body with red and yellow highlights and "Space" in script lettering over grip, grip shows rocket shooting toward planets, circular San/Japan logo behind trigger	1955	30	50	75
	Shawnee				
Tomi Space Gun	solid red plastic with yellow barrel plug, modelled after modern .45 caliber pistol with rounded reservoir lined with 2 horizontal fins over grip; embossed logo and circular Shawnee logos on grip	1950s	50	80	125
	Shudo/Japan				
Astro Ray Gun	9" long, friction spark action, tin litho body with clear red plastic barrel, red on yellow "ASTRO RAY GUN" lettering	1968	25	40	65
Astro Ray Gun	5-7/8" long, silver finish body with red, yellow and black detailing, friction sparkling action, single large spark window near muzzle, prominent "Astro Ray Gun" in center of body	1960s	15	30	45
Flash X-1	4" long, tin litho, friction sparkling action, red body with blue and yellow inset and grips, large white "Flash X-1" on body, four red tinted plastic sparkling windows	1967	15	25	35
Space Gun	4" tin litho, friction sparkling action, red body with blue inset and grips, yellow block letter "SPACE GUN," large yellow and white vertical painted fins, six red tinted plastic sparkling windows, oval Shudo logo by grip	1967	20	35	55
	Stevens				
Atomic Jet Gun	8-1/2" long, gold chromed die cast metal, cap shooting, "Atomic Jet" and large circular "S" logo on grip	1954	90	145	225
Jet Jr. Cap Gun	6-1/2" long, fires roll caps, side loading door, silver finish, rear jet "Blast Off Fins"	1950s	135	225	350
Space Police Neutron Blaster Cap Pistol	7-3/4" die cast, cap firing ray gun, lock mechanism pulls out through the top of the gun, silver finish	1949	185	310	475

Space

NAME	COMPANY/DESCRIPTION	YEAR	GOOD	EX	MIP
	T. Cohn				
Space Target Game	24" tall, metal target with rubber tipped darts and dartgun to shoot down all the jet rockets and missiles	1952	40	65	100
	T/Japan				
Space Atomic Gun	4" long, tin litho, friction sparkling action, silver gray finish with yellow and red trim, with "SPACE" on body in white small all caps and large yellow lower caps "atomic gun", small "T/Made in JAPAN" logo above trigger	1960	25	40	65
Universe Gun	4" long, blue, yellow and red tin litho gun with friction sparkling action, large all caps italic "Universe" on body side, sold in bag with header card	1960s	15	25	35
	Tarrson				
Ray Dart Gun	9-1/2" long, blue plastic body with yellow muzzle, with 3 darts, storage compartment in red handle base	1968	15	25	35
	TN/Japan				
Space Gun	8" long, battery operated, reciprocating barrel shaft has red and blue lenses that flash when fired, makes rat-a-tat noise, large circular "8" over handgrip, winged eagle over trigger, large block letter "SPACE GUN" on barrel	1960s	70	115	175
	TNT/Hong Kong				
Robot Raiders Space Signal Gun	6" long flashlight gun with interchangeable lenses and click sound	1980s	6	10	15
	U.S. Plastics				
Rocket Jet Water Pistol	5" long, red, orange or yellow clear plastic body, fill plug at top of gun, large integral gunsight fin at rear, small sight fin at front	1957	12	20	30
Space Patrol Rocket Gun	black or red plastic pistol body with red trigger, grip embossed with vertically printed "Space Patrol" in irregular oval grip design showing rocket, stars and ringed planet, shoots rubber tipped darts	1954	145	250	375
	Unknown				
Atom Bubble Gun	red tubular barrel with handle attached, two sets of silver finish fins--at barrel base and muzzle, wire loop projects from muzzle for bubble blowing, handle embossed "Atom Trade	1940s	75	130	200
Batman Ray Gun	cap pistol with bat symbol for the sight	1960s	35	55	85
Bicycle Water Cannon Ray Gun	10", red plastic, swivel mount attached to bicycle handles, fired by lever	1950s	30	50	75
Buck Rogers Rubber Band Gun	cut-out paper gun, on card, advertising premium item	1930s	30	50	75
Clicker Ray Gun	5" red, blue or gray hard plastic, no boxes, sold loose	1950s	15	25	35
Clicker Whistle Ray Gun	5" plastic, blue/green or olive/green swirl plastic, imprinted spacemen & rocket ships, back of gun is a whistle	1950s	15	25	35
Dan Dare & the Aliens Ray Gun	21" color tin litho gun	1950s	90	150	235

Space

NAME	COMPANY/DESCRIPTION	YEAR	GOOD	EX	MIP

Unknown

NAME	COMPANY/DESCRIPTION	YEAR	GOOD	EX	MIP
Lost In Space Laser Water Pistol	5" long, first season pistol style		30	50	75
Planet Patrol Saucer Gun	with spaceman motif	1950s	40	65	100
Radar Gun	5-1/2" long, mauve or silver/gray swirl plastic body with green or yellow spaceman sight and trigger, Saturn and star embossed above grip and "Radar Gun" embossed above that	1956	20	35	55
Rocket Pop Gun	wood, green and red horizontal striped body with black tri-fin pump base, cork and string stopper in nose, pump fins into body to make it pop	1955	25	40	65
Secret Squirrel Ray Gun		1960s	25	40	60
Signal Flash Gun	6" long, plastic flashlight, black body with translucent white plastic light housing at muzzle and pearl finish plastic grip plates, modern missile type sight on top of barrel, large "SIGNAL FLASH" above trigger	1957	20	35	55
Space Atomic Gun	4" silver, orange/red tin litho, sparking action	1960s	25	40	65
Space Control Ray Gun	5-1/2" long, red plastic with yellow trigger, clicks	1956	25	40	65
Space Dart Gun	6" long, gun has one white side and one black side, both with star and lightning motif, eight thin cooling fins on barrel	1950s	30	50	75
Space Patrol Cosmic Smoke Gun	solid color red or green plastic with "Space Patrol" on body above grip, TV show tie-in, shoots baking powder	1950s	135	225	350
Space Patrol Hydrogen Ray Gun Ring	glow-in-the-dark cap firing ring	1950s	70	115	175
Space Patrol Rocket Gun and Holster Set	with darts	1950s	185	310	475
Space Patrol Rocket Gun Set	with darts, sold without holster	1950s	105	180	275
Space Pistol	plastic ray gun that shoots rubber tip darts	1954	20	35	50
Star Trek Phaser Flashlight	battery operated, small phaser shape	1976	8	13	20
Superior Rocket Gun	8" long, dark gray plastic, embossed "Superior Rocket Gun" on grip	1956	30	50	75
Tom Corbett Flash X-1 Space Gun	5" long	1950s	55	95	145
Tom Corbett Space Cadet Atomic Rifle	well-designed silver/gray finish long rifle with futuristic styling and embossed logo on stock and above front tommy gun-styled handgrip	1950s	135	225	350

Webb Electric

NAME	COMPANY/DESCRIPTION	YEAR	GOOD	EX	MIP
Atom Buster Mystery Gun	11" long yellow plastic gun with inner bladder, fires blast of air at tissue paper atomic mushroom target, with instructions, atomic explosion cover art on box	1950s	105	180	275

Wham-O

NAME	COMPANY/DESCRIPTION	YEAR	GOOD	EX	MIP
Wham-O Air Blaster	10" long plastic gun uses rubber diaphragm to shoot air; styling is reminiscent of Budson Flash Gordon Air Ray Gun	1960s	70	115	175

Space

NAME	COMPANY/DESCRIPTION	YEAR	GOOD	EX	MIP
	Wyandotte				
Pop Ray Gun	red pressed steel body with five widely spaced vertical round fins, unpainted trigger and muzzle with large gunsight, rod connects body to pop mechanism in muzzle	1930s	70	115	175
Ray Gun	7" stamped metal pop gun that uses a captive cork to make the pop, red body, unpainted muzzle, with connecting rod from body to barrel tip	1936	35	55	85
	Yoshiya/Japan				
Space Gun	7" long, tin with sparkling action, shows a realistic white rocket blasting off over lunar terrain on side of body and atomic symbol on grip center with diamond-shaped "SY" logo and "Made in Japan" at bottom of grip	1957	40	65	100

Western

NAME	COMPANY/DESCRIPTION	YEAR	GOOD	EX	MIP
	Actoy				
Pony Cap Pistol	single shot, all-metal, nickel finish with eagle on grip	1950s	25	45	65
Wells Fargo Buntline Cap Pistol	11" long barrel, break-to-front, cream plastic stag grips	1950s	85	135	195
Wyatt Earp Buntline Special	11" barrel, die cast, friction break-to-front, white plastic grips, nickel finish	1950s	95	135	175
	Buzz-Henry				
Lone Rider Cap Pistol	8" die cast, white plastic inset rearing horse grips	1950s	35	50	90
	Carnell				
Maverick Cap Pistol	9" break-to-front, lever release, nickel finish, Maverick on sides, cream & brown swirl colored grips features notch bar with extra set of black plastic grips	1960	40	85	135
Maverick Two Gun Holster Set	9" break-to-front, lever release, nickel finish, Maverick on sides, cream/brown swirl grips features notch bar, black leather dbl. holster set with silver plates, studs & white trim, six loops, buckle	1960	175	250	365
	Classy				
Dale Evans Holster Set	brown & yellow leather, white fringe on holsters, stylized blue butterflies are also "DE" logo, if buckled in front, holsters are backwards; holsters only, no guns		55	100	150
Double Holster Set	imitation alligator-texture brown leather, steer-head conches on holsters, lots of studs, yellow felt backing, holsters only, no guns	1950s	100	150	225
Rebel Holster & Pistol	12" die cast long barrel pistol, brown plastic grips, black leather single holster left side, Rebel insignia on holster flap,	1960s	195	350	550
Roy Rogers Double Gun & Holster Set	die cast two 8-1/2" nickel finish pistols with copper figural grips, holster is brown & black leather with raised detail, plastic play bullets & leather tie-downs	1950s	275	525	800

Detective Shell Firing Pistol, 1950s, Nichols

Flash-O-Matic, The Safe Gun, 1950s, Royal Plastics

Western

NAME	COMPANY/DESCRIPTION	YEAR	GOOD	EX	MIP
Classy					
Roy Rogers Double Holster Set	10" guns with plain nickel finish & copper grips, lever release, brown & cream leather set, silver studs, gold fleck jewels & four wooden bullets		225	425	650
Roy Rogers Double Holster Set	black & white leather set, silver studs & conches, 9" Roy Rogers pistols with plain nickel finish & copper figural grips, friction release	1950s	240	450	695
Daisy					
760 Rapid Fire Shotgun Air Rifle	31" pump shotgun, gray metal one piece frame, brown plastic stock & slider grip, fires blast of air	1960s	65	95	135
Red Ryder BB Rifle	carved wooden stock	1980	25	40	60
Spittin Image Peacemaker BB Pistol	10-1/2" die cast, spring fired, single action, BBs load into spring fed magazine under barrel		25	50	75
Edison					
Susanna 90 12 Shot Cap Pistol	9" uses ring caps, wind out cylinder, black finish, plastic wood grips	1980s	10	15	25
Esquire Novelty					
Authentic Derringer	die cast, 2" cap firing, copper finish, twin swivel barrel	1960	15	25	45
Authentic Derringer	classic miniature Series #10, 2" cap firing, copper finish. twin swivel barrel	1960	15	25	45
Johnny Ringo, Adventure of, Gun & Holster	10-3/4" long barrel Actoy, friction break, black/gold plastic stag grips, black leather two-gun holster, felt backing, loops hold four to six bullets, silver buckle	1960	200	350	485
Pony Boy Double Holster Set	brown leather double holster with bucking broncs & studs, cuffs, spurs & spur leathers, guns are Actoy "Spitfires", die cast 8-1/2" copper finish, white plastic grips	1950s	100	195	275
Haig					
Western Buntline Pistol	13", pistol fires single caps and/or BBs, BBs are propelled down barrel sleeve by cap explosion	1963	75	120	165
Hubley					
2 Guns in 1 Cap Pistol	die cast, 8" with long barrel, twist-off barrels to change from long to short, side loading, white plastic grips	1950s	75	125	200
Colt .38 Detective Special	4-1/2" Colt .38 pistol single shot caps & loads six play bullets with suspenders chest holster	1959	35	75	125
Colt .45 Cap Pistol	die cast, 13", revolving cylinder produced with open or closed chamber ends, loads six two-piece cap-firing bullets, white plastic grips, red felt box	1959	100	150	300
Cowboy Cap Pistol	12", die cast, swing-out revolving cylinder, release on barrel, nickel finish, black plastic steer grips	1950s	100	160	250
Cowboy Cap Pistol	8" friction break-to-front nickel finish cast iron, rose swirl plastic grips w/Colt logo	1940	85	150	200
Cowboy Cap Pistol	die cast, 12" swing out revolving cyclinder, release on barrel, nickel or aluminum finish, white plastic steer grips with black steer head	1950s	95	165	225
Cowboy Jr. Cap Pistol No. 225	9" die cast, revolving cylinder, side loading, release on barrel, silver finish, white plastic cow grips, lanyard ring & cord	1950s	65	125	185

Western

Hubley

NAME	COMPANY/DESCRIPTION	YEAR	GOOD	EX	MIP
Dagger Derringer	7", unusual over & under pistol has hidden red plastic dagger that slides out from between barrels, rotating barrels load & fire two-pc. bullets	1958	55	100	145
Davy Crockett Buffalo Rifle	25" die cast & plastic, unusual flintlock style, fires single cap under pan cover, brown plastic stock, ammo storage door in stock	1950s	75	145	200
Deputy Cap Pistol	10" die cast, front breaking, release on barrel, ornate scroll work, nickel finish	1950s	45	75	115
Flintlock Jr. Cap Pistol	7-1/2" single shot, double action, brown swirl plastic stock	1955	10	20	35
Flintlock Pistol	9-1/4", two shot cap shooting single action double barrel, over & under style, brown swirl plastic stock, nickel finish	1954	50	95	145
Frontier Repeating Cap Rifle	35-1/4" rifle nickel finish with brown plastic stock & forestock, blue metal barrel, red plastic choke & front sight, pop down magazine, released by catch in front of trigger, scroll	1950s	75	165	210
Lone Ranger Rifle	29" long	1973	55	100	150
Marshal Cap Pistol	9-3/4" die cast, side loading, nickel finish with scrollwork, with brown and white plastic stag grips with a clip on left grip	1960	35	65	95
Model 1860 Cal .44 Cap Pistol	13", revolving cylinder with closed chamber ends, six 2-pc. bullets, flat aluminum finish, white plastic grips, complete with wooden display plaque	1959	125	250	350
Panther Pistol	die cast, 4" derringer style pistol snaps out from secret spring-loaded wrist holster	1958	85	130	185
Remington .36 Cap Pistol	8" long, nickel finish, black plastic grips, revolving cylinder chambers two-piece bullets	1950s	75	145	225
Rex Trailer Two Gun & Holster Set	9-1/2" side loading, nickel finish, stag plastic grips, brown textured tooled leather with white holsters & trim, six bullet loops with plastic silver bullets, plain buckle	1960	90	165	250
Ric-O-Shay .45 Cap Pistol	13" die cast, nickel chrome finish, large frame, revolving cylinder, fires rolled caps, chambers brass bullets and fires caps, black plastic grips, makes a twang sound when fired	1959	85	125	175
Ric-O-Shay .45 Cap Pistol Holster Set	13", die cast, revolving cylinder swings out to chamber six brass bullets, six loaded in gun and 12 on holster, 1" flake in nickel finish at heel, holster black leather with separate belt, horse-head emblem, rawhide tiedown	200	350	500	
Rifleman Flip Special Cap Rifle	3' long rifle, resembles classic Winchester with ring lever, brown plastic stock, pop down cap magazine	1959	125	225	275
Rodeo Cap Pistol	7-1/2" single shot, white plastic steer grips	1950s	20	40	60
Roy Rogers Tiny Tots Double Holster Set			55	100	150
Texan .38 Cap Pistol	10" long, revolving cylinder gun chambers six solid brass bullets (caps go into cylinder first), top release front break automatically ejects shells, plastic steer grips	1950s	100	165	225
Texan Cap Pistol	die cast, nickel finish, white plastic steer grips, star logo on grips	1950s	75	145	200
Texan Cap Pistol	9-1/4" cast iron revolving cylinder lever release, white plastic steer grips, nickel finish, Colt rearing horse logo on grips	1940	90	175	275

Western

NAME	COMPANY/DESCRIPTION	YEAR	GOOD	EX	MIP
	Hubley				
Texan Cap Pistol No. 285	9-1/4", cast iron, revolving cylinder, lever release, white plastic steer grips, nickel finish with star logo in grip	1940	85	185	300
Texan Dummy Cap Pistol	9-1/4" revolving cylinder, lever release, white plastic steer grips, nickel finish, Colt rearing horse logo	1940	65	110	165
Texan Dummy Cap Pistol	9-1/4", revolving cylinder, lever release, white plastic steer grips, nickel finish, star logo on grip	1950s	90	165	225
Texan Jr. Cap Pistol	10", spring button release on side of cylinder, break-to-front, nickel finish white plastic grips with black steers	1950s	65	90	125
Texan Jr. Cap Pistol	die cast, release under cylinder, nickel finish, white plastic Longhorn grips	1954	50	75	110
Texan Jr. Gold Plated Cap Pistol	9", gold finish with black longhorn steer grips, break-to-front release from cylinder	1950s	85	130	185
Western Cap Pistol	9" die cast, friction break, nickel finish with white plastic steer grips with black steer	1950s	40	55	80
Wyatt Earp Double Holster Set	black & white leather holster with silk screened "Marshal Wyatt Earp" logo, two No. 247 Hubley Wyatt Earp Buntline Specials, 10-3/4" nickel finish, purple swirl grips	1950s	175	275	385
	Ideal				
Yo Gun	7-1/4" red plastic gun releases yellow plastic ball which snaps back when trigger is pulled, functions like a yo-yo	1960s	30	45	60
	John Henry Products				
Matt Dillon Marshal Set	gun & holster set with jail keys, handcuffs and badge		55	100	150
	Kenton				
Gene Autry Cap Pistol	6-1/2" cast iron, nickel finish, red plastic grips with etched signature (Best/Logan G3.2.1)	1940	125	175	250
Gene Autry Cap Pistol	8-3/8" cast iron, long barrel, dark gray gunmetal finish with white plastic grips with signature, (Best/Logan G3.1.1)	1939	175	250	350
Gene Autry Dummy Cap Pistol	cast iron, 8-3/8", long barrel, dark gray, gunmetal finish, white plastic grips	1939	70	110	185
Lawmaker Cap Pistol	8-3/8", break-to-front friction break, unusual dark gray gunmetal finish, white plastic raised grips	1941	100	175	250
	Kilgore				
Big Horn Cap Pistol	7" all metal revolving cylinder, break-to-front, disc caps, silver finish	1950s	85	125	175
Bronco Cap Pistol	8-1/2", revolving swing-out cylinder fires Kilgore disc caps, silver finish, black plastic "Bronco" grips	1950s	50	100	175
Champion Quick Draw Timer Cap Pistol	silver finish, side loading, wind up mechanism in grip records elapsed time of draw, black plastic grips	1959	65	150	225
Cheyenne Cap Pistol	9-3/4" side loading, "Sure-K" plastic stag grips, silver finish	1974	10	15	25
Fastest Gun Electronic Draw Game	die cast, wire plug into "Rangers" gun grips, gun that shoots first lights eye of plastic battery operated steer head, red & blue plastic holsters w/matching cowboy gun grips, plastic	1958	80	145	250

NAME	COMPANY/DESCRIPTION	YEAR	GOOD	EX	MIP
	Kilgore				
Grizzly Cap Pistol	10" revolving cylinder fires disc caps, swing out cylinder, black plastic grips with grizzly bear	1950s	95	165	230
Hawkeye Cap Pistol	4-1/4" all metal, automatic style, side loading, silver finish	1950s	20	35	55
Lone Ranger Cap Pistol	8-1/2" cast iron, large hammer, nickel finish, spring release on side for break, red-brown, Hi-Yo Silver grips	1940	85	145	225
Lone Ranger Cap Pistol	8-1/4" cast iron, small hammer, nickel finish, friction break, purple plastic "Hi-Yo Silver" grips	1938	95	175	275
Long Boy Cap Gun	11-1/2" long, cast iron		65	125	175
Mustang Cap Pistol	9-1/2" chrome finish with "stag" plastic grips	1960s	20	35	55
Ranger Cap Pistol	8-1/2" nickel finish, brown swirl plastic grips, spring release on right side, break-to-front	1940	90	125	165
Roy Rogers Cap Pistol	10" revolving cylinder swings out to load, fires disc caps, white plastic horse-head grips with "RR" logo	1950s	125	225	350
	Langson				
Cody Colt Paper Buster Gun	7-3/4", paper popper, nickel finish, white plastic steer grips fires Cody Colt ammunition	1950s	35	65	90
	Leslie-Henry				
Gene Autry 44 Cap Pistol	11" lever release, side loading, long barrel, nickel finish, white plastic horse-head grips	1950s	75	135	275
Gene Autry 44 Cap Pistol	11" lever release, side loading, long barrel, loads solid metal bullets, nickel finish, brown translucent plastic horse-head grips	1950s	85	175	285
Gene Autry Cap Pistol	9" break-to-front lever release, copper finish, white plastic horse-head grips	1950s	65	155	250
Gene Autry Cap Pistol	9", break-to-front lever release, nickel finish, white plastic horse-head grips	1950s	90	145	250
Gene Autry Cap Pistol	9" break-to-front, lever release, nickel finish, black plastic horse-head grips	1950s	65	145	250
Gene Autry Cap Pistol	7-3/4" die cast, small size, lever release, break-to-front, nickel finish with extension scroll work, black plastic horse-head grips	1950s	65	145	250
Gunsmoke Double Holster Set	with copper clad grips		200	350	500
Longhorn Cap Pistol	10" die cast, release in front of trigger guard, scroll work, white plastic horse head grips, unusual pop-up cap magazine	1950s	85	125	175
Marshal Cap Pistol	10" revolving cylinder chambers, Nichols-style bullets, white plastic grips with star ovals	1950s	35	75	110
Marshal Matt Dillon "Gunsmoke" Cap Pistol	10" pop-up cap magazine, release in front of trigger guard, scroll work, bronze steer-head grips	1950s	50	80	135
Maverick Derringer	3-1/4" with removeable cap-shooting bullets, tan vinyl holster with two bullets	1958	35	55	85
Ranger Cap Pistol	7-3/4" derringer with removeable cap, shooting bullets, tan vinyl holster with two bullets	1950s	85	110	150
Texas Cap Pistol	9" die cast, break-to-front, lever release, nickel finish, plastic horse-head grips	1950s	65	95	125
Texas Ranger Cap Pistol	8-1/4" die cast, lever release break to front, nickel finish, scroll work, vasoline colored plastic grips		65	95	125

NAME	COMPANY/DESCRIPTION	YEAR	GOOD	EX	MIP

Leslie-Henry

NAME	COMPANY/DESCRIPTION	YEAR	GOOD	EX	MIP
Wagon Train Complete Western Cowboy Outfit	plastic flip ring lever rifle & wagon train pistol (late model L-H pistol) & leather holster	1960	75	145	200
Wild Bill Hickok 44 Cap Pistol Set	11" nickel finish, swing-out side loading action, revolving cylinder chambers six metal bullets, amber plastic horse-head grips, single holster black & brown leather w/silver studs, diamond conches	1950s	200	325	475
Wild Bill Hickok Cap Pistol	10" pop-up cap magazine, release in front of trigger guard, scroll work, translucent brown plastic grips with oval star inserts	1950s	90	135	190
Wild Bill Hickok Gun & Holster	single gun & holster set		100	200	300
Young Buffalo Bill Cowboy Outfit	black & white leather holster set with pistol, white grip, holster bands read Texas Ranger		100	175	225

Lone Star

NAME	COMPANY/DESCRIPTION	YEAR	GOOD	EX	MIP
Gunfighter Holster Set	9" Frontier Ace, lever release, break-to-front, silver finish, brown plastic grips, white & red leather "Laramie" single holster with separate belt	1960s	55	85	145
Pecos Kid Cap Pistol	9" silver chrome finish, brown plastic grips, lever release	1970s	10	15	30
Pepperbox Derringer Cap Pistol	die cast, 6-1/4" rotating barrel holds four cap loads, silver finish with black plastic grips	1960	75	110	185

Long Island Die Casting

NAME	COMPANY/DESCRIPTION	YEAR	GOOD	EX	MIP
Texas Cap Pistol	die cast, 8-1/2" friction break, Circle "T" logo, scroll work on barrel	1950s	45	85	110

Marx

NAME	COMPANY/DESCRIPTION	YEAR	GOOD	EX	MIP
Bonanza Guns Outfit	25" cap firing saddle rifle, magazine pulls down to load, 9-1/2" Western pistol fires two-piece Marx shooting bullets, wood plastic stocks & gun metal gray plastic body, tan vinyl	1960s	75	145	225
Buffalo	50 shooter repeater		70	100	140
Centennial Rifle	with big sound		25	35	65
Cork-Shooting Rifle			25	40	60
Double Holster Set	10" two pistols similar to 1860s Remington, fires roll caps by use of a lanyard that is pulled from the bottom of the grip, internal hammer, white plastic horse & steer grips, silver, brown vinyl holster	1960	70	120	175
Double-Barrel Pop Gun Rifle	28" long	1935	50	85	125
Double-Barrel Pop Gun Rifle	22" long	1935	45	75	100
Hi-Yo Silver Lone Ranger Pistol	tin gun		45	65	90
Historic Guns Derringer	Marx Historic Guns series, derringer with plastic presentation case, 4-1/2" long, on card	1974	15	25	30
How the West Was Won Gun Rifle	deep gray Winchester model with tan stock, in box		55	85	145
Johnny Ringo Gun & Holster Set	die cast gun, white plastic head grips, vinyl quick draw holster has rawhide tie, gun is fired by lanyard which passes through grip butt & attaches to belt, when pulled lanyard trips internal hammer	1960	80	130	185

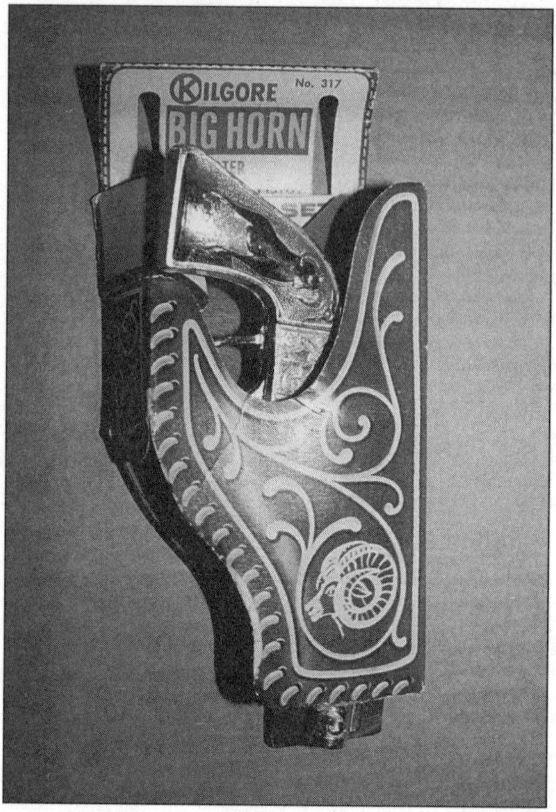

Big Horn Cap Pistol, 1950s, Kilgore

Wanted Dead or Alive Miniature Mare's Laig, 1959, Marx

Western

NAME	COMPANY/DESCRIPTION	YEAR	GOOD	EX	MIP
	Marx				
Lone Ranger 45 Flasher Flashlight Pistol			25	50	85
Lone Ranger Carbine	26" gray plastic repeater-style rifle has pull down cap magazine, western trim & Lone Ranger signature on stock	1950s	65	100	165
Lone Ranger Clicker Pistol	8", nickel finish, red jewels, inlaid white plastic grip with the Lone Ranger, Hi-Yo Silver & LR head embossed, brown leather holster	1938	60	100	150
Lone Ranger Double Target Set	9-1/2" square stand up target, tin litho, wire frame holds target upright, backed with bulls-eye target, 8" metal dart gun fires wooden shaft dart	1939	95	165	245
Lone Ranger Sparking Pop Pistol	tin litho		45	65	90
Mare's Laig Rifle Pistol	13-1/2" brown plastic, black plastic body, pull down magazine		60	95	135
Official Wanted Dead or Alive Mare's Laig Rifle	19" bullet loading, cap firing saddle rifle-pistol ejects plastic bullets, with holster		100	195	300
Ranch Rifle	plastic, repeater		30	45	60
Roy Rogers Carbine	26" gray plastic repeater-style rifle has pull down cap magazine, western trim & Roy Rogers signature on stock	1950s	80	120	155
Side-By Double Barrel Pop Gun Rifle	9" long		20	35	45
Tales of Wells Fargo Double Barrel Shotgun	26" long double barrel shotgun, two toy shotgun shells, decal on the butt of the gun	1950s	100	200	300
Thundergun Cap Pistol	12-1/2" single action,"Thundercaps" perforated roll cap system, silver finish, brown plastic grips	1950s	100	175	250
Wanted Dead or Alive Miniature Mare's Laig	Marx Miniatures series miniature cap rifle on "wood frame" card	1959	20	35	50
Wild West Rifle	30" long with sight, cap rifle		40	75	125
Zorro Flintock Pistol			30	45	75
Zorro Rifle			60	90	120
	Mattel				
Fanner 50 "Swivelshot Trick Holster" Set	die cast bullet loading Fanner 50, leather swivel style holster, attaches to any belt, gun fires in holster when swiveled, string included for last ditch draw	1958	85	165	250
Fanner 50 Cap Pistol	11" fanner non-revolving cylinder, stag plastic grips, nickel finish, black vinyl "Durahyde" holster	1960s	50	75	110
Fanner 50 Cap Pistol	later version 11" fires perforated roll caps, black finish, white plastic antelope grips	1960s	30	65	95
Fanner 50 Smoking Cap Pistol	10-1/2" with revolving cylinder, chambers six metal play bullets, die cast	1958	55	120	200
Fanner 50 Smoking Cap Pistol	10-1/2" with revolving cylinder, first version with grapefruit cylinder does not chamber bullets	1957	75	150	250
Shootin' Shell .45 Fanner Cap Pistol	11" revolving cylinder pistol shoots Mattel Shootin' Shell cartridges, shell ejector	1959	250	400	650
Shootin' Shell Buckle Gun	cap & bullet shooting copy of Remington Derringer pops out from belt buckle, two brass cartridges & six bullets	1958	45	80	125
Shootin' Shell Fanner	9" die cast chrome finish, revolving cylinder chambers six Shootin' Shell bullets	1958	65	125	225

Western

NAME	COMPANY/DESCRIPTION	YEAR	GOOD	EX	MIP
	Mattel				
Shootin' Shell Fanner & Derringer Set	small size Shooting Shell Fanner w/chrome finish, revolving cylinder, chambers six Shootin' Shell bullets, brown leather holster	1958	75	145	250
Shootin' Shell Fanner Single Holster	cowhide holster takes small size Shootin' Shell Fanner with six brass play bullets & tie-downs	1959	80	145	250
Shootin' Shell Indian Scout Rifle	29-1/2" plastic/metal Sharps rolling block rifle, chambers two-pc. Shootin' Shell bullets, secret compartment in stock for ammo storage, plastic stock & metal barrel	1958	100	145	250
Shootin' Shell Potshot Remington Derringer	3" derringer, on card	1959	35	60	90
Shootin' Shell Winchester Rifle	26" long		55	100	150
Showdown Set with Three Shootin' Shell Guns	30" single shot rifle with metal barrel, die cast w/ plastic stock, Shootin' Shell Fanner sm. size, revolving cylinder, chrome finish & imitation stag plastic grips, tan holster w/bullet loops	1958	300	600	950
Winchester Saddle Gun Rifle	33" die cast & plastic, perforated roll caps & chambers eight play bullets loaded through side door	1959	45	95	350
	National Metal & Plastic Toy				
The Plainsman Cap Pistol	10-1/2" revolving cylinder loads Kilgore-style disc caps, lever holds cylinder forward for loading, scroll work, white plastic grips	1950s	125	175	250
	Nichols				
Cap Gun Store Display	24" x 14" wood board, derringer & two strips of Nichols bullets	1950s	300	600	950
Dyna-Mite Derringer	3-1/4" die cast, loads single cap cartridge, silver finish, white plastic grips	1955	15	30	45
Dyna-Mite Derringer in Clip	3-1/2" die cast, fires single cap in Nichols cartridge, nickel finish, white plastic grips with small leather holster		25	45	65
Model 95 Shell Firing Rifle	35-1/2" rifle uses shell firing cartridges, holds five in removeable magazine & one chamber, lever action ejects cartridges, open frame box holds six bullets & 12 additional red bullet heads	1961	200	350	550
Pinto Cap Pistol	3-1/2", chrome finish, black plastic grips, flip out cylinder, white plastic "Pinto" holster in leather holster clip	1950s	20	35	50
Silver Pony Cap Pistol	7-1/2" single shot, silver metal grip & one replacement black plastic grip, silver finish	1950s	30	45	70
Spitfire Hip Gun No. 100	9" cap cartridge loading mini rifle, chrome finish, tan plastic stock	1950s	15	20	35
Spitfire with Clip	9" mini rifle, chrome finish, plastic stock, plastic holders with 2 extra cartridges	1950s	12	25	35
Stallion .22 Cap Pistol	7" revolving cylinder chambers five two-piece cartridges, single action, black plastic stag grips, never came in box	1950s	30	55	85
Stallion .22 Double Action Cap Pistol	7" double action, pull trigger to fire, white plastic grips, nickel finish, cylinder revolves	1950s	45	80	125
Stallion .38 Cap Pistol	9-1/2", chambers six two-pc. cap cartridges, nickel finish, white plastic grips	1950s	75	115	175
Stallion .45 MK I Cap Pistol	die cast, 12" chrome finish, revolving cylinder, chambers six two-pc. bullets, shell ejector, white "pearlescent" plastic grips with rearing stallion, red jewels & 6 bullets & Stallion caps	1950	100	250	375

NAME	COMPANY/DESCRIPTION	YEAR	GOOD	EX	MIP
	Nichols				
Stallion .45 MK II Cap Pistol	12" pistol, chrome finish, revolving cylinder, chambers six two-pc. bullets, shell ejector, extra set of white grips to replace black grips on gun & box of Stallion caps	1956	100	185	275
Stallion 32 Six Shooter	8" revolving cylinder chambers six two-pc. cartridges, nickel finish, black plastic grips	1955	75	110	165
Stallion 41-40 Cap Pistol	10-1/2" revolving cylinder chrome finish pistol, swing out cylinder that chambers six two-pc. cap cartridges, shell ejector, scroll work on frame, cream-purple swirl colored plastic	1950s	150	250	345
Tophand 250 Cap Pistol	9-1/2" break-to-front, lever release, black finish, brown plastic grips with a roll of "Tophand 250" caps	1960	45	85	135
	Ohio Art				
Sheriff's Derringer Pocket Pistol	3-1/4" silver finish derringer chambers two-pc., Nichols style cartridge, red plastic grips with an "A" logo, on card	1960s	10	20	25
	Pilgrim Leather				
Ruff Rider Western Holster Set	brown leather double holster, variety of studs & red jewels, 12 plastic silver bullets, tie-downs		100	150	200
	Product Engineering				
45 Smoker	10" single cap, shoots talcum-like powder by use of bellows when trigger is pulled, aluminum finish	1950s	45	75	115
Frontier Smoker	9-1/2" cap pistol, die cast, pop up magazine shoots white powder from internal bellows, all metal, black grips, silver finish, gold magazine, hammer & trigger		85	135	200
	Ralston-Purina				
Tom Mix Wooden Gun	three all-wood versions with leather holster, came in mailer, each	1930s	125	250	350
	Schmidt				
Buck 'n Bronc Marshal Cap Pistol	10" long barrel revolver style, lever release, break-to-front, plain silver finish, copper color metal grips	1950s	90	150	225
Hopalong Cassidy Cap Pistol	9" pull hammer to release, scroll work, nickel finish with black plastic grips with white bust of Hopalong Cassidy	1950	150	300	450
	Smart Style				
Real Texan Outfit with Nichols Stallion .22	brown/white leather double holsters have silver conches with red reflectors, silver horses at top of holster, belt with three bullet loops, guns are a pair of double action .22s	1950s	95	175	265
	Stevens				
49er Cap Pistol	cast iron, 9", unusual internal hammer with revolving steel cylinder, nickel finish, white plastic figural grips	1940	100	200	350
Billy The Kid Cap Pistol	8" long, cap pistol		55	100	150
Buffalo Bill Cap Pistol	7-3/4", silver nickel finish, side loading magazine door, white "tenite" plastic horse & cowboy grips, red jewels	1940	65	110	175

NAME	COMPANY/DESCRIPTION	YEAR	GOOD	EX	MIP
	Stevens				
Colt Cap Pistol	6-1/2", revolver style double action	1935	15	25	50
Cowboy Cap Pistol	3-1/2" cast iron, single shot single action, sold loose	1935	20	35	50
Cowboy King Cap Pistol	9" break-to-front release, gold finish cast iron, black plastic grips, yellow jewels	1940	85	175	225
	Topper/Deluxe				
Johnny Eagle Red River Bullet Firing	over 12" double action revolving cylinder pistol, die cast hammer, trigger, blue plastic overall with wood plastic grips with gold horse, side loading, shell ejector, fires two-piece plastic bullets	1965	65	95	150
	Unknown				
Davy Crockett Frontier Fighter Cork Gun	21" pop gun shoots cork on string & has cigarette flint mechanism at muzzle that makes sparks when fired, wood stock, leather sling	1950s	65	135	195
Gene Autry Champion Single Holster Set	leather & cardboard, red, yellow & green "jewels," four white wooden bullets, silver buckle	1940s	125	175	250
Lone Ranger Holster	9", leather/pressboard, Hi-Yo Silver & Lone Ranger printed, red jewel, belt loop		35	45	60
Tom Mix Gun & Holster Outfit	10" x 5" x 2" box, gun and holster unknown	1930s	55	85	125
Wagon Train Gun & Holster Set	two 5" plastic guns, vinyl holster with plastic bullets and metal badge	1950s	25	35	65
Wyatt Earp Double Holster Set	med. size, reflectors, black leather with brown rawhide fringe, holsters only	1950s	50	70	95
	USA				
Bobcat Saddle Gun	5-1/2" black finish, imprinted with "Official Wanted Dead or Alive "Mare's Laig" logo, brown plastic stock	1960s	10	20	30
	Wyandotte				
Red Ranger Jr. Cap Pistol	7-1/2" lever release, break-to-front, silver finish, white plastic horse grips	1950s	55	85	120
	Young Premiums				
Official Wyatt Earp Buntline Clicker Pistol	18-1/2" plastic		35	60	90

Lunch Boxes

Remember carrying your lunch to school in a colorful metal lunch box?

Lunch kits have been manufactured since the 1920s, but the boxes that are most collectible today are those bearing illustrations of popular licensed characters.

It took the power of television to launch the lunch box industry out of the domed steel domain of workmen into the colorful art boxes generations of children carried to school each day.

As World War II ended, Aladdin Industries returned to providing millions of workmen with sturdy, if uninspired, lunch kits designed to take the beating of the workplace. Great change occurred in 1950 when Aladdin released a pair of rectangular steel boxes, one red and one blue, sporting scalloped color decals of the TV western hero of the day, Hopalong Cassidy. Soon, 600,000 Hoppy boxes were being carried to school by proud young owners. The youth market had been found, and it would never be ignored again.

The envious classmates of those first Hoppy boxers would not be denied. American Thermos, Aladdin's chief competitor, went one up on Aladdin by introducing the 1953 Roy Rogers box in full-color lithography. Aladdin responded by issuing a new 1954 Hoppy box in full color litho, and the lunch box era officially began.

Throughout the latter 1950s, the box wars were fought in earnest between Aladdin and American Thermos, with occasional challenges by Adco Liberty, Ohio Art and Okay Industries.

The smaller firms produced some classic boxes, notably Mickey Mouse and Donald Duck (1954), Howdy Doody (1954) and Davy Crockett (1955) from Adco Liberty; and Captain Astro (1966), Bond XX (1967), Snow White (1980) and Pit Stop (1968) from Ohio Art. Okay Industries weighed in briefly later on with the now highly prized Wake Up America (1973) and Underdog (1974) boxes, but from the beginning it had always been a two-horse race.

The popular boxes of each year mirrored the stars, heroes, and interests of the times. From the Westerns and space explorations of the late 1950s through the 1960s, Americans enjoyed a golden age of cartoon and film heroes such as the Flintstones (1962), Dudley Do-Right (1962), Bullwinkle and Rocky (1962) and Mary Poppins (1965). As the decade progressed, America grew more aggressive, turning toward such violent heroes as the Man From U.N.C.L.E. (1966) and G.I. Joe (1967) before Vietnam changed the national consciousness.

The early 1970s brought us such innocuous role models as H.R. Pufnstuf (1970), The Partridge Family (1971) and Bobby Sherman (1972), and by decade's end we were greeting both the promise and the threat from beyond in Close Encounters (1978) and Star Wars (1978).

The metal box reigned supreme through the mid-1980s when parental groups began calling for a ban on metal boxes as "deadly weapons." The industry capitulated, and by 1986, both Aladdin and American Thermos were producing all their boxes in plastic.

The switch to plastic was not nearly as abrupt as might be expected. Aladdin and Thermos had been making plastic and vinyl boxes since the late 1950s. These included many character boxes that had no counterparts in metal, which is presently their major saving grace in the collector market.

Vinyl boxes were made of lower cost materials, consisting basically of cardboard sheathed in thin vinyl. They were not as popular as metal boxes, and their poor construction combined with lower unit sales have resulted in a field with higher rarity factors than the metal box arena. Additionally, vinyl was more affordable to small companies, which produced numerous limited-run boxes for sale or use as premiums.

Vinyl box collecting is an emerging field with few firmly established prices compared to the relative maturity of the metal box market, so any price guide such as this will be more open to debate. As the field matures, the pricing precedents of sales and time will

Munsters Thermos bottle, 1965, King Seeley Thermos

build into a stronger body of knowledge. In this book, for ease of searching, boxes are listed alphabetically by box composition — plastic, steel and vinyl.

Trends

Lunch boxes are still visible at toy and collectors' shows, but generalized dealers don't feature them quite as much. Specialized dealers still exist, however, dealing in good numbers. Character related boxes — from Superman to Western heroes to the Munsters — remain popular, rounding out character toy collections. Boxes aren't made like they used to be, and vintage ones hold a true nostalgia and make great display items.

Editor's note: A zero in the bottle column means the box did not come with a bottle.

The Top 10 Lunch Boxes
(in Near Mint condition)

1. 240 Robert, steel, Aladdin, 1978 .... $2,500
2. Toppie Elephant, steel, American Thermos, 1957 1,600
3. Home Town Airport Dome, steel, King Seeley Thermos, 1960 1,000
4. Underdog, steel, Okay Industries, 1974 900
5. Knight in Armor, steel, Universal, 1959 825
6. Ballerina, vinyl, Universal, 1960s 800
7. Superman, steel, Universal, 1954 800
8. Dudley Do-Right, steel, Universal, 1962 800
9. Bullwinkle & Rocky, steel, Universal, 1962 800
10. Little Friends, vinyl, Aladdin, 1982 760

LUNCH BOXES

Plastic

NAME, YEAR, COMPANY, DESCRIPTION	BOX NM	BOTTLE NM
101 Dalmatians, 1990, Aladdin	18	8
18 Wheeler, 1978, Aladdin	30	10
ALF, red plastic	20	0
Animalympics Dome, 1979, Thermos	35	10
Astronauts, 1986, Thermos	30	15
Atari Missile Command Dome, 1983, Aladdin	35	10
Back to School, 1980, Aladdin	60	20
Back to the Future, 1989, Thermos	30	12
Bang Bang, 1982, Thermos	45	0
Barbie with Hologram Mirror, 1990, Thermos	25	8
Batman (dark blue), 1989, Thermos	20	10
Batman (light blue), 1989, Thermos	40	10
Batman Returns, 1991, Thermos	15	5
Beach Bronto, 1984, Aladdin, no bottle	40	0
Beach Party (blue/pink), 1988, Deka, with generic plastic bottle	15	5
Bear with Heart (3-D), 1987, Servo	12	0
Beauty & the Beast, 1991, Aladdin	20	5
Bee Gees, 1978, Thermos	40	20
Beetlejuice, 1980, Thermos	10	4
Big Jim, 1976, Thermos	80	30
Bozostuffs, 1988, Deka	25	10
C.B. Bears, 1977, Thermos	20	0
Care Bears, 1986, Aladdin	10	5
Centurions, 1986, Thermos	15	8
Chiclets, 1987, Thermos, no bottle	40	0
Chipmunks, Alvin and the, 1983, Thermos	20	10
CHiPs, 1977, Thermos	45	15
Cinderella, 1992, Aladdin	25	10
Civil War, The, 1961, Universal, generic "Thermax" bottle	200	25
Colonial Bread Van, 1984, Moldmark Industries	60	20
Crestman Tubular!, 1980, Taiwan	50	20
Days of Thunder, 1988, Thermos	30	10
Deka 4 X 4, 1988, Deka, generic plastic bottle	25	5
Dick Tracy, 1989, Aladdin	20	10
Dino Riders, 1988, Aladdin	20	10
Dinobeasties, 1988, Thermos	15	0
Dinorocker with Radio & Headset, 1986, Fundes	45	0
Disney on Parade, 1970, Aladdin, plastic bottle, glass liner	30	15
Disney's Little Mermaid, 1989, Thermos, with generic plastic bottle	10	5
Duck Tales (4 X 4/Game), 1986, Aladdin	15	5
Dukes of Hazzard Dome, 1981, Aladdin	45	10
Dukes of Hazzard, 1981, Aladdin	45	10
Dune, 1984, Aladdin	45	20
Dunkin Munchkins, 1972, Thermos	25	15
Ecology Dome, 1980, Thermos	45	20
Ed Grimley, 1988, Aladdin	20	5
Entenmann's, 1989, Thermos	15	0
Ewoks, 1983, Thermos	20	5
Fame, 1972, Thermos	35	15
Fievel Goes West, 1991, Aladdin	10	4
Fire Engine Co. 7, 1985, D.A.S., with generic plastic bottle	30	5

Plastic

NAME, YEAR, COMPANY, DESCRIPTION	BOX NM	BOTTLE NM
Fisher-Price Mini Lunch Box, 1962, Fisher-Price, red with barnyard scenes,	20	5
Flash Gordon Dome, 1979, Aladdin	60	20
Flintstones Kids, 1987, Thermos	40	10
Flintstones, premium, Denny's Restaurants	30	0
Food Fighters, 1988, Aladdin	20	10
Fraggle Rock, 1987, Thermos	15	5
Frito Lay's, 1982, Thermos, no bottle	50	0
G.I. Joe (Space Mission), 1989, Aladdin	25	10
G.I. Joe, Live the Adventure, 1986, Aladdin	25	10
Garfield (food fight), 1979, Thermos	25	10
Garfield (lunch), 1977, Thermos	20	10
Geoffrey, 1981, Aladdin	30	10
Get Along Gang, 1983, Aladdin	10	5
Ghostbusters, 1986, Deka	15	5
Go Bots, 1984, Thermos	10	5
Golden Girls, 1984, Thermos	10	5
Goonies, 1985, Aladdin	20	5
Gumby, 1986, Thermos	60	20
Hot Wheels, 1984, Thermos	50	20
Howdy Doody Dome, 1977, Thermos	80	35
Incredible Hulk Dome, 1980, Aladdin	30	10
Inspector Gadget, 1983, Thermos	20	8
It's Not Just the Bus - Greyhound, 1980, Aladdin	60	20
Jabber Jaw, 1977, Thermos	50	20
Jetsons (paper picture), 1987, Servo	110	30
Jetsons 3-D, 1987, Servo	75	30
Jetsons, The Movie, 1990, Aladdin	30	15
Kermit the Frog, Lunch With, 1988, Thermos	18	5
Kermit's Frog Scout Van, 1989, Superseal, no bottle	15	0
Kool-Aid Man, 1986, Thermos	20	10
Lisa Frank, 1980, Thermos	10	5
Little Orphan Annie, 1973, Thermos	50	20
Looney Tunes Birthday Party, 1989, Thermos, blue or red	20	10
Looney Tunes Dancing, 1977, Thermos	20	10
Looney Tunes Playing Drums, 1978, Thermos	20	10
Looney Tunes Tasmanian Devil, 1988, Thermos, with generic plastic bottle	15	10
Los Angeles Olympics, 1984, Aladdin	20	5
Lucy's Luncheonette, 1981, Thermos, Peanuts characters	15	5
Lunch `N Tunes Safari, 1986, Fun Design, with built-in radio, no bottle	35	0
Lunch `N Tunes Singing Sandwich, 1986, Fun Design, with built-in radio, no bottle	35	0
Lunch Man with Radio, 1986, Fun Design, with built-in radio, no bottle	35	0
Lunch Time with Snoopy Dome, 1981, Thermos	15	5
Mad Balls, 1986, Aladdin	25	10
Marvel Super Heroes, 1990, Thermos	20	10
Max Headroom (Coca-Cola), 1985, Aladdin	50	25
McDonald's Happy Meal, 1986, Fisher-Price	15	0
Menudo, 1984, Thermos	12	5
Mickey & Minnie Mouse in Pink Car, 1988, Aladdin	10	5
Mickey Mouse & Donald Duck See-Saw, 1986, Aladdin	10	5
Mickey Mouse & Donald Duck, 1984, Aladdin	10	5
Mickey Mouse at City Zoo, 1985, Aladdin	10	5
Mickey Mouse Head, 1989, Aladdin	20	5
Mickey on Swinging Bridge, 1987, Aladdin	10	5
Mickey Skateboarding, 1980, Aladdin	18	5

Jabber Jaw plastic box, 1977, Thermos

Geoffrey plastic box, 1981, Aladdin

The Pussycats vinyl box, 1968, Aladdin

Donny & Marie, 1978, Aladdin, short hair version

Charlie's Angels Brunch Bag, 1978, Aladdin

Plastic

NAME, YEAR, COMPANY, DESCRIPTION	BOX NM	BOTTLE NM
Mighty Mouse, 1979, Thermos	35	10
Miss Piggy's Safari Van, 1989, Superseal, no bottle	15	0
Monster in My Pocket, 1990, Aladdin	30	5
Movie Monsters, 1979, Universal	35	12
Mr. T, 1984, Aladdin	20	10
Munchie Tunes Bear with Radio, 1986, Fun Design, with built-in radio	35	5
Munchie Tunes Punchie Pup w/Radio, 1986, Fun Design, with built-in radio	35	5
Munchie Tunes Robot with Radio, 1986, Fun Design, with built-in radio	35	5
Muppets (blue), 1982, Thermos	12	5
Muppets Dome, 1981, Thermos, plastic red box with matching bottle	20	5
New Kids on the Block (pink/orange), 1990, Thermos	10	5
Nosy Bears, 1988, Aladdin	12	5
Official Lunch Football, 1974, football shaped box, red or brown	100	0
Peanuts, Wienie Roast, 1985, Thermos	10	4
Pee Wee's Playhouse, 1987, Thermos, with generic plastic bottle	20	5
Peter Pan Peanut Butter, 1984, Taiwan	85	20
Pickle, 1972, Fesco, no bottle	140	0
Popeye & Son 3-D, 1987, Servo, plastic box, red or yellow, with matching bottle	50	12
Popeye & Son, 1987, Servo, plastic red box, flat paper label, with matching bottle	65	12
Popeye Dome, 1979, Aladdin	35	15
Popeye, Truant Officer, 1964, King Seeley Thermos, plastic red box, matching metal bottle (Canada)	150	35
Punky Brewster, 1984, Deka	20	10
Q-Bert, 1983, Thermos	15	12
Race Cars, 1987, Servo	20	0
Raggedy Ann & Andy, 1988, Aladdin	45	20
Rainbow Bread Van, 1984, Moldmark Industries	60	20
Rainbow Brite, 1983, Thermos	10	5
Robot Man and Friends, 1984, Thermos	20	10
Rocketeer, 1990, Aladdin	10	5
Rocky Roughneck, 1977, Thermos	25	10
Roller Games, 1989, Thermos	25	10
S.W.A.T. Dome, 1975, Thermos	45	15
Scooby Doo, 1973, Thermos	30	20
Scooby Doo, 1984, Aladdin	40	20
Scooby-Doo, A Pup Named, 1988, Aladdin	20	10
Sesame Street, 1985, Aladdin/Canada	10	5
Shirt Tales, 1981, Thermos	10	5
Sky Commanders, 1987, Thermos, generic plastic bottle	15	5
Smurfette, 1984, Thermos	10	5
Smurfs Dome, 1981, Thermos	20	5
Smurfs Fishing, 1984, Thermos	15	5
Smurfs, 1984, Thermos	15	5
Snak Shot Camera, 1987, Hummer, camera-shaped box, blue or green, with generic plastic bottle	30	2
Snoopy Dome, 1978, Thermos	20	5
Snorks, 1984, Thermos	12	5
Snow White, 1980, Aladdin	40	15
Spare Parts, 1982, Aladdin, with generic plastic bottle	35	10
Sport Billy, 1982, Thermos	20	10
Sport Goofy, 1986, Aladdin	30	10
Star Com. U.S. Space Force, 1987, Thermos	20	10
Star Trek Next Generation, 1988, Thermos, blue box, group picture, matching	35	10
Star Trek Next Generation, 1989, Thermos, red box, Picard, Data, Wesley,	50	20
Star Wars, Droids, 1985, Thermos	30	10

Plastic

NAME, YEAR, COMPANY, DESCRIPTION	BOX NM	BOTTLE NM
Strawberry Shortcake, 1980, Aladdin	10	5
Superman II Dome, 1986, Aladdin	40	20
Superman, This is a Job For, 1980, Aladdin, no bottle	25	0
Tail Spin, 1986, Aladdin	10	5
Tang Trio, 1988, Thermos, red or yellow box with generic plastic bottle	35	5
Teenage Mutant Ninja Turtles, 1990, Thermos, with generic plastic bottle	12	5
Thundarr the Barbarian Dome, 1981, Aladdin, plastic dome box with matching	25	10
Timeless Tales, 1989, Aladdin	10	5
Tiny Toon Adventures, 1990, Thermos	10	5
Tom & Jerry, 1989, Aladdin	30	10
Transformers Dome, 1986, Aladdin/Canada, dome box, generic plastic bottle	35	8
Transformers, 1985, Aladdin	15	5
Tweety & Sylvester, 1986, Thermos	45	20
Wayne Gretzky Dome, 1980, Aladdin	120	30
Wayne Gretzky, 1980, Aladdin	100	30
Where's Waldo, 1990, Thermos	10	5
Who Framed Roger Rabbit, 1987, Thermos, red or yellow, with matching bottle	20	10
Wild Fire, 1986, Aladdin	15	8
Wizard of Oz, 50th Anniversary, 1989, Aladdin	60	20
Woody Woodpecker, 1972, Aladdin, yellow box, red bottle	50	40
World Wrestling Federation, 1986, Thermos	10	5
Wrinkles, 1984, Thermos	10	5
Wuzzles, 1985, Aladdin	10	5
Yogi's Treasure Hunt 3-D, 1987, Servo, 3-D box, green or pink, with matching	55	30
Yogi's Treasure Hunt, 1987, Servo, flat paper label, with matching bottle	25	30

Steel

240 Robert, 1978, Aladdin	2500	300
A-Team, 1985, King Seeley Thermos, plastic bottle	20	15
Action Jackson, 1973, Okay Industries, matching steel bottle	600	200
Adam-12, 1973, Aladdin, matching plastic bottle	60	20
Addams Family, 1974, King Seeley Thermos, matching plastic bottle	100	30
Airline, 1969, Ohio Art, no bottle	80	0
All American, 1954, Universal, steel/glass bottle	350	65
America on Parade, 1976, Aladdin, matching plastic bottle	40	20
Americana, 1958, King Seeley Thermos, steel/glass bottle	325	125
Animal Friends, 1978, Ohio Art, yellow or red background behind name	35	0
Annie Oakley & Tagg, 1955, Aladdin, matching steel bottle	300	110
Annie, The Movie, 1982, Aladdin, plastic bottle	25	15
Apple's Way, 1975, King Seeley Thermos, plastic bottle	75	20
Archies, 1969, Aladdin, matching plastic bottle	70	30
Astronaut Dome, 1960, King Seeley Thermos, steel/glass bottle	250	60
Astronauts, 1969, Aladdin, matching plastic bottle	65	40
Atom Ant/Secret Squirrel, 1966, King Seeley Thermos, matching steel bottle	200	110
Auto Race, 1967, King Seeley Thermos, matching steel bottle	60	30
Back in `76, 1975, Aladdin, plastic bottle	55	25
Barbie Lunch Kit, 1962, King Seeley Thermos, tall steel/glass bottle	250	90
Basketweave, 1968, Ohio Art, no bottle	60	0
Batman, 1966, Aladdin, matching steel bottle	150	80
Battle Kit, 1965, King Seeley Thermos, matching steel bottle	85	50
Battle of the Planets, 1979, King Seeley Thermos, matching plastic bottle	45	25
Battlestar Galactica, 1978, Aladdin, matching plastic bottle	45	15
Beatles, 1966, Aladdin, blue, matching bottle	400	150

Gunsmoke, Marshal Matt Dillon steel box, 1962, Aladdin

Drag Strip steel box, 1975, Aladdin

Steel boxes: Disney's Magic Kingdom, 1980, Aladdin; Annie, The Movie, 1982 Aladdin; Dukes of Hazzard, 1983, Aladdin

Steel

NAME, YEAR, COMPANY, DESCRIPTION	BOX NM	BOTTLE NM
Bedknobs & Broomsticks, 1972, Aladdin, plastic bottle	40	25
Bee Gees, 1978, King Seeley Thermos, Barry on back, matching plastic bottle	35	15
Bee Gees, 1978, King Seeley Thermos, Maurice on back, matching plastic bottle	35	15
Bee Gees, 1978, King Seeley Thermos, Robin on back, matching plastic bottle	35	15
Berenstain Bears, 1983, American Thermos, matching plastic bottle	35	10
Beverly Hillbillies, 1963, Aladdin, matching steel bottle	180	80
Bionic Woman, with Car, 1977, Aladdin, plastic bottle	35	20
Bionic Woman, with Dog, 1978, Aladdin, matching plastic bottle	35	25
Black Hole, 1979, Aladdin, matching plastic bottle	60	30
Blondie, 1969, King Seeley Thermos, matching steel bottle	150	50
Boating, 1959, American Thermos, matching steel bottle	400	125
Bobby Sherman, 1972, King Seeley Thermos, matching steel bottle	75	50
Bonanza, 1963, Aladdin, green rim box, steel bottle	140	60
Bonanza, 1965, Aladdin, brown rim box, steel bottle	120	60
Bonanza, 1968, Aladdin, black rim box, steel bottle	160	90
Bond XX, 1967, Ohio Art, no bottle	150	0
Boston Bruins, 1973, Okay Industries, steel/glass bottle	525	250
Bozo the Clown Dome, Aladdin, steel bottle	280	100
Brady Bunch, 1970, King Seeley Thermos, matching steel bottle	250	75
Brave Eagle, 1957, American Thermos, red, blue, gray or green band, matching	220	120
Bread Box Dome, 1968, Aladdin, Campbell's Soup bottle	250	160
Buccaneer Dome, 1957, Aladdin, matching bottle	200	125
Buck Rogers, 1979, Aladdin, matching plastic bottle	35	20
Bugaloos, 1971, Aladdin, matching plastic bottle	125	45
Bullwinkle & Rocky, 1962, Universal, blue box, steel bottle	800	220
Cabbage Patch Kids, 1984, King Seeley Thermos, matching plastic bottle	15	5
Cable Car Dome, 1962, Aladdin, steel/glass bottle	600	125
Campbell's Kids, 1973, Okay, matching steel bottle	180	100
Campus Queen, 1967, King Seeley Thermos, matching steel bottle	50	25
Canadian Pacific Railroad, 1970, Ohio Art, no bottle	60	0
Captain Astro, 1966, Ohio Art, no bottle	325	0
Care Bear Cousins, 1985, Aladdin, matching plastic bottle	10	5
Care Bears, 1984, Aladdin, plastic bottle	10	5
Carnival, 1959, Universal, matching steel bottle	550	250
Cartoon Zoo Lunch Chest, 1962, Universal, steel/glass bottle	325	125
Casey Jones, 1960, Universal, steel dome box, steel/glass bottle	650	125
Chan Clan, The, 1973, King Seeley Thermos, plastic bottle	110	35
Charlie's Angels, 1978, Aladdin, matching plastic bottle	35	10
Chavo, 1979, Aladdin, matching plastic bottle	140	50
Children's, 1984, Ohio Art, no bottle	60	0
Children, Blue, 1974, Okay Industries, plastic bottle	160	40
Children, Yellow, 1974, Okay Industries, plastic bottle	210	40
Chitty Chitty Bang Bang, 1969, King Seeley Thermos, matching steel bottle	125	60
Chuck Wagon Dome, 1958, Aladdin, matching bottle	180	90
Circus Wagon Dome, 1958, King Seeley Thermos, steel/glass bottle	350	150
Clash of the Titans, 1981, King Seeley Thermos, matching plastic bottle	30	12
Close Encounters of the Third Kind, 1978, King Seeley Thermos, plastic bottle	80	20
Color Me Happy, 1984, Ohio Art, no bottle	110	0
Corsage, 1958, American Thermos, matching steel bottle	50	20
Cowboy in Africa, Chuck Connors, 1968, King Seeley Thermos, matching steel	190	75
Cracker Jack, 1969, Aladdin, matching plastic bottle	55	20
Curiosity Shop, 1972, King Seeley Thermos, matching steel bottle	45	30
Cyclist Dirt Bike, 1979, Aladdin, plastic bottle	40	20
Daniel Boone, 1955, Aladdin, matching steel bottle	350	110

Steel

NAME, YEAR, COMPANY, DESCRIPTION	BOX NM	BOTTLE NM
Daniel Boone, 1965, Aladdin, matching steel bottle	140	90
Dark Crystal, 1982, King Seeley Thermos, matching plastic bottle	20	10
Davy Crockett, 1955, Holtemp, matching steel bottle	140	75
Davy Crockett, 1955, Kruger, no bottle	350	0
Davy Crockett/Kit Carson, 1955, Adco Liberty	200	0
Debutante, 1958, Aladdin, matching steel bottle	110	75
Denim Diner Dome, 1975, Aladdin, matching plastic bottle	60	20
Dick Tracy, 1967, Aladdin, matching steel bottle	150	80
Disco Fever, 1980, Aladdin, matching plastic bottle	55	15
Disco, 1979, Aladdin, matching plastic bottle	50	15
Disney Express, 1979, Aladdin, matching plastic bottle	10	5
Disney Fire Fighters Dome, 1974, Aladdin, matching plastic bottle	150	60
Disney School Bus Dome, 1968, Aladdin, steel/glass bottle	75	30
Disney World, 1972, Aladdin, matching plastic bottle	30	10
Disney's Magic Kingdom, 1980, Aladdin, plastic bottle	15	10
Disney's Rescuers, The, 1977, Aladdin, plastic bottle	35	20
Disney's Robin Hood, 1974, Aladdin, plastic bottle	55	25
Disney, Wonderful World of, 1982, Aladdin, plastic bottle	15	10
Disneyland (Castle), 1957, Aladdin, matching steel bottle	160	115
Disneyland (Monorail), 1968, Aladdin, matching steel bottle	200	115
Donald Duck, 1980, Cheinco, no bottle	30	0
Double Decker, 1970, Aladdin, matching plastic bottle	60	40
Dr. Dolittle, Aladdin, steel/glass bottle	100	50
Dr. Seuss, 1970, Aladdin, matching plastic bottle	150	50
Drag Strip, 1975, Aladdin, matching plastic bottle	45	25
Dragon's Lair, 1983, Aladdin, matching plastic bottle	25	15
Duchess, 1960, Aladdin, steel/glass bottle	85	30
Dudley Do-Right, 1962, Universal, matching steel bottle	800	350
Dukes of Hazzard, 1983, Aladdin, matching plastic bottle	45	20
Dutch Cottage Dome, 1958, King Seeley Thermos, steel/glass bottle	450	150
Dyno Mutt, 1977, King Seeley Thermos, plastic bottle	45	20
E.T., The Extra-Terrestrial, 1982, Aladdin, matching plastic bottle	30	10
Early West Indian Territory, 1982, Ohio Art, no bottle	60	0
Early West Oregon Trail, 1982, Ohio Art, no bottle	60	0
Early West Pony Express, 1982, Ohio Art, no bottle	60	0
Emergency! Dome, 1977, Aladdin, plastic bottle	160	30
Emergency!, 1973, Aladdin, plastic bottle	50	30
Evel Knievel, 1974, Aladdin, plastic bottle	60	25
Exciting World of Metrics, The, 1976, King Seeley Thermos, plastic bottle	40	25
Fall Guy, 1981, Aladdin, matching plastic bottle	20	15
Family Affair, 1969, King Seeley Thermos, matching steel bottle	70	30
Fat Albert and the Cosby Kids, 1973, King Seeley Thermos, plastic bottle	45	20
Fess Parker, 1965, King Seeley Thermos, matching steel bottle	160	90
Fireball XL5, 1964, King Seeley Thermos, steel/glass bottle	185	85
Firehouse Dome, 1959, American Thermos, steel/glass bottle	350	150
Flag-O-Rama, 1954, Universal, steel/glass bottle	475	110
Flintstones, 1962, Aladdin, orange, 1st issue, matching bottle	150	80
Flintstones, 1963, Aladdin, yellow, 2nd issue, matching bottle	160	80
Flintstones, 1973, Aladdin, matching plastic bottle	130	50
Flipper, 1966, King Seeley Thermos, matching steel bottle	160	75
Floral, 1970, Ohio Art, no bottle	40	0
Flying Nun, 1968, Aladdin, matching steel bottle	150	80
Fonz, The, 1978, King Seeley Thermos, plastic bottle	40	20
Fox and the Hound, 1981, Aladdin, plastic bottle	30	10

Batman, 1966, Aladdin, steel Thermos bottle

Cowboy in Africa, Chuck Connors steel box, 1968, King Seeley Thermos

Steel

NAME, YEAR, COMPANY, DESCRIPTION	BOX NM	BOTTLE NM
Fraggle Rock, 1984, King Seeley Thermos, matching plastic bottle	15	5
Fritos, 1975, King Seeley Thermos, generic bottle	90	5
Frontier Days, 1957, Ohio Art, no bottle	210	0
Frost Flowers, 1962, Ohio Art, no bottle	70	0
Fruit Basket, 1975, Ohio Art, no bottle	35	0
Funtastic World of Hanna-Barbera, 1977, King Seeley Thermos, Huck Hound, plastic bottle	65	30
Funtastic World of Hanna-Barbera, 1978, King Seeley Thermos, Flintstones & Yogi, plastic bottle	75	30
G.I. Joe, 1967, King Seeley Thermos, steel/glass bottle	100	60
G.I. Joe, 1982, King Seeley Thermos, plastic bottle	25	15
Gene Autry, 1954, Universal, steel/glass bottle	425	125
Gentle Ben, 1968, Aladdin, plastic bottle, glass liner	95	30
Get Smart, 1966, King Seeley Thermos, steel/glass bottle	175	85
Ghostland, 1977, Ohio Art, spinner game, no bottle	35	0
Globe-Trotter Dome, 1959, Aladdin, steel dome box, matching steel/glass bottle	240	120
Gomer Pyle USMC, 1966, Aladdin, matching steel bottle	145	90
Goober and the Ghostchasers / Inch High, 1974, King Seeley Thermos, matching plastic bottle	35	15
Great Wild West, 1959, Universal, matching steel bottle	425	180
Green Hornet, 1967, King Seeley Thermos, matching steel bottle	375	150
Gremlins, 1984, Aladdin, matching plastic bottle	20	5
Grizzly Adams Dome, 1977, Aladdin, plastic bottle	80	25
Guns of Will Sonnett, The, 1968, King Seeley Thermos, steel/glass bottle	160	90
Gunsmoke, 1959, Aladdin, plastic bottle	160	80
Gunsmoke, 1972, Aladdin, mule splashing box with matching bottle	130	55
Gunsmoke, 1973, Aladdin, stagecoach box, matching bottle	130	55
Gunsmoke, Double L Version, 1959, Aladdin, double L error version, matching	550	80
Gunsmoke, Marshal Matt Dillon, 1962, Aladdin, matching steel bottle	180	75
H.R. Pufnstuf, 1970, Aladdin, matching plastic bottle	100	50
Hair Bear Bunch, The, 1972, King Seeley Thermos, plastic bottle	45	35
Hansel and Gretel, 1982, Ohio Art, no bottle	80	0
Happy Days, 1977, American Thermos, matching plastic bottle	40	20
Hardy Boys Mysteries, 1977, King Seeley Thermos, matching plastic bottle	40	30
Harlem Globetrotters, 1971, King Seeley Thermos, steel bottle, blue or purple	50	35
Have Gun, Will Travel, 1960, Aladdin, matching bottle	250	150
He-Man & Masters of the Universe, 1984, Aladdin, matching plastic bottle	5	5
Heathcliff, 1982, Aladdin, matching plastic bottle	20	10
Hector Heathcote, 1964, Aladdin, matching steel bottle	250	90
Hee Haw, 1971, King Seeley Thermos, matching steel bottle	70	50
Highway Signs, 1972, Ohio Art, no bottle	70	0
Hogan's Heroes Dome, 1966, Aladdin, steel/glass bottle	280	110
Holly Hobbie, 1968, Aladdin, red rim, matching plastic bottle	15	5
Holly Hobbie, 1973, Aladdin, matching plastic bottle	15	10
Holly Hobbie, 1979, Aladdin, matching plastic bottle	15	5
Home Town Airport Dome, 1960, King Seeley Thermos, steel/glass bottle	1000	275
Hong Kong Phooey, 1975, King Seeley Thermos, steel/glass bottle	40	15
Hopalong Cassidy, 1950, Aladdin, red or blue, steel/glass bottle	160	75
Hopalong Cassidy, 1952, Aladdin, full litho, matching steel bottle	210	70
Hopalong Cassidy, 1954, Aladdin, black rim, steel/glass bottle	250	125
Hot Wheels, 1969, King Seeley Thermos, matching steel bottle	70	25
How the West Was Won, 1979, King Seeley Thermos, matching plastic bottle	45	35
Howdy Doody, 1954, Adco Liberty	500	0
Huckleberry Hound, 1961, Aladdin, steel/glass bottle	175	60
Incredible Hulk, The, 1978, Aladdin, plastic bottle	30	10

Steel

NAME, YEAR, COMPANY, DESCRIPTION	BOX NM	BOTTLE NM
Indiana Jones Temple of Doom, 1984, King Seeley Thermos, matching plastic	20	15
Indiana Jones, 1984, King Seeley Thermos, matching plastic bottle	20	15
It's About Time Dome, 1967, Aladdin, matching bottle	220	125
Jack and Jill, 1982, Ohio Art	400	0
James Bond 007, 1966, Aladdin, matching steel bottle	260	135
Jet Patrol, 1957, Aladdin, matching steel bottle	360	150
Jetsons Dome, 1963, Aladdin, matching bottle	675	175
Joe Palooka, 1949, Continental Can, no bottle	90	0
Johnny Lightning, 1970, Aladdin, plastic bottle	75	30
Jonathan Livingston Seagull, 1973, Aladdin, matching plastic bottle	50	25
Julia, 1969, King Seeley Thermos, matching steel bottle	110	45
Jungle Book, 1968, Aladdin, matching steel bottle	65	60
Junior Miss, 1978, Aladdin, matching plastic bottle	30	20
Kellogg's, 1969, Aladdin, plastic bottle	150	60
King Kong, 1977, King Seeley Thermos, plastic bottle	55	25
KISS, 1977, King Seeley Thermos, plastic bottle	100	30
Knight in Armor, 1959, Universal, matching steel bottle	825	250
Knight Rider, 1984, King Seeley Thermos, matching plastic bottle	20	10
Korg, 1975, King Seeley Thermos, matching plastic bottle	50	30
Krofft Supershow, 1976, Aladdin, matching plastic bottle	85	35
Kung Fu, 1974, King Seeley Thermos, matching plastic bottle	65	25
Lance Link, Secret Chimp, 1971, King Seeley Thermos, matching steel bottle	110	65
Land of the Giants, 1968, Aladdin, plastic bottle	150	50
Land of the Lost, 1975, Aladdin, matching plastic bottle	80	35
Laugh-In (Helmet), 1969, Aladdin, helmet on back, matching plastic bottle	90	40
Laugh-In (Tricycle), 1969, Aladdin, trike on back, matching plastic bottle	140	40
Lawman, 1961, King Seeley Thermos, generic bottle	140	60
Legend of the Lone Ranger, 1980, Aladdin, plastic bottle	45	20
Lidsville, 1971, Aladdin, matching plastic bottle	90	45
Little Dutch Miss, 1959, Universal, matching steel bottle	110	60
Little Friends, 1982, Aladdin, matching plastic bottle	760	260
Little House on the Prairie, 1978, King Seeley Thermos, matching plastic bottle	80	35
Little Red Riding Hood, 1982, Ohio Art, no bottle	25	0
Lone Ranger, 1955, Adco Liberty, red rim, blue band, no bottle	450	0
Looney Tunes TV Set, 1959, King Seeley Thermos, steel/glass bottle	220	120
Lost in Space Dome, 1967, King Seeley Thermos, steel/glass bottle	600	60
Ludwig Von Drake, 1962, Aladdin, steel/glass bottle	190	90
Luggage Plaid, 1955, Adco Liberty, no bottle	75	0
Luggage Plaid, 1957, Ohio Art, no bottle	50	22
Magic of Lassie, 1978, King Seeley Thermos, matching plastic bottle	65	30
Major League Baseball, 1968, King Seeley Thermos, matching bottle	70	30
Man from U.N.C.L.E., 1966, King Seeley Thermos, matching steel bottle	140	90
Marvel Super Heroes, 1976, Aladdin, black rim, matching plastic bottle	35	5
Mary Poppins, 1965, Aladdin, steel/glass bottle	75	50
Masters of the Universe, 1983, Aladdin, matching plastic bottle	10	5
Mickey Mouse & Donald Duck, 1954, Adco Liberty, matching steel bottle	280	200
Mickey Mouse Club, 1963, Aladdin, yellow, steel/glass bottle	65	30
Mickey Mouse Club, 1976, Aladdin, white, matching steel bottle	65	30
Mickey Mouse Club, 1977, Aladdin, red rim, sky boat, matching bottle	35	15
Miss America, 1972, Aladdin, matching plastic bottle	55	30
Mod Floral Dome, 1975, Okay Industries, matching steel bottle	250	220
Monroes, 1967, Aladdin, matching steel bottle	160	110
Mork & Mindy, 1979, American Thermos, matching plastic bottle	25	15
Mr. Merlin, 1982, King Seeley Thermos, matching plastic bottle	25	10

Wild Bill Hickok steel box, 1955, Aladdin

Robin Hood steel box, 1956, Aladdin

Munsters steel box, 1956, King Seeley Thermos

Flipper steel box, 1966, King Seeley Thermos

Steel

NAME, YEAR, COMPANY, DESCRIPTION	BOX NM	BOTTLE NM
Munsters, 1965, King Seeley Thermos, matching steel bottle	275	125
Muppet Babies, 1985, King Seeley Thermos, matching plastic bottle	12	5
Muppet Movie, 1979, King Seeley Thermos, plastic bottle	30	10
Muppet Show, 1978, King Seeley Thermos, plastic bottle	20	10
Muppets, 1979, King Seeley Thermos, back shows Animal, Fozzie or Kermit, matching plastic bottle	20	10
My Lunch, 1976, Ohio Art, no bottle	30	0
Nancy Drew, 1978, King Seeley Thermos, plastic bottle	40	20
NFL Quarterback, 1964, Aladdin, matching steel bottle	150	60
NFL, 1972, Okay, black rim, steel/glass bottle	160	130
NFL, 1975, King Seeley Thermos, yellow rim, plastic bottle	30	15
NFL, 1976, King Seeley Thermos, red rim, matching plastic bottle	30	15
NFL, 1978, King Seeley Thermos, blue rim, matching plastic bottle	25	15
NHL, 1970, Okay Industries, plastic bottle	525	225
Orbit, 1963, King Seeley Thermos, matching steel bottle	250	90
Osmonds, The, 1973, Aladdin, matching plastic bottle	70	30
Our Friends, 1982, Aladdin, matching plastic bottle	575	180
Pac-Man, 1980, Aladdin, matching plastic bottle	10	5
Para-Medic, 1978, Ohio Art, no bottle	65	0
Partridge Family, 1971, King Seeley Thermos, plastic or steel bottle	50	30
Pathfinder, 1959, Universal, matching steel bottle	525	180
Patriotic, 1974, Ohio Art, no bottle	45	0
Peanuts, 1966, King Seeley Thermos, orange rim, matching steel bottle	35	15
Peanuts, 1973, King Seeley Thermos, red rim psychiatric box, plastic bottle	40	10
Peanuts, 1976, King Seeley Thermos, red pitching box, plastic bottle	25	5
Peanuts, 1980, King Seeley Thermos, pitching box, yellow face, green band, matching bottle	20	5
Pebbles & Bamm-Bamm, 1971, Aladdin, matching plastic bottle	55	40
Pele, 1975, King Seeley Thermos, matching plastic bottle	80	40
Pennant, 1950, Ohio Art, basket type box, no bottle	30	0
Pete's Dragon, 1978, Aladdin, matching plastic bottle	35	20
Peter Pan, 1969, Aladdin, matching plastic bottle, Disney	80	30
Pets `n Pals, 1961, King Seeley Thermos, matching steel bottle	75	30
Pigs In Space, 1977, King Seeley Thermos, matching plastic bottle	20	10
Pink Gingham, 1976, King Seeley Thermos, matching plastic bottle	45	20
Pink Panther & Sons, 1984, King Seeley Thermos, matching plastic bottle	45	20
Pinocchio, 1938, square	150	0
Pinocchio, 1938, steel round tin with handle	150	0
Pinocchio, 1971, Aladdin, plastic bottle	80	50
Pit Stop, 1968, Ohio Art	175	0
Planet of the Apes, 1974, Aladdin, matching plastic bottle	85	45
Play Ball, 1969, King Seeley Thermos, game on back, steel bottle	70	35
Police Patrol, 1978, Aladdin, plastic bottle	140	30
Polly Pal, 1975, King Seeley Thermos, matching plastic bottle	20	10
Pony Express, 1982, Ohio Art	90	0
Popeye, 1962, Universal, "Popeye socks Bluto" box, matching bottle	450	300
Popeye, 1964, King Seeley Thermos, "Popeye in boat" box with matching steel	120	80
Popeye, 1980, Aladdin, "arm wrestling" box, plastic bottle	40	25
Popples, 1986, Aladdin, plastic bottle	10	5
Porky's Lunch Wagon Dome, 1959, King Seeley Thermos, steel/glass bottle	375	110
Pro Sports, 1974, Ohio Art, no bottle	50	0
Psychedelic Dome, 1969, Aladdin, plastic bottle	335	85
Racing Wheels, 1977, King Seeley Thermos, plastic bottle	40	25
Raggedy Ann & Andy, 1973, Aladdin, plastic bottle	30	10
Rambo, 1985, King Seeley Thermos, matching plastic bottle	15	5

Steel

NAME, YEAR, COMPANY, DESCRIPTION	BOX NM	BOTTLE NM
Rat Patrol, 1967, Aladdin, steel/glass bottle	100	60
Red Barn Dome, 1957, King Seeley Thermos, closed door version, plain Holtemp	80	20
Red Barn Dome, 1958, King Seeley Thermos, open door version, matching steel	75	30
Red Barn Dome, 1972, Thermos, matching steel/glass bottle	65	45
Rifleman, The, 1961, Aladdin, steel/glass bottle	330	140
Road Runner, 1970, King Seeley Thermos, lavender or purple rim, steel or plastic	65	35
Robin Hood, 1956, Aladdin, matching bottle	190	120
Ronald McDonald, Sheriff, 1982, Aladdin, plastic bottle	30	10
Rose Petal Place, 1983, Aladdin, plastic bottle	15	5
Rough Rider, 1973, Aladdin, plastic bottle	55	30
Roy Rogers & Dale Dbl R Bar Ranch, 1953, King Seeley Thermos, steel/glass	130	75
Roy Rogers & Dale Dbl R Bar Ranch, 1954, American Thermos, blue or red band, woodgrain tall bottle	150	75
Roy Rogers & Dale Dbl R Bar Ranch, 1955, American Thermos, eight-scene box, red or blue band, matching bottle	150	85
Roy Rogers & Dale Evans, 1955, American Thermos, cowhide back box, red or blue band, matching bottle	160	80
Roy Rogers & Dale on Rail, 1957, American Thermos, red or blue band, matching	180	80
Roy Rogers Chow Wagon Dome, 1958, King Seeley Thermos, steel/glass bottle	250	80
Saddlebag, 1977, King Seeley Thermos, generic plastic bottle	120	30
Satellite, 1958, American Thermos, matching bottle	110	60
Satellite, 1960, King Seeley Thermos, steel bottle	110	60
Scooby Doo, 1973, King Seeley Thermos, yellow or orange rim, plastic bottle	45	25
Secret Agent T, 1968, King Seeley Thermos, matching bottle	95	45
Secret of NIMH, 1982, Aladdin, plastic bottle	30	10
Secret Wars, 1984, Aladdin, plastic bottle	45	20
See America, 1972, Ohio Art, no bottle	50	0
Sesame Street, 1983, Aladdin, yellow rim, plastic bottle	10	5
Sigmund and the Sea Monsters, 1974, Aladdin, plastic bottle	95	40
Six Million Dollar Man, 1974, Aladdin, plastic bottle	40	25
Six Million Dollar Man, 1978, Aladdin, plastic bottle	40	25
Skateboarder, 1977, Aladdin, plastic bottle	40	20
Sleeping Beauty, 1960, General Steel Ware/Canada, generic steel bottle	450	55
Smokey Bear, 1975, Okay Industries, plastic bottle	350	200
Smurfs, 1983, King Seeley Thermos, blue box, plastic bottle	45	20
Snoopy Dome, 1968, King Seeley Thermos, yellow, "Have Lunch With Snoopy", matching bottle	60	25
Snow White, Disney, 1975, Aladdin, orange rim, plastic bottle	55	25
Snow White, with Game, 1980, Ohio Art, no bottle	45	0
Space Explorer Ed McCauley, 1960, Aladdin, matching steel bottle	250	110
Space Ship, 1950, Decoware, dark blue square	250	0
Space Shuttle Orbiter Enterprise, 1977, King Seeley Thermos, plastic bottle	80	45
Space:1999, 1976, King Seeley Thermos, plastic bottle	55	25
Speed Buggy, 1974, King Seeley Thermos, red rim, plastic bottle	40	15
Spider-Man & Hulk, 1980, Aladdin, Captain America on back, plastic bottle	30	10
Sport Goofy, 1983, Aladdin, yellow rim, plastic bottle	25	10
Sport Skwirts, 1982, Ohio Art	35	0
Sports Afield, 1957, Ohio Art, no bottle	130	0
Star Trek Dome, 1968, Aladdin, matching bottle	700	375
Star Trek, The Motion Picture, 1980, King Seeley Thermos, matching bottle	80	40
Star Wars, 1978, King Seeley Thermos, cast or stars on band, matching plastic	60	30
Star Wars, Empire Strikes Back, 1980, King Seeley Thermos, plastic bottle	40	10
Star Wars, Return of the Jedi, 1983, King Seeley Thermos, plastic bottle	40	10
Stars and Stripes Dome, 1970, King Seeley Thermos, matching plastic bottle	90	30
Steve Canyon, 1959, Aladdin, steel/glass bottle	260	150

Steel

NAME, YEAR, COMPANY, DESCRIPTION	BOX NM	BOTTLE NM
Strawberry Shortcake, 1980, Aladdin, plastic bottle	10	5
Strawberry Shortcake, 1981, Aladdin, plastic bottle	10	5
Street Hawk, 1985, Aladdin, plastic bottle	160	90
Submarine, 1960, King Seeley Thermos, steel/glass bottle	110	60
Super Friends, 1976, Aladdin, matching plastic bottle	55	35
Super Powers, 1983, Aladdin, plastic bottle	65	30
Supercar, 1962, Universal, steel/glass bottle	325	150
Superman, 1954, Universal, blue rim	800	0
Superman, 1967, King Seeley Thermos, red rim, "under fire" art on back, matching steel/glass bottle	155	85
Superman, 1978, Aladdin, red rim, Daily Planet Office on back, matching bottle	35	20
Tapestry, 1963, Ohio Art, no bottle	60	0
Tarzan, 1966, Aladdin, steel/glass bottle	100	40
Teenager Dome, 1957, King Seeley Thermos, generic bottle	140	10
Teenager, 1957, King Seeley Thermos, generic bottle	85	10
Three Little Pigs, 1982, Ohio Art, red rim, generic/plastic bottle	70	0
Thundercats, 1985, Aladdin, plastic bottle	20	5
Tom Corbett Space Cadet, 1952, Aladdin, blue or red paper decal box, steel/glass	250	95
Tom Corbett Space Cadet, 1954, Aladdin, full litho, matching bottle	475	95
Toppie Elephant, 1957, American Thermos, yellow, matching bottle	1600	800
Track King, 1975, Okay Industries, matching steel bottle	260	180
Train, 1971, Ohio Art, no bottle	60	0
Transformers, 1986, Aladdin, red box, matching plastic bottle	10	5
Traveler, 1962, Ohio Art, no bottle	85	0
Trigger, 1956, King Seeley Thermos, no bottle	225	0
U.S. Mail Dome, 1969, Aladdin, plastic bottle	65	20
U.S. Space Corps, 1961, Universal, plastic rocket bottle	350	110
UFO, 1973, King Seeley Thermos, plastic bottle	90	30
Underdog, 1974, Okay Industries, plastic bottle	900	350
Universal's Movie Monsters, 1980, Aladdin, plastic bottle	70	25
Voyage to the Bottom of the Sea, 1967, Aladdin, steel/glass bottle	300	140
VW Bus Dome, 1960, Omni, plastic bottle	500	220
Wagon Train, 1964, King Seeley Thermos, matching steel bottle	160	50
Wags `n Whiskers, 1978, King Seeley Thermos, matching plastic bottle	30	10
Wake Up America, 1973, Okay Industries, matching steel bottle	600	250
Waltons, The, 1973, Aladdin, plastic bottle	50	20
Washington Redskins, 1970, Okay Industries, steel bottle	260	140
Wee Pals Kid Power, 1974, American Thermos, matching plastic bottle	40	10
Welcome Back Kotter, 1977, Aladdin, flat or embossed face, red rim, matching plastic bottle	40	15
Western, 1963, King Seeley Thermos, tan or red rim, steel/glass bottle	140	55
Wild Bill Hickok, 1955, Aladdin, steel/glass bottle	165	80
Wild Frontier, 1977, Ohio Art, spinner game on back, no bottle	45	0
Wild, Wild West, 1969, Aladdin, plastic bottle	165	80
Winnie the Pooh, 1976, Aladdin, blue rim, plastic bottle	190	70
Yankee Doodles, 1975, King Seeley Thermos, plastic bottle	40	20
Yellow Submarine, 1968, King Seeley Thermos, steel/glass bottle	375	175
Yogi Bear & Friends, 1961, Aladdin, black rim, matching steel bottle	140	100
Yogi Bear, 1974, Aladdin	95	45
Zorro, 1958, Aladdin, black band, steel/glass bottle	170	120
Zorro, 1966, Aladdin, red band, steel/glass bottle	200	120

Vinyl

Alice in Wonderland, 1972, Aladdin, matching plastic bottle	200	45

Superman steel box, 1967, King Seeley Thermos

H.R. Pufnstuf steel box, 1970, Aladdin

The Astronauts steel box, 1969, Aladdin

Vinyl

NAME, YEAR, COMPANY, DESCRIPTION	BOX NM	BOTTLE NM
All American, 1976, Bayville, Styrofoam bottle	160	20
All Dressed Up, 1970s, Bayville, Styrofoam bottle	90	20
All Star, 1960, Aladdin	450	60
Alvin and the Chipmunks, 1963, King Seeley Thermos, matching plastic bottle	400	140
Annie 1, 1981, Aladdin, matching plastic bottle	75	20
Bach's Lunch, 1975, Volkwein Bros., red Styrofoam bottle	130	20
Ballerina, 1960s, Universal, black, Thermax bottle	800	150
Ballerina, 1962, Aladdin, pink, steel/glass bottle	200	60
Ballet, 1961, Universal, red, plastic generic bottle	500	20
Banana Splits, 1969, King Seeley Thermos, matching steel/glass bottle	500	150
Barbarino Brunch Bag, 1977, Aladdin, zippered bag, plastic bottle	250	30
Barbie & Francie, 1965, King Seeley Thermos, black, matching steel/glass bottle	120	65
Barbie & Midge Dome, 1964, King Seeley Thermos, matching glass/steel bottle	530	65
Barbie & Midge, 1965, King Seeley Thermos, black, matching steel/glass bottle	110	65
Barbie Softy, 1988, King Seeley Thermos, pink, generic plastic bottle	45	15
Barbie, World of, 1971, King Seeley Thermos, blue box, matching steel/glass	90	25
Barbie, World of, 1971, King Seeley Thermos, pink box, matching steel/glass	75	25
Barnum's Animals, 1978, Adco Liberty, no bottle	60	0
Beany & Cecil, 1963, King Seeley Thermos, steel/glass bottle	560	150
Beatles Brunch Bag, 1966, Aladdin, zippered bag, matching bottle	650	150
Beatles Kaboodles Kit, 1965, Standard Plastic Products, no bottle	600	0
Beatles, 1965, Air Flite, no bottle	525	0
Betsey Clark Munchies Bag, 1977, King Seeley Thermos, zippered bag, plastic	90	10
Betsey Clark, 1977, King Seeley Thermos, yellow box, matching plastic bottle	110	15
Blue Gingham Brunch Bag, 1975, Aladdin, zippered box and plastic bottle	45	30
Bobby Soxer, 1959, Aladdin	300	0
Boston Red Sox, 1960s, Universal	30	20
Boy on the Swing, Abeama Industries	80	20
Buick 1910, 1974, Bayville, Styrofoam bottle	90	20
Bullwinkle, 1963, King Seeley Thermos, blue steel/glass bottle	650	200
Bullwinkle, 1963, King Seeley Thermos, yellow, generic steel bottle	450	60
Calico Brunch Bag, 1980, Aladdin, zippered bag, plastic bottle	70	30
Captain Kangaroo, King Seeley Thermos, steel/glass bottle	500	150
Carousel, 1962, Aladdin, matching steel/glass bottle	425	130
Cars, 1960, Universal	140	0
Casper the Friendly Ghost, 1966, King Seeley Thermos, blue box, orange steel	550	150
Challenger, Space Shuttle, 1986, Babcock, puffy box, no bottle	175	0
Charlie's Angels Brunch Bag, 1978, Aladdin, zippered bag, plastic bottle	170	30
Coca-Cola, 1947, Aladdin, Styrofoam bottle	160	20
Coco the Clown, 1970s, Gary, Styrofoam bottle	110	20
Combo Brunch Bag, 1967, Aladdin, zippered bag, steel/glass bottle	180	80
Corsage, 1970, King Seeley Thermos, steel/glass bottle	120	30
Cottage, 1974, King Seeley Thermos	130	0
Cowboy, 1960, Universal, plain plastic bottle	170	20
Dateline Lunch Kit, 1960, Hasbro, blue/pink, no bottle	250	0
Dawn Brunch Bag, 1971, Aladdin, zippered bag, plastic bottle	180	35
Dawn, 1971, Aladdin, matching plastic bottle	140	35
Dawn, 1972, Aladdin, matching plastic bottle	140	35
Denim Brunch Bag, 1980, Aladdin, zippered bag, plastic bottle	80	15
Deputy Dawg, 1964, Thermos, no bottle	550	0
Deputy Dawg, King Seeley Thermos, steel/glass bottle	550	120
Donny & Marie Brunch Bag, 1977, Aladdin, zippered bag, plastic bottle	125	20
Donny & Marie, 1977, Aladdin, long hair version, matching plastic bottle	110	20
Donny & Marie, 1978, Aladdin, short hair version, matching plastic bottle	120	20

Vinyl

NAME, YEAR, COMPANY, DESCRIPTION	BOX NM	BOTTLE NM
Dr. Seuss, 1970, Aladdin, plastic bottle	575	60
Dream Boat, 1960, Feldco, Styrofoam bottle	275	20
Eats `n Treats, King Seeley Thermos, blue steel/glass bottle	200	40
Fess Parker Kaboodle Kit, 1960s, Aladdin, matching steel bottle	425	90
Fishing, 1970, Universal, Styrofoam bottle	120	20
Frog Flutist, 1975, Aladdin, matching plastic bottle	75	20
Fun to See'n Keep Tiger, 1960, no bottle	175	0
G.I. Joe, 1989, King Seeley Thermos, generic plastic bottle	55	10
Gigi, 1962, Aladdin, matching steel/glass bottle	280	80
Girl & Poodle, 1960, Universal, Styrofoam bottle	140	20
Glamour Gal, 1960, Aladdin, steel/glass bottle	150	35
Go-Go Brunch Bag, 1966, Aladdin, plastic bottle	245	60
Goat Butt Mountain, 1960, Universal, Styrofoam bottle	140	20
Happy Powwow, 1970s, Bayville, red or blue, with Styrofoam bottle	60	20
Highway Signs Snap Pack, 1988, Avon	20	0
Holly Hobbie, 1972, Aladdin, white bag, matching plastic bottle	85	10
I Love a Parade, 1970, Universal, Styrofoam bottle	130	20
Ice Cream Cone, 1975, Aladdin, matching plastic bottle	55	20
It's a Small World, 1968, Aladdin, matching steel/glass bottle	250	110
Jonathan Livingston Seagull, 1974, Aladdin, matching plastic bottle	160	35
Junior Deb, 1960, Aladdin, steel/glass bottle	175	50
Junior Miss Safari, 1962, Prepac, no bottle	140	0
Junior Nurse, 1963, King Seeley Thermos, steel/glass bottle	320	90
Kaboodle Kit, 1960s, Aladdin, pink or white, no bottle	160	0
Kewtie Pie, Aladdin, steel/glass bottle	125	60
Kodak Gold, 1970s, Aladdin	85	20
Kodak II, 1970s, Aladdin	85	20
L'il Jodie (Puffy), 1985, Babcock	90	0
Lassie, 1960s, Universal, Styrofoam bottle	120	20
Liddle Kiddles, 1969, King Seeley Thermos, matching steel/glass bottle	250	60
Linus the Lion-Hearted, 1965, Aladdin, steel/glass bottle	550	110
Little Ballerina, 1975, Bayville, Styrofoam bottle	75	20
Little Old Schoolhouse, 1974, Dart	80	0
Love, 1972, Aladdin, matching plastic bottle	160	45
Lunch `n Munch, 1959, American Thermos, corsage bottle	400	75
Lunch `n Munch, 1959, King Seeley Thermos, steel/glass bottle	450	50
Mam'zelle, 1971, Aladdin, plastic bottle	180	60
Mardi-Gras, 1971, Aladdin, matching plastic bottle	80	20
Mary Ann Lunch `N Bag, 1960, Universal, no bottle	110	0
Mary Ann, 1960, Aladdin, matching steel/glass bottle	75	25
Mary Poppins Brunch Bag, 1966, Aladdin, steel/glass bottle	150	50
Mary Poppins, 1973, Aladdin, matching plastic bottle	90	50
Mod Miss Brunch Bag, 1969, Aladdin, plastic bottle	110	30
Monkees, 1967, King Seeley Thermos, matching steel/glass bottle	380	125
Moon Landing, 1960, Universal, Styrofoam bottle	180	20
Mr. Peanut Snap Pack, 1979, Dart, snap close bag, no bottle	110	0
Mushrooms, 1972, Aladdin, matching plastic bottle	125	45
New Zoo Revue, 1975, Aladdin, plastic bottle	210	60
Pac-Man (Puffy), 1985, Aladdin	65	0
Peanuts, 1967, King Seeley Thermos, red "kite" box, steel/glass bottle	90	30
Peanuts, 1969, King Seeley Thermos, red "baseball" box, steel bottle	90	30
Peanuts, 1971, King Seeley Thermos, green "baseball" box, steel bottle	150	30
Peanuts, 1973, King Seeley Thermos, white "piano" box, steel bottle	90	30
Pebbles & Bamm-Bamm, 1973, Gary, matching plastic bottle	250	55

Voyage to the Bottom of the Sea steel box, 1967, Aladdin

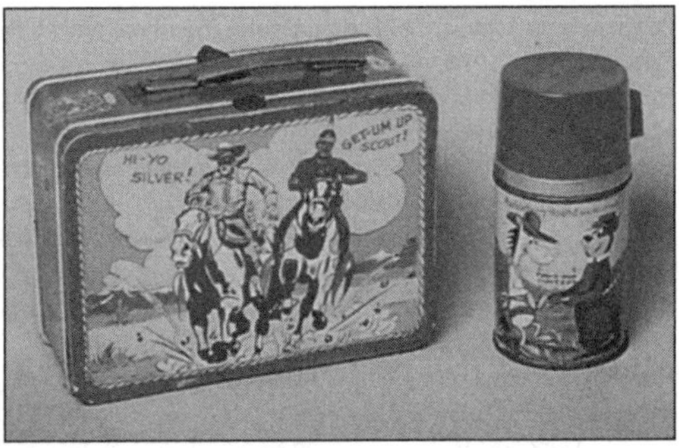

Lone Ranger steel lunch box, Adco Liberty; Huckleberry Hound bottle, Aladdin

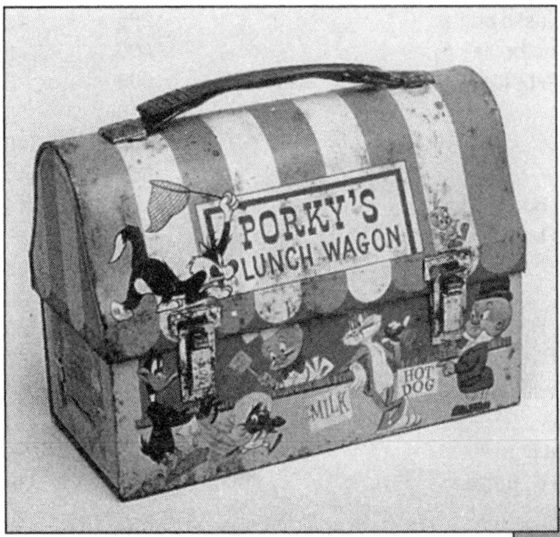

Porky's Lunch Wagon steel dome, 1959, King Seeley Thermos

Disney School Bus Dome steel box and bottle, 1968, Aladdin

Vinyl

NAME, YEAR, COMPANY, DESCRIPTION	BOX NM	BOTTLE NM
Penelope & Penny, 1970s, Gary, yellow box with Styrofoam bottle	120	20
Peter Pan, 1969, Aladdin, white box, matching plastic bottle	210	65
Pink Panther, 1980, Aladdin, matching plastic bottle	95	20
Pony Tail Tid-Bit-Kit, 1962, King Seeley Thermos, steel/glass satellite bottle	200	30
Pony Tail, 1960s, Thermos, white box, original art with gray border added, no	200	0
Pony Tail, 1965, King Seeley Thermos, white box, fold over lid, steel/glass bottle	200	30
Ponytails Poodle Kit, 1960, King Seeley Thermos, steel/glass bottle	150	20
Princess, 1963, Aladdin, steel/glass bottle	190	55
Psychedelic, 1969, Aladdin, yellow, matching steel/glass bottle	150	30
Pussycats, The, 1968, Aladdin, plastic bottle	220	80
Ringling Bros. Circus, 1970, King Seeley Thermos, orange box with matching steel/glass bottle	425	140
Ringling Bros. Circus, 1971, King Seeley Thermos, puffy blue box, steel/glass	110	40
Robo Warriors, 1970, no bottle	35	0
Roy Rogers Saddlebag, 1960, King Seeley Thermos, brown, steel/glass bottle	250	95
Roy Rogers Saddlebag, 1960, King Seeley Thermos, cream, steel/glass bottle	650	95
Sabrina, 1972, Aladdin, yellow box with matching plastic bottle	230	85
Sesame Street, 1979, Aladdin, orange, matching plastic bottle	35	10
Sesame Street, 1981, Aladdin, yellow, matching plastic bottle	85	15
Shari Lewis, 1963, Aladdin, matching steel/glass bottle	470	120
Sizzlers, Hot Wheels, 1971, King Seeley Thermos, matching steel/glass bottle	225	60
Skipper, 1965, King Seeley Thermos, steel/glass bottle	220	60
Sleeping Beauty, Disney, 1970, Aladdin, white box, matching plastic bottle	240	80
Smokey the Bear, 1965, King Seeley Thermos, steel/glass bottle	450	110
Snoopy Munchies Bag, 1977, King Seeley Thermos, plastic bottle	45	10
Snoopy Softy, 1988, King Seeley Thermos, matching plastic bottle	20	10
Snow White, 1975, Aladdin, white box with matching plastic bottle	285	45
Snow White, Disney, 1967, fold-over lid, tapered box, no bottle	400	0
Soupy Sales, 1966, King Seeley Thermos, blue box, no bottle	600	0
Spirit of `76, red	110	0
Sports Kit, 1960, Universal	350	40
Stewardess, 1962, Aladdin, steel/glass bottle	650	110
Strawberry Shortcake, 1980, Aladdin, matching plastic bottle	40	15
Tammy & Pepper, 1965, Aladdin, matching steel/glass bottle	240	85
Tammy, 1964, Aladdin, matching steel/glass bottle	240	85
Tinker Bell, Disney, 1969, Aladdin, plastic bottle	260	90
Twiggy, 1967, Aladdin, matching steel/glass bottle	225	80
Twiggy, 1967, King Seeley Thermos, steel/glass bottle	225	80
U.S. Mail Brunch Bag, 1971, Aladdin, zippered bag, plastic bottle	160	80
Winnie the Pooh, Aladdin, steel/glass bottle	450	110
Wonder Woman (blue), 1977, Aladdin, matching plastic bottle	150	35
Wonder Woman (yellow), 1978, Aladdin, matching plastic bottle	200	35
Wrangler, 1982, Aladdin, steel/glass bottle	325	95
Yosemite Sam, 1971, King Seeley Thermos, matching steel/glass bottle	560	140
Ziggy's Munch Box, 1979, Aladdin, plastic bottle	140	40

Marx Play Sets

For all practical purposes, play sets could have been invented by Louis Marx . . . at least as far as boys growing up in the 1950s and 1960s were concerned.

The words "Marx" and "play set" just went together, and they still go together today for many dedicated collectors.

A typical Marx play set included buildings, figures and lots of realistic accessories that helped bring the miniature world to life. The Fort Apache Stockade, for example, came with a hard plastic log fort, a colorful lithographed tin cabin, and, of course, pioneers and Indians locked in deadly combat. It was no wonder millions of kids had a burning desire for these toys. The play scenarios were almost endless.

This modern version of an age-old toy was a tribute to the marketing and manufacturing talents and whimsical genius of Louis Marx, the modern-day king of toys.

Not only was he responsible for developing the play set, but he popularized the yo-yo and produced some of the most innovative tin wind-ups, guns, dolls, trains, trikes, trucks and other toys that were commercially feasible. In 1955, Marx sold more than $30 million worth of toys, easily making it the largest toy manufacturer in the world.

What makes Marx's domination even more impressive was the fact that he rose from humble beginnings. He was born in Brooklyn in 1896 and didn't learn to speak English until he started school. At age 16, Marx went to work for Ferdinand Strauss, a toy manufacturer who produced items for Abraham & Strauss Department Stores. By the age of 20, Marx was managing the company's New Jersey factory.

After being fired by Strauss, Marx contracted with manufacturers to produce toys he designed. By the mid-1920s, Marx had three plants in the United States. By 1955, there were more than 5,000 items in the Marx toy line with plants worldwide.

Mass production and mass marketing through stores such as Sears and Montgomery Ward allowed Marx to keep prices low and quality high. Marx was also a master at producing new toys from the same basic components. Existing elements could be modified slightly, and new lithography would produce a new building from standard stock.

Part of Marx's repackaging genius included using popular TV or movie tie-ins to breathe new life into existing products. The Rifleman Ranch, Roy Rogers Ranch, Wyatt Earp and Wagon Train play sets were examples of repackaging existing parts to capture the fad of the day.

Gunsmoke Dodge City, 1960

Marx sold his company to the Quaker Oats Company in 1972 for $31 million. Quaker Oats sold the company four years later for $15 million after losing money every year of its ownership.

The Marx Toy Company is in existence once again making favorite Marx toys from original molds.

With the passing of a few short decades, once affordable children's toys have become highly prized collectibles. Play sets are among the price leaders in today's market for childhood treasures. And the figures that accompanied the play sets are also highly desired for their craftsmanship and detail.

A play set listed as MIB (Mint In Box) should be untouched and unassembled in the original box. Excellent condition means a complete set, but the buildings are assembled and the box may be worn or damaged. Good condition means the play set shows wear and may have a few minor pieces missing.

Trends

The play set market continues on a strong course, with several sets now commanding prices as high as some of their "classic" Marx tin wind-up counterparts. As it becomes increasingly difficult to find complete sets, individual pieces should sell well as collectors try to complete sets.

The Top 10 Marx Play Sets (in Mint in Box condition)

1. Johnny Ringo Western Frontier Set, 1959, #4784 .... $2,500
2. Civil War Centennial, 1961, #5929 2,000
3. Fire House, #4820 2,000
4. Gunsmoke Dodge City, 1960, #4268 2,000
5. Johnny Tremain Revolutionary War, 1957, #3402 2,000
6. Ben Hur, #4701 1,800
7. Sears Store, 1961,#5490 1,800
8. Custer's Last Stand, 1963 #4670 1,800
9. Wagon Train, #4888 1,800
10. World War II European Theatre, #5949 1,500

Ben Hur

MARX PLAY SETS

Miniature Play Sets

NAME	DESCRIPTION	NO.	GOOD	EX	MIB
101 Dalmatians	1961, "The Barn Scene"		75	300	450
101 Dalmatians	1961, "The Wedding Scene"		75	300	450
20 Minutes to Berlin	1964, 174 pieces		100	300	500
Alice in Wonderland	New series, 1961		100	225	350
Attack on Fort Apache	stable, cowboys, Indians	HK-8078	85	225	500
Babes In Toyland	six different scenes, each		25	65	100
Battleground	1963, 170 pieces	HK-6111	20	60	200
Blue and Gray	1960s, 101 individual pieces	HK-6109	90	175	325
Border Battle	Mexican-American War		145	350	700
Charge of the Bengal Lancers	British/Turks		125	325	500
Charge of the Light Brigade	smaller version, Lancers/Russians		75	225	325
Charge of the Light Brigade	2nd version, photo box art, Lancers/Turks		110	300	400
Charge of the Light Brigade	Sears, 216 pieces, Lancers/Cossacks		175	325	400
Cinderella	New series		100	225	350
Covered Wagon Attack			85	200	400
Custer's Last Stand	1964, 181 pieces		125	325	600
Disney 3-in-1 Set	original series		75	225	350
Disney Circus Parade	Super Circus performers, Disneykins		85	225	350
Disney See and Play Castle	1st and 2nd series Disneykins	48-24388	150	350	450
Disney See and Play Doll House	1st series Disneykins		100	265	350
Donald Duck	original series; Donald, Daisy, Louie, Goofy		45	100	150
Dumbo's Circus	original series		50	100	150
Fairykin	six different, each		30	80	125
Fairykin TV Scenes	12 different, each		8	20	30
Fairykin TV Scenes Gift Set	two different, each with six scenes, each		65	165	250
Fairykins 3-in-1 Diorama Set			100	265	400
Fairykins Gift Set	34 in window box		40	175	250
Fairykins TV Scenes Boxed Set of Eight			45	200	250
Fort Apache	large set, HQ bldg., cavalry/cowboys/ Indians		115	295	375
Fort Apache	1963, 90 pieces, Indians	HK-7526	55	80	165
Guerrilla Warfare	1960s, Viet Cong		275	350	450
Huckleberry Hound Presents	two different, each		75	115	175
Invasion Day	1964, 304 pieces		65	200	400
Jungle	smaller than Jungle Safari		50	85	175
Jungle Safari	260 pieces, hunters/natives		55	100	200
Knights and Castle	1963, 132 pieces	HK-7563	130	200	300
Knights and Castle	1964, 64 pieces	HK-7562	95	175	275
Knights and Vikings	1964, 143 pieces		145	275	425
Lady and the Tramp	New series, 1961		100	225	350
Lost Boys	New series		100	225	350
Lost Boys	second series, 1961		40	120	200
Ludwig Von Drake	1962, "The Nearsighted Professor"		50	100	150
Ludwig Von Drake	1962, "The Professor Misses"		50	100	150
Ludwig Von Drake	RCA premium set		65	130	200
Mickey Mouse and Friends	original series, display box		50	100	150
Munchville	vegetable characters		65	165	250
Noah's Ark	Ward's version, soft plastic figures		20	50	75

Fighting Knights Carry All, 1960s

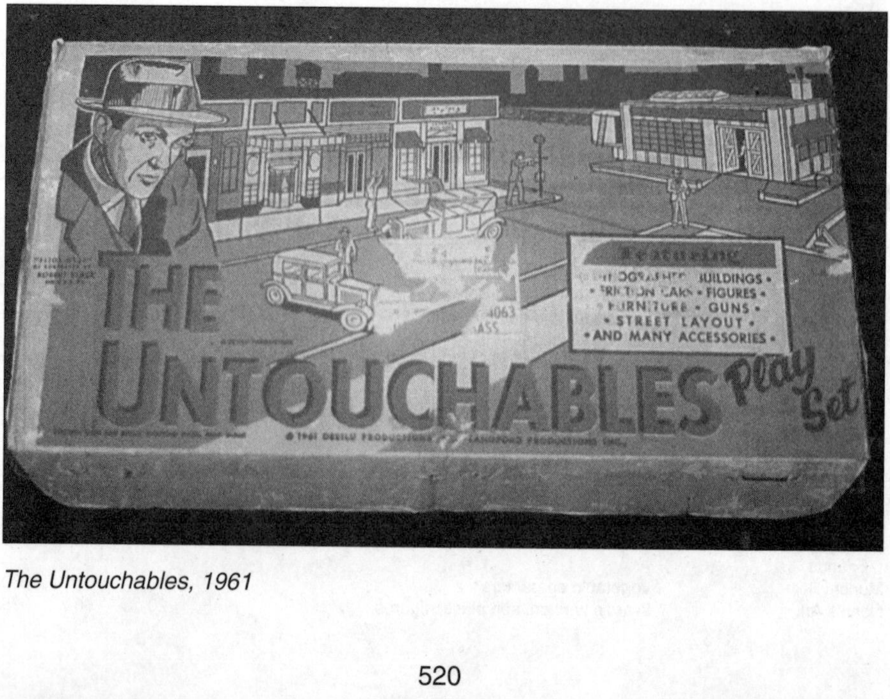

The Untouchables, 1961

Miniature Play Sets

NAME	DESCRIPTION	NO.	GOOD	EX	MIB
Noah's Ark	1968, 100 pieces		25	65	100
Over The Top	WWI, Germans/Doughboys		200	600	950
Panchito Western	original series, display box		50	100	150
Pinocchio	six different sets, each original series, display box		65	165	250
Pinocchio 3-in-1 Set			115	295	450
Quick Draw McGraw	two different, each		75	115	200
Revolutionary War	British/Colonials		95	250	500
Sands of Iwo Jima	1963, 205 pieces		115	210	325
Sands of Iwo Jima	1964, 296 pieces		150	295	450
Sands of Iwo Jima	1963, 88 pieces		75	145	225
See and Play Dollhouse	American Beauties/Campus Cuties		75	175	350
Sleeping Beauty	new series, 1961		75	175	275
Snow White and the Seven Dwarfs	original series, display box		50	100	150
Sunshine Farm Set	farmers and animals		45	115	175
Sword in the Stone	British only Disney release		300	1000	1500
Ten Commandments	Montgomery Ward		150	395	600
Three Little Pigs	new series		100	225	350
Tiger Town	ENCO-like tigers, 1960s		75	175	300
Top Cat	three different, each		75	115	200
Troll Village			80	300	350
TV-Tinykins Gift Set	set of 34 figures		115	350	550
TV-Tinykins TV Scenes	12 different, each		12	35	50
Western Town	over 170 pieces	48-24398	50	150	250
Wooden Horse of Troy	British only issue		125	600	800

Play Sets

NAME	DESCRIPTION	NO.	GOOD	EX	MIB
Adventures of Robin Hood	1956, Richard Greene TV series	4722	250	750	1250
Alamo	only two cannons	3546	100	300	500
Alamo	1960, for 54mm figures	3534	140	250	400
Alaska Frontier	1959, 100 pieces	3708	275	525	800
American Airlines Astro Jet Port		4822	150	250	450
American Airlines International Jet Port	1962, 98 pieces	4810	150	250	450
Arctic Explorer	1960, Series 2000	3702	250	450	700
Army Combat Set	Sears, 411 pieces	6019	100	300	500
Army Combat Training Center		2654	20	55	90
Babyland Nursery		3379	125	225	350
Bar-M Ranch		3956	50	100	150
Battle of Iwo Jima	1964, 128 pieces	6057	35	105	175
Battle of Iwo Jima	1964, 247 pieces	4147	80	240	400
Battle of Little Big Horn	1972	4679MO	125	250	400
Battle of the Blue & Gray	Series 2000, large set	4658	250	700	1200
Battle of the Blue & Gray	Series 1000, small set, no house	2646	80	240	400
Battle of the Blue & Gray	1959, Series 2000, 54mm	4745	175	375	600
Battle of the Blue & Gray	1963, Centennial edition	4744	200	700	1200
Battlefield	1958, Series 5000	4756	25	95	150
Battleground	1970s	4756	40	125	250
Battleground	1962, 200 pieces	4754	35	110	185
Battleground	U.S. and Nazi troops	4169	30	90	150
Battleground	1971, Montgomery Ward	4752	90	275	450
Battleground	1963, Sears, 160 pieces		70	210	350
Battleground	1958, largest of military sets	4750	130	395	650
Battleground	1959, 180 pieces	4751	35	110	185
Battleground	1963, Montgomery Ward	3745	80	240	400
Beach Head Landing Set	U.S. and Nazi Troops	4939	15	65	100
Ben Hur	1959, 132 pieces	4696	170	510	850

U.S. Army Training Center, 1950s

Battlegrouind #4752

Play Sets

NAME	DESCRIPTION	NO.	GOOD	EX	MIB
Ben Hur	1959, Series 2000, medium set	4702	250	750	1250
Ben Hur	Series 5000, large set	4701	350	1075	1800
Ben Hur	blister card	2648	25	95	150
Big Inch Pipeline	1963, 200 pieces	6008	80	240	400
Big Top Circus	1952	4310	80	325	500
Boot Camp	tin box set	4645	30	130	200
Boy Scout			115	600	900
Boys Camp	1956	4103	130	395	650
Cape Canaveral	1960	4524	85	195	300
Cape Canaveral	1959, Sears set	5963	80	325	500
Cape Canaveral Missile Center		4525	50	195	300
Cape Canaveral Missile Center	1959	2656	50	150	250
Cape Canaveral Missile Center	1959	4528	80	240	400
Cape Canaveral Missile Set	1958	4526	55	225	350
Cape Kennedy Carry All	1968, tin box set	4625	35	45	75
Captain Gallant of the Foreign Legion	1956	4729/ 4730	200	600	1000
Captain Space Solar Academy		7026	80	325	500
Captain Space Solar Academy	1954	7018	65	260	400
Castle and Moat Set	Sears exclusive	4734	65	260	400
Cattle Drive	mid-1970s	3983	60	245	375
Civil War Centennial	1961	5929	400	1200	2000
Comanche Pass	1976	3416	30	130	200
Complete Happitime Dairy Farm	Sears	5957	80	325	500
Complete U.S. Army Training Center	1954	4145	70	210	350
Construction Camp	1954	4439	90	275	450
Construction Camp	1956, 54mm, Series 1000	4442	110	325	550
Cowboy And Indian Camp	1953	3950	90	275	450
Custer's Last Stand	1956, Series 500	4779	80	325	500
Custer's Last Stand	1963, Sears, 187 pieces	4670	195	1200	1800
D-Day Army Set	U.S. and Nazi troops	6027	100	300	500
D.E.W. Defense Line Arctic Satellite Base		4802	100	300	500
Daktari	1967, 110 pieces	3717	100	300	500
Daktari	1967, 140 pieces	3720	130	395	650
Daktari		3718	80	325	500
Daniel Boone Frontier		1393	60	230	350
Daniel Boone Wilderness Scout	1964	0670	75	225	375
Daniel Boone Wilderness Scout	1964	0631	75	225	375
Daniel Boone Wilderness Scout	1964	2640	120	360	600
Davy Crockett at the Alamo		3442	120	360	600
Davy Crockett at the Alamo	1955, official Walt Disney, 100 pieces, first set	3530	65	260	400
Davy Crockett at the Alamo	1955, official Walt Disney, biggest set	3544	160	500	800
Desert Fox	1966, 244 pieces	4177	90	275	450
Desert Patrol	1967, U.S., Nazi troops	4174	60	175	300
Farm Set	Lazy Day, 1960, 100 pieces	3945	55	165	275
Farm Set	1958, 100 pieces, Series 2000	3948	80	250	400
Farm Set		5942	40	160	250

Western Town miniature set

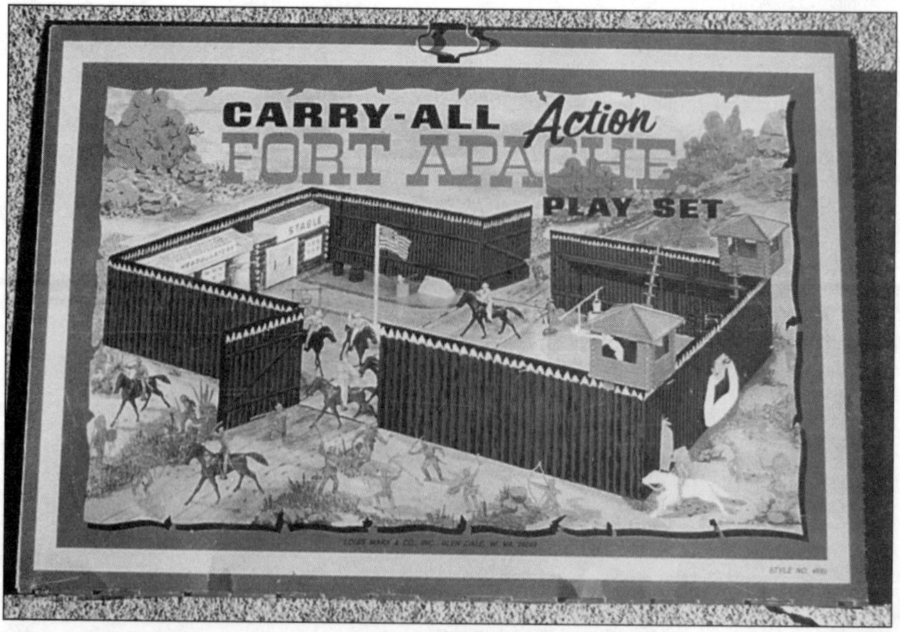

Fort Apache Carry All

Play Sets

NAME	DESCRIPTION	NO.	GOOD	EX	MIB
Farm Set		6006	50	195	300
Farm Set	deluxe, 1969	3953	75	225	375
Farm Set		6050	45	180	275
Fighting Knights Carry All	1966-68	4635	45	135	225
Fire House		4819	180	715	1100
Fire House	with two friction vehicles	4820	500	1500	2000
Fort Apache		3681A	30	90	150
Fort Apache	1965, Sears, 147 pieces		40	120	200
Fort Apache	1972, Sears, over 100 pieces	59093C	30	90	150
Fort Apache		6068	35	100	165
Fort Apache	1965, Sears, 335 pieces	6063	105	315	525
Fort Apache	Sears	6059	11	35	55
Fort Apache	1970s	4202	15	50	80
Fort Apache	giant set	3685	140	425	700
Fort Apache		3682	15	50	85
Fort Apache		3616	30	90	150
Fort Apache	1976	3681	40	120	200
Fort Apache	1967	3681	45	135	225
Fort Apache Carry All		4685	15	45	75
Fort Apache Rin Tin Tin	mixed scale set	3957	90	275	350
Fort Apache Rin Tin Tin	54mm	3658	90	275	350
Fort Apache Rin Tin Tin	early, 60mm	3627	100	300	450
Fort Apache Stockade	1953	3612	50	155	255
Fort Apache Stockade	1960, Series 2000, 60mm figures	3660	75	225	375
Fort Apache Stockade	1961, Series 5000		55	165	275
Fort Apache Stockade	1951	3610	70	210	350
Fort Apache with Famous Americans		3636	55	165	270
Fort Dearborn	larger set	3514	20	60	100
Fort Dearborn	with plastic walls	3688	80	240	400
Fort Dearborn	1952, with metal walls	3510	85	255	375
Fort Mohawk	British, Colonials, Indians, 54mm	3751	80	325	500
Fort Pitt	1959, Series 750, 54mm	3741	65	260	400
Fort Pitt	1959, Series 1000, 54mm	3742	70	290	450
Four-Level Allstate Service Station		6004	80	325	500
Four-Level Parking Garage		3502	40	120	200
Four-Level Parking Garage		3511	40	200	300
Freight Trucking Terminal	plastic trucks	5220	30	90	150
Freight Trucking Terminal	friction trucks	5422	30	90	150
Galaxy Command	1976	4206	10	30	50
Gallant Men	official set from TV series	4634	70	290	450
Gallant Men Army	U.S. troops	4632	65	260	400
Gunsmoke Dodge City	1960, official, Series 2000, 80 pieces	4268	245	1200	2000
Happitime Army and Air Force Training	1954, 147 pieces	4159	50	150	250
Happitime Civil War Centennial	1962, Sears	5929	115	455	700
Happitime Farm Set		3480	25	95	150
Happitime Roy Rogers Rodeo Ranch	1953	3990	60	180	300
Heritage Battle of the Alamo	1972, Heritage Series	59091	80	240	400
History in the Pacific	1972	4164	90	275	450
Holiday Turnpike	battery-operated with HO scale vehicles	5230	10	30	45
I.G.Y. Arctic Satellite Base	1959, Series 1000	4800	235	700	1175
Indian Warfare	Series 2000	4778	65	260	400

Pet Shop

Prehistoric Times Series 1000, 1957

Project Apollo Moon Landing

Play Sets

NAME	DESCRIPTION	NO.	GOOD	EX	MIB
Irrigated Farm Set	working pump	6021	7	20	35
Johnny Apollo Moon Launch Center	1970	4630	45	135	225
Johnny Ringo Western Frontier Set	1959, Series 2000	4784	1200	1800	2500
Johnny Tremain Revolutionary War	1957, official Walt Disney, Series 1000	3402	400	1200	2000
Jungle	1960, 48 pieces, Sears, large animals	3716	25	95	150
Jungle	metal trading post, Series 500	3705	110	325	550
Jungle Jim	1957, official, Series 1000	3706	280	850	1400
Knights and Vikings	1972	4743	50	150	250
Knights and Vikings	1973	4733	50	150	250
Knights and Vikings		4773	30	90	150
Little Red School House	1956	3381	90	270	450
Lone Ranger Ranch	1957, Series 500	3969	85	250	425
Lone Ranger Rodeo Set	1953	3696	30	90	150
Medieval Castle	with knights and Vikings	4707	35	110	180
Medieval Castle	1959, Sears, Series 2000	4708	120	350	600
Medieval Castle	1954	4709	25	95	150
Medieval Castle	with knights and Vikings	4733	30	130	200
Medieval Castle	Sears, with knights and Vikings	4734	75	290	450
Medieval Castle	1960, metallic knights	4700	75	290	450
Medieval Castle	1964, gold knights, moat	4704	30	90	150
Medieval Castle Fort	1953	4710	30	130	200
Midtown Service Station	1960	3420	50	150	250
Midtown Shopping Center		2644	30	90	150
Military Academy		4718	90	350	500
Modern Farm Set	1951, 54mm	3931	50	150	250
Modern Farm Set	1967	3932	60	180	300
Modern Service Center	1962	3471	70	210	350
Modern Service Station	1966	6044	35	105	175
Navarone Mountain Battleground Set	1976	3412	40	115	195
New Car Sales and Service		3466, 3465	75	290	450
One Million, B.C.	1970s		40	115	195
Operation Moon Base	1962	4654	90	270	450
Pet Shop		4209	60	230	350
Pet Shop	1953	4210	60	230	350
Prehistoric	1969	3398	35	105	175
Prehistoric Dinosaur	1978	4208	35	105	175
Prehistoric Times	Series 500	3389	35	105	175
Prehistoric Times		3391	20	55	95
Prehistoric Times	1957, Series 1000, big set	3390	60	230	350
Prehistoric Times		3388	25	75	125
Prehistoric Times		2650	30	130	200
Prince Valiant Castle	1955	4705	90	270	450
Prince Valiant Castle	1955, has figures	4706	100	300	500
Project Apollo Cape Kennedy		4523	25	75	125
Project Apollo Moon Landing		4646	50	150	250
Project Mercury Cape Canaveral	1959	4524	90	270	450
Raytheon Missile Test Center	1961	603-A	60	180	300
Real Life Western Wagon		4998	15	45	75
Red River Gang	1970s, mini set with cowboys	4104	35	105	175
Revolutionary War	Series 1000	3404	120	490	750
Revolutionary War	1959, 80 pieces, Sears	3408	100	390	600
Revolutionary War	1957, Series 500	3401	200	500	1000

Lone Ranger Ranch, Series 500, 1957

Medieval Castle, Series 2000, 1959

Construction Camp, 1954

Play Sets

NAME	DESCRIPTION	NO.	GOOD	EX	MIB
Rex Mars Planet Patrol		7040	100	300	500
Rex Mars Space Drome	1954	7016	130	395	650
Rifleman Ranch, The	1959	3998	130	395	650
Rifleman Ranch, The	1959	3997	115	455	700
Rin Tin Tin at Fort Apache	1956, Series 5000	3686R	240	725	1200
Rin Tin Tin at Fort Apache	1956, Series 500, 60mm	3628	160	475	800
Robin Hood Castle	60mm	4717	120	360	600
Robin Hood Castle	1958, 54mm	4718	80	325	500
Roy Rogers Double R Bar Ranch	1962	3982	100	300	500
Roy Rogers Mineral City	1958, 95 pieces	4227	100	300	500
Roy Rogers Ranch	with ranch kids	3980	200	300	500
Roy Rogers Rodeo		3689	20	60	100
Roy Rogers Rodeo Ranch	1952	3979	55	165	275
Roy Rogers Rodeo Ranch	1958	3986R	250	750	1250
Roy Rogers Rodeo Ranch	54mm	3988	65	195	325
Roy Rogers Rodeo Ranch	Series 2000	3996	130	395	650
Roy Rogers Rodeo Ranch	1952, 60mm	3985	45	135	225
Roy Rogers Western Town		4216	80	240	400
Roy Rogers Western Town	official, Series 5000	4259	80	235	395
Roy Rogers Western Town	1952, large set	4258	160	475	800
Sears Store	1961, Allstate box	5490	350	1200	1800
Service Station		5459	15	45	75
Service Station	with parking garage	3485	105	315	525
Service Station	with elevator	3495	30	90	150
Service Station	deluxe	3501	50	150	250
Shopping Center		3755	40	120	200
Silver City Western Town	has Custer, Boone, Carson, Buffalo Bill, Sitting Bull	4220	50	150	250
Skyscraper	working elevator and light	5450	155	800	1200
Skyscraper	working elevator	5449	155	800	1200
Sons of Liberty	Sears	4170	50	150	250
Star Station Seven	1970s		10	30	50
Strategic Air Command		6013	130	520	800
Super Circus	1952, with character figures	4320	75	290	450
Super Circus	1952, over 70 pieces	4319	80	240	400
Tactical Air Command	1970s	4106	15	40	65
Tales of Wells Fargo		4263	80	240	400
Tales of Wells Fargo	Series 1000	4264	150	450	750
Tales of Wells Fargo		4262	150	450	750
Tales of Wells Fargo Train Set	1959, with electric train	54752	240	600	800
Tank Battle	Sears, U.S., Nazi troops	6056	40	120	200
Tank Battle	U.S., Nazi troops	6060	40	120	200
Turnpike Service Center	1961	3460	100	300	500
U.S. Air Force		4807	30	90	160
U.S. Armed Forces		4151	70	210	350
U.S. Armed Forces Training Center	1955, Series 500	4149	85	255	425
U.S. Armed Forces Training Center	Marines, soldiers, sailors, airmen, tin litho building	4144	40	160	250
U.S. Armed Forces Training Center	1956	4158	110	330	550

Battle of the Little Big Horn, 1972

Troll Village miniature set

Jungle Jim, Series 1000, 1957

Play Sets

NAME	DESCRIPTION	NO.	GOOD	EX	MIB
U.S. Armed Forces Training Center		4150	50	150	250
U.S. Army Mobile Set	1956, flat figures	3655	20	60	100
U.S. Army Training Center		4122	20	60	100
U.S. Army Training Center		4123	25	75	125
U.S. Army Training Center		3146	20	55	95
U.S. Army Training Center		4153	15	55	85
U.S. Army Training Center		3378	20	55	95
Untouchables	1961, 90 pieces	4676	245	975	1500
Vikings and Knights		6053	60	180	300
Wagon Train	Series 1000, X Team	4805	160	480	800
Wagon Train	official, Series 5000	4888	245	975	1500
Wagon Train	official, Series 2000	4788	120	360	600
Walt Disney Television Playhouse	1953	4352	120	360	600
Walt Disney Television Playhouse	1953	4350	100	300	500
Walt Disney Television Playhouse	1953, Peter Pan figures	4352	105	420	650
Walt Disney's Zorro	1958, official	3758	240	725	1200
Walt Disney's Zorro	1958, official, Series 500	3753	230	695	1150
Walt Disney's Zorro	1958, official, Series 1000	3754	260	775	1300
Walt Disney's Zorro	1972, official, Series 1000	3758	160	500	800
Ward's Service Station	1959	3488	80	240	400
Western Frontier Set			90	275	450
Western Mining Town	1950s	4266	135	405	675
Western Mining Town	1950s	4265	135	400	675
Western Ranch Set		3954	35	105	175
Western Ranch Set		3980	35	105	175
Western Stagecoach	1965	1395	20	60	100
Western Town	1952, bi-level town	4229	120	490	650
Western Town	single level	2652	60	180	400
Westgate Auto Center	1968		40	120	200
White House	house with eight figures		15	40	70
White House & Presidents	house & figures, 1/48 scale presidents	3920	15	40	70
White House & Presidents	house & figures	3921	15	40	70
Wild Animal Jungle	large animals	3716	10	30	50
World War II Battleground	1970s	4204	30	90	150
World War II European Theatre	rare big set	5949	245	1200	1500
World War II European Theatre	Sears	5939	155	465	775
World War II Set	U.S., Nazi troops	5938	25	75	125
Wyatt Earp Dodge City Western Town	1957, Series 1000	4228	150	450	750

Model Kits

Model kits have always been popular toys for boys, and in recent years the kits have found a new following among older collectors, primarily men recapturing a part of their youth.

Plastic model kits were first produced shortly before World War II, but it wasn't until after the war that plastic kit building really began to take off. Automobiles, aircraft and ships all became subject matter for the miniature replicas popularized by such companies as Aurora, Revell, Monogram and Lindberg.

Each type of model kit has its own enthusiastic following, but probably the most collectible kits today are the figure and character kits produced in the 1960s. These kits have seen dramatic increases in collector values over the past 10 years.

The company that did the most to popularize the figure kit was Aurora, with its early 1960s kits representing the monsters from Universal Pictures. Aurora had been producing figure kits prior to that, but the monster craze of the period was responsible for a highly successful line of kits. These kits have become the backbone of the hobby, as our Top 10 listing at the end of this article shows.

Starting with the Frankenstein monster in 1961, Aurora produced kits of many memorable movie monsters before moving into more general monstrosities, such as its famous working guillotine kit. Such toys offended the sensibilities of some groups, who brought about political pressure that spelled the end of these kits. The firm also produced kits based on popular television shows, comic characters and sports celebrities. Some of the kits originally made by Aurora were later reissued by Monogram and Revell. Resin copies of the harder-to-find Aurora kits are still being produced and sold today by independent garage kit makers.

Other popular monster kits were also a fad in the 1960s. These weren't the traditional movie monsters, but rather an assortment of strange characters that often came in wild hot rods. Among the more popular were Revell kits based on Ed "Big Daddy" Roth's Rat Fink concept. Other firms, most notably Hawk, also produced kits of this new type of monster.

Even popular celebrities of the day became the subject of model kits. Revell, for example, issued figure kits of each of the four Beatles.

Figure kits began to enjoy new popularity in the 1980s as new large-scale kits of rather limited production runs were being made in vinyl and resin. Billiken, a Japanese company, produced vinyl kits of the classic movie monsters, some of which have become highly collectible. Screamin' and Horizon are two leaders in a burgeoning "garage kit" field of large-

Batman, Aurora, 1964

scale vinyl and resin kits of movie, monster, and comic book characters. This area bears watching as limited runs of these garage kits will no doubt translate into collectibility in the future.

The prices indicated provide general guidelines as to what these kits would sell for today at retail. MIB refers to a kit that is Mint in Box, in original condition in the original Mint box with instructions. The box may not be in the original factory seal, but if the kit pieces were contained in bags inside the box, the bags have not been opened. Kits that remain in pristine condition in factory seals may command a slight premium. Near Mint (NM) refers to a kit that is like new, complete and unassembled. The box may show some shelf wear and the interior bags may have been opened. B/U refers to a kit that has been assembled or built up. These price guidelines assume a neatly built, cleanly painted, complete kit.

Trends

The model kit market may have softened somewhat, but collectors are now seeking to upgrade their collections with Mint in Box kits, especially those that are still factory sealed. Aurora figure kits from the 1960s continue to be the cream of the figure model crop.

Figural kits from the 1960s are so popular that reissues have become popular.

The top end of the market remains firm, as do licensed kits like Star Trek. The true Aurora rarities like Godzilla's Go-Cart and those kits with cult and crossover followings like Lost In Space, continue to post steadily increasing values.

Godzilla, Aurora, 1964

The Top 10 Figural Model Kits (in Mint in Box condition)

1. Godzilla's Go-Cart, Aurora, 1966 . . . . . $3,200
2. Lost In Space, large kit w/chariot, Aurora, 1966 1,350
3. Frankenstein, Gigantic 1:5 scale, Aurora, 1964 1,300
4. Munsters Living Room, Aurora, 1964 1,300
5. King Kong's Thronester, Aurora, 1966 1,250
6. Lost In Space, small kit, Aurora, 1966 920
7. Lost In Space, The Robot, Aurora, 1968 800
8. Addams Family Haunted House, Aurora, 1964 800
9. Bride of Frankenstein, Aurora, 1965 750
10. Penguin (Batman), Aurora, 1967 525

MODEL KITS

Addar

NO.	NAME	YEAR	B/U	NM	MIP
106	Caesar, Planet of the Apes	1974	15	40	50
101	Cornelius, Planet of the Apes	1974	12	30	50
216	Cornfield Roundup, Planet of the Apes	1975	15	40	50
102	Dr. Zaius, Planet of the Apes	1974	10	25	40
105	Dr. Zira, Planet of the Apes	1974	10	25	40
152	Evil Knievel	1974	12	25	50
154	Evil Knievel's Sky Cycle	1974	12	25	50
103	Gen. Ursus, Planet of the Apes		12	30	35
104	Gen. Aldo, Planet of the Apes		10	25	30
217	Jail Wagon, Planet of the Apes	1975	15	40	45
270	Jaws	1975	35	55	100
270	Jaws in a Bottle	1975	20	50	60
227	Spirit in a Bottle	1975	10	20	45
107	Stallion & Soldier, Planet of the Apes	1974	25	75	100
215	Tree House, Planet of the Apes	1975	15	40	45

Airfix

3542	Anne Boleyn	1974	7	15	20
2502	Black Prince	1973	10	25	30
212	Boy Scout	1965	7	15	20
211	Charles I	1965	10	20	25
2501	Henry VIII	1973	4	8	10
823	James Bond's Aston Martin DB-5	1965	60	200	225
M401F	James Bond and Odd Job		30	65	200
2504	Julius Caesar	1973	10	25	30
2508	Napoleon	1978	4	8	10
3544	Queen Victoria	1976	7	15	20
3546	Queen Elizabeth I	1980	7	15	20
203	Richard I	1965	10	25	30
2507	Yeoman of the Guard	1978	4	8	10

AMT

7701	Bigfoot	1978	20	60	75
611	Brute Farce	1960s	5	10	15
610	Cliff Hanger	1960s	5	10	15
905	Drag-U-La, Munsters	1965	40	175	250
495	Flintstones Sports Car	1974	20	55	65
497	Flintstones Rock Crusher	1974	20	50	60
913	Girl From U.N.C.L.E. Car	1974	75	250	300
309	Graveyard Ghoul Duo (Munsters cars)	1970	50	100	150
2501	KISS Custom Chevy Van	1977	20	50	75
461	Laurel & Hardy `27 T Touring Car	1976	20	50	60
462	Laurel & Hardy `27 T Roadster	1976	20	50	60
912	Man From U.N.C.L.E. Car	1966	75	175	225
956	Mr. Spock, large box	1973	15	75	175
	Mr. Spock, small box	1973	20	100	150
901	Munster Koach	1964	50	100	200
904	My Mother The Car	1965	15	35	40
	Sonny & Cher Mustang		75	250	300
612	Threw'd Dude	1960s	5	10	15
614	Touchdown?	1960s	5	10	15
	UFO Mystery Ship		15	60	75
921-200	USS Enterprise w/lights, Star Trek	1967	40	200	250

AMT

NO.	NAME	YEAR	B/U	NM	MIP
950	USS Enterprise Bridge, Star Trek	1975	10	25	35
951-250	USS Enterprise, Star Trek	1966	40	125	150

Aurora

805	Addams Family Haunted House	1964	300	600	800
409	American Astronaut	1967	15	60	75
402	American Buffalo	1964	10	20	25
402	American Buffalo, reissue	1972	8	12	15
401	Apache Warrior on Horse	1960	175	300	450
K-10	Aramis, Three Musketeers	1958	20	75	100
582	Archie's Car	1969	25	85	100
819	Aston Martin Super Spy Car		40	150	200
K-8	Athos, Three Musketeers	1958	20	75	100
832	Banana Splits Banana Buggy	1969	150	400	500
811	Batboat	1968	150	300	450
810	Batcycle	1967	125	250	400
467	Batman	1964	15	75	250
187	Batman, Comic Scenes	1974	15	40	60
486	Batmobile	1966	100	195	325
487	Batplane	1967	50	100	250
407	Black Bear and Cubs	1962	15	30	40
407	Black Bear and Cubs, reissue	1969	15	20	25
400	Black Fury	1958	10	25	30
400	Black Fury, reissue	1969	10	13	15
K-3	Black Knight	1956	10	30	35
473	Black Knight, reissue	1963	10	13	15
463	Blackbeard	1965	75	200	225
K-2	Blue Knight	1956	10	35	50
472	Blue Knight, reissue	1963	10	17	20
414	Bond, James	1966	250	300	450
482	Bride of Frankenstein	1965	300	500	750
863	Brown, Jimmy	1965	75	150	175
409	Canyon, Steve	1958	75	175	250
480	Captain Action	1966	100	275	300
476	Captain America	1966	85	190	300
192	Captain America, Comic Scenes	1974	30	80	125
464	Captain Kidd	1965	25	50	80
738	Cave Bear	1971	15	25	40
416	Chinese Girl	1957	10	20	25
415	Chinese Mandarin	1957	12	25	30
213	Chinese Mandarin & Girl	1957	75	100	300
828	Chitty Chitty Bang Bang	1968	30	85	150
402	Confederate Raider	1959	150	200	350
426	Creature From The Black Lagoon	1963	65	300	425
483	Creature From The Black Lagoon, Glow Kit	1969	65	100	225
483	Creature From The Black Lagoon, Glow Kit	1972	65	100	200
653	Creature, Monsters of Movies	1975	75	100	225
730	Cro-Magnon Man	1971	10	30	45
731	Cro-Magnon Woman	1971	7	25	35
K-7	Crusader	1959	75	150	200
410	D'Artagnan, Three Musketeers	1966	50	150	175
861	Dempsey vs Firpo	1965	20	75	75
631	Dr. Deadly	1971	25	70	80
632	Dr. Deadly's Daughter	1971	25	65	75
460	Dr. Jekyll as Mr. Hyde	1964	45	250	350
482	Dr. Jekyll, Glow Kit	1969	45	100	175

Robin, Aurora, 1966

Lost in Space large kit w/Chariot, Aurora, 1966

Tarzan, Comic Scenes, Aurora, 1974

Hunchback of Notre Dame, Aurora, 1960s

Aurora

NO.	NAME	YEAR	B/U	NM	MIP
482	Dr. Jekyll, Glow Kit	1972	45	65	80
462	Dr. Jekyll, Monster Scenes	1971	40	90	125
654	Dr. Jekyll, Monsters of Movies	1975	25	60	70
424	Dracula	1962	25	225	300
454	Dracula, Frightning Lightning	1969	30	300	500
454	Dracula, Glow Kit	1969	20	75	150
454	Dracula, Glow Kit	1972	20	60	75
641	Dracula, Monster Scenes	1971	75	100	200
656	Dracula, Monsters of Movies	1975	75	100	250
466	Dracula's Dragster	1966	125	300	400
413	Dutch Boy	1957	10	25	30
209	Dutch Boy & Girl	1957	75	200	300
414	Dutch Girl	1957	10	20	25
817	Flying Sub	1968	35	175	200
254	Flying Sub, reissue	1975	35	85	100
422	Forgotten Prisoner	1966	65	350	400
453	Forgotten Prisoner, Frightning Lightning	1969	65	325	450
453	Forgotten Prisoner, Glow Kit	1972	65	150	175
453	Forgotten Prisoner, Glow Kit	1969	65	175	200
423	Frankenstein	1961	30	200	300
449	Frankenstein, Frightning Lightning	1969	30	300	400
470	Frankenstein, Gigantic 1/5 scale	1964	300	900	1300
449	Frankenstein, Glow Kit	1972	20	50	75
449	Frankenstein, Glow Kit	1969	20	65	150
633	Frankenstein, Monster Scenes	1971	50	75	100
651	Frankenstein, Monsters of Movies	1975	100	200	250
465	Frankie's Flivver	1964	150	350	400
451	Frog, Castle Creatures	1966	75	200	250
658	Ghidrah	1975	95	260	300
643	Giant Insect, Monster Scene	1971	95	350	400
469	Godzilla	1964	85	400	525
466	Godzilla, Glow Kit	1969	75	225	325
466	Godzilla, Glow Kit	1972	75	150	175
485	Godzilla's Go-Cart	1966	650	2200	3200
475	Gold Knight of Nice	1965	125	250	275
475	Gold Knight of Nice	1957	125	250	300
413	Green Beret	1966	75	150	175
489	Green Hornet Black Beauty	1966	125	350	500
634	Gruesome Goodies	1971	25	80	100
800	Guillotine	1964	125	350	400
637	Hanging Cage	1971	20	80	100
481	Hercules	1965	100	200	275
184	Hulk, Comic Scenes	1974	25	75	85
421	Hulk, Original	1966	75	250	300
460	Hunchback of Notre Dame	1964	45	200	300
481	Hunchback of Notre Dame, Glow Kit	1969	45	90	150
481	Hunchback of Notre Dame, Glow Kit	1972	45	65	75
417	Indian Chief	1957	40	90	100
212	Indian Chief & Squaw	1957	60	125	150
418	Indian Squaw	1957	15	38	45
411	Infantryman	1957	20	75	100
256	Invaders UFO	1975	25	65	75
813	Invaders UFO	1968	35	85	100
853	Iwo Jima	1966	75	175	200
408	Jesse James	1966	60	100	200
851	Kennedy, John F.	1965	50	70	150

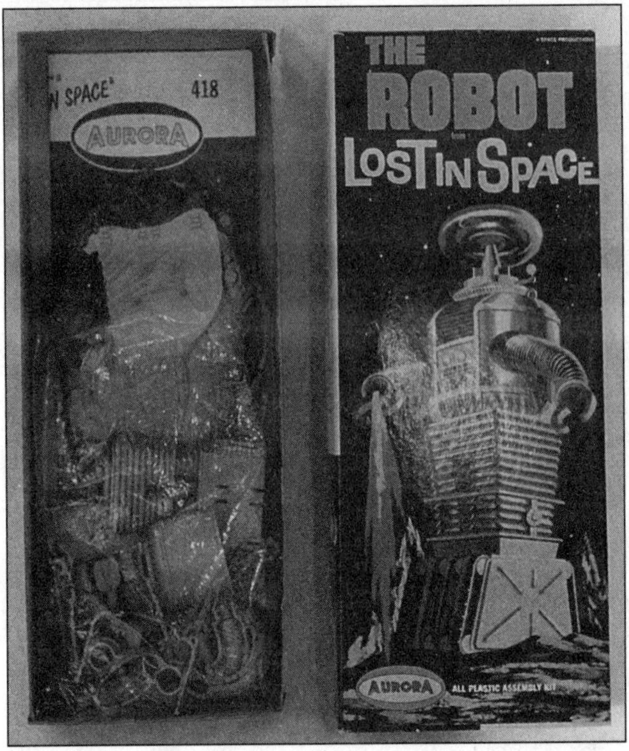

Lost in Space, The Robot, Aurora, 1968

Munster Koach, AMT, 1964

NO.	NAME	YEAR	B/U	NM	MIP
885	King Arthur	1973	50	100	200
825	King Arthur of Camelot	1967	30	65	75
468	King Kong	1964	75	300	450
468	King Kong, Glow Kit	1969	75	175	250
468	King Kong, Glow Kit	1972	75	125	175
484	King Kong's Thronester	1966	370	800	1250
830	Land of the Giants Space Ship	1968	150	290	400
816	Land of the Giants, Diorama	1968	150	350	450
808	Lone Ranger	1967	75	100	175
188	Lone Ranger, Comic Scenes	1974	20	40	50
420	Lost In Space, Large kit w/chariot	1966	450	1000	1350
419	Lost In Space, Small kit	1966	300	700	920
418	Lost In Space, The Robot	1968	250	600	800
455	Mad Barber	1972	45	100	150
457	Mad Dentist	1972	45	100	150
456	Mad Doctor	1972	45	100	150
412	Man From U.N.C.L.E., Illya Kuryakin	1966	75	100	200
411	Man From U.N.C.L.E., Napoleon Solo	1966	75	225	260
412	Marine	1959	20	80	100
860	Mays, Willie	1965	100	200	300
421	Mexican Caballero	1957	75	90	150
422	Mexican Senorita	1957	50	90	150
583	Mod Squad Wagon	1970	35	125	175
463	Monster Customizing Kit #1	1964	35	110	125
464	Monster Customizing Kit #2	1964	65	150	175
828	Moon Bus from 2001	1968	100	275	300
655	Mr. Hyde, Monsters of Movies	1975	25	65	75
922	Mr. Spock	1972	25	100	135
427	Mummy	1963	30	250	325
452	Mummy, Frightning Lightning	1969	30	200	400
452	Mummy, Glow Kit	1972	20	40	60
452	Mummy, Glow Kit	1969	20	90	175
459	Mummy's Chariot	1965	200	325	500
804	Munsters, Living Room	1964	400	900	1300
729	Neanderthal Man	1971	15	40	50
802	Neuman, Alfred E.	1965	100	275	400
806	Nutty Nose Nipper	1965	45	175	200
415	Odd Job	1966	200	350	450
635	Pain Parlor	1971	25	100	125
636	Pendulum	1971	25	50	75
416	Penguin	1967	200	400	525
428	Phantom of the Opera	1963	30	250	300
451	Phantom of the Opera, Frightning Lightning	1969	20	250	350
451	Phantom of the Opera, Glow Kit	1969	20	90	150
451	Phantom of the Opera, Glow Kit	1972	20	60	80
409	Pilot USAF	1957	75	145	175
K-9	Porthos, Three Musketeers	1958	25	75	100
814	Pushmi-Pullyu, Dr. Dolittle	1968	30	75	85
340	Rat Patrol	1967	30	75	95
474	Red Knight	1963	15	40	50
K-4	Red Knight	1957	15	75	100
488	Robin	1966	40	75	100
193	Robin, Comic Scenes	1974	20	70	90
657	Rodan	1975	125	200	350
405	Roman Gladiator with sword	1959	75	125	175
406	Roman Gladiator with Trident	1964	75	130	175

Aurora

NO.	NAME	YEAR	B/U	NM	MIP
216	Roman Gladiators	1959	100	200	250
862	Ruth, Babe	1965	100	225	325
419	Scotch Lad	1957	10	25	30
214	Scotch Lad & Lassie	1957	60	85	100
420	Scotch Lassie	1957	10	20	25
253	Seaview, Voyage to the Bottom of Sea	1975	100	160	190
707	Seaview, Voyage to the Bottom of Sea	1966	100	200	300
471	Silver Knight	1963	15	20	30
K-1	Silver Knight	1956	15	40	50
881	Sir Galahad	1973	15	40	50
826	Sir Galahad of Camelot	1967	25	90	175
882	Sir Kay	1973	20	40	50
883	Sir Lancelot	1973	20	40	50
827	Sir Lancelot of Camelot	1967	25	90	125
884	Sir Percival	1973	20	45	50
405	Spartacus (Gladiator/sword reissue)	1964	85	150	250
477	Spider-Man	1966	85	200	300
182	Spider-Man, Comic Scenes	1974	50	70	85
923	Star Trek, Klingon Cruiser	1972	20	65	75
921	Star Trek, USS Enterprise	1972	20	85	100
478	Superboy	1964	75	200	250
186	Superboy, Comic Scenes	1974	35	50	60
462	Superman	1963	25	275	340
185	Superman, Comic Scenes	1974	20	40	60
735	Tarpit	1972	50	90	125
820	Tarzan	1967	25	100	200
181	Tarzan, Comic Scenes	1974	15	25	35
207	Three Knights Set	1959	50	100	175
398	Three Musketeers Set	1958	95	200	350
809	Tonto	1967	20	100	200
183	Tonto, Comic Scenes	1974	15	20	30
818	Tracy, Dick	1968	60	100	250
819	Tracy, Dick, Space Coupe	1968	50	100	150
408	U.S. Marshal	1958	50	70	100
864	Unitas, Johnny	1965	75	125	175
410	United States Sailor	1957	10	25	30
452	Vampire, Castle Creatures	1966	60	175	250
638	Vampirella	1971	75	125	200
632	Victim	1971	20	65	75
K-6	Viking	1959	65	100	250
831	Voyager, Fantastic Voyage	1969	100	200	450
807	Wacky Back-Whacker	1965	50	100	250
852	Washington, George	1965	25	50	75
865	West, Jerry	1965	50	100	150
401	White Stallion	1964	10	25	30
401	White Stallion, reissue	1969	10	15	20
403	White-tailed Deer	1962	10	25	30
403	White-tailed Deer, reissue	1969	10	15	20
204	Whoozis, Alfalfa	1966	25	50	75
203	Whoozis, Denty	1966	25	50	75
202	Whoozis, Esmerelda	1966	25	50	75
205	Whoozis, Kitty	1966	25	50	75
206	Whoozis, Snuffy	1966	25	50	75
201	Whoozis, Susie	1966	25	50	75
483	Witch	1965	75	200	300
470	Witch, Glow Kit	1969	50	100	200

Penguin, Aurora, 1967

Odd Job, Aurora, 1966

Mummy's Chariot, Aurora, 1965

Aurora

NO.	NAME	YEAR	B/U	NM	MIP
470	Witch, Glow Kit	1972	75	100	125
425	Wolfman	1962	35	200	300
450	Wolfman, Frightning Lightning	1969	35	300	400
450	Wolfman, Glow Kit	1969	20	50	150
450	Wolfman, Glow Kit	1972	20	50	75
652	Wolfman, Monsters of Movies	1975	100	200	250
458	Wolfman's Wagon	1965	175	300	425
479	Wonder Woman	1965	150	300	500
801	Zorro	1965	125	200	300

Billiken

	Batman, type A	1989	35	85	100
	Batman, type B	1989	35	90	125
	Bride of Frankenstein		75	100	225
	Colossal Beast	1986	20	30	40
	Creature From Black Lagoon	1991	50	85	125
	Cyclops		75	100	200
	Dracula		60	100	150
	Frankenstein		60	90	125
	Joker	1989	40	90	150
	Mummy	1990	60	100	150
	Phantom of the Opera		100	200	275
	Predator		25	50	65
	Saucer Man		20	30	40
	She-Creature		25	40	50
	Syngenor		100	150	250
	The Thing		150	200	300
	Ultraman		20	30	40

Hawk

542	Beach Bunny	1964	25	65	75
532	Daddy the Way-Out Suburbanite	1963	30	75	90
531	Davy the Way-Out Cyclist	1963	30	75	90
530	Digger and Dragster	1963	30	75	90
	Drag Hag	1963	30	75	90
537	Endsville Eddie	1963	20	50	75
535	Francis The Foul	1963	15	35	45
548	Frantic Banana	1965	20	80	125
550	Frantic Cats	1965	20	70	100
547	Frantics Steel Pluckers	1965	20	70	100
549	Frantics Totally Fab	1965	20	80	100
533	Freddy Flameout	1963	20	65	80
543	Hidad Silly Surfer	1964	20	65	80
541	Hot Dogger Hangin' Ten	1964	20	65	80
538	Huey's Hut Rod	1963	20	45	55
	Killer McBash	1963	40	125	150
534	Leaky Boat Louie	1963	25	80	95
	Riding Tandem		25	65	75
637	Sling Rave Curvette	1964	15	25	45
547	Steel Pluckers	1965	20	65	85
550	Totally Fab	1965	25	75	100
636	Wade A Minute	1963	15	25	35
	Weird-Oh Customizing Kit	1964	75	250	300
545	Wild Woodie Car		20	50	60
540	Woodie on a Surfari	1964	25	85	100

Lindberg

NO.	NAME	YEAR	B/U	NM	MIP
6422	Bert's Bucket	1971	30	80	100
277	Big Wheeler	1965	30	80	100
280	Blurp	1964	10	20	45
273	Creeping Crusher	1965	20	40	55
6420	Fat Max	1971	30	80	90
281	Glob	1964	10	20	50
274	Green Ghoul	1965	20	35	50
272	Krimson Terror	1965	20	40	55
275	Mad Mangler	1965	20	40	55
276	Road Hog	1964	30	80	90
	Satan's Crate	1964	75	125	150
	Scuttle Bucket	1964	30	80	90
6421	Sick Cycle	1971	30	80	90
283	Voop	1964	10	20	50
282	Zopp	1964	10	20	50

Monogram

NO.	NAME	YEAR	B/U	NM	MIP
6028	Battlestar Galactica	1979	15	35	40
6008	Dracula	1983	20	25	40
105	Flip Out	1965	50	150	200
6007	Frankenstein	1983	20	30	40
6300	Godzilla	1978	40	65	80
6010	Mummy	1983	20	30	40
	Snoopy & Motorcycle	1971	15	25	30
6779	Snoopy & Sopwith Camel	1971	20	30	35
	Snoopy as Joe Cool	1971	25	50	100
MM106	Speed Shift	1965	70	100	200
	Super Fuzz	1965	80	100	225
6301	Superman	1978	20	30	40
6012	UFO, The Invaders	1979	15	35	40
6009	Wolfman	1983	20	30	35

MPC

NO.	NAME	YEAR	B/U	NM	MIP
1-1961	Alien	1979	25	75	100
0303	Ape Man Haunted Glow Head	1975	10	30	50
	AT-AT, Empire Strikes Back	1980	15	20	35
	Barnabas Vampire Van		75	100	250
1550	Barnabas, Dark Shadows	1968	100	325	400
1702	Batman	1984	20	30	35
612	Beverly Hillbillies Truck	1968	60	100	200
0609	Bionic Bustout, Six Million Dollar Man	1975	15	25	35
0610	Bionic Repair, Bionic Woman	1976	15	25	35
5003	Condemned to Chains Forever	1974	20	40	55
103	Curl's Gurl	1960s	25	65	80
	Darth Vader Bust	1977	20	40	55
	Darth Vader with Light Saber	1977	15	30	40
5005	Dead Man's Raft	1974	20	80	100
5001	Dead Men Tell No Tales	1974	20	40	50
1983	Encounter With Yoda Diorama	1981	15	20	35
5053	Escape From the Crypt	1974	20	40	50
604	Evil Rider, Six Million Dollar Man	1975	15	20	40
5004	Fate of the Mutineers	1974	20	40	50
602	Fight for Survival, Six Million Dollar Man	1975	15	20	40
0635	Fonzie & Dream Rod	1976	10	25	35
0634	Fonzie & Motorcycle		10	15	25

Mod Squad Wagon, Aurora, 1970

King Kong, Aurora, 1965

Evel Knievel, Addar, 1974

Rat Fink,
Revell, 1963

MPC

NO.	NAME	YEAR	B/U	NM	MIP
5007	Freed in the Nick of Time	1974	20	60	75
5006	Ghost of the Treasure Guard	1974	20	35	50
5051	Grave Robber's Reward	1974	20	40	50
402	Hogan's Heroes Jeep	1968	30	75	125
5002	Hoist High the Jolly Roger	1974	20	40	50
101	Hot Curl	1960s	20	40	50
	Hot Shot		20	40	50
1932	Hulk	1978	20	30	40
	Jabba's Throne Room	1983	20	30	45
603	Jaws of Doom, Six Million Dollar Man	1975	15	20	40
1925	Millennium Falcon with Light	1977	35	85	125
605	Monkeemobile	1967	70	175	225
1-0702	Muldowney, Shirley, Drag Kit		20	50	65
304	Mummy Haunted Glow Head	1975	20	30	50
	Night Crawler Wolfman Car	1971	45	75	125
622	Paul Revere & The Raiders Coach	1970	40	70	125
5052	Play It Again Sam	1974	35	80	100
1906	Raiders of the Lost Ark Chase Scene	1982	15	30	40
	Road Runner Beep Beep		20	50	75
1931	Spider-Man	1978	20	25	35
0902	Strange Changing Mummy	1974	15	30	40
0903	Strange Changing Time Machine	1974	20	40	50
0901	Strange Changing Vampire	1974	20	40	50
100	Stroker McGurk & Surf Rod	1960s	30	80	125
102	Stroker McGurk Tall T		30	80	125
1701	Superman	1984	15	20	25
641	Sweathog Dream Machine	1976	10	15	30
301	Vampire Haunted Glow Head	1975	15	20	45
5050	Vampire's Midnight Madness	1974	20	45	50
302	Werewolf Haunted Glow Head	1975	20	35	45
1552	Werewolf, Dark Shadows	1969	75	190	250
2651	Wile E. Coyote		20	50	65
617	Yellow Submarine	1968	70	170	250

Multiple Toymakers

955	Automatic Baby Feeder	1965	25	60	75
958	Back Scrubber	1965	25	60	75
981	Iron Maiden	1966	35	100	150
956	Painless Tooth Extractor	1965	25	60	75
957	Signal for Shipwrecked Sailors	1965	25	60	75
980	Torture Chair	1966	35	90	150
979	Torture Wheel	1966	35	90	150

Park

803	Castro, Born Losers	1965	35	65	125
802	Hitler, Born Losers	1965	35	65	175
801	Napoleon, Born Losers	1965	35	65	125

Precision

402	Cap'n Kidd the Pirate	1959	25	60	75
501	Crucifix		20	40	50

Pyro

NO.	NAME	YEAR	B/U	NM	MIP
166	Der-Baron	1958	50	75	100
175	Gladiator Show Cycle		20	40	50
281	Indian Chief		20	50	60
282	Indian Medicine Man		20	50	60
283	Indian Warrior	1960	20	50	60
168	Li'l Corporal	1970	25	65	75
276	Rawhide, Gil Favor	1958	20	50	60
277	Restless Gun Deputy	1959	20	50	60
176	Surf's Up	1970	15	35	40
286	U.S. Marshal		20	50	60
278	Wyatt Earp		20	50	60

Remco

452	Flintstones Motorized Paddy Wagon	1961	30	100	200
450	Flintstones Motorized Sports Car & Trailer	1961	30	100	200
451	Flintstones Motorized Yacht	1961	30	100	200

Revell

1307	Angel Fink	1965	40	100	140
1353	Beatles, George Harrison	1965	75	150	250
1352	Beatles, John Lennon	1965	75	150	250
1350	Beatles, Paul McCartney	1965	75	150	200
1351	Beatles, Ringo Starr	1965	75	150	200
	Birthday Bird	1959	40	150	250
1931	Bonanza	1965	50	100	150
1304	Brother Rat Fink	1963	20	50	65
	Busby the Tasselated Afghan Spaniel Yak	1960	50	125	250
2000	Cat in the Hat	1960	45	100	150
2050	Cat in the Hat with Thing 1 and Thing 2	1960s	45	100	225
1397	Charlie's Angels Van	1977	10	20	30
	Dr. Seuss Zoo Kit #1	1959	50	225	400
	Dr. Seuss Zoo Kit #2	1960	50	350	500
1303	Drag Nut	1963	20	50	65
1310	Fink-Eliminator	1965	30	100	200
1450	Flash Gordon & Martian	1965	60	100	150
1930	Flipper	1965	75	100	150
	Gowdy the Dowdy Grackle	1959	40	75	175
	Grickily the Gractus	1960	50	125	250
	Horton the Elephant	1960	150	200	400
323	McHale's Navy PT-73	1965	25	60	75
1302	Mother's Worry	1963	20	60	100
1301	Mr. Gasser	1963	30	75	100
3181	Mr. Gasser BMR Racer	1964	30	75	100
	Norval the Bashful Blinket	1959	40	75	175
1451	Phantom & Voodoo Witch Doctor	1965	50	100	200
1305	Rat Fink	1963	25	60	75
	Rat Fink Lotus Racer	1964	25	60	75
	Robbin' Hood Fink	1965	200	300	400
2004	Roscoe the Many-Footed Lion	1960s	30	75	165
1309	Scuz-Fink with Dingbat	1965	275	350	450
	Superfink	1964	150	250	350
1306	Surfink	1965	35	85	100
	Tingo the Noodle-Topped Stroodle	1960s	50	75	175
1271	Tweedy Pie with Boss-Fink	1965	200	300	400

Battlestar Galactica Cylon Base Star, Monogram, 1978

Angel Fink, Revell, 1965

Toy Biz

NO.	NAME	YEAR	B/U	NM	MIP
48660	Ghost Rider	1996	7	12	25
48656	Hulk	1996	5	10	20
48640	Onslaught	1997	7	12	25
48653	Silver Surfer	1996	5	10	20
48651	Spider-Man	1996	5	10	20
48658	Spider-Man, with wall	1996	5	10	20
48659	Storm	1996	5	10	20
48652	Thing	1996	5	10	20
48654	Venom	1996	5	10	20
48657	Wolverine	1996	5	10	20

Figural model kits are worth the most still boxed and unbuilt. Built kits, if built and painted well, still have collectible value. Pictured is Revell's Mr. Gasser.

PEZ Candy Dispensers

In 1952, Americans saw the inauspicious introduction of an Austrian mint in a handy dispenser. Long popular in the homeland, the pocket candy lost something in the translation. The marketing cure for this was successful beyond all expectations.

PEZ was created in 1927 as a peppermint candy and breath mint sold in a clever package which dispensed the candies one at a time. Highly successful in Europe, it became the fashionable adult candy of its time. But its launch in America found a disinterested public.

Soon PEZ would be reinvented for the American market as a children's candy with fruit-flavored candies replacing the staid pfefferminz of old. The dispensers were redesigned and given colorful heads in the shapes of popular cartoon characters, and American children quickly claimed the new candy as their own.

Today PEZ is available everywhere from the local Kmart to the corner convenience store, and few Americans can see a dispenser without it evoking a few childhood memories. This ability to reconnect us, either with our own childhoods or with our national past, is central to collectibility in any field, and a PEZ dispenser holds a rich postwar legacy in its tiny plastic container.

PEZ collectors nationwide have formed clubs, published newsletters and held national conventions. Long ago, kids threw away the dispensers, but these once lowly candy holders have grown in popularity and respect to the point that rare dispensers are now highly-prized collectibles and have even been sold by internationally-known auction houses like Christie's.

The PEZ market has developed some noteworthy variations on standard collecting procedures. In many fields, a toy still in the original package commands a premium over the same toy with no package. This is not usually the case in PEZ collecting. Before blister cards, PEZ dispensers were packaged in boxes or cellophane bags, which did not allow for either display or handling of the toys themselves.

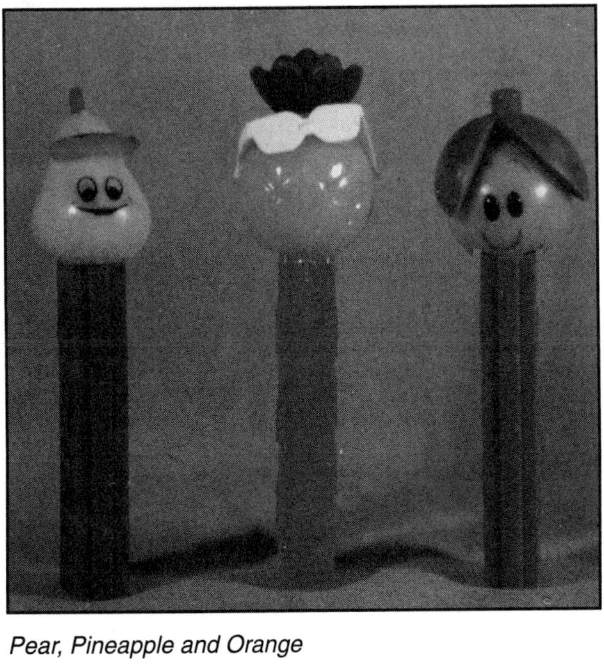

Pear, Pineapple and Orange

Since dispenser stems are easily interchangeable, PEZ authorities hold that only variations in head configuration or coloring affect value. There is no difference in value between dispensers with different colored stems but the same head.

Finally, there is the matter of feet and no feet. This refers to the presence or absence of a flattened rounded base on the stem resembling flat shoes. PEZ dispensers released in America before 1987 were all of the no-feet variety, so a dispenser with feet was made after that year. However, certain older molds continued to be produced with no feet after 1987 as well, but these are common dispensers with little variance in value between feet and no-feet varieties. The major difference in value here applies to older, no-feet dispensers that were discontinued and perhaps reissued after 1987 with feet.

Trends

PEZ dispensers have seen mixed performance at shows, but values are on the rise, with the most sought after dispensers being the Bridge, Mueslix, Pineapple, Vucko and Make-A-Face. Two PEZ items fast on the rise in popularity are Vucko from the 1984 Sarajevo Olympics and advertising regulars. The latter look like a regular non-headed PEZ with a company's name on the side.

Low-end common dispensers have been selling in large quantities, but many new collectors are hedging at the price points of dispensers in the $75 and up range, indicating this is still a strongly collector-based market. The top of the market appears healthy, with reports of dispensers achieving two and three times prior estimates at auction.

The Top 10 PEZ Dispensers (in Mint condition)

1. Make-A-Face, American card $3,300
2. Elephant 3,000
3. Witch Regular 3,000
4. Make-A-Face, German with Super Spiel card 3,000
5. Mueslix, European 2,900
6. Pineapple 2,750
7. Lion's Club Lion 2,500
8. Make-A-Face, no card 2,200
9. Space Trooper, gold 1,900
10. Bride 1,900

Contributor to this section: John Devlin, 5541 Oakville Center #119, St. Louis, MO 63129-3554.

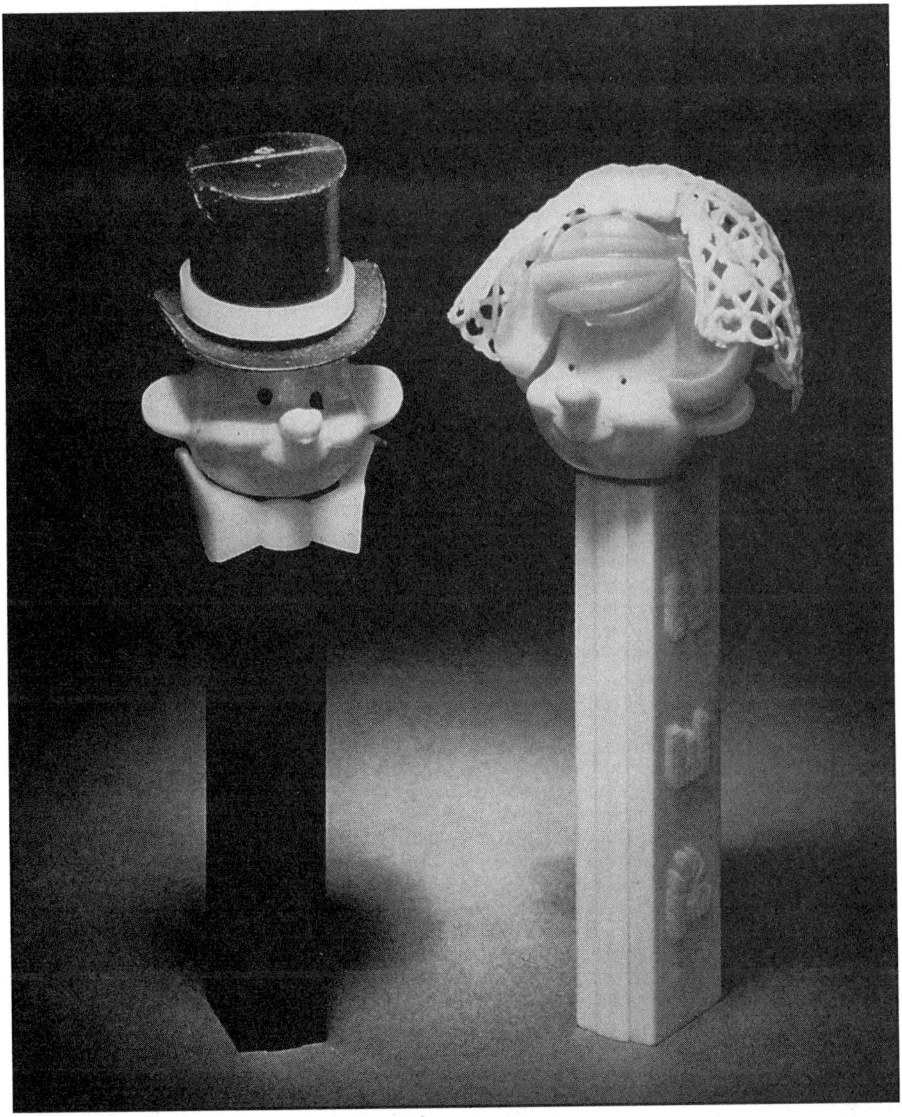

The Bride and Groom are two of the rarest dispensers. This set was offered several years ago at a Christie's auction.

PEZ

NAME	DESCRIPTION	EX	MINT
Advertising Regular	no feet, no head, advertising printed on side MUST be complete	600	1150
Air Spirit	no feet, reddish triangular fish face	175	250
Alpine	1972 Munich Olympics, no feet, green hat with beige plume, black mustache	650	850
Angel A	no feet, hair and halo	50	70
Arithmetic Regular	no feet, headless dispenser with white top, side of body has openings with columns of numbers	400	500
Arlene	with feet, pink head, Garfield's girlfriend	2	5
Asterix	no feet, blue hat with wings, yellow mustache, European	750	1200
Asterix (1998 foreign issue)	head different than old	3	5
Astronaut A	no feet, helmet, yellow visor, small head	400	550
Astronaut B	no feet, green stem, white helmet, yellow visor, large head	120	150
Baloo	no feet, Jungle Book	20	40
Baloo	with feet, Jungle Book	10	15
Bambi	with feet	20	35
Barney Bear	no feet, brown head, white cheeks and snout, black nose	25	45
Barney Bear	with feet, brown head, white cheeks and snout, black nose	15	25
Barney Rubble	Flintstones	1	2
Baseball Dispenser Set	no feet, baseball glove with ball, bat, white home plate marked "PEZ"	450	750
Baseball Glove	no feet, brown baseball glove with white ball	190	230
Batgirl	Soft Head Superhero, no feet, blue mask, black hair	125	150
Batman	no feet, blue cape, mask and hat	100	150
Batman	Soft Head Superhero, no feet, blue mask	145	175
Batman	with feet, blue	2	5
Betsy Ross	no feet, dark hair and white hat, Bicentennial issue	125	150
Big Top Elephant	no feet, orange head with blue pointed hat	60	85
Big Top Elephant	no feet, gray-green head, red flat hat	60	85
Big Top Elephant With Hair	no feet, yellow head and red hair	200	300
Bouncer Beagle		5	6
Boy	no feet, brown hair	15	25
Boy with Cap	no feet, white hair, blue cap	75	125
Bozo the Clown	die-cut Bozo and Butch on stem, no feet, white face, red hair and nose	125	170
Bride	no feet, white veil, light brown, blond or red hair	900	1900
Brutus	no feet, black beard and hair	160	250
Bugs Bunny	no feet, gray head with white cheeks	5	10
Bugs Bunny	with feet, gray head with white cheeks	2	4
Bullwinkle	no feet, brown head, yellow antlers	200	250
Bunny 1990	with feet, long ears, white face	1	3
Bunny Original A	no feet, narrow head and tall ears	300	375
Bunny Original B	no feet, tall ears and full face, smiling buck teeth	300	375
Bunny w/Fat Ears	no feet, wide ear version, yellow	15	20
Bunny w/Fat Ears	no feet, wide ear version, pink	20	25
C-3PO		1	3
Camel	with feet, brown face with red fez hat	45	60
Candy Shooter	red body, white grip, with German license and double PEZ candy	50	75
Candy Shooter	black gun with PEZ monogram on stock	85	150
Candy Shooter on Card		275	450
Captain (Paul Revere)	no feet, blue hat, Bicentennial issue	125	160
Captain America	blue mask	65	80
Captain America	no feet, blue cowl, black mask with white letter A	95	150
Captain Hook	no feet, black hair, flesh face winking with right eye open	40	80
Casper The Friendly Ghost	no feet, white face	90	150
Casper The Friendly Ghost	die-cut	150	200

NAME	DESCRIPTION	EX	MINT
Cat with Derby		45	70
Charlie Brown	with feet, blue cap, smile with red tongue at corner	15	10
Charlie Brown	with feet, crooked smile, blue cap	1	2
Charlie Brown	with feet, blue cap, eyes closed	20	35
Chewbacca		1	3
Chick in Egg	no feet, yellow chick in egg shell, red hat	15	25
Chick in Egg	no feet, yellow chick in egg shell, no hat	75	120
Chip	no feet, black top hat, tan head with white sideburns, brown nose, foreign issue	75	95
Clown	with feet, green hat, foreign issue (Merry Music Makers)	3	6
Clown with Chin	no feet, long chin, hat and hair	40	60
Clown with Collar	no feet, yellow collar, red hair, green hat	40	60
Cockatoo	no feet, yellow beak and green head, red head feathers	40	60
Cocoa Marsh Spaceman	no feet, clear helmet on small male head, with Cocoa Marsh embossed on side	145	210
Cool Cat	with feet, orange head, blue snout, black ears	20	35
Cow A	no feet, cow head, separate nose	50	70
Cow B	no feet, blue head, separate snout, horns, ears and eyes	75	120
Cowboy	no feet, human head, brown hat	225	300
Creature From Black Lagoon	no feet, green head and matching stem, with copyright	245	320
Crocodile	no feet, dark green head with red eyes	70	115
Daffy Duck A	no feet, black head, yellow beak, removable white eyes	10	15
Daffy Duck B	with feet, black head, yellow beak	3	5
Dalmatian Pup	with feet, white head with left ear cocked, foreign issue	30	50
Daniel Boone	no feet, light brown hair under dark brown hat, Bicentennial issue	145	200
Darth Vader		1	3
Dewey	no feet, blue hat, white head, yellow beak, small black eyes	20	35
Diabolic	no feet, soft orange monster head	175	250
Dino the Dinosaur		1	2
Dinosaurs	PEZasaurs line, four different dinosaurs, each	1	3
Doctor	no feet, white hair and mustache, gray reflector on white band, black stethoscope	125	195
Dog	no feet, orange dog face with black ears, foreign issue	10	35
Donald Duck	die-cut	125	170
Donald Duck A	no feet, blue hat, one-piece head and bill, open mouth	12	25
Donald Duck B	with feet, blue hat, white head and hair with large eyes, removable beak	5	10
Donkey	with feet, gray head with pink nose	5	10
Donkey Kong Jr.	no feet, blond monkey face, dark hair, white cap with "J" on it, with box	350	650
Dopey	no feet, flesh colored die cut face with wide ears, orange cap	160	200
Dr. Skull B	no feet, black collar	10	15
Droopy Dog	with feet	5	6
Droopy Dog A	with feet, white face, flesh snout, black movable ears and red hair	5	30
Droopy Dog B	with feet	5	6
Duck	with feet, brown head with yellow beak	20	30
Duck	no feet, brown head with yellow beak	25	40
Duckie with Flower	no feet, flower, duck head with beak	60	80
Dumbo	no feet, gray head with large ears, red hat	50	85
Dumbo	with feet, blue head with large ears, yellow hat	20	30
Easter Bunny Die Cut	no feet, die cut	425	550
Elephant	gold or black	2400	3000
Engineer	no feet, blue hat	90	150
Fireman	no feet, black mustache, red hat with gray #1 insignia	60	80

Bullwinkle

Psychedelic Eye (left) and Psychedelic Flower (right)

NAME	DESCRIPTION	EX	MINT
Foghorn Leghorn	no feet, brown head, yellow beak, red wattle	75	90
Foghorn Leghorn	with feet, brown head, yellow beak, red wattle	60	75
Football Player	no feet, white stem, red helmet with white stripe	110	140
Fozzie Bear	with feet, brown head, bow tie, small brown hat	1	2
Frankenstein	no feet, black hair, gray head	240	275
Fred Flintstone	Flintstones	1	2
Frog	no feet, yellow and green head with black eyes, foreign issue	40	50
Frog	with feet, yellow and green head with black eyes, foreign issue	35	45
Garfield	with feet, orange head	1	2
Garfield w/teeth	with feet, orange head, wide painted toothy grin	1	2
Garfield w/visor	with feet, orange face, green visor	1	2
Giraffe	no feet, orange head with horns, black eyes	100	150
Girl	with feet, pigtails	1	3
Girl	no feet, blond pigtails	15	25
Golden Glow	no feet, no head, gold stem and top	80	150
Gonzo	with feet, blue head, yellow eyelids, bow tie	1	3
Goofy B	no feet, red hat, painted nose, removable white teeth	35	50
Goofy C	same as version B except teeth are part of head	10	20
Goofy D	with feet, beige snout, green hat	1	3
Gorilla	no feet, black head with red eyes and white teeth	50	70
Green Hornet	no feet, green mask and hat; two hat styles exist	175	250
Groom	no feet, black top hat, white bow tie	225	375
Gyro Gearloose		5	6
Henry Hawk	no feet, light brown head, yellow beak	60	90
Henry Hawk	with feet	50	65
Hippo	no feet, green stem with "Hippo" printed on side, foreign issue	750	950
Huey	no feet, red hat, white head, yellow beak, small black eyes	20	35
Huey, Dewey or Louie Duck	with feet, red, blue, or green stem and matching cap, white head and orange beak	4	6
Hulk A	no feet, dark green head, black hair	25	40
Hulk B	with feet, light green head, tall dark green hair	3	6
Hulk B	no feet, light green head, dark green hair	20	25
Indian	with feet, black hair and green headband with feather, foreign issue (Merry Music Makers)	10	15
Indian Brave	no feet, small human head, Indian headband with one feather, Bicentennial issue	145	200
Indian Chief	no feet, war bonnet, Bicentennial issue	90	110
Indian Chief	no feet, marbleized headress	100	150
Indian Woman	no feet, black hair in braids with headband	95	140
Jerry (Tom & Jerry)	no feet, brown face, pink lining in ears	100	200
Jerry (Tom & Jerry)	with feet, brown face, multiple piece head	8	10
Jiminy Cricket	no feet, green hatband and collar, flesh face, black top hat	175	250
Joker	Soft Head Superhero, no feet, green painted hair	125	150
Kermit the Frog	with feet, green head	1	2
King Louie	with feet, brown hair and 'sideburns' over light brown head	10	15
Knight	no feet, gray helmet with plume	175	275
Koala	with feet, brown head with a black nose, foreign issue	20	25
Lamb	no feet, white head with a pink bow	5	20
Lamb	no feet, pink stem, white head, whistle	10	25
Li'l Bad Wolf	no feet, black ears, white face, red tongue	15	25
Li'l Bad Wolf	with feet, black ears, white face, red tongue	12	20
Lion with Crown	no feet, black mane, green head with yellow cheeks and red crown	40	70
Lion's Club Lion	no feet, stem imprinted "1962 Lion's Club Inter'l Convention," yellow roaring lion head	1600	2500

NAME	DESCRIPTION	EX	MINT
Little Lion	no feet, yellow head with brown mane	35	50
Little Orphan Annie	no feet, light brown hair, flesh face with black painted features	100	145
Louie	no feet, green hat, white head, yellow beak, small black eyes	5	15
Lucy	with feet, black hair	1	2
Maharajah	no feet, green turban with red inset	35	50
Make-A-Face	no card, feet, oversized head with 18 facial parts	1900	2200
Make-A-Face	American card	2300	3300
Make-A-Face	German "Super Spiel" card	2000	3000
Mary Poppins	no feet, flesh face, reddish hair, lavender hat	900	1000
Merlin Mouse	no feet, gray head with flesh cheeks, green hat	15	20
Merlin Mouse	with feet, gray head with flesh cheeks, green hat	10	15
Mexican	no feet, yellow sombrero, black beard and mustache, two earrings	125	155
Mickey Mouse A	no feet, black head and ears, pink face, mask with cut out eyes and mouth, nose pokes through	50	75
Mickey Mouse B	no feet, painted face, non-painted black eyes and mouth	75	125
Mickey Mouse C	no feet, flesh face, removable nose, painted eyes	5	10
Mickey Mouse D	no feet, flesh face, mask embossed white and black eyes	3	5
Mickey Mouse Die-Cut	no feet, die-cut stem with Minnie, die cut face mask	100	150
Mickey Mouse Die-Cut	no feet, die-cut stem with Minnie, painted face	100	200
Mickey Mouse E	with feet, flesh face, bulging black and white eyes, oval nose	1	3
Mimic the Monkey	no feet, wearing baseball cap, white eyes	25	35
Miraculix (1998 foreign issue)		3	5
Miss Piggy	with feet, pink face, yellow hair	1	2
Miss Piggy	with eyelashes	10	15
Monkey	with feet, tan monkey face in brown head, foreign issue	10	25
Monkey Sailor	no feet, cream face, brown hair, white sailor cap	40	50
Mowgli	with feet, black hair over brown head, Jungle Book	5	20
Mowgli	no feet, black hair over brown head, Jungle Book	15	35
Mr. Ugly	no feet, black hair, many color variations	40	80
Mueslix	no feet, white beard, moustache and eyebrows, European	2750	2900
Nermal	with feet, gray stem and head	1	2
Nurse	no feet, hair color variations, white nurse's cap	90	145
Obelix	no feet, red mustache and hair, blue hat, European	750	1200
Obelix (1998 foreign issue)	head different than old	3	5
Octopus	no feet; black; orange or black head	30	65
Octopus	no feet, red head	65	90
Olive Oyl	no feet, black hair and flesh painted face	100	200
One-Eyed Monster	no feet, gorilla head with one eye missing	75	95
Orange	no feet, orange (fruit) head with face and leaves on top	175	250
Panda A	no feet, yellow head with black eyes and ears	150	250
Panda A	no feet, white head with black eyes and ears	10	20
Panda B	with feet, white head with black eyes and ears	1	3
Panther	no feet, blue head with pink nose	120	165
Papa Smurf	with feet, red hat, white beard, blue face	3	6
Parrot	with feet, red hair, yellow beak and green eyes	5	6
Pear	no feet, yellow pear face, green visor	700	850
Pebbles Flintstone	Flintstones	1	2
Penguin	with feet, penguin head with yellow beak and red hat, foreign issue	5	6
Penguin (Batman villain)	Soft Head Superhero, no feet, yellow top hat, black painted monocle	125	150
Personalized Regular	no feet, no head, paper label for monogramming	175	250
Peter Pan	no feet, green hat, flesh face, orange hair	125	175

One of the numerous versioins of Make-A-Face, all of which are valuable.

From left, Indian Woman, Indian Chief, Indian Brave, and Pilgrim

PEZ

NAME	DESCRIPTION	EX	MINT
Peter PEZ	1990s remake with feet	1	2
Peter PEZ	no feet, blue top hat that says "PEZ," white face, yellow hair	50	65
Petunia Pig	with or without feet, black hair in pigtails	20	30
Pig	with feet, pink head	25	35
Pilgrim	no feet, pilgrim hat, blond hair, hat band	120	145
Pilot	no feet, blue hat, gray headphones	125	175
Pineapple	no feet, pineapple head with greenery and sunglasses	2250	2750
Pinocchio A	no feet, red or yellow cap, pink face, black painted hair	140	175
Pinocchio B	no feet, black hair, red hat	100	125
Pirate	no feet, red cap, patch over right eye	40	55
Pluto A	no feet, yellow head, long black ears, small painted eyes	5	20
Pluto C	with feet, yellow head, long painted black ears, large white and black decal eyes	1	5
Policeman	no feet, blue hat with gray badge	30	40
Pony-Go-Round	no feet, orange head, white harness, blue hair	65	90
Popeye A	no feet, yellow face, painted hat	90	120
Popeye B	no feet, removable white sailor cap	60	80
Popeye C	no feet, one eye painted, removeable pipe and cap	65	75
Practical Pig A	no feet, rounded blue hat, round snout	20	35
Practical Pig B	no feet, crooked blue hat	20	35
Psychedelic Eye	no feet, decal design on stem, beige or black hands; many color variations	400	525
Psychedelic Eye (1998 limited-edition)	limited-edition; has 1967 copyright date	10	20
Psychedelic Flower	no feet, stem with decal on side, many color variations	400	550
Pumpkin A	no feet, green stem, carved face	3	15
Pumpkin B	black facial features	1	3
Pussy Cat	no feet, cat head with hat	35	65
Raven	no feet, black head, beak, and glasses	25	40
Regular	no feet, no head, stem with top only; assorted colors; 1990s issues	2	3
Regular	no feet, no head, stem with top only; 1950s	100	150
Rhino	with feet, green head, red horn, foreign issue	5	6
Ringmaster	no feet, white bow tie, white hat with red hatband, and black handlebar moustache	200	300
Road Runner A	no feet, purple head, yellow beak	20	35
Road Runner B	with feet, purple head, yellow beak	8	15
Roman Soldier (1998 foreign issue)		5	7
Rooster	no feet, yellow head, comb and wattle	25	40
Rooster	no feet, white head, comb	20	35
Rudolph	no feet, brown deer head, red nose	30	50
Sailor	no feet, blue hat, white beard	100	150
Santa	full body stem with painted Santa suit and hat	135	190
Santa Claus A	no feet, ivory head with painted hat	80	125
Santa Claus B	no feet, small head with flesh painted face, black eyes, red hat	85	150
Santa Claus C	no feet, large head with white beard, flesh face, red open mouth and hat	1	5
Santa Claus C	with feet, removable red hat, white beard	1	2
Scarewolf	no feet, soft head with orange hair and ears	175	250
Scrooge McDuck A	no feet, white head, yellow beak, black top hat and glasses, white sideburns	10	25
Scrooge McDuck B	with feet, white head, removable yellow beak, tall black top hat and glasses, large eyes	5	6
Sheik	no feet, white head drape, headband	30	50
Sheriff	no feet, brown hat with badge	100	150

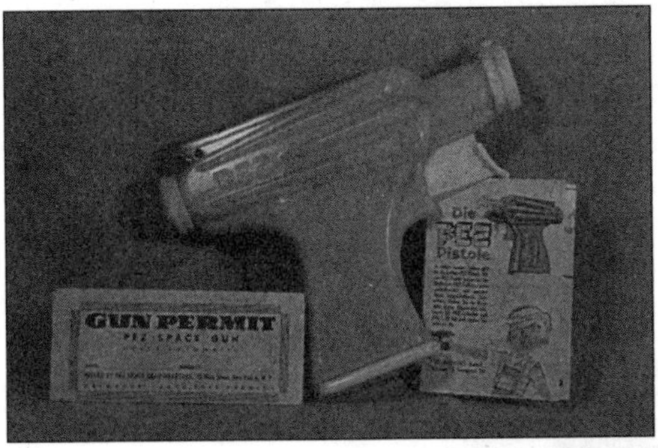

Space Gun, 1950s

Popular and valuable Olympics dispensers include, left to right, variations of Vucko Wolf and Winter Olympics Snowman.

Among the most unique, and valuable, PEZ dispensers, are the vintage dispensers with soft rubber heads. These include the Eeerie Spectres series.

Some of the spookiest PEZ dispensers are these Universal Monsters including, left to right, Wolf Man, Creature from the Black Lagoon and Frankenstein.

NAME	DESCRIPTION	EX	MINT
Smurf	no feet, blue face, white hat	3	6
Smurfette	with feet, blue face, yellow hair, white hat	3	8
Snoopy	with feet, with white head and black ears	1	2
Snow White	no feet, flesh face, black hair with ribbon and matching collar	150	200
Snowman	no feet, black hat, white face	5	10
Space Gun 1950s	various colors	250	375
Space Gun 1980s	red or silver, loose	65	90
Space Gun 1980s	red or silver, on blister pack	125	150
Space Trooper	gold	1200	1900
Space Trooper	full bodied robot with backpack; blue, red, or yellow	250	375
Spaceman	no feet, clear helmet over flesh-color head	90	150
Sparefroh (foreign issue)	no feet, green stem, red triangle hat, coin glued on stem	575	900
Speedy Gonzales	with feet, brown head, yellow sombrero	10	15
Speedy Gonzales	no feet, brown head, yellow sombrero	20	35
Spider-Man A	no feet, red head with black eyes	10	15
Spider-Man B	with feet, bigger head	1	2
Spike	with feet, brown face, pink snout	5	6
Spook	blue head with horns, soft head	175	250
Stand By Me	dispenser packed with mini film poster and candy	125	175
Stewardess	no feet, light blue flight cap, blond hair	125	175
Stormtrooper		1	3
Sylvester	no feet, black head, white whiskers, red nose	10	15
Sylvester	with feet, black head, white whiskers, red nose	1	3
Teenage Mutant Ninja Turtles	four different characters	1	3
Thor	no feet, yellow hair, gray winged helmet	200	300
Thumper	no feet, orange face, with logo	200	300
Thumper	with feet, no Disney logo	25	45
Tiger	with feet, tiger head with white snout	5	6
Tinkerbelle	no feet, pale pink stem, white hair, flesh face with blue and white eyes	175	225
Tom A (Tom & Jerry)	no feet, gray cat head with painted black features	20	35
Tom B (Tom & Jerry)	with feet, gray cat head with removable facial features	10	15
Truck A	cab, stem body, single rear axle	50	100
Truck B	cab, stem body, dual rear axle and dual arch fenders	50	75
Truck C	cab, stem body, dual rear wheels with single arch fender, movable wheels	10	25
Truck D	cab, stem body, dual rear wheels, single arch fender, nonmovable wheels	1	2
Tweety Bird	no feet, yellow head	10	20
Tweety Bird	with feet, yellow head	1	3
Tyke	with feet, brown head	10	25
Uncle Sam	no feet, stars and stripes on hat band, white hair and beard, Bicentennial issue	125	225
Vamp	no feet, light gray head on black collar, green hair and face, red teeth	200	275
Vucko Wolf	1984 Sarajevo Olympics issue, with feet, gray or brown face with ski hat	600	750
Vucko Wolf	1984 Sarajevo Olympics issue, with feet, gray or brown face with bobsled helmet	600	750
Vucko Wolf	1984 Sarajevo Olympics issue, with feet, gray or brown face	600	775
Whistles	no feet; police whistles on top	20	35
Wile E. Coyote	no feet, brown head	50	60
Wile E. Coyote	with feet, brown head	20	40
Winnie the Pooh	with feet, yellow head	50	60

NAME	DESCRIPTION	EX	MINT
Winter Olympics Snowman	1976 Innsbruck, red nose and hat, white head with arms extended, black eyes, blue smile	350	400
Witch 1 Piece	no feet, black stem with witch embossed on stem, orange one-piece head	175	275
Witch 3 Piece A	with feet, red head and hair, green mask, black hat	3	7
Witch 3 Piece B	no feet, chartreuse face, black hair, orange hat	60	100
Witch Regular	orange stem w/black witch graphics, no head	1500	3000
Wolf Man	no feet, black stem, gray head	200	265
Wonder Woman	with feet, red stem	1	3
Wonder Woman	no feet, black hair and yellow band with red star	5	10
Wonder Woman	Soft Head Superhero, no feet, black hair and yellow band with star	100	150
Woodstock	with painted feathers	5	6
Woodstock	with feet, yellow head	1	2
Wounded Soldier	no feet, white bandage, brown hair, Bicentennial issue	125	150
Yappy Dog	no feet, black floppy ears and nose, green or orange head	50	65
Yoda		1	3
Zombie	no feet, burgundy and black soft head	175	250
Zorro	no feet, black mask and hat	50	60
Zorro with Logo	no feet, black mask and hat, says "Zorro" on stem	60	80

Most modern PEZ dispensers are cheap, fun and colorful — but not valuable to collectors.

Restaurant Premiums

Collecting restaurant premiums — fast food toys — is a recent but already highly-developed area for collectors, with many clubs and conventions in existence. It offers collectors an entry into the field at relatively inexpensive prices, compared to other collectible toys.

The typical child's meal customer is under age 12, an age range not known for gentle play habits. Thus, condition of toys is critical to value. Only very rare toys hold any value at all if found in less than perfect condition. This price guide lists two grading classifications — Excellent and Mint. Excellent toys are rarely played with and exhibit no scratches or other imperfections.

Rarity is also a primary factor in determining value. The age of a toy plays a role in this, as does popularity and cross collectibility. Modern Disney movie tie-in toys are frequently worth more than older toys because of the strong Disneyana collector market. The same holds true for popular cartoon or comic character items.

Another factor affecting rarity is distribution. Some restaurant premiums were offered only in certain regions of the country. The toys of these campaigns are known as "regionals" and command higher than average prices due to their limited release areas.

Several years back some toys had to be recalled, resulting in the design of special one-piece toys for younger children, commonly called "Under 3" toys. Under 3 toys are not produced for each campaign, and are not normally advertised in the in-store displays. Lower numbers of these toys are released, again resulting in premiums typically 20 percent higher than regular toys of the same campaign.

One more factor deserves mention — international toys. Major film and comic character toy campaigns sometimes run worldwide with little or no changes from country to country. Sometimes only the package printing is changed. But occasionally foreign market toys are never released in America. These toys are highly valued by some collectors simply due to their foreign status. Other collectors also consider aspects such as popular character affiliation when calculating the value of these toys. Again, as the market matures, these values and item inventories will establish a track record.

Toy Story Puppets, 1995, Burger King

The 10-year-old McDonald's customer of 1979 will not turn 30 years old until 1999, but it seems likely that nostalgia will accompany following generations into their middle years, as it has for baby boomers. As McDonald's is the first global restaurant, and popular film and TV tie-ins are now the rule of the day, the future of this field looks secure.

Trends

McDonald's continues to lead the way. Across fast food companies, character toys lead all others, with certain series standing heads above their peers. Characters from Disney films continue to lead the pack. Over all, the restaurant premium market shows no sights of fatigue and continues on a strong course.

Note: Unless otherwise noted, prices listed are for each piece in a set, not the entire set.

The Top 10 Restaurant Premiums (in Mint condition)

1. Big Boy Nodder, Big Boy, 1960s $1,500
2. Black History Coloring Books, McDonald's, 1988 400
3. Big Boy Bank, large, Big Boy, 1960s 300
4. Big Boy Bank, medium, Big Boy, 1960s 165
5. Colonel Sanders Nodder, Kentucky Fried Chicken, 1960s 150
6. Transformers, McDonald's, 1985 140
7. Big Boy Board Game, Big Boy, 1960s 120
8. Big Boy Bank, small, Big Boy, 1960s 100
9. Triple Play Funmeal Boxes, Burger Chef, 1977 90
10. My Little Pony, regional promo, McDonald's, 1985 80

Contributor to this section: Jeff Escue, 164 Larchmont Ln., Bloomingdale, IL 60108-1412.

RESTAURANT PREMIUMS

Arby's

NAME	DESCRIPTION	YEAR	EX	MINT
Babar at the Beach Summer Sippers	set of three squeeze bottles: orange, yellow or purple top	1991	1	5
Babar Figures	set of four: Babar with sunglasses, elephant with binoculars, Babar with monkey, and with camera	1990	1	3
Babar License Plates	set of four: Paris, Brazil, USA, North Pole	1990	1	2
Babar Puzzles	set of four: Cousin Arthur's New Camera, Babar's Gondola Ride, Babar and the Haunted Castle, Babar's Trip to Greece	1990	2	4
Babar Squirters	set of three		1	3
Babar Stampers	set of three: Babar, Flora, Arthur	1990	2	4
Babar Storybooks	set of three: Read Get Ready, Set, Go, Calendar-Read and Have Fun-Read and Grow and Grow	1991	1	2
Babar World Tour Vehicles	set of three vehicles; Babar in helicopter, Arthur on trike, Zephyr in car	1990	2	4
Little Miss Figures	set of eight: Giggles, Helpful, Shy, Splendid, Late, Naughty, Star, Sunshine	1981	3	10
Looney Tunes Car-Tunes	set of six: Sylvester's Cat-illac, Daffy's Dragster, Yosemite Sam's Rackin Frackin Wagon, Taz's Slush Musher, Bugs' Buggy, Road Runner's Racer	1990	3	6
Looney Tunes Characters	set of three: Tazmanian Devil as pilot, Daffy as student, Sylvester as fireman	1989	3	6
Looney Tunes Christmas Ornament	Bugs, Porky Pig, Tweety	1989	3	8
Looney Tunes Figures	figures on oval base: Tasmanian Devil, Tweetie, Porky, Bugs, Yosemite Sam, Sylvester, Pepe Le Pew	1987	4	10
Looney Tunes Figures	Bugs, Daffy, Taz, Elmer, Road Runner, and Wile E. Coyote	1988	3	6
Looney Tunes Pencil Toppers	Sylvester, Yosemite, Porky, Bugs, Taz, Daffy, Tweety	1988	5	10
Looney Tunes Rings	set of four flicker rings: Bugs, Yosemite Sam, Porky, Daffy	1987	20	40
Megaphone, Minnesota Twins 25th Anniversary		1986	1	2
Mr. Men Figures	set of 10	1981	3	10
Polar Swirl Penguins	set of four	1987	15	40

Big Boy

NAME	DESCRIPTION	YEAR	EX	MINT
Action Figures	complete set of four: skater, pitcher, surfer, race driver	1990	2	4
Big Boy Bank, Large	produced from 1966-1976, 18" tall, full color	1960s	125	300
Big Boy Bank, Medium	produced from 1966-1976, 9" tall, brown	1960s	50	165
Big Boy Bank, Small	produced from 1966-1976, 7" tall, painted red/white	1960s	40	100
Big Boy Board Game		1960s	50	120
Big Boy Kite	kite with image of Big Boy	1960s	10	25
Big Boy Nodder	papier mache	1960s	900	1500
Big Boy Playing Cards	produced in four designs	1960s	15	45
Big Boy Stuffed Dolls	set of three: Big Boy, girlfriend Dolly, both 12" tall, and dog Nuggets, 7" tall	1960s	40	80
Helicopters	set of plastic vehicles; Ambulance, Police, Fire Department	1991	1	3
Monster In My Pocket	various secret monster packs	1991	2	4

Big Boy

NAME	DESCRIPTION	YEAR	EX	MINT
Sport Poses	set of four: surfing, baseball, racing, roller skating		3	5
Vehicles	set of three race cars: yellow, orange, purple	1992	2	5

Burger Chef

Funmeal Boxes	set of 24 featuring village buildings	1975	20	30
Funmeal Boxes	set of 24 featuring Burger Chef characters	1973-7	25	60
Star Wars Funmeal Boxes	set of seven	1978	25	40
Triple Play Funmeal Boxes	set of 24 featuring Major League Baseball teams	1977	20	90
Vacuform Race Cars		1981	20	30

Burger King

Adventure Kits	set of four activity kits with crayons; Passport, African Adventure, European Escapades, Worldwide Treasure Hunt	1991	1	3
Aladdin	set of five figures: Jafar and Iago, Genie in lamp, Jasmine and Rajah, Abu, Aladdin and the Magic Carpet	1992	1	3
Alf	joke & riddle disc, door knob card, sand mold, refrigerator magnet	1987	1	2
Alf Puppets	puppets with records; Sporting with Alf, Cooking with Alf, Born to Rock, Surfing with Alf	1987	3	8
Alvin and the Chipmunks	set of three toys; super ball, stickers, pencil topper	1987	1	2
Animal Boxes	set of four activity booklets; bear, hippo, lion, one more	1986	1	2
Aquaman Tub Toy	green	1987	3	5
Archie Cars	set of four: Archie in red car, Betty in aqua car, Jughead in green car, Veronica in purple car	1991	1	3
Barnyard Commandos	set of four: Major Legger Mutton in boat, Sgt. Shoat & Sweet in plane, Sgt. Wooley Pullover in sub, Pvt. Side O'Bacon in truck	1991	1	2
Batman Toothbrush Holder		1987	4	10
Beauty & the Beast	set of four PVC figures; Belle, Beast, Chip, Cogsworth	1991	2	5
Beetlejuice	set of six figures: Uneasy Chair, Head Over Heels, Ghost to Ghost TV, Charmer, Ghost Post, Peek A Boo Doo	1990	1	2
Bicycle License Plate			1	2
Bicycle Safety Fun Booklet			1	2
BK Kids Club Action Figures	set of four: Boomer, I.Q., Jaws, Kid Vid	1991	1	3
BK Kids Club Coolers	five colors	1995	2	4
BK Kids Club Glo Force	five different toys	1996	1	2
BK Kids Club Planet Patrol	Space Commander Jaws, I.Q.'s Planet Pacer, Boomer's Lightspeed Spacetop, Kid Vid's Glo Chopper, J.D. Shuttle Launch	1997	2	1
Bone Age Skeleton Kit	set of four dinos; T-Rex, Dimetron, Mastadon, Similodon	1989	3	6
Bone Age Skeleton Kit Boxes	The Past is a Blast, The Greatest Mystery in History	1989	1	2
Burger King Clubhouse	full size for kids to play in		15	35
Burger King Socks	rhinestone accents	2	5	
Calendar "20 Magical Years" Walt Disney World		1992	2	4

Beauty and the Beast, 1991, Burger King

DinoMotion Dinosaurs, 1993, McDonald's

NAME	DESCRIPTION	YEAR	EX	MINT
Capitol Critters	set of four: Hemmet for Prez in White House, Max at Jefferson Memorial, Muggle at Lincoln Memorial, Presidential Cat	1992	1	2
Capitol Critters Cartons	punch out masks: dog, chicken, duck, panda, rabbit, tiger, turtle	1992	2	4
Captain Planet	set of four flip over vehicles: Captain Planet & Hoggish Greedily, Linka, Ma-Ti & Dr. Blight ecomobile, Verminous Skumm & Kwane helicopter, Wheeler and Duke Nukem snowmobile	1991	1	2
Captain Planet Cartons	containers: Powerbase Spaceship, Biodread Patroller Spaceship, Powerjet XT-7	1991	2	3
Cartoon King Doll	16" tall; cloth	1972	20	30
Cartoon Network Racing Team	Speeding Bomber, Jeff Gordon Car, Scooby Doo Car, Burger King Race Car, Stoneage Rocker	1997	1	2
Christmas Cassette Tapes	set of three Christmas sing-a-long tapes: Joy to the World/ Silent Night, We Three Kings/ O Holy Night, Deck the Halls/ Night Before Christmas	1989	2	4
Christmas Crayola Bear Plush Toys	red, yellow, blue or purple	1986	3	6
Coloring Book	Keep Your World Beautiful		2	4
Crayola Coloring Books	set of six books: Boomer's Color Chase, I.Q.'s Computer Code, Kids Club Poster, Jaws' Colorful Clue, Snaps' Photo Power, Kid Vid's Video Vision	1990	2	4
Dino Meals	punch-out sheets: Stegosaurus, Woolly Mammoth, T-Rex, Triceratops	1987	3	6
Disney 20th Anniversary Figures	set of four wind-up vehicles with connecting track: Minnie, Donald, Roger Rabbit, Mickey	1992	3	5
Disney Collector Series Tumblers	Snow White, Jungle Book, Lion King, Peter Pan, Beauty and the Beast, Pinocchio, Dumbo, Aladdin	1994	2	4
Fairy Tale Cassette Tapes	set of four fairy tale cassettes: Goldilocks, Jack and the Beanstalk, Three Little Pigs, Hansel and Gretel	1989	1	3
Flying Disc	small yellow or orange, embossed Burger King		1	2
Freaky Fellas	set of four: blue, green, red, yellow, each with a roll of Life Savers candy	1992	2	4
Gargoyles	five different toys	1995	2	3
Go-Go Gadget Gizmos	set of four Inspector Gadget toys: gray copter, black inflatable, orange scuba, green surfer	1991	2	4
Golden Junior Classic Books	set of four: Roundabout Train, The Circus Train, Train to Timbucktoo, My Little Book of Trains		1	2
Goof Troop Bowlers	set of four: Goofy, Max, Pete, P.J.	1992	1	2
Hunchback of Notre Dame	four finger puppets: Quasimodo, Esmeralda, Gargoyle, Jester	1996	2	4
Hunchback of Notre Dame	eight PVC figures, each	1996	1	2
It's Magic	set of four: Magic Trunk, Disappearing Food, Magic Frame, Remote Control	1992	1	2
Kid Transformers	set of six: Bloomer w/Super Show, Kid Vid w/ SEGA Gamestar, I.Q. w/World Book Mobile, Snaps w/Camera Car, Jaws w/Burger Racer and Wheels w/Turbo Wheelchair	1990	1	3
Lickety Splits	plastic wheeled food items: Apple Pie Man, Flame Broiler Buggy, Drink Man, French Fry Man, Croissant Man, Chicken Tenders, French Toast Man	1990	1	2
Lion King Finger Puppets	Mufasa, Simba, Nala, Rafiki, Pumbaa and Timon, Ed the Hyena, Scar	1994	1	2
Lion King Timon & Pumbaa	Timon, Pumbaa, compass, Pumbaa/Timon together	1996	1	2

Burger King

NAME	DESCRIPTION	YEAR	EX	MINT
Lunch Box	blue plastic box embossed with BK logo		2	5
M & M's Figures	various colors	1997	1	2
Masters of the Universe Cups	Thunder Punch He-Man Saves the Day, He-Man and Roboto to the Rescue, He-Man Takes on the Evil Horde, Skeletor	1985	2	4
Matchbox Boxes (Buildings)	punch-out buildings: apartment, fire station, engine and firemen, restaurant, barn horse whirl and wheel game	1989	2	3
Matchbox Cars	set of four vehicles: blue Mountain Man 4 X 4, yellow Corvette, red Ferrari, black & white police car	1989	3	5
Mealbots	paper masks with 3-D lenses: red Broil Master, blue Winter Wizard, gray Beta Burger, yellow Galactic Guardians	1986	4	8
Nerfuls	rubber characters, interchangeable, set of four: Bitsy Ball, Fetch, Officer Bob, Scratch	1985	2	4
Oliver & Company	Sneak-A-Peek Oliver, Dashing Dodger, Surprise Attack DeSoto, Skateaway Tito	1996	1	2
Pilot Paks	set of four Styrofoam airplanes: two-seater, sunburst, lightning, one more	1988	4	8
Pinocchio	set of five toys, pail and following inflatables: Beach ball, Figaro, Jiminy Cricket, Monstro	1992	1	2
Pocahontas Figures	Meeko, Governor Radcliffe, Pocahontas, Flit, Captain John Smith, Grandmother Willow, Chief Powhatan	1995	1	2
Pocahontas Pop-Up Puppets	Peek-a-Boo Pocahontas, Meeko's Hideout, Pampered Percy, Ruthless Radcliffe, John Smith's Lookout, Busy Body Flit	1996	1	2
Pocahontas Tumblers	four versions	1995	3	5
Purrtenders	set of four: Free Wheeling Cheese Rider, Flip-Top Car, Radio Bank, Storybook	1988	2	5
Purrtenders	set of four plush toys: Hop-purr, Flop-purr, Scamp-purr, Romp-purr	1988	2	4
Record Breakers	set of six cars: Aero, Indy, Dominator, Accelerator, Fastland, Shockwave	1989	3	6
Rodney & Friends Reindeer	plush toys with holiday fun booklets: Ramona Holiday Sweets and Treats, Rodney Holiday Fun and Games Box, Rhonda Holiday decorating box	1987	5	10
Scooby Doo	Scooby, Mystery Machine Van, Scooby and Shaggy, Scrappy, Coffin	1996	1	3
Sea Creatures	terrycloth wash mitts: Stella Starfish, Dolly Dolphin, Sammy Seahorse, Ozzie Octopus	1989	2	3
Simpsons Cups	set of four	1991	1	2
Simpsons Dolls	set of five soft plastic dolls: Bart, Homer, Lisa, Marge, Maggie	1991	2	4
Simpsons Figures	set of five figures: Bart with backpack, Homer with skunk, Lisa with sax, Maggie with turtle, Marge with birds	1991	2	5
Spacebase Racers	set of five plastic vehicles: Moon Man Rover, Skylab Cruiser, Starship Viking, Super Shuttle, Cosmic Copter	1989	2	4
Sports Watches	white or yellow		2	5
Super Bowl Poster	1992		2	4
Super Heroes Cups	set of four cups with figural handles: Batman, Wonder Woman, Darkseid, Superman	1984	4	8
Superman Coin	flicker center	1987	5	15
Teenage Mutant Ninja Turtles Badges	six different: Michaelangelo, Leonardo, Raphael, Donatello, Heroes In a Half Shell, Shredder	1990	2	4
Teenage Mutant Ninja Turtles Poster		1991	2	4

RESTAURANT PREMIUMS

Star Wars toys, 1997, Taco Bell

California Raisins, Hardee's

Gargoyles, 1995, Burger King

Burger King

NAME	DESCRIPTION	YEAR	EX	MINT
Thundercats	set of four toys: cup/bank, Snarf strawholder, light switch plate, secret message ring	1986	3	8
Top Kids	set of four spinning tops with figural heads: Wheels, Kid Vid, two more	1992	1	2
Toy Story	Hopping Mr. Potato Head, Woody, Action Wing Buzz, Racing R.C. Car, Squash 'N Go Rex, Green Army Men Recon Squad	1995	2	4
Toy Story	Stroll 'n Scope Lenny, Jawbreaker Scud, Speedy Deposit Hamm, Round 'em Up Woody, Spin-Top Bo Peep, Blast-away Buzz, Spaced Out Alien, Slinky Dog	1996	2	3
Toy Story Puppets	Woody, Buzz Lightyear, Rex, or Hamm	1995	2	3
Toy Story Talking Puppets	R.C. Racer, Talking Woody, Talking Buzz	1996	3	5
Tricky Treaters Boxes	Monster Manor, Creepy Castle, Haunted House	1989	1	3
Tricky Treaters Figures	set of three PVC figures: Frankie Steen, Gourdy Goblin, Zelda Zoom Broom	1989	3	10
Trolls	set of four w/neon hair	1993	1	2
Universal Monsters	set of four: Wolf Man, Frankenstein, Dracula, Creature	1997	2	4
Water Club Mates	set of four: Lingo's Jet Ski, Snaps in Boat, Wheels on raft, I.Q. on dolphin	1991	1	2

Denny's

Dino-Makers	set of six: including blue dino, purple elephant, orange bird	1991	1	2
Flintstones Dino Racers	set of six: Fred, Bamm-Bamm, Dino, Pebbles, Barney, Wilma	1991	3	6
Flintstones Fun Squirters	set of six: Fred w/telephone, Wilma w/camera, Dino w/flowers, Bam Bam w/soda, Barney, Pebbles	1991	2	4
Flintstones Glacier Gliders	set of six: Bamm-Bamm, Barney, Fred, Dino, Hoppy, Pebbles	1990	2	4
Flintstones Mini Plush	in packages of two: Fred/Wilma, Betty/Barney, Dino/Hoppy, Pebbles/Bamm-Bamm, each 4" tall	1989	2	4
Flintstones Rock & Rollers	set of six: Fred w/guitar, Barney w/sax, Bam Bam, Dino w/piano, Elephant, Pebbles	1990	3	6
Flintstones Stone-Age Cruisers	set of six: Fred in green car, Wilma in red car, Dino in blue car, Pebbles in in purple bird, Bam Bam in orange car, Barney in yellow car w/sidecar	1991	2	4
Flintstones Vehicles	set of six	1990	2	4
Jetsons Game Packs	set of six: George, Elroy, Judy, Astro, Rosie, Jane	1992	2	4
Jetsons Go Back to School	set of six school tools: mini dictionary, folder, message board, pencil & topper, pencil box, triangle & curve	1992	2	4
Jetsons Puzzle Ornaments	set of six	1992	2	4
Jetsons Space Balls (Planets)	set of six: Jupiter, Neptune, Earth, Saturn, Mars, glow-in-the-dark Moon	1992	2	4
Jetsons Space Cards	set of five: Spacecraft, Phenomenon, Astronomers, Constellations, Planets	1992	2	4
Jetsons Space Travel Coloring Books	set of six books: each with four crayons	1992	2	4

Domino's Pizza

Noids	set of seven figures: Boxer, Clown, He Man, Holding Bomb, Holding Jack Hammer, Hunchback, Magician	1987	2	4

Domino's Pizza

NAME	DESCRIPTION	YEAR	EX	MINT
Quarterback Challenge Cards	pack of four cards	1991	1	2

Hardee's

NAME	DESCRIPTION	YEAR	EX	MINT
Backpack	orange		2	4
Beach Bunnies	set of four: girl with ball, boy with skateboard, girl with skates and boy with frisbee	1989	2	4
California Raisins	first set: dancer w/blue & white shoes, singer w/ mike, sax player, fourth with sunglasses	1987	2	8
California Raisins	second set: w/guitar in orange sneakers, on rollerskate w/yellow sneakers, with radio w/ yellow sneakers, surfer w/red sneakers, trumpet player w/blue sneakers	1988	2	4
California Raisins	third set: Alotta Stile in pink boots, Anita Break w/ package under her arm, Benny bowling, Buster w/skateboard	1991	2	6
California Raisins Plush	6" tall: lady in yellow shoes, dancer in yellow hat, with mike in white shoes, in sunglasses w/ orange hat	1988	3	6
Days of Thunder Racers	set of four cars: Mello Yellow #51, Hardee's #18 orange, City Chevy #46, Superflo in pink/white	1990	3	6
Disney's Animated Classics Plush Toys	Pinocchio, Bambi		3	5
Ertl Camaro	Hardee's Road Runner	1990	3	6
Fender Bender 500 Racers	set of five: Quick Draw/Baba Looey's Covered Wagon, Huckleberry/Snagglepuss's truck, Wally Gator/Magilla's Toilet, Dick Dastardly/ Mutley's rocket racer, Yogi/Boo Boo's basket	1990	2	3
Finger Crayons	set of four: not marked Hardee's		2	3
Flintstones First 30 Years	set of five: Fred w/TV, Barney w/grill, Pebbles w/ phone, Dino w/jukebox, Bamm Bamm w/pinball	1991	3	6
Flying Disc	4-3/4" wide, white with yellow imprinting		1	4
Food Squirters	set of four: cheeseburger, hot dog, shake, fries	1990	2	4
Ghostbusters Headquarters Posters			1	3
Gremlin Adventures	set of five book and record sets: Gift of the Mogwai, Gismo & the Gremlins, Escape from the Gremlins, Gremlins Trapped, The Late Gremlin	1989	2	4
Halloween Hideaways	set of four: goblin in blue cauldron, ghost in yellow bag, cat in pumpkin, bat in stump	1989	3	6
Home Alone 2	set of four	1992	1	2
Kazoo Crew Sailors	set of four: brown bear, monkey, rabbit, rhino	1991	1	2
Little Golden Books	set of four: The Little Red Caboose, The Three Bears, Old MacDonald Had a Farm, Three Little Kittens		2	4
Pound Puppies	set of six: white w/black spots, black, brown, gray w/brown spots, gray w/black ears, tan w/black ears	1991	3	6
Shirt Tales Plush Dolls	set of five, each 7" tall: Bogey, Pammy, Tyg, Digger, Rick		3	6
Smurfin' Smurfs	set of four surfers: Papa w/red board, boy w/ orange board, girl w/purple board, dog w/blue board	1990	2	4
Smurfs Box	Smurf Solution, Smurf Angle, Smurf Surprise, Stop & Smurf the Flowers	1990	1	2
Super Bowl Cloisonne Pins	25 NFL Super Bowl pins	1991	2	6
Super Heroes	Marvel figures in vehicles: She Hulk, Hulk, Captain America, Spider-Man	1990	3	6
Tang Mouth Figures	set of four: Lance, Tag, Flap, Annie	1989	2	4

Disney Collector Series Tumblers, 1994, Burger King

Pocahontas Tumblers, 1995, Burger King

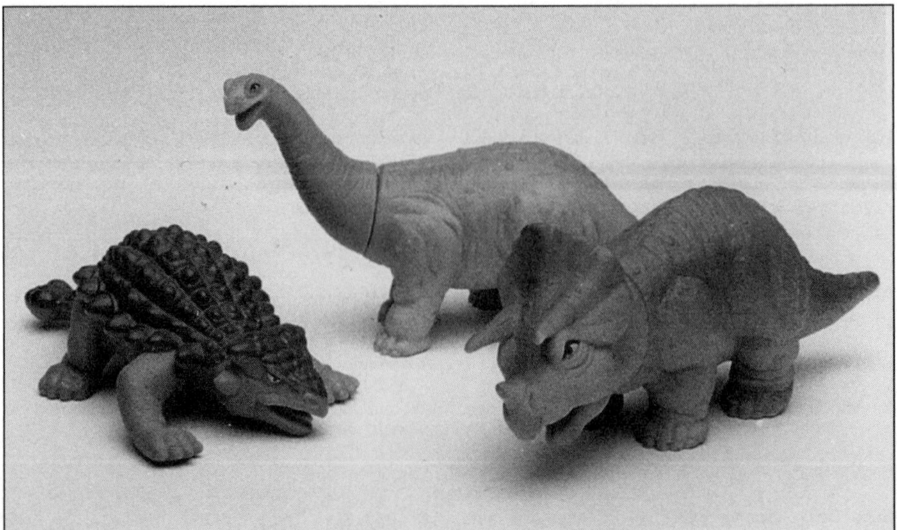

Definitely Dinosaurs, 1988-1989, Wendy's

Hardee's

NAME	DESCRIPTION	YEAR	EX	MINT
Waldo and Friends Holiday Ornaments			2	4
Waldo's Straw Buddies	set of four	1990	2	3
Waldo's Travel Adventure	set of four		2	3

Kentucky Fried Chicken

Alvin and the Chipmunks	Canadian issues; Alvin and Theodore	1991	3	5
Alvin and the Chipmunks	Canadian issues; Alvin and Simon	1992	2	4
Colonel Sanders Figure	9" tall	1960s	35	50
Colonel Sanders Nodder	7" tall; plastic	1960s	15	40
Colonel Sanders Nodder	7" tall; papier mache	1960s	50	150
Garfield Racers	set of six	1996	2	4
WWF Stampers	set of four: Canadian issues		2	4

Lee's Fried Chicken

Cartoon Viewers	set of six: Mighty Mouse, Woody Woodpecker, Popeye, Superman, Bugs Bunny, Porky Pig	1980s	20	50

Long John Silver's

Adventure on Volcano Island	paint with water activity book	1991	2	4
Fish Cars	red and yellow cars shaped like fish, w/stickers	1989	2	4
Sea Walkers	set of four packaged with string; Parrot, Penguin, Turtle and Sylvia	1990	3	8
Sea Walkers	set of four packaged w/o string; Parrot, Penguin, Turtle and Sylvia	1990	3	5
Sea Watchers Kaleidoscopes	set of three: orange, yellow, pink	1991	2	4
Treasure Trolls		1992	1	3
Water Blasters	set of four: Billy Bones, Captain Flint, Ophelia Octopus, Parrot	1990	3	5

McDonald's

101 Dalmatians	101 different PVC dogs	1996	2	4
101 Dalmatians	set of four PVC figures, Lucky, Pongo, Sergeant Tibbs, Cruella	1991	1	2
3-D Happy Meal	set of four cartons with 3-D designs with 3-D glasses inside: Bugsville, High Jinx, Loco Motion, Space Follies	1981	20	40
Adventures of Ronald McDonald	set of seven rubber figures: Ronald, Birdie, Big Mac, Captain Crook, Mayor McCheese, Hamburglar, Grimace	1981	5	10
Airport	Birdie Bentwing Blazer, Fry Guy Flyer, Grimace Bi-Plane, Big Mac Helicopter (green), Ronald Sea Plane	1986	3	5
Aladdin		1996	1	2
Alvin & The Chipmunks	set of four figures: Simon, Theodore, Brittany and Alvin	1991	3	6
An American Tail	set of four books: Fievel and Tiger, Fievel's Friends, Fievel's Boat Trip, Tony and Fievel	1986	2	3

McDonald's

NAME	DESCRIPTION	YEAR	EX	MINT
Animal Riddles	eight different rubber figures 2" to 2-1/2" tall: condor, snail, turtle, mouse, anteater, alligator, pelican, dragon, in various colors	1979	2	4
Astronauts	four different: Lunar Rover, Satellite dish, Command Module & Space Shuttle	1991	2	4
Astronauts	under 3 toy: Ronald McDonald in Lunar Rover	1991	2	4
Astrosnicks 1	eight different 3" rubber space creatures: Scout, Thirsty, Robo, Laser, Snickapotomus, Sport, Ice Skater, Astralia	1983	4	20
Astrosnicks 2	six different rubber space creatures: Copter, Drill, Ski, Racing, Perfido, Commander	1984	4	20
Astrosnicks 3	14 figures, many same as 1984 series, but without "M" logo: Commander, Robo, Perfido, Galaxo, Laser, Copter, Scout, Snikapotamus, boy, Pyramido, Racer	1985	5	40
Astrosnicks Rocket	9-1/2" rocket ship; coupon with Happy Meal	1984	30	60
Babe	plush toys	1996	1	2
Back to the Future	four figures: Marty, Einstein, Verne, Doc	1992	1	2
Bambi	set of four: Owl, Flower, Thumper, Bambi	1988	2	4
Barbie	set of eight dolls: Sparkle Eyes, Roller Blade, Rappin Rocking, My First Ballerina, Snap-On, Sun Sensation, Birthday Surprises, Rose Bride	1992	1	2
Barbie	eight dolls: My First Ballerina, Birthday Party, Western Stamping, Romantic Bride, Hollywood Hair, Paint `n Dazzle, Twinkle Lights, Secret Hearts	1993	1	2
Barbie	under 3 toy: Costume Ball Barbie	1991	2	4
Barbie	eight plastic dolls: Ice Capades, All American, Lights & Lace, Hawaiian Fun, Happy Birthday, Costume Ball, Wedding Day Midge, My First Barbie	1991	2	3
Barbie / Hot Wheels		1996	1	2
Baseball Cards	Donruss 1992, set of 32 cards with checklist	1992	10	25
Baseball Cards	Topps 1991, set of four cards	1991	15	35
Basketball Cards	Hoops 1992, set of 62 cards	1992	12	25
Basketball Cards	Hoops 1992, set of 70 cards including eight extra Chicago Bulls cards	1992	20	35
Batman Cups w/flying lids	set of six	1992	1	2
Batman Returns Press & Go Vehicles	set of four: Catwoman, Batman, Penguin in vehicles and Batmobile	1992	1	2
Beach Ball	set of three with Olympic theme: Grimace in kayak, Ronald, Birdie on sailboat	1984	3	6
Beach Ball	set of three inflatables: red Ronald, blue Birdie, yellow Grimace	1986	2	4
Beach Toys	set of eight: four inflatables: Fry Kid Super Sailer, Birdie Seaside Submarine, Grimace Bouncin' Beach Ball, Ronald Fun Flyer; four sand toys; two buckets, shovel and rake	1989	2	4
Beachcomber	white pail with blue lid and shovel	1986	5	10
Bedtime	set of four items with boxed tube of Crest toothpaste: Ronald toothbrush, Ronald bath mitt, Ronald glow-in-the-dark star figure, Ronald cup	1989	4	8
Berenstain Bears II	set of four figures: Papa with wheelbarrow, Mama with shopping cart, Brother with scooter, Sister with wagon	1987	2	4
Berenstain Bears Story Books	set of four: Attic Treasure, Substitute Teacher, Eager Beavers, Life With Papa	1990	2	4
Berenstain Bears Test Market Set	set of four Christmas figures: Papa with wheelbarrow, Mama with shopping cart, Brother on scooter, Sister on sled	1986	30	75

Lion King, 1996, Burger King

Berenstain Bears, McDonald's

Peanuts Vehicles, 1990, McDonald's

Super Heroes, 1990, Hardee's

McDonald's

NAME	DESCRIPTION	YEAR	EX	MINT
Bigfoot	set of four Ford trucks with two wheel sizes: Bigfoot, Ms. Bigfoot, Shuttle, Bronco	1987	3	5
Birdie Bike Horn	Japanese		4	9
Birdie Magic Trick	green or orange		2	5
Black History	two coloring books; sold in Detroit only	1988	200	400
Boats `n Floats	set of four plastic container boats: Chicken McNugget lifeboat, Birdie float, Fry Kids raft, Grimace power boat	1987	5	10
Bobby's World	set of four: Wagon-Race Car, Innertube-Submarine, Three Wheeler-Space Ship, Skates-Roller Coaster	1994	1	2
Bobby's World	under 3 toy: inner tube	1994	2	3
Cabbage Patch	under 3 toy: Anne Louise "Ribbons & Bows"	1992	1	2
Cabbage Patch	set of five Christmas dolls: Tiny Dancer, Holiday Pageant, Holiday Dreamer, Fun On Ice, All Dressed Up	1992	2	3
Camp McDonaldland	set of four: utensils, Birdie camper mess kit, collapsible cup, canteen	1990	1	2
Captain Crook Bike Reflector	blue plastic, Canada	1988	1	3
Castle Maker/Sand Castle	set of four molds: dome, square, cylinder, rectangle	1987	20	40
Changeables	set of six figures that change into robots: Big Mac, Shake, Egg McMuffin, Quarter Pounder, French Fries, Chicken McNuggets	1987	2	5
Changeables	set of eight figures that turn into robots: Large Fries, Quarter Pounder, Hot Cakes, Big Mac, Cheeseburger, Shake, Soft Serve Cone, Small Fries	1989	2	4
Changeables	under 3 toy: Character Changer Cube	1989	3	5
Chip `N Dale's Rescue Rangers	set of four figures in vehicles: Monterey Jack in Propelaphone, Gadget in Rescue Racer, Chip in Whirlicopter, Dale in Rotoroadster	1989	2	4
Chip `N Dale's Rescue Rangers	under 3 toy: Gadget Rider & Chip's Rockin Racer	1989	2	4
Christmas Ornaments	Fry Guy and Fry Girl, cloth, 3-1/2" tall	1987	3	6
Christmas Stocking	plastic, "Merry Christmas to My Pal"	1981	3	6
Circus	set of eight: fun house mirror, acrobat Ronald, French Fry Faller, Strong Gong, two punchout sheets, and fun house and puppet show background	1983	4	8
Circus Parade	set of four: Ringmaster Donald, Birdie Rider, Grimace caliope, Fry Guy and elephant	1991	2	4
Circus Wagon	set of four rubber toys: poodle, chimp, clown, horse	1979	2	3
Colorful Puzzles-Japan	Dumbo, Mickey & Minnie, Dumbo & Train		5	10
Coloring Stand-Ups	characters and backgrounds to color, punch out and stand	1978	4	8
Combs	Capt. Crook-red; Grimace-yellow; Ronald-yellow, blue or purple; Grimace Groomer-green	1988	1	2
Commandrons	set of four robots on blister cards: Solardyn, Magna, Motron, Velocitor	1985	5	15
Construx	set of four pieces to make a spaceship: axle, wing, body cylinder, canopy	1986	10	25
CosMc Crayola	set of five coloring kits: Crayolas, two with markers, chalk sticks, paint set	1988	2	4
CosMc Crayola	under 3 toy	1988	3	6
Crayola Magic	set of three stencil kits with crayons or markers: triangles with marker, rectangles with crayons, circles with crayons	1986	4	7
Crayola Squeeze Bottle, Kay Bee	regional set of four		3	6

McDonald's

NAME	DESCRIPTION	YEAR	EX	MINT
Crazy Vehicles	set of four: Ronald's dune buggy, Hamburglar's train, Birdie's airplane, Grimace's car	1991	3	5
Design-O-Saurs	set of four plastic interchangeable parts: Ronald on Tyrannosaurus, Grimace-Pterodactyl, Fry Guy-Brontosaurus, Hamburglar-Triceratops	1987	5	10
Dink the Dinosaur	set of six dino finger puppets: each packed with diorama and description; Dink, Flapper, Amber, Crusty, Scat, Shyler	1990	4	8
Dino-Motion Dinosaurs	under 3 toy: Baby Sinclair	1993	2	4
Dino-Motion Dinosaurs	set of six Sinclairs: Charlene, Earl, Grandma Ethyl, Fran, Baby, Robbie	1993	1	2
Dinosaur Days	set of six rubber dinos: Pteranodon, Triceratops, Stegosaurus, Dimetrodon, T-Rex, Ankylosaurus	1981	1	2
Discover the Rain Forest	set of four activity books with punch out figures: Sticker Safari, Wonders in the Wild, Paint It Wild, Ronald and the Jewel of the Amazon Kingdom	1991	2	4
Disney Favorites	set of four activity books: Lady and the Tramp, Dumbo, Cinderella, The Sword in the Stone	1987	3	4
Disney Masterpiece Collection	figures in video-shaped box	1996	1	2
Double Bell Alarm Clock	wind-up alarm clock with silver bells, hammer ringer, silver feet, image of Ronald on face with head tilted over folded hands, as if asleep		20	40
Duck Tales I	set of four toys: telescope, duck code quacker, magnifying glass, wrist decoder	1988	2	4
Duck Tales II	set of four toys: Uncle Scrooge in red car, Launchpad in plane, Huey, Dewey and Louie on jet ski, Webby on blue trike	1988	3	6
Dukes of Hazzard	set of five container vehicles, regional: white Caddy, Jeep, police car and pickup, orange General Lee Charger	1982	20	50
Dukes of Hazzard	set of six white plastic cups, national: Luke, Boss Hogg, Bo, Sheriff Roscoe, Daisy, Uncle Jesse	1982	3	6
E.T. Posters	set of four posters, 17" x 24": Boy and ET on bike	1985	5	11
E.T. Posters	Boy and ET finger touch or ET and radio	1985	3	8
E.T. Posters	ET waving	1985	5	10
Earth Days	set of four: birdfeeder, globe terrarium, binoculars, tool carrier with shovel	1994	1	2
Earth Days	under 3 toy: tool carrier with shovel	1994	1	2
Eric Carle Finger Puppets		1996	1	2
Fast Macs	set of four pull-back action cars: white Big Mac police car, yellow Ronald jeep, red Hamburglar racer, pink Birdie convertible	1984	3	5
Favorite Friends	set of seven character punch-out cards	1978	2	5
Feeling Good	set of five grooming toys: Grimace soap dish, Fry Guy sponge, Birdie mirror, Ronald or Hamburglar toothbrush, Captain Crook comb	1985	1	3
Field Trip	set of four: kaleidoscope, leaf printer, nature viewer, explorer bag	1993	1	2
Field Trip	under 3 toy: nature viewer	1993	1	2
Fitness Fun	set of eight toys	1992	2	4
Flintstone Kids	set of four figures in animal vehicles: Betty, Barney, Fred, Wilma	1987	4	8
Food Fundamentals	under 3 toy: Dunkan	1993	1	2
Food Fundamentals	set of four: Slugger, Otis, Milly & Ruby	1993	1	2
Fraggle Rock	set of four: Gobo in carrot car, Red in radish car, Mokey in eggplant car, Wembly and Boober in pickle car	1988	1	2
Fraggle Rock Doozers	set of two: Cotterpin in forklift and Bulldozer in bulldozer	1988	20	40

Food Squirters, 1990, Hardee's

101 Dalmatians, 1991, McDonald's

Tonka loader, 1992, McDonald's

McDonald's

NAME	DESCRIPTION	YEAR	EX	MINT
French Fry Radio	large red fry container with fries	1977	12	25
Friendship Spaceship Ring		1985	3	4
Frisbee - Canada	3" diameter, blue, early Ronald image		1	2
Fry Benders	set of four bendable fry figures with accessories: Grand Slam, Froggy, Roadie, Freestyle	1990	3	5
Fun To Go	set of seven cartons with games and activities	1977	2	4
Fun with Food	set of four multi-piece toys with faces: hamburger, fries, soft drink, McNuggets	1989	3	6
Funny Fry Friends	under 3 toy: two different, Lil' Chief & Little Darling	1990	3	5
Funny Fry Friends	set of eight: Too Tall, Tracker, Rollin' Rocker, Sweet Cuddles, ZZZ's, Gadzooks, Matey, Hoops	1990	2	4
Garfield Figures	set of four: on scooter, on skateboard, in jeep, with Odie on motorscooter	1989	3	6
Ghostbusters	set of five school tools: Slimer pencil, Stay Puft pad and eraser, containment chamber pencil case, Stay Puft pencil sharpener, Ghostbusters ruler	1987	3	7
Glow in the Dark Yo-Yo	no markings or dates	1978	2	5
Golf Ball	McDonald's logo		1	3
Good Morning	set of four grooming items: Ronald toothbrush, McDonaldland comb, Ronald play clock, white plastic cup	1991	1	2
Good Times Great Taste Record			2	4
Gravedale High	set of four mechanical Halloween figures: Cleofatra, Frankentyke, Vinnie Stoker, Sid the Invisible Kid	1991	2	3
Grimace Bank	purple ceramic, 9" tall	1985	10	20
Grimace Miniature Golf		1986	3	5
Grimace Pin	enamel		6	12
Grimace Ring		1970	8	15
Grimace Sponge	Grimace, Grimace Car Wash		2	4
Halloween Boo Bags	set of three glow-in-the-dark vinyl bags: Witch, Ghost and Frankie	1991	3	5
Halloween Buckets	set of three lidded pails with black plastic strap handles, safety stickers attached: orange pumpkin, white glow-in-dark ghost, Green witch	1990	2	4
Halloween Buckets	Ghost, Witch, Pumpkin	1992	1	2
Halloween Buckets	set of three pumpkin-shaped buckets: McGoblin, McPumpkin, McBoo	1986	2	4
Halloween Certificate Book w/Roger Rabbit Puffy Sticker		1988	1	3
Halloween McNugget Buddies	set of six: Pumpkin, McBoo, Monster, McNuggula, Witchie, Mummie	1993	1	2
Halloween McNugget Buddies	under 3 toy: McBoo McNugget	1993	2	3
Halloween Pumpkin Ring	orange pumpkin face		1	3
Hamburglar Doll	7" stuffed doll by Remco; one of set of seven, sold on blister card	1976	12	25
Hamburglar Hockey		1979	2	4
Happy Pails, Olympics	set of four with shovels: swimming, cycling, track, Olympic Games	1984	1	3
High Flying	set of three kites: Ronald, Birdie and Hamburglar	1987	3	8
Honey, I Shrunk the Kids Cups	set of three white 20 oz. plastic cups: giant bee, on the dog's nose, riding the ant	1988	1	2
Hook Figures	set of four: Peter Pan, Mermaid, Rufio, Hook	1991	1	2

NAME	DESCRIPTION	YEAR	EX	MINT
Hot Wheels	set of eight: Quaker State Racer #62, McDonald's Dragster, McDonald's Thunderbird #23, Hot Wheels Dragster, McDonald's Funny Car, Hot Wheels Funny Car, Hot Wheels Camaro #1, Duracell Racer #88	1993	2	3
Hot Wheels	under 3 toy: wrench and hammer	1993	3	6
Hot Wheels	under 3 toy: wrench and hammer	1991	2	3
Hot Wheels	split promo with Barbie, set of eight cars: purple or orange Z28 Camaro, white or yellow '55 Chevy, green or black '63 Corvette, turquoise or red '57 T-Bird	1991	2	3
Hot Wheels	set of eight cars	1983	3	7
Hot Wheels Mini-Streex	set of eight: Flame-Out, Quick Flash, Turbo Flyer, Black Arrow, Hot Shock, Racer Tracer, Night Shadow, Blade Burner	1992	2	4
Hot Wheels Mini-Streex	under 3 toy: orange arrow	1992	3	6
Hunchback of Notre Dame		1996	1	2
I Like Bikes	set of four bike accessories: Ronald Basket, Grimace mirror, Birdie spinner, Fry Guy Horn	1990	10	20
Jungle Book	set of four wind-up figures: Baloo the bear, Shere Kahn the tiger, King Louie the orangutan, Kaa the snake	1990	1	3
Jungle Book	under 3 toy, two different: Junior & Mowgli	1990	3	6
Kissyfur	furry figures	1987	8	25
Kissyfur	rubber figures	1987	3	6
Lego Building Sets	under 3 toys by Duplo: bird or boat	1986	4	8
Lego Building Sets	set of four Duplo kits: airplane, ship, truck, helicopter	1984	5	10
Lego Building Sets	under 3 Duplo toys: animal or building	1984	4	8
Lego Little Travelers Building Sets	four different sets: blue tanker boat, green airplane, red roadster, yellow helicopter	1986	5	10
Lego Motion	eight different kits: Gyro Bird, Lightning Striker, Land Laser, Sea Eagle, Wind Whirler, Sea Skimmer, Turbo Force, Swamp Stinger	1989	3	6
Linkables	set of four: Birdie on tricycle, Ronald in soap-box racer, Grimace in wagon, Hamburglar in airplane	1993	2	4
Lion Circus	set of four rubber figures: bear, elephant, hippo, lion	1979	2	3
Little Engineer	set of five train engines: Birdie Bright Lite, Fry Girl's Express, Fry Guy's Flyer, Grimace Streak, Ronald Rider	1987	4	8
Little Gardener	set of four tools and seed packets: Ronald Water Can, Birdie Shovel, Grimace Rake, Fry Guy Pail	1989	1	2
Little Golden Books	set of five books: Country Mouse and City Mouse, Tom & Jerry, Pokey Little Puppy, Benji, Monster at the End of This Block	1982	2	4
Little Mermaid	set of four figures: Flounder, Ursula, Prince Eric, Ariel with Sebastian	1989	1	2
Littlest Pet Shop Transformers		1996	1	2
Looney Tunes Christmas Dolls-Canada	set of four: Sylvester in nightgown & cap, Tasmanian Devil in Santa hat, Bugs in winter scarf and Tweetie dressed as Elf		3	6
Looney Tunes Quack Up Cars	Taz Tornado Tracker, Porky Ghost Catcher, Bugs Super Stretch Limo, Daffy Splittin' Sportster	1993	1	2
Looney Tunes Quack Up Cars	under 3 toy: Swingin' Sedan	1993	2	3
Lost Arches	periscope, flashlight, phone, camera	1991	1	2
Luggage Tags	set of four: Birdie, Grimace, Ronald, Hamburglar		2	4

Back to the Future, 1992, McDonald's

Gravedale High, 1991, McDonald's

McDonald's

NAME	DESCRIPTION	YEAR	EX	MINT
M-Squad	under 3 toy: Spytracker	1993	1	2
M-Squad	set of four: Spystamper, Spytracker, Spy-Nocular, Spycoder	1993	2	4
Mac Tonight	under 3 toy: Mac on skateboard	1990	3	6
Mac Tonight	set of six figures in vehicles: sports car, off-roader, motorcycle, scooter, jet ski, airplane	1990	3	6
Mac Tonight Pin	Moonface and slogan enamel pin	1988	2	4
Mac Tonight Puppet	Fingertronic foam puppet	1988	6	15
Mac Tonight Sunglasses	adult size	1988	2	5
Magic Show	set of four tricks: string pull, disappearing hamburger patch, magic tablet, magic picture	1985	3	6
Makin' Movies	set of four: sound effects machine, camera, clapboard with chalk, director megaphone	1994	1	2
Makin' Movies	under 3 toy: sound effects machine	1994	1	2
Marvel Superheroes		1996	1	2
Matchbox Mini-Flexies	eight rubber cars including Cosmobile, Hairy Hustler, Planet Scout, Hi-Tailer, Datsun, Beach Hopper, Baja Buggy	1979	2	4
McBunny Easter Pails	set of three: Pinky, Fluffy, Whiskers	1989	3	6
McDino Changeables	under 3 toy: Bronto cheeseburger, small fry, Ceratops	1991	2	3
McDino Changeables	set of eight dinosaurs: Happy Mealodon, Quarter Pounder Cheesosaur, Big Macosaurus Rex, McNuggetosaurus, Hotcakesodactyl, Large Fryosaur, Trishakatops, McDino cone	1991	2	3
McDonald's All-Star Race Team (MAXX) `91	complete set of cards	1991	4	5
McDonald's All-Star Race Team (MAXX) `92	complete set of 36 cards	1992	4	5
McDonald's Playing Cards	two decks to a set		2	5
McDonald's Spinner Top-Holland			2	4
McDonaldland Band	set of eight music toys: Grimace saxophone, Fry Guy trumpet and whistle, Ronald harmonica, whistle and pan pipes, kazoo, Hamburglar whistle	1987	3	5
McDonaldland Carnival	set of four toys: Birdie on swing, Grimace in turn-around, Hamburglar on ferris wheel, Ronald on carousel	1990	3	5
McDonaldland Connectables	set of four toys that connect to form a train of vehicles: Grimace in wagon, Birdie on a trike, Hamburglar in airplane, Ronald in a race car	1991	3	5
McDonaldland Express	set of four train car containers: Ronald engine, caboose, freight car, coach car	1982	25	50
McDonaldland Junction	train set of four snap together cars: yellow Birdie's Parlor car, red or blue Ronald Engine, purple Grimace caboose, green or white Hamburglar flat car	1983	5	10
McDonaldland Play-Doh	set of eight colors	1986	2	4
McDrive thru Crew	set of four vehicles: fries in potato roadster, shake in milk carton, McNugget in egg roadster, hamburger in ketchup bottle	1990	20	40
McNugget Buddies		1996	1	2
McNugget Buddies	set of 10 rubber figures and accessories: Sparky, Volley, Corny, Drummer, Cowpoke, Sarge, Snorkel, First Class, Rocker, Boomerang	1989	1	2
McNugget Buddies	under 3 toy: boy-Slugger, girl-Daisy	1989	1	2

McDonald's

NAME	DESCRIPTION	YEAR	EX	MINT
Michael Jordan Fitness	set of eight toys: soccer ball, squeeze bottle, stopwatch, basketball, football, baseball, jump rope, flying disc	1992	2	4
Mickey's Birthdayland	set of four Under 3 vehicles: Mickey's convertible, Goofy's car, Minnie's convertible, Donald's Jeep	1989	5	10
Mickey's Birthdayland	set of five characters in vehicles: Minnie's convertible, Donald's train, Goofy's Jalopy, Mickey's roadster, Pluto's rumbler	1989	1	3
Mighty Mini 4 X 4s	set of four big wheeled vehicles: Cargo Climber, Dune Buster, L'il Classic, Pocket Pickup	1991	1	2
Minnesota Twins Baseball Glove	Twins logo on side, Coca-Cola inside glove, McDonald's satin logo on back, given to the first 100 kids at 1984 game	1984	40	75
Minute Maid Juice Bottles	mini squeeze bottles	1991	1	2
Mix'em Up Monsters	set of four monsters with interchangeable parts: Corkle, Thugger, Gropple, Blibble	1990	2	4
Moveables	set of six vinyl bendies: Birdie, Captain Crook, Fry Girl, Hamburglar, Professor, Ronald	1988	4	12
Muppet Babies	four different: Kermit on red skateboard, Fozzie on hobby horse with wheels, Gonzo on tricycle with red wheels, Piggy in pink convertible	1987	2	4
Muppet Babies	four different: Piggy on trike, Gonzo in airplane, Fozzie in wagon, Kermit on soapbox car	1991	2	4
Muppet Treasure Island		1996	2	3
My Little Pony	regional promo with Transformers, set of six: Minty, Snuzzle, Blossom, Cotton Candy, Blue Belle, Butterscotch	1985	20	80
Mystery of the Lost Arches	set of four: mini-cassette, phone, telescope, camera	1992	1	2
Nature's Helpers	set of five garden tools with seeds: hinged trowel, rake, water can, terrarium, bird feeder	1991	1	2
Nature's Watch	set of four: double shovel/rake, greenhouse with seeds, bird buddy, water pail (no under 3 toy made)	1991	1	2
New Archies	set of six figures in bumper cars: Moose, Reggie, Archie, Veronica, Betty, Jughead	1988	5	15
New Food Changeables	set of eight: Krypto Cup, Fry Bot, Turbo Cone, Macro Mac, Gallacta Burger, Robo Cakes, C-2 Cheeseburger, Fry Force	1989	2	4
Nickelodeon	set of four: Blimp game, Loud-Mouth Mike, Gotcha Gusher, Applause Paws	1993	1	2
Nickelodeon	under 3 toy: Blimp Squirt toy	1993	1	2
Norman Rockwell Brass Ornament	50th Annivesary Norman Rockwell design, gift packaged with McDonald & Coca-Cola logos	1983	3	7
Norman Rockwell Ornament	clear acrylic, "Christmas Trio", gift boxed	1978	3	7
Old McDonald's Farm	set of six figures: farmer, wife, rooster, pig, sheep, cow	1986	5	10
Old West	set of six rubber figures: cowboy, frontiersman, lady, Indian, Indian woman, sheriff	1981	7	14
Oliver and Company	set of four: Oliver, Georgette, Francis and Dodger	1988	1	2
On the Go Games	set of five: stoplight bead game, Ronald slate board lift pad, Hamburglar lift pad, stop and go bead game, decal transfer	1985	3	8
On The Go Lunch Box	green, red or blue with arches on handles, stickers and embossed McDonaldland characters	1988	2	4
Paint with Water	paintless coloring board with self contained frame and easel	1978	5	10
Peanuts	under 3 toys: Charlie Brown's egg basket or Snoopy's potato sack	1990	2	4
Peanuts	set of four: Charlie, Snoopy, Lucy, Linus	1990	2	4

Jungle Book, 1990, McDonald's

Space Raiders, 1979, McDonald's

McDonald's

NAME	DESCRIPTION	YEAR	EX	MINT
Pencil Puppets	six different pencil toppers in shapes of McDonaldland characters	1978	2	4
Piggsburg Pigs	set of four figures on vehicles: Rembrandt in hotrod, Huff & Puff wolves on catapult, Piggy & Crackers on crate car, Portly & Pighead on motorcycle	1991	3	5
Playmobile	set of five toys and accesories: farmer, sheriff, Indian, umbrella girl, horse and saddle	1982	10	20
Popoids	set of four interconnecting constructor kits: cylinder, triangle, sphere, cube	1985	6	12
Potato Head Kids	set of eight	1992	3	6
Punkin' Makins	character cutoutss to decorate pumpkins: Ronald, Goblin, Grimace	1977	7	15
Raggedy Ann & Andy	set of four toys: Andy on slide, Grouchy on carousel, Ann with swing, camel on seesaw	1990	5	10
Records	set of four 45 RPM records in sleeves with different songs and colored labels	1985	3	6
Rescuers Down Under	set of four slide viewing movie camera toys: Jake, Wilbur, Bernard and Bianca, Cody	1990	2	4
Rescuers Down Under	under 3 toy: Bernard	1990	3	6
Rescuers Down Under Christmas Ornament	Miss Bianca, Bernard	1990	3	6
Rings	set of five rings with character heads: Big Mac, Captain Crook, Grimace, Hamburglar, Ronald	1977	5	10
Roger Rabbit Scarf-Japan	McDonald's logo, Japanese writing on scarf	1988	10	20
Ronald McDonald Bank	Ronald sitting with legs crossed, 7-1/2" tall		5	12
Ronald McDonald Bike Seat Pad-Japan			4	9
Ronald McDonald Cookie Cutter	green with balloons	1987	2	3
Ronald McDonald Doll	7" doll by Remco		18	25
Ronald McDonald Doll	14" vinyl head with a soft body by Dakin		15	35
Ronald McDonald Inflatable	12" with weighted base	1990	2	5
Ronald McDonald LCD Clock-Japan	foldable		4	9
Ronald McDonald Magic Tablet			1	3
Ronald McDonald Maze	lift up mystery game	1979	4	10
Ronald McDonald Pin	enamel, Ronald in Christmas wreath		6	12
Ronald McDonald Plastic Flyers	Ronald with legs and arms extended, red or yellow		1	3
Ronald McDonald Popsicle Maker-Canada	green or yellow	1984	1	3
Ronald McDonald Shoe & Sock Game-Japan	plastic with ball and string - in Japanese writing		5	10
Ronald McDonald Tote Bag-Japan	writing in Japanese		5	10
Ronald McDonald Wallet - Canada	yellow, attaches to shoes with laces	1987	1	2
Runaway Robots	set of six: dark blue Skull, green Coil, red Flame, purple Bolt, blue Beak and yellow Jab	1988	7	14
Safari Adventure	six different rubber animals: alligator, monkey, gorilla, tiger, hippo, rhino	1980	2	4
Sailors	set of four floating toys: Hamburglar Sailboat, Ronald Airboat, Grimace Sub and Fry Kids Ferry	1987	5	10
Santa Claus: the Movie Reindeer Ornament		1985	1	3

McDonald's

NAME	DESCRIPTION	YEAR	EX	MINT
School Days	set of five school tools: pencils (Ronald, Grimace, Hamburglar) erasers (Ronald, Grimace, Hamburglar, Captain Crook, Birdie) pencil sharpener, ruler, pencil case	1984	2	4
Serving Trays	set of six white plastic wedge shaped trays: Ronald, Big Mac, Mayor McCheese, Hamburglar, Grimace, Captain Crook		3	7
Ship Shape 1	set of four boat containers with stickers: Hamburglar, Ronald, Grimace, Captain Crook	1983	10	30
Ship Shape 2	same as 1983 set	1985	10	20
Sindy Doll	dressed in older McDonald's uniform	1970	4	8
Singing Wastebasket Bank	5-1/8" white plastic basket with coin slot in top		6	10
Sky-Busters	set of six rubber airplanes: Skyhawk AAF, Phantom, Mirage F1, United DC-10, MIG-21, Tornado	1982	3	5
Smart Duck	set of six rubber figures: duck, cat, donkey, chipmunk, two rabbits	1979	2	3
Snow White	under 3 toy: Dopey & Sneezy		2	4
Snow White Figures	Prince, Snow White, Dock, Bashful, Sleepy, Queen, Witch		2	4
Sonic 3 The Hedgehog	under 3 toy: Sonic Ball	1994	3	6
Sonic 3 The Hedgehog	set of four: Sonic the Hedgehog, Miles "Tails" Power, Knuckles & Dr. Ivo Robotnik	1994	2	4
Space Aliens	set of eight rubber monsters: lizard man, vampire bat, gill face, tree monster, winged fish, cyclops, veined brain, insectman	1979	2	3
Space Jam	interlocking pieces	1996	2	4
Space Raiders	set of eight rubber aliens: Drak, Dard, flying saucer, Rocket Kryoo-5, Horta, Zama, Rocket Ceti-3, Rocket Altair-2	1979	2	3
Speedee "Touch of Service" Pin	enamel		6	12
Spinner Baseball Game	green plastic with four characters	1983	2	4
Sports Balls	set of four: basketball, baseball, football and tennis ball	1988	20	40
Sports Balls	set of four: baseball, football, basketball, soccer	1990	2	4
Star Trek	rings: Kirk, Spock, Starfleet insignia, Enterprise	1979	15	20
Star Trek	Navigation bracelet with decals	1979	10	25
Star Trek	set of four glitter iron-ons: Kirk, Spock, McCoy, Ilia, packaged in pairs	1979	10	20
Star Trek	Starfleet game	1979	10	15
Star Trek	set of five video viewers, each with different story	1979	15	30
Sticker Club	set of five different sticker sheets: reflectors, scratch and sniff, color designs, action stickers, puffy designs	1985	1	3
Stomper Mini 4X4	set of eight big wheeled cars: Tercel, AMC Eagle, Chevy S-10 pickup, Chevy van, Chevy Blazer, Ford Ranger, Jeep Renegade, Dodge Ram,	1986	5	10
Sunglasses	Hamburglar, Ronald w/yellow lenses or Ronald McDonald on stem		2	4
Super Looney Tunes	set of four figures with costumes: Super Bugs, Bat Duck, Taz Flash, Wonder Pig	1991	2	4
Super Looney Tunes	under 3 toy: Bat Duck	1991	2	4
Super Mario Brothers	under 3 toy: Super Mario	1990	2	4
Super Mario Brothers	set of four action figures: Mario, Luigi, Little Goomba, Koopa	1990	2	4
Super Summer	Sand Castle Pail with shovel, Sand Pail with rake, Fish Sand Mold, Sailboat, Beach Ball	1988	2	4

Rescuers Down Under, 1990, McDonald's

Fast Food Racers, 1990, Wendy's

McDonald's

NAME	DESCRIPTION	YEAR	EX	MINT
Tailspin	set of four characters in airplanes: Molly, Balloo, Kit, Wildcat	1990	2	4
Tailspin	under 3 toys: Baloo's seaplane or Wildcat's jet	1990	2	4
Tic Tac Mac Game	yellow base, Grimace is X, Ronald is O	1981	2	5
Tinosaurs	set of eight figures: Link the Elf, Baby Jad, Merry Bones, Dinah, Time Traveller Fern, Tiny, Grumpy Spell, Kave Kolt Kobby	1986	5	10
Tiny Toons Flip Cars	under 3 toy: Sweetie	1991	1	2
Tiny Toons Flip Cars	set of four cars, each with two characters: Montana Max/Gobo Dodo, Babs/Plucky Duck, Hampton/Devil, Elmyra/Buster Bunny	1991	1	2
Tom & Jerry Band	under 3 toy: Droopy	1990	5	8
Tom & Jerry Band	set of four characters with instruments: Tom at keyboard, Jerry on drums, Spike on bass, Droopy at the mike	1990	4	8
Tonka	set of five: fire truck, loader, cement mixer, dump truck, backhoe	1992	1	2
Tonka	under 3 toy: dump truck	1992	1	2
Tootler Harmonica		1985	2	4
Tops	red, blue and green	1978	3	7
Totally Toy Holiday	under 3 toy: Keyforce car	1993	1	2
Totally Toy Holiday	set of nine: Attack pack, Keyforce truck, Keyforce charm Magic Nursery, Polly Pocket, Sally Secrets (Caucasion or African-American), Lil' Miss Candistripe, Tattoo Machine, Mighty May	1993	1	2
Totally Toy Holiday	under 3 toy: Magic Nursery (candy canes)	1993	1	2
Totally Toy Holiday	under 3 toy: Magic Nursery (Holly)	1993	1	2
Transformers	set of 24; non-transforming plastic figures, each	1985	50	140
Turbo Macs	set of four pull-back action cars with characters driving and large "M" on hood: Ronald, Grimace, Birdie, Hamburglar	1990	3	6
Under Sea	set of six cartons with undersea art: alligator, dolphin, hammerhead shark, sea turtle, seal, walrus	1980	2	3
United Airlines Friendly Skies	set of two airplanes with United markings: either Ronald or Grimace flying	1991	5	10
VR Troopers		1996	1	2
Walt Disney Video Viewer-Cinderella-Japan			5	10
What is It?	set of six rubber animals: skunk, squirrel, bear, owl, baboon, snake	1979	1	3
Who Framed Roger Rabbit Cups	Roger Rabbit-Hollywood, Benny the Cab, Roger Being Chased	1988	2	4
Wild Animal Toy Books	under 3 toy: Giant Panda	1991	3	6
Wild Animal Toy Books	complete set of four plus under three	1991	2	4
Winter World	set of five flat vinyl tree ornaments: Ronald, Hamburglar, Grimace, Mayor McCheese, Birdie	1983	5	10
Wrist Wallets	set of four watch-type bands with coin-holding dial: Ronald, Captain Crook, Big Mac, Hamburglar	1977	5	10
Yo Yogi	set of four characters	1991	3	5
Yo-Yo	half red, half yellow	1979	2	5
Young Astronauts	set of four snap-together models: Apollo Command Module, Argo Land Shuttle, Space Shuttle, Cirrus Vtol	1986	5	20
Young Astronauts	set of four vehicles	1992	2	4
Zoo Face	set of four rubber noses and makeup kits: alligator, monkey, tiger, toucan	1988	2	4

Pizza Hut

NAME	DESCRIPTION	YEAR	EX	MINT
Beauty & the Beast Puppets	set of four: Belle, Beast, Chip, Cogsworth	1992	4	8
Eureeka's Castle Puppets	set of three: Batly, Eureeka, Magellan	1991	3	6
Land Before Time Puppets	set of six: Spike, Sharptooth, Pteri, Little Foot, Cera, Ducky	1988	4	8
Universal Monster Cups	set of three holographic cards w/ 3-D cups		10	20

Roy Rogers

Critters	set of eight: blue eyes-yellow, blue eyes-orange, blue eyes-purple, blue eyes-red, yellow eyes-orange, yellow eyes-yellow, pink eyes, orange, pink eyes-purple		1	2
Gator Tales	set of four		1	2
Gumby	Gumby, blue girl		4	8
Ickky Stickky Bugs	set of sixteen		2	3
Skateboard Kids Figures	with yellow, orange, purple, or red knee pads		3	5
Star Searchers	saucer w/green top and orange pilot, sled w/ orange top and purple bottom, sled w/purple top and orange bottom, sled w/green top and orange bottom, robot w/orange head, green front		2	3

Taco Bell

Happy Talk Sprites	yellow Spark, white Twink		2	4
Hugga Bunch Plush Dolls			2	4
Star Wars	Death Star Spinner, Darth Vader/Yoda 3-D Cube, multi-scene cube, Bespin Cloud City, balancing Boba Fett, Millennium Falcon, R2-D2; Yoda under 3 toy	1997	3	5

Tastee Freeze

Roy Campanella Figure			20	35

Wendy's

Alf Tales	set of six: Sleeping Alf, Alf Hood, Little Red Riding Alf, Alf of Arabia, Three Little Pigs, Sir Gordon of Melmac	1990	2	4
Alien Mix-Ups	set of six: Crimsonoid, Bluezoid, Limetoid, Spotasoid, Yellowboid, Purpapoid	1990	1	2
All Dogs Go To Heaven	set of six: Anne Marie, Car Face, Charlie, Flo, Itchy, King Gator	1989	2	4
Definitely Dinosaurs	set of four: blue Apatosaurus, gray T-Rex, yellow Anatosaurus, green Triceratops	1988	3	6
Definitely Dinosaurs	set of five: green Ankylosaurus, blue Parasaurolophus, green Ceratosaurus, yellow Stegosaurus, pink Apatosaurus	1989	2	4
Fast Food Racers	set of five: hamburger, fries, shake, salad, kid's meal	1990	2	4
Fun Flyers	3-1/2" wide in red, yellow or blue		1	3
Furskins Plush Dolls	set of three: 7" tall: Boone in plaid shirt and red pants, Farrell in plaid shirt and blue jeans, Hattie in pink and white dress	1988	4	8

Jetsons, The Movie Space Gliders, 1990, Wendy's

Goof Troop Bowlers, 1992, Burger King

Sky-Busters, 1982, McDonald's

Wendy's

NAME	DESCRIPTION	YEAR	EX	MINT
Glass Hangers	yellow turtle, yellow frog, yellow penguin and purple gator		3	5
Glo Friends	set of 12: Book Bug, Bop Bug, Butterfly, Clutter Bug, Cricket, Doodle Bug, Globug, Granny Bug, Skunk Bug, Snail Bug, Snug Bug	1988	2	4
Good Stuff Gang	Cool Stuff, Cat, Hot Stuff, Overstuffed, Bear, Penguin	1985	2	4
Jetsons Figures	set of six figures in spaceships: George, Judy, Jane, Elroy, Astro, Spacely	1989	2	4
Jetsons: The Movie Space Gliders	set of six PVC figures on wheeled bases: Astro, Elroy, Judy, Fergie, Grunchee, George	1990	2	4
Micro Machines Super Sky Carriers	set of six kits that connect to form Super Sky Carrier	1990	2	4
Mighty Mouse	set of six: Bat Bat, Cow, Mighty Mouse, Pearl Pureheart, Petey, Scrappy	1989	3	5
Play-Doh Fingles	set of three finger puppet molding kits: green dough with black mold, blue dough with green mold, yellow dough with white mold	1989	3	6
Potato Head Kids	set of six: Captain Kid, Daisy, Nurse, Policeman, Slugger, Sparky	1987	4	8
Speed Writers	set of six car-shaped pens: black, blue, fuchsia, green, orange, red	1991	2	4
Summer Fun	float pouch, sky saucer	1991	1	2
Teddy Ruxpin	set of five: Professor Newton Gimmick, Teddy, Wolly Whats-It, Fob, Grubby Worm	1987	3	8
Too Kool for School	set of five		2	4
Tricky Tints	set of four		2	4
Wacky Wind-Ups	set of five: Milk Shake, Biggie French Fry, Stuff Potato, Hamburger, Hamburger in box	1991	2	3
Where's the Beef Stickers	set of six	1984	1	3
World Wildlife Foundation	set of four books: All About Koalas, All About Tigers, All About Snow Leopards, All About Pandas	1988	2	4
World Wildlife Foundation	set of four plush toys: panda, snow leopard, koala, tiger	1988	5	10
Yogi Bear & Friends	set of six: Ranger Smith in kayak, Boo Boo on skateboard, Yogi on skates, Cindy on red scooter, Huckleberry in inner tube, Snagglepuss with surfboard	1990	2	4

White Castle

Ballerina's Tiara			3	6
Camp White Castle			2	4
Camp White Castle Bowls	orange plastic		3	6
Castle Creatures			2	5
Castle Friends Bubble Makers	set of four		3	5
Castle Meal Friends	set of six	1989-9	4	10
Castle Meal Friends	set of five	1992-9	3	7
Castleburger Dudes Wind-Up Toys	set of four		2	4
Cosby Kids	set of four		8	15
Easter Pals	rabbit with carrot, rabbit with purse		3	5
Glow in the Dark Pull-Apart Monsters	set of three		3	6
Godzilla Squirter			3	6
Holiday Huggables	Candy Canine, Kitty Lights, Holly Hog		2	5

White Castle

NAME	DESCRIPTION	YEAR	EX	MINT
Nestle's Quik Plush Bunny			3	7
Push `N GO GO GO!	set of three		2	4
Silly Putty	set of three		2	4
Stunt Grip Geckos	set of four		2	4
Tootsie Roll Express	set of four		3	5
Totally U Back To School	pencil, pencil case		2	4
Willis the Dragon	Christmas giveaway		3	6
Willis the Dragon Sunglasses			2	4

M&Ms figures, 1997, Burger King

Rock and Roll

It's true that rock and roll is here to stay. That's easily evidenced by the number of toys and memorabilia avidly sought by collectors today.

The beginnings of rock and roll took root with the rhythm and blues music popularized after World War II. In the 1950s, Elvis Presley revolutionized the rock sound. But it wasn't until Liverpool's four young men swept America in the early 1960s that the rock and roll boom began. The Beatles' impact was so strong, in fact, it resonated not only in the music world, but in the merchandising and collectibles arena as well.

Because of the group's charm and popularity, companies soon learned the marketing potential of John, Paul, George and Ringo. Millions of young fans swooned over the Beatles, so it was only natural that licensees would immediately attach themselves to their success. Soon, and for decades following their debut, images of the Fab Four would appear on lunch boxes, wallets, clothing, toiletries, knickknacks, toys and more.

Not only were the Beatles' images used to endorse products, but many toys were made to help kids recreate the frenzy at home. Toy guitars, drums and dolls brought the Beatles to life to a younger audience. Today, collectors are thrilled to find such remnants of rock and roll history. And while Beatles collectibles aren't particularly difficult to find, those in the best condition are treasured.

Beatles Guitar, Selcol, 1960s

After the success companies enjoyed marketing Beatles toys, rock and roll continued to be a successful font for merchandisers. Although their musical talent was scrutinized, both the Monkees and the Partridge Family enjoyed added exposure on television. That exposure created much interest in related memorabilia, especially for young fans.

In the 1970s, KISS stormed onto the rock scene in wild face makeup and costumes unlike anything previously seen. Armies of KISS fans rocked and rolled all night, and in 1979, Mego made 12-inch dolls of band members Paul Stanley, Gene Simmons, Ace Frehley and Peter Criss. Dozens of KISS toys followed to the delight of eager fans. McFarlane Toys continued the KISS craze in 1997 with new action figures (see **Action Figures** chapter for values).

But KISS was one of the last rock and roll groups featured on numerous toys and memorabilia. While 1980s and 1990s icons like Michael Jackson and Madonna have been the subjects of some toys and collectibles, the trend has clearly ended.

Why? In their heyday, the Beatles were a clean-cut group with non-threatening personas. The Monkees and the Partridge Family were rather wholesome, humorous groups. KISS flaunted their distinctive look and image. But today's rockers (and rappers) do not lend themselves as well to having their images represented on toys. Their lyrics, images and attitudes are often sexual, violent or anti-establishment — not exactly the sweet, non-threatening images of the past.

Trends

Beatles toys lead the pack of rock and roll memorabilia. The market remains active with growth each year, particularly with renewed interest in the group since the release of the 1996 Beatles Anthology albums.

Non-toy items such as ticket stubs, recordings and posters are consistent sellers, but toys bearing the images of revered rockers seem to always be in vogue.

These licenses, although 30 years old, haven't lost their power. In fact, many new limited-edition collectibles (like plates, dolls and ornaments) have continued to satisfy the public's hunger for Beatles and Elvis items. Even the new toys of 1998 embraced these vintage rockers. New toys based on Elvis and KISS were represented at 1998's Toy Fair.

Toys featuring minor or fleeting rock phenomenons (like New Kids on the Block, MC Hammer or others) will likely see no future increase in collectibility. But collectors can't really go wrong picking up items featuring major players in rock and roll history.

One caveat, however. Reproductions and unlicensed items are rampant, so collectors should educate themselves about items before purchasing.

The Top 10 Rock and Roll Toys (in Mint in Box condition)

1.	Beatles Banjo, Mastro, 1964	$800
2.	Beatles Bobbin' Head Dolls, 1960s	750
3.	Beatles Guitar, Selcol, 1960s	550
4.	Beatles Magic Slate, Merit	500
5.	Beatles Cartoon Kit, Colorforms, 1966	500
6.	Beatles Jr. Guitar, Mastro, 1960s	425
7.	Beatles Costume, Ben Cooper, 1960s	400
8.	Paint Your Own Beatle Kit, Artistic Creations, 1960s	400
9.	Yellow Submarine Halloween Costume, Collegeville, 1960s	375
10.	Partridge Family Bus, Remco, 1973	300

ROCK AND ROLL

Beatles

NAME	COMPANY	YEAR	DESCRIPTION	GOOD	EX	MIB
Beatles Banjo	Mastro	1964		200	400	800
Beatles Cartoon Kit	Colorforms	1966		150	300	500
Beatles Coloring Book	Saalfield	1964		25	60	110
Beatles Costume	Ben Cooper	1960s	child's costume and mask; John, Paul, George, or Ringo, each	100	200	400
Beatles Forever Cloth Dolls	Applause	1987	22" tall, each	55	75	100
Beatles Guitar	Selcol	1960s	23", plastic	200	300	550
Beatles in Pepperland Puzzles	Jaymar		many variations and sizes, each	25	60	90
Beatles Jr. Guitar	Mastro	1960s	14" red/pink plastic guitar w/Beatles graphics	100	225	425
Beatles Magic Slate	Merit		British	150	300	500
Beatles Notebook Binder		1960s	binder from Beatles Fan Club	30	75	125
Beatles Toy Watches		1960s	four, tin w/plastic bands, on card	30	75	150
Beatles Wig	Lowell Toy			65	90	130
Bobbin' Head Dolls		1960s	8" tall, ceramic, four in set	300	500	750
Disk Go Case	Charter	1966	45 rpm carrying case; several colors	70	100	200
George Harrison Doll	Remco	1964		50	100	250
Harmonica	Hohner		in Beatles box	35	50	75
Inflatable Dolls		1966	Lux promotional; 13" tall; each	25	35	50
John Lennon Doll	Remco	1964		50	100	250
Paint Your Own Beatle Kit	Artistic Creations	1960s	oil painting kit; John, Paul, George, or Ringo, each	150	300	400
Paul McCartney Doll	Remco	1964		50	100	250
Paul McCartney Soaky	Colgate	1965	red plastic	70	120	250
Ringo Starr Doll	Remco	1964		50	100	250
Ringo Starr Soaky	Colgate	1965	blue plastic	70	120	250
Yellow Submarine Halloween Costume	Collegeville	1960s	Blue Meanie costume and mask	100	200	375
Yellow Submarine Water Color Set	Craft Master	1960s	pictures and paints	35	75	140

Elvis Presley

NAME	COMPANY	YEAR	DESCRIPTION	GOOD	EX	MIB
Elvis Presley Doll	Eugene	1984	12" tall	20	40	80
Elvis Presley Doll	World Dolls	1984	21" tall	50	100	200
Elvis Presley Wristwatch	Bradley	1983	white plastic case, quartz, stainless back, face shows a young Elvis, white vinyl band	20	35	75
Jigsaw Puzzle	Milton Bradley	1992	Elvis postage stamp	5	10	20

KISS

NAME	COMPANY	YEAR	DESCRIPTION	GOOD	EX	MIB
Jigsaw Puzzles	Milton Bradley	1970s		10	20	35
KISS Rub n' Play Magic Transfer Set	Colorforms	1979		20	40	80

Partridge Family Paper Dolls, Saalfield, 1970s

Beatles Dolls, Remco, 1964

Beatles Bobbin Head Dolls, Car Mascot, 1964

KISS

NAME	COMPANY	YEAR	DESCRIPTION	GOOD	EX	MIB
KISS Van Model Kit	AMT	1977		35	70	140
KISS Your Face Make-Up Kit	Remco	1978		30	65	115
Trading Cards Set	Donruss	1978	set of 132 cards	25	35	70
View-Master Set	GAF	1978		10	20	35
Wastebasket		1978	metal, cylindrical	20	45	65

Michael Jackson

NAME	COMPANY	YEAR	DESCRIPTION	GOOD	EX	MIB
Michael Jackson AM Radio	Ertl	1984		12	25	50
Michael Jackson Doll	LJN	1984	several styles	12	25	60
Michael Jackson Dress-Up Set	Colorforms	1984		10	15	25
Michael Jackson Microphone	LJN	1984	cordless, electronic	10	20	35
Michael's [Jackson] Pets	Ideal	1987	plush animals, 10 kinds, each	7	15	25

Miscellaneous

NAME	COMPANY	YEAR	DESCRIPTION	GOOD	EX	MIB
Andy Gibb Doll	Ideal	1979	7" tall	15	30	60
Boy George Doll	LJN	1980s	12", poseable in alphabet shirt	30	60	120
Boy George Doll	LJN	1980s	15", polka dot shirt	20	40	75
Cher Doll	Mego	1976	12", w/growing hair	35	75	140
Cher Doll	Mego	1976	12"	25	45	80
Debby Boone Doll	Mattel		10" tall	20	40	80
Dolly Parton Doll	Goldberger	1970s	12" tall	20	40	80
Pat Boone Paper Dolls	Whitman	1959	two cardboard dolls plus clothes	20	45	70
Pinky Lee Costume		1950s	hat, pants, and shirt	35	75	145
Pinky Lee Xylophone	Emenee			35	70	120
Rick Nelson Paper Dolls	Whitman	1959		20	45	70
Sonny & Cher Play Set	Mego	1976		20	45	80
Toni Tennile Doll			12" tall	8	15	30

Monkees

NAME	COMPANY	YEAR	DESCRIPTION	GOOD	EX	MIB
Flip Movies	Topps	1967	each	5	10	20
Halloween Costumes	Bland Charnas	1967	each	60	125	250
Jigsaw Puzzle	Fairchild	1967		15	25	45
Monkees Dolls	Remco	1967	4", rubber, each	35	80	175
Monkees Finger Puppets	Remco	1969		15	25	45
Talking Hand Puppet	Mattel	1966	cloth w/heads of Monkees on fingertips	50	100	200
Tambourine	Raybert	1967		45	100	200
Toy Guitar	Mattel	1966	14", wind-up crank	40	90	175
Toy Guitar	Mattel	1966	20"	60	125	250

Osmonds

NAME	COMPANY	YEAR	DESCRIPTION	GOOD	EX	MIB
Donny & Marie TV Show Play Set	Mattel	1976		15	35	55
Donny Osmond Doll	Mattel	1976	12" tall	15	25	45
Donny Osmond String Puppet	Madison	1978		20	35	60
Marie Osmond Doll	Mattel	1976	12" tall	10	25	40
Marie Osmond Modeling Doll	Mattel	1976	30" tall	20	40	80
Marie Osmond String Puppet	Madison	1978		25	35	65
Osmonds Colorforms Dress-Up Set	Colorforms	1976		10	25	45

Partridge Family

NAME	COMPANY	YEAR	DESCRIPTION	GOOD	EX	MIB
David Cassidy Dress-Up Kit	Colorforms	1972		20	40	75
Laurie Partridge Doll	Remco	1973	20" tall	55	120	225
Partridge Family Bus	Remco	1973	plastic, 14" long	55	160	300
Partridge Family Guitar	Carnival	1970s	19" plastic, decal of David Cassidy on body	35	75	150
Partridge Family Paper Dolls	Saalfield	1970s	several styles	20	40	75
Patti Partridge Doll	Ideal	1971		50	110	200

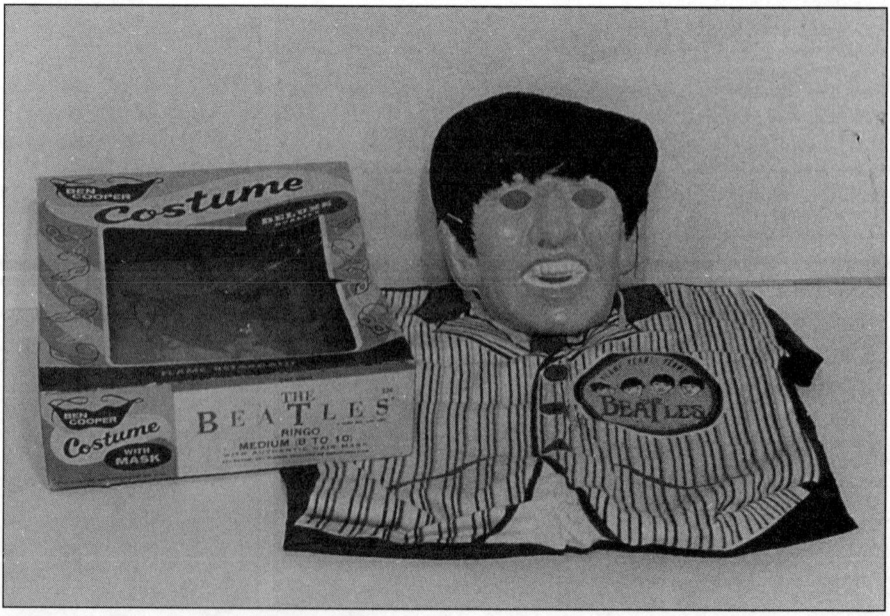

Beatles costume, 1960s, Ben Cooper

Space/Science Fiction Toys

Some would trace the modern age of science fiction to 1956 and *Forbidden Planet.* Undoubtedly, the toy world would be poorer for the lack of the movie's Robby the Robot. But through one medium or another, science fiction has enthralled millions for many years, back to Jonathan Swift's *Gulliver's Travels.*

The first universally acclaimed work of science fiction was Mary Shelley's *Frankenstein* or *The Modern Prometheus,* and her vision made way for Verne, Wells, Burroughs, Lovecraft, Heinlein, Asimov, Clarke and a host of others, whose collective imaginations led us up to today and through tomorrow.

Even with its classical pedigree, science fiction is almost exclusively a product of the 20th century, as the hard foundation of science had to exist before fiction writers could extrapolate upon it. In particular, science fiction is a phenomenon of the atomic age. World War II, more than any other event this century, opened our eyes to the wondrous and horrific potential of applied science.

Just as science fiction has captivated readers of all ages, so have toys. Buck Rogers made his first appearance in 1928, but in 1929, Buck Rogers went from pulp to newsprint, becoming the first science fiction comic strip. Flash Gordon followed Buck Rogers into print in 1934 and was an immediate success. Within two years, Flash was on the silver screen, portrayed by Buster Crabbe. Buck Rogers finally made it to the screen in 1939, also played by Crabbe.

During this period, Marx produced numerous toys in support of each character, including two ships that have become classics of the space toy field. Opinions vary as to which wind-up is better executed, Buck Rogers' 25th Century Rocket Ship or Flash Gordon's Rocket Fighter. Both are considered superb examples of tin character space toys.

From Ray Guns to Star Wars

No discussion of space toys would be complete without mention of ray guns. Here again Marx is a major player, producing numerous generic and character space guns.

Space Commander Walkie Talkies, 1968, Remco

Rocket Lite Flashlight, Rayovac, 1950s

Daisy, Hubley and Wyandotte, among others, all made memorable contributions as well.

Space toys have been made continuously for most of the 20th century. The 1930s and 1940s saw Buck Rogers and Flash Gordon. The 1950s saw fiction become reality with the growth of television. *Captain Video* was the first space series on TV, appearing in the summer of 1949. *Buzz Correy and his Space Patrol* and *Tom Corbett, Space Cadet* would feed the appetite for adventure until 1956 when the heavens took on a visual scale and grandeur never seen before — in the panoramic wonder of *Forbidden Planet.*

In 1966, when the low budget Star Trek went on the air, few dreamed that for millions of people, life would never be the same. Even though the original show ran only three seasons, its impact and legacy are undeniable. The phenomenon of *Star Trek* has grown far beyond cult status, and the extraordinary success of *Star Trek: The Next Generation* has only broadened its reach.

Star Trek may be big; however, the king of space toys has to be *Star Wars.* The array of books, models, figures, play sets and other items released since its debut in 1977 continued unabated until 1988. The license gained a new lease in 1987 with the opening of Star Tours at Disneyland and Disney World, generating still more new merchandise.

In terms of diversity of toys, the universe of *Star Wars* is easily the most fully realized and diversely populated in all science fiction. *Star Wars* figures, vehicles and play sets are the most widely-traded science fiction toys on the market today.

Today, store shelves are again seeing new *Star Wars* toys thanks to the highly successful 1997 re-release of the Star Wars trilogy and the forthcoming prequels. *Star Wars'* longevity and international name recognition are excellent assurances of the continuing popularity of its toys.

Trends

In general, the field of science fiction toys is one with particular growth potential, given the prevalence of science fiction in today's culture and the exceptional strength of franchises like *Star Trek* and *Star Wars.* Recent generations have been weaned on Luke Skywalker, Han Solo, Mr. Spock and Captain Picard. Newer incarnations such as *Star Trek: Voyager* only fuel the series' popularity. Toys from older, nostalgic series such as *Lost in Space* and *Buck Rogers in the 25th Century* continue to top lists of the most sought-after space/science fiction toys, and their values are soaring, particularly at auction.

The Top 10 Space / Science Fiction Toys (in Mint in Package condition)

1. Lost in Space Doll Set, Marusan/Japanese $7,000
2. Buck Rogers Solar Scouts Patch, Cream of Wheat, 1936 5,500
3. Buck Rogers Cut-Out Adventure Book, 1933 4,500
4. Space Patrol Monorail Set, Toys of Tomorrow, 1950s 4,000
5. Buck Rogers Roller Skates, Marx, 1935 3,500
6. Space Patrol Lunar Fleet Base, 1950s 2,600
7. Buck Rogers Pocket Watch, E. Ingraham, 1935 2,500
8. Lost in Space Switch-and-Go Set, Mattel, 1966 2,300
9. Buck Rogers Costume, Sackman Bros., 1934, ... 2,100
10. Lost in Space Roto-Jet Gun Set, Mattel, 1966 2,000

Editor's Note: Several abbreviations are used in this section to denote Star Wars and Star Trek movies. Key is as follows: ESB: The Empire Strikes Back; POTF: Power of the Force; ROTJ: Return of the Jedi; ST:TMP: Star Trek, The Motion Picture; ST:TNG: Star Trek, The Next Generation.

SPACE/SCIENCE FICTION TOYS

Alien/Aliens

NAME	COMPANY	YEAR	DESCRIPTION	GOOD	EX	MIB
Alien Blaster Target Set	HG Toys		larger set	75	170	250
Alien Chase Target Set	HG Toys		dart pistol, cardboard target	50	165	200
Alien Costume	Ben Cooper		black/white	50	65	100
Alien Figure	Kenner	1979	18" tall	75	200	400
Alien Model Kit	Tsukuda	1980s	vinyl, 1/6 scale	60	225	350
Alien Warrior Model Kit	Halcyon		base and egg	20	30	50
Aliens Colorforms Set	Colorforms			10	20	40
Aliens Computer Game	Commodore	1985		5	15	30
Glow Putty	Laramie		unlicensed art, carded	10	15	20
Movie Viewer	Kenner		"Alien Terror" film clip	40	75	100

Battlestar Galactica

NAME	COMPANY	YEAR	DESCRIPTION	GOOD	EX	MIB
Battlestar Galactica Model Kit	Monogram	1979	#6028	15	25	40
Colonial Viper Model Kit	Monogram	1979		15	25	40
Colorforms Adventure Set	Colorforms	1978		12	20	30
Cylon Base Star Model Kit	Monogram	1979	silver plastic	15	30	45
Cylon Helmet Radio		1979		10	15	25
Cylon Raider Model Kit	Monogram	1979		15	25	40
Cylon Warrior Costume		1978	boxed	10	15	25
Galactic Cruiser	Larami	1978	die-cast	5	8	12
Game of Starfighter Combat	FASA	1978	role playing game	10	15	25
L.E.M. Lander	Larami	1978	die-cast	5	8	12
Lasermatic Pistol	Mattel	1978		15	30	45
Lasermatic Rifle	Mattel	1978		25	40	65
Poster Art Set	Craft Master	1978		6	10	15
Puzzles	Parker Brothers	1978	The Rag-Tag Fleet, Starbuck, Interstellar Battle, each	6	10	15
Space Alert Game	Mattel	1978	hand-held electronic game	7	12	30

Buck Rogers

NAME	COMPANY	YEAR	DESCRIPTION	GOOD	EX	MIB
25th Century Police Patrol Rocket	Marx	1935	tin wind-up, 12" long	100	600	1200
25th Century Scientific Laboratory	Porter Chemical	1934	with three manuals	500	1000	2000
Adventures of Buck Rogers Book	Whitman	1934	All Pictures Comics edition, Big Big Book	20	80	125
Battle Cruiser Rocket	Tootsietoy	1937	two grooved wheels to run on string	60	180	275
Battle for the 25th Century Game	TSR	1988	role playing board game	15	25	35
Buck and Wilma Masks	Einson-Freeman	1933	paper litho, each	115	195	300
Buck Rogers 25th Century Rocket	Marx	1939	Buck and Wilma in window, 12" long, tin wind-up	200	500	1000

Buck Rogers

NAME	COMPANY	YEAR	DESCRIPTION	GOOD	EX	MIB
Buck Rogers and the Children of Hopetown Book	Golden Press	1980s	Little Golden Book	4	7	10
Buck Rogers and the Depth Men of Jupiter Book	Whitman	1935	Big Little Book	50	80	125
Buck Rogers and the Doom Comet Book	Whitman	1935	Big Little Book	50	80	125
Buck Rogers and the Overturned World Book	Whitman	1941	Big Little Book	45	70	110
Buck Rogers and the Planetoid Plot Book	Whitman	1936	Big Little Book	50	80	125
Buck Rogers and the Super Dwarf of Space Book	Whitman	1943	Big Little Book	45	70	110
Buck Rogers Figure	Tootsietoy	1937	1-3/4" tall, cast, gray	90	145	225
Buck Rogers Films	Irwin	1936	set of six	105	180	275
Buck Rogers in the 25th Century Book	Whitman	1933	Big Little Book, Cocomalt premium	50	80	125
Buck Rogers in the 25th Century Book	Whitman	1933	Big Little Book	70	115	175
Buck Rogers in the 25th Century Button		1935	pinback, color Buck bust profile on blue background, with small ray gun and rocket ship at his shoulders	25	75	125
Buck Rogers in the City Below The Sea Book	Whitman	1934	Big Little Book	75	130	200
Buck Rogers in the City of Floating Globes Book	Whitman	1935	Cocomalt premium, paperback Big Little Book	115	250	400
Buck Rogers in the War With the Planet Venus Book	Whitman	1938	Big Little Book	45	70	110
Buck Rogers on the Moons of Saturn Book	Whitman	1934	premium, paperback Big Little Book	75	130	200
Buck Rogers Pocket Watch	E. Ingraham	1935		400	1600	2500
Buck Rogers vs. The Fiend of Space Book	Whitman	1940	Big Little Book	45	70	110
Buck Rogers Wristwatch	E. Ingraham	1935		390	650	1000
Buck Rogers Wristwatch	Huckleberry Time	1970s		40	100	150
Century of Progess Medallion		1934	metal, reverse shows Buck silhouette profile	100	165	250
Century of Progress Button		1934	pinback, I Saw Buck Rogers 25th Century Show, color litho	135	225	350
Chemistry Set	Grooper	1937	advanced	250	700	1200
Chemistry Set	Grooper	1937	beginners	250	500	1000
Chief Explorer Badge		1936	red enamel	85	200	350
Chief Explorer Badge		1936	gold	150	350	600
Chief Explorer Folder		1936		70	115	175
Clock	Huckleberry Time	1970s		30	50	75
Colorforms Set	Colorforms	1979		10	15	35
Comet Socker Paddle Ball	Lee-Tex	1935		35	60	95
Communicator Set		1970s	with silver Twiki figure	10	15	25
Costume	Sackman Bros.	1934		825	1375	2100
Cut-Out Adventure Book		1933	Cocomalt premium	1000	2500	4500
Electric Caster Rocket	Marx	1930s		125	210	325

Men into Space Astronaut Space Helmet, 1960s, Ideal

Non-Fall Space Patrol X-16, 1955-1960, Masudaya

Space: 1999 Eagle One Spaceship, 1976, Mattel

Buck Rogers

NAME	COMPANY	YEAR	DESCRIPTION	GOOD	EX	MIB
Flash Blast Attack Ship Rocket	Tootsietoy	1937	Flash Blast Attack Ship 4-1/2", Venus Duo-Destroyer with two grooved wheels to run on string	90	145	225
Galactic Play Set	HG Toys	1980s		17	30	45
Helmet and Rocket Pistol Set	Einson-Freeman	1933	set of paper partial-face "helmet" mask and paper pop gun, in envelope	115	250	450
Helmet XZ-34	Daisy	1935	leather	285	475	725
Interplanetary Games Set		1934	three game boards in box: Cosmic Rocket Wars, Secrets of Atlantis, Siege of Gigantica, set	235	390	600
Interplanetary Space Fleet Model Kit		1935	six different kits, including instructions and poster, in box, each	100	165	250
Lite-Blaster Flashlight		1936		155	260	400
Martian Wars Game	TSR	1980s	role playing game	14	25	35
Official Utility Belt	Remco	1970s	in window box, with decoder glasses, wristwatch, disk-shooting ray gun, intruder detection badge, city decoder map, secret message	15	25	50
Paint By Number Set	Craft Master	1980s		8	13	20
Pencil Box	American Pencil	1930s		70	115	175
Pendant Watch	Huckleberry Time	1970s		105	180	275
Pocket Knife	Adolph Kastor	1934	red, green, blue	300	800	1600
Pocket Watch	E. Ingraham	1935	round, face shows Buck and Wilma, lightning bolt hands	400	1500	2500
Pocket Watch	Huckleberry Time	1970s		90	145	225
Punching Bag	Morton Salt	1942	balloon with characters	40	65	100
Puzzle	Puzzle Craft	1945	Buck Rogers and His Atomic Bomber	60	100	150
Puzzle	Milton Bradley	1979	two versions showing TV scenes, each	6	10	15
Puzzle	Milton Bradley	1952	space station scene, 14" x 10"	20	35	50
Puzzle	Milton Bradley	1950		20	35	50
Repeller Ray Ring			brass with inset green stone	200	600	925
Rocket Rangers Iron-On Transfers		1940s	set of three	30	50	75
Rocket Rangers Membership Card				45	75	120
Rocket Ship	Marx	1934	12" tall, wind-up	245	400	625
Roller Skates	Marx	1935		1350	2300	3500
Rubber Band Gun		1930s	cut-out paper gun, on card	30	50	75
Satellite Pioneers Button		1950s	green or blue	20	35	50
Satellite Pioneers Map of Solar System		1958		20	35	50
Satellite Pioneers Membership Card		1950s		30	50	75
Satellite Pioneers Starfinder		1950s	paper	20	35	50
Saturn Ring	Post Corn Toasties	1946	red stone, glow-in-the-dark white plastic on crocodile base	300	500	800
School Bag				60	100	150
Solar Scouts Member Badge		1935	gold	60	100	150

Buck Rogers

NAME	COMPANY	YEAR	DESCRIPTION	GOOD	EX	MIB
Solar Scouts Patch	Cream of Wheat	1936	three colors	1200	4000	5500
Solar Scouts Radio Club Manual		1936		70	200	350
Space Glasses	Norton-Honer	1955		40	65	100
Space Ranger Halolight Ring	Sylvania	1953		400	800	1500
Space Ranger Kit	Sylvania	1952	11" x 15" premium, envelope with six punch-out sheets	50	80	150
Spaceship Commander		1930s	stationary	50	80	125
Spaceship Commander Banner		1936		75	125	195
Spaceship Commander Whistling Badge		1930s		50	80	125
Star Fighter		1979	vehicle for 3-3/4" figures	20	35	50
Strange Adventures in the Spider Ship Pop-Up Book		1935		105	180	275
Strato-Kite	Aero-Kite	1946		17	30	45
Super Foto Camera	Norton-Honer	1955		40	65	100
Super Scope Telescope	Norton-Honer	1955	9" plastic telescope	40	65	100
Superdreadnought SD51X Model Kit		1936	6-1/2" long, balsa wood, one of Interplanetary Space Fleet kit set	100	165	250
Toy Watch	GLJ Toys	1978		10	16	25
Two-Way Transceiver	DA Myco	1948		80	130	200
View-Master Set	View-Master	1979	three-reel set, in envelope or on blister card	4	7	10
Walkie Talkies	Remco	1950s		60	100	150
Wilma Deering Figure	Tootsietoy	1937	1-3/4" tall, cast, gold	70	115	175

Captain Midnight

NAME	COMPANY	YEAR	DESCRIPTION	GOOD	EX	MIB
Air Heroes Stamp Album		1930s	12 stamps	30	50	100
Captain Midnight Medal		1930s	gold medal pin with centered wings and words "Flight Commander"; Capt. is embossed on top with medal dangling beneath	60	100	150
Cup	Ovaltine		plastic, 4" tall, "Ovaltine-The Heart of a Hearty Breakfast"	25	40	65
Membership Manual		1930s	Secret Squadron official code and manual guide	30	50	75
Secret Society Decoder		1949		22	35	55

Captain Video

NAME	COMPANY	YEAR	DESCRIPTION	GOOD	EX	MIB
Captain Video and Ranger Photo		1950s	premium	20	40	60
Captain Video Game	Milton Bradley	1952		75	125	200
Flying Saucer Ring		1950s	with two saucers	500	1000	1500
Galaxy Spaceship Riding Toy		1950s		250	425	650
Interplantary Space Men Figures		1950s	in die cut box	50	80	125
Mysto-Coder		1950s	with photo	60	150	300
Rite-O-Lite Flashlight Gun	Power House Candy	1950s		35	75	150

Star Trek: The Motion Picture Space Design Center, 1979, Avalon

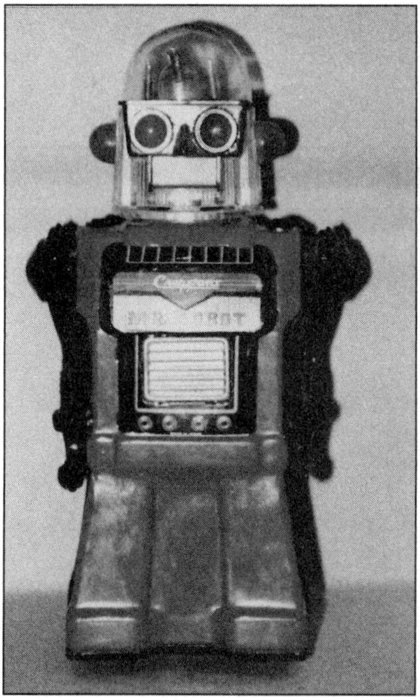

Cragstan's Mr. Robot, 1960s, Crangstan

Star Trek: The Motion Picture Kirk and Spock dolls, 1979, Avalon

Sky Patrol Jet, 1960s, Ideal

Captain Video

NAME	COMPANY	YEAR	DESCRIPTION	GOOD	EX	MIB
Rocket Tank	Lido	1952		55	95	145
Secret Seal Ring		1950s	with initials CV, gold or copper	250	400	600
Space Port Play Set	Superior	1950s		250	425	650
Troop Transport Ship	Lido	1950s	in box	55	95	145

Defenders of the Earth

NAME	COMPANY	YEAR	DESCRIPTION	GOOD	EX	MIB
Defenders Claw Copter	Galoob	1985		11	16	25
Flash Gordon Battle Action Figure	Galoob	1985	5-1/2" tall	7	10	15
Flash Swordship	Galoob	1985		11	16	25
Garax Battle Action Figure	Galoob	1985	5-1/2" tall	11	16	25
Garax Swordship	Galoob	1985		11	16	25
Gripjaw Vehicle	Galoob	1985		11	16	25
Lothar Battle Action Figure	Galoob	1985	5-1/2" tall	8	11	17
Mandrake the Magician Battle Action Figure	Galoob	1985	5-1/2" tall	7	10	15
Ming the Merciless Battle Action Figure	Galoob	1985	5-1/2" tall	8	11	17
Mongor Figure	Galoob	1985		16	23	35
Phantom Battle Action Figure	Galoob	1985	5-1/2" tall	7	10	15
Phantom Skull Copter	Galoob	1985		7	10	15
Puzzle			frame tray	9	13	20

Defenders of the Universe

NAME	COMPANY	YEAR	DESCRIPTION	GOOD	EX	MIB
Battling Black Lion Voltron Vehicle	LJN	1986		9	13	20
Coffin of Darkness Voltron Vehicle	LJN	1986		7	10	15
Doom Blaster Voltron Vehicle	LJN	1986	mysterious flying machine	9	13	20
Doom Commander Figure	Matchbox	1985		5	7	10
Green Lion Voltron Vehicle	LJN	1986		9	13	20
Hagar Figure	Matchbox	1985		5	7	10
Hunk Figure	Matchbox	1985		5	7	10
Keith Figure	Matchbox	1985		5	7	10
King Zarkon Figure	Matchbox	1985		5	7	10
Lance Figure	Matchbox	1985		5	7	10
Motorized Lion Force Voltron Vehicle Set	LJN	1986	black lion with blazing sound	9	13	20
Pidge Figure	Matchbox	1985		5	7	10
Prince Lothar Figure	Matchbox	1985		5	7	10
Princess Allura Figure	Matchbox	1985		5	7	10
Robeast Mutilor Figure	Matchbox	1985		5	7	10
Robeast Scorpious Figure	Matchbox	1985		5	7	10
Skull Tank Voltron Vehicle	LJN	1986		9	13	20
Vehicle Team Assembler	LJN	1986	forms Voltron	9	13	20

Defenders of the Universe

NAME	COMPANY	YEAR	DESCRIPTION	GOOD	EX	MIB
Voltron Lion Force & Vehicle Team Assemblers Gift Set	LJN	1986		9	13	20
Voltron Motorized Giant Commander	LJN	1984	plastic 36", multicolor body with movable head, arms and wings, wire remote control, battery operated	15	25	35
Zarkon Zapper Voltron Vehicle	LJN	1986	with galactic sound	11	16	25

Doctor Who

NAME	COMPANY	YEAR	DESCRIPTION	GOOD	EX	MIB
Ace Figure	Dapol	1986		11	16	25
Anniversary Set	Dapol	1986	Dr. Who, Melanie, K-9, Tardis, base and five-sided console	275	390	600
Cyberman	Dapol	1986		11	16	25
Cyberman Robot Doll	Denys Fisher	1970s	10"	250	350	550
Dalek Army Gift Set	Denys Fisher	1976	seven color varieties of Dapol Dalek plus Dalek Leader, Davros	46	60	95
Dalek Bagatelle	Denys Fisher	1976		70	100	150
Dalek Shooting Game	Marx	1965	8" x 20", four-color tin litho stand up target & generic cork rifle	225	325	500
Dalek's Oracle Question & Answer Board Game		1965	magnetized Dalek that spins	115	165	250
Davros Figure	Dapol	1986	villain with left arm	11	16	40
Doctor Who Card Set	Denys Fisher	1976	24 cards	18	26	40
Doctor Who Card Set		1970s	12 octagon cards	14	20	30
Doctor Who Doll	Denys Fisher	1976	10" tall with scarf and screwdriver	90	130	200
Doctor Who Tardis Play Set	Denys Fisher	1970s		205	295	450
Doctor Who Trump Card Game		1970s		9	13	20
Doctor Who...Dodge the Daleks Board Game		1965		115	165	250
Ice Warrior	Dapol	1986		9	13	20
K-9 Figure	Dapol	1986	the Doctor's dog	7	10	15
Mel Figure	Denys Fisher	1976	pink or blue jacket	9	13	20
Seventh Doctor Figure	Denys Fisher	1976	gray or brown jacket	9	13	20
Tardis Figure	Denys Fisher	1976	the Doctor's transporter	225	325	500

Flash Gordon

NAME	COMPANY	YEAR	DESCRIPTION	GOOD	EX	MIB
Adventure on the Moons of Mongo Game	House of Games	1977		15	25	35
Arak Figure	Mattel	1979	3-3/4", carded	17	30	45
Battle Rocket with Space Probing Action		1976		6	10	15
Beastman Figure	Mattel	1979	3-3/4", carded	15	25	40
Book Bag		1950s	12" wide, three-color art on flap	17	30	45
Candy Box		1970s	eight illustrated boxes, each	4	7	10
Captain Action Outfit	Ideal	1966	with space suit, helmet, mask, belt with ray gun, air tank, boots	165	275	425
Captain Action Outfit	Ideal	1967	all accessories and videomatic ring	185	310	475

Lost in Space 3-D Action Fun Game, 1966, Remco

Land of the Giants Colored Pencil Set, 1969, Hasbro

Flash Gordon

NAME	COMPANY	YEAR	DESCRIPTION	GOOD	EX	MIB
Dr. Zarkov Figure	Mattel	1979	3-3/4" figure, on card	15	25	40
Flash and Ming Button		1970s	shows Flash and Ming crossing swords	4	7	10
Flash Figure	Galoob	1986	Defenders of the Earth series	6	10	15
Flash Figure	Mattel	1979	3-3/4" figure, on card	10	16	25
Flash Gordon Air Ray Gun	Budson	1950s	10" unusual air blaster, handle on top cocks mechanism, pressed steel	215	350	550
Flash Gordon and Alien Model Kit	Revell	1965	#1450	60	100	150
Flash Gordon and the Ape Men of Mor Book	Dell	1942	196 pages, Fast Action Story	70	115	175
Flash Gordon and the Fiery Desert of Mongo Book	Whitman	1948	Big Little Book	30	50	80
Flash Gordon and the Monsters of Mongo Book	Whitman	1935	hardback Big Little Book	50	80	125
Flash Gordon and the Perils of Mongo Book	Whitman	1940	Big Little Book	35	60	90
Flash Gordon and the Power Men of Mongo Book	Whitman	1943	Big Little Book	35	55	85
Flash Gordon and the Red Sword Invaders Book	Whitman	1945	Big Little Book	30	50	80
Flash Gordon and the Tournaments of Mongo Book	Whitman	1935	paperback Big Little Book	45	70	110
Flash Gordon and the Tyrant of Mongo Book	Whitman	1941	Big Little Book, with flip pictures	35	60	95
Flash Gordon and the Witch Queen of Mongo Book	Whitman	1936	Big Little Book	45	70	110
Flash Gordon Arresting Ray Gun	Marx	1939	picture of Flash on handle, 12" long	175	295	450
Flash Gordon Costume	Esquire Novelty	1951		90	145	225
Flash Gordon Figure		1944	wood composition, 5" tall	115	195	300
Flash Gordon Game	House of Games	1970s		15	25	35
Flash Gordon Hand Puppet		1950s	rubber head	90	145	225
Flash Gordon in the Forest Kingdom of Mongo Book	Whitman	1938	Big Little Book	40	65	100
Flash Gordon in the Ice World of Mongo Book	Whitman	1942	Big Little Book, with flip pictures	35	60	90
Flash Gordon in the Jungles of Mongo Book	Whitman	1947	Big Little Book	35	55	85
Flash Gordon in the Water World of Mongo Book	Whitman	1937	Big Little Book	35	60	95
Flash Gordon Kite		1950s	21" x 17", paper	55	90	135
Flash Gordon on the Planet Mongo Book	Whitman	1934	Big Little Book	55	95	145
Flash Gordon Paint Book		1930s		60	100	150
Flash Gordon Radio Repeater Clicker Pistol	Marx		10" long, 1930s	215	350	550

Flash Gordon

NAME	COMPANY	YEAR	DESCRIPTION	GOOD	EX	MIB
Flash Gordon Signal Pistol	Marx	1930s	7", siren sounds when trigger is pulled, tin/pressed steel, green with red trim	195	325	500
Flash Gordon vs. the Emperor of Mongo Book	Dell	1936	244 pages, Fast Action Story	70	115	175
Flash Gordon Water Pistol	Marx	1940s	plastic with whistle in handle, 7-1/2" long	80	130	200
Flash Gordon Wristwatch	Bradley	1979	medium chrome case, back and sweep seconds, Flash in foreground with city behind	70	115	175
Flash Gordon, The Movie Buttons		1980	set of five, each	2	3	5
Home Foundry Casting Set		1935	lead casting set with molds of Flash and other characters	575	975	1500
Lizard Woman Figure	Mattel	1979	3-3/4", carded	15	25	35
Medals and Insignia	Larami	1978	set of five on blister card	3	5	8
Ming Figure	Mattel	1979	3-3/4", carded	12	20	30
Ming's Space Shuttle	Mattel			15	25	35
Pencil Box		1951		70	120	185
Puzzle	Milton Bradley	1951	frame tray	45	80	120
Puzzle		1930s	Featured Funnies	55	95	145
Puzzles	Milton Bradley	1951	set of three	105	180	275
Rocket Fighter	Marx	1939	tin wind-up, 12" long	175	295	450
Rocket Ship	Mattel	1979	inflatable, 3' long, with plastic nose, rocket and gondola attachments	20	35	50
Rocket Ship		1975	3" die cast metal	10	16	25
Solar Commando Set	Premier Products	1950s		65	105	165
Space Compass		1950s	ornately housed compass on illustrated watchband	25	40	65
Space Water Gun	Nasta	1976	water ray gun on illustrated card	6	10	15
Sunglasses	Ja-Ru	1981	plastic with emblem on bridge, carded	3	5	8
Three-Color Ray Gun	Nasta	1976		8	13	20
Thun, Lion Man Figure	Mattel	1979	3-3/4", carded	15	25	40
Two-Way Telephone	Marx	1940s		60	100	175
View-Master Set	View-Master	1963	three reels in envelope	20	35	50
View-Master Set	View-Master	1976	three reels, In the Planet Mongo	6	10	15
Vultan Figure	Mattel	1979	3-3/4", carded	15	30	45
Wallet		1949	with zipper	70	115	175
Water Pistol	Marx	1950s	7-1/2" plastic	155	260	425

Land of the Giants

NAME	COMPANY	YEAR	DESCRIPTION	GOOD	EX	MIB
Annual Book	World Dist./ UK	1969	two volumes, set	30	50	75
Colored Pencil Set	Hasbro	1969		60	100	150
Colorforms Set	Colorforms	1968		30	50	75
Costumes	Ben Cooper	1968	Steve Burton, Giant Witch, or Scientist, each	35	60	150
Double Action Bagatelle Game	Hasbro	1969	pinball game, cardboard back	35	65	150
Flight of Fear Book	Whitman		hardcover	8	15	25
Flying Saucer	Remco	1968	flying disk	60	100	150
Land of the Giants Book	Pyramid		paperback by Murray Leinster	8	13	20
Land of the Giants Coloring Book	Whitman	1968		20	35	50

Land of Giants Spindrift Toothpick Kit, 1968, Remco

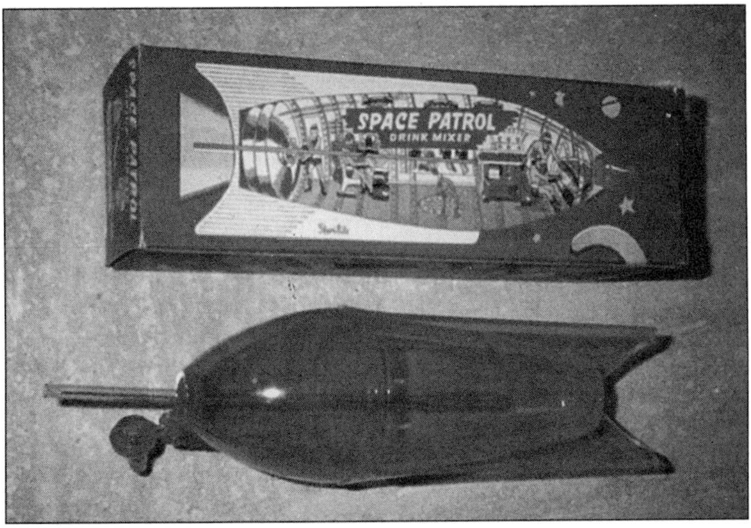

Space Patrol Drink Mixer, 1950s

Land of the Giants

NAME	COMPANY	YEAR	DESCRIPTION	GOOD	EX	MIB
Land of the Giants Comic Book #1	Gold Key	1968		10	15	25
Land of the Giants Comic Books #2-#5	Gold Key	1968	each	8	13	20
Motorized Flying Rocket	Remco	1968	plastic airplane with motor, LOTG logo on wings	80	130	200
Movie Viewer	Acme	1968	film strip viewer, on card	30	45	70
Painting Set	Hasbro	1969		40	65	100
Puzzle	Whitman	1968	round floor puzzle with cartoon illustration	35	55	85
Rub-Ons	Hasbro	1969		30	50	75
Shoot & Stick Target Rifle Set	Remco	1968	western rifle with logo decals	90	145	225
Signal Ray Space Gun	Remco	1968	ray gun with logo decals	70	115	175
Space Sled	Remco	1968	Supercar refitted with LOTG decals--Mike Mercury still sits behind the wheel	195	325	500
Spaceship Control Panel	Remco	1968	Firebird 99 dashboard with a cardboard cut-out of logo on top	195	325	500
Spindrift Interior Model Kit	Lunar Models	1989	#Sf029, interior for 16" model shell	35	55	85
Spindrift Model Kit	Aurora	1968	box shows model under a branch with logo on the side	275	450	700
Spindrift Model Kit	Aurora	1975	reissue, box shows ship in space & features the words Rocket Transport	115	195	300
Spindrift Toothpick Kit	Remco	1968	box of toothpicks with a few cardboard pieces to build ship	30	50	80
Target Set	Hasbro	1969	small guns with darts	60	100	150
The Hot Spot Book	Pyramid		paperback, #2 in series, by Leinster	12	20	30
Trading Card Wrapper	Topps/A & BC	1968		60	100	150
Trading Cards	Topps USA	1968	55 cards	275	450	700
Trading Cards	A & BC/ England	1968	55 cards	275	450	700
Trading Cards Box	Topps/A & BC	1968	display box only	395	650	1000
Unknown Danger Book	Pyramid		paperback #3 by Leinster	12	20	30
View-Master Set	GAF	1968	three reels, first episode	20	35	50
Walkie Talkies	Remco	1968	generic walkie talkies with LOTG decals added	80	130	200
Wrist Flashlight	Bantam Lite	1968		30	50	75

Lost in Space

NAME	COMPANY	YEAR	DESCRIPTION	GOOD	EX	MIB
3-D Fun Set	Remco	1966	three levels with small cardboard figures	500	775	1200
Chariot Model Kit	Lunar Models	1987	#SF009, 1/35 scale, with clear vacuform canopy and dome, plastic body, treads, roof rack	35	50	80
Chariot Model Kit	Marusan/ Japanese		figures and motor	625	975	1500
Costume	Ben Cooper	1965	silver spacesuit with logo	85	130	200
Diorama Model Kit	Aurora	1966	figures, Cyclops, mountain and boulders	425	650	1000
Doll Set	Marusan/ Japanese		dressed in spacesuits with their own freezing tubes with a cardboard insert with color photos and description	2900	4500	7000
Fan Cards		1960s	promo cards mailed to fans; black/white photo	15	25	35
Fan Cards		1960s	promo cards mailed to fans; color photo	20	35	50

Lost in Space

NAME	COMPANY	YEAR	DESCRIPTION	GOOD	EX	MIB
Helmet and Gun Set	Remco	1967	child size helmet with blue flashing light and logo decals, blue/red molded gun	325	525	800
Jupiter Model Kit	Marusan/ Japanese	1966	large version	425	650	1000
Jupiter-2 Model Kit	Comet/ England		2" diameter, solid metal	8	13	20
Jupiter-2 Model Kit	Marusan/ Japanese	1966	6" molded in green plstic with wheels and wind-up motor	425	650	1000
Laser Water Pistol			5" long, first season pistol style	30	50	75
Note Pad			June Lockhart on front	25	40	65
Puzzles	Milton Bradley	1966	frame tray; three poses with Cyclops	40	65	100
Robot	Aurora	1966	6" high with base	425	650	1000
Robot	AHI	1977	10" tall, plastic, stop-and-go action, blinking lights	20	35	50
Robot	Remco	1960s	10" tall, plastic, stop-and-go action, blinking lights	200	310	475
Robot	Remco	1966	12" high, motorized with blinking lights	295	450	700
Robot	Ahi/K-Mart	1977	green bubble and separated legs	40	65	100
Robot YM-3	Masudaya	1986	16" high, speaks English and Japanese	85	130	200
Robot YM-3	Masudaya	1985	4" high, wind-up	20	30	45
Roto-Jet Gun Set	Mattel	1966	gun can be turned into different variations of weapons that shot off small round discs	850	1300	2000
Saucer Gun	AHI	1977	toy gun with discs to shoot	30	50	75
Space Family Robinson Comic Book	Gold Key	1960s		15	25	40
Switch-and-Go Set	Mattel	1966	figures, Jupiter and chariot that ran around track	975	1500	2300
Trading Cards	Topps	1966	55 black and white cards, no wrappers or box	175	260	400
Tru-Vue Magic Eyes Set	GAF	1967	rectangular reels	30	50	75
View-Master Set	GAF	1967	Condemned of Space	25	40	60
Walkie Talkies	AHI	1977	small card	30	50	75

Miscellaneous

NAME	COMPANY	YEAR	DESCRIPTION	GOOD	EX	MIB
Astro Base	Ideal	1960	22" tall, red/white astronaut base, control panel opens lock door, extends crane & lowers astronaut in scout car	225	325	500
Astro Boy Mask/ Glasses		1960s	blue glasses with Astro boy hair on top	20	45	65
Astro-Ray Space Gun			10"	20	30	45
Astronaut Costume	Ben Cooper	1962		18	25	40
Astronaut Costume	Collegeville	1960		18	25	40
Astronaut Space Commander Play Suit	Yankeeboy	1950s	green outfit & cap (military style) with gold piping on collar & pants	35	50	80
Cherilea Space Gun	Marx		miniature scale, die cast	27	40	60
Dan Dare & the Aliens Ray Gun		1950s	21", tin litho gun	105	155	235
Martian Bobbing Head		1960s	7" tall, blue vinyl martian with bobbing eyes and exposed brain	23	35	50
Men into Space Astronaut Space Helmet	Ideal	1960s	plastic helmet with visor	35	50	75

Super Apollo Space Capsule, 1960s

Battlestar Galactica Adventure Set, 1978, Colorforms

Miscellaneous

NAME	COMPANY	YEAR	DESCRIPTION	GOOD	EX	MIB
Puzzle	Selchow & Righter	1970	10" x 14", picture of the moon's surface	14	20	30
Rex Mars Atomix Pistol Flashlight	Marx	1950s	plastic	50	75	100
Rocket Gun	Jak-Pak	1958	7" hard yellow/green plastic with spring loaded plunger that shoots corks up to 50 feet	9	13	20
Space Safari Planetary Play Set		1969	four battery operated space vehicles, 3" tall astronaut figures in silver plastic, 2" hard plastic aliens	45	65	95
Space Water Pistol	Nasta	1976		7	10	15
TV Space Riders Coloring Book	Abbott	1952	14" X 15"	7	10	15
V-Enemy Visitor Doll	LJN	1984	12"	16	23	35
Voyage to the Bottom of the Sea Play Set	Remco	1960s		45	195	350

Monsters

NAME	COMPANY	YEAR	DESCRIPTION	GOOD	EX	MIB
Creature from the Black Lagoon Figure	AHI		carded	295	425	700
Creature from the Black Lagoon Figure	Remco	1980	8", official Universal Studios figure	90	130	200
Creature from the Lagoon Glow-in-the-Dark Mini	Remco			27	40	60
MonstersDeadly Grell Figure	LJN	1983	bendable	5	7	10
Dracula Action Figure	AHI		with Aurora head	65	125	200
Dracula Figure	Remco	1980	8", official Universal Studios figure	30	45	70
Dracula Glow-in-the-Dark Mini Monsters	Remco			23	30	45
Dwarves of the Mountain Human/Monster Figure	LJN	1983		5	7	10
Evil Monster Figure Bugbear & Goblin	LJN	1983	Orcs of the Broken Bone	5	7	10
Frankenstein Figure	AHI		with Aurora head	65	125	200
Frankenstein Figure	Remco	1978	poseable, glow-in-the-dark features and removable cloth costumes	23	35	50
Frankenstein Figure	Remco	1980	8", official Universal Studios figure	27	40	60
Frankenstein Glow-in-the-Dark Mini Monsters	Remco			20	30	45
Godzilla Combat Joe Set		1984	vinyl, with 12" tall combat Joe figure, poseable	450	650	1000
Godzilla Figure	Imperial	1985	6-1/2" tall, arms, legs and tail movable	5	10	15
Godzilla Figure	Imperial	1985	13" tall, arms, legs and tail movable	15	23	35
Godzilla Figure	Mattel	1977	19" tall	20	30	55

Moon McDare

NAME	COMPANY	YEAR	DESCRIPTION	GOOD	EX	MIB
Action Communication Set	Gilbert	1966		25	35	55
Moon Explorer Set	Gilbert	1966		35	50	75

Moon McDare

NAME	COMPANY	YEAR	DESCRIPTION	GOOD	EX	MIB
Moon McDare Figure	Gilbert	1966	12" tall astronaut with blue jumpsuit	55	80	125
Space Accessory Pack	Gilbert	1966		25	35	55
Space Gun Set	Gilbert	1966		30	40	65
Space Mutt Set	Gilbert	1966		30	45	70

Other Worlds, The

NAME	COMPANY	YEAR	DESCRIPTION	GOOD	EX	MIB
Castle Zendo	Arco	1983		20	30	45
Fighting Glowgons Figure Set	Arco	1983		18	25	40
Fighting Terrans Figure Set	Arco	1983		20	30	45
Kamaro Figure	Arco	1983		8	12	18
Sharkoss Figure	Arco	1983		8	12	18

Planet of the Apes

NAME	COMPANY	YEAR	DESCRIPTION	GOOD	EX	MIB
Color-Vue Set	Hasbro	1970s	eight pencils and nine 12" x 13" pictures to color	30	45	65
Dr. Zaius Bank	Play Pal	1967	figural, vinyl, 11"	20	25	45
Fun-Doh Modeling Molds	Chemtoy	1974	molds of Zira, Cornelius, Zaius, and Aldo	20	30	45
Galen Bank	Play Pal	1960s		25	35	55
Planet of the Apes Activity Book	Saalfield	1974	#C3031	15	30	40
Planet of the Apes Coloring Book	Saalfield	1974	#C1531	15	30	40
Planet of the Apes Coloring Book	Artcraft	1974	#C1837	15	30	40
Puzzles	H.G. Toys		96-piece canister puzzles, each	7	10	20
View-Master Set	GAF	1967	#B507, three reels	10	25	40
Wagon	AHI		friction powered prison wagon	20	45	65
Wastebasket	Chein	1967	oval, tin	25	35	60

Robots

NAME	COMPANY	YEAR	DESCRIPTION	GOOD	EX	MIB
Answer Game Machine			battery-operated robot performs math tricks	300	650	900
Attacking Martian	S.H.	1960s	10"	45	65	100
B.O. Robot		1950s	7-1/2", electric remote control	325	450	700
Big Max & His Electronic Conveyor	Remco		9"	50	70	110
Captain Astro		1970	6", wind-up	40	55	85
Chief Robotman	KO	1965	12"	400	850	1200
Countdown-Y	Cragstan	1960s	9"	100	145	225
Cragstan's Mr. Robot	Cragstan	1960s	10-1/2" tin, red body, clear dome head	275	400	750
Electric Robot	Marx		plastic, 15" tall	125	200	275
Forbidden Planet Robby	Masudaya		16" tall, talking	80	115	200
Forbidden Planet Robby	Masudaya		5" tall, wind-up	16	23	35
Frankenstein Robot			tin and plastic, battery operated, wired remote control	550	900	1400
Laughing Robot	Marx			50	70	110
Launching Robot	S.H.	1975	10"	25	35	55

Doctor Who... Dodge the Daleks Board Game, 1965

Attacking Martian robot, 1960s, Japanese

Robots

NAME	COMPANY	YEAR	DESCRIPTION	GOOD	EX	MIB
Lunar Robot			wind-up, companion to Thunder Robot	225	450	650
Lunar Spaceman		1978	12", battery operated	20	30	45
Magnor	Cragstan	1975	9"	23	35	50
Mechanical Interplanetary Explorer		1950s	8", wind-up	180	260	400
Mechanical Moon Creature	Marx	1960	6", wind-up	90	130	200
Mechanical Television Spaceman	ALPS	1965	7", wind-up	45	60	95
Mechanized Robot		1960s	13", battery operated	425	600	925
Moon Creature	Marx	1960s	mechanical, wind-up, 5-1/2" tall	115	170	260
Myrobo		1970s	9", battery operated	25	35	55
Outer Space Ape Man Robot	Ilco	1970s		14	20	30
Outer Space Robot		1979	10", battery operated	20	30	45
Plastic Spaceman	Irwin	1950	wind-up	45	65	100
Radio Control Robot	Bilko	1970s		25	35	55
Red, Blue & Silver Robot		1970s	plastic with antenna, metal key	11	16	25
Rendezvous 7.8	Yanoman		15"	170	245	375
Ro-Gun "It's A Robot"	Arco	1984	robot changes into a rifle	11	16	25
Robert the Wonder Toy	Ideal	1960	14"	90	130	200
Robot	Marx		plastic and metal, 12" tall, 1967	115	165	250
Robot 2500	Durham	1979		14	20	30
Robot Commando	Ideal	1960s	19"	60	85	200
Robot Tank II		1965	10", battery operated	150	210	325
Robot YM-3	Masudaya	1985	5" tall	14	20	30
Sky Robot	S.H.	1970s	8"	20	30	45
Space Guard Pilot	Asak	1975	8"	20	30	45
Sparky Robot	KO	1960	7"	45	65	100
Super Astronaut	S.J.M.	1981		11	16	25
TR-2 Robotank		1975	5", battery operated	35	50	75
Zerak the Blue Destroyer Zeroid	Ideal	1968	6" metal/blue plastic with motor	45	65	175

Rocky Jones, Space Ranger

NAME	COMPANY	YEAR	DESCRIPTION	GOOD	EX	MIB
Rocky Jones, Space Ranger Coloring Book	Whitman	1951	14" x 16"	25	40	60
Space Ranger Button		1954		17	30	45
Wings Pin		1954		17	30	45
Wristwatch		1954	in illustrated box	80	130	200

Space Patrol

NAME	COMPANY	YEAR	DESCRIPTION	GOOD	EX	MIB
Atomic Pistol Flashlight Gun	Marx	1950s	plastic	80	130	200
Cosmic Cap		1950s		115	195	300
Cosmic Rocket Launcher Set		1950s		295	475	750
Cosmic Smoke Gun		1950s	red	100	165	275
Cosmic Smoke Gun		1950s	green	100	300	600
Drink Mixer		1950s	boxed	60	100	150
Emergency Kit		1950s		400	800	1500

Space Patrol

NAME	COMPANY	YEAR	DESCRIPTION	GOOD	EX	MIB
Handbook		1950s		55	95	145
Hydrogen Ray Gun Ring		1950s	glow-in-the-dark, fires caps	70	115	175
Interplanetary Space Patrol Credits Coins			different denominations and colors: Terra, Moon and Saturn, each	10	16	25
Jet Glow Code Belt		1950s	gold-finish metal, spaceship-shaped buckle, decoder ring behind buckle	105	180	275
Lunar Fleet Base		1950s	premium punch-outs in mailing envelope	300	1500	2600
Man From Mars Totem Head Mask		1950s	paper, several styles	65	105	165
Monorail Set	Toys of Tomorrow	1950s		1550	2600	4000
Non-Fall Space Patrol X-16	Matsudaya	1950s		45	100	250
Outer Space Helmet Mask		1950s	paper helmet with plastic one-way visor	100	165	250
Project-O-Scope		1950s	rocket-shaped film viewer with filmstrips	175	295	600
Puzzle	Milton Bradley	1950s	frame tray	35	60	95
Rocket Gun and Holster Set		1950s	with darts	175	295	450
Rocket Gun Set		1950s	with darts, without holster	105	180	275
Rocket Lite Flashlight	Rayovac	1950s	in box	135	225	350
Rocket Port Set	Marx	1950s		115	195	300
Rocket-Shaped Pen		1950s		105	180	275
Space Binoculars		1950s	green plastic, large logo on top	105	180	275
Space Binoculars		1950s	black plastic, logo on sides	70	115	175
Space Patrol Badge		1950s	metal oval	100	300	500
Space Patrol Badge		1950s	plastic, with ship and crest	60	100	180
Space Patrol Cadet Membership Card		1950s		20	35	50
Space Patrol Commander Helmet		1950s	plastic, in box	135	225	350
Space Patrol Periscope		1950s	paper with mirrors	60	125	250
Space Patrol Wristwatch		1950s	illustrated box with "Terra" compass	250	425	650
Space-A-Phones		1950s		145	300	500

Space:1999

NAME	COMPANY	YEAR	DESCRIPTION	GOOD	EX	MIB
Adventure Play Set	Amsco/Milton	1976		30	50	75
Bradley	Astro Popper Gun	1976	on card	6	10	35
Colorforms Adventure Set	Colorforms	1975		10	16	25
Commander Koenig Figure	Mattel	1976		17	30	45
Cut and Color Book	Saalfield	1975		6	10	40
Dr. Russell Figure	Mattel	1976		17	30	45
Eagle Freighter	Dinky	1975	#360, die cast	25	50	150
Eagle One Model Kit	MPC	1976		50	80	100
Eagle One Spaceship	Mattel	1976		100	300	500
Eagle Transport	Dinky	1975	#359, die cast	25	50	100
Eagle Transporter Model Kit	Airfix	1976		12	20	75
Film Viewer TV Set		1976		8	20	55
Galaxy Time Meter		1976		6	10	15

Super Astronaut, 1981, S.J.M.

Tom Corbett Push-Outs Book, 1952, Saalfield

Space:1999

NAME	COMPANY	YEAR	DESCRIPTION	GOOD	EX	MIB
Moon Base Alpha Model Kit	MPC	1976		15	30	50
Moonbase Alpha Play Set	Mattel	1976		30	50	150
Professor Bergman Figure	Mattel	1976		17	30	45
Puzzle	HG Toys	1976		6	10	25
Space Expedition Dart Set		1976	carded	6	10	15
Stamping Set		1976		8	13	20
Superscope		1976		6	10	15
Talking View-Master Set	View-Master	1975	three reels	6	20	40
Utility Belt Set	Remco	1976		12	40	100
View-Master Set	View-Master	1975	three reels	10	16	25
Walking Spaceman		1975		50	150	400
Zython Figure	Mattel	1976		20	35	50

Spaceships

NAME	COMPANY	YEAR	DESCRIPTION	GOOD	EX	MIB
Eagle Lunar Module		1960s	9"	80	115	175
Friendship 7			9-1/2", friction	35	50	75
Inter-Planet Toy Rocketank Patrol	Macrey	1950	10"	30	45	70
Jupiter Space Station	TN/Japan	1960s	8"	90	125	195
Moon-Rider Spaceship	Marx	1930s	tin wind-up	125	200	250
Mystery Spaceship	Marx	1960s	35mm astronauts and moonmen, rockets, launchers	50	75	100
Rocket Fighter	Marx	1950s	with tail fin and sparking action, tin wind-up	250	375	500
Rocket Fighter Spaceship	Marx	1930s	celluloid window, tin wind-up, 12" long	125	200	250
Satellite X-107	Cragstan	1965	9"	90	130	200
Sky Patrol Jet	TN/Japan	1960s	5" x 13" x 5", battery operated, working taillights	295	425	650
Solar-X Space Rocket	TN/Japan		15"	45	65	100
Space Bus			tin helicopter, battery operated with wired remote	350	500	750
Space Pacer		1978	7", battery operated	20	29	45
Space Survey X-09			battery operated, tin and plastic flying saucer with clear bubble	175	350	525
Space Train		1950s	9" long, engine & three metallic cars	18	26	40
Spaceship	Marx		bronze plastic	40	60	90
Super Space Capsule		1960s	9-1/2"	70	100	150
X-3 Rocket Gyro		1950s		25	35	50

Star Trek

NAME	COMPANY	YEAR	DESCRIPTION	GOOD	EX	MIB
Action Toy Book	Random House	1976		7	10	15
Beanbag Chair, ST:TMP				25	35	55
Bowl, ST:TMP	Deka	1979	plastic	3	4	10
Bridge Punch-Out Book, ST:TMP	Wanderer	1979		7	10	15
Bulletin Board, ST:TMP	Milton Bradley	1979	with four pens	6	8	12

Star Trek

NAME	COMPANY	YEAR	DESCRIPTION	GOOD	EX	MIB
Clock		1986	white wall clock, red 20th anniversary logo on face, Official Star Trek Fan Club	14	20	30
Clock		1989	Enterprise orbiting planet, rectangular	23	33	50
Colorforms Set	Colorforms	1975		15	20	35
Comb & Brush Set		1977	6" x 3", blue, oval brush	14	20	30
Command Bridge Model Kit	AMT	1975	#S950-601	40	50	75
Communicators	Mego	1976	blue plastic walkie talkies	70	100	155
Communicators	McNerney	1989	black plastic walkie talkies	35	50	75
Communicators, ST:TMP	Mego	1980	plastic wristband walkie talkies belt pack, battery operated	90	130	200
Controlled Space Flight	Remco	1976	plastic Enterprise, battery operated	80	115	200
Digital Travel Alarm	Lincoln Enterprises			15	20	30
Dinnerware Set, ST:TMP	Deka	1979	plate, bowl, glass and cup	15	25	35
Enterprise Make-A-Model, ST:TNG	Chatham River Press	1990		4	5	8
Enterprise Model Kit	Mego/Grand Toys	1980	#91232/B, Canadian issue, ST:TMP	90	100	130
Enterprise Model Kit	AMT	1966	#S921-200, 15" x 10" box, lights	200	350	500
Enterprise Punch-Out Book, ST:TMP	Wanderer	1979		9	13	20
Enterprise Wristwatch, ST:TMP	Bradley			20	30	45
Enterprise Wristwatch, ST:TMP	Rarities Mint	1989	gold-plated silver	55	80	125
Enterprise, ST:III	Ertl	1984	4" long, die cast with black plastic stand	10	12	25
Enterprise, ST:IV	Sterling	1986	24", silver plastic, inflatable	20	30	45
Enterprise, ST:TMP	South Bend	1979	20" long, white plastic, battery powered lights & sound with stand	80	115	175
Excelsior, ST:III	Ertl	1984	4" long, die cast with black plastic stand	7	10	25
Ferengi Costume, ST:TNG	Ben Cooper	1988		7	10	20
Figurine Paint Set	Milton Bradley	1979		14	20	30
Flashlight		1976	battery operated, small phaser shape	6	8	12
Flashlight, ST:TMP	Larami	1979		6	8	12
Galileo Shuttle Model Kit	AMT	1974	#S959-602	80	95	175
Giant in the Universe Pop-Up Book	Random House	1977		14	20	30
Golden Trivia Game	Western	1985		20	30	45
Helmet	Remco	1976	plastic, with sound and red lights	55	80	130
Kirk & Spock Wristwatch, ST:TMP	Bradley		LCD rectangular face display, Enterprise on blue face with Kirk & Spock	25	35	55
Kirk Bank	Play Pal	1975	12" plastic	25	35	55
Kirk Costume	Ben Cooper	1975	plastic mask, one-piece jumpsuit	9	13	20
Kirk Doll, ST:TMP	Knickerbocker	1979	13" tall, soft body with plastic head	16	23	35
Kirk or Spock Costumes	Ben Cooper	1967	tie-on jumpsuit, mask	11	16	25
Kirk Puzzle, ST:TMP	Larami	1979	15-piece sliding puzzle	5	7	10
Kite	Hi-Flyer	1975	TV Enterprise or Spock	14	20	30
Kite, ST:III	Lever Bros.	1984	pictures Enterprise	14	20	30

Doctor Who Cyberman, 1986, Dapol

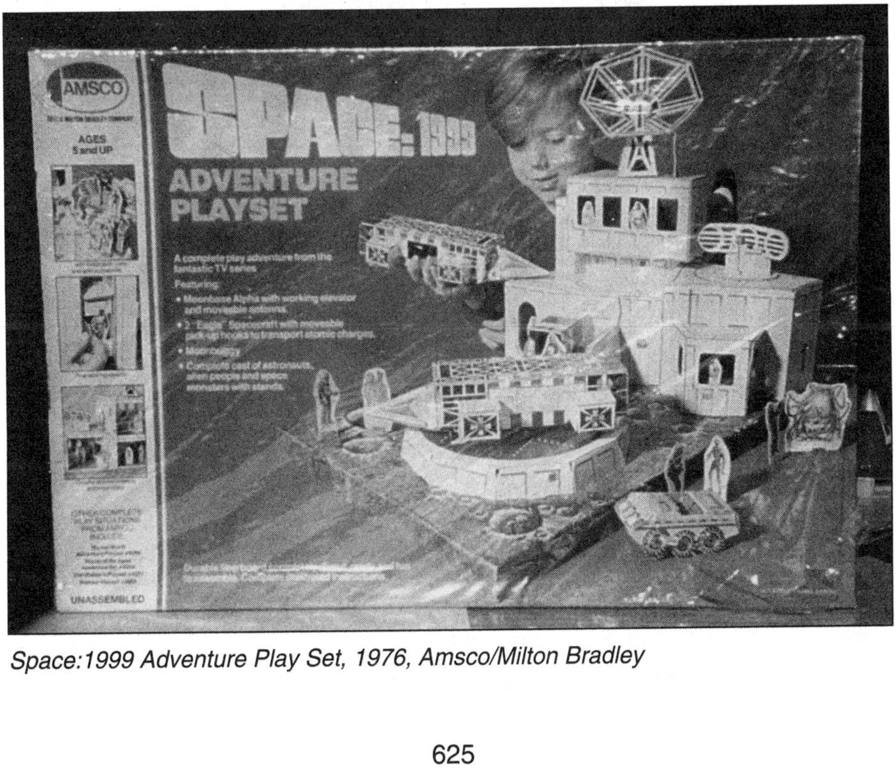

Space:1999 Adventure Play Set, 1976, Amsco/Milton Bradley

Star Trek

NAME	COMPANY	YEAR	DESCRIPTION	GOOD	EX	MIB
Kite, ST:TMP	Aviva	1976	picture of Spock	11	16	25
Klingon Bird of Prey, ST:III	Ertl	1984	3-1/2", die cast with black plastic stand	7	10	25
Klingon Costume	Ben Cooper	1975	plastic mask, one piece jumpsuit	9	13	25
Klingon Costume, ST:TNG	Ben Cooper	1988		7	10	15
Klingon Cruiser Model Kit	Matchbox/ AMT		#PK-5111, ST:TMP	45	60	80
Klingon Cruiser Model Kit	AMT	1968	#S952-802	135	155	225
Light Switch Cover	American Tack &	1985	ST:TMP	6	8	12
Hardware Magic Slates	Whitman	1979	four designs: Spock, Kirk, Kirk & Spock	7	10	15
Make-a-Game Book	Wanderer	1979		7	10	15
Metal Detector	Jetco	1976	U.S.S. Enterprise decal	100	145	225
Mirror		1966	2" x 3" metal, with black & white photo of crew	2	4	6
Mix `N Mold		1975	Kirk, Spock or McCoy, molding compound, paint & brush	35	50	75
Movie Viewer	Chemtoy	1967	3" red & black plastic	10	16	30
Needlepoint Kit	Arista	1980	Kirk	16	23	35
Needlepoint Kit	Arista	1980	14" x 18", "Live Long & Prosper"	16	23	35
Paint-By-Numbers Set	Hasbro	1972	small	20	35	50
Paint-By-Numbers Set	Hasbro	1972	large	35	50	75
Pen & Poster Kit	Open Door	1976	four versions; each	11	16	25
Pen & Poster Kit, ST:III	Placo	1984	3-D poster "Search for Spock" with overlay, 3-D glasses and four felt tip pens	9	13	20
Pennant	Image Products	1982	12" x 30" triangle, The Wrath of Khan	6	8	12
Pennant	Universal Studios	1988	Paramount Pictures Adventure	5	7	10
Pennant	Image Products	1982	12" x 30" triangular, black, yellow and red on white with "Spock Lives"	6	8	12
Phaser	Remco	1975	black plastic, shaped like pistol, electronic sound, flashlight projects target	35	50	80
Phaser Battle Game	Mego	1976	black plastic, 13" high battery operated electronic target game, LED scoring lights, sound effects & adjustable controls	195	275	450
Phaser Gun	Remco	1967	Astro Buzz-Ray Gun with three-color flash beam	80	115	200
Phaser Gun, ST:III	Daisy	1984	white & blue plastic gun with light & sound effects	35	50	80
Phaser Gun, ST:TNG	Galoob	1988	gray plastic light & sound hand phaser	14	20	35
Pinball Game, ST:TMP	Azrak-Hamway		12", plastic, Kirk or Spock	23	35	75
Pinball Game, ST:TMP	Bally	1979	electronic	200	295	600
Pocket Flix	Ideal	1978	battery operated movie viewer & film cartridge	18	25	40
Pop-Up Book, ST:TMP	Wanderer	1980		11	16	30
Puzzle	H.G. Toys	1976	150 pieces; Kirk, Spock, and McCoy	4	6	10
Puzzle	H.G. Toys	1974	150 pieces, Battle on the Planet Romulon	5	7	10
Puzzle	Whitman	1978	8-1/2" x 11" tray, Spock in spacesuit	2	3	8
Puzzle	Aviva	1979	551 pieces	9	13	20

Star Trek

NAME	COMPANY	YEAR	DESCRIPTION	GOOD	EX	MIB
Puzzle	H.G. Toys	1974	150 pieces, Attempted Hijacking of U.S.S. Enterprise	6	8	12
Puzzle	H.G. Toys	1974	150 pieces, Battle on the Planet Klingon	5	7	10
Puzzle	H.G. Toys	1974	150 pieces, Kirk and officers beaming down	5	7	10
Puzzle	H.G. Toys	1976	150 pieces, "Force Field Capture"	4	6	10
Puzzle	Milton Bradley	1979	ST:TMP, 50 pieces	5	7	10
Puzzle	Larami	1979	ST:TMP, 15-piece sliding puzzle	5	7	10
Puzzle	Mind's Eye Press	1986	551 pieces, ST:IV, "The Voyage Home"	14	20	30
Role Playing Game, 2001 Deluxe Edition	FASA		Star Trek Basic Set & the Star Trek III Combat Game	20	30	45
Role Playing Game, 2004 Basic Set	FASA		three books outlining Star Trek Universe	7	10	15
Role Playing Game, Second Deluxe Edition	FASA			14	20	30
Romulan Bird of Prey Model Kit	AMT	1975	#S957-601	100	120	130
Shuttlecraft Galileo, ST:TNG	Galoob	1989	white plastic with movable doors & sensor unit	16	23	40
Space Design Center, ST:TMP	Avalon	1979	blue plastic tray, paints, pens, crayons, project book & crew member cut-outs	70	100	150
Spock & Enterprise Wristwatch, ST:TMP	Lewco	1986	20th anniversary, digital	9	13	30
Spock Bank	Play Pal	1975	12" plastic	25	35	55
Spock Bop Bag		1975	plastic, inflatable	55	80	125
Spock Chair, ST:TMP		1979	inflatable	16	23	35
Spock Costume	Ben Cooper	1973	plastic mask, one-piece jumpsuit	11	16	30
Spock Doll, ST:TMP	Knickerbocker	1979	13" tall, soft body, plastic head	16	23	35
Spock Ears, ST:TMP	Aviva	1979		7	10	15
Spock Model Kit	AMT	1973	#S956-601, Spock with snake	120	130	175
Spock Tray	Aviva	1979	17-1/2" metal lap tray	9	13	20
Spock Wristwatch	Bradley		ST:TMP	20	30	45
Star Trek Cartoon Puzzle	Whitman	1978		4	5	8
Star Trek Color & Activity Book	Whitman	1979		4	5	8
Star Trek Coloring Book	Saalfield	1979		7	10	15
Star Trek Costume	Collegeville	1979	one-piece outfit, Spock, Kirk, Ilia or Klingon, each	11	16	25
Star Trek II U.S.S. Enterprise Ship	Corgi	1982	3" die cast	9	13	20
Star Trek Wristwatch	Bradley	1979	Spock on dial with revolving Enterprise and Shuttle craft hands	45	65	100
Telescreen	Mego	1976	plastic, battery operated target game with light & sound effects	70	100	155
Tracer Gun	Rayline	1966	plastic pistol with colored plastic discs	45	65	125
Tricorder	Mego	1976	blue plastic tape recorder, battery operated with shoulder strap	70	100	150
Trillions of Trilligs Pop-Up Book	Random House	1977		16	25	35
U.S.S. Enterprise Action Play Set	Mego	1975	8" dolls, stools, console, captain's chair, three scenes with blue plastic fold-out with picture of U.S.S. Enterprise	125	180	300

Yoda hand puppet, Kenner

C-3PO Cookie Jar, 1977, Roman

Star Trek

NAME	COMPANY	YEAR	DESCRIPTION	GOOD	EX	MIB
U.S.S. Enterprise Bridge, ST:TMP	Mego	1980	white plastic	70	100	200
Utility Belt	Remco	1975	black plastic phaser miniature, tricorder, communicator & belt with Star Trek buckle	45	65	120
View-Master Gift Pak, ST:TMP	GAF	1979	viewer, three reels, 3-D poster, glasses	35	50	75
View-Master Set	GAF	1974	three reels, "Mr. Spock's Time Trek"	11	16	25
View-Master Set	GAF	1968	booklet, three reels, "Omega Glory"	7	10	25
View-Master Set, ST:II	View-Master	1982	three reels	7	10	20
View-Master Set, ST:TMP	GAF	1979	three reels	7	10	20
View-Master Set, ST:TMP	GAF	1981	double plastic cassette with two filmstrips	11	16	25
Vulcan Shuttle Model Kit	Ertl	1984	#6679, ST:III	20	25	40
Vulcan Shuttle Model Kit	Ertl	1984	#6679, ST:TMP	18	20	30
Vulcan Shuttle Model Kit	Mego/Grand Toy	1980	#91231, ST:TMP	100	120	130
Wastebasket	Chein	1977	black metal	35	50	80
Wastebasket, ST:TMP	Chein	1979	13" high, metal rainbow painting with photograph of Enterprise surrounded by smaller pictures	11	16	35
Water Pistol	Azrak-Hamway	1976	white plastic, shaped like U.S.S. Enterprise	20	30	45
Water Pistol, ST:TMP	Aviva	1979	gray plastic, early phaser	11	16	25
Writing Tablet		1967	8" x 10"	11	16	25
Yo-Yo	Aviva	1979	ST:TMP, blue sparkle plastic	7	10	20

Star Wars

NAME	COMPANY	YEAR	DESCRIPTION	GOOD	EX	MIB
Admiral Ackbar Figurine Paint Set	Craft Master			9	13	20
Ben Kenobi/Darth Vader Poster	Proctor & Gamble	1978		9	13	20
Boba Fett Figure	Towle/Sigma	1983	bisque	30	45	65
Burger Chef Fun Book		1978		6	8	12
C-3PO & R2-D2 Alarm Clock	Bradley	1980		15	25	75
C-3PO & R2-D2 Wristwatch	Bradley	1970s	vinyl band, photo	30	45	65
C-3PO & R2-D2 Wristwatch	Bradley	1970s	digital, round, musical	70	100	150
C-3PO & R2-D2 Wristwatch	Bradley	1970s	digital, rectangular	30	45	65
C-3PO & R2-D2 Wristwatch	Bradley	1970s	white border, photo	45	65	95
C-3PO & R2-D2 Wristwatch	Bradley	1970s	vinyl band	45	60	95
C-3PO & R2-D2 Wristwatch	Bradley	1970s	digital	55	80	125
C-3PO & R2-D2 Wristwatch	Bradley	1970s	digital, round face	45	65	95
C-3PO Bank	Roman	1977	ceramic	25	50	75
C-3PO Cookie Jar	Roman	1977	ceramic	250	350	400
C-3PO Figurine Paint Set	Craft Master			9	13	20
C-3PO Mug	Sigma		ceramic	16	23	35
C-3PO Pencil Tray	Sigma			23	35	50

Star Wars

NAME	COMPANY	YEAR	DESCRIPTION	GOOD	EX	MIB
C-3PO Tape Dispenser	Sigma			23	35	50
Chewbacca & C-3PO Coloring Book	Kenner			5	7	12
Chewbacca & Leia Coloring Book	Kenner			5	7	12
Chewbacca Bank	Sigma	1983		25	35	55
Chewbacca Birthday Candle	Wilton			6	8	12
Chewbacca Medal	W. Berrie	1980		6	8	12
Chewbacca Mug	Sigma		ceramic	18	25	40
Chewbacca Poster	Burger King	1978		5	7	10
Chewbacca Punching Bag	Kenner	1977	50"	45	65	100
Chewbacca's Activity Book	Random House			3	5	10
Chewbacca, Han, Leia & Lando Coloring Book	Kenner			5	10	15
Chewbacca/Darth Vader Bookends	Sigma			25	35	55
Clock	Bradley	1982	3-D electronic, quartz	16	23	55
Darth Vader & Stormtroopers Coloring Book	Kenner			5	7	12
Darth Vader Bank	Roman	1977	ceramic	50	75	95
Darth Vader Bank	Leonard Silver	1981	silver plated	45	65	95
Darth Vader Bank	Adam Joseph	1983		9	13	20
Darth Vader Belt Buckle	Leather Shop	1977		14	20	30
Darth Vader Birthday Candle	Wilton			6	8	12
Darth Vader Duty Roster			school supplies	5	7	10
Darth Vader Figure	Towle/Sigma	1983	bisque	30	45	65
Darth Vader Mug	Sigma		ceramic	16	23	35
Darth Vader Paint Set			glow-in-the-dark	9	13	20
Darth Vader Picture Frame	Sigma			30	45	65
Darth Vader Pillow		1983		9	13	20
Darth Vader Poster	Burger King	1978		5	7	10
Darth Vader Poster	Nestea	1980		10	15	25
Darth Vader Poster	Proctor & Gamble	1980		5	8	12
Darth Vader Punching Bag	Kenner	1977	50"	25	40	60
Darth Vader Ring	W. Berrie	1980		6	8	12
Darth Vader Speaker Phone	ATC	1983		55	80	100
Darth Vader SSP Van	Kenner	1978	black	20	25	50
Darth Vader Wristwatch	Bradley	1970s	vinyl band	30	45	65
Darth Vader Wristwatch	Bradley	1970s	digital	30	45	65
Darth Vader Wristwatch	Bradley	1970s	star & planet on face	45	65	95
Darth Vader, R2-D2 & C-3PO Cookie Jar	Sigma		hexagon	55	80	165
Death Star Poster	Proctor & Gamble	1978		10	15	25
Degobah Poster	Burger King	1980		5	10	15
Droid Clock	Bradley		wall clock	20	30	45
Droids Wristwatch	Bradley	1970s	digital	30	45	65
Duel Racing Set	Lionel	1978		60	85	150

Buck Rogers Star Fighter, 1979, Mego

Space: 1999 figures, 1976, Mattel

Star Wars

NAME	COMPANY	YEAR	DESCRIPTION	GOOD	EX	MIB
Electric Toothbrush	Kenner	1978		16	23	35
Emperor's Royal Guard Bank	Adam Joseph	1983		9	13	20
Empire Strikes Back Clock	Bradley		wall clock	20	30	45
Empire Strikes Back Coloring Book		1980		5	7	15
Empire Strikes Back Dinnerware Set				15	25	35
Empire Strikes Back Panorama Book	Random House			15	20	30
Empire Strikes Back Pop-Up Book	Random House	1980		10	15	25
Empire Strikes Back Poster Album Vol. 1				10	15	25
Empire Strikes Back Radio Program Poster				15	20	30
Empire Strikes Back Sketchbook	Ballantine	1980		14	20	30
Escape from the Monster Ship Book	Random House	1985	book	6	8	12
Ewoks Coloring Book	Kenner	1983		7	10	15
Ewoks Play-Doh Set				11	16	30
Ewoks Wristwatch	Bradley	1970s	vinyl band	30	45	65
Flying R2-D2 Rocket Kit	Estes	1978		9	13	20
Fuzzy as an Ewok Book	Random House	1985		6	8	12
Galactic Emperor Figure	Towle/Sigma	1983	bisque	20	30	40
Gammorean Guard Figure	Towle/Sigma	1983	bisque	20	30	40
Gammorean Guard Mug	Sigma		ceramic	10	20	30
Gamorrean Guard Bank	Adam Joseph	1983		30	50	100
Give-A-Show Projector	Kenner	1979	with filmstrips	25	35	65
Han Solo Figure	Towle/Sigma	1983	bisque	25	35	50
Han Solo Figurine Paint Set	Craft Master			16	23	35
Han Solo Mug	Sigma		ceramic	23	35	50
Hoth Poster	Burger King	1980		5	7	10
How the Ewoks Saved the Trees Book	Random House	1985		6	8	12
Ice Planet Hoth Play-Doh Set				16	23	35
Inflatable Lightsaber	Kenner	1977		45	60	150
Intergalactic Passport & Stickers	Ballantine	1983		7	10	15
Jabba the Hutt Bank	Sigma	1983		25	35	55
Jabba the Hutt Play-Doh Set				11	16	30
Jabba the Hutt Puzzle	Craft Master	1983		4	7	10
Jabba the Hutt Wristwatch	Bradley	1970s	digital	30	45	65
Jawa Punching Bag	Kenner	1977	50"	45	65	100
Jedi Master's Quiz Book	Random House	1985		6	8	12
Kneesaa Bank	Adam Joseph	1983		7	10	20
Lando Calrissian Figure	Towle/Sigma	1983	bisque	30	45	65
Lando Calrissian Mug	Sigma		ceramic	16	23	35

Star Wars

NAME	COMPANY	YEAR	DESCRIPTION	GOOD	EX	MIB
Lando Fighting Skiff Guard Coloring Book	Kenner	1983		7	10	15
Lando in Falcon Cockpit Coloring Book	Kenner	1983		7	10	15
Learn-to-Read Activity Book	Random House	1985		6	8	12
Leia & Han Solo Glow-in-the-Dark Paint Set				9	13	20
Leia Figurine Paint Set	Craft Master			11	16	25
Leia Mug	Sigma		ceramic	20	30	45
Luke & Tauntaun Figurine Paint Set	Craft Master			14	20	30
Luke Skywalker AM Headset Radio				95	150	500
Luke Skywalker Coloring Book	Kenner	1983		7	10	15
Luke Skywalker Figure	Towle/Sigma	1983	bisque	30	45	65
Luke Skywalker Mug	Sigma		ceramic	20	30	45
Luke Skywalker Paint Set			glow-in-the-dark	9	13	20
Luke Skywalker Poster	Proctor & Gamble	1980		5	10	15
Luke Skywalker Poster	Nestea	1980		10	15	20
Luke Skywalker Poster	Burger King	1978		5	7	10
Max Rebo Coloring Book	Kenner	1983		7	10	15
Movie Viewer	Kenner	1978		15	20	40
My Jedi Journal Book	Ballantine			7	10	15
Original Fan Club Kit		1977		10	15	30
Original Press Kit		1977		25	35	55
Portable Clock/Radio	Bradley	1984		11	16	30
Puzzle	Kenner		500 pieces, Luke Skywalker	11	16	25
Puzzle	Kenner		140 pieces, Stormtroopers stop the Landspeeder	5	7	10
Puzzle	Kenner		140 pieces, Han Solo and Chewbacca	6	8	12
Puzzle	Kenner		500 pieces, Luke and Leia leap for their lives	6	8	12
Puzzle	Kenner		140 pieces, Attack of the Sand People	5	7	10
Puzzle	Kenner		140 pieces, Jawas capture R2-D2	4	6	10
Puzzle	Kenner		500 pieces, Space Battle	7	10	15
Puzzle	Kenner		500 pieces, Cantina Band	6	8	12
Puzzle	Kenner		140 pieces, Trapped in the Trash Compactor	5	7	10
Puzzle	Kenner		500 pieces, Victory Celebration	6	8	12
Puzzle	Kenner		500 pieces, X-Wing Fighters Prepare to Attack	6	8	12
R2-D2 & C-3PO Belt Buckles	Leather Shop	1977		14	20	30
R2-D2 & C-3PO Poster	Proctor & Gamble	1980		4	5	10
R2-D2 Bank	Roman	1977	ceramic	25	35	55
R2-D2 Bank	Adam Joseph	1983		7	10	20
R2-D2 Belt Buckle	Leather Shop	1977		14	20	30
R2-D2 Birthday Candle	Wilton			6	8	12

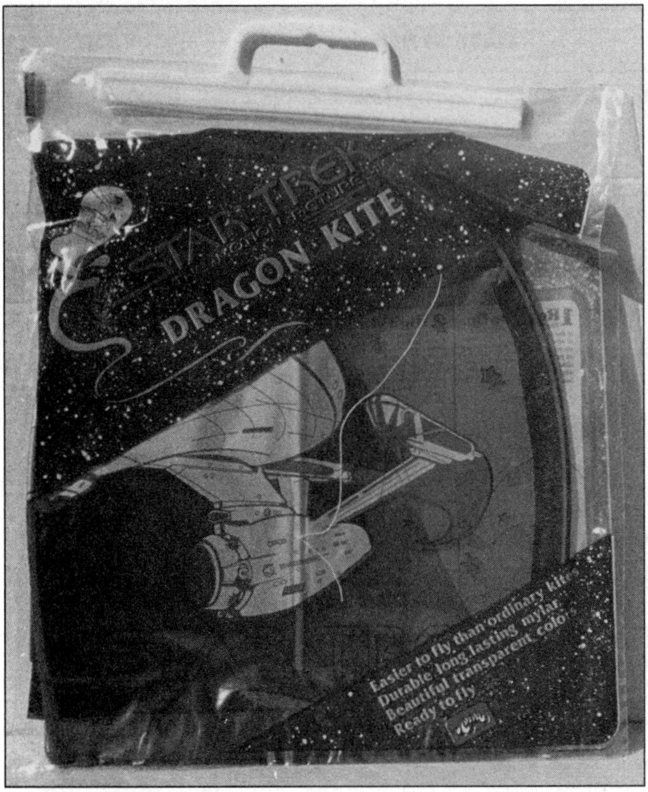

Star Trek: The Motion Picture kite

Lost in Space trading cards, 1966, Topps

Star Wars

NAME	COMPANY	YEAR	DESCRIPTION	GOOD	EX	MIB
R2-D2 Coloring Book	Kenner			5	7	12
R2-D2 Cookie Jar	Roman	1977	ceramic	90	130	200
R2-D2 Picture Frame	Sigma			30	45	65
R2-D2 Poster	Burger King	1978		5	7	10
R2-D2 Punching Bag	Kenner	1977	50"	25	40	60
R2-D2 String Dispenser with Scissors	Sigma			20	30	45
Return of the Jedi Activity Book	Happy House	1983		5	7	12
Return of the Jedi Belt	Leather Shop	1977		5	7	10
Return of the Jedi Candy Containers	Topps	1983	figural, set of 18	45	65	100
Return of the Jedi Coloring Book		1984		5	7	15
Return of the Jedi Maze Book	Happy House	1983		5	7	10
Return of the Jedi Monster Activity Book	Happy House	1983		5	7	10
Return of the Jedi Picture Puzzle Book	Happy House	1983		5	7	10
Return of the Jedi Pop-Up Book	Random House	1983		15	20	25
Return of the Jedi Punch-Out Book	Random House			14	20	30
Return of the Jedi Sketchbook	Ballantine	1983		11	16	25
Return of the Jedi Word Puzzle Book	Happy House	1983		5	7	10
Snow Speeder Toothbrush Holder	Sigma			25	35	55
Star Destroyer Poster	General Mills	1978		9	13	20
Star Wars Action Play-Doh Set				20	30	45
Star Wars Dinnerware Set				20	30	45
Star Wars Pop-Up Book	Random House	1978		11	16	25
Star Wars Poster Art Coloring Set	Craft Master	1978		9	13	20
Star Wars Questions & Answers About Space Book	Random House	1979		5	7	10
Star Wars Radio Program Poster	Golden			23	35	50
Star Wars Sketchbook	Ballantine	1977		14	20	30
Sticker Book	Panini	1977	256 stickers	23	35	50
Sticker Set & Album	Burger King			7	10	15
Stickers			cereal premium (several brands), set of four	3	4	6
TIE Fighter & X-Wing Poster	General Mills	1978		10	15	20
TIE Fighter Rocket Kit	Estes	1978		11	16	25
Wicket Bank	Adam Joseph	1983		7	10	20
Wicket Figurine Paint Set	Craft Master			9	13	20
Wicket the Ewok Wristwatch	Bradley	1970s		30	45	65
Wicket Toothbrush		1984	battery operated	9	13	20
X-Wing Fighter Rocket Kit	Estes	1978		11	16	25
X-Wing Medal	W. Berrie	1980		6	8	12
X-Wing with Maxi-Brutel Rocket Kit	Estes	1978		18	25	40

Star Wars

NAME	COMPANY	YEAR	DESCRIPTION	GOOD	EX	MIB
Yoda Backpack	Sigma			14	20	30
Yoda Bank			lithographed tin with combination dials	11	16	25
Yoda Bank	Sigma	1983		25	35	55
Yoda Coloring Book	Kenner			5	10	15
Yoda Figurine Paint Set	Craft Master			9	13	20
Yoda Hand Puppet				14	20	40
Yoda Jedi Master Fortune Teller Ball				35	50	70
Yoda Paint Set			glow-in-the-dark	9	13	20
Yoda Sleeping Bag				16	23	35
Yoda Tumbler/Pencil Cup	Sigma			20	30	45
Yoda Wristwatch	Bradley	1970s		30	45	65

Tom Corbett

NAME	COMPANY	YEAR	DESCRIPTION	GOOD	EX	MIB
Binoculars				60	100	150
Flash X-1 Space Gun			5" long	55	95	145
Model Craft Molding Super Set	Kay Standley	1950s	various characters	100	165	250
Official Space Pistol	Marx	1950s		50	80	125
Official Sparking Space Gun	Marx		21" long, with numerous apparatus on body	80	130	200
Polaris Wind-Up Spaceship	Marx	1952	12" long, 1952	195	325	500
Push-Outs Book	Saalfield	1952		30	50	75
Puzzles	Saalfield	1950s	frame tray, three versions, each	17	30	45
Rocket Scout Ring		1950s		10	16	25
Signal Siren Flashlight	Usalite	1950s		70	115	175
Space Academy Play Set	Marx	1952	#7010, 45mm figures	185	310	475
Space Academy Play Set	Marx	1950s	#7020	185	310	475
Space Cadet 2 Spaceship	Marx	1930s	tin wind-up, 12" long	225	375	575
Space Cadet Atomic Rifle		1950s		100	165	250
Space Cadet Belt		1950s		65	105	165
Space Gun		1950s	9-1/2" long light blue and black sparking	70	120	185
Space Suit Ring	1950s		10	16	25	
Tom Corbett Coloring Book	Saalfield	1950s	two versions, each	25	40	70
Tom Corbett Portrait Ring		1950s		20	35	50
Tom Corbett Wristwatch	Ingraham	1950s	round dial, embossed band with ship and planets, on illustrated rocket shaped card	250	425	650
View-Master Set	Sawyer's	1950s	three reels, Tom Corbett Secret from Space	30	50	75

Classic Tin Toys

Today's toys may be durable, but most are made of plastic, which just doesn't hold the charm and nostalgia of yesterday's tin creations. Those tin toys, made in large numbers mainly before World War II, are among the priciest collectible toys today.

Metal toys produced before World War I can be considered true works of art, especially since tin toys were often painstakingly hand painted.

But the advent of chromolithography changed the way most toys were produced. Chromolithography was actually developed late in the 19th century. The technique allowed multicolor illustrations to be printed on flat tin plates which were molded into toys.

Starting in the 1920s, lithographed tin toys began to dramatically change toy production. American manufacturers could produce these colorful toys more inexpensively than the classic European toys that had dominated the toy market until this time.

With mass production came mass appeal. New tin mechanical toys were based on the characters and celebrities that were popular at the time. Newspaper comic strips and Walt Disney movies provided already popular subject matter for toy marketers.

Among the most well-known makers of mechanical tin toys were Marx, Chein, Lehmann and Strauss. Others included Courtland, Girard, Ohio Art, Schuco, Unique Art and Wolverine.

Many of these manufacturers had business relationships with each other. Over the years, some would be found working together, producing toys for others, distributing others' toys or being absorbed by other companies. There even appeared to be some pilfering and reproducing others' ideas.

One of the advantages of lithography was that it allowed old toys to be recycled in many ways. When a character's public appeal began to wane, a new image could be printed on the same body to produce a new toy. Or when a toy company was absorbed by another, older models could be dusted off and dressed up with new lithography. Many of the mechanical tin wind-up toys show up in surprisingly similar versions with another manufacturer's name on them.

Of the companies listed here, Marx was no doubt the most prolific. The company's founder, Louis Marx, at one time was employed by another leading toy maker, Ferdinand Strauss. He left Strauss in 1918 to start his own company. Some of his first successes were new versions of old Strauss toys like the Climbing Monkey and Alabama Coon Jigger.

Many of the popular Marx tin wind-ups were based on popular characters. One of the most sought-after is the Merrymakers Band, a group of mouse musicians. Some of the other highly valued character toys are the Amos 'N Andy Walkers, the Donald Duck Duet, Popeye the Champ, Li'l Abner and his Dogpatch Band and the Superman Rollover Airplane.

While Marx went on to produce many different kinds of toys, other companies, such as Chein, specialized in inexpensive lithographed tin. And like Marx, Chein also capitalized on popular cartoon characters, producing several Popeye toys, among others. J. Chein and Company, which was founded in 1903, was best known for its carnival-themed mechanical toys. Its Ferris wheel is fairly well known among toy collectors and was made in several lithographed versions, including one with a Disneyland theme. Chein also produced a number of affordable tin banks.

Girard was founded shortly after Chein, but didn't start producing toys until 1918. It subcontracted toys for Marx and Strauss in the 1920s. In fact, several Girard and Marx toys are identical, having been produced in the same plant with different names on them. Marx later took over the company in the 1930s.

New Jersey-based Unique Art isn't known for an extensive line of toys, but it produced some that are favorites among tin toy collectors. It, too, reportedly was acquired by Marx at some point.

There are many other companies that produced lithographed tin toys not included in this section, particularly German and Japanese companies. Lehmann and Schuco, both German firms, are the only non-American toy makers listed in this guide. More lithographed tin toys can be found in the vehicles section of this book.

Prices listed are for toys in Good, Excellent and Mint conditions. Toys will usually command a premium over the listed price if they are in their original boxes.

Dancing Sailor, Lehmann

Trends

Older tin litho toy values are holding their own. However, due to the scarcity that has been created by collectors building and maintaining their collections, the values will continue to rise during 1998.

Remember, the better the condition, the better the value, especially with tin. Any character toys in Excellent condition or better are certain winners. Those with the original box are true treasures.

Tin robots have virtually become an endangered species. Whether wind-up or battery-operated, they are now quite difficult to find. Those that are on the market generally command premium prices.

The Top 20 Tin Toys (in Mint condition)

1. Popeye the Heavy Hitter, Chein $6,500
2. Popeye Acrobat, Marx 5,500
3. Popeye the Champ, Marx 4,600
4. Mikado Family, Lehmann 3,900
5. Red the Iceman, Marx 3,500
6. Popeye with Punching Bag, Chein 2,500
7. Lehmann's Autobus, Lehmann 2,500
8. Mortimer Snerd Hometown Band, Marx 2,500
9. Popeye Express, Marx 2,400
10. Hott and Trott, Unique Art 2,300
11. Superman Holding Airplane, Marx 2,250
12. Mortimer Snerd Bass Drummer, Marx 2,200
13. Tut-Tut Car, Lehmann 2,200
14. Ajax Acrobat, Lehmann 2,200
15. Hey Hey the Chicken Snatcher, Marx 2,200
16. Zig-Zag, Lehmann 2,000
17. Popeye Handcar, Marx 2,000
18. Howdy Doody & Buffalo Bob at Piano, Unique Art 2,000
19. KADI, Lehmann 2,000
20. Merrymakers Band, Marx 1,900

CLASSIC TIN

Chein

NAME	DESCRIPTION	GOOD	EX	MINT
Banks				
2nd National Duck Bank	1954, 3-1/2" high, Disney characters	90	145	225
Cash Box	1930, 2" high, round trap	30	50	75
Child's Safe Bank	1900's, 5-1/2" high	40	65	100
Child's Safe Bank	1910, 4" high, sailboat on front of door	35	60	90
Child's Safe Bank	1910, 3" high, dog on front of door	35	60	95
Church	1930s, 4" high	35	60	90
Church	1954, 3-1/2" high	70	115	175
Clown	1931, 5" high	70	115	175
Clown	1949, 5" high, says bank on front	35	60	95
Drum	1930s, 2-1/2" high	35	60	95
Elephant	1950s, 5" high	55	90	135
God Bless America	1930s, 2-1/2" high, drum shaped	30	50	75
Happy Days Cash Register	1930s, 4" high	45	80	120
Humpty Dumpty	1934, 5-1/4" high	60	100	150
Log Cabin	1930s, 3" high	80	130	200
Mascot Safe	1914, 5" high	35	60	95
Mascot Safe	1914, 4" high	35	55	85
Monkey	1950s, 5-1/4" high	55	90	135
New Deal	1930s, 3-1/4" high	50	80	125
Prosperity Bank	1930s, 2-1/4" high, pail shaped, with band	35	60	95
Prosperity Bank	1930s, 2-1/4" high, pail shaped, without band	30	50	75
Roly Poly	1940s, 6" high	135	230	350
Scout	1931, 3-1/4" high, cylinder	100	165	250
Three Little Pigs	1930s, 3" high	60	100	150
Treasure Chest	1930s, 2" high	35	60	90
Uncle Sam	1934, 4" high, hat shaped	35	60	95
Uncle Wiggly	1950s, 5" high	40	65	100
Miscellaneous Toys				
Army Drummer	1930s, 7" high, plunger-activated	70	115	175
Disneyland Tea Set	15-piece set featuring Disney characters	100	175	500
Dolly's Washer	washing machine	65	100	300
Easter Basket	nursery rhyme figures	35	55	100
Easter Egg	1938, 5-1/2", tin, chicken on top, opens to hold candy	35	55	85
Helicopter, Toy Town Airways	1950s, 13" long, friction drive	55	90	135
Indian in Headdress	1930s, 5-1/2" high	70	115	175
Marine	hand on belt	60	100	150
Melody Organ Player		60	100	150
Musical Top Clown	1950s, 7" high, clown head handle	75	125	195
Player Piano	eight rolls	195	325	500
Sand Toy	7" high, monkey bends and twists	30	50	75
Sand Toy Set	duck mold, sifter, frog on card	30	45	70
Scuba Diver	10" long	70	120	185
See-Saw Sand Toy	1930s, bright colors, boy and girl on see-saw move	55	90	135
See-Saw Sand Toy	1930s, pastel colors, boy and girl on see-saw move	70	120	185
Space Ride	tin litho, boxed, lever action with music	135	230	350
Sparkler Toy	5", on original card	30	50	75
Wind-up Toys				
Airplane, square-winged	early tin, 7" wingspan	90	145	225
Army Cargo Truck	1920s, 8" long	235	390	600
Army Plane	11" wingspan	135	230	350
Army Sergeant		70	120	185
Army Truck	8-1/2" long, open bed	40	65	100

Charlie McCarthy Benzine Buggy, 1930s, Marx

Hoky Poky, 1930s, Wyandotte

Chein

NAME	DESCRIPTION	GOOD	EX	MINT
	Wind-up Toys			
Army Truck	8-1/2" long, cannon on back	50	80	125
Barnacle Bill	1930s, looks like Popeye, waddles	330	550	850
Bear	1938, with hat, pants, shirt, bow tie	55	90	135
Bunny	1940s	35	60	100
Cabin Cruiser	1940s, 9" long	35	55	85
Cat	with wood wheels	40	65	100
Chick	4" high, bright colored clothes, polka dot bow tie	35	55	85
Chicken Pushing Wheelbarrow	1930s	35	60	100
China Clipper	10" long	135	225	350
Clown Boxing	8" tall, tin	235	390	600
Clown in Barrel	1930s, 8" high, waddles	175	295	450
Clown with Parasol	1920s, 8" tall	105	180	275
Dan-Dee Dump Truck		115	195	300
Disneyland Ferris Wheel	1940s	400	650	1200
Disneyland Roller Coaster		400	620	1200
Doughboy	1920s, 6" high, tin litho, WWI soldier with rifle	145	245	375
Drummer Boy	1930s, 9" high, with shako	90	145	225
Duck	1930, 4" high, waddles	35	60	100
Duck	1930, 6" high, waddles	50	80	150
Ferris Wheel	1930s, 16-1/2" high, six compartments, ringing bell	250	425	1000
Ferris Wheel, The Giant Ride	16" high	100	250	500
Greyhound Bus	6" long, wood tires	90	155	235
Handstand Clown		75	125	200
Handstand Clown	1930s, 5" tall	50	80	150
Happy Hooligan	1932, 6" high, tin litho	195	325	500
Hercules Ferris Wheel		145	245	400
Indian	4" high, red with headdress	50	80	125
Jumping Rabbit	1925	100	165	250
Junior Bus	9" long, yellow	70	115	175
Mack Hercules Motor Express	19-1/2" long, tin litho	235	385	595
Mack Hercules Truck	7-1/2" long	165	275	425
Mark 1 Cabin Cruiser	1957, 9" long	35	60	95
Mechanical Aquaplane, No. 39	1932, 8-1/2" long, boat-like pontoons	100	165	250
Mechanical Fish	1940s, 11" long	30	50	75
Merry-Go-Round	11" with swan chairs	475	775	1200
Motorboat	1950s, 9" long	40	65	100
Motorboat	1950s, 7" long, crank action	35	60	90
Musical Aero Swing	1940s, 10" high	295	490	750
Musical Merry-Go-Round	small version	155	260	400
Musical Toy Church	1937, crank music box	90	145	225
Peggy Jane Speedboat		35	55	85
Pelican		100	165	250
Penguin in Tuxedo	1940s	50	80	125
Pig		35	60	95
Playland Merry-Go-Round	1930s, 9-1/2" high	475	775	1200
Playland Whip, No. 340	four bump cars, driver's head wobbles	490	810	1250
Popeye in Barrel		380	625	1000
Popeye the Heavy Hitter	bell and mallet	2550	4225	6500
Popeye with Punching Bag		975	1625	2500
Ride-A-Rocket Carnival Ride	1950s, 19" high, four rockets	510	845	1300
Roadster	1925, 8-1/2" long	50	80	125
Roller Coaster	1938, includes two cars	255	425	650
Roller Coaster	1950s, includes two cars	225	375	575

Daredevil Motor Cop, Unique Art

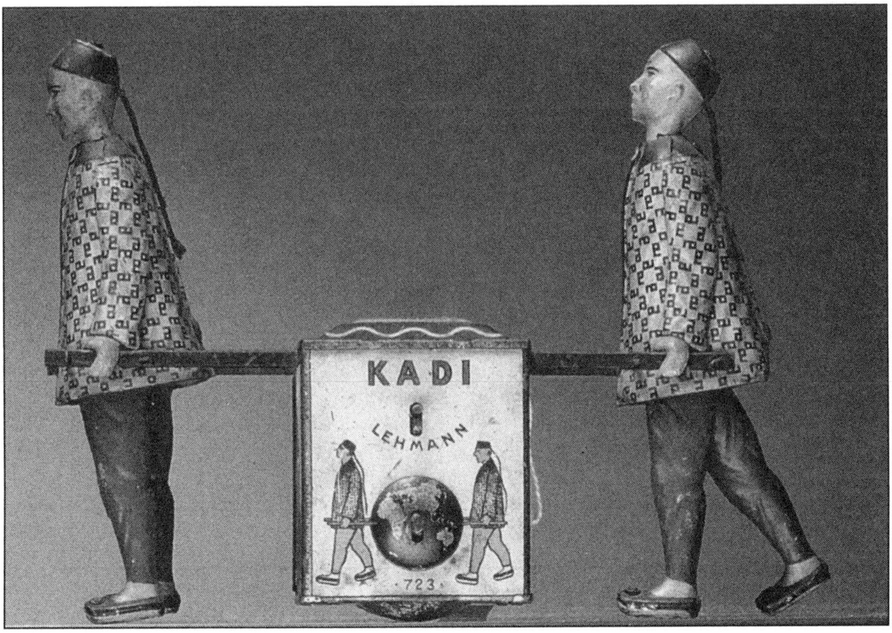

KADI, Lehmann

CLASSIC TIN

Chein

NAME	DESCRIPTION	GOOD	EX	MINT

Wind-up Toys

NAME	DESCRIPTION	GOOD	EX	MINT
Royal Blue Line Coast to Coast Service		410	685	1050
Sandmill	beach scene on side	90	145	225
Santa's Elf	1925, 6" high, boxed	235	390	600
Sea Plane	1930s, silver, red, and blue	115	195	300
Seal	balancing barbells	90	145	225
Ski-Boy	1930s, 6" long	165	275	425
Ski-Boy	1930s, 8" long	165	275	425
Speedboat	14" long	90	145	225
Touring Car	7" long	60	100	150
Turtle with Native on Its Back	1940s	165	275	425
Woody Car	1940s, 5" long, red	100	165	250
Yellow Cab	7" long	100	165	250
Yellow Taxi	6" long, orange and black	135	225	350

Lehmann

NAME	DESCRIPTION	GOOD	EX	MINT
Aha Truck		470	780	1200
Ajax Acrobat	does somersaults, 10" tall	850	1425	2200
Alabama Jigger	wind-up tap dancer on square base, 1920s	490	825	1500
Auton Boy & Cart		195	325	495
Captain of Kopenick	early 1900s	625	1050	1600
Crocodile	walks, mouth opens	285	475	725
Dancing Sailor		585	975	1500
Delivery Van	"Huntley & Palmers Biscuits"	650	1075	1650
Express Man & Cart		335	550	850
Flying Bird		295	475	750
Galop Race Car	1920s	235	390	600
Gustav The Climbing Miller		390	650	1000
Ito Sedan and Driver		525	875	1350
KADI	Chinese men carrying box	600	975	2000
Lehmann's Autobus		975	1625	2500
Li-La Car	driver in rear, women passengers	725	1200	1850
Mikado Family		1525	2550	3900
Minstrel Man	early 1900s	335	550	1000
New Century Cycle	driver and black man with umbrella	700	1175	1800
Ostrich Cart		380	650	975
Paddy Riding Pig		700	1175	1800
Quack Quack	duck pulling babies	295	495	750
Rooster and Rabbit	rooster pulls rabbit on cart	380	625	975
Sea Lion		145	250	375
Sedan and Garage		335	550	850
Shenandoah Zeppelin		155	255	395
Skier	wind-up, 1920s	510	850	1300
Taxi	10" long, 1920s	450	750	1150
Tut-Tut Car	driver has horn	850	1430	2200
Wild West Bucking Bronco		525	875	1350
Zebra Cart "Dare Devil"	1920s	255	425	650
Zig-Zag	handcar-type vehicle on oversized wheels	780	1300	2000

Marx

Buildings and Rooms

NAME	DESCRIPTION	GOOD	EX	MINT
Airport	1930s	115	195	300
Automatic Car Wash		135	225	350
Automatic Firehouse with Fire Chief Car	friction car, firehouse with plastic doors, 1940s	115	195	300
Automatic Garage	family car	145	245	375

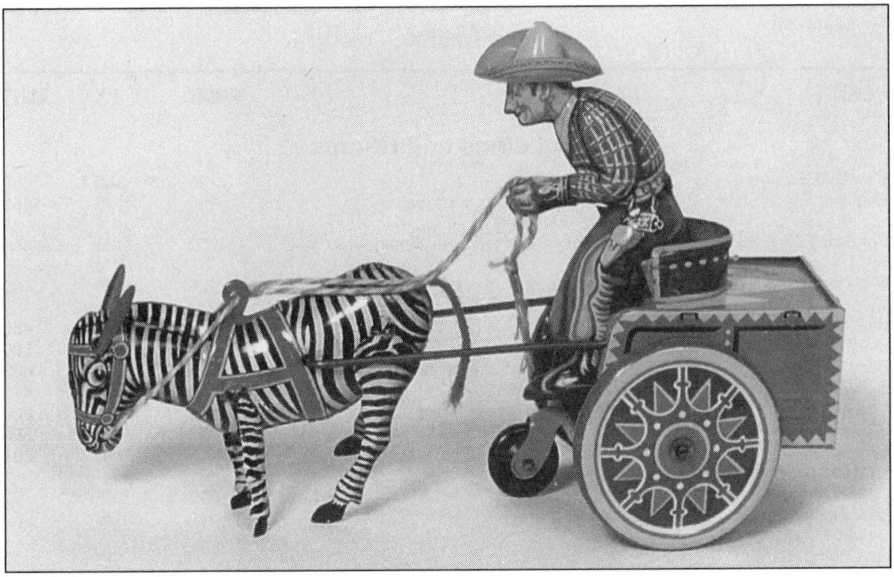

Zebra Cart, 1920s, Lehmann

Little Orphan Annie Skipping Rope, Marx

Marx

NAME	DESCRIPTION	GOOD	EX	MINT
	Buildings and Rooms			
Blue Bird Garage	1937	135	225	350
Brightlite Filling Station	bottle-shaped gas pumps, battery-operated, 1930s	255	425	650
Brightlite Filling Station	pump with round top says "Fresh Air," 1930s	250	425	650
Brightlite Filling Station	rectangular shaped pumps, battery-operated, late 1930s	255	425	650
Bus Terminal	1937	175	295	450
Busy Airport Garage	1936	195	325	500
Busy Parking Lot	five heavy gauge streamline autos, 1937	235	390	600
Busy Street	six vehicles, 1935	145	310	475
City Airport	with two metal planes, 1938	115	195	300
Crossing Gate House		115	195	300
Crossover Speedway	litho buildings on bridge, two cars, litho drivers, 1938	115	190	295
Crossover Speedway	1941	100	165	250
Dick Tracy Automatic Police Station	station and car	375	625	950
Gas Pump Island		115	195	300
General Alarm Fire House	wind-up alarm bell, steel chief car and patrol truck, 1938	175	295	450
Greyhound Bus Terminal	1938	115	195	300
Gull Service Station	1940s	185	310	475
Hollywood Bungalow House	1935	165	275	425
Home Town Drug Store	1930s	175	295	450
Home Town Favorite Store	1930s	175	295	450
Home Town Fire House	1930s	175	295	450
Home Town Grocery Store	1930s	165	275	425
Home Town Meat Market	1930s	180	300	465
Home Town Movie Theatre	1930s	145	245	375
Home Town Police Station	1930s	145	245	375
Home Town Savings Bank	1930s	145	245	375
Honeymoon Garage	1935	165	275	425
Lincoln Highway Set	pumps, oil-grease rack, traffic light and car, 1933	350	585	900
Loop-the-Loop Auto Racer	1-3/4", 1931	155	260	400
Magic Garage	litho garage, wind-up car, 1934	145	250	375
Magic Garage	with friction town car, 1934	145	250	375
Main Street Station	litho garage, 4" wind-up steel vehicles	165	275	425
Metal Service Station	1949-50	165	275	425
Military Airport		105	180	275
Model School House	1960s	50	80	125
Mot-O-Run 4 Lane Hi-Way	cars, trucks, buses move on 27" electric track, 1949	105	180	275
New York World's Fair Speedway	litho track, two red cars, 1939	295	495	750
Newlyweds' Bathroom	1920s	100	165	500
Newlyweds' Bedroom	1920s	100	165	500
Newlyweds' Dining Room	1920s	100	165	500
Newlyweds' Kitchen	1920s	100	165	500
Newlyweds' Library	1920s	100	165	500
Roadside Rest Service Station	Laurel and Hardy at counter, with stools in front, 1935	625	1050	1600
Roadside Rest Service Station	Laurel and Hardy at counter, no stool in front, 1938	550	950	1450
Service Station	two pumps, two friction vehicles, 1929	295	495	750

Butter and Egg Man, Marx

Red the Iceman, Marx

Marx

NAME	DESCRIPTION	GOOD	EX	MINT
	Buildings and Rooms			
Service Station Gas Pumps	wind-up, 9"	125	210	325
Sky Hawk Flyer	wind-up, two planes, tower, 7-1/2" tall	125	210	325
Stunt Auto Racer	two blue racers, 1931	135	225	350
Sunnyside Garage	1935	255	425	650
TV and Radio Station		135	225	350
Universal Motor Repair Shop	tin, 1938	255	425	650
Used Car Market	base, several vehicles and signs, 1939	255	425	650
Whee-Whiz Auto Racer	four 2" multicolored racers with litho driver, 1925	295	495	750
	Miscellaneous Toys			
Army Code Sender	pressed steel	20	35	50
Baby Grand Piano	with piano-shaped music books	50	100	250
Big Shot		50	85	130
Cat Pushing Ball	lever action, wood ball, 1938	50	80	125
Hopping Rabbit	metal and plastic, 4" tall, 1950s	40	65	100
Jumping Frog		40	65	100
Jungle Man Spear		80	130	200
King Kong	on wheels, with spring-loaded arms, 6-1/2" tall	35	60	95
Mysterious Woodpecker		50	80	125
Pathe Movie Camera	6" tall, 1930s	55	90	135
Rooster		60	100	150
Searchlight	3-1/2" tall	40	65	100
Toto the Acrobat		100	165	250
	Trains			
Commodore Vanderbilt Train	track, wind-up	75	125	190
Crazy Express Train	plastic and litho, wind-up, 12" long, 1960s	115	195	300
Disneyland Express	locomotive and three tin cars, wind-up, 21-1/2" long, 1950s	255	425	650
Disneyland Express, Casey Jr. Circus	wind-up, 12" long	100	165	250
Disneyland Train	Goofy drives locomotive with three tin cars, wind-up, 1950	135	225	350
Engine Train	ten cars, no track, HO, 1960s	65	110	170
Flintstones Choo Choo Train "Bedrock Express"	wind-up, 13" long, 1950s	235	390	600
Glendale Depot Railroad Station Train	1930s	235	390	600
Mickey Mouse Express Train Set	1952	235	390	900
Mickey Mouse Meteor Train	four cars/engine, wind-up, 1950s	255	425	650
Musical Choo-Choo	1966	35	60	90
Mystery Tunnel	wind-up	90	145	225
New York Central Engine Train	four cars	195	325	500
New York Circular with Train, with	wind-up, 1928	490	825	1250
New York Circular with Train, without airplane	wind-up, 1928	410	695	1050
Popeye Express	1936, version of Honeymoon Express, w/ airplane	925	1550	2400
Railroad Watch Tower	electric light, 9" tall	35	60	90
Scenic Express Train Set	wind-up, 1950s	60	105	160
Subway Express	with plastic tunnel, 1954	185	310	475

Flippo the Jumping Dog, 1940, Marx

Indian in Headdress, 1930s, Chein

Marx

Trains

NAME	DESCRIPTION	GOOD	EX	MINT
Train Set	plastic locomotive, tin cars, wind-up, 6" long, 1950s	60	100	150
Trolley No. 200	headlight, bell, tin wind-up, 9" long, 1920s	195	325	500

Wagons and Carts

NAME	DESCRIPTION	GOOD	EX	MINT
Bluto, Brutus, Horse and Cart	celluloid figure, metal, 1938	350	585	900
Busy Delivery	open three-wheel cart, wind-up, 9" long, 1939	295	475	750
Farm Wagon	horse pulling wagon, 10" long, 1940s	60	100	150
Horse and Cart	wind-up, 7" long, 1934	80	130	200
Horse and Cart	with driver, 9-1/2" long, 1950s	40	65	100
Horse and Cart with Clown Driver	wind-up, 7-5/8" long, 1923	135	225	350
Pinocchio Busy Delivery	on unicycle facing two-wheel cart, wind-up, 7-3/4" long, 1939	275	475	750
Popeye Horse and Cart	wind-up	335	550	850
Rooster Pulling Wagon	1930s	135	225	350
Toyland's Farm Products Milk Wagon	wind-up, 10-1/2" long, 1930s	145	245	375
Two Donkeys Pulling Cart	with driver, wind-up, 10-1/4" long, 1940s	105	180	275
Wagon with Two-Horse Team	late 1940s, wind-up	90	145	225

Wind-up Toys

NAME	DESCRIPTION	GOOD	EX	MINT
Acrobatic Marvel Monkey	balances on two chairs, 1930s	100	165	250
Amos 'n Andy Walkers	11" tall, 1930, each	650	1075	1800
B.O. Plenty Walker	8-1/2" tall, holds Sparkle Plenty, 1940s	185	310	475
Balky Mule	8-3/4" long, 1948	90	155	235
Ballerina	6" tall	90	145	225
Barney Rubble Riding Dino	8" long, 1960s	185	310	475
Bear Cyclist	5-3/4" tall, 1934	70	115	175
Bear Waddler	4" tall, 1960s	50	80	125
Beat! The Komikal Kop	1930s	125	210	325
Big Parade	1928	410	685	1050
Big Three Aerial Acrobats	1920	235	390	600
Boy on Trapeze		90	145	225
Busy Bridge	vehicles on bridge, 24" long, 1937	350	575	895
Busy Miners	miner's car, 16-1/2" long, 1930s	175	295	450
Butter and Egg Man	wind-up walker	700	1175	1800
Captain America	5" tall, 1968	55	90	140
Carter Climbing Monkey	8-1/2" tall, 1921	115	190	290
Charleston Trio	man, boy and dog dancers on roof, 9" tall, 1926	525	875	1350
Charlie McCarthy Bass Drummer	1939	450	750	1150
Charlie McCarthy Walker	1930s	335	550	850
Chipmunk		50	80	125
Chompy the Beetle	with action and sound, 6" tall, 1960s	60	100	150
Clancy	11" tall wind-up walker, 1931	195	325	500
Climbing Fireman	tin and plastic, 1950s	135	225	350
Coast Defense Revolving Airplane	circular with three cannons, 1929	390	650	1000
Cowboy on Horse	6" tall, 1925	70	115	175
Cowboy Rider	black horse version, 7", 1930s	165	275	425
Cowboy Rider	with lariat on black horse, 1941	145	245	375
Crazy Dora		165	275	425
Dapper Dan Coon Jigger	10" tall, 1922	375	625	950
Dapper Dan the Jigger Porter	9-1/2" tall, 1924	375	625	950
Dippy Dumper		135	225	350
Donald Duck and Scooter	1960s	90	145	225

Tom Tom Jungle Boy, Marx

One of numerous Honeymoon Express toys, Marx

Marx

NAME	DESCRIPTION	GOOD	EX	MINT

Wind-up Toys

NAME	DESCRIPTION	GOOD	EX	MINT
Donald Duck Duet	Donald and Goofy, 10-1/2" tall, 1946	470	780	1200
Donald Duck Walker	with three nephews	100	165	250
Donald the Drummer	10" tall, 1940s	135	225	350
Donald the Skier	plastic, metal skis, 10-1/2" tall, 1940s	165	275	425
Dopey	walker, 8" tall, 1938	310	525	1000
Doughboy Walker		215	360	550
Drummer Boy	1939	275	475	725
Dumbo	rollover action, 4" tall, 1941	175	295	450
Easter Rabbit	holds Easter basket, 5" tall	70	115	175
Ferdinand the Bull	tail spins, 4" tall, 1938	205	340	525
Ferdinand the Bull and the Matador	5-1/2" tall, 1938	295	475	750
Figaro (Pinocchio)	rollover action, 5" long, 1940	135	225	350
Fireman on Ladder	24" tall	165	275	425
Flipping Monkey		80	130	200
Flippo the Jumping Dog	3" tall, 1940	165	275	425
Flutterfly	3" long, 1929	90	145	225
George the Drummer Boy	moving eyes, 9" tall, 1930s	195	325	500
George the Drummer Boy	stationary eyes, 9" tall, 1930s	175	295	450
Gobbling Goose	lays golden eggs, 1940s	125	210	325
Golden Pecking Goose	9-1/2" long, 1924	135	225	350
Goofy	tail spins, plastic, 9" tall, 1950s	125	210	325
Goofy the Walking Gardener	holds a wheelbarrow, 9" tall, 1960	250	425	650
Hap/Hop Ramp Walker	2-1/2" tall, 1950s	40	65	100
Harold Lloyd Funny Face	walker, 11" tall, 1928	275	475	725
Hey Hey the Chicken Snatcher	1926, 8-1/2", black man with chicken	850	1450	2200
Hi-Yo Silver and the Lone Ranger	8" tall, 1938	195	325	650
Honeymoon Cottage, Honeymoon Express 7	square base	115	195	300
Honeymoon Express	1927, old-fashioned train on circular track	225	375	575
Honeymoon Express	1947, streamlined train on circular track	185	310	475
Honeymoon Express	1940s, circular train and plane	215	360	550
Hoppo the Monkey	plays cymbals, 8" tall, 1925	100	165	250
Howdy Doody	plays banjo and moves head, 5" tall, 1950	250	425	800
Howdy Doody	does jig and Clarabell sits at piano, 5-1/2" tall, 1950	525	875	1350
Jazzbo Jim	9" tall, 1920s	375	625	1200
Jetsons Figure	4" tall, 1960s	75	130	200
Jiminy Cricket Pushing Bass Fiddle	walker	85	145	220
Jiving Jigger	1950	135	225	350
Joe Penner and His Duck Goo-Goo	7-1/2" tall, 1934	375	625	1000
Jumbo The Climbing Monkey	9-3/4" tall, 1923	135	225	425
Knockout Champs Boxing Toy	1930s	175	295	450
Leopard	growls and walks, 1950	65	105	165
Little King Walkers	3" tall, 1963	50	80	125
Little Orphan Annie and Sandy	1930s	275	475	725
Little Orphan Annie Skipping Rope		165	275	425
Little Orphan Annie's Dog Sandy	tin litho	105	180	275
Mad Russian Drummer	7" tall	225	375	575
Main Street	street scene with moving cars, traffic cop, 1927	340	575	875
Mammy's Boy	wind-up walker, 11" tall, 1929	310	525	795
Merrymakers Band	with marquee, mouse band, 1931	700	1200	1900

G.I. Joe and His Jouncing Jeep, 1940s, Unique Art

Merrymakers Band, Marx

Marx

NAME	DESCRIPTION	GOOD	EX	MINT

Wind-up Toys

NAME	DESCRIPTION	GOOD	EX	MINT
Merrymakers Band	without marquee, mouse band, 1931	600	1000	1600
Mickey Mouse	7" tall	200	400	600
Minnie Mouse	7" tall	200	400	600
Minnie Mouse in Rocker	1950s	275	450	695
Minstrel Figure	11" tall	175	295	450
Monkey Cyclist	9-3/4" tall, 1923	105	180	275
Moon Creature	5-1/2" tall	135	225	350
Moon Mullins and Kayo on Handcar	6" long, 1930s	335	550	850
Mortimer Snerd Bass Drummer	1939	850	1450	2200
Mortimer Snerd Hometown Band	1935	925	1550	2500
Mortimer Snerd Walker	1939	185	310	475
Mother Goose	7-1/2" tall, 1920s	150	250	500
Mother Penguin with Baby Penguin on	3" long, 1950s	35	55	85
Musical Circus Horse	pull toy, metal drum rolls with chimes, 10-1/2" long, 1939	80	130	200
Mystery Cat	8-1/2" long, 1931	100	165	250
Mystery Pluto	8" long, 1948	100	165	250
Nodding Goose		30	50	75
Pecos Bill	twirls rope, plastic, 10" tall, 1950s	80	130	200
Pikes Peak Mountain Climber	vehicle on track	325	525	800
Pinched	square based, open circular track, 1927	275	450	695
Pinocchio	9" tall, 1938	275	450	700
Pinocchio	5" tall, 1950s	250	425	650
Pinocchio the Acrobat	16" tall, 1939	335	550	850
Pinocchio Walker	stationary eyes, 1930s	250	425	650
Pinocchio Walker	animated eyes, 8-1/2" tall, 1930s	275	450	700
Pluto Drum Major		235	390	595
Pluto Watch Me Roll-Over	8" long, 1939	165	275	425
Poor Fish	8-1/2" long, 1936	70	115	175
Popeye Acrobat		2150	3600	5500
Popeye and Olive Oyl Jiggers	10" tall, 1936	700	1175	1800
Popeye Express	with trunk and wheelbarrow	200	500	800
Popeye Handcar	1935	775	1300	2000
Popeye the Champ	tin and celluloid, 7" long, 1936	1800	3000	4600
Popeye the Pilot		100	300	500
Porky Pig Cowboy with Lariat	8" tall, 1949	280	475	725
Porky Pig with Rotating Umbrella	8" tall, 1939, with or without top hat	310	525	800
Red Cap Porter		375	625	950
Red the Iceman		1350	2275	3500
Ride `Em Cowboy		90	145	225
Ring-A-Ling Circus	7-1/2" diameter base, 1925	475	775	1200
Rodeo Joe	1933	165	275	425
Running Scottie	12-1/2" long, 1938	70	115	175
Smitty Riding a Scooter	8" tall, 1932	500	850	1300
Smokey Joe the Climbing Fireman	1930s, 7-1/2" tall	225	375	575
Smokey Sam the World Fireman	7" tall, 1950s	125	210	325
Snappy the Miracle Dog	with dog house, 3-1/2" long, 1931	80	130	200
Spic and Span	10" tall, black drummer and dancer, 1924	500	850	1300
Stop, Look and Listen	1927, circular track toy	295	490	750
Streamline Speedway	two racers on track	80	130	200
Subway Express	1950s	65	110	170
Superman Holding Airplane	6" wingspan on airplane, 1940	875	1450	2250

B.O. Plenty Walker, 1940s, Marx

Clown in Barrel, 1930s, Chein

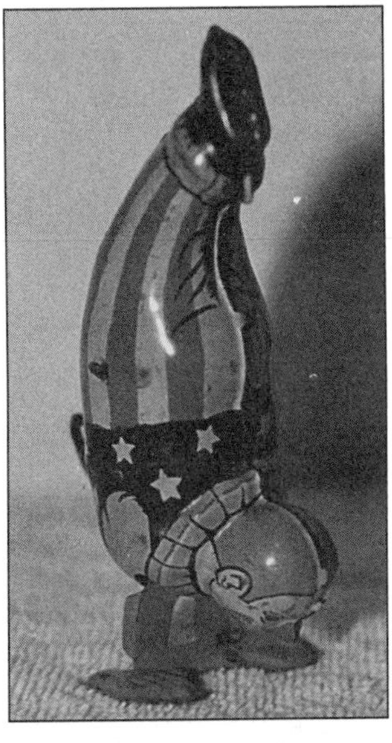

Handstand Clown, Chein

Marx

NAME	DESCRIPTION	GOOD	EX	MINT
Wind-up Toys				
Tidy Tim Streetcleaner	pushes wagon, 8" tall, 1933	175	295	450
Tom Tom Jungle Boy	7" tall	40	65	100
Tumbling Monkey	4-1/2" tall, 1942	75	125	190
Tumbling Monkey and Trapeze	5-3/4" tall, 1932	135	225	350
Walking Popeye	carrying parrots in cages, 8-1/4" tall, 1932	375	625	950
Walking Porter	carries two suitcases, 8" tall, 1930s	205	340	525
Wee Running Scottie	5-1/2" long, 1930s	65	110	165
Wee Running Scottie	5-1/2" long, 1952	35	50	100
Wise Pluto		100	165	250
WWI Soldier	prone position with rifle	50	80	125
Xylophonist	5" long	20	35	50
Zippo Monkey	9-1/2", 1938	70	115	175

Mattel

NAME	DESCRIPTION	GOOD	EX	MINT
Jack in the Boxes				
Flipper		45	75	125
Mother Goose		10	15	25
Porky Pig		40	75	150
Super Chief		50	200	300
Tom & Jerry		25	45	100
Woody Woodpecker		40	75	150
Musical Toys				
Hickory Dickory Dock Clock	crank; mouse climbs clock	10	25	40
Man on the Flying Trapeze	tin base, two metal rods holding trapeze man on top	75	100	145
Sing a Song of Sixpence	pie-shaped tin music box	20	35	65

Ohio Art

NAME	DESCRIPTION	GOOD	EX	MINT
Automatic Airport	two planes circle tower	195	325	500
Circus Shooting Gallery	1960s; with gun and darts	50	100	200
Coast Guard Plane		50	75	100
Coney Island Roller Coaster	1950s	100	165	250
Doll Stroller	teddy bear design	40	60	80
Donald Duck Carpet Sweeper	red w/Disney litho	60	100	250
Jungle Eyes Shooting Gallery	1960s; with gun and darts	40	60	125
Little Red Riding Hood Tea Set	seven-piece set	100	200	350
Mexican Boy Tea Set	nine-piece set	55	100	200
Mickey Mouse Tray		50	100	200
Mother Goose Tea Set	seven-piece set	90	150	250
Ten Little Indians Spinning Top		15	25	35
Three Little Pigs Spinning Top		15	25	35
Watering Cans	many variations, each	25	30	40

Ham and Sam, Strauss

Gobbling Goose, 1940s, Marx

Schuco

NAME	DESCRIPTION	GOOD	EX	MINT
1917 Ford		50	80	125
Airplane and Pilot	friction toy, oversized pilot, 1930s	135	225	350
Bavarian Boy	tin and cloth boy with beer mug, 5" tall, 1950s	115	195	295
Bavarian Dancing Couple	tin & cloth, 5" high	105	180	275
Black Man	tin & cloth, 5" high	250	425	650
Clown Playing Violin	tin and cloth, 4-1/2" tall, 1950s	135	225	350
Combinato Convertible	7-1/2" long, 1950s	100	165	250
Curvo Motorcycle	5" long, 1950s	135	225	350
Dancing Boy and Girl	tin and cloth, 1930s	115	195	300
Dancing Mice	large and small mouse, tin and cloth, 1950s	135	225	350
Dancing Monkey With Mouse	tin and cloth, 1950s	125	210	325
Drummer	tin and cloth, 5" tall, 1930s	125	210	325
Examico 4001 Convertible	maroon tin wind-up, 5-1/2"	145	250	375
Flic 4520	traffic cop type figure	135	225	350
Fox And Goose	tin and cloth, fox holding goose in cage, 1950s	145	250	375
Juggling Clown	tin and cloth, 4-1/2" tall	135	225	350
Mauswagen	tin & cloth mice and wagon	195	325	500
Mercer Car #1225	7-1/2" long, 1950s	100	150	225
Mickey & Minnie Dancing	tin & cloth	675	1150	1750
Monk Drinking Beer	tin & cloth, 5" high	115	195	300
Monkey Drummer	tin and cloth, 1950s	115	195	300
Monkey in Car	1930s	335	550	850
Monkey on Scooter	tin and cloth, 1930s	115	195	300
Monkey Playing Violin	tin and cloth, 1950s	125	210	325
Schuco Turn Monkey on Suitcase	tin and cloth, 1950s	115	195	300
Studio #1050 Race Car		90	145	225
Tumbling Boy	tin and cloth, 1950s	100	165	250
Yes-No Monkey		165	275	425

Unique Art

Artie the Clown in his Crazy Car		235	390	595
Bombo the Monk		115	195	295
Butter and Egg Man		475	825	1250
Capitol Hill Racer		125	210	325
Casey The Cop		350	575	895
Dandy Jim		295	490	750
Daredevil Motor Cop		205	350	525
Finnegan the Porter		135	225	350
Flying Circus		490	825	1250
G.I. Joe and His Jouncing Jeep	1940s, wind-up, 7"	125	210	325
G.I. Joe and His K-9 Pups		105	180	275
Gertie the Galloping Goose		115	190	295
Hee Haw	donkey pulling milk cart	115	190	295
Hillbilly Express		135	225	350
Hobo Train		175	295	450
Hott and Trott		900	1500	2300
Howdy Doody & Buffalo Bob at Piano		750	1300	2000
Kid-Go-Round		155	260	400
Kiddy Go-Round		155	260	400
Krazy Kar		200	335	515
Li'l Abner and His Dogpatch Band		350	650	1100
Lincoln Tunnel		235	375	595

Disneyland Ferris Wheel, 1940s, Chein

Acrobatic Marvel Monkey, 1930s, Marx

Unique Art

NAME	DESCRIPTION	GOOD	EX	MINT
Motorcycle Cop		75	125	195
Musical Sail-Way Carousel		175	295	450
Pecking Goose, Witch and Cat		175	295	450
Rodeo Joe Crazy Car		145	250	375
Rollover Motorcycle Cop		135	225	350
Sky Rangers		250	400	625

Wind-up Toys

Jazzbo Jim		375	625	1200

Wolverine

Arithmetic Quiz Toy	1950s; math quiz machine	30	50	75
Battleship	14" long, 1930s	70	120	185
Crane	red and blue, 18" high	40	65	100
Drum Major	round base, 13"	100	165	250
Drum Major		100	150	300
Express Bus		125	210	325
Jet Roller Coaster	21" long	105	180	275
Merry-Go-Round		235	390	595
Mystery Car		100	165	250
Sandy Andy Fullback	kicking fullback, 8" tall	175	295	450
Snow White Stove		20	35	50
Submarine	13" long	100	165	250
Sunny and Tank	yellow and green, 14-1/2" long	75	130	200
Sunny Suzy Deluxe Washing Machine		50	95	200
Yellow Taxi	13" long, 1940s	135	225	350
Zilotone	clown on xylophone, with musical discs, 1920s	450	750	1150

Wyandotte

Airplane Carousel	1930s	100	200	375
Carnival Set	1930s; diorama w/several rides on tin base	300	500	1000
Hoky Poky	1930s; clowns on railcar	95	200	300
Model Shooting Gallery	1930s	85	200	300
Mother Duck with Baby Ducks	two baby ducks on wheels pulled behind mother ducks	50	90	125
Trapeze Artist	1930s	95	200	325

Vehicle Toys

Modern man has always had a love affair with machines that move. Partial evidence of this is the amazing number of toy vehicles that have been produced in the 20th century. In fact, it could be reasonably argued that toy vehicles are collected more than any other type of toy.

With the dawn of the modern industrial age, the mass production of full-size automobiles and their toy counterparts seemed to go hand-in-hand. As cars rolled off assembly lines, their miniature replicas were not far behind.

Cars and trucks weren't the only toy vehicles, however. Any sort of vehicle — including boats, planes and horse-drawn wagons — was a natural for miniaturization. The types and manufacturers of toy vehicles were as varied as the real things too. Toy makers crafted them from everything from cast iron and tin to wood and plastic.

The earliest toy automobiles came along soon after their big daddy originals in the late 19th century and were produced in cast iron. But it wasn't until after World War I that toy automobile production really began to hit its stride.

The Early Days

Firms such as Arcade and Hubley are among the most well-known and sought-after manufacturers of early cast-iron vehicles.

Cars, trucks, and buses produced by Arcade Manufacturing of Freeport, Ill., are highly valued to toy vehicle collectors. Arcade actually began producing toys in the late 1800s, but it wasn't until around 1920 when the company reportedly issued its first toy vehicle, a replica of a Chicago Yellow Cab. After that came more realistic models of actual cars, trucks, and buses. The company's slogan was "They Look Real."

Hubley is another name associated with quality toy vehicles. This Pennsylvania company began manufacturing cast-iron toys in the 1890s, mostly horse-drawn wagons, trains, and guns. By the 1930s, Hubley was producing the cast-iron cars that became their most well-known products. Many were patterned after actual automobiles of

Lamborghini Marzel, 1969, Matchbox

VEHICLES

the day, while others were apparently looser interpretations of reality. Some of the Hubley vehicles also included company names, and some of the most interesting pieces had separate nickel-plated grilles.

One of the more skilled makers of smaller scale cast-iron vehicles was A.C. Williams. The Ohio company began producing toys in the late 1800s. The smaller cars and airplanes produced by A.C. Williams were intended for the five-and-dime market of the time. Williams' toys are difficult for the novice collector to identify since the toys bear no markings.

While heavy cast-iron toys ruled at the turn of the century, lithographed tin toys began stealing a large part of the market in the 1920s. One of the world's leading producers of these toys was Louis Marx. Over the years, Marx produced an extensive line of toy cars, trucks, airplanes, and farm equipment, not only in tin, but also in steel and later in plastic.

Marx capitalized on the popularity of certain celebrities and comic strip characters, incorporating them into its toy vehicles. With lithography, it was easy to put a new character into a car and create a brand new toy ready for market. Characters such as Mickey Mouse, Donald Duck, Dick Tracy, Blondie and Dagwood, Charlie McCarthy, Amos 'N Andy and Milton Berle show up in Marx cars.

Steel Takes Over

One of the most famous manufacturers of toy cars and trucks was Buddy L. These large pressed steel toys were not the kind of toys bought for display or quiet play on the living room floor. These were big trucks (around two feet long) designed for tough play.

Buddy L toys grew out of the Moline Pressed Steel Company of Moline, IL. The company was named for the son of the company's owner, reportedly for whom the first toys were produced. The Buddy L toys most sought by collectors were produced in the 1920s and 1930s and were of very heavy-duty construction. Starting in the early 1930s, the company began to use lighter-weight materials.

The Buddy L name has remained, but its post-World War II toys are not considered in the same league as its early issues, which command high collector prices today.

Buddy L is best remembered for its heavy duty trucks, but another name that was synonymous with trucks was Smith-Miller. Founded by Bob Smith and Matt Miller, the company specialized in "famous trucks in miniature." Smith-Miller was later known as Miller-Ironson Corporation, but is more commonly referred to as Smitty Toys. It produced large cast-metal and aluminum trucks.

Because of their outstanding quality, some of the Mack trucks made by Smith-Miller are very highly regarded among toy collectors. The Smith-Miller name continues today, with new limited edition trucks produced for collectors.

Wyandotte is another company associated with pressed steel vehicles. Known as both Wyandotte Toys or All Metal Products, this Michigan company produced several large steel vehicles with baked enamel finishes in the 1930s. Not all Wyandotte toys are marked, which tends to cause some confusion among collectors, but the vehicles can often be identified by their art deco-type styling and wooden wheels.

Another company that produced large steel toys was Structo. The company originally produced metal construction sets, but developed a line of vehicles in the 1920s.

While major toy companies were producing vehicles in cast iron, tin, and steel, others made toys in rubber. Probably the best-known manufacturer of rubber toys is the Auburn Rubber Company of Auburn, Ind. From the mid-1930s into the 1950s, Auburn produced rubber cars, trucks, tractors, motorcycles, airplanes, trains and boats.

Other popularly-collected vehicles are smaller die-cast models, generally three to six inches long. Probably the leading producer of this type of vehicle was Tootsietoy.

The company dates back to before the turn of the century to Samuel Dowst of Chicago. The trade name, which would eventually become the firm's mainstay, originates with Dowst's daughter, Tootsie. Although a few toy cars were produced by the firm before 1920, it was during the Roaring Twenties when the name Tootsietoy began to appear regularly. By the 1930s, the company was producing a wide line of toys, many of which are highly prized by collectors today. Tootsietoy's Federal vans from the 1920s are among the most sought-after toys, particularly those with company logos.

Being mass produced and economically priced, Tootsietoys were widely available in the five-and-dime arena. The success of these products no doubt led to several competitors.

One of those competitors was Barclay, which also manufactured die-cast vehicles, although most were generally considered of lesser quality than Tootsietoys. The first Barclay vehicles had metal tires, but in the mid-1930s, white rubber tires on wooden axles were introduced. Metal axles soon replaced the wood, and black tires replaced white after World War II.

Another competitor soon emerged from Europe — Dinky Toys were manufactured from 1933 through the 1970s in England and France. Their vehicles were high quality die cast, at least until the mid-1960s, generally in 1:43 scale.

Pictured are some of the most valuable Hot Wheels. At top is the Volkswagen Beach Bomb with surfboards in rear. It can sell for $4,000 or more in Near Mint condition. The Volkswagen at front is a prototype that was sold several years ago at a Christie's auction with a price estimate of $400-$600.

Another competitor in this classification of small die-cast vehicles is Corgi, which came on the scene in the late 1950s. Corgi was the trade name for the die-cast toys which were produced by England's Mettoy Company.

One of the best known series of toy cars today is Matchbox. These die-cast beauties are roughly three inches in length. However Lesney, the company that produced them, did produce several larger cars before it began the Matchbox line. Some of these early Lesney vehicles are valued at up to $2,000 each.

The Lesney company originated in England after World War II. The name came from the combination of parts of the two founders' first names, LESlie and RodNEY Smith. After tinkering with several products, including a few larger vehicles, the company hit paydirt with small cars that would eventually be dubbed "Matchbox." Soon the company began issuing 75 models each year, called the 1-75 Series. Lesney also issued a larger series of four- to six-inch cars called Models of Yesteryear.

Matchbox vehicles were immensely popular, so much so that in the United States, Mattel introduced a similar line called Hot Wheels. The California-based company gave its cars a California-type appeal, focusing on colorful hot rods that appealed to youngsters.

In the head-to-head battle that followed, Lesney at one time was producing 5.5 million toys a week. Eventually, Lesney lost the battle and went into receivership. Matchbox was restructured and sold twice, eventually landing with Tyco Toys. In 1996, Mattel purchased Tyco, bringing Matchbox cars to their company.

There are many other toy cars collected today than are listed in this book. For example, a wide variety of tin toy cars came out of Japan after World War II. Many Japanese toy companies emerged during that period; Bandai is listed separately. Other Japanese tin vehicles are listed under a general "Japanese" heading including numerous, and sometimes unmarked, manufacturers.

Tips on Grading

Since the universe of vehicle toys is diverse, grading condition must also take into consideration such differences. For example, when grading a much older cast-iron vehicle such as a Hubley or Arcade, one must consider the age and sturdiness of the vehicle. While any imperfections or scratches always lessen a vehicle's value, such things are more likely to deflate prices on vehicles such as Hot Wheels or Matchbox cars. On the latter two, any defect will bring the value down about 25 percent.

Editor's Note: Several vehicle listings may feature a zero in the MIP column. This means either the car did not come in a package, or it is ultimately too rare to be found in original packaging.

The Top 10 Vehicle Toys (in Mint condition)

1. Packard Straight 8, Hubley $15,000
2. White Dump Truck, Arcade 15,000
3. Checker Cab, Arcade 15,000
4. White Moving Van, Arcade 13,500
5. Elgin Street Sweeper, Hubley 11,500
6. World's Greatest Circus Truck, Keystone, 1930s 10,000
7. Ingersoll-Rand Compressor, Hubley 10,000
8. Motorized Sidecar Motorcycle, Hubley 10,000
9. Yellow Cab, Arcade 10,000
10. Ahrens-Fox Fire Engine, Hubley 8,000

The Top 10 Mattel Hot Wheels (in Mint in Package condition)

1. Volkswagen Beach Bomb, surf boards in rear window, 1969 . . . . $4,500
2. Custom Camaro, 1968, white enamel 2,000
3. Snake, 1973, white/yellow 1,500
4. Mongoose, 1973, red/blue 1,400
5. Mustang Stocker, 1975, white 1,200
6. Carabo, 1974, yellow 1,200
7. Custom Mustang, 1968 1,200
8. Mercedes C-111, 1973 1,200
9. Superfine Turbine, 1973 1,100
10. Ferrari 312P, 1973 1,100

Contributors to this section: **Majorette,** Dana Johnson, 1347 NW Albany Ave., Bend, OR 97701-3160; **Hot Wheels,** Jon Brecka, Krause Publications.

Carabo, 1970, Hot Wheels

VEHICLES

Arcade

NAME	DESCRIPTION	GOOD	EX	MINT
Airplanes				
Airplane	cast iron, 6" long	90	135	180
Boeing United Airplane	white rubber wheels, two propellors, cast iron, 5" wingspan, 1936	95	145	200
Boats and Ships				
Battleship New York	cast iron, 20" long, 1912	800	1200	1600
Showboat	cast iron, 10-3/4" long, 1934	475	715	950
Buses				
ACF Coach	dual rear wheels, front door opens and closes, cast iron, 11-1/2" long, 1925	1200	1600	2200
Bus	cast iron, 8" long	400	600	800
Century of Progress Bus	cast iron, 12" long, 1933	165	235	350
Century of Progress Bus	cast iron, 6" long, 1933	100	150	200
Century of Progress Bus	cast iron, 14-1/2" long, 1933	175	250	500
Century of Progress Bus	cast iron, 10-1/2" long, 1933	145	215	300
Century of Progress Bus	cast iron, 7-5/8" long, 1933	125	165	265
Double Deck Bus	green, white rubber wheels with red centers, "Chicago Motor Coach" on side, cast iron, 8-1/4" long, 1936	425	650	850
Double Deck Bus	rubber wheels, cast iron, 8" long, 1936	100	150	200
Double Deck Bus	cast iron, 8" long, 1938	350	525	700
Double Deck Yellow Coach Bus	nickel-plated driver, rubber balloon tires, cast iron, 13-1/2" long, 1926	800	1200	1500
Fageol Bus	cast iron, 13" long	335	500	675
Fageol Safety Bus	bright enamel colors, cast iron, 12" long, 1925	350	525	700
Fageol Safety Bus	nickel-plated wheels, with or without driver, cast iron, 7-3/4" long, 1926	200	300	400
Great Lakes Expo Bus	large, white rubber wheels with blue centers, cast iron, 11-1/4" long, 1936	300	575	1000
Great Lakes Expo Bus	small, white rubber wheels, cast iron, 7-1/4" long, 1936	225	385	695
Greyhound Cross Country Bus	"Greyhound Lines GMC" on top, white rubber wheels, cast iron, 7-3/4" long, 1936	125	215	300
Greyhound Trailer Bus	white with blue cab, "GMC Greyhound Lines" on top, blue centered rubber wheels, cast iron, 10-1/2" long	125	175	265
Sightseeing Bus	1933	125	200	300
Cars				
Andy Gump and Old 348	bright red car, green trim, green disc wheels with red hubcaps, cast iron, 7-1/4" long, 1923	800	1350	1800
Auto Racer	cast iron, 7-3/4" long, 1926	200	295	375
Boattail Racer	nickel plated wheels, cast iron, 5" long	75	115	150
Buick Coupe	green body, cast iron, 8-1/2" long, 1927	3000	5000	7500
Buick Sedan	green body, cast iron, 8-1/2" long, 1927	3000	5000	7500
Checker Cab	yellow body with black roof, cast iron, 9-1/4" long, 1932	5000	10000	15000

Arcade

NAME	DESCRIPTION	GOOD	EX	MINT
Checker Cab	deep green, cast iron, 9" long	500	750	1200
Checker Cab	plain two row checker, cast iron, 8" long	750	1000	1500
Chevrolet	white tires, 8" long, 1928	400	750	1000
Chevrolet Cab	metal tires, cast iron, 8" long, 1920s	500	850	1200
Chevrolet Coupe	Arizona gray body, spare wheel and tire on rear of car, with or without rubber balloon tires, cast iron, 8-1/4" long, 1927	1100	1800	3200
Chevrolet Coupe	cast iron, 8" long, 1929	400	750	1000
Chevrolet Sedan	Algerian blue body, spare wheel and tire on rear of car, with or without rubber balloon tires, cast iron, 8-1/4" long, 1927	1000	1700	2900
Chevrolet Superior Roadster	cast iron, 7" long, 1924	800	1250	1700
Chevrolet Superior Sedan	cast iron, 7" long, 1924			
Chevrolet Superior Touring Car	cast iron, 7" long, 1923	850	1300	2000
Chevrolet Utility Coupe	cast iron, 7" long, 1924	650	1000	1300
Coupe	solid wheels, no spokes, cast iron, 6-3/4" long, 1920s	125	185	250
Coupe	solid wheels, 6-1/2" long, 1920s	425	650	850
Coupe	two-toned, metal wheels, 5" long, 1920s	80	125	165
Coupe	red painted cast iron, 5" long, 1928	90	135	180
Coupe	with or without rubber tires, removable driver, cast iron, 9" long, 1922	375	575	750
DeSoto Sedan	white rubber wheels, cast iron, 4" long, 1936	75	150	250
Ford Coupe	nickel-plated wheels, no driver, cast iron, 5" long, 1926	150	225	325
Ford Coupe	with or without rubber tires, cast iron, 6-1/2" long, 1927	225	350	450
Ford Fordor Sedan	with or without rubber tires, removable driver, 6-1/2" long, 1920s	850	1275	1700
Ford Fordor Sedan	nickel-plated wheels, no driver, cast iron, 5" long, 1926	650	950	1200
Ford Sedan and Covered Wagon	cast iron, sedan 5-1/2" long, trailer 6-1/2" long, 1937	850	1200	1800
Ford Touring Car	with or without rubber tires, removable driver, cast iron, 6-1/2" long, 1926	450	650	900
Ford Tudor Sedan	with or without rubber tires, removable driver, cast iron, 6-1/2" long, 1920s	400	600	850
Ford Tudor Sedan	visor over front windshield, with or without rubber tires, driver, cast iron, 6-1/2" long, 1926	450	600	865
Ford With Rumble Seat In Back		40	60	80
Ford Yellow Cab	special edition for Chicago World's Fair, cast iron, 6-7/8" long, 1933	700	1000	1500
Limousine Yellow Cab	yellow with black body stripe, nickel-plated driver, spare tire at rear, cast iron, 8-1/2" long, 1930	650	850	1300
Model A	cast iron, 8-1/2" long	225	325	450
Model A	cast iron, 6-3/4" long, 1928	175	225	300
Model A Coupe		425	650	850
Model A Ford	white rubber tires, cast iron, 1929	225	350	450
Model A Sedan	orange, cast iron, 6-3/4" long, 1928	700	1050	1400
Model T	rubber tires, 6" long	125	185	250
Model T Sedan, center door	cast iron, 6-1/2" long, 1923	100	150	200

Arcade

NAME	DESCRIPTION	GOOD	EX	MINT
Pierce Arrow Coupe	"Silver Arrow" on sides, cast iron, 7-1/4" long, 1936	150	225	425
Plymouth Sedan	cast iron, 4-3/4" long, 1933	100	185	275
Pontiac Sedan	white rubber wheels, cast iron, 4-1/4" long, 1936	85	145	185
Racer	driver's head and number highlighted with gold bronze, cast iron, 8" long, 1936			
Racer	white rubber tires, cast iron, 5-3/4" long, 1936	50	75	100
Red Top Cab	cast iron, 8" long, 1924	700	1200	2500
Reo Coupe	cast iron, 9-3/8" long, 1931	100	2000	5000
Runabout Auto	wood, pressed steel, steel, cast iron, 8-1/2" long, 1908			
Sedan	cast iron, 5" long, 1920s	100	150	200
Sedan	cast iron, 8" long, late 1930s	125	185	250
Sedan with Red Cap Trailer	cast iron, sedan 5-7/8" long, trailer 2-1/2" long, 1939	250	400	650
Yellow Cab	yellow with "Yellow Cab" in black on top, cast iron, 4-1/4" long, 1936, "Darmalee" Cab	5000	7500	10000
Yellow Cab	bright yellow body, black top, white rubber wheels with black centers, cast iron, 8-1/4" long, 1936	1250	2000	3000
Yellow Cab	rubber tires, cast iron, 5-1/4" long, 1925	475	650	850
Yellow Cab	rubber tires, cast iron, 8" long, 1924	500	750	1000
Yellow Cab	rubber tires, cast iron, 9" long, 1925	600	825	1200

Emergency Vehicles

NAME	DESCRIPTION	GOOD	EX	MINT
Ambulance	cast iron, 6" long, 1932	500	750	1000
Fire Engine	red enamel with gold striping, cast iron, 7-1/2" long, 1926	750	1000	1500
Fire Engine	red, white rubber tires with green centers, cast iron, 4-1/2" long, 1936	125	175	250
Fire Engine	red trimmed in gold bronze, white rubber wheels with blue centers, cast iron, 6-1/4" long, 1936	150	225	300
Fire Engine	red trimmed in gold bronze, white rubber wheels with blue centers, cast iron, 9" long, 1936	300	450	750
Fire Ladder Truck	red trimmed in gold bronze, white rubber wheels with blue centers, cast iron, 7" long, 1936	400	600	800
Fire Ladder Truck	red trimmed in gold bronze, ladders yellow, white rubber wheels blue centers, cast iron, 12-1/2" long with ladders, 1936	475	750	950
Fire Pumper	with six firemen, cast iron, 13-1/4" long, 1938	500	800	1000
Fire Trailer Truck	red trimmed in gold bronze, two-piece fire engine, cast iron, truck 16-1/4" long, with ladders 20" long, 1936	500	775	1000
Fire Truck	cast iron, 15" long	175	265	350
Hook and Ladder Fire Truck	cast iron, 16" long	275	415	550
Mack Fire Apparatus Truck	bright red truck, hose reel, removable extention ladders, bell that rings, cast iron, 21" long, 1925	1000	1500	3000
Mack Fire Apparatus Truck	red trimmed in gold, yellow extension ladders, imitation hose, nickeled driver, nickeled bell that rings, 21" long, 1936	1000	1500	3000
Pontiac Boiler Fire Truck	cast iron	200	315	400
Pontiac Ladder Fire Truck	cast iron, 7" long	210	325	425

Ice Wagon, Arcade

Showboat, Arcade

White Moving Van, 1928, Arcade

Arcade

NAME	DESCRIPTION	GOOD	EX	MINT

Farm and Construction Equipment

NAME	DESCRIPTION	GOOD	EX	MINT
"Ten" Caterpillar Tractor	cast iron, 7-1/2" long, 1929	550	850	1200
Allis-Chalmers Tractor	cast iron, separate plated driver, 7" long, 1940	225	350	600
Allis-Chalmers Tractor	cast iron, cast in driver, 7" long, 1940	200	300	500
Allis-Chalmers Tractor	cast iron, 3" long, 1934	65	95	130
Allis-Chalmers Tractor	cast iron, 6" long, 1940	175	250	400
Allis-Chalmers Tractor Trailer	red tractor, trimmed gold bronze, green trailer, cast iron, 13" long, 1936	250	400	750
Allis-Chalmers Tractor with Earth Mover	cast iron, 5" long, 1934	95	145	190
Avery Tractor	gray frame with gold striping, red wheels, flat radiator, cast iron, 4-1/2" long, 1929	100	150	200
Caterpillar Tractor	yellow trimmed in black, black wheels, nickeled steel track and driver, cast iron, 7-3/4" long, 1936	500	850	1000
Caterpillar, No. 270	steel, 8-1/2" long, 1920s	750	1000	1500
Crawler	driver, chain tracks, cast iron, 3-7/8" long, 1930	160	245	325
Crawler	nickel-plated driver and tracks, cast iron, 7-1/2" long, 1930	625	950	1250
Crawler	nickel-plated driver and tracks, cast iron, 6 5/8" long, 1930	425	625	850
Crawler	driver, chain tracks, cast iron, 3" long, 1930			
Crawler	driver, chain tracks, cast iron, 5-5/8" long, 1930	325	450	650
Fairbanks-Moorse	small portable Z-engine, no wheels, cast iron, 3-1/2" long, 1930	200	350	500
Ford Tractor	cast iron, 1/12 scale, 1941	100	150	200
Ford Tractor	cast iron, 3-1/4" long, 1926	100	150	200
Ford Tractor	cast iron, cast in driver, 1/25 scale, 1940	100	150	200
Ford Tractor	cast iron, cast in driver, with plow, 1/25 scale, 1940	200	300	400
Fordson F/Loader Tractor	rear crank operated loader, cast iron, 1/16 scale	200	275	350
Fordson Tractor	cast iron, 4-3/4" long, 1926	75	100	150
Fordson Tractor	cast iron, 5-3/4" long, 1932	100	150	200
Fordson Tractor	disk rubber wheels, cast iron, 3-1/2" long, 1936	75	100	150
Fordson Tractor	with or without lugs on rear wheels, cast iron, 6" long, 1926	275	400	575
International-Harvester A Tractor	cast iron, 1/12 scale, 1941	500	750	900
International-Harvester Crawler	cast iron, 1/16 scale	375	565	750
International-Harvester Crawler	cast iron, plated driver, 1/16 scale, 1941	500	1000	2000
International-Harvester Crawler	cast iron, plated driver, 1/16 scale, 1936	800	1600	3000
International-Harvester M Tractor	rubber wheels, cast iron, 5-1/4" long, 1941	300	400	550
International-Harvester M Tractor	rubber wheels, cast iron, 7" long, 1940	450	700	800

Arcade

NAME	DESCRIPTION	GOOD	EX	MINT
John Deere Open Flywheel Tractor	cast iron, 1/16 scale, 1941	175	265	350
John Deere Wagon	wooden box, iron running gear, cast iron, 1/16 scale, 1940	150	225	300
McCormick-Deering Farmall Tractor	gray body trimmed in gold, red wheels, driver, cast iron, 6-1/4" long, 1936	300	450	600
McCormick-Deering M Tractor	wood wheels, cast iron, 4-1/4" long, 1942	175	250	400
McCormick-Deering Plow	red frame, yellow wheels, aluminum bronze plow, shares and disks, cast iron, 7-3/4" long, 1926	250	400	600
McCormick-Deering Plow	red frame, cream wheels, aluminum plow, shares and disks, cast iron, 7-3/4" long, 1936	175	250	325
McCormick-Deering Thresher	assorted colors with gold striping, red grain pipe, 9-1/2" long, 1936	150	250	400
McCormick-Deering Thresher	gray, red trim, cream colored wheels, cast iron, 12" long, 1929	300	600	1200
McCormick-Deering Tractor	belt pulley, cast iron, 1/16 scale, 1925	450	650	850
McCormick-Deering Tractor	assorted colors trimmed in gold, nickeled driver, cast iron, 7-1/2" long, 1936	300	400	500
Oliver Planter	red or green, 1/25 scale, cast iron, 1950	35	50	75
Oliver Plow	red cast iron, 1/16 scale, 1940	75	115	150
Oliver Plow	red or green cast iron, 1/25 scale, 1940	25	35	50
Oliver Plow	red enamel finish, blades striped with aluminum bronze, 6-1/2" long, 1926	225	350	475
Oliver Plow	red with nickel-plated wheels, aluminum bronze plow shares, 6-3/4" long, 1926	225	350	475
Oliver Spreader	yellow, cast iron, 1/16 scale, 1940	350	650	1200
Oliver Tractor	red or green with rubber tires, cast iron, 5-1/4" long, 1946	75	100	150
Oliver Tractor	red or green, cast iron, 7-1/2" long, 1940	300	600	1200
Threshing Machine	gray and white, red trim, cast iron, 10" long	160	245	325

Tanks

NAME	DESCRIPTION	GOOD	EX	MINT
Army Tank with Gun	shoots steel balls, cast iron, 8" long, 1940	200	350	600

Trucks

NAME	DESCRIPTION	GOOD	EX	MINT
"Yellow Baby" Dump Truck	cast iron, 10-3/4" long, 1935	2000	3500	5500
Auto Express 548 Truck	flatbed, cast iron, 9" long	225	350	450
Borden's Milk Bottle Truck	"Borden's" cast on side, cast iron, 6-1/4" long, 1936	1500	2750	8000
Carry-Car Truck and Trailer Set	red truck, green trailer, with four vehicles, cast iron, 28" long, 1936	1500	3000	4500
Carry-Car Truck and Trailer Set	red truck, green trailer, with three 3-3/4" Austin vehicles, cast iron, 14-1/4" long, 1936	650	1150	1700
Carry-Car Truck Trailer	cast iron, 24-1/2" long, 1928	750	1300	1900
Chevrolet Panel Delivery Van	white rubber tires with colored centers, cast iron, 4" long, 1936	125	200	300
Chevrolet Stake Truck		90	165	225
Chevrolet Utility Express Truck	cast iron, 9-1/4" long, 1923	450	675	1200
Chevrolet Wrecker		90	165	225
Chrome Wheeled Truck	cast iron, 7" long	225	350	450

NAME	DESCRIPTION	GOOD	EX	MINT
Delivery Truck	yellow, cast iron, 8-1/4" long, 1926	800	1200	1500
Dump Truck	spoked wheels, low bed, cast iron, 6" long	175	350	500
Dump Truck	cast iron, 6-1/2" long, 1941	175	350	500
Dump Truck	red chassis, green dump body, white rubber tires with green centers, cast iron, 4-1/2" long, 1936	75	100	150
Ford Anthony Dump Truck	nickel-plated spoked wheels, black enamel finish with gray dump body, cast iron, 8-1/2" long, 1926	1250	2000	2500
Ford Weaver Wrecker	cast iron, 8-1/4" long, 1928	700	950	1400
Ford Weaver Wrecker, Model A	cast iron, 11" long, 1926	550	800	1200
Ford Weaver Wrecker, Model T	cast iron, 11" long, 1926	600	875	1300
Ford Wrecker	white rubber wheels, cast iron, 7" long, 1936	200	400	600
Gasoline Truck	cast iron, 13" long, 1920s	600	900	1200
Ice Truck	red, cast iron, 7" long, 1930s	175	225	350
International Delivery Truck	cast iron, 9-1/2" long, 1936	1250	1875	2500
International Dump Truck	cast iron, 10-3/4" long, 1930	1200	1500	2000
International Dump Truck	white rubber wheels with red centers, cast iron, 10-1/2" long, 1935	1200	1500	2000
International Dump Truck	red cab, green dump, cast iron, 9-1/2" long, 1940	600	950	1250
International Harvester Dump Truck	cast iron, 11" long, 1941	300	465	675
International Harvester Pickup	cast iron, 9-1/2" long, 1941	300	465	675
International Stake Truck	white rubber wheels with red centers, cast iron, 12" long, 1935	1200	1500	2000
International Stake Truck	yellow, cast iron, 9-1/2" long, 1940	600	750	1250
International Wrecker Truck	cast iron, 11" long	1250	1875	2500
Mack Dump Truck	assorted colors trimmed in gold bronze, "Mack" decals on doors, dual rear wheels, cast iron, 12-1/4" long, 1936	750	1125	1500
Mack Dump Truck	cast iron, 13" long	375	565	750
Mack Dump Truck	light grey or blue, gold trim, white tires, cast iron, 12" long, 1925	700	1050	1400
Mack Gasoline Truck	cast iron, 13-1/4" long, 1925	750	1500	2200
Mack High Dump	cast iron, 12-3/8" long, 1931	1000	2000	3000
Mack High Dump	nickel-plated levers mechanically raise the dump bed, cast iron, 8-1/2" long, 1930	750	1500	2200
Mack Ice Truck	ice blocks and tongs, cast iron, 8-1/2" long, 1931	350	525	700
Mack Ice Truck	ice blocks and tongs, cast iron, 10-3/4" long, 1931	550	750	950
Mack Lubrite Tank Truck	cast iron, 13-1/4" long, 1925	1200	2000	3000
Mack Oil Truck	cast iron, 10" long, 1920s	350	525	700
Mack Tank Truck	cast iron with tin tank, 12-3/4" long, 1929	1200	1800	2900
Mack Tank Truck	assorted colors with gold trim, nickeled driver, dual rear wheels, tank holds water, cast iron, 13" long, 1936	1100	1800	2500
Mack Tank Truck	nickel-plated driver, cast iron, 13-1/4" long, 1926	1000	1500	2000

REO Coupe, 1930s, Arcade

McCormick-Deering Tractor, Arcade

Chevrolet Panel Delivery Van, 1930s, Arcade

Arcade

NAME	DESCRIPTION	GOOD	EX	MINT
Mack Wrecker Truck	white rubber tires, cast iron, 11" long	500	750	1000
Model-A Stakebody Truck	iron wheels, cast iron, 7-1/2" long, 1920s	150	225	300
Plymouth Wrecker		100	185	275
Pontiac Stake Truck	white rubber tires, cast iron, 6-1/4" long, 1936	225	350	500
Pontiac Stake Truck	white rubber tires, cast iron, 4-1/4" long, 1936	65	95	165
Pontiac Wrecker	white rubber tires, cast iron, 4-1/4" long, 1936	70	100	175
Red Baby Dump Truck	bright red truck, white enameled tires, crank dump, cast iron, 10-3/4" long, 1924	1200	1800	2400
Red Baby Truck	bright red truck, white enameled tires, cast iron, 10-1/4" long, 1924	1000	1500	2000
Semi Truck	cast iron, 1920s	500	750	1000
Stake Truck	cast iron, 7" long	110	175	225
Tow Truck	cast iron, 4" long	55	85	110
Transport	double-deck semi-trailer with four sedans, cast iron and pressed steel, 18-1/2" long, sedans 4-3/4" long, 1938	750	1000	2000
White Delivery Truck	cast iron, 8-1/2" long, 1931	1250	2100	3300
White Dump Truck	cast iron, 11-1/2" long	7500	10000	15000
White Moving Van	cast iron, 13-1/2" long, 1928	6000	8500	13500

Vehicle Banks

NAME	DESCRIPTION	GOOD	EX	MINT
Ford Touring Car Bank	removable driver, cast iron, 6-1/2" long, 1925	650	975	1300
Mack Dump Truck Bank	cast iron, 13" long	375	565	750
Yellow Cab Bank	cast iron, 8" long, 1924	475	775	1000
Yellow Cab Bank	with or without rubber tires, cast iron, 9" long, 1926	600	900	1200
Yellow Cab Bank	cast iron, 8-1/2" long, 1930	650	1400	2000

Wagons, Carts and Trailers

NAME	DESCRIPTION	GOOD	EX	MINT
Auto Dump Wagon	red and gold, cast iron, 7" long, 1920	200	400	600
Circus Wagon	circus wagon with driver and two horses, "Big Six Circus & Wild West" on side of wagon, 14-1/2" long, 1936	300	450	600
Ice Wagon	with black horses trimmed in gold, cast iron, 11-3/4" long, 1926	575	875	1300
McCormick-Deering Weber Wagon	removable wagon seat and box, with two horses, cast iron, overall length 12-1/8" long, 1925	200	400	600
Panama Dump Wagon	gray, with two horses, cast iron, nickeled pick and shovel, 14-1/4" long, 1926	300	450	600
Panama Dump Wagon	gray, nickeled pick and shovel, cast iron, 12-3/4" long, 1926	200	400	600
Whitehead & Kales Truck Trailer	sides and end gates removable by sections, cast iron, 8-1/2" long, 1926	150	250	400

Auburn

Airplanes

NAME	DESCRIPTION	GOOD	EX	MINT
Clipper Plane	rubber, 7" wingspan, 1941	10	20	25
Dive Bomber	rubber, 4" wingspan, 1937	10	20	25
Pursuit Ship Plane	4" wingspan, 1941	10	20	25
Two-Engine Transport Plane	1937	10	20	27

Auburn

NAME	DESCRIPTION	GOOD	EX	MINT

Boats and Ships

NAME	DESCRIPTION	GOOD	EX	MINT
Battleship	rubber, 8-1/4" long, 1941	15	25	30
Cruiser	rubber	15	25	30
Freighter	rubber, 8" long, 1941	15	25	30
Submarine	rubber, 6-1/2" long, 1941	10	20	27

Cars

NAME	DESCRIPTION	GOOD	EX	MINT
1947 Buick Coupe	#100 on license plate, 7" long	75	115	150
Airport Limousine	rubber, 8" long	25	40	50
Fire Chief's Car	red, yellow wheels	7	10	15
Ford	rubber, 1930s	10	20	25
Race Car	rubber, 6" long, 1930s	15	25	30
Race Car	red, rubber, 6" long	40	65	85
Race Car With Goggled Driver	rubber, 10" long	50	75	100
Racer	rubber	20	30	40
Racer	red vinyl with white plastic tires	25	40	50
Sedan	green, rubber, license #500R	15	25	30
Sedan	cast iron driver	25	40	50

Emergency Vehicles

NAME	DESCRIPTION	GOOD	EX	MINT
Fire Engine	red, rubber, 8" long	15	25	30
Fire Truck	black rubber wheels	7	10	15
Rescue Truck	dark army green	15	25	30

Farm and Construction Equipment

NAME	DESCRIPTION	GOOD	EX	MINT
Allis-Chalmers Tractor	red and silver plastic, 1/16 scale, 1950	25	40	50
Earthmover	red front, yellow back, plastic wheels	15	25	30
Giant Tractor	red tractor, silver motor, black tires, 7" long, 1950s	30	45	60
John Deere Tractor	plastic, 1/20 scale	25	40	50
Minneapolis Moline Tractor	red, large rubber tires, rubber, 1/16 scale, 1950	25	40	50
Tractor and Wagon	orange tractor, silver motor, black tires, red spreader wagon, yellow spoke tires	75	115	150

Motorcycles

NAME	DESCRIPTION	GOOD	EX	MINT
Motorcycle		20	30	40
Police Cycle	red rubber, drive chain, 6" long, 1950s	50	75	100

Trucks

NAME	DESCRIPTION	GOOD	EX	MINT
2-1/2 Ton Truck		20	30	40
Army Jeep	olive drab	5	7	10
Army Recon Half Truck	bright green	10	15	20
Stake Truck	rubber	15	25	30
Telephone Truck	6-1/2" long	35	55	75

Bandai

NAME	DESCRIPTION	GOOD	EX	MINT
	Buses			
Volkswagen Bus	red and white, battery-operated, 9-1/2" long, 1960s	65	150	275
	Cars			
1915 Ford Touring Car	7" long	90	135	180
1955 MG TF Convertible	8-1/2" aqua blue, aqua green, or black, friction	85	195	275
1956 Buick	6" green, battery-operated	35	60	85
1956 Chevrolet Convertible	10" cream, friction	135	315	450
1958 Chevrolet Convertible	8" light blue, friction	65	140	200
1958 Chrysler Imperial Convertible	8-1/2" maroon, friction	195	245	350
1958 Chrysler Imperial HT	8-1/2" two-tone green, friction	90	125	175
1958 Ford Station Wagon	8" two-tone green, friction	60	85	125
1958 Plymouth Station Wagon	8-1/2" green/white, friction	55	85	125
1960 Rolls-Royce Silver Cloud	blue body, white top, electric lights, 12" long	350	525	700
1961 Buick Station Wagon	blue, friction	50	85	125
1961 Plymouth Valiant	8" blue	45	105	150
1964 Chevrolet Malibu Fire Chief	8" red, friction	45	55	75
1964 Ford Fairlane	8" metallic red, friction	35	50	85
1965 Chrysler Four-Door HT	8-1/2" two-tone green, friction	40	55	75
1965 Ford Mustang	11" cream/black, friction	50	90	125
1967 Chevrolet Camaro	13-1/2" red battery-operated with lighted engine and turning fan	85	100	140
Cadillac	gold, hardtop, friction powered, 17" long, 1960s	195	350	550
Cadillac	copper, hardtop, friction powered, 11" long, 1960	75	125	175
Cadillac	white, hardtop, friction powered, 11" long, 1959	100	175	275
Cadillac	gold fins, black top, tin, 11-1/2" long, 1959	150	225	300
Cadillac	8" long	100	175	275
Cadillac Convertible	red, green interior, friction powered, 11" long, 1959	175	300	400
Cadillac Convertible	black, friction powered, 11" long, 1960	180	275	380
Cadillac Convertible	white, red interior, friction powered, 17" long, 1963	200	400	600
Chevrolet Corvette	white and black, battery-operated, 8" long, 1962	75	125	175
Chevrolet Corvette	red, friction powered, 8" long, 1963	75	125	175
Chevrolet Impala Convertible	white, friction powered, 11" long, 1961	250	400	550
Chevrolet Impala Sedan	cream, friction powered, 11" long, 1961	225	350	500
Citroen	blue and white, friction powered, 12" long, 1958	300	650	1000
Citroen DS-19 Sedan	8-1/2" metallic red/white, friction, 1950s	130	300	425
Corvair Bertone	white, battery-operated, tin, 12" long, 1963	70	120	175

Bandai

NAME	DESCRIPTION	GOOD	EX	MINT
Cougar	white, battery-operated, tin	100	150	200
D.K.W. Sedan	8-1/2" grey/black, friction, 1950s	175	245	350
Excalibur Roadster	white body, red fenders, black top, battery-operated, rubber wheels, motor sparks, 11" long	135	200	275
Ferrari	silver with red interior, battery-operated, tin, gearshift on floor, working lights, horn, and engine noise, 11" long, 1958	300	550	850
Fiat 600	7" blue, friction, opening sun roof, 1950s	35	70	100
Ford Convertible	red/black or two-tone green, friction powered, 12" long, 1957	150	275	375
Ford Convertible	green, friction powered, 12" long, 1955	350	650	900
Ford Country Sedan	blue and white, friction powered, 10-1/2" long, 1961	65	100	140
Ford F.B.I. Mustang	black and white, friction powered, 11" long, 1965	75	100	140
Ford Flower Delivery Wagon	blue, friction powered, 12" long, 1955	275	400	650
Ford GT	red, battery-operated, 10" long, 1960s	100	150	200
Ford Mustang	red, battery-operated, 11" long, 1965	125	200	275
Ford Mustang	red, battery-operated, 13" long, 1967	80	125	175
Ford Mustang	silver and black, battery-operated, 11" long, 1965	65	100	145
Ford Ranchero	two-tone blue, friction powered, 12"long, 1955	175	250	350
Ford Ranchero	black and red, friction powered, 12" long, 1957	175	250	350
Ford Standard Fresh Coffee Wagon	black and orange, friction powered, 12" long, 1955	350	625	975
Ford Station Wagon	cream and black, two-tone green, or red and black, friction powered, 12" long, 1955	75	125	200
Ford Thunderbird	red and black, friction powered, 10-3/4" long, 1965	100	175	225
Ford Thunderbird	red or red and black, friction powered, 8" long, 1962	65	100	150
Ford Wagon	green body, black top, 12" long	300	450	600
Ford Wagon	blue and white, friction powered, 12" long, 1957	90	150	200
GT-40	blue, hood and trunk open, rubber tires, tin, battery-operated, 11" long	100	150	200
Isetta	white and two-tone green, three wheels, friction powered, 6-1/2" long, 1950s	200	300	400
Jaguar 3.4 Sedan	8-1/2" light green, friction, 1960s	85	125	175
Jaguar XK 140 Convertible	various colors, 9-1/2" long, 1950s	125	200	275
Jaguar XK-E	red, battery-operated, 10" long, 1960s	100	150	250
Lincoln Continental	turquoise and white, matching interior, 1958	150	250	375
Lincoln Continental Convertible	white, red interior, 1958	200	300	425
Lincoln Mark III	turquoise and white, friction powered, 11" long, 1958	150	275	375
Lotus	blue, battery-operated, 9-1/2" long	110	175	225
Lotus Elite	red and black, friction powered, 8-1/2" long, 1950s	75	100	150
Mazda 360 Coupe	blue, friction powered, 7" long, 1960	65	100	125
Mercedes-Benz 219 Convertible	8" blue, wrap around windshield, friction	85	125	175
Mercedes-Benz 220 Sedan	"Scale Model Series", 10-1/2", red, battery-operated with working headlights	85	125	175
Mercedes-Benz 300 SL Coupe	silver and black, friction powered, 8" long, 1950s	160	245	325
Mercedes-Benz Taxi	black, battery-operated, 10" long, 1960s	125	250	350

Bandai

NAME	DESCRIPTION	GOOD	EX	MINT
MG 1600 Mark II	red, friction powered, 8-1/2" long, 1950s	75	125	175
MG TF	green, friction powered, 8" long, 1955	75	125	175
MGA-1600 Coupe	8-1/2" silver/red, friction, 1950s	85	125	175
Old Timer Police Car	battery-operated, 8" long	60	95	125
Olds Toronado	gold, battery-operated, 11" long, 1966	30	50	65
Oldsmobile	surrey top, friction, 1900's	75	115	150
Plymouth Valiant	blue, wind-up, 8" long	50	75	100
Pontiac Firebird	red, battery-operated, 9-1/2" long, 1967	75	125	175
Porsche 911	white, battery-operated, 10" long, 1960s	90	150	200
Racer with Hand Control		85	130	175
Rolls-Royce Convertible	several colors available, 12" long, friction, 1950s	200	300	450
Rolls-Royce Hardtop Sedan	blue, black, white, rare version with working headlights, battery-operated, 12" long, 1950s	275	400	600
Subaru 360	red, friction powered, 8" long, 1959	75	125	175
Taxi	friction, 1950s	75	115	150
Triumph TR-4	8" red/white battery-operated with signal lights, 1950s	50	100	140
Volkswagen	blue, battery-operated, 10-1/2" long, 1960s	75	125	175
Volkswagen	red, battery-operated, 15" long, 1960s	95	150	225
Volkswagen	red, with sun roof, battery-operated, 15" long, 1960s	90	150	225
Volkswagen	red, battery-operated, 8" long, 1960s	35	75	100
Volkswagen Convertible	white, battery-operated, 7-1/2" long, 1960s	50	95	125

Emergency Vehicles

NAME	DESCRIPTION	GOOD	EX	MINT
Plymouth Ambulance	white, red cross on doors, friction powered, 12" long, 1961	35	50	75
Rambler Ambulance	white, friction powered, 11" long, 1962	40	75	100

Motorcycles

NAME	DESCRIPTION	GOOD	EX	MINT
Police Auto Cycle	battery-operated, hard plastic, 10" long, 1970s	100	150	200

Sets

NAME	DESCRIPTION	GOOD	EX	MINT
Ferrari and Speed Boat	white car, red and white speed boat with white trailer, tin, overall 23" long, 1958	350	650	900
Lincoln Continental and Cabin	turquoise and white car and cruiser, red car interior and cruiser trailer, overall 23" long, 1958	300	550	800
Rambler Wagon and Cabin Cruiser	green and white Rambler wagon, red trailer, friction powered rambler, electric boat motor, overall 23" long, 1959	200	400	600
Rambler Wagon and Shasta Trailer	green and white Rambler wagon, yellow and white trailer, 11" long wagon, 12" long trailer, 1959	200	400	600
Rambler, Trailer and Cabin Cruiser	turquoise and white Rambler, red trailer, cruiser color varies, overall 35" long, 1959	350	650	1000

Trucks

NAME	DESCRIPTION	GOOD	EX	MINT
1958 Ford Ranchero Pickup Truck	8" long	110	175	225
Land Rover	maroon, 8" long	125	185	250
Land Rover	red, friction powered, 7-1/2" long, 1960	75	100	150
Volkswagen Truck	blue, open flatbed cargo section, battery-operated, 8" long, 1960s	90	150	200

Barclay

NAME	DESCRIPTION	GOOD	EX	MINT

Airplanes

NAME	DESCRIPTION	GOOD	EX	MINT
Dirigible Plane	4-3/8" long, 1930s	7	10	15
Lindy-Type Plane	4" wingspan, 1930s	7	10	15
Monoplane	single engine plane	15	28	35
Monoplane	single engine plane, red propeller, red metal wheels	15	30	40
U.S. Army Single Engine Transport Plane	white rubber wheels, 1940	10	15	20
U.S. Army Small Pursuit Plane	lead, rubber wheels, 1941	10	15	20

Buses

NAME	DESCRIPTION	GOOD	EX	MINT
Double-Decker Bus	4" long	20	30	40

Cannons

NAME	DESCRIPTION	GOOD	EX	MINT
Cannon	barrel elevated, 2-1/2" long	15	25	35
Cannon	1931	20	30	40
Cannon	very large wheels, 4" long	15	20	25
Cannon	silver with black rubber wheels, 7-3/4" long	20	30	40
Cannon	spoked wheels, 3" long	10	15	20
Coast Defense Rifle Cannon	4-1/2" long	25	40	50
Howitzer Cannon	horizontal loop hitch, four wheels, 3" long	15	25	35
Howitzer Cannon	vertical loop hitch, four wheels, 3" long	15	25	35
Mortar Cannon	swivels on base, 3" long	20	30	40
Spring-Firing Cannon	spoked wheels, 4" long	20	25	35

Cars

NAME	DESCRIPTION	GOOD	EX	MINT
1930s Sedan and Tourist Trailer	white rubber tires, 6-1/2"	35	55	75
Armoured Car	1937	15	25	35
Car Carrier	with two cars	25	40	55
Coupe	3" long, 1930s	15	25	35
Race Car	white tires, 4" long	25	40	55
Tow Car #205	white rubber tires, red wood wheels, 3"	25	35	50

Farm Equipment

NAME	DESCRIPTION	GOOD	EX	MINT
Fordson Tractor #203	white rubber tires, red wood wheels, 2-1/8"	15	20	25

Trucks

NAME	DESCRIPTION	GOOD	EX	MINT
Beer Truck	slush metal, 4" long	15	25	35
Mack Pickup Truck	3-1/2" long	25	40	55
Milk Truck #377	with milk cans	25	40	55
Open Truck	3-1/2" long	10	20	25
Stake Truck	slush metal, 5" long	15	25	35
White Horse Milk Van Delivery	2-7/8", white rubber tires	25	35	50

Brooklin

NAME	DESCRIPTION	GOOD	EX	MINT

Cars

NAME	DESCRIPTION	GOOD	EX	MINT
1932 Packard Light 8 Coupe		35	50	70
1933 Pierce Arrow	silver	35	50	70
1934 Chrysler Airflow Sedan	four door	35	50	70
1935 Dodge "City Ice Delivery" Van		40	60	80
1935 Dodge "Dr. Pepper" Van		45	70	90
1935 Dodge "Sears Roebuck" Van		40	60	80
1940 Ford Sedan Delivery "Ford Service"		40	60	80
1941 Packard Clipper" Van		40	60	80
1948 Tucker Torpedo		35	50	70
1949 Buick Roadmaster		35	55	75
1949 Mercury Coupe	two door	30	45	65
1952 Hudson Hornet Convertible	1:43 scale	30	45	65
1952 Studebaker Champion Starlight Coupe		40	60	80
1953 Buick Skylark		35	50	70
1953 Pontiac Sedan Delivery Gulf Oil Truck	1:43 scale	30	45	60
1953 Pontiac Sedan Delivery Mobil Oil Truck	1:43 scale	30	45	60
1953 Studebaker Commander	1:43 scale	30	45	60
1953 Studebaker Indiana State Police	1:43 scale	50	75	100
1954 Dodge 500 Indy Pace Car	1:43 scale	50	75	100
1955 Chrysler 300		30	45	60
1956 Ford Fairline Victoria	two door	25	40	55
1956 Ford Thunderbird 500	hardtop	30	45	65
1956 Lincoln Continental		30	45	60
1956 Lincoln Continental Mark II Coupe		30	45	65
1956 Lincoln Continental MKII		30	45	65
1957 Ford Fairlane Skyliner Police	1:43 scale	30	45	60
1958 Edsel Citation	two-door hardtop	30	45	30
1958 Pontiac Bonneville		30	45	65
1960 Ford Sunliner Convertible	1:43 scale	30	45	60

Brooklin

NAME	DESCRIPTION	GOOD	EX	MINT
1963 Chevrolet Corvette Stingray Coupe		35	55	75
1968 Shelby Mustang GT 500		35	55	75
Lincoln Mark	1:43 scale	30	45	60
Mini Marquee Packard Convertible	1:43 scale	50	75	100
Tucker	1:43 scale	30	45	60

Buddy L

Airplanes

NAME	DESCRIPTION	GOOD	EX	MINT
5000 Monocoupe "The Lone Eagle"	orange wing, black fuselage and tail with tailskid, all steel high wind cabin monoplane, 9-7/8" wingspan, 1929	250	300	400
Army Tank Transport Plane	low-wing monoplane, two small four-wheel tanks that clip beneath wings, 27" wingspan, 1941	250	350	400
Catapult Airplane and Hangar	5000 Monocoupe with tailwheel, 9-7/8" wingspan, olive/gray hangar, black twin-spring catapult, 1930	950	1200	1500
Four Motor Air Cruiser	white, red engine cowlings, yellow fuselage and twin tails, four engine monoplane, 27" wingspan, 1952	200	300	350
Four-Engine Transport	green wings, white engine cowlings, yellow fuselage and twin tails, four engine monoplane, 27" wingspan, 1949	200	300	400
Hangar and Three 5000 Monocoupes	olive/gray hangar, windows outlined in red or orange, planes 9-7/8" wingspan, all steel high wing cabin monoplanes, 1930	750	1000	2000
Transport Airplane	white wings and engine cowlings, red fuselage and twin tails, four engine monoplane, 27" wingspan, 1946	200	300	400

Cars

NAME	DESCRIPTION	GOOD	EX	MINT
Army Staff Car	olive drab body, 15-3/4" long, 1964	100	150	200
Bloomin' Bus	chartreuse body, white roof and supports, similar to VW minibus, 10-3/4" long, 1969	90	135	180
Buddywagon	red body with white roof, 10-3/4" long, 1966	100	150	200
Buddywagon	red body with white roof, no chrome on front, 10-3/4" long, 1967	95	145	190
Colt Sportsliner	red open body, white hardtop, off-white seats and interior, 10-1/4" long, 1967	35	50	70
Colt Sportsliner	light blue-green open body, white hardtop, pale tan seats and interior, 10-1/4" long, 1968	30	45	65
Colt Utility Car	red open body, white plstic seats, floor and luggage space, 10-1/4" long, 1967	35	50	70
Colt Utility Car	light orange body, tan interior, 10-1/4" long, 1968	30	45	65
Country Squire Wagon	off-white hood fenders, end gate and roof, brown woodgrain side panels, 15-1/2" long, 1963	85	130	175
Country Squire Wagon	red hood fenders, end gate and roof, brown woodgrain side panels, 15" long, 1965	75	115	150
Deluxe Convertible Coupe	metallic blue enamel front, sides and deck, cream top retracts into rumble seat, 19" long, 1949	300	450	600

Buddy L

NAME	DESCRIPTION	GOOD	EX	MINT
Desert Rats Command Car	light tan open body, light beige interior, black .50-caliber machine gun swivels on post between seats, 10-1/4" long, 1967	50	75	100
Desert Rats Command Car	light tan open body, light beige interior, black .50-caliber machine gun swivels on post between seats, 10-1/4" long, blackwall tires, 1968	45	65	90
Flivver Coupe	black with red eight-spoke wheels, black hubs, aluminum tires, flat, hardtop roof on enclosed glass-window-style body, 11" long, 1924	775	1100	1550
Flivver Roadster	black with red eight-spoke wheels, black hubs, aluminum tires, simulated soft, folding top, 11" long, 1924	750	1000	1500
Jr. Camaro	metallic blue body, white racing stripes across hood nose, 9" long, 1968	50	75	100
Jr. Flower Power Sportster	purple hood, fenders and body, white roof and supports, white plastic seats, lavender and orange five-petal blossom decals on hood top, roof, and sides, 6" long, 1969	35	55	75
Jr. Sportster	blue hood and open body, white hardtop and upper sides, 6" long, 1968	35	55	75
Mechanical Scarab Automobile	red radically streamlined body, bright metal front and rear bumpers, 10-1/2" long, 1936	200	300	500
Police Colt	deep blue open body, white hardtop, "POLICE" across top of hood, "POLICE 1" on sides, 10-1/4" long, 1968	50	75	100
Ski Bus	white body and roof, similar to VW minibus, 10-3/4" long, 1967	75	115	150
Station Wagon	light blue/green body and roof, 15-1/2" long, 1963	75	115	150
Streamline Scarab	red, radically streamlined body, non-mechanical, 10-1/2" long, 1941	145	225	290
Suburban Wagon	powder blue or white body and roof, 15-1/2" long, 1963	75	115	150
Suburban Wagon	gray/green body and roof, 15-3/4" long, 1964	70	100	140
Town and Country Convertible	maroon front, hood, rear deck and fenders, gray top retracts into rumble seat, 19" long, 1947	300	450	600
Travel Trailer and Station Wagon	red station wagon, two-wheel trailer with red lower body and white steel camper-style upper body, 27-1/4" long, 1965	150	225	300
Yellow Taxi with Skyview	yellow hood, roof and body, red radiator front and fenders, 18-1/2" long, 1948	200	400	600

Emergency Vehicles

NAME	DESCRIPTION	GOOD	EX	MINT
Aerial Ladder and Emergency Truck	red with white ladders, bumper and steel disc wheels, three 8-rung steel ladders, 22-1/4" long, 1952	200	300	400
Aerial Ladder and Emergency Truck	red with white ladders, bumper and steel disc wheels, three 8-rung steel ladders, no rear step, no siren or SIREN decal, 22-1/4" long, 1953	225	345	450
Aerial Ladder Fire Engine	snub-nose red tractor and semi-trailer, white swivel-mounted aerial ladder with side rails, two white 13-rung sectional ladders, 27-1/2" long, 1970	100	150	200
Aerial Ladder Fire Engine	red tractor and semi-trailer, white plastic bumper with integral grille guard, two aluminum 13-rung extension ladders on sides, swivel-base aluminum central ladder, 26-1/2" long, 1961	125	185	250

NAME	DESCRIPTION	GOOD	EX	MINT
Aerial Ladder Fire Engine	red tractor, wraparound bumpevr and semi-trailer, two aluminum 13-rung extension ladders on sides, swivel-base aluminum central ladder, 26-1/2" long,	125	185	250
Aerial Ladder Fire Engine	red cab-over-engine tractor and semi-trailer units, two 13-rung white sectional ladders and swivel-mounted aerial ladder with side rails, 25-1/2" long, 1968	100	150	200
Aerial Ladder Fire Engine	red tractor and semi-trailer, chrome one-piece wraparound bumper, slotted grille, two aluminum 13-rung extension ladders on sides, swivel-base aluminum central ladder, 26-1/2" long, 1966	125	185	250
Aerial Truck	red with nickel ladders, black hand wheel, brass bell, and black hubs, 39" long with ladder down, 1925	850	1300	1700
American LaFrance Aero-Chief	red cab-over-engine and body, white underbody, rear step and simulated hose reels, black extension ladders on right side, 25-1/2" long, 1972	90	150	200
Brute Fire Pumper	red cab-over-engine body and frame, two yellow 5-rung sectional ladders on sides of open body, 5-1/4" long, 1969	30	50	75
Brute Hook-N-Ladder	red cab-over-engine tractor and detachable semi-trailer, white elevating, swveling aerial ladder with side rails, 10" long, 1969	30	40	55
Extension Ladder Fire Truck	red with silver ladders and yellow removable rider seat, enclosed cab, 35" long, 1945	200	300	400
Extension Ladder Rider Fire Truck	duo-tone slant design, tractor has white front, lower hood sides and lower doors, red hood top, cab and frame, red semi-trailer, white 10-rung and 8-rung ladders, 32-1/2" long, 1949	150	225	300
Extension Ladder Trailer Fire Truck	red tractor with enclosed cab, boxy fenders, red semi-trailer with fenders, two white 8-rung side ladders, 10-rung central extension ladder, 29-1/2" long,	200	300	400
Extension Ladder Trailer Fire Truck	red tractor unit and semi-trailer, enclosed cab, two white 13-rung side extension ladders on sides, white central ladder on swivel base, 29-1/2" long, 1956	125	185	250
Fire and Chemical Truck	duo-tone slant design, white front, lower hood sides and lower doors, rest is red, bright-metal or white eight-rung ladder on sides, 25" long, 1949	125	185	250
Fire Department Emergency Truck	red streamlined body, enclosed cab, chrome one-piece grille, bumper, and headlights, 12-3/4" long, 1953	100	150	200
Fire Engine	red with nickel rim flywheels on dummy pump, brass bell, dim-or-bright electric headlights, 25-1/2" long, 1933	500	750	1500
Fire Engine	red with nickel-plated upright broiler, nickel rims and flywheels on dummy water pump, brass bell, 23-1/4" long, 1925-29	500	800	1600
Fire Hose and Water Pumper	red with two white five-rung ladders, two removable fire extinguishers, enclosed cab, 12-1/2" long, 1950	100	150	200
Fire Hose and Water Pumper	red with two white five-rung ladders, one red/white removable fire extinguisher, enclosed cab, 12-1/2" long, 1952	100	150	200
Fire Pumper	red cab-over-engine and open body, 11-rung white 10" ladder on each side, 16-1/4" long, 1968	100	150	200

NAME	DESCRIPTION	GOOD	EX	MINT
Fire Pumper with Action Hydrant	red wraparound bumper, hood cab and cargo section, aluminum nine-rung ladders, white hose reel, 15" long, 1960	75	115	150
Fire Truck	red with white ladders, black rubber wheels, enclosed cab, 12" long, 1945	75	115	150
Fire Truck	red with nickel or white ladders, bright-metal radiator grille and black removable rider saddle, 25-1/2" long, 1935	250	375	500
Fire Truck	duo-tone slant design, yellow front, bumper, hood sides and skirted fenders, rest is red, nickel ladders, 28-1/2" long, 1939	500	700	1000
Fire Truck	red with two white ladders, enclosed cab, bright metal grille and headlights, 25" long, 1948	125	200	250
Fire Truck	duo-tone slant design, yellow front, single-bar bumper, hood sides and removable rider seat, rest is red, 25-1/2" long, 1936	170	250	500
Fire Truck	bright red, red floor, open driver's seat, 26" long, 1928	450	800	1200
Fire Truck	bright red with black inverted L-shaped crane mounted in socket on seat back, red floor, open driver's seat, 26" long, 1925	450	800	1600
Fire Truck	duo-tone slant design, tractor has white front, lower hood sides and lower doors, red hood top, cab and frame, red semi-trailer, rubber wheels with black tires, 32-1/2" long, 1953	200	300	400
Fire Truck	bright red with black inverted L-shaped crane mounted in socket on seat back, open driver's seat, 26" long, 1924	500	900	1850
Fire Truck	red with black solid-rubber Firestone tires on red seven-spoke embossed metal wheels, two 18-1/2" red steel sectional ladders, 26" long, 1930	800	1000	1600
GMC Deluxe Aerial Ladder Fire Engine	white tractor and semi-trailer units, golden 13-rung extension ladder on sides, golden central aerial ladder, black and white DANGER battery case with two flashing lights, 28" long, 1959	225	345	450
GMC Extension Ladder Trailer Fire Engine	red tractor with chrome GMC bar grille, red semi-trailer, white 13-rung extension ladders on sides, white swiveling central ladder with side rails, 27-1/4"	100	250	400
GMC Fire Pumper with Horn	red with aluminum-finish 11-rung side ladders and white reel of black plastic hose in open cargo section, chrome GMC bar grille, 15" long, 1958	150	200	300
GMC Hydraulic Aerial Ladder Fire Engine	red tractor unit with chrome GMC bar grille, red semi-trailer, white 13-rung extension ladders on sides, white swiveling central ladder, 26-1/2" long, 1958	125	185	250
GMC Red Cross Ambulance	all white, removable fabric canopy with a red cross and "Ambulance" in red, 14-1/2" long, 1960	150	250	400
Hook & Ladder Fire Truck	medium-dark red, with black inverted L-shaped crane mounted in socket on seat back, open driver's seat, 26" long, 1923	1200	1800	2400
Hose Truck	red with two white hose pipes, white cord hose on reeland brass nozzle, electric headlights with red bulbs, 21 3/4" long, 1933	500	750	1000
Hydraulic Aerial Truck	duo-tone slant design, yellow front, single-bar bumper, chassis, radiator, front fender, lower sides and removable rider saddle, rest is red, 40" long with ladders down, 1936	550	825	1100

Buddy L

NAME	DESCRIPTION	GOOD	EX	MINT
Hydraulic Aerial Truck	red with brass bell on cowl, nickel ladders mounted on 5-1/2" turntable rotated by black hand wheel, 39" long, 1927	850	1300	1700
Hydraulic Snorkel Fire Pumper	red cab-over-engine and open rear body, white 11-rung 10" ladder on each side, snorkel pod with solid sides, 21" long, 1969	100	150	200
Hydraulic Water Tower Truck	red with nickel water tower, dim/bright electric headlights, brass bell, 44-7/8" long with tower down, 1933	800	1000	1700
Hydraulic Water Tower Truck	red with nickel water tower, dim/bright electric headlights, brass bell, added-on bright-metal grille, 44-7/8" long with tower down, 1935	600	900	1500
Hydraulic Water Tower Truck	duo-tone slant design, yellow bumper, hood sides, front fenders, rest is red, electric headlights, added-on bright-metal grille, 44-7/8" long with tower down,	600	1000	1500
Hydraulic Water Tower Truck	duo-tone slant design, yellow front, single-bar bumper and hood sides, red hood top, enclosed cab and water tank, brass bell, nickel water tower, 46" long with tower down, 1939	900	1200	1700
Jr. Fire Emergency Truck	red cab-over-engine and body, one-piece chrome wraparound narrow bumper and 24-hole grille with plastic vertical-pair headlights, 6-3/4" long, 1968	50	75	100
Jr. Fire Emergency Truck	red cab-over-engine and body, wider one-piece chrome wraparound narrow bumper and four-slot grille with two square plastic headlights, 6-3/4" long, 1969	50	75	100
Jr. Fire Snorkel Truck	red cab-over-engine and body, full-width chrome one-piece bumper and four-slot grille with two square plastic headlights, 11" long, 1969	60	150	200
Jr. Fire Snorkel Truck	red cab-over-engine and body, chrome one-piece narrow wraparound bumper and 24-hole grille with plastic vertical-pair headlights, 11-1/2" long, 1968	100	150	200
Jr. Hook-n-Ladder Aerial Truck	red cab-over-engine tractor and semi-trailer, white high-sides ladder, chrome one-piece wraparound bumper and 24-hole grille, plastic veritcal-pair headlights, 17" long, 1967	100	150	200
Jr. Hook-n-Ladder Aerial Truck	red cab-over-engine tractor and semi-trailer, white high-sides ladder, one-piece chrome four-slot grille, two square plastic headlights, 17" long, 1969	75	115	150
Jr. Hook-n-Ladder Aerial Truck	red cab-over-engine tractor with one-piece four-slot grille and two square plastic headlights, red semi-trailer, white high-sides ladder, plastic vertical-pair headlights, 17" long, 1969	75	115	150
Ladder Fire Truck	red with bright-metal V-nose radiator, headlights and ladder, black wooden wheels, 12" long, 1941	125	200	250
Ladder Truck	red with two yellow ladders, enclosed square cab with sharply protruding visor, bright-metal radiator front, 22-3/4" long, 1935	250	375	500
Ladder Truck	red with two white ladders, bright-metal radiator grille and headlights, 24" long, 1941	125	200	250
Ladder Truck	modified duo-tone slant design, white front, front fenders and lower doors, white ladders, bright-metal grille, no bumper, 17-1/2" long, 1941	200	300	400

Buddy L

NAME	DESCRIPTION	GOOD	EX	MINT
Ladder Truck	red with yellow severely streamlined, skirted fenders and lower doors, white ladders, bright-metal grille, no bumper, 17-1/2" long, 1940	200	300	400
Ladder Truck	red with bright-metal grille and headlights, two white ladders, 24" long, 1939	100	200	300
Ladder Truck	duo-tone slant design, white front, hood sides, fenders and two ladders, rest is red, square enclosed cab with sharply protruding visor, 22-3/4" long, 1936	150	225	300
Ladder Truck	red with two yellow ladders, enclosed square cab with sharply protruding visor, 22-3/4" long, 1934	200	300	400
Ladder Truck	red with two yellow sectional ladders, enclosed square cab, 22-3/4" long, 1933	120	250	500
Ladder Truck	duo-tone slant design, white front, hood sides, fenders and two ladders, rest is red, square enclosed cab with sharply protruding visor, no headlights, 22-3/4" long, 1937	135	200	275
Police Squad Truck	yellow front and front fenders, dark blue-green body, yellow fire extinguisher, 21-1/2" long over ladders, 1947	150	300	550
Pumping Fire Engine	red with nickel stack on boiler, nickel rims on pump flywheels, nickel-rim headlights and searchlight, 23-1/2" long, 1929	3000	3500	4000
Rear Steer Trailer Fire Truck	red with two white 10-rung ladders, chrome one-piece grille, headlights and bumper, 20" long, 1952	65	100	200
Red Cross Ambulance	all white, removable fabric canopy with a red cross and "Ambulance" in red, 14-1/2" long, 1958	60	95	125
Suburban Pumper	red station wagon body, white plastic wraparound bumpers, one-piece grille and double headlights, 15" long, 1964	100	150	175
Texaco Fire Chief American	promotional piece, red rounded-front enclosed cab and body, white one-pice underbody, running boards and rear step, 25" long, 1962	200	300	400
LaFrance Pumper				
Trailer Ladder Truck	duo-tone slant design, tractor unit has yellow front, lower hood sides and lower doors, red hood top, enclosed cab and semi-trailer, nickel 10-rung ladders, 30" long with ladders, 1940	200	300	400
Trailer Ladder Truck	all red with cream removable rider saddle, three bright metal 10-rung ladders, 20" long over ladders, 1941	150	225	300
Water Tower Truck	red with nickel two-bar front bumper, red nickel-rim headlights plus searchllight on cowl, nickel latticework water tower, 45-1/2" long with tower down,	3000	4500	6000

Farm and Construction Equipment

NAME	DESCRIPTION	GOOD	EX	MINT
Aerial Tower Tramway	two tapering dark green 33-1/2" tall towers and 12" square bases, black hand crank, 1928	1000	1300	2700
Big Derrick	red mast and 20" boom, black base, 24" tall, 1921	600	900	1200
Brute Articulated Scooper	yellow front-loading scoop, cab, articulated frame and rear power unit, black radiator, exhaust, steering wheel and driver's seat, 5-1/2" long, 1970	50	75	100
Brute Double Dump Train	yellow hood, fenders and back on tractor unit, yellow coupled bottom-dumping earth carriers, 9-1/2" long, 1969	50	75	100

Buddy L

NAME	DESCRIPTION	GOOD	EX	MINT
Brute Dumping Scraper	yellow hood, fenders and back on two-wheel tractor unit, yellow scraper-dump unit, 7" long, 1970	50	75	100
Brute Farm Tractor-n-Cart	bright blue tractor body and rear fenders, green plastic radiator, engine, exhaust and driver's seat, bright blue detachable, square, two-wheel open cart, 6-1/4" long, 1969	30	40	55
Brute Road Grader	yellow hood, cab, frame and adjustable blade, black radiator, driver's seat and steering wheel, 6-1/2" long, 1970	50	75	100
Cement Mixer on Treads	medium gray with black treads and water tank, 16" tall, 1929-31	2500	3500	4500
Cement Mixer on Wheels	medium gray with black cast steel wheels and water tank, 14-1/2" tall, 1926-29	700	900	1500
Concrete Mixer	medium gray with black cast-steel wheels, black water tank, with wood-handle, steel-blade scoop shovel, 17-3/4" long with tow bar up, 1926	500	700	1000
Concrete Mixer	green with black cast-steel wheels, crank, gears, and band mixing drum, 10-1/2" long, 1930	175	265	350
Concrete Mixer	yellow/orange frame and base, red hopper and drum, black crank handle, 10-1/2" long, 1936	110	175	225
Concrete Mixer	red frame and base, cream/yellow hopper, drum and crank handle, 10-1/2" long, 1941	125	185	250
Concrete Mixer	green frame, base, crank, crank handle and bottom of mixing drum, gray hopper and top of drum, 9-5/8" long, 1949	100	150	200
Concrete Mixer with Motor Sound	green frame, base, crank, crank handle and bottom of mixing drum, gray hopper and top of drum, with sound when crank rotates drum, 9 5/8" long, 1950	75	115	150
Dandy Digger	yellow seat lower control lever and main boom, black underframe, skids, shovel and arm, 38-1/2" long with shovel arm extended, 1953	75	115	150
Dandy Digger	yellow seat, lower control lever and main boom, black underframe, skids, shovel, arm and control lever, 38-1/2" long with shovel arm extended, 1953	75	115	150
Dandy Digger	yellow main frame, operators, seat and boom, brown shovel, arm, under frame and twin skids, 27" long, 1941	95	145	195
Dandy Digger	yellow main frame, operators, seat and boom, green shovel, arm, under frame and twin skids, 27" long, 1936	85	130	175
Dandy Digger	red main frame, operators, seat and boom, black shovel, arm, under frame and twin skids, 27" long, 1931	100	160	215
Digger	red main frame, operators, seat and boom, black shovel, arm, lower frame and twin skids, curved connecting rod, boom tilts down for digging, 11-1/2" long with shovel arm extended, 1935	100	150	200
Dredge	red corrugated roof and base with four wide black wheels, red hubs, black boiler, floor, frame boom and clamshell bucket, 19" long, 1924	750	1000	1500
Giant Digger	red main frame, operators, seat and boom, black shovel, arm, lower frame and twin skids, boom tilts down for digging, 42" long with shovel arm extended,	275	415	550

Buddy L

NAME	DESCRIPTION	GOOD	EX	MINT
Giant Digger	red main frame, operators, seat and boom, black shovel, arm, lower frame and twin skids, curved connecting rod, boom tilts, 31" long with shovel arm extended, 1933	265	395	525
Giant Digger	yellow main frame, operators, seat and boom, green shovel, arm, lower frame and twin skids, boom tilts, 11-1/2" long with shovel arm extended, 1936	85	130	175
Giant Digger	yellow main frame, operators, seat and boom, brown shovel, arm, lower frame and twin skids, boom tilts, 11-1/2" long with shovel arm extended, 1941	75	115	155
Gradall	bright yellow truck and superstructure, black plastic bumper and radiator, 32" long with digging arm extended, 1965	350	750	1000
Hauling Rig with Construction	duo-tone slant design tractor, yellow bumper, lower hood and cab sides, white upper hood and cab, white trailer with yellow loading ramp, overall 38-1/2" long, 1953	175	250	400
Hauling Rig with Construction	yellow tractor unit, green semi-trailer, winch on front of trailer makes sound, 36-3/4" long, 1954	110	175	225
Hoisting Tower	dark green, hoist tower and three distribution chutes, 29" tall, 1928-31	1500	2000	2500
Husky Tractor	bright yellow body and large rear fenders, black engine block, exhaust, steering wheel and driver's seat, 13" long, 1966	50	75	100
Husky Tractor	bright blue body and large rear fenders, black engine block, exhaust, steering wheel and driver's seat, 13" long, 1969	40	60	80
Husky Tractor	bright yellow body, red large rear fenders and wheels, black engine block, exhaust, steering wheel and driver's seat, 13" long, 1970	30	45	65
Improved Steam Shovel	black with red roof and base, 14" tall, 1927-29	100	150	200
Junior Excavator	red shovel, arm, underframe, control lever and twin skids, yellow boom, rear lever, frame and seat, 28" long, 1945	75	115	150
Mechanical Crane	orange removable roof, boom and wheels in black cleated rubber crawler treads, olive green enclosed cab and base, hand crank with rat-tat motor noise, 20" tall, 1950	175	265	350
Mechanical Crane	orange removable roof, boom, yellow wheels in white rubber crawler treads, olive green enclosed cab and base, hand crank with rat-tat motor noise, 20"	150	225	300
Mobile Construction Derrick	orange laticework main mast, swiveling base, yellow latticework boom, green clamshell bucket and main platform base, 25-1/2" long with boom lowered,	150	250	350
Mobile Construction Derrick	orange laticework main mast, swiveling base, yellow latticework boom, gray clamshell bucket, orange main platform base, 25-1/2" long with boom	150	250	350
Mobile Construction Derrick	orange laticework main mast, swiveling base, yellow latticework boom, gray clamshell bucket, green main platform base, 25-1/2" long with boom lowered,	150	250	350
Mobile Power Digger Unit	clamshell dredge mounted on 10-wheel truck, orange truck, yellow dredge cab on swivel base, 31-3/4" long with boom lowered, 1955	125	185	250

VEHICLES

Buddy L

NAME	DESCRIPTION	GOOD	EX	MINT
Mobile Power Digger Unit	clamshell dredge mounted on six-wheel truck, orange truck, yellow dredge cab on swivel base, 31-3/4" long with boom lowered, 1956	100	150	250
Overhead Crane	black folding end frames and legs, braces, red crossbeams and platform, 45" long, 1924	1500	2000	3000
Pile Driver on Wheels	black with red roof and base, 22-1/2" tall, 1924-27	800	1800	2600
Polysteel Farm Tractor	orange molded plastic four-wheel tractor, silver radiator front, headlights, and motor parts, 12" long, 1961	75	115	150
Pull-n-Ride Horse-Drawn Farm	red four-wheel steel hopper-body wagon, detailed litho horse, 22-3/4" long, 1952	150	225	300
Road Roller	dark green with red roof and rollers, nickel plated steam cylinders, 20" long, 1929-31	3000	4000	5000
Ruff-n-Tuff Tractor	yellow grille, hood and frame, black plastic engine block and driver's seat, 10-1/2" long, 1971	50	75	100
Sand Loader	warm gray with 12 black buckets, 21" long, 18" high, 1924	150	200	450
Sand Loader	warm gray with 12 black buckets, chain-tension adjusting device at bottom of elevator side frames, 21" long, 1929	175	300	500
Sand Loader	yellow with 12 black buckets, chain-tension adjusting device at bottom of elevator side frames, 21" long, 1931	200	250	350
Scoop-n-Load Conveyor	cream body frame, green loading scoop, black circular crank operates black rubber cleated conveyor belt, 18" long, 1953	75	115	150
Scoop-n-Load Conveyor	cream body frame, green loading scoop, black circular crank operates black rubber cleated conveyor belt, "PORTABLE" decal in red, 18" long, 1954	65	100	135
Scoop-n-Load Conveyor	cream body frame, red loading scoop and chute, bright-plated circular crank operates black rubber cleated conveyor belt, "PORTABLE" decal in white, 18" long, 1955	60	95	125
Scoop-n-Load Conveyor	cream body frame, red loading scoop and chute, bright-plated circular crank operates black rubber cleated conveyor belt, "PORTABLE" decal in yellow, 18" long, 1956	55	85	115
Side Conveyor Load-n-Dump	yellow plastic front end including cab, yellow steel bumper, green frame and dump body, red conveyor frame with chute, 20-1/2" long, 1953	70	125	145
Side Conveyor Load-n-Dump	all steel yellow cab, hood, bumper and frame, white dump body and tailgate, red conveyor frame with chute, 21-1/4" long, 1954	65	100	135
Side Conveyor Load-n-Dump	all steel yellow cab, hood, bumper and frame, deep blue dump body, white tailgate, red conveyor frame with chute, 21-1/4" long, 1955	60	95	125
Small Derrick	red 20" movable boom and three angle-iron braces, black base and vertical mast, 21-1/2" tall, 1921	500	750	1000
Steam Shovel	black with red roof and base, 25-1/2" tall, 1921-22	125	250	500
Traveling Crane	red crane, carriage, and long cross beams, hand wheel rotates crane boom, 46" long, 1928	1275	1900	2550
Trench Digger	yellow main frame, base, and motor housing, red elevator and conveyor frame and track frames, 20" tall, 1928-31	2000	3500	5000

NAME	DESCRIPTION	GOOD	EX	MINT

Sets

NAME	DESCRIPTION	GOOD	EX	MINT
Army Combination Set	searchlight repair-it truck, transport truck and howitzer, ammunition conveyor, stake delivery truck, ammo, soldiers, 1956	300	400	500
Army Commando Set	14-1/2" truck, searchlight unit, two-wheel howitzer, soldiers, 1957	125	185	250
Big Brute 3-Piece Highway Set	bulldozer, dump truck, yellow four-wheel trailer, 1971	100	150	200
Big Brute 3-Piece Road Set	cement mixer truck, scooper, dump truck, 1971	125	185	250
Big Brute 4-Piece Freeway Set	scraper, grader, scooper and dump truck, 1971	125	185	250
Brute Fire Department Set	semi-trailer aerial ladder truck, fire pumper, fire wrecker, brute tow truck, 1970	75	115	150
Brute Five-Piece Highway Set	bulldozer, grader, scraper, dumping scraper and double dump train, 1970	50	75	100
Brute Fleet Set	car carrier with two plastic coupes, dump truck, pickup truck, cement mixer truck, tow truck, 1969	85	130	175
Delivery Set Combination	16-1/2" long wrigley express truck, 15" long sand and stone dump truck, 14-1/4" long freight conveyor and 14-1/4" long stake delivery truck, 1955	175	265	350
Family Camping Set	Camper/cruiser truck, 15-1/2" long maroon suburban wagon, and brown/light gray/beige folding teepee camping trailer, 1963	60	95	125
Family Camping Set	blue camping trailer and suburban wagon, blue camper-n-cruiser, 1964	50	75	100
Farm Combination Set	cattle transport stake truck with six plastic steers, hydraulic farm supplies trailer dump truck, trailer and three farm machines and farm machinery trailer hauler truck, 1956	100	150	200
Fire Department Set	aerial ladder fire engine, fire pumper with action hydrant that squirts water, two plastic hoses, two plastic firemen, fire chief's badge, 1960	250	375	500
Freight Conveyor and Stake Delivery Truck	blue frame 14-1/4" long conveyor, red, white and yellow body 14-3/4" long truck, 1955	125	185	250
GMC Air Defense Set	15" long, GMC army searchlight truck, 15" long, GMC signal corps truck, two four-wheel trailers, plastic soliders, 1957	300	500	700
GMC Brinks Bank Set	silver gray, barred windows on sides and in double doors, coin slot and hole in roof, brass padlock with two keys, pouch, play money, two gray plastic guard figures, 16" long, 1959	300	350	450
GMC Fire Department Set	red GMC extension ladder trailer and GMC pumper with ladders and hose reel, four-wheel red electric searchlight trailer, warning barrier, red plastic helmet, firemen, policeman, 1958	300	500	750
GMC Highway Maintenance Fleet	orange maintenance truck with trailer, sand & stone dump truck, scoop-n-load conveyor, sand hopper, steel scoop shovel, four white steel road barriers,	300	500	700
GMC Livestock Set	red fenders, hood, cab and frame, white flatbed cargo section , six sections of brown plastic rail fencing, five black plastic steers, 14-1/2" long, 1958	300	400	500

Buddy L

NAME	DESCRIPTION	GOOD	EX	MINT
GMC Western Roundup Set	blue fenders, hood, cab and frame, white flatbed cargo section, plastic six sections of rail fencing with swinging gate, rearing and standing horse, cowboys, calf, steer, 1959	300	400	500
Highway Construction Set	orange and black bulldozer and driver, truck with orange pickup body, orange dump truck, 1962	200	300	400
Highway Maintenance Mechanical Truck & Concrete Mixer	20" truck plus movable ramp, with duo-tone slant design, blue lower hood sides, yellow hood top and cab, 10-3/4" blue and yellow mixer, overall 36" long,	160	245	325
Interstate Highway Set	orange, parks department dumper, landscape truck, telephone truck, accessories include trees, drums, workmen and traffic cones, scoop shovel, 1959	250	400	500
Interstate Highway Set	orange, husky dumper, contractor's truck and ladder, utility truck, plastic pickaxe, spade, shovel, nail keg, 1960	250	350	500
Jr. Animal Farm Set	6-1/2" long Jr. Giraffe Truck, 6-1/4" long Jr. Kitty Kennel, 11-1/4" long Jr. Pony Trailer with Sportster, 1968	125	185	250
Jr. Fire Department	17" long Jr. hook-n-ladder aerial truck, 11-1/2" long Jr. fire snorkel, 6-3/4" long truck, all have 24-hole chrome grilles, 1968	125	185	250
Jr. Fire Department	17" long Jr. hook-n-ladder aerial truck, 11-1/2" long Jr. fire snorkel, 6-3/4" long truck, all have four-slot grilles and two square plastic headlights, 1969	125	185	250
Jr. Highway Set	yellow and black Jr. scooper tractor, yellow and white Jr. cement mixer truck, yellow Jr. dump truck, 1969	200	300	400
Jr. Sportsman Set	Jr. camper pickup with red cab and body and yellow camper, towing 6" plastic runabout on yellow two-wheel boat trailer, 1971	50	75	100
Loader, Dump Truck, and Shovel Set	conveyor, blue body sand and gravel dump truck, 8-3/4" long blue-enameled steel scoop shovel, 1955	85	130	175
Loader, Dump Truck, and Shovel Set	conveyor, green body sand and gravel dump truck, 8-3/4" long green-enameled steel scoop shovel, 1954	100	150	200
Mechanical Hauling Truck and	truck with duo-tone slant design, red-orange lower hood sides, dark green upper hood, cab, trailer and ramp, 9-5/8" green mixer, gray hopper, 38" long	160	245	325
Concrete Mixer	with ramps, 1950			
Mechanical Hauling Truck and Concrete Mixer	truck with duo-tone slant design, red/orange lower hood sides, dark green upper hood, cab, ramp, yellow trailer, 9-5/8" green mixer, gray hopper, 38" long with ramps, 1951	150	225	300
Polysteel Farm Set	blue milkman truck with rack and nine milk bottles, red and gray milk tanker, orange farm tractor, 1961	75	115	150
Road Builder Set	green/white cement mixer truck, yellow/black bulldozer, red dump truck, husky dumper, 1963	200	300	400
Truck with Concrete Mixer Trailer	22" truck with duo-tone slant design, green fenders and lower sides, yellow squarish cab and body, 10" mixer with yellow frame and red hopper, overall 34-1/2" long, 1937	175	265	350
Truck with Concrete Mixer Trailer	22" truck with duo-tone slant design, green front and lower hood sides, yellow upper hood, cab and body, 10" mixer with yellow frame and red hopper, overall 32-1/2" long, 1938	165	250	330

Buddy L

NAME	DESCRIPTION	GOOD	EX	MINT
Warehouse Set	Coca-Cola truck, two hand trucks, eight cases Coke bottles, store-door delivery truck, lumber, sign, two barrels, forklift, 1958	175	265	350
Warehouse Set	Coca-Cola truck, two hand trucks, eight cases Coke bottles, store-door delivery truck, sign, two barrels, forklift, 1959	150	225	300
Western Roundup Set	turquoise fenders, hood, cab and frame, white flatbed cargo section, six sections of rail fencing with swinging gate, rearing and standing horse, cowboys, calf, steer, 1960	175	250	400

Trucks

NAME	DESCRIPTION	GOOD	EX	MINT
Air Force Supply Transport	blue with blue removable fabric canopy, rubber wheels, decals on cab doors, 14-1/2" long, 1957	125	250	350
Air Mail Truck	black front, hood fenders, enclosed cab and opening doors, red enclosed body and chassis, 24" long, 1930	675	1000	1400
Allied Moving Van	tractor and semi-trailer van, duo-tone slant design, black front and lower sides, orange hood top, cab and van body, 29-1/2" long, 1941	600	900	1200
Army Electric Searchlight Unit	shiny olive drab flatbed truck, battery-operated searchlight, 14-3/4" long, 1957	125	225	325
Army Half-Track and Howitzer	olive drab with olive drab carriage, 12-1/2" truck, 9-3/4" gun, overall 22-1/2" long, 1953	100	150	200
Army Half-Track with Howitzer	olive drab steel, red firing knob on gun, 17" truck, 9-3/4" gun, overall 27" long, 1955	100	150	200
Army Medical Corps Truck	white, black rubber tires on white steel disc wheels, 29-1/2" long, 1941	125	185	250
Army Searchlight Repair-It Truck	shiny olive drab truck and flatbed cargo section, 15" long, 1956	125	175	225
Army Supply Truck	shiny olive drab truck and removable fabric cover, 14-1/2" long, 1956	100	150	175
Army Transport Truck and Trailer	olive drab truck, 20-1/2" long, trailer 34-1/2" long, 1940	250	350	450
Army Transport with Howitzer	olive drab steel, re-firing knob on gun, 17" truck, 9-3/4" gun, overall 27" long, 1954	115	175	230
Army Transport with Howitzer	olive drab, 12" truck, 9-3/4" gun, overall 28" long, 1953	100	150	200
Army Transport with Howitzer	olive drab steel, 17" truck, 9-3/4" gun, overall 27" long, 1955	150	250	350
Army Transport with Tank	olive drab, 15-1/2" long truck, 11-1/2" long detachable two-wheel trailer, overall 26-1/2" long, 7-1/2" long tank, 1959	100	150	200
Army Troop Transport with	dark forest green truck and gun, canopy mixture of greens, 14" long truck, 12" long, gun, overall 25-3/4" long, 1965	100	150	200
Army Truck	olive drab, 20-1/2" long, 1939	110	175	225
Army Truck	olive drab, 17" long, 1940	150	200	250
Atlas Van Lines	green tractor unit, chrome one-piece toothed grille and headlights, green lower half of semi-trailer van body, cream upper half, silvery roof, 29" long,	200	300	400
Auto Hauler	yellow cab-over-engine tractor unit and double-deck semi-trailer, three 8" long vehicles, 25-1/2" long, 1968	75	115	150
Auto Hauler	snub-nose medium blue tractor unit and double-deck semi-truck trailer, three plastic coupes, overall 27-1/2" long, 1970	65	95	130

Buddy L

NAME	DESCRIPTION	GOOD	EX	MINT
Baggage Rider	duo-tone horizontal design, green bumper, fenders and lower half of truck, white upper half, 28" long, 1950	250	175	500
Baggage Truck	green front, hood, and fenders, non-open doors, yellow cargo section slat or solid sides, metal grille, 26-1/2" long, 1935	350	500	650
Baggage Truck	green hood, fenders, and cab, yellow cargo section, no bumper, 17-1/2" long, 1945	175	265	350
Baggage Truck	black front, hood, and fenders, enclosed cab with opening doors, nickel-rim, red-shell headlights, yellow stake body, 26-1/2" long, 1930-32	3000	5000	7000
Baggage Truck	black front, hood, and fenders, doorless cab, yellow four-post stake sides, two chains across back, 26-1/2" long, 1927	800	1000	2000
Baggage Truck	duo-tone slant design, yellow skirted fenders and cargo section, green hood top, enclosed cab, 27-3/4" long, 1938	300	600	1000
Baggage Truck	duo-tone slant design, yellow fenders, green hood top, cab, and removable rider seat, 26-1/2" long, 1936	350	750	1000
Big Brute Dumper	yellow cab-over-engine, frame and tiltback dump section with cab shield, striped black and yellow bumper, black grille, 8" long, 1971	50	75	100
Big Brute Mixer Truck	yellow cab-over-engine, body and frame, white plastic mixing drum, white plastic seats, 7" long, 1971	35	50	70
Big Fella Hydraulic Rider Dumper	duo-tone slant design, yellow front and lower hood, red upper cab, dump body and upper hood, rider seat has large yellow sunburst-style decal, 26-1/2" long,	110	175	225
Big Mack Dumper	yellow front, hood cab, chassis and tiltback dump section, black plastic bumper, heavy-duty black balloon tires on yellow plastic five-spoke wheels, 20-1/2" long, 1971	60	90	120
Big Mack Dumper	off-white front, hood cab and chassis, blue-green tiltback dump section, white plastic bumper, 20-1/2" long, 1964	75	115	150
Big Mack Dumper	yellow front, hood cab, chassis and tiltback dump section, black plastic bumper, 20-1/2" long, 1967	70	100	140
Big Mack Dumper	yellow front, hood cab, chassis and tiltback dump section, black plastic bumper, single rear wheels, 20-1/2" long, 1968	65	95	130
Big Mack Hydraulic Dumper	red hood, cab and tiltback dump section with cab shield, dump body sides have a large circular back, white plastic bumper, short step ladder on each side, 20-1/2" long, 1970	40	60	80
Big Mack Hydraulic Dumper	red hood, cab and tiltback dump section with cab shield, white plastic bumper, short step ladder on each side, 20-1/2" long, 1968	50	75	100
Big Mack Hydraulic Dumper	white hood, cab and tiltback dump section with cab shield, white plastic bumper, short step ladder on each side, 20-1/2" long, 1969	45	65	90
Boat Transport	blue flatbed truck carrying 8" litho metal boat, boat deck white, hull red, truck 15" long, 1959	300	550	750
Borden's Milk Delivery Van	white upper cab-over-engine van body and sliding side doors, yellow lower body, metal-handle yellow plastic tray and six white milk bottle with yellow caps, 11-1/2" long, 1965	125	200	275
Brute Car Carrier	bright blue cab-over-engine tractor unit and detachable double-deck semi-trailer, two plastic cars, 10" long, 1969	60	95	125

Buddy L

NAME	DESCRIPTION	GOOD	EX	MINT
Brute Cement Mixer Truck	sand-beige cab-over-engine body and frame, white plastic mixing drum, white plastic seats, 5-1/4" long, 1968	35	55	75
Brute Cement Mixer Truck	blue cab-over-engine body and frame, white plastic mixing drum, white plastic seats, white-handled crank rotates drum, 5-1/4" long, 1969	30	45	65
Brute Dumper	red cab-over-engine body and cab shield on tiltback dump section, wide chrome wraparound bumper, 5" long, 1968	35	55	75
Brute Monkey House	yellow cab-over-engine body, red and white awning roof, cage on back, two plastic monkeys, 5" long, 1969	40	60	80
Brute Monkey House	yellow cab-over-engine body, striped orange and white awning roof, cage on back, two plastic monkeys, 5" long, 1968	50	75	100
Brute Sanitation Truck	lime green cab-over-engine and frame, white open-top body, wide chrome wraparound bumper, 5-1/4" long, 1969	50	75	100
Buddy L Milk Farms Truck	white body, black roof, short hood with black wooden headlights, 13-1/2" long, 1945	150	300	450
Buddy L Milk Farms Truck	light cream body, red roof, nickel glide headlights, sliding doors, 13" long, 1949	170	300	500
Camper	bright medium blue steel truck and camper body, 14-1/2" long, 1964	60	95	125
Camper	medium blue truck and back door, white camper body, 14-1/2" long, 1965	50	75	100
Camper-N-Cruiser	powder blue pickup truck and trailer, pale blue camper body, 24-1/2" long, 1963	60	95	125
Camper-N-Cruiser	bright medium blue camper with matching boat trailer and 8-1/2" long plastic sport cruiser, overall 27" long, 1964	50	75	100
Campers Truck	turquoise pickup truck, pale turquoise plastic camper, 14-1/2" long, 1961	55	85	110
Campers Truck with Boat	green/turquoise pickup truck, lime green camper body, red plastic runabout boat on camper roof, 14-1/2" long, 1962	50	100	150
Campers Truck with Boat	green/turquoise pickup, no side mirror, lime green camper body with red plastic runabout boat on top, 14-1/2" long, 1963	50	100	150
Camping Trailer and Wagon	bright medium blue suburban wagon, matching teepee trailer, overall 24-1/2" long, 1964	60	95	125
Cattle Transport Truck	red with yellow stake sides, 15" long, 1956	75	115	150
Cattle Transport Truck	green and white with white stake sides, 15" long, 1957	75	115	150
Cement Mixer Truck	turquoise body, tank ends, and chute, white side ladder, water tank, mixing drum and loading hopper, 16-1/2" long, 1964	60	95	125
Cement Mixer Truck	red body, tank ends, and chute, white side ladder, water tank, mixing drum and loading hopper, 15-1/2" long, 1965	75	115	150
Cement Mixer Truck	red body, tank ends, and chute, white water tank, mixing drum and loading hopper, black wall tires, 15-1/2" long, 1967	60	95	125
Cement Mixer Truck	red body, tank ends, and chute, white water tank, mixing drum and loading hopper, whitewall tires, 15-1/2" long, 1968	50	75	100
Cement Mixer Truck	snub-nosed yellow body, cab, frame and chute, white plastic mixing drum, loading hopper and water tank with yellow ends, 16" long, 1970	35	50	70

Buddy L

NAME	DESCRIPTION	GOOD	EX	MINT
Charles Chip Delivery Truck Van	tan/beige body, decal on sides has brown irregular center resembling a large potato chip, 1966	125	200	275
City Baggage Dray	duo-tone slant design, green front and fenders, yellow hood top and cargo section, 19" long, 1936	150	300	400
City Baggage Dray	green front, hood, and fenders, non-open doors, yellow stake-side cargo section, 19" long, 1934	150	300	400
City Baggage Dray	duo-tone slant design, green front and fenders, yellow hood top and cargo section, dummy headlights, 19" long, 1937	150	300	400
City Baggage Dray	duo-tone slant design, green front and skirted fenders, yellow hood top, enclosed cab and cargo section, 20-3/4" long, 1938	150	200	350
City Baggage Dray	light green with aluminum-finish grille, no bumper, black rubber wheels, 20-3/4" long, 1939	95	250	300
City Baggage Dray	cream with aluminum-finish grille, no bumper, black rubber wheels, 20-3/4" long, 1940	100	200	300
City Baggage Dray	green front, hood, and fenders, non-open doors, yellow stake-side cargo section, bright metal grille, 19" long, 1935	150	300	400
Coal Truck	black hopper body and fully enclosed cab with opening doors, red wheels, 25" long, 1930	900	1500	2000
Coal Truck	black front, hood, fenders, doorless cab, red chassis and disc wheels, 25" long, 1926	700	1000	2000
Coal Truck	black front, hood, fenders, sliding discharge door on each side of hopper body, red chassis and disc wheels, 25" long, 1927	800	1000	2000
Coca-Cola Bottling Route Truck	bright yellow, with two small metal hand trucks and eight yellow cases of miniature green Coke bottles, 14-3/4" long, 1957	110	175	225
Coca-Cola Bottling Route Truck	bright yellow, with small metal hand truck, six or eight yellow cases of miniature green Coke bottles, 14-3/4" long, 1955	125	175	250
Coca-Cola Delivery Truck	red lowercab-over-engine and van body, white upper cab, left side of van lifts to reveal 10 miniature bottle cases, 9-1/2" long, 1971	25	40	55
Coca-Cola Delivery Truck	orange/yellow cab and double-deck, open-side cargo, two small hand trucks, four red and four green cases of bottles, 15" long, 1964	60	95	125
Coca-Cola Delivery Truck	orange/yellow cab and double-deck, open-side cargo, two small hand trucks, four red and four green cases of bottles, 15" long, 1960	100	150	200
Coca-Cola Delivery Truck	orange/yellow cab and double-deck, open-side cargo, two small hand trucks, four red and four green cases of bottles, 15" long, 1963	75	100	150
Coke Coffee Co. Delivery Truck Van	black lower half of body, orange upper half, roof and sliding side doors, 1966	85	130	175
Colt Vacationer	blue/white Colt sportsliner with trailer carrying 8-1/2" long red/white plastic sport cruiser, overall 22-1/2" long, 1967	60	95	125
Curtiss Candy Trailer Van	blue tractor and bumper, white semi-trailer van, blue roof, chrome one-piece toothed grille and headlights, white drop-down rear door, 32-3/4" long with tailgate/ramp lowered, 1955	250	400	500
Dairy Transport Truck	duo-tone slant design, red front and lower hood sides, white hood top, cab and semi-trailer tank body, tank opens in back, 26" long, 1939	150	225	300
Deluxe Auto Carrier	turquoise tractor unit, aluminum loading ramps, three plastic cars, overall 34" long including, 1962	100	175	250

Buddy L

NAME	DESCRIPTION	GOOD	EX	MINT
Deluxe Camping Outfit	turquoise pickup truck and camper, and 8-1/2" long plastic boat on pale turquoise boat trailer, overall 24" long, 1961	60	95	125
Deluxe Hydraulic Rider Dump Truck	duo-tone slant design, red front and lower hood sides, white upper cab, dump body and chassis, red or black removable rider saddle, 26" long, 1948	175	265	350
Deluxe Motor Market	duo-tone slant design, red front, curved bumper, lower hood and cab sides, white hood top, body and cab, 22-1/4" long, 1950	250	350	500
Deluxe Rider Delivery Truck	duo-tone horizontal design, deep blue lower half, gray upper half, red rubber disc wheels, black barrel skid, 22-3/4" long, 1945	135	200	270
Deluxe Rider Delivery Truck	duo-tone horizontal design, gray lower half, blue upper half, red rubber disc wheels, black barrel skid, 22-3/4" long, 1945	135	200	270
Deluxe Rider Dump Truck	various colors, dual rear wheels, no bumper, 25-1/2" long, 1945	75	115	150
Double Hydraulic	green front loading scoop with yellow arms attached to cab sides, yellow hood and enclosed cab, orange frame and wide dump body, 29" long with scoop lowered, 1956	85	130	175
Double Tandem Hydraulic Dump and Trailer	truck has red bumper, hood, cab and frame, four-wheel trailer with red tow and frame, both with white tiltback dump bodies, 38" long, 1957	85	130	175
Double-Deck Boat Transport	light blue steel flatbed truck carrying three 8" white plastic boats with red decks, truck 15" long, 1960	150	250	400
Dr. Pepper Delivery Truck Van	red, white, and blue, 1966	85	130	175
Dump Body Truck	black front, hood, open driver's seat and dump section, red chassis, chain drive dump mechanism, 25" long, 1923	1200	1800	2500
Dump Body Truck	black front, hood, open driver's seat and dump section, red chassis, crank windlass with ratchet raises dump bed, 25" iong, 1921	800	1400	2000
Dump Truck	duo-tone slant design, red lower cab, lower hood, front and dump body, yellow upper hood, upper cab and chassis, no bumper, 22-1/4" long, 1939	250	375	500
Dump Truck	various colors, black rubber wheels, 12" long, 1945	50	75	100
Dump Truck	duo-tone slant design, red front, fenders and dump body, yellow hood top, upper sides, upper cab and chassis, no bumper, 22-1/2" long, 1948	125	185	250
Dump Truck	red hood top and cab, white or cream dump body and frame, no bumper, 17-1/2" long, 1945	75	115	150
Dump Truck	white upper hood, enclosed cab, wide-skirt fenders and open-frame chassis, orange dump body, bright-metal grille, no bumper, 17-3/8" long, 1941	85	130	175
Dump Truck	green with cream hood top and upper enclosed cab, no bumper, bright-metal headlights and grille, 22-1/4" long, 1941	85	130	175
Dump Truck	duo-tone slant design, red front, fenders and lower doors, white upper, bright radiator grille and headlights, no bumper, 22-1/4" long, 1939	250	375	500

Buddy L

NAME	DESCRIPTION	GOOD	EX	MINT
Dump Truck	duo-tone slant design, yellow enclosed cab and hood, red front and dump body, no bumper, bright-metal headlights, 20" long, 1936	250	375	500
Dump Truck	yellow enclosed cab, front and hood, red dump section, no bumper, bright-metal radiator, 20" long, 1935	325	485	650
Dump Truck	yellow enclosed cab, front and hood, red dump section, no bumper, 20" long, 1934	275	415	550
Dump Truck	black enclosed cab and opening doors, front and hood, red dump body and chassis, simple lever arrangement lifts dump bed, 24" long, 1930	650	975	1300
Dump Truck	yellow upper hood and enclosed cab, red wide-skirt fenders and open-frame chassis, blue dump body, no bumper, 17-1/4" long, 1940	85	130	175
Dump Truck	black enclosed cab and opening doors, front and hood, red dump body and chassis, crank handle lifts dump bed, 24" long, 1931	750	1125	1500
Dump Truck	duo-tone slant design, yellow enclosed cab and hood, red front and dump body, no bumper, dummy headlights, 20" long, 1937	250	375	500
Dump Truck-Economy Line	dark blue dump body, remainder is yellow, bright-metal grille and headlights, no bumper or running boards, 12" long, 1941	75	115	150
Dump-n-Dozer	orange husky dumper truck and orange flatbed four-wheel trailer carrying orange bulldozer, 23" long including trailer, 1962	75	115	150

Hydraulic Dump Truck, Buddy L

NAME	DESCRIPTION	GOOD	EX	MINT
Dumper with Shovel	medium green body, frame and dump section, white one-piece bumper and grille guard, no side mirror, large white steel scoop shovel, spring suspension on front axle only, 15" long, 1964	75	115	150
Dumper with Shovel	orange body, frame and dump section, chrome one-piece grille, no bumper guard, no side mirror, large white steel scoop shovel, no spring suspension, 15-3/4" long, 1965	75	115	150
Dumper with Shovel	medium green body, frame and dump section, white one-piece bumper and grille guard, no side mirror, large white steel scoop shovel, 15" long, 1963	75	115	150
Dumper with Shovel	turquoise body, frame and dump section, white one-piece bumper and grille guard, large white steel scoop shovel, 15" long, 1962	75	115	150
Express Trailer Truck	red tractor unit, hood, fenders and enclosed cab, green semi-trailer van with removable roof and drop-down rear door, bright-metal dummy headlights, 23-3/4" long, 1934	350	525	700
Express Trailer Truck	red tractor unit, hood, fenders and enclosed cab, green semi-trailer van with removable roof and drop-down rear door, 23-3/4" long, 1933	350	525	700
Express Truck	all black except red frame, enclosed cab with opening doors, nickel-rim, red-shell headlights, six rubber tires, double bar front bumper, 24-1/2" long,	3000	4500	6000
Farm Machinery Hauler Trailer	blue tractor unit, yellow flatbed semi-trailer, 31-1/2" long, 1956	125	185	250
Farm Supplies Automatic Dump	duo-tone slant design, blue curved bumper, front, lower hood sides and cab, yellow upper hood, cab and rest of body, 22-1/2" long, 1950	125	185	250
Farm Supplies Dump Truck	duo-tone slant design, red front, fenders and lower hood sides, yellow upper hood, cab and body, 22-3/4" long, 1949	125	185	250
Farm Supplies Hydraulic Dump	green tractor unit, long cream body on semi-trailer, 14 rubber wheels, 26-1/2" long, 1956	100	150	200
Fast Delivery Pickup	yellow hood and cab, red open cargo body, removable chain across open back, 13-1/2" long, 1949	100	150	200
Finger-Tip Steering Hydraulic	powder blue bumper, fenders, hood, cab and frame, white tiltback dump body, 22" long, 1959	75	115	150
Fisherman	pale blue/green station wagon with four-wheel boat trailer carrying plastic 8-1/2" long boat, overall 27-1/2" long, 1963	75	115	150
Fisherman	metallic sage green pickup truck with boat trailer carrying plastic 8-1/2" long sport cruiser, overall 25" long, 1964	70	100	140
Fisherman	sage gray/green and white pickup truck with steel trailer carrying plastic 8-1/2" long sport cruiser, overall 25" long, 1965	65	95	130
Fisherman	light tan pickup truck with tan steel trailer carrying plastic 8-1/2" long sport crusier, overall 24-1/4" long, 1962	80	120	160
Flivver Dump Truck	black with red eight-spoke wheels, black hubs, aluminum tires, flat, open dump section with squared-off back with latching, drop-down endgate, 11" long,	700	1000	1500
Flivver Scoop Dump Truck	black with red eight-spoke wheels, 12-1/2" long, 1926-27, 1929-30	1500	2500	3500

Buddy L

NAME	DESCRIPTION	GOOD	EX	MINT
Flivver Truck	black with red eight-spoke wheels with aluminum tires, black hubs, 12" long, 1924	800	1000	1500
Ford Flivver Dump Cart	black with red eight-spoke wheels, black hubs, aluminum tires, flat, short open dump section tapers to point on each side, 12-1/2" long, 1926	1500	2500	3500
Frederick & Nelson Delivery Truck Van	medium green body, roof and sliding side doors, 1966	125	200	275
Freight Delivery Stake Truck	red hood, bumper, cab and frame, white cargo section, yellow three-post, three-slat removable stake sides, 14-3/4" long, 1955	75	125	150
Front Loader Hi-Lift Dump Truck	red scoop and arms attached to white truck at rear fenders, green dump body, 17-3/4" long with scoop down and dump body raised, 1955	85	130	175
Giant Hydraulic Dumper	red bumper, frame, hood and cab, light tan tiltback dump body and cab shield 23-3/4" long, 1960	125	185	250
Giant Hydraulic Dumper	overall color turquoise, dump lever has a red plastic tip, 22-3/4" long, 1961	135	200	275
Giraffe Truck	powder blue hood, white cab roof, high-sided open-top cargo section, two orange/yellow plastic giraffes, 13-1/4" long,1968	60	95	125
GMC Air Force Electric Searchlight Unit	all blue flatbed, off white battery-operated searchlight swivel mount, decals on cab doors, 14-3/4" long, 1958	200	300	400
GMC Airway Express Van	green hood, cab and van body, latching double rear doors, shiny metal drum coin bank and metal hand truck, 17-1/2" long with rear doors open, 1957	250	350	450
GMC Anti-Aircraft Unit with Searchlight	15" truck with four-wheel trailer, battery-operated, over 25-1/4" long, 1957	250	350	450
GMC Army Hauler with Jeep	shiny olive drab tractor unit and flatbed trailer, 10" long jeep, overall 31-1/2" long, 1958	200	300	400
GMC Army Transport with Howitzer	shiny olive drab, 14-1/2" long, truck, overall with gun 22-1/2" long, 1957	200	300	400
GMC Brinks Armored Truck Van	silver gray, barred windows on sides and in double doors, coin slot and hole in roof, brass padlock with two keys, pouch, play money, three gray plastic guard figures, 16" long, 1958	300	350	450
GMC Coca-Cola Route Truck	lime/yellow, with small metal hand truck and eight cases of miniature green Coke bottles, 14-1/8" long, 1957	200	300	400
GMC Coca-Cola Route Truck	orange/yellow, with two small metal hand trucks and eight cases of miniature green Coke bottles, 14-1/8" long, 1958	200	300	400
GMC Construction Company	pastel blue including control lever on left and dump section with cab shield, hinged tailgate, chrome GMC bar grille, six wheels, 16" long, 1958	200	300	400
GMC Construction Company	pastel blue including control lever on left and dump section with cab shield, hinged tailgate, chrome GMC bar grille, four wheels, 16" long, 1959	150	250	350
GMC Highway Giant Trailer	blue tractor, blue and white van, chrome GMC bar grille and headlights, blue roof on semi-trailer, white tailgate doubles as loading ramp, 18-wheeler, 31-1/4" long, 1957	250	350	450
GMC Highway Giant Trailer Truck	blue tractor, blue and white van, chrome GMC bar grille and headlights, blue roof on semi-trailer, white tailgate doubles as loading ramp, 14-wheeler, 30-3/4" long, 1958	200	300	400

NAME	DESCRIPTION	GOOD	EX	MINT
GMC Husky Dumper	red hood, bumper, cab and chassis, chrome GMC bar grille and nose emblem, white oversize dump body, red control lever on right side, 17-1/2" long, 1957	150	250	350
GMC Self-Loading Auto Carrier	yellow tractor and double-deck semi trailer, three plastic cars, overall 33-1/4" long, 1959	200	300	400
GMC Signal Corps Unit	both olive drab, 14-1/4" long truck with removable fabric canopy, 8" long four-wheel trailer, 1957	150	200	250
Grocery Motor Market Truck	duo-tone slant design, yellow front, lower hood sides, fenders and lower doors, white hood top, enclosed cab and body, no bumper, 20-1/2" long, 1937	275	415	550
Grocery Motor Market Truck	duo-tone slant design, yellow front, lower hood sides, skirted fenders and lower doors, white hood top, cab and body, no bumper, 21-1/2" long, 1938	275	415	550
Heavy Hauling Dumper	red hood, bumper, cab and frame, cream tiltback dump body, 20-1/2" long, 1955	75	125	150
Heavy Hauling Dumper	red hood, bumper, cab and frame, cream oversize dump body, hinged tailgate, 21-1/2" long, 1956	70	125	140
Heavy Hauling Hydraulic Dumper	green hood, cab and frame, cream tiltback dump body, and cab shield, raising dump body almost to vertical, 23" long, 1956	70	105	140
Hertz Auto Hauler	bright yellow tractor and double-deck semi-trailer, three plastic vehicles, 27" long, 1965	100	150	200
Hi-Lift Farm Supplies Dump	all steel, red front end including hood and enclosed cab, yellow dump body, cab shield and hinged tailgate, 23-1/2" long, 1954	100	175	225
Hi-Lift Farm Supplies Dump	red plastic front end including hood and enclosed cab, yellow dump body, cab shield and hinged tailgate, 21-1/2" long, 1953	100	175	225
Hi-Lift Scoop-n-Dump Truck	blue hood, fenders and cab, yellow front loading scoop and arms attached to fenders, white frame, dump body, cab shield, and running boards, 17-3/4" long,	65	100	135
Hi-Lift Scoop-n-Dump Truck	orange hood, fenders and cab, yellow front loading scoop and arms attached to fenders, white frame, dump body and cab shield, 17-3/4" long, 1956	70	125	145
Hi-Lift Scoop-n-Dump Truck	orange truck with deeply fluted sides, dark green scoop on front rises to empty load into deep hi-lift slightly orange dump body, 16" long, 1955	75	115	155
Hi-Lift Scoop-n-Dump Truck	orange truck with deeply fluted sides, dark green scoop on front rises to empty load into hi-lift cream/yellow dump body, 16" long, 1952	85	130	175
Hi-Lift Scoop-n-Dump Truck	orange truck with deeply fluted sides, dark green scoop on front rises to empty load into hi-lift light cream dump body, 16" long, 1953	80	125	165
Hi-Tip Hydraulic Dumper	orange hood, cab and frame, cream tiltback dump body, and cab shield, raising dump body almost to vertical, 23" long, 1957	75	115	150
Highway Hawk Trailer Van	bronze cab tractor, chrome metallized plastic bumper, grille, air cleaner and exhaust, 19-3/4" long, 1985	50	75	100
Highway Maintenance Truck with Trailer	orange with black rack of four simulated floodlights behind cab, 19-1/2" long including small two-wheel trailer, 1957	100	150	200
Husky Dumper	bright yellow, chrome one-piece bumper, slotted rectangular grille and double headlights, 14-1/2" long, 1966,	75	115	150

Buddy L

NAME	DESCRIPTION	GOOD	EX	MINT
Husky Dumper	snub-nose red body, tiltback dump section snda cab shield, full-width chrome bumperless grille, white-tipped dump-control lever on left, deep-tread whitewall tires, 14-1/2" long, 1971	40	60	80
Husky Dumper	snub-nose red body, tiltback dump section, cab shield, full-width chrome bumperless grille, deep-tread whitewall tires, 14-1/2" long, 1970	45	70	90
Husky Dumper	white plastic wraparound bumper, tan body, frame and dump section, hinged tailgate, plated dump lever on left side, 15-1/4" long, 1961	70	125	140
Husky Dumper	orange wraparound bumper, body, frame and dump section, hinged tailgate, plated dump lever on left side, 15-1/4" long, 1960	75	115	150
Husky Dumper	red hood, cab, chassis and dump section, chrome one-piece bumper and slotted grille with double headlights, 14-1/2" long, 1968	60	95	125
Husky Dumper	yellow hood, cab, fram and tiltback dump section with cab shield, crome one-piece wraparound bumper, 14-1/2" long, 1969	50	75	100
Hydraulic Auto Hauler with Four GMC Cars	powder blue GMC tractor, 7" long plastic cars, overall 33-1/2" long including loading ramp, 1958	250	350	450
Hydraulic Construction Dumper	red front, cab and chassis, large green dump section with cab shield, 15-1/4" long, 1962	65	100	135
Hydraulic Construction Dumper	tan/beige front, cab and chassis, large green dump section with cab shield, 15-1/4" long, 1963	60	95	125
Hydraulic Construction Dumper	bright blue front, cab and chassis, large green dump section with cab shield, 15-1/2" long, 1964	50	75	100
Hydraulic Construction Dumper	bright green front, cab and chassis, large green dump section with cab shield, 14" long, 1965	50	75	100
Hydraulic Construction Dumper	medium blue front, cab and chassis, large green dump section with cab shield, 15-1/4" long, 1967	50	75	100
Hydraulic Dump Truck	black front, hood, fenders, open seat, and dump body, red chassis and disc wheels with aluminum tires, 25" long, 1926	500	800	1000
Hydraulic Dump Truck	duo-tone slant design, red hood sides, dump body and chassis, white upper hood, cab and removable rider seat, electric headlights, 24-3/4" long, 1936	300	500	700
Hydraulic Dump Truck	black front, hood, fenders, dark reddish maroon dump body, red chassis and disc wheels with seven embossed spokes, black hubs, 25" long, 1931	675	900	1300
Hydraulic Dump Truck	black front, hood, fenders and enclosed cab, red dump body, chassis and wheels with six embossed spokes, bright hubs, 24-3/4" long, 1933	325	485	650
Hydraulic Dumper	green, plated dump lever on left side, large hooks on left side hold yellow or off-white steel scoop shovel, white plastic side mirro and grille guard, 17"	125	185	250
Hydraulic Dumper with Shovel	green, plated dump lever on left side, large hooks on left side hold yellow or off-white steel scoop shovel, 17" long, 1960	125	185	250
Hydraulic Hi-Lift Dumper	duo-tone slant design, green hood nose and lower cab sides, remainder white with chrome grille, enclosed cab, 24" long, 1953	75	115	150

Buddy L

NAME	DESCRIPTION	GOOD	EX	MINT
Hydraulic Hi-Lift Dumper	green hood, fenders, cab, and dump-body supports, white dump body with cab shield, 22-1/2" long, 1954	85	130	175
Hydraulic Hi-Lift Dumper	blue hood, fenders, cab, and dump-body supports, white dump body with cab shield, 22-1/2" long, 1955	75	115	150
Hydraulic Highway Dumper	orange with row of black square across scraper edges, one-piece chrome eight-hole grille and double headlights, no scraper blade, 17-3/4" long over blade and raised dump body, 1959	50	75	100
Hydraulic Highway Dumper with Scraper Blade	orange with row of black square across scraper edges, one-piece chrome eight-hole grille and double headlights, 17-3/4" long over blade and raised dump body, 1958	75	115	150
Hydraulic Husky Dumper	red body, frame, dump section and cab shield, 15-1/4" long, 1962	65	100	135
Hydraulic Husky Dumper	red body, white one-piece bumper and grille guard, heavy side braces on dump section, 14" long, 1963	50	75	100
Hydraulic Rider Dumper	duo-tone slant design, yellow front and lower hood, red upper cab, dump body and upper hood, 26-1/2" long, 1949	175	265	350
Hydraulic Sturdy Dumper	lime green hood, cab, frame and tiltback dump section, green lever on left side controls hydraulic dumping, 14-1/2" long, 1969	50	75	100
Hydraulic Sturdy Dumper	yellow hood, cab, fram and tiltback dump section, green lever on left side controls hydraulic dumping, 14-1/2" long, 1969	50	75	100
Hydraulic Sturdy Dumper	snub-nose green/yellow body, cab and tiltback dump section, white plastic seats, 14-1/2" long, 1970	45	60	80
Ice Truck	black front, hood, fenders and doorless cab, yellow open cargo section, canvas sliding cover, 26-1/2" long, 1926	700	900	1600
Ice Truck	black front, hood, fenders and enclosed cab, yellow open cargo section, canvas, ice cakes, miniature tongs, 26-1/2" long, 1930	700	900	1500
Ice Truck	black front, hood, fenders and enclosed cab, yellow ice compartment, canvas, ice cakes, tongs, 26-1/2" long, 1933-34	700	900	1300
Ice Truck	black front, hood, fenders and enclosed cab, yellow ice compartment, 26-1/2" long, 1933	700	900	1500
IHC "Red Baby" Express Truck	red with black hubs and aluminum tires, 24-1/4" long, 1929	800	1300	1900
IHC "Red Baby" Express Truck	red doorless roofed cab, open pickup body, chassis and fenders, 24-1/4" long, 1928	750	1000	2000
Insurance Patrol	red with open driver's seat and body, brass bell on cowl and full-length handrails, 27" long, 1925	650	1000	1300
Insurance Patrol	red with open driver's seat and body, brass bell on cowl and full-length handrails, no CFD decal, 27" long, 1928	625	950	1250
International Delivery Truck	red with removable black rider saddle, black-edged yellow horizontal strip on cargo body, 24-1/2" long, 1935	225	350	450
International Delivery Truck	duo-tone slant design, red front, bumper and lower hood sides, yellow hood top, upper sides, cab and open cargo body, 24-1/2" long, 1936	200	300	400

NAME	DESCRIPTION	GOOD	EX	MINT
International Delivery Truck	duo-tone slant design, red front, bumper and lower hood sides, yellow hood top, upper sides, cab and open cargo body, bright metal dummy headlights, 24-1/2" long, 1938	150	225	300
International Dump Truck	red with bright-metal radiator grille, and black removable rider saddle, 25-3/4" long, 1935	325	485	650
International Dump Truck	duo-tone slant design, yellow radiator, fenders, lower hood and detachable rider seat, rest of truck is red, 25-3/4" long, 1936	315	475	630
International Dump Truck	red, with red headlights on radiator, black removable rider saddle, 25-3/4" long, 1938	125	185	250
International Railway Express Truck	duo-tone slant design, yellow front, lower hood sides and removable top, green hood top, enclosed cab and van body, electric headlights, 25" long, 1937	350	525	700
International Railway Express Truck	duo-tone slant design, yellow front, lower hood sides and removable top, green hood top, enclosed cab and van body, dummy headlights, 25" long, 1938	345	525	690
International Wrecker Truck	duo-tone slant design, yellow upper cab, hood, and boom, red lower cab, fenders, grille and body, rubber tires, removable rider seat, 32" long, 1938	600	900	1800
Jewel Home Service Truck Van	dark brown body and sliding side doors, 1967	125	200	275
Jewel Home Shopping Truck Van	pale mint green upper body and roof, darker mint green lower half, no sliding doors, 1968	125	200	275
Jolly Joe Ice Cream Truck	white with black roof, black tires and wooden wheels, 17-1/2" long, 1947	225	350	450
Jolly Joe Popsicle Truck	white with black roof, black tires and wooden wheels, 17-1/2" long, 1948	275	425	550
Jr. Animal Ark	fuschia lapstrake hull, four black tires, 10 pairs of plastic animals, 5" long, 1970	40	60	80
Jr. Auto Carrier	yellow cab-over-engine tractor unit and double-deck semi-trailer, two red plastic cars, 15-1/2" long, 1967	50	75	100
Jr. Auto Carrier	bright blue cab-over-engine tractor unit and double-deck semi-trailer, two plastic cars, 17-1/4" long, 1969	60	95	125
Jr. Beach Buggy	yellow hood, fenders and topless jeep body, red plastic seats, white plastic surfboard that clips to roll bar and windshield, truck 6" long, 1969	45	65	90
Jr. Beach Buggy	lime green hood, fenders and topless jeep body, red plastic seats, lime green plastic surfboard that clips to roll bar and windshield, truck 6" long, 1971	35	50	70
Jr. Buggy Hauler	fuschia jeep body with orange seats, orange two-wheel trailer tilts to unload sandpiper beach buggy, 12" long including jeep and trailer, 1970	35	55	75
Jr. Camper	red cab and pickup body wih yellow camper body, 7" long, 1971	50	75	100
Jr. Canada Dry Delivery Truck	green/lime cab-over-engine body, hand truck, 10 cases of green bottles, 9-1/2" long, 1968	100	150	200
Jr. Canada Dry Delivery Truck	green/lime cab-over-engine body, hand truck, 10 cases of green bottles, 9-1/2" long, 1969	85	130	170
Jr. Cement Mixer Truck	blue cab-over-engine body, frame and hopper, white plastic mixing drum, white plastic seats, 7-1/2" long, 1968	50	75	100

Buddy L

NAME	DESCRIPTION	GOOD	EX	MINT
Jr. Cement Mixer Truck	blue cab-over-engine body, frame and hopper, white plastic mixing drum, white plastic seats, wide one-piece chrome bumper, 7-1/2" long, 1969	35	50	70
Jr. Dump Truck	red cab-over-engine, frame and tiltback dump section, plastic vertical headlights, 7-1/2" long, 1967	50	75	100
Jr. Dumper	avocado cab-over-engine, frame and tiltback dump section with cab shield, one-piece chrome bumper and four-slot grille, 7-1/2" long, 1969	335	55	75
Jr. Giraffe Truck	turquoise cab-over-engine body, white cab roof, plastic giraffe, 6-1/2" long, 1968	50	75	100
Jr. Giraffe Truck	turquoise cab-over-engine body, white cab roof, plastic giraffe, 6-1/4" long, 1969	40	60	80
Jr. Kitty Kennel	pink cab-over-engine body, white cab roof, four white plastic cats, 6-1/4" long, 1969	55	85	115
Jr. Kitty Kennel	pink cab-over-engine body, white cab roof, four colored plastic cats, 6-1/4" long, 1968	60	95	125
Jr. Sanitation Truck	blue cab-over-engine, white frame, refuse body and loading hopper, 10" long, 1968	75	115	150
Jr. Sanitation Truck	yellow cab-over-engine and underframe, refuse body and loading hopper, full width bumper and grille, 10" long, 1969	75	115	150
Junior Line Air Mail Truck	black enclosed cab, red chassis and body, headlights and double bar bumper, six rubber tires, 24" long, 1930-32	400	650	1200
Kennel Truck	medium blue pickup body and cab, clear plastic 12-section kennel with 12 plastic dogs fits in cargo box, 13-1/2" long, 1964	60	90	120
Kennel Truck	turquoise pickup body and cab, clear plastic 12-section kennel with 12 plastic dogs fits in cargo box, 13-1/2" long, 1965	60	95	125
Kennel Truck	bright blue pickup body and cab, clear plastic 12-section kennel with 12 plastic dogs fits in cargo box, 13-1/4" long, 1966	95	145	190
Kennel Truck	bright blue pickup body and cab, clear plastic 12-section kennel with 12 plastic dogs fits in cargo box, 13-1/4" long, 1967	85	130	175
Kennel Truck	cream/yellow pickup body and cab, clear plastic 12-section kennel with 12 plastic dogs fits in cargo box, 13-1/4" long, 1968	80	120	160
Kennel Truck	red/orange pickup body and cab, yellow roof, six section kennel with six plastic dogs fits in cargo box, 13-1/4" long, 1969	65	100	135
Kennel Truck	snub-nosed red/orange body and cab, plastic kennel section in back, six kennels with six plastic dogs, 13-1/4" long, 1970	60	95	125
Lumber Truck	black front, hood, fenders, cabless open seat and low-sides cargo bed, red bumper, chassis and a pair of removable solid stake sides, load of lumber pieces, 24" long, 1924	750	1000	1500
Lumber Truck	black front, hood, fenders, doorless cab and low-sides cargo bed, red bumper, chassis and a pair of removable solid stake sides, load of lumber, 25-1/2" long,	650	1300	2700
Mack Hydraulic Dumper	red front, hood, cab, chassis and tiltback dump section with cab shield, white plastic bumper, short step ladder on each side, 20-1/2" long, 1967	50	75	100
Mack Hydraulic Dumper	red front, hood, cab, chassis and tiltback dump section with cab shield, white plastic bumper, 20-1/2" long, 1965	60	95	125

Buddy L

NAME	DESCRIPTION	GOOD	EX	MINT
Mack Quarry Dumper	orange front, hood cab and chassis, blue-green tiltback dump section, white plastic bumper, 20-1/2" long, 1965	75	115	150
Mammoth Hydraulic Quarry Dumper	deep green hood, cab and chassis, red tiltback dump section, black plastic bumper, 23" long, 1962	50	100	175
Mammoth Hydraulic Quarry Dumper	deep green hood, red cab, chassis and tiltback dump section, black plastic bumper, 22-1/2" long, 1963	60	90	125
Marshall Field's Delivery Truck Van	hunter green body, sliding doors and roof, 1966	125	200	275
Milkman Truck	medium blue hood, cab and flatbed body, white side rails, eight 3" white plastic milk bottles, 14-1/4" long, 1961	110	175	225
Milkman Truck	lime yellow hood, cab and flatbed body, white side rails, 14 3" white plastic milk bottles, 14-1/4" long, 1964	75	115	150
Milkman Truck	light blue hood, cab and flatbed body, white side rails, 14 3" white plastic milk bottles, 14-1/4" long, 1963	85	130	175
Milkman Truck	light yellow hood, cab and flatbed body, white side rails, 14 3" white plastic milk bottles, 14-1/4" long, 1964	75	115	150
Milkman Truck	light blue hood, cab and flatbed body, white side rails, 14 3" white plastic milk bottles, 14-1/4" long, 1963	85	130	175
Milkman Truck	deep cream hood, cab and flatbed body, white side rails, fourteen 3" white plastic milk bottles with red caps, 14-1/4" long, 1962	50	100	200
Milkman Truck	medium blue hood, cab and flatbed body, white side rails, eight 3" white plastic milk bottles, 14-1/4" long, 1961	50	100	200
Milkman Truck	deep cream hood, cab and flatbed body, white side rails, 14 3" white plastic milk bottles with red caps, 14-1/4" long, 1962	100	150	200
Mister Buddy Ice Cream Truck	white cab-over-engine van body, pale blue or off-white plastic underbody and floor, 11-1/2" long, 1964	45	60	100
Mister Buddy Ice Cream Truck	white cab-over-engine van body, red plastic underbody and floor, 11-1/2" long, 1966	45	75	100
Mister Buddy Ice Cream Truck	white cab-over-engine van body, red plastic underbody and floor, red bell knob, 11-1/2" long, 1967	55	85	115
Model T Flivver Truck	black with red eight-spoke wheels with aluminum tires, black hubs, 12" long, 1924	1000	1500	2000
Motor Market Truck	duo-tone horizontal design, white hood top, upper cab and high partition in cargo section, yellow-orange grille, fenders, lower hood and cab sides, 21-1/2" long, 1941	200	350	550
Moving Van	black front, hood and seat, red chassis and disc wheels with black hubs, green van body, roof extends forward above open driver's seat, 25" long, 1924	1200	2000	3000
Overland Trailer Truck	yellow tractor, enclosed cab, red semi-trailer and four-wheel full trailer with removable roofs, 39-3/4" long, 1935	350	525	700
Overland Trailer Truck	duo-tone horizontal design, red and white tractor has red front, lower half chassis, chassis, enclosed cab, 40", 1939	350	550	700
Overland Trailer Truck	duo-tone slant design, green and yellow tractor unit with yellow cab, red semi-trailer and four-wheel full trailer with yellow removable roofs, 39-3/4" long,	325	485	650

NAME	DESCRIPTION	GOOD	EX	MINT
Overland Trailer Truck	duo-tone slant design, green and yellow semi-streamlined tractor and green hood sides, yellow hood, chassis, enclosed cab, 40", 1939	350	550	700
Pepsi Delivery Truck	powder blue hood and lower cab, white upper cab and double-deck cargo section, two hand trucks, four blue cases of red bottles, four red cases of blue bottles, 15" long, 1970	60	95	125
Polysteel Boat Transport	medium blue soft plastic body, steel flatbed carrying 8" white plastic runabout boat with red deck, truck 12-1/2" long, 1960	75	115	150
Polysteel Coca-Cola Delivery Truck	yellow plastic truck, slanted bottle racks, eight red Coke cases with green bottles, small metal hand truck, 12-1/2" long, 1961	50	75	100
Polysteel Coca-Cola Delivery Truck	yellow plastic truck, slanted bottle racks, eight green Coke cases with red bottles, small metal hand truck, 12-1/4" long, 1962	60	90	120
Polysteel Dumper	orange plastic body and tiltback dump section with cab shield, no "Come-Back Motor", no door decals, 13-1/2" long, 1962	60	95	125
Polysteel Dumper	green soft molded plastic front, cab and frame, yellow steel dump body with sides rounded at back, hinged tailgate, 13" long, 1959	100	150	200
Polysteel Dumper	medium blue soft molded plastic front, cab and frame, off-white steel dump body with sides rounded at back, hinged tailgate, 13" long, 1960	87	130	175
Polysteel Dumper	orange plastic body and tiltback dump section with cab shield, "Come-Back Motor", 13" long, 1961	75	115	150
Polysteel Highway Transport	red soft plastic tractor, cab roof lights, double horn, radio antenna and side fuel tanks, white steel semi-trailer van, 20-1/2" long, 1960	100	150	200
Polysteel Hydraulic Dumper	red soft molded-plastic front, cab and frame, light green steel dump section with sides rounded at rear, 13" long, 1960	80	120	160
Polysteel Hydraulic Dumper	yellow soft plastic body, frame and tiltback ribbed dump section with cab shield, 13" long, 1961	75	115	150
Polysteel Hydraulic Dumper	red soft plastic body, frame and tiltback ribbed dump section with cab shield, 13" long, 1962	65	100	130
Polysteel Hydraulic Dumper	beige soft molded-plastic front, cab and frame, off-white steel dump section with sides rounded at rear, 13" long, 1959	60	95	125
Polysteel Milk Tanker	red soft plastic tractor unit, light blue/gray semi-trailer tank with red ladders and five dooms, 22" long, 1961	60	95	125
Polysteel Milk Tanker	turquoise soft plastic tractor unit, light blue/gray semi-trailer tank with red ladders and five dooms, 22" long, 1961	60	95	125
Polysteel Milkman Truck	light blue soft plastic front, cab and frame, light yellow steel open cargo section with nine oversized white plastic milk bottles, 11-3/4" long, 1960	65	100	130
Polysteel Milkman Truck	light blue soft plastic front, cab and frame, light blue steel open cargo section with nine oversized white plastic milk bottles, 11-3/4" long, 1961	60	95	125
Polysteel Milkman Truck	turquoise soft plastic front, cab and frame, light blue steel open cargo section with nine oversized white plastic milk bottles with red caps, 11-3/4" long,	35	50	70

Buddy L

NAME	DESCRIPTION	GOOD	EX	MINT
Polysteel Supermarket Delivery	medium blue soft molded-plastic front, hood, cab and frame, steel off-white open cargo section, 13" long, 1959	75	115	150
Pull-N-Ride Baggage Truck	duo-tone horizontal design, light cream upper half, off-white lower half and bumper, 24-1/4" long, 1953	150	225	300
R E A Express Truck	dark green cab-over-engine van body, sliding side doors, double rear doors, white plastic one-piece bumper, no spring suspension, 11-1/2" long, 1965	130	195	260
R E A Express Truck	dark green cab-over-engine van body, sliding side doors, double rear doors, white plastic one-piece bumper, no spring suspension, side doors are embossed "BUDDY L", 11-1/2" long, 1966	125	185	250
R E A Express Truck	dark green cab-over-engine van body, sliding side doors, double rear doors, white plastic one-piece bumper, 11-1/2" long, 1964	200	300	400
Railroad Transfer Rider Delivery	duo-tone horizontal design, yellow upper half, hood top, cab and slatted caro sides, green lower half, small hand truck, two milk cans with removable lids, 23-1/4" long, 1949	70	100	140
Railroad Transfer Store Door	duo-tone horizontal design, yellow hood top, cab and upper body, red lower half of hood and body, small hand truck, two metal drums with coin slots, 23-1/4" long, 1950	90	135	180
Railway Express Truck	black front hood, fenders, seat and low body sides, dark green van body, red chassis, 25" long, 1926	400	800	1600
Railway Express Truck	green plastic hood and cab, green steel high-sides open body, frame and bumper, small two-wheel hand truck, steel four-rung barrel skid, 20-3/4" long, 1953	125	185	250
Railway Express Truck	deep green plastic "Diamond T" hood and cab, deep green steel frame and van body with removable silvery roof, small two-wheel hand truck, steel four-rung barrel skid, 21" long, 1952	200	300	400
Railway Express Truck	duo-tone horizontal design, tractor unit has yellow front, lower door and chassis, green hood top and enclosed upper cab, semi-trailer has yellow lower sides, 25" long, 1941	325	485	650
Railway Express Truck	green all-steel hood, cab, frame and high-sides open bady, sides have three horizontal slots in upper back corners, 22" long, 1954	75	115	150
Railway Express Truck	duo-tone slant design, tractor has silvery and hood sides, green hood top, enclosed cab, green semi-trailer, "Wrigley's Spearmint Gum" poster on trailer sides, 23" long, 1935	400	600	800
Railway Express Truck	duo-tone slant design, tractor unit has white skirted fenders and hood sides, green hood top, enclosed cab and chassis, green semi-trailer with white removable roof, 25" long, 1939	350	475	700
Railway Express Truck	red tractor unit, enclosed square cab, green 12-1/4" long two-wheel semi-trailer van with removable roof, "Wrigley's Spearmint Gum" poster on trailer sides, 23" long, 1935	375	565	750
Railway Express Truck	yellow and green tractor unit has white skirted fenders and hood sides, green hood top, enclosed cab and chassis, green semi-trailer with yellow removable roof, 25" long, 1940	330	495	660

Buddy L

NAME	DESCRIPTION	GOOD	EX	MINT
Railway Express Truck	dark green or light green screen body, double-bar nickel front bumper, brass radiator knob, red wheels, 25" long, 1930	500	1000	2000
Ranchero Stake Truck	medium green, white plastic one-piece bumper and grille guard, four-post, four-slat fixed stake sides and cargo section, 14" long, 1963	50	75	100
Rider Dump Truck	duo-tone horizontal design, yellow hood top, upper cab and upper dump body, red front, hood sides, lower doors and lower dump body, no bumper, 21-1/2" long, 1945	160	245	325
Rider Dump Truck	duo-tone horizontal design, yellow hood top, upper cab and upper dump body, red front, hood sides, lower doors and lower dump body, no bumper, 23"	75	115	150
Rival Dog Food Delivery Van	cream front, cab and boxy van body, metal drum coin bank with "RIVAL DOG FOOD" label in blue, red, white and yellow, 16-1/2" long, 1956	160	245	325
Robotoy	black fenders and chassis, red hood and enclosed cab with small visor, green dump body's front and back are higher than sides, 21-5/8" long, 1932	575	900	1200
Rockin' Giraffe Truck	powder blue hood, cab, and high-sided open-top cargo section, two orange and yellow plastic giraffes, 13-1/4" long, 1967	75	115	150
Ruff-n-Tuff Cement Mixer Truck	yellow snub-nosed cab-over-engine body, frame and water-tank ends, white plastic water tank and mixing drum, white seats, 16" long, 1971	35	55	75
Ruff-n-Tuff Log Truck	yellow snub-nose cab-over-engine, frame and shallow truck bed, black full-width grille, 16" long, 1971	50	75	100
Ryder City Special Delivery Truck	duo-tone horizontal design, yellow upper half including hood top and cab, brown removable van roof, warm brown front and lower half of van body, 24-1/2" long, 1949	150	225	300
Ryder Van Lines Trailer	duo-tone slant design, black front and lower hood sides and doors, deep red hood top, enclosed cab and chassis, 35-1/2" long, 1949	350	525	700
Saddle Dump Truck	duo-tone slant design, yellow front, fenders and removable rider seat, red enclosed square cab and dump body, no bumper, 19-1/2" long, 1937	200	300	400
Saddle Dump Truck	duo-tone slant design, yellow front, fenders, lower hood and cab, and removable rider seat, rest of body red, no bumper, 21-1/2" long, 1939	125	185	250
Saddle Dump Truck	duo-tone horizontal design, deep blue hood top, upper cab and upper dump body, orange fenders radiator front lower two-thirds of cab and lower half of dump body, 21-1/2" long, 1941	85	130	175
Sand and Gravel Rider Dump Truck	duo-tone horizontal design, blue lower half, yellow upper half including hoop top and enclosed cab, 24" long, 1950	350	525	700
Sand and Gravel Truck	duo-tone horizontal design, red front, bumper, lower hood, cab sides, chassis and lower dump body sides, white hood top, enclosed cab and upper dump body, 23-3/4" long, 1949	350	525	700
Sand and Gravel Truck	black with red chassis and wheels, nickel-rim, red-shell headlights, enclosed cab with opening doors, 25-1/2" long, 1930-32	2000	3000	5000

Buddy L

NAME	DESCRIPTION	GOOD	EX	MINT
Sand and Gravel Truck	black body, doorless roofed cab and steering wheel, red chassis and disc wheels with black hubs, 25-1/2" long, 1926	600	1000	1200
Sand and Gravel Truck	dark or medium green hood, cab, roof lights and skirted body, white or cream dump section, 13-1/2" long, 1949	100	150	200
Sand Loader and Dump Truck	duo-tone horizontal design, yellow hood top and upper dump blue cab sides, frame and lower dump body, red loader on dump with black rubber conveyor belt, 24-1/2" long, 1950	175	265	350
Sand Loader and Dump Truck	duo-tone horizontal design, yellow hood top and upper dump blue cab sides, frame and lower dump body, red loader on dump with black rubber conveyor belt, 24-1/2" long, 1952	60	95	125
Sanitation Service Truck	blue front fenders, hood, cab and chassis, white encllosed dump section and hinged loading hopper, one-piece chrome bumper, plastic windows in garbage section, 16-1/2" long, 1967	75	150	200
Sanitation Service Truck	blue front fenders, hood, cab and chassis, white encllosed dump section and hinged loading hopper, one-piece chrome bumper, no plastic windows in garbage section, 16-1/2" long, 1968	75	115	150
Sanitation Service Truck	blue snub-nose hood, cab and frame, white cargo dump body and rear loading unit, two round plastic headlights, 17" long, 1972	75	115	150
Sears Roebuck Delivery Truck Van	gray/green and off-white, no side doors, 1967	125	200	275
Self-Loading Auto Carrier	medium tan tractor unit, three plastic cars, overall 34" long including loading ramp, 1960	85	130	175
Self-Loading Boat Hauler	pastel blue tractor and semi-trailer with three 8-1/2" long boats, no side mirror on truck, overall 26-1/2" long, 1963	150	225	350
Self-Loading Boat Hauler	pastel blue tractor and semi-trailer with three 8-1/2" long boats, overall 26-1/2" long, 1962	150	225	350
Self-Loading Car Carrier	lime green tractor unit, three plastic cars, overall 33-1/2" long including, 1963	75	115	150
Self-Loading Car Carrier	beige/yellow tractor unit, three plastic cars, overall 33-1/2" long including, 1964	60	95	125
Shell Pickup and Delivery	reddish orange hood and body, open cargo section with solid sides, chain across back, red coin-slot oil drum with Shell emblem and lettering, 13-1/4" long,	135	200	275
Shell Pickup and Delivery	yellow/orange hood and body, open cargo section, three curved slots toward rear in sides, chains across back, red coin-slot oil drum with Shell emblem and lettering, 13-1/4" long, 1952	125	185	250
Shell Pickup and Delivery	yellow/orange hood and body, open cargo section with three curved slots toward rear in sides, red coin-slot oil drum with Shell emblem and lettering, 13-1/4" long, 1953	110	175	225
Smoke Patrol	lemon yellow body, six wheels, garden hose attaches and water squirts through large chrome swivel-mount water cannon on rear deck, 7" long, 1970	50	75	100
Sprinkler Truck	black front, hood, fenders and cabless open driver's seat, red bumper and chassis, bluish/gray/green water tank, 25" long, 1929	800	1500	2000

Buddy L

NAME	DESCRIPTION	GOOD	EX	MINT
Stake Body Truck	black cabless open driver's seat, hood, front fenders and flatbed body, red chassis and five removable stake sections, 25" long, 1921	1000	1500	2000
Stake Body Truck	black cabless open driver's seat, hood, front fenders and flatbed body, red chassis and five removable stake sections, cargo bed with low sidesboards, drop-down tailgate, 25" long, 1924	1000	1500	2000
Standard Coffee Co. Delivery Truck Van	1966	125	200	275
Standard Oil Tank Truck	duo-tone slant design, white upper cab and hood, red lower cab, grille, fenders and tank, rubber wheels, electric headlights, 26" long, 1936-37	350	500	1000
Stor-Dor Delivery	red hood and body, open cargo body with four horizontal slots in sides, plated chains across open back, 14-1/2" long, 1955	65	100	175
Street Sprinkler Truck	black front, hood, front fenders and cabless open driver's seat, red bumper and chassis, bluish/gray/green water tank, 25" long, 1929	700	900	1800
Street Sprinkler Truck	black front, hood, and fenders, open cab, nickel-rim, red-shell headlights, double bar front bumper, bluish/gray/green water tank, six rubber tires, 25" long,	800	1000	1900
Sunshine Delivery Truck Van	bright, yellow cab-over-engine van body and opening double rear doors, off-white plastic bumper and under body, 11-1/2" long, 1967	125	200	275
Super Motor Market	duo-tone horizontal design, white hood top, upper cab and high partition in cargo section, yellow/orange lower hood and cab sides, semi-trailer carrying supplies, 21-1/2" long, 1942	300	500	700
Supermarket Delivery	all white with rubber wheels, enclosed cab, pointed nose, bright metal one-piece grille, 13-3/4" long, 1950	125	185	250
Supermarket Delivery	blue bumper, front, hood, cab and frame, one-piece chrome four-hole grille and headlights, 14-1/2" long, 1956	75	115	150
Tank and Sprinkler Truck	black front, hood, fenders, doorless cab and seat, dark green tank and side racks, black or dark green sprinkler attachment, 26-1/4" long with sprinkler attachment, 1924	700	900	1700
Teepee Camping Trailer and Wagon	maroon suburban wagon, two-wheel teepee trailer and its beige plastic folding tent, overall 24-1/2" long, 1963	75	175	250
Texaco Tank Truck	red steel GMC 550-series blunt-nose tractor and semi-trailer tank, 25" long, 1959	175	250	400
Tom's Toasted Peanuts Delivery Truck Van	light tan/beige body, no seat or sliding doors, blue bumpers, floor and underbody, 11-1/2" long, 1973	125	200	275
Trail Boss	red, square-corner body with sloping sides, open cockpit, white plastic seat, 7" long, 1970	40	60	80
Trail Boss	lime green, square-corner body with sloping sides, open cockpit, yellow plastic seat, 7" long, 1971	35	55	75
Trailer Dump Truck	cream tractor unit with enclosed cab, dark blue semi-trailer dump body with high sides and top-hinged opening endgate, no bumper, 20-3/4" long, 1941	75	115	155
Trailer Van Truck	red tractor and van roof, blue bumper, white semi-trailer van, chrome one-piece toothed grille and headlights, white drop-down rear door, 29" long with tailgate/ramp lowered, 1956	150	225	300

Buddy L

NAME	DESCRIPTION	GOOD	EX	MINT
Trailer Van with Tailgate Loader	green high-impacted styrene plastic tractor on steel frame, cream steel detachable semi-trailer van with green roof and crank operated tailgate, 33" long with tailgate lowered, 1953	125	185	250
Trailer Van with Tailgate Loader	green steel tractor, bumper, chrome one-piece toothed grille and headlights, cream van with green roof and tailgate loader, 31-3/4" long, with tailgate	125	185	250
Traveling Zoo	red high side pickup with yellow plastic triple-cage unit, six compartments with plastic animals, 13-1/4" long, 1965	85	130	175
Traveling Zoo	red high side pickup with yellow plastic triple-cage unit, six compartments with plastic animals, 13-1/4" long, 1967	75	115	150
Traveling Zoo	yellow high side pickup with red plastic triple-cage unit, six compartments with plastic animals, 13-1/4" long, 1969	65	95	130
Traveling Zoo	snub-nosed yellow body and cab, six red plastic cages with six plastic zoo animals, 13-1/4" long, 1970	60	95	125
U.S. Army Half-Track and Howitzer	olive drab, 12-1/2" truck, 9-3/4" gun, overall 22-1/2" long, 1952	125	200	275
U.S. Mail Delivery Truck	white upper cab-over-engine, sliding side doors and double rear doors, red belt-line stripe on sides and front, blue lower body, 11-1/2" long, 1964	65	100	200
U.S. Mail Delivery Truck	blue cab, hood, bumper, frame and removable roof on white van body, 23-1/4" long, 1956	150	300	500
U.S. Mail Truck	shiny olive green body and bumper, yellow-cream removable van roof, enclosed cab, 22-1/2" long, 1953	225	400	575
United Parcel Delivery Van	duo-tone horizontal design, deep cream upper half with brown removable roof, chocolate brown front and lower half, 25" long, 1941	250	450	650
Utility Delivery Truck	duo-tone slant design, blue front and lower hood sides, gray hood top, cab and open body with red and yellow horizontal stripe, 22-3/4" long, 1940	250	450	650
Utility Delivery Truck	duo-tone horizontal design, green upper half including hood top, dark cream lower half, green wheels, red and yellow horizontal stripe, 22-3/4" long, 1941	125	185	250
Utility Dump Truck	duo-tone slant design, red front, lower doord and fenders, gray chassis and enclosed upper cab, royal blue dump body, yellow removable rider seat, 25-1/2" long, 1940	125	185	250
Utility Dump Truck	duo-tone slant design, red front, lower door and fenders, gray chassis, red upper hood, upper enclosed cab and removable rider seat, yellow body, 25-1/2"	85	130	175
Van Freight Carriers Trailer	red streamlined tractor, bright blue enclosed cab, light cream/white semi-trailer van with removable white roof, 22" long, 1953	125	185	250
Van Freight Carriers Trailer	bright blue streanlined tractor and enclosed cab, cream/yellow semi-trailer van, removable silvery roof, 22" long, 1949	65	100	135
Van Freight Carriers Trailer	red streamlined tractor, bright blue enclosed cab, cream/yellow semi-trailer van, white removable van roof, 22" long, 1952	55	85	115
Wild Animal Circus	red tractor unit and semi-trailer, trailer cage doors lighter than body, 26" long, 1970	100	150	200
Wild Animal Circus	red tractor unit and semi-trailer, three cages with plastic elephant, lion, tiger, 26" long, 1966	150	225	300

Buddy L

NAME	DESCRIPTION	GOOD	EX	MINT
Wild Animal Circus	red tractor unit and semi-trailer, three cages with six plastic animals, 26" long, 1967	110	175	225
Wrecker Truck	black front, hood, and fenders, open cab, four rubber tires, red wrecker body, 26-1/2" long, 1930	800	1000	2500
Wrecker Truck	duo-tone slant design, red upper cab, hood, and boom, white lower cab, grille, fenders, body, rubber wheels, electric headlights, removable rider seat, 31"	250	450	700
Wrecker Truck	black open cab, red chassis and bed, disc wheels, 26-1/2" long, 1928-29	700	1000	2000
Wrigley Express Truck	forest green with chrome one-piece, three-bar grille and headlights, "Wrigley's Spearmint Gum" poster on sides, 16-1/2" long, 1955	135	200	275
Zoo-A-Rama	lime green Colt Sportsliner with four-wheel trailer cage, cage contains plastic tree, monkeys and bears, 20-3/4" long, 1967	100	150	200
Zoo-A-Rama	sand yellow four-wheel trailer cage, matching Colt Sportsliner with white top, three plastic animals, 20-3/4" long, 1968	100	150	200
Zoo-A-Rama	greenish yellow four-wheel trailer cage, matching Colt Sportsliner with white top, three plastic animals, 20-3/4" long, 1969	85	130	175

Corgi

Adams Drag-Star	4-3/8" long, orange body, red nose, gold engines, chrome pipes and hood panels	20	30	45
Adams Probe 16	3-5/8" long, one-piece body, blue sliding canopy; metallic burgundy, or metallic lime/gold with and without racing stripes	15	25	40
Agricultural Set	#69 Massey Ferguson tractor, #62 trailer, #438 Land Rover, #484 Farm Truck, #71 harrow, #1490 skip, and accessories	120	180	350
Agricultural Set	two versions: with #55 tractor and yellow trailer, 1962-64; with #60 tractor and red trailer, 1965-66	280	420	750
Agricultural Set	with mustard yellow conveyor	60	90	150
Alfa Romeo P33 Pininfarina	3-5/8" long; gold or black spoiler	16	24	40
Alfa Romeo P33 Pininfarina	3-5/8" long; gold or black spoiler	16	24	40
All Winners Set	five vehicle set: white Mustang and Marcos, red Ferrari, silver Corvette and gold Jaguar, 1966 only; prior vehicles in other colors; or Toronado, Ferrari, MGB, Corvette and Jaguar	160	240	450
Allis-Chalmers AFC 60 Fork Lift	4-3/8" long, yellow body, white engine hood, with driver	15	20	35
AMC Pacer	4-3/4" metallic red body, white Pacer X decals, working hatch, clear windows, light yellow interior, chrome bumpers and wheels	15	20	35
AMC Pacer Rescue Car	4-7/8" long, chrome roll bars and red roof lights, white with black engine hood; with or without Secours decal	16	25	40
American La France Ladder Truck	11-1/8" long, red working cab, trailer and ladder rack; red/chrome body with red wheels or red/white body with unpainted wheels	60	90	150
AMX 30D Recovery Tank	6-7/8" long, olive body with black plastic turret and gun, accessories and three figures	35	50	80

Corgi

NAME	DESCRIPTION	GOOD	EX	MINT
Army Equipment Transporter	9-1/2" long, olive cab and trailer with white U.S. Army decals	70	105	175
Army Troop Transporter	5-1/2" long, olive with white U.S. Army decals	70	105	175
Aston Martin DB4	3-3/4" long, red or yellow body with working hood, detailed engine, clear windows, plastic interior, silver lights, grille, license plate & bumpers, red taillights, rubber tires, working scoop on early models	45	65	110
Aston Martin DB4	3-3/4" white top w/aqua green sides, yellow plastic interior, racing #1, 3 or 7	50	75	125
Austin A40	3-1/8" long, one-piece light blue or red body with clear windows, smooth wheels, rubber tires	35	50	85
Austin A40-Mechanical	3-1/8" long, friction motor, red body w/black roof	55	75	130
Austin A60 Driving School	3-3/4" medium blue body with silver trim, left hand drive steering wheel	45	65	110
Austin A60 Motor School	3-3/4" long, light blue body with silver trim, red interior, single body casting, right hand drive steering wheel, two figures	45	65	110
Austin Cambridge	3-1/2" long, available in gray, green/gray, silver/green, aqua	40	60	100
Austin Cambridge-Mechanical	3-1/2" long, fly-wheel motor, available in orange, cream, light or dark gray	50	75	125
Austin London Taxi	3-7/8" long, black body w/yellow plastic interior, with or without driver, smooth or rubber wheels	36	55	90
Austin London Taxi	4-5/8" long, black body w/two working doors, light brown interior	15	20	35
Austin London Taxi/Reissue	3-7/8" long, updated version with Whizz Wheels, black or maroon body	15	20	35
Austin Mini Countryman	3-1/8" long, turquoise body with two working doors; three versions: one with shaped wheels, the others with cast wheels and with or without aluminum parts	55	80	150
Austin Mini Van	3-1/8" long, with two working doors, clear windows, metallic deep green body	40	60	100
Austin Mini-Metro	3-1/2", three versions with plastic interior, working rear hatch and doors, clear windows, folding seats, chrome headlights, or ange taillights, black plastic base, grille, bumpers, Whizz Wheels	18	27	45
Austin Mini-Metro Datapost	3-1/2", white body, blue roof, hood & trim, red plastic interior, hepolite & #77 decals, working hatch & doors, clear windows, folding seats, chrome headlights, orange taillights, Whizz Wheels	15	18	30
Austin Police Mini Van	3-1/8" long, dark blue body with policeman and dog figures, white police decals	50	75	175
Austin Seven Mini	2-3/4" red or yellow body, yellow interior, silver bumpers, grille & headlights, orange taillights	50	75	125
Austin-Healey	3-1/4" cream body w/red seats or red body w/cream seats	50	75	125
Avengers Set	white Lotus, red or green Bentley; Jonathan Steed and Emma Peel figures with umbrellas	260	390	800
Basil Brush's Car	3-5/8" long, red body, dark yellow chassis, gold lamps and dash, Basil Brush figure, red plastic wheels, plastic tires	70	105	175
Batbike	4-1/4" black body, one-piece body, black & red plastic parts, gold engine & exhaust pipes, clear windshield, chrome stand, black plastic five-spoked wheels, Batman and decals	40	60	125
Batboat	5-1/8" long, black plastic boat, red seats, fin and jet, blue windshield, Batman and Robin figures, gold cast trailer, large black/yellow decals on fin, cast wheels, plastic tires	60	90	175

Corgi

NAME	DESCRIPTION	GOOD	EX	MINT
Batboat	5-1/8" long, black plastic boat with Batman and Robin figures, small decals on fin, Whizz Wheels on trailer	30	45	100
Batcopter	5-1/2" long, black body with yellow/red/black decals, red rotors, Batman figure	26	39	95
Batman Set	three vehicle set: Batmobile, Batboat with trailer and Batcopter, Whizz Wheels on trailer	90	135	300
Batmobile	5" long, gold hubs, bat logos on door, maroon interior, black body, plastic rockets, gold headlights & rocket control, tinted canopy, working front chain cutter, no tow hook	200	300	500
Batmobile	5" long, chrome hubs with red bat logos on door, maroon interior, red plastic tires, gold tow hook, plastic rockets, gold headlight & rocket control, tinted canopy with chrome support, chain cutter	140	200	350
Batmobile	5" long, chrome hubs with red bat logos on door, light red interior, regular wheels, gold tow hook, plastic rockets, gold headlights and rocket control, tinted canopy with chrome support	80	120	200
Batmobile	5" long, gold hubs, bat logos on door, maroon interior, black body, plastic rockets, gold headlights & rocket control, tinted canopy, working front chain cutter, with two hooks	180	270	450
Batmobile, Batboat & Trailer	four versions: red bat hubs on wheels, 1967-72; red tires and chrome wheels, 1973; black tires, big decals on boat, 1974-76; chrome wheels, boat decals, Whizz Wheels on trailer, 1977-81 each set	240	360	650
Beach Buggy & Sailboat	purple buggy, yellow trailer and red/white boat	20	30	50
Beast Carrier Trailer	4-1/2" long, red chassis, yellow body and tailgate, four plastic calves, red plastic wheels, black rubber tires	24	36	60
Beatles' Yellow Submarine	5" long, yellow and white body, working hatches with two Beatles in each	180	270	450
Bedford AA Road Service Van	3-5/8" long, dark yellow body in two versions: divided windshield, 1957-59; single windshield, 1960-62	50	75	125
Bedford Army Tanker	7-3/8" long, olive cab and tanker, with white U.S. Army decals	140	210	350
Bedford Articulated Horse Box	10" long, cast cab, lower body and three working ramps, yellow interior, plastic upper body, with horse and Newmarket Racing Stables decals, dark metallic green or light green body with orange or yellow upper	32	48	80
Bedford Car Transporter	10-1/4" black die cast cab base with blue cab, yellow semi trailer, blue lettering decals and/or red cab, pale green upper & blue lower semi-trailer, white decals, lower tailgate, clear windshield	70	105	175
Bedford Car Transporter	10-5/8" long, red cab with blue lower and light green upper trailer, working ramp, yellow interior, clear windows, white wording and Corgi dog decals	60	90	150
Bedford Carrimore Low Loader	8-1/2" red or yellow cab, metallic blue semi trailer & tailgate; smooth and/or shaped wheels	60	90	150
Bedford Carrimore Low Loader	9-1/2" long, yellow cab and working tailgate, red trailer, clear windows, red interior, suspension, shaped wheels, rubber tires	56	84	140
Bedford Corgi Toys Van	3-1/4" long, Corgi Toys decals, with either yellow body/blue roof or yellow upper/blue lower body	60	90	175
Bedford Daily Express Van	3-1/4" long, dark blue body with white Daily Express decals, divided windshield, smooth wheels, rubber tires	60	90	150
Bedford Dormobile	3-1/4" long, two versions and several colors: divided windshield with cream, green or metallic maroon body; or single windshield with yellow body/blue roof w/shaped or smooth wheels	50	75	125
Bedford Dormobile-Mechanical	3-1/4" long, friction motor, dark metallic red or turquoise body	60	90	150

Corgi

NAME	DESCRIPTION	GOOD	EX	MINT
Bedford Evening Standard Van	3-1/4" long, black body/silver roof or black lower body/silver upper body and roof, Evening Standard decals	55	80	130
Bedford Fire Tender	3-1/4" long, single windshield version, red body with either black ladders and smooth wheels or unpainted ladders and shaped wheels	60	90	150
Bedford Fire Tender	3-1/4" long, divided windshield, red or green body, each with different decals	60	90	150
Bedford Fire Tender-Mechanical	3-1/4" long, friction motor, red body with Fire Dept. decals	70	105	175
Bedford Giraffe Transporter	red Bedford truck with blue giraffe box with Chipperfield decal, three giraffes	60	90	175
Bedford KLG Van-Mechanical	3-1/4" long, with friction motor, in either red body with KLG Spark Plugs decals, or dark blue body with Daily Express decals	70	105	175
Bedford Military Ambulance	3-1/4" long, clear front and white rear windows, olive body with Red Cross decals, with or without suspension	56	84	140
Bedford Milk Tanker	7-3/4" long, light blue cab and lower semi, white upper tank with blue/white milk decals	110	165	275
Bedford Milk Tanker	7-1/2" long, light blue cab and lower semi, white upper tank, with blue/white milk decals, shaped wheels, rubber tires	100	150	250
Bedford Mobilgas Tanker	7-3/4" long, red cab and tanker with red/white/blue Mobilgas decals, shaped wheels, rubber tires	100	150	250
Bedford Mobilgas Tanker	7-5/8" long, red or blue cab with Mobilgas decals, shaped wheels, rubber tires	100	150	250
Bedford Tanker	7-1/2" long, red cab with black chassis, plastic tank with chrome catwalk, Corgi Chemco decals	15	20	35
Bedford TK Tipper Truck	4-1/8" long, four color variations	26	39	65
Bedford Utilecon Ambulance	3-1/4" long, divided windshield, cream body with red/white/blue decals, smooth wheels	50	75	125
Beep Beep London Bus	4-3/4" long, battery-operated working horn, red body, black windows, BTA decals	26	39	65
Belgian Police Range Rover	4" long, white body, working doors, red interior, Belgian Police decals	22	33	55
Bell Army Helicopter	5-1/4" long, two-piece olive/tan camo body, clear canopy, olive green rotors, U.S. Army decals	24	36	60
Bell Rescue Helicopter	5-3/4" long, two-piece blue body with working doors, red interior, yellow plastic floats, black rotors, white N428 decals	20	30	50
Bentley Continental	4-1/4" four different versions; red interior, clear windows, chrome grille & bumpers, jewel headlights, red jeweled taillights, suspension, shaped wheels, gray rubber tires	45	65	110
Bentley T Series	4-1/2" red body, cream interior; working hood, trunk & doors; clear windows, folding seats, chrome bumper/grille, jewel headlights	36	55	90
Berliet Articulated Horse Box	10-7/8" long, bronze cab and lower semi body, cream chassis, white upper body, black interior, three working ramps, National Racing Stables decals, horse figures	30	45	75
Berliet Container Truck	US Cines	30	45	75
Berliet Dolphinarium Truck	56	84	175	
Berliet Fruehauf Dumper	11-1/4" yellow cab, fenders & dumper; black cab & semi chassis; plastic orange dumper body; or dark orange, black interior	30	45	75
Berliet Holmes Wrecker	5" red cab & bed, blue rear body, white chassis, black interior, two gold booms & hooks, yellow dome light, driver, amber lenses & red/white/blue stripes	30	45	75

Corgi

NAME	DESCRIPTION	GOOD	EX	MINT
Bertone Barchetta Runabout	3-1/4" long, black interior, amber windows, die cast air foil, suspension, red/yellow Runabout decals, Whizz Wheels	15	18	30
Bertone Shake Buggy	3-3/8" long, clear windows, green interior, gold engine, four variations: yellow upper/white lower body or metallic mauve upper/white lower body with spoked or solid chrome wheels	15	18	30
-BL Roadtrain & Trailers		16	24	40
Bloodhound & Launching Platform		110	165	275
Bloodhound Launching Ramp		34	51	85
Bloodhound Loading Trolley		40	60	100
Bloodhound Missile		70	105	175
Bloodhound Missile on Trolley		120	180	300
BMC Mini-Cooper	3" white body, black working hood, trunk, two doors, red interior, clear windows, orange/black stripes & #177 decals, suspension, Whizz Wheels	30	45	75
BMC Mini-Cooper Magnifique	2-7/8" long, metallic blue or olive green body with working doors, hood and trunk, clear windows and sunroof, cream interior with folding seats, jewel headlights, cast wheels, plastic tires	34	51	85
BMC Mini-Cooper S	3" bright yellow body, red plastic interior, chrome plastic roof rack with two spare wheels, clear windshield, one-piece body silver grille, bumpers, headlights, red taillights, suspension, Whizz Wheels	45	65	110
BMC Mini-Cooper S Rally	2-7/8" long, red body, white roof, chrome roof rack with two spare tires, Monte Carlo Rally and #177 decals, with shaped wheels/rubber tires or cast wheels/plastic tires	40	60	100
BMC Mini-Cooper S Rally Car	2-7/8" long, red body, white roof, five jewel headlights, Monte Carlo Rally decals with either #52 (1965) or #2 (1966) with drivers' autographs on roof	60	90	150
BMC Mini-Cooper S Rally Car	2-7/8" long, red body, white roof with six jewel headlights, RAC Rally and #21 decals	60	90	150
BMW M1	5" yellow body, black plastic base, rear panel & interior, white seats, clear windshield, multicolored stripes, lettering & #25 decal	15	20	35
BMW M1 BASF	4-7/8" long, red body, white trim with black/white BASF and #80 decals	15	18	30
Breakdown Truck	3-7/8" long, red body, black plastic boom with gold hook, yellow interior, amber windows, black/yellow decals, Whizz Wheels	15	18	30
British Leyland Mini 1000	3-3/8" long, metallic blue body, working doors, black base, clear windows, white interior, silver lights, grille and bumper, Union Jack decal on roof, Whizz Wheels	18	27	45
British Leyland Mini 1000	3-1/4" long, red interior, chrome lights, grille and bumper, #8 decal; three variations: silver body with decals, 1978-82; silver body, no decals; orange body with extra hood stripes, 1983	16	24	40
British Racing Cars	set of three cars, three versions: blue Lotus, green BRM, green Vanwall, all with smooth wheels, 1959; same cars with shaped wheels, 1960-61; red Vanwall, green BRM and blue Lotus, 1963, each set	140	210	375
BRM Racing Car	3-1/2" long, silver seat, dash and pipes, smooth wheels, rubber tires, in three versions: dark green body, 1958-60; light green body w/driver and various number decals 1961-65; light green body, no driver	50	75	125

VEHICLES

Corgi

NAME	DESCRIPTION	GOOD	EX	MINT
Buck Rogers Starfighter	6-1/2" long, white body with yellow plastic wings, amber windows, blue jets, color decal, Buck and Wilma figures	32	48	90
Buick & Cabin Cruiser	three versions: light blue, dark metallic blue or gold metallic Buick	80	120	200
Buick Police Car	4-1/8" long, metallic blue body with white stripes and Police decals, chrome light bar with red lights, orange taillights, chrome spoke wheels	18	27	45
Buick Riviera	4-1/4" long, metallic gold or dark blue, pale blue or bronze body, red interior, gray steering wheel, & tow hook, clear windshield, chrome grille & bumpers, suspension, Tan-o-lite tail & headlights, spoked wheels & rubber tires	30	45	75
Cadillac Superior Ambulance	4-1/2" long, battery-operated, in two versions: red lower/cream upper body, or white lower/blue upper body	60	90	150
Cafe Racer Motorcycle		15	18	30
Campbell Bluebird	5-1/8" long, blue body, red exhaust, clear windshield, driver, in two versions: black plastic wheels, 1960; metal wheels and rubber tires	56	84	175
Canadian Mounted Police Set	blue Land Rover with Police sign on roof and RCMP decals, plus mounted Policeman	30	45	75
Captain America Jetmobile	6" white body, metallic blue chassis, black nose cone, red shield & jet, red-white-blue Captain America decals, light blue seats & driver, chrome wheels, red tires	24	36	60
Captain Marvel Porsche	4-3/4" white body, gold parts, red seat, driver, red/yellow/blue Captain Marvel decals, black plastic base, gold wheels	20	30	50
Car Transporter & Cars	Scammell transporter with five cars: Ford Capri, the Saint's Volvo, Pontiac Firebird, Lancia Fulvia, MGC GT, Marcos 3 Litre, set	200	300	600
Car Transporter & Four Cars	two versions: with Fiat 1800, Renault Floride, Mercedes 230SE and Ford Consul, 1963-65; with Chevy Corvair, VW Ghia, Volvo P-1800 and Rover 2000, 1966 only, each set	200	300	600
Carrimore & Six Cars	sold by mail order only	240	360	700

Hardy Boys' Rolls Royce, Corgi

Corgi

NAME	DESCRIPTION	GOOD	EX	MINT
Carrimore Car Transporter	three versions: transporter with Riley, Jaguar, Austin Healey and Triumph, 1957-60; with four American cars, 1959; with Triumph, Mini, Citroen and Plymouth, 1961-62, each set	300	450	800
Caterpillar Tractor	4-1/4", lime green body with black or white rubber treads, gray plastic seat, driver figure, controls, stacks	70	105	175
Centurion Mark III Tank		30	45	75
Centurion Tank & Transporter	olive tank and transport	55	80	130
Chevrolet Astro I	4-1/8" long, dark metallic green/blue body with working rear door, cream interior with two passengers, in two versions: gold wheels with red plastic hubs or Whizz wheels	18	27	45
Chevrolet Camaro SS	4" long, blue or turquoise body with white stripe, cream interior, working doors, white plastic top, clear windshield, folding seats, silver air intakes, red taillights, black grille & headlights, suspension, Whizz Wheels	30	45	95
Chevrolet Camaro SS	4" long, metallic lime/gold body with two working doors, black roof and stripes, red interior, cast wheels, plastic tires	30	45	75
Chevrolet Caprice Classic	5-7/8" long, working doors and trunk, whitewall tires, two versions: light metallic green body with green interior or silver on blue body with brown interior	24	36	60
Chevrolet Caprice Classic	6" long, white upper body, red sides with red/white/blue stripes and #43 decals, tan interior	24	36	60
Chevrolet Caprice Fire Chief Car	5-3/4" red body, red-white-orange decals, chrome roof bar, opaque black windows, red dome light, chrome bumpers, grille & headlights, orange taillights, Fire Dept. & Fire Chief decals, chrome wheels	28	42	70
Chevrolet Caprice Police Car	5-7/8" long, black body with white roof, doors and trunk, red interior, silver light bar, Police decals	20	30	50
Chevrolet Caprice Taxi	5-7/8" long, orange body with red interior, white roof sign, Taxi and TWA decals	20	30	50
Chevrolet Charlie's Angels Van	4-5/8" long, light rose-mauve body with Charlie's Angels decals, in two versions: either solid or spoked chrome wheels	10	15	40
Chevrolet Coca-Cola Van	4-5/8" long, red body, white trim, with Coca Cola logos	15	20	35
Chevrolet Corvair	3-3/4" three body versions with yellow interior & working hood, detailed engine, clear windows, silver bumpers, headlights & trim, red taillights, rear window blind, shaped wheels, rubber tires	36	55	90
Chevrolet Impala	4-1/4" long, pink body, yellow plastic interior, clear windows, silver headlights, bumpers, grille and trim, suspension, die cast base with rubber tires	50	75	125
Chevrolet Impala	4-1/4" tan body, cream interior, gray steering wheel, clear windshields, chrome bumpers, grille, headlights, suspension, red taillights shaped wheels & rubber tires	50	75	125
Chevrolet Impala Fire Chief	4-1/8" long, red body, yellow interior, in three versions: with four white doors, with round shield decals on two doors, or with white decals on doors	55	80	130
Chevrolet Impala Fire Chief	4" long, with Fire Chief decal on hood, yellow interior with driver, in two versions: either all red body or red on white body	55	80	130
Chevrolet Impala Police Car	4" long, black lower body and roof, white upper body, yellow interior with driver, Police and Police Patrol decals on doors and hood	55	80	130

Corgi

NAME	DESCRIPTION	GOOD	EX	MINT
Chevrolet Impala Taxi	4-1/4" light orange body, base with hexagonal panel under rear axle & smooth wheels, or two raised lines & shaped wheels, one-piece body, clear windows, plastic interior, silver grille, headlights & bumpers, rubber tires	50	75	125
Chevrolet Impala Yellow Cab	4" long, red lower body, yellow upper, red interior with driver, white roof sign, red decals	80	120	200
Chevrolet Kennel Club Van	4" long, white upper, red lower body, working tailgate and rear windows, green interior, dog figures, kennel club decals, cast wheels, rubber tires	56	84	140
Chevrolet Performing Poodles Van	4" blue upper body & tailgate, red lower body & base, clear windshield, pale blue interior with poodles in back & ring of poodles & trainer, plastic tires	160	240	450
Chevrolet Rough Rider Van	4-5/8" long, yellow body with working rear doors, cream interior, amber windows, Rough Rider decals	15	18	30
Chevrolet Spider-Van	4-5/8" long, dark blue body with Spider-Man decals, in two versions: with either spoke or solid wheels	26	39	65
Chevrolet State Patrol Car	4" black body, State Patrol decals, smooth wheels with hexagonal panel or raised lines & shaped wheels, yellow plastic interior, gray antenna, clear windows, silver bumpers, grille, headlights & trim, rubber tires	50	75	125
Chevrolet Superior Ambulance	4-3/4" long, white body, orange roof and stripes, two working doors, clear windows, red interior with patient on stretcher and attendant, Red Cross decals	30	45	75
Chevrolet Vanatic Van	4-5/8" long, off white body with Vanatic decals	10	15	25
Chevrolet Vantastic Van	4-5/8" long, black body with Vantastic decals	10	15	25
Chieftain Medium Tank		30	45	75
Chitty Chitty Bang Bang	6-1/4" metallic copper body, dark red interior & spoked wheels, figures, black chassis with silver running boards, silver hood, horn, brake, dash, tail & headlights, gold radiator, red & orange wings, handbrake operates	180	270	425
Chopper Squad Helicopter		20	30	50
Chopper Squad Rescue Set	blue Jeep with Chopper Squad decal and red/white boat with Surf Rescue decal	40	60	100
Chrysler Imperial Convertible	4-1/4" red or blue-green body with gray base, working hood, trunk & doors, golf bag in trunk, detailed engine, clear windshield, aqua interior, driver, chrome bumpers	45	65	110
Chubb Pathfinder Crash Tender		45	65	110
Chubb Pathfinder Crash Truck	9-1/2" red body with either "Airport Fire Brigade" or "New York Airport" decals, upper & lower body, gold water cannon unpainted & sirens, clear windshield, yellow interior, black steering wheel, chrome plastic deck, silver lights, plastic	60	90	150
Circus Cage Wagon		56	84	140
Circus Crane & Cage	red and blue trailer	400	600	1000
Circus Crane & Cage Wagon	crane truck, cage wagon and accessories	150	225	375
Circus Crane Truck		80	120	200
Circus Horse Transporter		80	120	200
Circus Human Cannonball Truck		30	45	75
Circus Land Rover & Elephant Cage	red Range Rover with blue canopy, Chipperfields Circus decal on canopy, burnt orange elephant cage on red bed trailer	90	135	250
Circus Land Rover & Trailer	yellow/red Land Rover with Pinder-Jean Richard decals	30	45	75

Corgi

NAME	DESCRIPTION	GOOD	EX	MINT
Circus Menagerie Transporter		120	180	350
Circus Set	vehicle and accessory set in two versions: with #426 Booking Office, 1963-65; with #503 Giraffe Truck, 1966, each set	340	510	1000
Citroen 2CV Charleston	4-1/8" long, yellow/black or maroon/black body versions	15	18	30
Citroen Alpine Rescue Safari	4" white body, light blue interior, red roof & rear hatch, yellow roof rack & skis, clear windshield, man & dog, gold die cast bobsled, Alpine Rescue decals	80	120	225
Citroen DS 19 Rally	4" long, light blue body, white roof, yellow interior, four jewel headlights, Monte Carlo Rally and #75 decals	70	105	175
Citroen DS19	4" long, one-piece body in several colors, clear windows, silver lights, grille and bumpers, smooth wheels, rubber tires: colors: cream, yellow/black, red, metallic green, yellow	56	84	140
Citroen Dyane	4-1/2" metallic yellow or green body, black roof & interior, working rear hatch, clear windows, black base & tow bar, silver bumpers, grille & headlights, red taillights, marching duck & French flag decals, suspension, chrome wheels	15	18	30
Citroen ID-19 Safari	4" long, orange body with red/brown or red/green luggage on roof rack, green/brown interior, working hatch, two passengers, Wildlife Preservation decals	40	60	100
Citroen Le Dandy Coupe	4" metallic maroon body & base, yellow interior, working trunk & two doors, clear windows, plastic interior, folding seats, chrome grille & bumpers, jewel headlights, red taillights, suspension, spoked wheels, rubber tires	50	75	125
Citroen Le Dandy Coupe	4" metallic dark blue hood, sides & base, plastic aqua interior, white roof & trunk lid, clear windows, folding seats, chrome grille & bumpers, jewel headlights, red taillights, suspension, spoked wheels, rubber tires	70	105	175
Citroen SM	4-3/16" metallic lime gold with chrome wheels or mauve body with spoked wheels, pale blue interior & lifting hatch cover, working rear hatch & two doors, chrome inner drs., window frames, bumpers, grille, amber headlights, red	16	24	40
Citroen Tour de France Car	4-1/4" red body, yellow interior & rear bed, clear windshield & headlights, driver, black plastic rack with four bicycle wheels, swiveling team manager figure with megaphone in back of car, Paramount & Tour de France decals, taillights, black Whizz Wheels	40	60	100
Citroen Winter Olympics Car	4-1/8" long, white body, blue roof and hatch, blue interior, red roof rack with yellow skis, gold sled with rider, skier, gold Grenoble Olympiade decals on car roof	70	105	200
Citroen Winter Sports Safari	4" long, white body in three versions: two with Corgi Ski Club decals and either with or without roof ski rack, or one with 1964 Winter Olympics decals	56	84	140
Coast Guard Jaguar XJ12C	3-1/4" long, olive drab body either with or without suspension, Red Cross decals, clear front windows, white rear windows	18	27	45
Combine, Tractor & Trailer	set of three: #1111 combine, #50 Massey Ferguson tractor, and #51 trailer	110	165	275
Commer 3/4 Ton Ambulance	3-1/2" long, in either white or cream body, red interior, blue dome light, red Ambulance decals	36	55	90
Commer 3/4 Ton Milk Float	3-1/2" long, white cab with either light or dark blue body, one version with CO-OP decals	32	48	80
Commer 3/4 Ton Pickup	3-1/2" long, either red cab with orange canopy or yellow cab with red canopy, yellow interior in both	30	45	75

Beatles' Yellow Submarine, Corgi

Chitty Chitty Bang Bang, Corgi

Corgi

NAME	DESCRIPTION	GOOD	EX	MINT
Commer 3/4 Ton Police Bus	3-1/2" long, battery operated working dome light, in several color combinations of dark or light metallic blue or green bodies	45	65	110
Commer 3/4 Ton Van	3-1/2" long, either dark blue body with Hammonds decals (1971) or white body with CO-OP decals (1970)	45	65	110
Commer 5 Ton Dropside Truck	4-5/8" long, either blue or red cab, both with cream rear body, sheet metal tow hook, smooth wheels, rubber tires	40	60	100
Commer 5 Ton Platform Truck	4-5/8" long, either yellow or metallic blue cab with silver body	40	60	100
Commer Holiday Mini Bus	3-1/2" white interior, clear windshield, silver bumpers, grille & headlights, Holiday Camp Special decal, roof rack, two working rear doors	30	45	75
Commer Military Ambulance	3-5/8" long, olive drab body, blue rear windows and dome light, driver, Red Cross decals	50	75	125
Commer Military Police Van	3-5/8" long, olive drab body, barred rear windows, white MP decals, driver	55	80	130
Commer Mobile Camera Van	3-1/2" long, metallic blue lower body and roof rack, white upper body, two working rear doors, black camera on gold tripod, cameraman	60	90	175
Commer Refrigerator Van	4-5/8" long, either light or dark blue cab, both with cream bodies and red/white/blue Wall's Ice Cream decals	80	120	225
Commuter Dragster	4-7/8" long, maroon body with Ford Commuter, Union Jack and #2 decals, cast silver engine, chrome plastic suspension and pipes, clear windshield, driver, spoke wheels	30	45	75
Concorde-First Issues	BOAC	20	30	50
Concorde-First Issues	Air France	20	30	50
Concorde-First Issues	Air Canada	80	120	200
Concorde-First Issues	Japan Airlines	280	420	700
Concorde-Second Issues		15	20	35
Construction Set	orange tractor and Mazda	32	48	80
Constructor Set	one each red and white cab bodies, with four different interchangeable rear units; van, pickup, milk truck, and ambulance	48	72	120
Cooper-Maserati Racing Car	3-3/8" long, blue body with red/white/blue Maserati and #7 decals, unpainted engine and suspension, chrome plastic steering wheel, roll bar, mirrors and pipes, driver, cast eight-spoke wheels, plastic tires	26	39	65
Cooper-Maserati Racing Car	3-3/8" long, yellow/white body with yellow/black stripe and #3 decals, driver tilts to steer car	18	27	45
Corgi Flying Club Set	blue/orange Land Rover with red dome light, blue trailer with either orange/yellow or orange/white plastic airplane	24	36	60
Corporal Missile & Erector	1959-62	240	360	600
Corporal Missile Launcher	1960-61	36	55	90
Corporal Missile on Launcher	1959-62	80	120	200
Corporal Missile Set	missile and ramp, erector vehicle and army truck	340	510	850
Corvette Sting Ray	4" metallic green or red body, yellow interior, black working hood, working headlights, clear windshield, amber roof panel, gold dash, chrome grille & bumpers, decals, gray die cast base, Golden jacks, cast wheels, plastic tires	40	60	100

Corgi

NAME	DESCRIPTION	GOOD	EX	MINT
Corvette Sting Ray	3-3/4" metallic silver/red body, two working headlights, clear windshield, yellow interior, silver hood panels, four jewel headlights, suspension, chrome bumpers, with spoked or shaped wheels, rubber tires	60	90	175
Corvette Sting Ray	3-3/4" long, yellow body, red interior, suspension, #13 decals	55	85	140
Corvette Sting Ray	3-5/8" long, metallic gray body with black hood, Whizz Wheels	50	75	125
Corvette Sting Ray	3-7/8" long, either dark metallic blue or metallic mauve-rose body, chrome dash, Whizz Wheels	50	75	125
Country Farm Set	#50 Massey Ferguson tractor, red hay trailer with load, fences, figures	30	45	75
Country Farm Set	same as 4-B but without hay load on trailer	30	45	75
Daily Planet Helicopter	1979-81	24	36	60
Daimler 38 1910	1964-69	20	30	50
Daktari Set	two versions: cast wheels, 1968-73; Whizz Wheels, 1974-75, each set	50	75	150
Datsun 240Z	3-5/8" long, red body with #11 and other decals, two working doors, white interior, orange roll bar and tire rack; one version also has East Africa Rally decals	15	20	35
Datsun 240Z	3-5/8" long, white body with red hood and roof, #46 and other decals	15	20	35
David Brown Combine		30	45	75
David Brown Tractor	4-1/8" long, white body with black/white David Brown #1412 decals, red chassis and plastic engine	15	18	30
David Brown Tractor & Trailer	two-piece set; #55 tractor and #56 trailer	30	45	75
De Tomaso Mangusta	3-7/8" long, metallic dark green body with gold stripes and logo on hood, silver lower body, clear front windows, cream interior, amber rear windows and headlights, gray antenna, spare wheel, Whizz Wheels	26	39	65
De Tomaso Mangusta	5" white upper/light blue lower body/base, black interior, clear windows, silver engine, black grille, amber headlights, red taillights, gray antenna, spare wheel, gold stripes & black logo decal on hood, suspension, removable gray chassisDecca Airfield Radar Van	32	48	80
Decca Radar Scanner	3-1/4" long, with either orange or custard colored scanner frame, silver scanner face, with gear on base for turning scanner	34	51	85
Dick Dastardly's Racing Car	5" long, dark blue body, yellow chassis, chrome engine, red wings, Dick and Muttley figures	40	60	150
Dodge Kew Fargo Tipper	5-1/4" long, white cab and working hood, blue tipper, red interior, clear windows, black hydraulic cylinders, cast wheels, plastic tires	34	51	85
Dodge Livestock Truck	5-3/8" long, tan cab and hood, green body, working tailgate and ramps, five pigs	34	51	85
Dolphin Cabin Cruiser	5-1/4" long, white hull, blue deck plastic boat with red/white stripe decals, driver, blue motor with white cover, gray prop, cast trailer with smooth wheels, rubber tires	24	36	60
Dougal's Magic Roundabout Car	4-1/2" long, yellow body, red interior, clear windows, dog and snail, red wheels with gold trim, Magic Roundabout decals	70	105	175
Drax Jet Helicopter	5-7/8" long, white body, yellow rotors and fins, yellow/black Drax decals	24	36	75
Dropside Trailer	4-3/8" long, cream body, red chassis in five versions: smooth wheels 1957-61; shaped wheels, 1962-1965; white body, cream or blue chassis; or silver gray body, blue chassis, each	10	15	25

Corgi

NAME	DESCRIPTION	GOOD	EX	MINT
Ecurie Ecosse Racing Set	transporter with three cars in two versions: RRM, Vanwall and Lotus XI, 1961-64; BRM, Vanwall and Ferrari, 1964-66, each set	140	210	450
Ecurie Ecosse Transporter	7-3/4" long, in dark blue body with either blue or yellow lettering, or light blue body with red or yellow lettering, working tailgate and sliding door, yellow interior, shaped wheels, rubber tires	70	105	200
Emergency Set	Land Rover Police Car and Police Helicopter with figures and accessories	40	60	100
Emergency Set	three-vehicle set with figures and accessories, Ford Cortina Police car, Police Helicopter, Range Rover Ambulance	40	60	100
ERF 44G Dropside Truck		36	55	90
ERF 44G Moorhouse Van	4-5/8" long, yellow cab, red body, Moorhouse Lemon Cheese decals	100	150	250
ERF 44G Platform Truck	4-5/8" long, light blue cab with either dark blue or white flatbed body	36	55	90
ERF Dropside Truck & Trailer	#456 truck and #101 trailer with #1488 cement sack load and #1485 plank load	60	90	150
ERF Neville Cement Tipper	3-3/4" long, yellow cab, gray tipper, cement decal, with either smooth or shaped wheels	32	48	80
ERG 64G Earth Dumper	4" long, red cab, yellow tipper, clear windows, unpainted hydraulic cylinder, spare tire, smooth wheels, rubber tires	30	45	75
Euclid Caterpillar Tractor	4-1/4", TC-12 lime green body with black or white rubber treads, gray plastic seat, driver figure, controls, stacks, silver grille, painted blue engine sides & Euclid decals			
Euclid TC-12 Bulldozer	5", yellow body with black treads or lime green body with white treads, silver blade surface, gray plastic seat controls & stacks, silver grille & lights, painted blue engine sides, sheet metal base, rubber treads & Euclid decals	80	120	200
Euclid TC-12 Bulldozer	6-1/8" long, red or green body, metal control rod, driver, black rubber treads	80	120	200
Ferrari 206 Dino	4-1/8" long, black interior and fins, in either red body with #30 and gold or Whizz Wheels, or yellow body with #23 and gold or Whizz Wheels	24	36	60
Ferrari 308GTS	4-5/8" long, red or black body with working rear hood, black interior with tan seats, movable chrome headlights, detailed engine	15	20	35
Ferrari 308GTS Magnum	4-5/8" long, red body with solid chrome wheels	24	36	60
Ferrari 312 B2 Racing Car	4" long, red body, white fin, gold engine, chrome suspension, mirrors and wheels, Ferrari and #5 decals	16	24	40
Ferrari Berlinetta 250LM	3-3/4" red body with yellow stripe, blue windshields, chrome interior, grille & exhaust pipes, detailed engine, #4 Ferrari logo & yellow stripe decals, spoked wheels & spare, rubber tires	30	45	75
Ferrari Daytona	5" apple green body, black tow hook, red-yellow-silver black Daytona #5 & other racing decals, amber windows, headlights, black plastic interior, base, four spoke chrome wheels	15	20	35
Ferrari Daytona	4-3/4" long, white body with red roof and trunk, black interior, two working doors, amber windows and headlights, #81 and other decals	15	20	35
Ferrari Daytona & Racing Car	blue/yellow Ferrari and Surtees on yellow trailer	15	18	30
Ferrari Daytona JCB	4-3/4" long, orange body with #33, Corgi and other decals, chrome spoked wheels	15	25	50

NAME	DESCRIPTION	GOOD	EX	MINT
Ferrari Racing Car	3-5/8" long, red body, chrome plastic engine, roll bar and dash, driver, silver cast base and exhaust, Ferrari and #36 decals	24	36	60
Fiat 1800	3-3/4" long, one-piece body in several colors, clear windows, plastic interior, silver lights, grille and bumpers, red taillights, smooth wheels, rubber tires, colors: blue body with light or bright yellow interior, mustard or cream body	24	36	60
Fiat 2100	3-3/4" long, light mauve body, yellow interior, purple roof, clear windows with rear blind, silver grille, license plates & bumpers, red taillights, shaped wheels, rubber tires	22	33	55
Fiat X 1/9 & Powerboat	green/white Fiat, with white/gold boat	30	45	75
Fiat X1/9	4-3/4" metallic blue body & base, white Fiat #3, multicolored lettering & stripe decals, black roof, trim, interior, rear panel, grille, bumpers & tow hook, chrome wheels & detailed engine	15	20	35
Fiat X1/9	4-1/2" metallic light green or silver body with black roof, trim & interior, two working doors, rear panel, grille, tow hook & bumpers, detailed engine, suspension, chrome wheels	15	20	35
Fire Bug	1972-73	20	30	50
Fire Engine	1975-78	16	24	40
Flying Club Set	green/white Jeep with Corgi Flying Club decals, green trailer, blue/white airplane	36	55	90
Ford 5000 Super Major Tractor	3-3/4" long, blue body/chassis with Ford Super Major 5000 decals, gray cast fenders and rear wheels, gray plastic front wheels, black plastic tires, driver	30	45	75
Ford 5000 Tractor with Scoop	3-1/8" long, blue body/chassis, gray fenders, yellow scoop arm and controls, chrome scoop, black control lines	55	80	130
Ford Aral Tank Truck		20	30	50
Ford Capri	4" orange-red or dark red body, gold wheels with red hubs, two working doors, clear windshield & headlights, black interior, folding seats, black grille, silver bumpers	15	20	35
Ford Capri 3 Litre GT		15	20	35
Ford Capri 30 S		15	20	35
Ford Capri S	4-3/4" white body, red lower body & base, red interior, clear windshield, black bumpers, grille & tow hook, chrome headlights & wheels, red taillights, #6 & other racing decals	15	20	35
Ford Capri Santa Pod Gloworm	4-3/8" long, white/blue body with red/white/blue lettering and flag decals, red chassis, amber windows, gold-based black engine, gold scoop, pipes and front suspension	18	27	45
Ford Car Transporter		20	30	50
Ford Car Transporter		20	30	50
Ford Cobra Mustang		15	18	30
Ford Consul	3-5/8" long, one-piece body in several colors, clear windows, silver grille, lights and bumpers, smooth wheels, rubber tires	45	65	110
Ford Consul Classic	3-3/4" long, cream or gold body & base, yellow interior, pink roof, clear windows, gray steering wheel, silver bumpers, grille	35	55	90
Ford Consul-Mechanical	same as model 200-A but with friction motor and blue or green body	55	85	140
Ford Cortina Estate Car	3-1/2" metallic dark blue body & base, brown & cream simulated wood panels, cream interior, chrome bumpers & grille, jewel headlights	35	55	90

Saint's Volvo P-1800, Corgi

Jeep FC-150 Pickup, Corgi

Corgi

NAME	DESCRIPTION	GOOD	EX	MINT
Ford Cortina Estate Car	3-3/4" red body & base or metallic charcoal gray body & base, cream interior, chrome bumpers & grille, jewel headlights	35	55	90
Ford Cortina GXL	4" tan or metallic silver blue body, black roof & stripes, red plastic interior, working doors, clear windshield	30	45	75
Ford Cortina Police Car	4" white body, red or pink & black stripe labels, red interior, folding seats, blue dome light, clear windows, chrome bumpers	15	18	30
Ford Covered Semi-Trailer		15	25	40
Ford Escort 13 GL		8	12	20
Ford Escort Police Car	4-3/16" blue body & base, tan interior, white doors, blue dome lights, red Police labels, black grille & bumpers	8	15	20
Ford Esso Tank Truck	1976-81	15	25	40
Ford Express Semi-Trailer	1965-70	60	90	150
Ford Exxon Tank Truck	1976-81	15	25	40
Ford GT 70	1972-73	10	15	25
Ford Guinness Tanker	1982	20	30	50
Ford Gulf Tank Truck	1976-78	15	25	40
Ford Holmes Wrecker	1967-74	60	90	200
Ford Michelin Container Truck	1981	15	25	40
Ford Mustang Fastback	1965-69	25	35	60
Ford Mustang Fastback	1965-66	30	45	95
Ford Mustang Mach 1	4-1/4" green upper body, white lower body & base, cream interior, folding seat backs, chrome headlights & rear bumper	25	35	60
Ford Sierra	5" many versions with plastic interior, working hatch & two doors, clear windows, folding seat back, lifting hatch cover	8	15	20
Ford Sierra and Caravan Trailer	blue #299 Sierra, two-tone blue/white #490 Caravan	15	20	35
Ford Sierra Taxi		8	15	20
Ford Thames Airborne Caravan	3-3/4" different versions of body & plastic interior with table, white blinds, silver bumpers, grille & headlights, two doors	35	55	95
Ford Thames Wall's Ice Cream Van	4" light blue body, cream pillar, chimes, chrome bumpers & grille, crank at rear to operate chimes, no figures	35	50	175
Ford Thunderbird 1957	5-3/16" cream body, dark brown, black or orange plastic hardtop, black interior, open hood & trunk, chrome bumpers	10	15	25
Ford Thunderbird 1957	5-1/4" white body, black interior & plastic top, amber windows, white seats, chrome bumpers, headlights & spare wheel cover	10	15	25
Ford Thunderbird Hardtop	4-1/8" long, clear windows, silver lights, grille and bumpers, red taillights, rubber tires; light green body 1959-61	50	80	130
Ford Thunderbird	4-1/8" long, same as 214-A but with friction motor and pink or light green body	70	105	175
Hardtop-MechanicalFord Thunderbird Roadster	4-1/8" long, clear windshield, silver seats, lights, grille and bumpers, red taillights, rubber tires, white body	50	75	125
Ford Torino Road Hog	5-3/4" orange-red body, yellow and gray chassis, gold lamps, chrome radiator shell, windows & bumpers, one-piece body	15	20	35
Ford Tractor & Conveyor	tractor, conveyor with trailer, figures and accessories	60	90	150

Corgi

NAME	DESCRIPTION	GOOD	EX	MINT
Ford Tractor and Beast Carrier		60	90	150
Ford Tractor with Trencher	5-5/8" long, blue body/chassis, gray fenders, cast yellow trencher arm and controls, chrome trencher, black control lines	50	75	125
Ford Transit Milk Float	5-1/2" white one-piece body, blue hood & roof, tan interior, chrome & red roof lights, open compartment door & milk cases	15	25	40
Ford Transit Tipper		10	15	25
Ford Transit Wrecker		25	35	60
Ford Wall's Ice Cream Van	3-1/4" light blue body, dark cream pillars, plastic striped rear canopy, white interior, silver bumpers, grille & headlights	50	75	150
Ford Zephyr Estate Car	3-7/8" light blue one-piece body, dark blue hood & stripes, red interior, silver bumpers, grille & headlights, red taillights	30	45	75
Ford Zephyr Patrol Car	3-3/4" white or cream body, blue & white Police/Politie/Rijkspolitie decals, red interior, blue dome light, silver bumpers	35	50	85
Fordson Power Major Halftrack Tractor	3-1/2" long, blue body/chassis, silver steering wheel, seat and grille, three versions: orange cast wheels, gray treads	90	135	225
Fordson Power Major Tractor	3-3/8" long, blue body with Fordson Power Major decals, driver, blue chassis and steering wheel, silver seat, hitch, exhaust	50	75	125
Fordson Power Major Tractor	3-1/4" long, blue body/chassis with Fordson Power Major decals, silver steering wheel, seat, exhaust, grille and lights	45	65	110
Fordson Tractor & Plow	#55 Fordson Tractor & 56 Four Furrow plow	55	85	140
Fordson Tractor & Plowc	tractor and four-furrow plow	55	85	140
Four Furrow Plow	3-5/8" long, red frame, yellow plastic parts	15	20	35
Four Furrow Plow	3-3/4" long, blue frame with chrome plastic parts	15	20	35
French Construction Set	1980	25	35	60
Futuristic Space Vehicle		15	18	30
Futuristic Space Vehicle		15	18	30
German Life Saving Set	red/white Land Rover and lifeboat, white trailer, German decals	30	45	75
Ghia L64 Chrysler V8	4-1/4", different color versions, plastic interior, hood, trunk & two doors working, detailed engine, clear windshield	25	40	65
Ghia-Fiat 600 Jolly	3-1/4", light or dark blue body, red & silver canopy, red seats, two figures, windshield, chrome dash, floor, steering wheels	45	65	110
Ghia-Fiat 600 Jolly	3-1/4" long, dark yellow body, red seats, two figures and a dog, clear windshield, silver bumpers and headlights, red taillights	60	90	150
Giant Daktari Set	black/green Land Rover, tan Giraffe truck, blue/brown Dodge Livestock truck, tan figures, set	225	350	650
Giant Tower Crane		35	50	85
Glider Set	two versions: white Honda, 1981-82; yellow Honda, 1983 on, each set	30	45	75
Golden Eagle Jeep	3-3/4" three different versions, tan plastic top, chrome plastic base, bumpers & steps, chrome wheels	8	15	20
Golden Guinea Set	three vehicle set, gold plated Bentley Continental, Chevy Corvair and Ford Consul	90	135	300
GP Beach Buggy		15	20	35
Grand Prix Racing Set	four vehicle set with accessories in two versions: with #330 Porsche, 1969; with Porsche #371, 1970-72, each set	135	210	400
Grand Prix Set	#30-A, sold by mail order only	50	75	125

NAME	DESCRIPTION	GOOD	EX	MINT
Green Hornet's Black Beauty	5" black body, green window/interior, two figures, working chrome grille & panels with weapons, green headlights, red taillights	175	275	500
Green Line Bus	4-7/8" green body, white interior & stripe, TDK labels, six spoked wheels	10	15	25
Half Track Rocket Launcher & Trailer	6-1/2" two rocket launchers & single trailer castings, gray plastic roll cage, man with machine gun, front wheels & hubs	20	35	55
Hardy Boys' Rolls-Royce	4-5/8" long, red body with yellow hood, roof and window frames, band figures on roof on removable green base	70	105	200
HDL Hovercraft SR-N1	blue superstructure, gray base and deck, clear canopy, red seats, yellow SR-N1 decals	60	90	150
Hesketh-Ford Racing Car	5-5/8" long, white body with red/white/blue Hesketh, stripe and #24 decals, chrome suspension, roll bar, mirrors and pipes	15	18	30
HGB-Angus Firestreak	6-1/4" long, chrome plastic spotlight and ladders, black hose reel, red dome light, white water cannon, in two interior versions	35	50	85
Hillman Hunter	4-1/4" blue body, gray interior, black hood, white roof, unpainted spotlights, clear windshield, red radiator screen, black equipment	45	65	125
Hillman Husky	3-1/2" long, one-piece tan or metallic blue/silver body, clear windows, silver lights, grille and bumpers, smooth wheels	40	60	100
Hillman Husky-Mechanical	3-1/2" long, same as 206-A but with friction motor, black base and dark blue, gray or cream body	50	75	125
Hillman Imp	3-1/4" metallic copper, blue, dark blue, gold & maroon one-piece bodies, with white/yellow interior, silver bumpers, headlights	30	45	75
Hillman Imp Rally	3-1/4" long, in various metallic body colors, with cream interior, Monte Carlo Rally and #107 decals	30	45	75
Honda Ballade Driving School	4-3/4" red body/base, tan interior, clear windows, tow hook, mirrors, bumpers	10	15	25
Honda Prelude	4-3/4" long, dark metallic blue body, tan interior, clear windows, folding seats, sunroof, chrome wheels	8	15	20
Hughes Police Helicopter	5-1/2" long, red interior, dark blue rotors, in several international imprints, Netherlands, German, Swiss, in white or yellow	20	30	50
Hyster 800 Stacatruck	8-1/2" long, clear windows, black interior with driver	35	50	85
Incredible Hulk Mazda Pickup	5" metallic light brown body, gray or red plastic cage, black interior, green & red figure, Hulk decal on hood, chrome wheels	20	30	75
Inter-City Mini Bus	4-3/16" long, orange body with brown interior, clear windows, green/yellow/black decals, Whizz Wheels	8	15	20
International 6x6 Army Truck	5-1/2" long, olive drab body with clear windows, red/blue decals, six cast olive wheels with rubber tires	70	105	175
Iso Grifo 7 Litre	4" metallic blue body, light blue interior, black hood & stripe, clear windshield, black dash, folding seats, chrome bumpers	15	18	30
Jaguar 1952 XK120 Rally	4-3/4" long, cream body with black top and trim, red interior, Rally des Alps and #414 decals	8	15	20
Jaguar 2.4 Litre	3-7/8" long, one-piece white body with no interior 1957-59, or yellow body with red interior 1960-63, clear windows	50	80	130
Jaguar 2.4 Litre Fire Chief's Car	3-3/4" long, red body with unpainted roof signal/siren, red/white fire and shield decals on doors, in two versions	60	90	150
Jaguar 2.4 Litre-Mechanical	3-7/8" long, same as 208-A but with friction motor and metallic blue body	60	90	150

NAME	DESCRIPTION	GOOD	EX	MINT
Jaguar E Type	3-3/4" maroon or metallic dark gray body, tan interior, red & clear plastic removeable hardtop, clear windshield, folded top	45	65	110
Jaguar E Type 2+2	4-3/16" long, working hood, doors and hatch, black interior with folding seats, copper engine, pipes and suspension, spoked wheels	40	60	100
Jaguar E Type 2+2	4-1/8" long, in five versions: red or yellow with nonworking doors; or with V-12 engine in yellow body or metallic yellow body	35	55	90
Jaguar E Type Competition	3-3/4" gold or chrome plated body, black interior, blue & white stripes & black #2 decals, no top, clear windshield & headlights	45	65	110
Jaguar Mark X Saloon	4-1/4", seven different versions with working front & rear hood castings, clear windshields, plastic interior, gray steering wheel	35	55	90
Jaguar XJ12C	5-1/4" five different metallic versions, working hood & two doors, clear windows, tow hook, chrome bumpers, grille & headlights	10	15	25
Jaguar XJ12C Police Car	5-1/8" long, white body with blue and pink stripes, light bar with blue dome light, tan interior, police decals	15	18	30
Jaguar XJS	5-3/4" long, metallic burgundy body, tan interior, clear windows, working doors, spoked chrome wheels	10	15	25
Jaguar XJS Motul	5-3/4" long, black body with red/white Motul and #4, chrome wheels	8	15	20
Jaguar XJS-HE Supercat	5-1/4" black body with silver stripes & trim, red interior, dark red taillights, light gray antenna, no tow hook, clear windshield	8	15	20
Jaguar XK120 Hardtop	4-3/4" long, red body, black hardtop, working hood and trunk, detailed engine, cream interior, clear windows, chrome wheels	8	15	20
James Bond Aston Martin	4" metallic silver body, red interior, two figures, working roof hatch, ejector seat, bullet shield and guns, chrome bumpers	100	150	275
James Bond Aston Martin	5" metallic silver body & die cast base, red interior, two figures, clear windows, passenger set raises to eject	30	45	75
James Bond Aston Martin DB5	3-3/4" metallic gold body, red interior, working roof hatch, clear windows, two figures, left seat ejects	70	105	250
James Bond Bobsled	2-7/8" long, yellow body, silver base, Bond figure, 007 decals, Whizz Wheels	60	90	175
James Bond Citroen 2CV6	4-1/4" dark yellow body & hood, red interior, clear windows, chrome headlights, red taillights, black plastic grille	15	25	50
James Bond Lotus Esprit	4-3/4" white body & base, black windshield, grille & hood panel, white plastic roof device that triggers fins & tail, rockets	30	45	95
James Bond Moon Buggy	4-3/8" long, white body with blue chassis, amber canopy, yellow tanks, red radar dish, arms and jaws, yellow wheels	175	275	525
James Bond Mustang Mach 1	4-3/8" long, red and white body with black hood	100	150	275
James Bond Set	set of three: Lotus Esprit, Space Shuttle and Aston Martin	80	120	225
James Bond Space Shuttle	5-7/8" long, white body with yellow/black Moonraker decals	30	45	75
James Bond SPECTRE Bobsled	2-7/8" long, orange body with wild boar decals	60	90	175
James Bond Toyota 2000GT	4" long, white body, black interior with Bond and passenger, working trunk and gun rack, spoked wheels, plastic tires	115	180	375
JCB 110B Crawler Loader	6-1/2" long, white cab, yellow body, working red shovel, red interior with driver, clear windows, black treads, JCB decals	20	30	50

Jaguar Mark X Saloon, Corgi

Popeye's Paddle Wagon, Stromberg Jet Ranger Helicopter, Corgi

Corgi

NAME	DESCRIPTION	GOOD	EX	MINT
Jean Richard Circus Set	yellow/red Land Rover and cage trailer with Pinder-Jean Richard decals, office van and trailer, Human Cannonball truck, ring	90	135	275
Jeep & Horse Box	metallic painted Jeep and trailer	15	25	40
Jeep & Motorcycle Trailer	red working Jeep with two blue/yellow bikes on trailer	15	20	35
Jeep CJ-5	4" long, dark metallic green body, removable white top, white plastic wheels, spare tire	8	15	20
Jeep FC-150 Covered Truck	four versions: blue body, rubber tires (1965-67), yellow/brown body, rubber tires (1965-67), blue body, plastic tires	30	45	75
Jeep FC-150 Pickup	3-1/2" long, blue body, clear windows, sheet metal tow hook, in two wheel versions: smooth or shaped wheels	35	55	90
Jeep FC-150 Pickup with Conveyor Belt	7-1/2" long, red body, yellow interior, orange grille, two rubber belts, shaped wheels, black rubber tires	45	65	110
Jeep FC-150 Tower Wagon	4-5/8" long, metallic green body, yellow interior and basket with workman figure, clear windows, with either rubber or plastic wheels	40	60	100
Jet Ranger Police Helicopter	5-7/8" long, white body with chrome interior, red floats and rotors, amber windows, Police decals	25	40	65
JPS Lotus Racing Car	10-1/2" long, black body, scoop and wings with gold John Player Special, Texaco and #1 decals, gold suspension, pipes and wheels	30	45	75
Karrier Bantam Two Ton Van	4" long, blue body, red chassis and bed, clear windows, smooth wheels, rubber tires	35	55	95
Karrier Butcher Shop	3-5/8" long, white body, blue roof, butcher shop interior, Home Service decals, in two versions: with or without suspension	65	100	165
Karrier Circus Booking Office	3-5/8" long, red body, light blue roof, clear windows, circus decals, shaped wheels, rubber tires	105	165	275
Karrier Dairy Van	4-1/8" long, light blue body with Drive Safely on Milk decals, white roof, with either smooth or shaped wheels	50	75	125
Karrier Field Kitchen	3-5/8" long, olive body, white decals	60	90	150
Karrier Ice Cream Truck	3-5/8" long, cream upper, blue lower body and interior, clear windows, sliding side windows, Mister Softee decals, figure inside	90	135	225
Karrier Lucozade Van	4-1/8" long, yellow body with gray rear door, Lucozade decals, rubber tires, with either smooth or shaped wheels	70	105	175
Karrier Mobile Canteen	3-5/8" long, blue body, white interior, amber windows, roof knob rotates figure, working side panel counter	60	90	150
Karrier Mobile Grocery	3-5/8" long, light green body, grocery store interior, red/white Home Service decals, friction motor, rubber tires	70	110	185
King Tiger Heavy Tank	6-1/8" long, tan and rust body, working turret and barrel, tan rollers and treads, German decals	30	45	75
Kojak's Buick Regal	5-3/4" metallic bronze brown body, off-white interior, two doors, clear windows, chrome bumpers, grille & headlights, red taillights	25	40	85
Lamborghini Miura	3-3/4" long, silver body, black interior, yellow/purple stripes and #7 decal	30	45	75
Lamborghini Miura P400	3-3/4" long, with red or yellow body, working hood, detailed engine, clear windows, jewel headlights, bull, Whizz Wheels	40	60	100
Lancia Fulvia Zagato	3-5/8" long, metallic blue body, light blue interior, working hood and doors, folding seats, amber lights, cast wheels	25	35	60
Lancia Fulvia Zagato	3-5/8" long, orange body, black working hood and interior, Whizz Wheels	15	25	40

NAME	DESCRIPTION	GOOD	EX	MINT
Land Rover & Ferrari Racer	red/tan Land Rover, yellow trailer	60	90	150
Land Rover & Horse Box	blue/white Land Rover with horse trailer in two versions: cast wheels, 1968-74; Whizz Wheels, 1975-77	50	75	125
Land Rover 109WB	5-1/4" long, working rear doors, tan interior, spare on hood, plastic tow hook	15	18	30
Land Rover and Pony Trailer	two versions: tan/cream Rover, 1958-62; tan/cream #438 Land Rover, 1963-68, each set	50	75	125
Land Rover Breakdown Truck	4-3/8" long, red body with silver boom and yellow canopy, revolving spotlight, Breakdown Service decals	35	55	90
Land Rover Breakdown Truck	4-3/8" long, red body, yellow canopy, chrome revolving spotlight, Breakdown Service decals	25	35	60
Land Rover Circus Vehicle	3-1/2" long, red body, yellow interior, blue rear and speakers, revolving clown, chimp figures, Chipperfield decals	60	90	150
Land Rover Pickup	3-3/4" long, yellow or metallic blue body, spare on hood, clear windows, sheet metal tow hook, rubber tires	45	70	120
Land Rover with Canopy	3-3/4" long, one-piece body with clear windows, plastic interior, spare on hood, issued in numerous colors	35	55	90
Lincoln Continental	5-3/4" metallic gold or blue body, black roof, maroon plastic interior, working hood, trunk & doors, clear windows, TV	60	90	150
Lions of Longleat	black/white Land Rover pickup with lion cages and accessories, two versions: cast wheels, 1969-73; Whizz Wheels, 1974, each	60	90	200
London Set	taxi and bus with policeman, in two versions: "Corgi Toys" on bus, 1964-66; "Outspan Oranges" on bus, 1967-68, each set	55	85	140
London Set	orange Mini, Policeman, London Taxi and Routemaster bus	50	75	125
London Set	London Taxi and Routemaster bus in two versions: with mounted Policeman, 1980-81; without Policeman, 1982 on, each set	25	35	60
London Transport Routemaster Bus	4-7/8" long, clear windows, interior, some models have driver and conductor, released with numerous advertiser logos	25	35	60
London Transport Routemaster Bus	4-1/2" long, clear windows with driver and conductor, released with numerous advertiser logos	35	50	85
Lotus Elan S2 Hardtop	2-1/4" long, cream interior with folding seats and tan dash, working hood, separate chrome chassis, issued in blue body	30	45	75
Lotus Elan S2 Roadster	3-3/8" long, working hood, plastic interior with folding seats, shaped wheels and rubber tires, issued in metallic blue or white	30	45	75
Lotus Eleven	2-1/4" long, clear windshield and plastic headlights, smooth wheels, rubber tires, racing decals, in several color variations	60	95	160
Lotus Elite	5-1/8" red body, white interior, two working doors, clear windshield, black dash, hood panel, grille, bumpers, base & tow hook	15	18	30
Lotus Elite 22	4-3/4" long, dark blue body with silver trim	15	18	30
Lotus Racing Car	5-5/8" long, black body and base, gold cast engine, roll bar, pipes, dash and mirrors, driver, gold cast wheels, in two versions	25	35	60
Lotus Racing Set	three versions: #3 on Elite and JPS on racer; #7 on Elite and JPS on racer; #7 on Elite and Texaco on racer, each set	30	45	75

NAME	DESCRIPTION	GOOD	EX	MINT
Lotus Racing Team Set	four vehicle set in two versions, with accessories: #319 Lotus has either red or black interior, each set	30	45	75
Lotus-Climax Racing Car	3-5/8" long, green body and base with black/white #1 and yellow racing stripe decals, unpainted engine and suspension	25	35	60
Lotus-Climax Racing Car	3-5/8" long, orange/white body with black/white stripe and #8 decals, unpainted cast rear wing, cast 8-spoke wheels	15	25	40
Lunar Bug	5" long, white body with red roof, blue interior and wings, clear and amber windows, red working ramp, Lunar Bug decals	25	40	85
M60 A1 Medium Tank	4-3/4" long, green/tan camo body, working turret and barrel, green rollers, white decals	30	45	75
Mack Container Truck	11-3/8" long, yellow cab, red interior, white engine, red suspension, white ACL decals	30	50	80
Mack Esso Tank Truck	10-3/4" long, white cab and tank with Esso decals, red tank chassis and fenders	20	30	50
Mack Exxon Tank Truck	10-3/4" long, white cab and tank, red tank chassis and fenders, red interior, chrome catwalk, Exxon decals	15	25	40
Mack Trans Continental Semi	10", orange cab body & semi chassis & fenders, metallic light blue semi body, unpainted trailer rests	35	55	90
Mack-Priestman Crane Truck	9" long, red truck, yellow crane cab, red interior, black engine, Hi Lift and Long Vehicle decals	50	75	125
Magic Roundabout Train	1973	70	105	225
Man From U.N.C.L.E. THRUSH-Buster	4-1/8" long, plastic interior, blue windows, two figures, two spotlights, dark metallic blue	80	120	250
Man From U.N.C.L.E. THRUSH-Buster	4-1/8" long, plastic interior, blue windows, two figures, two spotlights, cream body	100	350	550
Marcos 3 Litre	3-3/8" long, working hood, detailed engine, black interior, Marcos decal, Whizz Wheels, issued in orange or metallic blue/green	20	30	50
Marcos Matis	4-1/4" metallic red body & doors, cream interior & headlights, silver gray lower body base, bumpers, hood panel	20	35	55
Marcos Volvo 1800 GT	3-5/8" long, issued with either white or blue body, plastic interior with driver, spoked wheels, rubber tires	25	35	60
Massey Ferguson 165 Tractor	3" long, gray engine and chassis, red hood and fenders with black/white Massey Ferguson 165 decals, white grille, red cast wheels	35	55	90
Massey Ferguson 165 Tractor with Saw	3-1/2" long, red hood and fenders, gray engine and seat, cast yellow arm and control, chrome circular saw	55	85	140
Massey Ferguson 165 Tractor with Shovel	5-1/8" long, gray chassis, red hood, fenders and shovel arms, unpainted shovel and cylinder, red cast wheels, black plastic tires	45	65	110
Massey Ferguson 50B Tractor	4" long, yellow body, black interior and roof, red plastic wheels with black plastic tires	15	18	30
Massey Ferguson 65 Tractor	3" long, silver steering wheel, seat and grille, red engine hood, red wheels with black rubber tires	40	60	100
Massey Ferguson 65 Tractor And Shovel	4-3/4" long, two versions: either cream or gray chassis, each	55	85	140
Massey Ferguson Combine	6-1/2" long, red body with yellow metal blades, metal tines, black/white decals, orange wheels	70	105	175
Massey Ferguson Combine	6-1/2" long, red body, plastic blades, red wheels	60	90	150
Massey Ferguson Tipping Trailer	3-5/8" long, two versions: either yellow or gray tipper and tailgate, each	10	15	25

NAME	DESCRIPTION	GOOD	EX	MINT
Massey Ferguson Tractor & Tipping Trailer		50	75	125
Massey Ferguson Tractor & Tipping Trailer	#50 MF tractor with driver, #51 trailer	50	75	125
Massey Ferguson Tractor with Fork	4-7/8" long, red cast body and shovel, arms, cream chassis, red plastic wheels, black rubber tires, Massey Ferguson 65 decals	60	90	150
Massey Ferguson Tractor with Shovel	6" long, two versions: either yellow and red or red and white body colors	20	30	50
Massey Ferguson Tractor with Shovel & Trailer	#35 MF tractor with driver and shovel, #62 trailer	30	45	75
Matra & Motorcycle Trailer	red Rancho with two yellow/blue bikes on trailer	15	20	35
Matra & Racing Car	black/yellow Rancho and yellow car with Team Corgi decals	15	25	40
Mazda 4X4 Open Truck	4-7/8" long, blue body, white roof, black windows, no interior, white plastic wheels	15	20	35
Mazda B-1600 Pickup Truck	4-7/8" long, issued in either blue and white or blue and silver bodies with working tailgate, black interior, chrome wheels	15	20	35
Mazda Camper Pickup	5-3/8" long, red truck and white camper with red interior and folding supports	15	25	40
Mazda Custom Pickup	4-7/8" long, orange body with red roof	15	18	30
Mazda Motorway Maintenance Truck	6-1/8" long, deep yellow body with red base, black interior and hydraulic cylinder, yellow basket with workman figure	18	25	45
Mazda Pickup & Dinghy	two versions: red Mazda with "Ford" decals; or with "Sea Spray" decals	25	35	60
McLaren M19A Racing Car	4-5/8" long, white body, orange stripes, chrome engine, exhaust and suspension, black mirrors, driver, Yardley McLaren #55 decals	15	25	40
McLaren M23 Racing Car	10-1/4" long, red/white body and wings with red/white/black Texaco-Marlboro #5 decals, chrome pipes, suspension and mirrors	30	45	75
Mercedes-Benz & Caracan	truck and trailer in two versions: with blue Mercedes truck, 1975-79; with brown Mercedes, 1980-81	15	25	40
Mercedes-Benz 220SE Coupe	3-3/4" long, cream, black or dark red body, red plastic interior, clear windows, working trunk, silver bumpers, grille & plate	40	60	100
Mercedes-Benz 220SE Coupe	4" metallic maroon body, yellow plastic interior, light gray base, clear windows, silver bumpers, headlights, grille & license	40	60	100
Mercedes-Benz 220SE Coupe	4" metallic dark blue body, cream plastic interior, medium gray base, clear windows, silver bumpers, headlights, grille & license	40	60	100
Mercedes-Benz 240D	5-1/4" three different versions, working trunk, two doors, clear windows, plastic interior, two hook, chrome bumpers, grille & headlights	10	15	25
Mercedes-Benz 240D Rally	5-1/8" cream or tan body, black, red & blue lettering & dirt, red plastic interior, clear windows, black radiator guard & roof	10	15	25
Mercedes-Benz 240D Taxi	5" orange body, orange interior, black roof sign with red and white Taxi labels, black on door	15	18	30
Mercedes-Benz 300SC Convertible	5" black body, black folded top, white interior, folding seat backs, detailed engine, chrome grille & wheels, lights, bumpers	8	12	20

Fire Engine, Corgi

James Bond Lotus Esprit, Corgi

NAME	DESCRIPTION	GOOD	EX	MINT
Mercedes-Benz 300SC Hardtop	5" maroon body, tan top & interior, open hood & trunk, clear windows, folding seat backs, top with chrome side irons	8	15	20
Mercedes-Benz 300SL	5" red body & base, tan interior, open hood & two gullwing doors, black dash, detailed engine, clear windows, chrome bumpers	8	15	20
Mercedes-Benz 300SL	4-3/4" silver body, tan interior, black dash, clear windows, open hood & two gullwing doors, detailed engine, chrome bumpers	8	15	20
Mercedes-Benz 300SL Coupe	3-3/4" different body & interior versions, hardtop, clear windows, `59-60 smooth wheels no suspension, `61-65 racing stripes	45	65	110
Mercedes-Benz 300SL Roadster	3-3/4" different body & interior versions, plastic interior, smooth, shaped or cast wheels, racing stripes & number, driver	45	65	110
Mercedes-Benz 350SL	3-3/4" white body, spoke wheels or metallic dark blue body solid wheels, pale blue interior, folding seats, detailed engine	15	25	40
Mercedes-Benz 600 Pullman	4-1/4" metallic maroon body, cream interior & steering wheel, clear windshields, chrome grille, trim & bumpers	40	60	100
Mercedes-Benz Ambulance	5-3/4" three different versions, white interior, open rear & two doors, blue windows & dome lights, chrome bumpers, grille & headlights	15	20	35
Mercedes-Benz Ambulance	5-3/4" white body & base, red stripes & taillights, Red Cross & black & white ambulance labels, open rear door, white interior	15	20	35
Mercedes-Benz C-111	4" orange main body with black lower & base, black interior, vents, front & rear grilles, silver headlights, red taillights	15	18	30
Mercedes-Benz Fire Chief	5" light red body, black base, tan plastic interior, blue dome light, white Notruf 112 decals, red taillights, no tow hook	15	25	40
Mercedes-Benz Police Car	5" white body with two different hood versions, brown interior, polizei or police lettering, blue dome light	15	18	30
Mercedes-Benz Refrigerator	8" yellow cab & tailgate, red semi-trailer, two-piece lowering tailgate & yellow spare wheel base, red interior, clear window	15	18	30
Mercedes-Benz Semi-Trailer		15	18	30
Mercedes-Benz Semi-Trailer Van	8-1/4" black cab & plastic semi trailer, white chassis & airscreen, red doors, red-blue & yellow stripes, white Corgi lettering	15	18	30
Mercedes-Benz Tanker	7-1/4" tan cab, plastic tank body, black chassis, black & red Guinness labels, with chrome or black plastic catwalk, clear windows	15	18	30
Mercedes-Benz Tanker	7-1/4" two different versions, cab & tank, chassis, chrome or black plastic catwalk, red/white/green 7-Up labels	15	18	30
Mercedes-Benz Unimog & Dumper	6-3/4" yellow cab & tipper, red fenders & tipper chassis, charcoal gray cab chassis, black plastic mirrors or without	25	35	60
Mercedes-Benz Unimog 406	3-3/4" yellow body, red front fenders & bumpers, metallic charcoal gray chassis with olive or tan rear plastic covers, red interior	18	25	45
Mercedes-Faun Street Sweeper	5" orange body with light orange or brown figure, red interior, black chassis & unpainted brushing housing & arm castings	15	25	40
Metropolis Police Car	6" metallic blue body, off white interior, white roof/stripes, two working doors, clear windows, chrome bumpers, grille & headlights	20	30	50

Corgi

NAME	DESCRIPTION	GOOD	EX	MINT
MG Maestro	4-1/2" yellow body, black trim, opaque black windows, black plastic grille, bumpers, spoiler, trim & battery hatch, clear headlights	15	20	35
MGA	3-3/4" metallic light brown body, all white interior, black dash, clear windshield, silver bumpers, grille & headlight decals	60	90	150
MGB GT	3-1/2" dark red body, pale blue interior, open hatch & two doors, jewel headlights, chrome grille & bumpers, orange taillights	50	75	125
MGC GT	3-1/2" bright yellow body & base, black interior, hood & hatch, folding seats, luggage, jewel headlights, red taillights	50	75	125
MGC GT	3-1/2" red body, black hood & base, black interior, open hatch & two doors, folding seat backs, luggage, orange taillights	50	75	125
Midland Red Express Coach	5-1/2" red one-piece body, black roof with shape or smooth wheels, yellow interior, clear windows, silver grille & headlights	70	105	175
Military Set	set of three, Tiger tank, Bell Helicopter, Saladin Armored Car	60	90	150
Milk Truck & Trailer	blue/white milk truck with trailer	60	90	150
Mini Camping Set	cream Mini, with red/blue tent, grille and figures	25	40	65
Mini-Marcos GT850	2-1/8" white body, red-white-blue racing stripe & #7 labels, clear headlights, Whizz Wheels	20	30	50
Mini-Marcos GT850	3-1/4" metallic maroon body, white name & trim decals, cream interior, open hood & doors, clear windows & headlights	30	45	75
Minissima	2-1/4" cream upper body, metallic lime green lower body with black stripe centered, black interior, clear windows, headlights	15	20	35
Monkeemobile	4-3/4" red body/base, white roof, yellow interior, clear windows, four figures, chrome grille, headlights, engine, orange taillights	145	225	375
Monte Carlo Rally Set	three vehicle set, Citroen, Mini and Rover rally cars	295	450	800
Morris Cowley	3-1/8" long, one-piece body in several colors, clear windows, silver lights, grille and bumper, smooth wheels, rubber tires	45	65	110
Morris Cowley-Mechanical	3-1/8" long, same as 202-A but with friction motor, available in off-white or green body	55	85	140
Morris Marina	3-3/4" metallic dark red or lime green body, cream interior, working hood & two doors, clear windshield, chrome grille & bumpers	15	25	40
Morris Mini-Cooper	2-3/4" yellow or blue body either body & base and/or hood, white roof and/or hood, two versions, red plastic interior, jewel headlights	25	40	70
Morris Mini-Cooper	2-7/8" red body & base, white roof, yellow interior, chrome spotlight, No. 37 & Monte Carlo Rally decals	55	85	140
Morris Mini-Cooper Deluxe	2-3/4" black body/base, red roof, yellow & black wicker work decals on sides & rear, yellow interior, gray steering wheel, jewel headlights	45	65	110
Morris Mini-Minor	2-7/8" long, one-piece body in several colors, plastic interior, silver lights, grille and bumpers, red taillights, Whizz Wheels	30	45	75
Morris Mini-Minor	2-3/4" three to four different versions with shaped and/or smooth wheels, plastic interior, silver bumpers, grille & headlights	40	60	100
Motorway Ambulance	4" white body, dark blue interior, red-white-black labels, dark blue windows, clear headlights, red die cast base & bumpers	10	15	25
Mr. McHenry's Trike	1972-74	70	105	175
Muppet Vehicles	#2030	15	18	30

Corgi

NAME	DESCRIPTION	GOOD	EX	MINT
Musical Carousel	1973	275	425	750
Mustang Organ Grinder Dragster	4" long, yellow body with green/yellow name, #39 and racing stripe decals, black base, green windshield, red interior, roll bar	20	30	50
NASA Space Shuttle	6" white body, two open hatches, black plastic interior, jets & base, unpainted retracting gear castings, black plastic wheels	30	45	75
National Express Bus		8	15	20
Noddy's Car	3-3/4" yellow body, red fenders & base, Chubby, Golliwogg & Noddy figures, chrome bumpers & grille castings, black grille	107	165	300
Noddy's Car	3-3/4" yellow body, red chassis, Chubby inside 1970; 3-1/2" 1975-1977 Noddy alone, closed trunk with spare tire	60	90	175
NSU Sport Prinz	3-1/4" metallic burgundy or maroon body, yellow interior, one-piece body, silver bumpers, headlights & trim, shaped wheels	30	45	75
Off Road Set	#5 decal on Jeep, blue boat	15	20	35
Olds Toronado & Speedboat	blue Toronado, blue/yellow boat with swordfish decals	60	90	150
Oldsmobile 88 Staff Car	4-1/4" drab olive body, four figures, white decals	50	75	125
Oldsmobile Sheriff's Car	4-1/4" long, black upper body with white sides, red interior with red dome light & County Sheriff decals on doors, single body casting	50	75	125
Oldsmobile Super 88	4-1/4" long, three versions: light blue, light or dark metallic blue body with white stripes, red interior, single body casting	40	60	100
Oldsmobile Toronado	4-3/16" metallic copper red one-piece body, cream interior, Golden jacks, gray tow hook, clear windows, bumpers, grille, headlights	35	55	90
Oldsmobile Toronado	4-1/8" metallic peacock blue body, cream interior, one-piece body, clear windshield, chrome bumpers, grille, headlight covers	35	55	90
Opel Senator Doctor's Car	1980-81	10	15	25
Open Top Disneyland Bus	4-3/4" yellow body, red interior & stripe, Disneyland labels, eight spoked wheels or orange body, white interior & stripe	30	50	95
OSI DAF City Car	2-3/4" orange/red body, light cream interior, textured black roof, sliding left door, working hood, hatch & two right doors	18	25	45
Penguinmobile	3-3/4" black & white lettering on orange-yellow-blue decals, gold body panels, seats, air scoop, chrome engine	20	30	65
Pennyburn Workmen's Trailer	3-1/8" long, blue body w/working lids, red plastic interior, chrome tools, red cast wheels, plastic tires	15	20	35
Peugeot 505 STI	4-7/8" cream body & base, red interior, blue-red-white Taxi labels, black grille, bumpers, tow hook, chrome headlights & wheels	8	15	20
Peugeot 505 Taxi	4-7/8" long, cream body, red interior, red/white/blue taxi decals	8	15	20
Platform Trailer	4-3/8" long, in five versions: silver body, blue chassis; silver body, yellow chassis; blue body, red chassis, blue body, yellow chassis	10	15	25
Playground	1973	295	450	750
Plymouth Sports Suburban	4-1/4" long, dark cream body, tan roof, red interior, die cast base, red axle, silver bumpers, trim and grille and rubber tires	40	60	100
Plymouth Sports Suburban	4-1/4" pale blue body with silver trim, red roof, yellow interior, gray die cast base without rear axle bulge, shaped wheels	40	60	100

Corgi

NAME	DESCRIPTION	GOOD	EX	MINT
Plymouth Suburban Mail Car	4-1/4" white upper, blue lower body with red stripes, gray die cast base without rear axle bulge, silver bumpers & grille	55	85	140
Police Land Rover	5" white body, red & blue police stripes, black lettering, open rear door, opaque black windows, blue dome light, roof light	15	25	40
Police Land Rover & Horse Box	white Land Rover with police decals and mounted policeman	30	45	75
Police Vigilant Range Rover	4" white body, red interior, black shutters, blue dome light, two chrome & amber spotlights, black grille, silver headlights	25	35	60
Pontiac Firebird	4" metallic silver body & base, red interior, black hood, stripes & convertible top, doors open, clear windows, folding seats	50	75	125
Pony Club Set	brown/white Land Rover with Corgi Pony Club decals, horse box, horse and rider	30	45	75
Pop Art Mini-Motest	2-3/4" light red body & base, yellow interior, jewel headlights, orange taillights, yellow-blue-purple pop art & "Motest" decals	100	150	250
Popeye's Paddle Wagon	4-7/8" yellow body, red chassis, blue rear fenders, bronze & yellow stacks, white plastic deck, blue lifeboat with Swee' Pea	195	300	525
Popeye's Paddle Wagon Jr.		70	105	200
Porsche 917	4-1/4" red or blue body, black or gray base, blue or amber tinted windows & headlights, open rear hood, headlights	15	20	35
Porsche 92 Turbo	4-1/2" black body with gold trim, yellow interior, four chrome headlights, clear windshield, taillight-license plate decal, black	15	20	35
Porsche 924	4-7/8" red or metallic light brown body, dark red interior, two doors open & rear window, chrome headlights, black plastic grille	10	15	25
Porsche 924	4-1/2" bright orange body, dark red interior, black plastic grille, multicolored stripes, swivel roof spotlight	10	15	25
Porsche 924 Police Car	4-1/4" white body with different hood & doors versions, blue & chrome light, Polizei white on green panels or Police labels	15	25	40
Porsche Carrera 6	3-7/8" white body, red or blue trim, blue or amber tinted engine covers, black interior, clear windshield & canopy, red jewel taillights	30	45	75
Porsche Carrera 6	3-3/4" white upper body, red front hood, doors, upper fins & base, black interior, purple rear window, tinted engine cover	25	35	60
Porsche Targa 911S	3-1/2" three different versions, black roof with or without stripe, orange interior, open hood & two doors, chrome engine & bumpers	25	35	60
Porsche Targa Police Car	3-1/2" white body & base, red doors & hood, black roof & plastic interior also comes with an orange interior, unpainted siren	25	35	60
Porsche-Audi 917	4-3/4" white body, red & black no. 6, L & M, Porsche Audi & stripe labels or orange body, orange two-tone green white no. 6	15	20	35
Powerboat Team	white/red Jaguar with red/white boat on silver trailer, Team Corgi Carlsberg, Union Jack and #1 decals on boat	25	35	60
Priestman Cub Crane	9" orange body, red chassis & two-piece bucket, unpainted bucket arms, lower boom, knobs, gears & drum castings, clear window	50	75	125

Chevrolet
Astro I, Corgi

Batman Set, Corgi

Circus Horse Transporter, Corgi

Corgi

NAME	DESCRIPTION	GOOD	EX	MINT
Priestman Cub Power Shovel	6" orange upper body & panel, yellow lower body, lock rod & chassis, rubber or plastic treads, pulley panel, gray boom	40	60	100
Priestman Shovel & Carrier	cub shovel & low loader machinery carrier	90	135	225
Professionals Ford Capri	5" metallic silver body & base, red interior, black spoiler, grille, bumpers, tow hook & trim, blue windows, chrome wheels	30	45	75
Psychedelic Ford Mustang	3-3/4" light blue body & base, aqua interior, red-orange-yellow No. 20 & flower decals, cast eight spoke wheels, plastic tire	30	45	75
Public Address Land Rover	4" green body, yellow plastic rear body & loudspeakers, red interior, clear windows, silver bumper, grille & headlights	50	75	125
Quartermaster Dragster	5-3/4" long, dark metallic green upper body with green/yellow/black #5 and Quartermaster decals, light green lower body	30	45	75
RAC Land Rover	3-3/4" three versions of body, plastic interior & rear cover, RAC & Radio Rescue decals	60	90	150
Radio Luxembourg Dragster	5-3/4" long, blue body with yellow/white/blue John Wolfe Racing, Radio Luxembourg and #5 decals, silver engine	30	45	75
Radio Roadshow Van	4-3/4" white body, red plastic roof & rear interior, opaque black windows, red-white-black Radio Tele Luxembourg labels, gray	25	35	60
RAF Land Rover	3-3/4" blue body & cover, one-piece body, sheet metal rear cover, RAF rondel decal, with or without suspension, silver bumper	60	90	150
RAF Land Rover & Bloodhound	set of three standard colored, Massey Ferguson Tractor, Bloodhound Missile, Ramp & Trolley	150	240	400
RAF Land Rover & Thunderbird	Standard colors, #350 Thunderbird Missileon Trolley & 351 R.A.F. Land Rover	100	150	250
Rambler Marlin Fastback	4-1/8" red or blue body, black roof & trim, cream interior, clear windshield, folding seats, chrome bumpers, grille & headlights	35	55	90
Rambler Marlin with Kayak & Trailer	blue Marlin with roof rack, blue/white trailer	100	150	250
Range Rover Ambulance	4" two different versions of body sides, red interior, raised roof, open upper & lower doors, black shutters, blue dome light	20	30	50
Raygo Rascal Roller	4-7/8" dark yellow body, base & mounting, green interior & engine, orange & silver roller mounting & castings, clear windshield	15	25	40
Red Wheelie Motorcycle	4" long, red plastic body and fender with black/white/yellow decals, black handlebars, kickstand and seat, chrome engine, pipes	10	15	25
Reliant Bond Rug 700 E.S.	2-1/2" bright orange or lime green body, off white seats, black trim, silver headlights, red taillights	15	25	40
Renault 11 GTL	4-1/4" light tan body & base, red interior, open doors & rear hatch, lifting hatch cover, folding seats, grille	15	25	40
Renault 16	3-3/4" metallic maroon body, dark yellow interior, chrome base, grille & bumpers, clear windows, hatch cover/Renault decal	25	35	60
Renault 16TS	3-7/8" long, metallic blue body with Renault decal on working hatch, clear windows, detailed engine, yellow interior	25	35	60
Renault 5 Police Car	3-7/8" white body, red interior, blue dome light, black hood, hatch & doors with white Police labels, orange taillights	15	20	35

NAME	DESCRIPTION	GOOD	EX	MINT
Renault 5 Turbo	3-3/4" bright yellow body, red plastic interior, black roof & hood, working hatch & two doors, black dash, chrome rear engine	15	18	30
Renault 5 Turbo	4" white body, red roof, red & blue trim painted on, No. 5 lettering, blue & white label on windshield	15	18	30
Renault 5TS	3-3/4" metallic golden orange body, black trim, tan plasic interior, working hatch & two doors, clear windows & headlights	15	18	30
Renault 5TS	3-3/4" light blue body, red plastic interior, dark blue roof, dome light, S.O.S. Medicine lettering, working hatch & two doors	15	18	30
Renault 5TS Fire Chief	3-3/4" red body, tan interior, amber headlights, gray antenna, black/white Sapeurs Pompiers decals, blue dome light	15	25	40
Renault Alpine 5TS	3-3/4" dark blue body, off white interior, red & chrome trim, clear windows & headlights, gray base & bumpers, black grille	15	25	40
Renault Floride	3-5/8" long, one-piece body, clear windows, silver bumper, grille, lights and plates, red taillights, rubber tires	35	55	95
Renegade Jeep	4" dark blue body with no top, white interior, base & bumper, white plastic wheels & rear mounted spare	8	15	20
Renegade Jeep with Hood	4" yellow body with removeable hood, red interior, base, bumper, white plastic wheels, side mounted spare, number 8	8	15	20
Rice Beaufort Double Horse Box	3-3/8" long, blue body and working gates, white roof, brown plastic interior, two horses, cast wheels, plastic tires	15	25	40
Rice Pony Trailer	3-3/8" long, cast body and chassis w/working tailgate, horse, in four versions: tan body, cream roof	20	30	50
Riley Pathfinder	4" long, red or dark blue one-piece body, clear windows, silver lights, grille and bumpers, smooth wheels, rubber tires	45	65	110
Riley Pathfinder Police Car	4" long, black body with blue/white Police lettering, unpainted roof sign, gray antenna	50	75	125
Riley Pathfinder-Mechanical	4" long with friction motor and either red or blue body	50	75	125
Riot Police Quad Tractor	3-3/4" white body & chassis, brown interior, red roof with white panel, gold water cannons, gold spotlight with amber lense	15	20	35
Road Repair Unit	10" dark yellow Land Rover with battery hatch & trailer with red plastic interior with sign & open panels, stripe & Roadwork	15	25	40
Rocket Age Set	set of eight standard including colored Thunderbird Missile on Trolley, R.A.F. Land Rover, R.A.F. Staff Car, Radar Scanner, Decca Radar	295	450	850
Rocket Launcher & Trailer	1975-80	25	35	60
Roger Clark's Capri	4" white body, black hood, grille & interior, open doors, folding seats, chrome bumpers, clear headlights, red taillights	15	18	30
Rolls-Royce Corniche	5-1/2" different versions with light brown interior, working hood, trunk & two doors, clear windows, folding seats, chrome bumpers	10	15	25
Rolls-Royce Silver Ghost	4-1/2" silver body/hood, charcoal & silver chassis, bronze interior, gold lights, box & tank, clear windows, dash lights, radiator	15	25	40
Rolls-Royce Silver Shadow	4-3/4" metallic white upper/dusty blue lower body, working hood, trunk & two doors, clear windows, folding seats, chrome bumpers	30	45	75

Corgi

NAME	DESCRIPTION	GOOD	EX	MINT
Rolls-Royce Silver Shadow	4-3/4" metallic silver upper/metallic blue lower body, light brown interior, hole in trunk for spare tire mounting	25	40	65
Rolls-Royce Silver Shadow	4-3/4" metallic silver upper/metallic blue lower body, light brown interior, no hole in trunk for spare tire	25	40	65
Rolls-Royce Silver Shadow	4-3/4" metallic blue body, bright blue interior, working hood, trunk & two doors, clear windows, folding seats, spare wheel	25	40	65
Routemaster Bus-Promotionals	4-7/8" different body & interior versions promotional	15	25	40
Rover 2000	3-3/4" metallic blue with red interior or maroon body with yellow interior, gray steering wheel, clear windshields	30	45	75
Rover 2000 Rally	3-3/4" two different versions, metallic dark red body, white roof, shaped wheels, No. 136 & Monte Carlo Rally decal	50	75	125
Rover 2000TC	3-3/4" metallic olive green or maroon one-piece body, light brown interior, chrome bumpers/grille, jewel headlights, red taillights	30	45	75
Rover 2000TC	3-3/4" metallic purple body, light orange interior, black grille, one-piece body, amber windows, chrome bumpers & headlights	25	35	60
Rover 3500	5-1/4" three different body & interior versions, plastic interior, open hood, hatch & two doors, lifting hatch cover	8	15	20
Rover 3500 Police Car	5-1/4" white body, light red interior, red stripes, white plastic roof sign, blue dome light, red & blue Police & badge label	8	15	20
Rover 3500 Triplex	5-1/4" white sides & hatch, blue roof & hood, red plastic interior & trim, detailed engine, red-white-black no. 1	8	15	20
Rover 90	3-7/8" long, one-piece body in several colors, silver headlights, grille and bumpers, smooth wheels, rubber tires; colors available	50	75	125
Rover 90-Mechanical	3-7/8" long with friction motor and red, green, gray or metallic green body	60	90	150
Safari Land Rover & Trailer	black/white Land Rover in two versions: with chrome wheels, 1976; with red wheels, 1977-80	20	30	50
Saint's Jaguar XJS	5-1/4" white body, red interior, black trim, Saint figure hood label, open doors, black grille, bumpers & tow hook, chrome headlights	30	45	75
Saint's Volvo P-1800	3-5/8" long, one-piece white body with red Saint decals on hood, gray base, clear windows, black interior with driver	55	85	175
Saint's Volvo P-1800	3-3/4" three versions of white body with silver trim & different colored Saint decals on hood, driver, one-piece body	55	85	175
Saladin Armored Car	3-1/4" drab olive body, swiveling turret & raising barrel castings, black plastic barrel end & tires, olive cast wheels	30	45	75
Scammell Carrimore Tri-deck Car Transporter	11" orange cab chassis & lower deck, white cab & middle deck, blue top deck (three decks), red interior, black hydraulic cylinders	35	55	95
Scammell Circus Crane Truck	8" red upper cab & silver rear body, light blue crane base & winch crank housing, red interior & tow hook, jewel headlights	175	275	450
Scammell Coop Semi-Trailer Truck	9" white cab & fenders, light blue semi-trailer, red interior, gray bumper base, jewel headlights, black hitch lever, spare wheel	135	210	350
Scammell Ferrymasters Semi-Trailer Truck	9-1/4" long, white cab, red interior, yellow chassis, black fenders, clear windows, jewel headlights, cast wheels, plastic tires	60	90	150

NAME	DESCRIPTION	GOOD	EX	MINT
Scania Bulk Carrier	5-5/8" long, white cab, blue and white silos, ladders and catwalk, amber windows, blue British Sugar decals, Whizz Wheels	6	9	15
Scania Bulk Carrier	5-5/8" long, white cab, orange and white silos, clear windows, orange screen, black/orange Spillers Flour decals	6	9	15
Scania Container Truck	5-1/2" long, yellow truck and box with red Ryder Truck rental decals, clear windows, black exhaust stack, red rear doors	6	9	15
Scania Container Truck	5-1/2" long, blue cab with blue and white box and rear doors, white deck, Securicor Parcels decals, in two rear door colors	6	9	15
Scania Container Truck	5-1/2" long, white cab and box with BRS Truck Rental decals, blue windows, red screen, roof and rear doors	6	9	15
Scania Dump Truck	5-3/4" long, white cab with green tipper, black/green Barratt decals, black exhaust and hydraulic cylinders, spoked Whizz Wheels	6	9	15
Scania Dump Truck	5-3/4" long, yellow truck and tipper with black Wimpey decals, in two versions: either clear or green windows	6	9	15
Security Van	4" long, black body, blue windows and dome light, yellow/black Security decals, Whizz Wheels	6	9	15
Service Ramp	accessory	30	45	75
Shadow-Ford Racing Car	5-5/8" long, white body, red stripes, driver, chrome plastic pipes, mirrors and steering wheel, in two versions	10	15	25
Shadow-Ford Racing Car	5-5/8" long, black body and base with white/black #17, UOP and American flag decals, cast chrome suspension and pipes	10	15	25
Shell or BP Garage	gas station/garage with pumps and other accessories in two versions: Shell or B.P., each set	295	450	750
Shelvoke & Drewry Garbage Truck	5-7/8" long, orange cab, silver body with City Sanitation decals, black interior, grille and bumpers, clear windows	15	25	40
Sikorsky Skycrane Army Helicopter	5-1/2" long, olive drab and yellow body with Red Cross and Army decals	15	20	35
Sikorsky Skycrane Casualty	6-1/8" long, red and white body, black rotors and wheels, orange pipes, working rear hatch, Red Cross decals	15	20	35
HelicopterSilo & Conveyor Belt	with yellow conveyor and Corgi Harvesting Co. decal on silo	35	50	85
Silver Jubilee Landau	Landua with four horses, two footmen, two riders, Queen and Prince figures, and Corgi dog, in two versions	15	25	40
Silver Jubilee London Transport Bus	4-7/8" long, silver body with red interior, no passengers, decals read "Woolworth Welcomes the World" and "The Queen's Silver	15	18	30
Silver Streak Jet Dragster	6-1/4" long, metallic blue body with sponsor and flag decals on tank, silver engine, orange plastic jet and nose cone	15	18	30
Silverstone Racing Layout	seven-vehicle set with accessories; Vanwall, Lotus IX, Aston Martin, Mercedes 300SL, BRM, Ford Thunderbird, Land Rover Truck	400	600	1200
Simca 1000	3-1/2" chrome plated body, #8 & red-white-blue stripe decals, one-piece body, clear windshield, red interior	30	45	75
Simon Snorkel Fire Engine	10-1/2" long, red body with yellow interior, blue windows and dome lights, chrome deck, black hose reels and hydraulic cylinders	30	45	75
Simon Snorkel Fire Engine	9-7/8" long, red body with yellow interior, two snorkle arms, rotating base, five firemen in cab and one in basket	35	55	90

Man From U.N.C.L.E. THRUSH-buster, Corgi

James Bond Toyota, Corgi

Ghia, Corgi

Corgi

NAME	DESCRIPTION	GOOD	EX	MINT
Skyscraper Tower Crane	9-1/8" tall, red body with yellow chassis and booms, gold hook, gray loads of block, black/white Skyscraper decals	30	45	75
Spider-Bike	4-1/2" medium blue body, one-piece body, dark blue plastic front body & seat, blue & red Spider-Man figure, amber windshield	40	60	100
Spider-Buggy	5-1/8" red body, blue hood, clear windows, dark blue dash, seat & crane, chrome base with bumper & steps. silver headlights	50	75	125
Spider-Copter	5-5/8" long, blue body with Spider-Man decals, red plastic legs, tongue and tail rotor, black windows and main rotor	30	45	85
Spider-Man Set	set of three: Spider-Bike, Spider-Copter and Spider-Buggy	80	120	225
Standard Vanguard	3-5/8" long, one-piece red and white body, clear windows, silver lights, grille and bumpers, smooth wheels, rubber tires	50	75	125
Standard Vanguard RAF Staff Car	3-3/4" long, blue body with friction motor, RAF decals	55	85	140
Standard Vanguard-Mechanical	3-5/8" long, with friction motor and red/off-white body with black or gray base, or red/gray body	55	85	140
Starsky & Hutch Ford Torino	5-3/4" red one-piece body, white trim, light yellow interior, clear windows, chrome bumpers, grille & headlights, orange taillights	35	55	100
STP Patrick Eagle Racing Car	5-5/8" long, red body with red/white/black STP and #20 decals, chrome lower engine and suspension, black plastic upper engine	20	30	50
Stromberg Jet Ranger Helicopter	5-5/8" long, black body with yellow trim and interior, clear windows, black plastic rotors, white/blue decals	30	45	85
Studebaker Golden Hawk	4-1/8" long, one-piece body in several colors, clear windows, silver lights, grille and bumpers, smooth wheels, rubber tires	55	85	140
Studebaker Golden	4-1/8" long, with friction motor and white body with gold trim	70	105	175
Hawk-MechanicalStunt Motorcycle	3" long, made for Corgi Rockets race track, gold cycle, blue rider with yellow helmet, clear windshield, plastic tires	70	105	175
SU-100 Medium Tank	5-5/8" long, olive and cream camo upper body, gray lower, working hatch and barrel, black treads, red star and #103 decals	30	50	80
Sunbeam Imp Police Car	3-1/4" white or light blue body, tan interior, driver, black or white hood & lower doors, dome light, Police decals, cast wheels	25	40	65
Sunbeam Imp Rally	3-3/8" long, metallic blue body with white stripes, Monte Carlo Rally and #77 decals, cast wheels	20	35	55
Super Karts	two karts, orange and blue, Whizz Wheels in front, slicks on rear, silver and gold drivers	15	18	30
Superman Set	set of three: Supermobile, Daily Planet Helicopter and Metropolis Police Car	70	120	225
Supermobile	5-1/2" blue body, red, chrome or gray fists, red interior, clear canopy, driver, chrome arms with removeable "striking fists"	30	45	75
Supervan	4-5/8" long, silver van with Superman decals, working rear doors, chrome spoked wheels	15	25	50
Surtees TS9 Racing Car	4-5/8" long, black upper engine, chrome lower engine, pipes and exhaust, driver, Brook Bond Oxo-Rob Walker decals	15	18	30
Surtees TS9B Racing Car	4-3/8" long, red body with white stripes and wing, black plastic lower engine, driver, chrome upper engine, pipes, suspension	15	18	30

Corgi

NAME	DESCRIPTION	GOOD	EX	MINT
Talbot-Matra Rancho	4-3/4" long, working tailgate and hatch, clear windows, plastic interior, black bumpers, grille and tow hook, in several colors	10	15	25
Tandem Disc Harrow	3-5/8" long, yellow main frame, red upper frame, working wheels linkage, unpainted linkage and cast discs, black plastic tires	15	20	35
Tarzan Set	metallic green Land Rover with trailer, cage and other accessories	100	150	275
Thunderbird Bermuda Taxi	4" white body with blue, yellow, green plastic canopy with red fringe, yellow interior, driver, yellow & black labels	50	75	125
Thunderbird Missile & Trolley	5-1/2" ice blue or silver missile, RAF blue trolley, red rubber nose cone, plastic tow bar, steering front & rear axles	55	85	165
Thwaites Tusker Skip Dumper	3-1/8" yellow body, chassis & tipper, driver & seat, hydraulic cylinder, red wheels, black tires two sizes, name labels	10	15	25
Tiger Mark I Tank	6" tan & green camouflage finish, German emblem, swiveling turret & raising barrel castings, black plastic barrel end, antenna	30	45	75
Tipping Farm Trailer	4-1/4" long, red working tipper and tailgates, yellow chassis, red plastic wheels, black tires	10	15	25
Tipping Farm Trailer	5-1/8" long, cast chassis and tailgate, red plastic tipper and wheels, black tires, in two versions	10	15	25
Tour de France Set	Renault with Paramount Film roof sign, rear platform with cameraman and black camera on tripod, plus bicycle and rider	60	90	200
Tour de France Set	with white Peugeot	25	40	75
Touring Caravan	4-3/4" white body with blue trim, white plastic open roof & door, pale blue interior, red plastic hitch & awning	15	25	40
Tower Wagon & Lamp Standard	red Jeep Tower wagon with yellow basket, workman figure	40	60	100
Toyota 2000 GT	4" metallic dark blue or purple body, cream interior, one-piece body, red gear shift & antenna, two red & two amber taillights	15	25	40
Tractor & Beast Carrier	Fordson tractor, figures & beast carrier	65	100	165
Tractor with Shovel & Trailer	standard colors, No. 69 Massey Ferguson Tractor & 62 Tipping Trailer	65	100	165
Tractor, Trailer & Field Gun	10-3/4" tractor body & chassis, trailer body, base & opening doors, gun chassis & raising barrel castings, brown plastic interior	30	50	80
Transporter & Six Cars	Ford transporter with six cars, Mini DeLuxe, Mini Rally, Mini, Rover, Sunbeam Imp, Ford Cortina Estate Car	225	365	700
Transporter & Six Cars	Scammell transporter with six cars, Mini DeLuxe, Mini, Mini Rally, The Saint's Volvo, Sunbeam Imp, MGC GT	250	395	750
Triumph Acclaim Driving School	4-3/4" yellow or red body/base, Corgi Motor School decals, black roof mounted steering wheel steers front wheels, clear windows	15	25	40
Triumph Acclaim Driving School	4-3/4" dark yellow body with black trim, black roof mounted steering wheel steers front wheels, clear windows, mirrors, bumpers	15	25	40
Triumph Acclaim HLS	4-3/4" metallic peacock blue body/base, black trim, light brown interior, clear windows, mirrors, bumpers, vents, tow hook	15	18	30
Triumph Herald Coupe	3-1/2" long, blue or gold top and lower body, white upper body, red interior, clear windows, silver bumpers, grille, headlights	35	50	85
Triumph TR2	3-1/4" cream body with red seats, light green body with white or cream seats, one-piece body, clear windshield, silver grille	70	105	175

Corgi

NAME	DESCRIPTION	GOOD	EX	MINT
Triumph TR3	2-1/4" metallic olive or cream body, red seats, one-piece body, clear windshield, silver grille, bumpers & headlights	60	90	150
Trojan Heinkel	2-1/2" long, issued in mauve, red or orange body, plastic interior, silver bumpers & headlights, red taillights, suspension	35	55	95
Twin Packs and 2601		6	9	15
Tyrrell P34 Racing Car	4-3/8" long, dark blue body and wings with yellow stripes, #4 and white Elf and Union Jack decals, chrome plastic engine	20	30	55
Tyrrell P34 Racing Car	without yellow decals	20	30	55
Tyrrell-Ford Racing Car	4-5/8" long, dark blue body with blue/black/white Elf and #1 decals, chrome suspension, pipes, mirrors, driver	18	25	45
U.S. Racing Buggy	3-3/4" long, white body with red/white/blue stars, stripes and USA #7 decals, red base, gold engine, red plastic panels	18	25	45
Unimog Dump Truck	3-3/4" blue cab, yellow tipper, fenders & bumpers, metallic charcoal gray chassis, red interior, black mirrors, gray tow hook	20	30	50
Unimog Dump Truck	4" yellow cab, chassis, rear frame & blue tipper, fenders & bumpers, red interior, no mirrors, gray tow hook, hydraulic cylinders	20	30	50
Unimog Dumper & Priestman Cub Shovel	standard colors, #1145 Mercedes-Benz unimog with Dumper & 1128 Priestman Cub Shovel	70	105	175
Unimog with Snowplow (Mercedes-Benz)	6" four different body versions, red interior, cab, rear body, fender-plow mounting, lower & charcoal upper chassis, rear fenders	30	45	75
Vanwall Racing Car	3-3/4" long, clear windshield, unpainted dash, silver pipes and decals, smooth wheels, rubber tires, in three versions: green body	35	55	90
Vauxhall Velox	3-3/4" long, one-piece body in several colors, clear windows, silver lights, grille and bumpers, smooth wheels, rubber tires	50	75	125
Vauxhall Velox-Mechanical	3-3/4" long, with friction motor; orange or red body	60	90	150
Vegas Ford Thunderbird	5-1/4" orange/red body & base, black interior & grille, open hood & trunk, amber windshield, white seats, driver, chrome bumper	25	40	65
VM Polo Mail Car		25	35	60
Volkswagen 1200	3-1/2" seven different versions, plastic interior, one-piece body, silver headlights, red taillights, die cast base & bumpers	20	30	50
Volkswagen 1200	3-1/2" dark yellow body, white roof, red interior & dome light, unpainted base & bumpers, black & white ADAC Strassenwacht	60	90	150
Volkswagen 1200 Driving School	3-1/2" metallic red or blue body, yellow interior, gold roof mounted steering wheel that steers, silver headlights, red taillights	25	35	60
Volkswagen 1200 Police Car	3-1/2" two different body versions made for Germany, Netherlands & Switzerland, blue dome light in chrome collar	40	60	100
Volkswagen 1200 Rally	3-1/2" light blue body, off-white plastic interior, silver headlights, red tailights, suspension, Whizz Wheels	20	30	50
Volkswagen Breakdown Van	4" tan or white body, red interior & equipment boxes, clear windshield, chrome tools, spare wheels, red VW emblem, no lettering	50	75	125
Volkswagen Delivery Van	3-1/4" white upper & red lower body, plastic red or yellow interior, silver bumpers & headlights, red VW emblem, shaped wheels	55	85	140

Corgi

NAME	DESCRIPTION	GOOD	EX	MINT
Volkswagen Driving School	3-1/2" metallic blue body, yellow interior, gold roof mounted steering wheel that steers, silver headlights, red taillights	25	40	70
Volkswagen East African Safari	3-1/2" light red body, brown interior, working front & rear hood, clear windows, spare wheel on roof steers front wheels, jewel headlights	50	80	225
Volkswagen Kombi Bus	3-3/4" off green upper & olive green lower body, red interior, silver bumpers & headlights, red VW emblem, shaped wheels	50	75	125
Volkswagen Military Personnel Carrier	3-1/2" drab olive body, white decals, driver	55	85	140
Volkswagen Pickup	3-1/2" dark yellow body, red interior & rear plastic cover, silver bumpers & headlights, red VW emblem, shaped wheels	45	65	110
Volkswagen Police Car/Foreign Issues	3-1/2" five different versions, one-piece body, red interior, dome light, silver headlights, red taillights, clear windows	60	90	150
Volkswagen Tobler Van	3-1/2" light blue body, plastic interior, silver bumpers, Trans-o-lite headlights & roof panel, shaped wheels, rubber tires	55	85	140
Volvo Concrete Mixer	8-1/4" yellow or orange cab, red or white mixer with yellow & black stripes, rear chassis, chrome chute & unpainted hitch casings	30	45	75
Volvo P-1800	3-1/2" one-piece body with six versions, clear windows, plastic interior, shaped wheels, rubber tires	40	60	100
VW 1500 Karmann-Ghia	3-1/2" three color versions, plastic interior and taillights, front & rear working hoods, clear windshields, silver bumpers	35	55	90
VW Polo	3-3/4" metallic light brown body, off-white interior, black dash, clear windows, silver bumpers, grille & headlights	15	18	30
VW Polo	3-3/4" apple green or bright yellow body, black DBP & posthorn (German Post Office) decals, off white interior, black dash	25	40	65
VW Polo Auto Club Car		15	25	40
VW Polo German Auto Club Car	3-1/2" yellow body, off-white interior, black dash, silver bumpers, grille & headlights, white roof, yellow dome light	25	35	60
VW Polo Police Car	3-1/2" white body, green hood & doors, black dash, silver bumpers, grille & headlights, white roof, blue dome light	15	25	40
VW Polo Turbo	3-3/4" cream body, red interior with red & orange trim, working hatch & two door castings, clear windshield, black plastic dash	15	18	30
VW Racing Tender & Cooper	white VW with racing decals, blue Cooper	50	75	125
VW Racing Tender & Cooper Maserati	two versions: tan or white VW truck, each set	50	75	125
Warner & Swasey Crane	8-1/2" yellow cab & body, blue chassis, blue/yellow stripe decals, red interior, black steering wheel, silver knob, gold hook	30	45	75
White Wheelie Motorcycle	4" long, white body with black/white police decals	15	20	35
Wild Honey Dragster	3" long, yellow body with red/yellow Wild Honey and Jaguar Powered decals, green windows and roof, black grille, driver, Whizz Wheels	25	40	65

Dinky

NAME	NO.	YEAR	GOOD	EX	MINT
	Aircraft				
A.W. Ensign	68a	1940	150	300	450
A.W. Ensign	62x	1945-49	95	275	350
A.W. Ensign	62P	1945-49	65	145	175
Airspeed Envoy	62m	1945-49	40	125	200
Airspeed Envoy	62m	1938-41	55	175	250
Airspeed Envoy	62k	1938-40	70	200	300
Amiot 370	64a	1939-48	70	130	200
Arc-en-Ciel	60a	1935-40	200	375	450
Armstrong Whitworth Ensign	62p	1938-41	90	175	250
Atalanta	66a	1940	200	450	650
Atalanta (Imperial Airways Liner)	60a	1934-41	150	375	500
Autogyro	60f	1934-41	90	180	250
Autogyro	66f	1940	125	250	350
Avro Vulcan Delta Wing Bomber	749/992	1955-56	800	1500	4500
Avro York	70a/704	1946-59	45	155	200
Beechcraft Baron	715	1968-76	25	60	90
Beechcraft Bonanza	710	1965-76	20	55	80
Beechcraft T42A	712	1972-77	30	75	110
Bell 47 Police Helicopter	732	1974-80	10	30	55
Bloch 220	64b	1939-48	75	150	250
Boeing 737	717	1970-75	20	45	90
Boeing Flying Fortress	62g	1945-48	50	125	225
Boeing Flying Fortress	62g	1939-41	100	200	300
Breguet Corsair	60d	1935-40	100	200	300
Bristol 173 Helicopter	715	1956-63	20	60	95
Bristol Blenheim	62b/62d	1940-41	60	140	175
Bristol Blenheim	62B	1945-48	35	95	120
Bristol Blenheim	62B		60	85	130
Bristol Brittania	998	1959-65	50	250	500
Cierva Autogyro	60f	1935-40	100	250	250
Clipper III Flying Boat	60w	1938-41	150	250	400
Clipper III Flying Boat	60w	1945-49	75	175	250
D.H. Albatross	62w	1939-41	75	275	400
D.H. Albatross	68b	1940	150	300	450
D.H. Albatross	62r	1939-41	75	275	400
D.H. Albatross	62R	1945-49	75	170	250
D.H. Comet Jetliner	702/999	1954-65	35	100	200
D.H. Comet Racer	60g	1946-49	50	110	155
D.H. Comet Racer	60g	1935-40	60	125	225
D.H. Sea Vixen	738	1960-65	30	65	110
Dewoitine 500	60e	1935-40	100	200	300
Dewoitine D338	61a/64	1937-46	225	350	500
Douglas DC3	60t	1938-41	125	300	650
Empire Flying Boat	60x	1938-40	350	600	1000
Empire Flying Boat	60r	1945-49	90	175	250

Dinky

NAME	NO.	YEAR	GOOD	EX	MINT
Empire Flying Boat	60r	1937-41	150	325	450
Empire Flying Boat	700	1937-41	150	275	350
Fairy Battle	60n	1937-41	40	100	140
Fairy Battle	60s	1937-41	50	125	200
General Monospar	60e	1934-41	75	225	300
General Monospar	66e	1940	90	300	350
Gloster Gladiator	60p	1937-40	90	175	225
Gloster Javelin	735	1956-65	15	50	90
Gloster Meteor	70e/732	1946-62	10	25	35
Hanriot 180M	61e	1937-40	90	150	225
Hawker Harrier	722	1970-80	25	65	95
Hawker Hunter	736	1955-63	15	45	85
Hawker Hurricane	62h	1939-41	30	70	115
Hawker Hurricane	62S	1945-49	40	95	130
Hawker Hurricane IIc	718	1972-75	50	95	155
Hawker Siddeley HS 125	723/728	1970-75	20	50	75
Hawker Tempest II	70b/730	1946-55	15	45	75
Henriot 180T	60c	1935-40	110	170	225
Junkers JU87B Stuka	721	1969-80	40	75	145
Junkers JU89	62y	1938-41	100	275	350
Junkers JU89	62Y	1945-49	70	160	225
Junkers JU89 Heavy Bomber	67a	1941	100	400	600
Junkers JU90 Airliner	62n	1938-40	90	250	325
Leopard Moth	66b	1940	70	225	300
Leopard Moth	60b	1934-41	50	140	180
Lockheed Constellation	60c/892	1956-63	90	200	350
Lockheed P-80 Shooting Star	701/733	1947-62	10	20	30
M.D. F-4 Phantom	725/727/73	1972-77	70	125	175
Mayo Composite	63	1939-41	200	450	750
ME BI 109	726	1972-76	50	100	175
Mercury Seaplane	700	1949-57	35	75	110
Mercury Seaplane	63b	1939-41	50	100	175
Mitsubishi A65M Zero	739	1975-78	75	150	225
MRCA Tornado	729	1974-76	35	90	155
Mystere IV	60a/800	1957-63	30	75	110
Nord Noratlas	804	1960-64	125	250	400
P1B Lightning	737	1959-69	20	55	100
Percival Gull	66c	1940	100	225	275
Percival Gull	66c	1945-48	65	125	150
Percival Gull	60c	1934-41	70	150	225
Potez 56	61b	1937-40	125	200	270
Potez 58	60b/61d	1935-40	100	260	350
Potez 58 Sanitaire	61d	1937-40	110	175	250
Potez 63	64c	1939-48	150	250	400
Potez 662	64d	1939-40	125	250	400
Republic P47 Thunderbolt	734	1975-78	75	175	250
S.E. Caravelle Airliner	997	1962-69	70	150	250

Range Rover Ambulance, 1970s, Dinky

Euclid Dump Truck, Dinky

Plymouth Yellow Cab, 1970s, Dinky

VEHICLES

Dinky

NAME	NO.	YEAR	GOOD	EX	MINT
Sea King Helicopter	724/736	1971-79	15	35	85
SEPCAT Jaguar	731	1973-76	25	70	115
Short Shetland Flying Boat	701	1947-49	200	550	750
Sikorsky S58 Helicopter	60d/802	1957-61	45	135	190
Singapore Flying Boat	60m	1936-41	125	400	550
Singapore Flying Boat	60h	1936-41	100	300	400
Spitfire	62A	1945-49	30	75	125
Spitfire	62e/62a	1940-41	35	150	200
Spitfire II	700	1979	65	165	250
Spitfire II	741	1978-80	40	80	135
Sud Est Caravelle	60f/891	1959-62	40	165	225
Supermarine Spitfire II	719	1969-78	60	95	175
Supermarine Swift	734	1955-63	10	45	85
Twin Engined Fighter	70d/731	1946-55	10	25	35
Vautour	60b/801	1957-63	30	80	125
Vickers Jockey	66d	1940	75	175	225
Vickers Jockey	60d	1934-41	75	110	150
Vickers Viking	70c/705	1947-63	15	40	65
Vickers Viscount	706	1956-65	40	125	200
Vickers Viscount	708	1956-65	40	125	200
Viscount	60e/803	1957-60	75	125	195
Vulcan Bomber	749/707/99	1955-56	500	1500	4000
Westland Sikorsky Helicopter	716	1957-63	30	75	125
Whitley Bomber	60v	1937-41	95	150	225
Whitley Bomber	62t	1937-41	95	300	350

Buses & Taxis

Austin Taxi	40H/254	1951-62	60	95	135
Austin/London Taxi	284	1972-79	25	35	75
Autobus Parisien	F29D	1948-51	80	135	200
Autocar Chausson	F29F/571	1956-60	70	120	200
B.O.A.C. Coach	283	1956-63	55	80	115
Continental Touring Coach	953	1963-66	135	200	350
Ford Vedette Taxi	F24XT	1956-59	60	95	150
Observation Coach	29F/280	1954-60	50	75	100
Peugeot 404 Taxi	F1400	1967-71	50	75	100
Plymouth Plaza Taxi	266	1960-67	60	95	150
Routemaster Bus, Tern Shirts	289	1964-80	75	100	150
Silver Jubilee Bus	297	1977	25	35	70

Cars

Armstrong Siddeley, blue or brown	36A	1937-40	85	130	225
Austin Atlantic Convertible, blue	106/140A	1954-58	60	95	150
Austin Mini-Moke	342	1967-75	20	30	55
Cadillac Eldorado	131	1956-62	60	95	135
Chrysler Airflow	32/30A	1935-40	130	250	450
Chrysler Saratoga	F550	1961-66	70	100	190

Dinky

NAME	NO.	YEAR	GOOD	EX	MINT
Citroen 2 CV	F535/24T	1959-63	50	70	90
Citroen DS-19	F522/24C	1959-68	60	90	135
DeSoto Diplomat, green	F545	1960-63	70	100	190
DeSoto Diplomat, orange	F545	1960-63	60	85	125
Dodge Royal	191	1959-64	75	115	150
Estate Car	27D/344	1954-61	45	70	115
Ford Cortina Rally Car	212	1967-69	35	55	75
Ford Fairlane, pale green	148	1962-66	30	55	80
Ford Fairlane, South African issue, bright blue	148	1962-66	150	300	700
Ford Thunderbird (Hong Kong)	57/005	1965-67	50	70	100
Ford Thunderbird, South African Issue, blue	F565		120	250	600
Jaguar D-Type	238	1957-65	60	86	125
Jaguar XK 120, green, yellow, red	157	1954-62	50	95	135
Jaguar XK 120, turquoise, cerise	157	1954-62	80	125	250
Jaguar XK 120, white	157	1954-62	120	200	400
Jaguar XK 120, yellow/gray	157	1954-62	80	125	250
Lotus Racing Car	241	1963-70	20	30	50
Maserati Race Car	231	1954-64	45	75	110
Mustang Fastback	161	1965-73	35	55	75
Panhard PL17	F547	1960-68	45	80	120
Peugeot 403 Sedan	F521/24B	1959-61	50	90	135
Plymouth Fury Sports	115	1965-69	35	55	80
Renault Dauphine	F524/24E	1959-62	50	80	125
Rolls-Royce	30B	1946-50	65	100	125
Rolls-Royce Phantom V	198	1962-69	50	75	100
Singer Vogue	145	1962-67	50	75	100
Standard Vanguard	153	1954-60	60	85	120
Studebaker Commander	F24Y/540	1959-61	65	90	150
Town Sedan	24C	1934-40	85	130	200
Triumph TR-2, gray	105	1957-60	60	85	135
Triumph TR-2, yellow	105	1957-60	75	120	200
Volkswagen 1300 Sedan	129	1965-76	20	35	75
VW Karmann-Ghia	187	1959-64	45	80	125

Emergency Vehicles

NAME	NO.	YEAR	GOOD	EX	MINT
Ambulance	30F		100	160	275
Bedford Fire Escape	956	1969-74	75	110	175
Citroen DS19 Police Car	F501	1967-70	75	95	175
Citroen Fire Van	F25D/562	1959-63	80	110	250
Commer Fire Engine	955	1955-69	60	85	135
Delahaye Fire Truck	F32D/899	1955-70	120	190	375
Fire Chief Land Rover			35	50	85
Ford Police Car	F551	1960s	50	100	150
Mersey Tunnel Police Land Rover	255	1955-61	60	85	135
Plymouth Police Car	244	1977-80	25	35	50
Range Rover Ambulance	268	1974-78	25	35	50
Streamlined Fire Engine	25H/25	1946-53	75	100	175

Bedford Van, 1950s, Dinky

Lady Penelope's Fab 1, Dinky

Foden Flat Truck, Dinky

NAME	NO.	YEAR	GOOD	EX	MINT
Superior Criterion Ambulance	263	1962-68	50	75	100
USA Police Car (Pontiac)	251		35	50	85
Vauxhall Victor Ambulance	278	1964-70	55	85	115

Farm & Construction

NAME	NO.	YEAR	GOOD	EX	MINT
Atlas Digger	984		30	45	70
Blaw Knox Bulldozer	561		45	75	115
David Brown Tractor	305	1966-75	35	50	75
Euclid Dump Truck			45	75	115
Field Marshall Tractor	37N/301	1954-65	60	85	150
Garden Roller	105A	15	25	35	
Hayrake	324	1954-71	30	40	60
Massey-Harris Tractor	27A/300		50	75	120
Moto-Cart	27G/342	1954-60	35	50	75
Muir Hill Two-Wheel Loader	437	1962-78	30	40	60
Richier Road Roller	F830	1959-69	75	100	150
Road Grader	963	1973-75	30	45	70
Salev Crane	F595	1959-61	65	100	175

Military

NAME	NO.	YEAR	GOOD	EX	MINT
10 Ton Army Truck	622	1954-63	30	50	75
5.5 Medium Gun	692		15	30	50
AEC with Helicopter	618		50	85	125
AMX Bridge Layer	F883		75	110	200
AMX Tank	F80C/817		50	75	100
Armoured Command Vehicle	677		50	85	125
Austin Covered Truck	30SM/625		85	135	275
Austin Paramoke	601		25	35	50
Bedford Military Truck	25WM/60		80	125	250
Berliet Missile Loader	620		75	100	175
Berliet Wrecker	F806		60	90	140
Centurian Tank	651		30	50	75
Commando Jeep	612		25	35	50
Daimler Ambulance	30HM/624		80	125	250
Dodge Command Car	F810		40	60	85
Ferret Armoured Car	630		25	35	50
GMC Tanker	F823		125	250	500
Jeep	F816		50	75	115
Military Ambulance	626		25	45	75
Military Ambulance	F80F/820		50	75	100
Missile Servicing Platform	667		80	115	225
Reconnaisance Car	152B		40	60	100
Recovery Tractor	661		60	90	140
Searchlight, prewar	161A		125	250	500
Tank Transporter	660		75	100	175
Three Ton Army Wagon	621	1954-63	50	85	125

NAME	NO.	YEAR	GOOD	EX	MINT
Motorcycles & Caravans					
A.A. Motorcycle Patrol	270/44B	1946-64	30	45	70
Caravan	190	1956-64	30	45	60
Caravan, postwar	30G		40	60	85
Caravan, prewar	30G		55	85	150
Caravane Caravelair			75	150	250
Police Motorcycle Patrol	42B	1936-40	50	75	125
Police Motorcycle Patrol	42B	1946-53	30	45	70
Police Motorcyclist	37B	1938-40	50	75	125
Police Motorcyclist	37B	1946-48	30	45	70
Touring Secours Motorcycle Patrol (Swiss Version)	271	1960s	70	110	200
Space Vehicles					
Galactic War Chariot	361	1979-80	30	45	70
Joe's Car	102	1969-75	60	90	140
Klingon Battle Cruiser	357	1976-79	30	45	70
Lady Penelope's Fab 1, pink version	100	1966-76	80	135	220
Lady Penelope's Fab 1, shocking pink version	100	1966-76	115	200	340
Pathe News Camera Car			65	100	175
Prisoner Mini-Moke			115	200	340
Renault Sinpar	F1406	1968-71	80	135	220
Santa Special Model T Ford	485	1964-68	65	100	150
Tiny's Mini-Moke	350	1970-73	60	85	120
U.S.S. Enterprise	371/803	1980	30	45	70
Trucks					
A.E.C. Hoynor Transporter	974	1969-75	60	90	130
Atco Delivery Van, type 2	28N	1935-40	200	375	850
Atco Delivery Van, type 3	28N	1935-40	135	200	350
Austin Van, Nestle's	471	1955-63	60	110	175
Austin Van, Raleigh	472	1957-60	60	110	175
Austin Van, Shell/BP	470	1954-56	60	110	175
B.E.V. Truck	14A/400	1954-60	15	30	70
Bedford Van, Dinky Toys	482	1956-58	60	115	200
Bedford Van, Heinz	923	1955-59	100	165	300
Berliet Transformer Carrier	F898	1961-65	100	200	450
Big Bedford (blue/yellow)	408/922		90	135	210
Big Bedford (maroon/fawn)	408/922		80	120	185
Chevrolet El Camino	449	1961-68	35	65	100
Citroen Cibie Delivery Van	F561	1960-63	90	150	350
Citroen Milk Truck	F586	1961-65	145	275	600
Citroen Wrecker	F35A/582	1959-71	75	120	250
Coles Mobile Crane	971	1955-66	40	70	110
Covered Wagon (green, gray)	25B		65	115	160
Covered Wagon, Carter Paterson	25B		150	300	750
Electric Articulated Vehicle	30W/421		60	85	120

Road Grader, 1970s, Dinky

Commer Fire Engine, Dinky

Dinky

NAME	NO.	YEAR	GOOD	EX	MINT
Ensign Delivery Van, type 1	28E	1934	300	500	1000
Foden Flat Truck w/ Tailboard 1, gray/blue	503/903		140	210	450
Foden Flat Truck w/ Tailboard 1, red/black	503/903		140	210	450
Foden Flat Truck w/ Tailboard 2, blue/orange	503/903		90	150	275
Foden Mobilgas Tanker	941	1954-57	145	350	750
Foden Regent Tanker	942		135	300	550
Ford Transit Van	417	1978-80	15	20	30
Forward Control Wagon	25R	1948-53	45	65	90
Guy Van, Lyons	514		275	550	1600
Guy Van, Spratts	514		135	300	575
Guy Warrior 4 Ton	431	1958-64	150	270	450
Johnston Road Sweeper	449/451	1970s	25	50	75
Leland Comet Cement Truck	419/933	1956-59	85	150	250
Leland Tanker, Corn Products			700	1200	3000
Leland Tanker, Shell/BP	944	1963-69	125	215	450
Market Gardeners Wagon (yellow)	25F		65	115	160
Midland Bank	280	1966-68	60	85	120
Mighty Antar with Propeller	986	1956-64	125	215	400
Mini Minor Van, Joseph Mason Paints	274		150	300	500
Mini Minor Van, R.A.C.	273	1960s	65	115	150
Motor Truck (red, green, blue)	22C		80	120	200
Motor Truck (red/blue)	22C		150	350	650
Panhard Esso Tanker	F32C	1954-59	75	120	170
Panhard Kodak Semi Trailer	F32AJ	1952-54	140	250	450
Panhard SNCF Semi Trailer	F32AB	1954-59	100	165	280
Petrol Wagon, Power	25D		150	300	500
Pickfords Delivery Van, type 1	28B	1934-35	300	500	1000
Pickfords Delivery Van, type 2	28B	1934-35	200	375	600
Pullmore Car Transporter	982	1954-63	75	125	175
Renault Estafette	F561		50	85	150
Royal Mail Van	260	1955-61	65	115	150
Saviem Race Horse Van	F571	1969-71	125	225	400
Simca Glass Truck (gray/green)	F33C/579		75	120	170
Simca Glass Truck (yellow/green)	F33C/579		100	150	250
Studebaker Mobilgas Tanker	440	1954-61	70	100	175
Thames Flat Truck	422/30R	1951-60	45	75	110
Trojan Dunlop Van	31B/451	1952-57	70	110	185
Unic Bucket Truck	F38A/895	1957-65	75	120	225
Willeme Log Truck	F36A/897	1956-71	75	120	200
Willeme Semi Trailer Truck	F36B/896	1959-71	85	130	225

Doepke

NAME	DESCRIPTION	GOOD	EX	MINT
Adams Road Grader	26" long, orange, #2006	115	190	290
Adams Road Grader	26" long, yellow, #2006	165	240	340
American LaFrance Aerial Ladder Truck	33-1/2" long, red, #2008	195	275	350
American LaFrance Improved Aerial Ladder Truck	33-1/2" long, red, with outriggers and cast aluminum ladder, #2014	225	310	385
American LaFrance Pumper	19" long, red, #2010	225	310	395
American LaFrance Searchlight	white with battery-operated search light, #2023	950	1200	1800
Barber-Greene Bucket Loader	on tracks, 18" tall, early model w/swivel chute, green, #2001	330	410	525
Barber-Greene Bucket Loader	on tracks, 18" tall, later model w/out swivel chute, green, #2001	225	300	425
Barber-Greene Bucket Loader	on tracks, 18" tall, later model w/out swivel chute, orange, #2001	275	350	475
Barber-Greene Bucket Loader	on wheels, 22" long, green, #2013	250	350	475
Bulldozer	15" long, yellow, #2012	375	450	550
Clark Airport Tractor and Trailers	26-1/2" long, three pieces: red tractor, green trailer, yellow trailer, #2015	325	400	500
Euclid Truck	27" long, olive green, #2009	225	300	395
Euclid Truck	27" long, forest green, #2009	200	275	350
Euclid Truck	27" long, orange, #2009	175	250	330
Heiliner Scraper	29" long, red, #2011	220	300	395
Jaeger Concrete Mixer	15" long, yellow w/black drum, #2002	225	310	450
Jaguar	18" long, kit or built, light blue or red, #2018	395	495	565
MG Auto	15-1/2" long, kit, aluminum body, red, yellow, or gray primer, #2017	325	410	495
Unit Crane	11" long, without boom, orange, with finger wheel, #2007	225	300	375
Unit Crane	11" long, without boom, orange, w/out finger wheel, #2007	195	275	350
Wooldridge Earth Hauler	25", yellow, #2000	165	230	350

Hess

Amerada Hess Tanker	1969	650	1500	3500
Car Hauler	w/race car, 1992	20	30	40
Car Hauler	w/race car, 1991	15	20	35
Car Hauler	w/race car, 1988	25	40	75
Fire Truck	reissue of 1970, 1971	75	300	600
Fire Truck	1970	100	350	700
Fire Truck	with ladder, 1986	25	40	85
Fire Truck	w/ladder, 1989	25	40	60
Fuel Oils Truck	reissue of 1975 w/unlabeled oil drums, 1976	100	200	400
Fuel Oils Truck	tractor trailer, 1975	75	250	400
Gasoline Tanker	1965	400	1000	2000
Gasoline Tanker	1990	20	30	50
Gasoline Tanker	reissue of 1968-1969, 1974	50	100	250
Gasoline Tanker	reissue of 1977, 1978	50	100	225
Gasoline Tanker	1977	50	100	225
Gasoline Tanker	reissue of 1968-1969, 1972	50	100	250
Gasoline Tanker Bank	reissue of 1977 as a bank, 1984	25	35	70
Helicopter Transporter	w/helicopter, 1995	25	35	60
Mack Gasoline Tanker	1967	400	900	1500
Mack Gasoline Tanker	reissue of 1968, 1969	50	250	500
Mack Gasoline Tanker	reissue of 1967, 1968	50	250	500
Oil Delivery Truck	1934 Chevy Oil Tanker, 1982	25	40	80
Oil Delivery Truck Bank	reissue of 1982 as bank, 1983	25	45	75
Oil Delivery Truck Bank	1985	25	40	70
Police Car	1993	15	25	35
Rescue Truck	1994	15	25	35
Tractor Trailer Truck	with three labeled barrels, 1987	25	40	75
Training Van	1980	100	200	400
Voyager Ship	1966	500	1300	2500

Hot Wheels — Vintage

NAME	NO	DESCRIPTION	YEAR	MNP	MIP
#43-STP		petty blue, gray rollbars, blackwalls	1992	10	30
`31 Doozie	9649	orange, redline	1977	15	60
`31 Doozie	9649	orange, blackwall	1977	8	15
`32 Ford Delivery		white/pink, Early Times logo, blackwalls	1993	15	30
`40 Ford Two-Door	4367	black with white hubs, Real Rider	1983	15	40
`55 Chevy		blue #92, Real Rider	1992	10	30
`55 Chevy		black #92, Real Rider	1992	10	30
`55 Nomad		purple, Real Rider	1993	10	30
`55 Nomad		purple, blackwalls	1992	10	30
`56 Hi Tail Hauler	9647	orange, blackwall	1977	10	35
`56 Hi Tail Hauler	9647	orange, redline	1977	15	60
`57 Chevy		white #22, Real Rider	1992	15	30
`57 Chevy	9638	red, redline	1977	20	85
`57 Chevy	9638	red, blackwall	1977	10	30
`57 T-Bird	9522	black with white hubs, Real Rider	1986	30	150
`59 Caddy		pink, Canadian, blackwalls	1990	10	30
`59 Caddy		gold, blackwalls	1993	10	30
3 Window `34	4352	black, Real Rider	1984	50	160
A-OK		red, Real Rider	1981	75	275
Alien		blue	1988	10	20
Alive `55	6968	blue	1974	90	350
Alive `55	6968	assorted	1973	75	500
Alive `55	6968	green	1974	50	110
Alive `55	9210	chrome, redline	1977	15	55
Alive `55	9210	chrome, blackwall	1977	15	30
Ambulance	6451	assorted	1970	30	50
American Hauler	9118	blue	1976	20	50
American Tipper	9089	red	1976	20	50
American Victory	7662	light blue	1975	20	60
AMX/2	6460	assorted	1971	30	100
Aw Shoot	9243	olive	1976	15	25
Backwoods Bomb	7670	light blue	1975	40	125
Backwoods Bomb	7670	green, redline or blackwall	1977	30	120
Baja Bruiser	8258	orange	1974	30	75
Baja Bruiser	8258	yellow, blue in tampo	1974	200	900
Baja Bruiser	8258	yellow, magenta in tampo	1974	200	900
Baja Bruiser	8258	blue, redline or blackwall	1977	25	85
Baja Bruiser	8258	light green	1976	300	1000
Beatnik Bandit	6217	assorted	1968	15	45
Black Passion		black	1990	15	45
Boss Hoss	6406	assorted	1971	75	175
Boss Hoss	6499	chrome, Club Kit	1970	50	160
Brabham-Repco F1	6264	assorted	1969	10	25

Hot Wheels — Vintage

NAME	NO	DESCRIPTION	YEAR	MNP	MIP
Bronco 4-Wheeler	1690	Toys R Us	1981	50	120
Bugeye	6178	assorted	1971	30	75
Buzz Off	6976	gold plated, redline or blackwall	1977	15	30
Buzz Off	6976	assorted	1973	75	400
Buzz Off	6976	blue	1974	30	90
Bye Focal	6187	assorted	1971	90	375
Bywayman	2509	Toys R Us	1979	50	100
Bywayman	2196	blue, red interior	1989	40	50
Cadillac Seville		gold, Mexican, Real Rider	1987	50	100
Cadillac Seville		gray, French, Real Rider	1983	40	100
Captain America	2879	white, Scene Machine	1979	40	100
Carabo	7617	yellow	1974	400	1200
Carabo	7617	light green	1974	25	70
Carabo	6420	assorted	1970	20	60
Cement Mixer	6452	assorted	1970	20	45
Chapparal 2G	6256	assorted	1969	15	35
Chevy Monza 2+2	9202	light green	1975	200	600
Chevy Monza 2+2	7671	orange	1975	40	110
Chief's Special Cruiser	7665	red	1975	30	75
Chief's Special Cruiser	7665	red, redline	1977	25	65
Chief's Special Cruiser	7665	red, blackwall	1977	10	20
Circus Cats	3303	white 60	1981	50	120
Classic `31 Ford Woody	6251	assorted	1969	20	70
Classic `32 Ford Vicky	6250	assorted	1969	20	50
Classic `36 Ford Coupe	6253	blue	1969	10	25
Classic `36 Ford Coupe	6253	assorted	1969	15	50
Classic `57 T-Bird	6252	assorted	1969	25	70
Classic Caddy	2529	red/white/blue, Museum Exhibit car	1992	15	35
Classic Cobra		blue with white hubs, Real Rider	1985	40	80
Classic Nomad	6404	assorted	1970	30	110
Cockney Cab	6466	assorted	1971	30	100
Continental Mark III	6266	assorted	1969	20	60
Cool One	9120	plum, blackwall	1977	20	40
Corvette Stingray	9506	chrome, blackwall set only	1977	55	0
Corvette Stingray	9241	red	1976	30	80
Corvette Stingray	9506	chrome	1976	20	50
Custom AMX	6267	assorted	1969	25	80
Custom Barracuda	6211	assorted	1968	35	325
Custom Camaro	6208	white enamel	1968	300	2000
Custom Camaro	6208	assorted	1968	50	375
Custom Charger	6268	assorted	1969	50	190
Custom Corvette	6215	assorted	1968	50	250
Custom Cougar	6205	assorted	1968	60	275
Custom El Dorado	6218	assorted	1968	25	100

Hot Wheels — Vintage

NAME	NO	DESCRIPTION	YEAR	MNP	MIP
Custom Firebird	6212	assorted	1968	45	220
Custom Fleetside	6213	assorted	1968	60	250
Custom Mustang	6206	assorted	1968	75	425
Custom Mustang	6206	assorted with open hood scoops or ribbed	1968	400	1200
Custom Police Cruiser	6269	assorted	1969	55	200
Custom T-Bird	6207	assorted	1968	50	165
Custom VW Bug	6220	assorted	1968	15	60
Datsun 200SX	3255	maroon, Canada	1982	75	175
Demon	6401	assorted	1970	15	35
Deora	6210	assorted	1968	60	375
Double Header	5880	assorted	1973	120	450
Double Vision	6975	assorted	1973	110	400
Dune Daddy	6967	assorted	1973	110	400
Dune Daddy	6967	light green	1975	25	75
Dune Daddy	6967	orange	1975	175	450
El Rey Special	8273	light green	1974	95	300
El Rey Special	8273	light blue	1974	200	650
El Rey Special	8273	dark blue	1974	225	450
El Rey Special	8273	green	1974	40	75
Emergency Squad	7650	red	1975	15	50
Evil Weevil	6471	assorted	1971	40	85
Ferrari 312P	6973	red	1974	40	80
Ferrari 312P	6417	assorted	1970	20	30
Ferrari 312P	6973	assorted	1973	300	1100
Ferrari 512-S	6021	assorted	1972	75	250
Fire Chief Cruiser	6469	red	1970	10	25
Fire Engine	6454	red	1970	25	60
Flat Out 442		green, Canada	1984	75	120
Ford J-Car	6214	assorted	1968	10	60
Ford MK IV	6257	assorted	1969	10	35
Formula 5000	9119	white	1976	20	45
Formula 5000	9511	chrome	1976	30	65
Fuel Tanker	6018	assorted	1971	60	175
Funny Money	7621	magenta	1974	30	80
Funny Money	7621	gray, blackwall	1977	20	65
Funny Money	7621	gray, redline	1977	25	75
Funny Money	6005	gray	1972	60	325
GMC Motorhome	9645	orange, redline	1977	200	600
GMC Motorhome	9645	orange, blackwall	1977	10	25
Gold Passion		gold, Toy Fair promo	1992	40	100
Good Humor Truck		white	1986	55	125
Goodyear Blimp		chrome, Mattel promo	1992	65	0
Grass Hopper	7622	light green, no engine	1975	90	350
Grass Hopper	7621	light green	1974	30	90

Classic '36 Ford Coupe, 1969, Hot Wheels

Olds 442, 1971, Hot Wheels

Shelby Turbine, 1969, Hot Wheels

Hot Wheels — Vintage

NAME	NO	DESCRIPTION	YEAR	MNP	MIP
Grass Hopper	6461	assorted	1971	25	55
Greased Gremlin		red, Mexican, Real Rider	1987	100	300
Gremlin Grinder	9201	chrome, blackwall	1977	15	30
Gremlin Grinder	7652	green	1975	25	60
GT Racer	1789	blue	1989	25	60
Gun Bucket	9090	olive	1976	25	60
Gun Bucket	9090	olive, blackwall	1977	25	60
Gun Slinger	7664	olive, blackwall	1976	15	30
Gun Slinger	7664	olive	1975	25	50
Hairy Hauler	6458	assorted	1971	20	50
Hammer Down		red set only	1980	125	0
Heavy Chevy	7619	light green	1974	200	750
Heavy Chevy	6189	chrome, Club Kit	1970	50	175
Heavy Chevy	7619	yellow	1974	75	175
Heavy Chevy	9212	chrome, redline or blackwall	1977	40	120
Heavy Chevy	6408	assorted	1970	25	55
Hiway Robber	6979	assorted	1973	75	250
Hood	6175	assorted	1971	15	90
Hot Bird		blue	1980	40	100
Hot Bird		brown	1980	45	120
Hot Heap	6219	assorted	1968	10	35
Human Torch	2881	black	1979	10	25
Ice T	6980	light green, blackwall	1977	20	35
Ice T	6184	yellow	1971	40	200
Ice T	6980	assorted	1973	200	650
Ice T	6980	yellow with hood tampo	1974	200	525
Ice T	6980	light green	1974	30	75
Incredible Hulk Van	2850	white, Scene Machine	1979	45	100
Indy Eagle	6263	gold	1969	50	200
Indy Eagle	6263	assorted	1969	10	25
Inferno	9186	yellow	1976	30	60
Jack Rabbit Special	6421	white	1970	10	55
Jack-in-the-Box Promotion	6421	white, Jack rabbit w/decals	1970	225	0
Jet Threat	6179	assorted	1971	45	160
Jet Threat II	8235	magenta	1976	35	80
Khaki Kooler	9183	olive	1976	15	30
King Kuda	6411	assorted	1970	25	100
King Kuda	6411	chrome, Club Kit	1970	30	120
Large Charge	8272	green	1975	25	60
Letter Getter	9643	white, redline	1977	175	550
Letter Getter	9643	white, blackwall	1977	5	10
Light My Firebird	6412	assorted	1970	20	55
Lola GT 70	6254	assorted	1969	10	30
Lotus Turbine	6262	assorted	1969	10	30

NAME	NO	DESCRIPTION	YEAR	MNP	MIP
Lowdown	9185	light blue	1976	30	75
Lowdown	9185	gold plated, redline or blackwall	1977	15	30
Mantis	6423	assorted	1970	15	40
Masterati Mistral	6277	assorted	1969	50	125
Maxi Taxi	9184	yellow, blackwall	1977	20	60
Maxi Taxi	9184	yellow	1976	25	60
McClaren M6A	6255	assorted	1969	10	40
Mercedes 280SL	6962	assorted	1973	100	450
Mercedes 280SL	6275	assorted	1969	10	40
Mercedes C-111	6978	assorted	1973	300	1200
Mercedes C-111	6978	red	1974	40	90
Mercedes C-111	6169	assorted	1972	80	250
Mighty Maverick	9209	light green	1975	200	400
Mighty Maverick	7653	blue	1975	30	65
Mighty Maverick	6414	assorted	1970	35	85
Mighty Maverick	9209	chrome, blackwall	1977	25	50
Mod-Quad	6456	assorted	1970	15	40
Mongoose	6970	red/blue	1973	400	1400
Mongoose Funny Car	6410	red	1970	50	160
Mongoose II	5954	metallic blue	1971	75	350
Mongoose Rail Dragster	5952	blue, two pack	1971	75	600
Monte Carlo Stocker	7660	yellow	1975	45	90
Monte Carlo Stocker	7660	yellow, blackwall	1977	20	50
Motocross I	7668	red	1975	50	160
Motorcross Team Van	2853	red, Scene Machine	1979	50	125
Movin' On		white set only	1980	125	0
Moving Van	6455	assorted	1970	50	125
Mustang Stocker	9203	chrome	1976	40	90
Mustang Stocker	9203	yellow with red in tampo	1975	300	900
Mustang Stocker	7664	yellow with magenta tampo	1975	90	300
Mustang Stocker	9203	chrome, redline or blackwall	1977	40	90
Mustang Stocker	7664	white	1975	400	1200
Mutt Mobile	5185	assorted	1971	75	175
NASCAR Stocker	3927	white, NASCAR/Mountain Dew base	1983	90	165
Neet Streeter	9244	blue	1976	20	60
Neet Streeter	9510	chrome, blackwall set only	1977	40	0
Neet Streeter	9510	chrome	1976	20	40
Neet Streeter	9244	blue, blackwall	1977	15	30
Nitty Gritty Kitty	6405	assorted	1970	25	65
Noodle Head	6000	assorted	1971	40	150
Odd Job	6981	assorted	1973	100	600
Odd Rod	9642	yellow, redline	1977	30	50
Odd Rod	9642	yellow, blackwall	1977	20	40
Odd Rod	9642	plum, blackwall or redline	1977	200	400

Hot Wheels — Vintage

NAME	NO	DESCRIPTION	YEAR	MNP	MIP
Old Number 5	1695	red, no louvers	1982	10	20
Olds 442	6467	assorted	1971	275	625
Open Fire	5881		1972	100	400
Paddy Wagon	6966	blue	1973	30	120
Paddy Wagon	6966	blue, blackwall	1977	10	20
Paddy Wagon	6402	blue	1970	15	30
Paramedic	7661	yellow	1976	25	45
Paramedic	7661	yellow, blackwall or redline	1977	25	45
Paramedic	7661	white	1975	25	55
Peepin' Bomb	6419	assorted	1970	10	25
Pepsi Challenger	2023	yellow funny car	1982	15	20
Pit Crew Car	6183	white	1971	50	450
Poison Pinto	9240	green, blackwall	1977	10	20
Poison Pinto	9508	chrome, blackwall set only	1977	50	0
Poison Pinto	9240	light green	1976	25	60
Poison Pinto	9508	chrome	1976	20	40
Police Cruiser	6963	white with blue light	1977	30	65
Police Cruiser	6963	white	1974	35	90
Police Cruiser	6963	white, blackwall	1977	25	45
Police Cruiser	6963	white	1973	250	600
Porsche 911	7648	yellow	1975	40	75
Porsche 911	7648	black, six pack blackwall	1977	175	350
Porsche 911	9206	chrome, redline or blackwall	1977	20	40
Porsche 911	6972	orange	1975	25	60
Porsche 917	6416	assorted	1970	15	40
Porsche 917	6972	red	1974	175	500
Porsche 917	6972	orange	1974	40	75
Porsche 917	6972	orange, blackwall	1977	15	25
Porsche 917	6972	assorted	1973	300	950
Power Pad	6459	assorted	1970	25	65
Prowler	9207	chrome, blackwall	1977	35	70
Prowler	6965	assorted	1973	200	1000
Prowler	6965	light green	1974	500	1000
Prowler	6965	orange	1974	35	75
Python	6216	assorted	1968	10	55
Race Ace	2620	white	1986	15	30
Racer Rig	6194	red/white	1971	100	375
Racing Team Van		yellow, Scene Machine	1981	40	60
Ramblin' Cruiser	7659	white without phone number	1977	15	20
Ramblin' Wrecker	7659	white	1975	20	45
Ramblin' Wrecker	7659	white, blackwall	1977	10	20
Ranger Rig	7666	green	1975	20	65
Rash I	7616	blue	1974	300	800
Rash I	7616	green	1974	35	65
Rear Engine Mongoose	5699	red	1972	200	600

NAME	NO	DESCRIPTION	YEAR	MNP	MIP
Rear Engine Snake	5856	yellow	1972	200	600
Red Baron	6964	red, blackwall	1977	10	20
Red Baron	6964	red	1973	30	200
Red Baron	6400	red	1970	15	40
Red Passion		red	1994	10	20
Rescue Squad	3304	red, Scene Machine	1982	50	90
Road King Truck	7615	yellow set only	1974	400	1000
Rock Buster	9507	chrome	1976	15	30
Rock Buster	9088	yellow, blackwall	1977	10	15
Rock Buster	9507	chrome, blackwall set only	1977	45	0
Rock Buster	9088	yellow	1976	20	35
Rocket Bye Baby	6186	assorted	1971	60	200
Rodger Dodger	8259	gold plated, blackwall or redline	1977	25	70
Rodger Dodger	8259	magenta	1974	40	90
Rodger Dodger	8259	blue	1974	200	550
Rolls-Royce Silver Shadow	6276	assorted	1969	25	45
Ruby Red Passion		red	1992	25	45
S'Cool Bus	6468	yellow	1971	175	750
S.W.A.T. Van	2854	blue, Scene Machine	1979	70	110
Sand Crab	6403	assorted	1970	10	40
Sand Drifter	7651	green	1975	150	375
Sand Drifter	7651	yellow	1975	20	50
Sand Witch	6974	assorted	1973	100	400
Scooper	6193	assorted	1971	100	325
Seasider	6413	assorted	1970	50	120
Second Wind	9644	white, blackwall or redline	1977	35	75
Shelby Turbine	6265	assorted	1969	10	25
Short Order	6176	assorted	1971	35	100
Show Hoss II	9646	yellow, redline	1977	300	600
Show Hoss II	9646	yellow, blackwall	1977	35	60
Show-Off	6982	assorted	1973	140	400
Sidekick	6022	assorted	1972	80	200
Silhouette	6209	assorted	1968	20	90
Simpsons Camper		blue, Scene Machine	1990	10	20
Simpsons Van		yellow, Scene Machine	1990	10	20
Sir Rodney Roadster	8261	yellow, blackwall	1977	25	40
Sir Sidney Roadster	8261	yellow	1974	25	65
Sir Sidney Roadster	8261	light green	1974	325	650
Sir Sidney Roadster	8261	orange/brown	1974	375	700
Six Shooter	6003	assorted	1971	75	225
Sky Show Fleetside (Aero Launcher)	6436	assorted	1970	400	850
Snake	6969	white/yellow	1973	600	1500
Snake Dragster	5951	white, two-pack	1971	75	0
Snake Funny Car	6409	assorted	1970	60	300

Racers Don "The Snake" Prudhomme and Tom "The Mongoose" McEwen were favorites with fans in the early 1970s. Mattel made several Hot Wheels versions of their vehicles including dragsters and funny cars.

NAME	NO	DESCRIPTION	YEAR	MNP	MIP
Snake II	5953	white	1971	60	275
Snorkel	6020	assorted	1971	60	150
Space Van	2855	gray, Scene Machine	1979	45	100
Special Delivery	6006	blue	1971	45	150
Spider-Man	2852	black	1979	10	30
Spider-Man Van	2852	white, Scene Machine	1979	50	125
Splittin' Image	6261	assorted	1969	10	35
Spoiler Sport	9641	light green, blackwall	1977	5	10
Spoiler Sport	9641	light green, redline	1977	15	25
Staff Car	9521	olive, blackwall	1977	500	750
Staff Car	9521	olive, six-pack only	1977	400	750
Steam Roller	9208	chrome, redline or blackwall	1977	20	40
Steam Roller	9208	chrome with seven stars	1977	100	300
Steam Roller	8260	white	1974	25	70
Steam Roller	8260	white with seven stars	1974	100	300
Street Eater	7669	black	1975	30	50
Street Rodder	9242	black, blackwall	1977	20	45
Street Rodder	9242	black	1976	40	85
Street Snorter	6971	assorted	1973	110	400
Strip Teaser	6188	assorted	1971	65	200
Sugar Caddy	6418	assorted	1971	20	70
Super Chromes	9505	chrome, blackwall six-pack	1977	375	0
Super Van	7649	Toys-R-Us	1975	100	350
Super Van	9205	chrome	1976	20	40
Super Van	7649	black, blackwall	1977	15	25
Super Van	7649	blue	1975	650	0
Super Van	7649	plum	1975	90	250
Superfine Turbine	6004	assorted	1973	300	1100
Sweet "16"	6007	assorted	1973	90	375
Swingin' Wing	6422	assorted	1970	15	40
T-4-2	6177	assorted	1971	35	165
T-Totaller	9648	brown, blackwall	1977	10	30
T-Totaller	9648	black, Red Line, six-pack only	1977	500	1000
T-Totaller	9648	black, blackwall	1977	10	30
Team Trailer	6019	white/red	1971	95	225
Thing, The	2882	dark blue	1979	15	35
Thor	2880	yellow	1979	10	30
Thrill Driver Torino	9793	red/white, blackwall, set of two	1977	275	0
TNT-Bird	6407	assorted	1970	25	70
Top Eliminator	7630	blue	1974	50	165
Top Eliminator	7630	gold plated, redline or blackwall	1977	25	45
Torero	6260	assorted	1969	10	60
Torino Stocker	7647	gold plated, redline or blackwall	1977	30	70
Torino Stocker	7647	red	1975	35	70
Tough Customer	7655	olive	1975	15	55

Hot Wheels — Vintage

NAME	NO	DESCRIPTION	YEAR	MNP	MIP
Tow Truck	6450	assorted	1970	30	80
Tri-Baby	6424	assorted	1970	15	40
Turbo Mustang		blue	1984	35	65
Turbofire	6259	assorted	1969	10	45
Twinmill	6258	assorted	1969	10	35
Twinmill II	9502	chrome, blackwall set only	1977	45	0
Twinmill II	8240	orange, blackwall	1977	10	25
Twinmill II	9509	chrome	1976	20	45
Twinmill II	8240	orange	1976	10	35
Vega Bomb	7658	green	1975	250	800
Vega Bomb	7654	orange, blackwall	1977	25	55
Vega Bomb	7658	orange	1975	40	85
Volkswagen	7620	orange with bug on roof	1974	30	60
Volkswagen	7620	orange with stripes on roof	1974	100	400
Volkswagen Beach Bomb	6274	surf boards on side raised panels	1969	50	275
Volkswagen Beach Bomb	6274	surf boards in rear window	1969	4500	0
VW Bug		pink, Real Rider	1993	15	35
Warpath	7654	white	1975	50	110
Waste Wagon	6192	assorted	1971	90	325
What-4	6001	assorted	1971	50	150
Whip Creamer	6457	assorted	1970	15	40
White Passion		white, in box	1990	10	30
Winnipeg	7618	yellow	1974	90	300
Xploder	6977	assorted	1973	100	500
Z Whiz	9639	white, redline	1977	1500	0
Z Whiz	9639	gray, blackwall	1977	10	20
Z Whiz	9639	gray, redline	1977	25	55
Z Whiz	9639	blue	1982	20	55

Hot Wheels - Numbered Packs

NO.	NAME	YEAR	DESCRIPTION	MNP	MIP
1	Old Number 5	1982	red	94	125
2	Sol Aire CX4	1989	black, Ultra Hots wheels	12	15
2	Sol Aire CX4	1989	black with yellow window, Ultra Hots wheels	15	20
2	Sol Aire CX4	1989	blue, Ultra Hots wheels	15	20
2	Sol Aire CX4	1989	black, Ultra Hots wheels, gold	68	90
2	Sol Aire CX4	1989	black, Ultra Hots wheels, silver	19	25
3	Wheel Loader	1990	yellow, construction tires, yellow	9	12
4	XT-PR	1989		9	12
5	Good Humor Truck	1996	white, 7-spoke wheel	7	10
5	Good Humor Truck	1988	white, small or large window	3	5
5	Good Humor Truck	1992	white	3	5
5	Good Humor Truck	1995	white, 5-spoke wheels	3	4
5	Good Humor Truck	1996	white, sawblade wheels	3	4
6	Blazer 4x4	1984	black, construction tires, silver	49	65
6	Blazer 4x4	1984	black, construction tires	41	55
6	Blazer 4x4	1991	blue, construction tires	9	12
6	Blazer 4x4	1991	blue, construction tires, silver	11	14
7	Troop Convoy		olive	83	110
8	Vampyra	1991	purple, with or without Ultra Hots wheels	7	10
9	Hiway Haul	1996	red, Real Rider wheels, gold	15	20
10	Baja Breaker	1989	all-terrain vehicle, white	34	45
11	`31 Doozie	1990	maroon, whitewalls	19	25
12	Roll Patroll		olive, construction tires, black	113	150
13	Delivery Van - Bob's	1996	red	12	15
14	Custom Corvette	1989		19	25
14	Custom Corvette	1989	red, Ultra Hots wheels	19	25
14	Custom Corvette	1989	red, clear window, Ultra Hots wheels	34	45
15	Peterbilt (Shell)	1989	yellow	19	25
16	CAT Earthmover	1988	yellow, construction tires, gold	49	65
16	CAT Earthmover	1988	yellow, construction tires, yellow	19	25
17	Suzuki Quadracer	1987	yellow, construction tires, yellow and silver	19	25
17	Suzuki Quadracer	1987	yellow, construction tires, silver and yellow	9	12
18	Mercedes 540K	1988	black	12	15
19	Shadow Jet	1988	maroon	49	65
19	Shadow Jet	1988	blue	9	12
20	Rocketank	1988	olive	12	15
21	Nissan Hardbody	1989	white, construction tires	25	30
21	Nissan Hardbody	1989	white, construction tires, silver	26	35
22	Talbot Largo	1990	white, whitewalls	12	15
23	`80s Firebird	1990	yellow, yellow or black taillights	7	10
23	`80s Firebird	1990	yellow	7	10
24	Hiway Hauler	1990	turquoise with Ocean Pacific tampo	9	12
24	Hiway Hauler	1990	red with Pepsi	9	12
26	`65 Mustang Conv	1987	turquoise	19	25

Hot Wheels - Numbered Packs

NO.	NAME	YEAR	DESCRIPTION	MNP	MIP
26	`65 Mustang Conv	1987	blue, whitewalls	12	15
26	`65 Mustang Conv	1987	white, whitewalls	83	110
26	`65 Mustang Conv	1987	whitewalls	19	25
27	Command Tank	1985	olive	17	22
28	`37 Bugatti	1989	yellow	49	65
28	`37 Bugatti	1989	yellow, whitewalls	25	30
28	`37 Bugatti	1991	blue, blue fenders, whitewalls	6	8
28	`37 Bugatti	1991	blue, yellow fender, whitewalls	9	12
28	`37 Bugatti	1995	blue, 7-spoke wheels	2	3
28	`37 Bugatti	1996	blue	30	40
29	Tail Gunner	1987	olive, construction tires	34	45
29	Tail Gunner	1987	olive, construction tires	49	65
30	`80s Corvette	1989	blue	49	65
31	Classic Cobra	1988	red	3	5
31	Cobra	1995	red, metal base, 7-spoke wheels	3	4
31	Cobra	1995	red, black plastic base, 7-spoke wheels	3	4
31	Cobra	1995	red, 7-spoke wheels	3	5
32	Sharkruiser	1987	gray, Ultra Hots wheels	7	10
32	Sharkruiser	1991	gray, Hot Ones wheels, silver	9	12
33	Camaro Z-28	1991	orange	7	10
33	Camaro Z-28	1995	red with black base, Ultra Hots wheels	3	4
33	Camaro Z-28	1991	orange, Ultra Hots wheels	57	75
33	Camaro Z-28	1986	red, metal base, Hot Ones wheels	9	12
33	Camaro Z-28	1986	red, plastic base, Ultra Hots wheels	7	10
33	Camaro Z-28	1986	red, plastic base, Hot Ones wheels	19	25
33	Camaro Z-28	1986	red, plastic base	6	8
33	Camaro Z-28	1991	purple	6	8
33	Camaro Z-28	1986	red, metal base, Ultra Hots wheels	9	12
34	Bulldozer	1980		9	12
35	Ferrari Testarossa	1989	red with black interior, Ultra Hots Wheels	4	6
35	Ferrari Testarossa	1996	black with black interior, lace wheels	27	35
35	Ferrari Testarossa	1995	black, lace wheels	2	3
35	Ferrari Testarossa	1994	black with tan interior, Ultra Hots wheels,	9	12
35	Ferrari Testarossa	1989	red with tan interior, Ultra Hots Wheels	7	9
35	Ferrari Testarossa	1995	black, 3-spoke wheels, gold	7	9
36	Baja Bug	1987	white	19	25
37	Hot Bird	1990	black with tan interior, Hot Ones wheels	38	50
37	Hot Bird	1990	white with red interior, Ultra Hots wheels	4	6
37	Hot Bird	1990	black with red interior, Hot Ones wheels	15	20
37	Hot Bird	1990	blue, Hot Ones wheels	15	20
37	Hot Bird	1990	blue, Ultra Hots wheels	15	20
37	Hot Bird	1990	white with red interior, Hot Ones wheels, chrome	12	15
37	Hot Bird	1990	white with tan interior, Ultra Hots wheels	26	35
37	Hot Bird	1990	white with red interior, Ultra Hots wheel, golds	113	150

Hot Wheels - Numbered Packs

NO.	NAME	YEAR	DESCRIPTION	MNP	MIP
38	Dump Truck	1992	yellow with metal bed, construction tires, yellow	10	13
38	Dump Truck	1992	yellow with pastic bed, construction tires, yellow	4	6
39	Monster Vette	1990	purple, construction tires	9	12
39	Monster Vette	1990	purple, construction tires, silver	12	15
39	Monster Vette	1987	yellow, construction tires	75	100
39	Monster Vette	1987	yellow, construction tires, silver	75	100
40	Power Plower	1989	black, construction tires	7	9
40	Power Plower	1989	black	19	25
42	Oshkosh Snowplow	1990	green with cab, construction tires, orange	11	14
42	Oshkosh Snowplow	1990	green with plastic cab, construction tires	9	12
42	Oshkosh Snowplow	1990	green with metal cab, construction tires	9	12
42	Oshkosh Snowplow	1990	green with metal cab, construction tires, silver	13	16
43	Tall Ryder	1987	gray, construction tires, silver	19	25
44	`35 Classic Caddy	1994	blue, whitewalls	7	10
44	`35 Classic Caddy	1994	blue, door handles, whitewalls	12	15
44	`35 Classic Caddy	1996	blue, 5-spoke wheels	2	3
45	Rescue Ranger	1988	red, blackwalls	12	15
46	Rig Wrecker	1990	white	25	30
47	`57 Chevy	1989	turquoise, Ultra Hots wheels	49	65
47	`57 Chevy	1989	turquoise, Hot Ones wheels	64	85
48	Pearl Passion	1995	white, 5-spoke wheels	19	25
49	Rolls Royce	1984	blue, whitewalls	94	125
49	Gulch Stepper	1987	red without Penzoil tampo, construction tires	12	15
49	Gulch Stepper	1987	red, construction tires	9	12
49	Gulch Stepper	1987	red, construction tires silver	12	15
51	`40s Woodie	1987	yellow	450	600
52	Delivery Van	1990	white	19	25
53	Zombot	1992	gold, Hot Ones wheels, silver	7	10
53	Zombot	1992	gold, Ultra Hots wheels	3	5
54	Nissan 300ZX	1989	white, Ultra Hots wheels	12	15
54	Nissan 300ZX	1989	white, Hot Ones wheels	34	45
54	Nissan 300ZX	1989	red, Ultra Hots wheels	9	12
55	BMW 850i	1995	blue, 3-spoke wheels	2	3
55	Road Roller	1987	yellow	7	10
56	Bronco 4-wheeler	1990	white, construction tires	34	45
56	Bronco 4-wheeler	1990	turquoise, construction tires	12	15
56	Bronco 4-wheeler	1990	white	34	45
57	3-Window `34	1989	purple	12	15
58	Blown Camaro Z28	1990	turquoise	34	45
58	Blown Camaro Z28	1990	turquoise, Ultra Hots wheels	19	25
58	Blown Camaro Z28	1990	turquoise, Hot Ones wheels	25	30
59	Sheriff Patrol	1990	blue	7	10
59	Sheriff Patrol	1990	black, 7-spoke wheels	2	3

Hot Wheels - Numbered Packs

NO.	NAME	YEAR	DESCRIPTION	MNP	MIP
59	Sheriff Patrol	1990	black	6	8
60	Lamborghini Countach	1989	white, Ultra Hots wheels	7	10
62	Alien	1988	red, Ultra Hots wheels	4	6
62	Alien	1994	red, 5-spoke wheels	7	10
63	Radar Ranger	1997	silver, sawblade wheels	2	3
63	Radar Ranger	1988	silver, construction tires, silver	7	10
63	Radar Ranger	1988	silver, construction tires, silver	12	15
63	Radar Ranger	1988	silver, construction tires, black	25	30
64	Ferrari F40	1995	red, 3-spoke wheels	2	3
65	VW Bug	1989	turquoise	19	25
65	VW Bug	1990	red	12	15
65	VW Bug	1990	red	12	15
66	Custom Corvette	1989	red, Ultra Hots wheels	13	16
66	Custom Corvette	1989	red	21	28
67	`32 Ford Delivery	1989	yellow	12	15
67	T-Bucket	1989	yellow	4	6
68	T-Bucket	1995	yellow, 5-spoke wheels	3	4
69	Ferrari F40	1995	red, 5-spoke wheels	3	4
69	Ferrari F40	1995	red, Ultra Hots wheels	7	9
69	Ferrari F40	1994	red, Ultra Hots wheels, gold	2	3
69	Ferrari F40	1995	red, 3-spoke wheels	2	3
69	Ferrari F40	1995	red, 5-spoke wheels, gold	3	4
70	Chevy Stocker	1989	black	12	15
70	Chevy Stocker	1989	black	14	18
71	Ambulance	1989	white	4	6
71	Ambulance	1996	white, 7-spoke wheels	15	20
71	Ambulance	1996	white, sawblade wheels	2	3
72	School Bus	1989	yellow	3	5
73	Street Roader	1989	white, construction tires	7	10
73	Street Roader	1989	white, construction tires, silver	38	50
74	GT Racer	1989	purple without Valvoline tampo, Ultra Hots wheels	9	12
74	GT Racer	1989	purple without Valvoline tampo, Ultra Hots wheels	6	8
74	GT Racer	1989	purple	9	12
75	Pontiac Banshee	1996	red, 7-spoke wheels	3	4
75	Pontiac Banshee	1990	red, Ultra Hots wheels	4	6
75	Pontiac Banshee	1990	red, Ultra Hots wheels, gold	113	150
75	Pontiac Banshee	1995	red, 5-spoke wheels	3	5
76	Kenworth Big Rig	1990	black	7	9
76	Kenworth Big Rig	1995	black, 7-spoke	3	5
77	Bywayman	1989	blue metal flake with red interior, construction tires	57	75
77	Bywayman	1989	maroon, construction tires	11	14
77	Bywayman	1989	maroon, construction tires, white	94	125
77	Bywayman	1990	black, construction tires	12	15

Pit Crew Car, 1971, Hot Wheels

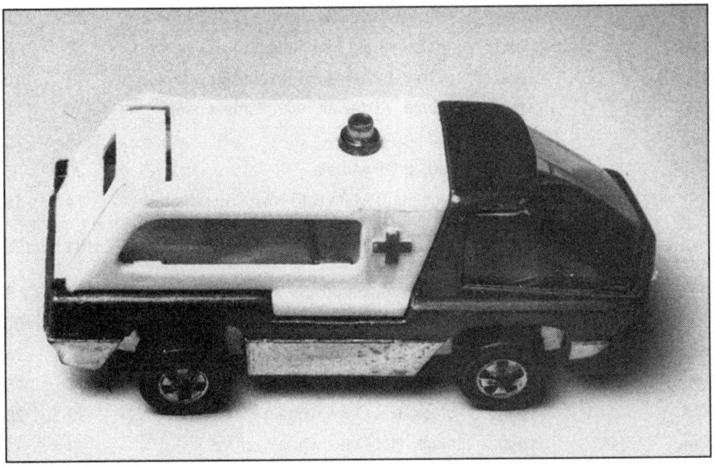

Ambulance, 1970, Hot Wheels

Volkswagen Beach Bomb, (side surf-boards), 1969, Hot Wheels

Hot Wheels - Numbered Packs

NO.	NAME	YEAR	DESCRIPTION	MNP	MIP
78	Peterbilt Cement	1985	red	19	25
79	Big Bertha	1985	olive	12	15
80	Porsche 959	1990	red, number 59, Ultra Hots wheels	9	12
80	Porsche 959	1990	red, number 59, Hot Ones wheels	11	14
80	Porsche 959	1990	red metal flake, Ultra Hots wheels	9	12
80	Porsche 959	1990	red, number 7, Ultra Hots wheels	9	12
81	Ratmobile	1992	white, Ultra Hots wheels	4	6
81	Ratmobile	1992	white, Hot Ones wheels, chrome	9	12
82	Fire-Eater	1982	red	7	10
83	Tank Gunner	1987	olive, blackwalls black	19	25
83	Tank Gunner	1987	tan, blackwalls black	19	25
84	Probe Funny Car	1984	red	19	25
86	Propper Chopper	1990	white with yellow triangle A	17	22
86	Propper Chopper	1990	white without triangle A	6	8
87	Purple Passion	1990	purple, whitewalls with scallop wheels	7	10
87	Purple Passion	1993	purple, whitewalls with flame wheels	6	8
88	T-Bird Stock	1990	red	15	20
88	T-Bird Stock	1990	red, small wheels	26	35
88	T-Bird Stock	1991	black and white with Havoline tampo, Ultra Hots wheels	14	18
88	T-Bird Stock	1991	black and white with Havoline tampo, blackwalls	131	175
89	Mini Truck	1990	turquoise, Ultra Hots wheels	4	6
89	Mini Truck	1990	turquoise, Hot Ones wheels, chrome	120	160
90	Mongoose Dragster	1995	blue, 5-spoke wheels	11	15
91	Snake Dragster	1995	yellow, 5-spoke wheels	11	15
92	Mercedes 380SL	1990	black, Hot Ones wheels	19	25
92	Mercedes 380SL	1990	black, Hot Ones wheels, silver	9	11
92	Mercedes 380SL	1990	black, Ultra Hots wheels	5	7
94	Auburn 852	1991	red, whitewalls	7	10
94	Auburn 852	1991	red, blackwalls	131	175
95	`55 Chevy	1991	yellow	11	15
95	`55 Chevy	1991	white	9	12
98	Nissan 300ZX	1992	red, Unltra Hots wheels	7	10
98	Nissan 300ZX	1992	red, Hot Ones wheels, silver	10	13
98	Nissan 300ZX	1992	red, Hot Ones wheels, gold	131	175
99	Ford Stake Bed	1991	blue	9	12
100	Peterbuilt Dump	1989	red, blackwalls, white	19	25
100	Peterbuilt Dump	1996	red, 7-spoke wheels	2	3
101	Split Window `63	1991	green	7	10
102	Surf Patrol	1991	yellow, construction tires	6	8
102	Surf Patrol	1991	yellow, construction tires, no logo	3	5
103	Range Rover	1990	white, construction tires	7	10
104	Turbo Streak	1991	pink	7	9
104	Turbo Streak	1991	pink with pink spoiler	49	65
105	Peugot 295	1980	white	25	30

Hot Wheels - Numbered Packs

NO.	NAME	YEAR	DESCRIPTION	MNP	MIP
106	VW Golf	1991	red	7	10
106	VW Golf	1991	white	15	20
108	Ramp Truck	1991	clear windows	15	20
108	Ramp Truck	1991	dark windows	10	15
110	Trailbuster	1991	turquoise, construction tires	6	8
110	Trailbuster	1991	turquoise, construction tires	12	16
111	Street Beast	1991	white and turquoise with whitewalls	4	6
112	Limozeen	1991	white with whitewalls	6	8
113	Speed Shark	1995	black	3	4
113	Speed Shark	1996	black, 5-spoke wheels	3	4
113	Speed Shark	1991	purple	3	4
113	Speed Shark	1991	maroon	26	35
114	Fiero 2M4	1987	red	64	85
114	Fiero 2M4	1987	red, Ultra Hots wheels	12	15
114	Fiero 2M4	1991	black	12	15
114	Fiero 2M4	1991	black with black base	4	6
114	Fiero 2M4	1991	black, Hot Ones wheels, silver	12	16
114	Fiero 2M4	1991	black, Ultra Hots wheels	8	11
115	Roll Patrol	1991	olive, construction tires	12	15
115	Roll Patrol	1991	olive, construction tires, black	6	8
116	Classic Ferrari	1991	yellow	12	15
116	Classic Ferrari	1995	yellow, 7-spoke wheels	3	4
116	Classic Ferrari	1991	yellow	3	5
116	Mazda Miata MX5	1991	red, no side	6	8
116	Classic Ferrari	1991	yellow with black pipes and yellow	9	12
116	Mazda Miata MX5	1991	red	6	8
118	Ferrari 348	1991	yellow, Ultra Hots wheels	4	6
118	Ferrari 348	1991	yellow, Hot Ones wheels, silver	9	12
122	Toyota MR2 Rally	1991	white, Hot Ones, silver	9	12
122	Toyota MR2 Rally	1991	white, Ultra Hots wheels	4	6
123	Lambo Diablo	1991	red, Ultra Hots wheels	7	9
125	Zender Fact 4	1991	Silver, no Fact 4	5	7
125	Zender Fact 4	1991	Silver, Hot Ones wheels, chrome	9	12
125	Zender Fact 4	1991	Silver, Ultra Hots wheels, gold	90	120
125	Zender Fact 4	1991	silver, Ultra Hots wheels	7	9
126	Chevy Lumina	1992	red, large wheels	12	15
126	Chevy Lumina	1992	red	2	3
127	Power Plower	1990	light purple, construction tires	6	8
127	Chevy Lumina	1992	red	2	3
127	Power Plower	1990	purple metal flake, construction tires	7	10
127	Power Plower	1990	purple, construction tires	7	10
129	Suziki Quadracer	1991	white, construction tires, yellow	12	15
129	Suziki Quadracer	1991	white, construction tires	15	20
131	Nissan Hardbody	1991	black with pink interior, construction tires	12	15

Hot Wheels - Numbered Packs

NO.	NAME	YEAR	DESCRIPTION	MNP	MIP
131	Nissan Hardbody	1991	black with yellow interior, construction tires	2	3
131	Nissan Hardbody	1992	white, construction tires	34	45
131	Nissan Hardbody	1997	red, no. 1	2	3
133	Shadow Jet PR90	1990	purple with green interior	3	5
133	Shadow Jet PR90	1990	purple with yellow interior	15	20
133	Shadow Jet PR90	1990	purple, small wheels	26	35
134	Mercedes 540K	1990	white	4	6
135	`32 Ford Delivery	1991	white/turquoise	4	6
135	`32 Ford Delivery	1995	purple, Real Riders, green	132	175
135	`32 Ford Delivery	1995	white/turquoise, 7-spoke wheels	3	4
135	`32 Ford Delivery	1995	purple, Real Rider wheels, green	23	30
136	`56 Flashsider	1992	turquoise, Ultra Hots wheels	2	3
136	`56 Flashsider	1992	turquoise, Hot Ones wheels, silver	38	50
136	`56 Flashsider	1996	turquoise, 5-dot wheels	1	2
136	`56 Flashsider	1996	turquoise, 5-spoke wheels	3	4
137	Goodyear Blimp	1992	gray	4	6
137	Goodyear Blimp	1992	gray, no. 1	4	6
140	Flashfire	1992	black, Ultra Hots wheels	4	6
140	Flashfire	1992	black, Hot Ones wheels, silver	8	11
140	Flashfire	1996	black, sawblade wheels	2	3
140	Flashfire	1996	black, 3-spoke wheels	2	3
141	Shock Factor	1995	black, red insert, construction tires	23	30
141	Shock Factor	1992	black, construction tires	2	3
142	Hiway Haul	1992	red, Kool-Aid	80	10
143	Recycling Truck	1992	orange	1	2
143	Recycling Truck	1995	orange, 7-spoke wheels	2	3
144	Oshkosh Cement	1995	blue and white	2	3
145	Tractor	1992	yellow, construction tires, yellow	8	10
145	Tractor	1992	red, construction tires	2	3
145	Tractor	1996	red, sawblade wheels	2	3
146	Bulldozer	1992	yellow	2	3
147	Tank Truck	1995	orange, Unical, 7-spoke wheels	2	3
147	Tank Truck	1996	orange	2	3
147	Tank Truck	1996	red	2	3
147	Tank Truck	1992	red, Unical tampo	4	5
147	Tank Truck	1995	red, 7-spoke wheels	4	5
148	Porsche 930	1995	red, 5-spoke wheels	2	3
148	Porsche 930	1991	purple	49	65
148	Porsche 930	1995	green	56	75
148	Porsche 930	1991	red	3	4
149	BMW 850i	1991	blue, Ultra Hots wheels	2	3
149	BMW 850i	1993	blue, yellow wheels	19	25
149	BMW 850i	1993	blue with green/yellow wheels	4	5
149	BMW 850i	1991	blue, Hot Ones wheels, silver	8	10

Hot Wheels - Numbered Packs

NO.	NAME	YEAR	DESCRIPTION	MNP	MIP
149	BMW 850i	1991	blue, Pow wheels	4	5
150	BMW 323 Cabriolet	1991	black, Ultra Hots wheels	5	7
150	BMW 323 Cabriolet	1991	black, blackwalls	11	15
151	Ford Aerostar	1991	purple	6	8
153	Thundersteak	1993	blue/green, without no. 1	8	10
153	Thundersteak	1993	blue/green, no. 1	9	12
153	Thundersteak	1992	yellow, pennzoil	6	8
154	`59 Cadillac	1991	white, whitewall	3	4
155	Turboa	1986	yellow, Ultra Hots wheels	3	4
155	Turboa	1986	yellow, Hot Ones wheels, silver	7	9
156	Rodzilla	1988	purple with red base, Ultra Hots wheels	6	8
156	Rodzilla	1988	purple with purple base, Ultra Hots wheels	5	7
156	Rodzilla	1988	purple with red eyes	49	65
157	`57 Chevy	1991	yellow, Ultra Hots wheels	8	11
157	`57 Chevy	1991	yellow, Hot Ones wheels, silver	15	20
158	Mercedes Unimog	1991	white, construction tires	6	8
159	Big Bertha	1992	gray	2	3
160	Command Tank	1992	white	4	6
161	Roll Patrol CJ-7	1992	tan, construction tires, black	5	7
161	Roll Patrol CJ-7	1992	tan, no star, construction tires, black	5	7
161	Roll Patrol CJ-7	1992	tan, blackwalls	5	7
161	Roll Patrol CJ-7	1992	tan, construction tires	11	14
161	Roll Patrol CJ-7	1992	tan, no star, blackwalls, black	5	7
162	`65 Mustang Convertible	1992	red, whitewalls	8	11
162	`65 Mustang Convertible	1995	red, 5-spoke wheels	2	3
162	`65 Mustang Convertible	1995	red, 7-spoke wheels	4	6
163	Talbot Lago	1992	red, whitewalls	4	6
164	Mercedes 540K	1993	purple, sawblade wheels	6	8
164	Mercedes 540K	1996	purple, 3-spoke wheels	2	3
164	Mercedes 540K	1996	purple, 5-spoke wheels	2	3
164	Mercedes 540K	1996	purple, 7-spoke wheels	2	3
164	Mercedes 540K	1996	purple, 5-dot wheels	2	3
165	Suzuki Quadracer	1992	pink, construction tires	2	3
166	Vamypra	1992	black	2	3
166	Vamypra	1995	black with yellow/green eye, Ultra Hots	2	3
166	Vamypra	1995	black, Hot Ones, silver	19	25
166	Vamypra	1995	black with yellow top, Ultra Hots wheels	3	4
166	Vamypra	1995	black, Ultra Hots wheels	2	3
167	`80s Firebird	1992	orange	5	7
167	`80s Firebird	1992	orange, extra bolt	9	12
168	GT Racer	1992	black	3	4
169	Sol-Aire CX4	1992	blue, Ultra Hots wheels	5	7
169	Sol-Aire CX4	1992	blue with plastic base, Ultra Hots wheels	5	7
170	Chevy Stocker	1990	orange	4	5

Ice T, 1971, Hot Wheels

Mutt Mobile, 1971, Hot Wheels

Snake and Mongoose Funny Cars, 1970, Hot Wheels

Hot Wheels - Numbered Packs

NO.	NAME	YEAR	DESCRIPTION	MNP	MIP
171	VW Bug	1991	purple	9	12
172	Mazda Miata	1996	red, Pow wheels	3	4
172	Mazda Miata	1991	yellow	1	2
172	Mazda Miata	1995	yellow with yellow/green wheels	4	6
172	Mazda Miata	1995	yellow, 7-spoke wheels	34	45
174	Limozeen	1992	blue, whitewalls	3	2
175	Ferrari 348	1992	white, Ultra Hots wheels	4	6
175	Ferrari 348	1992	white, Hot Ones wheels, chrome	10	8
176	Lamborghini Diablo	1992	blue, Ultra Hots wheels	5	7
177	Zender fact 4	1993	purple, Hot Ones wheels, silver	6	8
177	Zender fact 4	1993	purple, Ultra Hots wheels	4	6
178	Hot Bird	1991	black, Ultra Hots wheels	5	7
179	Porsche 959	1992	purple, Hot Ones wheels, silver	7	9
179	Porsche 959	1992	purple, Ultra Hotss wheels	3	4
181	Fiero 2M4	1993	green, Ultra Hots wheels	6	8
181	Fiero 2M4	1992	green, Hot Ones, silver	8	10
182	Shadow Jet	1996	green, 5-spoke wheels	2	3
182	Shadow Jet	1992	light green	4	5
182	Shadow Jet	1992	green	3	4
183	VW Golf	1992	green metal flake	18	24
183	VW Golf	1992	pink metal flake, silver plastic base	9	12
183	VW Golf	1992	pink metal flake, black plastic base	6	8
184	Mercedes 380SEL	1992	blue, Ultra Hots wheels	4	6
184	Mercedes 380SEL	1992	blue, Hot Ones wheels, silver	7	9
185	Propper Chopper	1992	white	6	8
185	Propper Chopper	1997	yellow	2	3
186	Ford Aerostar	1991	white	8	10
187	Ramp truck	1992	yellow	8	10
187	Ramp truck	1996	yellow, 7-spoke wheels	2	3
188	Hummer	1995	brown	3	2
188	Hummer	1992	tan, construction tires	6	7
188	Hummer	1993	brown, construction tires	4	5
188	Hummer	1995	brown, plain hood, construction tires	2	3
189	Gleamer Patrol	1992	silver, tan interior	4	5
189	Gleamer Patrol	1992	chrome, black interior	4	5
189	Gleamer Patrol	1992	chrome, tan interior	8	10
190	`57 T-Bird	1992	gold	4	5
191	Aeroflash	1992	purple, Ultra Hots wheels	4	5
191	Aeroflash	1992	purple, Ultra Hots wheels, gold	9	12
192	Corvette Stingray	1992	green	11	15
192	Corvette Stingray	1992	chrome	6	8
192	Corvette Stingray	1992	chrome, Ultra Hots wheels	6	7
193	Porsche 959	1992	pink chrome, Ultra Hots wheels	3	4
193	Porsche 959	1992	silver, Ultra Hots wheels	6	8
194	Goodyear Blimp	1992	gray, no. 1	3	4

Hot Wheels - Numbered Packs

NO.	NAME	YEAR	DESCRIPTION	MNP	MIP
195	Troop Convoy	1993	tan, blackwall, black	4	5
196	3-window `34	1993	white with purple fenders	4	6
196	3-window `34	1993	white with pink fenders	30	40
196	3-window `34	1993	yellow/green	90	120
197	Corvette Split Window	1993	blue	18	24
197	Corvette Split Window	1993	blue with gray plastic base, whitewalls	3	4
197	Corvette Split Window	1995	blue, 5-spoke wheels	4	5
198	Pathbeater	1989	black	26	35
198	Pathbeater	1995	lime green, construction tires	2	3
198	Pathbeater	1995	lime green, construction tires, black	5	7
199	Double Demon	1995	yellow, Ultra Hots wheels	3	4
199	Double Demon	1994	lime, Ultra Hots wheels	3	4
200	Custom Corvette	1997	purple, silver lace wheels	3	4
200	Custom Corvette	1993	white, Ultra Hots wheels	1	2
200	Custom Corvette	1995	purple, Ultra Hots wheels	3	4
200	Custom Corvette	1995	purple, 5-spoke or 7-spoke wheels	2	3
201	Oshkosh Snowplow	1996	orange, directional wheels, orange	2	3
201	Oshkosh Snowplow	1991	orange, construction tires, orange	2	3
201	Oshkosh Snowplow	1991	orange, construction tires, black	7	9
202	`93 Camaro	1993	purple, gray interior, Ultra Hots wheels or Ultra Hots wheels, gold	3	4
202	`93 Camaro	1993	purple, white interior, Ultra Hots wheels or Ultra Hots wheels, gold	6	8
203	Jaguar XJ220	1994	silver, Ultra Hots wheels	2	3
203	Jaguar XJ220	1994	silver with open spoiler, Ultra Hots wheels	4	5
203	Jaguar XJ220	1994	silver, Ultra Hots wheels, gold	9	12
203	Jaguar XJ220	1994	blue, Ultra Hots wheels, gold	15	20
204	Oscar Mayer	1993	yellow and red	3	5
204	Oscar Mayer	1993	yellow and red	2	3
204	Oscar Mayer	1993	yellow and red, 7-spoke wheels	3	4
205	Treadator	1995	green	3	4
205	Treadator	1993	red	2	3
206	Pipe Jammer	1993	yellow, Ultra Hots wheels	1	2
207	Vector Avtech	1993	lavender, Ultra Hots wheels	1	2
207	Vector Avtech	1995	lavender, 5-spoke wheels	3	4
208	Avus Quattro	1995	silver, 5-spoke wheels	3	5
209	Avus Quattro	1992	silver, Ultra Hots wheels	2	3
209	Lexus SC400	1993	black, Ultra Hots wheels	2	3
210	Dodge Viper	1993	red, Ultra Hots wheels	6	8
210	Dodge Viper	1995	red, Ultra Hots wheels, gold	9	12
210	Dodge Viper	1995	red, 5-spoke wheels	71	95
210	Dodge Viper	1995	green, 3-spoke wheels, gold	3	4
210	Dodge Viper	1995	green, 3-spoke wheels, gold	3	5
210	Dodge Viper	1995	yellow, no logo, Ultra Hots wheels, gold	9	12
210	Dodge Viper	1995	yellow, window logo, Ultra Hots wheels, gold	3	5

Hot Wheels - Numbered Packs

NO.	NAME	YEAR	DESCRIPTION	MNP	MIP
211	Twinmill II	1993	yellow with charcoal base, Ultra Hots wheels	8	10
211	Twinmill II	1993	yellow with black base, Ultra Hots Wheels	34	45
211	Twinmill II	1993	yellow with gray base, Ultra Hots wheels	5	7
212	Silhouette II	1993	purple with charcoal or gray base, Ultra Hots wheels	2	1
213	`57 Chevy	1993	turquoise, bow tie, Ultra Hots wheels	8	10
213	`57 Chevy	1989	turquoise, no bow tie, Ultra Hots wheels	15	20
213	`57 Chevy	1994	turquoise, no bow tie, Ultra Hots wheels, gold	8	10
214	Swingfire	1993	blue, whitewalls	1	2
214	Swingfire	1995	blue, 7-spoke	2	3
215	Auburn	1993	red, whitewalls	4	6
216	Fat Fendered `40	1993	purple	8	10
217	`40s Woodie	1993	turquoise	8	10
217	`40s Woodie	1995	turquoise, 5-spoke wheels	3	4
217	`40s Woodie	1994	red	17	22
219	Gulch Stepper	1993	lime, construction tires	5	7
220	Bywayman	1996	white	2	3
220	Bywayman	1993	white with white or black interior, construction tires	2	3
221	Range Rover	1996	black	2	3
221	Range Rover	1993	black, construction tires	3	5
222	Blazer 4x4	1993	blue	4	6
222	Blazer 4x4	1993	blue, construction tires, black	7	9
223	Baja Bug	1992	red	15	20
224	Zombot	1995	gray chrome, purple gun	8	10
224	Zombot	1994	blue chrome, Ultra Hots wheels	2	3
224	Zombot	1995	gray chrome, orange gun, Ultra Hots wheels	3	5
224	Zombot	1995	gray chrome, orange gun, 7-spoke wheels	2	3
224	Zombot	1995	gray chrome, purple gun, 7-spoke wheels	8	10
225	Limozeen	1993	black, whitewalls	5	7
226	Ferrari 348	1995	pink, 5-spoke wheels	2	3
226	Ferrari 348	1993	red, Ultra Hots wheels	3	5
226	Ferrari 348	1992	pink, Ultra Hots wheels	2	3
226	Ferrari 348	1995	red or black, 5-spoke or 7-spoke wheels	3	4
227	Lamborghini Diablo	1995	purple, 5-spoke wheels	5	7
227	Lamborghini Diablo	1995	yellow, 5-spoke wheels	8	10
227	Lamborghini Diablo	1993	yellow, Ultra Hots wheels	2	3
228	Zender Fact 4	1993	blue, Ultra Hots wheels	2	3
228	Zender Fact 4	1994	green, Ultra Hots wheels	2	3
228	Zender Fact 4	1994	green with yellow and clear windows, Ultra Hots wheels	4	6
228	Zender Fact 4	1994	green with clear windows, Ultra Hots wheels	12	15
228	Zender Fact 4	1993	blue with clear window, Ultra Hots wheels	14	18

Hot Wheels - Numbered Packs

NO.	NAME	YEAR	DESCRIPTION	MNP	MIP
229	Mercedes 380SEL	1993	pink, Ultra Hots wheels	6	8
230	XT-3	1993	white with red dome	3	4
230	XT-3	1993	blue, 5-spoke wheels	2	3
230	XT-3	1993	white with blue dome	34	45
230	XT-3	1993	blue	3	5
231	Mini Truck	1993	orange, Ultra Hots wheels	2	3
231	Mini Truck	1993	orange, big Ultra Hots wheels	15	20
231	Mini Truck	1996	orange, 5-dot wheels, white	1	2
231	Mini Truck	1993	orange, 5-dot wheels	2	3
232	Lamborghini Countach	1989	red, Ultra Hots wheels	4	6
232	Lamborghini Countach	1994	red, Ultra Hots wheels, gold	5	7
233	Toyota MR2 Rally	1996	black, 3-spoke wheels	19	25
233	Toyota MR2 Rally	1997	white, no. 2, silver lace wheels	3	4
233	Toyota MR2 Rally	1996	white, Ultra Hots wheels	68	90
233	Toyota MR2 Rally	1996	black, 5-spoke wheels	38	50
233	Toyota MR2 Rally	1996	white, 3-spoke wheels	2	3
233	Toyota MR2 Rally	1991	white, Ultra Hots wheels	1	2
233	Toyota MR2 Rally	1996	white, 3-spoke wheels	2	3
234	Nissan Custom Z	1993	purple, Ultra Hots wheels	2	3
234	Nissan Custom Z	1994	purple, Ultra Hots wheels, gold	2	3
234	Nissan Custom Z	1995	purple, 3-spoke wheels	6	8
234	Nissan Custom Z	1995	purple, 5-spoke wheels	3	5
235	Turbo Streak	1993	yellow	2	3
235	Turbo Streak	1993	yellow, HiPipes	45	60
236	Ford Aerostar	1993	black	3	4
237	Ford Stake Bed	1996	red, 7-spoke wheels	4	6
237	Ford Stake Bed	1992	red	4	6
238	Hiway Hauler	1994	purple, Hot Wheels tampo	7	9
239	Mercedes Unimog	1992	tan, construction wheels, white	3	4
239	Mercedes Unimog	1992	tan, construction wheels, black	9	12
239	Mercedes Unimog	1996	tan, construction wheels, white	2	3
239	Mercedes Unimog	1992	tan, construction wheels	8	10
242	`93 Camaro	1993	blue, enamel, Ultra Hots wheels	101	135
242	`93 Camaro	1993	blue enamel, blackwalls	18	25
242	`93 Camaro	1993	blue	2	3
242	`93 Camaro	1993	blue, Ultra Hots wheels	132	175
244	No Fera Race Car	1993	black	4	6
244	No Fera Race Car	1993	black, no trademark on card	8	10
245	Driven to Max	1994	orange	4	6
246	Shadow Jet II	1994	black, assorted wheels types	2	3
247	Rigor Motor	1994	red	2	3
248	Spittin' Image II	1995	purple, pink window, Ultra Hots wheels	6	8
248	Spittin' Image II	1995	purple, Ultra Hots wheels	8	10
248	Spittin' Image II	1994	purple, Ultra Hots wheels	2	3
249	Fuji Buimp	1994	white	2	3

Tow Truck 1970, Hot Wheels

Snorkel, 1971, Hot Wheels

Hot Wheels - Numbered Packs

NO.	NAME	YEAR	DESCRIPTION	MNP	MIP
250	Talbot Lago	1995	black, 7-spoke wheels	2	3
250	Talbot Lago	1994	black, whitewalls	3	4
251	Gulch Stepper	1995	black, construction tires	2	3
251	Gulch Stepper	1996	black	30	40
251	Gulch Stepper	1994	red, construction tires	3	5
252	Street Roader	1996	white	2	3
252	Street Roader	1995	white, construction tires	2	3
253	Mercedes 380	1994	red, Ultra Hots wheels	3	4
253	Mercedes 380	1995	red, 5-spoke or 7-spoke wheels	2	3
253	Mercedes 380	1996	red	4	6
254	Sol-Aire CX4	1995	blue, 7-spoke wheels	2	3
254	Sol-Aire CX4	1995	blue, Ultra Hots wheels	2	3
254	Sol-Aire CX4	1995	blue, 5-spoke wheels	6	8
255	BMW 850i	1994	blue, clear windows, Ultra Hots wheels	6	8
255	BMW 850i	1994	blue, dark windows, Ultra Hots wheels	3	5
256	`80s Firebird	1994	pink	12	15
257	3-Window `34	1994	silver	90	120
257	3-Window `34	1995	silver, 7-spoke wheels	8	10
258	Blazer 4x4	1994	blue, construction tires	2	3
258	Blazer 4x4	1994	blue, construction tires, black	11	14
259	Lumina Minivan	1995	yellow, 5-spoke wheels	3	5
259	Lumina Minivan	1994	yellow, blackwalls	2	3
260	Twin Mill II BK94	1996	black with red windows, 5-dot wheels	2	3
260	Twin Mill II BK94	1994	black with red windows, Ultra Hots wheels	3	4
260	Twin Mill II BK94	1994	blue, Ultra Hots wheels	4	6
261	Cyber Cruiser	1995	purple, Ultra Hots wheels	2	3
261	Cyber Cruiser	1995	purple, Ultra Hots wheels	19	25
262	`93 Camaro	1994	red with tan interior, Ultra Hots wheels	4	5
262	`93 Camaro	1995	red, 7-spoke wheels	26	35
262	`93 Camaro	1996	blue with gray base, 7-spoke wheels	12	15
262	`93 Camaro	1995	blue, Ultra Hots wheels, blue	3	5
262	`93 Camaro	1995	dark blue, Ultra Hots wheels, gold	3	5
262	`93 Camaro	1994	red with white interior, Ultra Hots wheels	2	3
262	`93 Camaro	1994	red with gray base, Ultra Hots wheels	11	15
262	`93 Camaro	1995	red, 5-spoke wheels	2	3
263	Mean Green Passion	1994	green, whitewalls	10	14
264	Lexus SC400	1995	red	5	7
264	Lexus SC400	1994	red, Ultra Hots wheels	3	5
264	Lexus SC400	1996	red, 3-spoke wheels	3	4
264	Lexus SC400	1995	red, 5-spoke wheels	3	4
264	Lexus SC400	1996	red, 5-dot wheels	2	3
264	Lexus SC400	1996	red, sawblade wheels	2	3
265	Olds Aurora	1994	green with dark windows	9	12
265	Olds Aurora	1994	green with opaque windows	5	7

Hot Wheels - Numbered Packs

NO.	NAME	YEAR	DESCRIPTION	MNP	MIP
266	`59 Caddy	1995	lavender, 5-spoke wheels	3	4
266	`59 Caddy		lavender, whitewalls	2	3
267	Olds 442 W-30	1994	yellow	3	4
267	Olds 442 W-30	1995	yellow, 7-spoke wheels	6	8
268	GM Lean Machine	1994	Ultra Hots wheels	3	5
268	GM Lean Machine	1995	5-spoke wheels	2	3
268	GM Lean Machine	1995	7-spoke wheels	2	3
269	Oshkosh Cement	1994	yellow	2	3
269	Oshkosh Cement	1995	yellow, 5-spoke	2	3
269	Oshkosh Cement	1995	yellow, 7-spoke	2	3
269	Oshkosh Cement	1996	yellow	2	3
270	Chevy Stocker	1994	gold, Ultra Hots wheels, gold	19	25
270	Chevy Stocker	1994	red with yellow interior, Ultra Hots wheels, gold	4	6
270	Chevy Stocker	1994	red with red interior, Ultra Hots wheels, gold	30	40
271	Dragster	1996	blue	9	12
273	Tail Gunner	1994	white, construction tires, white	4	6
273	Tail Gunner	1994	white, construction tires	5	7
274	Super Cannon	1994	olive, blackwalls, white	3	5
274	Super Cannon	1994	olive, 5-spoke wheels, white	4	6
275	Lumina Stocker	1995	blue	3	5
275	Lumina Stocker	1995	blue with Pontiac front, Ultra Hots wheels	94	125
275	Lumina Stocker	1995	dark blue, Ultra Hots wheels	3	5
275	Lumina Stocker	1995	dark blue, 7-spoke wheels	2	3
276	Hot Wheels 500	1995	blue	7	9
276	Hot Wheels 500	1995	blue, Ultra Hots wheels	15	20
276	Hot Wheels 500	1995	blue, 7-spoke wheels	2	3
276	Hot Wheels 500	1995	blue metal flake	2	3
276	Hot Wheels 500	1995	blue metal flake	45	60
277	Side Splitter	1995	blue metal flake	3	5
277	Side Splitter	1995	blue	3	5
277	Side Splitter	1995	blue, 5-spoke wheels	5	7
278	Dragster	1995	blue metal flake	5	7
278	Dragster	1995	dark blue, 5-spoke wheels	4	6
278	Dragster	1995	dark blue, 5-spoke wheels	2	3
278	Dragster	1995	dark blue	3	4
280	Sharkruiser	1995	blue, 7-spoke wheels	3	4
280	Sharkruiser	1995	blue with painted base, Ultra Hots wheels	2	3
281	Turboa	1995	purple, 7-spoke wheels	2	3
281	Turboa	1995	purple, Ultra Hots wheels	3	4
282	`63 Split Window	1995	green	2	3
282	`63 Split Window	1995	green, 5-spoke wheels	3	4
282	`63 Split Window	1995	green, 7-spoke wheels	4	6
284	Flashfire	1995	purple with orange dome, 5-spoke wheels	3	5

Hot Wheels - Numbered Packs

NO.	NAME	YEAR	DESCRIPTION	MNP	MIP
284	Flashfire	1995	purple, 5-spoke wheels	2	3
284	Flashfire	1995	purple, Ultra Hots wheels	4	5
285	Steel Passion	1995	black, whitewalls	3	5
285	Steel Passion	1995	black, 7-spoke wheels	3	4
287	Zender Fact 4	1995	purple, 3-spoke wheels	2	3
287	Zender Fact 4	1995	purple, 5-spoke wheels	2	3
287	Zender Fact 4	1995	purple, Ultra Hots wheels	3	5
287	Zender Fact 4	1995	purple with clear windows, 5-spoke wheels	12	15
289	`56 Flashsider	1995	maroon, Ultra Hots wheels	3	5
289	`56 Flashsider	1995	maroon, 5-spoke wheels	9	12
290	`57 Chevy	1995	blue with black interior, 5-spoke wheels	15	20
290	`57 Chevy	1995	blue, 7-spoke wheels	3	4
290	`57 Chevy	1995	blue, 5-spoke wheels	3	4
290	`57 Chevy	1995	blue, Ultra Hots wheels	3	4
292	Pearl Passion	1995	lavender, 7-spoke wheels	2	3
292	Pearl Passion	1995	lavender, whitewalls	5	7
293	VW Bug	1995	pink	3	4
293	VW Bug	1995	pink, 5-spoke wheels	14	18
293	VW Bug	1995	pink, 7-spoke wheels	2	3
295	Talbot Lago	1995	blue, whitewalls	2	3
295	Talbot Lago	1995	blue, 7-spoke wheels	3	4
296	Jaguar XJ220	1995	white, Ultra Hots wheels	3	4
296	Jaguar XJ220	1995	white, 7-spoke wheels	8	10
296	Jaguar XJ220	1995	white, 5-spoke wheels	2	3
296	Jaguar XJ220	1995	white, 3-spoke wheels	3	4
297	Splittin Image	1995	black, 6-spoke wheels	4	6
297	Splittin Image	1995	black, 7-spoke wheels	2	3
298	Twin Mill II	1995	black, 6-spoke wheels	3	5
298	Twin Mill II	1995	black, 7-spoke wheels	2	3
299	Silhouette II	1995	black, 6-spoke wheels	2	3
299	Silhouette II	1995	black with silver base, 6-spoke wheels	2	3
299	Silhouette II	1995	black, 7-spoke wheels	2	3
300	Rigor Motor	1995	black, 5-spoke wheels	2	3
300	Rigor Motor	1995	black, 6-spoke wheels	5	7
300	Rigor Motor	1995	black, 7-spoke wheels	3	4
302	Mini Truck	1995	tan, 7-spoke wheels	7	10
302	Mini Truck	1995	tan, Ultra Hots wheels	4	6
302	Mini Truck	1995	tan, 5-spoke wheels	2	3
302	Mini Truck	1995	tan with orange base, 5-spoke wheels	38	50
303	Street Rodder	1995	orange, construction tires	26	35
303	Street Rodder	1995	orange, construction tires, orange	2	3
304	Roll Patrol	1995	white, construction tires	23	30
304	Roll Patrol	1995	white, construction tires, yellow	2	3
305	Classic Cobra	1995	lime green, metal base, 7-spoke wheels	3	5

Hot Wheels - Numbered Packs

NO.	NAME	YEAR	DESCRIPTION	MNP	MIP
305	Classic Cobra	1995	lime green, black metal pase, 7-spoke wheels	6	8
305	Classic Cobra	1995	lime green, black plastic base, 7-spoke wheels	2	3
307	Cyber Cruiser	1995	red, 3-spoke wheels	2	3
308	Vampyra	1995	purple, 3-spoke wheels, gold	2	3
308	Vampyra	1995	purple, 6-spoke wheels, gold	2	3
310	Shadow Jet	1995	green, 3-spoke	2	3
311	Suzuki Quad Racer	1995	yellow, 3-spoke wheels, yellow	2	3
311	Suzuki Quad Racer	1995	lime green, 3-spoke wheels, yellow	2	3
311	Suzuki Quad Racer	1995	yellow with blue seat, 3-spoke wheels, yellow	2	3
312	3-Window `34	1995	7-spoke wheels	2	3
313	T-Bucket	1995	purple with clear windows, 5-spoke wheels	30	40
313	T-Bucket	1995	purple, 5-spoke wheels	2	3
313	T-Bucket	1995	purple, 7-spoke wheels	2	3
315	Ratmobile	1995	black, Ultra Hots wheels	2	3
315	Ratmobile	1995	black, 7-spoke wheels	2	3
316	Limozeen	1995	gold, whitewalls	4	6
316	Limozeen	1995	gold, 7-spoke wheels	2	3
317	Dump Truck	1995	yellow, Real Rider wheels, yellow	12	15
318	Mercedes Unimog	1995	gray, Real Rider wheels, orange	14	18
320	`59 Caddy	1995	pink, Real Rider wheels, silver	19	18
321	Corvette Stingray	1995	green, Real Rider wheels, green	30	40
322	Fire Eater	1995	chrome, silver base, 7-spoke wheels	6	8
322	Fire Eater	1995	chrome	2	3
322	Fire Eater	1995	chrome, 5-spoke wheels	2	3
322	Fire Eater	1995	chrome, Ultra Hots wheels	2	3
322	Fire Eater	1995	chrome, 7-spoke wheels	2	3
322	Fire Eater	1995	chrome with silver base, Ultra Hots wheels	6	8
325	Shopper Chopper	1995		2	3
328	School Bus	1995	clear windows, blackwalls	6	8
328	School Bus	1995	dark windows, blackwalls	15	20
328	School Bus	1995	dark windows, 7-spoke wheels	2	3
328	School Bus	1995	clear windows, 7-spoke wheels	2	3
328	School Bus	1995	clear windows, 5-spoke wheels	2	3
328	School Bus	1995	clear windows, 7-spoke wheels	2	3
331	Aerostar	1995	white, 7-spoke	2	3
332	Flyin' Aces Blimp	1995	white, 7-spoke	2	3
333	Tank Truck	1995	blue, 7-spoke wheels	2	3
335	Hiway Hauler	1995	green, 7-spoke wheels	2	3
336	Race Truck	1995	black, no JT tampo, construction tires	2	3
336	Race Truck	1995	black, JT tampo, construction tires	2	3
337	Ramp Truck	1995	purple, 7-spoke	2	3
337	Ramp Truck	1995	blue, 7-spoke wheels	6	8

T-4-2, 1971, Hot Wheels

Maserati Mistral, 1969, Hot Wheels

Silhouette, 1968, Hot Wheels

Hot Wheels - Numbered Packs

NO.	NAME	YEAR	DESCRIPTION	MNP	MIP
338	Camaro Racer	1995	Blue, 5-spoke wheels	9	12
338	Camaro Racer	1995	no name roof, 5-spoke wheels	4	6
340	Dragster	1995	blue, 5-spoke wheels	2	3
341	`58 Vette Coupe	1995	pink, 5-spoke wheels	4	6
341	`58 Vette Coupe	1995	pink, 7-spoke wheels	3	4
341	`58 Vette Coupe	1996	pink, 5-spoke wheels	3	5
341	`58 Vette Coupe	1996	pink, 7-spoke wheels	7	10
342	Mercedes SL	1996	red with dark red base, Ultra Hots wheels	5	7
342	Mercedes SL	1996	black, 7-spoke wheels	2	3
342	Mercedes SL	1996	black with tan interior, 5-spoke wheels	30	40
342	Mercedes SL	1996	black, 5-spoke wheels	3	4
342	Mercedes SL	1995	red, 5-spoke wheels	3	4
342	Mercedes SL	1995	red, Ultra Hots wheels	3	4
342	Mercedes SL	1996	red with dark red base, 5-spoke wheels	3	4
343	Speed Blaster	1995	green, Ultra Hots wheels	4	6
343	Speed Blaster	1995	green, 5-dot wheels	3	4
343	Speed Blaster	1995	green, 7-spoke wheels	23	30
343	Speed Blaster	1995	green, sawblade wheels	2	3
343	Speed Blaster	1995	green, 3-spoke wheels	2	3
343	Speed Blaster	1995	blue with purple base, 5-spoke wheels	3	4
343	Speed Blaster	1995	blue with charteuse base, Ultra Hots wheels	6	8
343	Speed Blaster	1995	blue with black base, Ultra Hots wheels	38	50
343	Speed Blaster	1995	blue with purple base, Ultra Hots wheels	8	10
343	Speed Blaster	1995	green, 5-spoke wheels	3	5
343	Speed Blaster	1995	blue with chartreuse base, 5-spoke wheels	3	4
344	Camaro Convertible	1995	green, Ultra Hots wheels	7	9
344	Camaro Convertible	1995	green, 5-spoke wheels	2	3
344	Camaro Convertible	1995	green, 3-spoke wheels	2	3
344	Camaro Convertible	1995	red, 3-spoke wheels	2	3
344	Camaro Convertible	1995	red, 5-spoke wheels	3	4
344	Camaro Convertible	1995	red, sawblade wheels	2	3
345	Speed-A-Saurus	1995	green, 5-spoke wheels	2	3
345	Speed-A-Saurus	1995	green	2	3
346	Hydro Plane	1995	blue	3	4
347	Power Pistons	1996	red with silver base, 3-spoke wheels	11	15
347	Power Pistons	1995	red, Ultra Hots wheels	11	15
347	Power Pistons	1995	red, 3-spoke wheels	3	5
347	Power Pistons	1995	red, 5-spoke wheels	3	5
347	Power Pistons	1996	red, 7-spoke wheels	9	12
347	Power Pistons	1995	red with purple package, 5-spoke wheels	3	5
348	Dodge Ram Truck	1995	green, 5-spoke wheels	3	5
348	Dodge Ram Truck	1997	green, 5-dot wheels	3	5
348	Dodge Ram Truck	1997	green, 5-dot wheels, white	3	5
349	Power Pipes	1995	purple, 3-spoke wheels	2	3

Hot Wheels - Numbered Packs

NO.	NAME	YEAR	DESCRIPTION	MNP	MIP
349	Power Pipes	1995	purple, 5-spoke wheels	7	9
349	Power Pipes	1995	purple, 7-spoke wheels	7	9
350	Ferrari 355	1996	yellow, sawblade wheels	2	3
350	Ferrari 355	1996	yellow, directional wheels	2	3
350	Ferrari 355	1996	yellow, 5-dot wheels, white	2	3
350	Ferrari 355	1996	yellow, 5-dot wheels	2	3
350	Ferrari 355	1995	yellow, 5-spoke	4	6
350	Ferrari 355	1995	yellow, painted 5-spoke wheels	3	4
350	Ferrari 355	1995	yellow, 3-spoke	3	4
351	Power Rocket	1995	purple with painted base, 5-spoke wheels	2	3
351	Power Rocket	1995	purple	3	5
351	Power Rocket	1995	purple, 5-spoke wheels	2	3
351	Power Rocket	1995	purple, 3-spoke wheels	2	3
351	Power Rocket	1995	purple	2	3
352	Big Chill	1995	white	2	3
352	Big Chill	1995	white without shredder tampo	8	10
352	Big Chill	1995	white with orange ski	7	9
352	Big Chill	1995	white with pink ski	3	5
352	Big Chill	1995	white with black ski	2	3
367	Chevy 1500	1996	gray, 7-spoke wheels, black	2	3
367	Chevy 1500	1996	gray, large 7-spoke wheels, black	5	7
368	Road Rocket	1996	green	2	3
369	Turbo Flame	1996	white, 5-spoke wheels	3	5
369	Sizzler	1996	white	3	4
370	Rail Rodder	1996	black	3	5
372	VW Bus	1996	blue, 5-spoke wheels	11	15
373	Street Cleaver	1996	yellow without flames, 5-spoke wheels	2	3
373	Street Cleaver	1996	yellow without flames	2	3
373	Street Cleaver	1996	yellow with flames, 5-spoke wheels	4	6
373	Street Cleaver	1996	yellow with flames, 3-spoke wheels	2	3
373	Street Cleaver	1996	yellow	2	3
374	Radio Flyer Wagon	1996	red, 5-spoke wheels	4	6
375	Dog Fighter	1996	red	3	4
375	Dog Fighter	1996	red, 5-spoke wheels	3	4
376	Twang Thang	1996	silver, 5-spoke wheels	2	3
377	Ferrari F50	1996	red, silver lace wheels	3	4
377	Ferrari F50	1996	red, lace wheels	3	4
378	Mustang GT	1996	red	2	3
378	Mustang GT	1996	red, 7-spoke wheels	2	3
378	Mustang GT	1996	red	2	3
378	Mustang GT	1996	red, 3-spoke wheels	2	3
378	Mustang GT	1996	red, 5-spoke wheels	2	3
380	Dodge Ram 1500	1996	red, 7-spoke wheels, black	2	3
380	Dodge Ram 1500	1996	red without hood, 7-spoke wheels, black	8	10

Hot Wheels - Numbered Packs

NO.	NAME	YEAR	DESCRIPTION	MNP	MIP
381	Kenworth	1996	silver, 7-spoke wheels, black	2	3
381	Ford LTL	1996	silver, 7-spoke wheels, black	7	9
382	`70 Dodge Daytona	1996	red, 7-spoke wheels, gold	5	7
382	`70 Dodge Daytona	1996	red	5	7
382	`56 Flashsider	1996	black, 7-spoke wheels, unpainted wheels	2	3
382	`56 Flashsider	1996	black with unpainted wheels, 7-spoke wheels	6	8
383	Nissan truck	1996	blue, construction tires	2	3
383	Nissan truck	1996	blue, yellow interior, construction tires	2	3
383	Nissan truck	1996	blue, construction tires, black	2	3
384	`57 T-Bird	1996	white with brown flame, 7-spoke wheels	5	7
384	`57 T-Bird	1996	white with pink flame, 7-spoke wheels	2	3
384	`57 T-Bird	1996	white with Hot Wheels logo, 7-spoke wheels	2	3
385	Hydroplane	1996	yellow	2	3
385	Hydroplane	1996	yellow with reverse flames	11	15
386	Range Rover	1996	red, construction tires	2	3
386	Range Rover	1996	red with Hot Wheels logo, construction tires	2	3
387	Oshkosh Snowplow	1995	black, construction tires	2	3
387	Oshkosh Snowplow	1995	black, construction tires, black	3	5
388	Radar Range	1996	white, construction tires	2	3
389	GM Lean Machine	1995	white and black, 5-spoke tires	2	3
390	Alien	1996	blue, 5-spoke wheels	2	3
391	Treadator	1996	blue chrome	2	3
391	Treadator	1996	white	2	3
392	Ramp Truck	1996	blue, 5-spoke wheels	2	3
392	Ramp Truck	1996	blue, 7-spoke wheels	11	15
393	Baja bug	1996	blue, 5-spoke wheels	3	5
394	`57 Chevy	1996	blue, 5-spoke wheels	2	3
395	Bywayman	1996	blue, construction wheels	2	3
395	Bywayman	1996	blue	2	3
396	Hummer	1996	pink, construction tires	2	3
396	Hummer	1996	pink	2	3
397	School bus	1995	blue, 7-spoke wheels	3	4
398	VW Bug	1996	blue, 7-spoke wheels	3	4
399	`67 Camaro	1996	green, 7-spoke wheels	3	5
399	`67 Camaro	1996	green, 5-spoke wheels	3	5
399	`67 Camaro	1996	green with green windows, 5-spoke wheels	3	5
400	Big Chill	1996	black	2	3
401	Streat Beast	401	black, 7-spoke wheels	2	3
402	Thunderstreak	401	black, 7-spoke wheels	2	3
403	Power Pistons	1996	chartreuse, 7-spoke wheels	2	3
404	Porsche 930	1996	silver, 7-spoke wheels	2	3
405	Custom Corvette	1996	purple, 7-spoke wheels	2	3

Hot Wheels - Numbered Packs

NO.	NAME	YEAR	DESCRIPTION	MNP	MIP
405	Custom Corvette	1996	purple with red interior, 7-spoke wheels	2	3
406	Classic Cobra 427	1996	white, 7-spoke wheels	3	5
406	Classic Cobra 427	1996	white, 5-spoke wheels	11	15
407	`59 Caddy	1996	black, 7-spoke wheels	2	3
408	Rescue Ranger	1996	orange, 5-spoke wheels	2	3
408	Rescue Ranger	1996	orange	3	5
409	Side Splitter	1996	white, 5-spoke wheels	2	3
410	`55 Chevy	1996	yellow Splatter paint, 5-spoke wheels	2	3
410	`55 Chevy	1996	yellow Splatter paint	2	3
411	`80s Camaro Z-28	1996	white Splatter Paint, 5-spoke wheels	2	3
412	Speed Machine	1996	green, 5-spoke	2	3
412	Speed Machine	1996	green, 7-spoke	2	3
413	Silhouette II	1996	purple, 5-spoke wheels	2	3
414	Proper Chopper	1996	blue	2	3
415	Roll Patrol	1996	tan, construction tires, yellow	2	3
415	Roll Patrol	1996	tan	2	3
416	pizza Vette	1996	white, 3-spoke	2	3
417	Pasta Pipes	1996	white, 3-spoke wheels	2	3
418	Sweet Stocker	1996	white, 3-spoke wheels	2	3
419	Crunch Chief	1996	white, 3-spoke wheels	2	3
420	Dump Truck	1996	chartreuse	2	3
421	`40s Woodie	1996	chartreuse, 7-spoke wheels	4	6
424	Ambulance	1996	green, 5-spoke wheels	2	3
424	Ambulance	1996	green, 7-spoke wheels	2	3
425	Rescue Ranger	1996	yellow, 5-spoke wheels	2	3
427	Fire Eater	1996	red, 5-spoke wheels	2	3
439	`37 Bugatti	1996	blue, 6-spoke wheels	14	18
440	Monte Carlo Stocker	1996	blue, 7-spoke wheels	3	4
440	Monte Carlo Stocker	1996	blue, 7-spoke wheels, black	3	4
441	Chevy Stocker	1996	black, 7-spoke wheels	2	3
442	Ferrari F40	1996	white, 5-spoke wheels	2	3
442	Ferrari F40	1996	white, 5-dot wheels, white	2	3
442	Ferrari F40	1996	white, 7-spoke wheels	2	3
443	Ferrari 348	1996	black, 7-spoke wheels	3	5
443	Ferrari 348	1996	black, 5-spoke wheels	8	10
444	Aeroflash	1996	white, 7-spoke wheels	3	5
444	Aeroflash	1996	white, 7-spoke wheels, gold	2	3
445	Jaguar 220	1996	green, 5-spoke wheels	2	3
445	Jaguar 220	1996	green, 7-spoke wheels	9	12
446	`32 Ford Delivery	1996	blue, 7-spoke wheels	2	3
447	`63 Split Window	1996	green, 3-spoke wheels	2	3
447	`63 Split Window	1996	green, 5-spoke wheels	7	8
447	`63 Split Window	1996	green, 7-spoke wheels	2	3
449	Camaro Z-28	1996	orange, 3-spoke wheels	3	4
449	Camaro Z-28	1996	orange, 5-spoke wheels	4	5

Hot Wheels - Numbered Packs

NO.	NAME	YEAR	DESCRIPTION	MNP	MIP
450	Corvette Stingray	1996	white, 5-spoke wheels	2	3
450	Corvette Stingray	1996	white, 7-spoke wheels	2	3
450	Corvette Stingray	1996	white, 3-spoke wheels	2	3
451	3-Window	1996	pink, 3-spoke wheels	4	5
451	3-Window	1996	pink, 7-spoke wheels	2	3
451	Ferrari 250	1996	green, 7-spoke wheels	2	3
453	Avus Quattro	1996	red, 7-spoke wheels	2	3
453	Avus Quattro	1996	red, 5-spoke wheels	8	10
454	Zender Fact	1996	white, 5-spoke wheels	4	6
454	Zender Fact	1996	white, 7-spoke wheels	2	3
455	`65 Mustang Convertible	1996	gold, 3-spoke wheels	2	3
455	`65 Mustang Convertible	1996	gold, 5-spoke wheels	4	6
455	`65 Mustang Convertible	1996	gold, 7-spoke wheels	2	3
457	Pontiac	1996	black, 5-spoke wheels	1	2
457	Pontiac	1996	black, silver lace wheels	1	2
460	Zombot	1996	gray and orange, 5-spoke wheels	1	2
461	Enforcer	1996	purple, 5-spoke wheels	1	2
462	`80s Firebird	1996	blue, 5-spoke wheels	2	3
462	`80s Firebird	1997	blue, silver lace wheels	1	2
463	Pontiac Fiero	1996	lime green, 5-spoke wheels	2	3
467	Peugot 405	1996	green, 5-spoke wheels	2	3
467	Peugot 405	1996	gray, 5-spoke wheels	2	3
468	GT Racer	1996	red, 5-spoke wheels	2	3
469	Hot Bird	1996	gold, 5-spoke wheels	2	3
469	Road Rocket	1996	green, silver lace wheels	4	5
470	Turbo Streak	1996	white and blue, 5-spoke wheels	2	3
471	Velociter	1996	blue or blue, 5-spoke wheels	2	3
472	Buick Stocker	1996	yellow, 5-spoke wheels	2	3
472	Buick Stocker	1996	yellow	3	4
473	Street Beast	1996	green or silver, 5-spoke wheels	2	3
474	Forklit	1996	yellow, 5-spoke wheels	2	3
474	VW Golf	1996	black, 5-spoke wheels	2	3
477	Double Demon	1996	green, 5-spoke wheels	2	3
478	Dragon Wagon	1996	green, 5-spoke wheels	2	3
479	Computer Warrior	1996	gray and orange, 5-spoke wheels	3	4
481	Tall Ryder	1996	green and yellow, construction tires, silver	2	3
482	Earthmover	1996	yellow, construction tires, silver	1	2
483	Thunder Roller	1996	red	2	3
484	Grizzlor	1996	white, 5-spoke tires	2	3
485	Evil Weevil	1996	orange, 5-spoke wheels	2	3
486	Command Tank	1996	black and magenta	2	3
487	Troop Convoy	1996	red and orange, 5-spoke	1	2
488	Sting Rod	1996	gray, construction tires, silver	2	3
489	Big Bertha	1996	gray and orange	1	2

Hot Wheels - Numbered Packs

NO.	NAME	YEAR	DESCRIPTION	MNP	MIP
491	Rocket Set	1996	gray and orange	1	2
492	Swingfire	1996	red or white, 5-spoke wheels	2	3
493	Porsche 911	1996	lime green, 5-spoke wheels	2	3
494	Mercedes 500	1996	gray, 5-spoke wheels	2	3
496	Ferrari 308GT	1996	red, 5-spoke wheels	3	4
498	BMW 850i	1996	silver, 5-spoke wheels	2	3
499	Corvete Coupe	1996	green, 5-spoke wheels	2	3
502	Chevy Nomad	1996	red, 7-spoke, gold	3	4
503	`80s Corvette	1996	red, 3-spoke wheels	2	3
504	Camaro Z-28	1996	white, 5-spoke wheels	3	4
504	Camaro Z-28	1997	white, 3-spoke wheels	2	3
505	1993 Camaro	1997	black	2	3
509	Firebird Funny Car	1996	blue, 5-spoke wheels	3	4
512	Excavator	1997	white	2	3
513	Ford F-150	1997	red, 5-spoke wheel	3	4
520	Saltflat Racer	1997	red, 5-spoke wheels	2	3
525	Trailbuster	1997	black	2	3
527	Second Wind	1997	white	4	5
528	Beach Blaster	1997	white	2	3
529	Power Rocket	1997	green, 3-spoke wheels	2	3
532	Power Pipes	1997	orange and yellow, 3-spoke wheels	2	3
533	Hummer	1997	blue, sawblade wheels	2	3
536	`80s Corvette	1997	blue, 5-dot wheels	2	3
537	Police Cruiser	1997	green, 5-spoke wheels	2	3
539	Peterbilt Tank	1997	dark red, 7-spoke wheels	2	3
577	Police Cruiser	1996	black, 7-spoke wheels, black	3	4
590	Porsche 911	1996	red, 5-spoke wheels	3	4
590	Porsche 911	1997	red, 5-dot wheels	2	3
591	Porsche 911	1997	blue/gray, 5-spoke wheels	3	4
597	Buick Wildcat	1997	red, 7-spoke wheels	2	3

Treasure Hunts

NO.	NAME	YEAR	DESCRIPTION	MNP	MIP
353	Olds 442	1995	limited to 10,000	40	50
354	Gold Passion	1995	metallic gold, limited to 10,000	50	70
355	`67 Camaro	1995	limited to 10,000	100	125
356	`57 T-Bird	1995	limited to 10,000	40	50
357	Volkswagen Bug	1995	flourescent lime green, limited to 10,000	35	44
358	`63 Split Window	1995	limited to 10,000	42	53
359	Stutz Blackhawk	1995	limited to 10,000	35	44
360	Rolls Royce Phanton II	1995	limited to 10,000	33	41
361	Classic Caddy	1995	limited to 10,000	40	50
362	Nomad	1995	limited to 10,000	45	57
363	Classic Cobra	1995	limited to 10,000	60	75
364	`31 Doozie	1995	limited to 10,000	35	44
428	1940s Woodie	1996	limited to 25,000	35	44

Hot Wheels - Numbered Packs

NO.	NAME	YEAR	DESCRIPTION	MNP	MIP
		Treasure Hunts			
429	Lamborghini Countach	1996	limited to 25,000	30	38
430	Ferrari 250	1996	limited to 25,000	30	39
431	Jaguar XJ220	1996	limited to 25,000	30	39
432	`59 Caddy	1996	limited to 25,000	35	44
433	Dodge Viper RT/10	1996	limited to 25,000	55	67
434	1957 Chevy	1996	limited to 25,000	45	56
435	Ferrari 355	1996	limited to 25,000	30	38
436	`58 Corvette	1996	limited to 25,000	40	50
437	Auburn 852	1996	limited to 25,000	28	35
438	Dodge Ram 1500	1996	limited to 25,000	35	44
439	1937 Bugatti	1996	limited to 25,000	25	31
579	Silouette	1997	limited to 25,000	14	18
580	Mercedes 500SL	1997	limited to 25,000	14	18
581	Street Cleaver	1997	limited to 25,000	14	18
582	Lean Machine	1997	limited to 25,000	14	18
583	Hot Rod Wagon	1997	limited to 25,000	20	25
584	Olds Aurora	1997	limited to 25,000	15	19
585	Dog Fighter	1997	limited to 25,000	15	19
586	Buick Wildcat	1997	limited to 25,000	10	13
587	Blimp	1997	limited to 25,000	10	13

King Kuda, 1970, Hot Wheels

VEHICLES

Hubley

NAME	DESCRIPTION	GOOD	EX	MINT
	Airplanes			
America Plane	trimotor, open cockpit, co-pilot, pilot, cast iron, 17" wingspan	2500	5000	7000
American Eagle Airplane	WWII Fighter, 11" wingspan	150	225	300
American Eagle Carrier Plane	cast metal, 11" wingspan, 1971	60	95	125
B-17 Bomber	15" wingspan	125	185	250
Bremen Junkers Monoplane	10" wingspan	1200	1750	2000
Corsair-Type Fighter Plane		30	45	65
Delta Wing Jet		60	95	125
DO-X Plane	six engines, 5-7/8" wingspan, 1935	165	200	285
Flying Circus	12" wingspan	45	70	90
Lindy Plane	cast iron, 10" wingspan	500	750	1000
Lindy Plane	cast iron, 13-1/4" long	1200	1750	2000
Navy WWII Fighter	folding wings and wheels	30	45	65
P-38 Fighter	camouflage paint, 12-1/2" wingspan	75	100	185
P-38 Plane	black rubber tires	70	95	165
Piper Cub Plane	pot metal	50	75	115
Sea Plane	orange/blue, two engines	35	55	75
Single Engine Fighter	3-1/2"	60	90	120
U.S. Air Force	12" wingspan	25	45	75
U.S. Army Monoplane	7-5/8" wingspan, 1941	50	75	100
U.S. Army Single Engine Fighter	black rubber tires	20	35	50
	Boats and Ships			
Penn Yan Motorboat	15" long	1700	2600	3500
	Buses			
School Bus	metal, wooden wheels	60	85	125
Service Coach	cast iron, 5" long	675	900	1500
	Cars			
Auto and Trailer	cast iron, 6-3/4" long, 1939	175	235	295
Auto and Trailer	cast iron, sedan 7-1/4" long, trailer 7-1/8" long, 1936	185	245	315
Buick Convertible	opening top, 6-1/2" long	50	65	90
Buick Convertible	top down, 6-1/2" long	45	60	80
Cadillac	black rubber tires, die cast with tin bottom plate	25	40	70
Car Carrier With Four Cars	cast iron, 10" long	300	475	675
Chrysler Airflow	6" long	145	200	350
Chrysler Airflow	battery-operated lights, cast iron, 1934	1250	1900	2750
Chrysler Airflow	cast iron, 4-1/2" long	110	195	325
Coupe	cast iron, 9-1/2" long, 1928	900	1400	1800
Coupe	cast iron, 8-1/2" long, 1928	600	1100	1500
Coupe	cast iron, 7" long, 1928	400	650	800
Ford Convertible	cast iron, V/8, 1930s	65	100	185
Ford Coupe	cast iron, V-8, 1930s	65	100	185
Ford Model-T	movable parts	100	150	200
Ford Sedan	cast iron, V-8, 1930s	65	100	185

Hubley

NAME	DESCRIPTION	GOOD	EX	MINT
Ford Town Car	cast iron, V-8, 1930s	65	100	185
Limousine	cast iron, 7" long, 1918	250	325	400
Lincoln Zephyr	1937	250	350	450
Mr. Magoo Car	old timer car, battery-operated, 9" long, 1961	75	115	150
Open Touring Car	cast iron, 7-1/2" long, 1911	675	900	1250
Packard Roadster	9-1/2" long, 1930	90	145	250
Packard Straight 8	hood raises, detailed cast motor, cast iron, 11" long, 1927	7500	10000	15000
Race Car #22	cast iron, 7-1/2" long	40	55	85
Race Car #2241	7" long, 1930s	50	75	100
Racer	white rubber wheels, cast iron, 7" long	200	300	400
Racer	cast iron, 10-3/4" long, 1931	75	150	250
Racer	nickel-plated driver, cast iron, 4-3/4" long, 1960s	75	115	150
Racer	red with black wheels, silver grille and driver, 7-1/2" long	35	55	75
Racer #12	die cast, prewar	60	95	125
Racer #629	7" long, 1939	50	75	135
Roadster	cast iron, 7-1/2" long, 1920	75	150	225
Sedan	cast iron, 7" long, 1920s	175	265	350
Service Car	5" long, 1930s	75	115	150
Speedster	cast iron, 7" long, 1911	125	225	350
Station Wagon	die cast	50	75	100
Streamlined Racer	cast iron, 5" long	70	125	140
Studebaker	take-apart, 5" long	400	600	800
Tinytown Station Wagon and Boat Trailer		45	70	90
Touring Car	woman and dog seated in back, driver in front, 10" long, 1920	650	1000	1450
Yellow Cab	with luggage rack, cast iron, 8" long, 1940	325	500	700

Emergency Vehicles

NAME	DESCRIPTION	GOOD	EX	MINT
Ahrens-Fox Fire Engine	cast iron, 11-1/2" long, 1932	5000	6500	8000
Auto Fire Engine	cast iron, 15" long, 1912	4200	5500	7000
Fire Engine	cast iron, 14-1/2" long, 1932	1575	2250	3000
Fire Engine	blue and green, large rear wheels with smaller front wheels, cast iron, 10-3/4" long, 1920"s	4500	5750	7500
Fire Truck	5" long, 1930s	75	100	150
Fire Truck No. 468		60	95	125
Hook and Ladder Fire Truck	rubber wheels, die cast, 18" long	150	250	500
Hook and Ladder Truck	cast iron, 8" long	100	125	200
Hook and Ladder Truck	cast iron, 23" long, 1912	1850	3000	4500
Hook and Ladder Truck	cast iron, 16-1/2" long, 1926	850	1200	1700
Ladder Fire Truck	cast iron, 5-1/2" long	40	60	80
Ladder Truck	14" long, 1940s	150	275	500
Police Patrol	with three policemen, cast iron, 11" long, 1919	900	1450	2500
Pumper Fire Truck	plastic, 1950s	25	45	65
Seven Man Fire Patrol	cast iron, 15" long, 1912	3575	5700	7500
Special Ladder Truck	cast iron, 13" long, 1938	465	575	975

Hubley

NAME	DESCRIPTION	GOOD	EX	MINT
	Farm and Construction Equipment			
Avery	round radiator, cast iron, 4-1/2" long, 1920	175	300	450
Diesel Road Roller	plastic, 1950s	15	25	40
Elgin Street Sweeper	brush sweeps dirt into a bin in the body, uniformed driver, cast iron, 8-1/2" long, 1930	4000	8000	11500
Farm Trailer	with gate, 8" long	30	45	60
Ford 4000	blue and gray, die cast, 1:12 scale	100	200	300
Ford 6000	blue and gray, die cast, 1:12 scale, 1963	125	250	400
Ford 961 Powermaster	red and gray, row crop, die cast, 1:12 scale, 1961	100	200	300
Ford 961 Powermaster	red and gray, die cast, 1:12 scale, 1961	125	250	400
Ford 961 Select-O-Speed	red and gray, die cast, 1:12 scale, 1962	125	250	400
Ford Commander 6000	blue and gray, die cast, 1:12 scale, 1963	125	225	300
Fordson	with loader, cast iron, 8-1/2" long, 1938	750	1250	1700
Fordson	cast iron, 5-1/2" long	150	225	300
Fordson F	with crank and driver, cast iron, 5-1/2" long	150	225	300
Huber Road Roller	large, with standing driver, cast iron, 15" long, 1927	1675	2250	3750
Huber Steam Roller	cast iron, 1:25 scale, 1929	365	450	600
Huber Steam Roller	cast iron, 3-1/4" long, 1929	100	150	215
Junior Tractor		60	95	125
Oliver 70 Orchard	fenders over rear wheels, cast iron, 5" long, 1938	175	300	575
Road Scraper	plastic, 1950s	30	45	60
Steam Shovel	red, nickel-plated boom, cast iron, 4-3/4" long	75	115	150
Tractor	yellow, 5-1/4" long	35	55	75
Tractor and Farmer	plastic	25	35	55
Tractor Shovel	cast iron, 8-1/2" long, 1933	1250	1650	2000
	Motorcycles			
Crash Car	motorcycle with cart on back, cast iron, 9" long, 1930s	1000	1650	2000
Harley-Davidson Motorcycle	cast iron, 7-1/2" long, 1932	300	450	800
Harley-Davidson Parcel Post	cast iron, 10" long, 1928	1500	2500	4000
Harley-Davidson Sidecar Motorcycle	cast iron, 9" long, 1930	900	1625	1900
Hill Climber	cast iron, 6-3/4" long, 1935	375	500	900
Indian Air Mail	cast iron, 9-1/4" long, 1929	1575	2650	3500
Indian Armored Car	motorcyle police, cast iron, 8-1/2" long, 1928	1750	3500	6000
Indian Four-Cylinder Motocycle	cast iron, 9" long, 1929	1700	2425	3000
Indian Motorcycle	cast iron, 9" long	600	850	1500
Marathon Rider	bicycle, cast iron	200	300	400
Motorcycle	three wheels, cast iron	400	600	800
Motorcycle Cop With Sidecar	Harley-Davidson, cast iron	700	1000	1500
Motorcycle Crash Car	with cart on back, cast iron, 5" long, 1930s	100	135	195
Motorized Sidecar Motorcycle	clockwork motor, cast iron, 8-1/2" long, 1932	4000	6250	10000
P.D. Motorcycle Cop	red plastic cycle	35	65	95

Hubley

NAME	DESCRIPTION	GOOD	EX	MINT
Patrol Motorcycle	green, 6-1/2" long	275	350	475
Popeye Patrol	cast iron, 9" long, 1938	425	600	950
Popeye Spinach Delivery	red motorcycle, cast iron, 6" long, 1938	375	500	750
Traffic Car	three-wheel transport vehicle, cast iron, 12" long, 1930	600	950	1500

Trucks

NAME	DESCRIPTION	GOOD	EX	MINT
Auto Dump Coal Wagon	cast iron, 16-1/4" long, 1920	800	1200	1500
Auto Express with Roof	cast iron, 9-1/2" long, 1910	500	875	1200
Auto Truck	spoke wheels, cast iron, 10" long, 1918	600	1200	1650
Auto Truck	five-ton truck, cast iron, 17-1/2" long, 1920	1000	1650	2250
Bell Telephone Truck	white tires, cast iron, 7" long, 1930	300	400	500
Bell Telephone Truck	12" long, 1940	50	75	100
Bell Telephone Truck	spoke wheels, cast iron, 5-1/2" long, 1930	225	335	450
Bell Telephone Truck	white tires, winch works, cast iron, 10" long, 1930	500	1000	1500
Bell Telephone Truck	no driver, white tires, cast iron, 3-3/4" long, 1930	150	225	300
Borden's Milk Truck	cast iron, 7-1/2" long, 1930	1650	2850	4250
Compressor Truck	1953 Ford	60	90	120
Delivery Van	cast iron, 4-1/2" long, 1932	365	475	675
Dump Truck	cast iron, 7-1/2" long	225	335	450
Dump Truck	white rubber tires, 4-1/2" long, 1930s	100	150	200
Dump Truck	plastic, 1950s	25	35	55
Gas Tanker	5-1/2" long	115	225	295
General Shovel Truck	dual rear wheels, cast iron, 10" long, 1931	500	750	1000
Ingersoll-Rand Compressor	cast iron, 8-1/4" long, 1933	3250	6500	10000
Lifesaver Truck	cast iron, 4-1/4" long	350	475	700
Long Bed Dump Truck	series 510, Ford, cast iron	110	175	300
Mack Dump Truck	cast iron, 11" long, 1928	600	800	1450
Merchants Delivery Truck	cast iron, 6-1/4" long, 1925	400	600	850
Milk Truck	cast iron, 3-3/4" long, 1930	115	200	295
Nucar Transport	with four vehicles, cast iron, 16" long, 1932	675	1200	1500
Open Bed Auto Express	cast iron, 9-1/2" long, 1910	725	1200	1700
Panama Shovel Truck	Mack truck, cast iron, 13" long, 1934	750	1500	2000
Railway Express Truck	cast iron	135	225	275
Shovel Truck	metal, 10" long	250	375	500
Stake Bed Truck	cast iron, 3" long	75	115	150
Stake Truck	white rubber tires, 7" long, 1930s	85	130	175
Stake Truck	white cab, blue stake bed, 12" long	100	150	200
Stake Truck	die cast, 7" long, 1950s	75	115	150
Stockyard Truck #851	with three pigs	60	95	165
Tanker	cast iron, 7" long, 1940s	85	130	175
Tow Truck	Ford, cast metal, 7" long, 1950s	50	65	165
Tow Truck	9" long	135	225	275

Hubley

NAME	DESCRIPTION	GOOD	EX	MINT
Truckmixer	Ford, mixer cylinder rotates when truck moves, cast iron, 8" long, 1932	50	85	185
Wrecker	whitewall tires, green/white, 11-1/2" long	25	50	75
Wrecker	cast iron, 5" long	60	95	125
Wrecker	6" long	70	125	145
Wrecker	red, die cast, 9-1/2" long, 1940"s	60	95	165

Wagons and Carts

Alphonse in Goat-Pulled Wagon	13-3/4" long, 1900's	100	150	210
Alphonse in Mule-Pulled Wagon	6-1/2" long	225	335	450

Japanese

Cars

1935 Pontiac Four-Door Sedan	8", maroon, friction	30	65	90
1950s Cunningham Roadster	7-1/2", light blue, friction	40	90	125
1950s DeSoto	6", green, friction	20	45	65
1950s DeSoto	Asahi Toy, 8", green, friction	40	90	125
1950s Jaguar XKE Convertible	Tomiyama, 12", white, friction	195	455	650
1950s Jeep Station Wagon	Yonezawa, 7-1/2", two-tone brown, friction	100	210	300
1950s Kaiser Darren Convertible	6-1/2", red, friction	300	70	100
1950s Mercedes Convertible	Alps, 9", red, friction	105	245	350
1950s Studebaker Lark	5-1/2", blue, friction	20	45	65
1950s Volvo	5-1/2", red, friction	20	38	55
1950s Volvo PV-544	HoKu, 7-1/2", black, friction	165	385	550
1950s VW Convertible	9-1/2", dark blue, maroon, or light metallic blue, friction with battery-operated engine light	85	195	275
1950s VW Sedan	7-1/2", gray, oval window, friction	45	105	150
1950s Zephyr Deluxe Convertible	11", maroon/yellow/blue, friction	120	280	400
1951 Cadillac Four-Door Sedan	Marusan, 12-1/2", gray, black, white, or red, friction	300	700	1000
1951 Cadillac Four-Door Sedan	Marusan, 12-1/2", gray, battery-operated, remote control, working headlights	480	1120	1600
1951 Ford Sedan	7", tan, battery-operated	30	70	100
1951 Futuristic Buick LeSabre	Yonezawa, 7-1/2", black, friction	240	560	800
1952 Ford Yellow Cab	Marusan, 10-1/2", yellow, friction, working money meter	165	385	550
1953 Chrysler Orion Convertible	6-1/2", blue/green, friction	25	60	85
1953 Studebaker Coupe	9", yellow, friction, working wipers	45	100	140
1954 Chevrolet Bel Air	Marusan and Linemar, 11", rare orange/yellow, friction	660	1550	2200
1954 Chevrolet Bel Air	Marusan and Linemar, 11", gray/black, friction	400	840	1200

1959 Oldsmobile Highway Patrol Car, Japanese

1960s Ferrari Berlinetta 250 LeMans, Japanese

Japanese

NAME	DESCRIPTION	GOOD	EX	MINT
1955 Buick Special	8-1/2", two-tone blue, battery-operated, working headlights	55	122	175
1955 Chevrolet Bel Air	Asahi Toy, 7" light green, friction	50	105	150
1955 Ford Convertible	Haji, 6-1/2", two-tone blue or red/white, friction	60	140	200
1955 Ford Thunderbird Convertible	8", orange, friction	45	100	140
1955 Mercedes 300 SL Coupe	9", metallic red, opening gull-wing doors, battery-operated	105	245	350
1956 Ford Two-Door HT	Ichiko, 10", two-tone blue or orange/white, friction	165	385	550
1956 Ford Two-Door Sedan	Marusan, 13", orange/white or blue/white	1050	2450	3500
1956 GM Gas Turbine Firebird II	8-1/2", red, friction	240	560	800
1956 Lincoln Premiere Two-Door	7-1/2", orange, friction	30	70	100
1956 Oldsmobile Super 88	Modern Toys, 14", orange, battery-operated, working headlights and signal lights	225	525	750
1956 Plymouth HT	Alps, 8-1/2", two-tone green, friction	165	385	550
1957 Chrysler New Yorker	6-1/2", red/black, friction	25	60	85
1957 Ford Convertible	HTC, 12", orange/pink, friction	115	265	375
1957 Ford Fairlane 500 HT	ToyMaster, 9-1/2", green/yellow, friction	45	100	140
1958 Dodge Four-Door HT	8-1/2", orange/white, friction	55	125	175
1958 Edsel Station Wagon	11", red/black, friction	225	525	750
1958 Ford HT Convertible	9-1/2", blue/white, battery-operated	85	195	275
1958 Ford HT Convertible	11", orange/white, battery-operated	60	140	200
1958 Oldsmobile	Asahi Toy, 12", gold/black, friction	540	1260	1800
1958 Oldsmobile Station Wagon	7-1/2", red/black, friction	30	70	100
1958 Pontiac Four-Door HT	Asahi Toy, 8", green/pink, friction	45	105	150
1959 Buick Convertible	11", orange/yellow, friction, dog and driver figures	115	245	350
1959 Buick HT Convertible	Linemar, 9-1/2" red/white, friction	50	105	150
1959 Buick Station Wagon	Yonezawa, 9", two-tone green, friction	50	105	150
1959 Chevrolet Highway Patrol Car	ASC, 10", black/white, friction	55	125	175
1959 Chevrolet HT	7", green, friction	30	70	100
1959 Dodge Two-Door HT	9", blue/white, friction	135	315	450
1959 Ford HT Convertible	11", blue/white, red/white, or green/white, battery-operated	70	160	225
1959 Ford Station Wagon	10-1/2", green/white, friction	40	90	125
1959 Oldsmobile Highway Patrol Car	Ichiko, 12-1/2", black/white, friction, working speed meter on trunk	75	175	250
1959 Oldsmobile Two-Door HT	Ichiko, 12-1/2", two-tone blue, two-tone green, or brown/white, friction	135	315	450

Excalibur Roadster, Japanese

1960s Porsche 911 Rally, Japanese

VEHICLES

Japanese

NAME	DESCRIPTION	GOOD	EX	MINT
1959 Plymouth Convertible	Asahi Toy, 11", red/white, friction	330	770	1100
1960 Cadillac Four-Door Sedan	Yonezawa, 18" black or maroon, friction	300	630	900
1960 Chevrolet Impala HT	Alps, 9", red/white, friction	105	245	350
1960 Ford Gyron	Ichida, 11", red/white, battery-operated	135	315	450
1960 Ford Gyron	Ichida, 11", red/black, remote control, battery-operated	85	195	275
1960 Ford Gyron	Ichida, red/black, friction	45	100	140
1960s BMW Coupe	Yonezawa, 11" tan, battery-operated	50	105	150
1960s Ferrari Berlinetta 250 LeMans	Asahi Toy, 11", red, friction	115	265	375
1960s Ford Falcon	Marusan, 9", red/white, friction	25	55	75
1960s Jaguar XKE Coupe	10-1/2", red, friction	55	125	175
1960s Mercedes Convertible	HTC, 8", red, opening door with swing-out driver, friction	45	105	150
1960s Porsche 911 Rally	11", red, friction	70	160	225
1960s Porsche 914 Rally	Daiya, 9", blue, battery-operated	25	55	75
1961 Buick Fire Department Car	16", red, friction, working wipers, revolving emergency light	50	105	150
1961 Mercedes 220-S	12", black, jack-up feature, friction	75	175	250
1961 Oldsmobile Rally Car	Asahi Toy, 15", red, friction	55	125	175
1962 Cadillac Polic Car	Ichiko, 6-1/2" black/white, friction with siren	30	65	90
1962 Ford Thunderbird HT	Yonezawa, 11-1/2", red, battery-operated	105	245	350
1963 Corvette Coupe	12", metallic red or white, battery-operated, working headlights	180	420	600
1963 Ford Fire Chief Car	Taiyo, 12-1/2", red, battery-operated	23	55	75
1963 Ford Stock Car	Taiyo, 10-1/2", red/silver/blue, friction	25	60	85
1964 Ford Thunderbird HT	Ichiko, 15-1/2", red, working side windows, friction	120	280	400
1964 Lincoln	11", burgundy, battery-operated	165	385	550
1965 Ford Country Squire Wagon	9", white, friction	30	70	100
1965 Ford Mustang GT	15-1/2", red, friction	85	195	275
1966 Dodge Charger Sonic Car	16", red, battery-operated	145	335	475
1970s VW Rabbit Rally Team Car	Asahi Toy, 8", yellow, battery-operated	20	45	65
Ford Model-T	9", black, open top, friction	25	55	75
Ford Model-T	9", red, hard top, friction	25	55	75

Johnny Lightning / Topper

NAME	DESCRIPTION	GOOD	EX	MINT
1932 Roadster		20	40	55
A.J. Foyt Indy Special		15	30	65
Al Unser Indy Special		15	30	65

Johnny Lightning / Topper

NAME	DESCRIPTION	GOOD	EX	MINT
Baja		20	45	125
Big Rig		45	75	200
Bubble		15	25	40
Bug Bomb		20	35	65
Condor		45	85	350
Custom Dragster		15	35	75
Custom El Camino		75	175	400
Custom Eldorado		75	125	225
Custom Ferrari	w/sealed doors	15	35	50
Custom Ferrari	w/opening doors	70	160	350
Custom GTO		75	250	450
Custom Mako Shark	w/opening doors	75	150	350
Custom Mako Shark	w/sealed doors	15	35	75
Custom Spoiler		10	25	50
Custom T-Bird		75	150	275
Custom Toronado		75	150	350
Custom Turbine	w/unpainted interior	15	30	45
Custom Turbine	w/painted interior	20	50	65
Custom XKE	w/sealed doors	15	35	60
Custom XKE	w/opening doors	55	150	350
Double Trouble		20	60	250
Flame Out		15	40	125
Flying Needle		15	35	55
Frantic Ferrari		15	35	45
Glasser		15	35	55
Hairy Hauler		30	75	200
Jumpin' Jag		15	30	50
Leapin' Limo		15	40	175
Mad Maverick		15	40	125
Monster		10	25	35
Movin' Van		10	25	45
Nucleon		15	40	95
Parnelli Jones Indy Special		15	30	70
Pipe Dream		30	75	200
Sand Stormer		10	20	40
Screamer		15	25	45
Sling Shot		10	30	65
Smuggler		15	30	40
Stiletto		10	20	55
T.N.T.		15	30	40
Triple Threat		15	30	60
Twin Blaster		30	75	200
Vicious Vette		20	40	75
Vulture		50	95	350
Wasp		20	45	125
Wedge		15	30	45
Whistler		30	55	225
Wild Winner		35	75	200

Keystone

NAME	DESCRIPTION	GOOD	EX	MINT
	Airplanes			
Air Mail Plane	25", olive green, with three propellors	1600	2400	2800
Air Mail Plane	25", olive green	1500	2200	3000
Rapid Fire Air Mail Plane	25", olive green	2000	3000	4200
Ride `Em Air Mail Plane	25", seat and handles allowed child to ride	1200	1700	2400
	Emergency Vehicles			
Aerial Ladder Truck	27-1/2"	800	1200	1500
Ambulance	27-1/2", with canvas cover and stretcher	1000	1500	2000
Fire Truck	27-1/2", red w/hose reel	700	1000	1600
Water Pump Tower Truck	29", red cab, rubber tires	850	1200	1600
	Miscellaneous Vehicles			
Coast to Coast Bus	31-1/4", blue	1200	1700	2500
Ride `Em Locomotive	28"	300	500	650
Ride `Em Pullman Car	25", silver or red passenger car	300	500	725
Steam Roller	20", black body, with whistle and bell	200	300	600
Steam Shovel	20-3/4"	75	175	225
Truck Loader	tan w/black base, 20-1/4"	175	375	500
	Trucks			
American Railway Express Truck	26-1/4", green body, screened back	600	1000	1600
Dump Truck	black w/red chassis; 26-1/4"; rubber tires	300	500	720
Koaster Truck	26", open cab w/flatbed, black cab, red body	700	1000	1900
Moving Van	26-1/4", black cab, red body, rubber tires	1000	1500	2000
Packard U.S. Mail Truck	26-1/4", black cab, tan body	400	1000	1700
Police Patrol Truck	27-1/2", Police Patrol decals	700	1200	1600
Ride `Em Dump Truck	26-1/2", green cab, red body	250	450	700
Standard Dump Truck	Packard decal; black w/red chassis, 26-1/4"	350	500	800
Tank Truck	26-1/4", black cab, dark green body	1200	1800	2200
U.S. Army Truck	26-1/4", with canvas cover	500	900	1300
World's Greatest Circus Truck	26-1/4", with six removable animal cages, w/circus lithography	2500	5000	10000
Wrecker	27-1/2", red	800	1000	1500

Majorette

NAME	NO.	DESCRIPTION	MINT
		100 Series	
BRM Formula 1 Racer	101-A	unchromed or chromed	18
Citroen DS 21	113-A		23
Ferrari LeMans	109-A	unchromed or chromed	18
Fiat Multi Benne Skip Truck	114-A	orange with red skip, 1:100-scale (later reissued as 222-A)	18
Hotchkiss Jeep with Cable Carrier	107-A		15
Peugeot 404 Saloon	116-A		23
Porsche Formula 1 Racer	102-A	uncromed or chromed	18
		200 Series	
`57 Chevy Hot Rod	223-D	assorted	4
`57 Chevy Hot Rod	223-D	Kool Kromes	5
`57 Chevy Hot Rod	223-D	Road Eaters assorted	5
1969 Plymouth Fury Police	216-B	metallic blue	8
20 Panel Truck	214-C		8
Airport Minibus	262-A	various, 1:87-scale	4
Airport Minibus	262-A	red and yellow, 1:87-scale	5
Airport Minibus	262-A	white, marked "Air France" or "TWA," 1:87-scale	8
Airport Minibus	262-A	red with Coca-Cola sun	7
Alfa 75	271-A	red with black accents with black interior, no trim on grille	8
Alfa 75	271-A	red with black accents with black interior, silver trim on grille	6
Alfa 75	271-A	red with tan or white interior	5
Alfa Romeo Giulietta	271-B	red with black and silver accents	5
Alfa Romeo Giulietta	271-B	blue Polizia with whit interior, 1:55-scale	6
Alpine A310 special Team Unit	264-A	various	10
Ambulance Truck	255-D	white with orange accents, 1:60-scale	4
Aston-Martin DB7	229-D	blue	5
ATF	256-A		14
Audi 90	259-D	yellow, 1:60-scale	4
Audi 90	239-D	assorted, 1:60-scale	4
Audi Quattro	221-C	assorted	5
Autobianchi A112	269-B	assorted, 1:53-scale	8
Bank Security Armored Truck	204-C	assorted, 1:57-scale	4
Bernard Cattle Carrier	206-C	1:50-scale	15
Bernard Circus Truck	212-A		18
Bernard Dump truck	205-A		10
Bernard Fire Engine	204-A		18

Majorette

NAME	NO.	DESCRIPTION	MINT
Bernard Flat truck with Racks	206-A	1:50-scale	15
Bernard Flat Truck with Scraper	206-B	1:50-scale	15
Bernard Sanitation Truck	218-A		14
Bernard Sanitation Truck	218-A		12
Bernard Snow Plow	208-A	1:50-scale	15
Bernard Stake Truck	219-A		9
Bernard Truck with Chalet	236-B		14
Bertone Camargue	221-B		8
BMW 2800CS Coupe	235-A		9
BMW 3.0	235-C	yellow, 1:60-scale	6
BMW 3258i	257-C	various, 1:56-scale	4
BMW 325i	229-C	assorted	4
BMW 733	256-B	various, 1:60-scale	7
BMW Turbo	217-B		8
British Bus	259-E	red with amber windows, with "British Airways" label, 1:125-scale	8
British Bus	286-B	assorted, 1:125-scale	8
BRM Fi Racer	228-A		12
Bulldozer	287-B	yellow with black cab	5
Bulldozer	255-A		6
BX4TC	225-B	white	7
Cadillac Allante	253-C	various	4
Cadillac Allante	253-C	silver with red roof	7
Camping Trailer	259-B		8
Chevrolet El Camino SS Pickup	296-A	Road Eaters fluorescent lime green, 1:59-scale	5
Chevrolet El Camino SS Pickup	296-A	Kool Kromes red with pink flames, 1:59-scale	5
Chevrolet El Camino SS Pickup	296-A	assorted, 1:59-scale	4
Chevrolet El Camino SS Pickup	296-A	Road Eaters dark blue, 1:59-scale	5
Chevrolet Blazer	236-E		4
Chevrolet Covered Truck	241-C	blue, marked "Cadbury Roses Chocolates"	8
Chevrolet Grand Prix Corvette	215-C	assorted, 1:57-scale	4
Chevrolet Grand Prix Corvette	215-C	metallic blue with gold abd black accents	6
Chevrolet Impala Police Car	240-B	black with white doors	4
Chevrolet Impala Police Car	240-B	various	5
Chevrolet Impala Taxi	219-D	assorted	5

NAME	NO.	DESCRIPTION	MINT
Chevrolet Impala Taxi	213-F	yellow, 1:69-scale	5
Chevrolet Impala Taxi	213-F	yellow and black, 1:69-scale	4
Chevrolet Pickup	217-D		3
Chevy Blazer Pickup 4x4	291-A	assorted	4
Chevy Blazer Pickup 4x4	291-A	blue with yellow interior or yellow with red interior	5
Chevy Blazer Wrecker	228-B	assorted	4
Chevy Blazer Wrecker (Depanneuse)	228-B	red without markings, 1:62-scale	5
Chrysler 180	208-B	metallic green, 1:60-scale	13
Citroen Ambulance	206-E		10
Citroen CX	265-A	burgundy, silver or metallic brown, 1:60-scale	8
Citroen Ds19	213-A	blue with red interior	14
Citroen DS21	213-B		12
Citroen DS21 with Boat	214-A		18
Citroen DS21 with Caravan	215-A		18
Citroen Dyane Raid	231-A		10
Citroen GS	201-B		8
Citroen Maserati SM	250-A		12
Citroen Visa Chrono Mille Pistes	201-C	yellow or white, 1:52-scale	4
Citroen XM	254-C	white	4
Citroen XM	254-C	metal flake silver	5
Container Truck	265-B	white with gold tampo, marked "Futura Motors Pty.Ltd.," Australian promotional issue, 1:100-scale	18
Container Truck	265-B	blue, marked "Cadbury Roses Chocolate" or "Cadbury Dairy Milk," 1:100-scale	8
Container Truck	265-B	red, labelled "Coca-Cola," 1:100-scale	8
Container Truck	265-B	white, labelled "Total," 1:100-scale	6
Container Truck	265-B	Road Eaters white, labelled "Diet Pepsi," 1:100-scale	6
Container Truck	265-B	various, 1:100-scale	5
Corvette Turbo Racer	268-B	assorted, 1:54-scale	5
Corvette Turbo Racer	268-B	black with silver and gold accents, 1:54-scale	6
Corvette Turbo Racer	268-B	white, 1:54-scale	6
Crane Truck	283-A	assorted, 1:100-scale	4
Crazy Car	267-B	black with white interior	5
Crazy Car 4x4	223-B		8
Custom Ford Transit Van Pickup	264-B	red with white interior	6
DAF 2600 Bucket Truck	246-C	1:100-scale	17

Majorette

NAME	NO.	DESCRIPTION	MINT
DAF 2600 Covered Platform Truck	241-A	1:100-scale	9
DAF 2600 Crane Truck	247-A	1:100-scale	14
DAF 2600 Tanker	245-A	marked "Shell," 1:100-scale	8
DAF Covered Trailer	243-A		14
Datsun 260Z	229-B	metallic turquoise or light green, 1:60-scale	7
Datsun 260Z	229-B	red, 1:60-scale	6
Datsun 260Z	229-B	yellow, 1:60-scale	5
Desert Raider 4x4	223-C	black	5
Dodge Camper	209-A		15
Dodge Fire Rescue	204-B		8
Dodge Safari Truck	225-A	assorted	12
Dodge Snow Top truck with Plow	242-A		8
Dodge Wrecker Tow Truck	212-B	white wit blue light	8
Dune Buggy	248-A	various colors, flowers and lightning decals on roof, 1:55-scale	8
Dune Buggy	232-C	hot pink, "Fun Buggy"	6
Dune Buggy	232-C	hot pink, "Ice Cream"	5
Dune Buggy	259-A		12
Dune Buggy	258-A	various, with awning and amber windshield, chrome spiral window wheel design	12
Dune Buggy Surfer JP4 with surfboards	252-A	various, 1:47-scale	6
Etalmobile Warehouse Vehicle	203-A	1:50-scale	18
Excalibur	267-A	silver or metallic brown, 1:56-scale	6
Excalibur	267-A	yellow with passenger, 1:56-scale	8
Excalibur	267-A	metallic light blue with white interior and black roof, 1:56-scale	7
Explorateur 4x4	260-B	various, 1:59-scale	7
Explorer	261-B		4
Extending Ladder Fire Truck	207-D	red with gold lettering, wheat laurel or four-petal insignia	4
Extending Ladder Fire Truck (Pompier)	207-D	red without lettering, 1:100-scale	6
F1 Racer	234-B		6
Farm Tractor	208-C	assorted	4
Farm Tractor	208-C	with cultivator	4
Ferrari 456 GT	204-D		4
Ferrari F50 Coupe	204-E		4
Ferrari FA0	280-B	Kool Kromes red or yellow	5
Ferrari FA0	280-B	red with or without red plastic window, 1:58-scale	4
Ferrari FI Racer	229-A	metallic purple	12
Ferrari GTO	211-E	assorted, 1:56-scale	4
Ferrari Testarossa	211-F		4
FI Ferrari	282-A	assorted	4
FI Ferrari	282-A	red and green marked "Benetton 8"	6
FI Ferrari	282-A	red and green marked "Benetton 23"	6
FI Racer	213-G		4

Mustang SVO, Majorette

Mercedes-Benz 450SL, Majorette

Peugeot 504, Majorette

Majorette

NAME	NO.	DESCRIPTION	MINT
Fiat 127	203-B	blue, metalillic blue or metallic yellow-green; doors open, dog on back seat looking out left rear window, 1:55-scale	8
Fiat 127	203-B	blue, metallic blue or metallic yellow green, dog on back seat, 1:55-scale	10
Fiat Ritmo/Strada	239-C	assorted	6
Fiat Ritmo/Strada	239-C	red without markings	7
Fiat Skip Truck Multi Benne	222-A	red with yellow skip, 1:100-scale	8
Fiat Skip Truck multi Benne	222-A	orange with blue skip, 1:100-scale	10
Fiat Tipo	286-B	assorted, 1:54-scale	5
Ford 5000 Farm Tractor	253-A	blue with white fenders, 1:55-scale	10
Ford Bronco 4x4	251-B	black with silver and gold accents, with and without sunroof, 1:56-scale	4
Ford Bronco 4x4	251-B	mustard yellow, 1:56-scale	6
Ford Capri	251-A	various	8
Ford Covered Truck	241-C	red with white cargo cover with Coca-Cola label, 1:100-scale	6
Ford Covered Truck	241-C	various, 1:100-scale	4
Ford Escort GT	275-B		4
Ford Escort XR3	212-C	assorted, 1:52-scale	4
Ford Model A Van	201-D	blue or metallic red, "Tea Shop," 1:60-scale	6
Ford Model A Van	201-D	blue, "Orange Company," 1:60-scale	4
Ford Model A Van	201-D	Road Eaters bright orange or red, 1:60-scale	5
Ford Model A Van	201-D	blue, 1:60-scale	8
Ford Mondeo	269-B		4
Ford Mustang GT Convertible	227-C	assorted, 1:59-scale	4
Ford Tanker	245-C	various	5
Ford Tempo/Sierra	272-A	assorted, 1:58-scale	4
Ford Thunderbird Turbo	217-C	Smelly Speeders, white with brown trim, 1:67-scale	5
Ford Thunderbird Turbo	217-C	assorted, 1:67-scale	4
Ford Thunderbird Turbo	217-C	blue with red trim, 1:67-scale	5
Ford Transit Custom Tow Truck	295-A	red with white bumpers, marked "Jack's Towing 24 Hr Service"	4
Ford Transit Custom Tow Truck	295-A	fluorescent orange, marked "Racing Service"	5
Ford Transit Custom Tow Truck	230-D	assorted	4
Ford Transit Van	243-C	various	4
Ford US Van	250-B	black, 1:65-scale	7
Formula 1 Brabham	232-B	assorted, 1:53-scale	4
Formula 1 Racer	238-B	green, red and white, "Benetton," 1:55-scale	8
Formula I Racer	238-B	assorted, 1:55-scale	4
Four Wheel Loader	211-B		10
Fourgon Commercial Van	234-C	red, "Coca-Cola"	6

Majorette

NAME	NO.	DESCRIPTION	MINT
Fourgon Commercial Van	234-C	assorted, 1:53-scale	4
Fourgon Commercial Van	234-C	blue, "Cadbury Dairy Milk Buttons"	8
Fourgon Ice Cream Van	259-C	white with red awning	6
Fourgon Ice Cream Van	259-C	yellow with pink or blue windows, white awning	8
Fourgon Ice Cream Van	259-C	various	7
Fourgon Motor Home	224-B	assorted, 1:67-scale	10
Fourgon Police Van	279-A	assorted, 1:65-scale	5
Front End Loader	263-A	yellow, 1:87-scale	4
Front End Loader	263-A	orange, 1:87-scale	5
GMC Jimmy	249-D		4
Honda Accord	219-C	assorted, 1:59-scale	5
Honda Accord NSX	220-C		4
Honda Prelude 4WD	252-B	red with black and silver accents, 1:58-scale	5
Hotchkiss Jeep with Cattle Trailer	211-A		18
Jaguar E	207-B		10
Jaguar J6	293-B	assorted, 1:65-scale	4
Jaguar J6 Police	205-D		5
Jaguar XKE 2+2 Coupe	207-A		10
Jeep 4x4	244-B	various	5
Jeep 4x4 Rallye	289-A	assorted, 1:62-scale	4
Jeep Cherokee	236-D	yellow with "Indian" on roof	4
Jeep Cherokee 4x4	236-D	flourescent green, no dog, 1:64-scale	4
Jeep Cherokee 4x4	236-D	Road Eaters, black, "Chee-tos Chester Cheetah," no dog, 1:64-scale	5
Jeep Cherokee 4x4	236-D	green	4
Jeep Cherokee 4x4	236-D	metallic brown, "Big Chief," with dog, ribbed roof, 1:64-scale	7
Jeep Cherokee 4x4	236-D	red, "Mad Bull," ribbed roof, 1:64-scale	6
Jeep Cherokee 4x4	236-D	light brown, "Big Chief," with dog, ribbed roof, 1:64-scale	5
Jeep Cherokee 4x4	236-D	Smelly Speeders, yellow, ribbed roof, 1:64-scale	4
Jeep Cherokee 4x4	236-D	white, "Coca-Cola" with polar bear	12
Jeep Cherokee 4x4	236-D	flourescent orange, ribbed roof, 1:64-scale	4
Jeep Cherokee 4x4	236-D	red with white and black, orange and bllue accents, with dog, 1:65-scale	5
Jeep Cherokee 4x4	236-D	red with white and black, orange and blue accents, no dog, 1:65-scale	4
Jeep Cherokee 4x4	236-D	yellow with red and black accents, ribbed roof, with dog, 1:64-scale	5
Jeep Cherokee Ambulance	269-A	white with no markings, 1:64-scale	6
Jeep Cherokee Ambulance	269-A	white with red or blue accents, 1:64-scale	5
Jeep Cherokee Limited	224-C	with surfboards on roof, assorted, 1:60-scale	4
Jeep Cherokee Sheriff	285-B	assorted	4

Majorette

NAME	NO.	DESCRIPTION	MINT
Jeep CJ	268-A	metallic brown, with conventional chassis, 1:54-scale	6
Jeep CJ	268-A	assorted, with conventional chassis, 1:54-scale	5
Lamborghini Diablo	219-E	Road Eaters with Pepsi logo, 1:58-scale	6
Lamborghini Diablo	219-E	assorted, 1:58-scale	5
Lancia Montecarlo	285-A	mustard yellow with black interior	6
Lancia Montecarlo	285-A	white with red accents	5
Land Rover 4x4	266-B	tan with zebra stripes, 1:60-scale	4
Land Rover 4x4	266-b	various, 1:60-scale	5
Locomotive	234-A		17
Lotus FI Racer	227-A		10
Mack Dump Truck	297-A	assorted, 1:100-scale	4
Mack Tow Truck	256-C	red, 1:100-scale	4
Magirus Beton Cement Mixer	227-B	1:100-scale	10
Matra Simca 670	239-B	metallic blue	7
Matra Simca Bagheera	219-B	blue, 1:55-scale	7
Mazda RX7 Daytona	257-B	orange with blue accents, narrow tires, 1:56-scale	6
Mazda RX7 Daytona	257-B	various, 1:56-scale	4
Mercedes-Benz 300TE Station Wagon	250-C	various, 1:63-scale	5
Mercedes Cattle Truck	254-B	yellow, with one white and one black steer, 1:100-scale	8
Mercedes Public Works Truck (Trax Publics)	233-B	metallic light blue, 1:70-scale	6
Mercedes Public Works Truck (Trax Publics)	233-B	orange, 1:70-scale	7
Mercedes-Benz 190E 2.3-16	231-B	assorted, 1:59-scale	5
Mercedes-Benz 2.3-16	231-B	white, 1:59-scale	4
Mercedes-Benz 350SL Convertible	213-C	silver with red-orange accents, 1:60-scale	6
Mercedes-Benz 350SL Convertible	213-C	yellow, 1:60-scale	8
Mercedes-Benz 450 SE	249-B	metallic gold or light green, 1:60-scale	5
Mercedes-Benz 450 SE	249-B	metallic silver, 1:60-scale	4
Mercedes-Benz 450SL	213-D		6
Mercedes-Benz 500 SL Roadster	260-C	Road Eaters bright red, marked "Willy Wonka NERDS," 1:58-sclae	5
Mercedes-Benz 500 SL Roadster	260-C	various, 1:58-scale	4
Mercedes-Benz 500 SL Roadster	260-C	dark red with maroon interior, 1:58-scale	5
Mercedes-Benz Fire Engine	258-B	red, 1:70-scale	8
Mercedes-Benz Sanitation Truck	218-B	assorted, 1:100-scale	6

Majorette

NAME	NO.	DESCRIPTION	MINT
Mercedes-Benz Sanitation Truck	218-B	orange and gray	8
Mercedes-Benz Stake Truck	213-E	red with yellow hay bales, 1:100-scale	8
Mercedes-Benz Stake Truck with Hay Load	254-A	red and brown, 1:100-scale	8
Mobile Home Camping Car	278-B	assorted	6
Mobile Home Camping Car	278-B	red with white camper	4
Mobile Office	223-A		10
Morgan	261-A	various, 1:50-scale	4
Morgan	261-A	Kool Kromes blue with white interior, 1:50-scale	5
Motorboat and Trailer	235-B		15
Mustang SVO	220-B	white with oversized tires, 1:59-scale	4
Mustang SVO	220-B	metallic periwinkle, narrow or oversize tires, 1:59-scale	6
Mustang SVO	220-B	metallic periwinkle with red and white accents, oversize tires, 1:59-scale	4
Mustang SVO	220-B	white with narrow tires, 1:59-scale	5
Mustang SVO	220-B	red with narrow tires	4
Nissan 300ZX Turbo T-Roof	214-D	assorted	4
Oldsmobile Omega	253-B	various, 1:75-scale	8
Panther Bertone Course Racer	233-A	assorted, 1:65-scale	12
Peugeot 204 Roadster	230-A		12
Peugeot 205 CTI/GTI Cabriolet	210-C	assorted, 1:53-scale	4
Peugeot 205 GTI/CTI Hardtop	210-D	1:53-scale	4
Peugeot 205 GTI/CTI Hardtop Sedan	281-A	white with red and blue accents, marked "Police," 1:53-scale	5
Peugeot 205 GTI/CTI Hardtop Sedan	281-A	assorted, 1:53-scale	4
Peugeot 404 Ambulance	206-D		15
Peugeot 404 Police	216-A		10
Peugeot 404 Saloon with Alpine	217-A		18
Peugeot 405 mi 16	218-C	assorted, 1:62-scale	4
Peugeot 405 T16	202-D	Kool Kromes orange or blue	5
Peugeot 405 T16	202-D	yellow or red	4
Peugeot 406	218-D		4
Peugeot 504	239-A		8
Peugeot 604	238-A	assorted	7
Peugeot 604	236-C		6
Police Motorcycle	203-C		23
Pontiac Fiero	206-F	assorted	4
Pontiac Firebird	212-D		4

Simca 1308, Majorette

Saviem Container Truck, metallic blue with office, Majorette

Dodge Snow Plow, Majorette

Majorette

NAME	NO.	DESCRIPTION	MINT
Pontiac Firebird Trans Am	248-B	various, 1:55-scale	4
Pontiac Firebird Turbo	293-A		5
Pontiac Trans Sport SE	268-C	red or black with white interior	4
Porsche 911 Turbo	209-B	yellow with orange and blue accents, narrow tires, 1:57-scale	8
Porsche 911 Turbo	209-B	assorted, wide tires, 1:57-scale	5
Porsche 911 Turbo	209-B	assorted	4
Porsche 924	247-B	various, amber windows, 1:60-scale	8
Porsche 924	247-B	metallic blue with orange accents, clear windows	6
Porsche LeMans Racer	232-A	metallic red, 1:65-scale	11
Power Shovel	242-B	yellow body with black base and shovel, 1:100-scale	4
Power Shovel	242-B	red body with yellow base and yellow shovel, 1:100-scale	5
Pro Stocker Firebird with Oversized Engine	258-C	various, 1:62-scale	4
Range Rover Rescue Unit	246-C	with closed rear section, red with black or white interior, 1:60-scale	4
Range Rover Rescue Unit	246-B	with closed rear section, 1:60-scale	6
Refuse Truck	247-C	vatious, 1:100-scale	4
Refuse TrucK	247-C	green body with gray container	7
Refuse Truck	247-C	green body, orange container, 1:100-scale	6
Refuse Truck	247-C	lime green body with orange container with hippo on sides, 1:100-scale	5
Renault 11 Encore	274-A	dark maroon with yellow interior, 1:56-scale	6
Renault 11 Encore	274-A	metallic green with black, orange and silver accents, 1:54-scale	7
Renault 11 Encore	274-A	assorted, 1:54-scale	5
Renault 17	260-A		8
Renault 18	266-A	metallic silver, 1:60-scale	8
Renault 18	266-A	various, 1:60-scale	6
Renault 18	266-A	yellow, no marking, with passenger, 1:60-scale	8
Renault 19 Convertible	225-C		4
Renault 25 V6	222-B	assorted, 1:63-scale	4
Renault 4L Delivery Van	230-B	assorted, 1:55-scale	10
Renault 5 LeCar	257-A	various, with antenna and rearview mirrors	10
Renault 5 LeCar	280-A	metallic olive green, without rearview mirrors or antenna, 1:51-scale	8
Renault Acadiane Service Van	235-D	assorted, 1:53-scale	6
Renault Clio	270-B		5
Renault Espace	272-B	white ambulance	4
Renault Express Van	233-C	red, "Avis"	6
Renault Express Van	233-C	orange or white, "europcar rentacar"	7
Renault Express Van	233-C	blue, "Satellite Service"	5

Majorette

NAME	NO.	DESCRIPTION	MINT
Renault Maxi 5 Turbo	255-C	blue, 1:53-scale	5
Renault R16	221-A		12
Renault R5 Turbo	255-B	red with yellow interior, 1:53-scale	6
Renault Safrane	221-E	assorted	4
Renault Super Cinq GT Turbo	205-C	black, 1:51-scale	5
Renault Super Cinq GT Turbo	205-C	blue, 1:51-scale	5
Renault Twingo Minivan	206-G		6
Renault X54	221-D		5
Repco FI Racer	226-A		12
Road Grader Shovel	211-D		6
Road Roller	226-C	assorted	4
Roadster	273-B	red, based on the Dodge Viper, 1:58-scale	4
Rock Motorcycle	207-C		15
Saab 900Turbo	284-A	assorted, 1:62-scale	5
Saab Scania Dump Truck	205-B	yellow or silver with red dumper	8
Saviem Canvas Platform Truck	241-B	various	7
Saviem Container Truck	214-B	Adidas logo	7
Saviem Container Truck	214-B	Pepsi logo	8
Saviem Tanker	245-B	red-orange with white tank, marked "Ewig Oil Co.," 1:100-scale	22
Saviem Tanker	245-B	marked "Texaco," 1:100-scale	26
Saviem Tanker	245-B	various, 1:100-scale	8
Service Broom Truck	251-C		4
Simca 1100	234-B		10
Simca 1308	240-A	dark silver or blue, 1:60-scale	8
Ski-Doo Nordic Snowmobile	249-A		16
Snowmobile	284-B	white with red skis	6
Snowmobile	284-B	white with silver skis	8
Sports Proto Racer	235-F		4
St. Tropez Travel Trailer	201-A	1:68-scale	12
Stock Car	279-A	assorted, 1:60-scale	4
Streckeman Lovely 400 Travel Trailer	236-A	1:65-scale	14
Super Dump Truck	274-A	yellow with silver dumper, 1:100-scale	4
Toyota 4x4	276-A	red with black interior	4
Toyota 4x4	276-A	metallic maroon	4
Toyota Celica 2.0 GT	249-C	various, 1:58-scale	4
Toyota Hi-Lux Pickup	292-A	hot pink or metallic blue, modified 4x4 chassis	4
Toyota Hi-Lux Pickup	287-A	assorted, conventional chassis, 1:56-scale	8
Toyota Hi-Lux Pickup	292-A	yellow with black interior, modified 4x4 chassis	6

Majorette

NAME	NO.	DESCRIPTION	MINT
Toyota Hi-Lux Pickup	292-A	assorted, modified 4x4 chassis	5
Toyota Landcruiser	277-A	beige with green accents and map of Africa on roof, conventional chassis, 1:53-scale	6
Toyota Landcruiser	277-A	black and white zebra stripes, conventional chassis, 1:53-scale	6
Toyota Landcruiser	277-A	metallic green, conventional chassis, 1:53-scale	7
Toyota Landcruiser	277-A	red with yellow interior, conventional chassis, 1:53-scale	5
Toyota Landcruiser 4x4	277-B	bright green with yellow interior or red with black and gold accents, modified chassis, 1:53-scale	5
Toyota Landcruiser 4x4	277-B	red and black or white and black zebra stripes, modified chassis, 1:53-scale	6
Toyota Landcruiser 4x4	277-B	white with red and gold accents, modified chassis, 1:53-scale	4
Toyota Landcruiser 4x4	277-B	black with gold accents and map of Africa on roof, modified chassis, 1:53-scale	6
Toyota Landcruiser 4x4	277-B	beige with red interior and map of Africa on roof, modified chassis, 1:53-scale	5
Toyota Lite Ace Van Wagon	216-C	assorted, 1:52-scale	5
Toyota Lite Ace Van Wagon	216-C	yellow with bluebird accents on side	6
Toyota Tercel 4WD	273-A	assorted, 1:55-scale	6
Tractor with Plow	211-C		8
Triumph TR7	202-C	red or orange	4
Unimog Snow Plow	224-A		10
Unimog with Fork Lift	215-B	blue and white	8
Volkswagen 1302	203-D	blue or lime green, trunk does not open (Coccinelle)	5
Volkswagen 1302	203-D	light blue, trunk doesn't open (Coccinelle)	6
Volkswagen Ambulance	244-A	metallic blue with white interior	10
Volkswagen Golf	210-B	assorted, 1:60-scale	6
Volkswagen Golf	264-C	red, silver or teal, 1:56-sclae	4
Volkswagen K70	210-A	assorted, 1:60-scale	10
Volkswagen Van	226-B		12
Volkwagen Ambulance	244-A	various	8
Volvo 245 DL Station Wagon	220-A	assorted, 1:60-scale	8
Volvo 760 GLE	230-C	assorted, 1:61-scale	5
VW 113	202-A		10
VW 1302 Beetle	202-B	red with yellow lightning bolt on trunk, opening trunk, 1:60-scale	8
VW 1302 Beetle	202-B	red with yellow flower on trunk, opening trunk, 1:60-scale	11
Western Train	278-A	metallic blue, 1:87-scale	12

Saab Scania Dump Truck, Majorette

Peugeot 604, Majorette

Bernard Cattle Carrier, Majorette

Marx

NAME	DESCRIPTION	GOOD	EX	MINT

Airplanes

NAME	DESCRIPTION	GOOD	EX	MINT
727 Riding Jet	jet engine sound	150	225	300
Air-Sea Power Bombing Set	12" wingspan, 1940s	325	450	650
Airmail Biplane	four engines, tin wind-up, 18" wingspan, 1936	225	325	450
Airmail Monoplane	two engines, tin wind-up, 1930	100	150	225
Airplane	monoplane, pressed steel, 9" wingspan, 1942	110	165	300
Airplane	medium fuselage, tin wind-up	125	150	250
Airplane	monoplane, adjustable rudder, tin wind-up, 9-1/4" wingspan	150	225	300
Airplane	adjustable rudder, tin wind-up, 10" wingspan, 1926	150	225	300
Airplane	tin wind-up, 9-1/4" wingspan, 1926	150	225	300
Airplane	mail biplane, tin wind-up, 9-3/4" wingspan, 1926	150	225	300
Airplane	two propellers, tin wind-up, 9 7/8" wingspan, 1927	200	300	400
Airplane	twin engine, tin wind-up, 9-1/2" wingspan	100	150	200
Airplane	no engines, tin wind-up, 9-1/2" wingspan	135	200	300
Airplane	light fuselage, tin wind-up	125	175	250
Airplane #90	tin wind-up, 5" wingspan, 1930	240	360	480
Airplane with Parachute	monoplane, tin wind-up, 13" wingspan, 1929	115	170	325
Airways Express Plane	tin wind-up, 13" wingspan, 1929	200	300	400
American Airlines Airplane	passenger plane, tin wind-up, 27" wingspan, 1940	130	190	400
American Airlines Flagship	pressed steel, wood wheels, 27" wingspan, 1940	200	300	500
Army Airplane	tin, mechanical fighter, 7" wingspan	125	170	230
Army Airplane	two engines, tin wind-up, 18" wingspan, 1938	125	190	250
Army Airplane	biplane, tin wind-up, 25-3/4" wingspan, 1930	225	340	450
Army Airplane	18" wingspan, 1951	150	225	300
Army Bomber	tri-motor, 25-1/2" wingspan, 1935	250	375	500
Army Bomber	two engines, tin wind-up, 18" wingspan, 1940s	250	375	500
Army Bomber	monoplane, litho machine gun & pilot, 25-1/2" wingspan, 1935	300	450	600
Army Bomber with Bombs	camouflage pattern, metal, wind-up, 12" wingspan, 1930s	100	150	250
Army Fighter Plane	tin wind-up, 5" wingspan, 1940s	100	150	250
Autogyro	tin wind-up, 27" wingspan, 1940s	150	225	375
Blue and Silver Bomber	two engines, tin wind-up, 18" wingspan, 1940	190	280	375
Bomber	four propellers, metal, wind-up, 14-1/2" wingspan	100	150	300
Bomber with Tricycle Landing Gear	four engine, tin wind-up, 18" wingspan, 1940	225	325	425
Camouflage Airplane	four engines, 18" wingspan, 1942	125	200	295
China Clipper	four engines, tin wind-up, 18-1/4" wingspan, 1938	100	150	250
City Airport	extra tower and planes, 1930s	125	190	250
Crash-Proof Airplane	monoplane, tin wind-up, 11-3/4" wingspan, 1933	100	150	200
Cross Country Flyer	19" tall, 1929	375	550	725
Dagwood's Solo Flight Airplane	wind-up, 9" wingspan, 1935	200	300	950
Daredevil Flyer	tin wind-up, 1929	115	170	325

Marx

NAME	DESCRIPTION	GOOD	EX	MINT
Daredevil Flyer	Zeppelin-shaped, 1928	225	350	475
DC-3 Airplane	aluminum, wind-up, 9-1/2" wingspan, 1930s	125	190	300
Eagle Air Scout	monoplane, tin wind-up, 26-1/2" wingspan, 1929	200	300	400
Fighter Jet, USAF	battery-operated, 7" wingspan	90	135	250
Fighter Plane	battery-operated, remote controlled, 1950s	90	135	250
Fix All Helicopter		275	400	600
Flip-Over Airplane	tin wind-up	200	300	450
Floor Zeppelin	9-1/2" long, 1931	225	340	500
Floor Zeppelin	16-1/2" long, 1931	350	525	750
Flying Fortress 2095	sparking, four engines, 1940	150	245	400
Flying Zeppelin	wind-up, 9" long, 1930	225	340	475
Flying Zeppelin	wind-up, 17" long, 1930	350	525	750
Flying Zeppelin	wind-up, 10" long	275	400	600
Four-Motor Transport Plane	friction, tin litho	120	180	325
Golden Tricky Airplane		75	115	225
Hangar with One Plane	1940s	150	225	500
International Airline Express	monoplane, tin wind-up, 17-1/2" wingspan, 1931	200	300	425
Jet Plane	friction, 6" wingspan, 1950s	65	90	195
Little Lindy Airplane	friction, 2-1/4" wingspan, 1930	200	300	500
Looping Plane	silver version, tin wind-up, 7" wingspan, 1941	225	325	525
Lucky Stunt Flyer	tin wind-up, 6" long, 1928	150	225	350
Mammoth Zeppelin, 1st Mammoth	pull toy, 28" long, 1930	400	600	900
Mammoth Zeppelin, 2nd Mammoth	pull toy, 28" long, 1930	375	575	775
Municipal Airport Hangar	1929	100	150	425
Overseas Biplane	three propellers, tin wind-up, 9-7/8" wingspan, 1928	150	275	395
PAA Clipper Plane	pressed steel, 27" wingspan, 1952	125	175	525
PAA Passenger Plane	tin litho, 14" wingspan, 1950s	120	175	275
Pan American	pressed steel, four motors, 27" wingspan, 1940	90	150	525
Piggy Back Plane	tin wind-up, 9" wingspan, 1939	100	150	350
Pioneer Air Express Monoplane	tin litho, pull toy, 25-1/2" wingspan	125	190	300
Popeye Flyer	Wimpy and Swee'Pea litho on tower, 1936	600	900	1600
Popeye Flyer	Popeye and Olive Oyl in plane, tin litho tower, wind-up, 1936	475	700	1250
Pursuit Planes	one propeller, 8" wingspan, 1930s	125	200	300
Rollover Airplane	tin wind-up, forward and reverse, 6" wingspan, 1947	200	300	575
Rollover Airplane	tin wind-up, 1920s	200	300	575
Rookie Pilot	tin litho, wind-up, 7" long, 1930s	225	340	550
Seversky P-35	single-engine plane, 16" wingspan, 1940s	125	200	350
Sky Bird Flyer	two planes, 9-1/2" tower, 1947	275	400	550
Sky Cruiser Two-Motored Transport Plane	18" wingspan, 1940s	125	175	325
Sky Flyer	biplane and Zeppelin, 8-1/2" tall tower, 1927	225	340	450
Sky Flyer	9" tall tower, 1937	150	225	375
Spirit of America	monoplane, tin wind-up, 17-1/2" wingspan, 1930	325	500	650

Marx

NAME	DESCRIPTION	GOOD	EX	MINT
Spirit of St. Louis	tin wind-up, 9-1/4" wingspan, 1929	150	225	595
Stunt Pilot	tin wind-up	175	250	425
Tower Flyers	1926	175	250	350
Trans-Atlantic Zeppelin	wind-up, 10" long, 1930	225	350	450
TWA Biplane	four-engine, 18" wingspan	225	350	450
U.S. Marines Plane	monoplane, tin wind-up, 17-7/8" wingspan, 1930	200	300	400
Zeppelin	all metal, pull toy, 28" long, 1929	400	600	800
Zeppelin	friction pull toy, steel, 6" long	100	200	250
Zeppelin	flies in circles, wind-up, 17" long, 1930	350	525	700

Boats and Ships

NAME	DESCRIPTION	GOOD	EX	MINT
Battleship USS Washington	friction, 14" long, 1950s	50	75	225
Caribbean Luxury Liner	sparkling, friction, 15" long	50	75	225
Luxury Liner Boat	tin, friction	100	150	250
Mosquito Fleet Putt Putt Boat		40	55	95
River Queen Paddle Wheel Station	plastic	50	75	150
Sparkling Warship	tin wind-up, 14" long	50	75	195
Tugboat	plastic, battery-operated, 6" long, 1966	50	75	195

Buses

NAME	DESCRIPTION	GOOD	EX	MINT
American Van Lines Bus	cream and red, tin wind-up, 13-1/2" long	65	100	130
Blue Line Tours Bus	tin litho, wind-up, 9-1/2" long, 1930s	150	225	425
Bus	red, 4" long, 1940	35	50	95
Coast to Coast Bus	tin litho, wind-up, 10" long, 1930s	125	200	325
Greyhound Bus	tin litho, wind-up, 6" long, 1930s	100	150	200
Liberty Bus	tin litho, wind-up, 5" long, 1931	75	125	200
Mystery Speedway Bus	tin litho, wind-up, 14" long, 1938	200	300	550
Royal Bus Lines Bus	tin litho, wind-up, 10-1/4" long, 1930s	135	200	450
Royal Van Co. Truck "We Haul Anywhere"	tin wind-up, 9" long, 1920s-30's	140	225	500
School Bus	steel body, wooden wheels, pull toy, 11-1/2" long	125	200	325

Cars

NAME	DESCRIPTION	GOOD	EX	MINT
Amos 'N' Andy Fresh Air Taxi Cab	tin litho, wind-up, 8" long, 1930	650	875	1600
Anti-Aircraft Gun on Car	5-1/4" long	50	75	225
Army Car	battery-operated	65	100	200
Army Staff Car	with flasher and siren, tin wind-up, 11" long, 1940s	75	125	400
Army Staff Car	litho steel, tin wind-up, 1930s	125	200	425
Big Lizzie Car	tin wind-up, 7-1/4" long, 1930s	75	125	235
Blondie's Jalopy	tin litho, 16" long, 1941	325	500	850
Boat Tail Racer #3	tin wind-up, 5" long, 1930s	40	55	200
Bouncing Benny Car	pull toy, 7" long, 1939	325	500	750
Bumper Auto	large bumpers front and rear, tin wind-up, 1939	60	100	225

NAME	DESCRIPTION	GOOD	EX	MINT
Cadillac Coupe	trunk w/tools on luggage carrier, tin wind-up, 11" long, 1931	200	300	525
Cadillac Coupe	8-1/2" long, 1931	175	275	400
Camera Car	heavy gauge steel car, 9-1/2" long, 1939	850	1300	1900
Careful Johnnie	plastic driver, 6-1/2" long, 1950s	100	150	350
Charlie McCarthy "Benzine Buggy" Car	with red wheels, tin wind-up, 7" long, 1938	600	900	1450
Charlie McCarthy "Benzine Buggy" Car	with white wheels, tin wind-up, 7" long, 1938	450	625	950
Charlie McCarthy and Mortimer Snerd Private Car	tin wind-up, 16" long, 1939	600	900	1500
Charlie McCarthy Private Car	wind-up, 1935	1450	2200	3500
College Boy Car	blue car with yellow trim, tin wind-up, 8" long, 1930s	300	450	525
Convertible Roadster	nickel-plated tin, 11" long, 1930s	175	275	395
Coo Coo Car	8" long, tin wind-up, 1931	375	575	825
Crazy Dan Car	tin wind-up, 6" long, 1930s	140	225	395
Dagwood the Driver	8" long, tin wind-up, 1941	200	300	900
Dan Dipsy Car	nodder, tin wind-up, 5-3/4" long, 1950s	250	375	450
Dick Tracy Police Car	9" long	150	225	350
Dick Tracy Police Station Riot Car	friction, sparkling, 7-1/2" long, 1946	130	200	375
Dick Tracy Squad Car	yellow flashing light, tin litho, wind-up, 11" long, 1940s	250	375	525
Dick Tracy Squad Car	battery-operated, tin litho, 11-1/4" long, 1949	170	275	475
Dick Tracy Squad Car	friction, 20" long, 1948	125	170	300
Dippy Dumper	Brutus or Popeye, celluloid figure, tin wind-up, 9", 1930s	350	525	900
Disney Parade Roadster	tin litho, wind-up, 1950s	100	150	900
Donald Duck Disney Dipsy Car	plastic Donald, tin wind-up, 5-3/4" long, 1953	425	650	895
Donald Duck Go-Kart	plastic and metal, friction, rubber tires, 1960s	75	130	325
Donald the Driver	plastic Donald, tin car, wind-up, 6-1/2" long, 1950s	200	300	595
Dora Dipsy Car	nodder, tin wind-up, 5-3/4" long, 1953	400	600	725
Dottie the Driver	nodder, tin wind-up, 6-1/2" long, 1950s	150	225	450
Drive-Up Self Car	turns left, right or straight, 1940	100	150	225
Driver Training Car	tin wind-up, 1930s	80	120	225
Electric Convertible	tin and plastic, 20" long	65	100	295
Falcon	plastic bubble top, black rubber tires	50	75	175
Funny Fire Fighters	7" long, tin wind-up, 1941	800	1200	1600
Funny Flivver Car	tin litho, wind-up, 7" long, 1926	275	425	675
G-Man Pursuit Car	sparks, 14-1/2" long, 1935	190	285	750
Gang Buster Car	tin wind-up, 14-1/2" long, 1938	200	300	800
Giant King Racer	dark blue, tin wind-up, 12-1/4" long, 1928	250	375	725
Hot Rod #23	friction motor, tin, 8" long, 1967	45	75	90
Huckleberry Hound Car	friction	125	200	250
International Agent Car	tin wind-up	30	55	125
International Agent Car	friction, tin litho, 1966	60	100	195

International Agent Car, friction, 1966, Marx

Rollover Airplane, 1920s, Marx

Old Jalopy, 1950, Marx

Marx

NAME	DESCRIPTION	GOOD	EX	MINT
Jaguar	battery-operated, 13" long	225	325	450
Jalopy	tin driver, friction, 1950s	125	200	250
Jalopy Car	tin driver, motor sparks, crank, wind-up	140	225	280
Jolly Joe Jeep	tin litho, 5-3/4" long, 1950s	150	225	375
Joy Riders Crazy Car	tin litho, wind-up, 8" long, 1928	340	500	675
Jumping Jeep	tin litho, 5-3/4" long, 1947	210	325	425
King Racer	yellow body, red trim, tin wind-up, 8-1/2" long, 1925	375	575	750
King Racer	yellow with black outlines, 8-1/2" long, 1925	250	430	575
Komical Kop	black car, tin litho, wind-up, 7-1/2" long, 1930s	450	675	900
Leaping Lizzie Car	tin wind-up, 7" long, 1927	250	375	500
Learn To Drive Car	wind-up	110	165	225
Lonesome Pine Trailer and Convertible Sedan	22" long, 1936	375	600	795
Machine Gun on Car	hand crank activation on gun, 3" long	75	130	225
Magic George and Car	litho, 1940s	170	250	350
Mechanical Speed Racer	tin wind-up, 12" long, 1948	125	200	275
Mickey Mouse Disney Dipsy Car	plastic Mickey, tin wind-up, 5-3/4" long, 1953	425	655	875
Mickey the Driver	plastic Mickey, tin car, wind-up, 6-1/2" long, 1950s	170	250	450
Midget Racer "Midget Special" #2	miniature car, clockwork-powered, 5" long, 1930s	125	200	250
Midget Racer "Midget Special" #7	miniature car, tin wind-up, 5" long, 1930s	125	200	250
Milton Berle Crazy Car	tin litho, wind-up, 6" long, 1950s	250	375	595
Mortimer Snerd's Tricky Auto	tin litho, wind-up, 7-1/2" long, 1939	400	600	750
Mystery Car	press down activation, 9" long, 1936	125	200	275
Mystery Taxi	press down activation, steel, 9" long, 1938	160	250	375
Nutty Mad Car	red tin car, vinyl driver, friction, 4" long, 1960s	100	150	225
Nutty Mad Car	with driver, battery-operated, 1960s	100	150	225
Nutty Mad Car	blue car with goggled driver, friction, hard plastic, 1960s	75	130	200
Old Jalopy	tin wind-up, driver, "Old Jalopy" on hood, 7" long, 1950	225	325	425
Parade Roadster	with Disney characters, tin litho, wind-up, 11" long, 1950	225	325	900
Peter Rabbit Eccentric Car	tin wind-up, 5-1/2" long, 1950s	250	375	500
Queen of the Campus	with four college students' heads, 1950	250	400	525
Race `N Road Speedway	HO scale racing set, 1950s	60	100	125
Racer #12	tin litho, wind-up, 16" long, 1942	225	325	625
Racer #3	miniature car, tin wind-up, 5" long	75	125	195
Racer #4	miniature car, tin wind-up, 5" long	75	125	195
Racer #5	miniature car, tin wind-up, 5" long, 1948	75	125	195
Racer #61	miniature car, tin wind-up, 4-3/4" long, 1930	75	125	195
Racer #7	miniature car, tin wind-up, 5" long, 1948	75	125	195
Racing Car	two man team, tin litho, wind-up, 12" long, 1940	125	200	425
Racing Car	plastic driver, tin wind-up, 27" long, 1950	100	175	450
Roadster	11-1/2" long, 1949	100	150	225

Marx

NAME	DESCRIPTION	GOOD	EX	MINT
Roadster and Cannon Ball Keeper	wind-up, 9" long	175	250	350
Roadster Convertible with Trailer and Racer	mechanical, 1950	125	200	275
Rocket Racer	tin litho, 1935	275	425	550
Rolls-Royce	black plastic, friction, 6" long, 1955	40	60	80
Royal Coupe	tin litho, wind-up, 9" long, 1930	175	275	375
Secret Sam Agent 012 Car	tin litho, friction, 5" long, 1960s	40	65	165
Sedan	battery-operated, plastic, 9-1/2" long	175	275	350
Sheriff Sam and His Whoopee Car	plastic, tin wind-up, 5-3/4" long, 1949	200	300	475
Siren Police Car	15" long, 1930s	75	125	395
Smokey Sam the Wild Fireman Car	6-1/2" long, 1950	125	200	395
Smokey Stover Whoopee Car	1940s	175	275	525
Snoopy Gus Wild Fireman	7" long, 1926	500	750	1150
Speed Cop	two 4" all tin wind-up cars, track, 1930s	175	275	795
Speed King Racer	tin litho, wind-up, 16" long, 1929	325	500	650
Speed Racer	13" long, 1937	250	375	500
Speedway Coupe	tin wind-up, battery-operated headlights, 8" long, 1938	200	300	400
Speedway Set	two wind-up sedans, figure eight track, 1937	250	375	500
Sports Coupe	tin, 15" long, 1930s	125	200	250
Station Wagon	litho family of four with dogs on back windows, 6-3/4" long	60	100	275
Station Wagon	light purple with woodgrain pattern, wind-up, 7-1/2" long	60	100	275
Station Wagon	friction, 11" long, 1950	125	200	325
Station Wagon	green with woodgrain pattern, wind-up, 7" long, 1950	50	75	250
Streamline Speedway	two tin wind-up racing cars, 1936	175	275	395
Stutz Roadster	driver, 15" long, wind-up, 1928	325	500	750
Super Hot Rod	"777" on rear door, 11" long, 1940s	200	300	400
Super Streamlined Racer	tin wind-up, 17" long, 1950s	125	200	400
The Marvel Car, Reversible Coupe	tin wind-up, 1938	125	200	400
Tricky Safety Car	6-1/2" long, 1950	100	150	200
Tricky Taxi	black/white version, tin wind-up, 4-1/2" long, 1935	160	250	375
Tricky Taxi	red, black, and white, tin wind-up, 4-1/2" long, 1940s	175	275	375
Uncle Wiggly, He Goes A Ridin' Car	rabbit driving, tin wind-up, 7-1/2" long, 1935	425	650	850
Walt Disney Television Car	friction, 7-1/2" long, 1950s	125	200	425
Western Auto Track	steel, 24" long	75	125	225
Whoopee Car	witty slogans, tin litho, wind-up, 7-1/2" long, 1930s	375	580	775
Whoopee Cowboy Car	bucking car, cowboy driver, tin wind-up, 7-1/2" long, 1930s	400	600	800
Woody Sedan	tin friction, 7-1/2" long	60	100	225
Yellow Taxi	wind-up, 7" long, 1927	275	425	575
Yogi Bear Car	friction, 1962	50	85	195

NAME	DESCRIPTION	GOOD	EX	MINT

Emergency Vehicles

NAME	DESCRIPTION	GOOD	EX	MINT
Ambulance	tin litho, 11" long	125	175	375
Ambulance	13-1/2" long, 1937	225	350	650
Ambulance with Siren	tin wind-up	100	150	400
Army Ambulance	13-1/2" long, 1930s	250	375	750
Boat Tail Racer #2	litho, 13" long, 1948	125	200	425
Chief-Fire Department No. 1 Truck	friction, 1948	60	90	195
Chrome Racer	miniature racer, 5" long, 1937	85	125	250
City Hospital Mack Ambulance	tin litho, wind-up, 10" long, 1927	190	280	500
Electric Car	runs on electric power, license #A7132, 1933	225	325	500
Electric Car	wind-up, 1933	175	250	475
Fire Chief Car	working lights, 16" long	125	200	295
Fire Chief Car	wind-up, 6-1/2" long, 1949	75	125	275
Fire Chief Car	battery-operated headlights, wind-up, 11" long, 1950	100	150	325
Fire Chief Car	friction, loud fire siren, 8" long, 1936	150	250	425
Fire Chief Car with Bell	10-1/2" long, 1940	175	250	450
Fire Engine	sheet iron, 9" long, 1920s	100	175	335
Fire Truck	battery-operated, two celluloid firemen, 12" long	50	75	300
Fire Truck	friction, all metal, 14" long, 1945	90	135	295
Giant King Racer	red, 13" long, 1941	200	300	550
Giant King Racer	pale yellow, tin wind-up, 12-1/2" long, 1928	225	325	575
Giant Mechanical Racer	tin litho, 12-3/4" long, 1948	100	175	350
H.Q. - Staff Car	14-1/2" long, 1930s	325	500	750
Hook and Ladder Fire Truck	three tin litho firemen, 13-1/2" long	90	135	225
Hook and Ladder Fire Truck	plastic ladder on top, 24" long, 1950	90	135	225
Plastic Racer	6" long, 1948	50	75	150
Racer with Plastic Driver	tin litho car, 16" long, 1950	150	225	375
Rocket-Shaped Racer #12	1930s	275	425	600
Siren Fire Chief Car	red car with siren, 1934	225	325	495
Siren Fire Chief Truck	battery-operated, 15" long, 1930s	100	175	325
Tricky Fire Chief Car	4-1/2" long, 1930s	250	375	625
V.F.D. Emergency Squad	with ladder, metal, electrically powered, 14" long, 1940s	70	125	200
V.F.D. Fire Engine	with hoses and siren, 14" long, 1940s	120	180	295
V.F.D. Hook and Ladder Fire Truck	33" long, 1950	140	225	325
War Department Ambulance	1930s	100	150	695

Farm and Construction Equipment

NAME	DESCRIPTION	GOOD	EX	MINT
Aluminum Bulldog Tractor Set	tin wind-up, 9-1/2" long tractor, 1940	250	375	500
American Tractor	with accessories, tin wind-up, 8" long, 1926	150	225	300

Marx

NAME	DESCRIPTION	GOOD	EX	MINT
Army Design Climbing Tractor	tin wind-up, 7-1/2" long, 1932	80	120	300
Automatic Steel Barn and Mechanical Plastic Tractor	tin wind-up, 7" long red tractor, 1950	85	125	475
Bulldozer Climbing Tractor	bumper auto, large bumpers, tin wind-up, 1939	50	75	325
Bulldozer Climbing Tractor	caterpillar type, tin wind-up, 10-1/2" long, 1950s	45	75	295
Caterpillar Climbing Tractor	yellow tractor, tin wind-up, 9-1/2" long, 1942	75	125	300
Caterpillar Climbing Tractor	orange tractor, tin wind-up, 9-1/2" long, 1942	100	175	325
Caterpillar Tractor and Hydraulic	tin wind-up, 1948	50	75	295
Climbing Tractor	with driver, tin wind-up, 1920s	50	100	275
Climbing Tractor	tin wind-up, 8-1/4" long, 1930	100	150	325
Climbing Tractor with Chain Pull	tin wind-up, 7-1/2" long, 1929	150	225	350
Co-Op Combine	tin friction, 6"	30	45	95
Construction Tractor	reversing, tin wind-up, 14" long, 1950s	100	150	465
Covered Wagon	friction, tin litho, 9" long	20	30	195
Crawler	with or without blades and drivers, litho, 1/25 scale, 1950	75	125	150
Crawler with Stake Bed	litho, 1/25 scale, 1950	75	130	175
Farm Tractor and Implement Set	tin wind-up, tractor mower, hayrake, three-gang plow, 1948	200	300	400
Farm Tractor Set	40 pieces, tin wind-up, 1939	325	500	650
Farm Tractor Set and Power Plant	32 pieces, tin wind-up, 1938	200	300	400
Hill Climbing Dump Truck	tin wind-up, 13-1/2" long, 1932	140	210	375
Industrial Tractor Set	orange and red heavy gauge plate tractor, 7-1/2" long, 1930	145	225	395
International Harvester Tractor	diesel, driver and set of tools, 1/12 scale, 1954	75	125	150
Magic Barn and Tractor	plastic tractor, tin litho barn, 1950s	70	100	375
Mechanical Tractor	tin wind-up, 5-1/2" long, 1942	125	200	275
Midget Climbing Tractor	tin wind-up, 5-1/4" long, 1935	125	200	250
Midget Road Building Set	tin wind-up, 5-1/2" long tractor, 1939	200	300	400
Midget Tractor	copper color metal, tin wind-up, 5-1/4" long, 1940	40	60	125
Midget Tractor	red metal, tin wind-up, 5-1/4" long, 1940	30	45	125
Midget Tractor and Plow	tin wind-up, 1937	70	100	225
Midget Tractor with Driver	red metal, tin wind-up, 5-1/4" long, 1940	40	60	220
No. 2 Tractor	red with black wheels, tin wind-up, 8-1/2" long, 1940	95	150	225
Plastic Sparkling Tractor Set	tin wind-up, 6-1/2" long tractor with 10-1/2" long wagon, 1950	40	60	195
Plastic Tractor with Scraper	tin wind-up, 8" long with road scraper, 1949	50	75	195
Power Grader	black or white wheels, 17-1/2" long	50	100	125
Power Shovel		50	75	100

Speed Cop, 1930s, Marx

PAA Airplane, 1950s, Marx

NAME	DESCRIPTION	GOOD	EX	MINT
Reversible Six-Wheel Farm Tractor-Truck	tin wind-up, 13-3/4" steel tractor, 7-1/2" stake truck, 1950	80	125	425
Reversible Six-Wheel Tractor	red steel tractor, tin wind-up, 11-3/4" long, 1940	200	300	475
Self-Reversing Tractor	tin wind-up, 10" long, 1936	125	200	350
Sparkling Climbing Tractor	tin wind-up, 8-1/2" long, 1950s	100	150	300
Sparkling Climbing Tractor	tin wind-up, 10" long, 1940s	175	275	350
Sparkling Heavy Duty Bulldog	with road scraper, tin wind-up, 11" long, 1950s	40	60	225
Sparkling Hi-Boy Climbing Tractor	10-1/2" long, 1950s	25	40	195
Sparkling Tractor	with driver and trailer, tin wind-up, 16" long, 1950s	60	90	295
Sparkling Tractor	with plow blade, tin wind-up, 1939	50	75	235
Sparkling Tractor and Trailer Set	"Marborook Farms," tin wind-up, 21" long, 1950s	55	75	255
Steel Farm Tractor and Implements	tin wind-up, 15" long steel bulldozer tractor, 1947	160	240	325
Super Power Reversing Tractor	tin wind-up, 12" long, 1931	100	150	325
Super Power Tractor and Trailer Set	tin wind-up, 8-1/2" tractor, 1937	125	200	375
Super-Power Bulldog Tractor with V-Shaped Plow	aluminum finish, tin wind-up, 1938	75	130	200
Super-Power Climbing Tractor and Nine-Piece Set	tin wind-up, 9-1/2" long tractor, 1942	160	240	395
Super-Power Giant Climbing Tractor	tin wind-up, 13" long, 1939	180	270	450
Tractor	red tractor, tin wind-up, 8-1/2" long, 1941	100	175	300
Tractor	tin wind-up, 8-1/2" long, 1941	100	175	315
Tractor and Equipment Set	five pieces, tin wind-up, 16" long tractor, 1949	160	240	320
Tractor and Mower	tin wind-up, 5" long litho steel tractor, 1948	50	75	215
Tractor and Six Implement Set	tin wind-up, 8-1/2" long aluminum tractor, 1948	160	240	425
Tractor and Trailer	tin wind-up, 16-1/2" long, 1950s	40	60	115
Tractor Road Construction Set	36 pieces, tin wind-up, 8-1/2" long tractor, 1938	200	325	595
Tractor Set	40 pieces, tin wind-up, 8-1/2" long, 1942	300	450	800
Tractor Set	seven pieces, tin wind-up, 8-1/2" long tractor, 1932	125	200	400
Tractor Set	five pieces, tin wind-up, 8-1/2" long tractor, 1935	70	100	220
Tractor Set	four pieces, tin wind-up, 8-1/2" long, 1936	100	150	325
Tractor Set	32 pieces, tin wind-up, 1937	180	270	450
Tractor Set	two-pieces, tin wind-up, 19" long steel tractor, 1950	65	100	215
Tractor Set	five pieces, tin wind-up, 8-1/2" tractor, 1938	90	135	295
Tractor Trailer and Scraper	tin wind-up, 8-1/2" long tractor, 1946	80	120	200
Tractor Train with Tractor Shed	tin wind-up, 8-1/2" long, 1936	100	150	220
Tractor with Airplane	wind-up, 5-1/2" long tractor, 27" wingspan on airplane, 1941	250	375	675

Marx

NAME	DESCRIPTION	GOOD	EX	MINT
Tractor with Driver	wind-up, 1940s	100	150	300
Tractor with Earth Grader	tin wind-up, mechanical, 21-1/2" long, 1950s	40	60	190
Tractor with Plow and Scraper	aluminum tractor, tin wind-up, 1938	70	125	295
Tractor with Plow and Wagon	tin wind-up, 1934	70	125	295
Tractor with Road Scraper	tin wind-up, 8-1/2" long climbing tractor, 1937	90	125	325
Tractor with Scraper	tin wind-up, 8-1/2" long, 1933	70	125	295
Tractor with Trailer and Plow	tin wind-up, 8-1/2" long, 1940	80	120	295
Tractor, Trailer, and V-Shaped Plow	tin wind-up, 8-1/2" steel tractor, 1939	85	125	325
Tractor-Trailer Set	tin wind-up, 8-1/2" long copper-colored tractor, 1939	70	125	295
Yellow and Green Tractor	tin wind-up, 8-1/2" long, 1930	90	135	335

Motorcycles

NAME	DESCRIPTION	GOOD	EX	MINT
Motorcycle Cop	tin litho, mechanical, siren, 8-1/4" long	75	125	235
Motorcycle Delivery Toy	"Speedy Boy Delivery Toy" on rear of cart, tin wind-up, 1932	175	275	500
Motorcycle Delivery Toy	"Speedy Boy Delivery Toy" on side of cart, tin wind-up, 1930s	175	280	525
Motorcycle Police	red uniform on cop, tin wind-up, 8" long, 1930s	125	200	275
Motorcycle Police #3	tin wind-up, 8-1/2" long	100	150	250
Motorcycle Policeman	orange/blue, tin wind-up, 8" long, 1920s	100	150	325
Motorcycle Trooper	tin litho, wind-up, 1935	80	120	275
Mystery Police Cycle	yellow, tin wind-up, 4-1/2" long, 1930s	75	125	225
Mystic Motorcycle	tin litho, wind-up, 4-1/4" long, 1936	75	130	235
P.D. Motorcyclist	tin wind-up, 4" long	40	60	195
Pinched Roadster Motorcycle Cop	in circular track, tin wind-up, 1927	125	200	365
Pluto Motorcycle with Siren	1930s	150	225	550
Police Motorcycle with Sidecar	tin wind-up, 3-1/2" long, 1930s	300	450	750
Police Motorcycle with Sidecar	tin wind-up, 8" long, 1930s	425	650	950
Police Motorcycle with Sidecar	tin litho, wind-up, 8" long, 1950	175	280	425
Police Patrol Motorcycle with	tin wind-up, 1935	110	165	400
Police Siren Motorcycle	tin litho, wind-up, 8" long, 1938	150	225	350
Police Squad Motorcycle Sidecar	tin litho, wind-up, 8" long, 1950	100	150	375
Police Tipover Motorcycle	tin litho, wind-up, 8" long, 1933	200	300	525
Rookie Cop	yellow with driver, tin litho, wind-up, 8" long, 1940	175	275	425
Sparkling Soldier Motorcycle	tin litho, wind-up, 8" long, 1940	175	275	425
Speeding Car and Motorcycle Policeman	tin litho, wind-up, 1939	90	150	350
Tricky Motorcycle	tin wind-up, 4-1/2" long, 1930s	100	150	250

Marx

NAME	DESCRIPTION	GOOD	EX	MINT
Tanks				
Anti-Aircraft Tank Outfit	four flat metal soldiers, tank, anti-aircraft gun, 1941	90	135	300
Anti-Aircraft Tank Outfit	three cardboard tanks	150	225	350
Army Tank	sparking climbing tank, tin wind-up, 1940s	150	225	375
Climbing Fighting Tank	tin wind-up	90	135	250
Climbing Tank	tin wind-up, 9-1/2" long, 1930	125	200	325
Doughboy Tank	doughboy pops out, tin litho, wind-up, 9-1/2" long, 1930	150	225	425
Doughboy Tank	sparking tank, tin wind-up, 10" long, 1937	125	200	400
Doughboy Tank	tin wind-up, 10" long, 1942	150	225	450
E12 Tank	green tank, 9-1/2" long, tin wind-up, 1942	150	225	375
M48T Tank	battery-operated, 1960s	50	100	225
Midget Climbing Fighting Tank	tin litho, wind-up, 5-1/4" long, 1931	90	135	235
Midget Climbing Fighting Tank	wide plastic wheels, supergrid tread, 5-1/4" long, 1951	100	150	300
Midget Climbing Fighting Tank	5-1/2" long, 1937	100	175	375
Refrew Tank	tin wind-up	75	125	200
Rex Mars Planet Patrol Tank	tin wind-up, 10" long, 1950s	150	225	375
Sparkling Army Tank	camouflage hull, two olive guns, tin wind-up, 5-1/2" long	100	150	275
Sparkling Army Tank	camouflage hull, two khaki guns, tin wind-up, 5-1/2" long	125	200	325
Sparkling Army Tank	tan or khaki hull, tin litho, wind-up, 1938	95	150	235
Sparkling Climbing Tank	tin wind-up, 10" long, 1939	110	165	310
Sparkling Space Tank	tin wind-up, 1950s	250	375	600
Sparkling Super Power Tank	tin wind-up, 9-1/2" long, 1950s	75	125	250
Sparkling Tank	tin wind-up, 4" long, 1948	75	130	195
Superman Turnover Tank	Superman lifts tank, 4" long, tin wind-up, 1940	250	375	600
Tank	pop-up army man shooting	150	225	400
Turnover Army Tank	camouflage, tan or khaki hull, tin wind-up, 1938	175	275	450
Turnover Army Tank	tin wind-up, 9" long, 1930	150	225	375
Turnover Tank	tin litho, wind-up, 4" long, 1942	100	150	275
Trucks				
A & P Supermarket Truck	pressed steel, rubber tires, litho, 19" long	50	100	275
Aero Oil Co. Mack Truck	tin litho, friction, 5-1/2" long, 1930	125	200	350
Air Force Truck	32" long	75	125	300
American Railroad Express Agency Inc. Truck	open cab, 7" long, 1930s	100	150	325
American Truck Co. Mack Truck	friction, 5" long	100	150	225

NAME	DESCRIPTION	GOOD	EX	MINT
Armored Trucking Co. Mack Truck	black cab, yellow printing, wind-up, 9-3/4" long	200	300	400
Armored Trucking Co. Truck	tin litho, wind-up, 10" long, 1927	100	150	325
Army Truck	tin, 12" long, 1950	60	90	150
Army Truck	20" long	50	75	125
Army Truck	canvas top, 20" long, 1940s	150	225	325
Army Truck	olive drab truck, 4-1/2" long, 1930s	125	200	350
Army Truck with Rear Benches and Canopy	olive drab paint, 10" long	75	125	225
Artillery Set	three-piece set, 1930	100	150	225
Auto Carrier	two yellow plastic cars, two ramp tracks, 14" long, 1950	125	200	295
Auto Mack Truck	yellow/red, 12" long, 1950	50	75	175
Auto Transport Mack Truck and	dark blue cab, dark green trailer, wind-up, 11-1/2" long, 1932	150	225	375
Auto Transport Mack Truck and	medium blue cab, friction, 11-1/2" long, 1932	150	225	425
Auto Transport Mack Truck and	dark blue cab, wind-up, 11-1/2" long, 1932	150	225	400
Auto Transport Truck	with three cars, 21" long, 1940	250	375	575
Auto Transport Truck	21" long, 1947	175	250	400
Auto Transport Truck	with dump truck, roadster and coupe, 30-1/2" long, 1938	275	425	575
Auto Transport Truck	double decker transport truck, 24-1/2" long, 1935	275	425	575
Auto Transport Truck	with three racing coupes, 22" long, 1933	250	375	525
Auto Transport Truck	with three wind-up cars, 22-3/4" long, 1931	250	375	525
Auto Transport Truck	with two tin litho cars, 34" long, 1950s	60	90	200
Auto Transport Truck	pressed steel, with two plastic cars, 14" long, 1940	75	125	225
Auto Transport Truck	with two plastic sedans, wooden wheels, 13-3/4" long, 1950	75	130	225
Auto Transwalk Truck	with three cars, 1930s	175	250	395
Bamberger Mack Truck	dark green, wind-up, 5" long, 1920s	200	300	525
Big Load Van Co. Hauler and Trailer	with little cartons of products, 12-3/4" long, 1927	200	300	525
Big Load Van Co. Mack Truck	wind-up, 13" long, 1928	225	350	600
Big Shot Cannon Truck	battery-operated, 23" long, 1960s	150	250	325
Cannon Army Mack Truck	9" long, 1930s	175	250	325
Carpenter's Truck	stake bed truck, pressed steel, 14" long, 1940s	175	250	350
Carrier with Three Racers	tin litho, wind-up, 22-3/4" long, 1930	150	225	425
Cement Mixer Truck	red cab, tin finish mixing barrel, 6" long, 1930s	100	150	300
City Coal Co. Mack Dump Truck	14" long, 1934	225	350	525
City Delivery Van	yellow steel truck, 11" long	140	210	280
City Sanitation Dept. "Help Keep Your City Clean" Truck	12-3/4" long, 1940	60	90	250
Coal Truck	battery-operated, automatic dump, forward and reverse, tin	75	125	175

Nutty Mad Car, Marx

A&P Super Market Truck, Marx

Marx

NAME	DESCRIPTION	GOOD	EX	MINT
Coal Truck	1st version, red cab, litho blue and yellow dumper, 12" long	140	225	280
Coal Truck	2nd version, light blue truck, 12" long	140	225	280
Coal Truck	3rd version, Lumar Co. truck, 10" long, 1939	150	250	310
Coca-Cola Truck	tin, 17" long, 1940s	125	200	250
Coca-Cola Truck	yellow, 20" long, 1950	150	225	300
Coca-Cola Truck	stamped steel, 20" long, 1940s	175	250	350
Coca-Cola Truck	red steel, 11-1/2" long, 1940s	100	175	250
Contractors and Builders Truck	10" long	125	200	250
Curtiss Candy Truck	red plastic truck, 10" long, 1950	125	200	250
Dairy Farm Pickup Truck	22" long	60	100	150
Delivery Truck	blue truck, 4" long, 1940	100	175	225
Deluxe Delivery Truck	with six delivery boxes, stamped steel, 13-1/4" long, 1948	100	150	200
Deluxe Trailer Truck	tin and plastic, 14" long, 1950s	70	100	145
Dodge Salerno Engineering Department Truck		600	900	1200
Dump Truck	4-1/2" long, 1930s	100	175	325
Dump Truck	motor, tin friction, 12" long, 1950s	50	75	100
Dump Truck	red cab, gray bumper, yellow bed, 18" long, 1950	100	150	200
Dump Truck	6" long, 1930s	100	150	250
Dump Truck	red cab, green body, 6-1/4" long	50	75	100
Dump Truck	yellow cab, blue bumper, red bed, 18" long, 1950	100	150	200
Emergency Service Truck	friction, tin	125	200	275
Emergency Service Truck	friction, tin, searchlight behind car and siren	150	225	300
Firestone Truck	metal, 14" long, 1950s	50	75	150
Ford Heavy Duty Express Truck	cab with canopy, 1950s	60	100	175
Gas Truck	green truck, 4" long, 1940	60	90	165
Giant Reversing Tractor Truck	with tools, tin wind-up, 14" long, 1950s	75	130	350
Gravel Truck	pressed steel cab, red tin dumper, 10" long, 1930	100	150	250
Gravel Truck	metal, 8-1/2" long, 1940s	75	125	200
Gravel Truck	metal with "Gravel Mixer" drum, 10" long, 1930s	100	150	250
Grocery Truck	cardboard boxes, tinplate and plastic, 14-1/2" long	90	150	225
Guided Missile Truck	blue, red and yellow body, friction, 16" long, 1958	75	125	225
Highway Express Truck	tin, tin tires, 16" long, 1940s	50	75	185
Highway Express Truck	"Nationwide Delivery," 1950s	100	150	225
Highway Express Truck	pressed steel	95	150	185
Highway Express Van Truck	metal, 15-1/2" long, 1940s	100	150	225
Jalopy Pickup Truck	tin wind-up, 7" long	60	90	175
Jeep	11" long, 1946-50	75	130	200
Jeepster	mechanical, plastic	110	165	250

NAME	DESCRIPTION	GOOD	EX	MINT
Lazy Day Dairy Farm Pickup Truck and Trailer	22" long	45	75	275
Lincoln Transfer and Storage Co. Mack Truck	wheels have cut-out spokes, tin litho, wind-up, 13" long, 1928	350	525	750
Lone Eagle Oil Co. Mack Truck	bright blue cab, green tank, wind-up, 12" long, 1930	225	350	550
Lumar Contractors Scoop/Dump	17-1/2" long, 1940s	75	125	200
Lumar Lines and Gasoline Set	1948	250	375	725
Lumar Lines Truck	red cab, aluminum finished trailer, 14" long	125	200	300
Lumar Motor Transport Truck	litho, 13" long, 1942	75	130	250
Machinery Moving Truck		60	90	200
Mack Army Truck	pressed steel, wind-up, 7-1/2" long	70	100	250
Mack Army Truck	friction, 5" long, 1930	125	200	350
Mack Army Truck	khaki brown body, wind-up, 10-1/2" long	125	225	400
Mack Army Truck	13-1/2" long, 1929	225	350	600
Mack Dump Truck	dark red cab, medium blue bed, wind-up, 10" long, 1928	150	225	475
Mack Dump Truck	medium blue truck, wind-up, 13" long, 1934	200	300	525
Mack Dump Truck	silver cab, medium blue dump, wind-up, 12-3/4" long, 1936	225	355	525
Mack Dump Truck	no driver, 19" long, 1930	300	450	650
Mack Dump Truck	tin litho, wind-up, 13-1/2" long, 1926	225	350	575
Mack Railroad Express Truck #7	tin, 1930s	75	125	450
Mack Towing Truck	dark green cab, wind-up, 8" long, 1926	175	275	450
Mack U.S. Mail Truck	black body, wind-up, 9-1/2" long	250	375	500
Magnetic Crane and Truck	1950	175	275	350
Mammoth Truck Train	truck with five trailers, 1930s	150	250	450
Meadowbrook Dairy Truck	14" long, 1940	150	275	375
Mechanical Sand Dump Truck	steel, 1940s	100	150	275
Medical Corps Ambulance Truck	olive drab paint, 1940s	100	150	275
Merchants Transfer Mack Truck	red open-stake truck, 10" long	225	350	450
Merchants Transfer Mack Truck	13-1/3" long, 1928	250	375	600
Military Cannon Truck	olive drab paint, cannon shoots marbles, 10" long, 1939	125	200	250
Milk Truck	white truck, 4" long, 1940	60	90	120
Miniature Mayflower Moving Van	operating lights	60	90	125
Motor Market Truck	10" long, 1939	100	175	225
Navy Jeep	wind-up	50	75	100
North American Van Lines Tractor Trailer	wind-up, 13" long, 1940s	100	175	225
Panel Wagon Truck		30	55	75
Pet Shop Truck	plastic, six compartments with six vinyl dogs, 11" long	125	200	250

Marx

NAME	DESCRIPTION	GOOD	EX	MINT
Pickup Truck	blue/yellow with wood tires, 9" long, 1940s	50	75	150
Polar Ice Co. Ice Truck	13" long, 1940s	100	175	275
Police Patrol Mack Truck	wind-up, 10" long	200	300	475
Popeye Dippy Dumper Truck	Popeye is celluloid, tin wind-up	325	500	800
Pure Milk Dairy Truck	glass bottles, pressed steel, 1940	55	80	250
Railway Express Agency Truck	green closed van truck, 1940s	90	135	300
Range Rider	tin wind-up, 1930s	250	375	500
RCA Television Service Truck	plastic Ford panel truck, 8-1/2" long, 1948-50	150	225	300
Reversing Road Roller	tin wind-up	60	100	125
Road Builder Tank	1950	125	200	250
Rocker Dump Truck	17-1/2" long	60	90	120
Roy Rogers and Trigger Cattle Truck	metal, 15" long, 1950s	60	100	225
Royal Oil Co. Mack Truck	dark red cab, medium green tank, wind-up, 8-1/4" long, 1927	200	300	500
Royal Van Co. Mack Truck	1927	250	375	600
Royal Van Co. Mack Truck	red cab, tin litho and paint, wind-up, 9" long, 1928	210	315	550
Run Right To Read's Truck	14" long, 1940	125	200	250
Sand and Gravel Truck	"Builder's Supply Co.," tin wind-up, 1920	150	225	400
Sand Truck	tin litho, 12-1/2" long, 1940s	125	200	275
Sand Truck	9" long, 1948	100	150	200
Sand-Gravel Dump Truck	tin litho, 12" long, 1950	150	225	300
Sand-Gravel Dump Truck	blue cab, yellow dump with "Gravel" on side, tin, 1930s	140	225	285
Sanitation Truck	1940s	135	200	270
Searchlight Truck	pressed steel, 9-3/4" long, 1930s	160	250	325
Side Dump Truck	1940	90	130	275
Side Dump Truck	10" long, 1930s	140	225	280
Side Dump Truck and Trailer	15" long, 1935	150	225	300
Sinclair Tanker	tin, 14" long, 1940s	150	225	300
Stake Bed Truck	red cab, blue stake bed, 6" long, 1930s	75	125	150
Stake Bed Truck	rubber stamped chicken on one side of truck, bunny on other	100	150	200
Stake Bed Truck	pressed steel, wooden wheels, 7" long, 1936	60	90	200
Stake Bed Truck	red cab, green stake bed, 20" long, 1947	100	150	200
Stake Bed Truck	medium blue cab, red stake bed, 10" long, 1940	75	130	185
Stake Bed Truck	red cab, yellow and red trailer, 14" long	100	150	200
Stake Bed Truck and Trailer	red truck, silver stake bed	125	200	250
Streamline Mechanical Hauler, Van, and Tank Truck Combo	heavy gauge steel, 10 3/8" long, 1936	170	275	350
Sunshine Fruit Growers Truck	red cab, yellow/white trailer with blue roof, 14" long	100	150	225

Marx

NAME	DESCRIPTION	GOOD	EX	MINT
Tipper Dump Truck	wind-up, 9-3/4" long, 1950	75	130	175
Tow Truck	aluminum finsih, tin litho wind-up, 6-1/4" long	75	125	150
Tow Truck	aluminum finish, wind-up, 6-1/4" long	95	150	190
Tow Truck	10" long, 1935	125	200	250
Tow Truck	red cab, yellow towing unit, 6" long, 1930s	137	200	275
Toyland Dairy Truck	10" long	140	210	280
Toyland's Farm Products Mack Milk Truck	with 12 wooden milk bottles, 10-1/4" long, 1931	200	300	400
Toytown Express Truck	plastic cab	45	70	95
Tractor Trailer with Dumpster	blue/yellow hauler, tan dumpster	125	200	250
Truck Train	stake hauler and five trailers, 41" long, 1933	250	400	525
Truck Train	stake hauler and four trailers, 41" long, 1938	350	550	725
Truck with Electric Lights	15" long, 1930s	110	165	300
Truck with Electric Lights	battery-operated lights, 10" long, 1935	125	200	375
Truck with Searchlight	toolbox behind cab, 10" long, 1930s	150	225	400
U.S. Air Force Willy's Jeep	tin body, plastic figures	70	100	140
U.S. Army Jeep with Trailer		100	150	200
U.S. Mail Truck	metal, 14" long, 1950s	225	350	450
U.S. Trucking Co. Mack Truck	dark maroon cab, friction, 5-1/2" long, 1930	100	150	200
Van Truck	plastic, 10" long, 1950s	40	60	80
Western Auto Truck	steel, 25" long	60	90	125
Willy's Jeep	steel, 12" long, 1938	125	200	250
Willy's Jeep and Trailer	1940s	100	160	215
Wrecker Truck	1930s	145	225	290

Matchbox

NO.	NAME	DESCRIPTION	GOOD	EX	MINT
	"B" Type London Bus	red w/black driver; black wheels; 8 over 4 side windows; Dewar's labels	45	60	85
43-F	0-4-0 Steam Loco	red cab, bin and tanks, black boiler, coal, and base, 3" long, 1978	4	5	7
47-A	1 Ton Trojan Van	red body, no windows, 2-1/4" long, 1958	25	35	45
73-A	10 Ton Pressure Refueller	bluish gray body, six gray plastic wheels, 2-5/8" long, 1959	20	25	35
4-H	1957 Chevy	metallic rose body, chrome interior, large five arch rear wheels, 2-15/16" long, 1979	2	8	12
42-G	1957 Ford T-Bird	red convertible, white interior, silver grille and trunk mounted spare, 1982	3	5	6
30-B	6-Wheel Crane Truck	silver body, orange crane, metal or plastic hook, gray wheels, 2-5/8" long, 1961	20	30	40
30-D	8-Wheel Crane Truck	red body, yellow plastic hook, five spoke thin wheels, 3" long, 1970	5	7	10
30-C	8-Wheel Crane Truck	green body, orange crane, red or yellow hook, eight black wheels on four axles, 3" long, 1965	15	20	25
51-D	8-Wheel Tipper	yellow cab, silver/gray tipper, blue windows, five spoke thin wheels, 3" long, 1970	4	6	8
51-C	8-Wheel Tipper	blue tinted windows, eight black plastic wheels, 3" long, 1969	7	10	15
65-E	Airport Coach	white top and roof, metallic blue bottom, amber windows, yellow interior, comes with varying airline logo decals, 3" long, 1977	5	10	15
51-A	Albion Chieftan	yellow body, tan and light tan bags, small round decal on doors, 2-1/2" long, 1958	10	20	25
75-D	Alfa Carabo	pink body, ivory interior, black trunk, five spoke wide wheels, 3" long, 1971	3	4	5
61-B	Alvis Stalwart	white body, yellow plastic removable canopy, green windows, six plastic wheels, 2-5/8" long, 1966	20	30	50
69-E	Armored Truck	red body, white plastic roof, silver/gray base and grille, "Wells Fargo" on sides, 2-13/16" long, 1978	3	5	8
30-G	Articulated Truck	blue cab, white grille, silver/gray dumper, five spoke accent wheels, 3" long, 1981	3	5	6
50-E	Articulated Truck	purple tinted windows, small wide wheels, five spoke wide wheels, 2-3/4" long, 1973	6	8	12
53-A	Aston Martin DB2 Saloon	metallic light green, 2-1/2" long, 1958	20	30	40
19-C	Aston Martin Racing Car	metallic green, metal steering wheel and wire wheels, black plastic tires, 2-1/2" long, 1961	10	30	45
15-B	Atlantic Tractor Super	orange body, tow hook, spare wheel behind cab on body, 2-5/8" long, 1959	15	30	40
16-A	Atlantic Trailer	tan body, six metal wheels, tan tow bar, 3-1/8" long, 1956	15	25	35
16-B	Atlantic Trailer	orange body, eight gray plastic wheels with knobby treads, 3-1/4" long, 1957	15	50	80
32-G	Atlas Extractor	red/orange body, gray platform, turret and treads, black wheels, 3" long 1981	2	4	6
23-E	Atlas Truck	metallic blue cab, silver interior, orange dumper, red and yellow labels on doors, 3" long, 1975	6	8	10
23-G	Audi Quattro	white body, red and black print sides, clear windows, "Audi Sport" on doors, 1982	2	3	5
71-A	Austin 200 Gallon Water Truck	olive green body, four black plastic wheels, 2-3/8" long, 1959	10	25	30
36-A	Austin A50	silver grille, with or without silver rear bumper, no windows, 2-3/8" long, 1957	15	20	30
29-B	Austin A55 Cambridge Sedan	two-tone green, light green roof and rear top half of body, dark metallic green hood and lower body, 2-3/4" long, 1961	10	20	25

Matchbox

NO.	NAME	DESCRIPTION	GOOD	EX	MINT
68-A	Austin MK II Radio Truck	olive green body, four black plastic wheels, 2-3/8" long, 1959	20	25	30
17-B	Austin Taxi Cab	maroon body, 2-1/4" long, 1960	35	50	70
1-D	Aveling Barford Road Roller	green body, canopy, tow hook, 2-5/8" long, 1962	7	10	15
43-B	Aveling Barford Tractor Shovel	yellow body, yellow or red driver, four large plastic wheels, 2-5/8" long, 1962	10	15	25
16-E	Badger Exploration Truck	metallic red body, silver grille, 2-1/4" long, 1974	3	4	6
13-F	Baja Dune Buggy	metallic green, orange interior, silver motor, 2-5/8" long, 1971	3	4	6
58-A	BEA Coach	blue body, four wheels with small knobby treads, 2-1/2" long, 1958	20	25	35
30-E	Beach Buggy	pink, yellow paint splatters, clear windows, 2-1/2" long, 1970	3	4	6
47-E	Beach Hopper	dark metallic blue body, hot pink splattered over body, bright orange interior, tan driver, 2-5/8" long, 1974	3	4	5
14-C	Bedford Ambulance	white body, silver trim, two rear doors open, 2-5/8" long, 1962	15	60	80
28-A	Bedford Compressor Truck	silver front and rear grilles, metal wheels, 1-3/4" long, 1956	15	35	45
42-A	Bedford Evening News Van	yellow/orange body, silver grille, 2-1/4" long, 1957	15	35	45
27-A	Bedford Low Loader	light blue cab, dark blue trailer, silver grille and side gas tanks, four metal wheels on cab, two metal wheels on trailer, 3-1/8" long, 1956	100	350	450
27-B	Bedford Low Loader	green cab, tan trailer, silver grille, four gray wheels on cab, two wheels on trailer, 3-3/4" long, 1959	20	55	75
29-A	Bedford Milk Delivery Van	tan body, white bottle load, 2-1/4" long, 1956	15	25	30
17-A	Bedford Removals Van	maroon body, peaked roof, gold grille, 2-1/8" long, 1956	35	115	150
40-A	Bedford Tipper Truck	red cab, silver grille, two front wheels, four dual rear wheels, 2-1/8" long, 1957	25	40	50
3-B	Bedford Ton Tipper	gray cab, gray wheels, dual rear wheels, 2-1/2" long, 1961	10	15	20
13-A	Bedford Wreck Truck	tan body, red metal crane and hook, 2" long, 1955	30	40	55
13-B	Bedford Wreck Truck	tan body, red metal crane and hook, crane attached to rear axle, 2-1/8" long, 1958	30	45	60
23-A	Berkeley Cavalier Trailer	decal on lower right rear of trailer, metal wheels, flat tow hook, 2-1/2" long, 1956	10	30	40
26-E	Big Banger	red body, blue windows, small front wheels, large rear wheels, 3" long, 1972	4	6	8
12-F	Big Bull	orange body, green plow blade, base and sides, chrome seat and engine, orange rollers, 2-1/2" long, 1975	4	6	8
22-F	Blaze Buster	red body, silver interior, yellow label, five spoke slicks, 3" long, 1975	4	6	8
61-C	Blue Shark	metallic dark blue, white driver, clear glass, four spoke wide wheels, 3" long, 1971	3	4	6
23-B	Bluebird Dauphine Trailer	green with gray plastic wheels, decal on lower right rear of trailer, door on left rear side opens, 2-1/2" long, 1960	45	135	200
56-C	BMC 1800 Pininfarina	clear windows, ivory interior, five spoke wheels, 2-3/4" long, 1970	7	10	15
45-D	BMW 3.0 CSL	orange body, yellow interior, 2-7/8" long, 1976	3	5	7

Matchbox

NO.	NAME	DESCRIPTION	GOOD	EX	MINT
53-B	BMW M1	silver gray metallic body with plastic hood, red interior, black stripes and "52" on sides, 2-15/16" long, 1981	2	4	5
9-D	Boat and Trailer	white hull, blue deck, clear windows, five spoke wheels on trailer, 3-1/4" long, 1970	3	5	20
9-C	Boat and Trailer	white hull, blue deck, clear windows, five spoke wheels on trailer, 3-1/4" long, 1966	4	6	15
72-E	Bomag Road Roller	yellow body, base and wheel hubs, black plastic roller, 2-15/16" long, 1979	2	4	6
44-E	Boss Mustang	yellow body, amber windows, silver interior, clover leaf wide wheels, 2-7/8" long, 1972	3	4	6
25-C	BP Petrol Tanker	yellow hinged cab, white tanker body, six black plastic wheels, 3" long, 1964	8	15	20
52-B	BRM Racing Car	blue or red body, white plastic driver, yellow wheels, 2-5/8" long, 1965	10	20	25
54-C	Cadillac Ambulance	white body, blue tinted windows, white interior, red cross labels on sides, 2-3/8" long, 1970	5	7	20
27-C	Cadillac Sedan	with or without silver grille, clear windows, white roof, silver wheels, 2-3/4" long, 1960	15	30	40
38-G	Camper	red body, off white camper, unpainted base, 3" long 1980	3	5	7
16-D	Case Tractor Bulldozer	red body, yellow base, motor and blade, black plastic rollers, 2-1/2" long, 1969	5	9	12
18-C	Caterpillar Bulldozer	yellow body with driver, green rubber treads, 2-1/4" long, 1961	10	15	25
18-B	Caterpillar Bulldozer	yellow body and driver, yellow blade, No. 18 cast on back of blade, metal rollers, 2" long, 1958	30	45	65
18-D	Caterpillar Crawler Bulldozer	yellow body, no driver, green rubber treads, 2-3/8" long, 1964	30	45	60
18-A	Caterpillar D8 Bulldozer	yellow body and driver, red blade and side supports, 1-7/8" long, 1956	25	35	45
8-A	Caterpillar Tractor	driver has same color hat as body, metal rollers, rubber treads, crimped axles, 1-1/2" long, 1955	40	60	80
8-D	Caterpillar Tractor	yellow body, no driver, plastic rollers, rubber treads, rounded axles, 2" long, 1964	10	15	20
8-C	Caterpillar Tractor	yellow body and driver, large smoke stack, metal rollers, rubber treads, rounded axles, 1-7/8" long, 1961	10	15	25
8-B	Caterpillar Tractor	yellow body and driver, large smoke stack, metal rollers, rubber treads, crimped axles, 1-5/8" long, 1959	20	35	45
37-E	Cattle Truck	gray plastic box, white plastic cattle inside, five spoke thin wheels, green tinted windows, 2-1/2" long, 1970	5	7	20
37-D	Cattle Truck	yellow body, gray plastic box with fold down rear door, black plastic wheels, green tinted windows, 2-1/4" long, 1966	7	10	15
71-F	Cattle Truck	metallic brown body, yellow/orange cattle carrier, 3" long, 1976	4	6	7
25-H	Celica GT	blue body, black base, white racing stripes and "78" on roof and doors, 2-15/16" long, 1978	3	7	12
25-J	Celica GT	yellow body, blue interior, red "Yellow Fever" on hood, side racing stripes, clear windows, large rear wheels, 1982	2	4	6
3-A	Cement Mixer	blue body and rotating barrel, orange metal wheels, 1-5/8" long, 1953	25	35	45
19-G	Cement Truck	red body, yellow plastic barrel with red stripes, large wide arch wheels, 3" long, 1976	4	5	7
41-F	Chevrolet Ambulance	white body, blue windows and dome light, gray interior, 2-15/16" long, 1978	4	7	10

NO.	NAME	DESCRIPTION	GOOD	EX	MINT
57-B	Chevrolet Impala	pale blue roof, metallic blue body, green tinted windows, 2-3/4" long, 1961	20	25	35
20-C	Chevrolet Impala Taxi Cab	orange/yellow or bright yellow body, ivory or red interior and driver, 3" long, 1965	10	15	20
44-G	Chevy 4x4 Van	green body and windows, white "Ridin High" with horse and fence on sides, 1982	2	3	5
34-G	Chevy Pro Stocker	white body, red interior, clear front and side windows, frosted rear window, 3" long, 1981	2	3	5
68-E	Chevy Van	orange body, unpainted base and grille, large rear wheels, 3" long, 1979	3	5	7
49-D	Chop Suey Motorcycle	metallic dark red body, yellow bull's head on front handle bars, 2-3/4" long, 1973	5	7	10
12-G	Citroen CX	metallic body, silver base and lights, blue plastic hatch door, 3" long, 1979	4	8	12
66-A	Citroen DS 19	light or dark yellow body, with or without silver grille, four plastic wheels, 2-1/2" long, 1959	10	30	40
51-E	Citroen SM	clear windows, frosted rear windows, five spoke wheels, 3" long, 1972	3	4	6
65-C	Claas Combine Harvester	red body, yellow plastic rotating blades and front wheels, black plastic front tires, solid rear wheels, 3" long, 1967	7	10	15
39-D	Clipper	metallic dark pink, amber windows, bright yellow interior, 3" long, 1973	3	4	5
11-H	Cobra Mustang	orange body, "The Boss" on doors, 1982	2	3	5
37-C	Coca-Cola Lorry	yellow body of various shades, even case load, silver wheels, black base, 2-1/4" long, 1960	25	55	125
37-B	Coca-Cola Lorry	orange/yellow body, uneven case load, open base, metal rear fenders, 2-1/4" long, 1957	25	75	140
51-F	Combine Harvester	red body, black painted base, yellow plastic grain chute, 2-3/4" long, 1978	3	5	8
69-A	Commer 30 CWT Van	silver grille, sliding left side door, four plastic wheels, yellow "Nestle's" decal on upper rear panel, 2-1/4" long, 1959	10	30	40
47-B	Commer Ice Cream Canteen	metallic blue body, cream or white plastic interior with man holding ice cream cone, black plastic wheels, 1963	60	95	100
21-C	Commer Milk Truck	pale green body, clear or green tinted windows, ivory or cream bottle load, 2-1/4" long, 1961	20	30	40
50-A	Commer Pickup Truck	with or without silver grille and bumpers, four wheels, 2-1/2" long, 1958	35	55	75
62-G	Corvette	metallic red body, unpainted base, gray interior, 1979	2	4	6
40-F	Corvette T Roof	white body and interior, black "09" on door, red and black racing stripes, 1982	2	3	5
26-G	Cosmic Blues	white body, blue "COSMIC BLUES" and stars on sides, 2-7/8" long, 1970	2	3	4
74-E	Cougar Village	metallic green body, yellow interior, unpainted base, 3-1/16" long, 1978	3	4	6
41-A	D-Type Jaguar	dark green body, tan driver, open air scoop, 2-13/16" long, 1957	15	30	55
41-B	D-Type Jaguar	dark green body, tan driver, silver wheels, open and closed air scoop, 2-7/16" long, 1960	50	100	150
58-C	D.A.F. Girder Truck	cream body shades, green tinted windows, six black wheels, red plastic girders, 3" long, 1968	7	10	15
58-D	D.A.F. Girder Truck	green windows, five spoke thin wheels, red plastic girders, 2-7/8" long, 1970	5	7	20
47-D	DAF Tipper Container Truck	silver cab, yellow tipper box, five spoke thin wheels, 3" long, 1970	4	6	20
47-C	DAF Tipper Container Truck	aqua or silver cab, yellow tipper box with light gray or dark gray plastic roof, 3" long, 1968	6	8	12

Coca-Cola Lorry 37-C, Matchbox

Jaguar XKE 32-B, 1962, Matchbox

NO.	NAME	DESCRIPTION	GOOD	EX	MINT
14-B	Daimler Ambulance	silver trim, "AMBULANCE" cast on sides, red cross on roof, 2-5/8" long, 1958	40	65	85
14-A	Daimler Ambulance	cream body, silver trim, no number cast on body, "AMBULANCE" cast on sides, 1-7/8" long, 1956	20	35	45
74-C	Daimler Bus	double deck, white plastic interior, five spoke thin wheels, 3" long, 1970	7	10	20
74-B	Daimler Bus	double deck, white plastic interior, four black plastic wheels, 3" long, 1966	10	15	20
67-E	Datsun 260Z 2+2	metallic burgundy body, black base and grille, yellow interior, 3" long, 1978	3	4	6
24-G	Datsun 280ZX	black body and base, clear windows, five spoke wheels, 2-7/8" long, 1979	2	3	5
33-E	Datsun or 126X	yellow body, amber windows, silver interior, 3" long, 1973	5	8	10
9-A	Dennis Fire Escape Engine	red body, metal wheels, no front bumper, 2-1/4" long, 1955	10	20	30
20-F	Desert Dawg Jeep 4x4	white body, red top and stripes, white "Jeep" and yellow, red and green, "Desert Dawg" decal, 1982	2	3	5
1-A	Diesel Road Roller	dark green body, flat canopy, tow hook, driver, 1-7/8" long, 1953	15	25	35
63-G	Dodge Challenger	green body, black base, bumpers and grille, clear windows, 2-7/8" long, 1980	3	5	7
1-H	Dodge Challenger	red body, white plastic top, silver interior, wide five spoke wheels, 2-15/16" long, 1976	3	5	8
52-C	Dodge Charger	clear windows, black interior, five spoke wide wheels, 2-7/8" long, 1970	4	6	8
63-C	Dodge Crane Truck	yellow body, green windows, six black plastic wheels, rotating crane cab, 3"long, 1968	7	10	15
63-D	Dodge Crane Truck	yellow body, green windows, four spoke wide wheels, yellow plastic hook, 2-3/4" long, 1970	5	7	20
70-D	Dodge Dragster	pink body, clear windows, silver interior, five spoke wide front wheels, 3" long, 1971	7	10	15
13-D	Dodge Wreck Truck	green cab and crane, yellow body, green windows, 3" long, 1965	300	500	700
13-E	Dodge Wreck Truck	yellow cab, rear body, red plastic hook, green windows, 3" long, 1970	10	15	40
43-E	Dragon Wheels Volkswagen	light green body, amber windows, silver interior, orange on black "Dragon Wheels" on sides, large rear wheels, 2-13/16" long, 1972	5	7	9
58-B	Drott Excavator	red or orange body, movable front shovel, green rubber treads, 2-5/8" long, 1962	35	50	70
2-A	Dumper	green body, red dumper, gold trim, thin driver, green painted wheels, 1-5/8" long, 1953	35	50	70
2-B	Dumper	green body, red dumper, no trim color, fat driver, 1-7/8" long, 1957	20	30	40
48-C	Dumper Truck	red body, green tinted windows, 3" long, 1966	10	20	25
48-D	Dumper Truck	bright blue cab, yellow body, green windows, 3" long, 1970	5	7	25
25-A	Dunlop Truck	dark blue body, silver grille, 2-1/8" long, 1956	10	20	25
57-E	Eccles Trailer Caravan	orange roof, green plastic interior, five spoke thin wheels, 3" long, 1970	5	7	10
20-B	ERF 686 Truck	dark blue body, silver radiator, eight plastic silver wheels, No. 20 cast on black base, 2-5/8" long, 1959	25	45	60
6-B	Euclid Quarry Truck	yellow body, four ribs on dumper sides, plastic wheels, 2-1/2" long, 1957	10	20	25
6-C	Euclid Quarry Truck	yellow body, three round axles, two front black plastic wheels, two solid rear dual wheels, 2-5/8" long, 1964	15	25	30

NO.	NAME	DESCRIPTION	GOOD	EX	MINT
35-D	Fandango	white body, red interior, chrome rear engine, large five spoke rear wheels, 3" long, 1975	3	5	8
58-F	Faun Dump Truck	yellow cab and dumper, black base, 2-7/8" long, 1976	5	10	15
70-F	Ferrari 308 GTB	red body and base, black plastic interior, side stripe, 2-15/16" long, 1981	2	3	5
75-B	Ferrari Berlinetta	metallic green body of various shades, ivory interior and tow hook, four wire or silver plastic wheels, 3" long, 1965	10	20	25
75-C	Ferrari Berlinetta	ivory interior, five spoke thin wheels, 2-3/4" long, 1970	5	7	25
73-B	Ferrari F1 Racing Car	light and dark red body, plastic driver, white and yellow "73" decal on sides, 2-5/8" long, 1962	15	25	40
61-A	Ferret Scout Car	olive green, tan driver faces front or back, four black plastic wheels, 2-1/4" long, 1959	10	20	25
56-B	Fiat 1500	silver grille, red interior and tow hook, brown or tan luggage on roof, 2-1/2" long, 1965	10	15	20
9-G	Fiat Abarth	white body, red interior, 1982	2	3	5
18-F	Field Car	yellow body, tan roof, red wheels, ivory interior and tow hook, 2-5/8" long, 1970	5	7	20
18-E	Field Car	yellow body, tan plastic roof, ivory interior and tow hook, green plastic tires, 2-5/8" long, 1969	50	145	200
29-C	Fire Pumper Truck	red body, metal grille, white plastic hose and ladders, 3" long, 1966	3	7	20
29-D	Fire Pumper Truck	red body, metal grille, white plastic hose and ladders, 3" long, 1970	2	5	20
53-G	Flareside Pick-up	blue body, white interior, grille and pipes, clear windshield, lettered with "326", "Baja Bouncer" and "B.F. Goodrich", 1982	2	4	5
11-F	Flying Bug	metallic red, gray windows, small five spoke front wheels, large five spoke rear wheels, 2-7/8" long, 1972	5	7	10
63-B	Foamite Fire Fighting Crash Tender	red body, six black plastic wheels, white plastic hose and ladder on roof, 2-1/4" long, 1964	10	15	40
21-D	Foden Concrete Truck	orange/yellow body and rotating barrel, green tinted windows, eight plastic wheels, 3" long, 1968	3	5	15
21-E	Foden Concrete Truck	red body, orange barrel, green base and windows five spoke wheels, 2-7/8" long, 1970	2	10	20
26-A	Foden Ready Mix Concrete Truck	orange body and rotating barrel, silver or gold grille, four silver plastic wheels, 1-3/4" long, 1956	50	100	140
26-B	Foden Ready Mix Concrete Truck	orange body, gray plastic rotating barrel, with or without silver grille, six gray wheels, 2-1/4" long, 1961	60	130	180
7-B	Ford Anglia	blue body, green tinted windows, 2-5/8" long, 1961	10	15	30
54-D	Ford Capri	ivory interior and tow hook, clear windows, five spoke wide wheels, 3" long, 1971	3	4	6
45-B	Ford Corsair with Boat	pale yellow body, red interior and tow hook, green roof rack with green plastic boat, 2-3/8" long, 1965	10	15	20
55-I	Ford Cortina	metallic tan body, yellow interior, blue racing stripes, 1982	2	4	6
55-H	Ford Cortina	metallic gold/green body, unpainted base and grille, wide multispoke wheels, 3-1/16" long, 1979	3	4	5
25-E	Ford Cortina	clear windows, ivory interior and tow hook, thin five spoke wheels, 2-3/4" long, 1970	3	6	10
25-D	Ford Cortina G.T.	light brown body in various shade, ivory interior and tow hook, 2-7/8" long, 1968	4	7	10

NO.	NAME	DESCRIPTION	GOOD	EX	MINT
31-A	Ford Customline Station Wagon	yellow body, no windows, with or without red painted tail lights, 2-5/8" long, 1957	20	30	40
9-F	Ford Escort RS2000	white body, black base and grille, tan interior, wide multispoke wheels, 3" long, 1978	3	5	7
59-B	Ford Fairlane Fire Chief's Car	red body, ivory interior, clear windows, four plastic wheels, 2-5/8" long, 1963	35	75	125
55-B	Ford Fairlane Police Car	silver grille, ivory interior, clear windows, four plastic wheels, 2-5/8" long, 1963	20	55	100
31-B	Ford Fairlane Station Wagon	green or clear windows, with or without red painted tail lights, 2-3/4" long, 1960	10	30	35
59-C	Ford Galaxie Fire Chief's Car	red body, ivory interior, driver and tow hook, clear windows, four black plastic wheels, 2-7/8" long, 1966	5	10	40
59-D	Ford Galaxie Fire Chief's Car	red body, ivory interior and tow hook, clear windows, four spoke thin wheels, 2-7/8" long, 1970	3	7	30
55-C	Ford Galaxie Police Car	white body, ivory interior, driver and tow hook, clear windows, 2-7/8" long, 1966	10	15	30
45-C	Ford Group 6	metallic green body, ivory interior, clear windows, five spoke wide wheels, 3" long, 1970	5	7	10
41-C	Ford GT	white or yellow body, red interior, clear windows, yellow or red plastic wheels, 2-5/8" long, 1965	20	30	40
41-D	Ford GT	white body, red interior, clear windows, five spoke wheels, 2-5/8" long, 1970	5	7	20
71-D	Ford Heavy Wreck Truck	red cab, white body, green windows and dome light, four spoke wide wheels, 3" long, 1970	15	25	30
71-C	Ford Heavy Wreck Truck	red cab, white bumper, amber or green windows, 3" long, 1968	25	75	100
8-F	Ford Mustang	wide five spoke wheels, interior and tow hook same color, 2-7/8" long, 1970	10	25	50
8-E	Ford Mustang Fastback	white body, red interior, clear windows, 2-7/8" long, 1966	7	10	30
6-E	Ford Pick-up	red body, white removable canopy, five spoke wheels, 2-3/4" long, 1970	5	7	15
6-D	Ford Pick-up	red body, white removable plastic canopy, four black plastic wheels, 2-3/4" long, 1968	10	15	20
30-A	Ford Prefect	blue body, metal wheels, silver grille, black tow hook, 2-1/4" long, 1956	20	60	90
7-D	Ford Refuse Truck	gray plastic body, silver metal dumper, 3" long, 1970	6	8	15
7-C	Ford Refuse Truck	orange cab, gray plastic dumper, silver metal loader, 3" long, 1966	7	10	15
63-A	Ford Service Ambulance	olive green body, four plastic wheels, round white circle on sides with red cross, 2-1/2" long, 1959	15	20	45
70-A	Ford Thames Estate Car	yellow upper, bluish/green lower, four plastic wheels, 2-1/8" long, 1959	10	15	30
59-A	Ford Thames Van	silver grille, four plastic knobby wheels, 2-1/8" long, 1958	35	75	110
75-A	Ford Thunderbird	cream top half, pink bottom half, green tinted windows, 2-5/8" long, 1960	25	35	45
39-C	Ford Tractor	blue body, black plastic steering wheel and tires, with or without yellow hood, 2-1/8" long, 1967	5	10	20
46-F	Ford Tractor	blue body, black base, large black plastic rear wheels, 2 3/16" long, 1987	3	5	8
66-F	Ford Transit	orange body, unpainted base, yellow interior, green windows, 2-3/4" long, 1977	2	3	4
61-D	Ford Wreck Truck	red body, black base and grille, frosted amber windows, 3" long, 1978	3	5	7
33-B	Ford Zephyr 6MKIII	blue/green body shades, clear windows, ivory interior, 2-5/8" long, 1963	15	20	30

NO.	NAME	DESCRIPTION	GOOD	EX	MINT
53-D	Ford Zodiac	clear windows, ivory interior, five spoke wheels, 2-3/4" long, 1970	5	7	15
39-A	Ford Zodiac Convertible	peach/pink body shades, tan driver, metal wheels, silver grille, 2-5/8" long, 1957	35	65	90
53-C	Ford Zodiac MK IV	metallic silver blue body, clear windows, ivory interior, four black plastic wheels, 2-3/4" long, 1968	7	10	15
33-A	Ford Zodiac MKII Sedan	with or without silver grille, with or without red painted tail lights, 2-5/8" long, 1957	20	30	45
72-A	Fordson Tractor	blue body with tow hook, 2" long, 1959	10	20	25
15-F	Fork Lift Truck	red body, yellow hoist, 2-1/2" long, 1972	3	4	6
34-E	Formula 1 Racing Car	metallic pink, white driver, clear glass, wide four spoke wheels, 2-7/8" long, 1971	7	10	15
36-F	Formula 5000	orange body, silver rear engine, large clover leaf rear slicks, 3" long, 1975	5	7	10
28-H	Formula Racing Car	gold body, silver engine and pipes, white driver and "Champion", black "8" on front and sides, large clover leaf rear wheels, 1982	2	3	5
22-E	Freeman Inter-City Commuter	clear windows, ivory interior, five spoke wide wheels, 3" long, 1970	5	7	10
63-E	Freeway Gas Truck	red cab, purple tinted windows, small wide wheels on front, clover leaf design, 3" long, 1973	10	15	20
62-A	General Service Lorry	olive green body, six black wheels, 2-5/8" long, 1959	15	25	30
44-D	GMC Refrigerator Truck	green windows, four spoke wheels, gray plastic rear door, 2-13/16" long, 1970	5	7	15
44-D	GMC Refrigerator Truck	red ribbed roof cab, turquoise box with gray plastic rear door that opens, green windows, 1967	7	10	20
26-D	GMC Tipper Truck	red cab, silver/gray tipper body, four spoke wide wheels, 2-1/2" long, 1970	10	15	20
26-C	GMC Tipper Truck	red tipping cab, silver tipper body with swinging door, four wheels, 2-5/8" long, 1968	5	7	20
66-C	Greyhound Bus	silver body, white plastic interior, clear or dark amber windows, six black plastic wheels, 3" long, 1967	25	35	45
66-D	Greyhound Bus	silver body, white interior, amber windows, five spoke thin wheels, 3" long, 1970	5	10	20
70-B	Grit Spreader Truck	dark red cab, four black plastic wheels, 2-5/8" long, 1966	5	7	15
70-C	Grit Spreader Truck	red cab, yellow body, green windows, gray plastic rear pull, 2-5/8" long, 1970	6	8	15
4-F	Gruesome Twosome	metallic gold body, wide five spoke wheels, 2-7/8" long, 1971	5	7	10
23-F	GT 350	white body, blue stripes on hood, roof and rear deck, 2-7/8" long, 1970	3	4	5
7-E	Hairy Hustler	metallic bronze, silver interior, five spoke front wheels, clover leaf rear wheels, 2-7/8" long, 1971	5	7	10
50-G	Harley-Davidson Motorcycle	silver/brown metallic frame and tank, chrome engine and pipes, brown rider, 2-11/16" long, 1980	2	3	7
66-B	Harley-Davidson Motorcycle/ Sidecar	metallic bronze body, three wire wheels, 2-5/8" long, 1962	20	45	75
69-B	Hatra Tractor Shovel	orange or yellow movable shovel arms, four plastic tires, 3" long, 1965	20	30	40
40-C	Hay Trailer	blue body with tow bar, yellow plastic racks, yellow plastic wheels, 3-3/4" long, 1967	5	10	15

Saracen Personnel Carrier, Matchbox

Rolls-Royce Silver Cloud, 1958, Matchbox

Fire Chief's Car, 1960s, Matchbox

Matchbox

NO.	NAME	DESCRIPTION	GOOD	EX	MINT
55-G	Hellraiser	white body, unpainted base and grille, silver rear engine, 3" long, 1975	3	5	7
15-G	Hi Ho Silver	metallic pearl gray body, 2-1/2" long, 1971	7	10	15
56-D	Hi-Tailer	white body, silver engine and windshield, wide five spoke front wheels, wide clover leaf rear wheels, 3" long, 1974	4	6	8
43-A	Hillman Minx	with or without silver grille, with or without red painted tail lights, 2-5/8" long, 1958	15	20	30
38-D	Honda Motorcycle and Trailer	yellow trailer with five spoke thin wheels, 2-7/8" long, 1970	5	7	20
38-C	Honda Motorcycle and Trailer	metallic blue/green cycle with wire wheels, black plastic tires, orange trailer, 2-7/8" long, 1967	10	15	20
18-G	Hondarora Motorcycle	red frame and fenders chrome bars, fork, engine, black seat, 2-3/8" long, 1975	5	15	25
40-E	Horse Box	orange cab, off white van with tan plastic door, small wheels, 2-13/16" long, 1977	4	5	7
17-E	Horse Box	blue tinted windows, five spoke thin wheels, white plastic horses inside box, 2-3/4" long, 1970	5	10	15
17-D	Horse Box, Ergomatic Cab	red cab, green plastic box, gray side door, 1969	10	15	20
7-A	Horse Drawn Milk Float	orange body, white driver and bottle load, brown horse with white mane and hoofs, 2-1/4" long, 1954	25	55	90
46-G	Hot Chocolate	metallic brown front lid and sides, black roof, 2-13/16" long, 1972	3	4	5
67-d	Hot Rocker	metallic lime/green body, white interior and tow hook, five spoke wide wheels, 3" long, 1973	3	5	7
36-E	Hot Rod Draguar	metallic red body, clear canopy, wide five spoke wheels, 2-13/16" long, 1970	4	6	8
72-D	Hovercraft	white body, black bottom and base, red props, 3" long, 1972	4	7	10
2-G	Hovercraft	metallic green top, tan base, silver engine, yellow windows,3-1/8" long, 1976	4	8	12
17-C	Hoveringham Tipper	red body, orange dumper, 2-7/8" long, 1963	7	10	25
42-D	Iron Fairy Crane	four spoke wheels, yellow plastic hook, 3" long, 1970	25	35	50
42-C	Iron Fairy Crane	red body, yellow/orange crane, black plastic wheels, yellow plastic single cable hook, 3" long, 1969	7	10	30
14-D	Iso Grifo	blue body, light blue interior and tow hook, clear windows, 3" long, 1968	5	7	15
14-E	Iso Grifo	five spoke wheels, clear windows, 3" long, 1969	5	10	15
65-A	Jaguar 3.4 Litre Saloon	silver grille, silver or black bumpers, four gray plastic wheels, 2-1/2" long, 1959	7	10	15
65-B	Jaguar 3.8 Litre Sedan	red body shades, green tinted windows, four plastic wheels, 2-5/8" long, 1962	5	7	30
28-C	Jaguar Mark 10	light brown body, off white interior, working hood, gray motor and wheels, 2-3/4" long, 1964	35	65	90
32-A	Jaguar XK 140 Coupe	with or without silver grille, with or without red painted tail lights, 2-3/8" long, 1957	20	30	40
32-B	Jaguar XKE	metallic red body, ivory interior, clear or tinted windows, 2-5/8" long, 1962	15	20	35
38-F	Jeep	olive green body, black base and interior, wide five spoke reverse accent wheels, no hubs, 2-3/8" long, 1976	5	8	12
5-H	Jeep 4x4 Golden Eagle	brown body, wide four spoke wheels, eagle decal on hood, 1982	2	5	8
72-B	Jeep CJ5	yellow body, red plastic interior/tow hook, four yellow wheels, black plastic tires, 2-3/8", 1966	10	15	20

NO.	NAME	DESCRIPTION	GOOD	EX	MINT
72-C	Jeep CJ5	red interior and tow hook, eight spoke wheels 2-3/8" long, 1970	5	7	10
53-F	Jeep CJ6	red body, unpainted base, bumper and winch, five spoke rear accent wheels, 2-15/16" long, 1977	2	3	5
71-B	Jeep Gladiator Pickup Truck	red body, clear windows, green or white interior, four black plastic wheels, fine treads, 2-5/8" long, 1964	18	25	30
2-F	Jeep Hot Rod	cream seats and tow hook, large wide four spoke wheels, 2-5/16" long, 1971	7	10	15
50-B	John Deere Tractor	green body and tow hook, yellow plastic wheels, 2-1/8" long, 1964	10	20	30
51-B	John Deere Trailer	green tipping body with tow bar, two small yellow wheels, three plastic barrels, 2-5/8" long, 1964	20	35	50
11-C	Jumbo Crane	yellow body, black plastic wheels, 3" long, 1965	5	10	15
71-E	Jumbo Jet Motorcycle	dark metallic blue body, red elephant head on handle bars, wide wheels, 2-3/4" long, 1973	4	6	8
38-A	Karrier Refuse Collector	silver grille headlights and bumper, 2-3/8" long, 1957	15	25	30
50-D	Kennel Truck	green windows, light blue tinted canopy, four plastic dogs, 2-3/4" long, 1970	5	15	45
50-C	Kennel Truck	metallic green body, clear or blue tinted canopy, four plastic dogs, 2-3/4" long, 1969	5	15	40
45-E	Kenworth Caboner Aerodyne	white body with blue and brown side stripes, silver grille, tanks and pipes, 1982	2	3	5
41-G	Kenworth Conventional Aerodyne	red cab and chassis, silver tanks and pipes, black and white stripes on cab, 1982	2	3	5
27-F	Lamborghini Countach	yellow body, silver interior and motor, five spoke wheels, 2-7/8" long, 1973	5	7	10
20-D	Lamborghini Marzel	amber windows, ivory interior, 2-3/4" long, 1969	15	20	35
33-C	Lamborghini Miura	metal grille, silver plastic wheels, red or white interior, clear or frosted back window, 2-3/4" long, 1969	10	15	20
33-D	Lamborghini Miura	clear windows, frosted rear window, five spoke wheels, 2-3/4" long, 1970	25	40	50
36-B	Lambretta TV 175 Motor Scooter and Sidecar	metallic green, three wheels, 2" long, 1961	25	35	45
12-A	Land Rover	olive green body, tan driver, metal wheels, 1-3/4" long, 1955	20	25	35
12-B	Land Rover	olive green body, no driver, tow hook, 2-1/4" long, 1959	35	55	75
57-C	Land Rover Fire Truck	red body, blue tinted windows, white plastic ladder on roof, 2-1/2" long, 1966	10	15	25
57-D	Land Rover Fire Truck	red body, blue tinted windows, white plastic removable ladder, 2-1/2" long, 1970	5	7	20
32-D	Leyland Petrol Tanker	green cab, white tank body, blue tinted windows, five spoke thin wheels, 3" long, 1970	10	20	50
32-C	Leyland Petrol Tanker	green cab, white tank body, blue tinted windows, eight plastic wheels, 3" long, 1968	25	40	50
40-B	Leyland Royal Tiger Coach	silver/gray body, green tinted windows four plastic wheels, 3" long, 1961	10	15	20
31-D	Lincoln Continental	clear windows, ivory interior, five spoke wheels, 2-3/4" long, 1970	5	10	30
31-C	Lincoln Continental	clear windows, ivory interior, black plastic wheels, 2-7/8" long, 1964	10	15	20
28-G	Lincoln Continental MK-V	red body, tan interior, 3" long, 1979	10	15	20

Matchbox

NO.	NAME	DESCRIPTION	GOOD	EX	MINT
5-C	London Bus	red body, silver grille and headlights, 2-9/16" long, 1961	10	20	25
5-A	London Bus	red body, gold grille, metal wheels, 2" long, 1954	20	45	65
5-D	London Bus	red body, white plastic seats, black plastic wheels, 2-3/4" long, 1965	5	10	30
5-B	London Bus	red body, 2-1/4" long, 1957	20	45	65
	London E Class Tramcar	red, cream roof; News of the World decals	25	60	90
56-A	London Trolley Bus	red body, two trolley poles on top of roof, six wheels, 2-5/8" long, 1958	45	70	90
17-F	Londoner Bus	red body, white interior, five spoke wide wheels, 3" long, 1972	10	15	20
21-B	Long Distance Coach	green body, black base, No. 21 cast on baseplate, "London to Glasgow" orange decal on sides, 2-5/8" long, 1958	20	60	45
21-A	Long Distance Coach	light green body, black base, "London to Glasgow" orange decal on sides, 2-1/4" long, 1956	10	20	25
5-E	Lotus Europa	metallic blue body, clear windows, ivory interior and tow hook, 2-7/8" long, 1969	5	7	10
19-E	Lotus Racing Car	metallic purple, white driver, five spoke wide wheels with clover leaf design, 2-3/4" long, 1970	5	10	15
19-D	Lotus Racing Car	white driver, large rear wheels, 2-3/4" long, 1966	10	15	20
60-D	Lotus Super Seven	butterscotch, clear windshield, black interior and trunk, four spoke wide wheels, 2-7/8" long,1971	5	7	10
49-A	M3 Army Personnel Carrier	olive green body, gray rubber treads, 2-1/2" long, 1958	15	25	30
28-D	Mack Dump Truck	orange body, green windows, four large plastic wheels, 2-5/8" long, 1968	4	7	15
28-E	Mack Dump Truck	pea green body, green windows, large ballon wheels with clover leaf design, 2-5/8" long, 1970	2	5	15
35-A	Marshall Horse Box	red cab, brown horse box, silver grille, three rear windows in box, 2" long, 1957	20	25	35
52-A	Maserati 4 Cl. T/1948	red or yellow body, cream or white driver with or without circle on left shoulder, 2-3/8" long, 1958	10	17	25
32-E	Maserati Bora	metallic burgundy, clear windows, bright yellow interior, wide five spoke wheels, 3" long, 1972	5	7	10
4-A	Massey Harris Tractor	red body with rear fenders, tan driver, four spoke metal front wheels, 1954	30	40	55
4-B	Massey Harris Tractor	red body, no fenders, tan driver, solid metal front wheels, hollow inside rear wheels, 1-5/8" long, 1957	25	40	50
72-F	Maxi Taxi	yellow body, black "MAXI TAXI" on roof, five spoke wheels, 3" long, 1973	2	3	4
66-E	Mazda RX 500	orange body, purple windows, silver rear engine, five spoke wide wheels, 3" long, 1971	3	4	5
31-G	Mazda RX-7	white body, black base, burgundy stripe, black "RX-7", 3" long, 1979	3	4	5
31-H	Mazda RX-7	gray body with sunroof, black interior, 1982	15	25	35
10-B	Mechanical Horse and Trailer	red cab, ribbed bed in trailer, metal front wheels on cab, 2-15/16" long, 1958	25	35	45
10-A	Mechanical Horse and Trailer	red cab with three metal wheels, gray trailer with two metal wheels, 2-3/8" long, 1955	15	25	35
53-B	Mercedes-Benz 220 SE	silver grille, clear windows, ivory interior, four wheels, 2-3/4" long, 1963	15	25	30
27-D	Mercedes-Benz 230 SL	unpainted metal grille, red plastic interior and tow hook, black plastic wheels, 3" long, 1966	5	10	20

NO.	NAME	DESCRIPTION	GOOD	EX	MINT
27-E	Mercedes-Benz 230 SL	metal grille, blue tinted windshield, five spoke wheels, 2-7/8" long, 1970	5	10	20
46-D	Mercedes-Benz 300 SE	clear windows, ivory interior, five spoke thin wheels, 2-7/8" long, 1970	5	10	20
46-C	Mercedes-Benz 300 SE	clear windows, ivory interior, black plastic wheels, 2-7/8" long, 1968	5	10	20
6-F	Mercedes-Benz 350 SL	orange body, black plastic convertible top, light yellow interior, 3" long, 1973	5	10	20
56-E	Mercedes-Benz 450 SEL	metallic blue body, unpainted base and grille, 3" long, 1979	3	4	5
3-C	Mercedes-Benz Ambulance	varying body colors, white interior and stretcher, blue windows and dome light, metal grille, black plastic wheels, 2-7/8" long, 1968	5	7	15
3-D	Mercedes-Benz Ambulance	ivory interior, Red Cross label on side doors, 2-7/8" long, 1970	5	10	15
68-B	Mercedes-Benz Coach	white plastic top half, white plastic interior, clear windows, four black plastic wheels, 2-7/8" long, 1965	30	40	55
42-F	Mercedes-Benz Container Truck	red body, black base and grille, removable ivory container with red top and back door, six wheels, 3" long, 1977	4	5	7
1-E	Mercedes-Benz Lorry	pale green body, removable orange plastic canopy, 3" long, 1967	5	7	15
1-F	Mercedes-Benz Lorry	metallic gold, removable orange or yellow canopy, 3" long, 1970	4	6	8
56-F	Mercedes-Benz Taxi	tan plastic interior, unpainted base, clear plastic windows, red "Taxi" sign on roof, 3" long, 1980	3	4	5
2-D	Mercedes-Benz Trailer	pale green body, removable orange canopy, tow hook, black plastic wheels, 3 1/2" long, 1968	5	7	15
2-E	Mercedes-Benz Trailer	metallic gold body, removable canopy, rotating tow bar, 3-1/4" long, 1970	3	5	15
49-B	Mercedes-Benz Umimog	silver grille, four black plastic tires, 2-1/2" long, 1967	10	15	20
62-C	Mercury Cougar	metallic lime green body shades, red plastic interior and tow hook, silver wheels, 3" long, 1968	7	10	15
62-D	Mercury Cougar	red interior and tow hook, five spoke thin wheels, 3" long, 1970	3	5	15
62-E	Mercury Cougar "Rat Rod"	red interior and tow hook, small five spoke front wheels, larger five spoke wide rear windows, 3" long, 1970	4	6	8
59-E	Mercury Fire Chief's Car	red body, ivory interior, two occupants, clear windows, five spoke wide wheels, 3" long, 1971	5	7	20
55-E	Mercury Police Car	white body, ivory interior, two occupants, five spoke thin wheels, 3" long, 1970	5	7	15
55-D	Mercury Police Car	white body, ivory interior with two figures, clear windows, four silver wheels with black plastic tires, 3" long, 1968	10	15	40
55-F	Mercury Police Station Wagon	white body, ivory interior, no occupants, five spoke wide wheels, 3" long, 1971	5	7	10
73-D	Mercury Station Wagon	red body, ribbed rear roof, ivory interior with two dogs, 3" long, 1970	3	5	10
73-E	Mercury Station Wagon	red body, ribbed rear roof, ivory interior with two dogs, 3" long, 1972	3	5	7
73-C	Mercury Station Wagon	metallic lime green body shades, ivory interior with dogs in rear, 3-1/8" long, 1968	7	10	15
35-C	Merryweather Fire Engine	metallic red body, blue windows, white removable ladder on roof, five spoke thin wheels, 3" long, 1969	5	7	10
48-A	Meteor Sports Boat and Trailer	metal boat with tan deck and blue hull, black metal trailer with tow bar, 2-3/8" long, 1958	30	45	60

Combine Harvester, Matchbox

Commer Milk Truck, 1961, Matchbox

Matchbox

NO.	NAME	DESCRIPTION	GOOD	EX	MINT
19-A	MG Midget	white body, tan driver, red seats, spare tire on trunk, 2" long, 1956	30	60	75
19-B	MG Midget	silver or gold grilles, tan driver, 2-1/4" long, 1958	20	45	75
64-B	MG-1100	green body, ivory interior, driver, dog and tow hook, clear windows, four black plastic wheels, 2-5/8" long, 1966	5	7	15
64-C	MG-1100	ivory interior and tow hook, one occupant and dog, clear windows, 2-5/8" long, 1970	7	10	15
19-B	MGA Sports Car	white body variation, silver wheels, tan driver, silver or gold grilles, 2-1/4" long, 1958	50	95	125
51-G	Midnight Magic	black body, silver stripes on hood, five spoke front wheels, clover leaf rear windows, 1972	2	3	4
14-F	Mini Haha	red body, pink driver, silver engine, large spoke rear slicks, 2-3/8" long, 1975	5	9	12
74-A	Mobile Refreshment Canteen	cream, white, or silver body, upper side door opens with interior utensils, "Refreshment" on front side, 2-5/8" long, 1959	15	40	65
1-G	Mod Rod	yellow body, tinted windows, red or black wheels, 2-7/8" long, 1971	10	15	20
25-F	Mod Tractor	metallic purple, orange/yellow seat and tow hook, 2-1/8" long, 1972	10	15	20
73-G	Model A Ford	off white body, black base, green fenders and running boards, 1979	2	3	5
3-E	Monteverdi Hai	dark orange body, blue tinted windows, ivory interior, 2-7/8" long, 1973	3	6	8
60-A	Morris J2 Pickup	blue body, open windshield and side door windows, four plastic wheels, 2-1/4" long, 1958	15	25	30
46-A	Morris Minor 1000	dark green body, metal wheels, no windows, 2" long, 1958	20	30	45
2-C	Muirhill Dumper	red cab, green dumper, black plastic wheels, 2-1/6" long, 1961	10	20	25
54-G	NASA Tracking Vehicle	white body, silver radar screen, red windows, blue "Space Shuttle Command Center", red "NASA" on roof, 1982	2	3	5
36-C	Opel Diplomat	metallic light gold body, white interior and tow hook, clear windows, black plastic wheels, 2-3/5" long, 1966	10	15	20
36-D	Opel Diplomat	ivory interior and tow hook, clear windows, five spoke thin wheels, 2-7/8" long, 1970	5	7	15
74-F	Orange Peel	white body, wide orange and black stripe and black "ORANGE PEEL" on each side, 3" long, 1971	3	4	5
47-F	Pannier Tank Loco	green body, black base and insert, six large plastic wheels, 3" long, 1979	3	5	7
8-H	Pantera	white body, blue base, red/brown interior, five spoke rear slicks, 3" long, 1975	35	45	60
54-E	Personnel Carrier	olive green body, green windows, black base and grille, tan men and benches, 3" long, 1976	4	5	7
43-G	Perterbilt Conventional	black cab and chassis, silver grille, fenders and tanks, red and white side stripes, six wheels, 3" long, 1982	2	3	5
19-H	Peterbilt Cement Truck	green body, orange barrel, "Big Pete" decal on hood, 1982	2	3	5
30-H	Peterbilt Quarry Truck	yellow body, gray dumper, silver tanks, "Dirty Dumper" on sides, six wheels, 1982	2	4	6
56-G	Peterbilt Tanker	blue cab, white tank with red "Milk's the One", silver tanks, grille, and pipes, 1982	15	25	40
48-E	Pi-Eyed Piper	metallic blue body, amber windows, small front wheels, large rear wheels, 2-7/8" long, 1972	5	7	10
46-B	Pickford Removal Van	green body, with or without silver grilles, 2-5/8" long, 1960	15	30	50

NO.	NAME	DESCRIPTION	GOOD	EX	MINT
10-D	Pipe Truck	red body, gray pipes, "Leyland" or "Ergomatic" on front base, eight black plastic wheels, 2-7/8", 1966	7	10	15
10-E	Pipe Truck	black pipe racks, eight five spoke thin wheels, 2-7/8" long, 1970	4	6	15
60-F	Piston Popper	yellow body, red windows, silver engine, labels top and sides, large rear wheels, 1982	2	3	5
10-F	Piston Popper	metallic blue body, white interior, 2-7/8" long, 1973	4	6	8
59-F	Planet Scout	metallic green top, green bottom and base, silver interior, grille and roof panels, large multispoke rear wheels, 2-3/4" long 1975	4	5	7
10-G	Plymouth Gran Fury Police Car	white body w/black detailing, "Police" on doors, white interior, 3" long, 1979	3	4	5
52-D	Police Launch	white deck, blue hull and men, 3" long, 1976	2	4	6
33-F	Police Motorcyclist	white frame, seat and bags, silver engine and pipes, wire wheels, 2-1/2" long, 1977	5	7	10
20-E	Police Patrol	white body, "Police" on orange side stripe, orange interior, 2-7/8" long, 1975	6	8	12
39-B	Pontiac Convertible	purple body, with or without silver grille, cream or ivory interior, silver wheels, 2-3/4" long, 1962	30	50	75
4-G	Pontiac Firebird	metallic blue body, silver interior, slick tires, 2-7/8" long, 1975	2	7	12
22-D	Pontiac Grand Prix Sports Coupe	light gray interior, clear windows, five spoke thin wheels, 3" long, 1970	2	4	6
22-C	Pontiac Grand Prix Sports Coupe	light gray interior and tow hook, clear windows, four black plastic wheels, 3" long, 1964	6	9	12
16-G	Pontiac Trans Am	white body, red interior, clear windows, blue eagle decal, 1982	2	3	4
35-F	Pontiac Trans Am T Roof	black body, red interior, yellow "Turbo" on doors, yellow eagle on hood, 1982	2	3	5
43-C	Pony Trailer	yellow body, clear windows, gray plastic rear fold-down door, four plastic wheels, 2-5/8" long, 1968	7	10	15
43-D	Pony Trailer	yellow body, clear windows, gray rear door, five spoke thin wheels, 2-5/8" long, 1970	3	5	15
68-C	Porsche 910	amber windows, ivory interior, five spoke wheels, 2-7/8" long, 1970	7	10	15
59-G	Porsche 928	metallic brown body, black base, wide five spoke wheels, 3" long, 1980	3	5	6
3-F	Porsche Turbo	metallic brown body, black base, yellow interior, wide five arch wheels, 3" long, 1978	4	7	10
15-A	Prime Mover	silver trim on grille and tank, tow hook same color as body, 2-1/8" long, 1956	20	35	50
6-A	Quarry Truck	orange cab, gray dumper with six vertical ribs, metal wheels, 2-1/8" long, 1954	15	30	50
29-E	Racing Mini	clear windows, five spoke wide wheels, 2-1/4" long, 1970	5	7	10
44-F	Railway Passenger Car	cream plastic upper and roof, red metal lower, black base, 3-1/16" long, 1978	3	5	7
14-G	Rallye Royal	metallic pearl gray body, black plastic interior, five spoke wide wheels, 2-7/8" long, 1973	3	4	5
48-G	Red Rider	red body, white "Red Rider" and flames on sides, 2-7/8" long, 1972	2	3	4
36-G	Refuse Truck	red metallic body, silver/gray base, orange plastic container, 3" long, 1980	2	3	4
15-C	Refuse Truck	blue body, gray dumper with opening door, 2-1/2" long, 1963	10	15	20
62-F	Renault 17TL	white interior, green tinted windows, green "9" in yellow and black circle, 3" long, 1974	5	7	10

Matchbox

NO.	NAME	DESCRIPTION	GOOD	EX	MINT
21-G	Renault 5TL	yellow body and removable rear hatch, tan interior, silver base and grille, 2 11/16" long, 1978	4	9	15
1-I	Revin' Rebel	orange body, blue top, black interior, large five spoke rear wheels, 1982	2	3	5
19-F	Road Dragster	ivory interior, silver plastic motor, 2-7/8" long, 1970	3	4	6
1-B	Road Roller	pale green body, canopy, tow hook, dark tan or light tan driver, 2-1/4" long, 1953	25	45	65
1-C	Road Roller	light green or dark green body, canopy, metal rollers, tow bar, driver, 2-3/8" long, 1958	20	25	35
21-F	Road Roller	yellow body, red seat, black plastic rollers, 2-5/8" long, 1973	7	10	15
11-B	Road Tanker	red body, gas tanks, "11" on baseplate, black plastic wheels, 2-1/2" long, 1958	20	55	80
11-A	Road Tanker	green body, flat base between cab and body, gold trim on front grille, gas tanks, metal wheels, no number cast, 2" long, 1955	60	265	360
44-B	Rolls-Royce Phantom V	clear windows, ivory interior, black plastic wheels, 2-7/8" long, 1964	15	20	30
44-A	Rolls-Royce Silver Cloud	metallic blue body, no windows, with or without silver grille, 2-5/8" long, 1958	15	20	25
24-C	Rolls-Royce Silver Shadow	metallic red body, ivory interior, clear windows, silver hub caps or solid silver wheels, 3" long, 1967	10	15	20
24-D	Rolls-Royce Silver Shadow	ivory interior, clear windows, five spoke wheels, 3" long, 1970	5	7	20
69-C	Rolls-Royce Silver Shadow Coupe	amber windshield, five spoke wheels, 3" long, 1969	5	7	10
39-E	Rolls-Royce Silver Shadow II	metallic silver gray body, red interior, clear windshield, 3-1/16" long, 1979	3	5	7
7-G	Rompin' Rabbit	white body, red windows, yellow lettered "Rompin Rabbit" on side, 1982	2	3	5
54-B	S & S Cadillac Ambulance	white body, blue tinted windows, white interior, red cross decal on front doors, 2-7/8" long, 1965	10	15	30
65-D	Saab Sonnet	metallic blue body, amber windows, light orange interior and hood, five spoke wide wheels, 2-3/4" long, 1973	5	7	10
12-D	Safari Land Rover	metallic gold, clear windows, tan luggage, five spoke thin wheels, 2-13/16" long, 1970	10	20	30
12-C	Safari Land Rover	clear windows, white plastic interior and tow hook, black plastic wheels, 2-1/3" long, 1965	7	10	20
67-A	Saladin Armoured Car	olive green body, rotating gun turret, six black plastic wheels, 2-1/2" long, 1959	15	20	25
48-F	Sambron Jacklift	yellow body, black base and insert, no window, orange and yellow fork and boom combinations, 3-1/16" long, 1977	4	7	10
54-A	Saracen Personnel Carrier	olive green body, six black plastic wheels, 2-1/4" long, 1958	10	17	25
11-D	Scaffolding Truck	silver body, green tinted windows, black plastic wheels, 2-1/2" long, 1969	4	7	25
11-E	Scaffolding Truck	silver/gray body, green tinted windows, yellow pipes, 2-7/8" long, 1969	5	10	25
64-A	Scammel Breakdown Truck	olive green, double cable hook, six black plastic wheels, 2-1/2" long, 1959	15	25	30
16-C	Scammel Mountaineer Dump Truck/Snow Plow	gray cab, orange dumper body, six plastic wheels, 3" long, 1964	10	20	25
5-F	Seafire Boat	white deck, blue hull, silver engine, red pipes, 2-15/16" long, 1975	6	8	10

Matchbox

NO.	NAME	DESCRIPTION	GOOD	EX	MINT
75-F	Seasprite Helicopter	white body, red base, black blades, 1977	3	5	7
12-E	Setra Coach	clear windows, ivory interior, five spoke thin wheels, 3" long, 1970	5	7	10
29-F	Shovel Nose Tractor	yellow body and base, red plastic shovel, silver engine, 2-7/8" long, 1976	8	15	20
24-F	Shunter	metallic green body, red base, tan instruments, no window, 3" long, 1978	3	5	7
26-F	Site Dumper	yellow body and dumper, black base, 2-5/8" long, 1976	2	3	5
60-C	Site Hut Truck	blue cab, blue windows, five spoke thin wheels, 2-1/2" long, 1970	5	7	15
60-B	Site Hut Truck	blue body, blue windows, four black plastic wheels, 2-1/2" long, 1966	7	10	15
41-E	Siva Spider	metallic red body, cream interior, clear windows, wide five spoke wheels, 3" long, 1972	5	7	10
37-G	Skip Truck	red body, yellow plastic bucket, light amber windows, silver interior, 2-11/16" long, 1976	3	5	7
64-D	Slingshot Dragster	pink body, white driver, five spoke thin front wheels, eight spoke wide rear wheels, 3" long, 1971	7	10	15
13-G	Snorkel Fire Engine	red body, yellow plastic snorkel and fireman, 3" long, 1977	3	5	7
35-B	Snowtrac Tractor	red body, silver painted grille, green windows, white rubber treads, 2-3/8" long, 1964	10	15	20
37-F	Soopa Coopa	metallic blue, amber windows, yellow interior, 2-7/8" long, 1972	3	4	5
48-B	Sports Boat and Trailer	plastic boat, red or white deck, hulls in red, white or cream, gold or silver motors, blue metal two-wheel trailers, boat 2-3/8" long, trailer 2-5/8" long, 1961	35	65	80
4-E	Stake Truck	cab colors vary, 2-7/8" long, 1970	5	7	15
4-D	Stake Truck	yellow cab, green tinted windows, 2-7/8" long, 1967	5	10	30
20-A	Stake Truck	gold trim on front grille and side gas tanks, ribbed bed, metal wheels, 2-3/8" long, 1956	50	75	100
38-E	Stingeroo Cycle	metallic purple body, ivory horse head at rear of seat, five spoke wide rear wheels, 3" long, 1973	4	6	8
46-E	Stretcha Fetcha	white body, blue windows, pale yellow interior, 2" long, 1972	6	8	12
28-F	Stroat Armored Truck	metallic gold body, brown plastic observer coming out of turret, five spoke wide wheels, 2-5/8" long, 1974	8	15	25
42-B	Studebaker Lark Wagonaire	blue body, sliding rear roof panel, white plastic interior and tow hook, 3" long, 1965	10	15	20
10-C	Sugar Container Truck	blue body, eight gray plastic wheels, "Tate & Lyle" decals on sides and rear, 2-5/8" long, 1961	30	55	75
37-H	Sun Burner	black body, red and yellow flames on hood and sides, 3" long, 1972	2	3	4
30-F	Swamp Rat	green deck, yellow plastic hull, tan soldier, black engine and prop, 3" long, 1976	2	4	6
27-G	Swing Wing Jet	red top and fins, white belly and retractable wings, 3" long, 1981	2	3	5
53-E	Tanzara	orange body, silver interior, small front wheels, larger rear wheels, 3" long, 1972	3	4	5
24-E	Team Matchbox	white driver, silver motor, wide clover leaf wheels, 2-7/8" long, 1973	15	20	25
62-B	Television Service Van	cream body, green tinted windows with roof window, four plastic wheels, 2-1/2" long, 1963	25	40	50

NO.	NAME	DESCRIPTION	GOOD	EX	MINT
28-B	Thames Trader Compressor Truck	yellow body, black wheels, 2-3/4" long, 1959	20	25	35
13-C	Thames Wreck Truck	red body, bumper and parking lights, 2-1/2" long, 1961	15	25	30
74-D	Toe Joe	metallic lime green body, yellow interior, five spoke wide wheels, 2-3/4" long, 1972	3	4	6
23-C	Trailer Caravan	yellow or pink body with white roof, blue removable interior, 2-7/8" long, 1965	4	7	15
4-C	Triumph Motorcycle and Sidecar	silver/blue body, wire wheels, 2-1/8" long, 1960	25	40	60
42-E	Tyre Fryer	metallic red body, cream interior, clear windows, wide five spoke wheels, 3" long, 1972	3	4	6
5-G	U.S. Mail Jeep	blue body, white base and bumpers, black plastic seat, white canopy, wide five-arch rear wheels, 2-3/8" long, 1978	5	10	15
34-F	Vantastic	orange body, white base and interior, silver engine, large rear slicks, 2-7/8" long, 1975	4	7	9
22-B	Vauxhall Cresta	with or without silver grille, tow hook, plastic wheels, 2-5/8" long, 1958	20	40	60
40-D	Vauxhall Guildsman	pink body, light green windows, light cream interior and tow hook, wide five spoke wheels, 3" long, 1971	3	4	5
22-A	Vauxhall Sedan	dark red body, cream or off white roof, tow hook, 2-1/2" long, 1956	15	30	45
38-B	Vauxhall Victor Estate Car	yellow body, red or green interior, clear windows, 2-5/8" long, 1963	10	18	25
45-A	Vauxhall Victor Saloon	yellow body, with or without green tinted windows, with or without silver grille, 2-3/8" long, 1958	10	15	25
31-E	Volks Dragon	red body, purple tinted windows, 2-1/2" long, 1971	3	4	5
25-B	Volkswagen 1200 Sedan	silver-blue body, clear or tinted windows, 2-1/2" long, 1960	20	40	55
15-D	Volkswagen 1500 Saloon	off white body and interior, clear windows, "137" on doors, 2-7/8" long, 1968	10	20	30
15-E	Volkswagen 1500 Saloon	clear windows, "137" on doors, red decal on front, 2-7/8" long, 1968	7	15	20
67-B	Volkswagen 1600 TL	ivory interior, four black plastic tires, 2-3/4" long, 1967	10	15	20
67-C	Volkswagen 1600 TL	ivory interior, clear windows, five spoke wheels, 2-5/8" long, 1970	5	10	20
23-D	Volkswagen Camper	orange top, clear windows, five spoke wheels, 2-1/8" long, 1970	5	7	10
34-C	Volkswagen Camper Car	silver body, orange interior, black plastic wheels, raised roof, six windows, 2-5/8" long, 1967	15	20	30
34-D	Volkswagen Camper Car	silver body, orange interior, black plastic wheels, short raised sun roof, 2-5/8" long, 1968	10	20	25
7-F	Volkswagen Golf	green body, black base and grille, 2-7/8" long, 1976	4	8	12
34-A	Volkswagen Microvan	blue body, gray wheels, "Matchbox International Express" on sides, 2-1/4" long, 1957	30	40	50
34-B	Volkswagen Microvan	light green body, dark green interior, flat roof window tinted green, 2 3/5" long, 1962	20	30	40
73-F	Weasel	metallic green body, large five spoke slicks, 2-7/8" long, 1974	3	4	6
24-B	Weatherhill Hydraulic Excavator	yellow body, small and medium front wheels, large rear wheels, 2-5/8" long, 1959	10	15	20

NO.	NAME	DESCRIPTION	GOOD	EX	MINT
24-A	Weatherhill Hydraulic Excavator	metal wheels, "Weatherhill Hydraulic" decal on rear, 2-3/8" long, 1956	20	25	35
57-F	Wild Life Truck	yellow body, red windows, light tinted blue canopy, 2-3/4" long, 1973	3	4	6
57-A	Wolseley 1500	with or without grilles, four wheels, 2-1/8" long, 1958	20	30	35
58-E	Woosh-n-Push	yellow body, red interior, large rear wheels, 2-7/8" long, 1972	3	4	5
35-E	Zoo Truck	1981	4	7	10

Snowtrac Tractor, Matchbox

D.A.F. Girder Truck 58-C, Matchbox

Nylint

NAME	DESCRIPTION	GOOD	EX	MINT

Cars

NAME	DESCRIPTION	GOOD	EX	MINT
Howdy Doody Pump Mobile	8-1/2" long	250	450	650

Emergency Vehicles

NAME	DESCRIPTION	GOOD	EX	MINT
Ladder Truck	post war, 30" long	100	175	250

Farm and Construction Equipment

NAME	DESCRIPTION	GOOD	EX	MINT
Michigan Shovel	bright yellow, bucket tips automatically when raised to boom, boom raises and lowers, 10 wheels, steerable front wheels	150	225	275
Payloader	bright red, 3-3/4" rubber tires, 18" long, 1955	125	187	250
Road Grader	sturdy blade can be raised, lowered, or tilted; tandem-pivoted rear wheels, 3-3/4" steel wheels, 19-1/4" long, 1955	100	175	225
Speed Swing Pettibone	orange, raise or lower bucket and tip to dump, steerable wheels, 3-3/4" rubber tires, "Pettibone" decal on sides, 19" long	200	300	400
Street Sweep	wind-up, 8-1/4" long	175	275	350
Tournahopper	huge hopper, pull lever at rear opens wide clamshell jaws for bottom dumping, 3-3/4" rubber-tired steel wheels, 22-1/2" long,	100	150	200
Tournarocker	oversize hopper, crank action hoist, 3-3/4" rubber-tired steel wheels, 18" long, 1955	75	125	175
Tournatractor	yellow, big powerful adjustable blade on front, pivoted tow-bar on rear, 14-3/4" long, 1955	100	150	200
Traveloader	orange, synchronized feeders, buckets, and rubber conveyor belt, hand crank, steel wheels with 3-3/4" rubber tires, 30" long,	200	300	400

Trucks

NAME	DESCRIPTION	GOOD	EX	MINT
Guided Missile Launcher	1957	75	125	175
Tournahauler	dark green, tractor with enclosed cab, platform trailer, slid-out ramps, 41-1/2" long with ramp extended, 1955	125	150	250
U-Haul Ford Truck and Trailer	with twin I-Beam suspension	125	187	250

Schuco

Cars

NAME	DESCRIPTION	GOOD	EX	MINT
1902 Mercedes Simplex 32PS	wind-up, 8-1/2" long	125	187	250
1913 Mercer	wind-up, 7-1/2" long	87	130	175
Renault 6CV Model 1911	open two-seater, 7" long	125	187	250
Sedan	blue, tin litho, wind-up, 4-1/2" long, 1950s	200	300	400

Sets

NAME	DESCRIPTION	GOOD	EX	MINT
Highway Patrol Official Squad Car Road Set	1958	100	150	200

Schuco

NAME	DESCRIPTION	GOOD	EX	MINT

Tanks

NAME	DESCRIPTION	GOOD	EX	MINT
Military Miniature Tank	keywind	37	55	75

Trucks

NAME	DESCRIPTION	GOOD	EX	MINT
Van	battery-operated, 4" long	75	112	150

Smith-Miller

Emergency Vehicles

NAME	DESCRIPTION	GOOD	EX	MINT
"L" Mack Aerial Ladder	red w/gold lettering; polished aluminum surface, SMFD decals on hood and trailer sides, six-wheeler, 1950	375	475	795

Trucks

NAME	DESCRIPTION	GOOD	EX	MINT
"B" Mack Associated Truck Lines	red cab, polished aluminum trailer, decals on trailer sides, six-wheel tractor, eight-wheel trailer, 1954	500	850	1200
"B" Mack Blue Diamond Dump	all white truck with blue decals, hydraulic piston, 10-wheeler, 1954	600	950	1300
"B" Mack Lumber Truck	yellow cab and timber deck, three rollers, loading bar and two chains, six-wheeler, load of nine timbers, 1954	450	650	1000
"B" Mack Orange Dump Truck	construction orange all over, no decals, hydraulic piston, 10-wheeler, 1954	650	1150	1650
"B" Mack P.I.E.	red cab, polished trailer, six-wheel tractor, eight-wheel trailer, 1954	375	600	850
"B" Mack Searchlight	dark red paint schemes, fully rotating and elevating searchlight, battery-operated, 1954	500	775	1100
"B" Mack Silver Streak	yellow cab, unpainted, unpolished trailer sides, "Silver Streak" decal on both sides, six-wheel tractor, eight-wheel trailer, 1954	450	775	1050
"B" Mack Watson Bros.	yellow cab, polished aluminum trailer, decals on trailer sides and cab doors, 10-wheel tractor, eight-wheel trailer, 1954	650	1100	1500
"L" Mack Army Materials Truck	Army green, flatbed with dark green canvas, 10-wheeler, load of three wood barrels, two boards, large and small crate, 1952	375	500	750
"L" Mack Army Personnel Carrier	all Army green, wood sides, Army seal on door panels, military star on roof, 10-wheeler, 1952	375	500	750
"L" Mack Bekins Van	white, covered with "Bekins" decals, six-wheel tractor, four-wheel trailer, 1953	1000	1650	2000
"L" Mack Blue Diamond Dump	white cab, white dump bed, blue fenders and chassis, hydraulically operated, 10-wheeler, 1952	425	750	1050
"L" Mack International Paper Co.	white tractor cab, "International Paper Co." decals, six wheel tractor, four wheel trailer, 1952	375	650	900
"L" Mack Lyon Van	silver gray cab, dark blue fenders and frame, silver gray van box with blue "Lyon" decal, six-wheeler, 1950	425	800	1100
"L" Mack Material Truck	light metallic green cab, dark green fenders and frame, wood flatbed, six-wheeler, load of two barrels and six timbers, 1950	400	600	875

1952 Ford Yellow Cab, Japanese

U-Haul Truck and Trailer, Nylint

Schuco

NAME	DESCRIPTION	GOOD	EX	MINT
"L" Mack Merchandise Van	red cab, black fenders and frame, "Smith-Miller" decals on both sides of van box, double rear doors, six-wheeler, 1951	425	695	1000
"L" Mack Mobil Tandem Tanker	all red cab, "Mobilgas" and "Mobiloil" decals on tank sides, six-wheel tractor, six-wheel trailer, 1952	450	725	1000
"L" Mack Orange Hydraulic Dump	orange cab, orange dump bed, hydraulic, 10-wheeler, may or may not have "Blue Diamond" decals, 1952	850	1500	1950
"L" Mack Orange Materials Truck	all orange, flatbed with canvas, 10-wheeler, load of three barrels, two boards, large and small crate, 1952	400	650	900
"L" Mack P.I.E.	all red tractor, polished aluminum trailer, "P.I.E." decals on sides and front, six wheel tractor, eight-wheel trailer, 1950	395	550	850
"L" Mack Sibley Van	dark green cab, black fenders and frame, dark green van box with "Sibley's" decal in yellow on both sides, six-wheeler, 1950	850	1375	1850
"L" Mack Tandem Timber	red/black cab, six-wheeler, load of six wood lumber rollers, two loading bars, four chains and 18 or 24 boards, 1950	400	550	725
"L" Mack Tandem Timber	two-tone green cab, six-wheeler, load of six wood lumber rollers, two loading bars, four chains, and 18 timbers, 1953	400	550	725
"L" Mack Telephone Truck	all dark or two-tone green truck, "Bell Telephone System" decals on truck sides, six-wheeler, 1952	475	750	975
"L" Mack West Coast Fast Freight	silver with red/black or silver cab and chassis, "West Coast-Fast Freight" decals on sides of box, six-wheeler, 1952	475	775	1000
Chevrolet Arden Milk Truck	red cab, white wood body, four-wheeler, 1945	275	465	800
Chevrolet Bekins Van	blue die cast cab, all white trailer, 14-wheeler, 1945	275	350	750
Chevrolet Coca-Cola Truck	red cab, wood body painted red, four-wheeler, 1945	300	600	850
Chevrolet Flatbed Tractor-Trailer	unpainted wood trailer, unpainted polished cab, 14-wheeler, 1945	250	300	500
Chevrolet Heinz Grocery Truck	yellow cab, load of four waxed cases, 1946	225	325	475
Chevrolet Livestock Truck	polished, unpainted tractor cab and trailer, 1946	175	275	375
Chevrolet Lumber	green cab, load of 60 polished boards and two chains, 1946	150	195	275
Chevrolet Lyons Van	blue cab, silver trailer, 1946	165	325	500
Chevrolet Material Truck	green cab, no side rails, load of three barrels, two cases and 18 boards, 1946	135	185	225
Chevrolet Stake	yellow tractor cab	185	250	425
Chevrolet Transcontinental Vanliner	blue tractor cab, white trailer, "Bekins" logos and decals on trailer sides, 1946	200	350	495
Chevrolet Union Ice Truck	blue cab, white body, load of eight waxed blocks of ice, 1946	300	495	800
Ford Bekins Van	red sand-cast tractor, gray sheet metal trailer, 14-wheeler, 1944	275	500	750
Ford Coca-Cola Truck	red sandcast cab, wood body painted red, four-wheeler, 1944	400	650	900
GMC Arden Milk Truck	red cab, white painted wood body with red stakes, four-wheeler, 1947	200	425	650
GMC Bank of America Truck	dark brownish green cab and box, `Bank of America' decal on box sides, four-wheeler, 1949	115	165	275

Schuco

NAME	DESCRIPTION	GOOD	EX	MINT
GMC Be Mac Tractor-Trailer	red cab, plain aluminum frame, "Be Mac Transport Co." in white letters on door panels, 14-wheeler, 1949	250	350	700
GMC Bekins Vanliner	blue cab, metal trailer painted white, 14-wheeler, 1947	175	275	425
GMC Coca-Cola Truck	red cab, yellow wood body, four-wheeler, load of 16 Coca-Cola cases, 1947	400	675	895
GMC Coca-Cola Truck	all yellow truck, red Coca-Cola decals, five spoke hubs, four-wheeler, load of six cases each with 24 plastic bottles, 1954	275	450	750
GMC Drive-O	red cab, red dump body, runs forward and backward with handturned control at end of 5-1/2 ft. cable, six-wheeler, 1949	175	300	450
GMC Dump Truck	all red truck, six-wheeler, 1950	150	200	285
GMC Emergency Tow Truck	white cab, red body and boom, `Emergency Towing Service' on body side panels, four-wheeler, 1953	185	250	400
GMC Furniture Mart	blue cab, off-white body, "Furniture Mart, Complete Home Furnishings" markings on body sides, four-wheeler, 1953	135	275	295
GMC Heinz Grocery Truck	yellow cab, wood body, six-wheeler, 1947	250	325	450
GMC Highway Freighter	red tractor cab, hardwood bed on trailer with full length wood fences, "Fruehauf" decal on trailer, 14-wheeler, 1948	150	210	325
Tractor-Trailer				
GMC Kraft Foods	yellow cab, yellow steel box, large "Kraft" decal on both sides, four-wheeler, 1948	200	300	450
GMC Lumber Tractor-Trailer	metallic blue cab and trailer, three rollers and two chains, 14-wheeler, 1949	185	250	350
GMC Lumber Truck	green cab, six-wheeler, 1947	165	215	300
GMC Lyons Van Tractor-Trailer	blue tractor cab, "Lyons Van" decals on both sides, fold down rear door, 14-wheeler, 1948	165	250	400
GMC Machinery Hauler	construction orange cab and lowboy trailer, "Fruehauf" decal on gooseneck, 13-wheeler, 1949	150	225	335
GMC Machinery Hauler	construction orange cab and lowboy trailer, "Fruehauf" decal on gooseneck, 13-wheeler, 1949	150	225	335
GMC Machinery Hauler	construction orange, two loading ramps, 10-wheeler, 1953	200	295	425
GMC Marshall Field's & Company Tractor-Trailer	dark green cab and trailer, double rear doors, never had Smith-Miller decals, 10-wheeler, 1949	295	395	500
GMC Material Truck	green cab, wood body, six-wheeler, load of three barrels, three cases and 18 boards, 1947	115	150	250
GMC Material Truck	yellow cab, natural finish hardwood bed and sides, four-wheeler, load of four barrels and two timbers, 1949	125	175	265
GMC Mobilgas Tanker	red cab and tanker trailer, large "Mobilgas", "Mobiloil" emblems on sides and rear panel of tanker, 14-wheeler, 1949	135	225	400
GMC Oil Truck	orange cab, rear body unpainted, six-wheeler, load of three barrels, 1947	115	185	265
GMC P.I.E.	red cab, polished aluminum box trailer, double rear doors, "P.I.E." decals on sides and front panels, 14-wheeler, 1949	150	265	350
GMC People's First National Bank and Trust Company	dark brownish green cab and box, "People's First National Bank & Trust Co." decals on box sides, 1951	165	250	385

Schuco

NAME	DESCRIPTION	GOOD	EX	MINT
GMC Rack Truck	red or yellow cab, natural finish wood deck, red stake sides, six-wheeler, 1948	135	200	325
GMC Redwood Logger	green or maroon cab, unpainted aluminum trailer with four hardwood stakes, load of three cardboard logs, 1948	365	585	700
GMC Rexall Drug Truck	orange cab and closed steel box body, "Rexall" logo on both sides and on front panel of box, four-wheeler, 1948	500	750	1000
GMC Scoop Dump	rack and pinion dump with a scoop, five spoke wheels, six-wheeler, 1954	275	350	575
GMC Searchlight Truck	four wheel truck pulling four wheel trailer, color schemes vary, "Hollywood Film Ad" on truck body side panels, 1953	300	415	695
GMC Silver Streak	unpainted polished cab and trailer, wrap around sides and shield, some had tail gate, 1950	140	200	300
GMC Sunkist Special Tractor-Trailer	cherry/maroon tractor cab, natural mahogany trailer bed, 14-wheeler, 1947	165	275	475
GMC Super Cargo Tractor-Trailer	silver gray tractor cab, hardwood bed on trailer with red wraparound side rails, 14-wheeler, load of 10 barrels, 1948	150	225	395
GMC Timber Giant	green or maroon cab, unpainted aluminum trailer with four hardwood stakes, load of three cardboard logs, 1948	175	285	495
GMC Tow Truck	white cab, red body and boom, five spoke cast hubs, "Emergency Towing Service" on body side panels, four-wheeler, 1954	95	135	200
GMC Transcontinental	red tractor cab, hardwood bed on trailer with full length wood fences, "Fruehauf" decal on trailer, 14-wheeler, 1948	150	210	325
GMC Triton Oil Truck	blue cab, mahogany body unpainted, six-wheeler, load of three Triton Oil drums (banks) and side chains, 1947	115	185	265
GMC U.S. Treasury Truck	gray cab and box, "U.S. Treasury" insignia and markings on box sides, four-wheeler, 1952	235	325	475

Structo

Cars

Auto Builder Auto	1918; red w/black fenders; 16"	550	1100	1500
Auto Builder Racing Auto	1919; 12-1/2"; red	600	900	1200
Bearcat Auto	1919; 16"	850	1400	2000
Roadster	1924; 10-1/2"; orange or red body	900	1500	2000

Miscellaneous

Caterpillar Tractor	1921; 9"	325	500	850
Climbing Military Tank	1929; green tank; chain link treads	450	700	1100
Climbing Tractor	1929; 12-1/2"; chain link treads	450	600	1100
Excavator	1931; 29"	325	600	950
Fire Insurance Patrol	1928; 18"	250	500	700
High Wheel Tractor/Trailer	1919; 22", green with red wheel guards	550	1000	1500
Hook and Ladder Truck	1930; 24"	300	450	690
Lift Crane	1924; 13"	100	150	225

Structo

NAME	DESCRIPTION	GOOD	EX	MINT
Lone Eagle Airplane	1928; 19-1/2"; monoplane w/spring-driven motor; blue/green with orange wing	600	800	1100
Moving Van	1928; 17"; orange with green wheels and black tires	350	650	950
Pile Driver	1924; blue crane with red roof; 10-3/4"	100	200	300
Police Patrol	1928; 16-1/2"; blue with decals; red wheels with white tires	350	650	950
Pumping Fire Engine	1928; 21"; red truck with orange water tank	600	1000	1400
Road Grader	1922; four-wheel grader; green; 16"	150	350	650
Sky King Airplane	1929; 21-1/2"; blue with gray wings	900	1000	1500
Whippet Military Tank	1920; 12"	425	650	1000
Yuba Tractor	1919; 20"	750	1000	1475

Trucks

NAME	DESCRIPTION	GOOD	EX	MINT
Ambulance Truck	1928; 17"; green with Ambulance markings	550	800	1350
Contractor Truck	1924; 12-1/2"; orange dump truck with black fenders	525	850	1000
Dump Truck	1924-1932; 14-1/2"; assorted colors and styles	200	300	450
Giant Truck Builder Dump Truck	1919; 18-1/2"	850	1300	1850
Motor Dispatch	1929; 24"; blue with decals; orange wheels and black tires	850	1000	1400
Red Rider Truck	1932; 17-5/8"; red	250	450	750
Screen Side Truck	1928; 17"; white with black wheels and white tires	350	650	1000
U.S. Mail Truck	1928; 17"; green with red wheels and white tires	450	700	1100

Road Grader, Structo

Sportsman Pickup Truck, Tonka

Car Hauler, 1960s, Tonka

Tonka

NAME	DESCRIPTION	GOOD	EX	MINT
	Construction Toys			
Back Hoe	1963	80	145	250
Bulldozer	1960; orange w/rubber treads	50	75	125
Crane and Clam	1949, yellow; with tracks	95	145	245
Crane and Clam	1947, yellow	95	145	245
Giant Bulldozer	1961; orange w/rubber treads	75	175	240
Hi-Way Mobile Clam	1961; orange	100	225	340
Hi-Way Mobile Dragline	1960; orange	125	225	345
Power Lift Truck and Trailer	1948; red	125	260	500
Road Grader	1953; orange	50	75	125
Shovel	1964	45	90	150
Steam Shovel	1953; red	95	125	245
Steam Shovel	1947, red	90	120	240
Steam Shovel	1949; red	90	130	210
Trencher	1963	50	100	200
	Emergency Vehicles			
Fire Department Rescue Van		60	150	275
Fire Jeep	1963-64	100	200	325
Hydraulic Aerial Ladder	1957	100	200	400
Ladder Truck	1959; white	175	325	600
Ladder Truck	1954	135	225	450
Suburban Pumper	1957	135	200	375
Suburban Pumper	1959; white w/red hydrant	150	250	425
T.F.D. Tanker	1958; white w/red decals	275	425	800
U.S.A.F. Ambulance		100	160	250
<sub-header>	Sets			
Bulk Storage Set	1961	250	400	850
Hi-Way Set	1956; four trucks plus road signs	500	750	1750
Hi-Way Sign Set	1959	150	250	350
Trailer Sales	1961	195	300	600
	Trucks			
Ace Hardware Semi Truck	1955; red w/decals	200	425	850
Airport Service Truck	1962	125	225	300
Allied Van Lines	1953; orange w/black lettering	100	200	350
Army Troop Carrier	1964	75	150	250
Big Mike with V-Plow	1957; orange with Hi-Way Dept. decals	300	500	850
Car Hauler	1961; yellow	150	250	400
Cargo King	1956	75	150	250
Carnation Milk Van	1955; white w/decals	125	225	400
Cement Truck	1960; red	100	200	300
Coast to Coast Stores Truck	red cab	100	170	295
Cross Country Freight Semi Truck	1955; white w/red logo	160	225	475
Deluxe Fisherman	1960; with boat and trailer	100	200	400
Deluxe Fisherman	1960; with houseboat and trailer	135	225	425
Deluxe Sportsman	1958; with boat and trailer	160	225	375
Dump Truck		85	135	200
Dump Truck	1955-57	75	125	250
Dump Truck and Sandloader	1961	100	200	300
Eibert Coffee Van	1954	150	340	575

Cement Truck, 1960s, Tonka

Log Hauler, 1955, Tonka

Tonka

NAME	DESCRIPTION	GOOD	EX	MINT
Gambles Pickup Truck	1955-63; white	100	200	300
Gambles Semi Truck	1956; white	190	270	450
Gasoline Tanker	1957; red	200	400	850
Gasoline Tanker	1958; orange	250	450	900
Grain Hauler	1952-53; red cab	65	175	250
Green Giant Transport Truck	1953; white w/green lettering	100	200	400
Green Giant Utility Truck	1953; white w/green decals	100	160	300
Hi-Way Dump Truck	1956	100	150	300
Hydraulic Dump Truck	1957	100	175	250
Hydraulic Land Rover	1959; orange	200	500	1000
Livestock Van	1952-53; red	75	150	275
Log Hauler	red cab w/logs	100	150	250
Minute Maid Box Van	1955; white w/decals	200	400	675
Parcel Delivery Van	1954; brown	100	200	400
Pickup Truck	1955	150	250	350
Pickup Truck w/Camper	1963	80	195	300
Pickup Truck/Stake Trailer	1957; with animal	150	225	400
Ramp Hoist	1963	175	250	550
Rescue Squad Van	1960-61; white w/red accents	100	250	500
Sanitary Service Truck	1959	190	350	750
Sportsman Pickup Truck	1959; tan w/boat on top	125	225	300
Standard Tanker	1961; "Standard" on side	200	450	700
Star-Kist Tuna Box Van	1954; red cab; blue body	200	500	700
Steel Carrier	1950-53; orange or yellow cab; green body	125	180	265
Thunderbird Express Semi Truck	1959; red or white	100	200	400
Tonka Air Express	1959; black w/decals	175	265	400
Tonka Express	1950; green cab; red box with decals	100	300	500
Tonka Farms Stock Rack Truck	1957	150	225	400
Tonka Marine Service	1961	125	350	500
Tonka Service Van	1961; light blue	95	200	280
Tonka Toy Transport Truck	1949; red w/silver roof	150	225	325
Tractor and Carry-All	1949-50	100	200	300
Utility Truck	1950-53; several colors	100	175	200
Wrecker Truck	1949-53; blue	100	175	225

Pickup Truck/Stake Trailer, Tonka

Tonka Farms Truck, 1950s, Tonka

Allied Van Lines, 1950s, Tonka

Hydraulic Dump Truck, Tonka

Pickup Truck, 1960s, Tonka

Tootsietoy

NAME	DESCRIPTION	GOOD	EX	MINT
	Airplanes			
Aero-Dawn	1928	20	35	65
Atlantic Clipper	2" long	5	10	20
Autogyro	white, 1934	25	40	90
Beechcraft Bonanza	orange, front propellor	6	10	35
Bi-Wing Seaplane	1926, yellow	30	40	65
Bleriot Plane	1910	25	50	100
Crusader		25	50	75
Curtis P-40	light green	120	250	525
Dirigible U.S.N. Los Angeles	silver	40	60	90
Douglas D-C 2 TWA	1935	20	45	85
F-94 Starfire	green, four engines, 1970s	10	15	35
F9F-2 Panther Shooting Star		25	35	70
Fly-N-Gyro	1938	50	100	310
KOP-1 USN		20	30	60
Low Wing Plane	miniature Piper Cub	20	30	50
Navion	red, front propellor	10	15	35
Navy Jet	red, 1970s	5	10	30
Navy Jet Cutlass	red with silver wings	10	15	30
P-38 Plane	9-3/4" wingspan	40	60	110
Piper Cub	blue, front propellor	10	15	30
S-58 Sikorsky Helicopter	1970s	15	30	50
Snow Skids Airplane	rotating prop, 4" wingspan	40	60	85
Stratocruiser		30	50	110
Supermainliner		20	30	50
Top Wing Plane	miniature	20	30	50
Transport Plane	1941, orange	40	75	110
Tri-Motor Plane	three propellors	50	85	150
TWA Electra	two engines, propellors	10	20	45
Twin Engine Airliner	10 windows, DC4	10	20	45
U.S. Army Plane	1936	15	25	50
UX214 Monoplane	4", 1930s	30	60	90
Waco Bomber	blue base/silver top or silver base	50	80	150
	Boats and Ships			
Battleship	silver with some red on top, 6" long, 1939	10	20	35
Carrier	silver	15	25	40
Cruiser	silver; some red on top, 6" long, 1939	10	15	25
Destroyer	4" long, 1939	10	20	40
Freighter	6" long, 1940	10	20	45
Submarine	4" long, 1939	10	15	30
Tanker	black, 6" long, 1940	10	20	25
Tender	4" long, 1940	10	15	25
Transport	6" long, 1939	15	20	30
Yacht	4" long, 1940	10	15	30
	Buses			
Cross Country Bus		30	45	65
Fageol Bus	1927-33	20	40	65
GMC Greyhound Bus	blue/silver, 6", 1948	20	35	55

Tootsietoy

NAME	DESCRIPTION	GOOD	EX	MINT
GMC Scenicruiser Bus	blue and silver, raised passenger roof with windows, 6" long, 1957	20	35	55
Greyhound Bus	blue, 1937-41	25	60	80
Overland Bus	1929-33	45	65	125
Twin Coach Bus	red with black tires, 3" long, 1950	20	35	55

Cannons and Tanks

NAME	DESCRIPTION	GOOD	EX	MINT
Army Tank	1931-41	35	50	100
Army Tank	miniature	5	10	25
Four Wheel Cannon	4" long, 1950s	10	20	45
Long Range Cannon		7	10	30
Six-Wheel Army Cannon	1950s	10	20	50

Cars

NAME	DESCRIPTION	GOOD	EX	MINT
Andy Gump Car	1932; Funnies series	75	300	500
Armored Car	"U.S. Army" on sides, camouflage, black tires, 1938-41	25	35	65
Auburn Roadster	red, white rubber wheels	15	35	50
Austin-Healy	light brown roadster; 6" long, 1956	20	30	60
Baggage Car		10	15	30
Bluebird Daytona Race Car		20	35	65
Boat Tail Roadster	red roadster, 6" long	20	35	55
Buick Brougham	tan/black	20	35	70
Buick Coupe	blue with white wheels, 1924	30	42	65
Buick Coupe	4"	20	35	50
Buick Estate Wagon	yellow and maroon with black wheels, 6" long, 1948	20	30	50
Buick Experimental Car	blue with black wheels, detailed tin bottom, 6" long, 1954	25	50	85
Buick LaSabre	red open top, black wheels, 6" long, 1951	25	45	70
Buick Roadmaster	blue with black wheels, four-door, 1949	25	40	65
Buick Roadster	yellow open top, black wheels, 4" long, 1938	25	45	65
Buick Sedan	6" long	25	35	55
Buick Special	4" long, 1947	15	25	50
Buick Station Wagon	green with yellow top, black wheels, 6" long, 1954	20	30	45
Buick Tourer	red with white wheels, 1925	25	45	70
Buick Touring Car		50	75	110
Cadillac	HO series, blue car/white top, 2" long, 1960	10	20	30
Cadillac 60	red-orange with black wheels, four-door, 1948	20	35	60
Cadillac 62	red-orange with white top, black wheels, four-door, 6" long, 1954	20	35	120
Cadillac Brougham		40	60	85
Cadillac Coupe	blue/tan, black wheels	40	60	85
Cadillac Sedan	white rubber wheels	40	60	85
Cadillac Touring Car	1926	50	90	120
Chevrolet Ambulance	4"	15	20	35
Chevrolet Bel Air	yellow with black wheels, 3" long, 1955	15	30	50
Chevrolet Brougham		40	60	85
Chevrolet Coupe		20	35	90
Chevrolet Coupe	green with black wheels, 3"	20	35	55
Chevrolet Fastback	blue with black wheels, 3" long, 1950	15	20	35
Chevrolet Roadster		20	35	90

Tootsietoy

NAME	DESCRIPTION	GOOD	EX	MINT
Chevrolet Sedan		20	35	75
Chevrolet Touring Car		60	150	200
Chrysler Convertible	blue-green with black wheels, 4" long, 1960	15	25	35
Chrysler Experimental Roadster	orange open top, black wheels	25	40	60
Chrysler New Yorker	blue with black wheels, four-door, 6" long, 1953	25	35	55
Chrysler Windsor Convertible	green with black wheels, 4" long, 1941	20	30	45
Chrysler Windsor Convertible	black wheels, 6" long, 1950	50	90	120
Classic Series 1906 Cadillac or Studebaker	green and black, spoke wheels	10	15	25
Classic Series 1907 Stanley Steamer	yellow and black, spoke wheels, 1960-65	10	15	25
Classic Series 1912 Ford Model T	black with red seats, spoke wheels	10	15	25
Classic Series 1919 Stutz Bearcat	black and red, solid wheels	10	15	25
Classic Series 1929 Ford Model A	blue and black, black tread wheels, 1960-65	10	15	25
Corvair	red, 4" long, 1960s	30	55	75
Corvette Roadster	blue open top, black wheels, 4" long, 1954-55	15	20	35
Coupe	miniature	20	25	40
DeSoto Airflow	green with white wheels	20	35	60
Doodlebug	same as Buick Special	50	75	100
Ferrari Racer	red with gold driver, black wheels, 6" long, 1956	30	40	65
Ford	red with open top, black wheels, 6" long, 1940	15	25	40
Ford B Hotrod	1931	15	20	40
Ford Convertible Coupe	1934	30	50	70
Ford Convertible Sedan	red with black wheels, 3" long, 1949	10	20	35
Ford Coupe	powder blue with tan top, white wheels, 1934	30	50	75
Ford Coupe	blue or red with white wheels, 1935	25	35	45
Ford Customline	blue with black wheels, 1955	15	20	30
Ford Fairlane 500 Convertible	red with black wheels, 3" long, 1957	10	15	30
Ford Falcon	red with black wheels, 3" long, 1960	10	15	30
Ford LTD	blue with black wheels, 4" long, 1969	15	20	35
Ford Mainliner	red with black wheels, four-door, 3" long, 1952	10	20	30
Ford Model A Coupe	blue with white wheels	25	35	50
Ford Model A Sedan	green with black wheels	25	35	50
Ford Ranch Wagon	green with yellow top, four-door, 4" long, 1954	15	25	35
Ford Ranch Wagon	red with yellow top, four-door, 3" long, 1954	15	20	35
Ford Roadster	powder blue with open top, white wheels	25	40	60
Ford Sedan	1934	35	50	70
Ford Sedan	powder blue with white solid wheels, 1935	25	35	50
Ford Sedan	lime green with black wheels, four-door, 3" long, 1949	15	20	35
Ford Station Wagon	powder blue with white top, black wheels, 6" long, 1959	15	20	35
Ford Station Wagon	blue with black wheels, 3" long, 1960	10	15	25
Ford Station Wagon	red with white top, black wheels, four-door, 6" long, 1962	25	40	55
Ford Tourer	open top, red with silver spoke wheels	20	30	45

NAME	DESCRIPTION	GOOD	EX	MINT
Ford V-8 Hotrod	red with open top, black wheels, open silver motor, 6" long, 1940	15	25	35
Ford w/Trailer	blue sedan, white rubber wheels	50	75	130
Graham Convertible Coupe	rear spare tire, rubber wheels, 1933-35	50	125	175
Graham Convertible Coupe	side spare tire, rubber wheels, 1933-35	50	125	175
Graham Convertible Sedan	rear spare tire, rubber wheels, 1933-35	50	125	175
Graham Convertible Sedan	side spare tire, rubber wheels, 1933-35	50	125	175
Graham Coupe	rear spare tire, rubber wheels, 1933-35	50	115	150
Graham Coupe	side spare tire, rubber wheels, 1933-35	50	115	150
Graham Roadster	rear spare tire, rubber wheels, 1933-35	50	115	150
Graham Roadster	side spare tire, rubber wheels, 1933-35	50	115	150
Graham Sedan	side spare tire, rubber wheels, 1933-35	50	115	150
Graham Sedan	rear spare tire, rubber wheels, 1933-35	50	115	150
Graham Towncar	rear spare tire, rubber wheels, 1933-35	50	115	150
Graham Towncar	side spare tire, 1933-35	50	115	140
Insurance Patrol	miniature	15	25	35
International Station Wagon	red/yellow, 3" long	15	25	50
International Station Wagon	4" long, rubber wheels, 1940s	25	45	50
International Station Wagon	red/yellow, white wheels, 1939-41	15	25	50
International Station Wagon	orange with black wheels, postwar	15	20	30
Jaguar Type D	green with black wheels, 3" long, 1957	10	15	25
Jaguar XK 120 Roadster	green open top, black wheels, 3" long	10	15	25
Jaguar XK 140 Coupe	blue with black wheels, 6" long	20	30	50
Kaiser Sedan	blue with black wheels, 6" long, 1947	25	45	55
KO Ice	1932; Funnies series	100	325	425
Lancia Racer	dark green with black wheels, 6" long, 1956	30	50	75
Large Bluebird Racer	green with yellow solid wheels	20	40	60
LaSalle Convertible	rubber wheels	80	200	300
LaSalle Convertible Sedan	rubber wheels	80	200	300
LaSalle Coupe	rubber wheels	100	200	300
LaSalle Sedan	red with black rubber wheels, 3" long	15	20	30
LaSalle Sedan	rubber wheels	140	200	275
Limousine	blue with silver spoke wheels; prewar	20	40	65
Lincoln	prewar; red with white rubber wheels, four-door	110	400	500
Lincoln Capri	red with yellow top, black wheels, two-door, 6" long	20	35	50
Mercedes 190 SL Coupe	powder blue with black wheels, 6" long, 1956	15	25	40
Mercury	red with black wheels, four-door, 4" long, 1952	15	20	40
Mercury Custom	blue with black wheels, four-door, 4" long, 1949	15	30	40
Mercury Fire Chief Car	red with black wheels, 4" long, 1949	25	35	50
MG TF Roadster	red open top, black wheels, 6" long, 1954	15	25	45
MG TF Roadster	blue open top, black wheels, 3" long, 1954	15	20	35
Model T Pick-up	3", black, spoked metal wheels	25	40	50
Moon Mullins Police Car	1932; Funnies series	130	200	425

Tootsietoy

NAME	DESCRIPTION	GOOD	EX	MINT
Nash Metropolitan Convertible	red with black tires, 1954	25	35	60
Observation Car		10	15	30
Offenhauser Racer	dark blue with black wheels, 4" long, 1947	10	20	35
Oldsmobile 88 Convertible	yellow with black wheels, 4" long, 1949	15	25	35
Oldsmobile 88 Convertible	bright green with black wheels, 6" long, 1959	20	25	40
Oldsmobile 98	white body with blue top, skirted fenders, black wheels, 4" long, 1955	20	25	40
Oldsmobile 98	red body with yellow top, open fenders, black wheels, 4" long, 1955	20	25	45
Oldsmobile 98 Staff Car		20	25	45
Oldsmobile Brougham		25	35	50
Oldsmobile Coupe		25	35	50
Oldsmobile Roadster	orange/black, white wheels, 1924	25	45	75
Oldsmobile Sedan		25	35	50
Oldsmobile Touring		25	35	50
Open Touring	green convertible, white wheels, 3"	25	35	45
Packard	white body with blue top, black wheels, four-door, 6" long, 1956	25	35	45
Plymouth	dark blue with black wheels, two-door, 3" long, 1957	10	15	20
Plymouth Sedan	blue with black wheels, four-door, 3" long, 1950	15	20	30
Pontiac Fire Chief	red with black wheels, 4" long, 1950	20	35	50
Pontiac Sedan	green with black wheels, two-door, 4" long, 1950	15	25	45
Pontiac Star Chief	red with black wheels, four-door, 4" long, 1959	15	25	40
Porsche Roadster	red with open top, black wheels, two-door, 6" long, 1956	20	25	40
Pullman Car		10	15	30
Racer	miniature	25	40	50
Racer	orange with black wheels, 3" long, 1950s	10	15	30
Rambler Wagon	dark green w/yellow top, black wheels, yellow interior, 1960s	15	25	35
Rambler Wagon	blue with black wheels, 4" long, 1960	15	25	35
Roadster	miniature	15	25	40
Roadster		50	100	140
Sedan		50	100	135
Sedan	miniature	20	25	40
Small Racer	blue with driver, white wheels, 1927	50	110	175
Smitty	1932; Funnies series	170	250	500
Station Wagon	red w/tan upper, 3"	15	35	55
Studebaker Coupe	green with black wheels, 3" long, 1947	25	35	50
Studebaker Lark Convertible	lime green with black wheels, 3" long, 1960	10	15	25
Tank Car	miniature	20	25	40
Thunderbird Coupe	blue with black wheels, 3" long, 1955	15	20	30
Thunderbird Coupe	powder blue with black wheels, 4" long, 1955	15	30	40
Torpedo Coupe	gray; prewar	20	25	45
Torpedo Sedan	red	20	30	50
Triumph TR 3 Roadster	black wheels, 3" long, 1956	10	20	35
Uncle Walt	1932; Funnies series	150	300	450
Uncle Willie	1932; Funnies series	150	300	450

LaSalle Coupe, Tootsietoy

A collection of Tootsietoy's Funnies series from 1932

NAME	DESCRIPTION	GOOD	EX	MINT
VW Bug	metallic gold with black tread wheels, 6" long, 1960	15	25	30
VW Bug	lime green with black tread wheels, 3" long, 1960	10	20	35
Yellow Cab Sedan	green with white wheels, 1921	10	20	30

Emergency Vehicles

NAME	DESCRIPTION	GOOD	EX	MINT
American LaFrance Pumper	red, 3" long, 1954	15	20	35
Chevrolet Ambulance	yellow, red cross on top, 4" long, 1950	15	25	40
Chevrolet Ambulance	army green, red cross on roof top, army star on top of hood, 4" long, 1950	15	25	50
Fire Hook and Ladder	red/blue with side ladders	25	40	50
Fire Water Tower Truck	blue/orange, red water tower	30	60	80
Ford Wrecker	3" long, 1935, brown, white rubber wheels	30	50	65
Graham Ambulance	white with red cross on sides	40	95	125
Graham Wrecker	red/black; rubber wheels	50	110	150
Hook and Ladder	#1040	20	25	40
Hook and Ladder	with driver; white rubber wheels, 1937-41	40	60	75
Hook and Ladder	red and silver; white rubber wheels	50	75	90
Hose Car	with driver and figure standing by water gun; 1937-41	35	50	80
Hose Wagon	red, black rubber wheels, postwar	20	25	40
Hose Wagon	red with silver hose, white rubber wheels, 3" long, prewar	25	30	45
Insurance Patrol	red, white wheels, prewar	25	30	45
Insurance Patrol	red, black rubber wheels, postwar	20	25	40
Insurance Patrol	with driver	25	40	60
Jumbo Wrecker	6" long, 1941	25	40	60
Lincoln Wrecker	sedan w/wrecker hook	200	425	600
Mack L-Line Fire Pumper	red with ladders on sides	35	65	75
Mack L-Line Hook and Ladder	red with silver ladder	35	65	75

Farm and Construction Equipment

NAME	DESCRIPTION	GOOD	EX	MINT
Caterpillar Bulldozer	yellow, 6" long	25	45	55
Caterpillar Scraper	yellow with black wheels, silver blade, 6" long, 1956	15	25	40
Caterpillar Tractor	miniature	15	25	35
Caterpillar Tractor	1931	20	30	50
D7 Crawler with Blade	1:50 scale, die cast, 1956	20	35	55
D8 Crawler with Blade	1:87 scale, die-cast	20	30	45
Farm Tractor	with driver	60	100	145
Ford Tractor	red with loader, die cast, 1/32 scale	30	40	65
Grader	1:50 scale, 1956, 6"	20	30	50
International Tractor		10	15	25
Steamroller	1931-34	75	150	200

Sets

NAME	DESCRIPTION	GOOD	EX	MINT
Box Trailer and Road Scraper Set	with driver on road scraper	75	130	250

Tootsietoy

NAME	DESCRIPTION	GOOD	EX	MINT
Contractor Set	pickup truck with three wagons	50	100	130
Four-Car Transport Set	flatbed trailer carries cars	50	90	135
Freight Train	five-piece set	35	60	80
Grand Prix #1687 Set	seven vehicles, 1969	50	95	150
Midget Series	single engine plane, St. Louis, bomber, Atlantic Clipper, 1" long, 1936-41	10	15	25
Midget Series	yellow stake truck, red limo, green doodlebug, yellow railcar, blue racer, red fire truck, 1" long, 1936-41	20	40	60
Midget Series	assorted ships, 1" long, 1936-41	20	40	60
Midget Series	green cannon, blue tank, green armored car, green tow truck, green camelback van, 1" long, 1936-41	20	40	60
Milk Trailer Set	tractor with three milk tankers	50	100	200
Passenger Train	five-piece set	40	60	100
Playtime Set	six cars, two trucks, two planes	200	500	850
Tractor with Scoop Shovel and	red with silver scoop, flatbed trailer, 1946-52	125	185	250

Space Vehicles

NAME	DESCRIPTION	GOOD	EX	MINT
Buck Rogers Battlecruiser	1937	75	110	165
Buck Rogers Flash Blast Attack	1937	50	90	165
Buck Rogers Venus Duo-Destroyer	1937	75	110	150

Trailers

NAME	DESCRIPTION	GOOD	EX	MINT
Boat Trailer	two-wheel	10	15	20
Horse Trailer	red with white top, two-wheel, black tread wheels	10	15	20
House Trailer	powder blue with black wheels, two-wheel, door opens	10	15	20
Restaurant Trailer	yellow with black tread wheels, two-wheel, open sides	30	40	65
Small House Trailer	two-wheel, three side windows, 1935	30	45	65
U-Haul Trailer	red with black tread wheels, two-wheel, U-Haul logo on sides	10	15	30

Trains

NAME	DESCRIPTION	GOOD	EX	MINT
Borden's Milk Tank Car	yellow; red base	15	20	35
Box Car	"Southern"	15	20	35
Caboose	red	10	15	25
Coal Car	red	10	15	25
Cracker Jack Railroad Car	embossed white metal, painted orange, black rubber tires, 3" long, 1930s	60	150	200
Fast Freight Set	five-piece set, 1940	35	60	80
Log Car	silver w/red wheels; logs chained on	10	15	25
Oil Tank Car	silver top, red base, "Sinclair"	15	20	35
Passenger Train Set	four-piece set, 1925	50	125	160
Pennsylvania Engine	red/silver	20	30	50
Refrigerator Car	yellow ochre, red base, black roof	10	15	30
Santa Fe Engine	black/silver	15	20	35
Stock Car	red	10	15	30
Tootsietoy Flyer	silver; three-piece set, 1937	35	60	90
Wrecking Crane	green crane on silver flatbed car w/red wheels	20	30	40
Zephyr Railcar	dark green, 4" long, 1935	35	60	90

Tootsietoy

NAME	DESCRIPTION	GOOD	EX	MINT

Trucks

NAME	DESCRIPTION	GOOD	EX	MINT
Army Half Truck	1941	30	55	75
Army Jeep	windshield up, 6" long, 1950s	15	25	40
Army Jeep CJ3	extended back, windshield down, 4" long, 1950	10	15	35
Army Jeep CJ3	no windshield, 3" long, 1950	10	15	40
Army Supply Truck	with driver	25	40	55
Box Truck	red w/white wheels, 3" long	15	20	30
Buick Delivery Van		25	35	50
Cadillac Delivery Van		25	35	50
Chevrolet Cameo Pickup	green with black wheels, 4" long, 1956	15	25	40
Chevrolet Delivery Van		25	35	50
Chevrolet El Camino	red	20	25	40
Chevrolet El Camino Camper and	blue body with red camper, black/white boat on top of camper	25	35	45
Chevrolet Panel Truck	green, front fenders opened, 3" long, 1950s	15	20	35
Chevrolet Panel Truck	light green with black wheels, 4" long, 1950	25	30	40
Chevrolet Panel Truck	green, 3" long, 1950	10	20	30
Civilian Jeep	burnt orange, open top, black wheels, 3" long, 1950	10	15	30
Civilian Jeep	red, open top, black wheels, 4" long, 1950	15	20	35
Civilian Jeep	blue with black tread wheels, 6" long, 1960	15	20	35
CJ3 Army Jeep	open top, no steering wheel cast on dashboard, 3" long, 1950	10	15	30
CJ5 Jeep	red with black tread wheels, windshield up, 6" long, 1960s	15	20	30
CJ5 Jeep	red with black tread wheels, windshield up, 6" long, 1950s	15	25	35
Coast to Coast Van	9" long	40	75	100
Commercial Tire Van	"Commercial Tire & Supply Co."; white rubber wheels; prewar	70	150	200
Diamond T K5 Dump Truck	yellow cab and chassis, green dump body, 6" long	25	35	50
Diamond T K5 Semi	red tractor and light green closed trailer	25	45	55
Diamond T K5 Stake Truck	orange, open sides, 6" long, 1940	25	35	55
Diamond T K5 Stake Truck	orange, closed sides, 6" long, 1940	25	35	55
Diamond T Metro Van	powder blue, 6" long	35	75	100
Diamond T Tow Truck	red with silver tow bar	25	35	55
Dodge D100 Panel	green and yellow, 6" long	25	40	55
Dodge Pickup	lime green, 4" long	20	30	40
Federal Bakery Van	black with cream wheels, 1924	50	85	110
Federal Florist Van	black with cream wheels, 1924	75	175	220
Federal Grocery Van	black with cream wheels, 1924	45	70	100
Federal Laundry Van	black with cream wheels, 1924	50	85	110
Federal Market Van	black with cream wheels, 1924	55	85	110
Federal Milk Van	black with cream wheels, 1924	55	85	110
Ford C600 Oil Tanker	bright yellow, 3" long	10	15	20
Ford C600 Oil Tanker	red, 6" long, 1962	15	30	40

Tootsietoy

NAME	DESCRIPTION	GOOD	EX	MINT
Ford Econoline Pickup	red, 1962	15	25	35
Ford F1 Pickup	orange, closed tailgate, 3" long, 1949	15	20	35
Ford F1 Pickup	orange, open tailgate, 3" long, 1949	15	20	35
Ford F6 Oil	orange, 4" long, 1949	10	15	25
Ford F6 Oil Tanker	red with Texaco, Sinclair, Shell or Standard on sides, 6" long, 1949	25	50	100
Ford F6 Pickup	red, 4" long, 1949	15	25	40
Ford F600 Army Anti-Aircraft Gun	tractor-trailer flatbed, guns on flatbed	20	30	45
Ford F600 Army Radar	tractor-trailer flatbed, yellow radar unit on flatbed, 6" long, 1955	20	30	45
Ford F600 Army Stake Truck	tractor-trailer box, army star on top of trailer box roof and "U.S. Army" on sides, 6" long, 1955	25	40	55
Ford F600 Stake Truck	light green, 6" long, 1955	15	25	35
Ford Pickup	3", beige; 1935; white rubber wheels	25	40	55
Ford Shell Oil Truck		45	60	85
Ford Styleside Pickup	orange, 3" long, 1957	10	15	30
Ford Texaco Oil Truck		45	60	85
Hudson Pickup	red, 4" long, 1947	25	40	55
International Bottle Truck	lime green	30	45	65
International Car Transport Truck	red tractor, orange double-deck trailer with cars	35	50	70
International Gooseneck Trailer	orange tractor and flatbed trailer	30	40	55
International K1 Oil Truck	green, comes with oil brands on sides, 6" long	20	30	45
International K1 Panel Truck	blue, 4" long	20	30	55
International RC180 Grain Semi	green tractor and red trailer	30	50	70
International Sinclair Oil Truck	6" long	35	75	100
International Standard Oil Truck	6" long	35	75	100
Jeepster	bright yellow with open top, black wheels, 3" long, 1947	10	15	35
Jumbo Pickup	6" long, green w/black wheels, 1936-41	25	35	50
Mack Anti-Aircraft Gun		25	40	55
Mack B-Line Cement Truck	red truck with yellow cement mixer, 1955	20	35	55
Mack B-Line Oil Tanker	red tractor and trailer, "Mobil"	20	35	55
Mack B-Line Stake Trailer	red tractor, orange closed trailer, 1955	20	35	55
Mack Coal Truck	"City Fuel Company," 10 wheels	60	120	175
Mack Coal Truck	orange cab with blue bed, four wheels, 1925	90	100	250
Mack Coal Truck	red cab with black bed, 1928	60	120	160
Mack Dairy Tanker	1930s; two-piece cab; Tootsietoy Dairies	75	100	225
Mack L-Line Dump Truck	yellow cab and chassis, light green dump body, 6" long, 1947	20	35	55
Mack L-Line Semi and Stake Trailer	red tractor and trailer	50	95	125

Tootsietoy

NAME	DESCRIPTION	GOOD	EX	MINT
Mack L-Line Semi-Trailer	red tractor cab, silver semi-trailer, "Gerard Motor Express" on sides	60	115	145
Mack L-Line Stake Truck	red with silver bed inside	25	35	55
Mack L-Line Tow Truck	red with silver tow bar	25	35	55
Mack Log Hauler	red cab, trailer with load of logs, 1940s	50	95	135
Mack Mail Truck	red cab with light brown box, "U.S. Mail Airmail Service" on sides, 3" long, 1920s	40	70	100
Mack Milk Truck	"Tootsietoy Dairy," one-piece cab	50	110	175
Mack Oil Tanker	"DOMACO" on side of tanker	60	100	155
Mack Oil Truck	red cab with orange tanker, 1925	25	40	55
Mack Railway Express	1930s; Wrigley's Gum	55	115	165
Mack Searchlight Truck	1931-41	25	40	55
Mack Stake Trailer-Truck	enclosed cab, open stake trailer, `Express' on sides of trailer	50	90	120
Mack Stake Truck	orange cab with red stake bed, 1925	25	40	60
Mack Trailer-Truck	open cab	50	85	110
Mack Transport	red, open cab with flatbed trailer	60	150	200
Mack Transport	yellow, 1941, with three cars at angle	150	500	700
Mack Van Trailer-Truck	enclosed cab and box trailer	50	100	140
Mack Wrigley's Spearmint Gum	4", green with white rubber wheels, red hubs	70	150	225
Model T Pickup	3" long, 1914	30	50	75
Oil Tanker	blue and silver, two caps on top of tanker, 3" long	20	30	40
Oil Tanker	orange, four caps on top of tanker, 3" long, postwar	20	25	40
Oil Tanker	blue, three caps on top, 2" long, 1932	20	25	50
Oil Tanker	green with white wheels; prewar	15	25	35
Oldsmobile Delivery Van		25	35	50
Shell Oil Truck	yellow/silver, 6", white rubber wheels	50	85	120
Sinclair Oil Truck	6" long; green/silver	50	85	120
Special Delivery	1936	20	25	40
Stake Truck	miniature	25	40	55
Standard Oil Truck	red/silver; 6", white rubber wheels	80	100	150
Texaco Oil Truck	red/silver; 6", white rubber wheels	55	85	120
Tootsietoy Dairy	semi trailer truck	75	110	140
Tootsietoy Oil Tanker	red cab, silver tanker, "Tootsietoy Line" on side, 1950s	60	95	125
Wrigley's Box Van	with or without decal, 1940s	45	60	75

Williams, A.C.

Cars

1930s Chrysler Convertible Coupe	6", operating rumble seat	300	400	500
1930s Chrysler Roadster	5"	100	200	300
1930s Coupe	5", rubber tires, turned metal wheels, twin sidemount spare tires	150	250	350

Williams, A.C.

NAME	DESCRIPTION	GOOD	EX	MINT
1930s Sedan	5", rubber tires, turned metal wheels, twin sidemount spare tires	150	250	350
1930s Sedan	6-1/2", rubber tires, turned metal wheels, twin sidemount spare tires	250	350	450
Dream Car	cast iron, 4-7/8" long, 1930	75	150	250
Ford Model A Fordor Sedan	6", nickel-plated cast spoke wheels	600	900	1200
Ford Roadster	1936	450	550	650
Lincoln Touring Cars	spoked wheels, cast iron, 8-3/4" long, 1924	400	600	800
Racer	yellow, cast iron, 8-1/2" long, 1932	300	450	600
Taxi	cast iron, 5-1/4" long, 1920	200	350	500
Touring Cars	solid wheels, cast iron, 11-3/4" long, 1917	500	750	1200
Touring Cars	disc wheels, cast iron, 9-1/8" long, 1922	400	600	800

Trucks

NAME	DESCRIPTION	GOOD	EX	MINT
Austin Transport Set	with three vehicles, cast iron, 12-1/2" long, 1930	500	850	1250
Car Carrier w/Three Austin Cars	12"	800	1200	1600
Coast to Coast Cartage Semi Stake Trailer	10", nickel plated stamped steel spoke wheels	400	600	800
Coast to Coast Cartage Semi Stake Trailer	6-1/2"	100	150	200
Interchangeable Delivery Truck	cast iron, 7-1/4" long, 1932	175	250	350
Mack Bulldog Gasoline Tank Truck	7", nickel plated cast spoke wheels	300	450	600
Mack Bulldog Gasoline Tank Truck	5", nickel plated stamped steel spoke wheels	125	175	225
Mack Bulldog Stake Truck	4-3/4", nickel plated stamped steel spoke wheels	100	150	200
Moving Van	cast iron, 4-3/4" long 1930	150	225	300
Pickup Truck	cast iron, 4-3/4" long, 1926	100	150	200

Winross

AACA Hershey Region Fall Meet	sleeper single axle (White 7000 cab) tanker, (incentive)	95	100	125
AACA Hershey Region Fall Meet	long nose single axle (White 9000 cab), stk., wind screen doubles (incentive)	95	100	125
AACA Library and Research Center	FL/T stk. aerodynamic wind screen	60	65	75
ACME Printing	long nose tandem axle (White 9000)	30	45	50
Adirondack Beverage Co.	Ford cab (Ford C1 9000), long nose tandem axle, stacks, wind screen	20	25	35
Almond Joy	Ford cab (Ford C1 9000), long nose tandem axle, stacks, other side Mounds	45	50	60
Alpo	long nose tandem axle (White 9000 cab)	30	45	50
Amana	sleeper tandem axle (White 7000 cab)	35	45	50
American Red Cross	International 8300/T stacks Hanover top logo not Winross 1/600	55	65	75
Amoco Mileage Caravan	sleeper single axle (White 7000 cab) blue cab swing dolly	75	80	90
Anderson Windows	Internationl 8300/T stacks	50	60	65

Winross

NAME	DESCRIPTION	GOOD	EX	MINT
Andes Candies	Mack cab (Mack Ultra-liner), sleeper tandem axle (White 7000 cab), wind screen, reefer	40	50	60
Antique Car Show (Hershey)	International 8300/T stacks	55	65	75
Antique Car Show (Hershey)	Ford cab (Ford C1 9000), long nose tandem axle stacks, drop bed	55	65	75
Avis Truck Rental	cab over single axle (White 5000 cab)	25	35	45
Bicentennial Trail Issue	long nose tandem axle (White 9000 cab)	40	50	60
Bon Ton Potato Chips	Ford cab (Ford C1 9000), long nose single axle, chrome tanks	35	45	55
Bon Ton Potato Chips	Ford cab (Ford C1 9000), long nose single axle, black tanks, old suspension	90	100	125
Borden	milk tanker, screw replaces the rivet in the ear of the floor trailer, with ladder	80	90	110
Borden	Credit & Sales sleeper tandem axle (White 7000 cab), plastic dolly	35	40	50
Bowman Trans.	long nose tandem axle (White 9000 cab), stacks, wind screen	90	100	125
Bubble Yum	cab over single axle (White 5000 cab), red, plastic dolly	55	65	75
Bud Light (Fox Dist.)	Ford cab (Ford C1 9000), long nose tandem axle, stacks	90	100	125
Budd Movers	sleeper tandem axle (White 7000 cab), stacks, drop bed	135	175	200
Busch	Ford cab (Ford C1 9000), long nose tandem axle, stacks, wind screen (Hauck & Sons)	80	90	100
Butternut Coffee	sleeper tandem axle (White 7000 cab), stacks	25	35	45
California Raisins	Ford cab (Ford C1 9000), long nose tandem axle, stacks	60	70	80
Campbell's Soup	Mack cab (Mack Ultra-liner), sleeper tandem axle (White 7000 cab), aerodynamic wind screen, tanker Tomato Juice	70	80	90
Cerro Cooper	sleeper tandem axle (White 7000 cab), stacks	30	40	50
Cherry Hill Orchard	Ford cab (Ford C1 9000), long nose tandem axle, stack, full fairing T-Bird reefer	45	55	65
Cherry Hill Orchard	Mack cab (Mack Ultra-liner), sleeper tandem axle (White 7000 cab), aerodynamic wind screen, full fairing T-Bird reefer	45	55	65
Cola-Cola	long nose single axle (White 9000 cab) unpainted doors	100	125	150
Cola-Cola	long nose single axle (White 9000 cab) red plain doors	100	125	150
Cola-Cola	Ford Aeromax 120 cab, aerodynamic conventional sleeper, tandem axle, stacks, Dearborn Convention	175	200	300
Coleman's Ice Cream	Ford cab (Ford C1 9000), long nose tandem axle, stack, aerodynamic wind screen tool box sleeper	40	50	60
Colorado Beef	long nose single axle and long nose tandem axle (White 9000 cab) and stacks	30	40	50
Coors	Ford cab (Ford C1 9000), long nose tandem axle, stack, full fairing drop bed "Bill Elliott"	185	200	250
Coors	Ford Aeromax 120 cab, an aerodynamic conventional sleeper, stack	55	65	75
Corning	sleeper tandem axle (white 7000 cab), "Lots for You" (both sides shown)	30	40	50
Cracker Jack	silk screen, plastic dolly	200	250	300
Dairymen	Ford cab (Ford C1 9000), stack, tanker	90	100	125
Dannon Yogurt	cab over single axle with Beatrice logo	25	30	35

NAME	DESCRIPTION	GOOD	EX	MINT
Diamond Crystal Salt	silk screen	50	55	60
Diefenbach's Potato Chips	Ford cab (Ford C1 9000), long nose tandem axle, stack, 25th Anniversary	40	50	60
Domino's	Kenworth T800, tandem axle, stack, T-Bird reefer	40	50	60
Downy's Honey Butter	Ford cab (Ford C1 9000), long nose tandem axle, stack, T-Bird reefer, chassis cylinder (fuel tank for a reefer)	40	50	60
Eastman Kodak	long nose tandem axle (White 9000 cab), metal dolly, Kodak logo	55	65	75
Eastman Kodak	Mack ultra-liner cab, sleeper tandem axle (White 7000 cab), stacks, drop bed, (racing team)	350	400	500
Eastwood Company	Ford Aeromax 120 cab, an aerodynamic conventional sleeper, long nose tandem axle, stack, wind screen, doubles	40	50	60
Eastwood Company	Ford cab (Ford C1 9000), long nose tandem axle, stack, turbo wind screen (Motor Sports)	55	65	75
Eastwood Company	Kenworth T800, tandem axle, stack, turbo wind screen, parabolic shape, straight truck	55	65	75
Emergency Fire	1500 cab light or dark red	100	125	150
Emergency Fire	3000 cab white	300	325	350
Evergreen Juice Co.	Mack cab (Mack Ultra-liner), sleeper tandem axle (White 7000 cab), stack	30	40	50
Firestone	International 8300, tandem axle	55	65	75
Florigold (Sealed Sweet)	sleeper tandem axle	55	65	75
Ford, Story of	Ford Aeromax 120 cab, tandem axle, stack, yellow cab (#1), 1905	150	175	200
Fourth of July	Ford cab (Ford C1 9000), long nose tandem axle, stack, aerodynamic wind screen	55	65	75
Foxx Paper (fictional company)	3000 cab, 32' flat bed with side boards and simulated paper roll load	20	25	30
Georgia Pacific	Mack cab (Mack Ultra-liner), sleeper tandem axle (White 7000 cab), stack, wind screen, drop bed	250	300	350
Girl Scout Cookies	Ford cab (Ford C1 9000), long nose tandem axle, stack	55	65	75
Glade Spinfresh	long nose single axle (White 9000 cab)	25	30	35
Good & Plenty Candy	International 8300, tandem axle, stack	30	40	50
Good Poultry Services	Ford cab (Ford C1 9000), long nose tandem axle, stack, wind screen, tanker	30	40	50
Goodwill	Ford cab (Ford C1 9000), long nose single axle, stack	30	40	50
Goodwrench	International 8300, tandem axle, stack, drop bed #3 Dale Earnhart with cars	100	125	150
Graebel	sleeper tandem axle (White 7000 cab), stack, wind screen, drop bed (Movers)	75	85	100
Graebel	sleeper tandem axle (White 7000 cab), stack, wind screen, drop bed (Van Line)	30	35	45
Great American Van Lines	single tandem axle (White 7000 cab), stack, drop bed	70	75	90
H & H Excavating	Ford cab (Ford C1 9000), long nose tandem axle, stack	30	40	50
H & R Block	Mack cab (Mack Ultra-liner), sleeper tandem axle (White 7000 cab), stack	30	40	50
Halls	long nose tandem axle, plastic dolly, smooth front trailer	90	100	125
Hanover Brands	silk screen, swing dolly (both sides)	125	150	200

NAME	DESCRIPTION	GOOD	EX	MINT
Hanover Transfer Co.	International 8300, tandem axle, stack	70	85	100
Hardee's	Mack cab (Mack Ultra-liner), sleeper tandem axle (White 7000 cab), stack, reefer, under chassis cylinder (fuel tank for a reefer)	65	70	80
Hawaiian Punch	sleeper tandem axle (White 7000 cab), wind screen	100	125	150
Hershey's Chocolate	Ford cab (Ford C1 9000), long nose tandem axle, stack, red cab foil tanker	100	125	150
Hershey's Chocolate	Kenworth T800, tandem axle, stack, Strawberry Syrup tanker	75	85	100
Hershey's Chocolate	sleeper single axle (White 7000 cab), foil tanker	100	125	150
Hershey's Chocolate	Ford cab (Ford C1 9000), long nose tandem axle, stack, aerodynamic wind screen, milk tanker	75	85	100
Hertz	long nose single axle (White 9000 cab), 32' wheel, dolly cast doors	75	85	100
Hess Mills (Purina Chows)	Ford cab (Ford C1 9000), long nose tandem axle, stack. vert. brush tanker	40	50	60
Hostess Cake	Ford cab (Ford C1 9000), long nose tandem axle, stack, wind screen, other side Wonder	60	70	80
Iceland Seafood	sleeper single axle (White 7000 cab)	70	80	90
Iola Car Show	long nose single axle (White 9000 cab)	20	25	30
Iola Car Show	cab over single axle (White 5000 cab)	20	25	30
James River Corp.	Mack cab (Mack Ultra-liner), sleeper tandem axle (White 7000 cab), stack, wind screen	35	45	55
Jeno's Pizza	sleeper single axle (White 7000 cab), pin, plastic dolly	50	55	60
Johnson Wax	sleeper tandem axle (White 7000 cab), vert. brush tanker "Innobulk"	25	30	40
Juice Bowl	sleeper tandem axle (White 7000 cab), wind screen	45	55	65
Kraft	International 8300, tandem axle, stack, T-Bird reefer, "America Spells Cheese" top logo	60	70	80
Lancaster Farm Toy Show	Mack cab (Ultra-liner), sleeper tandem axle, brown cab, flatbed with J.D. farm equipment	150	175	200
Lancaster Farm Toy Show	Mack cab (Ultra-liner), sleeper tandem axle, blue cab, flatbed with International farm tractor	150	175	200
Lancaster Farm Toy Show	Mack cab (Ultra-liner), sleeper tandem axle, red cab, flatbed with Ford farm tractor	150	175	200
Lancaster Farm Toy Show	Ford cab (Ford C1 9000), long nose tandem axle, stack, met, brown, flatbed with J.D. farm equipment	150	175	200
Lancaster Farm Toy Show	Ford cab (Ford C1 9000), long nose tandem axle, stack, met, slate blue, flatbed with J.D. farm equipment	150	175	200
Lancaster Farm Toy Show	Ford cab (Ford C1 9000), long nose tandem axle, stack, met, maroon, flatbed with J.D. farm equipment	150	175	200
Lea & Perrins	sleeper tandem axle (White 7000 cab)	25	30	40
Leinenkugel Brewery	Ford cab (Ford C1 9000), long nose tandem axle, stack	35	45	55
Londonderry Fire Co.	Ford cab (Ford C1 9000), long nose tandem axle, stack, vert. brush tanker	40	50	60
Lysol	long nose single axle (White 9000 cab)	25	30	40
Mack Trucks "Story of Mack Trucks Set #1"	Mack cab (Mack Ultra-liner), sleeper tandem axle (White 7000 cab), stack 1893 new & old suspension	125	150	175

Winross

NAME	DESCRIPTION	GOOD	EX	MINT
Mack Trucks "Story of Mack Trucks Set #1"	Mack cab (Mack Ultra-liner). sleeper tandem axle (White 7000 cab), stack 1905 new & old suspension	125	150	175
Mack Trucks "Story of Mack Trucks Set #1"	Mack cab (Mack Ultra-liner). sleeper tandem axle (White 7000 cab), stack 1909 new & old suspension	125	150	175
Martin's Potato Chips	Ford cab (Ford C1 9000), long nose single axle, stack, aerodynamic wind screen, white cab	35	45	55
Maxwell House (Sterling Martin)	Ford Aeromax 120 cab, an aerodynamic conventional sleeper	55	65	75
McDonald's	sleeper tandem axle (White 7000 cab), Martin Brower, T-Bird reefer	40	50	60
Michelob Fox District	Ford cab (Ford C1 9000), long nose tandem axle, stack, tanker	100	110	125
Monfort	sleeper tandem axle (White 7000 cab)	100	110	125
Morton Salt	sleeper single axle (White 7000 cab)	70	80	90
Mountain Dew	International 8300, tandem axle, stack	35	45	55
Mrs. Paul's	International 8300, tandem axle, stack, tanker	60	70	80
Mt. Joy Co-op	Ford cab (Ford C1 9000), long nost tandem axle, foil tanker with graphics	70	85	100
Nabisco	sleeper single axle (White 7000 cab), Jr. Mints-Chuckles both sides shown	85	95	110
National Private Trucking	Ford cab (Ford C1 9000), long nose tandem axle, stack	20	25	30
National Toy Show	long nose single axle, Ford "F" Series, Ford cab with pop-up hood, two tractors both sides and cabs	75	85	100
Nestle's Quik	sleeper tandem axle (White 7000 cab), vert. brush	70	80	90
Old Milwaukee	long nose single axle (White 9000 cab), metal dolly, black chassis	70	80	90
Old Style (Heileman Brewery)	sleeper tandem axle (White 7000 cab), plastic dolly, one shield	55	65	75
Old Toyland Shows	long nose single axle (White 9000 cab), stacks, white cab both sides shown	25	30	35
Owens Corning Fiberglass	sleeper single axle (White 7000 cab), vert. brush tanker	40	50	60
P.I.E. Nationwide	long nose tandem axle (White 9000 cab), wind screen	55	65	75
P.I.E. Nationwide	Ford cab, long nose single axle, wind screen, stacks, doubles, with Olympic rings	300	350	400
Pennsylvania Pump Primers	Ford cab (Ford C1 9000), long nose tandem axle, stack, tool box #1	30	40	50
Pepsi	screw replaces the rivet in the rear of the floor of the trailer, plastic dolly	150	175	200
Pepsi	International 8300, tandem axle, stack, special edition	150	175	200
Pillsbury	sleeper tandem axle (White 7000 cab), Hungry Jack/Crescent Rolls	80	90	100
Prince Spaghetti	long nose single axle (White 9000 cab)	55	65	75
Quaker Oats	Mack cab (Mack Ultra-liner), sleeper tandem axle (White 7000 cab), stack, Kankakee Distribution Center	30	40	50
Quaker State	sleeper single axle (White 7000 cab), swing dolly	100	150	200
RCA	long nose tandem axle (White 9000 cab), stack, "Home Video"	80	85	100
Reading Railroad	long nose tandem axle (White 9000 cab), no stack	65	70	80

Winross

NAME	DESCRIPTION	GOOD	EX	MINT
Reading Railroad	long nose tandem axle (White 9000 cab), one stack	85	90	100
Red Ball Movers	sleeper tandem axle (White 7000 cab), van red, metal dolly	80	90	100
Red Hawk Racing	Ford Aeromax 120 cab, an aerodynamic conventional sleeper, stack, tandem axle, double bed, Jeff McClure	90	100	120
Reese's	long nose tandem axle (White 9000 cab), cab logo, foil tanker, (peanut butter cups)	35	45	55
Rochester Smelting	long nose tandem axle (White 9000 cab), metal dolly, flat bed with block load	150	175	200
Sakrete	screw replaces the rivet in the rear of the floor of the trailer	85	95	100
Schmidt's Beer	long nose tandem axle (White 9000 cab)	100	125	150
Seven Up	sleeper single axle (White 7000 cab), red wheels, red metal dolly	90	100	125
Shasta	sleeper single axle (White 7000 cab)	30	35	45
Silver Spring Fire Co.	Mack cab (Mack Ultra-liner), sleeper tandem axle (White 7000 cab), stack, tanker	40	50	60
Simon Candy	Mack cab (Mack Ultra-liner), sleeper tandem axle (White 7000 cab), stack, full fairing clear sided doubles; with candy	55	65	75
Snyder's of Hanover	Mack cab (Mack Ultra-liner), sleeper single axle (White 7000 cab), stack	90	100	110
Sony	Mack cab (Mack Ultra-liner), sleeper tandem axle (White 7000 cab), stack, wind screen	60	65	75
Spickler's	Ford cab (Ford C1 9000), long nose tandem axle, stack, vert. brush, tanker	45	55	65
Stephens Boat Works (fictional company)	long nose single axle (White 9000 cab), blue 32'	150	175	200
Sunoco	Ford Aeromax 120 cab, an aerodynamic conventional sleeper, tandem axle, stack, drop bed, Terry Labonte	200	250	300
Sunoco	Ford cab (Ford C1 9000), long nose tandem axle, stack, full fairing drop bed "Ultra Racing Team" (Marlin) plain	125	150	175
SuperAmerica	long nose tandem axle (White 9000 cab), tanker	30	35	45
Superbubble	sleeper tandem axle (White 7000 cab)	25	30	35
Timberline	sleeper tandem axle (White 7000 cab), stack	35	45	55
TMI (Three Mile Island)	sleeper tandem axle (White 7000 cab), stack, wind screen, flat bed, nuc. waste load, each numbered	100	125	150
Toledo Toy Show	International 8300, tandem axle, stack	25	30	40
Totinos	sleeper tandem axle (White 7000 cab), stack, T-Bird reefer	35	40	50
Transport for Christ (Mobile Chapel)	Mack cab (Mack Ultra-liner), sleeper tandem axle (White 7000 cab), stack, wind screen	35	40	50
Transport Topics	long nose tandem axle (White 9000 cab), tanker, swing dolly	90	100	125
Tyson Foods	sleeper tandem axle (White 7000 cab), wind screen, reefer, "America's Choice"	80	90	110
U.S. Brands	sleeper tandem axle (White 7000 cab), wind screen	25	35	45
U.S. Gypsum	sleeper tandem axle (White 7000 cab), blue with red letters	55	65	75
U.S. Mail	cab over single axle (White 5000 cab), "Zip" blue metal dolly	55	65	75

Winross

NAME	DESCRIPTION	GOOD	EX	MINT
U.S. Steel	long nose tandem axle (White 9000 cab), long wheel base cab, green, 32' flat bed, silver I-beam	90	100	125
Union Carbide	sleeper tandem axle (White 7000 cab), vert. brush, tanker	55	65	75
Unique Garden Center	sleeper tandem axle (White 7000 cab), stack, reefer	25	30	40
United Auto Workers (UAW)	Ford cab (Ford C1 9000), long nose tandem axle, stack (America Works)	65	70	75
United Way	Ford cab (Ford C1 9000), long nose tandem axle, stack, tanker (Collector Model 1 of 500)	30	40	50
Warner-Lambert	Ford cab (Ford C1 9000), long nose tandem axle, aerodynamic wind screen, "Efferdent" on side	25	30	35
Watergate	cab over single axle (White 5000 cab)	20	25	30
Weaver Chicken	sleeper tandem axle (White 7000 cab), white metal dolly "Country Style"	80	90	110
Westmans 32' Transport Tanker	3000 cab with wheels, blue with red, logo on trailer, wheel dolly	175	200	250
Westmans 32' Transport Tanker	long nose single axle (White 9000 cab), white trailer, red letters, wheel dolly	150	175	200
White Oak Mills	sleeper tandem axle (White 7000 cab), stack, tanker with catwalk	90	100	110
Wilbur Chocolate	Ford cab (Ford C1 9000), long nose tandem axle, stack, aerodynamic wind screen, T-Bird reefer, "Wilbur Buds"	60	65	75
Winross at Dyersville	Ford Aeromax 120 cab, an aerodynamic conventional sleeper, stack, Erie Canal	65	70	75
Winross at Hershey	Ford Aeromax 120 cab, an aerodynamic conventional sleeper, tandem axle, stack, quilts, different shade pink wind screen (Collectors Series)	40	45	50
Winross at Hershey	Kenworth T800, tandem axle, tent scene	30	40	50
Winross at Hershey	Mack cab (Mack Ultra-liner), tandem axle, gray stacks, aerodynamic wind screen, car restoration, Collector Series	30	40	50
Winross at Hershey	Mack cab (Mack Ultra-liner), sleeper tandem axle (White 7000 cab), aerodynamic wind screen, green cab	30	40	50
Winross at Hershey	Mack cab (Mack Ultra-liner), tandem axle, chrome stacks, aerodynamic wind screen, car restoration	30	40	50
Winross at Hershey	Ford Aeromax 120 cab, an aerodynamic conventional sleeper, tandem axle	75	80	90
Winross at Hershey	Ford Aeromax 120 cab, an aerodynamic conventional sleeper, stack, tandem axle, quilts	75	80	90
Winross at Hershey	International 8300, tandem axle, Hershey Commemorative	40	45	50
Winross at Hershey	Ford cab (Ford C1 9000), long nose tandem axle, stack, aerodynamic wind screen, blue cab	30	40	50
Winross at Hershey	Mack cab (Mack Ultra-liner), sleeper tandem axle (White 7000 cab), aerodynamic wind screen, farm scene	30	40	50
Winross at Rochester	Kenworth T800, tandem axle, stack, Erie Canal	30	40	50
Winross Hospitality Day	Mack cab (Mack Ultra-liner), tandem axle, stack, wind screen, blue cab, "You've Got a Friend in PA"	50	55	60
Winston Motor Sports	Ford Aeromax 120 cab, an aerodynamic conventional sleeper, stack, 20-year anniversary	45	50	55

Winross

NAME	DESCRIPTION	GOOD	EX	MINT
Wonder Bread	Ford cab (C1 9000), long nose tandem axle, stack, wind screen, other side "Hostess"	70	75	80
Wyler's	sleeper single axle (White 7000 cab), pin, plastic dolly, "Realemon" on both sides	25	30	40
Y & S Candies	Mack cab (Mack Ultra-liner), sleeper tandem axle (White 7000 cab), stack, wind screen	55	65	75
Yellow Freight System	screw replaces the rivet in the rear of the floor of the trailer, rib trailer, plastic dolly	25	30	40
Yoplait Yogurt	sleeper tandem axle (White 7000 cab), stack, wind screen, reefer, trailer edged white	30	40	50
Zeager Bros. Inc.	Ford cab (C1 9000), long nose tandem axle, stack, flat bed with two lumber stacks	100	125	150
Zembo Temple	Ford cab (C1 9000), long nose tandem axle, stack, wind screen, white wheels	85	95	100

Wyandotte

Airplanes

Army Bombing Plane	8-1/2" wingspan; rubber wheels	25	65	85
Autogyro Plane	9-1/4" long; rubber wheels	75	190	300
China Clipper Airplane	9-1/4" long; rubber wheels	90	200	300
Defense Bomber Airplane	9-1/4" long; "U.S. Army" marked on wing	70	150	225
Stratoship Mystery Plane	4-1/4" long	10	25	40

Boats

Battleship	6" long, rubber wheels	35	55	70
S.S. America Boat	12-1/2" long, wood wheels	40	75	100

Cars

Air Speed Coupe	5-7/8" long, rubber wheels	40	75	95
Coupe	6-1/2" long, wood or rubber wheels	40	75	90
Rocket Racer	6" long	50	75	100
Soap Box Derby Racer	6-1/4"	45	100	150
Station Wagon	woodgrain and passenger lithography; 21" long	200	300	450
Zephyr Racer	10" long	75	125	150
Zephyr Roadster	13-3/8" long; rubber wheels	400	600	750

Emergency Vehicles

Ambulance	6", wood wheels	45	70	90
Ambulance	11-1/4" long; Red Cross and "Wyandotte Toys" on side	75	150	185

Miscellaneous

Auto Hauler Set	trailer, sedan, dump truck, wooden wheels; 18-1/2" long	175	325	375
Coast to Coast Bus Lines	21" long; rubber wheels	200	300	450
Flash Strat-O-Wagon	6" long; rubber wheels	35	65	90
Gasoline and Service Station	Shell station; w/two vehicles	250	550	800

Sedan, Wyandotte

Stake Truck, 1930s, Wyandotte

Wyandotte

NAME	DESCRIPTION	GOOD	EX	MINT
Land Cruiser Auto Set	sedan w/trailer; 11-3/4" long	50	100	150
Sand Hopper Set	7" high; w/shovel	20	35	50
Streamlined Wagon	5-1/4" long, rubber wheels	25	35	50

Trucks

NAME	DESCRIPTION	GOOD	EX	MINT
Circus Truck and Wagon	19" long; red/yellow; w/cardboard animals	650	950	1400
Contractors' Truck	11-1/4" long; metal wheels; w/miniature wheelbarrow	70	125	175
Engineer Corps Truck	17-1/2" long; wood wheels; marked "Army Engineer Corps"	75	130	190
Gasoline Truck	21" long; rubber wheels	100	200	300
Hook `n Ladder Truck	10-1/4" long, detachable ladders, rubber wheels	90	130	160
Ice Truck	11-1/2" long, wood wheels	100	190	250
Medical Corps Truck	11-3/4" long; metal wheels; "U.S.A. Medical Corps" on side	90	175	250
Milk Truck	11-1/2" long, wood wheels	100	190	250
Semi-Trailer Dump Truck	17-3/8" long; rubber wheels	60	90	150
Stake Truck	6-3/4" long, rubber or wood wheels	35	65	80

View-Master Reels

View-Master, the popular 3-D collectible, was first introduced at the 1939 New York World's Fair. Originally considered a "photographic" souvenir, it has since become a favorite toy of young and old alike.

During the 1940s and 1950s, the company mainly produced reels for various national parks and other scenic attractions. During World War II, View-Master produced millions of reels for the U.S. government to aid in airplane and ship identification and range estimation.

The company began producing reels for children with the development of their Fairy Tale series in the late 1940s. They went on to produce hundreds of other favorite cartoon characters. In 1951, View-Master purchased the competition, the Tru-Vue Company. Since Tru-Vue held the license to use Disney characters, this acquisition was a coup for View-Master.

Since 1939, the style and construction of the View-Master viewer has greatly changed. Originally made of black Bakelite, the first models were round. The Model "C," which debuted in 1946, was the first square viewer and became the model for all viewers produced since. This model is easily-found today and very affordable. Character viewers were added to the line in 1989 and have since become the mainstay of the View-Master product line.

A new 3-D product created for 1997 distribution is a talking View-Master, which will feature scenes and sound from several motion pictures.

Tyco Toys purchased View-Master in 1989. In 1996, Mattel Toys acquired Tyco, bringing the View-Master name under its umbrella.

In pricing reels, the important thing to remember is condition. Three-reel packets consist of the reels, a book (if indicated on the packet reverse side), and the outer envelope with full-color picture. If any element is missing, the overall price drops dramatically. Also, if any part is torn or damaged in any way, the price should be adjusted accordingly.

Prices listed are for reels Mint in Package (MIP). Those in Excellent condition command about 85 percent of the MIP price. Note: All packages listed have three reels unless otherwise noted.

Trends

Prices for View-Master reels have risen consistently over the past several years, but the market appears to have topped off. Scenic titles continue to be modestly priced, but the demand for character reels, mostly by crossover character toy collectors, has resulted in a higher demand for these items. World's Fair titles are also very popular.

Bullwinkle

The Top 10 View-Master Reel Sets
(Mint in package; not including Movie Preview Reels)

1.	The Munsters	$200
2.	Joe 90	95
3.	The Addams Family	85
4.	World Bobsled Championships (ABC's WW/Sports)	80
5.	NCAA Track & Field Championships (ABC's WW/Sports)	75
6.	Cartoon Carnival with Supercar	75
7.	Steve Canyon	75
8.	Dr. Who	75
9.	Davy Crockett	75
10.	International Swimming & Diving Meet (ABC's WW/Sports)	70

Contributors to this section: Mary Ann and Wolfgang Sell, 3752 Broadview Dr., Cincinnati, OH 45208.

Buck Rogers BL015

VIEW-MASTER REELS

NO.	NAME	MNP	MIP
B506	$1,000,000 Duck	7	10
3014	101 Dalmatians	3	5
B532	101 Dalmatians	3	5
86	1939 New York World's Fair (single reel)	10	25
87	1939 New York World's Fair (single reel)	10	25
89	1939 New York World's Fair (single reel)	5	10
88	1939 New York World's Fair (single reel)	5	8
59	1940 Golden Gate International Expo (single reel)	5	8
58	1940 Golden Gate International Expo (single reel)	5	8
57	1940 Golden Gate International Expo (single reel)	10	20
56	1940 Golden Gate International Expo (single reel)	10	20
A272	1962 Seattle World's Fair (four reels)	20	30
A2726	1962 Seattle World's Fair (single reel)	7	10
A2725	1962 Space Needle U.S.A. (single reel)	7	10
B937	1970s America's Cup (ABC's WW/Sports)	45	40
B370	20,000 Leagues Under the Sea	7	15
4045	A-Team	6	12
BD199	Adam & the Ants	15	20
B593	Adam-12	10	12
B486	Addams Family	60	85
BD205	Adventures of Morph	7	8
3088	Aladdin	3	4
BD265	Alex	9	10
4082	Alf	5	10
4111	An American Tail II	5	6
N3	Annie	5	6
B470	Annie Oakley	12	25
B558	Apple's Way	14	16
B574	Archie	7	8
	Arena (Movie Preview Reel)	75	125
B365	Aristocats, The	5	6
B457	Astrix & Cleopatra	21	25
B948	Auto Racing, Phoenix 200 (ABC's WW/Sports)	35	50
B375	Babes in Toyland	21	25
H77	Bad News Bears in "Breaking Training"	7	8
B502	Banana Splits	10	20
BD239	Bananaman	9	10
BB492	Batman	13	15
B492	Batman	9	10
1086	Batman	4	5
3086	Batman - The Animated Series	2	3
4137	Batman Returns	7	8
4011	Batman, The Perfect Crime	7	8
L16	Battle Beyond the Stars	12	20
BD185	Battle of the Planets	13	15
3079	Beauty & the Beast	2	3
B366	Bedknobs & Broomsticks	5	10
1074	Beetlejuice	5	6
J51	Benji's Very Own Christmas	7	8
4018	Benji, Superstar	5	6
BD259	Bertha	7	8
B570	Beverly Hillbillies	12	25
B587	Big Blue Marble	10	12
D135	Black Beauty	9	10
BK035	Black Hole	7	10
K35	Black Hole	15	18
BD207	Bollie & Billie	7	8
BB487	Bonanza	10	20
B471	Bonanza	10	20
B487	Bonanza (w/o Pernell Roberts)	30	35
BD1484	Bozo	13	15
B568	Brady Bunch	17	20
B466	Brave Eagle	21	25
BD272	Bravestar	9	10
L15	Buck Rogers	7	8
4056	Buckaroo Banzai	10	12
965abc	Buffalo Bill, Jr.	21	25
B464	Buffalo Bill, Jr.	21	25
B531	Bugs Bunny	9	10
1077	Bugs Bunny & Tweety	2	3

View-Master Reels

NO.	NAME	MNP	MIP
800	Bugs Bunny and Elmer Fudd (one reel)	7	8
B549	Bugs Bunny, Big Top Bunny	5	6
M10	Bugs Bunny/Road Runner Show	3	4
B515	Bullwinkle	12	15
BD212	Button Moon	7	8
L1	Can't Stop the Music	17	20
H43	Captain America	4	5
755abc	Captain Kangaroo	14	16
B560	Captain Kangaroo	10	12
B565	Captain Kangaroo Show	10	12
BD264	Care Bears	5	6
B521	Cartoon Carnival with Supercar	50	75
BD171	Casimir Costureiro	7	8
BB533	Casper The Friendly Ghost	5	6
B533	Casper The Friendly Ghost	5	6
J22	Cat from Outer Space	7	8
1057	Centurions	3	4
	Charge at Feather River, The (Movie Preview Reel)	128	150
L2	Charlie Brown, Bon Voyage	4	5
B556	Charlie Brown, It's a Bird	4	5
1039	Charlie Brown, It's Your First Kiss	9	10
B321	Charlotte's Web	5	6
3075	Chip 'n Dale Rescue Rangers	4	5
L14	CHiPs	13	15
FT5	Cinderella (one reel)	1	2
960	Cisco Kid (one reel)	2	3
B496	City Beneath the Sea	15	30
J47	Close Encounters of the Third Kind	10	15
B461	Cowboy Stars	21	25
B564	Curiosity Shop	10	12
B498	Daktari	13	15
B463	Dale Evans	23	27
944abc	Dale Evans	26	30
BD214	Danger Mouse	15	18
	Dangerous Mission (Movie Preview Reel)	128	150
B479	Daniel Boone	15	20
4036	Dark Crystal	7	8
B503	Dark Shadows	30	60
935abc	Davy Crockett	64	75
BD244	Dempsey & Makepeace	7	8
B539	Dennis the Menace	3	4
1065	Dennis the Menace	2	3
B519	Deputy Dawg	30	50
	Devil's Canyon (Movie Preview Reel)	128	150
4105	Dick Tracy	7	8
BD188	Dick Turpin	9	10
4138	Dinosaurs	4	5
A177	Disneyland, Adventureland	6	20
A178	Disneyland, Fantasyland	6	20
A176	Disneyland, Frontierland	6	20
A175	Disneyland, Main Street U.S.A.	6	20
A180	Disneyland, New Orleans Square	6	20
A179	Disneyland, Tomorrowland	6	20
B525	Donald Duck	7	8
H2	Dr. Shrinker & Wonderbug	10	14
BD216	Dr. Who	35	75
BD187	Dr. Who	35	75
B324	Dracula	13	15
	Drums of Tahiti (Movie Preview Reel)	128	150
3055	Duck Tales	4	5
L17	Dukes of Hazzard	7	8
M19	Dukes of Hazzard #2	6	7
4000	Dukes of Hazzard 2	6	7
J60	Dumbo	7	8
BD1474	Dumbo	9	10
4058	Dune	7	12
4117	E.T. (reissued)	3	5
N7	E.T. The Extra-Terrestrial	15	18

Toby Tyler

Legend of the Lone Ranger

VIEW- MASTER

View-Master Reels

NO.	NAME	MNP	MIP
4001	E.T., More Scenes from	15	18
K76	Eight is Enough	13	15
H3	Electra Woman & Dyna Girl	7	8
4125	Elmo Wants to Play	2	3
B597	Emergency	10	12
BD122	Emil	10	12
A071	Expo 67 Montreal	12	25
A074	Expo 67 Montreal	12	25
A073	Expo 67 Montreal	12	25
	Expo 70 Osaka	20	40
	Expo 74 Spokane	5	10
BD251	Fabeltjes Krant	10	12
B571	Family Affair	21	25
4118	Family Matters	4	5
K66	Fang Face	5	6
K36	Fantastic Four	9	10
B546	Fantastic Voyage	10	12
B554	Fat Albert & Cosby Kids	5	6
BD269	Ferdy	9	10
B390	Fiddler on the Roof	21	25
	Flash Gordon in the Planet Mongo	21	25
	Flight to Tangier (Movie Preview Reel)	128	150
1066	Flintstone Kids	2	3
1080	Flintstones	2	3
L6	Flintstones	10	12
BB480	Flipper	10	12
B485	Flipper	10	12
BD189	Flying Kiwi	10	12
B495	Flying Nun	21	25
BJ013	Fonz, The	7	8
H54	For the Love of Benji	7	8
	Fort Ti (Movie Preview Reel)	128	150
L29	Fox & Hound	7	8
3000	Fox & the Hound, The (Disney)	4	5
1067	Fraggle Rock	4	5
4053	Fraggle Rock	4	5
B323	Frankenstein	13	15
	French Line, The (Movie Preview Reel)	128	150
4119	Full House	3	4

NO.	NAME	MNP	MIP
B585	G.I. Joe	13	15
L28	Garfield	3	4
950	Gene Autry (one reel)	2	3
951	Gene Autry, "The Kidnapping" (one reel)	2	3
1062	Ghostbusters, The Real	4	5
BD225	Gil & Julie	7	8
	Glass Web (Movie Preview Reel)	128	150
J23	Godzilla	13	15
B945	Gold Cup Hydroplane Races (ABC's WW/Sports)	34	40
FT6	Goldilocks and the Three Bears (one reel)	1	2
4064	Goonies	7	8
M7	Great Muppet Caper	4	5
B488	Green Hornet	30	60
4055	Gremlins	7	10
J10	Grizzly Adams	9	10
	Gun Fury (Movie Preview Reel)	128	150
B589	Gunsmoke	21	25
B552	Hair Bear Bunch	7	8
1081	Hammerman	3	4
	Hannah Lee (Movie Preview Reel)	128	150
BB586	Happy Days	7	8
J13	Happy Days	7	8
B586	Happy Days	9	10
B547	Hardy Boys	9	10
B590	Hawaii Five-O	17	20
B578	Herbie Rides Again	7	8
B588	Here's Lucy	43	50
956	Hopalong Cassidy (one reel)	2	3
955	Hopalong Cassidy (one reel)	2	3
	House of Wax (Movie Preview Reel)	213	250
4073	Howard the Duck	7	8
B343	Huckleberry Finn	5	8
B512	Huckleberry Hound & Yogi Bear	4	5
L32	I Go Pogo	7	15

NO.	NAME	MNP	MIP
	Inferno (Movie Preview Reel)	128	150
BD232	Inspector Gadget	7	8
B946	International Moto-Cross (ABC's WW/Sports)	15	30
B936	International Swimming & Diving Meet (ABC's WW/Sports)	60	70
H44	Ironman	2	3
T100	Isis	10	12
B367	Island at Top of the World	21	25
	It Came From Outer Space (Movie Preview Reel)	213	250
FT3	Jack and the Beanstalk (one reel)	1	2
BB393	James Bond, Live & Let Die	17	20
B393	James Bond, Live & Let Die	12	20
K68	James Bond, Moonraker	13	15
4041	Jaws 3-D	4	5
1059	Jem	6	7
	Jesse James vs. The Daltons (Movie Preview Reel)	128	150
L27	Jetsons	5	6
K27	Jim Henson's Muppet Movie	4	5
BD261	Jimbo and the Jet Set	7	8
B456	Joe 90	70	95
BB454	Joe Forrester	9	10
B468	Johnny Mocassin	21	25
937abc	Johnny Mocassin	21	25
B572	Julia	15	25
4150	Jurassic Park	6	8
B392	King Kong	7	8
	Kiss Me Kate (Movie Preview Reel)	128	150
4054	Knight Rider	5	6
B557	Korg 70,000 B.C.	10	12
B598	Kung Fu	13	15
B504	Lancelot Link Secret Chimp	21	25
B494	Land of the Giants	25	50
B579	Land of the Lost	9	15
H1	Land of the Lost 2	9	15
BD190	Larry the Lamb	7	8
B472	Lassie & Timmy	13	15
B480	Lassie Look Homeward	9	10
B489	Lassie Rides the Log Flume	13	15
4057	Last Starfighter, The	5	6
B497	Laugh-In	13	15
J20	Laverne & Shirley	5	6
4092	Legend of Indiana Jones	5	6
4033	Legend of the Lone Ranger	5	6
L26	Legend of the Lone Ranger	7	8
BD203	Les Maitres Du Temps	7	8
B940	Little League World Series (ABC's WW/Sports)	43	50
3078	Little Mermaid	2	3
3089	Little Mermaid - TV Show	2	3
B319	Little Red Hen/ Thumbelina/Pied Piper	7	8
B465	Lone Ranger	21	25
962abc	Lone Ranger, The	21	25
B482	Lost in Space	40	60
	Lost Treasures of the Amazon (Movie Preview Reel)	128	150
B501	Love Bug, The	9	10
B455	Lucky Luke vs. The Daltons	21	25
BJ011	M*A*S*H	5	6
J11	M*A*S*H	9	10
B441	Magic Roundabout, The	13	15
BD182	Maja the Bee	7	8
B484	Man from U.N.C.L.E.	25	35
BB450	Mannix	21	25
B376	Mary Poppins	7	8
BB372	Mary Poppins	7	8
1056	Mask	4	5
	Maze, The (Movie Preview Reel)	128	150
BD217	Metal Mickey	7	8

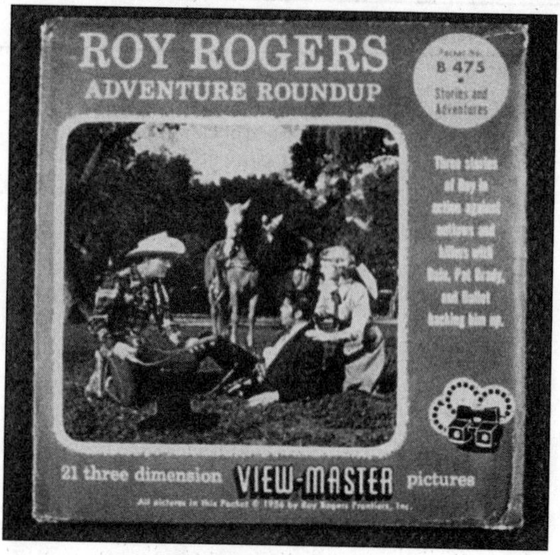

One of numerous Roy Rogers sets.

Zorro

View-Master Reels

NO.	NAME	MNP	MIP
K46	Meteor	9	10
D122	Michael	26	30
4047	Michael Jackson's Thriller	4	5
B528	Mickey Mouse	5	6
B551	Mickey Mouse - Clock Cleaners	9	10
865abc	Mickey Mouse Club	21	25
B524	Mickey Mouse Club Mouseketeers	21	25
J29	Mickey Mouse Jubilee	9	10
B526	Mighty Mouse	17	20
BB526	Mighty Mouse	4	5
	Miss Sadie Thompson (Movie Preview Reel)	128	150
B505	Mission Impossible	15	18
B478	Mod Squad	15	18
	Money From Home (Movie Preview Reel)	128	150
B493	Monkees	12	25
K67	Mork & Mindy	7	8
740	Movie Stars I (one reel)	13	15
741	Movie Stars II (one reel)	13	15
742	Movie Stars III (one reel)	13	15
H56	Mr. Magoo	5	6
BD197	Munch Bunch	7	8
B481	Munsters	105	200
4005	Muppet Movie, Scenes From The	4	5
L25	Muppets Go Hawaiian, The	4	5
K26	Muppets, Meet Jim Henson's	4	5
BK026	Muppets, The	4	5
B573	Nanny & The Professor	30	35
B935	NCAA Track & Field Championships (ABC's WW/Sports)	50	75
	Nebraskan, The (Movie Preview Reel)	128	150
H9	New Mickey Mouse Club	5	6
B566	New Zoo Revue	13	15
B567	New Zoo Revue 2	13	15
B443	Old Surehand	43	50
B377	One of Our Dinosaurs Is Missing	10	12
A0250	Oregon Centennial Exposition (1959)	30	50
BD266	Orm & Cheap	10	12
B592	Partridge Family	21	25
BB5924	Partridge Family	17	20
B569	Partridge Family	17	20
4074	Pee-Wee's Playhouse	9	10
B943	Pendelton Round-Up (ABC's WW/Sports)	30	35
BD184	Perishers, The	4	5
H38	Pete's Dragon	7	8
	Peter Pan, Disney's	5	6
J12	Pink Panther	5	6
BJ012	Pink Panther	4	5
750	Pinky Lee's 7 Days (one reel)	21	25
B322	Pippi Longstocking	13	15
D113	Pippi Longstocking	13	15
B507	Planet of the Apes	30	35
BB507	Planet of the Apes	30	35
3013	Pluto	4	5
BB529	Pluto	4	5
B529	Pluto	9	10
B442	Polly in Portugal	26	30
D100	Polly in Venice	17	20
B516	Popeye	5	6
	Popeye Talking View-Master Set	9	10
B527	Popeye's Fun	7	8
BD226	Portland Bill	10	12
B391	Poseidon Adventure	21	25
BD218	Postman Pat	7	8
BD220	Pumcki	7	8
4068	Punky Brewster	7	8
4003	Puppets Audition Night	4	5
L9	Puppets Audition Night, The	4	5
FT1	Red Riding Hood (one reel)	2	3
BH026	Rescuers, The	5	6
H26	Rescuers, The	5	6
J25	Return to Witch Mountain	5	6
B467	Rin-Tin-Tin	13	15
930abc	Rin-Tin-Tin	13	15
	Robin Hood	21	25

View-Master Reels

NO.	NAME	MNP	MIP
B373	Robin Hood Meets Friar Tuck	17	20
4115	Rocketeer, The	7	8
BD240	Roland Rat Superstar	7	8
K20	Romper Room	7	8
BB452	Rookies, The	13	15
B475	Roy Rogers	21	25
948abc	Roy Rogers	21	25
B462	Roy Rogers	21	25
945	Roy Rogers (one reel)	2	3
B594	Run Joe Run	13	15
BD109	Rupert the Bear	10	12
BB453	S.W.A.T.	10	12
	Sangaree (Movie Preview Reel)	128	150
B553	Scooby Doo	4	5
1079	Scooby Doo	2	3
B591	Search	17	20
B452	Sebastian	26	30
D101	Sebastian	26	30
	Second Chance (Movie Preview Reel)	128	150
B535	Secret Squirrel & Atom Ant	9	10
BD208	Secret Valley	10	12
4066	Sesame Street - Follow That Bird	4	5
M12	Sesame Street - People in Your Neighborhood	4	5
4049	Sesame Street - People in Your Neighborhood	4	5
4051	Sesame Street Alphabet	2	3
4072	Sesame Street Baby Animals	4	5
4097	Sesame Street Circus Fun	4	5
4050	Sesame Street Counting	4	5
4077	Sesame Street Goes on Vacation	4	5
4085	Sesame Street Goes Western	4	5
4083	Sesame Street Nursery Rhymes	4	5
4052	Sesame Street Shapes, Colors	4	5
4017	Sesame Street Visits the Zoo	4	5
b368	Shaggy D.A.	10	12
B550	Shazam	4	5
BD270	Shoe People, The	10	12
B595	Sigmund & the Sea Monsters (correct issue numbers)	15	18
B559	Sigmund & the Sea Monsters, (wrong issue number)	15	18
1058	Silverhawks	4	5
B556	Six Million Dollar Man	13	15
B308	Sleeping Beauty, Disney's	4	5
B490	Smith Family, The	10	30
BD194	Smuggler	7	8
BD246	Smurf, Baby	7	8
N1	Smurf, Flying	4	5
N2	Smurf, Traveling	4	5
BD172	Smurfs	5	6
B544	Snoopy and the Red Baron	7	8
BD250	Snorkes	7	8
K69	Snow White & the Seven Dwarfs	7	8
FT4	Snow White (one reel)	3	4
BD262	Snowman	7	8
	Son of Sinbad (Movie Preview Reel)	128	150
B509	Space Mouse	10	20
BD150	Space: 1999	21	25
BB451	Space: 1999	21	25
K31	Spider-Man	9	10
H11	Spider-Man	10	12
BH011	Spider-Man	7	8
B555	Star Trek (Cartoon Series)	9	10
B499	Star Trek (TV Series), "Omega Glory"	21	25
K57	Star Trek - The Motion Picture	10	12
4095	Star Trek - The Next Generation	7	8
M38	Star Trek - Wrath of Khan	10	12
B582	Steve Canyon	64	75
	Stranger Wore a Gun, The (Movie Preview Reel)	128	150
1064	Superman	4	5
BJ78	Superman	5	6
B584	Superman (cartoon)	4	5

NO.	NAME	MNP	MIP
J78	Superman - The Movie	10	20
L46	Superman II	17	25
4044	Superman III	7	15
3081	Tailspin	4	5
B580	Tarzan	9	10
975	Tarzan (one reel)	4	5
976A	Tarzan Finds a Son (one reel)	3	4
976abc	Tarzan of the Apes	21	25
	Taza, Son of Cochise (Movie Preview Reel)	128	150
1073	Teenage Mutant Ninja Turtles	2	3
4114	Teenage Mutant Ninja Turtles - Movie II	3	4
4149	Teenage Mutant Ninja Turtles - Movie III	2	3
4109	Teenage Mutant Ninja Turtles - The Movie	3	4
BD243	Telecat	7	8
BD230	Terrahawks	10	20
	They Called Him Hondo (Movie Preview Reel)	128	150
BD238	Thomas the Tank Engine	6	10
H39	Thor	2	3
	Those Redheads from Seattle (Movie Preview Reel)	128	150
B453	Thunderbirds	43	50
B491	Time Tunnel	15	30
1076	Tiny Toon Adventures	2	3
BD205	Tiswas	7	8
B476	Toby Tyler	30	35
810	Tom & Jerry (one reel)	2	3
B511	Tom & Jerry, Two Musketeers	9	10
B581	Tom Corbett, Secret from Space	21	25
970abc	Tom Corbett, Space Cadet	21	25
B340	Tom Sawyer	7	8
D123	Tom Thumb	13	15
BB513	Top Cat	13	15
B513	Top Cat	13	15
B947	Tournament of Thrills (ABC's WW/Sports)	30	35
BD242	Tripods, The	10	12
M37	Tron	7	8
B477	TV Shows at Universal Studios	26	30
745	TV Stars I (one reel)	17	20
746	TV Stars II (one reel)	17	20
747	TV Stars III (one reel)	17	20
J28	Tweety & Sylvester	3	4
BD1161	Tweety & Sylvester	7	8
4043	Twice Upon a Time	3	4
B417	U.F.O.	38	45
BD198	Ulysses 31	7	8
BD224	Victor & Maria	7	8
B483	Voyage to the Bottom of the Sea	10	12
B596	Waltons, The	10	12
BB596	Waltons, The	10	12
J19	Welcome Back Kotter	10	12
4086	Who Framed Roger Rabbit	8	16
B473	Wild Bill Hickcock & Jingles	26	30
BD215	Willo the Wisp	7	8
BD231	Wind in the Willows	7	8
4084	Wind in the Willows	7	8
	Wings of the Hawk (Movie Preview Reel)	128	150
BB7284	Winnetou	21	25
B728	Winnetou	26	30
BB731	Winnetou	21	25
B728	Winnetou & Halfblood Apache	26	30
K37	Winnie the Pooh & The Blustery Day	7	8
J14	Wiz, The	17	20
FT45abc	Wizard of Oz	13	15
BD267	Wizard of Oz	7	8
BD131	Wombles, The	7	8
D131	Wombles, The	13	15
B522	Woody Woodpecker	13	15
820	Woody Woodpecker Pony Express Ride (one reel)	3	4
B949	World Bobsled Championships (ABC's WW/Sports)	60	80
BD185	Worzel Gummidge	7	8
4067	Wrestling Superstars	7	8
4140	Young Indiana Jones Chronicles	6	8
B469	Zorro	34	40

Lassie & Timmy

James Bond Live and Let Die

A look at how View-Master viewers have changed throughout the years. Most vintage viewers sell in the $15-$35 or more range. Newer plastic viewers have little value, although some character viewers may be sought by collectors.

Western Toys

A quarter doesn't buy what it used to. In the 1940s and 1950s, American cowgirls and cowboys could plunk down their hard-earned silver for genuine articles endorsed (or at least bearing the names of) their Western heroes.

Heroes were aplenty — Gene Autry, Hopalong Cassidy, the Lone Ranger, and, of course, Roy Rogers and Dale Evans.

Today, grownup cowpokes drive hard bargains to get their hands on well-preserved toys to evoke well-preserved memories. That's because they don't make toys, or heroes, like they used to.

Remembering those days is only a toy away.

In the 1940s, straight shooters like Buck Jones, Red Ryder and Tom Mix were everywhere — not only were they peering down from the movie screen, but their images also graced cereal boxes, comic books and toy rifles.

By the time Hopalong Cassidy, Gene Autry and Roy Rogers became household words, merchandising of cowboy toys was everywhere. Soon every child could wear, eat, play with, read or decorate his room with cowboy memorabilia.

The heyday of cowboy fever blossomed in the 1940s and 1950s. Stores like Sears Roebuck had entire sections dedicated to cowboy merchandise.

Then, toys ranged from 25 cents to $5. There were six-guns of every size, both in sets and individually. Generic guns, hats and other pieces can still be widely found today at reasonable prices. Toys featuring the "Big Four" (Hoppy, Lone Ranger, Gene

Hopalong Cassidy Mechanical Shooting Gallery, Automatic Toy

and Roy) command higher premiums. But other cowboy hereos enjoyed successful merchandising as well, including Rex Allen, Wild Bill Elliott, Tom Mix, The Cisco Kid and Tex Ritter.

Trends

Western toys, particularly those from the pre-World War II era, should only increase in collector interest and value. Paper ephemera, such as lobby cards, advertisements, food wrappers or club membership items, are especially valuable because they were often discarded and, ultimately, hard to find today.

Premium rings often bore the likenesses of cowboy heroes. These are among some of the most valuable items as well. Other premiums, offered by companies like Ovaltine, Kellogg's and Ralston, were available only by mail, adding to their frequent scarcity.

Beware of reproductions, however. In recent years, many unlicensed reproduction and fake items bearing the Lone Ranger, Roy Rogers or Hopalong Cassidy names have been encountered.

Roy Rogers' death in 1998 may result in a renewed interest — and increased value — in related memorabilia.

Editor's Note: Most Western toy guns are listed in the Guns section of this book.

The Top 10 Western Toys (in Mint condition)

1. Hopalong Cassidy Roller Skates, Rollfast $1,000
2. Hopalong Cassidy Radio, Arvin, 1950s 600
3. Hopalong Cassidy Cap Gun, Wyandotte, 1950s 600
4. Buck Jones Rangers Cowboy Suit, Yankiboy, 1930s 500
5. Roy Rogers Toy Chest, 1950s 500
6. Hopalong Cassidy Western Series Figures, Timpo, 1950s 475
7. Lone Ranger Record Player, Dekka, 1940s 450
8. Big Little Book Picture Puzzles, 1930s 425
9. Tom Mix Rodeorope, Mordt, 1930s 425
10. Roy Rogers Play Set, Amsco, 1950s 400

WESTERN TOYS

Annie Oakley

NAME	COMPANY	YEAR	DESCRIPTION	GOOD	EX	MIP
Annie Oakley Cut-Out Dolls	Watkins-Strathmore	1956	paper dolls	20	35	50
Annie Oakley Cut-Out Dolls	Whitman	1958	paper dolls	20	35	50
Annie Oakley Sewing Set	Pressman	1950s		20	35	50
Annie Oakley Sharpshooter Book	Golden	1956	Little Golden Book	3	5	10
Annie Oakley Sparkle PIcture Craft	Gabriel	1950s		20	30	40
Annie Oakley with Tagg and Lofty Cut-Out Dolls	Whitman	1955	paper dolls	20	35	50
Holster Set	Daisy	1950s	leather holsters and cap pistols	60	100	200
Puzzle	Milton Bradley	1955	boxed jigsaw puzle	5	10	25

Bonanza

NAME	COMPANY	YEAR	DESCRIPTION	GOOD	EX	MIP
Ben Cartwright and his Horse	American Character	1966	plastic figures	60	140	290
Bonanza Four-in-One Wagon	American Character	1960s		75	150	300
Foto Fantastiks Photo Coloring Set	Eberhard Faber	1960s		40	75	150
Hoss Cartwright and his Horse	American Character	1966	plastic figures	60	140	290
Hoss Range Pistol	Marx	1960s	carded plastic gun	35	65	125
Little Joe Cartwright Figure	American Character	1966		60	140	290
Mustang	American Character		the outlaw's horse	25	50	100
Outlaw Figure	American Character	1966		40	80	160
Palomino	American Character		Ben's horse	25	50	100
Pinto	American Character		Little Joe's horse	25	50	100
Puzzle	Saalfield	1960s	frame tray	10	20	40
Puzzle	Milton Bradley	1964		20	40	75
Stallion	American Character		Hoss's horse	25	50	100
Woodburning Set	ATF Toys	1960s		65	125	250

Daniel Boone

NAME	COMPANY	YEAR	DESCRIPTION	GOOD	EX	MIP
Daniel Boone Card Game	Ed-U-Cards	1965		7	10	15
Daniel Boone Figure	Remco	1964	5" tall, hard plastic body, vinyl head, cloth coonskin cap and long rifle	35	75	160
Daniel Boone Film Viewer	Acme	1964	w/two films	7	10	20
Daniel Boone Inflatable Toy	Multiple Toymakers	1965		25	50	75

Davy Crockett

NAME	COMPANY	YEAR	DESCRIPTION	GOOD	EX	MIP
3-D Moving Picture and Viewer	Armour-Cloverbloom Margarine	1950s	premium, cardboard viewer with 3-D cards	40	80	150
Alamo Construction Set	Practi-Cole	1950s		40	75	150
Auto-Magic Picture Gun	Stephens	1950s	projection gun w/films	40	75	150
Baby Davy Crockett Doll	Spunky	1950s	8" tall vinyl squeeze toy	20	40	75
Davy Crockett and His Horse Figures	Ideal	1950s	plastic	30	50	100
Davy Crockett Doll	Exel	1973	Legends of the West series, 9-1/2"	30	55	95
Davy Crockett Doll		1950s	8", hard plastic, sleepy eyes, leather clothes, two rifles, hat	60	130	250
Davy Crockett Frontier Target Game	American Toys		tin litho dartboard, darts	35	75	130
Davy Crockett Guitar	Peter Puppet Playthings	1950s	24" long, wood	80	175	350
Davy Crockett Push Puppet	Kohner Bros.	1950s		30	65	100
Davy Crockett Ring		1950s	bronze-colored	40	50	60
Dispatch Case	Neptune Plastics	1950s	vinyl pouch, yellow graphics	15	30	50
Frontier Rifle	Marx	1950s	32" rifle	40	75	150
Frontierland Pencil Case	Hassenfeld Bros.	1950s	cardboard	45	75	100
Frontierland Pioneer Scout Outfit	Eddy	1950s	costume set	50	100	250
Iron-Ons	Vogart	1950s	three transfers on sheet	15	35	50
Magic Paint with Water Pictures	Artcraft	1950s	construction and art set with Frontierland fort	75	150	275
Offical Davy Crockett Color TV Set	Lido		plastic viewer, four films	75	150	275
Official Davy Crockett Belt			on card	35	80	150
Official Davy Crockett Tool Kit	Liberty Steel Chest		with tools	75	175	250
Pocket Knife	Disney	1950s	single blade; 2", yellow	20	40	75
Pony Express Bank	Randing	1950s	brown/white cloth pouch with lock and key	15	30	50
Puzzle	Marx		Siege on the Fort	15	40	75
Puzzle	Whitman	1955	11-1/4" x 15"	15	40	75
Ride-On Bouncing Horse	Rich Toys	1950s	37" x 31", white plastic with black/yellow saddle	75	150	275
Sand Pail	Ohio Art	1950s	tin litho	20	30	50
Suspenders		1950s	tan with Crockett graphics	25	50	90
Travel Bag	Neevel	1950s	6-1/2" x 12" x 10" heavy cardboard with brass hinges & plastic handle	30	75	130
Wallet	Walt Disney	1950s	red vinyl with faux fur	20	45	80

Gene Autry

NAME	COMPANY	YEAR	DESCRIPTION	GOOD	EX	MIP
Frame Tray Puzzle	Whitman	1950s	wood	15	20	35
Gene Autry Drum Set	Colmor	1940s		50	100	250
Gene Autry Guitar	Emenee	1950s	plastic	55	100	200
Gene Autry Jump-Up Book	Adprint Limited London	1955		30	50	75

Roy Rogers
Nodder,
Japanese

Roy Rogers
Ranch Lantern,
Ohio Art

Gene Autry

NAME	COMPANY	YEAR	DESCRIPTION	GOOD	EX	MIP
Gene Autry Ranch Outfit	Henry	1940s	dress-up kit w/holster and gun	55	100	200
Gene Autry's Champion Slate	Lowe	1950s		15	25	45
Gene Autry's Stencil Book		1950s		15	30	50
Spurs	Leslie-Henry	1960s		25	55	160
Stringless Marionette	National Mask & Puppet	1950s	14-1/2" tall	45	75	150

Gunsmoke

Gunsmoke Puzzle	Whitman	1950s	boxed	19	34	60
Gunsmoke Puzzle	Whitman	1958	frame tray	18	33	55

Hartland Figures

Annie Oakley and Target	Hartland	1950s	9" plastic rider w/horse	70	140	250
Brave Eagle and White Cloud	Hartland	1950s	9" plastic rider w/horse	75	150	200
Bret Maverick	Hartland	1960s	5" plastic rider w/horse	20	40	80
Bret Maverick	Hartland	1950s	9" plastic rider w/horse	50	100	150
Buffalo Bill	Hartland	1950s	9" plastic rider w/horse	70	140	250
Cheyenne	Hartland	1960s	5" plastic rider w/horse	30	50	100
Cheyenne	Hartland	1950s	9" plastic rider w/horse	40	80	150
Cochise	Hartland	1950s	9" plastic rider w/horse	45	90	150
Dale Evans	Hartland	1950s	9" plastic rider w/horse	40	80	200
Davy Crockett	Hartland	1950s	9" plastic rider w/horse	45	90	150
Jim Bowie	Hartland	1950s	9" plastic rider w/horse	70	140	250
Lone Ranger	Hartland	1950s	9" plastic rider w/horse	45	90	150
Lone Ranger	Hartland	1960s	5" plastic rider w/horse	15	30	60
Marshal Matt Dillon	Hartland	1950s	9" plastic rider w/horse	50	100	175
Marshal Matt Dillon	Hartland	1960s	5" plastic rider w/horse	15	30	60
Paladin	Hartland	1950s	9" plastic rider w/horse	45	90	150
Paladin	Hartland	1960s	5" plastic rider w/horse	15	30	60
Roy Rogers and Trigger	Hartland	1960s	5" plastic rider w/horse	50	100	200
Roy Rogers and Trigger	Hartland	1950s	9" plastic rider w/horse	50	100	175
Tom Jeffords	Hartland	1950s	9" plastic rider w/horse	70	140	250
Tonto	Hartland	1960s	5" plastic rider w/horse	30	60	120
Tonto	Hartland	1950s	9" plastic rider w/horse	45	90	150
Wyatt Earp	Hartland	1950s	9" plastic rider w/horse	45	90	150
Wyatt Earp	Hartland	1960s	5" plastic rider w/horse	15	30	60

Have Gun, Will Travel

Have Gun, Will Travel Play Set	Prestige	1958		50	100	200
Have Gun, Will Travel Slate		1960s		10	25	50
Holster Set	Halco	1960s	holsters, belt, canteen	40	100	200
Paladin Checkers	Ideal	1960s	carded	25	50	75
Paladin Western Outfit	Ben Cooper	1959	mask, vest	30	65	150

Hopalong Cassidy

NAME	COMPANY	YEAR	DESCRIPTION	GOOD	EX	MIP
Automatic Television Set	Automatic Toy	1950s		60	100	200
Bar 20 Ranch Badge		1950s		25	50	75
Bar Twenty Shooting Game	Chad Valley	1950s		50	100	250
Bread Wrapper	Butternut	1950s	14" x 23"	50	70	95
Canvas School Bag		1950s		25	50	150
Cap Gun	Wyandotte	1950s	7", silver metal with white handles, Hoppy pictured on sides	150	300	600
Coloring Outfit	Transogram	1950s		35	75	150
Compass Hat Ring		1950s	brass, features "HC" on one side and "20" on other; compass on top	195	250	300
Crayon and Stencil Set	Transogram	1950s		35	75	150
Frame Tray Puzzle	Whitman	1950s		10	20	35
Hop-A-Long Cassidy Tin Toy	Marx	1950s	wind-up rocker toy	90	150	325
Hopalong Canasta	Pacific Playing Card	1950s		55	100	250
Hopalong Cassidy Chinese Checkers	Milton Bradley	1950		60	130	225
Hopalong Cassidy Figure	Ideal	1950s	plastic figure of Hoppy with Topper	40	65	150
Hopalong Cassidy Lasso Game	Transogram	1950s		35	75	150
Hopalong Cassidy Western Series	Timpo	1950s	British set of seven metal figures	100	250	475
Mechanical Shooting Gallery	Automatic Toy	1950s		75	150	350
Original Hopalong Pogo Stick		1950s		50	100	225
Picture Gun and Theatre	Stephens Products	1950s	projects film, Hoppy decals on sides, includes film, gun, cardboard stage	75	150	225
Pony Express Toss Game	Transogram	1950s		30	45	70
Puzzles	Milton Bradley	1950s	boxed set of three	30	50	100
Radio	Arvin	1950s	red metal case, embossed image of Hoppy and Topper on front	200	450	600
Roller Skates	Rollfast			200	500	1000
Stationary	Whitman	1950s		20	50	100
Topper Rocking Horse	Rich Toys	1950s	plastic/wood	65	125	250
Wrist Cuffs	1950s			55	100	250

How the West Was Won

NAME	COMPANY	YEAR	DESCRIPTION	GOOD	EX	MIP
Dakota Figure	Mattel	1978		10	20	35
How the West Was Won Action Play Set	Timpo	1977		15	25	50
Lone Wolf Figure	Mattel	1978		10	20	35
Puzzle	HG Toys	1978	boxed jigsaw puzzle	5	7	15
Zeb Macahan Figure	Mattel	1978		10	20	35

Lone Ranger Sheriff Jail Keys, Esquire Novelty, 1945

Roy Rogers & Dale Evans Western Dinner Set, Ideal

Lone Ranger

NAME	COMPANY	YEAR	DESCRIPTION	GOOD	EX	MIP
Electric Drawing Set	Lakeside	1960s		5	10	20
Filmstrip Saddle Ring	General Mills	1950s	gold ring holds silver finish viewer with adjustable focus	75	150	300
Flashlight Ring		1948	premium	45	90	175
Hand Puppet		1940s	cloth body, blue/white polka dot shirt with bells in both hands	40	90	160
Lone Ranger and Tonto Target Set	Multiple Toymakers	1970s	3" figures with horses, guns, darts	15	35	65
Lone Ranger Coloring Book	Whitman	1975	#1010	8	15	25
Lone Ranger Record Player	Dekka	1940s	12" x 10" x 6" wooden box with burned in illustrations, leather strap	100	225	450
Lone Ranger Sheriff Jail Keys	Esquire Novelty	1945	5", on ring, came on card with cut-out Sheriff card	30	63	125
Pencil Box	American Pencil	1940s		20	40	75
Tonto Indian Outfit	Esquire	1950s	costume set	50	100	350

Miscellaneous Characters

NAME	COMPANY	YEAR	DESCRIPTION	GOOD	EX	MIP
Bat Masterson Holster Set	Carnell	1950s	holster, belt, cane, yellow/black vest	75	200	300
Buck Jones Rangers Cowboy Suit	Yankiboy	1930s	costume set	60	200	500
Buffalo Bill Puzzle	Built-Rite	1956	frame tray	10	20	35
Cheyenne Book	Whitman	1958	Little Golden Book	6	13	24
Cheyenne Little Golden Record		1950s	45 rpm	7	15	25
Cheyenne Puzzle	Milton Bradley	1957	frame tray	12	25	45
Cisco Kid Puzzle Set	Saalfield	1950s	set of three puzzles	30	50	70
Johnny Moccasin View-Master Reels	Sawyers	1957	set of three reels, booklet	20	45	60
Johnny Ringo Hand Puppet	Tops in Toys	1950s	15" full body puppet, vinyl head	65	100	200
Laramie Cowboy Holster Set	Clarke Bros.	1960s		65	100	225
Rin Tin Tin Magic Slate	Whitman	1950s		10	20	30
Rin Tin Tin Rusty Costume	Ben Cooper	1950s		25	40	65
Straight Arrow Target Game	Novel Novelties	1949	tin litho target w/crossbow	40	75	150
Wild Bill Hickok and Jingles Puzzle Set	Built-Rite	1950s	set of four	25	50	75
Wild Bill Hickok Treasure Map/Guide	Kellogg's	1950s	cereal premium	40	65	90

Red Ryder

NAME	COMPANY	YEAR	DESCRIPTION	GOOD	EX	MIP
Frame Tray Puzzle	Jaymar	1951	Red Ryder or Little Beaver	5	10	20
Little Beaver Archery Set		1951	cardboard target	10	20	30
Little Beaver Coloring Book	Whitman	1956		5	10	25

Red Ryder

NAME	COMPANY	YEAR	DESCRIPTION	GOOD	EX	MIP
Red Ryder Corral Bagatelle Game	Gotham	1940s		25	45	75
Red Ryder Pop-Um Shooting Game	Daisy	1940s		30	50	100
Red Ryder Target Game	Whitman	1939		25	50	100
Red Ryder Whirli-Crow Game	Daisy	1940s		30	50	100

Roy Rogers

NAME	COMPANY	YEAR	DESCRIPTION	GOOD	EX	MIP
Dale Evans Wristwatch	Ingraham	1951	Dale inside upright horseshoe, tan background, chrome case, black leather band	60	150	300
Give-A-Show Projector	Kenner	1960s	projector w/slides	25	50	75
Horsedrawn Wagon Pull Toy	Hill	1950s	18", wood w/paper litho	55	100	250
Ranch Lantern	Ohio Art	1950s	tin litho, battery-operated	55	100	250
Roy Rogers & Dale Evans Coloring Book	Whitman	1975		10	20	30
Roy Rogers & Dale Evans Paper Dolls	Whitman	1954		35	55	80
Roy Rogers & Dale Evans Western Dinner Set	Ideal	1950s	utensils in 14" x 24" box	35	65	95
Roy Rogers Alarm Clock	Ingraham	1950s	4", square ivory-color case, full-color graphics	150	300	400
Roy Rogers Crayon Set	Standard Toykraft	1950s		35	50	100
Roy Rogers Fix-It Chuck Wagon and Jeep	Ideal	1950s	set with two horses, four figures, Nellybelle, accessories	70	150	300
Roy Rogers Horeshoe Set	Ohio Art		tin litho	40	100	250
Roy Rogers Horseshoe Set		1950s	rubber horseshoes and pegs	30	50	130
Roy Rogers Nodder	Japanese		composition; blue shirt, white hat and pants and red bandana and boots	90	165	250
Roy Rogers Play Set	Amsco	1950s	cardboard w/magnetic figures	50	200	400
Roy Rogers Puzzle		1950s	frame tray	20	35	50
Roy Rogers Rodeo Board Game	Rogden	1949	four games in one	100	140	200
Roy Rogers Rodeo Sticker Fun Book	Whitman	1953		35	55	110
Roy Rogers Stagecoach	Ideal	1950s	14", plastic with two harnessed horses, Roy, and accessories	40	65	90
Roy Rogers Toy Chest		1950s	17" x 17" x 12" w/Roy and Bullet graphics	175	300	500
Roy Rogers Toy Football		1950s	white vinyl w/logo	20	40	60
Roy Rogers Trigger Trotter		1950s	pogo stick	150	225	300
Roy Rogers Truck	Marx	1950s	14", tin litho, red/yellow/blue	100	150	250
Trigger Rocking Horse	Bell Toys	1950s	wood w/metal seat	50	100	200

Hopalong Cassidy Coloring Outfit, Transogram, 1950s

Wyatt Earp figure, Hartland

Tom Mix

NAME	COMPANY	YEAR	DESCRIPTION	GOOD	EX	MIP
Big Little Book Picture Puzzles		1930s	boxed set of two	100	225	425
Bullet Flashlight		1930s	3" long	50	75	130
Riding Horse	Mengel	1930s	wood	75	150	365
Tom Mix Rodeorope	Mordt	1930s		75	250	425
Tom Mix Shooting Gallery	Parker Brothers	1935		75	150	320

Zorro

NAME	COMPANY	YEAR	DESCRIPTION	GOOD	EX	MIP
Dart Rifle Target Set		1960s		40	70	175
Hand Puppet	Gund	1958	vinyl and fabric	25	50	75
Official Zorro Action Set	Marx	1950s	costume and weapons set	75	150	350
Paint by Number Set	Hasbro		canvas and paints	35	60	80
Pencil by Number Set		1960s		35	60	80
Puzzle	Jaymar	1960s	boxed jigsaw puzzle	7	20	35
Target Shoot		1950s	gun w/soldier targets	40	75	150
Walt Disney's The Adventures of Zorro	Golden	1958	Big Golden Book	10	20	30
Walt Disney's Zorro		1958	Little Golden Book	5	7	12
Walt Disney's Zorro	Golden	1958	Golden Book	5	10	20
Zorro Dominoes	Halsam	1950s		25	45	65
Zorro Fencing Set		1950s	swords, masks, face guards	45	75	150
Zorro Spring Action Target	Knickerbocker	1950s		45	75	150

Toy Manufacturers Directory

Have you ever wanted to reach a toy company but didn't know how? The following is a representative, but not all-inclusive, list of most of the current, notable toy manufacturers.

Remember: Most manufacturers are not able to answer questions about secondary market value of their toys. And many companies do not maintain comprehensive archives of past products, so they may not be able to answer questions about past toy lines. The addresses and phone numbers may be helpful, however, when trying to reach the company with a customer service question or product concerns.

Aladdin Industries (lunch kits)
703 Murfreesboro Rd.
Nashville, TN 37210
615-748-3000

Alexander Doll Company
615 W. 131st St.
New York, NY 10027-7982
212-283-5900

Applause (plush, plastic)
6101 Variel Ave.
Woodland Hills, CA 91365-4183
818-992-6000

Bachmann Industries (trains)
1400 E. Erie Ave.
Philadelphia, PA 19124
215-533-1600

Bandai America (action figures)
12851 E. 166th St.
Cerritos, CA 90701
310-926-0947

Binney & Smith (Crayola crayons)
1100 Church La., P.O. Box 431
Easton, PA 19042
610-253-6271

Cadaco (games)
4300 W. 47th St.
Chicago, IL 60632-4477
312-927-1500

Duncan Toys (yo-yos)
15981 Valplast Rd.
Middlefield, OH 44062
216-632-1631

Ertl (farm toys, die-cast)
Highways 136 & 20
P.O. Box 500
Dyersville, IA 52040-0500
319-875-5607

Fisher-Price (preschool toys)
636 Girard Ave.
East Aurora, NY 14052
716-687-3449

Galoob Toys, Lewis (action figures)
500 Forbes Blvd. S.
San Francisco, CA 94080
415-873-0680

Gund (plush)
1 Runyons Ln.
P.O. Box H
Edison, NJ 08818
908-248-1500

Hasbro
(Milton Bradley, Tonka, Kenner, Parker Brothers and Playskool)
1027 Newport Ave.
Pawtucket, RI 02862-1059
401-727-5582

James Industries (Slinky)
P.O. Box 407
Hollidaysburg, PA 16648
814-695-5681

Kenner Products (G.I. Joe)
615 Elsinore Pl.
Cincinnati, OH 45202
513-579-4927

Larami Corp. (miscellaneous)
340 N. 12th St.
Philadelphia, PA 19107-1123
215-923-4900

LEGO Systems (construction toys)
555 Taylor Rd.
P.O. Box 1600
Enfield, CT 06083-1600
203-763-6731

Lionel Trains
50625 Richard W. Blvd.
Chesterfield, MI 48051
810-949-4100

Little Tykes (preschool toys)
2180 Barlow Rd.
Hudson, OH 44236
216-650-3000

Mattel (Barbie, Hot Wheels, Matchbox)
333 Continental Blvd.
El Segundo, CA 90245-5012
310-252-2000

McFarlane Toys (Spawn toys)
15155 Fogg St.
Plymouth, MI 48170
313-414-3500

Meccano-Erector (Erector sets)
1675 Broadway, 31st Floor
New York, NY 10019
212-397-0711

Milton Bradley (games)
443 Shaker Rd. E.
Longmeadow, MA 01028-3149
413-525-6411

Nintendo (video games)
4820 150th Ave. NE
Redmond, WA 98052-5111
206-882-2040

Nylint (vehicle toys)
1800 Sixteenth Ave.
Rockford, IL 61104-5491
815-397-2880

Ohio Art (Etch-a-Sketch)
One Toy Street
Bryan, OH 43506
419-636-3141

Parker Brothers (games)
50 Dunham Rd.
Beverly, MA 01915
617-927-7600

PEZ Candy
35 Prindle Hill Rd.
Orange, CT 06477
203-795-0531

Playing Mantis (die-cast cars)
3600 McGill St., Suite 300
P.O. Box 3688
South Bend, IN 46619-3688
219-232-0300

Playmates (action figures)
611 Anton Blvd. #600
Costa Mesa, CA 92626
714-428-2000

Playmobil (figures)
11-E Nichols Ct.
Dayton, NJ 08810
908-274-0101

Reeves International (distributor of Corgi, Breyer)
14 Industrial Rd.
Pequannock, NJ 07440
201-694-5006

Revell-Monogram (model kits)
8601 Waukegan Rd.
Morton Grove, IL 60053
708-966-3500

Sega of America (video games)
255 Shoreline Dr.
Redwood City, CA 94065
415-508-2800

Smith-Miller (vehicle toys)
P.O. Box 139
Canoga Park, CA 91305
818-703-8588

Steiff USA (teddy bears)
200 Fifth Ave., Suite 1205
New York, NY 10010
212-675-2727

Thermos Co. (lunch kits)
Rt. 75 East
Freeport, IL 61032
815-232-2111

Thinkway Toys (Toy Story, etc.)
8885 Woodbine Ave.
Markham, Ontario
Canada L3R 5G9
905-470-8883

Today's Kids (miscellaneous)
13630 Neutron Rd.
Dallas, TX 75244
214-404-9335

Tootsietoy (vehicle toys)
600 N. Pulaski Rd.
Chicago, IL 60624
312-638-1000

Toy Biz (action figures)
333 East 38th St.
New York, NY 10016
212-682-4700

Trendmasters (action figures)
611 North 10th St., Suite 555
St. Louis, MO 63101
314-231-2250

Ty, Inc. (Beanie Babies)
P.O. Box 5377
Oakbrook, IL 60522
630-920-1515

Winross (die-cast trucks)
Box 23860
Rochester, NY 14692
716-381-5638

Directory of Auction Houses

The following is a representative, but not all-inclusive, list of nationwide auction houses. Most hold general antiques and collectibles auction; some may specialize in toys. For information on upcoming auctions, contact the auction house directly.

Noel Barrett Antiques & Auctions
P.O. Box 300
6193 Carversville Rd.
Carversville, PA 18913
215-297-5109

Bill Bertoia Auctions
2413 Madison Ave.
Vineland, NJ 08360
609- 692-1881

Block's Box
P.O. Box 51
Trumbull, CT 06611
206-926-8448

Butterfield & Butterfield
7601 Sunset Blvd.
Los Angeles, CA 90046
213-850-7500

Christie's
502 Park Ave.
New York, NY 10022
212-546-1000

Christie's East
219 East 67th St.
New York, NY 10021
212-606-0400

Continental Auctions
P.O. Box 193
Sheboygan, WI 53082
920-693-3371

William Doyle Galleries
175 E. 87th St.
New York, NY 10128-2205
212- 427-2730

Dumouchelle Art Galleries
409 East Jefferson Ave.
Detroit, MI 48226
313-963-6255 / 313-963-0248

Eldred's
P.O. Box 796
East Dennis, MA 02641-0796
508- 385-3116

Guernsey's
108 East 73rd St.
New York, NY 10021
212-794-2280

Hake's Americana & Collectibles
P.O. Box 1444
York, PA 17405-1444
717-848-1333

Randy Inman Auctions
40 College Ave.
P.O. Box 726
Waterville, ME 04901
207-872-6900

James D. Julia Auctioneers Inc.
Rte. 201, Skowhegan Rd.
P.O. Box 830
Fairfield, ME 04937
207-453-7125

Just Kids
310 New York Ave.
Huntington, NY 11743
516-423-8449

Manion's Auction House
PO Box 12214
Kansas City, KS 66112-0214
913-299-6692

Mapes Auctioneers & Appraisers
1600 Vestal Parkway West
Vestal, NY 13850
607-754-9193

Ted Maurer
1003 Brookwood Dr.
Pottstown, PA 19646
215-323-1573

McMasters Doll Auctions
P.O. Box 1755
Cambridge, OH 43725
614-432-4419

Mid-Hudson Auction Galleries
One Idlewild Ave.
Croton-On-Hudson, NY 12520

New England Auction Gallery
Box 2273-T
West Peabody, MA 01960
508-535-3140

Richard Opfer Auctioneering Inc.
1919 Greenspring Dr.
Lutherville Timonium, MD 21093-4113
410-252-5035

Phillips Fine Art & Auctioneers
406 East 79th St.
New York, NY 10022
212-570-4830

Lloyd W. Ralston
173 Post Rd.
Fairfield, CT 06430
203-255-1233

Skinner Inc.
357 Main St.
Bolton, MA 01740-1104
508-779-6241/ 617-350-5400

Smith House
P.O. Box 336
Eliot, ME 03903
207-439-4614

Sotheby's
1334 York Ave. at 72nd St.
New York, NY 10021
212- 606-7370 / 212- 606-7000

Theriault's
P.O. Box 151
Annapolis, MD 21404
800-638-0422

Toy Scouts
137 Casterton Ave.
Akron, OH 44303
330-836-0668

Withington Inc.
RD2 Box 440
Hillsboro, NH 03244
603-464-3232